BUILDING CONSTRUCTION HANDBOOK

Seventh edition

R. Chudley
MCIOB

and

R. Greeno
BA (Hons) FCIOB FIPHE FRSA

ELSEVIER

AMSTERDAM • BOSTON • HEIDELBERG • LONDON • NEW YORK • OXFORD
PARIS • SAN DIEGO • SAN FRANCISCO • SINGAPORE • SYDNEY • TOKYO

Butterworth-Heinemann is an imprint of Elsevier

Butterworth-Heinemann is an imprint of Elsevier
Linacre House, Jordan Hill, Oxford OX2 8DP, UK
30 Corporate Drive, Suite 400, Burlington, MA 01803, USA

First edition 1988
Second edition 1995
Third edition 1998
Reprinted 1999, 2000
Fourth edition 2001
Fifth edition 2004
Sixth edition 2006
Reprinted 2007
Seventh edition 2008

British Library Cataloguing in Publication Data
A catalogue record for this book is available from the British Library

Library of Congress Cataloging-in-Publication Data Control Number: 2005938728

ISBN: 978-0-7506-86228

For information on all Butterworth-Heinemann publications
visit our website at books.elsevier.com

Typeset by Integra Software Services
Printed and bound in Great Britain
08 09 10 11   11 10 9 8 7 6 5 4 3 2

Working together to grow
libraries in developing countries

www.elsevier.com | www.bookaid.org | www.sabre.org

ELSEVIER   BOOK AID
             International   Sabre Foundation

CONTENTS

Contents

Contents

Part Eight   Domestic Services

PREFACE TO SEVENTH EDITION

The presentation of this seventh edition continues the familiar and unique format of clear illustrations supplemented with comprehensive notes throughout. The benefit of data accumulated from the numerous previous editions, permits traditional construction techniques to be retained alongside contemporary and developing practice. Established procedures are purposely retained with regard to maintenance and refurbishment of existing building stock.

Progressive development, new initiatives and government directives to reduce fuel energy consumption in buildings by incorporating sustainable and energy efficient features is included. In support of these environmental issues, the companion volume *Building Services Handbook* should be consulted for applications to energy consuming systems, their design and incorporation within the structure.

The diverse nature of modern construction practice, techniques and developments with new and synthetic materials cannot be contained in this volume alone. The content is therefore intended as representative and not prescriptive. Further reading of specific topics is encouraged, especially through professional journals, trade and manufacturers' literature, illustrative guides to the Building Regulations and the supplementary references given hereinafter.

R.G.

1 GENERAL

BUILT ENVIRONMENT
THE STRUCTURE
PRIMARY AND SECONDARY ELEMENTS
CONSTRUCTION ACTIVITIES
CONSTRUCTION DOCUMENTS
CONSTRUCTION DRAWINGS
BUILDING SURVEY
HIPS/EPCs
PLANNING APPLICATION
MODULAR COORDINATION
CONSTRUCTION REGULATIONS
CDM REGULATIONS
SAFETY SIGNS AND SYMBOLS
BUILDING REGULATIONS
CODE FOR SUSTAINABLE HOMES
BRITISH STANDARDS
EUROPEAN STANDARDS
CPI SYSTEM OF CODING
CI/SFB SYSTEM OF CODING

Environment = surroundings which can be natural, man-made or a combination of these.

Built Environment = created by man with or without the aid of the natural environment.

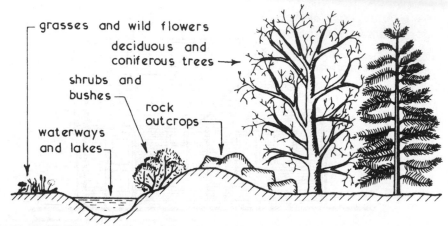

grasses and wild flowers

deciduous and coniferous trees →

shrubs and bushes

rock outcrops

waterways and lakes

ELEMENTS of the NATURAL ENVIRONMENT

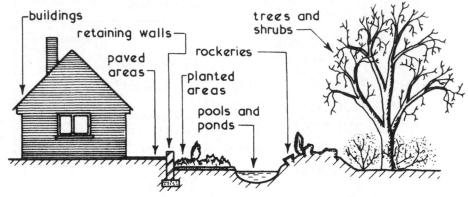

buildings

retaining walls

paved areas

rockeries

planted areas

pools and ponds

trees and shrubs

ELEMENTS of the BUILT ENVIRONMENT (EXTERNAL)

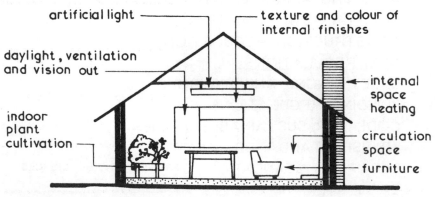

artificial light

texture and colour of internal finishes

daylight, ventilation and vision out

internal space heating

indoor plant cultivation

circulation space

furniture

ELEMENTS of the BUILT ENVIRONMENT (INTERNAL)

Environmental Considerations
1. Planning requirements.
2. Building Regulations.
3. Land restrictions by vendor or lessor.
4. Availability of services.
5. Local amenities including transport.
6. Subsoil conditions.
7. Levels and topography of land.
8. Adjoining buildings or land.
9. Use of building.
10. Daylight and view aspects.

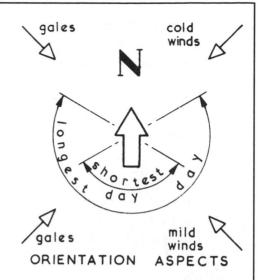

ORIENTATION ASPECTS

Examples:~

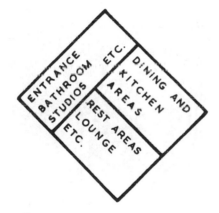

HOUSES

SCHOOLS

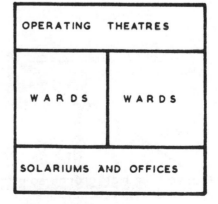

FACTORIES

HOSPITALS

Physical considerations

1. Natural contours of land.
2. Natural vegetation and trees.
3. Size of land and/or proposed building.
4. Shape of land and/or proposed building.
5. Approach and access roads and footpaths.
6. Services available.
7. Natural waterways, lakes and ponds.
8. Restrictions such as rights of way; tree preservation and ancient buildings.
9. Climatic conditions created by surrounding properties, land or activities.
10. Proposed future developments.

Examples:~

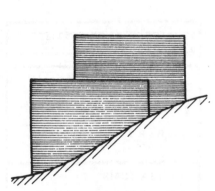

Split level construction to form economic shape.

Shape determined by existing trees.

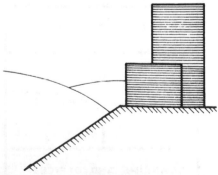

Plateau or high ground solution giving dry site conditions on sloping sites.

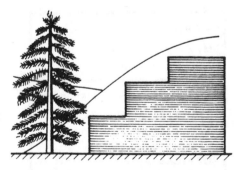

Stepped elevation or similar treatment to blend with the natural environment.

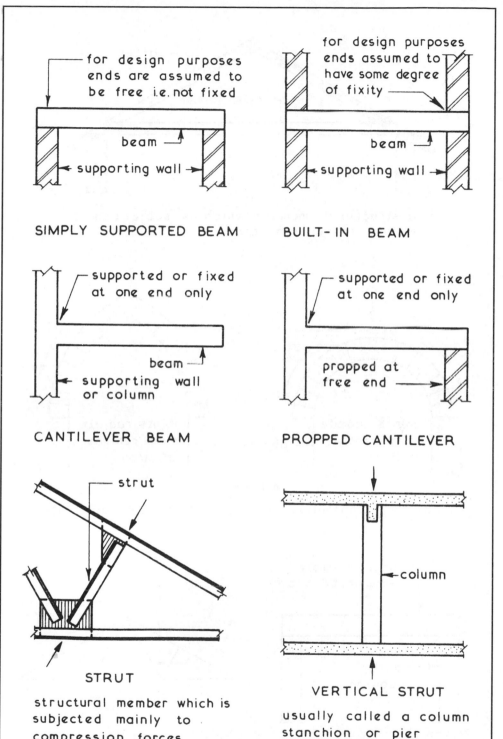

for design purposes ends are assumed to be free i.e. not fixed

beam

supporting wall

SIMPLY SUPPORTED BEAM

for design purposes ends assumed to have some degree of fixity

beam

supporting wall

BUILT-IN BEAM

supported or fixed at one end only

beam

supporting wall or column

CANTILEVER BEAM

supported or fixed at one end only

propped at free end

PROPPED CANTILEVER

strut

STRUT

structural member which is subjected mainly to compression forces

column

VERTICAL STRUT

usually called a column stanchion or pier

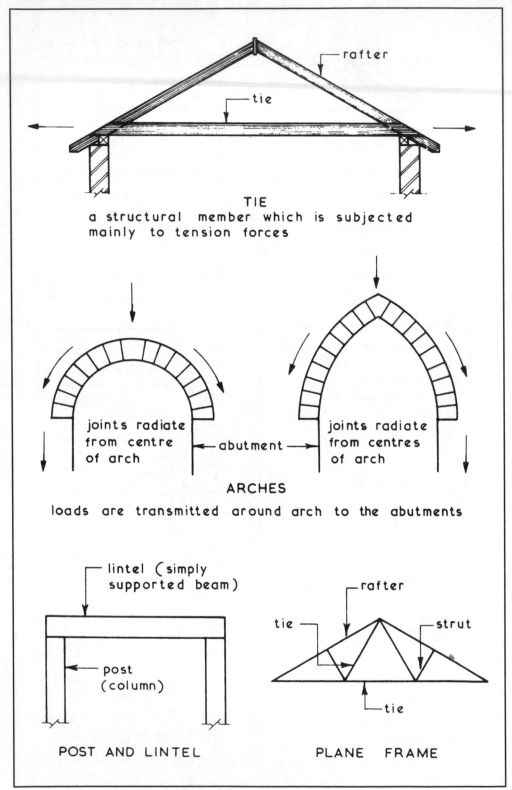

TIE

a structural member which is subjected
mainly to tension forces

joints radiate
from centre
of arch

←— abutment —→

joints radiate
from centres
of arch

ARCHES

loads are transmitted around arch to the abutments

lintel (simply
supported beam)

post
(column)

rafter

tie

strut

tie

POST AND LINTEL

PLANE FRAME

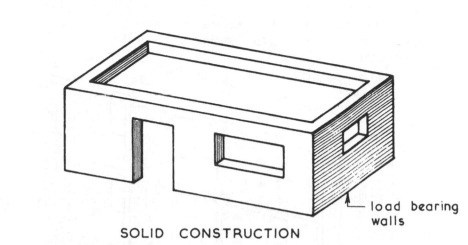

load bearing walls

SOLID CONSTRUCTION

structurally limited confined usually to buildings of low height and short spans

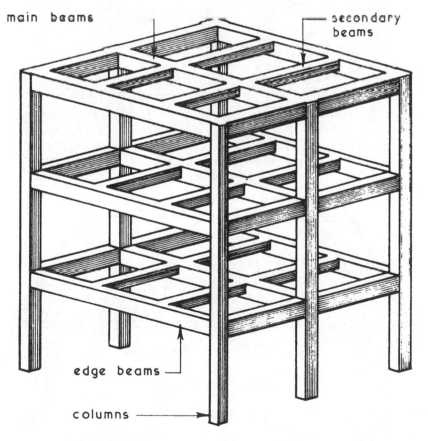

main beams

secondary beams

edge beams

columns

FRAMED OR SKELETAL CONSTRUCTION

structure consists of a series
of interconnected plates forming
structural walls and floors

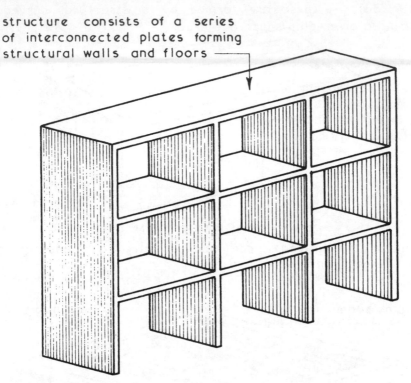

PANEL OR BOX CONSTRUCTION

flat slab folded so that roof will
behave as a beam spanning along fold

diaphragms

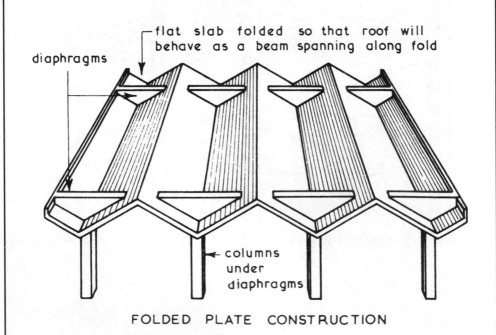

← columns
under
diaphragms

FOLDED PLATE CONSTRUCTION

Shell Roofs ~ these are formed by a structural curved skin covering a given plan shape and area.

Examples ~

double curvature shell formed by rotating a plain curved shape about a vertical axis

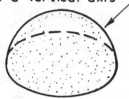
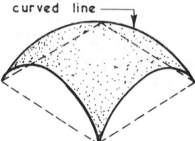

DOME OR ROTATIONAL SHELL

hemispherical rotational dome

vertical cut plane

pendentive

inscribed polygon

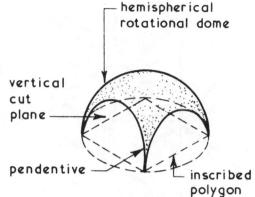

PENDENTIVE DOME

formed by a curved line moving over another curved line

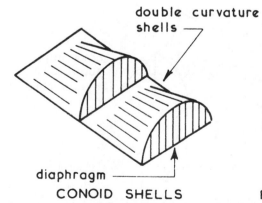

TRANSLATIONAL DOME

cut cylinder giving a single curvature shell

diaphragm

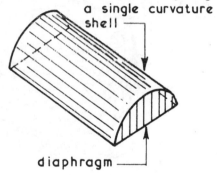

BARREL VAULT

double curvature shells

diaphragm

CONOID SHELLS

double curvature saddle shaped shell

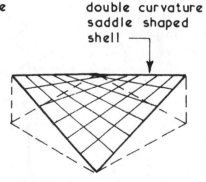

HYPERBOLIC PARABOLOID

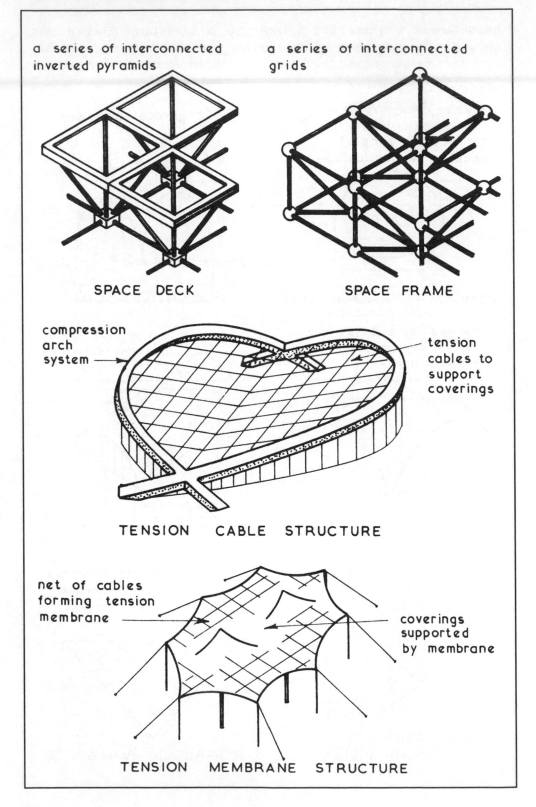

a series of interconnected inverted pyramids

a series of interconnected grids

SPACE DECK

SPACE FRAME

compression arch system

tension cables to support coverings

TENSION CABLE STRUCTURE

net of cables forming tension membrane

coverings supported by membrane

TENSION MEMBRANE STRUCTURE

Substructure ~ can be defined as all structure below the superstructure which in general terms is considered to include all structure below ground level but including the ground floor bed.

Typical Examples~

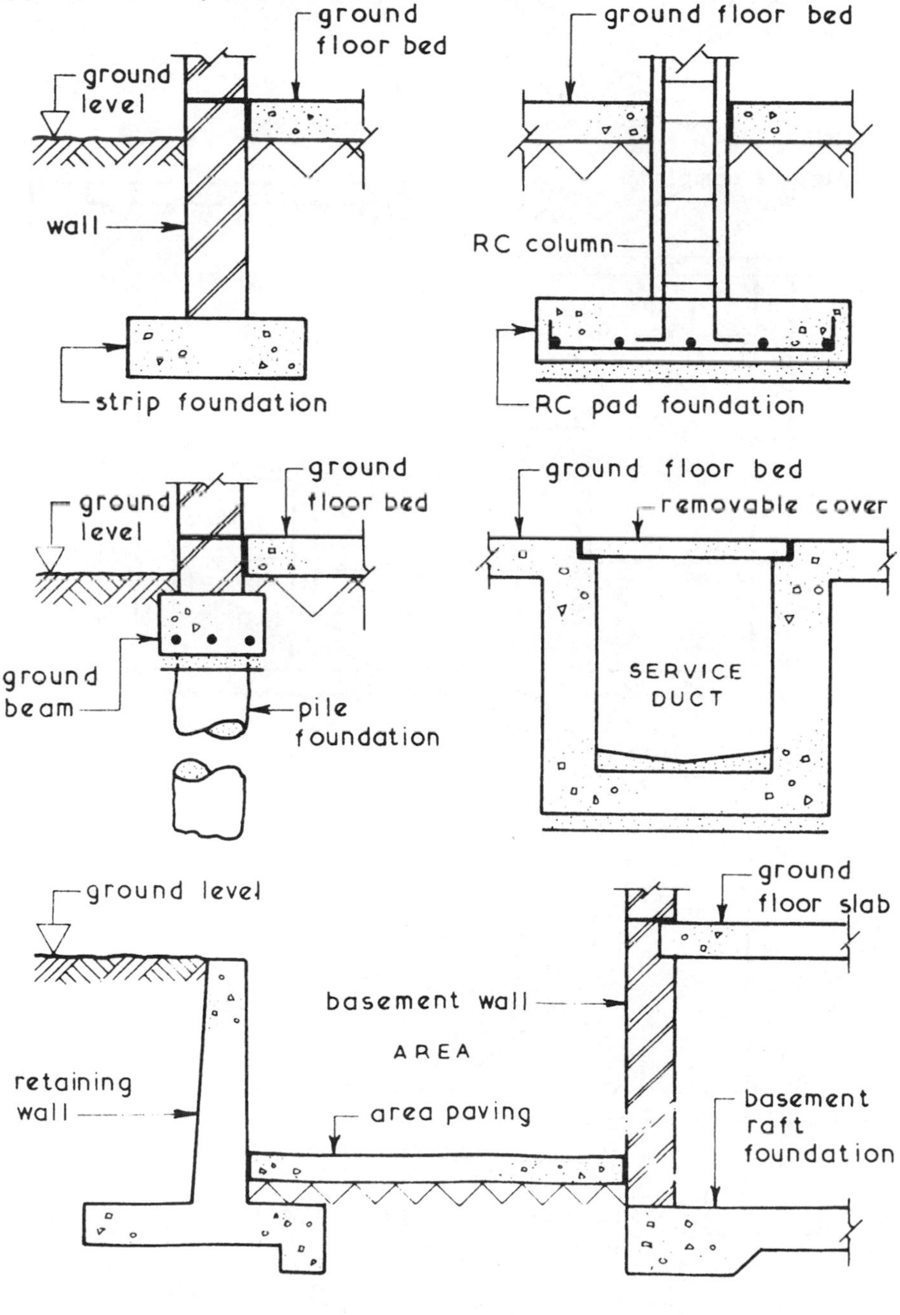

strip foundation

RC pad foundation

pile foundation

SERVICE DUCT

basement raft foundation

Superstructure ~ can be defined as all structure above substructure both internally and externally.

Primary Elements ~ basically components of the building carcass above the substructure excluding secondary elements, finishes, services and fittings.

Typical Examples~

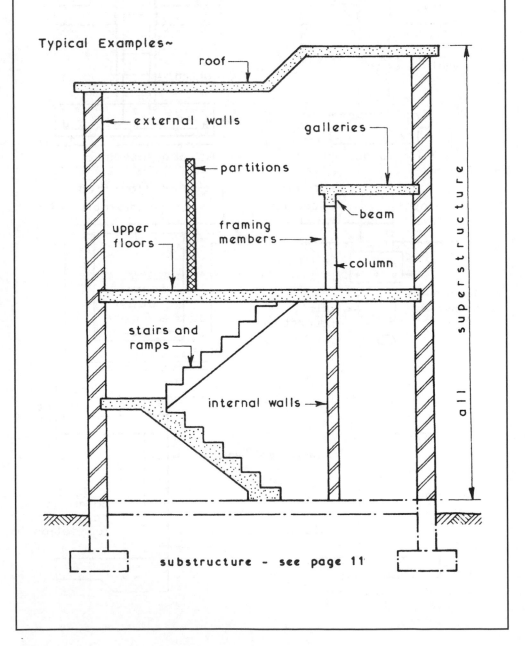

Secondary Elements ~ completion of the structure including completion around and within openings in primary elements.

Typical Examples~

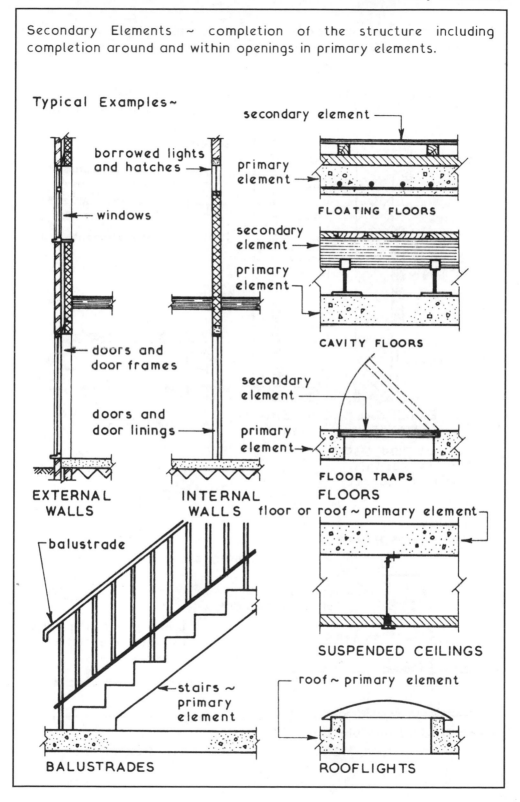

borrowed lights and hatches →

windows

doors and door frames

doors and door linings →

EXTERNAL WALLS

INTERNAL WALLS

secondary element

primary element →

FLOATING FLOORS

secondary element →

primary element →

CAVITY FLOORS

secondary element

primary element →

FLOOR TRAPS

FLOORS

floor or roof ~ primary element

SUSPENDED CEILINGS

balustrade

stairs ~ primary element

BALUSTRADES

roof ~ primary element

ROOFLIGHTS

Finish ~ the final surface which can be self finished as with a trowelled concrete surface or an applied finish such as floor tiles.

Typical Examples~

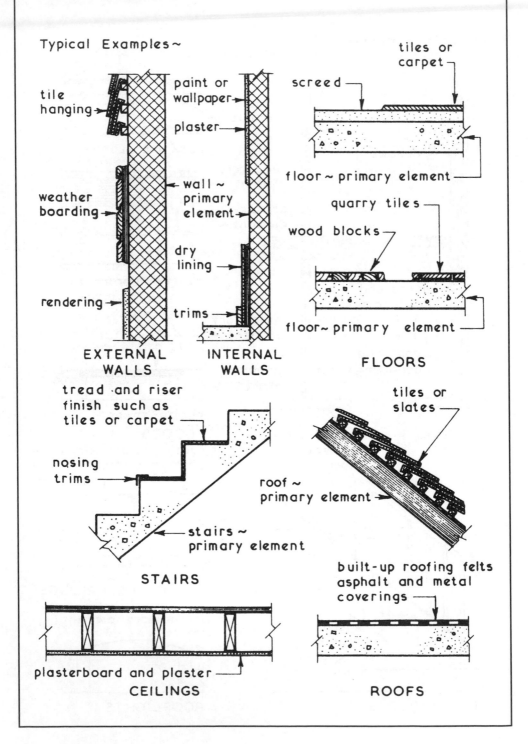

tile hanging

paint or wallpaper

plaster

weather boarding

wall ~ primary element

dry lining

rendering

trims

EXTERNAL WALLS

INTERNAL WALLS

tiles or carpet

screed

floor ~ primary element

quarry tiles

wood blocks

floor ~ primary element

FLOORS

tread and riser finish such as tiles or carpet

nosing trims

stairs ~ primary element

STAIRS

tiles or slates

roof ~ primary element

built-up roofing felts asphalt and metal coverings

plasterboard and plaster
CEILINGS

ROOFS

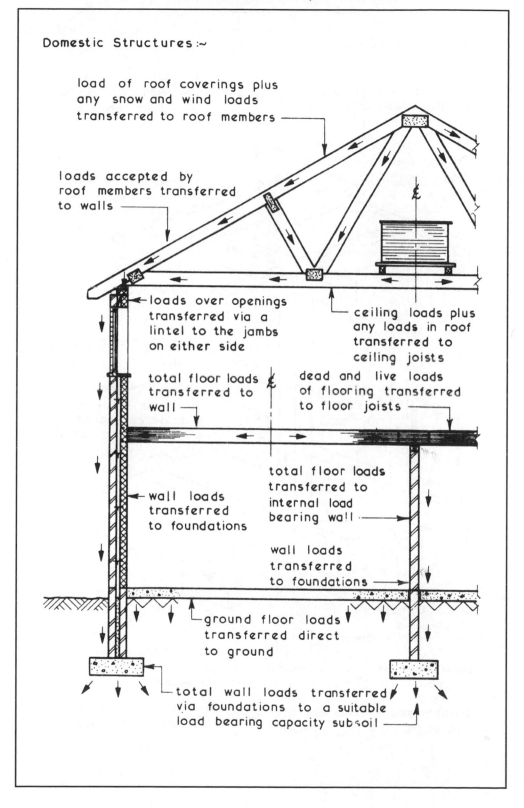

Domestic Structures :~

load of roof coverings plus any snow and wind loads transferred to roof members

loads accepted by roof members transferred to walls

loads over openings transferred via a lintel to the jambs on either side

ceiling loads plus any loads in roof transferred to ceiling joists

total floor loads transferred to wall

dead and live loads of flooring transferred to floor joists

wall loads transferred to foundations

total floor loads transferred to internal load bearing wall

wall loads transferred to foundations

ground floor loads transferred direct to ground

total wall loads transferred via foundations to a suitable load bearing capacity subsoil

Framed Structures:~

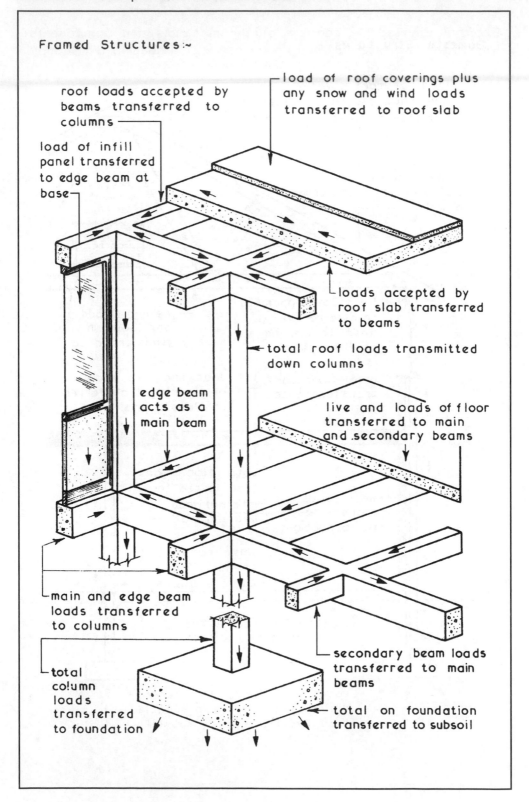

roof loads accepted by beams transferred to columns

load of infill panel transferred to edge beam at base

load of roof coverings plus any snow and wind loads transferred to roof slab

loads accepted by roof slab transferred to beams

total roof loads transmitted down columns

edge beam acts as a main beam

live and loads of floor transferred to main and secondary beams

main and edge beam loads transferred to columns

secondary beam loads transferred to main beams

total column loads transferred to foundation

total on foundation transferred to subsoil

External Envelope ~ consists of the materials and components which form the external shell or enclosure of a building. These may be load bearing or non-load bearing according to the structural form of the building.

Primary Functions :~

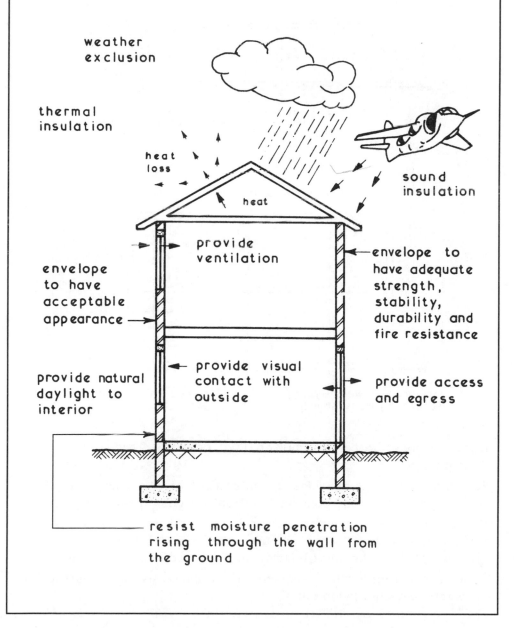

weather exclusion

thermal insulation

heat loss

heat

sound insulation

provide ventilation

envelope to have acceptable appearance

envelope to have adequate strength, stability, durability and fire resistance

provide natural daylight to interior

provide visual contact with outside

provide access and egress

resist moisture penetration rising through the wall from the ground

Internal Separation and Compartmentation

Dwelling houses ~

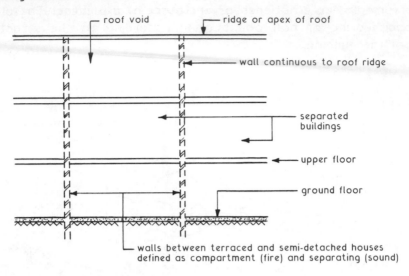

- roof void
- ridge or apex of roof
- wall continuous to roof ridge
- separated buildings
- upper floor
- ground floor
- walls between terraced and semi-detached houses defined as compartment (fire) and separating (sound)

Flats ~

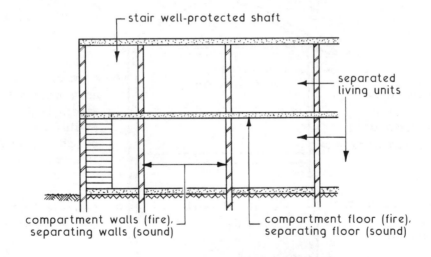

- stair well-protected shaft
- separated living units
- compartment walls (fire), separating walls (sound)
- compartment floor (fire), separating floor (sound)

Note: Floors within a maisonette are not required to be ''compartment''.

For non-residential buildings, compartment size is limited by floor area depending on the building function (purpose group) and height.

Compartment ~ a building or part of a building with walls and floors constructed to contain fire and to prevent it spreading to another part of the same building or to an adjoining building.

Separating floor/wall ~ element of sound resisting construction between individual living units.

A Building or Construction Site can be considered as a temporary factory employing the necessary resources to successfully fulfil a contract.

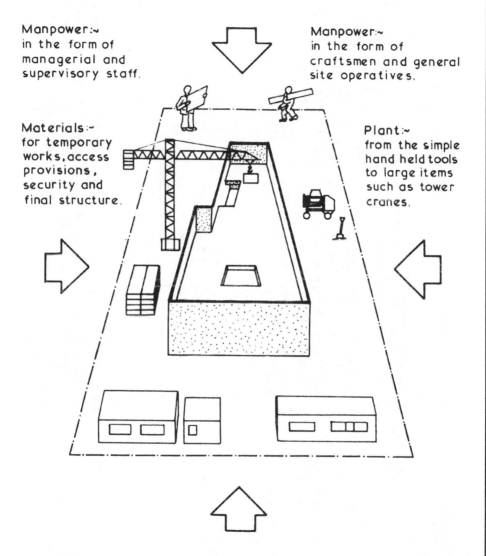

Manpower:~ in the form of managerial and supervisory staff.

Manpower:~ in the form of craftsmen and general site operatives.

Materials:~ for temporary works, access provisions, security and final structure.

Plant:~ from the simple hand held tools to large items such as tower cranes.

Money :~ in the form of capital investment from the building owner to pay for the land, design team fees and a building contractor who uses his money to buy materials, buy or hire plant and hire labour to enable the project to be realised.

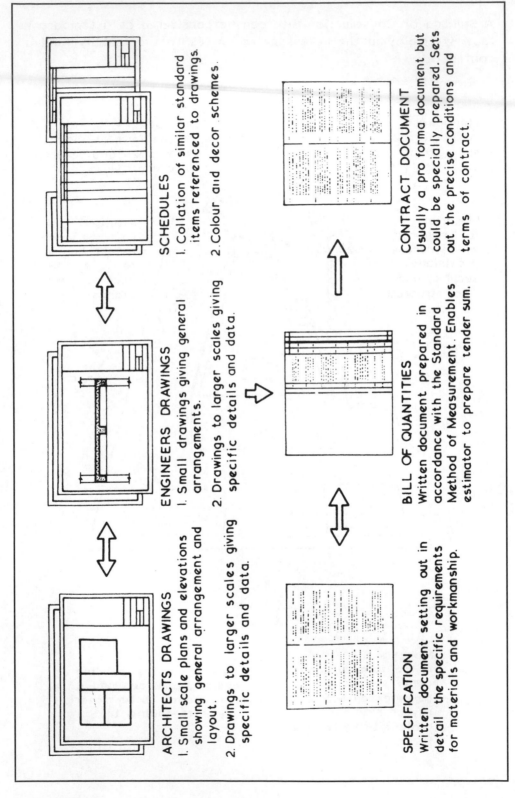

ARCHITECTS DRAWINGS
1. Small scale plans and elevations showing general arrangement and layout.
2. Drawings to larger scales giving specific details and data.

ENGINEERS DRAWINGS
1. Small drawings giving general arrangements.
2. Drawings to larger scales giving specific details and data.

SCHEDULES
1. Collation of similar standard items referenced to drawings.
2. Colour and decor schemes.

SPECIFICATION
Written document setting out in detail the specific requirements for materials and workmanship.

BILL OF QUANTITIES
Written document prepared in accordance with the Standard Method of Measurement. Enables estimator to prepare tender sum.

CONTRACT DOCUMENT
Usually a pro forma document but could be specially prepared. Sets out the precise conditions and terms of contract.

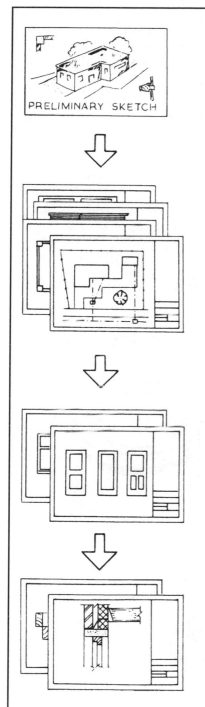

PRELIMINARY SKETCH

Location Drawings ~

Site Plans – used to locate site, buildings, define site levels, indicate services to buildings, identify parts of site such as roads, footpaths and boundaries and to give setting out dimensions for the site and buildings as a whole. Suitable scale not less than 1:2500

Floor Plans – used to identify and set out parts of the building such as rooms, corridors, doors, windows, etc., Suitable scale not less than 1:100

Elevations – used to show external appearance of all faces and to identify doors and windows. Suitable scale not less than 1:100

Sections – used to provide vertical views through the building to show method of construction. Suitable scale not less than 1:50

Component Drawings ~

used to identify and supply data for components to be supplied by a manufacturer or for components not completely covered by assembly drawings. Suitable scale range 1:100 to 1:1

Assembly Drawings ~

used to show how items fit together or are assembled to form elements. Suitable scale range 1:20 to 1:5

All drawings should be fully annotated, fully dimensioned and cross referenced.

Ref. BS EN ISO 7519: Technical drawings. Construction drawings. General principles of presentation for general arrangement and assembly drawings.

Sketch ~ this can be defined as a draft or rough outline of an idea, it can be a means of depicting a three-dimensional form in a two-dimensional guise. Sketches can be produced free-hand or using rules and set squares to give basic guide lines.

All sketches should be clear, show all the necessary detail and above all be in the correct proportions.

Sketches can be drawn by observing a solid object or they can be produced from conventional orthographic views but in all cases can usually be successfully drawn by starting with an outline 'box' format giving length, width and height proportions and then building up the sketch within the outline box.

Example~ Square Based Chimney Pot.

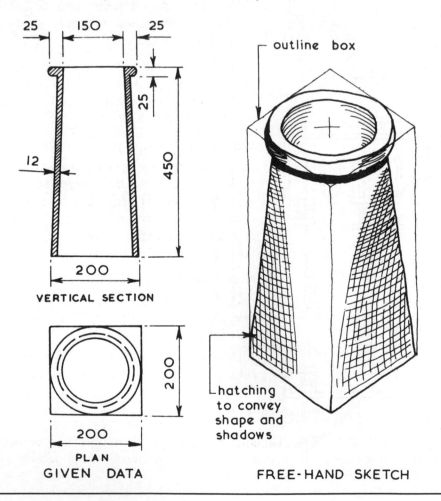

GIVEN DATA   FREE-HAND SKETCH

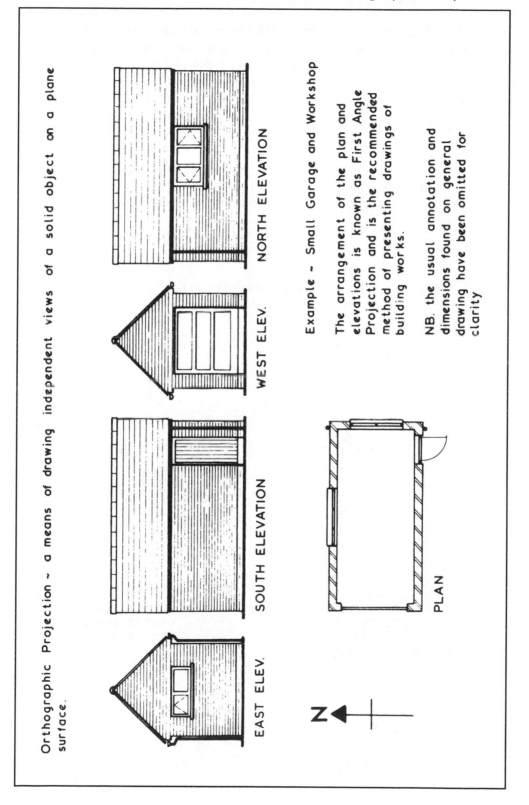

Orthographic Projection ~ a means of drawing independent views of a solid object on a plane surface.

NORTH ELEVATION

WEST ELEV.

SOUTH ELEVATION

EAST ELEV.

PLAN

N

Example ~ Small Garage and Workshop

The arrangement of the plan and elevations is known as First Angle Projection and is the recommended method of presenting drawings of building works.

NB. the usual annotation and dimensions found on general drawing have been omitted for clarity

Isometric Projections ~ a pictorial projection of a solid object on a plane surface drawn so that all vertical lines remain vertical and of true scale length, all horizontal lines are drawn at an angle of 30° and are of true scale length therefore scale measurements can be taken on the vertical and 30° lines but cannot be taken on any other inclined line.

A similar drawing can be produced using an angle of 45° for all horizontal lines and is called an Axonometric Projection

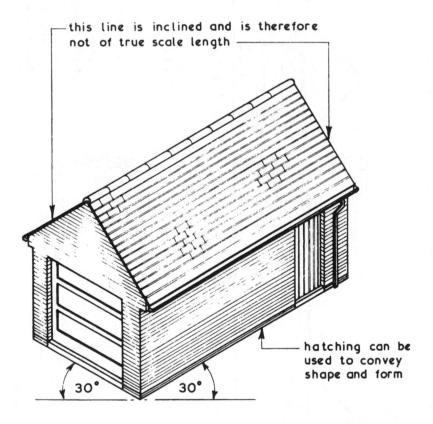

this line is inclined and is therefore not of true scale length

hatching can be used to convey shape and form

30°   30°

ISOMETRIC PROJECTION SHOWING SOUTH AND WEST ELEVATIONS OF SMALL GARAGE AND WORKSHOP ILLUSTRATED ON PAGE 23

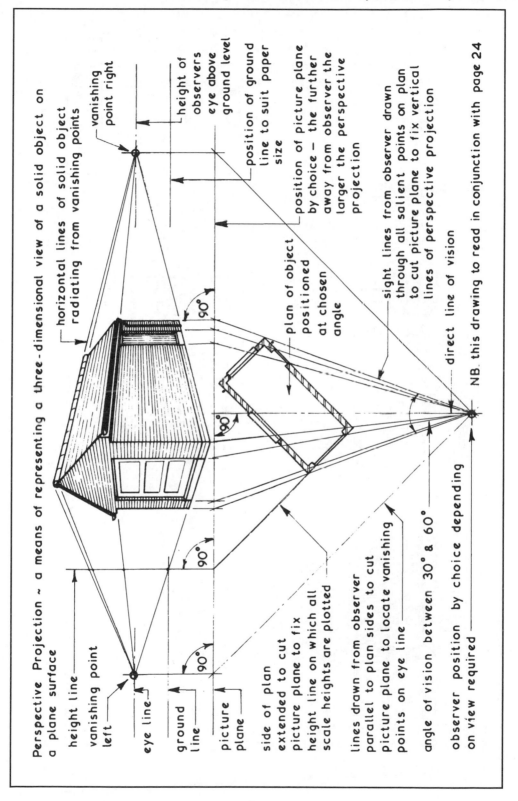

Perspective Projection ~ a means of representing a three-dimensional view of a solid object on a plane surface

horizontal lines of solid object radiating from vanishing points

vanishing point right

height of observers eye above ground level

position of ground line to suit paper size

position of picture plane by choice – the further away from observer the larger the perspective projection

plan of object positioned at chosen angle

sight lines from observer drawn through all salient points on plan to cut picture plane to fix vertical lines of perspective projection

direct line of vision

NB. this drawing to read in conjunction with page 24

90°

90°

90°

90°

height line

vanishing point left

eye line

ground line

picture plane

side of plan extended to cut picture plane to fix height line on which all scale heights are plotted

lines drawn from observer parallel to plan sides to cut picture plane to locate vanishing points on eye line

angle of vision between 30° & 60°

observer position by choice depending on view required

25

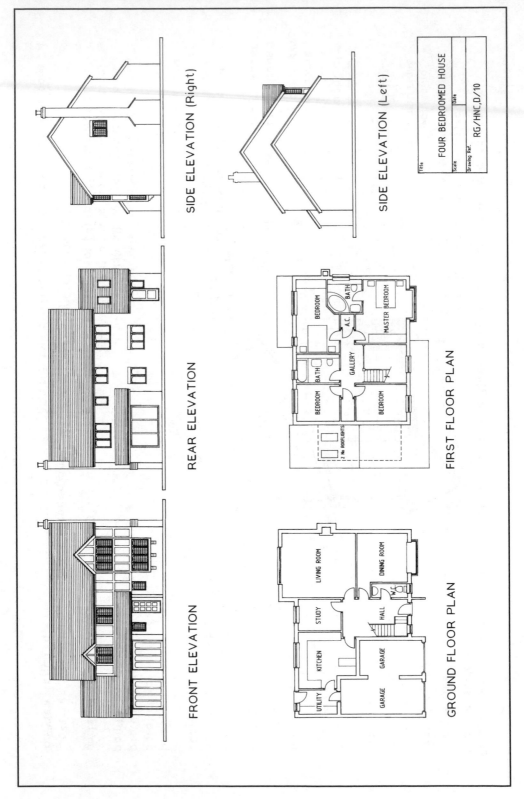

SIDE ELEVATION (Right)

SIDE ELEVATION (Left)

FOUR BEDROOMED HOUSE

Title

Scale | Date

Drawing Ref. | RG/HNC,D/10

REAR ELEVATION

FRONT ELEVATION

FIRST FLOOR PLAN

GROUND FLOOR PLAN

BEDROOM

BATH

MASTER BEDROOM

A.C.

GALLERY

BATH

BEDROOM

BEDROOM

2 No ROOFLIGHTS

LIVING ROOM

DINING ROOM

STUDY

HALL

W.C.

KITCHEN

GARAGE

UTILITY

GARAGE

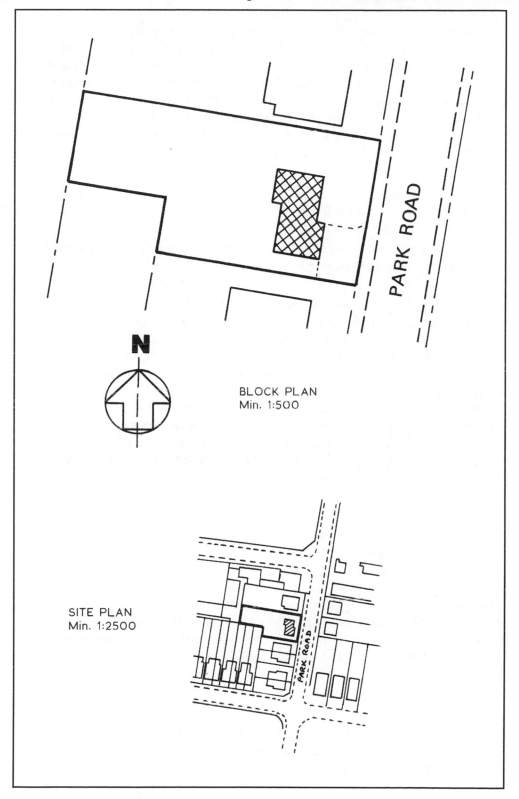

PARK ROAD

N

BLOCK PLAN
Min. 1:500

SITE PLAN
Min. 1:2500

PARK ROAD

Construction Defects – correct application of materials produced to the recommendations of British, European and International Standards authorities, in accordance with local building regulations, by laws and the rules of building guarantee companies, i.e. National House Building Council (NHBC) and Zurich Insurance, should ensure a sound and functional structure. However, these controls can be seriously undermined if the human factor of quality workmanship is not fulfilled. The following guidance is designed to promote quality controls:

BS 8000: Workmanship on building sites.

Building Regulations, Approved Document to support Regulation 7 – materials and workmanship.

No matter how good the materials, the workmanship and supervision, the unforeseen may still affect a building. This may materialise several years after construction. Some examples of these latent defects include: woodworm emerging from untreated timber, electrolytic decomposition of dissimilar metals inadvertently in contact, and chemical decomposition of concrete. Generally, the older a building the more opportunity there is for its components and systems to have deteriorated and malfunctioned. Hence the need for regular inspection and maintenance. The profession of facilities management has evolved for this purpose and is represented by the British Institute of Facilities Management (BIFM).

Property values, repairs and replacements are of sufficient magnitude for potential purchasers to engage the professional services of a building surveyor. Surveyors are usually members of the Royal Institution of Chartered Surveyors (RICS). The extent of survey can vary, depending on a client's requirements. This may be no more than a market valuation to secure financial backing, to a full structural survey incorporating specialist reports on electrical installations, drains, heating systems, etc.

Further reading:

BRE Digest No. 268 – Common defects in low-rise traditional housing. Available from Building Research Establishment Bookshop – www.brebookshop.com.

Established Procedure – the interested purchaser engages a building surveyor.

UK Government Requirements – the seller to provide a property/home information pack (HIP) which can include 'A survey report on the condition of the property, including requirements for urgent or significant repairs...'.

Survey document preliminaries:

* Title and address of property

* Client's name, address and contacts

* Survey date and time

* Property status – freehold, leasehold or commonhold

* Occupancy – occupied or vacant. If vacant, source of keys

* Extent of survey, e.g. full structural + services reports

* Specialists in attendance, e.g. electrician, heating engineer, etc.

* Age of property (approx. if very dated or no records)

* Disposition of rooms, i.e. number of bedrooms, etc.

* Floor plans and elevations if available

* Elevation (flooding potential) and orientation (solar effect)

* Estate/garden area and disposition if appropriate

* Means of access – roads, pedestrian only, rights of way

Survey tools and equipment:

* Drawings + estate agent's particulars if available

* Notebook and pencil/pen

* Binoculars and a camera with flash facility

* Tape measure, spirit level and plumb line

* Other useful tools, to include small hammer, torch, screwdriver and manhole lifting irons

* Moisture meter

* Ladders – eaves access and loft access

* Sealable bags for taking samples, e.g. wood rot, asbestos, etc.

Estate and garden:

* Location and establishment of boundaries
* Fences, gates and hedges – material, condition and suitability
* Trees – type and height, proximity to building
* Pathways and drives – material and condition
* Outbuildings – garages, sheds, greenhouses, barns, etc.
* Proximity of water courses

Roof:

* Tile type, treatment at ridge, hips, verge and valleys
* Age of covering, repairs, replacements, renewals, general condition, defects and growths
* Eaves finish, type and condition
* Gutters – material, size, condition, evidence of leakage
* Rainwater downpipes as above
* Chimney – dpcs, flashings, flaunching, pointing, signs of movement
* Flat roofs – materials, repairs, abutments, flashings and drainage

Walls:

* Materials – type of brick, rendering, cladding, etc., condition and evidence of repairs
* Solid or cavity construction, if cavity extent of insulation and type
* Pointing of masonry, painting of rendering and cladding
* Air brick location, function and suitability
* Dpc, material and condition, position relative to ground level
* Windows and doors, material, signs of rot or damage, original or replacement, frame seal
* Settlement – signs of cracking, distortion of window and door frames – specialist report

Drainage:

A building surveyor may provide a general report on the condition of the drainage and sanitation installation. However, a full test for leakage and determination of self-cleansing and flow conditions to include fibre-optic scope examination is undertaken as a specialist survey.

Roof space:
* Access to all parts, construction type – traditional or trussed
* Evidence of moisture due to condensation – ventilation at eaves, ridge, etc.
* Evidence of water penetration – chimney flashings, abutments and valleys
* Insulation – type and quantity
* Party wall in semi-detached and terraced dwellings – suitability as fire barrier
* Plumbing – adequacy of storage cistern, insulation, overflow function

Floors:
* Construction – timber, pre-cast or cast in-situ concrete? Finish condition?
* Timber ground floor – evidence of dampness, rot, woodworm, ventilation, dpcs
* Timber upper floor stability, ie. wall fixing, strutting, joist size, woodworm, span and loading

Stairs:
* Type of construction and method of fixing – built in-situ or preformed
* Soffit, re. fire protection (plasterboard?)
* Balustrading – suitability and stability
* Safety – adequate screening, balusters, handrail, pitch angle, open tread, tread wear

Finishes:
* General décor, i.e. paint and wallpaper condition – damaged, faded
* Woodwork/joinery – condition, defects, damage, paintwork
* Plaster – ceiling (plasterboard or lath and plaster?) – condition and stability
* Plaster – walls – render and plaster or plasterboard, damage and quality of finish
* Staining – plumbing leaks (ceiling), moisture penetration (wall openings), rising damp
* Fittings and ironmongery – adequacy and function, weather exclusion and security

Supplementary enquiries should determine the extent of additional building work, particularly since the planning threshold of 1948. Check for planning approvals, permitted development and Building Regulation approvals, exemptions and completion certificates.

Services – apart from a cursory inspection to ascertain location and suitability of system controls, these areas are highly specialised and should be surveyed by those appropriately qualified.

Home Information Packs ~ otherwise known as HIPS or "seller's packs". A HIP is provided as supplementary data to the estate agent's sales particulars by home sellers when marketing a house. The packs place emphasis on an energy use assessment and contain some contract preliminaries such as evidence of ownership. Property developers are required to provide a HIP as part of their sales literature. Preparation is by a surveyor, specifically trained in energy performance assessment.

Compulsory Content ~
• Index
• Energy performance certificate
• Sales statement
• Standard searches, e.g. LA enquiries, planning consents, drainage arrangements, utilities providers
• Evidence of title (ownership)
• Leasehold and commonhold details (generally flats and maisonettes)

Optional Content ~
• Home condition report (general survey)
• Legal summary – terms of sale
• Home use and contents form (fixtures and fittings)
• Guarantees and warrantees
• Other relevant information, e.g. access over ancillary land

Energy Performance Certificate (EPC) ~ provides a rating between A and G. A is the highest possible grade for energy efficiency and lowest impact on environmental damage in terms of CO_2 emissions. The certificate is similar to the EU energy label (see page 445 as applied to windows) and it relates to SAP numerical ratings (see page 442). The certificate is an asset rating based on a building's performance relating to its age, location/exposure, size, appliance efficiency e.g. boiler, glazing type, construction, insulation and general condition.

EPC rating (SAP rating) ~

A (92–100)	B (81–91)	C (69–80)	D (55–68)
E (39–54)	F (21–38)	G (1–20)	

Ref. The Home Information Pack Regulations 2006.

A method statement precedes preparation of the project programme and contains the detail necessary for construction of each element of a building. It is prepared from information contained in the contract documents – see page 20. It also functions as a brief for site staff and operatives in sequencing activities, indicating resource requirements and determining the duration of each element of construction. It complements construction programming by providing detailed analysis of each activity.

A typical example for foundation excavation could take the following format:

Activity	Quantity	Method	Output/hour	Labour	Plant	Days
Strip site for excavation	300 m²	Exc. to reduced level over construction area – JCB-4CX face shovel/loader. Topsoil retained on site.	50 m²/hr	Exc. driver +2 labourers	JCB-4CX backhoe/loader	0·75
Excavate for foundations	60 m³	Excavate foundation trench to required depth – JCB-4CX backhoe. Surplus spoil removed from site.	15 m³/hr	Exc. driver +2 labourers. Truck driver.	JCB-4CX backhoe/loader. Tipper truck.	0·50

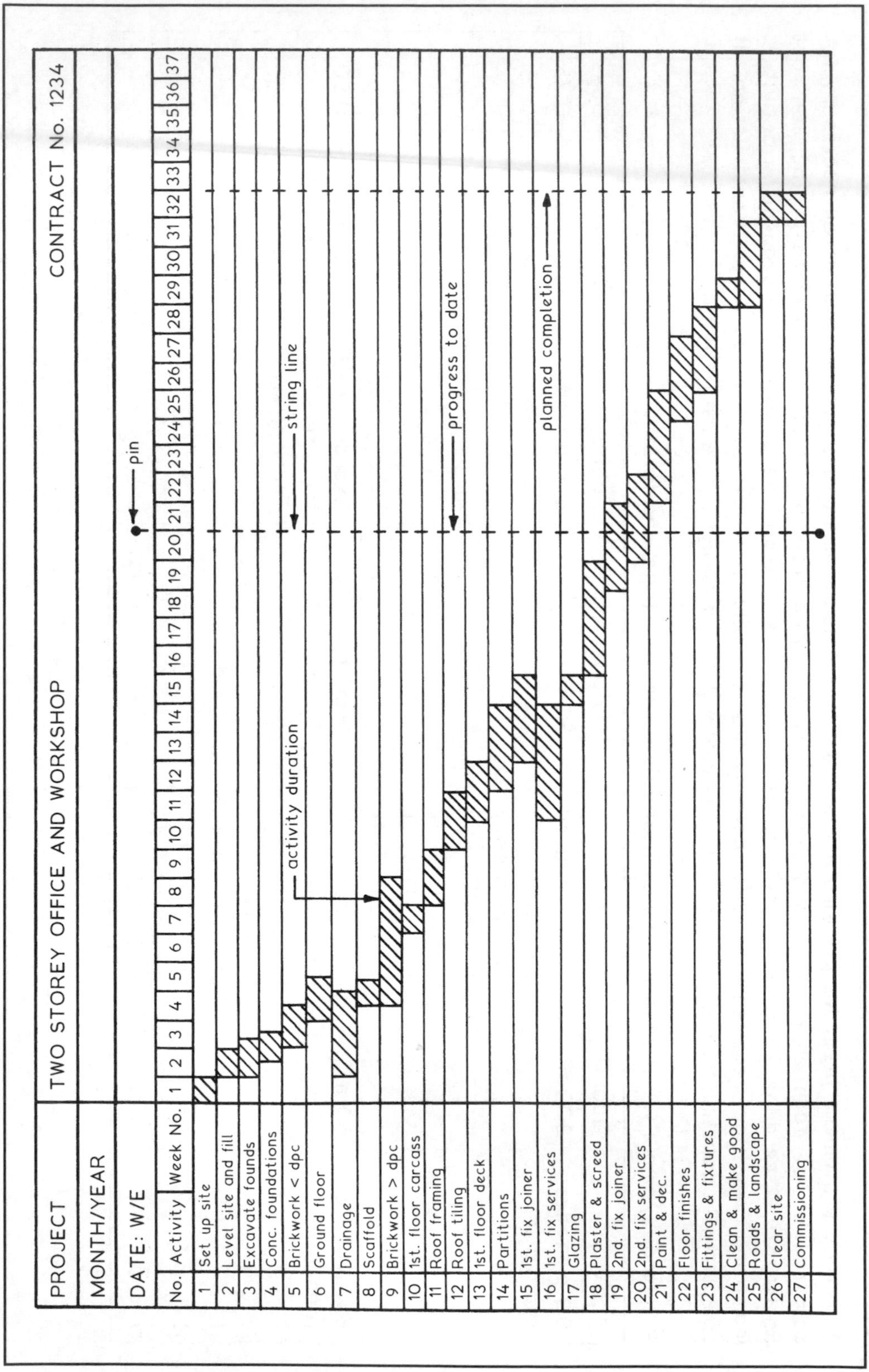

CONTRACT No. 1234

PROJECT TWO STOREY OFFICE AND WORKSHOP

MONTH/YEAR

DATE: W/E

pin — string line — progress to date — planned completion — activity duration

No.	Activity	Week No.
1	Set up site	
2	Level site and fill	
3	Excavate founds	
4	Conc. foundations	
5	Brickwork < dpc	
6	Ground floor	
7	Drainage	
8	Scaffold	
9	Brickwork > dpc	
10	1st. floor carcass	
11	Roof framing	
12	Roof tiling	
13	1st. floor deck	
14	Partitions	
15	1st. fix joiner	
16	1st. fix services	
17	Glazing	
18	Plaster & screed	
19	2nd. fix joiner	
20	2nd. fix services	
21	Paint & dec.	
22	Floor finishes	
23	Fittings & fixtures	
24	Clean & make good	
25	Roads & landscape	
26	Clear site	
27	Commissioning	

Week numbers: 1 2 3 4 5 6 7 8 9 10 11 12 13 14 15 16 17 18 19 20 21 22 23 24 25 26 27 28 29 30 31 32 33 34 35 36 37

Material	Weight (kg/m²)
BRICKS, BLOCKS and PAVING –	
Clay brickwork – 102.5 mm	
low density	205
medium density	221
high density	238
Calcium silicate brickwork – 102.5 mm	205
Concrete blockwork, aerated	78
.. .. lightweight aggregate	129
Concrete flagstones (50 mm)	115
Glass blocks (100 mm thick) 150 × 150	98
.. ..200 × 200	83
ROOFING –	
Thatching (300 mm thick)	40·00
Tiles – plain clay	63·50
.. – plain concrete	93·00
.. single lap, concrete	49·00
Tile battens (50 × 25) and felt underlay	7·70
Bituminous felt underlay	1·00
Bituminous felt, sanded topcoat	2·70
3 layers bituminous felt	4·80
HD/PE breather membrane underlay	0·20
SHEET MATERIALS –	
Aluminium (0·9 mm)	2·50
Copper (0·9 mm)	4·88
Cork board (standard) per 25 mm thickness	4·33
.. .. (compressed)	9·65
Hardboard (3·2 mm)	3·40
Glass (3 mm)	7·30
Lead (1·32 mm – code 3)	14·97
.. .. (3·15 mm – code 7)	35·72
Particle board/chipboard (12 mm)	9·26
.. .. (22 mm)	16·82
Planking, softwood strip flooring (ex 25 mm)	11·20
.. .. hardwood	16·10
Plasterboard (9·5 mm)	8·30
.. .. (12·5 mm)	11·00
.. .. (19 mm)	17·00
Plywood per 25 mm	1·75
PVC floor tiling (2·5 mm)	3·90
Strawboard (25 mm)	9·80
Weatherboarding (20 mm)	7·68
Woodwool (25 mm)	14·50

Typical Weights of Building Materials and Densities

Material	Weight (kg/m^2)
INSULATION	
Glass fibre thermal (100 mm)	2·00
.. .. acoustic	4·00
APPLIED MATERIALS -	
Asphalte (18 mm)	42
Plaster, 2 coat work	22
STRUCTURAL TIMBER -	
Rafters and Joists (100 × 50 @ 400 c/c)	5·87
Floor joists (225 × 50 @ 400 c/c)	14·93

Densities -

Material	Approx. Density (kg/m^3)
Cement	1440
Concrete (aerated)	640
.. .. (broken brick)	2000
.. .. (natural aggregates)	2300
.. .. (no-fines)	1760
.. .. (reinforced)	2400
Metals -	
Aluminium	2770
Copper	8730
Lead	11325
Steel	7849
Timber (softwood/pine)	480 (average)
.. .. .. (hardwood, eg. maple, teak, oak)	720 .. .. ..
Water	1000

Refs. BS 648: Schedule of Weights of Building Materials.
BS 6399-1: Loadings for buildings. Code of Practice
for Dead and Imposed Loads.

Drawings ~ these are the major means of communication between the designer and the contractor as to what, where and how the proposed project is to be constructed.

Drawings should therefore be clear, accurate, contain all the necessary information and be capable of being easily read.

To achieve these objectives most designers use the symbols and notations recommended in BS 1192-5 and BS EN ISO 7519 to which readers should refer for full information.

Typical Examples~

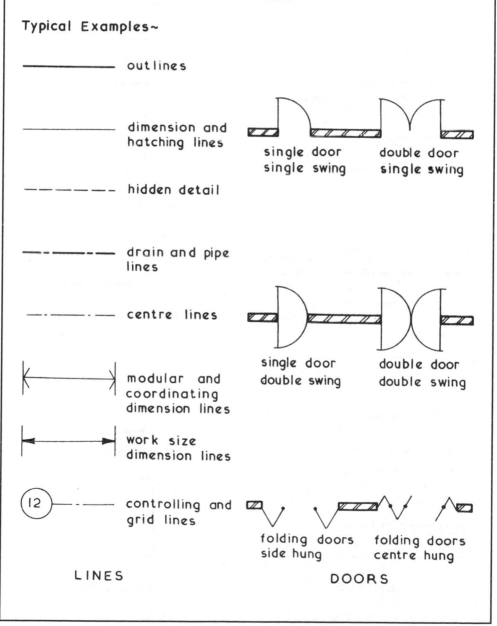

————————— outlines

——————— dimension and hatching lines

single door single swing

double door single swing

— — — — — hidden detail

—·——·— drain and pipe lines

—·—·—·— centre lines

single door double swing

double door double swing

modular and coordinating dimension lines

work size dimension lines

(12)—·—·— controlling and grid lines

folding doors side hung

folding doors centre hung

LINES

DOORS

37

Hatchings ~ the main objective is to differentiate between the materials being used thus enabling rapid recognition and location. Whichever hatchings are chosen they must be used consistently throughout the whole set of drawings. In large areas it is not always necessary to hatch the whole area.

Symbols ~ these are graphical representations and should wherever possible be drawn to scale but above all they must be consistent for the whole set of drawings and clearly drawn.

Typical Examples~

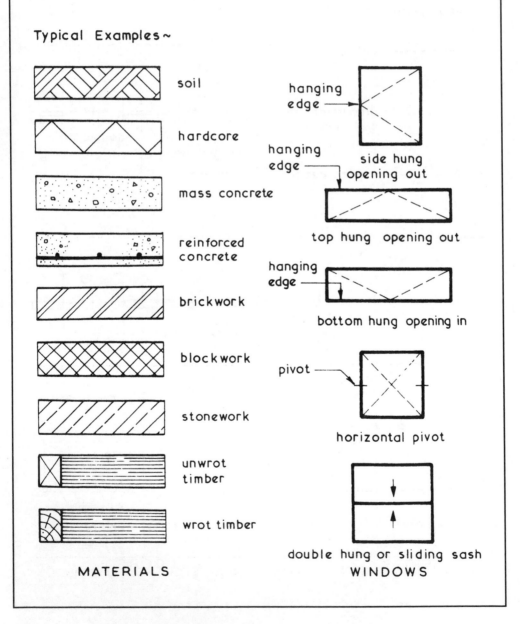

MATERIALS

WINDOWS

Name	Symbol	Name	Symbol
Rainwater pipe	○ R W P	Distribution board	□
Gully	□ G	Electricity meter	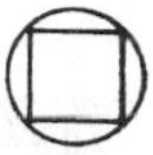 ◉
Inspection chambers	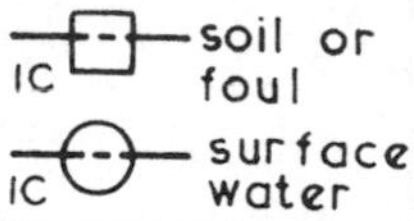 IC —□— soil or foul IC —⊙— surface water	Switched socket outlet	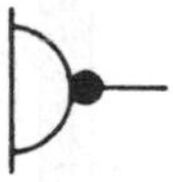 ◗—•
Boiler	□ B	Switch	 ⟋•
Sink	 [S •]	Two way switch	 ⟋•
Bath	 [•]	Pendant switch	 ⟋•
Wash basin	 [W•B]	Filament lamp	○
Shower unit	 [S]	Fluorescent lamp	 ▬○▬
Urinal	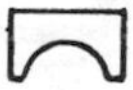 ⌂ stall  ⌓ bowl	Bed	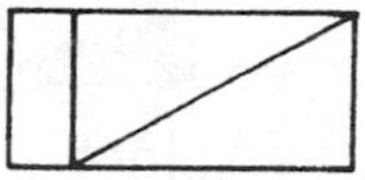 ▱
Water closet	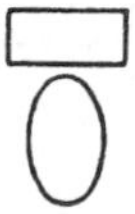 ▭ ⬭	Table and chairs	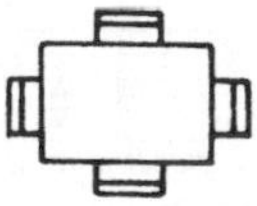 ⊞

TYPICAL COMPONENT, FITMENT AND ELECTRICAL SYMBOLS

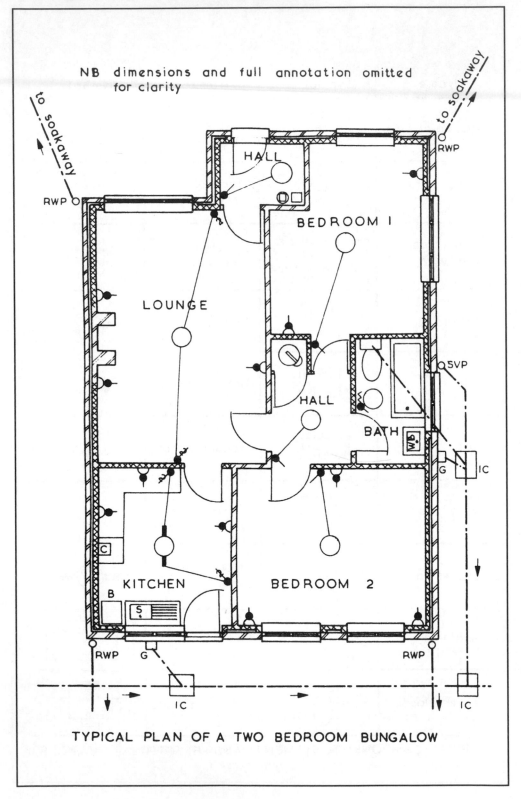

NB dimensions and full annotation omitted for clarity

TYPICAL PLAN OF A TWO BEDROOM BUNGALOW

Principal legislation: ~'

The Town & Country Planning Act 1990 – Effects control over volume of development, appearance and layout of buildings. The Public Health Acts 1936 to 1961 – Limits development with regard to emission of noise, pollution and public nuisance. The Highways Act 1980 – Determines layout and construction of roads and pavements.

The Building Act 1984 – Effects the Building Regulations 2000, which enforce minimum material and design standards. The Civic Amenities Act 1967 – Establishes conservation areas, providing local authorities with greater control of development. The Town & Country Amenities Act 1974 – Local authorities empowered to prevent demolition of buildings and tree felling.

Procedure: ~

Outline Planning Application – This is necessary for permission to develop a proposed site. The application should contain:
An application form describing the work.
A site plan showing adjacent roads and buildings (1:2500).
A block plan showing the plot, access and siting (1:500).
A certificate of land ownership.

Detail or Full Planning Application – This follows outline permission and is also used for proposed alterations to existing buildings.

It should contain: details of the proposal, to include trees, materials, drainage and any demolition.

Site and block plans (as above). A certificate of land ownership. Building drawings showing elevations, sections, plans, material specifications, access, landscaping, boundaries and relationship with adjacent properties (1:100).

Permitted Developments – Small developments may be exempt from formal application. These include house extensions <15% of the original volume (vol. in 1948 for older houses), <10% for terraced properties. Other exceptions include porches <2m^2 floor area, <3m in height and over 2m from the boundary.

Note: All developments are subject to Building Regulation approval.

Certificates of ownership – Article 7 of the Town & Country Planning (General Development Procedure) Order 1995:
Cert. A – States the applicant is sole site freeholder.
Cert. B – States the applicant is part freeholder or prospective purchaser and all owners of the site know of the application.
Cert. C – As Cert. B, but the applicant is only able to ascertain some of the other land owners.
Cert. D – As Cert. B, but the applicant cannot ascertain any owners of the site other than him/herself.

PLANNING APPLICATION

APPLICATION No

Use this form to apply for Planning Permission for:-
• an Extension   • a High Wall or Fence
• a Loft Conversion   • a Garage or Outbuilding
• a New or Altered Access • a Satellite Dish

Please return:- • 6 copies of the Form
                 • 6 copies of the Plans
                 • a Certificate under
                   Article 7
                 • the correct fee

DATE RECEIVED

1. NAME AND ADDRESS OF APPLICANT

Post Code _____

Tel. No. _____

2. NAME AND ADDRESS OF AGENT (If Used)

Post Code _____

Tel. No. _____

3. ADDRESS OF PROPERTY TO BE ALTERED OR EXTENDED

4. OWNERSHIP

Please indicate applicants interest in the property and complete the appropriate Certificate under Article 7.

Freeholder ☐   Other ☐

Leaseholder ☐

Purchaser ☐

5. BRIEF DESCRIPTION OF WORKS (include any demolition work)

6. DESCRIPTION OF EXTERNAL MATERIALS

7. ACCESS AND PARKING

Will your proposal affect? Please tick appropriate boxes

Vehicular Access   Yes ☐   No ☐
A Public Right of Way   Yes ☐   No ☐
Existing Parking   Yes ☐   No ☐

8. DRAINAGE

a. Please indicate method of Surface Water Disposal

b. Please indicate method of Foul Water Disposal
Please tick one box

Mains Sewer ☐   Septic Tank ☐

Cesspit ☐   Other ☐

9. TREES

Does the proposal involve the felling of any trees?

Please tick box   Yes ☐   No ☐

If yes, please show details on plans

10. PLEASE SIGN AND DATE THIS FORM BEFORE SUBMITTING

I/We hereby apply for Full Planning Permission for the development described above and shown on the accompanying plans.

Signed _____   Date _____

Date
On behalf of (if agent) _____

Use this form to apply for **Planning Permission for:-**
Outline Permission
Full Permission
Approval of Reserved Matters
Renewal of Temporary Permission
Change of Use

Please return:-
  * 6 copies of the Form
  * 6 copies of the Plans
  * a Certificate under
    Article 7
  * the correct fee

DATE RECEIVED

DATE VALID

1. NAME AND ADDRESS OF APPLICANT

_____
_____
_____
_____
_____

                    Post Code _____

Day Tel. No. _____ Fax No. _____

Email: _____

2. NAME AND ADDRESS OF AGENT (If Used)

_____
_____
_____
_____
_____

                    Post Code _____

Tel. No. _____ Fax No. _____

Email: _____

3. ADDRESS OR LOCATION OF LAND TO WHICH APPLICATION RELATES

_____
_____
_____

State Site Area _____ Hectares
This must be shown edged in Red on the site plan

4. OWNERSHIP

Please indicate applicants interest in the property and complete the appropriate Certificate under Article 7.

Freeholder ☐   Other ☐

Leaseholder ☐   Purchaser ☐

Any adjoining land owned or controlled and not part of application must be edged Blue on the site plan

5. WHAT ARE YOU APPLYING FOR?   Please tick one box and then answer relevant questions.

☐ **Outline Planning Permission**   Which of the following are to be considered?

☐ Siting ☐ Design ☐ Appearance ☐ Access ☐ Landscaping

☐ **Full Planning Permission/Change of use**

☐ **Approval of Reserved Matters following Outline Permission.**

O/P No. _____ Date_____ No. of Condition this application refers to: _____

☐ **Continuance of Use without complying with a condition of previous permission**

P/P No. _____ Date_____ No. of Condition this application relates to: _____

☐ **Permission for Retention of works.**

Date of Use of land or when buildings or works were constructed: _____ Length of temporary permission: _____

Is the use temporary or permanent? _____ No. of previous temporary permission if applicable: _____

6. BRIEF DESCRIPTION OF PROPOSED DEVELOPMENT

Please indicate the purpose for which the land or buildings are to be used. _____
_____
_____

7. NEW RESIDENTIAL DEVELOPMENTS. Please answer the following if appropriate:

What type of building is proposed? _____

No. of dwellings: _____ No. of storeys: _____ No. of Habitable rooms: _____

No. of Garages: _____ No. of Parking Spaces: _____ Total Grass Area of all buildings: _____

How will surface water be disposed of? _____

How will foul sewage be dealt with? _____

8. ACCESS

Does the proposed development involve any of the following? Please tick the appropriate boxes.

New access to a highway	☐	Pedestrian	☐	Vehicular
Alteration of an existing highway	☐	Pedestrian	☐	Vehicular
The felling of any trees	☐	Yes	☐	No

If you answer Yes to any of the above, they should be clearly indicated on all plans submitted.

9. BUILDING DETAIL

Please give details of all external materials to be used, if you are submitting them at this stage for approval.

_____

_____

List any samples that are being submitted for consideration. _____

_____

10. LISTED BUILDINGS OR CONSERVATION AREA

Are any Listed buildings to be demolished or altered? ☐ Yes ☐ No

If Yes, then Listed Building Consent will be required and a separate application should be submitted.

Are any non-listed buildings within a Conservation Area to be demolished? ☐ Yes ☐ No

If Yes, then Conservation Area consent will be required to demolish. Again, a separate application should be submitted.

11. NOTES

A special Planning Application Form should be completed for all applications involving Industrial, Warehousing, Storage, or Shopping development.
An appropriate Certificate must accompany this application unless you are seeking approval to Reserved Matters.
A separate application for Building Regulation approval is also required.
Separate applications may also be required if the proposals relate to a Listed Building or non-listed building within a Conservation Area.

12. PLEASE SIGN AND DATE THIS FORM BEFORE SUBMITTING
I/We hereby apply for Planning Permission for the development described above and shown on the accompanying plans.

Signed _____

TOWN AND COUNTRY PLANNING ACT

TOWN AND COUNTRY PLANNING (General Development Procedure) ORDER
Certificates under Article 7 of the Order

CERTIFICATE A   **For Freehold Owner (or his/her Agent)**

I hereby certify that:-

1.   No person other than the applicant was an owner of any part of the land to which the application relates at the beginning of the period of 21 days before the date of the accompanying application.

2.   ***Either (i)** None of the land to which the application relates constitutes or forms part of an agricultural holding:

       *or (ii) *(I have) (the applicant has) given the requisite notice to every person other than *(myself) (himself) (herself) who, 21 days before the date of the application, was a tenant of any agricultural holding any part of which was comprised in the land to which the application relates, viz:-

Name and Address of Tenant...

....

.... Signed ....,,,,,, ... Date.........................

Date of Service of Notice... *On Behalf of ....

CERTIFICATE B   **For Part Freehold Owner or Prospective Purchaser (or his/her Agent) able to ascertain all the owners of the land**

I hereby certify that:-

1.   *(I have) (the applicant has) given the requisite notice to all persons other than (myself) (the applicant) who, 21 days before the date of the accompanying application were owners of any part of the land to which the application relates, viz:-

Name and Address of Owner ....

....

.... Date of Service of Notice ....

2.   ***Either (i)** None of the land to which the application relates constitutes or forms part of an agricultural holding;

       *or (ii) *(I have) (the applicant has) given the requisite notice to every person other than *(myself) (himself) (herself) who, 21 days before the date of the application, was a tenant of any agricultural holding any part of which was comprised in the land to which the application relates, viz:-

Name and Address of Tenant...

....

.... Signed .... Date.........................

Modular Coordination ~ a module can be defined as a basic dimension which could for example form the basis of a planning grid in terms of multiples and submultiples of the standard module.

Typical Modular Coordinated Planning Grid ~

Let M = the standard module

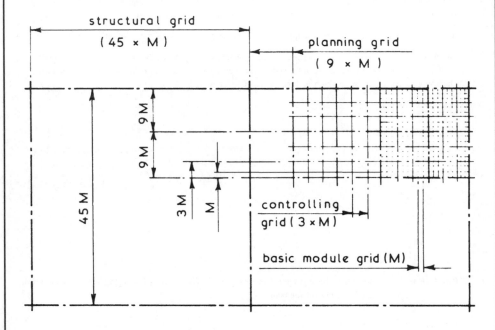

Structural Grid ~ used to locate structural components such as beams and columns.

Planning Grid ~ based on any convenient modular multiple for regulating space requirements such as rooms.

Controlling Grid ~ based on any convenient modular multiple for location of internal walls, partitions etc.

Basic Module Grid ~ used for detail location of components and fittings.

All the above grids, being based on a basic module, are contained one within the other and are therefore interrelated. These grids can be used in both the horizontal and vertical planes thus forming a three dimensional grid system. If a first preference numerical value is given to M dimensional coordination is established − see next page.

Ref. BS 6750: Specification for modular coordination in building.

Dimensional Coordination ~ the practical aims of this concept are to:-

1. Size components so as to avoid the wasteful process of cutting and fitting on site.
2. Obtain maximum economy in the production of components.
3. Reduce the need for the manufacture of special sizes.
4. Increase the effective choice of components by the promotion of interchangeability.

BS 6750 specifies the increments of size for coordinating dimensions of building components thus:-

Preference	1st	2nd	3rd	4th
Size (mm)	300	100	50	25

the 3rd and 4th preferences having a maximum of 300mm

Dimensional Grids – the modular grid network as shown on page 46 defines the space into which dimensionally coordinated components must fit. An important factor is that the component must always be undersized to allow for the joint which is sized by the obtainable degree of tolerance and site assembly:-

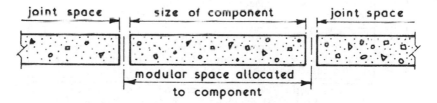

Controlling Lines, Zones and Controlling Dimensions – these terms can best be defined by example:-

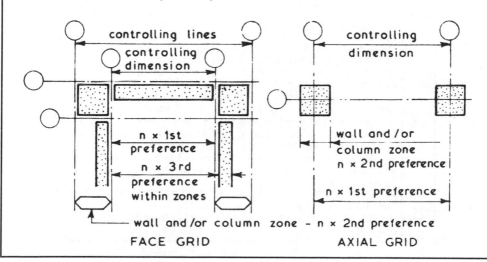

FACE GRID   AXIAL GRID

Construction Regulations ~ these are Statutory Instruments made under the Factories Acts of 1937 and 1961 and come under the umbrella of the Health and Safety at Work etc., Act 1974. They set out the minimum legal requirements for construction works and relate primarily to the health, safety and welfare of the work force. The requirements contained within these documents must therefore be taken into account when planning construction operations and during the actual construction period. Reference should be made to the relevant document for specific requirements but the broad areas covered can be shown thus:-

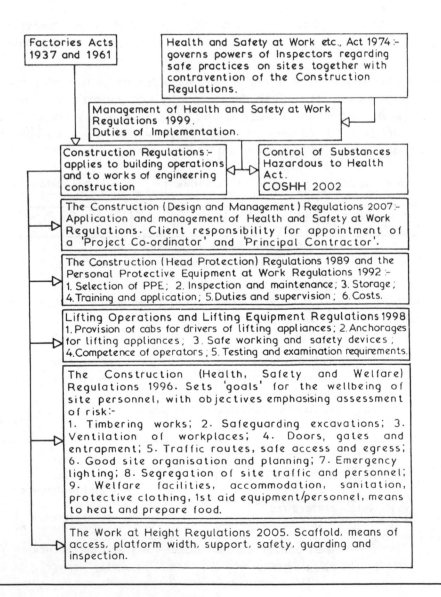

Factories Acts 1937 and 1961

Health and Safety at Work etc., Act 1974 :- governs powers of Inspectors regarding safe practices on sites together with contravention of the Construction Regulations.

Management of Health and Safety at Work Regulations 1999. Duties of Implementation.

Construction Regulations :- applies to building operations and to works of engineering construction

Control of Substances Hazardous to Health Act. COSHH 2002

The Construction (Design and Management) Regulations 2007 :- Application and management of Health and Safety at Work Regulations. Client responsibility for appointment of a 'Project Co-ordinator' and 'Principal Contractor'.

The Construction (Head Protection) Regulations 1989 and the Personal Protective Equipment at Work Regulations 1992 :- 1. Selection of PPE; 2. Inspection and maintenance; 3. Storage; 4. Training and application; 5. Duties and supervision; 6. Costs.

Lifting Operations and Lifting Equipment Regulations 1998 1. Provision of cabs for drivers of lifting appliances; 2. Anchorages for lifting appliances; 3. Safe working and safety devices; 4. Competence of operators; 5. Testing and examination requirements.

The Construction (Health, Safety and Welfare) Regulations 1996. Sets 'goals' for the wellbeing of site personnel, with objectives emphasising assessment of risk:-
1. Timbering works; 2. Safeguarding excavations; 3. Ventilation of workplaces; 4. Doors, gates and entrapment; 5. Traffic routes, safe access and egress; 6. Good site organisation and planning; 7. Emergency lighting; 8. Segregation of site traffic and personnel; 9. Welfare facilities, accommodation, sanitation, protective clothing, 1st aid equipment/personnel, means to heat and prepare food.

The Work at Height Regulations 2005. Scaffold, means of access, platform width, support, safety, guarding and inspection.

Objective – To create an all-party integrated and planned approach to health and safety throughout the duration of a construction project.

Administering Body – The Health and Safety Executive (HSE).

Scope – The CDM Regulations are intended to embrace all aspects of construction, with the exception of very minor works.

Responsibilities – The CDM Regulations apportion responsibility to everyone involved in a project to cooperate with others and for health and safety issues to all parties involved in the construction process, i.e. client, designer, project coordinator and principal contractor.

Client – Appoints a project coordinator and the principal contractor. Provides the project coordinator with information on health and safety matters and ensures that the principal contractor has prepared an acceptable construction phase plan for the conduct of work. Ensures adequate provision for welfare and that a health and safety file is available.

Designer – Establishes that the client is aware of their duties. Considers the design implications with regard to health and safety issues, including an assessment of any perceived risks. Coordinates the work of the project coordinator and other members of the design team.

Project Coordinator – Ensures that:
* a pre-tender, construction phase plan is prepared.
* the HSE are informed of the work.
* designers are liaising and conforming with their health and safety obligations.
* a health and safety file is prepared.
* contractors are of adequate competence with regard to health and safety matters and advises the client and principal contractor accordingly.

Principal Contractor – Develops a construction phase plan, collates relevant information and maintains it as the work proceeds. Administers day-to-day health and safety issues. Co-operates with the project coordinator, designers and site operatives preparing risk assessments as required.

Note: The CDM Regulations include requirements defined under The Construction (Health, Safety and Welfare) Regulations.

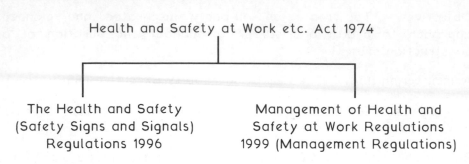

Under these regulations, employers are required to provide and maintain health and safety signs conforming to European Directive 92/58 EEC:

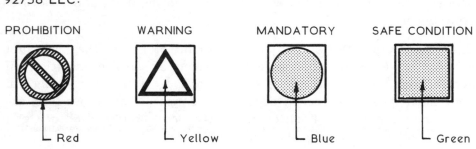

In addition, employers obligations include the need to provide:

Risk Assessment – provide and maintain safety signs where there is a risk to health and safety, eg. obstacles. Train staff to comprehend safety signs.

Pictorial Symbols – pictograms alone are acceptable but supplementary text, eg. FIRE EXIT, is recommended.

Fire/Emergency Escape Signs – conformity to BS 5499-1:2002. A green square or rectangular symbol.

Positioning of signs – primarily for location of fire exits, fire equipment, alarms, assembly points, etc. Not to be located where they could be obscured.

Marking of Hazardous Areas – to identify designated areas for storing dangerous substances: Dangerous Substances (Notification and Marking of Sites) Regulations 1990. Yellow triangular symbol.

Pipeline Identification – pipes conveying dangerous substances to be labelled with a pictogram on a coloured background conforming to BS 1710:1984 and BS 4800:1989. Non-dangerous substances should also be labelled for easy identification.

Typical Examples on Building Sites ~

PROHIBITION (Red)

| Authorised personnel only | Children must not play on this site | Smoking prohibited | Access not permitted |

WARNING (Yellow)

| Dangerous substance | Flammable liquid | Danger of electric shock | Compressed gas |

MANDATORY (Blue)

| Safety helmets must be worn | Protective footwear must be worn | Use ear protectors | Protective clothing must be worn |

SAFE CONDITIONS (Green)

| Emergency escapes | Treatment area | Safe area |

The Building Regulations ~ this is a Statutory Instrument which sets out the minimum performance standards for the design and construction of buildings and where applicable to the extension of buildings. The regulations are supported by other documents which generally give guidance on how to achieve the required performance standards. The relationship of these and other documents is set out below:-

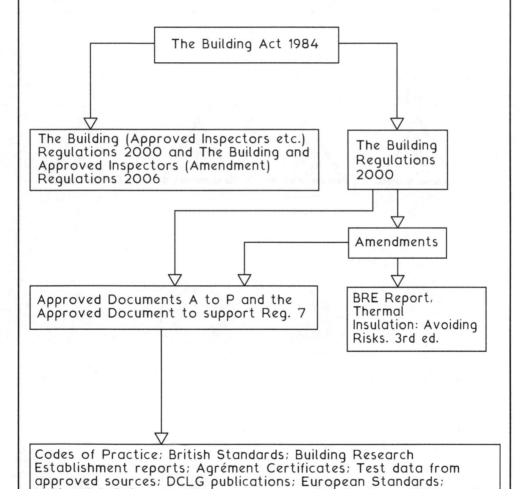

NB. The Building Regulations apply to England and Wales but not to Scotland and Northern Ireland which have separate systems of control.

Approved Documents ~ these are non-statutory publications supporting the Building Regulations prepared by the Department for Communities and Local Government approved by the Secretary of State and issued by The Stationery Office. The Approved Documents (ADs) have been compiled to give practical guidance to comply with the performance standards set out in the various regulations. They are not mandatory but in the event of a dispute they will be seen as tending to show compliance with the requirements of the Building Regulations. If other solutions are used to satisfy the requirements of the Regulations the burden of proving compliance rests with the applicant or designer.

Approved Document A — STRUCTURE

Approved Document B — FIRE SAFETY
Volume 1 — Dwelling houses
Volume 2 — Buildings other than dwelling houses

Approved Document C — SITE PREPARATION AND RESISTANCE TO CONTAMINANTS AND MOISTURE

Approved Document D — TOXIC SUBSTANCES

Approved Document E — RESISTANCE TO THE PASSAGE OF SOUND

Approved Document F — VENTILATION

Approved Document G — HYGIENE

Approved Document H — DRAINAGE AND WASTE DISPOSAL

Approved Document J — COMBUSTION APPLIANCES AND FUEL STORAGE SYSTEMS

Approved Document K — PROTECTION FROM FALLING, COLLISION AND IMPACT

Approved Document L — CONSERVATION OF FUEL AND POWER
L1A — New dwellings
L1B — Existing dwellings
L2A — New buildings other than dwellings
L2B — Existing buildings other than dwellings

Approved Document M — ACCESS TO AND USE OF BUILDINGS

Approved Document N — GLAZING — SAFETY IN RELATION TO IMPACT, OPENING AND CLEANING

Approved Document P — ELECTRICAL SAFETY

Approved Document to support Regulation 7 MATERIALS AND WORKMANSHIP

Example in the Use of Approved Documents

Problem:- the sizing of suspended upper floor joists to be spaced at 400mm centres with a clear span of 3·600m for use in a two storey domestic dwelling.

Building Regulation A1:- states that the building shall be constructed so that the combined dead, imposed and wind loads are sustained and transmitted by it to the ground –

(a) safely, and

(b) without causing such deflection or deformation of any part of the building, or such movement of the ground, as will impair the stability of any part of another building.

Approved Document A:- guidance on sizing floor joists can be found in 'Span Tables for Solid Timber Members in Dwellings', published by the Timber Research And Development Association (TRADA), and BS5268-2: Structural use of timber. Code of practice for permissible stress design, materials and workmanship.

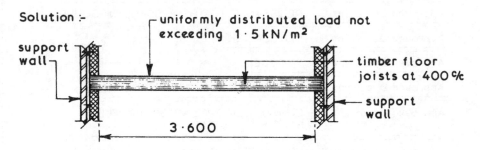

Dead load (kN/m²) supported by joist excluding mass of joist:-

Floor finish — carpet	— 0·03	weights of
Flooring — 20 mm thick particle board	— 0·15	materials
Ceiling — 9·5 mm thick plasterboard	0·08	from BS648
Ceiling finish — 3 mm thick plaster	— 0·04	
total dead load —	0·30 kN/m³	

Dead loading is therefore in the 0·25 to 0·50kN/m² band
From table on page xxx suitable joist sizes are:- 38 × 200, 50 × 175, 63 × 175 and 75 × 150.

Final choice of section to be used will depend upon cost; availability; practical considerations and/or personal preference.

Building Control ~ unless the applicant has opted for control by a private approved inspector under The Building (Approved Inspectors etc.) Regulations 2000 the control of building works in the context of the Building Regulations is vested in the Local Authority. There are two systems of control namely the Building Notice and the Deposit of Plans. The sequence of systems is shown below:-

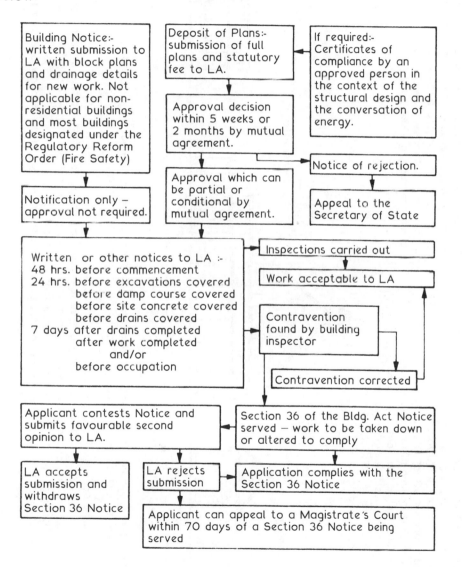

NB. In some stages of the above sequence statutory fees are payable as set out in The Building (Local Authority Charges) Regulations 1998.

small detached buildings:

1. floor area < 15 m² not containing sleeping accommodation, or
2. floor area < 30 m² not containing sleeping accommodation, and either:
   • constructed substantially from non-combustible materials, or
   • located in excess of 1 m from the boundary

The guidance shown indicates the categories of buildings that do not normally require submission of a Building Notice or Deposit of Plans for approval by the Building Control Section of the Local Authority. However, they may still require planning permission – see page 41

greenhouse, unless for commercial use, ie. retailing, packing or exhibiting

conservatory*

carport*

open sides

porch*

boundary

see note 2 above

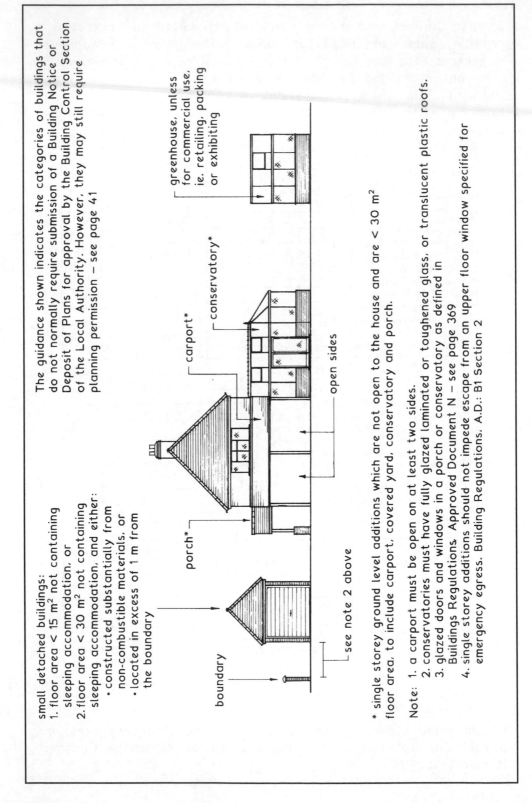

* single storey ground level additions which are not open to the house and are < 30 m² floor area, to include carport, covered yard, conservatory and porch.

Note:
1. a carport must be open on at least two sides.
2. conservatories must have fully glazed laminated or toughened glass, or translucent plastic roofs.
3. glazed doors and windows in a porch or conservatory as defined in Buildings Regulations, Approved Document N – see page 369
4. single storey additions should not impede escape from an upper floor window specified for emergency egress. Building Regulations, A.D.: B1 Section 2

BUILDING REGULATIONS APPLICATION

Use this form to give notice of intention to erect, extend, or alter a building, install fittings or make a material change of use of the building.	Unless specified differently overleaf, Please return:- • 2 copies of the Form • 4 copies of the Plans • the correct fee

APPLICATION No

DATE RECEIVED

1. NAME AND ADDRESS OF APPLICANT
Applicant will be invoiced on commencement of work.

Post Code _____

Tel. No. _____

2. NAME AND ADDRESS OF AGENT (If Used)

Post Code _____

Tel. No. _____

3. ADDRESS OR LOCATION OF PROPOSED WORK

4. DESCRIPTION OF PROPOSED WORKS

5. IF NEW BUILDING OR EXTENSION PLEASE STATE PROPOSED USE

6. IF EXISTING BUILDING PLEASE STATE PRESENT USE

7. DRAINAGE

Please state means of:-

Water Supply _____

Foul Water Disposal _____

Storm Water Disposal _____

8. CONDITIONS

Do you consent to the Plans being passed subject to conditions where appropriate?   Yes ☐   No ☐

Do you agree to an extension of time if this is required by the Council?   Yes ☐   No ☐

9. COMPLETION CERTIFICATE

Do you wish the Council to issue a Completion Certificate upon satisfactory completion of the work?

Yes ☐   No ☐

10. REGULATORY REFORM ORDER (Fire Safety) 2005

Is the building intended for any other purpose than occupation as a domestic living unit by one family group?

Yes ☐   No ☐

11. FEE
Please state estimated cost of the work (at current market value) £........................ Amount of Fee submitted £........................

Has Planning Permission been sought?   Yes ☐   No ☐   If Yes, please give Application No _____

12. PLEASE SIGN AND DATE THIS FORM BEFORE SUBMITTING

I/We hereby give notice of intention to carry out the work set out above and deposit the attached drawings and documents in accordance with the requirements of Regulations 11 (1) (b). Also enclosed is the appropriate Plan Fee and I understand that a further Fee will be payable when the first inspection of work on site is made by the Local Authority.

Signed _____ Date _____ On behalf of (if agent) _____

Published ~ 2006 by the Department for Communities and Local Government (DCLG) in response to the damaging effects of climate change. The code promotes awareness and need for new energy conservation initiatives in the design of new dwellings.

Objective ~ to significantly reduce the 27% of UK CO^2 emissions that are produced by 21 million homes. This is to be a gradual process, with the target of reducing CO^2 emissions from all UK sources by 60% by 2050.

Sustainability ~ measured in terms of a quality standard designed to provide new homes with a factor of environmental performance. This measure is applied primarily to categories of thermal energy, use of water, material resources, surface water run-off and management of waste.

Measurement ~ a 'green' star rating that indicates environmental performance ranging from one to six stars. Shown below is the star rating criteria applied specifically to use of thermal energy. A home with a six star rating is also regarded as a zero carbon home.

Proposed Progression ~

Percentage Improvement compared with AD L 2006	Year	Star rating
10	–	1
18	–	2
25	2010	3
44	2013	4
100	2016	5 and 6

Zero Carbon Home ~ zero net emissions of CO^2 from all energy use in the home. This incorporates insulation of the building fabric, heating equipment, hot water systems, cooling, washing appliances, lighting and other electrical/electronic facilities. Net zero emissions can be measured by comparing the carbon emissions produced in consuming on- or off-site fossil fuel energy use in the home, with the amount of on-site renewable energy produced. Means for producing low or zero carbon energy include micro combined heat and power units, photovoltaic (solar) panels, wind generators and ground energy heat pumps, (see Building Services Handbook).

British Standards ~ these are publications issued by the British Standards Institution which give recommended minimum standards for materials, components, design and construction practices. These recommendations are not legally enforceable but some of the Building Regulations refer directly to specific British Standards and accept them as deemed to satisfy provisions. All materials and components complying with a particular British Standards are marked with the British Standards kitemark thus:- together with the appropriate BS number.

This symbol assures the user that the product so marked has been produced and tested in accordance with the recommendations set out in that specific standard. Full details of BS products and services can be obtained from, Customer Services, BSI, 389 Chiswick High Road, London, W4 4AL. Standards applicable to building may be purchased individually or in modules, GBM 48, 49 and 50; Construction in General, Building Materials and Components and Building Installations and Finishing, respectively. British Standards are constantly under review and are amended, revised and rewritten as necessary, therefore a check should always be made to ensure that any standard being used is the current issue. There are over 1500 British Standards which are directly related to the construction industry and these are prepared in four formats:-

1. British Standards – these give recommendations for the minimum standard of quality and testing for materials and components. Each standard number is prefixed BS.

2. Codes of Practice – these give recommendations for good practice relative to design, manufacture, construction, installation and maintenance with the main objectives of safety, quality, economy and fitness for the intended purpose. Each code of practice number is prefixed CP or BS.

3. Draft for Development – these are issued instead of a British Standard or Code of Practice when there is insufficient data or information to make firm or positive recommendations. Each draft number is prefixed DD. Sometimes given a BS number and suffixed DC, ie. Draft for public Comment.

4. Published Document – these are publications which cannot be placed into any one of the above categories. Each published document is numbered and prefixed PD.

European Standards – since joining the European Union (EU), trade and tariff barriers have been lifted. This has opened up the market for manufacturers of construction-related products, from all EU and European Economic Area (EEA) member states. Before 2004, the EU was composed of 15 countries: Austria, Belgium, Denmark, Finland, France, Germany, Greece, Ireland, Italy, Luxemburg, Netherlands, Portugal, Spain, Sweden and the United Kingdom. It now includes Bulgaria, Cyprus, the Czech Republic, Estonia, Hungary, Latvia, Lithuania, Malta, Poland, Romania, Slovakia and Slovenia. The EEA extends to: Iceland, Liechtenstein and Norway. Nevertheless, the wider market is not so easily satisfied, as regional variations exist. This can create difficulties where product dimensions and performance standards differ. For example, thermal insulation standards for masonry walls in Mediterranean regions need not be the same as those in the UK. Also, preferred dimensions differ across Europe in items such as bricks, timber, tiles and pipes.

European Standards are prepared under the auspices of Comité Européen de Normalisation (CEN), of which the BSI is a member. European Standards that the BSI have not recognised or adopted, are prefixed EN. These are EuroNorms and will need revision for national acceptance.

For the time being, British Standards will continue and where similarity with other countries' standards and ENs can be identified, they will run side by side until harmonisation is complete and approved by CEN.

eg. BS EN 295, complements the previous national standard:
    BS 65 – Vitrefied clay pipes . . . . . for drains and sewers.

European Pre-standards are similar to BS Drafts for Development. These are known as ENVs.

Some products which satisfy the European requirements for safety, durability and energy efficiency, carry the CE mark. This is not to be assumed a mark of performance and is not intended to show equivalence to the BS kitemark. However, the BSI is recognised as a Notified Body by the EU and as such is authorised to provide testing and certification in support of the CE mark.

International Standards – these are prepared by the International Organisation for Standardisation and are prefixed ISO. Many are compatible with and complement BSs, e.g. the ISO 9000 Quality Management series and BS 5750: Quality systems.

For manufacturers' products to be compatible and uniformly acceptable in the European market, there exists a process for harmonising technical specifications. These specifications are known as harmonised European product standards (hENs), produced and administered by the Comité Européen de Normalisation (CEN). European Technical Approvals (ETAs) are also acceptable where issued by the European Organisation for Technical Approvals (EOTA). These standards are not a harmonisation of regulations. Whether or not the technical specification satisfies regional and national regulations is for local determination. However, for commercial purposes a technical specification should cover the performance characteristics required by regulations established by any member state in the European Economic Area (EEA).

CPD harmonises:
* methods and criteria for testing
* methods for declaring product performance
* methods and measures of conformity assessment

UK attestation accredited bodies include: BBA, BRE and BSI.

CE mark – a marking or labelling for conforming products. A 'passport' permitting a product to be legally marketed in any EEA. It is not a quality mark, e.g. BS Kitemark, but where appropriate this may appear with the CE marking.

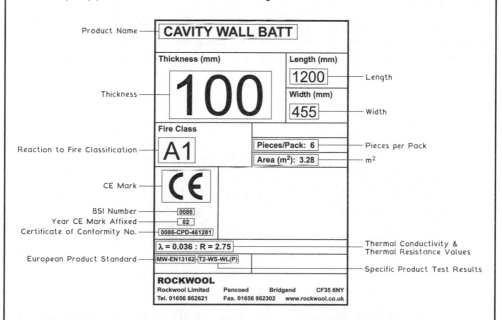

CE marking – reproduced with kind permission of Rockwool Ltd.

Building Research Establishment ~ The BRE was founded as a UK Government agency in 1921 and was known until the early 1970s as the Building Research Station.

In addition to UK Government funding, some financial support is now provided by the European Union. Additional funding is derived from a variety of sources, including commercial services for private industry and from publications. The latter includes the BRE's well known regular issue of research information products, i.e. Digests, Information Papers, Good Building Guides and Good Repair Guides.

UK Government support is principally through the Department of Trade and Industry (DTI) and the Department for Communities and Local Government (DCLG). The DCLG works with the BRE in formulating specific aspects of the Approved Documents to the Building Regulations. Commissioned research is funded by BRE Trust.

The BRE incorporates and works with other specialised research and material testing organisations, e.g. see LPCB, below. It is accredited under the United Kingdom Accreditation Service (UKAS) as a testing laboratory authorised to issue approvals and certifications such as CE product marking (see pages 60 and 61). Certification of products, materials and applications is effected through BRE Certification Ltd.

Loss Prevention Certification Board (LPCB) ~ The origins of this organisation date back to the latter part of the 19th century, when it was established by a group of building insurers as the Fire Offices' Committee (FOC).

Through a subdivision known as the Loss Prevention Council (LPC), the FOC produced a number of technical papers and specifications relating to standards of building construction and fire control installations. These became the industry standards that were, and continue to be, frequently used by building insurers as supplementary to local byelaws and latterly the Building Regulation Approved Documents.

In the late 1980s the LPC was renamed as the LPCB as a result of reorganisation within the insurance profession. At this time the former LPC guidance documents became established in the current format of Loss Prevention Standards.

In 2000 the LCPB became part of the BRE and now publishes its Standards under BRE Certification Ltd.

CPI System of Coding ~ the Co-ordinated Project Information initiative originated in the 1970s in response to the need to establish a common arrangement of document and language communication, across the varied trades and professions of the construction industry.

However, it has only been effective in recent years with the publication of the Standard Method of Measurement 7th edition (SMM 7), the National Building Specification (NBS) and the Drawings Code. (Note: The NBS is also produced in CI/SfB format.)

The arrangement in all documents is a coordination of alphabetic sections, corresponding to elements of work, the purpose being to avoid mistakes, omissions and other errors which have in the past occurred between drawings, specification and bill of quantities descriptions.

The coding is a combination of letters and numbers, spanning 3 levels:-

Level 1 has 24 headings from A to Z (omitting I and O). Each heading relates to part of the construction process, such as groundwork (D), Joinery (L), surface finishes (M), etc.

Level 2 is a sub-heading, which in turn is sub-grouped numerically into different categories. So for example, Surface Finishes is sub-headed; Plaster, Screeds, Painting, etc. These sub-headings are then extended further, thus Plaster becomes; Plastered/Rendered Coatings, Insulated Finishes, Sprayed Coatings etc.

Level 3 is the work section sub-grouped from level 2, to include a summary of inclusions and omissions.

As an example, an item of work coded M21 signifies:-

> M – Surface finishes
>
> 2 – Plastered coatings
>
> 1 – Insulation with rendered finish

The coding may be used to:-
(a) simplify specification writing
(b) reduce annotation on drawings
(c) rationalise traditional taking-off methods

CI/SfB System ~ this is a coded filing system for the classification and storing of building information and data. It was created in Sweden under the title of Samarbetskommittën för Byggnadsfrågor and was introduced into this country in 1961 by the RIBA. In 1968 the CI (Construction Index) was added to the system which is used nationally and recognised throughout the construction industry. The system consists of 5 sections called tables which are subdivided by a series of letters or numbers and these are listed in the CI/SfB index book to which reference should always be made in the first instance to enable an item to be correctly filed or retrieved.

Table 0 – Physical Environment

This table contains ten sections 0 to 9 and deals mainly with the end product (i.e. the type of building.) Each section can be further subdivided (e.g. 21, 22, et seq.) as required.

Table 1 – Elements

This table contains ten sections numbered (--) to (9-) and covers all parts of the structure such as walls, floors and services. Each section can be further subdivided (e.g. 31, 32 et seq.) as required.

Table 2 – Construction Form

This table contains twenty five sections lettered A to Z (O being omitted) and covers construction forms such as excavation work, blockwork, cast in-situ work etc., and is not subdivided but used in conjunction with Table 3.

Table 3 – Materials

This table contains twenty five sections lettered a to z (l being omitted) and covers the actual materials used in the construction form such as metal, timber, glass etc., and can be subdivided (e.g. n1, n2 et seq.) as required.

Table 4 – Activities and Requirements

This table contains twenty five sections lettered (A) to (Z), (O being omitted) and covers anything which results from the building process such as shape, heat, sound, etc. Each section can be further subdivided ((M1), (M2) et seq.) as required.

2 SITE WORKS

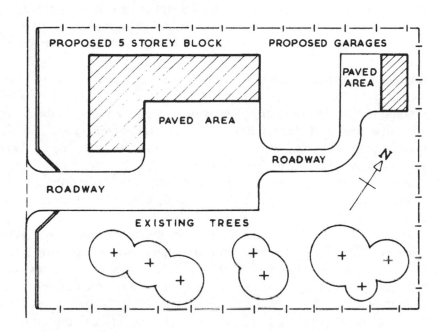

PROPOSED 5 STOREY BLOCK   PROPOSED GARAGES

PAVED AREA

PAVED AREA

ROADWAY

ROADWAY

N

EXISTING TREES

SITE INVESTIGATIONS
SOIL INVESTIGATION
SOIL ASSESSMENT AND TESTING
SITE LAYOUT CONSIDERATIONS
SITE SECURITY
SITE LIGHTING AND ELECTRICAL SUPPLY
SITE OFFICE ACCOMMODATION
MATERIALS STORAGE
MATERIALS TESTING
SETTING OUT
LEVELS AND ANGLES
ROAD CONSTRUCTION
TUBULAR SCAFFOLDING AND SCAFFOLDING SYSTEMS
SHORING SYSTEMS
DEMOLITION

Site Analysis – prior to purchasing a building site it is essential to conduct a thorough survey to ascertain whether the site characteristics suit the development concept. The following guidance forms a basic checklist:

* Refer to Ordnance Survey maps to determine adjacent features, location, roads, facilities, footpaths and rights of way.
* Conduct a measurement survey to establish site dimensions and levels.
* Observe surface characteristics, i.e. trees, steep slopes, existing buildings, rock outcrops, wells.
* Inquire of local authority whether preservation orders affect the site and if it forms part of a conservation area.
* Investigate subsoil. Use trial holes and borings to determine soil quality and water table level.
* Consider flood potential, possibilities for drainage of water table, capping of springs, filling of ponds, diversion of streams and rivers.
* Consult local utilities providers for underground and overhead services, proximity to site and whether they cross the site.
* Note suspicious factors such as filled ground, cracks in the ground, subsidence due to mining and any cracks in existing buildings.
* Regard neighbourhood scale and character of buildings with respect to proposed new development.
* Decide on best location for building (if space permits) with regard to 'cut and fill', land slope, exposure to sun and prevailing conditions, practical use and access.

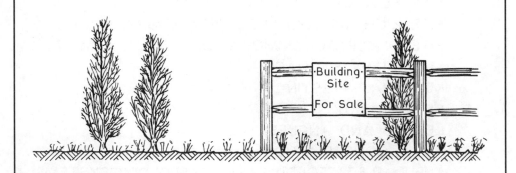

See also, desk and field studies on page 68

Site Investigation For New Works ~ the basic objective of this form of site investigation is to collect systematically and record all the necessary data which will be needed or will help in the design and construction processes of the proposed work. The collected data should be presented in the form of fully annotated and dimensioned plans and sections. Anything on adjacent sites which may affect the proposed works or conversely anything appertaining to the proposed works which may affect an adjacent site should also be recorded.

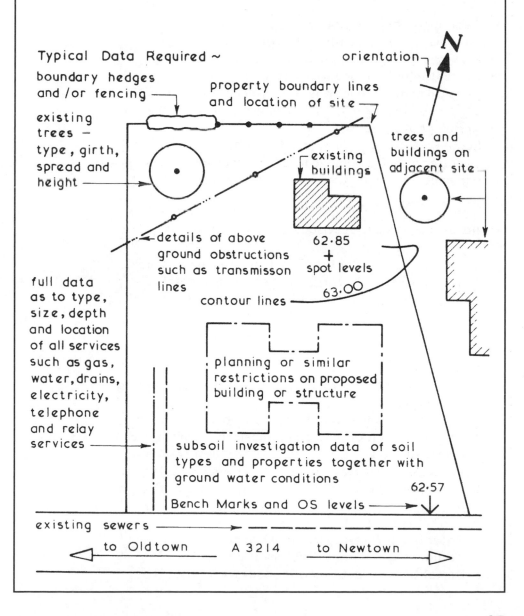

Typical Data Required ~

boundary hedges and /or fencing

property boundary lines and location of site

orientation

N

existing trees — type, girth, spread and height

trees and buildings on adjacent site

existing buildings

details of above ground obstructions such as transmisson lines

62·85
+
spot levels

contour lines

63·00

full data as to type, size, depth and location of all services such as gas, water, drains, electricity, telephone and relay services

planning or similar restrictions on proposed building or structure

subsoil investigation data of soil types and properties together with ground water conditions

62·57

Bench Marks and OS levels

existing sewers

to Oldtown   A 3214   to Newtown

Procedures ~

1. Desk study
2. Field study or walk-over survey
3. Laboratory analysis (see pages 77–78 and 81–83)

Desk Study ~ collection of known data, to include:

- Ordnance Survey maps – historical and modern, note grid reference.
- Geological maps – subsoil types, radon risk.
- Site history – green-field/brown-field.
- Previous planning applications/approvals.
- Current planning applications in the area.
- Development restrictions – conservation orders.
- Utilities – location of services on and near the site.
- Aerial photographs.
- Ecology factors – protected wildlife.
- Local knowledge – anecdotal information/rights of way.
- Proximity of local land fill sites – methane risk.

Field Study ~ intrusive visual and physical activity to:

- Establish site characteristics from the desk study.
- Assess potential hazards to health and safety.
- Appraise surface conditions:
  * Trees – preservation orders.
  * Topography and geomorphological mapping.
- Appraise ground conditions:
  * Water table.
  * Flood potential – local water courses and springs.
  * Soil types.
  * Contamination – vegetation die-back.
  * Engineering risks – ground subsidence, mining, old fuel tanks.
  * Financial risks – potential for the unforeseen.
- Take subsoil samples and conduct in-situ tests.
- Consider the need for subsoil exploration, trial pits and bore holes.
- Appraise existing structures:
  * Potential for re-use/refurbishment.
  * Archaeological value/preservation orders.
  * Demolition – costs, health issues e.g. asbestos.

Purpose ~ primarily to obtain subsoil samples for identification, classification and ascertaining the subsoil's characteristics and properties. Trial pits and augered holes may also be used to establish the presence of any geological faults and the upper or lower limits of the water table.

Typical Details ~

minimum plan size to provide access for operatives 1·200 x 1·200

diameter range 50 to 150mm

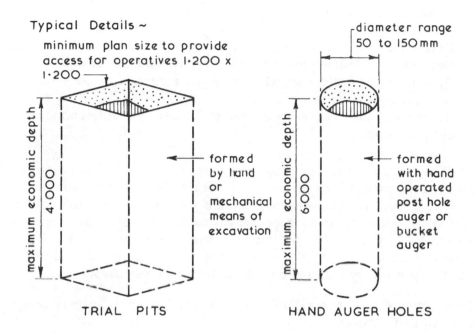

maximum economic depth 4·000

formed by hand or mechanical means of excavation

maximum economic depth 6·000

formed with hand operated post hole auger or bucket auger

TRIAL PITS

HAND AUGER HOLES

General use ~

dry ground which requires little or no temporary support to sides of excavation.

Subsidiary use~
to expose and/or locate underground services.

Advantages ~
subsoil can be visually examined in-situ — both disturbed and undisturbed samples can be obtained.

General use ~

dry ground but liner tubes could be used if required to extract subsoil samples at a depth beyond the economic limit of trial holes.

Advantages ~
generally a cheaper and simpler method of obtaining subsoil samples than the trial pit method.

Trial pits and holes should be sited so that the subsoil samples will be representative but not interfering with works.

Site Investigation ~ this is an all embracing term covering every aspect of the site under investigation.

Soil Investigation ~ specifically related to the subsoil beneath the site under investigation and could be part of or separate from the site investigation.

Purpose of Soil Investigation ~

1. Determine the suitability of the site for the proposed project.
2. Determine an adequate and economic foundation design.
3. Determine the difficulties which may arise during the construction process and period.
4. Determine the occurrence and/or cause of all changes in subsoil conditions.

The above purposes can usually be assessed by establishing the physical, chemical and general characteristics of the subsoil by obtaining subsoil samples which should be taken from positions on the site which are truly representative of the area but are not taken from the actual position of the proposed foundations. A series of samples extracted at the intersection points of a 20·000 square grid pattern should be adequate for most cases.

Soil Samples ~ these can be obtained as disturbed or as undisturbed samples.

Disturbed Soil Samples ~ these are soil samples obtained from bore holes and trial pits. The method of extraction disturbs the natural structure of the subsoil but such samples are suitable for visual grading, establishing the moisture content and some laboratory tests. Disturbed soil samples should be stored in labelled airtight jars.

Undisturbed Soil Samples ~ these are soil samples obtained using coring tools which preserve the natural structure and properties of the subsoil. The extracted undisturbed soil samples are labelled and laid in wooden boxes for dispatch to a laboratory for testing. This method of obtaining soil samples is suitable for rock and clay subsoils but difficulties can be experienced in trying to obtain undisturbed soil samples in other types of subsoil.

The test results of soil samples are usually shown on a drawing which gives the location of each sample and the test results in the form of a hatched legend or section.

Depth of Soil Investigation ~ before determining the actual method of obtaining the required subsoil samples the depth to which the soil investigation should be carried out must be established. This is usually based on the following factors –

1. Proposed foundation type.
2. Pressure bulb of proposed foundation.
3. Relationship of proposed foundation to other foundations.

Typical Examples ~

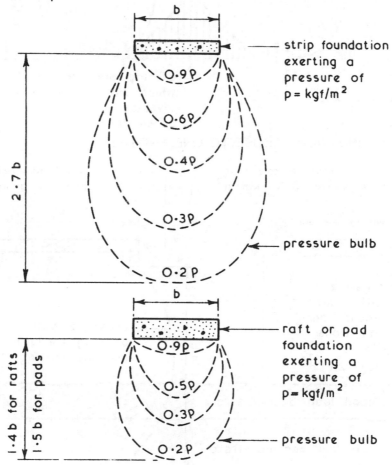

Pressure bulbs of less than 20% of original loading at foundation level can be ignored – this applies to all foundation types.

For further examples see next page.

Typical Examples of Depth of Soil Investigation Considerations ~

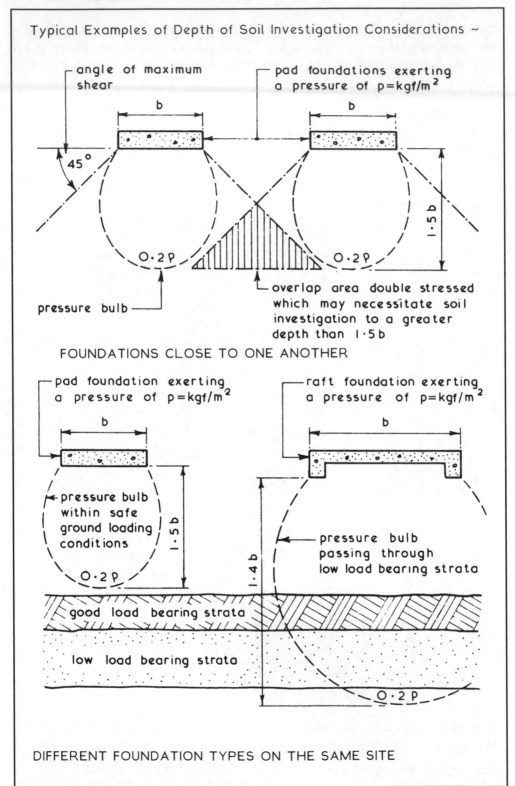

angle of maximum shear

pad foundations exerting a pressure of $p = kgf/m^2$

b

b

$45°$

$1.5 b$

O·2P

O·2P

pressure bulb

overlap area double stressed which may necessitate soil investigation to a greater depth than $1.5 b$

FOUNDATIONS CLOSE TO ONE ANOTHER

pad foundation exerting a pressure of $p = kgf/m^2$

raft foundation exerting a pressure of $p = kgf/m^2$

b

b

pressure bulb within safe ground loading conditions

$1.5 b$

O·2P

$1.4 b$

pressure bulb passing through low load bearing strata

good load bearing strata

low load bearing strata

O·2P

DIFFERENT FOUNDATION TYPES ON THE SAME SITE

Soil Investigation Methods ~ method chosen will depend on several factors -

1. Size of contract.
2. Type of proposed foundation.
3. Type of sample required.
4. Type of subsoils which may be encountered.

As a general guide the most suitable methods in terms of investigation depth are -

1. Foundations up to   3·000   deep - trial pits.
2. Foundations up to 30·000   deep - borings.
3. Foundations over 30·000 deep - deep   borings   and   in-situ examinations from tunnels and/or deep pits.

Typical Trail Pit Details ~

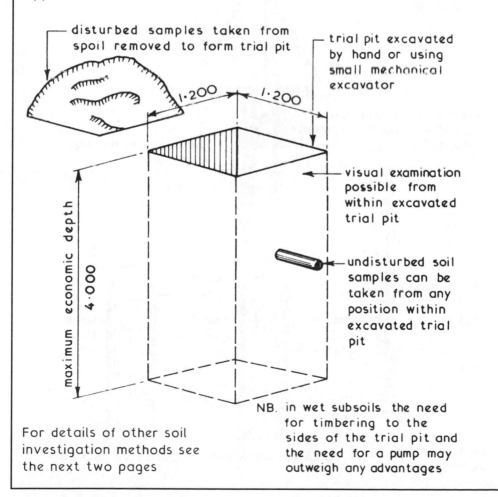

disturbed samples taken from spoil removed to form trial pit

trial pit excavated by hand or using small mechanical excavator

1·200   1·200

maximum economic depth   4·000

visual examination possible from within excavated trial pit

undisturbed soil samples can be taken from any position within excavated trial pit

NB. in wet subsoils the need for timbering to the sides of the trial pit and the need for a pump may outweigh any advantages

For details of other soil investigation methods see the next two pages

Boring Methods to Obtain Disturbed Soil Samples ~

1. Hand or Mechanical Auger – suitable for depths up to 3·000 using a 150 or 200mm diameter flight auger.
2. Mechanical Auger – suitable for depths over 3·000 using a flight or Cheshire auger – a liner or casing is required for most granular soils and may be required for other types of subsoil.
3. Sampling Shells – suitable for shallow to medium depth borings in all subsoils except rock.

Typical Details ~

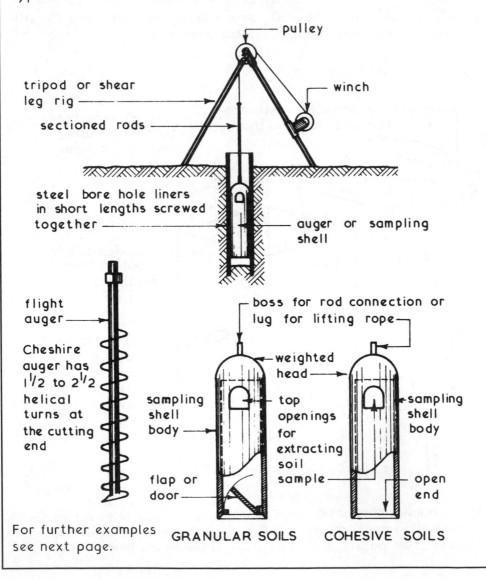

pulley

tripod or shear leg rig

winch

sectioned rods

steel bore hole liners in short lengths screwed together

auger or sampling shell

flight auger

Cheshire auger has 1¹/2 to 2¹/2 helical turns at the cutting end

boss for rod connection or lug for lifting rope

weighted head

sampling shell body

top openings for extracting soil sample

sampling shell body

flap or door

open end

For further examples see next page.

GRANULAR SOILS   COHESIVE SOILS

Wash Boring ~ this is a method of removing loosened soil from a bore hole using a strong jet of water or bentonite which is a controlled mixture of fullers earth and water. The jetting tube is worked up and down inside the bore hole, the jetting liquid disintegrates the subsoil which is carried in suspension up the annular space to a settling tank. The settled subsoil particles can be dried for testing and classification. This method has the advantage of producing subsoil samples which have not been disturbed by the impact of sampling shells however it is not suitable for large gravel subsoils or subsoils which contain boulders.

Typical Wash Boring Arrangement ~

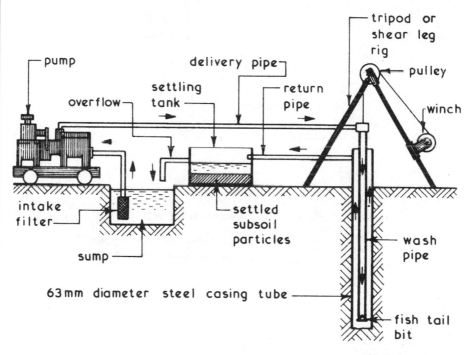

Mud-rotary Drilling ~ this is a method which can be used for rock investigations where bentonite is pumped in a continuous flow down hollow drilling rods to a rotating bit. The cutting bit is kept in contact with the bore face and the debris is carried up the annular space by the circulating fluid. Core samples can be obtained using coring tools.

Core Drilling ~ water or compressed air is jetted down the bore hole through a hollow tube and returns via the annular space. Coring tools extract continuous cores of rock samples which are sent in wooden boxes for laboratory testing.

Bore Hole Data ~ the information obtained from trial pits or bore holes can be recorded on a pro forma sheet or on a drawing showing the position and data from each trial pit or bore hole thus:-

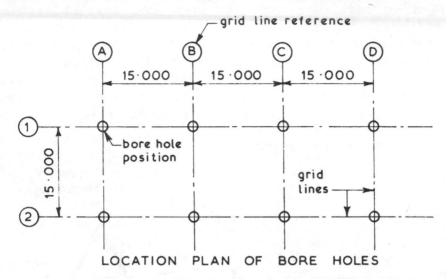

LOCATION PLAN OF BORE HOLES

Bore holes can be taken on a 15·000 to 20·000 grid covering the whole site or in isolated positions relevant to the proposed foundation(s)

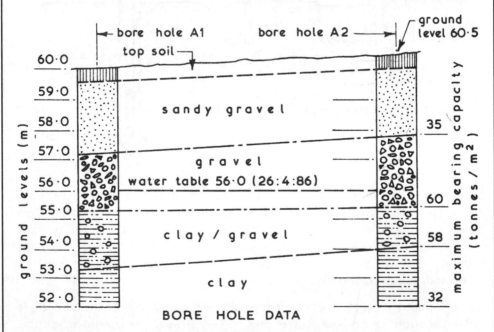

BORE HOLE DATA

As a general guide the cost of site and soil investigations should not exceed 1% of estimated project costs

Soil Assessment ~ prior to designing the foundations for a building or structure the properties of the subsoil(s) must be assessed. These processes can also be carried out to confirm the suitability of the proposed foundations. Soil assessment can include classification, grading, tests to establish shear strength and consolidation. The full range of methods for testing soils is given in BS 1377: Methods of test for soils for civil engineering purposes.

Classification ~ soils may be classified in many ways such as geological origin, physical properties, chemical composition and particle size. It has been found that the ⌐particle size and physical properties of a soil are closely linked and are therefore of particular importance and interest to a designer.

Particle Size Distribution ~ this is the percentages of the various particle sizes present in a soil sample as determined by sieving or sedimentation. BS 1377 divides particle sizes into groups as follows:-

Gravel particles – over 2mm
Sand particles – between 2mm and 0·06mm
Silt particles – between 0·06mm and 0·002mm
Clay particles – less than 0·002mm

The sand and silt classifications can be further divided thus:-

CLAY	SILT			SAND			GRAVEL
	fine	medium	coarse	fine	medium	coarse	
0·002	0·006	0·02	0·06	0·2	0·6	2	

The results of a sieve analysis can be plotted as a grading curve thus:-

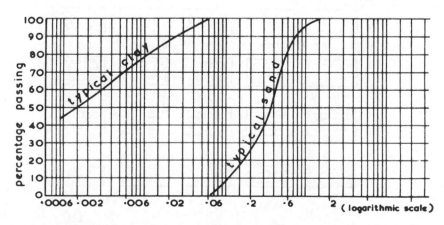

77

Triangular Chart ~ this provides a general classification of soils composed predominantly from clay, sand and silt. Each side of the triangle represents a percentage of material component. Following laboratory analysis, a sample's properties can be graphically plotted on the chart and classed accordingly.

e.g. Sand – 70%. Clay – 10% and Silt – 20% = Sandy Loam.

Note:

Silt is very fine particles of sand, easily suspended in water.
Loam is very fine particles of clay, easily dissolved in water.

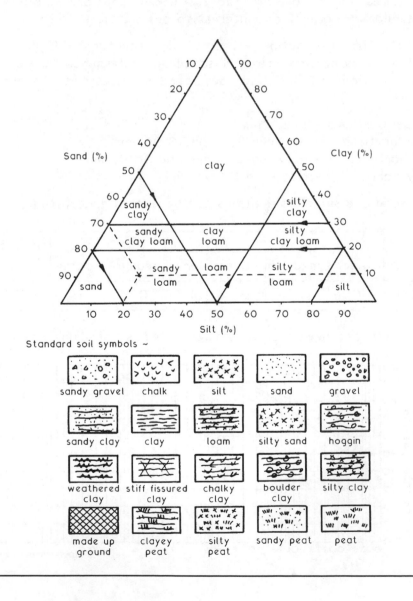

Standard soil symbols ~

sandy gravel	chalk	silt	sand	gravel
sandy clay	clay	loam	silty sand	hoggin
weathered clay	stiff fissured clay	chalky clay	boulder clay	silty clay
made up ground	clayey peat	silty peat	sandy peat	peat

Site Soil Tests ~ these tests are designed to evaluate the density or shear strength of soils and are very valuable since they do not disturb the soil under test. Three such tests are the standard penetration test, the vane test and the unconfined compression test all of which are fully described in BS 1377; Methods of test for soils for civil engineering purposes.

Standard Penetration Test ~ this test measures the resistance of a soil to the penetration of a split spoon or split barrel sampler driven into the bottom of a bore hole. The sampler is driven into the soil to a depth of 150 mm by a falling standard weight of 65 kg falling through a distance of 760 mm. The sampler is then driven into the soil a further 300 mm and the number of blows counted up to a maximum of 50 blows. This test establishes the relative density of the soil.

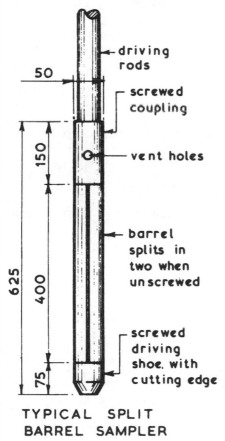

TYPICAL SPLIT BARREL SAMPLER

driving rods — screwed coupling — vent holes — barrel splits in two when unscrewed — screwed driving shoe, with cutting edge

TYPICAL RESULTS

Non-cohesive soils:-

No. of Blows	Relative Density
0 to 4	very loose
4 to 10	loose
10 to 30	medium
30 to 50	dense
50+	very dense

Cohesive soils:-

No. of Blows	Relative Density
0 to 2	very soft
2 to 4	soft
4 to 8	medium
8 to 15	stiff
15 to 30	very stiff
30+	hard

The results of this test in terms of number of blows and amounts of penetration will need expert interpretation.

Vane Test ~ this test measures the shear strength of soft cohesive soils. The steel vane is pushed into the soft clay soil and rotated by hand at a constant rate. The amount of torque necessary for rotation is measured and the soil shear strength calculated as shown below.

This test can be carried out within a lined bore hole where the vane is pushed into the soil below the base of the bore hole for a distance equal to three times the vane diameter before rotation commences. Alternatively the vane can be driven or jacked to the required depth, the vane being protected within a special protection shoe, the vane is then driven or jacked a further 500mm before rotation commences.

Calculation of Shear Strength –

Formula:- $S = \dfrac{M}{K}$

where S = shear value in kN/m^2

M = torque required to shear soil

K = constant for vane

= $3 \cdot 66 \, D^3 \times 10^{-6}$

D = vane diameter

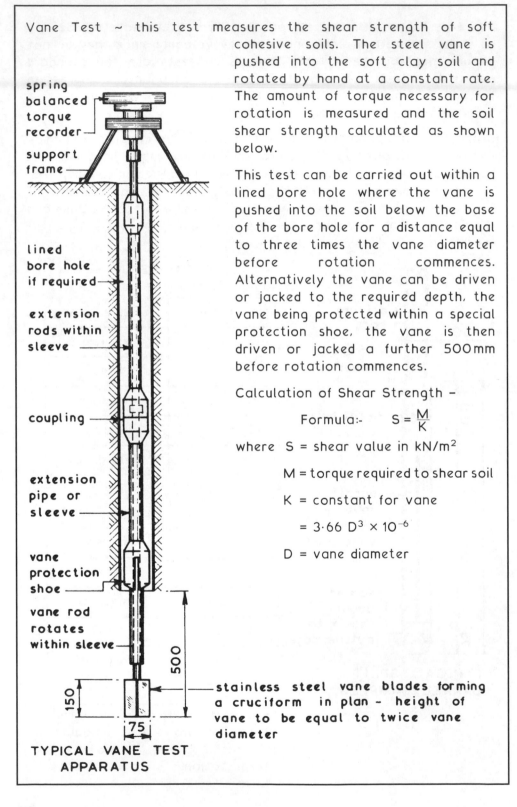

spring balanced torque recorder

support frame

lined bore hole if required

extension rods within sleeve

coupling

extension pipe or sleeve

vane protection shoe

vane rod rotates within sleeve

500

150

75

stainless steel vane blades forming a cruciform in plan - height of vane to be equal to twice vane diameter

TYPICAL VANE TEST APPARATUS

Unconfined Compression Test ~ this test can be used to establish the shear strength of a non-fissured cohesive soil sample using portable apparatus either on site or in a laboratory. The 75mm long × 38mm diameter soil sample is placed in the apparatus and loaded in compression until failure occurs by shearing or lateral bulging. For accurate reading of the trace on the recording chart a transparent viewfoil is placed over the trace on the chart.

Typical Apparatus Details~

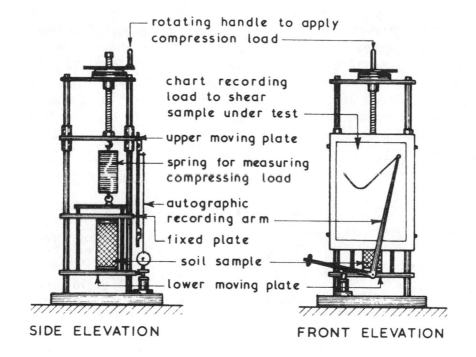

rotating handle to apply
compression load

chart recording
load to shear
sample under test

upper moving plate

spring for measuring
compressing load

autographic
recording arm

fixed plate

soil sample

lower moving plate

SIDE ELEVATION   FRONT ELEVATION

Typical Results ~ showing compression strengths of clays:-

Very soft clay   – less than 25kN/m²

Soft clay   – 25 to 50kN/m²

Medium clay   – 50 to 100kN/m²

Stiff clay   – 100 to 200kN/m²

Very stiff clay   – 200 to 400kN/m²

Hard clay   – more than 400kN/m²

NB. The shear strength of clay soils is only half of the compression strength values given above.

Laboratory Testing ~ tests for identifying and classifying soils with regard to moisture content, liquid limit, plastic limit, particle size distribution and bulk density are given in BS 1377.

Bulk Density ~ this is the mass per unit volume which includes mass of air or water in the voids and is essential information required for the design of retaining structures where the weight of the retained earth is an important factor.

Shear Strength ~ this soil property can be used to establish its bearing capacity and also the pressure being exerted on the supports in an excavation. The most popular method to establish the shear strength of cohesive soils is the Triaxial Compression Test. In principle this test consists of subjecting a cylindrical sample of undisturbed soil (75mm long × 38mm diameter) to a lateral hydraulic pressure in addition to a vertical load. Three tests are carried out on three samples (all cut from the same large sample) each being subjected to a higher hydraulic pressure before axial loading is applied. The results are plotted in the form of Mohr's circles.

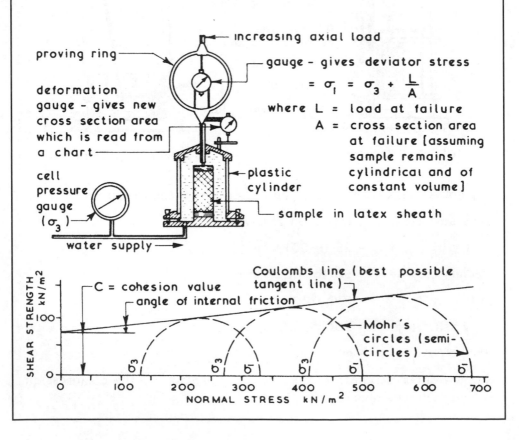

proving ring

deformation gauge - gives new cross section area which is read from a chart

cell pressure gauge (σ_3)

water supply →

increasing axial load

gauge - gives deviator stress

$$= \sigma_1 = \sigma_3 + \frac{L}{A}$$

where L = load at failure
A = cross section area at failure [assuming sample remains cylindrical and of constant volume]

plastic cylinder

sample in latex sheath

Coulombs line (best possible tangent line)

C = cohesion value
angle of internal friction

Mohr's circles (semi-circles)

SHEAR STRENGTH kN/m²

NORMAL STRESS kN/m²

Shear Strength ~ this can be defined as the resistance offered by a soil to the sliding of one particle over another. A simple method of establishing this property is the Shear Box Test in which the apparatus consists of two bottomless boxes which are filled with the soil sample to be tested. A horizontal shearing force (S) is applied against a vertical load (W) causing the soil sample to shear along a line between the two boxes.

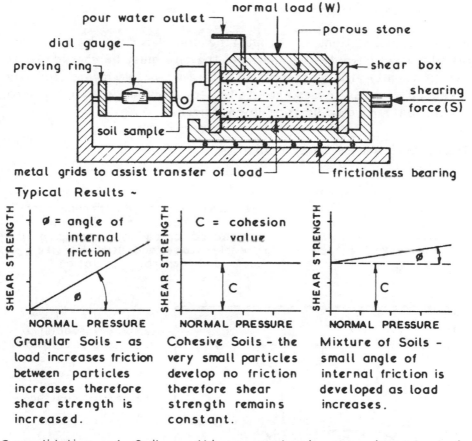

Typical Results ~

Granular Soils - as load increases friction between particles increases therefore shear strength is increased.

Cohesive Soils - the very small particles develop no friction therefore shear strength remains constant.

Mixture of Soils - small angle of internal friction is developed as load increases.

Consolidation of Soil ~ this property is very important in calculating the movement of a soil under a foundation. The laboratory testing apparatus is called an Oedometer.

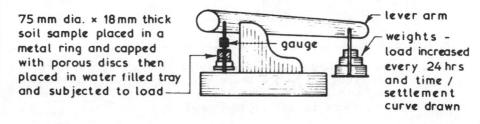

75 mm dia. × 18 mm thick soil sample placed in a metal ring and capped with porous discs then placed in water filled tray and subjected to load

lever arm

weights - load increased every 24 hrs and time / settlement curve drawn

General Considerations ~ before any specific considerations and decisions can be made regarding site layout a general appreciation should be obtained by conducting a thorough site investigation at the pre-tender stage and examining in detail the drawings, specification and Bill of Quantities to formulate proposals of how the contract will be carried out if the tender is successful. This will involve a preliminary assessment of plant, materials and manpower requirements plotted against the proposed time scale in the form of a bar chart (see page 34).

Access Considerations ~ this must be considered for both on- and off-site access. Routes to and from the site must be checked as to the suitability for transporting all the requirements for the proposed works. Access on site for deliveries and general circulation must also be carefully considered.

Typical Site Access Considerations ~

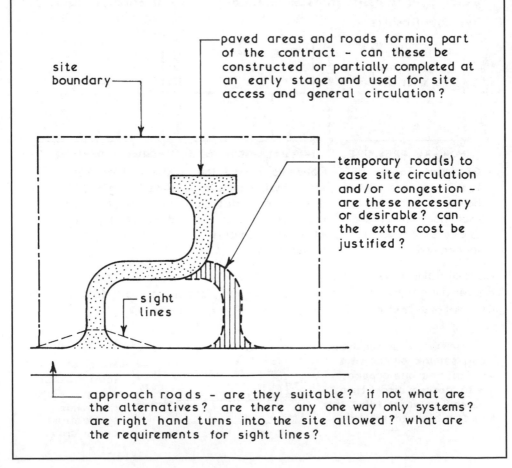

site boundary

paved areas and roads forming part of the contract - can these be constructed or partially completed at an early stage and used for site access and general circulation?

temporary road(s) to ease site circulation and/or congestion - are these necessary or desirable? can the extra cost be justified?

sight lines

approach roads - are they suitable? if not what are the alternatives? are there any one way only systems? are right hand turns into the site allowed? what are the requirements for sight lines?

Storage Considerations ~ amount and types of material to be stored, security and weather protection requirements, allocation of adequate areas for storing materials and allocating adequate working space around storage areas as required, siting of storage areas to reduce double handling to a minimum without impeding the general site circulation and/or works in progress.

Accommodation Considerations ~ number and type of site staff anticipated, calculate size and select units of accommodation and check to ensure compliance with the minimum requirements of the Construction (Health, Safety and Welfare) Regulations 1996, select siting for offices to give easy and quick access for visitors but at the same time giving a reasonable view of the site, select siting for messroom and toilets to reduce walking time to a minimum without impeding the general site circulation and/or works in progress.

Temporary Services Considerations ~ what, when and where are they required? Possibility of having permanent services installed at an early stage and making temporary connections for site use during the construction period, coordination with the various service undertakings is essential.

Plant Considerations ~ what plant, when and where is it required? static or mobile plant? If static select the most appropriate position and provide any necessary hard standing, if mobile check on circulation routes for optimum efficiency and suitability, provision of space and hard standing for on-site plant maintenance if required.

Fencing and Hoarding Considerations ~ what is mandatory and what is desirable? Local vandalism record, type or types of fence and/or hoarding required, possibility of using fencing which is part of the contract by erecting this at an early stage in the contract.

Safety and Health Considerations ~ check to ensure that all the above conclusions from the considerations comply with the minimum requirements set out in the various Construction Regulations and in the Health and Safety at Work etc., Act 1974.

For a typical site layout example see next page.

Typical Site Layout Example ~

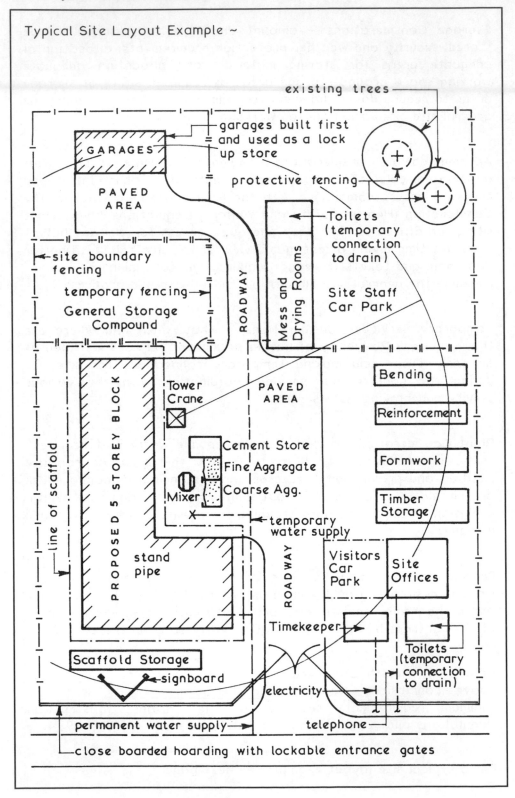

existing trees

garages built first
and used as a lock
up store

GARAGES

PAVED
AREA

protective fencing

Toilets
(temporary
connection
to drain)

site boundary
fencing

temporary fencing

General Storage
Compound

ROADWAY

Mess and
Drying Rooms

Site Staff
Car Park

Tower
Crane

PAVED
AREA

Bending

Reinforcement

Cement Store

Fine Aggregate

Coarse Agg.

Mixer

Formwork

Timber
Storage

line of scaffold

PROPOSED 5 STOREY BLOCK

stand
pipe

temporary
water supply

Visitors
Car
Park

Site
Offices

ROADWAY

Timekeeper

Scaffold Storage

signboard

electricity

Toilets
(temporary
connection
to drain)

permanent water supply

telephone

close boarded hoarding with lockable entrance gates

Site Security ~ the primary objectives of site security are -

1. Security against theft.
2. Security from vandals.
3. Protection from innocent trespassers.

The need for and type of security required will vary from site to site according to the neighbourhood, local vandalism record and the value of goods stored on site. Perimeter fencing, internal site protection and night security may all be necessary.

Typical Site Security Provisions ~

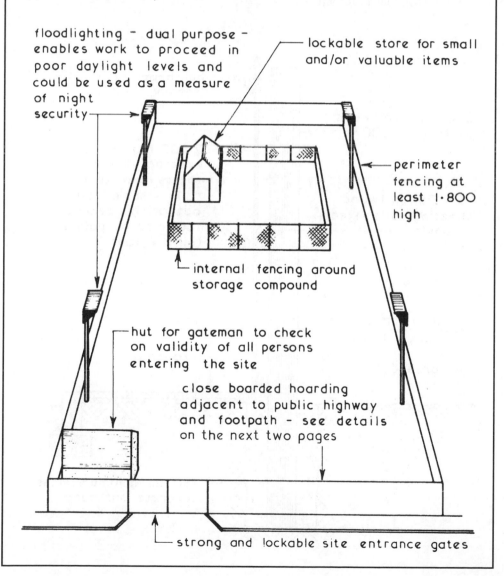

floodlighting - dual purpose - enables work to proceed in poor daylight levels and could be used as a measure of night security

lockable store for small and/or valuable items

perimeter fencing at least 1·800 high

internal fencing around storage compound

hut for gateman to check on validity of all persons entering the site

close boarded hoarding adjacent to public highway and footpath - see details on the next two pages

strong and lockable site entrance gates

Hoardings ~ under the Highways Act 1980 a close boarded fence hoarding must be erected prior to the commencement of building operations if such operations are adjacent to a public footpath or highway. The hoarding needs to be adequately constructed to provide protection for the public, resist impact damage, resist anticipated wind pressures and adequately lit at night. Before a hoarding can be erected a licence or permit must be obtained from the local authority who will usually require 10 to 20 days notice. The licence will set out the minimum local authority requirements for hoardings and define the time limit period of the licence.

Typical Hoarding Details ~

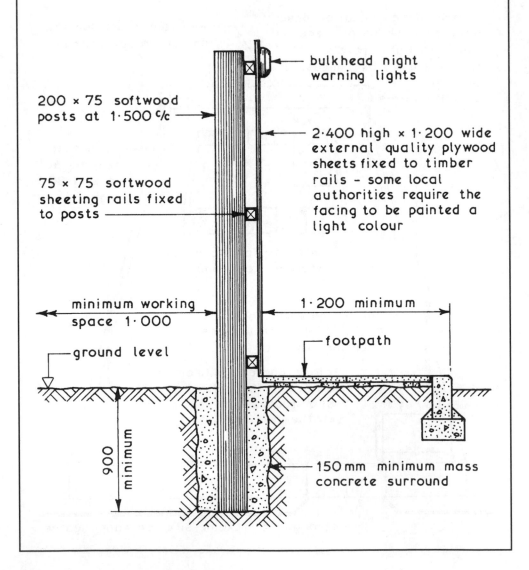

200 × 75 softwood posts at 1·500 c/c

75 × 75 softwood sheeting rails fixed to posts

bulkhead night warning lights

2·400 high × 1·200 wide external quality plywood sheets fixed to timber rails - some local authorities require the facing to be painted a light colour

minimum working space 1·000

1·200 minimum

ground level

footpath

900 minimum

150 mm minimum mass concrete surround

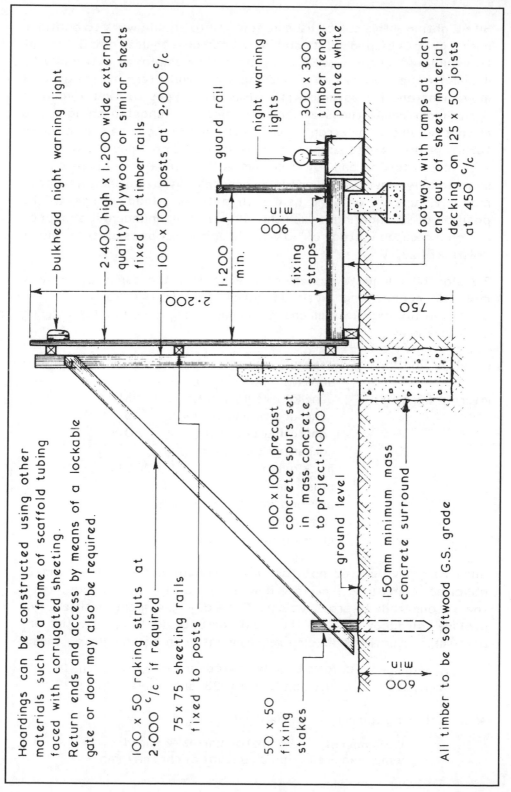

Hoardings can be constructed using other materials such as a frame of scaffold tubing faced with corrugated sheeting.
Return ends and access by means of a lockable gate or door may also be required.

100 x 50 raking struts at 2·000 c/c if required

75 x 75 sheeting rails fixed to posts

50 x 50 fixing stakes

100 x 100 precast concrete spurs set in mass concrete to project 1·000

ground level

150mm minimum mass concrete surround

All timber to be softwood G.S. grade

bulkhead night warning light

2·400 high x 1·200 wide external quality plywood or similar sheets fixed to timber rails

100 x 100 posts at 2·000 c/c

guard rail

night warning lights

300 x 300 timber fender painted white

fixing straps

footway with ramps at each end and out of sheet material decking on 125 x 50 joists at 450 c/c

2·200

1·200 min.

900 min.

750

600 min.

Site Lighting ~ this can be used effectively to enable work to continue during periods of inadequate daylight. It can also be used as a deterrent to would-be trespassers. Site lighting can be employed externally to illuminate the storage and circulation areas and internally for general movement and for specific work tasks. The types of lamp available range from simple tungsten filament lamps to tungsten halogen and discharge lamps. The arrangement of site lighting can be static where the lamps are fixed to support poles or mounted on items of fixed plant such as scaffolding and tower cranes. Alternatively the lamps can be sited locally where the work is in progress by being mounted on a movable support or hand held with a trailing lead. Whenever the position of site lighting is such that it can be manhandled it should be run on a reduced voltage of 110 V single phase as opposed to the mains voltage of 230 V.

To plan an adequate system of site lighting the types of activity must be defined and given an illumination target value which is quoted in lux (lx). Recommended minimum target values for building activities are:-

External lighting   –   general circulation ⎫   10 lx
                             materials handling ⎭

Internal lighting   –   general circulation   5 lx
                             general working areas   15 lx
                             concreting activities   50 lx
                             carpentry and joinery ⎫
                             bricklaying   ⎬   100 lx
                             plastering   ⎭
                             painting and decorating ⎫
                             site offices   ⎬ 200 lx
                             drawing board positions   300 lx

Such target values do not take into account deterioration, dirt or abnormal conditions therefore it is usual to plan for at least twice the recommended target values. Generally the manufacturers will provide guidance as to the best arrangement to use in any particular situation but lamp requirements can be calculated thus:-

$$\frac{\text{Total lumens}}{\text{required}} = \frac{\text{area to be illluminated (m}^2) \times \text{target value (lx)}}{\text{utilisation factor 0·23 [dispersive lights 0·27]}}$$

After choosing lamp type to be used:-

$$\frac{\text{Number of}}{\text{lamps required}} = \frac{\text{total lumens required}}{\text{lumen output of chosen lamp}}$$

Typical Site Lighting Arrangement:-

Area lighting using high mounted lamps ~

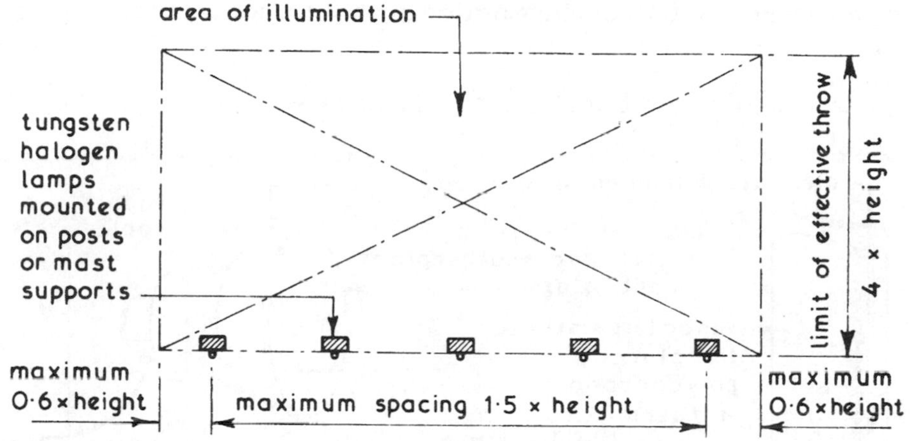

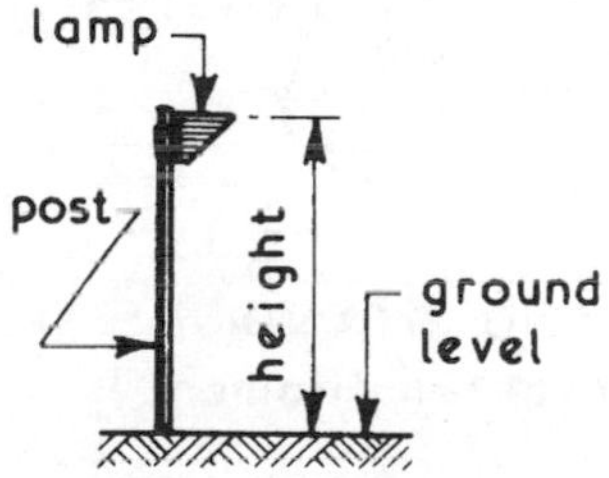

area of illumination

tungsten halogen lamps mounted on posts or mast supports

limit of effective throw

4 × height

maximum 0·6×height

maximum spacing 1·5 × height

maximum 0·6×height

lamp

post

height

ground level

Typical minimum heights for tungsten halogen lamps :-

500 watts - 7·500 metres
1000 watts - 9·000 metres
2000 watts - 15·000 metres

Area lighting using overhead dispersive lights suspended from a grid or from the structure ~

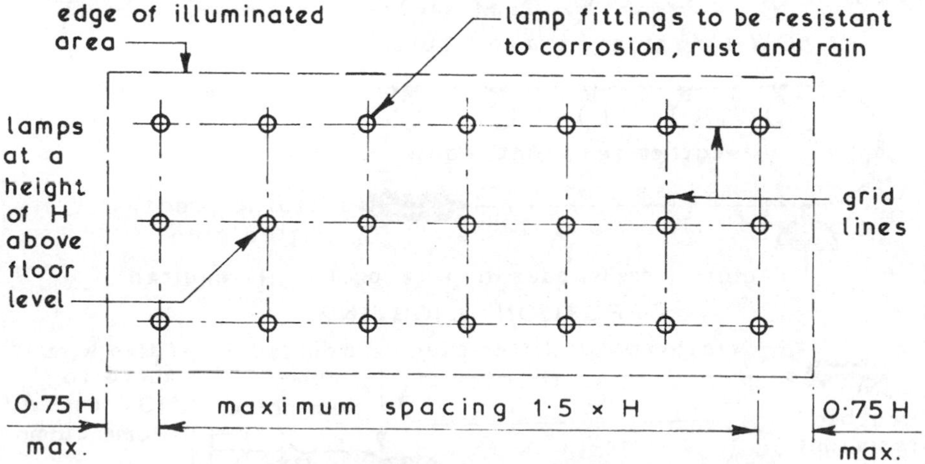

edge of illuminated area

lamp fittings to be resistant to corrosion, rust and rain

lamps at a height of H above floor level

grid lines

0·75 H max.

maximum spacing 1·5 × H

0·75 H max.

Typical minimum heights for dispersive lamps:

Fluorescent 40 to 125 W – 2·500 m; Tungsten filament 300 W – 3·000 m

91

Walkway and Local Lighting ~ to illuminate the general circulation routes bulkhead and/or festoon lighting could be used either on a standard mains voltage of 230V or on a reduced voltage of 110V. For local lighting at the place of work hand lamps with trailing leads or lamp fittings on stands can be used and positioned to give the maximum amount of illumination without unacceptable shadow cast.

Typical Walkway and Local Lighting Fittings ~

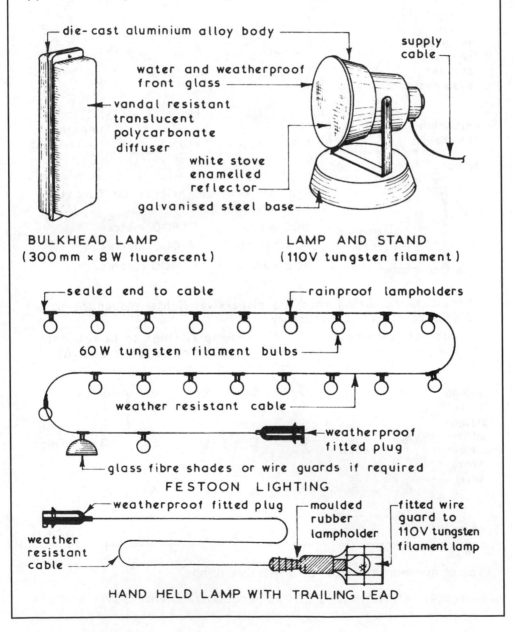

die-cast aluminium alloy body

supply cable

water and weatherproof front glass

vandal resistant translucent polycarbonate diffuser

white stove enamelled reflector

galvanised steel base

BULKHEAD LAMP
(300 mm × 8 W fluorescent)

LAMP AND STAND
(110V tungsten filament)

sealed end to cable — rainproof lampholders

60 W tungsten filament bulbs

weather resistant cable

weatherproof fitted plug

glass fibre shades or wire guards if required

FESTOON LIGHTING

weatherproof fitted plug — moulded rubber lampholder — fitted wire guard to 110V tungsten filament lamp

weather resistant cable

HAND HELD LAMP WITH TRAILING LEAD

Electrical Supply to Building Sites ~ a supply of electricity is usually required at an early stage in the contract to provide light and power to the units of accommodation. As the work progresses power could also be required for site lighting, hand held power tools and large items of plant. The supply of electricity to a building site is the subject of a contract between the contractor and the local area electricity company who will want to know the date when supply is required; site address together with a block plan of the site; final load demand of proposed building and an estimate of the maximum load demand in kilowatts for the construction period. The latter can be estimated by allowing 10 W/m² of the total floor area of the proposed building plus an allowance for high load equipment such as cranes. The installation should be undertaken by a competent electrical contractor to ensure that it complies with all the statutory rules and regulations for the supply of electricity to building sites.

Typical Supply and Distribution Equipment ~

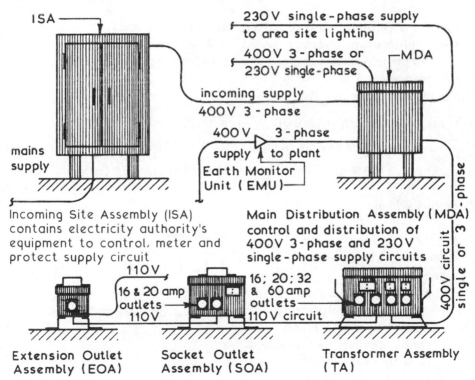

ISA

230 V single-phase supply to area site lighting

400 V 3-phase or 230 V single-phase

incoming supply

400 V 3-phase

—MDA

400 V 3-phase supply to plant

Earth Monitor Unit (EMU)

mains supply

Incoming Site Assembly (ISA) contains electricity authority's equipment to control, meter and protect supply circuit

Main Distribution Assembly (MDA) control and distribution of 400 V 3-phase and 230 V single-phase supply circuits

400 V circuit single or 3-phase

110 V

16 & 20 amp outlets

110 V

16; 20; 32 & 60 amp outlets

110 V circuit

Extension Outlet Assembly (EOA)

Socket Outlet Assembly (SOA)

Transformer Assembly (TA)

The units must be strong, durable and resistant to rain penetration with adequate weather seals to all access panels and doors. All plug and socket outlets should be colour coded :- 400 V - red; 230 V - blue; 110 V - yellow

Office Accommodation ~ the arrangements for office accommodation to be provided on site is a matter of choice for each individual contractor. Generally separate offices would be provided for site agent, clerk of works, administrative staff, site surveyors and sales staff.

The minimum requirements of such accommodation is governed by the Offices, Shops and Railway Premises Act 1963 unless they are ~

1. Mobile units in use for not more then 6 months.
2. Fixed units in use for not more than 6 weeks.
3. Any type of unit in use for not more than 21 man hours per week.
4. Office for exclusive use of self employed person.
5. Office used by family only staff.

Sizing Example ~

Office for site agent and assistant plus an allowance for 3 visitors.
Assume an internal average height of 2·400.
Allow 3·7m² minimum per person and 11·5m³ minimum per person.
Minimum area = 5 × 3·7 = 18·5m²
Minimum volume = 5 × 11·5 = 57·5m³

Assume office width of 3·000 then minimum length required is
$= \dfrac{57·5}{3 \times 2·4} = \dfrac{57·5}{7·2} = 7·986$ say 8·000
Area check 3 × 8 = 24m² which is > 18·5m² ∴ satisfactory

Typical Example ~

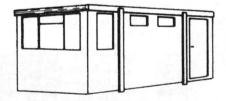

Portable cabin with four adjustable steel legs with attachments for stacking. Panelling of galvanised steel sheet and rigid insulation core. Plasterboard inner lining to walls and ceiling. Pyro-shield windows with steel shutters and a high security steel door.

Ref. Fire prevention on construction sites – the joint code of practice on protection from fire of construction sites and buildings undergoing renovation. Published by Construction Confederation and The Fire Protection Association.

The requirements for health and wellbeing of persons on construction sites are enforced by the Health and Safety Executive, through the Health and Safety at Work etc. Act 1974 and the Construction (Health, Safety and Welfare) Regulations 1996. The following minimum requirements apply and the numbers of persons on site were established by the Construction Regulations of 1966.

Provision	Requirement	No of persons employed on site
FIRST AID	Box to be distinctively marked and in charge of responsible person.	5 to 50 – first aid boxes 50+ first aid box and a person trained in first aid
AMBULANCES	Stretcher(s) in charge of responsible person	25+ notify ambulance authority of site details within 24 hours of employing more than 25 persons
FIRST AID ROOM	Used only for rest or treatment and in charge of trained person	If more than 250 persons employed on site each employer of more than 40 persons to provide a first aid room
SHELTER AND ACCOMMODATION FOR CLOTHING	All persons on site to have shelter and a place for changing, drying and depositing clothes. Separate facilities for male and female staff.	Up to 5 where possible a means of warming themselves and drying wet clothes 5+ adequate means of warming themselves and drying wet clothing
REST ROOM	Drinking water, means of boiling water, preparing and eating meals for all persons on site. Arrangements to protect non-smokers from tobacco smoke.	10+ facilities for heating food if hot meals are not available on site
WASHING FACILITIES	Washing facilities to be provided for all persons on site for more than 4 hours. Ventilated and lit. Separate facilities for male and female staff.	20 to 100 if work is to last more than 6 weeks – hot and cold or warm water, soap and towel. 100+ work lasting more than 12 months – 4 wash places+1 for every 35 persons over 100
SANITARY FACILITIES	To be maintained, lit, ventilated and kept clean. Separate facilities for male and female staff	Up to 100 – 1 convenience for every 25 persons 100+ convenience for every 35 persons

Site Storage ~ materials stored on site prior to being used or fixed may require protection for security reasons or against the adverse effects which can be caused by exposure to the elements.

Small and Valuable Items ~ these should be kept in a secure and lockable store. Similar items should be stored together in a rack or bin system and only issued against an authorised requisition.

Large or Bulk Storage Items ~ for security protection these items can be stored within a lockable fenced compound. The form of fencing chosen may give visual security by being of an open nature but these are generally easier to climb than the close boarded type of fence which lacks the visual security property.

Typical Storage Compound Fencing ~

Close boarded fences can be constructed on the same methods used for hoardings – see pages 86 & 87.

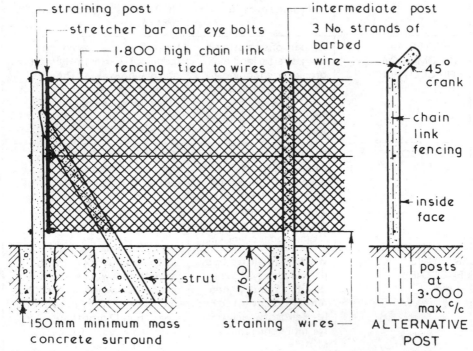

straining post
stretcher bar and eye bolts
1·800 high chain link fencing tied to wires
intermediate post
3 No. strands of barbed wire
45° crank
chain link fencing
inside face
strut
760
150mm minimum mass concrete surround
straining wires
posts at 3·000 max. ᶜ/c
ALTERNATIVE POST

CHAIN LINK FENCING WITH PRECAST CONCRETE POSTS

Alternative Fence Types ~ woven wire fence, strained wire fence, cleft chestnut pale fence, wooden palisade fence, wooden post and rail fence and metal fences – see BS 1722: Fences, for details.

Storage of Materials ~ this can be defined as the provision of adequate space, protection and control for building materials and components held on site during the construction process. The actual requirements for specific items should be familiar to students who have completed studies in construction technology at an introductory level but the need for storage and control of materials held on site can be analysed further:-

1. Physical Properties – size, shape, weight and mode of delivery will assist in determining the safe handling and stacking method(s) to be employed on site, which in turn will enable handling and storage costs to be estimated.

2. Organisation – this is the planning process of ensuring that all the materials required are delivered to site at the correct time, in sufficient quantity, of the right quality, the means of unloading is available and that adequate space for storage or stacking has been allocated.

3. Protection – building materials and components can be classified as durable or non-durable, the latter will usually require some form of weather protection to prevent deterioration whilst in store.

4 Security – many building materials have a high resale and/or usage value to persons other than those for whom they were ordered and unless site security is adequate material losses can become unacceptable.

5. Costs – to achieve an economic balance of how much expenditure can be allocated to site storage facilities the following should be taken into account:-

   a. Storage areas, fencing, racks, bins, etc.
   b. Protection requirements.
   c. Handling, transporting and stacking requirements.
   d. Salaries and wages of staff involved in storage of materials and components.
   e. Heating and/or lighting if required.
   f. Allowance for losses due to wastage, deterioration, vandalism and theft.
   g. Facilities to be provided for subcontractors.

6. Control – checking quality and quantity of materials at delivery and during storage period, recording delivery and issue of materials and monitoring stock holdings.

Site Storage Space ~ the location and size(s) of space to be allocated for any particular material should be planned by calculating the area(s) required and by taking into account all the relevant factors before selecting the most appropriate position on site in terms of handling, storage and convenience. Failure to carry out this simple planning exercise can result in chaos on site or having on site more materials than there is storage space available.

Calculation of Storage Space Requirements ~ each site will present its own problems since a certain amount of site space must be allocated to the units of accommodation, car parking, circulation and working areas, therefore the amount of space available for materials storage may be limited. The size of the materials or component being ordered must be known together with the proposed method of storage and this may vary between different sites of similar building activities. There are therefore no standard solutions for allocating site storage space and each site must be considered separately to suit its own requirements.

Typical Examples ~

Bricks – quantity = 15,200 to be delivered in strapped packs of 380 bricks per pack each being 1100 mm wide × 670 mm long × 850 mm high. Unloading and stacking to be by forklift truck to form 2 rows 2 packs high.

Area required :- number of packs per row = $\dfrac{15,200}{380 \times 2}$ = 20

length of row = 10 × 670 = 6·700
width of row = 2 × 1100 = 2·200

allowance for forklift approach in front of stack = 5·000 ∴ minimum brick storage area = 6·700 long × 7·200 wide

Timber – to be stored in open sided top covered racks constructed of standard scaffold tubes. Maximum length of timber ordered = 5·600. Allow for rack to accept at least 4 No. 300 mm wide timbers placed side by side then minimum width required = 4 × 300 = 1·200
Minimum plan area for timber storage rack = 5·600 × 1·200
Allow for end loading of rack equal to length of rack
∴ minimum timber storage area = 11·200 long × 1·200 wide
Height of rack to be not more than 3 × width = 3·600

Areas for other materials stored on site can be calculated using the basic principles contained in the examples above.

Site Allocation for Materials Storage ~ the area and type of storage required can be determined as shown on pages 96 to 98, but the allocation of an actual position on site will depend on:-

1. Space available after areas for units of accommodation have been allocated.
2. Access facilities on site for delivery, vehicles.
3. Relationship of storage area(s) to activity area(s) – the distance between them needs to be kept as short as possible to reduce transportation needs in terms of time and costs to the minimum. Alternatively storage areas and work areas need to be sited within the reach of any static transport plant such as a tower crane.
4. Security – needs to be considered in the context of site operations, vandalism and theft.
5. Stock holding policy – too little storage could result in delays awaiting for materials to be delivered, too much storage can be expensive in terms of weather and security protection requirements apart from the capital used to purchase the materials stored on site.

Typical Example ~

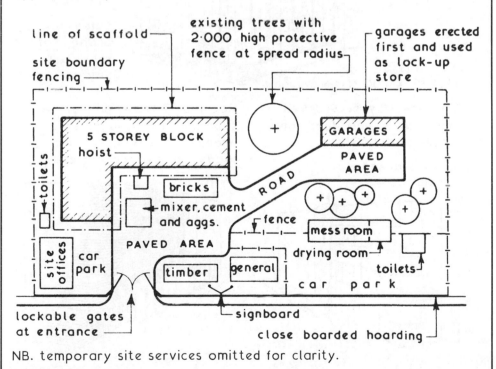

NB. temporary site services omitted for clarity.

Bricks ~ may be supplied loose or strapped in unit loads and stored on timber pallets

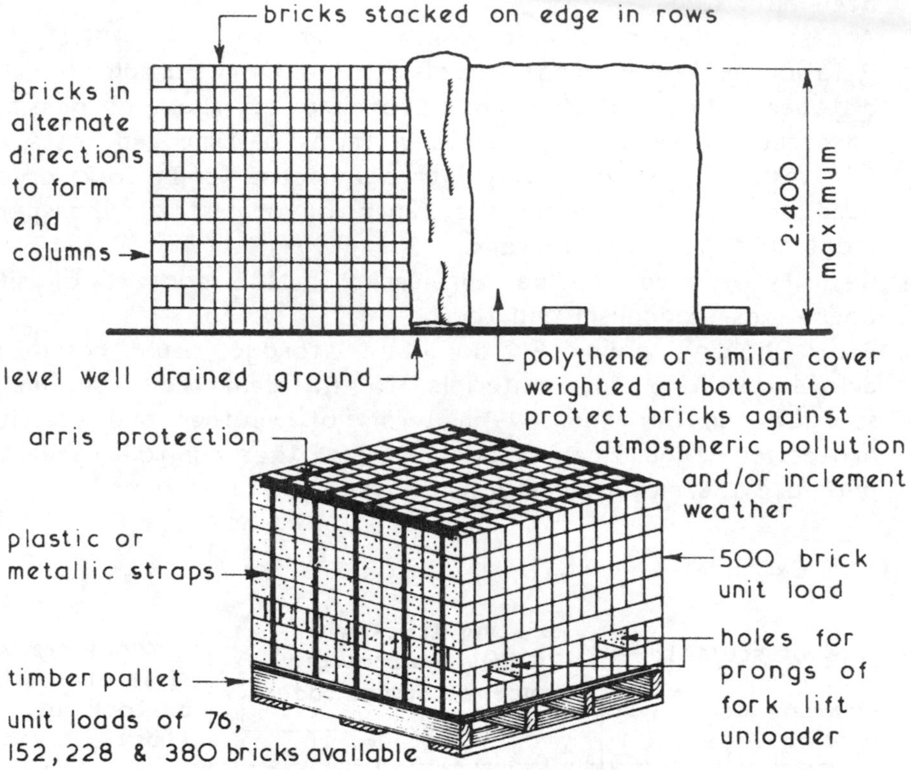

bricks stacked on edge in rows

bricks in alternate directions to form end columns →

2·400 maximum

level well drained ground →

polythene or similar cover weighted at bottom to protect bricks against atmospheric pollution and/or inclement weather

arris protection

plastic or metallic straps

500 brick unit load

timber pallet

holes for prongs of fork lift unloader

unit loads of 76, 152, 228 & 380 bricks available

Blocks ~ may be supplied loose or in unit loads on timber pallets

Roofing Tiles ~ may be supplied loose, in plastic wrapped packs or in unit loads on timber pallets

blocks stacked in 'columns'

protective cover

8 courses maximum

6 rows maximum

ridge tiles stored on ends

end laid flat and staggered

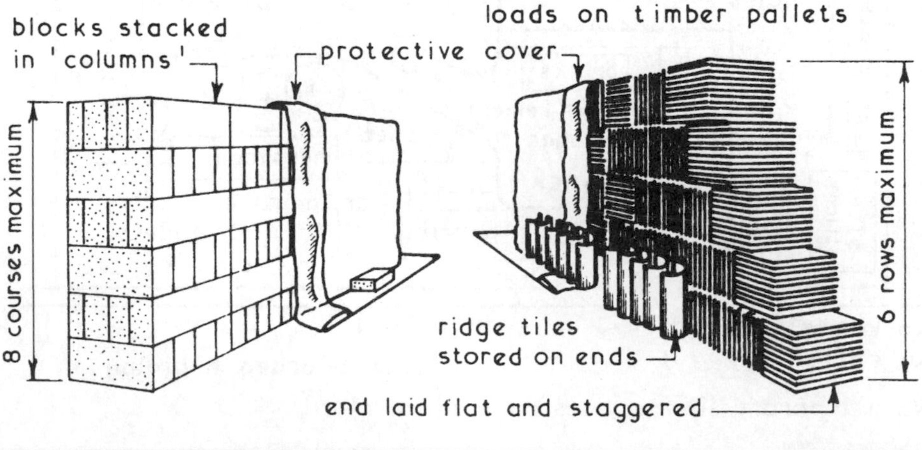

Drainage Pipes ~ supplied loose or strapped together on timber pallets

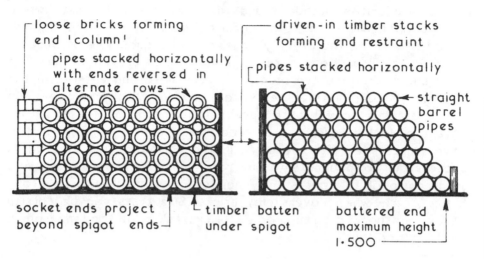

loose bricks forming end 'column'

pipes stacked horizontally with ends reversed in alternate rows

driven-in timber stacks forming end restraint

pipes stacked horizontally

straight barrel pipes

socket ends project beyond spigot ends

timber batten under spigot

battered end maximum height 1·500

Gullies etc., should be stored upside down and supported to remain level

Baths ~ stacked or nested vertically or horizontally on timber battens

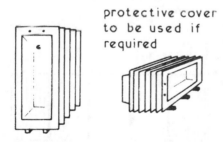

protective cover to be used if required

Basins ~ stored similar to baths but not more than four high if nested one on top of another

Corrugated and Similar Sheet Materials ~ stored flat on a level surface and covered with a protective polythene or similar sheet material

Timber and Joinery Items ~ should be stored horizontally and covered but with provison for free air flow

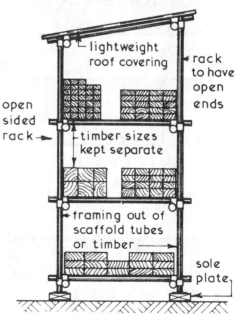

lightweight roof covering

rack to have open ends

open sided rack

timber sizes kept separate

framing out of scaffold tubes or timber

sole plate

Cement, Sand and Aggregates ~ for supply and storage details see pages 267 & 271

Site Tests ~ the majority of materials and components arriving on site will conform to the minimum recommendations of the appropriate British Standard and therefore the only tests which need be applied are those of checking quantity received against amount stated on the delivery note, ensuring quality is as ordered and a visual inspection to reject damaged or broken goods. The latter should be recorded on the delivery note and entered in the site records. Certain site tests can however be carried out on some materials to establish specific data such as the moisture content of timber which can be read direct from a moisture meter. Other simple site tests are given in the various British Standards to ascertain compliance with the recommendations, such as tests for dimensional tolerances and changes given in BS EN 771-1 and BS EN 772-16 which covers clay bricks. Site tests can be carried out by measuring a sample of 24 bricks taken at random from a delivered load thus:-

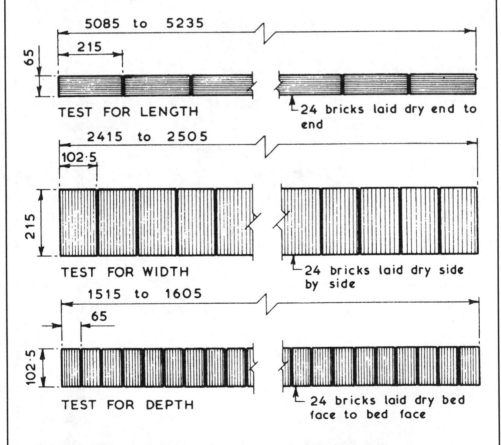

5085 to 5235
215
65
TEST FOR LENGTH
24 bricks laid dry end to end

2415 to 2505
102·5
215
TEST FOR WIDTH
24 bricks laid dry side by side

1515 to 1605
65
102·5
TEST FOR DEPTH
24 bricks laid dry bed face to bed face

Refs. BS EN 772-16: Methods of test for masonry units.
BS EN 771-1: Specification for masonry units.

Site Test ~ apart from the test outlined on page 102 site tests on materials which are to be combined to form another material such as concrete can also be tested to establish certain properties which if not known could affect the consistency and/or quality of the final material.

Typical Example ~ Testing Sand for Bulking ~

this data is required when batching concrete by volume – test made at commencement of mixing and if change in weather

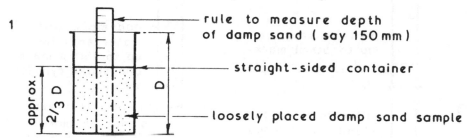

1 — rule to measure depth of damp sand (say 150 mm)

— straight-sided container

— loosely placed damp sand sample

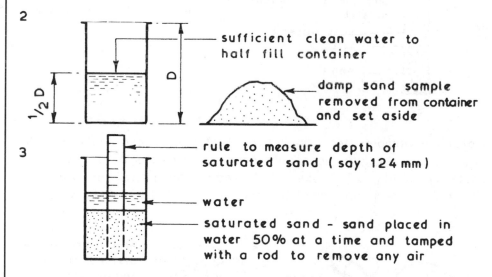

2 — sufficient clean water to half fill container

— damp sand sample removed from container and set aside

3 — rule to measure depth of saturated sand (say 124 mm)

— water

— saturated sand – sand placed in water 50% at a time and tamped with a rod to remove any air

4 Calculation :-

$$bulking = \frac{\text{difference in height between damp \& saturated sand}}{\text{depth of saturated sand}}$$

$$\% \text{ bulking} = \frac{150 - 124}{124} \times 100 = \frac{26}{124} \times 100 = 20 \cdot 96774 \%$$

therefore volume of sand should be increased by 21% over that quoted in the specification

NB. a given weight of saturated sand will occupy the same space as when dry but more space when damp

Silt Test for Sand ~ the object of this test is to ascertain the cleanliness of sand by establishing the percentage of silt present in a natural sand since too much silt will weaken the concrete

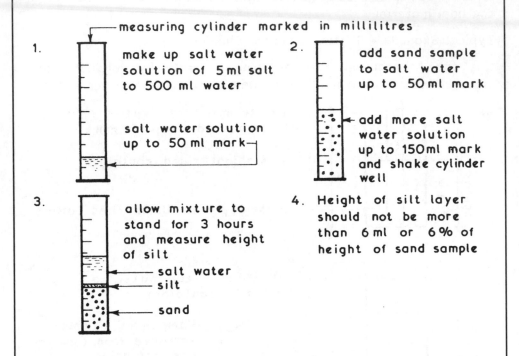

measuring cylinder marked in millilitres

1. make up salt water solution of 5 ml salt to 500 ml water

salt water solution up to 50 ml mark

2. add sand sample to salt water up to 50 ml mark

add more salt water solution up to 150 ml mark and shake cylinder well

3. allow mixture to stand for 3 hours and measure height of silt
 - salt water
 - silt
 - sand

4. Height of silt layer should not be more than 6 ml or 6% of height of sand sample

Obtaining Samples for Laboratory Testing ~ these tests may be required for checking aggregate grading by means of a sieve test, checking quality or checking for organic impurities but whatever the reason the sample must be truly representative of the whole:-

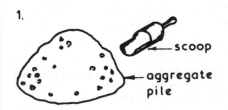

1.
 - scoop
 - aggregate pile

samples extracted by means of a scoop from at least ten different positions in the pile

sample required :-
fine aggregate - 50 kg
coarse aggregate - 200 kg

2. well mixed sample divided into four equal parts - opposite quarters are discarded -

remainder of sample remixed and quartered - whole process is repeated until required size of sample is left.
samples required :-
fine aggregate ✈ 6 mm - 3 kg
              ✈ 10 mm - 6 kg
coarse aggregate ✈ 20 mm - 25 kg
              ✈ 32 mm - 50 kg

Concrete requires monitoring by means of tests to ensure that subsequent mixes are of the same consistency and this can be carried out on site by means of the slump test and in a laboratory by crushing test cubes to check that the cured concrete has obtained the required designed strength.

Slump Test ~

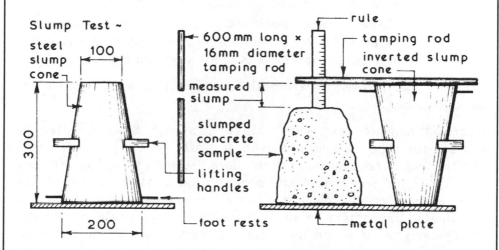

The slump cone is filled to a quarter depth and tamped 25 times – filling and tamping is repeated three more times until the cone is full and the top smoothed off. The cone is removed and the slump measured, for consistent mixes the slump should remain the same for all samples tested. Usual specification 50 mm or 75 mm slump.

Test Cubes – these are required for laboratory strength tests~

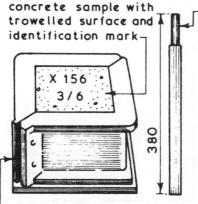

concrete sample with trowelled surface and identification mark

X 156
3/6

150 × 150 × 150 standard steel test cube mould thinly coated inside with mould oil

25 × 25 mm square end tamping bar

1. Sample taken from discharge outlet of mixer or from point of placing using random selection by means of a scoop.

2. Mould filled in three equal layers each layer well tamped with at least 35 strokes from the tamping bar.

3. Sample left in mould for 24 hours and covered with a damp sack or similar at a temperature of 4·4 to 21°C

4. Remove sample from mould and store in water at temperature of 10 to 21°C until required for testing

Refs. BS EN 12350-2 (Slump) and BS EN 12390-1 (Cubes)

Non destructive testing of concrete. Also known as in-place or in-situ tests.

Changes over time and in different exposures can be monitored.

References: BS 6089: Guide to assessment of concrete strength in existing structures;
BS 1881: Testing concrete.
BS EN 13791: Assessment of in-situ compressive strength in structures and pre-cast concrete components.

Provides information on: strength in-situ, voids, flaws, cracks and deterioration.

Rebound hammer test – attributed to Ernst Schmidt after he devised the impact hammer in 1948. It works on the principle of an elastic mass rebounding off a hard surface. Varying surface densities will affect impact and propagation of stress waves. These can be recorded on a numerical scale known as rebound numbers. It has limited application to smooth surfaces of concrete only. False results may occur where there are local variations in the concrete, such as a large piece of aggregate immediately below the impact surface. Rebound numbers can be graphically plotted to correspond with compressive strength.

Ref: BS EN 12504-2: Testing concrete in structures.

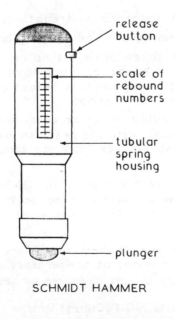

SCHMIDT HAMMER

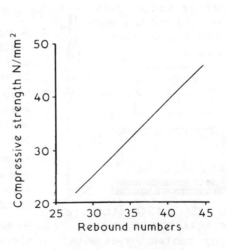

CONVERSION CHART
(illustration only)

Penetration or Windsor probe test ~ there are various interpretations of this test. It is a measure of the penetration of a steel alloy rod, fired by a predetermined amount of energy into concrete. In principle, the depth of penetration is inversely proportional to the concrete compressive strength. Several recordings are necessary to obtain a fair assessment and some can be discarded particularly where the probe cannot penetrate some dense aggregates. The advantage over the rebound hammer is provision of test results at a greater depth (up to 50mm).

Pull out test ~ this is not entirely non destructive as there will be some surface damage, albeit easily repaired. A number of circular bars of steel with enlarged ends are cast into the concrete as work proceeds. This requires careful planning and location of bars with corresponding voids provided in the formwork. At the appropriate time, the bar and a piece of concrete are pulled out by tension jack. Although the concrete fails in tension and shear, the pull out force can be correlated to the compressive strength of the concrete.

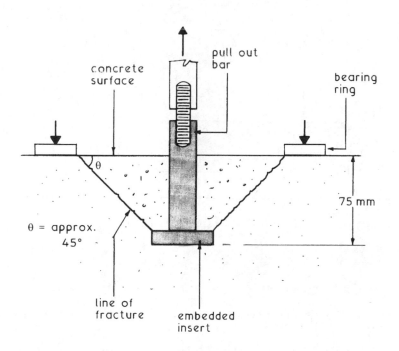

Ref: BS 1881-207: Testing concrete. Recommendations for the assessment of concrete strength by near-to-surface tests.

Vibration test ~ a number of electronic tests have been devised, which include measurement of ultrasonic pulse velocity through concrete. This applies the principle of recording a pulse at predetermined frequencies over a given distance. The apparatus includes transducers in contact with the concrete, pulse generator, amplifier, and time measurement to digital display circuit. For converting the data to concrete compressive strength, see BS EN 12504-4: Testing concrete. Determination of ultrasonic pulse velocity.

A variation, using resonant frequency, measures vibrations produced at one end of a concrete sample against a receiver or pick up at the other. The driving unit or exciter is activated by a variable frequency oscillator to generate vibrations varying in resonance, depending on the concrete quality. The calculation of compressive strength by conversion of amplified vibration data is by formulae found in BS 1881-209: Testing concrete. Recommendations for the measurement of dynamic modulus of elasticity.

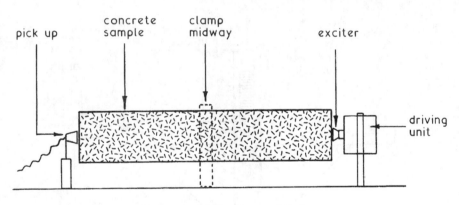

RESONANT FREQUENCY TEST

Other relevant standards:-

BS 1881-122: Testing concrete. Method for determination of water absorption.
BS 1881-124: Testing concrete. Methods for analysis of hardened concrete.
BS EN 12390-7: Testing hardened concrete. Density of hardened concrete.

Trees ~ these are part of our national heritage and are also the source of timber – to maintain this source a control over tree felling has been established under the Forestry Act 1967 which places the control responsibility on the Forestry Commission. Local planning authorities also have powers under the Town and Country Planning Act 1990 and the Town and Country Amenities Act 1974 to protect trees by making tree preservation orders. Contravention of such an order can lead to a substantial fine and a compulsion to replace any protected tree which has been removed or destroyed. Trees on building sites which are covered by a tree preservation order should be protected by a suitable fence.

tree covered by a tree preservation order

cleft chestnut or similar fencing at least 1·200 high erected to the full spread and completely encircling the tree

Trees, shrubs, bushes and tree roots which are to be removed from site can usually be grubbed out using hand held tools such as saws, picks and spades. Where whole trees are to be removed for relocation special labour and equipment is required to ensure that the roots, root earth ball and bark are not damaged.

Structures ~ buildings which are considered to be of historic or architectural interest can be protected under the Town and Country Acts provisions. The Department of the Environment lists buildings according to age, architectural, historical and/or intrinsic value. It is an offence to demolish or alter a listed building without first obtaining 'listed building consent' from the local planning authority. Contravention is punishable by a fine and/or imprisonment. It is also an offence to demolish a listed building without giving notice to the Royal Commission on Historic Monuments, this is to enable them to note and record details of the building.

Services which may be encountered on construction sites and the authority responsible are:-

Water – Local Water Company

Electricity – transmission ~ RWE npower, BNFL and E-on.

distribution ~ Area Electricity Companies in England and Wales. Scottish Power and Scottish Hydro-Electric, EDF Energy.

Gas – Local gas or energy service providers, e.g. British Gas.

Telephones – National Telecommunications Companys, eg. BT, C&W, etc.

Drainage – Local Authority unless a private drain or sewer when owner(s) is responsible.

All the above authorities must be notified of any proposed new services and alterations or terminations to existing services before any work is carried out.

Locating Existing Services on Site ~

Method 1 – By reference to maps and plans prepared and issued by the respective responsible authority.

Method 2 – Using visual indicators ~

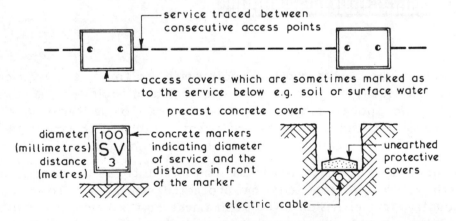

service traced between consecutive access points

access covers which are sometimes marked as to the service below e.g. soil or surface water

precast concrete cover

diameter (millimetres)
distance (metres)

100
S V
3

concrete markers indicating diameter of service and the distance in front of the marker

unearthed protective covers

electric cable

Method 3 – Detection specialist contractor employed to trace all forms of underground services using electronic subsurface survey equipment.

Once located, position and type of service can be plotted on a map or plan, marked with special paint on hard surfaces and marked with wood pegs with indentification data on earth surfaces.

Setting Out the Building Outline ~ this task is usually undertaken once the site has been cleared of any debris or obstructions and any reduced level excavation work is finished. It is usually the responsibility of the contractor to set out the building(s) using the information provided by the designer or architect. Accurate setting out is of paramount importance and should therefore only be carried out by competent persons and all their work thoroughly checked, preferably by different personnel and by a different method.

The first task in setting out the building is to establish a base line to which all the setting out can be related. The base line very often coincides with the building line which is a line, whose position on site is given by the local authority in front of which no development is permitted.

Typical Setting Out Example ~

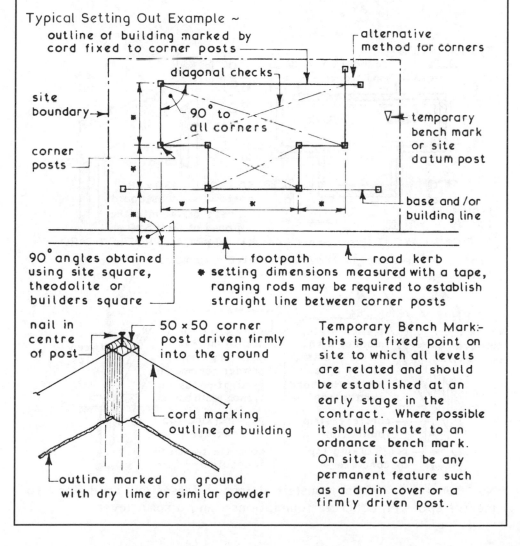

outline of building marked by cord fixed to corner posts — alternative method for corners

diagonal checks

site boundary→

90° to all corners

corner posts

temporary bench mark or site datum post

base and/or building line

90° angles obtained using site square, theodolite or builders square

footpath — road kerb

* setting dimensions measured with a tape, ranging rods may be required to establish straight line between corner posts

nail in centre of post — 50 × 50 corner post driven firmly into the ground

cord marking outline of building

outline marked on ground with dry lime or similar powder

Temporary Bench Mark:- this is a fixed point on site to which all levels are related and should be established at an early stage in the contract. Where possible it should relate to an ordnance bench mark. On site it can be any permanent feature such as a drain cover or a firmly driven post.

Setting Out Trenches ~ the objective of this task is twofold. Firstly it must establish the excavation size, shape and direction and secondly it must establish the width and position of the walls. The outline of building will have been set out and using this outline profile boards can be set up to control the position, width and possibly the depth of the proposed trenches. Profile boards should be set up at least 2·000 clear of trench positions so they do not obstruct the excavation work. The level of the profile crossboard should be related to the site datum and fixed at a convenient height above ground level if a traveller is to be used to control the depth of the trench. Alternatively the trench depth can be controlled using a level and staff related to site datum. The trench width can be marked on the profile with either nails or sawcuts and with a painted band if required for identification.

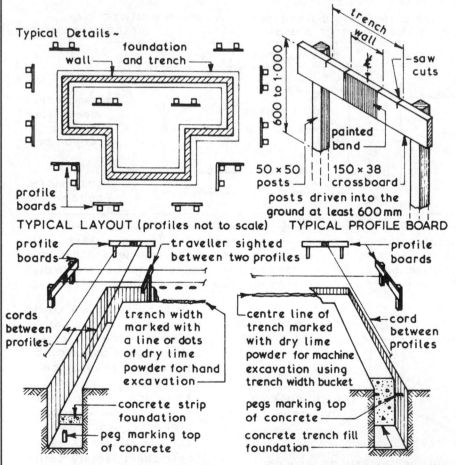

Typical Details ~

wall — foundation and trench

profile boards

TYPICAL LAYOUT (profiles not to scale)

trench wall

saw cuts

painted band

50 × 50 posts — 150 × 38 crossboard

posts driven into the ground at least 600 mm

600 to 1·000

TYPICAL PROFILE BOARD

profile boards — traveller sighted between two profiles

profile boards

cords between profiles

trench width marked with a line or dots of dry lime powder for hand excavation

centre line of trench marked with dry lime powder for machine excavation using trench width bucket

cord between profiles

concrete strip foundation

peg marking top of concrete

pegs marking top of concrete

concrete trench fill foundation

NB. Corners of walls transferred from intersecting cord lines to mortar spots on concrete foundations using a spirit level

Setting Out a Framed Building ~ framed buildings are u̶
related to a grid, the intersections of the grid lines being ̶
centre point of an isolated or pad foundation. The grid is usually
set out from a base line which does not always form part of the
grid. Setting out dimensions for locating the grid can either be
given on a drawing or they will have to be accurately scaled off a
general layout plan. The grid is established using a theodolite and
marking the grid line intersections with stout pegs. Once the grid
has been set out offset pegs or profiles can be fixed clear of any
subsequent excavation work. Control of excavation depth can be
by means of a traveller sighted between sight rails or by level and
staff related to site datum.

Typical Details ~

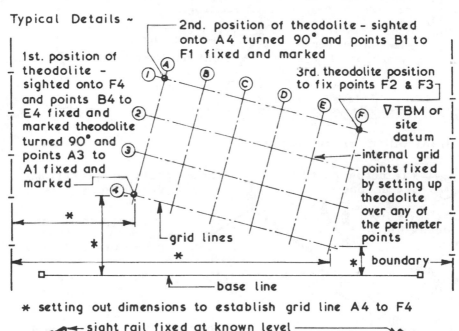

2nd. position of theodolite - sighted
onto A4 turned 90° and points B1 to
F1 fixed and marked

1st. position of
theodolite -
sighted onto F4
and points B4 to
E4 fixed and
marked theodolite
turned 90° and
points A3 to
A1 fixed and
marked

3rd. theodolite position
to fix points F2 & F3

▽ TBM or
site
datum

internal grid
points fixed
by setting up
theodolite
over any of
the perimeter
points

grid lines

boundary

base line

* setting out dimensions to establish grid line A4 to F4

sight rail fixed at known level
fenced peg - alternative to
profile

profile
board

cords

profile
board

excavation

traveller
sighted
between sight
rails to
control depth

pad template

grid setting out peg

1. Pad template positioned with
   cords between profiles and
   pad outline marked with dry
   lime or similar powder.

2. Pad pits excavated using
   traveller sighted between
   sight rails fixed at a level
   related to site datum.

113

...educed Level Excavations ~ the overall outline of ...vel area can be set out using a theodolite, ranging ...d pegs working from a base line. To control the depth ...., sight rails are set up at a convenient height and at ...ch will enable a traveller to be used.

Typical Details ~

1. Setting up sight rails :-

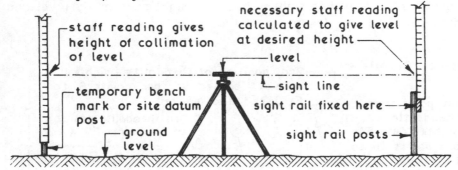

staff reading gives height of collimation of level

temporary bench mark or site datum post

ground level

necessary staff reading calculated to give level at desired height

level

sight line

sight rail fixed here

sight rail posts

2. Controlling excavation depth :-

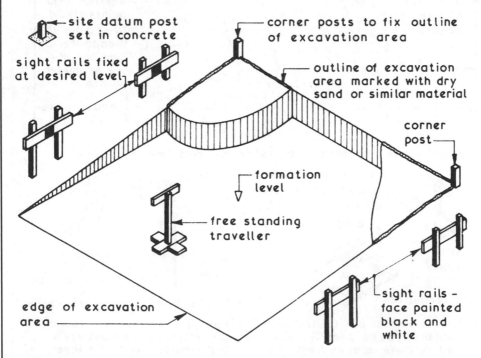

site datum post set in concrete

sight rails fixed at desired level

corner posts to fix outline of excavation area

outline of excavation area marked with dry sand or similar material

corner post

formation level

free standing traveller

edge of excavation area

sight rails - face painted black and white

height of traveller = desired level of sight rail - formation level

Levelling ~ the process of establishing height dimensions, relative to a fixed point or datum. Datum is mean sea level, which varies between different countries. For UK purposes this is established at Newlyn in Cornwall, from tide data recorded between May 1915 and April 1921. Relative levels defined by benchmarks are located throughout the country. The most common, identified as carved arrows, can be found cut into walls of stable structures. Reference to Ordnance Survey maps of an area will indicate benchmark positions and their height above sea level, hence the name Ordnance Datum (OD).

On site it is usual to measure levels from a temporary benchmark (TBM), i.e. a manhole cover or other permanent fixture, as an OD may be some distance away.

Instruments consist of a level (tilting or automatic) and a staff. A tilting level is basically a telescope mounted on a tripod for stability. Correcting screws establish accuracy in the horizontal plane by air bubble in a vial and focus is by adjustable lens. Cross hairs of horizontal and vertical lines indicate image sharpness on an extending staff of 3, 4 or 5m length. Staff graduations are in 10mm intervals, with estimates taken to the nearest millimetre. An automatic level is much simpler to use, eliminating the need for manual adjustment. It is approximately levelled by centre bulb bubble. A compensator within the telescope effects fine adjustment.

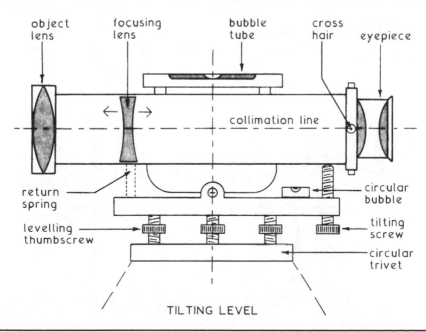

TILTING LEVEL

Application ~ methods to determine differences in ground levels for calculation of site excavation volumes and costs.

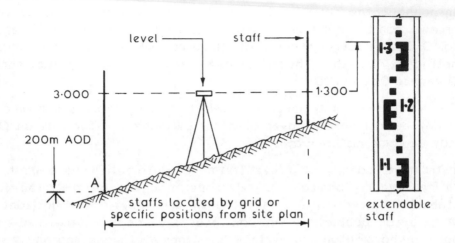

3·000

200m AOD

A

level

staff

B

1·300

staffs located by grid or specific positions from site plan

extendable staff

Rise and fall:

Staff reading at A = 3·00 m, B = 1·30 m
Ground level at A = 200 m above ordnance datum (AOD)
Therefore level at B = 200 m + rise (– fall if declining)
So level at B = 200 + (3·00 – 1·30) = 201·7 m

Height of collimation (HC):

HC at A = Reduced level (RL) + staff reading
= 200 m + 3·00 m = 203 m AOD
Level at B = HC at A – staff reading at B
= 203 – 1·30 = 201·7 m

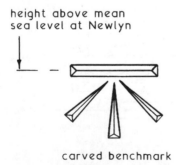

height above mean sea level at Newlyn

carved benchmark

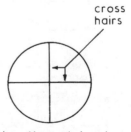

cross hairs

view through level

Theodolite – a tripod mounted instrument designed to measure angles in the horizontal or vertical plane.

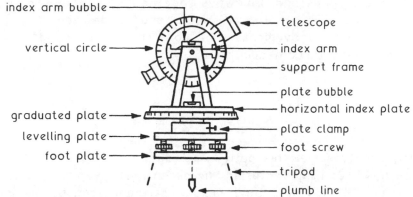

The theodolite in principle

Measurement – a telescope provides for focal location between instrument and subject. Position of the scope is defined by an index of angles. The scale and presentation of angles varies from traditional micrometer readings to computer compatible crystal displays. Angles are measured in degrees, minutes and seconds, e.g. 165° 53′ 30″.

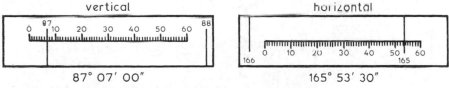

vertical	horizontal
87° 07′ 00″	165° 53′ 30″

Direct reading micrometer scale

Application – at least two sightings are taken and the readings averaged. After the first sighting, the horizontal plate is rotated through 180° and the scope also rotated 180° through the vertical to return the instrument to its original alignment for the second reading. This process will move the vertical circle from right face to left face, or vice-versa. It is important to note the readings against the facing – see below.

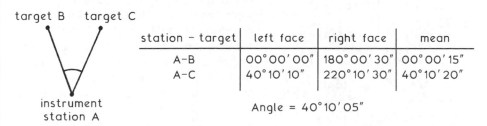

station – target	left face	right face	mean
A–B	00° 00′ 00″	180° 00′ 30″	00° 00′ 15″
A–C	40° 10′ 10″	220° 10′ 30″	40° 10′ 20″

Angle = 40° 10′ 05″

Defining an angle

Road Construction ~ within the context of building operations roadworks usually consist of the construction of small estate roads, access roads and driveways together with temporary roads laid to define site circulation routes and/or provide a suitable surface for plant movements. The construction of roads can be considered under three headings:-

1. Setting out.
2. Earthworks (see page 119).
3. Paving Construction (see pages 120 & 121).

Setting Out Roads ~ this activity is usually carried out after the topsoil has been removed using the dimensions given on the layout drawing(s). The layout could include straight lengths junctions, hammer heads, turning bays and intersecting curves.

Straight Road Lengths – these are usually set out from centre lines which have been established by traditional means

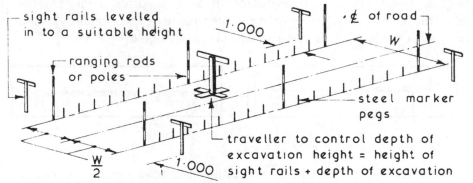

NB. curve road lengths set out in a similar manner

Junctions and Hammer Heads –

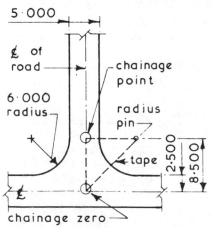

Centre lines fixed by traditional methods. Tape hooked over pin at chainage zero and passed around chainage point pin at 8·500 then returned to chainage zero with a tape length of 29·021. Radius pin held tape length 17·000 and tape is moved until tight between all pins. Radius pin is driven and a 6·000 tape length is swung from the pin to trace out curve which is marked with pegs or pins.

Tape length $= 17 + \sqrt{2} \times 8·5$
$\qquad\qquad\quad = 29·021$

Earthworks ~ this will involve the removal of topsoil together with any vegetation, scraping and grading the required area down to formation level plus the formation of any cuttings or embankments. Suitable plant for these operations would be tractor shovels fitted with a 4 in 1 bucket (page 158): graders (page 157) and bulldozers (page 155). The soil immediately below the formation level is called the subgrade whose strength will generally decrease as its moisture content rises therefore if it is to be left exposed for any length of time protection may be required. Subgrade protection may take the form of a covering of medium gauge plastic sheeting with 300mm laps or alternatively a covering of sprayed bituminous binder with a sand topping applied at a rate of 1 litre per m². To preserve the strength and durability of the subgrade it may be necessary to install cut off subsoil drains alongside the proposed road (see Road Drainage on page 688).

Paving Construction ~ once the subgrade has been prepared and any drainage or other buried services installed the construction of the paving can be undertaken. Paved surfaces can be either flexible or rigid in format. Flexible or bound surfaces are formed of materials applied in layers directly over the subgrade whereas rigid pavings consist of a concrete slab resting on a granular base (see pages 120 & 121).

Typical Flexible Paving Details ~

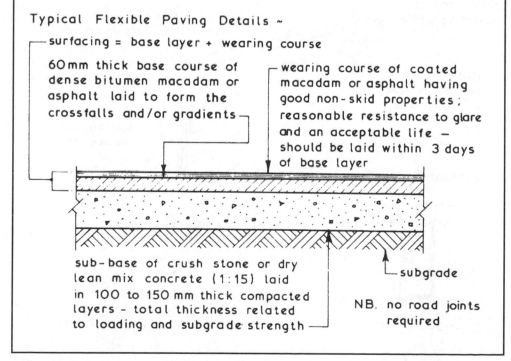

surfacing = base layer + wearing course

60mm thick base course of dense bitumen macadam or asphalt laid to form the crossfalls and/or gradients

wearing course of coated macadam or asphalt having good non-skid properties; reasonable resistance to glare and an acceptable life — should be laid within 3 days of base layer

sub-base of crush stone or dry lean mix concrete (1:15) laid in 100 to 150 mm thick compacted layers - total thickness related to loading and subgrade strength

subgrade

NB. no road joints required

119

Rigid Pavings ~ these consist of a reinforced or unreinforced in-situ concrete slab laid over a base course of crushed stone or similar material which has been blinded to receive a polythene sheet slip membrane. The primary objective of this membrane is to prevent grout loss from the in-situ slab.

Typical Rigid Paving Details ~

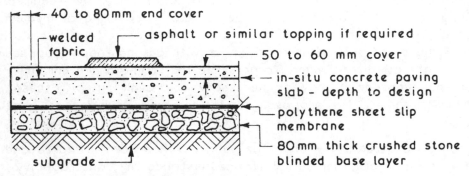

The paving can be laid between metal road forms or timber edge formwork. Alternatively the kerb stones could be laid first to act as permanent formwork.

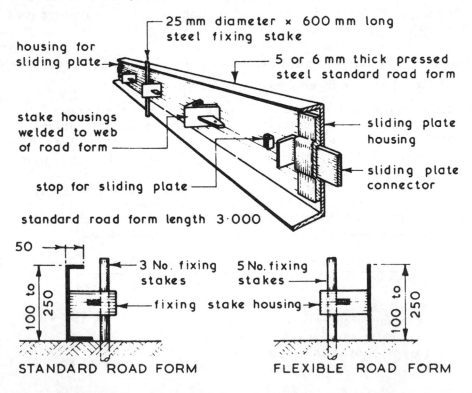

Joints in Rigid Pavings ~ longitudinal and transverse joints are required in rigid pavings to:-

1. Limit size of slab.
2. Limit stresses due to subgrade restraint.
3. Provide for expansion and contraction movements.

The main joints used are classified as expansion, contraction or longitudinal, the latter being the same in detail as the contraction joint differing only in direction. The spacing of road joints is determined by:-

1. Slab thickness.
2. Whether slab is reinforced or unreinforced.
3. Anticipated traffic load and flow rate.
4. Temperature at which concrete is laid.

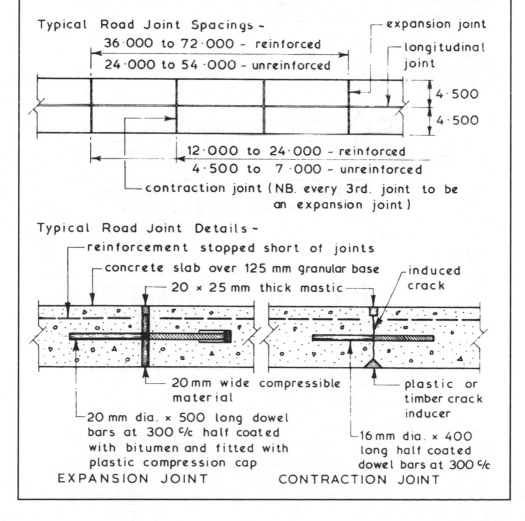

Typical Road Joint Spacings ~

36·000 to 72·000 – reinforced
24·000 to 54·000 – unreinforced

expansion joint
longitudinal joint

4·500
4·500

12·000 to 24·000 – reinforced
4·500 to 7·000 – unreinforced

contraction joint (NB. every 3rd. joint to be an expansion joint)

Typical Road Joint Details ~

reinforcement stopped short of joints
concrete slab over 125 mm granular base
20 × 25 mm thick mastic
induced crack

20 mm wide compressible material
plastic or timber crack inducer

20 mm dia. × 500 long dowel bars at 300 c/c half coated with bitumen and fitted with plastic compression cap

16 mm dia. × 400 long half coated dowel bars at 300 c/c

EXPANSION JOINT   CONTRACTION JOINT

Typical Examples ~

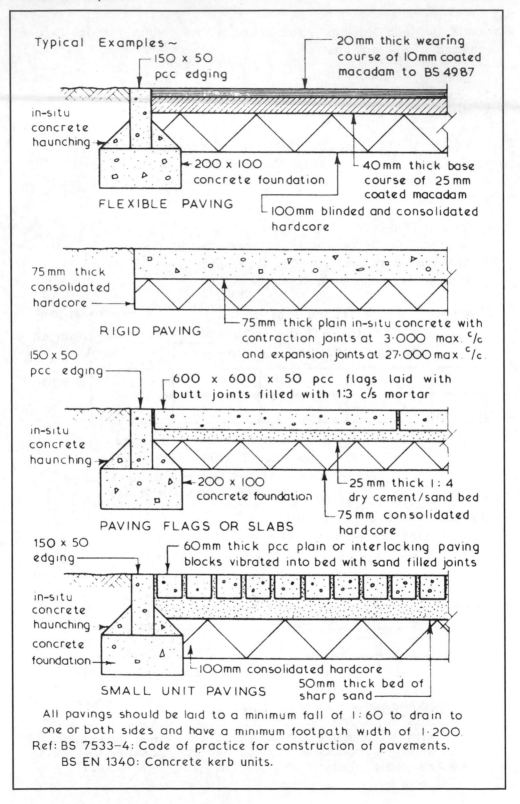

150 x 50 pcc edging

20mm thick wearing course of 10mm coated macadam to BS 4987

in-situ concrete haunching

200 x 100 concrete foundation

40mm thick base course of 25mm coated macadam

100mm blinded and consolidated hardcore

FLEXIBLE PAVING

75mm thick consolidated hardcore

75mm thick plain in-situ concrete with contraction joints at 3·000 max. c/c and expansion joints at 27·000 max. c/c.

RIGID PAVING

150 x 50 pcc edging

600 x 600 x 50 pcc flags laid with butt joints filled with 1:3 c/s mortar

in-situ concrete haunching

200 x 100 concrete foundation

25mm thick 1:4 dry cement/sand bed

75mm consolidated hardcore

PAVING FLAGS OR SLABS

150 x 50 edging

60mm thick pcc plain or interlocking paving blocks vibrated into bed with sand filled joints

in-situ concrete haunching

concrete foundation

100mm consolidated hardcore

50mm thick bed of sharp sand

SMALL UNIT PAVINGS

All pavings should be laid to a minimum fall of 1:60 to drain to one or both sides and have a minimum footpath width of 1·200.
Ref: BS 7533–4: Code of practice for construction of pavements.
BS EN 1340: Concrete kerb units.

Available sections ~ manufactured in 915mm lengths
from silver/grey aggregate concrete.

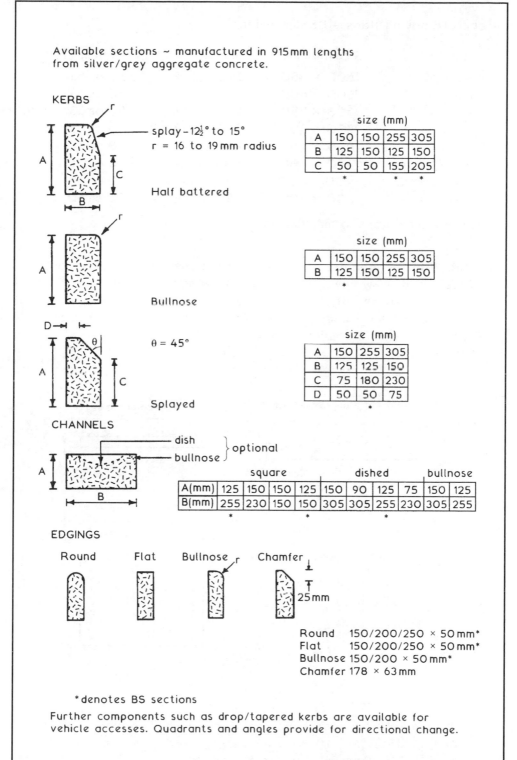

KERBS

splay – 12½° to 15°
r = 16 to 19 mm radius

Half battered

size (mm)

A	150	150	255	305
B	125	150	125	150
C	50	50	155	205
	*		*	*

Bullnose

size (mm)

A	150	150	255	305
B	125	150	125	150
	*			

θ = 45°

Splayed

size (mm)

A	150	255	305
B	125	125	150
C	75	180	230
D	50	50	75
		*	

CHANNELS

dish
bullnose } optional

	square				dished			bullnose	
A(mm)	125	150	150	125	150	90	125	75	150	125
B(mm)	255	230	150	150	305	305	255	230	305	255
	*				*			*		

EDGINGS

Round   Flat   Bullnose r   Chamfer

25mm

Round   150/200/250 × 50 mm*
Flat   150/200/250 × 50 mm*
Bullnose 150/200 × 50 mm*
Chamfer 178 × 63 mm

*denotes BS sections

Further components such as drop/tapered kerbs are available for
vehicle accesses. Quadrants and angles provide for directional change.

Concrete paving flags – BS dimensions:

Type	Size (nominal)	Size (work)	Thickness (T)
A – plain	600 × 450	598 × 448	50 or 63
B – plain	600 × 600	598 × 598	50 or 63
C – plain	600 × 750	598 × 748	50 or 63
D – plain	600 × 900	598 × 898	50 or 63
E – plain	450 × 450	448 × 448	50 or 70
TA/E – tactile	450 × 450	448 × 448	50 or 70
TA/F – tactile	400 × 400	398 × 398	50 or 65
TA/G – tactile	300 × 300	298 × 298	50 or 60

Note: All dimensions in millimetres.

Tactile flags – manufactured with a blistered (shown) or ribbed surface. Used in walkways to provide warning of hazards or to enable recognition of locations for people whose visibility is impaired. See also, Department of Transport Disability Circular DU 1/86[1], for uses and applications.

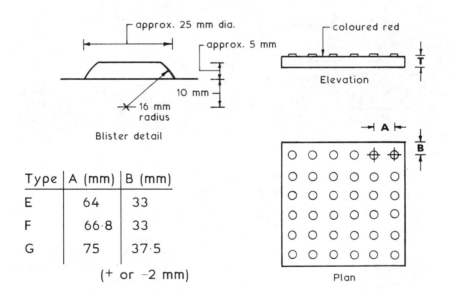

Blister detail

Type	A (mm)	B (mm)
E	64	33
F	66·8	33
G	75	37·5

(+ or −2 mm)

Plan

Ref. BS 7263–1: Pre-cast concrete flags, kerbs, channels, edgings and quadrants.
BS EN 1339: Concrete paving flags.

Landscaping ~ in the context of building works this would involve reinstatement of the site as a preparation to the landscaping in the form of lawns, paths, pavings, flower and shrub beds and tree planting. The actual planning, lawn laying and planting activities are normally undertaken by a landscape subcontractor. The main contractor's work would involve clearing away all waste and unwanted materials, breaking up and levelling surface areas, removing all unwanted vegetation, preparing the subsoil for and spreading topsoil to a depth of at least 150mm.

Services ~ the actual position and laying of services is the responsibility of the various service boards and undertakings. The best method is to use the common trench approach, avoid as far as practicable laying services under the highway.

Typical Common Trench Details ~

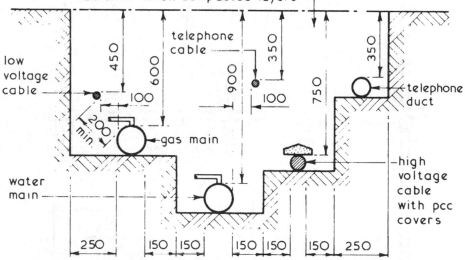

Road Signs ~ these can range from markings painted on roads to define traffic lanes, rights of way and warnings of hazards to signs mounted above the road level to give information, warning or directives, the latter being obligatory.

Typical Examples ~

INFORMATION   WARNING – ROAD WORKS   DIRECTIVE – NO LEFT TURN

Scaffolds ~ these are temporary working platforms erected around the perimeter of a building or structure to provide a safe working place at a convenient height. They are usually required when the working height or level is 1·500 or more above the ground level. All scaffolds must comply with the minimum requirements and objectives of the Work at Height Regulations 2005.

Component Parts of a Tubular Scaffold ~

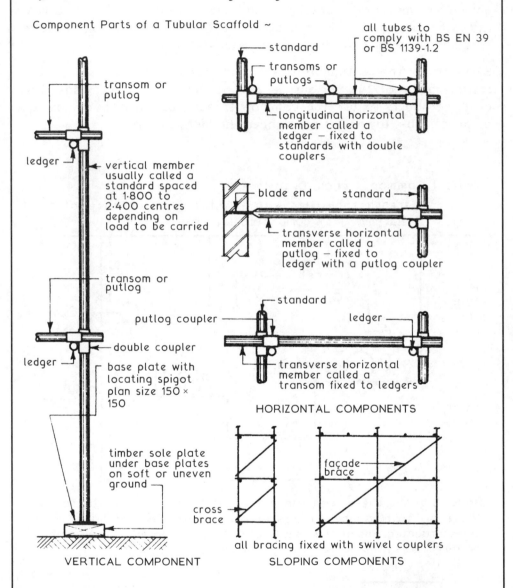

all tubes to comply with BS EN 39 or BS 1139-1.2

standard

transoms or putlogs

longitudinal horizontal member called a ledger — fixed to standards with double couplers

blade end   standard

transverse horizontal member called a putlog — fixed to ledger with a putlog coupler

standard

putlog coupler   ledger

transverse horizontal member called a transom fixed to ledgers

HORIZONTAL COMPONENTS

transom or putlog

ledger

vertical member usually called a standard spaced at 1·800 to 2·400 centres depending on load to be carried

transom or putlog

ledger

double coupler

base plate with locating spigot plan size 150 × 150

timber sole plate under base plates on soft or uneven ground

façade brace

cross brace

all bracing fixed with swivel couplers

VERTICAL COMPONENT   SLOPING COMPONENTS

Refs. BS EN 39: Loose steel tubes for tube and coupler scaffolds.
BS 1139-1.2: Metal scaffolding. Tubes. Specification for aluminium tube.

Putlog Scaffolds ~ these are scaffolds which have an outer row of standards joined together by ledgers which in turn support the transverse putlogs which are built into the bed joints or perpends as the work proceeds, they are therefore only suitable for new work in bricks or blocks.

Typical Details ~

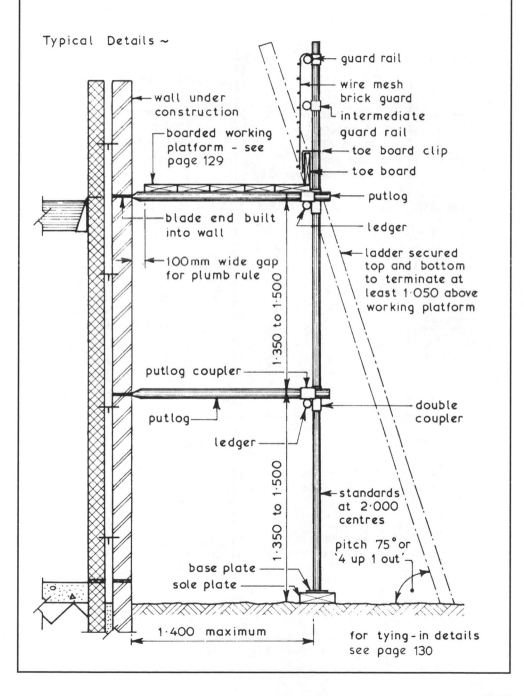

- wall under construction
- boarded working platform – see page 129
- blade end built into wall
- 100 mm wide gap for plumb rule
- putlog coupler
- putlog
- ledger
- base plate
- sole plate
- 1·400 maximum

1·350 to 1·500

1·350 to 1·500

- guard rail
- wire mesh brick guard
- intermediate guard rail
- toe board clip
- toe board
- putlog
- ledger
- ladder secured top and bottom to terminate at least 1·050 above working platform
- double coupler
- standards at 2·000 centres
- pitch 75° or '4 up 1 out'
- for tying-in details see page 130

Independent Scaffolds ~ these are scaffolds which have two rows of standards each row joined together with ledgers which in turn support the transverse transoms. The scaffold is erected clear of the existing or proposed building but is tied to the building or structure at suitable intervals – see page 130

Typical Details ~

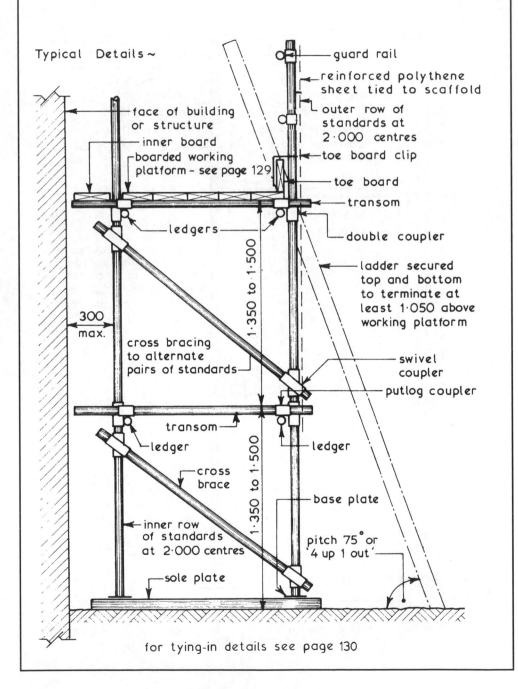

guard rail

reinforced polythene sheet tied to scaffold

outer row of standards at 2·000 centres

face of building or structure

inner board

boarded working platform – see page 129

toe board clip

toe board

transom

ledgers

double coupler

1·350 to 1·500

ladder secured top and bottom to terminate at least 1·050 above working platform

300 max.

cross bracing to alternate pairs of standards

swivel coupler

putlog coupler

transom

ledger

ledger

cross brace

1·350 to 1·500

base plate

inner row of standards at 2·000 centres

pitch 75° or '4 up 1 out'

sole plate

for tying-in details see page 130

Working Platforms ~ these are close boarded or plated level surfaces at a height at which work is being carried out and they must provide a safe working place of sufficient strength to support the imposed loads of operatives and/or materials. All working platforms above the ground level must be fitted with a toe board and a guard rail.

Typical Details ~

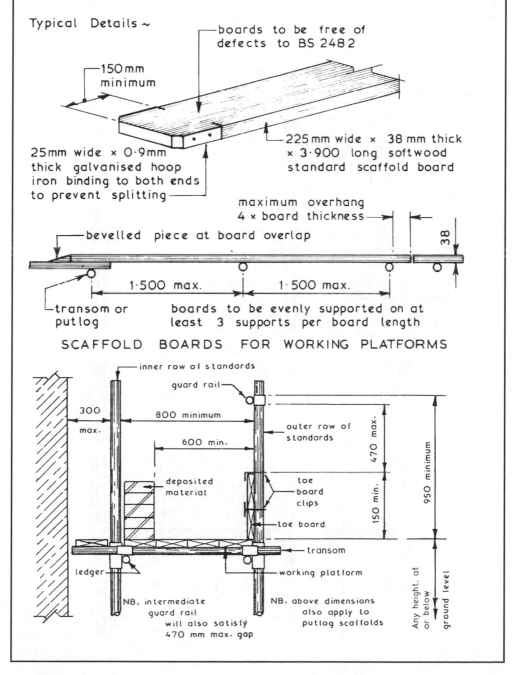

SCAFFOLD BOARDS FOR WORKING PLATFORMS

129

Tying-in ~ all putlog and independent scaffolds should be tied securely to the building or structure at alternate lift heights vertically and at not more than 6·000 centres horizontally. Putlogs should not be classified as ties.

Suitable tying-in methods include connecting to tubes fitted between sides of window openings or to internal tubes fitted across window openings, the former method should not be used for more than 50% of the total number of ties. If there is an insufficient number of window openings for the required number of ties external rakers should be used.

Typical Details ~

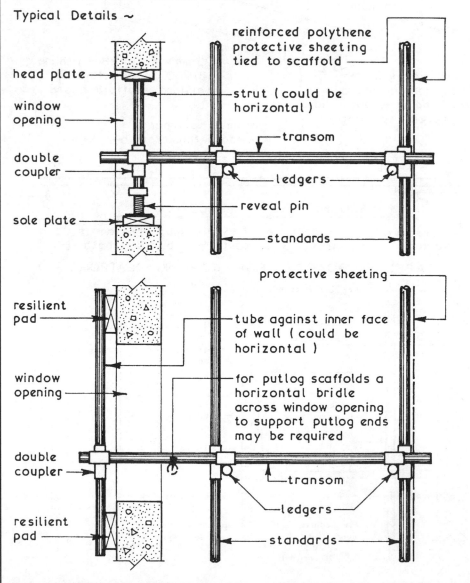

Mobile Scaffolds ~ otherwise known as mobile tower scaffolds. They can be assembled from pre-formed framing components or from standard scaffold tube and fittings. Used mainly for property maintenance. Must not be moved whilst occupied by persons or equipment.

Typical detail ~

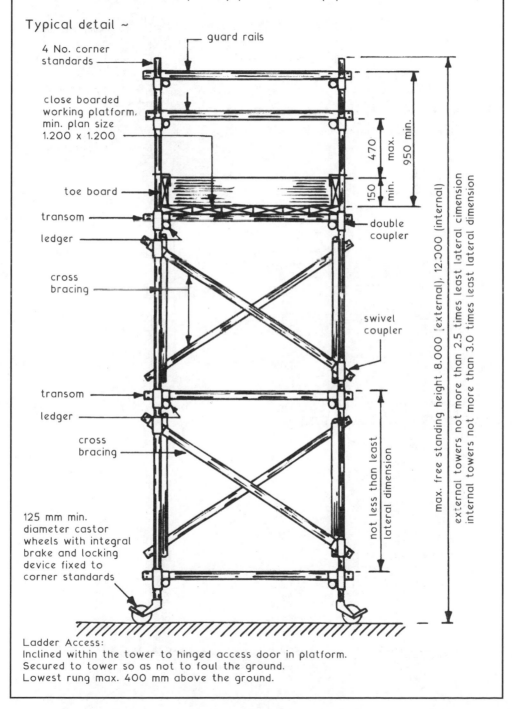

guard rails

4 No. corner standards

close boarded working platform, min. plan size 1.200 x 1.200

toe board

transom

ledger

cross bracing

transom

ledger

cross bracing

125 mm min. diameter castor wheels with integral brake and locking device fixed to corner standards

470 max. 150 min. 950 min.

double coupler

swivel coupler

not less than least lateral dimension

max. free standing height 8.000 (external). 12.000 (internal)

external towers not more than 2.5 times least lateral dimension
internal towers not more than 3.0 times least lateral dimension

Ladder Access:
Inclined within the tower to hinged access door in platform.
Secured to tower so as not to foul the ground.
Lowest rung max. 400 mm above the ground.

Some basic fittings ~

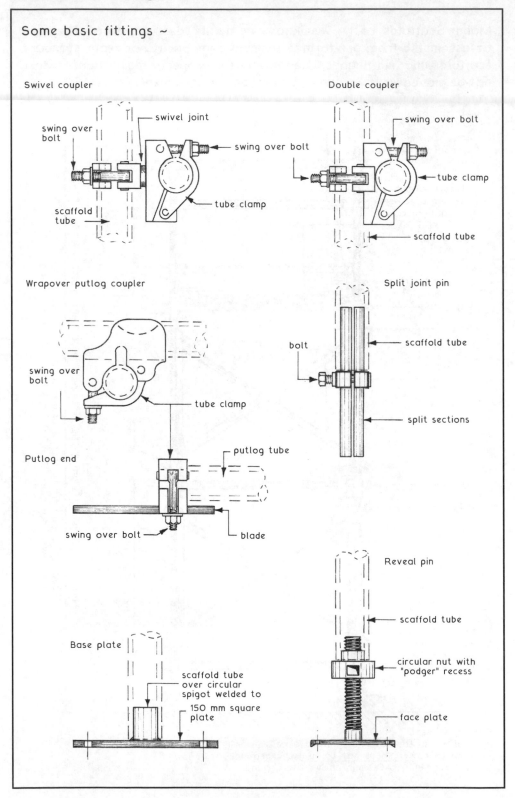

Swivel coupler

swivel joint

swing over bolt

swing over bolt

tube clamp

scaffold tube

Double coupler

swing over bolt

tube clamp

scaffold tube

Wrapover putlog coupler

swing over bolt

tube clamp

Putlog end

putlog tube

swing over bolt

blade

Split joint pin

bolt

scaffold tube

split sections

Base plate

scaffold tube over circular spigot welded to 150 mm square plate

Reveal pin

scaffold tube

circular nut with "podger" recess

face plate

Patent Scaffolding ~ these are systems based on an independent scaffold format in which the members are connected together using an integral locking device instead of conventional clips and couplers used with traditional tubular scaffolding. They have the advantages of being easy to assemble and take down using semi-skilled labour and should automatically comply with the requirements set out in the Work at Height Regulations 2005. Generally cross bracing is not required with these systems but façade bracing can be fitted if necessary. Although simple in concept patent systems of scaffolding can lack the flexibility of traditional tubular scaffolds in complex layout situations.

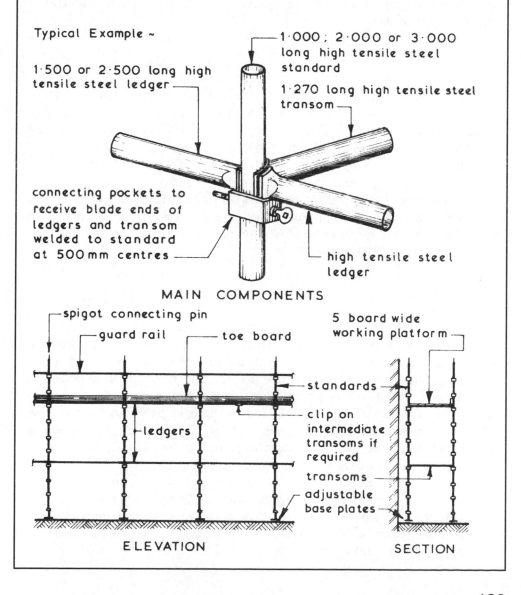

Typical Example ~

1·500 or 2·500 long high tensile steel ledger

1·000; 2·000 or 3·000 long high tensile steel standard

1·270 long high tensile steel transom

connecting pockets to receive blade ends of ledgers and transom welded to standard at 500mm centres

high tensile steel ledger

MAIN COMPONENTS

spigot connecting pin
guard rail
toe board

5 board wide working platform

standards

clip on intermediate transoms if required

ledgers

transoms

adjustable base plates

ELEVATION

SECTION

Scaffolding Systems ~ these are temporary stagings to provide safe access to and egress from a working platform. The traditional putlog and independent scaffolds have been covered on pages 126 to 130 inclusive. The minimum legal requirements contained in the Construction (Health Safety and Welfare) Regulations 1996 applicable to traditional scaffolds apply equally to special scaffolds. Special scaffolds are designed to fulfil a specific function or to provide access to areas where it is not possible and or economic to use traditional formats. They can be constructed from standard tubes or patent systems, the latter complying with most regulation requirements are easy and quick to assemble but lack the complete flexibility of the traditional tubular scaffolds.

Birdcage Scaffolds ~ these are a form of independent scaffold normally used for internal work in large buildings such as public halls and churches to provide access to ceilings and soffits for light maintenance work like painting and cleaning. They consist of parallel rows of standards connected by leaders in both directions, the whole arrangement being firmly braced in all directions. The whole birdcage scaffold assembly is designed to support a single working platform which should be double planked or underlined with polythene or similar sheeting as a means of restricting the amount of dust reaching the floor level.

Slung Scaffolds ~ these are a form of scaffold which is suspended from the main structure by means of wire ropes or steel chains and is not provided with a means of being raised or lowered. Each working platform of a slung scaffold consists of a supporting framework of ledgers and transoms which should not create a plan size in excess of 2·500 × 2·500 and be held in position by not less than six evenly spaced wire ropes or steel chains securely anchored at both ends. The working platform should be double planked or underlined with polythene or similar sheeting to restrict the amount of dust reaching the floor level. Slung scaffolds are an alternative to birdcage scaffolds and although more difficult to erect have the advantage of leaving a clear space beneath the working platform which makes them suitable for cinemas, theatres and high ceiling banking halls.

Suspended Scaffolds ~ these consist of a working platform in the form of a cradle which is suspended from cantilever beams or outriggers from the roof of a tall building to give access to the façade for carrying out light maintenance work and cleaning activities. The cradles can have manual or power control and be in single units or grouped together to form a continuous working platform. If grouped together they are connected to one another at their abutment ends with hinges to form a gap of not more than 25mm wide. Many high rise buildings have a permanent cradle system installed at roof level and this is recommended for all buildings over 30·000 high.

Typical Example ~

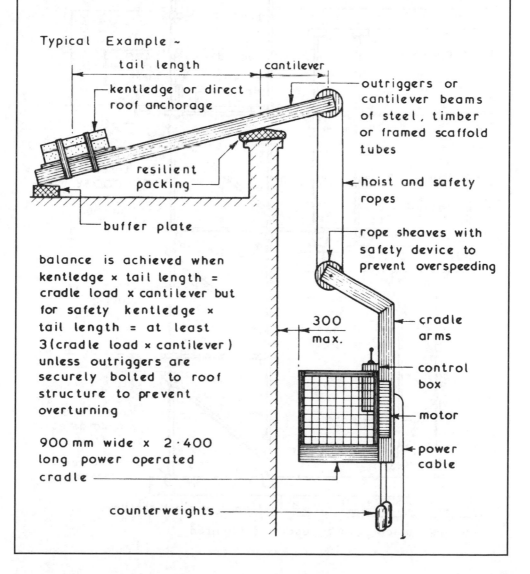

tail length

cantilever

kentledge or direct roof anchorage

outriggers or cantilever beams of steel, timber or framed scaffold tubes

resilient packing

hoist and safety ropes

buffer plate

rope sheaves with safety device to prevent overspeeding

balance is achieved when kentledge × tail length = cradle load × cantilever but for safety kentledge × tail length = at least 3(cradle load × cantilever) unless outriggers are securely bolted to roof structure to prevent overturning

300 max.

cradle arms

control box

motor

900 mm wide × 2·400 long power operated cradle

power cable

counterweights

Cantilever Scaffolds ~ these are a form of independent tied scaffold erected on cantilever beams and used where it is impracticable, undesirable or uneconomic to use a traditional scaffold raised from ground level. The assembly of a cantilever scaffold requires special skills and should therefore always be carried out by trained and experienced personnel.

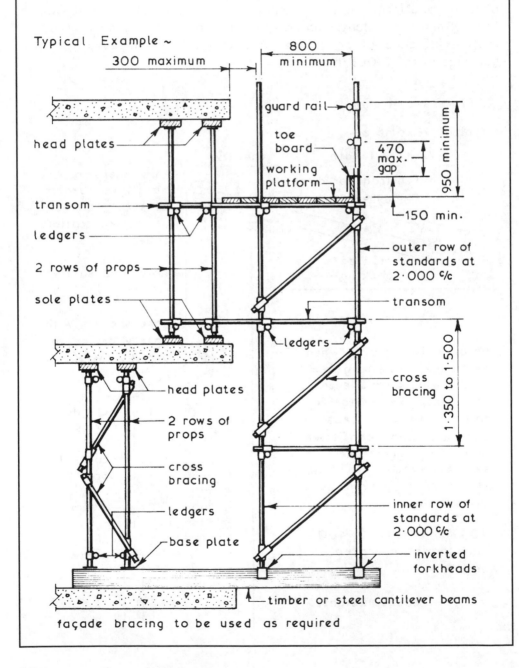

Typical Example ~

300 maximum

800 minimum

head plates

transom

ledgers

2 rows of props

sole plates

head plates

2 rows of props

cross bracing

ledgers

base plate

guard rail

toe board

working platform

470 max. gap

950 minimum

150 min.

outer row of standards at 2·000 ℅

transom

1·350 to 1·500

ledgers

cross bracing

inner row of standards at 2·000 ℅

inverted forkheads

timber or steel cantilever beams

façade bracing to be used as required

Truss-out Scaffold ~ this is a form of independent tied scaffold used where it is impracticable, undesirable or uneconomic to build a scaffold from ground level. The supporting scaffold structure is known as the truss-out. The assembly of this form of scaffold requires special skills and should therefore be carried out by trained and experienced personnel.

Typical Example ~

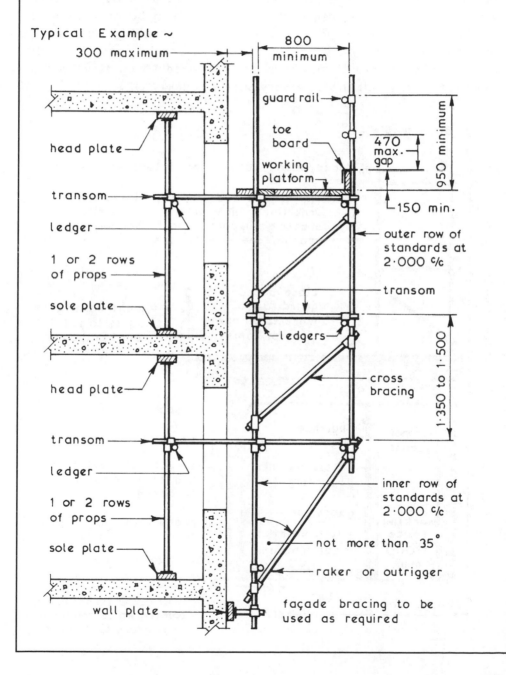

Gantries ~ these are elevated platforms used when the building being maintained or under construction is adjacent to a public footpath. A gantry over a footpath can be used for storage of materials, housing units of accommodation and supporting an independent scaffold. Local authority permission will be required before a gantry can be erected and they have the power to set out the conditions regarding minimum sizes to be used for public walkways and lighting requirements. It may also be necessary to comply with police restrictions regarding the loading and unloading of vehicles at the gantry position. A gantry can be constructed of any suitable structural material and may need to be structurally designed to meet all the necessary safety requirements.

Typical Example ~

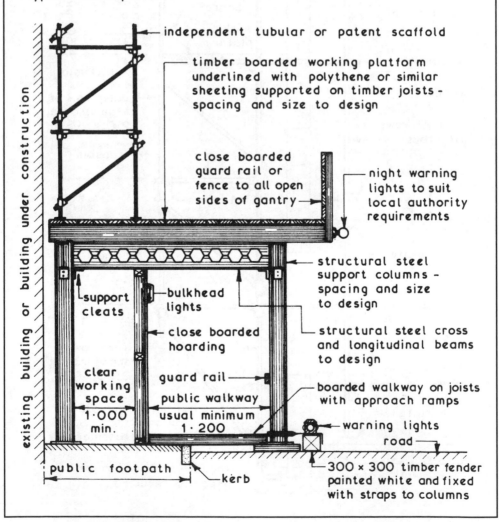

Shoring ~ this is a form of temporary support which can be given to existing buildings with the primary function of providing the necessary precautions to avoid damage to any person from collapse of structure as required by the Construction (Health, Safety and Welfare) Regulations 1996.

Shoring Systems ~ there are three basic systems of shoring which can be used separately or in combination with one another to provide the support(s) and these are namely:-

1. Dead Shoring – used primarily to carry vertical loadings.
2. Raking Shoring – used to support a combination of vertical and horizontal loadings.
3. Flying Shoring – an alternative to raking shoring to give a clear working space at ground level.

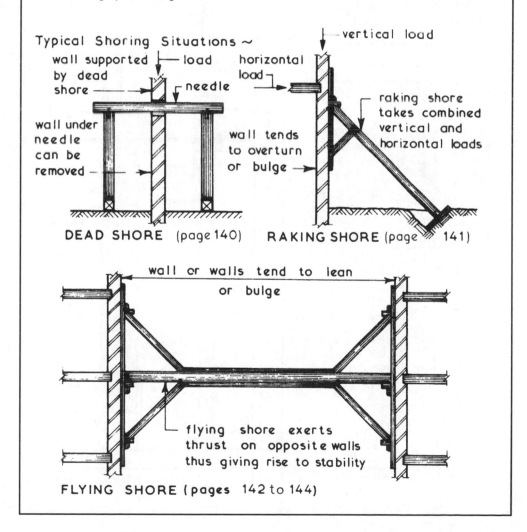

Typical Shoring Situations ~

DEAD SHORE (page 140)   RAKING SHORE (page 141)

FLYING SHORE (pages 142 to 144)

Dead Shores ~ these shores should be placed at approximately 2·000 c/c and positioned under the piers between the windows, any windows in the vicinity of the shores being strutted to prevent distortion of the openings. A survey should be carried out to establish the location of any underground services so that they can be protected as necessary. The sizes shown in the detail below are typical, actual sizes should be obtained from tables or calculated from first principles. Any suitable structural material such as steel can be substituted for the timber members shown.

Typical Detail ~

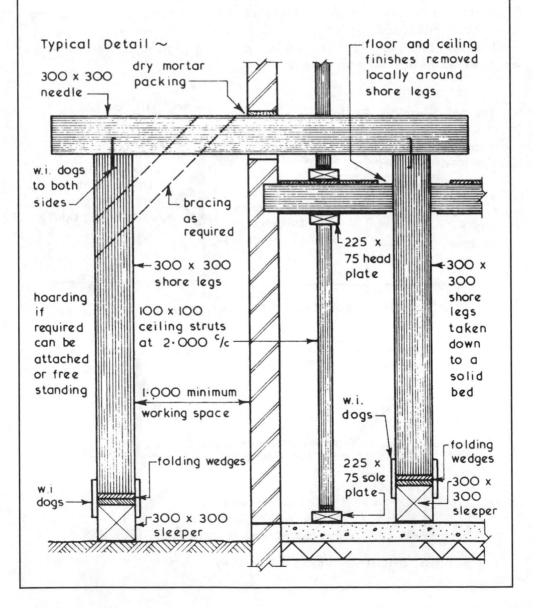

300 x 300 needle

dry mortar packing

floor and ceiling finishes removed locally around shore legs

w.i. dogs to both sides

bracing as required

300 x 300 shore legs

225 x 75 head plate

300 x 300 shore legs taken down to a solid bed

hoarding if required can be attached or free standing

100 x 100 ceiling struts at 2·000 c/c

1·000 minimum working space

w.i. dogs

folding wedges

w.i dogs

folding wedges

225 x 75 sole plate

300 x 300 sleeper

300 x 300 sleeper

Raking Shoring ~ these are placed at 3·000 to 4·500 c/c and can be of single, double, triple or multiple raker format. Suitable materials are timber structural steel and framed tubular scaffolding.

Typical Multiple Raking Shore Detail ~

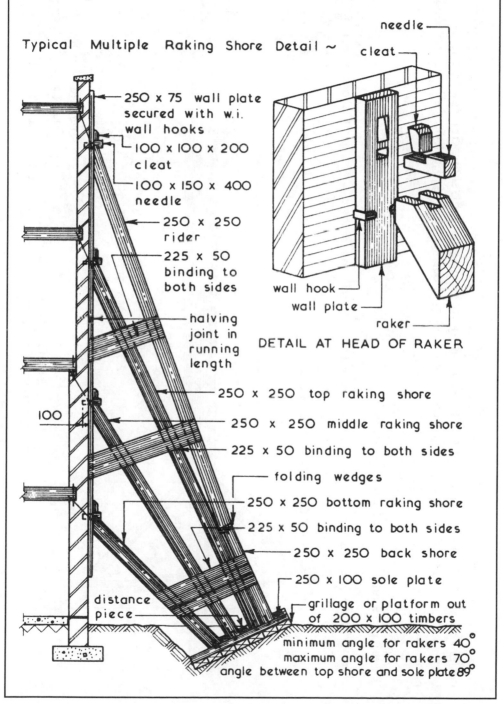

250 x 75 wall plate secured with w.i. wall hooks

100 x 100 x 200 cleat

100 x 150 x 400 needle

250 x 250 rider

225 x 50 binding to both sides

halving joint in running length

100

distance piece

needle

cleat

wall hook

wall plate

raker

DETAIL AT HEAD OF RAKER

250 x 250 top raking shore

250 x 250 middle raking shore

225 x 50 binding to both sides

folding wedges

250 x 250 bottom raking shore

225 x 50 binding to both sides

250 x 250 back shore

250 x 100 sole plate

grillage or platform out of 200 x 100 timbers

minimum angle for rakers 40°
maximum angle for rakers 70°
angle between top shore and sole plate 89°

141

Flying Shores ~ these are placed at 3.000 to 4.500 c/c and can be of a single or double format. They are designed, detailed and constructed to the same basic principles as that shown for raking shores on page 141. Unsymmetrical arrangements are possible providing the basic principles for flying shores are applied – see page 144.

Typical Single Flying Shore Detail ~

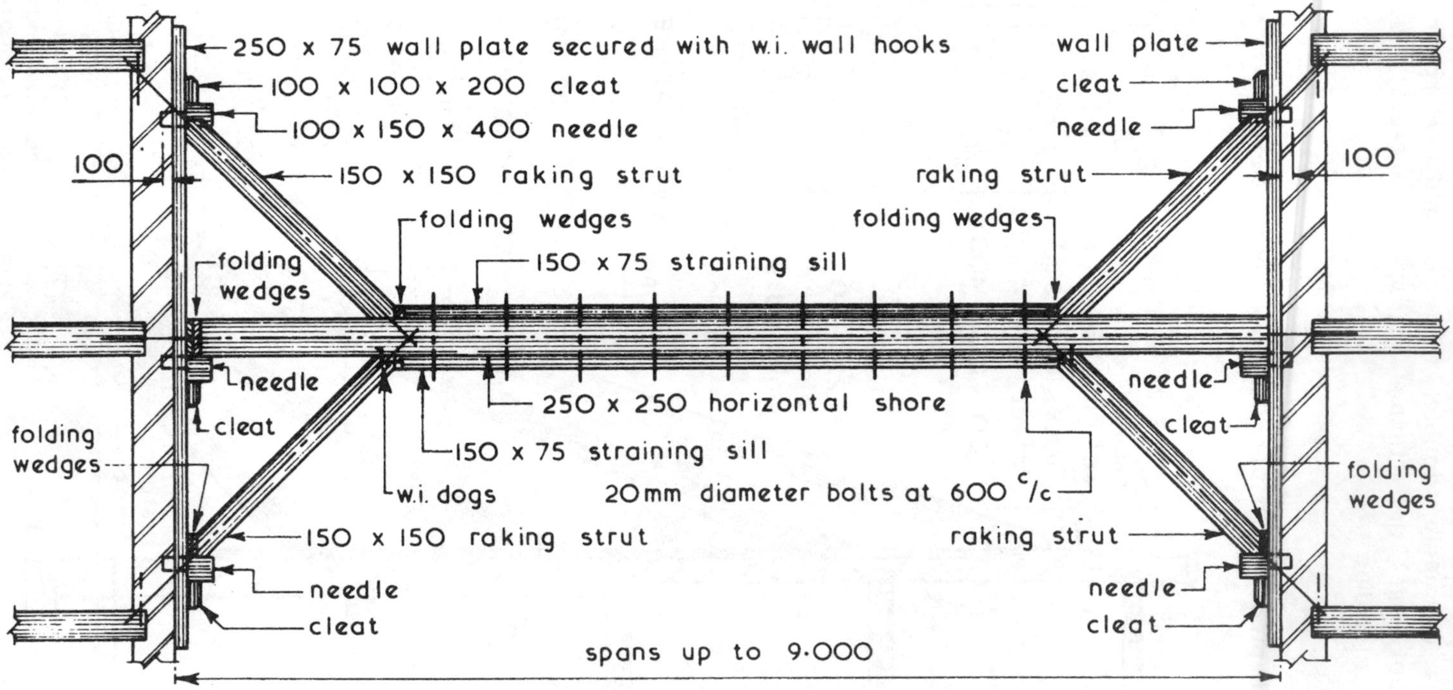

- 250 x 75 wall plate secured with w.i. wall hooks
- 100 x 100 x 200 cleat
- 100 x 150 x 400 needle
- 150 x 150 raking strut
- folding wedges
- 150 x 75 straining sill
- 150 x 75 straining sill
- w.i. dogs
- 150 x 150 raking strut
- needle
- cleat
- folding wedges
- needle
- cleat
- wall plate
- cleat
- needle
- raking strut
- folding wedges
- 250 x 250 horizontal shore
- 20mm diameter bolts at 600 c/c
- raking strut
- needle
- cleat
- needle
- cleat
- folding wedges
- 100
- 100
- spans up to 9.000

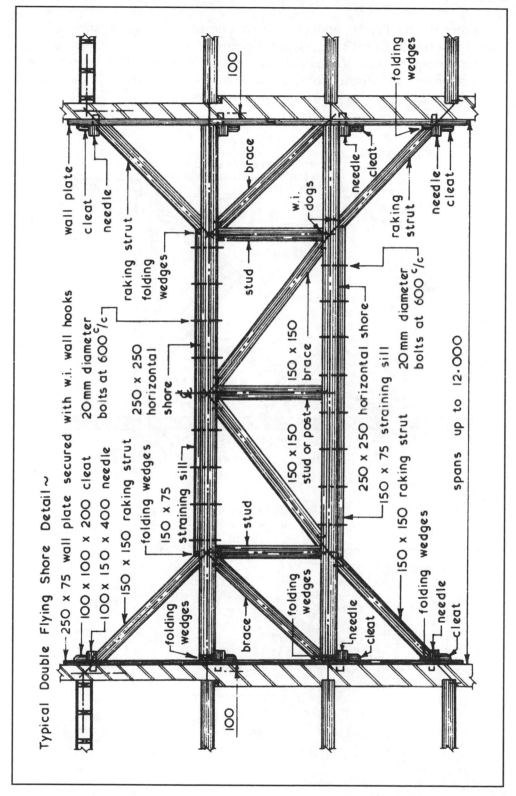

Typical Double Flying Shore Detail ~

wall plate
cleat
needle

raking strut
folding wedges

250 x 250 horizontal shore

150 x 75 straining sill

250 x 75 wall plate secured with w.i. wall hooks

100 x 100 x 200 cleat

100 x 150 x 400 needle

150 x 150 raking strut
folding wedges

150 x 75 straining sill

folding wedges

brace

stud

folding wedges

needle
cleat

brace

w.i. dogs

150 x 150 brace

150 x 150 stud or post

250 x 250 horizontal shore

150 x 75 raking strut

20mm diameter bolts at 600 c/c

raking strut

needle
cleat

folding wedges

needle
cleat

folding wedges

stud

20mm diameter bolts at 600 c/c

spans up to 12.000

needle
cleat

100

100

Unsymmetrical Flying Shores ~ arrangements of flying shores for unsymmetrical situations can be devised if the basic principles for symmetrical shores is applied (see page 142). In some cases the arrangement will consist of a combination of both raking and flying shore principles.

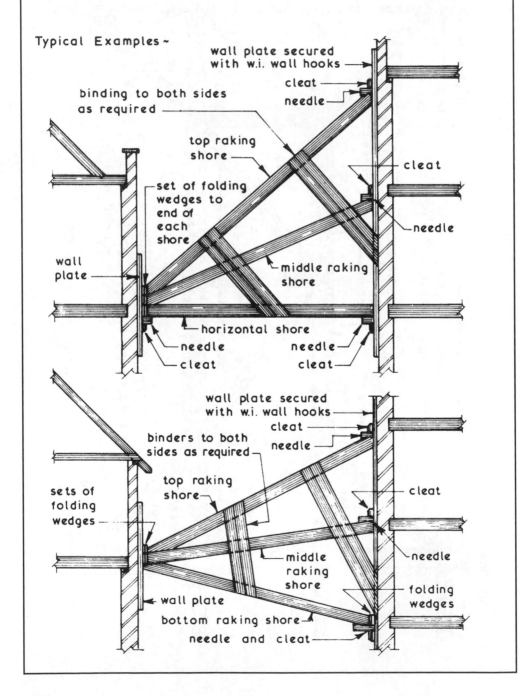

Typical Examples ~

wall plate secured with w.i. wall hooks

cleat

needle

binding to both sides as required

top raking shore

cleat

set of folding wedges to end of each shore

needle

wall plate

middle raking shore

horizontal shore

needle

cleat

needle

cleat

wall plate secured with w.i. wall hooks

cleat

needle

binders to both sides as required

top raking shore

cleat

sets of folding wedges

middle raking shore

needle

wall plate

folding wedges

bottom raking shore

needle and cleat

Temporary Support Determination ~ the basic sizing of most temporary supports follows the principles of elementary structural design. Readers with this basic knowledge should be able to calculate such support members which are required, particularly those used in the context of the maintenance and adaptation of buildings such as a dead shoring system.

Typical Example ~

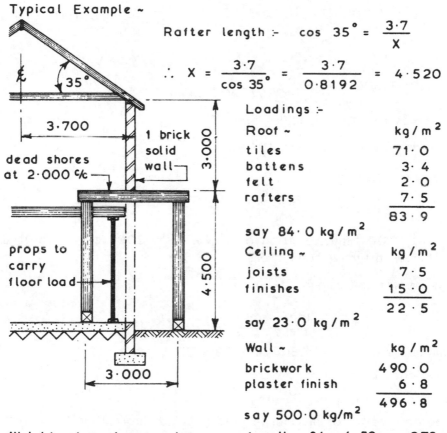

Rafter length :- $\cos 35° = \dfrac{3 \cdot 7}{X}$

$\therefore X = \dfrac{3 \cdot 7}{\cos 35°} = \dfrac{3 \cdot 7}{0 \cdot 8192} = 4 \cdot 520$

Loadings :-

Roof ~	kg/m²
tiles	71·0
battens	3·4
felt	2·0
rafters	7·5
	83·9

say 84·0 kg/m²

Ceiling ~	kg/m²
joists	7·5
finishes	15·0
	22·5

say 23·0 kg/m²

Wall ~	kg/m²
brickwork	490·0
plaster finish	6·8
	496·8

say 500·0 kg/m²

Weight of roof per metre run of wall = 84 × 4·52 = 379·68
Weight of ceiling per metre run of wall = 23 × 3·70 = 85·10
Weight of wall per metre run of wall = 500 × 3·00 = 1500·00

Total weight of wall per meter run = 1964·78

Total weight supported by needle = 1964·78 × shore centres
= 1964·78 × 2·000
= 3929·56 kg
say 3930 kg

For design calculations see next page
Note: for other typical weights of building materials
see pages 35 & 36

145

Design calculations reference previous page.

Needle Design :-

$W = 3930\,kg \qquad R_A = R_B = \dfrac{W}{2}$

hence force

$= 3930 \times 9.81 \doteq 39300\,N \qquad = \dfrac{39300}{2}$

$L = 3.000$

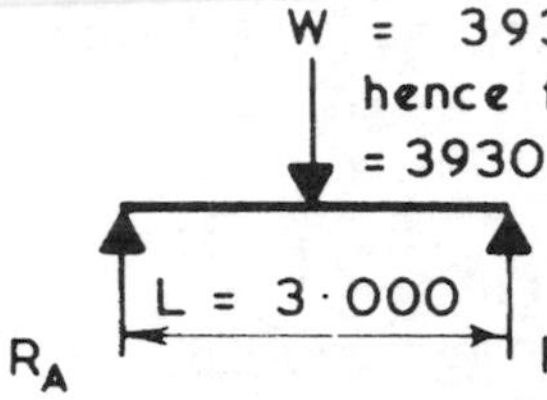

$R_A \qquad\qquad R_B \qquad\qquad\qquad = 19650\,N$

$BM = \dfrac{WL}{4} = \dfrac{39300 \times 3000}{4} = 29475000\,N/mm$

$MR = stress \times section\,modulus = fZ = f\dfrac{bd^2}{6}$

assume $b = 300\,mm$ and $f = 7\,N/mm^2$

then $29475000 = \dfrac{7 \times 300 \times d^2}{6}$

$d = \sqrt{\dfrac{29475000 \times 6}{7 \times 300}} = 290.2\,mm$

use 300×300 timber section or 2 No. 150×300 sections bolted together with timber connectors.

Props to Needle Design:-

$$area = \dfrac{load}{stress} = \dfrac{19650}{7} = 2807.143\,mm^2$$

$\therefore$ minimum timber size $= \sqrt{2807.143} = 53 \times 53$

check slenderness ratio:

slenderness ratio $= \dfrac{l}{b} = \dfrac{4500}{53} = 84.9$

slenderness ratio for medium term load is not more than 17·3 (from CP 112 – now BS 5268: Structural use of timber)

$\therefore$ minimum timber prop size $= \dfrac{l}{sr} = \dfrac{4500}{17.3} = 260.12\,mm$

for practical reasons use 300×300 prop $\therefore$ new sr $= 15$

Check crushing at point of loading on needle:-

wall loading on needle $= 3930\,kg = 39300\,N = 39.3\,kN$

area of contact $=$ width of wall $\times$ width of needle

$= 215 \times 300 = 64500\,mm^2$

safe compressive stress perpendicular to grain $= 1.72\,N/mm^2$

$\therefore$ safe load $= \dfrac{64500 \times 1.72}{1000} = 110.94\,kN$ which is $> 39.3\,kN$

Town and Country Planning Act ~ demolition is generally not regarded as development, but planning permission will be required if the site is to have a change of use. Attitudes to demolition can vary between local planning authorities and consultation should be sought.

Planning (Listed Buildings and Conservation Areas) Act ~ listed buildings and those in conservation areas will require local authority approval for any alterations. Consent for change may be limited to partial demolition, particularly where it is necessary to preserve a building frontage for historic reasons. See the arrangements for temporary shoring on the preceding pages.

Building Act ~ intention to demolish a building requires six weeks written notice of intent. The next page shows the typical outline of a standard form for submission to the building control department of the local authority, along with location plans. Notice must also be given to utilities providers and adjoining/ adjacent building owners, particularly where party walls are involved. Small buildings of volume less than 50 m^3 are generally exempt. Within six weeks of the notice being submitted, the local authority will specify their requirements for shoring, protection of adjacent buildings, debris disposal and general safety requirements under the HSE.

Public Health Act ~ the local authority can issue a demolition enforcement order to a building owner, where a building is considered to be insecure, a danger to the general public and detrimental to amenities.

Highways Act ~ concerns the protection of the general public using a thoroughfare in or near to an area affected by demolition work. The building owner and demolition contractor are required to ensure that debris and other materials are not deposited in the street unless in a suitable receptacle (skip) and the local authority highways department and police are in agreement with its location. Temporary road works require protective fencing and site hoardings must be robust and secure. All supplementary provisions such as hoardings and skips may also require adequate illumination. Provision must be made for immediate removal of poisonous and hazardous waste.

Anytown Borough Council
Building Control Section
Anytown   Tel: . . . .
UK   Fax: . . . .
                                                 Email: . . . .

NOTICE TO LOCAL AUTHORITY TO CARRY OUT DEMOLITION WORKS

THE BUILDING ACT 1984 – SECTION 80

It is my intention to commence demolition of:

. . . .

. . . .

. . . .

As shown on the attached site plan, on the . . . .
(date)

This date is at least six weeks from the date of this notice. Under section 81 of the Building Act, I anticipate your notification within six weeks.

Copies of this notice have been sent to:

• The occupants/owners of any/all buildings adjacent to the proposed demolition.
• The public services/utilities supply companies.

Signed . . . .   Date . . . .

Company name and address . . . .

. . . .

. . . .

Demolition ~ skilled and potentially dangerous work that should only be undertaken by experienced contractors.

Types of demolition ~ partial or complete removal. Partial is less dynamic than complete removal, requiring temporary support to the remaining structure. This may involve window strutting, floor props and shoring. The execution of work is likely to be limited to manual handling with minimal use of powered equipment.

Preliminaries ~ a detailed survey should include:

* an assessment of condition of the structure and the impact of removing parts on the remainder.
* the effect demolition will have on adjacent properties.
* photographic records, particularly of any noticeable defects on adjacent buildings.
* neighbourhood impact, ie. disruption, disturbance, protection.
* the need for hoardings, see pages 85 to 89.
* potential for salvaging/recycling/re-use of materials.
* extent of basements and tunnels.
* services – need to terminate and protect for future reconnections.
* means for selective removal of hazardous materials.

Insurance ~ general builders are unlikely to find demolition cover in their standard policies. All risks indemnity should be considered to cover claims from site personnel and others accessing the site. Additional third party cover will be required for claims for loss or damage to other property, occupied areas, business, utilities, private and public roads.

Salvage ~ salvaged materials and components can be valuable, bricks, tiles, slates, steel sections and timber are all marketable. Architectural features such as fireplaces and stairs will command a good price. Reclamation costs will be balanced against the financial gain.

Asbestos ~ this banned material has been used in a variety of applications including pipe insulation, fire protection, sheet claddings, linings and roofing. Samples should be taken for laboratory analysis and if necessary, specialist contractors engaged to remove material before demolition commences.

Generally ~ the reverse order of construction to gradually reduce the height. Where space in not confined, overturning or explosives may be considered.

Piecemeal ~ use of hand held equipment such as pneumatic breakers, oxy-acetylene cutters, picks and hammers. Care should be taken when salvaging materials and other reusable components. Chutes should be used to direct debris to a suitable place of collection (see page 169).

Pusher Arm ~ usually attached to a long reach articulated boom fitted to a tracked chassis. Hydraulic movement is controlled from a robust cab structure mounted above the tracks.

Wrecking Ball ~ largely confined to history, as even with safety features such as anti-spin devices, limited control over a heavy weight swinging and slewing from a crane jib will be considered unsafe in many situations.

Impact Hammer ~ otherwise known as a "pecker". Basically a large chisel operated by pneumatic power and fitted to the end of an articulated boom on a tracked chassis.

Nibbler ~ a hydraulically operated grip fitted as above that can be rotated to break brittle materials such as concrete.

Overturning ~ steel wire ropes of at least 38 mm diameter attached at high level and to an anchored winch or heavy vehicle. May be considered where controlled collapse is encouraged by initial removal of key elements of structure, typical of steel framed buildings. Alternative methods should be given preference.

Explosives ~ demolition is specialised work and the use of explosives in demolition is a further specialised practice limited to very few licensed operators. Charges are set to fire in a sequence that weakens the building to a controlled internal collapse.

Some additional references ~

BS 6187: Code of practice for demolition.

The Construction (Health, Safety and Welfare) Regulations.

The Management of Health and Safety at Work Regulations.

3 BUILDERS PLANT

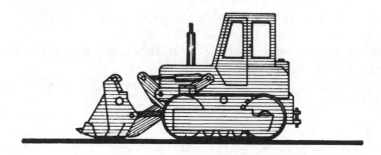

GENERAL CONSIDERATIONS
BULLDOZERS
SCRAPERS
GRADERS
TRACTOR SHOVELS
EXCAVATORS
TRANSPORT VEHICLES
HOISTS
RUBBLE CHUTES AND SKIPS
CRANES
CONCRETING PLANT

General Considerations ~ items of builders plant ranging from small hand held power tools to larger pieces of plant such as mechanical excavators and tower cranes can be considered for use for one or more of the following reasons:-

1. Increased production.
2. Reduction in overall construction costs.
3. Carry out activities which cannot be carried out by the traditional manual methods in the context of economics.
4. Eliminate heavy manual work thus reducing fatigue and as a consequence increasing productivity.
5. Replacing labour where there is a shortage of personnel with the necessary skills.
6. Maintain the high standards required particularly in the context of structural engineering works.

Economic Considerations ~ the introduction of plant does not always result in economic savings since extra temporary site works such as roadworks, hardstandings, foundations and anchorages may have to be provided at a cost which is in excess of the savings made by using the plant. The site layout and circulation may have to be planned around plant positions and movements rather than around personnel and material movements and accommodation. To be economic plant must be fully utilised and not left standing idle since plant, whether hired or owned, will have to be paid for even if it is non-productive. Full utilisation of plant is usually considered to be in the region of 85% of on site time, thus making an allowance for routine, daily and planned maintenance which needs to be carried out to avoid as far as practicable plant breakdowns which could disrupt the construction programme. Many pieces of plant work in conjunction with other items of plant such as excavators and their attendant haulage vehicles therefore a correct balance of such plant items must be obtained to achieve an economic result.

Maintenance Considerations ~ on large contracts where a number of plant items are to be used it may be advantageous to employ a skilled mechanic to be on site to carry out all the necessary daily, preventive and planned maintenance tasks together with any running repairs which could be carried out on site.

Plant Costing ~ with the exception of small pieces of plant, which are usually purchased, items of plant can be bought or hired or where there are a number of similar items a combination of buying and hiring could be considered. The choice will be governed by economic factors and the possibility of using the plant on future sites thus enabling the costs to be apportioned over several contracts.

Advantages of Hiring Plant:-

1. Plant can be hired for short periods.
2. Repairs and replacements are usually the responsibility of the hire company.
3. Plant is returned to the hire company after use thus relieving the building contractor of the problem of disposal or finding more work for the plant to justify its purchase or retention.
4. Plant can be hired with the operator, fuel and oil included in the hire rate.

Advantages of Buying Plant:-

1. Plant availability is totally within the control of the contractor.
2. Hourly cost of plant is generally less than hired plant.
3. Owner has choice of costing method used.

Typical Costing Methods ~

1. Straight Line — simple method

Capital Cost = £ 100 000
Anticipated life = 5 years
Year's working = 1500 hrs
Resale or scrap value = £9000
Annual depreciation ~

$$= \frac{100\,000 - 9000}{5} = £\,18\,200$$

Hourly depreciation ~

$$= \frac{18200}{1500} = 12 \cdot 13$$

Add 2% insurance = 0·27
10% maintenance = 1·33
Hourly rate = £13·73

2. Interest on Capital Outlay- widely used more accurate method

Capital Cost = £ 100 000
C.I. on capital
(8% for 5 yrs) = 46 930
146 930
Deduct resale value 9 000
137 930
+ Insurance at 2% = 2 000
+ Maintenance at 10% = 10 000
149 930

Hourly rate ~

$$= \frac{149\,930}{5 \times 1500} = £\,20 \cdot 00$$

N.B. add to hourly rate running costs

Output and Cycle Times ~ all items of plant have optimum output and cycle times which can be used as a basis for estimating anticipated productivity taking into account the task involved, task efficiency of the machine, operator's efficiency and in the case of excavators the type of soil. Data for the factors to be taken into consideration can be obtained from timed observations, feedback information or published tables contained in manufacturer's literature or reliable textbooks.

Typical Example ~

Backacter with $1m^3$ capacity bucket engaged in normal trench excavation in a clayey soil and discharging directly into an attendant haulage vehicle.

Optimum output	= 60 bucket loads per hour
Task efficiency factor	= 0·8 (from tables)
Operator efficiency factor	= 75% (typical figure)
∴ Anticipated output	= 60 × 0·8 × 0·75
	= 36 bucket loads per hour
	= 36 × 1 = 36 m^3 per hour

An allowance should be made for the bulking or swell of the solid material due to the introduction of air or voids during the excavation process

∴ Net output allowing for a 30% swell = 36 – (36 × 0·3)

$$= say\ 25\ m^3\ per\ hr.$$

If the Bill of Quantities gives a total net excavation of $950\,m^3$

time required $= \dfrac{950}{25} = \underline{38\ hours}$

or assuming an 8 hour day–1/2 hour maintenance time in

days $= \dfrac{38}{7·5} =$ say $\underline{5\ days}$

Haulage vehicles required $= 1 + \dfrac{\text{round trip time of vehicle}}{\text{loading time of vehicle}}$

If round trip time = 30 minutes and loading time = 10 mins.

number of haulage vehicles required $= 1 + \dfrac{30}{10} = 4$

This gives a vehicle waiting overlap ensuring excavator is fully utilised which is economically desirable.

Bulldozers ~ these machines consist of a track or wheel mounted power unit with a mould blade at the front which is controlled by hydraulic rams. Many bulldozers have the capacity to adjust the mould blade to form an angledozer and the capacity to tilt the mould blade about a central swivel point. Some bulldozers can also be fitted with rear attachments such as rollers and scarifiers.

The main functions of a bulldozer are:-

1. Shallow excavations up to 300 m deep either on level ground or sidehill cutting.
2. Clearance of shrubs and small trees.
3. Clearance of trees by using raised mould blade as a pusher arm.
4. Acting as a towing tractor.
5. Acting as a pusher to scraper machines (see next page).

NB. Bulldozers push earth in front of the mould blade with some side spillage whereas angledozers push and cast the spoil to one side of the mould blade.

Typical Bulldozer Details ~

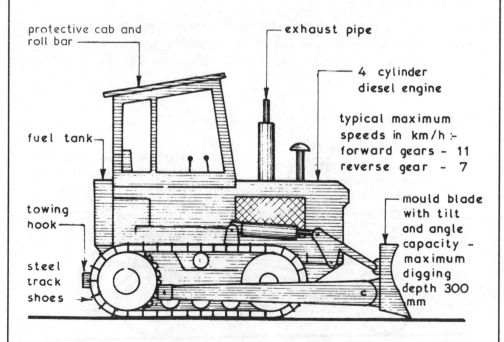

protective cab and roll bar

exhaust pipe

4 cylinder diesel engine

typical maximum speeds in km/h :-
forward gears – 11
reverse gear – 7

fuel tank

mould blade with tilt and angle capacity – maximum digging depth 300 mm

towing hook

steel track shoes

Note: Protective cab/roll bar to be fitted before use.

Scrapers ~ these machines consist of a scraper bowl which is lowered to cut and collect soil where site stripping and levelling operations are required involving large volume of earth. When the scraper bowl is full the apron at the cutting edge is closed to retain the earth and the bowl is raised for travelling to the disposal area. On arrival the bowl is lowered, the apron opened and the spoil pushed out by the tailgate as the machine moves forwards. Scrapers are available in three basic formats:-

1. Towed Scrapers – these consist of a four wheeled scraper bowl which is towed behind a power unit such as a crawler tractor. They tend to be slower than other forms of scraper but are useful for small capacities with haul distances up to 300·00.
2. Two Axle Scrapers – these have a two wheeled scraper bowl with an attached two wheeled power unit. They are very manoeuvrable with a low rolling resistance and very good traction.
3. Three Axle Scrapers – these consist of a two wheeled scraper bowl which may have a rear engine to assist the four wheeled traction engine which makes up the complement. Generally these machines have a greater capacity potential than their counterparts, are easier to control and have a faster cycle time.

To obtain maximum efficiency scrapers should operate downhill if possible, have smooth haul roads, hard surfaces broken up before scraping and be assisted over the last few metres by a pushing vehicle such as a bulldozer.

Typical Scraper Details ~

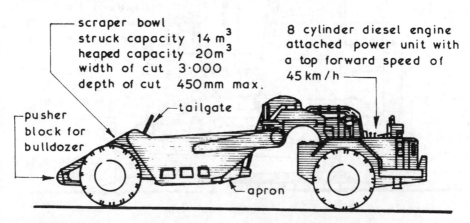

scraper bowl
struck capacity 14 m^3
heaped capacity 20m^3
width of cut 3·000
depth of cut 450mm max.

8 cylinder diesel engine attached power unit with a top forward speed of 45 km/h

pusher block for bulldozer

tailgate

apron

Note: Protective cab/roll bar to be fitted before use.

Graders ~ these machines are similar in concept to bulldozers in that they have a long slender adjustable mould blade, which is usually slung under the centre of the machine. A grader's main function is to finish or grade the upper surface of a large area usually as a follow up operation to scraping or bulldozing. They can produce a fine and accurate finish but do not have the power of a bulldozer therefore they are not suitable for oversite excavation work. The mould blade can be adjusted in both the horizontal and vertical planes through an angle of 300° the latter enabling it to be used for grading sloping banks.

Two basic formats of grader are available:-

1. Four Wheeled – all wheels are driven and steered which gives the machine the ability to offset and crab along its direction of travel.
2. Six Wheeled – this machine has 4 wheels in tandem drive at the rear and 2 front tilting idler wheels giving it the ability to counteract side thrust.

Typical Grader Details ~

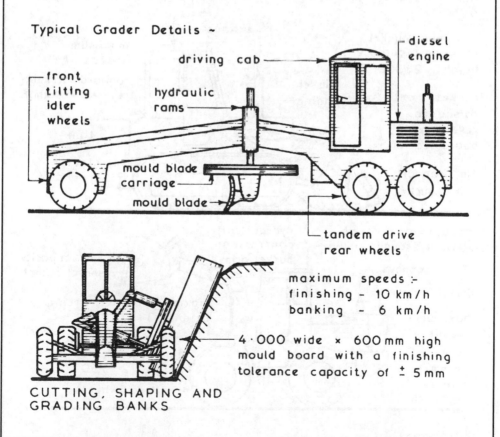

driving cab

diesel engine

front tilting idler wheels

hydraulic rams

mould blade carriage

mould blade

tandem drive rear wheels

maximum speeds :-
finishing – 10 km/h
banking – 6 km/h

4·000 wide × 600 mm high mould board with a finishing tolerance capacity of ± 5 mm

CUTTING, SHAPING AND GRADING BANKS

Tractor Shovels ~ these machines are sometimes called loaders or loader shovels and primary function is to scoop up loose materials in the front mounted bucket, elevate the bucket and manoeuvre into a position to deposit the loose material into an attendant transport vehicle. Tractor shovels are driven towards the pile of loose material with the bucket lowered, the speed and power of the machine will enable the bucket to be filled. Both tracked and wheeled versions are available, the tracked format being more suitable for wet and uneven ground conditions than the wheeled tractor shovel which has greater speed and manoeuvring capabilities. To increase their versatility tractor shovels can be fitted with a 4 in 1 bucket enabling them to carry out bulldozing, excavating, clam lifting and loading activities.

Typical Tractor Shovel Details ~

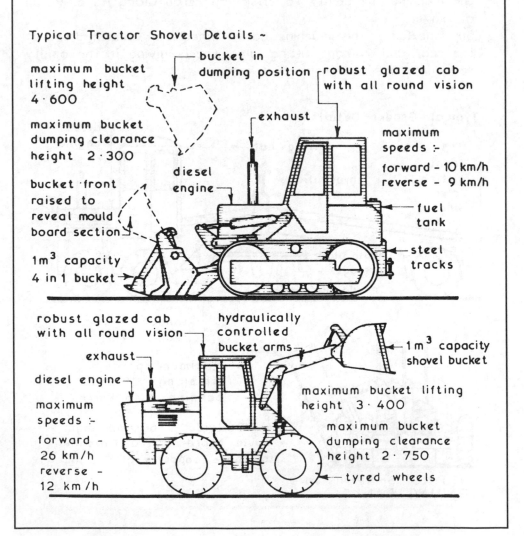

Excavating Machines ~ these are one of the major items of builders plant and are used primarily to excavate and load most types of soil. Excavating machines come in a wide variety of designs and sizes but all of them can be placed within one of three categories:-

1. Universal Excavators – this category covers most forms of excavators all of which have a common factor the power unit. The universal power unit is a tracked based machine with a slewing capacity of 360° and by altering the boom arrangement and bucket type different excavating functions can be obtained. These machines are selected for high output requirements and are rope controlled.

2. Purpose Designed Excavators – these are machines which have been designed specifically to carry out one mode of excavation and they usually have smaller bucket capacities than universal excavators; they are hydraulically controlled with a shorter cycle time.

3. Multi-purpose Excavators – these machines can perform several excavating functions having both front and rear attachments. They are designed to carry out small excavation operations of low output quickly and efficiently. Multi-purpose excavators can be obtained with a wheeled or tracked base and are ideally suited for a small building firm with low excavation plant utilisation requirements.

Skimmers ~ these excavators are rigged using a universal power unit for surface stripping and shallow excavation work up to 300 mm deep where a high degree of accuracy is required. They usually require attendant haulage vehicles to remove the spoil and need to be transported between sites on a low-loader. Because of their limitations and the alternative machines available they are seldom used today.

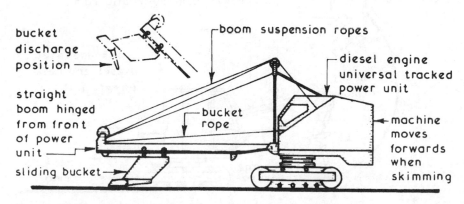

bucket discharge position ~

straight boom hinged from front of power unit

sliding bucket

boom suspension ropes

bucket rope

diesel engine universal tracked power unit

machine moves forwards when skimming

Face Shovels ~ the primary function of this piece of plant is to excavate above its own track or wheel level. They are available as a universal power unit based machine or as a hydraulic purpose designed unit. These machines can usually excavate any type of soil except rock which needs to be loosened, usually by blasting, prior to excavation. Face shovels generally require attendant haulage vehicles for the removal of spoil and a low loader transport lorry for travel between sites. Most of these machines have a limited capacity of between 300 and 400 mm for excavation below their own track or wheel level.

Typical Face Shovel Details ~

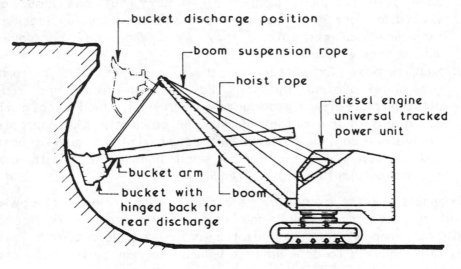

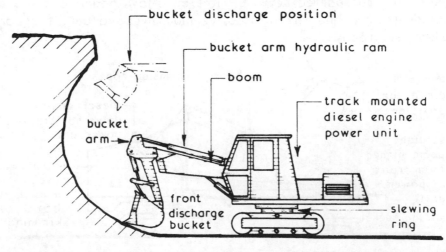

Backacters ~ these machines are suitable for trench, foundation and basement excavations and are available as a universal power unit base machine or as a purpose designed hydraulic unit. They can be used with or without attendant haulage vehicles since the spoil can be placed alongside the excavation for use in backfilling. These machines will require a low loader transport vehicle for travel between sites. Backacters used in trenching operations with a bucket width equal to the trench width can be very accurate with a high output rating.

Typical Backacter Details ~

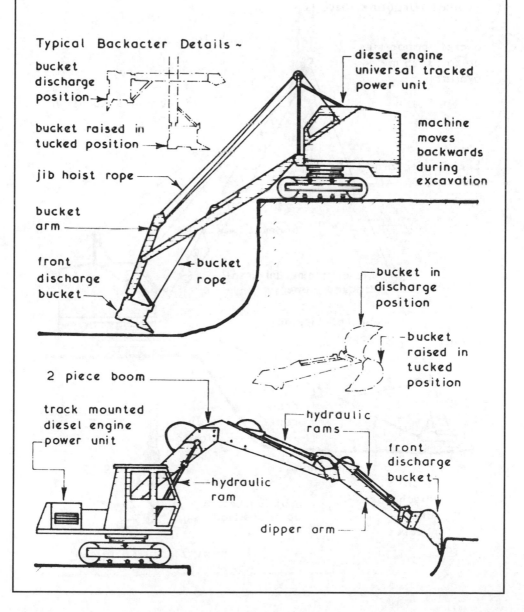

bucket discharge position

bucket raised in tucked position

jib hoist rope

bucket arm

front discharge bucket

bucket rope

diesel engine universal tracked power unit

machine moves backwards during excavation

bucket in discharge position

bucket raised in tucked position

2 piece boom

track mounted diesel engine power unit

hydraulic ram

hydraulic rams

front discharge bucket

dipper arm

Draglines ~ these machines are based on the universal power unit with basic crane rigging to which is attached a drag bucket. The machine is primarily designed for bulk excavation in loose soils up to 3·000 below its own track level by swinging the bucket out to the excavation position and hauling or dragging it back towards the power unit. Dragline machines can also be fitted with a grab or clamshell bucket for excavating in very loose soils.

Typical Dragline Details ~

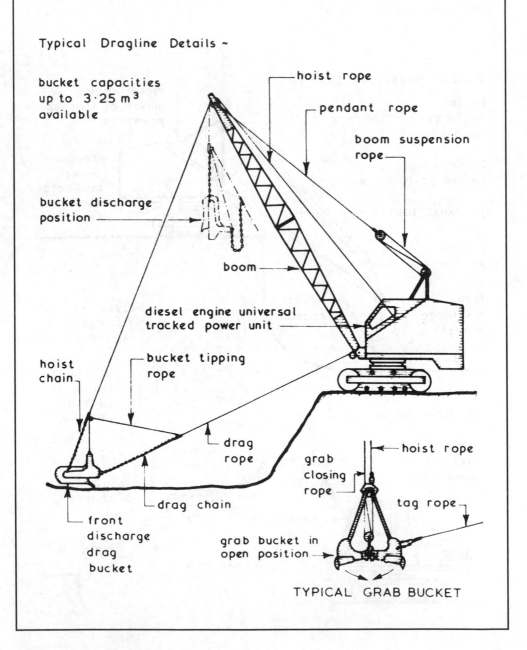

bucket capacities
up to 3·25 m³
available

hoist rope

pendant rope

boom suspension rope

bucket discharge position

boom

diesel engine universal tracked power unit

bucket tipping rope

hoist chain

drag rope

drag chain

front discharge drag bucket

grab closing rope

hoist rope

tag rope

grab bucket in open position

TYPICAL GRAB BUCKET

Multi-purpose Excavators ~ these machines are usually based on the agricultural tractor with 2 or 4 wheel drive and are intended mainly for use in conjunction with small excavation works such as those encountered by the small to medium sized building contractor. Most multi-purpose excavators are fitted with a loading/excavating front bucket and a rear backacter bucket both being hydraulically controlled. When in operation using the backacter bucket the machine is raised off its axles by rear mounted hydraulic outriggers or jacks and in some models by placing the front bucket on the ground. Most machines can be fitted with a variety of bucket widths and various attachments such as bulldozer blades, scarifiers, grab buckets and post hole auger borers.

Typical Multi-purpose Excavator Details ~

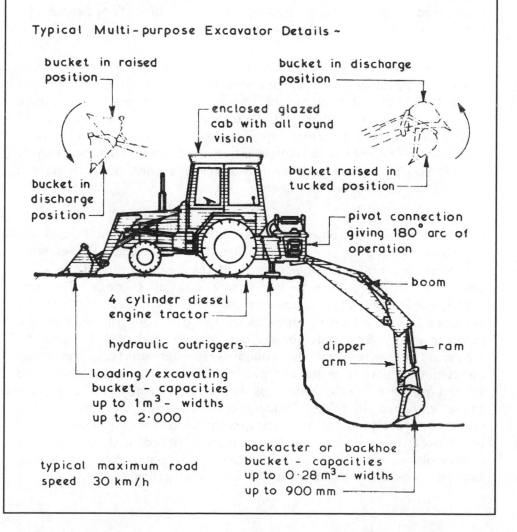

bucket in raised position

bucket in discharge position

enclosed glazed cab with all round vision

bucket in discharge position

bucket raised in tucked position

pivot connection giving 180° arc of operation

boom

4 cylinder diesel engine tractor

hydraulic outriggers

dipper arm

ram

loading/excavating bucket - capacities up to 1 m³ - widths up to 2·000

typical maximum road speed 30 km/h

backacter or backhoe bucket - capacities up to 0·28 m³ - widths up to 900 mm

Transport Vehicles ~ these can be defined as vehicles whose primary function is to convey passengers and/or materials between and around building sites. The types available range from the conventional saloon car to the large low loader lorries designed to transport other items of builders plant between construction sites and the plant yard or depot.

Vans – these transport vehicles range from the small two person plus a limited amount of materials to the large vans with purpose designed bodies such as those built to carry large sheets of glass. Most small vans are usually fitted with a petrol engine and are based on the manufacturer's standard car range whereas the larger vans are purpose designed with either petrol or diesel engines. These basic designs can usually be supplied with an uncovered tipping or non-tipping container mounted behind the passenger cab for use as a `pick-up´ truck.

Passenger Vehicles – these can range from a simple framed cabin which can be placed in the container of a small lorry or `pick-up´ truck to a conventional bus or coach. Vans can also be designed to carry a limited number of seated passengers by having fixed or removable seating together with windows fitted in the van sides thus giving the vehicle a dual function. The number of passengers carried can be limited so that the driver does not have to hold a PSV (public service vehicle) licence.

Lorries – these are sometimes referred to as haul vehicles and are available as road or site only vehicles. Road haulage vehicles have to comply with all the requirements of the Road Traffic Acts which among other requirements limits size and axle loads. The off-highway or site only lorries are not so restricted and can be designed to carry two to three times the axle load allowed on the public highway. Site only lorries are usually specially designed to traverse and withstand the rough terrain encountered on many construction sites. Lorries are available as non-tipping, tipping and special purpose carriers such as those with removable skips and those equipped with self loading and unloading devices. Lorries specifically designed for the transportation of large items of plant are called low loaders and are usually fitted with integral or removable ramps to facilitate loading and some have a winching system to haul the plant onto the carrier platform.

Dumpers ~ these are used for the horizontal transportation of materials on and off construction sites generally by means of an integral tipping skip. Highway dumpers are of a similar but larger design and can be used to carry materials such as excavated spoil along the roads. A wide range of dumpers are available of various carrying capacities and options for gravity or hydraulic discharge control with front tipping, side tipping or elevated tipping facilities. Special format dumpers fitted with flat platforms, rigs to carry materials skips and rigs for concrete skips for crane hoisting are also obtainable. These machines are designed to traverse rough terrain but they are not designed to carry passengers and this misuse is the cause of many accidents involving dumpers.

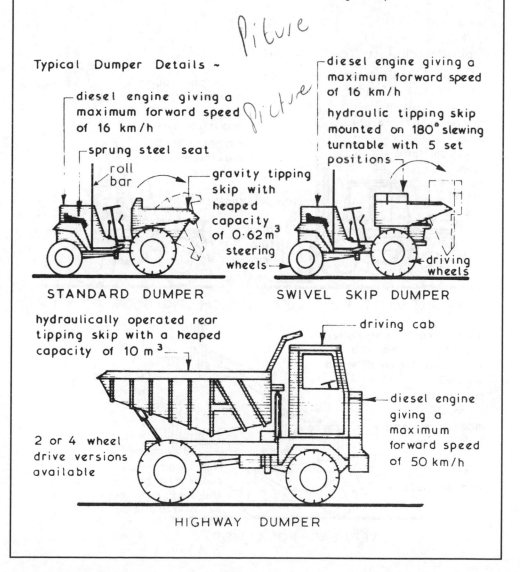

Typical Dumper Details ~

diesel engine giving a maximum forward speed of 16 km/h

sprung steel seat

roll bar

gravity tipping skip with heaped capacity of 0·62 m³

steering wheels

STANDARD DUMPER

diesel engine giving a maximum forward speed of 16 km/h

hydraulic tipping skip mounted on 180° slewing turntable with 5 set positions

driving wheels

SWIVEL SKIP DUMPER

hydraulically operated rear tipping skip with a heaped capacity of 10 m³

driving cab

diesel engine giving a maximum forward speed of 50 km/h

2 or 4 wheel drive versions available

HIGHWAY DUMPER

Fork Lift Trucks ~ these are used for the horizontal and limited vertical transportation of materials positioned on pallets or banded together such as brick packs. They are generally suitable for construction sites where the building height does not exceed three storeys. Although designed to negotiate rough terrain site fork lift trucks have a higher productivity on firm and level soils. Three basic fork lift truck formats are available, namely straight mast, overhead and telescopic boom with various height, reach and lifting capacities. Scaffolds onto which the load(s) are to be placed should be strengthened locally or a specially constructed loading tower could be built as an attachment to or as an integral part of the main scaffold.

Typical Fork Lift Truck Details ~

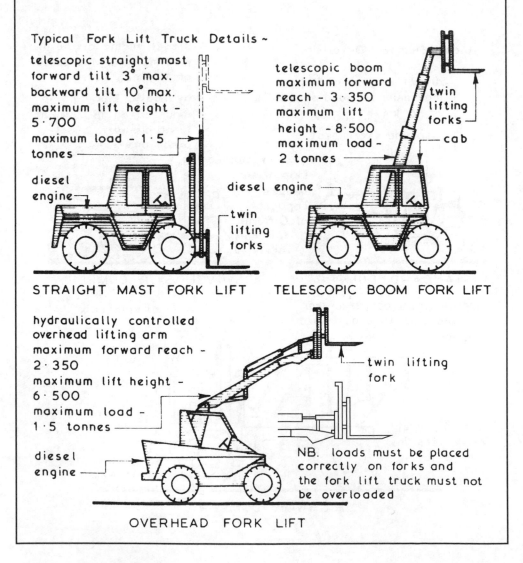

telescopic straight mast
forward tilt 3° max.
backward tilt 10° max.
maximum lift height –
5·700
maximum load – 1·5
tonnes

diesel
engine

twin
lifting
forks

STRAIGHT MAST FORK LIFT

telescopic boom
maximum forward
reach – 3·350
maximum lift
height – 8·500
maximum load –
2 tonnes

twin
lifting
forks

cab

diesel engine

TELESCOPIC BOOM FORK LIFT

hydraulically controlled
overhead lifting arm
maximum forward reach –
2·350
maximum lift height –
6·500
maximum load –
1·5 tonnes

diesel
engine

twin lifting
fork

NB. loads must be placed
correctly on forks and
the fork lift truck must not
be overloaded

OVERHEAD FORK LIFT

Hoists ~ these are designed for the vertical transportation of materials, passengers or materials and passengers (see page 168). Materials hoists are designed for one specific use (i.e. the vertical transportation of materials) and under no circumstances should they be used to transport passengers. Most material hoists are of a mobile format which can be dismantled, folded onto the chassis and moved to another position or site under their own power or lowed by a haulage vehicle. When in use material hoists need to be stabilised and/or tied to the structure and enclosed with a protective screen.

Typical Materials Hoist Details ~

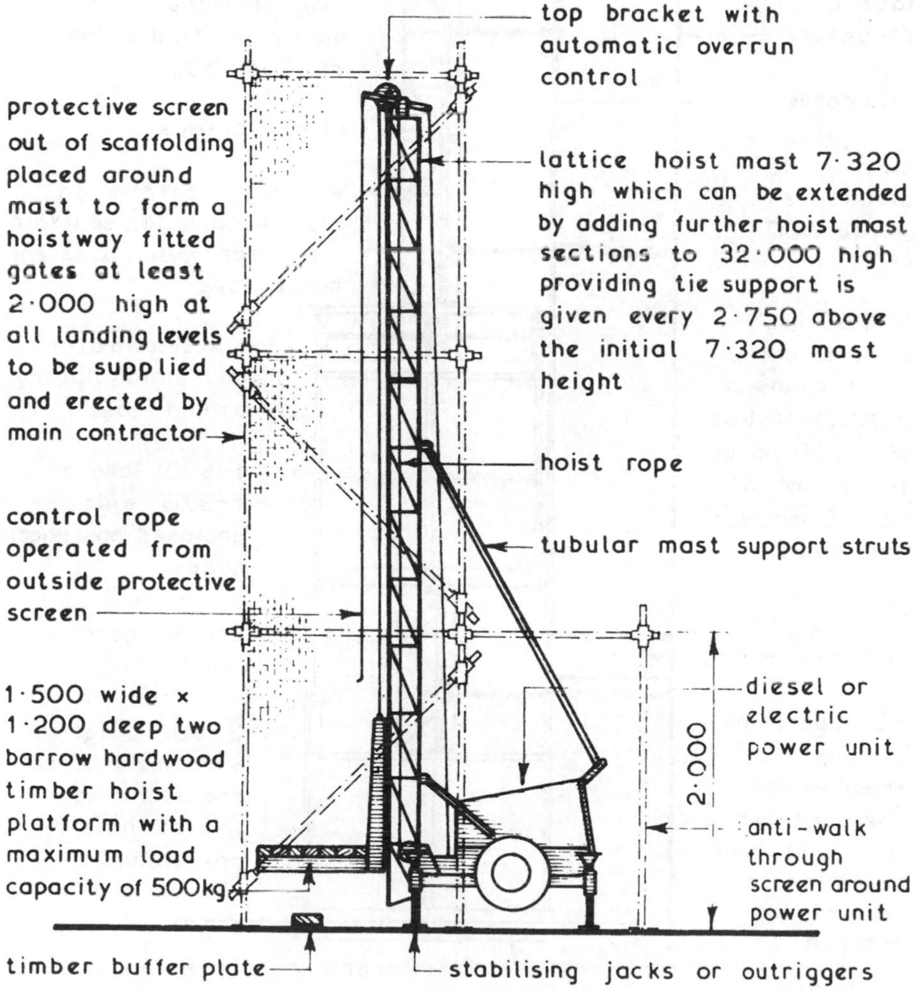

top bracket with automatic overrun control

protective screen out of scaffolding placed around mast to form a hoistway fitted gates at least 2·000 high at all landing levels to be supplied and erected by main contractor

lattice hoist mast 7·320 high which can be extended by adding further hoist mast sections to 32·000 high providing tie support is given every 2·750 above the initial 7·320 mast height

hoist rope

control rope operated from outside protective screen

tubular mast support struts

1·500 wide × 1·200 deep two barrow hardwood timber hoist platform with a maximum load capacity of 500kg

diesel or electric power unit

2·000

anti-walk through screen around power unit

timber buffer plate

stabilising jacks or outriggers

Passenger Hoists ~ these are designed to carry passengers although most are capable of transporting a combined load of materials and passengers within the lifting capacity of the hoist. A wide selection of hoists are available ranging from a single cage with rope suspension to twin cages with rack and pinion operation mounted on two sides of a static tower.

Typical Passenger Hoist Details ~

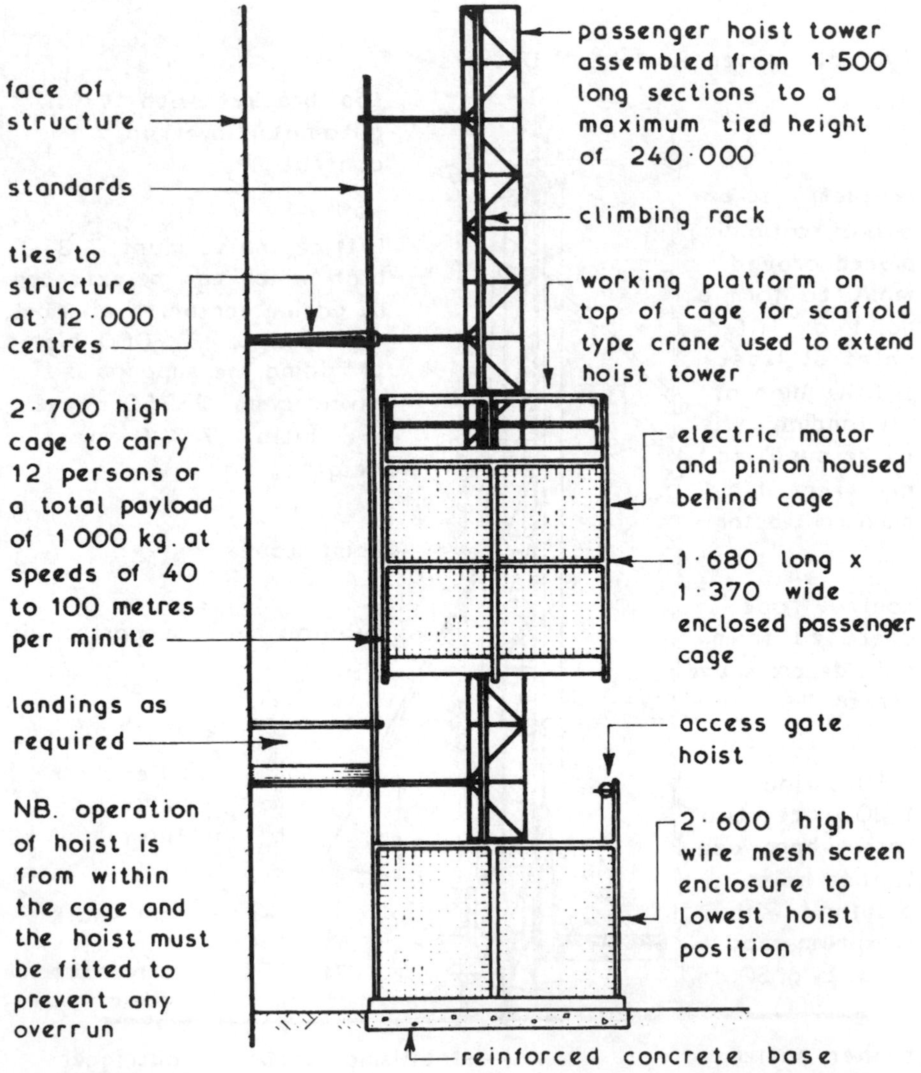

face of structure

standards

ties to structure at 12·000 centres

2·700 high cage to carry 12 persons or a total payload of 1 000 kg. at speeds of 40 to 100 metres per minute

landings as required

NB. operation of hoist is from within the cage and the hoist must be fitted to prevent any overrun

passenger hoist tower assembled from 1·500 long sections to a maximum tied height of 240·000

climbing rack

working platform on top of cage for scaffold type crane used to extend hoist tower

electric motor and pinion housed behind cage

1·680 long x 1·370 wide enclosed passenger cage

access gate hoist

2·600 high wire mesh screen enclosure to lowest hoist position

reinforced concrete base

Rubble Chutes ~ these apply to contracts involving demolition, repair, maintenance and refurbishment. The simple concept of connecting several perforated dustbins is reputed to have been conceived by an ingenious site operative for the expedient and safe conveyance of materials.

In purpose designed format, the tapered cylinders are produced from reinforced rubber with chain linkage for continuity. Overall unit lengths are generally 1100 mm, providing an effective length of 1 m. Hoppers and side entry units are made for special applications.

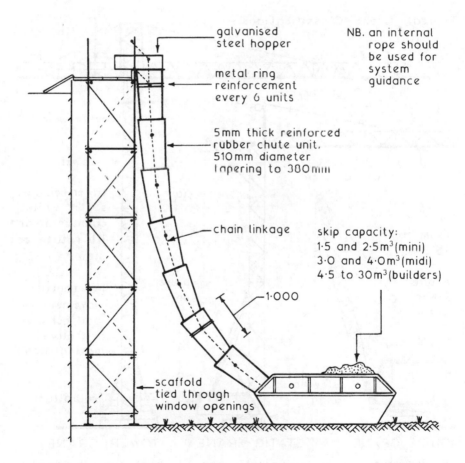

galvanised steel hopper

metal ring reinforcement every 6 units

5mm thick reinforced rubber chute unit, 510mm diameter tapering to 300mm

chain linkage

1·000

NB. an internal rope should be used for system guidance

skip capacity:
1·5 and 2·5m^3 (mini)
3·0 and 4·0m^3 (midi)
4·5 to 30m^3 (builders)

scaffold tied through window openings

Ref. Highways Act – written permit (license) must be obtained from the local authority highways department for use of a skip on a public thoroughfare. It will have to be illuminated at night and may require a temporary traffic light system to regulate vehicles.

169

Cranes ~ these are lifting devices designed to raise materials by means of rope operation and move the load horizontally within the limitations of any particular machine. The range of cranes available is very wide and therefore choice must be based on the loads to be lifted, height and horizontal distance to be covered, time period(s) of lifting operations, utilisation factors and degree of mobility required. Crane types can range from a simple rope and pulley or gin wheel to a complex tower crane but most can be placed within 1 of 3 groups, namely mobile, static and tower cranes.

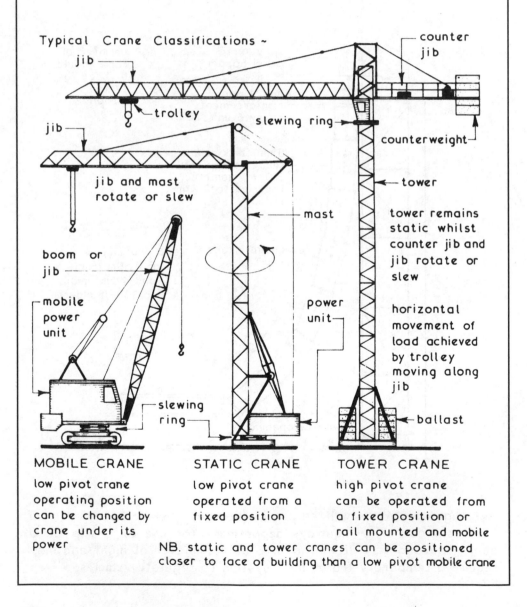

Typical Crane Classifications ~

jib

counter jib

trolley

slewing ring

counterweight

jib

jib and mast rotate or slew

tower

boom or jib

mast

tower remains static whilst counter jib and jib rotate or slew

mobile power unit

power unit

horizontal movement of load achieved by trolley moving along jib

slewing ring

ballast

MOBILE CRANE
low pivot crane operating position can be changed by crane under its power

STATIC CRANE
low pivot crane operated from a fixed position

TOWER CRANE
high pivot crane can be operated from a fixed position or rail mounted and mobile

NB. static and tower cranes can be positioned closer to face of building than a low pivot mobile crane

Self Propelled Cranes ~ these are mobile cranes mounted on a wheeled chassis and have only one operator position from which the crane is controlled and the vehicle driven. The road speed of this type of crane is generally low, usually not exceeding 30 km p.h. A variety of self propelled crane formats are available ranging from short height lifting strut booms of fixed length to variable length lattice booms with a fly jib attachment.

Typical Self Propelled Crane Details ~

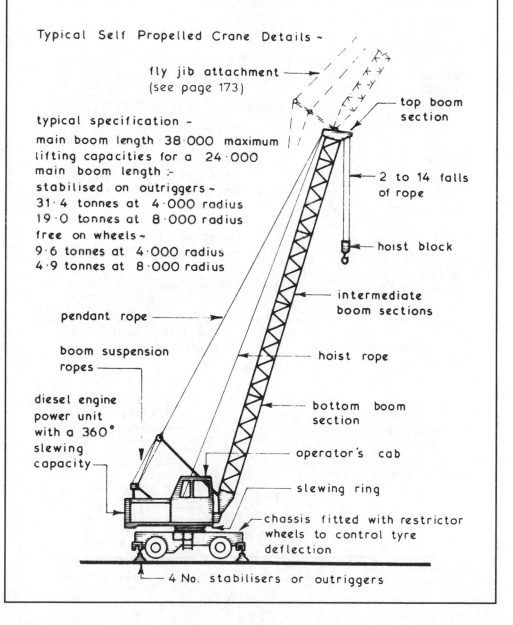

fly jib attachment ⟶
(see page 173)

top boom
section

typical specification ~
main boom length 38·000 maximum
lifting capacities for a 24·000
main boom length :-
stabilised on outriggers ~
31·4 tonnes at 4·000 radius
19·0 tonnes at 8·000 radius
free on wheels ~
9·6 tonnes at 4·000 radius
4·9 tonnes at 8·000 radius

2 to 14 falls
of rope

hoist block

intermediate
boom sections

pendant rope ⟶

hoist rope

boom suspension
ropes

diesel engine
power unit
with a 360°
slewing
capacity

bottom boom
section

operator's cab

slewing ring

chassis fitted with restrictor
wheels to control tyre
deflection

4 No. stabilisers or outriggers

171

Lorry Mounted Cranes ~ these mobile cranes consist of a lattice or telescopic boom mounted on a specially adapted truck or lorry. They have two operating positions: the lorry being driven from a conventional front cab and the crane being controlled from a different location. The lifting capacity of these cranes can be increased by using outrigger stabilising jacks and the approach distance to the face of building decreased by using a fly jib. Lorry mounted telescopic cranes require a firm surface from which to operate and because of their short site preparation time they are ideally suited for short hire periods.

Typical Lorry Mounted Telescopic Crane Details ~

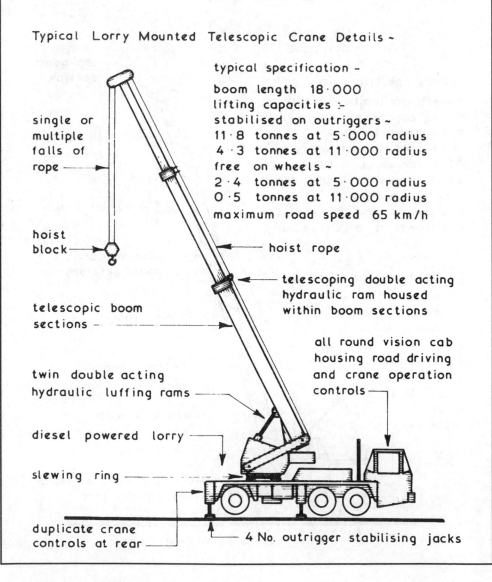

single or multiple falls of rope

hoist block

telescopic boom sections

twin double acting hydraulic luffing rams

diesel powered lorry

slewing ring

duplicate crane controls at rear

typical specification –

boom length 18·000
lifting capacities :-
stabilised on outriggers ~
11·8 tonnes at 5·000 radius
4·3 tonnes at 11·000 radius
free on wheels ~
2·4 tonnes at 5·000 radius
0·5 tonnes at 11·000 radius
maximum road speed 65 km/h

hoist rope

telescoping double acting hydraulic ram housed within boom sections

all round vision cab housing road driving and crane operation controls

4 No. outrigger stabilising jacks

Lorry Mounted Lattice Jib Cranes ~ these cranes follow the same basic principles as the lorry mounted telescopic cranes but they have a lattice boom and are designed as heavy duty cranes with lifting capacities in excess of 100 tonnes. These cranes will require a firm level surface from which to operate and can have a folding or sectional jib which will require the crane to be rigged on site before use.

Typical Lorry Mounted Lattice Jib Crane Details ~

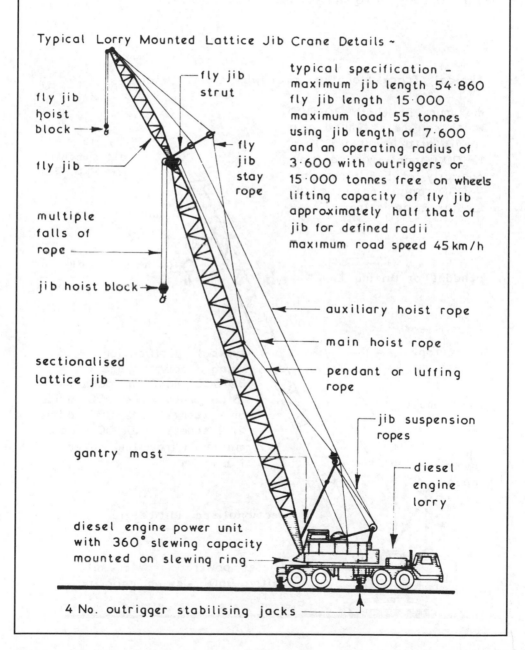

fly jib strut

fly jib hoist block

fly jib

fly jib stay rope

typical specification –
maximum jib length 54·860
fly jib length 15·000
maximum load 55 tonnes
using jib length of 7·600
and an operating radius of
3·600 with outriggers or
15·000 tonnes free on wheels
lifting capacity of fly jib
approximately half that of
jib for defined radii
maximum road speed 45 km/h

multiple falls of rope

jib hoist block

auxiliary hoist rope

main hoist rope

pendant or luffing rope

sectionalised lattice jib

jib suspension ropes

gantry mast

diesel engine lorry

diesel engine power unit with 360° slewing capacity mounted on slewing ring

4 No. outrigger stabilising jacks

Track Mounted Cranes ~ these machines can be a universal power unit rigged as a crane (see page 162) or a purpose designed track mounted crane with or without a fly jib attachment. The latter type are usually more powerful with lifting capacities up to 45 tonnes. Track mounted cranes can travel and carry out lifting operations on most sites without the need for special road and hardstand provisions but they have to be rigged on arrival after being transported to site on a low loader lorry.

Typical Track Mounted or Crawler Crane Details ~

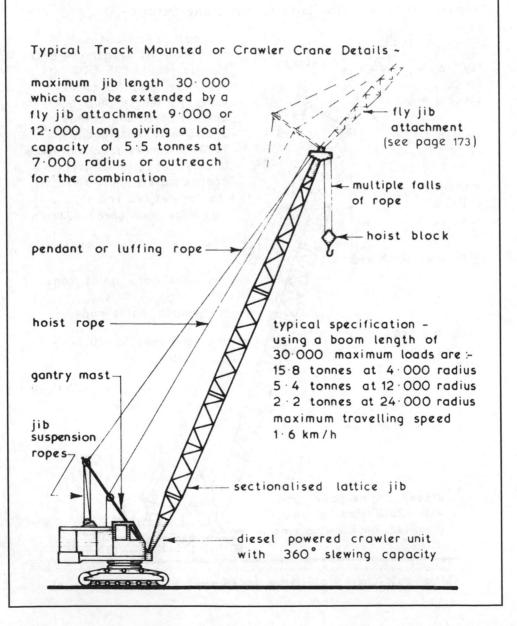

maximum jib length 30·000 which can be extended by a fly jib attachment 9·000 or 12·000 long giving a load capacity of 5·5 tonnes at 7·000 radius or outreach for the combination

fly jib attachment (see page 173)

multiple falls of rope

hoist block

pendant or luffing rope →

hoist rope →

typical specification - using a boom length of 30·000 maximum loads are :-
15·8 tonnes at 4·000 radius
5·4 tonnes at 12·000 radius
2·2 tonnes at 24·000 radius
maximum travelling speed 1·6 km/h

gantry mast

jib suspension ropes

sectionalised lattice jib

diesel powered crawler unit with 360° slewing capacity

Gantry Cranes ~ these are sometimes called portal cranes and consist basically of two 'A' frames joined together with a cross member on which transverses the lifting appliance. In small gantry cranes (up to 10 tonnes lifting capacity) the 'A' frames are usually wheel mounted and manually propelled whereas in the large gantry cranes (up to 100 tonnes lifting capacity) the 'A' frames are mounted on powered bogies running on rail tracks with the driving cab and lifting gear mounted on the cross beam or gantry. Small gantry cranes are used primarily for loading and unloading activities in stock yards whereas the medium and large gantry cranes are used to straddle the work area such as in power station construction or in repetitive low to medium rise developments. All gantry cranes have the advantage of three direction movement –

1. Transverse by moving along the cross beam.
2. Vertical by raising and lowering the hoist block.
3. Horizontal by forward and reverse movements of the whole gantry crane.

Typical Gantry Crane Details ~

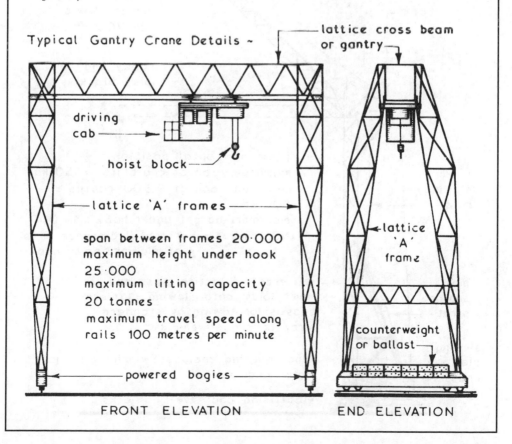

lattice cross beam or gantry

driving cab

hoist block

lattice 'A' frames

span between frames 20·000
maximum height under hook 25·000
maximum lifting capacity 20 tonnes
maximum travel speed along rails 100 metres per minute

powered bogies

lattice 'A' frame

counterweight or ballast

FRONT ELEVATION   END ELEVATION

Mast Cranes ~ these are similar in appearance to the familiar tower cranes but they have one major difference in that the mast or tower is mounted on the slewing ring and thus rotates whereas a tower crane has the slewing ring at the top of the tower and therefore only the jib portion rotates. Mast cranes are often mobile, self erecting, of relatively low lifting capacity and are usually fitted with a luffing jib. A wide variety of models are available and have the advantage over most mobile low pivot cranes of a closer approach to the face of the building.

Typical Mast Crane Details ~

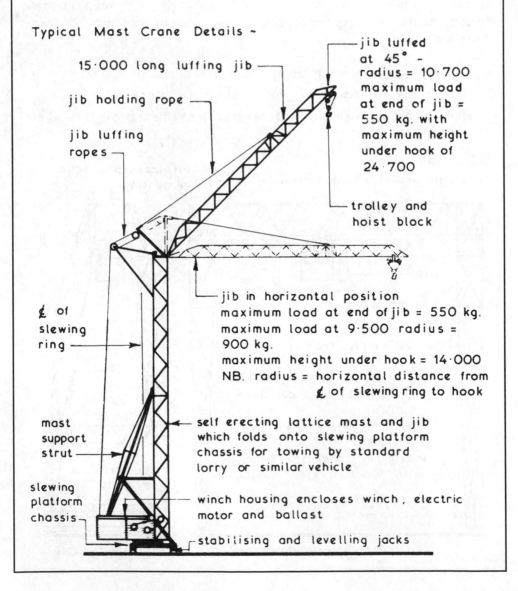

15·000 long luffing jib

jib holding rope

jib luffing ropes

jib luffed at 45° - radius = 10·700 maximum load at end of jib = 550 kg. with maximum height under hook of 24·700

trolley and hoist block

£ of slewing ring

jib in horizontal position
maximum load at end of jib = 550 kg.
maximum load at 9·500 radius = 900 kg.
maximum height under hook = 14·000
NB. radius = horizontal distance from
£ of slewing ring to hook

mast support strut

self erecting lattice mast and jib which folds onto slewing platform chassis for towing by standard lorry or similar vehicle

slewing platform chassis

winch housing encloses winch, electric motor and ballast

stabilising and levelling jacks

Tower Cranes ~ most tower cranes have to be assembled and erected on site prior to use and can be equipped with a horizontal or luffing jib. The wide range of models available often make it difficult to choose a crane suitable for any particular site but most tower cranes can be classified into one of four basic groups thus:-

1. Self Supporting Static Tower Cranes - high lifting capacity with the mast or tower fixed to a foundation base - they are suitable for confined and open sites. (see page 178)

2. Supported Static Tower Cranes - similar in concept to self supporting cranes and are used where high lifts are required, the mast or tower being tied at suitable intervals to the structure to give extra stability. (see page 179)

3. Travelling Tower Cranes - these are tower cranes mounted on power bogies running on a wide gauge railway track to give greater site coverage - only slight gradients can be accommodated therefore a reasonably level site or specially constructed railway support trestle is required. (see page 180)

4. Climbing Cranes - these are used in conjunction with tall buildings and structures. The climbing mast or tower is housed within the structure and raised as the height of the structure is increased. Upon completion the crane is dismantled into small sections and lowered down the face of the building. (see page 181)

All tower cranes should be left in an 'out of service' condition when unattended and in high wind conditions, the latter varying with different models but generally wind speeds in excess of 60 km p.h. would require the crane to be placed in an out of service condition thus:-

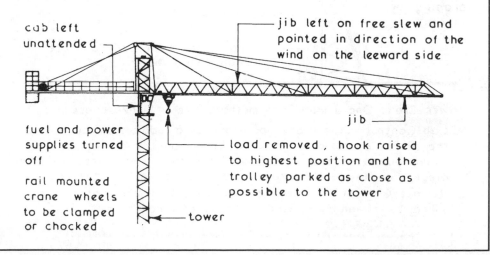

cab left unattended

jib left on free slew and pointed in direction of the wind on the leeward side

jib

fuel and power supplies turned off

rail mounted crane wheels to be clamped or chocked

load removed, hook raised to highest position and the trolley parked as close as possible to the tower

tower

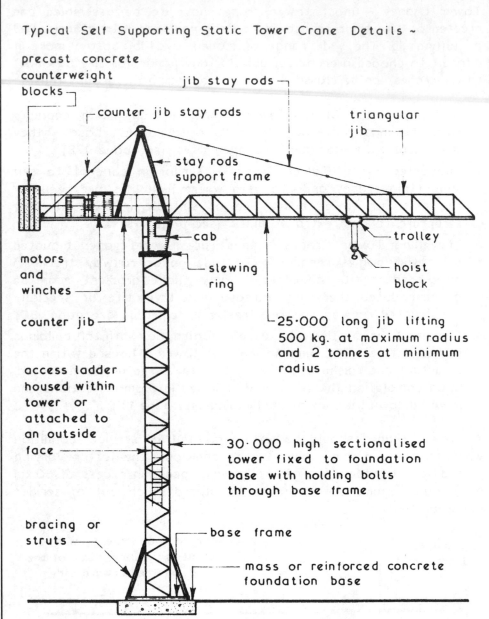

Typical Self Supporting Static Tower Crane Details ~

precast concrete counterweight blocks

jib stay rods

counter jib stay rods

triangular jib

stay rods support frame

cab

slewing ring

motors and winches

counter jib

trolley

hoist block

25·000 long jib lifting 500 kg. at maximum radius and 2 tonnes at minimum radius

access ladder housed within tower or attached to an outside face

30·000 high sectionalised tower fixed to foundation base with holding bolts through base frame

bracing or struts

base frame

mass or reinforced concrete foundation base

Tower Crane Operation ~ two methods are in general use :-

1 Cab Control - the crane operator has a good view of most of the lifting operations from the cab mounted at top of the tower but a second person or banksman is required to give clear signals to the crane operator and to load the crane

2 Remote Control - the crane operator carries a control box linked by a wandering lead to the crane controls.

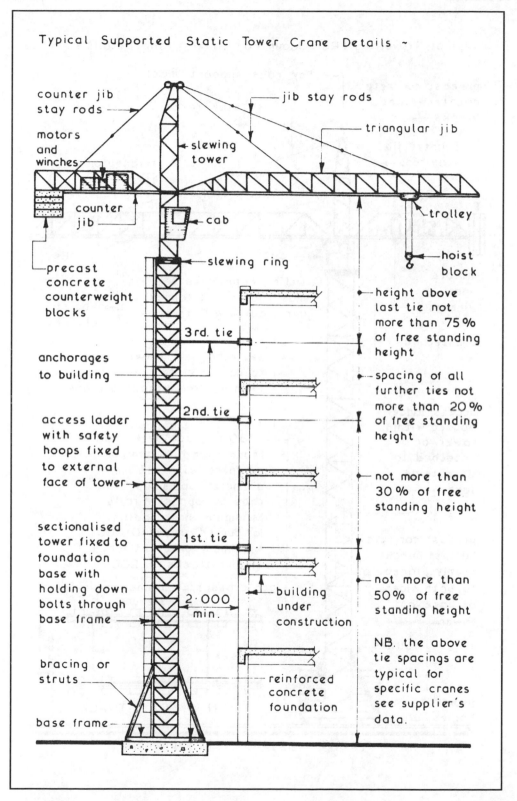

Typical Supported Static Tower Crane Details ~

counter jib stay rods

jib stay rods

triangular jib

motors and winches

slewing tower

counter jib

cab

precast concrete counterweight blocks

slewing ring

trolley

hoist block

height above last tie not more than 75% of free standing height

3rd. tie

anchorages to building

spacing of all further ties not more than 20% of free standing height

2nd. tie

access ladder with safety hoops fixed to external face of tower

not more than 30% of free standing height

sectionalised tower fixed to foundation base with holding down bolts through base frame

1st. tie

2·000 min.

building under construction

not more than 50% of free standing height

NB. the above tie spacings are typical for specific cranes see supplier's data.

bracing or struts

reinforced concrete foundation

base frame

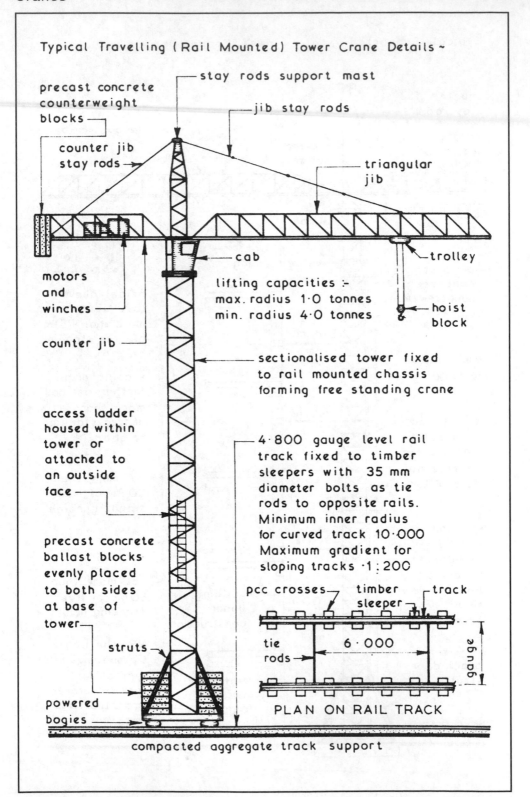

Typical Travelling (Rail Mounted) Tower Crane Details ~

stay rods support mast

precast concrete counterweight blocks

jib stay rods

counter jib stay rods

triangular jib

cab

trolley

motors and winches

lifting capacities :-
max. radius 1·0 tonnes
min. radius 4·0 tonnes

hoist block

counter jib

sectionalised tower fixed to rail mounted chassis forming free standing crane

access ladder housed within tower or attached to an outside face

4·800 gauge level rail track fixed to timber sleepers with 35 mm diameter bolts as tie rods to opposite rails. Minimum inner radius for curved track 10·000 Maximum gradient for sloping tracks ·1:200

precast concrete ballast blocks evenly placed to both sides at base of tower

pcc crosses

timber sleeper

track

tie rods

6·000

gauge

struts

powered bogies

PLAN ON RAIL TRACK

compacted aggregate track support

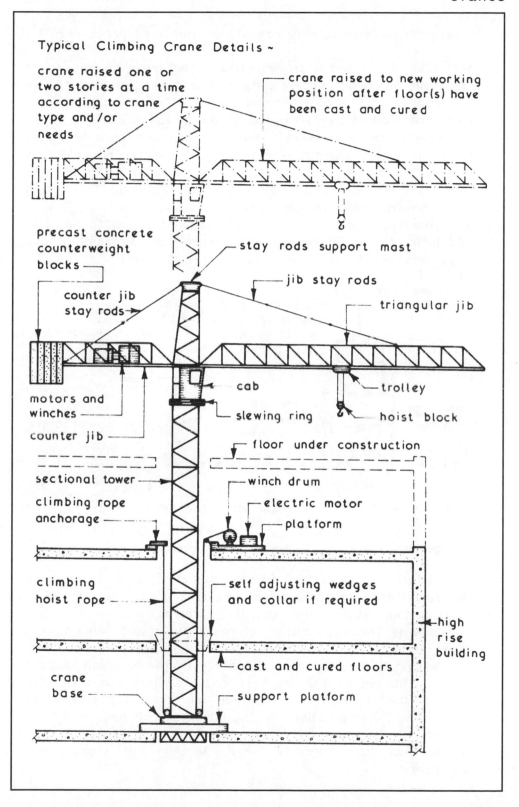

Typical Climbing Crane Details ~

crane raised one or two stories at a time according to crane type and /or needs

crane raised to new working position after floor(s) have been cast and cured

precast concrete counterweight blocks

stay rods support mast

jib stay rods

counter jib stay rods

triangular jib

motors and winches

cab

trolley

counter jib

slewing ring

hoist block

floor under construction

sectional tower

winch drum

climbing rope anchorage

electric motor

platform

climbing hoist rope

self adjusting wedges and collar if required

high rise building

crane base

cast and cured floors

support platform

Concreting ~ this site activity consists of four basic procedures –

1. Material Supply and Storage – this is the receiving on site of the basic materials namely cement, fine aggregate and coarse aggregate and storing them under satisfactory conditions. (see Concrete Production – Materials on pages 266 & 267)
2. Mixing – carried out in small batches this requires only simple hand held tools whereas when demand for increased output is required mixers or ready mixed supplies could be used. (see Concrete Production on pages 268 to 271 and Concreting Plant on pages 183 to 188)
3. Transporting – this can range from a simple bucket to barrows and dumpers for small amounts. For larger loads, especially those required at high level, crane skips could be used:-

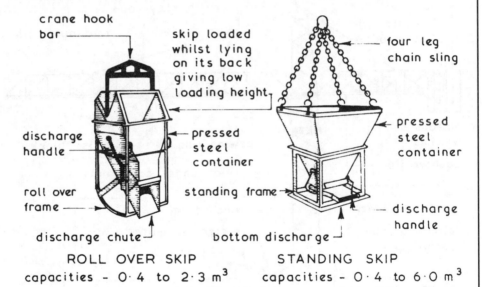

crane hook bar

skip loaded whilst lying on its back giving low loading height

four leg chain sling

discharge handle

pressed steel container

pressed steel container

roll over frame

standing frame

discharge chute

bottom discharge

discharge handle

ROLL OVER SKIP
capacities - 0·4 to 2·3 m^3

STANDING SKIP
capacities - 0·4 to 6·0 m^3

For the transportation of large volumes of concrete over a limited distance concrete pumps could be used. (see page 186)
4. Placing Concrete – this activity involves placing the wet concrete in the excavation, formwork or mould; working the concrete between and around any reinforcement; vibrating and/or tamping and curing in accordance with the recommendations of BS 8110: Structural use of concrete. This standard also covers the striking or removal of the formwork. (see Concreting Plant on page 187 and Formwork on page 478)
Further ref. BS 8000-2.1: Workmanship on building sites. Codes of practice for concrete work. Mixing and transporting concrete.

Concrete Mixers ~ apart from the very large output mixers most concrete mixers in general use have a rotating drum designed to produce a concrete without segregation of the mix.

Concreting Plant ~ the selection of concreting plant can be considered under three activity headings –
1. Mixing. 2. Transporting. 3. Placing.

Choice of Mixer ~ the factors to be taken into consideration when selecting the type of concrete mixer required are –

1. Maximum output required (m³/hour).
2. Total output required (m³).
3. Type or method of transporting the mixed concrete.
4. Discharge height of mixer (compatibility with transporting method).

Concrete mixer types are generally related to their designed output performance, therefore when the answer to the question 'How much concrete can be placed in a given time period?' or alternatively 'What mixing and placing methods are to be employed to mix and place a certain amount of concrete in a given time period?' has been found the actual mixer can be selected. Generally a batch mixing time of 5 minutes per cycle or 12 batches per hour can be assumed as a reasonable basis for assessing mixer output.

Small Batch Mixers ~ these mixers have outputs of up to 200 litres per batch with wheelbarrow transportation an hourly placing rate of 2 to 3 m³ can be achieved. Most small batch mixers are of the tilting drum type. Generally these mixers are hand loaded which makes the quality control of successive mixes difficult to regulate.

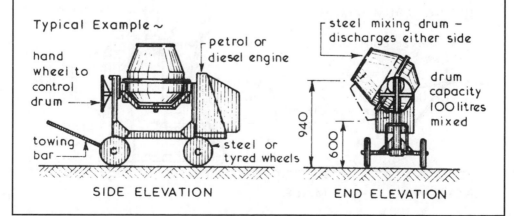

Typical Example ~

hand wheel to control drum

towing bar

petrol or diesel engine

steel or tyred wheels

SIDE ELEVATION

steel mixing drum – discharges either side

drum capacity 100 litres mixed

940

600

END ELEVATION

Medium Batch Mixers ~ outputs of these mixers range from 200 to 750 litres and can be obtained at the lower end of the range as a tilting drum mixer or over the complete range as a non-tilting drum mixer with either reversing drum or chute discharge. The latter usually having a lower discharge height. These mixers usually have integral weight batching loading hoppers, scraper shovels and water tanks thus giving better quality control than the small batch mixers. Generally they are unsuitable for wheelbarrow transportation because of their high output.

Typical Examples ~

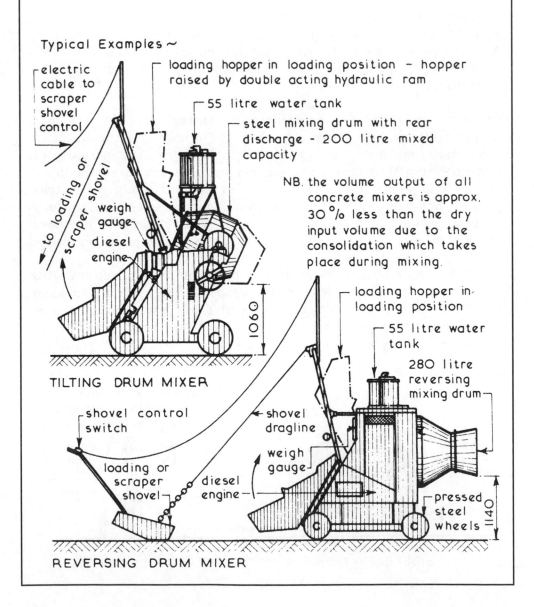

electric cable to scraper shovel control

loading hopper in loading position - hopper raised by double acting hydraulic ram

55 litre water tank

steel mixing drum with rear discharge - 200 litre mixed capacity

to loading or scraper shovel

weigh gauge

diesel engine

NB. the volume output of all concrete mixers is approx. 30 % less than the dry input volume due to the consolidation which takes place during mixing.

1060

TILTING DRUM MIXER

shovel control switch

shovel dragline

loading or scraper shovel

diesel engine

weigh gauge

loading hopper in loading position

55 litre water tank

280 litre reversing mixing drum

pressed steel wheels

1140

REVERSING DRUM MIXER

Transporting Concrete ~ the usual means of transporting mixed concrete produced in a small capacity mixer is by wheelbarrow. The run between the mixing and placing positions should be kept to a minimum and as smooth as possible by using planks or similar materials to prevent segregation of the mix within the wheelbarrow.

Dumpers ~ these can be used for transporting mixed concrete from mixers up to 600 litre capacity when fitted with an integral skip and for lower capacities when designed to take a crane skip.

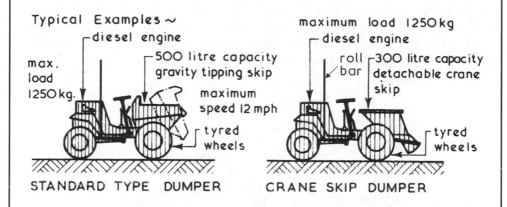

Typical Examples ~

diesel engine

max. load 1250 kg.

500 litre capacity gravity tipping skip

maximum speed 12 mph

tyred wheels

STANDARD TYPE DUMPER

maximum load 1250 kg

diesel engine

roll bar

300 litre capacity detachable crane skip

tyred wheels

CRANE SKIP DUMPER

Ready Mixed Concrete Trucks ~ these are used to transport mixed concrete from a mixing plant or depot to the site. Usual capacity range of ready mixed concrete trucks is 4 to 6 m³. Discharge can be direct into placing position via a chute or into some form of site transport such as a dumper, crane skip or concrete pump.

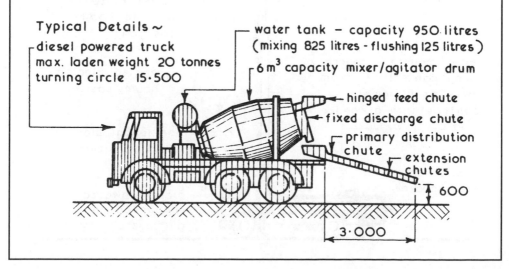

Typical Details ~

diesel powered truck
max. laden weight 20 tonnes
turning circle 15·500

water tank – capacity 950 litres
(mixing 825 litres - flushing 125 litres)

6 m³ capacity mixer/agitator drum

hinged feed chute

fixed discharge chute

primary distribution chute

extension chutes

600

3·000

Concrete Pumps ~ these are used to transport large volumes of concrete in a short time period (up to 100 m³ per hour) in both the vertical and horizontal directions from the pump position to the point of placing. Concrete pumps can be trailer or lorry mounted and are usually of a twin cylinder hydraulically driven format with a small bore pipeline (100 mm diameter) with pumping ranges of up to 85·000 vertically and 200·000 horizontally depending on the pump model and the combination of vertical and horizontal distances. It generally requires about 45 minutes to set up a concrete pump on site including coating the bore of the pipeline with a cement grout prior to pumping the special concrete mix. The pump is supplied with pumpable concrete by means of a constant flow of ready mixed concrete lorries throughout the pumping period after which the pipeline is cleared and cleaned. Usually a concrete pump and its operator(s) are hired for the period required.

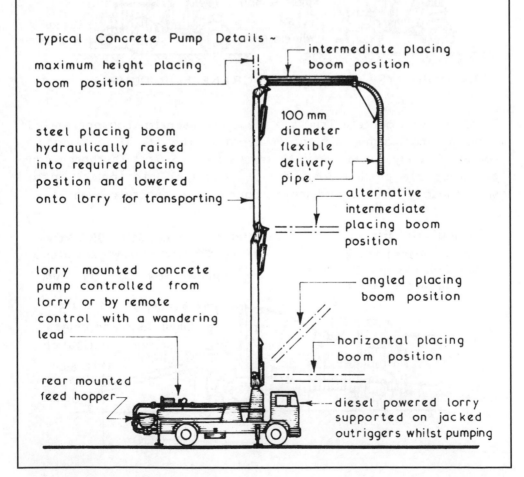

Typical Concrete Pump Details ~

maximum height placing boom position

intermediate placing boom position

steel placing boom hydraulically raised into required placing position and lowered onto lorry for transporting

100 mm diameter flexible delivery pipe

alternative intermediate placing boom position

lorry mounted concrete pump controlled from lorry or by remote control with a wandering lead

angled placing boom position

horizontal placing boom position

rear mounted feed hopper

diesel powered lorry supported on jacked outriggers whilst pumping

Placing Concrete ~ this activity is usually carried out by hand with the objectives of filling the mould, formwork or excavated area to the correct depth, working the concrete around any inserts or reinforcement and finally compacting the concrete to the required consolidation. The compaction of concrete can be carried out using simple tamping rods or boards or alternatively it can be carried out with the aid of plant such as vibrators.

Poker Vibrators ~ these consist of a hollow steel tube casing in which is a rotating impellor which generates vibrations as its head comes into contact with the casing –

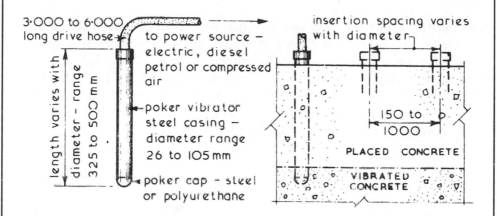

Poker vibrators should be inserted vertically and allowed to penetrate 75 mm into any previously vibrated concrete.

Clamp or Tamping Board Vibrators ~ clamp vibrators are powered either by compressed air or electricity whereas tamping board vibrators are usually petrol driven –

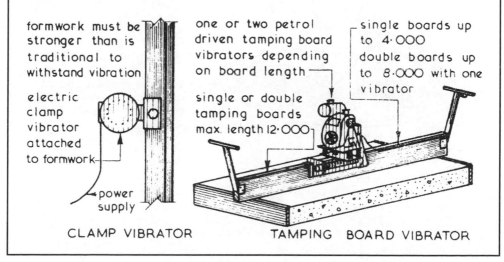

Power Float – a hand-operated electric motor or petrol engine, surmounted over a mechanical surface skimmer. Machines are provided with an interchangeable revolving disc and a set of blades. These are used in combination to produce a smooth, dense and level surface finish to in-situ concrete beds.

The advantages offset against the cost of plant hire are:

* Eliminates the time and materials needed to apply a finishing screed.
* A quicker process and less labour-intensive than hand troweling.

Application – after transverse tamping, the concrete is left to partially set for a few hours. Amount of setting time will depend on a number of variables, including air temperature and humidity, mix specification and machine weight. As a rough guide, walking on the concrete will leave indentations of about 3–4 mm. A surfacing disc is used initially to remove high tamping lines, before two passes with blades to finish and polish the surface.

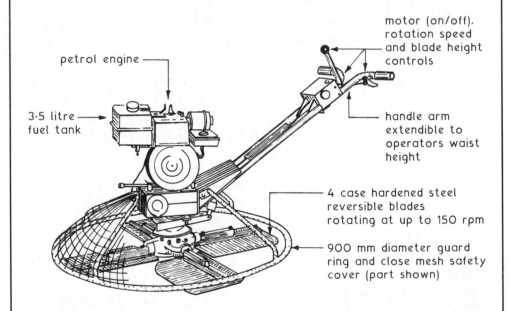

motor (on/off), rotation speed and blade height controls

petrol engine

3·5 litre fuel tank

handle arm extendible to operators waist height

4 case hardened steel reversible blades rotating at up to 150 rpm

900 mm diameter guard ring and close mesh safety cover (part shown)

Power or mechanical float

4 SUBSTRUCTURE

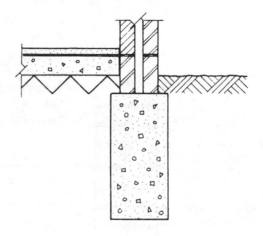

Foundations ~ the function of any foundation is to safely sustain and transmit to the ground on which it rests the combined dead, imposed and wind loads in such a manner as not to cause any settlement or other movement which would impair the stability or cause damage to any part of the building.

Example ~

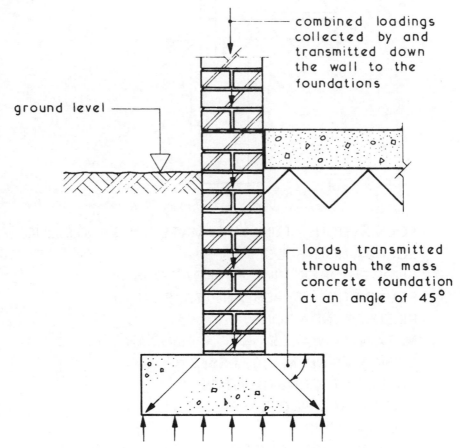

combined loadings collected by and transmitted down the wall to the foundations

ground level

loads transmitted through the mass concrete foundation at an angle of 45°

Subsoil beneath foundation is compressed and reacts by exerting an upward pressure to resist foundation loading. If foundation load exceeds maximum passive pressure of ground (i.e. bearing capacity) a downward movement of the foundation could occur. Remedy is to increase plan size of foundation to reduce the load per unit area or alternatively reduce the loadings being carried by the foundations.

Subsoil Movements ~ these are due primarily to changes in volume when the subsoil becomes wet or dry and occurs near the upper surface of the soil. Compact granular soils such as gravel suffer very little movement whereas cohesive soils such as clay do suffer volume changes near the upper surface. Similar volume changes can occur due to water held in the subsoil freezing and expanding – this is called Frost Heave.

Typical Examples ~

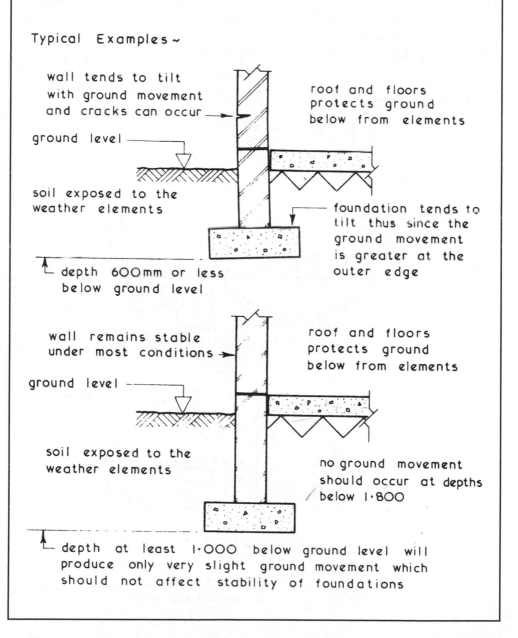

wall tends to tilt with ground movement and cracks can occur →

ground level

soil exposed to the weather elements

roof and floors protects ground below from elements

foundation tends to tilt thus since the ground movement is greater at the outer edge

depth 600mm or less below ground level

wall remains stable under most conditions →

ground level

soil exposed to the weather elements

roof and floors protects ground below from elements

no ground movement should occur at depths below 1·800

depth at least 1·000 below ground level will produce only very slight ground movement which should not affect stability of foundations

191

Trees ~ damage to foundations. Substructural damage to buildings can occur with direct physical contact by tree roots. More common is the indirect effect of moisture shrinkage or heave, particularly apparent in clay subsoils.

Shrinkage is most evident in long periods of dry weather, compounded by moisture abstraction from vegetation. Notably broad leaved trees such as oak, elm and poplar in addition to the thirsty willow species. Heave is the opposite. It occurs during wet weather and is compounded by previous removal of moisture-dependent trees that would otherwise effect some drainage and balance to subsoil conditions.

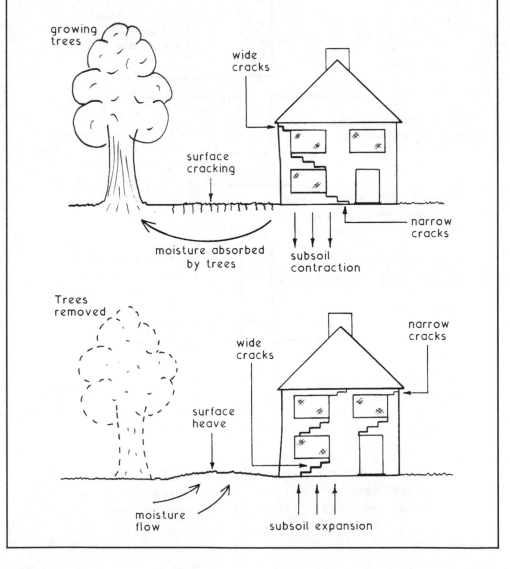

Trees ~ effect on foundations. Trees up to 30 m distance may have an effect on foundations, therefore reference to local authority building control policy should be undertaken before specifying construction techniques.

Traditional strip foundations are practically unsuited, but at excavation depths up to 2·5 or 3·0 m, deep strip or trench fill (preferably reinforced) may be appropriate. Short bored pile foundations are likely to be more economical and particularly suited to depths exceeding 3·0 m.

For guidance only, the illustration and table provide an indication of foundation depths in shrinkable subsoils.

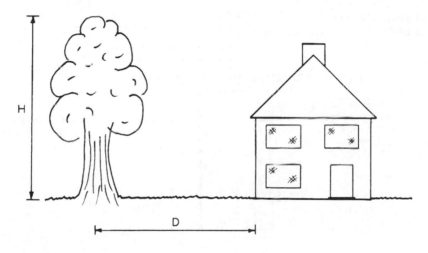

H = Mature height of tree
D = Distance to centre of tree

D/H – Distance from tree/Height of tree

Tree species	0·10	0·25	0·33	0·50	0·66	0·75	1·00
Oak, elm, poplar and willow	3·00	2·80	2·60	2·30	2·10	1·90	1·50
All others	2·80	2·40	2·10	1·80	1·50	1·20	1·00

Minimum foundation depth (m)

Trees ~ preservation orders (see page 109) may be waived by the local planning authority. Permission for tree felling is by formal application and will be considered if the proposed development is in the economic and business interests of the community. However, tree removal is only likely to be acceptable if there is an agreement for replacement stock being provided elsewhere on the site.

In these circumstances, there is potential for ground heave within the 'footprint' of felled trees. To resist this movement, foundations must incorporate an absorbing layer or compressible filler with ground floor suspended above the soil.

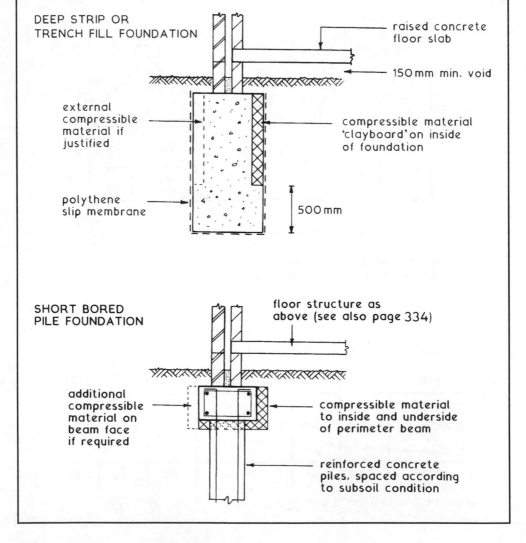

DEEP STRIP OR
TRENCH FILL FOUNDATION

raised concrete
floor slab

150 mm min. void

external
compressible
material if
justified

compressible material
'clayboard' on inside
of foundation

polythene
slip membrane

500 mm

SHORT BORED
PILE FOUNDATION

floor structure as
above (see also page 334)

additional
compressible
material on
beam face
if required

compressible material
to inside and underside
of perimeter beam

reinforced concrete
piles, spaced according
to subsoil condition

Cracking in Walls ~ cracks are caused by applied forces which exceed those that the building can withstand. Most cracking is superficial, occurring as materials dry out and subsequently shrink to reveal minor surface fractures of < 2 mm. These insignificant cracks can be made good with proprietary fillers.

Severe cracking in walls may result from foundation failure, due to inadequate design or physical damage. Further problems could include:

* Structural instability
* Air infiltration
* Sound insulation reduction
* Rain penetration
* Heat loss
* Visual depreciation

A survey should be undertaken to determine:
1. The cause of cracking, i.e.
    * Loads applied externally (tree roots, subsoil movement).
    * Climate/temperature changes (thermal movement).
    * Moisture content change (faulty dpc, building leakage).
    * Vibration (adjacent work, traffic).
    * Changes in physical composition (salt or ice formation).
    * Chemical change (corrosion, sulphate attack).
    * Biological change (timber decay).
2. The effect on a building's performance (structural and environmental).
3. The nature of movement – completed, ongoing or intermittent (seasonal).

Observations over a period of several months, preferably over a full year, will determine whether the cracking is new or established and whether it is progressing.

Simple method for monitoring cracks –

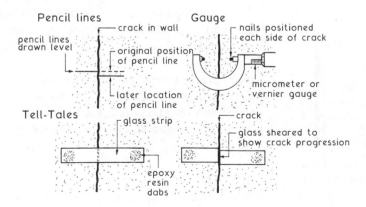

Further reading – BRE Digest 251: Assessment of damage in low rise buildings.

195

Foundation Materials ~ from page 190 one of the functions of a foundation can be seen to be the ability to spread its load evenly over the ground on which it rests. It must of course be constructed of a durable material of adequate strength. Experience has shown that the most suitable material is concrete.

Concrete is a mixture of cement + aggregates + water in controlled proportions.

CEMENT

Manufactured from clay and chalk and is the matrix or binder of the concrete mix. Cement powder can be supplied in bags or bulk —

Bags ~

25 kg.

air-tight sealed bags requiring a dry damp free store.

Bulk ~

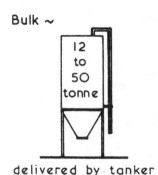

12 to 50 tonne

delivered by tanker and pumped into storage silo.

AGGREGATES

Coarse aggregate is defined as a material which is retained on a 5mm sieve.

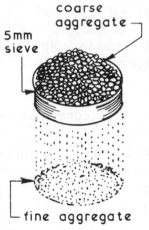

coarse aggregate

5mm sieve

fine aggregate

Fine aggregate is defined as a material which passes a 5mm sieve.

Aggregates can be either natural rock which has disintegrated or crushed stone or gravel.

WATER

Must be of a quality fit for drinking.

MIXES

These are expressed as a ratio thus:~
1 : 3 : 6 / 20mm
which means —
1 part cement.
3 parts of fine aggregate.
6 parts of coarse aggregate
20mm — maximum size of coarse aggregate for the mix.

Water is added to start the chemical reaction and to give the mix workability ~ the amount used is called the — Water / Cement Ratio and is usually about 0·4 to 0·5.

Too much water will produce a weak concrete of low strength whereas too little water will produce a concrete mix of low and inadequate work-ability.

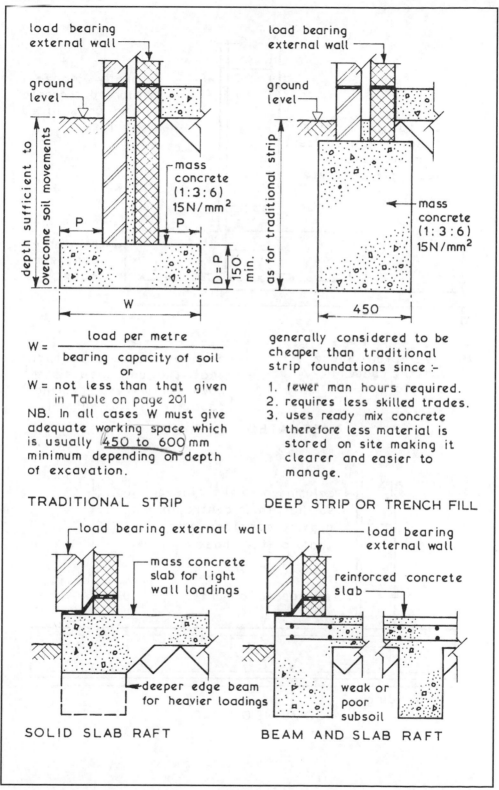

load bearing external wall

ground level

depth sufficient to overcome soil movements

mass concrete (1:3:6) 15 N/mm²

$$W = \frac{\text{load per metre}}{\text{bearing capacity of soil}}$$

or

W = not less than that given in Table on page 201

NB. In all cases W must give adequate working space which is usually 450 to 600 mm minimum depending on depth of excavation.

TRADITIONAL STRIP

load bearing external wall

ground level

as for traditional strip

mass concrete (1:3:6) 15 N/mm²

450

generally considered to be cheaper than traditional strip foundations since :-

1. fewer man hours required.
2. requires less skilled trades.
3. uses ready mix concrete therefore less material is stored on site making it clearer and easier to manage.

DEEP STRIP OR TRENCH FILL

load bearing external wall

mass concrete slab for light wall loadings

deeper edge beam for heavier loadings

SOLID SLAB RAFT

load bearing external wall

reinforced concrete slab

weak or poor subsoil

BEAM AND SLAB RAFT

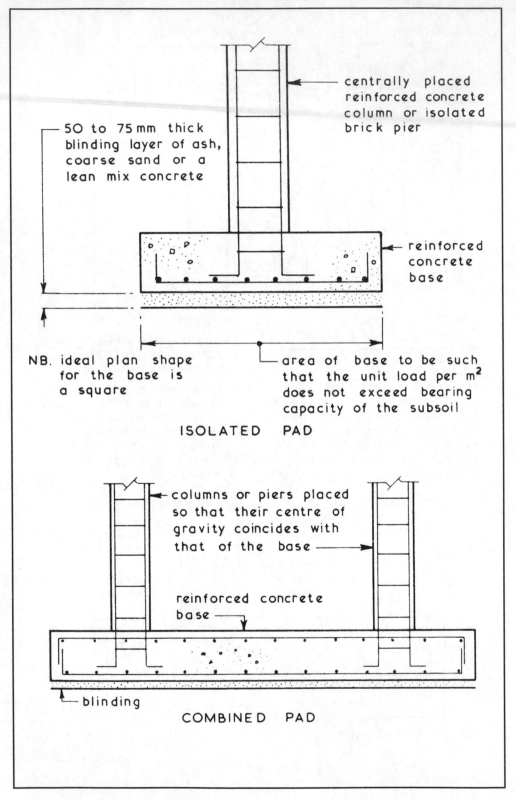

50 to 75 mm thick blinding layer of ash, coarse sand or a lean mix concrete

centrally placed reinforced concrete column or isolated brick pier

reinforced concrete base

NB. ideal plan shape for the base is a square

area of base to be such that the unit load per m² does not exceed bearing capacity of the subsoil

ISOLATED PAD

columns or piers placed so that their centre of gravity coincides with that of the base

reinforced concrete base

blinding

COMBINED PAD

Bed ~ a concrete slab resting on and supported by the subsoil, usually forming the ground floor surface. Beds (sometimes called oversite concrete) are usually cast on a layer of hardcore which is used to make up the reduced level excavation and thus raise the level of the concrete bed to a position above ground level.

Typical Example ~

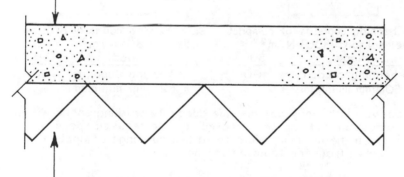

mass concrete bed (1 : 3 : 6 / 20 mm mix 15 N / mm^2). Thickness for domestic work is usually 100 to 150 mm and the bed is constructed so as to prevent the passage of moisture from the ground to the upper surface of the floor – this is usually achieved by incorporating into the design a damp-proof membrane ~ for details see page 596

100 to 150 mm thick layer of hardcore ~ material used should be inert and not affected by water. Suitable materials are gravel; crushed rock; quarry waste; concrete rubble; brick or tile rubble; blast furnace slag and pulverised fuel ash. The hardcore material should be laid evenly and well compacted with the upper surface blinded with fine grade material as required.

199

Basic Sizing ~ the size of a foundation is basically dependent on two factors –

1. Load being transmitted, max 70 kN/m (dwellings up to 3 storeys).
2. Bearing capacity of subsoil under proposed foundation.

Bearing capacities for different types of subsoils may be obtained from tables such as those in BS 8004: Code of practice for foundations and BS 8103-1: Structural design of low rise buildings. Also, directly from soil investigation results.

Typical   Examples ~

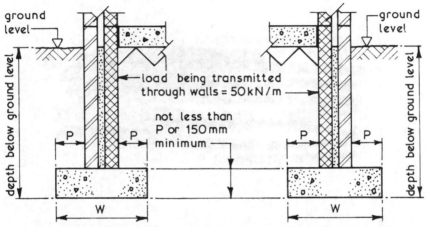

safe bearing capacity of compact gravel subsoil = $100 \, kN/m^2$

$$W = \frac{load}{bearing\ capacity} = \frac{50}{100}$$
$$= 500\,mm \ minimum$$

safe bearing capacity of clay subsoil = $80 \, kN/m^2$

$$W = \frac{load}{bearing\ capacity} = \frac{50}{80}$$
$$= 625\,mm \ minimum$$

The above widths may not provide adequate working space within the excavation and can be increased to give required space. Guidance on the minimum width for a limited range of applications can be taken from the table on the next page.

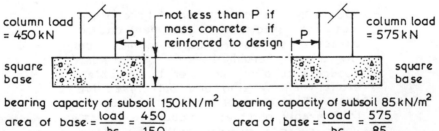

bearing capacity of subsoil $150 \, kN/m^2$

$$area\ of\ base = \frac{load}{bc} = \frac{450}{150}$$
$$= 3\,m^2 \ \therefore \ side = \sqrt{3}$$
$$= 1 \cdot 732 \ min.$$

bearing capacity of subsoil $85 \, kN/m^2$

$$area\ of\ base = \frac{load}{bc} = \frac{575}{85}$$
$$= 6 \cdot 765\,m^2 \ \therefore \ side = \sqrt{6 \cdot 765}$$
$$= 2 \cdot 6 \ min.$$

Ground type	Ground condition	Field test	Max. total load on load bearing wall (kN/m)					
			20	30	40	50	60	70
			Minimum width (mm)					
Rock	Not inferior to sandstone, limestone or firm chalk.	Requires a mechanical device to excavate.	At least equal to the width of the wall					
Gravel Sand	Medium density Compact	Pick required to excavate. 50 mm square peg hard to drive beyond 150 mm.	250	300	400	500	600	650
Clay Sandy clay	Stiff Stiff	Requires pick or mechanical device to aid removal. Can be indented slightly with thumb.	250	300	400	500	600	650
Clay Sandy clay	Firm Firm	Can be moulded under substantial pressure by fingers.	300	350	450	600	750	850
Sand Silty sand Clayey sand	Loose Loose Loose	Can be excavated by spade. 50 mm square peg easily driven.	400	600	Conventional strip foundations unsuitable for a total load exceeding 30 kN/m.			
Silt Clay Sandy clay Silty clay	Soft Soft Soft Soft	Finger pushed in up to 10 mm. Easily moulded with fingers.	450	650				
Silt Clay Sandy clay Silty clay	Very soft Very soft Very soft Very soft	Finger easily pushed in up to 25 mm. Wet sample exudes between fingers when squeezed.	Conventional strip inappropriate. Steel reinforced wide strip, deep strip or piled foundation selected subject to specialist advice.					

Adapted from Table 10 in the Bldg. Regs., A.D: A – Structure.

Typical procedure (for guidance only) –

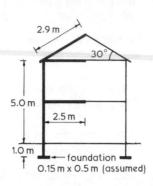

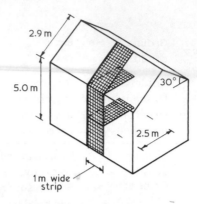

foundation
0.15 m x 0.5 m (assumed)

1 m wide strip

Dead load per m run (see pages 35 and 36)

Substructure brickwork, 1 m × 1 m × 476 kg/m²	=	476 kg
……… cavity conc. (50 mm), 1 m × 1 m × 2300 kg/m³	=	115 kg
Foundation concrete, 0·15 m × 1 m × 0·5 m × 2300 kg/m³	=	173 kg
Superstructure brickwork, 5 m × 1 m × 221 kg/m²	=	1105 kg
………… blockwork & ins., 5 m × 1 m × 79 kg/m²	=	395 kg
………… 2 coat plasterwork, 5 m × 1 m × 22 kg/m²	=	110 kg
Floor joists/boards/plstrbrd., 2·5 m × 1 m × 42·75 kg/m²	=	107 kg
Ceiling joists/plstrbrd/ins., 2·5 m × 1 m × 19·87 kg/m²	=	50 kg
Rafters, battens & felt, 2·9 m × 1 m × 12·10 kg/m²	=	35 kg
Single lap tiling, 2·9 m × 1 m × 49 kg/m²	=	142 kg
		2708 kg

Note: kg × 9·81 = Newtons

Therefore: 2708 kg × 9·81 = 26565 N or 26·56 kN

Imposed load per m run (see BS 6399-1: Code of practice for dead and imposed loads) –

Floor, 2·5 m × 1 m × 1·5 kN/m²   = 3·75 kN

Roof, 2·9 m × 1 m × 1·5 kN/m² (snow) = 4·05 kN

                                7·80 kN

Note: For roof pitch >30°, snow load = 0·75 kN/m²

Dead + imposed load is, 26·56 kN + 7·80 kN = 34·36 kN

Given that the subsoil has a safe bearing capacity of 75 kN/m²,

W = load ÷ bearing capacity = 34·36 ÷ 75 = 0·458 m or 458 mm

   Therefore a foundation width of 500 mm is adequate.

Note: This example assumes the site is sheltered. If it is necessary to make allowance for wind loading, reference should be made to BS 6399-2: Code of practice for wind loads.

Stepped Foundations ~ these are usually considered in the context of strip foundations and are used mainly on sloping sites to reduce the amount of excavation and materials required to produce an adequate foundation.

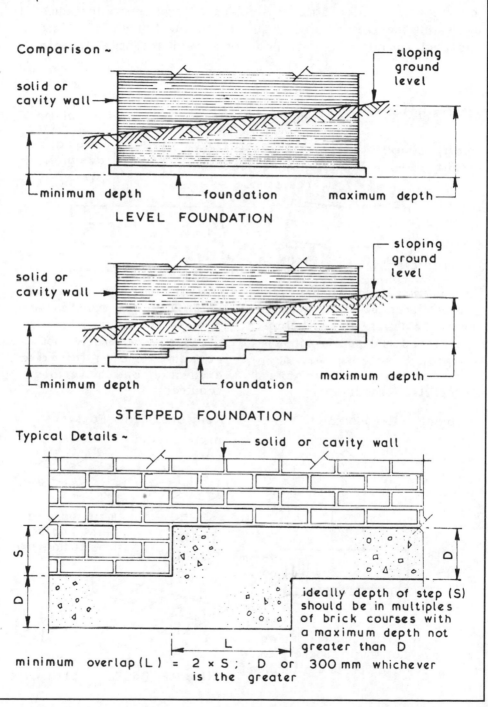

Comparison ~

solid or cavity wall →

sloping ground level

minimum depth   foundation   maximum depth

LEVEL FOUNDATION

solid or cavity wall →

sloping ground level

minimum depth   foundation   maximum depth

STEPPED FOUNDATION

Typical Details ~   solid or cavity wall

ideally depth of step (S) should be in multiples of brick courses with a maximum depth not greater than D

minimum overlap (L) = 2 × S; D or 300 mm whichever is the greater

Concrete Foundations ~ concrete is a material which is strong in compression but weak in tension. If its tensile strength is exceeded cracks will occur resulting in a weak and unsuitable foundation. One method of providing tensile resistance is to include in the concrete foundation bars of steel as a form of reinforcement to resist all the tensile forces induced into the foundation. Steel is a material which is readily available and has high tensile strength.

Comparisons~

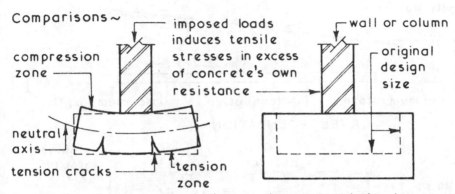

foundation tends to bend, the upper fibres being compressed and the lower fibres being stretched and put in tension- remedies increase size of base or design as a reinforced concrete foundation

size of foundation increased to provide the resistance against the induced tensile stresses - generally not economic due to the extra excavation and materials required

Typical RC Foundation

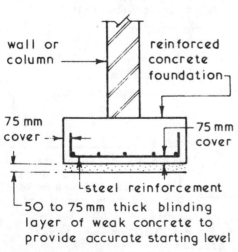

Reinforcement Patterns

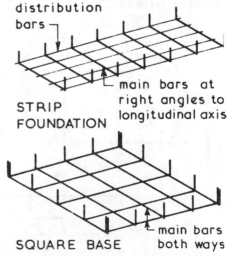

Short Bored Piles ~ these are a form of foundation which are suitable for domestic loadings and clay subsoils where ground movements can occur below the 1·000 depth associated with traditional strip and trench fill foundations. They can be used where trees are planted close to a new building since the trees may eventually cause damaging ground movements due to extracting water from the subsoil and root growth. Conversely where trees have been removed this may lead to ground swelling.

Typical Details ~

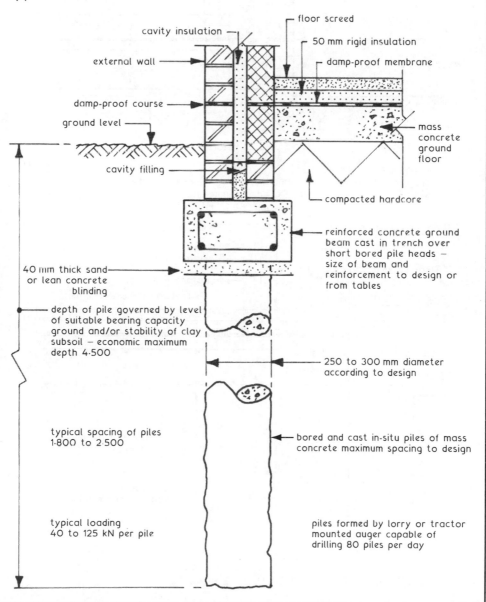

cavity insulation

floor screed

50 mm rigid insulation

external wall

damp-proof membrane

damp-proof course

ground level

mass concrete ground floor

cavity filling

compacted hardcore

reinforced concrete ground beam cast in trench over short bored pile heads — size of beam and reinforcement to design or from tables

40 mm thick sand or lean concrete blinding

depth of pile governed by level of suitable bearing capacity ground and/or stability of clay subsoil — economic maximum depth 4·500

250 to 300 mm diameter according to design

typical spacing of piles 1·800 to 2·500

bored and cast in-situ piles of mass concrete maximum spacing to design

typical loading 40 to 125 kN per pile

piles formed by lorry or tractor mounted auger capable of drilling 80 piles per day

Simple Raft Foundations ~ these can be used for lightly loaded buildings on poor soils or where the top 450 to 600 mm of soil is overlaying a poor quality substrata.

Typical Details ~

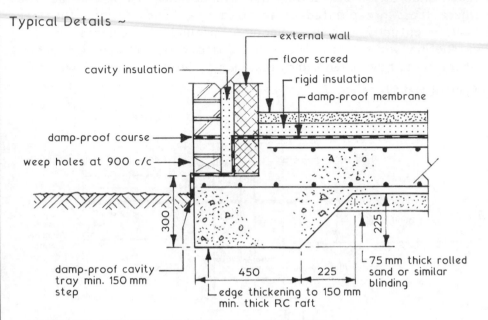

REINFORCED CONCRETE RAFT WITH EDGE THICKENING

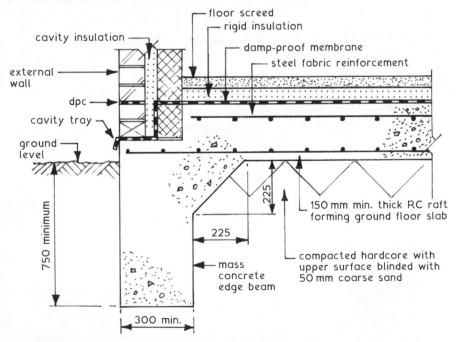

REINFORCED CONCRETE RAFT WITH EDGE BEAM

Foundation Design Principles ~ the main objectives of foundation design are to ensure that the structural loads are transmitted to the subsoil(s) safely, economically and without any unacceptable movement during the construction period and throughout the anticipated life of the building or structure.

Basic Design Procedure ~ this can be considered as a series of steps or stages –

1. Assessment of site conditions in the context of the site and soil investigation report.

2. Calculation of anticipated structural loading(s).

3. Choosing the foundation type taking into consideration –

    a. Soil conditions;
    b. Type of structure;
    c. Structural loading(s);
    d. Economic factors;
    e. Time factors relative to the proposed contract period;
    f. Construction problems.

4. Sizing the chosen foundation in the context of loading(s), ground bearing capacity and any likely future movements of the building or structure.

Foundation Types ~ apart from simple domestic foundations most foundation types are constructed in reinforced concrete and may be considered as being shallow or deep. Most shallow types of foundation are constructed within 2·000 of the ground level but in some circumstances it may be necessary to take the whole or part of the foundations down to a depth of 2·000 to 5·000 as in the case of a deep basement where the structural elements of the basement are to carry the superstructure loads. Generally foundations which need to be taken below 5·000 deep are cheaper when designed and constructed as piled foundations and such foundations are classified as deep foundations. (For piled foundation details see pages 212 to 229)

Foundations are usually classified by their type such as strips, pads, rafts and piles. It is also possible to combine foundation types such as strip foundations connected by beams to and working in conjunction with pad foundations.

Strip Foundations ~ these are suitable for most subsoils and light structural loadings such as those encountered in low to medium rise domestic dwellings where mass concrete can be used. Reinforced concrete is usually required for all other situations.

Typical Strip Foundation Types ~

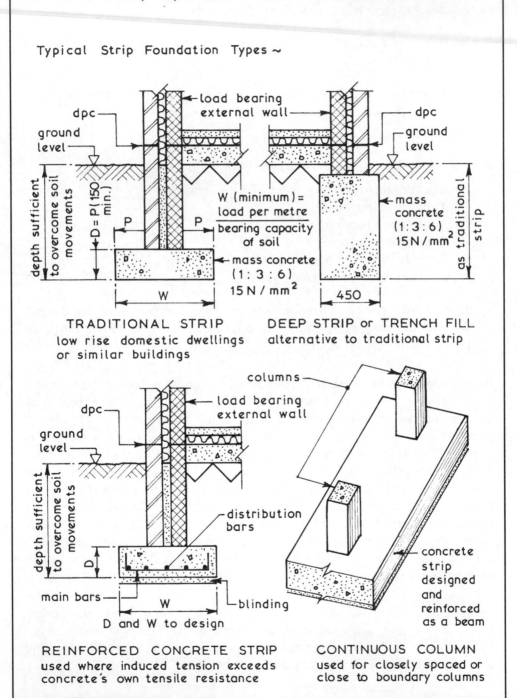

TRADITIONAL STRIP
low rise domestic dwellings
or similar buildings

DEEP STRIP or TRENCH FILL
alternative to traditional strip

$$W \text{ (minimum)} = \frac{\text{load per metre}}{\text{bearing capacity of soil}}$$

REINFORCED CONCRETE STRIP
used where induced tension exceeds
concrete's own tensile resistance

CONTINUOUS COLUMN
used for closely spaced or
close to boundary columns

Pad Foundations ~ suitable for most subsoils except loose sands, loose gravels and filled areas. Pad foundations are usually constructed of reinforced concrete and where possible are square in plan.

Typical Pad Foundation Types ~

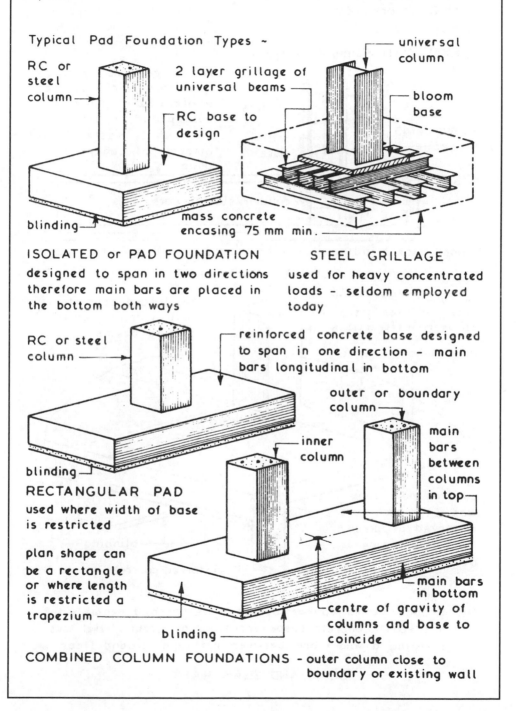

RC or steel column

2 layer grillage of universal beams

RC base to design

blinding

mass concrete encasing 75 mm min.

universal column

bloom base

ISOLATED or PAD FOUNDATION
designed to span in two directions therefore main bars are placed in the bottom both ways

STEEL GRILLAGE
used for heavy concentrated loads - seldom employed today

RC or steel column

reinforced concrete base designed to span in one direction - main bars longitudinal in bottom

blinding

RECTANGULAR PAD
used where width of base is restricted

plan shape can be a rectangle or where length is restricted a trapezium

outer or boundary column

inner column

main bars between columns in top

main bars in bottom

centre of gravity of columns and base to coincide

blinding

COMBINED COLUMN FOUNDATIONS - outer column close to boundary or existing wall

Raft Foundations ~ these are used to spread the load of the superstructure over a large base to reduce the load per unit area being imposed on the ground and this is particularly useful where low bearing capacity soils are encountered and where individual column loads are heavy.

Typical Raft Foundation Types ~

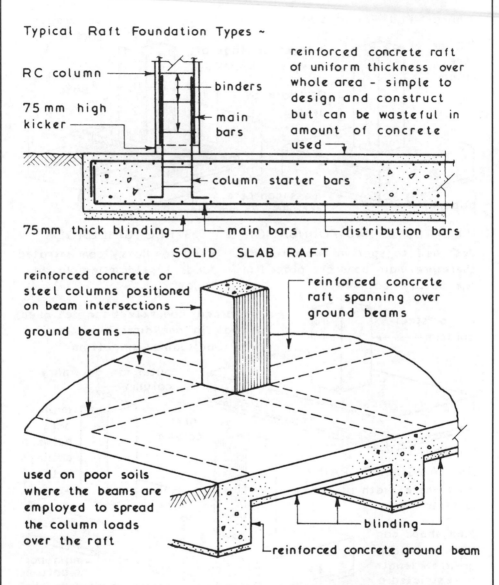

RC column

75 mm high kicker

binders

main bars

reinforced concrete raft of uniform thickness over whole area - simple to design and construct but can be wasteful in amount of concrete used

column starter bars

75 mm thick blinding   main bars   distribution bars

SOLID SLAB RAFT

reinforced concrete or steel columns positioned on beam intersections

reinforced concrete raft spanning over ground beams

ground beams

used on poor soils where the beams are employed to spread the column loads over the raft

blinding

reinforced concrete ground beam

NB. Ground beams can be designed as upstand beams with a precast concrete suspended floor at ground level thus creating a void space between raft and ground floor.

BEAM AND SLAB RAFT

Cantilever Foundations ~ these can be used where it is necessary to avoid imposing any pressure on an adjacent foundation or underground service.

Typical Cantilever Foundation Types ~

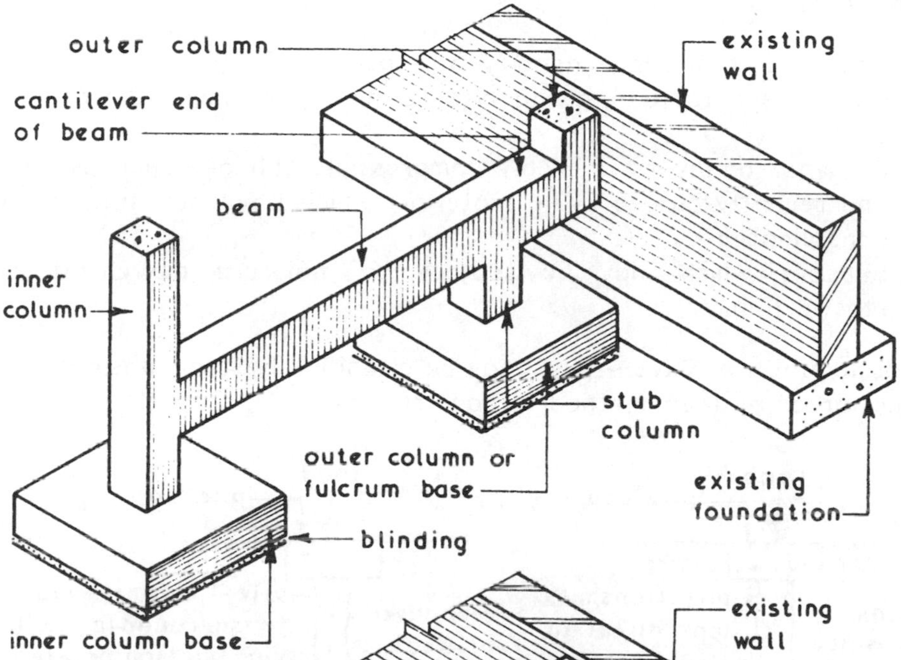

outer column

cantilever end of beam

beam

inner column

outer column or fulcrum base

blinding

inner column base

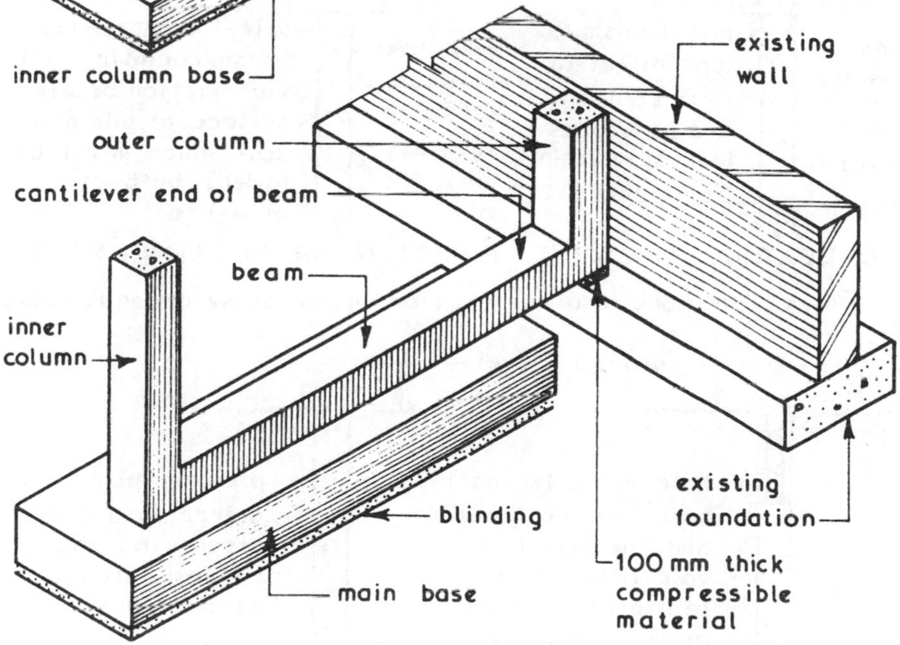

existing wall

stub column

existing foundation

existing wall

outer column

cantilever end of beam

beam

inner column

blinding

main base

existing foundation

100 mm thick compressible material

Cantilever foundations designed and constructed in reinforced concrete

211

Piled Foundations ~ these can be defined as a series of columns constructed or inserted into the ground to transmit the load(s) of a structure to a lower level of subsoil. Piled foundations can be used when suitable foundation conditions are not present at or near ground level making the use of deep traditional foundations uneconomic. The lack of suitable foundation conditions may be caused by:-

1. Natural low bearing capacity of subsoil.
2. High water table – giving rise to high permanent dewatering costs.
3. Presence of layers of highly compressible subsoils such as peat and recently placed filling materials which have not sufficiently consolidated.
4. Subsoils which may be subject to moisture movement or plastic failure.

Classification of Piles ~ piles may be classified by their basic design function or by their method of construction:-

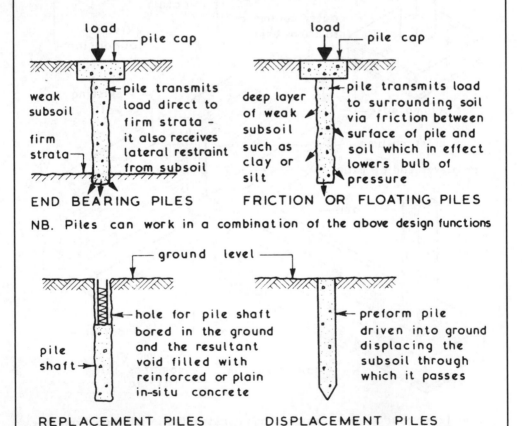

load ── pile cap

weak subsoil
firm strata ── pile transmits load direct to firm strata – it also receives lateral restraint from subsoil

END BEARING PILES

load ── pile cap

deep layer of weak subsoil such as clay or silt ── pile transmits load to surrounding soil via friction between surface of pile and soil which in effect lowers bulb of pressure

FRICTION OR FLOATING PILES

NB. Piles can work in a combination of the above design functions

── ground level ──

pile shaft ── hole for pile shaft bored in the ground and the resultant void filled with reinforced or plain in-situ concrete

REPLACEMENT PILES

── preform pile driven into ground displacing the subsoil through which it passes

DISPLACEMENT PILES

Replacement Piles ~ these are often called bored piles since the removal of the spoil to form the hole for the pile is always carried out by a boring technique. They are used primarily in cohesive subsoils for the formation of friction piles and when forming pile foundations close to existing buildings where the allowable amount of noise and/or vibration is limited.

Replacement Pile Types ~

PERCUSSION BORED

small or medium size contracts with up to 300 piles

load range - 300 to 1300 kN

length range - up to 24·000

diameter range - 300 to 900

may have to be formed as a pressure pile in waterlogged subsoils - see page 214

FLUSH BORED

large projects - these are basically a rotary bored pile using bentonite as a drilling fluid

load range - 1000 to 5000 kN

length range - up to 30·000

diameter range - 600 to 1500

see page 215

ROTARY BORED

Small Diameter - < 600 mm

light loadings - can also be used in groups or clusters with a common pile cap to receive heavy loads

load range - 50 to 400 kN

length range - up to 15·000

diameter range - 240 to 600

see page 216

Large Diameter - > 600 mm

heavy concentrated loadings - may have an underreamed or belled toe

load range - 800 to 15000 kN

length range - up to 60·000

diameter range - 600 to 2400

see page 217

NB. The above given data depicts typical economic ranges. More than one pile type can be used on a single contract.

Percussion Bored Piles

Typical Details ~

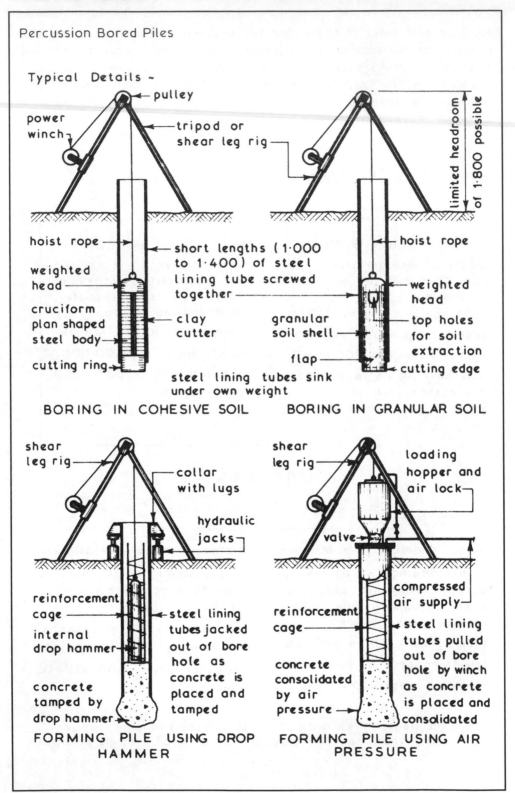

power winch

pulley

tripod or shear leg rig

limited headroom of 1·800 possible

hoist rope

weighted head

cruciform plan shaped steel body

cutting ring

short lengths (1·000 to 1·400) of steel lining tube screwed together

clay cutter

steel lining tubes sink under own weight

BORING IN COHESIVE SOIL

hoist rope

weighted head

granular soil shell

flap

top holes for soil extraction

cutting edge

BORING IN GRANULAR SOIL

shear leg rig

collar with lugs

hydraulic jacks

reinforcement cage

internal drop hammer

concrete tamped by drop hammer

steel lining tubes jacked out of bore hole as concrete is placed and tamped

FORMING PILE USING DROP HAMMER

shear leg rig

loading hopper and air lock

valve

reinforcement cage

concrete consolidated by air pressure

compressed air supply

steel lining tubes pulled out of bore hole by winch as concrete is placed and consolidated

FORMING PILE USING AIR PRESSURE

Flush Bored Piles

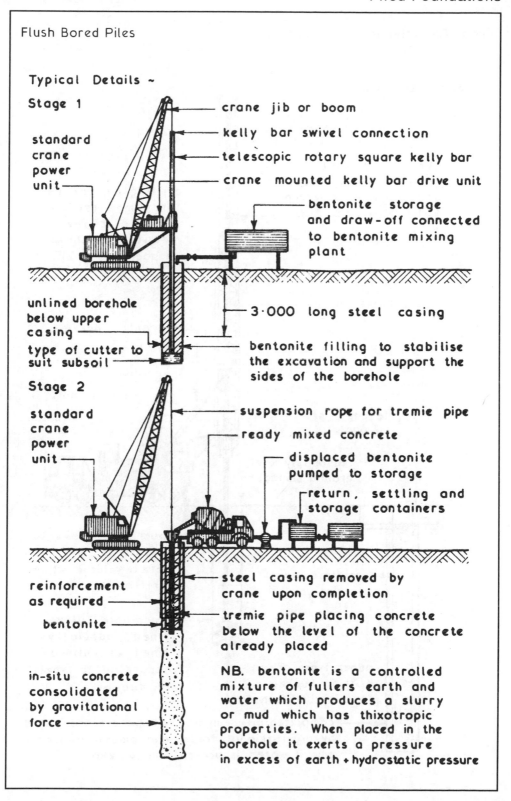

Typical Details ~

Stage 1

standard crane power unit

crane jib or boom

kelly bar swivel connection

telescopic rotary square kelly bar

crane mounted kelly bar drive unit

bentonite storage and draw-off connected to bentonite mixing plant

unlined borehole below upper casing

3·000 long steel casing

type of cutter to suit subsoil

bentonite filling to stabilise the excavation and support the sides of the borehole

Stage 2

standard crane power unit

suspension rope for tremie pipe

ready mixed concrete

displaced bentonite pumped to storage

return, settling and storage containers

reinforcement as required

bentonite

steel casing removed by crane upon completion

tremie pipe placing concrete below the level of the concrete already placed

in-situ concrete consolidated by gravitational force

NB. bentonite is a controlled mixture of fullers earth and water which produces a slurry or mud which has thixotropic properties. When placed in the borehole it exerts a pressure in excess of earth + hydrostatic pressure

215

Small Diameter Rotary Bored Piles

Typical Details ~

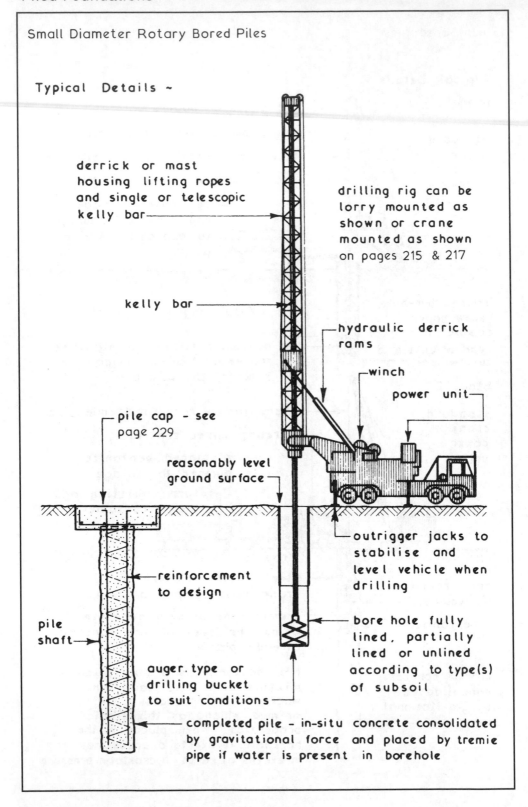

derrick or mast housing lifting ropes and single or telescopic kelly bar

drilling rig can be lorry mounted as shown or crane mounted as shown on pages 215 & 217

kelly bar

hydraulic derrick rams

winch

power unit

pile cap - see page 229

reasonably level ground surface

outrigger jacks to stabilise and level vehicle when drilling

reinforcement to design

pile shaft

bore hole fully lined, partially lined or unlined according to type(s) of subsoil

auger type or drilling bucket to suit conditions

completed pile - in-situ concrete consolidated by gravitational force and placed by tremie pipe if water is present in borehole

Large Diameter Rotary Bored Piles

Typical Details ~

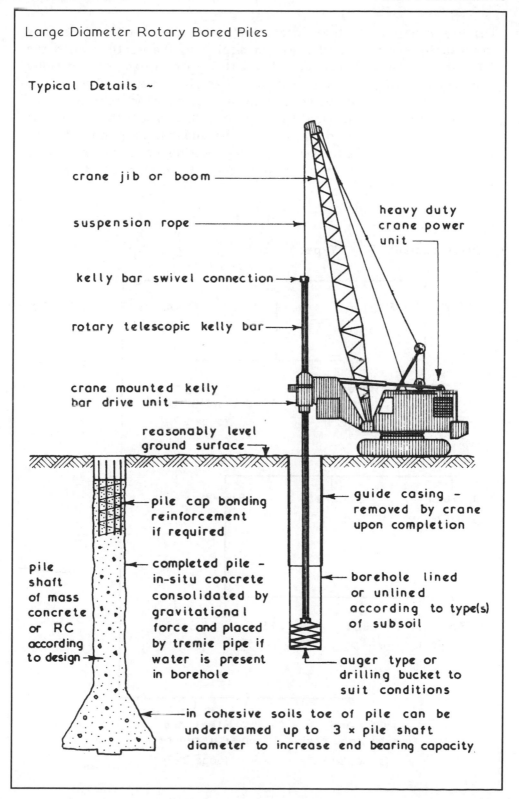

crane jib or boom

suspension rope

kelly bar swivel connection

rotary telescopic kelly bar

crane mounted kelly bar drive unit

reasonably level ground surface

heavy duty crane power unit

pile cap bonding reinforcement if required

completed pile - in-situ concrete consolidated by gravitational force and placed by tremie pipe if water is present in borehole

pile shaft of mass concrete or RC according to design

guide casing - removed by crane upon completion

borehole lined or unlined according to type(s) of subsoil

auger type or drilling bucket to suit conditions

in cohesive soils toe of pile can be underreamed up to 3 × pile shaft diameter to increase end bearing capacity.

Displacement Piles ~ these are often called driven piles since they are usually driven into the ground displacing the earth around the pile shaft. These piles can be either preformed or partially preformed if they are not cast in-situ and are available in a wide variety of types and materials. The pile or forming tube is driven into the required position to a predetermined depth or to the required `set´ which is a measure of the subsoils resistance to the penetration of the pile and hence its bearing capacity by noting the amount of penetration obtained by a fixed number of hammer blows.

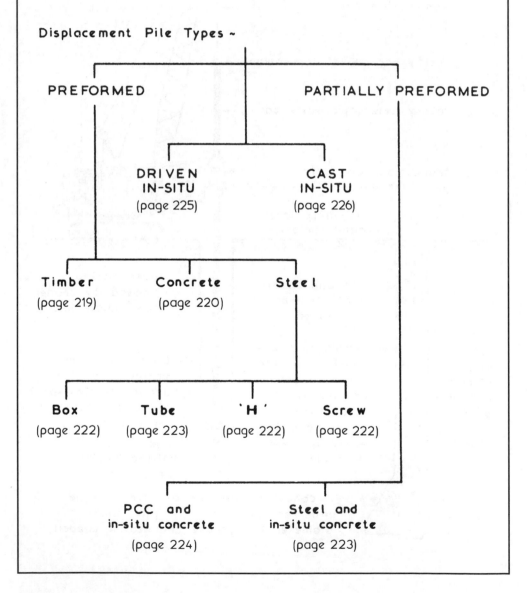

Displacement Pile Types ~

PREFORMED   PARTIALLY PREFORMED

DRIVEN   CAST
IN-SITU   IN-SITU
(page 225)   (page 226)

Timber   Concrete   Steel
(page 219)   (page 220)

Box   Tube   `H´   Screw
(page 222)   (page 223)   (page 222)   (page 222)

PCC and   Steel and
in-situ concrete   in-situ concrete
(page 224)   (page 223)

Timber Piles ~ these are usually square sawn and can be used for small contracts on sites with shallow alluvial deposits overlying a suitable bearing strata (e.g. river banks and estuaries.) Timber piles are percussion driven.

Typical Example ~

mild
steel
band

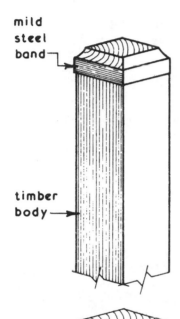

timber
body

Typical Data :-

load range - 50 to 350 kN

length range - up to 12·000
without splicing

size range - 225 × 225
300 × 300 *
350 × 350 *
400 × 400 *
450 × 450
600 × 600

* common sizes

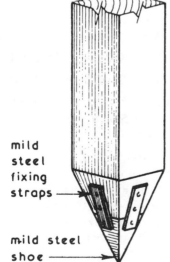

mild
steel
fixing
straps

mild steel
shoe

NB. timber piles are not easy to splice and are liable to attack by marine borers when set in water therefore such piles should always be treated with a suitable preservative before being driven.

Preformed Concrete Piles ~ variety of types available which are generally used on medium to large contracts of not less than one hundred piles where soft soil deposits overlie a firmer strata. These piles are percussion driven using a drop or single acting hammer.

Typical Example [West's Hardrive Precast Modular Pile] ~

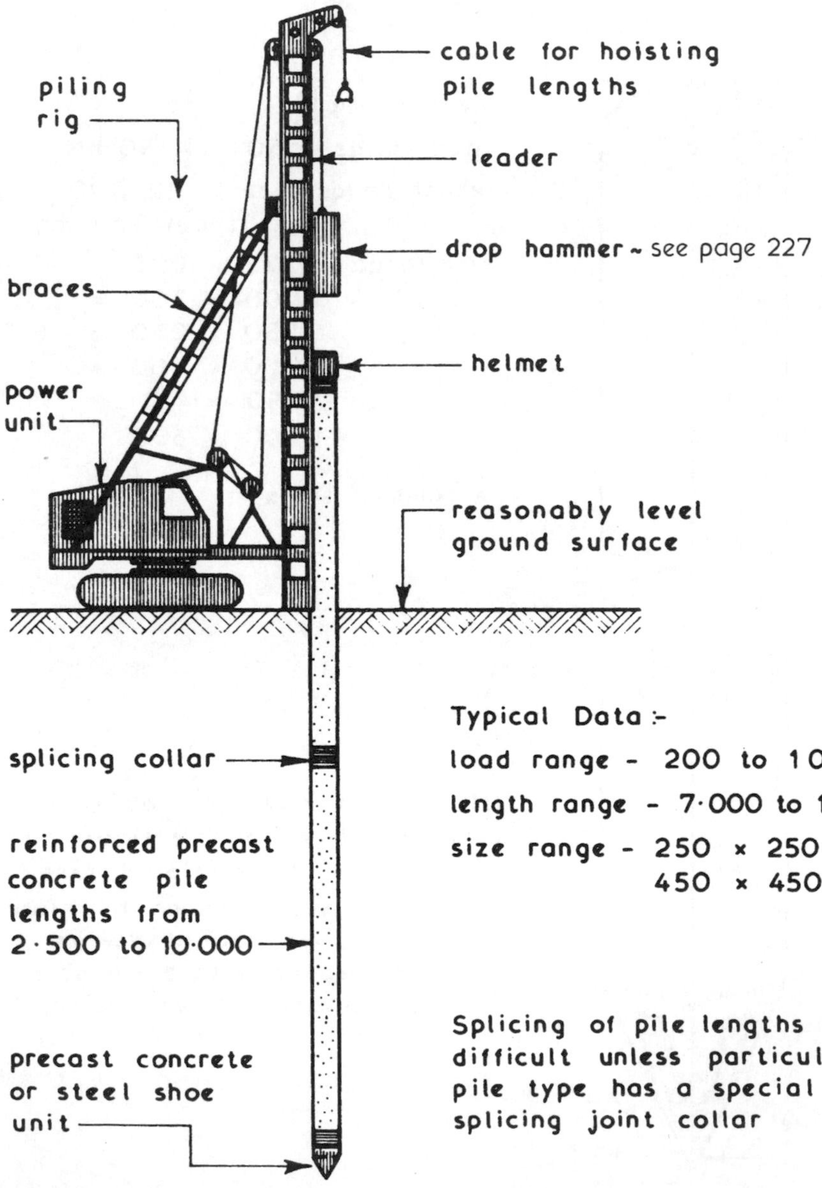

cable for hoisting pile lengths

piling rig

leader

drop hammer ~ see page 227

braces

helmet

power unit

reasonably level ground surface

splicing collar

Typical Data :-
load range - 200 to 1 000 kN
length range - 7·000 to 18·000
size range - 250 × 250 up to
450 × 450

reinforced precast concrete pile lengths from 2·500 to 10·000

precast concrete or steel shoe unit

Splicing of pile lengths is difficult unless particular pile type has a special splicing joint collar

Preformed Concrete Piles – jointing with a peripheral steel splicing collar as shown on the preceding page is adequate for most concentrically or directly loaded situations. Where very long piles are to be used and/or high stresses due to compression, tension and bending from the superstructure or the ground conditions are anticipated, the 4 or 8 lock pile joint [AARSLEFF PILING] may be considered.

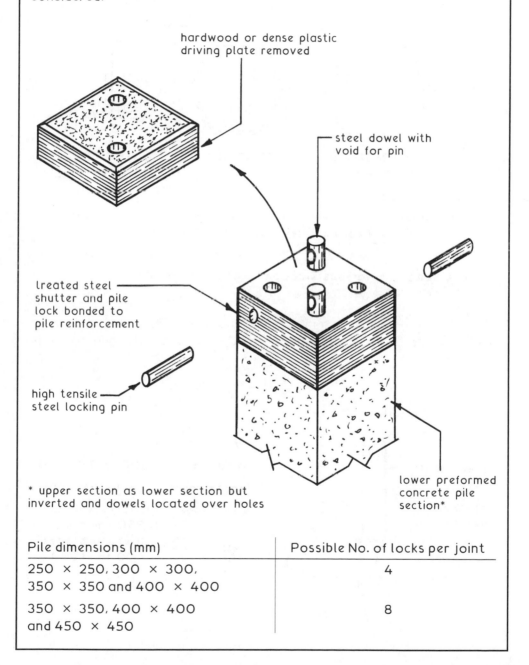

hardwood or dense plastic driving plate removed

steel dowel with void for pin

treated steel shutter and pile lock bonded to pile reinforcement

high tensile steel locking pin

lower preformed concrete pile section*

* upper section as lower section but inverted and dowels located over holes

Pile dimensions (mm)	Possible No. of locks per joint
250 × 250, 300 × 300, 350 × 350 and 400 × 400	4
350 × 350, 400 × 400 and 450 × 450	8

Steel Box and 'H' Sections ~ standard steel sheet pile sections can be used to form box section piles whereas the 'H' section piles are cut from standard rolled sections. These piles are percussion driven and are used mainly in connection with marine structures.

Typical Examples ~

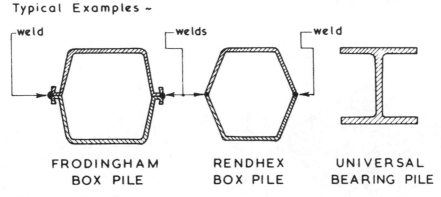

FRODINGHAM
BOX PILE

RENDHEX
BOX PILE

UNIVERSAL
BEARING PILE

Typical Data :-

load range - box piles 300 to 1500 kN
bearing piles 300 to 1700 kN

length range - all types up to 36·000

size range - various sizes and profiles available

Steel Screw Piles ~ rotary driven and used for dock and jetty works where support at shallow depths in soft silts and sands is required.

Typical Example ~

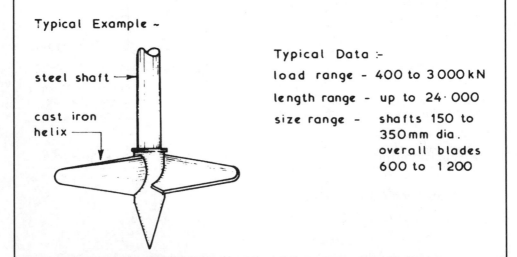

steel shaft →

cast iron
helix →

Typical Data :-

load range - 400 to 3000 kN

length range - up to 24·000

size range -   shafts 150 to
350mm dia.
overall blades
600 to 1200

Steel Tube Piles ~ used on small to medium size contracts for marine structures and foundations in soft subsoils over a suitable bearing strata. Tube piles are usually bottom driven with an internal drop hammer. The loading can be carried by the tube alone but it is usual to fill the tube with mass concrete to form a composite pile. Reinforcement, except for pile cap bonding bars, is not normally required.

Typical Example [BSP Cased Pile] ~

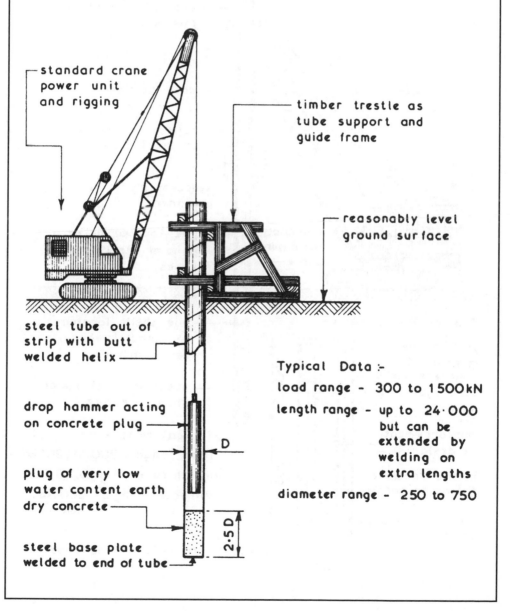

standard crane power unit and rigging

timber trestle as tube support and guide frame

reasonably level ground surface

steel tube out of strip with butt welded helix

drop hammer acting on concrete plug

plug of very low water content earth dry concrete

steel base plate welded to end of tube

D

2·5 D

Typical Data :-

load range - 300 to 1500kN

length range - up to 24·000 but can be extended by welding on extra lengths

diameter range - 250 to 750

Partially Preformed Piles ~ these are composite piles of precast concrete and in-situ concrete or steel and in-situ concrete (see page 223). These percussion driven piles are used on medium to large contracts where bored piles would not be suitable owing to running water or very loose soils.

Typical Example [West's Shell Pile] ~

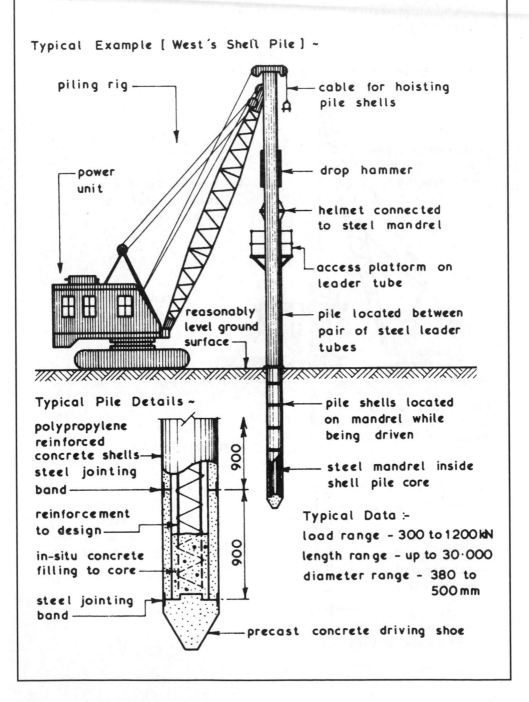

piling rig

cable for hoisting pile shells

power unit

drop hammer

helmet connected to steel mandrel

access platform on leader tube

reasonably level ground surface

pile located between pair of steel leader tubes

Typical Pile Details ~

pile shells located on mandrel while being driven

polypropylene reinforced concrete shells

steel jointing band

900

steel mandrel inside shell pile core

reinforcement to design

900

Typical Data :-
load range - 300 to 1200 kN
length range - up to 30·000
diameter range - 380 to 500 mm

in-situ concrete filling to core

steel jointing band

precast concrete driving shoe

Driven In-situ Piles ~ used on medium to large contracts as an alternative to preformed piles particularly where final length of pile is a variable to be determined on site.

Typical Example [Franki Driven Insitu Pile] ~

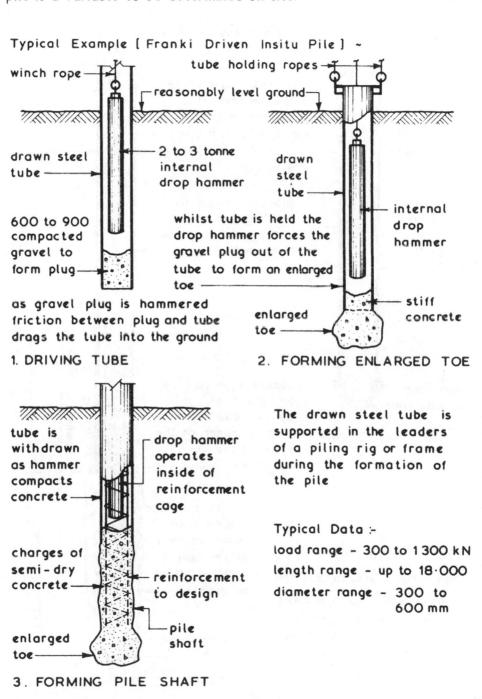

winch rope

tube holding ropes

reasonably level ground

drawn steel tube

2 to 3 tonne internal drop hammer

drawn steel tube

internal drop hammer

600 to 900 compacted gravel to form plug

whilst tube is held the drop hammer forces the gravel plug out of the tube to form an enlarged toe

as gravel plug is hammered friction between plug and tube drags the tube into the ground

stiff concrete

enlarged toe

1. DRIVING TUBE

2. FORMING ENLARGED TOE

tube is withdrawn as hammer compacts concrete

drop hammer operates inside of reinforcement cage

charges of semi-dry concrete

reinforcement to design

enlarged toe

pile shaft

3. FORMING PILE SHAFT

The drawn steel tube is supported in the leaders of a piling rig or frame during the formation of the pile

Typical Data :-

load range - 300 to 1300 kN

length range - up to 18.000

diameter range - 300 to 600 mm

Cast In-situ Piles ~ an alternative to the driven in-situ piles (see page 225)

Typical Example [Vibro Cast Insitu Pile] ~

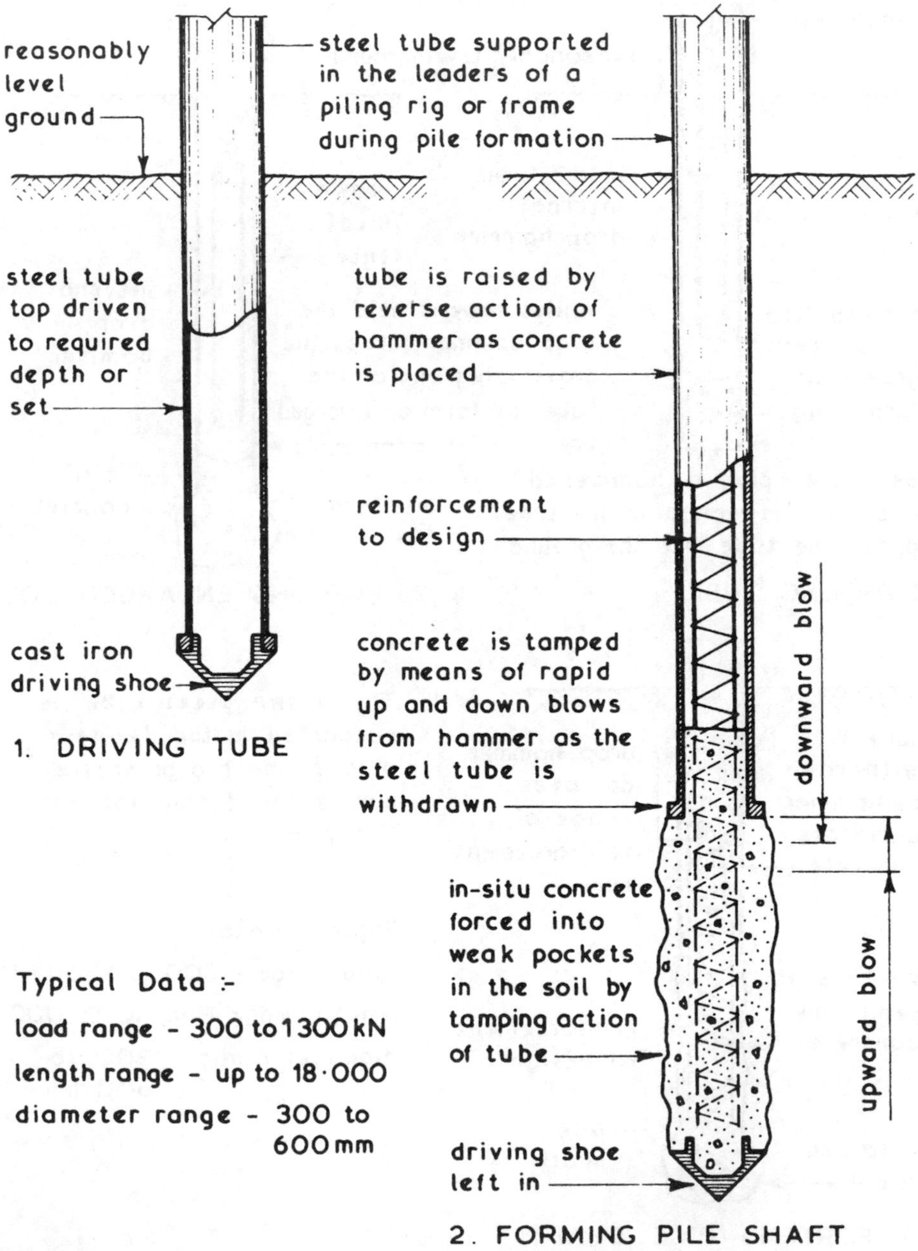

reasonably level ground ──

steel tube supported in the leaders of a piling rig or frame during pile formation ──→

steel tube top driven to required depth or set ──

tube is raised by reverse action of hammer as concrete is placed ──→

reinforcement to design ──→

cast iron driving shoe ──→

concrete is tamped by means of rapid up and down blows from hammer as the steel tube is withdrawn ──→

downward blow

1. DRIVING TUBE

in-situ concrete forced into weak pockets in the soil by tamping action of tube ──→

upward blow

Typical Data :-

load range - 300 to 1300 kN

length range - up to 18·000

diameter range - 300 to 600 mm

driving shoe left in ──→

2. FORMING PILE SHAFT

Piling Hammers ~ these are designed to deliver an impact blow to the top of the pile to be driven. The hammer weight and drop height is chosen to suit the pile type and nature of subsoil(s) through which it will be driven. The head of the pile being driven is protected against damage with a steel helmet which is padded with a sand bed or similar material and is cushioned with a plastic or hardwood block called a dolly.

Drop Hammers ~ these are blocks of iron with a rear lug(s) which locate in the piling rig guides or leaders and have a top eye for attachment of the winch rope. The number of blows which can be delivered with a free fall of 1·200 to 1·500 ranges from 10 to 20 per minute. The weight of the hammer should be not less than 50% of the concrete or steel pile weight and 1 to 1·5 times the weight of a timber pile.

Single Acting Hammers ~ these consist of a heavy falling cylinder raised by steam or compressed air sliding up and down a fixed piston. Guide lugs or rollers are located in the piling frame leaders to maintain the hammer position relative to the pile head. The number of blows delivered ranges from 36 to 75 per minute with a total hammer weight range of 2 to 15 tonnes.

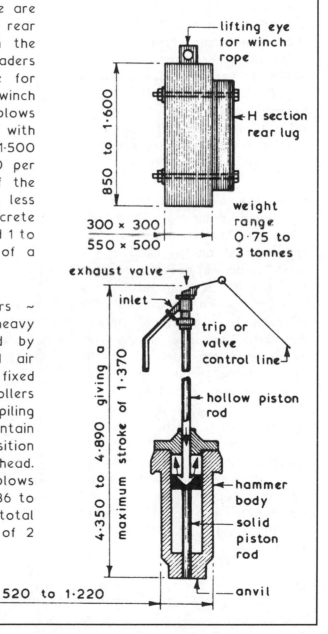

Double Acting Hammers ~ these consist of a cast iron cylinder which remains stationary on the pile head whilst a ram powered by steam or compressed air for both up and down strokes delivers a series of rapid blows which tends to keep the pile on the move during driving. The blow delivered is a smaller force than that from a drop or single acting hammer. The number of blows delivered ranges from 95 to 300 per minute with a total hammer weight range of 0·7 to 6·5 tonnes. Diesel powered double acting hammers are also available.

Diesel Hammers ~ these are self contained hammers which are located in the leaders of a piling rig and rest on the head of the pile. The driving action is started by raising the ram within the cylinder which activates the injection of a measured amount of fuel. The free falling ram compresses the fuel above the anvil causing the fuel to explode and expand resulting in a downward force on the anvil and upward force which raises the ram to recommence the cycle which is repeated until the fuel is cut off. The number of blows delivered ranges from 40 to 60 per minute with a total hammer weight range of 1·0 to 4·5 tonnes.

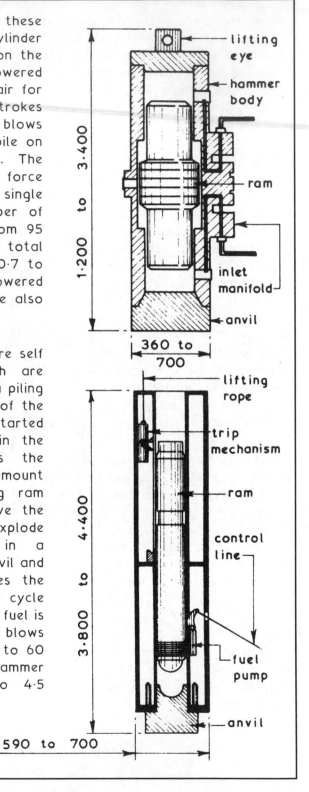

Pile Caps ~ piles can be used singly to support the load but often it is more economical to use piles in groups or clusters linked together with a reinforced concrete cap. The pile caps can also be linked together with reinforced concrete ground beams.

The usual minimum spacing for piles is:-

1. Friction Piles – 1·100 or not less than 3 × pile diameter, whichever is the greater.
2. Bearing Piles – 750 mm or not less than 2 × pile diameter, whichever is the greater.

Typical Examples ~

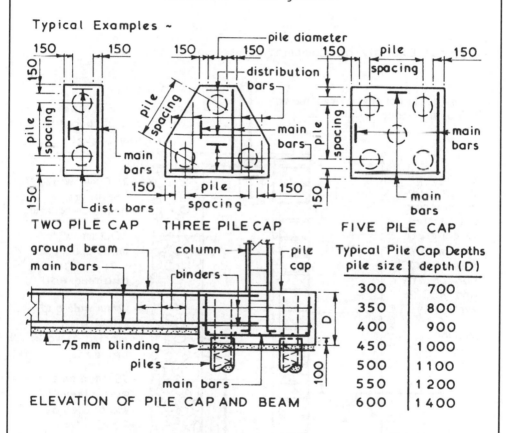

TWO PILE CAP   THREE PILE CAP   FIVE PILE CAP

ELEVATION OF PILE CAP AND BEAM

Typical Pile Cap Depths	
pile size	depth (D)
300	700
350	800
400	900
450	1000
500	1100
550	1200
600	1400

Pile Testing ~ it is advisable to test load at least one pile per scheme. The test pile should be overloaded by at least 50% of its working load and this load should be held for 24 hours. The test pile should not form part of the actual foundations. Suitable testing methods are:-

1. Jacking against kentledge placed over test pile.
2. Jacking against a beam fixed to anchor piles driven in on two sides of the test pile.

Retaining Walls ~ the major function of any retaining wall is to act as on earth retaining structure for the whole or part of its height on one face, the other being exposed to the elements. Most small height retaining walls are built entirely of brickwork or a combination of brick facing and blockwork or mass concrete backing. To reduce hydrostatic pressure on the wall from ground water an adequate drainage system in the form of weep holes should be used, alternatively subsoil drainage behind the wall could be employed.

Typical Example of Combination Retaining Wall ~

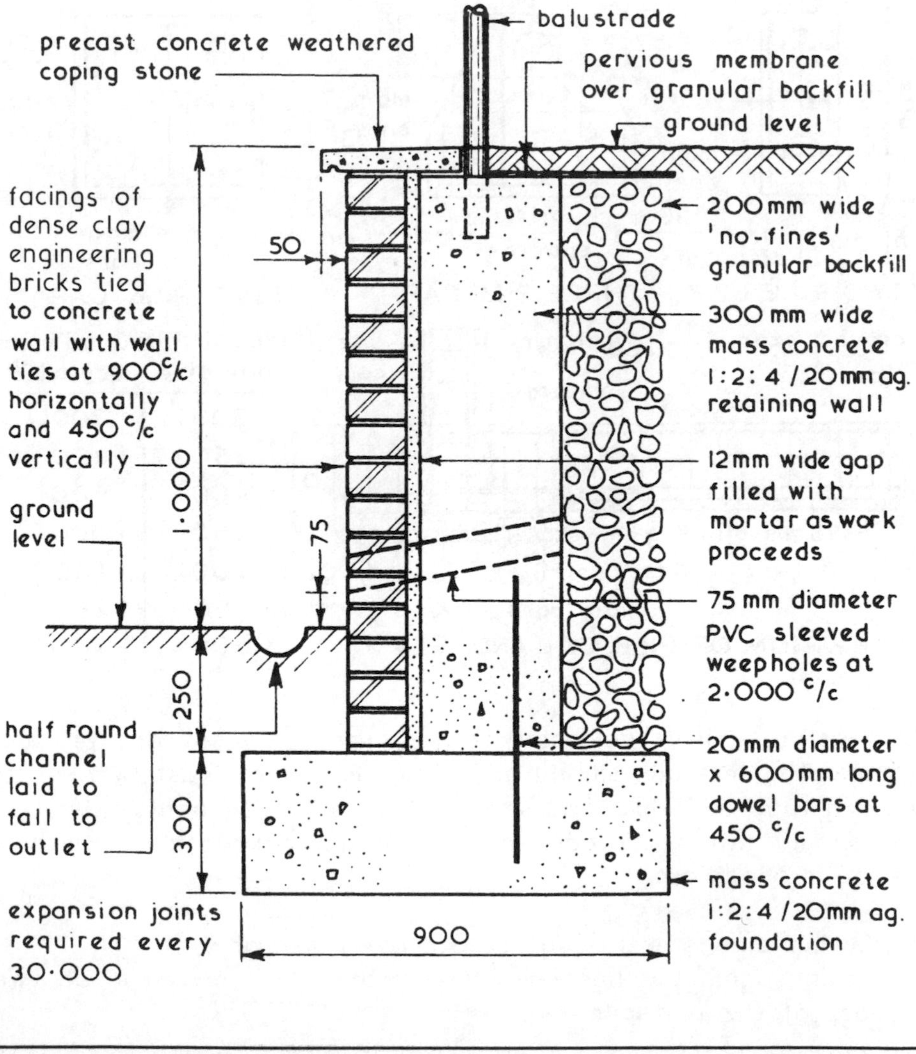

balustrade

precast concrete weathered coping stone

pervious membrane over granular backfill

ground level

facings of dense clay engineering bricks tied to concrete wall with wall ties at 900°/c horizontally and 450°/c vertically

50

200 mm wide 'no-fines' granular backfill

300 mm wide mass concrete 1:2:4 /20mm ag. retaining wall

1·000

ground level

12mm wide gap filled with mortar as work proceeds

75

75 mm diameter PVC sleeved weepholes at 2·000 °/c

half round channel laid to fall to outlet

250

20 mm diameter x 600mm long dowel bars at 450 °/c

300

mass concrete 1:2:4 /20mm ag. foundation

expansion joints required every 30·000

900

Small Height Retaining Walls ~ retaining walls must be stable and the usual rule of thumb for small height brick retaining walls is for the height to lie between 2 and 4 times the wall thickness. Stability can be checked by applying the middle third rule –

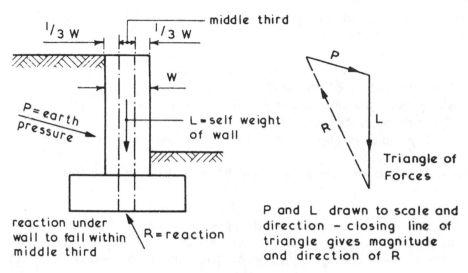

reaction under wall to fall within middle third

P and L drawn to scale and direction – closing line of triangle gives magnitude and direction of R

Typical Example of Brick Retaining Wall ~

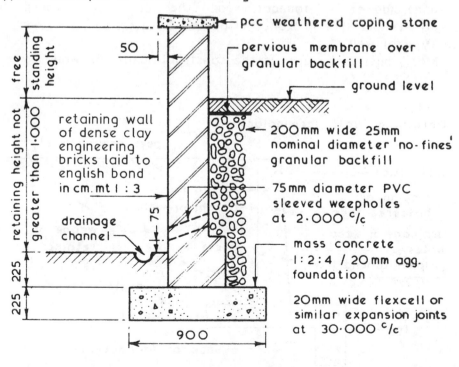

pcc weathered coping stone

pervious membrane over granular backfill

ground level

200mm wide 25mm nominal diameter 'no-fines' granular backfill

75mm diameter PVC sleeved weepholes at 2·000 c/c

mass concrete 1:2:4 / 20mm agg. foundation

20mm wide flexcell or similar expansion joints at 30·000 c/c

retaining wall of dense clay engineering bricks laid to english bond in cm. mt 1 : 3

drainage channel

Retaining Walls up to 6·000 high ~ these can be classified as medium height retaining walls and have the primary function of retaining soils at an angle in excess of the soil's natural angle of repose. Walls within this height range are designed to provide the necessary resistance by either their own mass or by the principles of leverage.

Design ~ the actual design calculations are usually carried out by a structural engineer who endeavours to ensure that:-

1. Overturning of the wall does not occur.
2. Forward sliding of the wall does not occur.
3. Materials used are suitable and not overstressed.
4. The subsoil is not overloaded.
5. In clay subsoils slip circle failure does not occur.

The factors which the designer will have to take into account:-

1. Nature and characteristics of the subsoil(s).
2. Height of water table – the presence of water can create hydrostatic pressure on the rear face of the wall, it can also affect the bearing capacity of the subsoil together with its shear strength, reduce the frictional resistance between the underside of the foundation and the subsoil and reduce the passive pressure in front of the toe of the wall.
3. Type of wall.
4. Material(s) to be used in the construction of the wall.

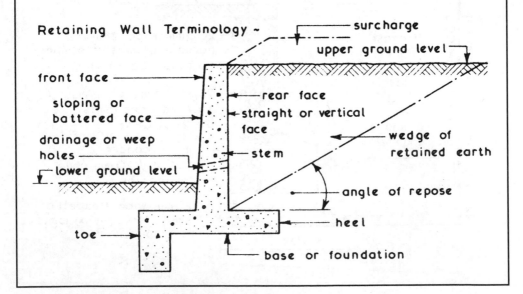

Retaining Wall Terminology ~

Earth Pressures ~ these can take one of two forms namely:-

1. Active Earth Pressures – these are those pressures which tend to move the wall at all times and consist of the wedge of earth retained plus any hydrostatic pressure. The latter can be reduced by including a subsoil drainage system behind and/or through the wall.
2. Passive Earth Pressures ~ these are a reaction of an equal and opposite force to any imposed pressure thus giving stability by resisting movement.

Typical Examples ~

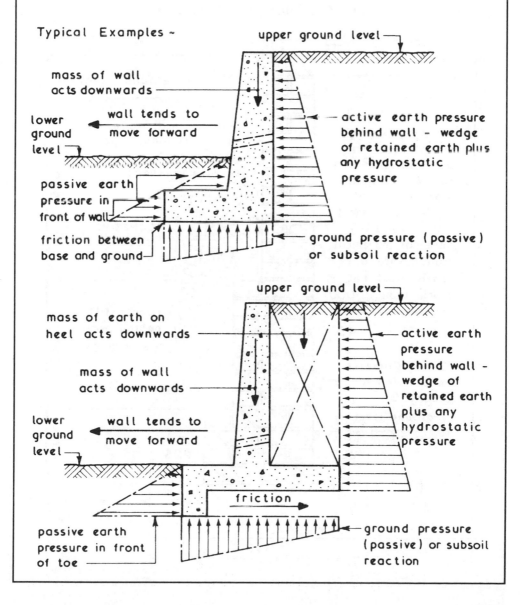

Mass Retaining Walls ~ these walls rely mainly on their own mass to overcome the tendency to slide forwards. Mass retaining walls are not generally considered to be economic over a height of 1·800 when constructed of brick or concrete and 1·000 high in the case of natural stonework. Any mass retaining wall can be faced with another material but generally any applied facing will not increase the strength of the wall and is therefore only used for aesthetic reasons.

Typical Brick Mass Retaining Wall Details ~

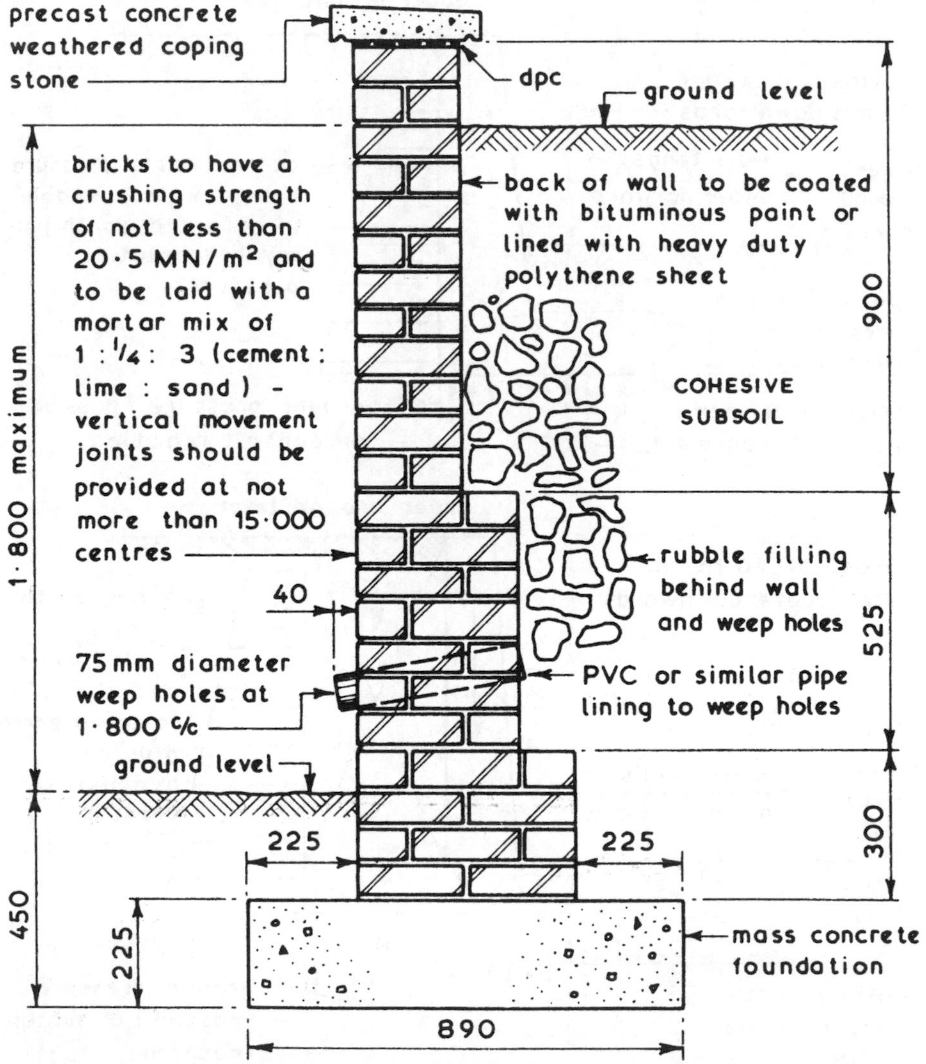

precast concrete weathered coping stone

dpc

ground level

bricks to have a crushing strength of not less than 20·5 MN/m² and to be laid with a mortar mix of 1 : ¼ : 3 (cement : lime : sand) – vertical movement joints should be provided at not more than 15·000 centres

back of wall to be coated with bituminous paint or lined with heavy duty polythene sheet

COHESIVE SUBSOIL

1·800 maximum

900

40

rubble filling behind wall and weep holes

525

75 mm diameter weep holes at 1·800 %

PVC or similar pipe lining to weep holes

ground level

225   225

300

450

225

mass concrete foundation

890

Typical Brick Faced Mass Concrete Retaining Wall Detail ~

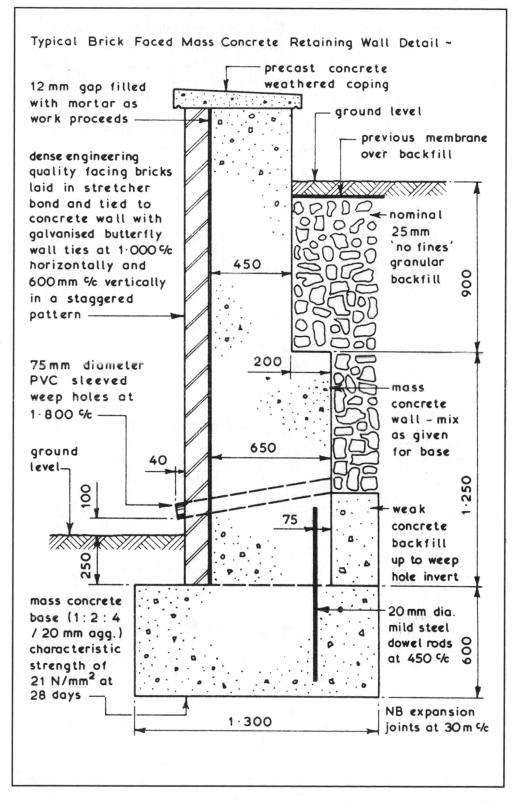

precast concrete
weathered coping

12 mm gap filled
with mortar as
work proceeds

ground level

previous membrane
over backfill

dense engineering
quality facing bricks
laid in stretcher
bond and tied to
concrete wall with
galvanised butterfly
wall ties at 1·000 c/c
horizontally and
600 mm c/c vertically
in a staggered
pattern

450

nominal
25 mm
'no fines'
granular
backfill

900

75 mm diameter
PVC sleeved
weep holes at
1·800 c/c

200

mass
concrete
wall – mix
as given
for base

ground
level

40

650

100

weak
concrete
backfill
up to weep
hole invert

75

1·250

250

mass concrete
base (1:2:4
/ 20 mm agg.)
characteristic
strength of
21 N/mm^2 at
28 days

20 mm dia.
mild steel
dowel rods
at 450 c/c

600

1·300

NB expansion
joints at 30 m c/c

Cantilever Retaining Walls ~ these are constructed of reinforced concrete with an economic height range of 1·200 to 6·000. They work on the principles of leverage where the stem is designed as a cantilever fixed at the base and base is designed as a cantilever fixed at the stem. Several formats are possible and in most cases a beam is placed below the base to increase the total passive resistance to sliding. Facing materials can be used in a similar manner to that shown on page 235.

Typical Formats ~

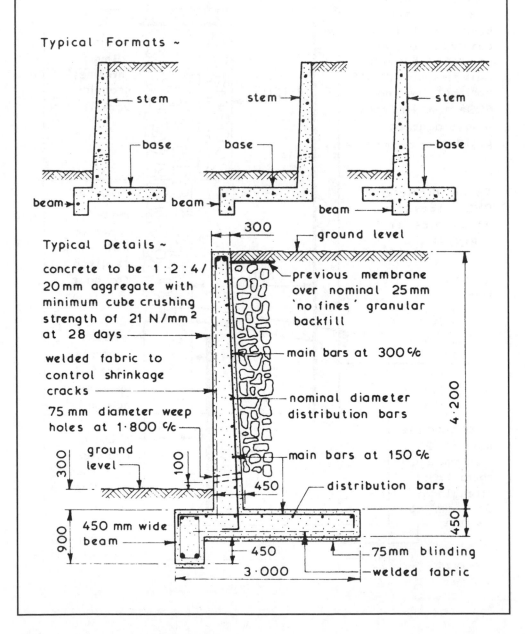

Typical Details ~

concrete to be 1:2:4/ 20 mm aggregate with minimum cube crushing strength of 21 N/mm² at 28 days

welded fabric to control shrinkage cracks

75 mm diameter weep holes at 1·800 ℅

ground level

450 mm wide beam

300

ground level

previous membrane over nominal 25mm 'no fines' granular backfill

main bars at 300 ℅

nominal diameter distribution bars

main bars at 150 ℅

distribution bars

4·200

450

75mm blinding

welded fabric

300

100

450

900

450

3·000

Formwork ~ concrete retaining walls can be cast in one of three ways – full height; climbing (page 238) or against earth face (page 239).

Full Height Casting ~ this can be carried out if the wall is to be cast as a freestanding wall and allowed to cure and gain strength before the earth to be retained is backfilled behind the wall. Considerations are the height of the wall, anticipated pressure of wet concrete, any strutting requirements and the availability of suitable materials to fabricate the formwork. As with all types of formwork a traditional timber format or a patent system using steel forms could be used.

Typical Details ~

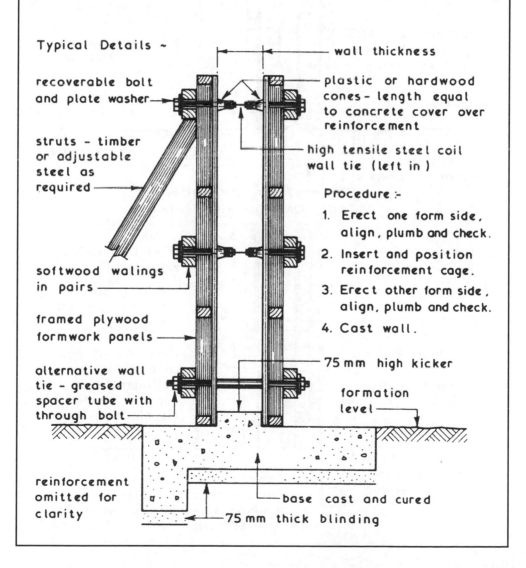

recoverable bolt and plate washer

plastic or hardwood cones – length equal to concrete cover over reinforcement

struts – timber or adjustable steel as required

high tensile steel coil wall tie (left in)

wall thickness

softwood walings in pairs

framed plywood formwork panels

alternative wall tie – greased spacer tube with through bolt

reinforcement omitted for clarity

Procedure :-

1. Erect one form side, align, plumb and check.

2. Insert and position reinforcement cage.

3. Erect other form side, align, plumb and check.

4. Cast wall.

75 mm high kicker

formation level

base cast and cured

75 mm thick blinding

237

Climbing Formwork or Lift Casting ~ this method can be employed on long walls, high walls or where the amount of concrete which can be placed in a shift is limited.

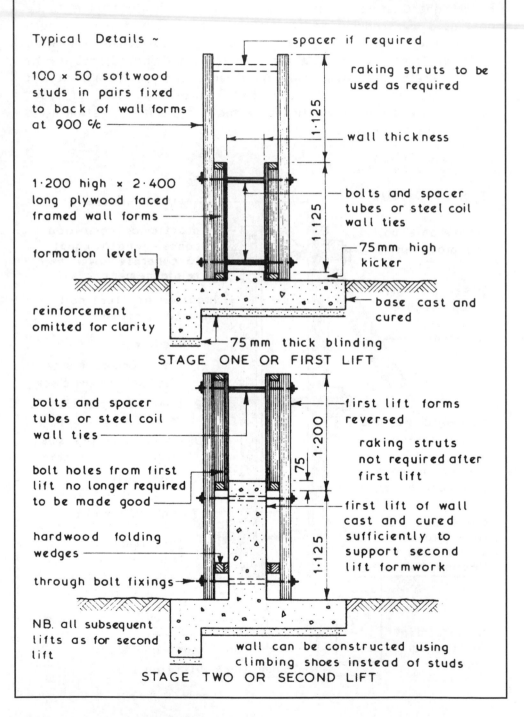

Typical Details ~

spacer if required

100 × 50 softwood studs in pairs fixed to back of wall forms at 900 c/c

raking struts to be used as required

1·125

wall thickness

1·200 high × 2·400 long plywood faced framed wall forms

bolts and spacer tubes or steel coil wall ties

1·125

formation level

75mm high kicker

reinforcement omitted for clarity

base cast and cured

75mm thick blinding

STAGE ONE OR FIRST LIFT

bolts and spacer tubes or steel coil wall ties

first lift forms reversed

1·200

raking struts not required after first lift

75

bolt holes from first lift no longer required to be made good

first lift of wall cast and cured sufficiently to support second lift formwork

hardwood folding wedges

1·125

through bolt fixings

NB. all subsequent lifts as for second lift

wall can be constructed using climbing shoes instead of studs

STAGE TWO OR SECOND LIFT

Casting Against Earth Face ~ this method can be an adaptation of the full height or climbing formwork systems. The latter uses a steel wire loop tie fixing to provide the support for the second and subsequent lifts.

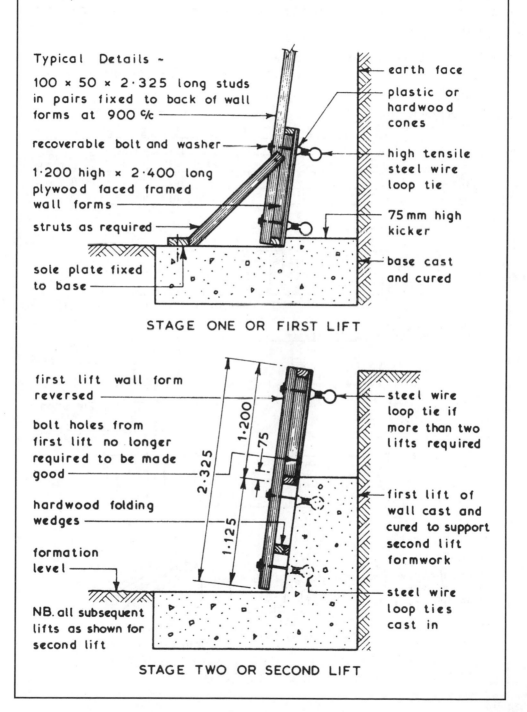

Typical Details ~

100 × 50 × 2·325 long studs in pairs fixed to back of wall forms at 900 ⁰/c

recoverable bolt and washer

1·200 high × 2·400 long plywood faced framed wall forms

struts as required

sole plate fixed to base

earth face

plastic or hardwood cones

high tensile steel wire loop tie

75 mm high kicker

base cast and cured

STAGE ONE OR FIRST LIFT

first lift wall form reversed

bolt holes from first lift no longer required to be made good

hardwood folding wedges

formation level

NB. all subsequent lifts as shown for second lift

steel wire loop tie if more than two lifts required

first lift of wall cast and cured to support second lift formwork

steel wire loop ties cast in

STAGE TWO OR SECOND LIFT

239

Masonry units – these are an option where it is impractical or cost-ineffective to use temporary formwork to in-situ concrete. Exposed brick or blockwork may also be a preferred finish. In addition to being a structural component, masonry units provide permanent formwork to reinforced concrete poured into the voids created by:

* Quetta bonded standard brick units, OR
* Stretcher bonded standard hollow dense concrete blocks.

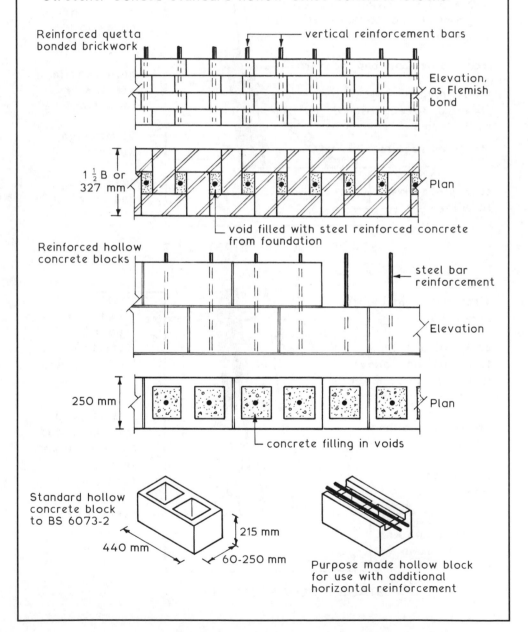

Reinforced quetta bonded brickwork

vertical reinforcement bars

Elevation, as Flemish bond

$1\frac{1}{2}$ B or 327 mm

Plan

void filled with steel reinforced concrete from foundation

Reinforced hollow concrete blocks

steel bar reinforcement

Elevation

250 mm

Plan

concrete filling in voids

Standard hollow concrete block to BS 6073-2

215 mm

440 mm

60-250 mm

Purpose made hollow block for use with additional horizontal reinforcement

Construction – a reinforced concrete base is cast with projecting steel bars accurately located for vertical continuity. The wall may be built solid, e.g. Quetta bond, with voids left around the bars for subsequent grouting. Alternatively, the wall may be of wide cavity construction, where the exposed reinforcement is wrapped in 'denso' grease tape for protection against corrosion. Steel bars are threaded at the top to take a tensioning nut over a bearing plate.

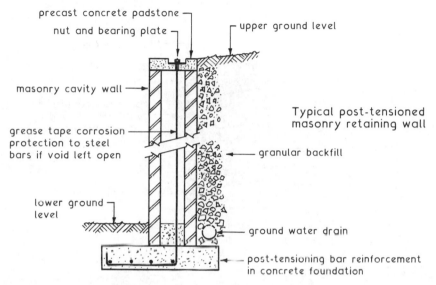

precast concrete padstone

nut and bearing plate

upper ground level

masonry cavity wall

Typical post-tensioned masonry retaining wall

grease tape corrosion protection to steel bars if void left open

granular backfill

lower ground level

ground water drain

post-tensioning bar reinforcement in concrete foundation

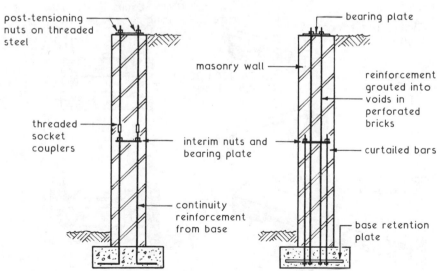

post-tensioning nuts on threaded steel

bearing plate

masonry wall

reinforcement grouted into voids in perforated bricks

threaded socket couplers

interim nuts and bearing plate

curtailed bars

continuity reinforcement from base

base retention plate

Staged post-tensioning to high masonry retaining walls

BS 5628-2: Code of practice for use of masonry. Structural use of reinforced and prestressed masonry.

Crib Retaining Walls – a system of pre-cast concrete or treated timber components comprising headers and stretchers which interlock to form a three-dimensional framework. During assembly the framework is filled with graded stone to create sufficient mass to withstand ground pressures.

Principle –

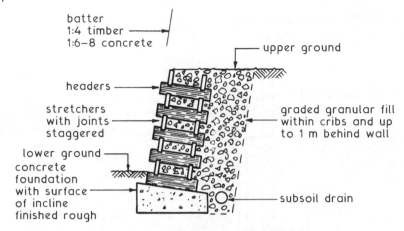

batter
1:4 timber
1:6–8 concrete

upper ground

headers

stretchers with joints staggered

graded granular fill within cribs and up to 1 m behind wall

lower ground
concrete foundation with surface of incline finished rough

subsoil drain

Note: height limited to 10 m with timber

Components –

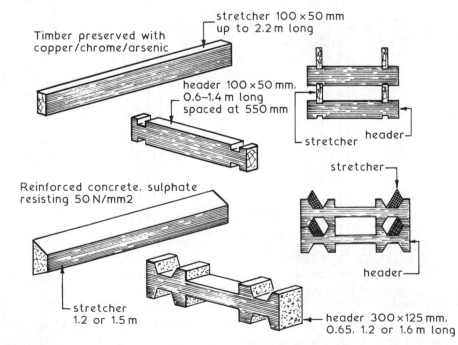

Timber preserved with copper/chrome/arsenic

stretcher 100 × 50 mm up to 2.2 m long

header 100 × 50 mm, 0.6–1.4 m long spaced at 550 mm

stretcher   header

Reinforced concrete, sulphate resisting 50 N/mm2

stretcher

header

stretcher 1.2 or 1.5 m

header 300 × 125 mm, 0.65, 1.2 or 1.6 m long

Soil Nailing ~ a cost effective geotechnic process used for retaining large soil slopes, notably highway and railway embankments.

Function ~ after excavating and removing the natural slope support, the remaining wedge of exposed unstable soil is pinned or nailed back with tendons into stable soil behind the potential slip plane.

Types of Soil Nails or Tendons ~

• Solid deformed steel rods up to 50 mm in diameter, located in bore holes up to 100 mm in diameter. Cement grout is pressurised into the void around the rods.
• Hollow steel, typically 100 mm diameter tubes with an expendable auger attached. Cement grout is injected into the tube during boring to be ejected through purpose-made holes in the auger.
• Solid glass reinforced plastic (GRP) with resin grouts.

Embankment Treatment ~ the exposed surface is faced with a plastic coated wire mesh to fit over the ends of the tendons. A steel head plate is fitted over and centrally bolted to each projecting tendon, followed by spray concreting to the whole face.

Typical Application ~

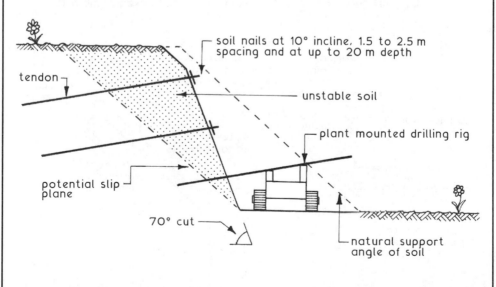

soil nails at 10° incline, 1.5 to 2.5 m spacing and at up to 20 m depth

tendon

unstable soil

plant mounted drilling rig

potential slip plane

70° cut

natural support angle of soil

Gabion ~ a type of retaining wall produced from individual rectangular boxes made from panels of wire mesh, divided internally and filled with stones. These units are stacked and overlapped (like stretcher bonded masonry) and applied in several layers or courses to retained earth situations. Typical sizes, 1·0 m long x 0·5 m wide x 0·5 m high, up to 4·0 m long x 1·0 m wide x 1·0 m high.

Mattress ~ unit fabrication is similar to a gabion but of less thickness, smaller mesh and stone size to provide some flexibility and shaping potential. Application is at a much lower incline. Generally used next to waterways for protection against land erosion where tidal movement and/or water level differentials could scour embankments. Typical sizes, 3·0 m long x 2·0 m wide x 0·15 m thick, up to 6·0 m long x 2·0 m wide x 0·3 m thick.

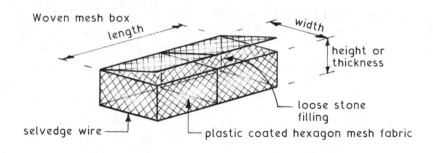

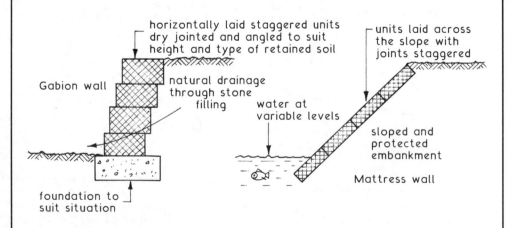

Design of Retaining Walls ~ this should allow for the effect of hydrostatics or water pressure behind the wall and the pressure created by the retained earth (see page 233). Calculations are based on a 1m unit length of wall, from which it is possible to ascertain:

1. The resultant thrust

2. The overturning or bending moment

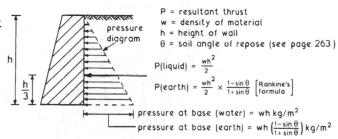

P = resultant thrust
w = density of material
h = height of wall
θ = soil angle of repose (see page 263)

$$P\text{(liquid)} = \frac{wh^2}{2}$$

$$P\text{(earth)} = \frac{wh^2}{2} \times \frac{1-\sin\theta}{1+\sin\theta} \left[\begin{array}{c}\text{Rankine's}\\\text{formula}\end{array}\right]$$

pressure at base (water) = wh kg/m^2

pressure at base (earth) = $wh\left(\frac{1-\sin\theta}{1+\sin\theta}\right)$ kg/m^2

P, the resultant thrust, will act through the centre of gravity of the pressure diagram, i.e. at h/3.

The overturning moment due to water is therefore:

$$\frac{wh^2}{2} \times \frac{h}{3} \text{ or } \frac{wh^3}{6}$$

and for earth:

$$\frac{wh^2}{2} \times \frac{1-\sin\theta}{1+\sin\theta} \times \frac{h}{3} \text{ or } \frac{wh^3}{6} \times \frac{1-\sin\theta}{1+\sin\theta}$$

Typical example ~

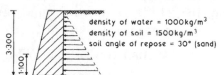

density of water = 1000kg/m^3
density of soil = 1500kg/m^3
soil angle of repose = 30° (sand)

For water:

$$p = \frac{wh^2}{2} = \frac{1000 \times (3\cdot3)^2}{2} = 5445 \text{ kg}$$

NB. kg × gravity = Newtons

Therefore, 5445 kg × 9·81 = 53·42 kN

The overturning or bending moment will be: P × h/3 = 53·42 kN x 1.1 m = 58·8 kNm

For earth:

$$p = \frac{wh^2}{2} \times \frac{1-\sin\theta}{1+\sin\theta}$$

$$p = \frac{1500 \times (3\cdot3)^2}{2} \times \frac{1-\sin 30°}{1+\sin 30°} = 2723 \text{ kg} \text{ or } 26\cdot7 \text{ kN}$$

The overturning or bending moment will be: P × h/3 = 26·7 kN × 1·1m = 29·4 kNm

A graphical design solution, to determine the earth thrust (P) behind a retaining wall. Data from previous page:

h = 3·300 m

θ = 30°

w = 1500 kg/m³

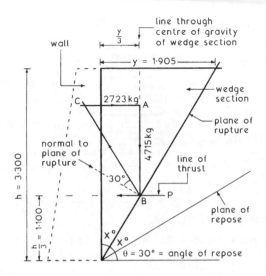

Wall height is drawn to scale and plane of repose plotted. The wedge section is obtained by drawing the plane of rupture through an angle bisecting the plane of repose and vertical back of the wall. Dimension 'y' can be scaled or calculated:

Tangent x = $\dfrac{y}{3\cdot3}$ x = 30°, and tan 30° = 0·5774

therefore, y = 3·3 × 0·5774 = 1·905 m

Area of wedge section = $\dfrac{3\cdot3}{2}$ × 1·905 m = 3·143 m²

Volume of wedge per metre run of wall = 3·143 × 1 = 3·143 m³

Weight .. .. = 3·143 × 1500 = 4715 kg

Vector line A – B is drawn to a scale through centre of gravity of wedge section, line of thrust and plane of rupture to represent 4715 kg.

Vector line B – C is drawn at the angle of earth friction (usually same as angle of repose, i.e. 30° in this case), to the normal to the plane of rupture until it meets the horizontal line C – A.

Triangle ABC represents the triangle of forces for the wedge section of earth, so C – A can be scaled at 2723 kg to represent (P), the earth thrust behind the retaining wall.

Open Excavations ~ one of the main problems which can be encountered with basement excavations is the need to provide temporary support or timbering to the sides of the excavation. This can be intrusive when the actual construction of the basement floor and walls is being carried out. One method is to use battered excavation sides cut back to a safe angle of repose thus eliminating the need for temporary support works to the sides of the excavation.

Typical Example of Open Basement Excavations ~

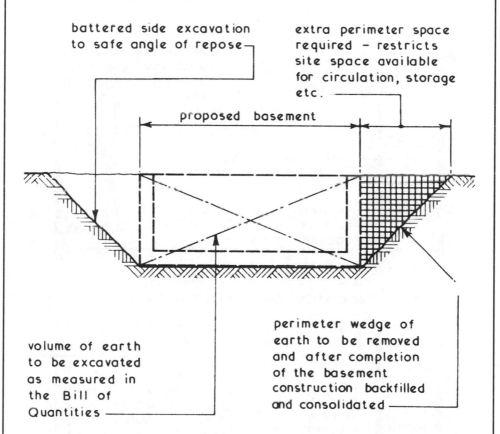

battered side excavation to safe angle of repose

extra perimeter space required – restricts site space available for circulation, storage etc.

proposed basement

volume of earth to be excavated as measured in the Bill of Quantities

perimeter wedge of earth to be removed and after completion of the basement construction backfilled and consolidated

In economic terms the costs of plant and manpower to cover the extra excavation, backfilling and consolidating must be offset by the savings made by omitting the temporary support works to the sides of the excavation. The main disadvantage of this method is the large amount of free site space required.

Perimeter Trench Excavations ~ in this method a trench wide enough for the basement walls to be constructed is excavated and supported with timbering as required. It may be necessary for runners or steel sheet piling to be driven ahead of the excavation work. This method can be used where weak subsoils are encountered so that the basement walls act as permanent timbering whilst the mound or dumpling is excavated and the base slab cast. Perimeter trench excavations can also be employed in firm subsoils when the mechanical plant required for excavating the dumpling is not available at the right time.

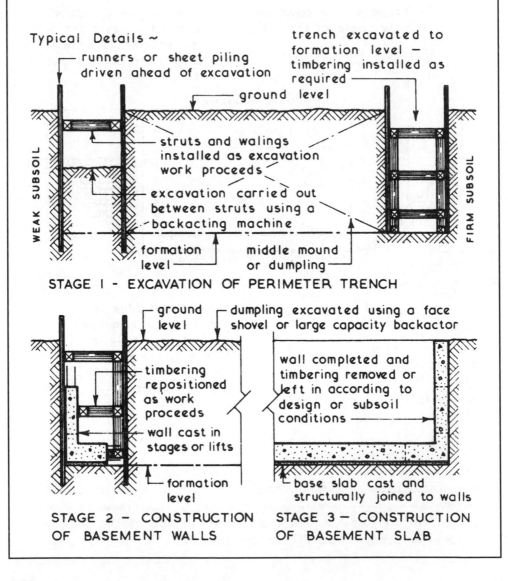

Typical Details ~

runners or sheet piling driven ahead of excavation

trench excavated to formation level — timbering installed as required

ground level

struts and walings installed as excavation work proceeds

excavation carried out between struts using a backacting machine

formation level

middle mound or dumpling

WEAK SUBSOIL

FIRM SUBSOIL

STAGE I - EXCAVATION OF PERIMETER TRENCH

ground level

dumpling excavated using a face shovel or large capacity backactor

timbering repositioned as work proceeds

wall cast in stages or lifts

formation level

wall completed and timbering removed or left in according to design or subsoil conditions

base slab cast and structurally joined to walls

STAGE 2 – CONSTRUCTION OF BASEMENT WALLS

STAGE 3 – CONSTRUCTION OF BASEMENT SLAB

Complete Excavation ~ this method can be used in firm subsoils where the centre of the proposed basement can be excavated first to enable the basement slab to be cast thus giving protection to the subsoil at formation level. The sides of excavation to the perimeter of the basement can be supported from the formation level using raking struts or by using raking struts pitched from the edge of the basement slab.

Typical Details ~

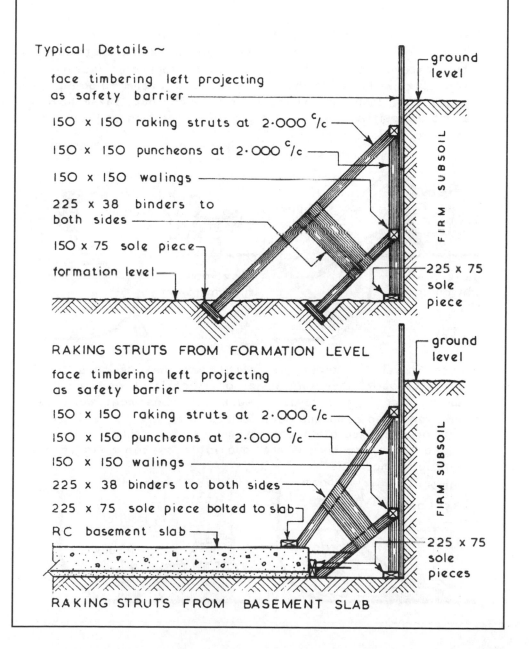

face timbering left projecting as safety barrier ———

150 x 150 raking struts at 2·000 ^{c}/c ———

150 x 150 puncheons at 2·000 ^{c}/c ———

150 x 150 walings ———

225 x 38 binders to both sides ———

150 x 75 sole piece ———

formation level ———

ground level

FIRM SUBSOIL

225 x 75 sole piece

RAKING STRUTS FROM FORMATION LEVEL

face timbering left projecting as safety barrier ———

150 x 150 raking struts at 2·000 ^{c}/c ———

150 x 150 puncheons at 2·000 ^{c}/c ———

150 x 150 walings ———

225 x 38 binders to both sides ———

225 x 75 sole piece bolted to slab ———

RC basement slab ———

ground level

FIRM SUBSOIL

225 x 75 sole pieces

RAKING STRUTS FROM BASEMENT SLAB

249

Excavating Plant ~ the choice of actual pieces of plant to be used in any construction activity is a complex matter taking into account many factors. Specific details of various types of excavators are given on pages 159 to 163. At this stage it is only necessary to consider basic types for particular operations. In the context of basement excavation two forms of excavator could be considered.

1. Backactors – these machines are available as cable rigged or hydraulic excavators suitable for trench and bulk excavating. Cable rigged backactors are usually available with larger bucket sizes and deeper digging capacities than the hydraulic machines but these have a more positive control and digging operation and are also easier to operate.

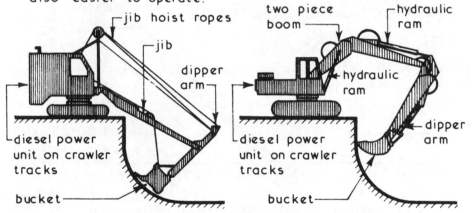

2. Face Shovels – these are robust machines designed to excavate above their own wheel or track level and are suitable for bulk excavation work. In basement work they will require a ramp approach unless they are to be lifted out of the excavation area by means of a crane. Like backactors face shovels are available as cable rigged or hydraulic machines.

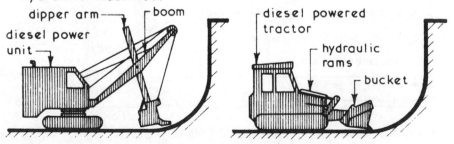

Basement Construction ~ in the general context of buildings a basement can be defined as a storey which is below the ground storey and is therefore constructed below ground level. Most basements can be classified into one of three groups:-

1. Retaining Wall and Raft Basements - this is the general format for basement construction and consists of a slab raft foundation which forms the basement floor and helps to distribute the structural loads transmitted down the retaining walls.

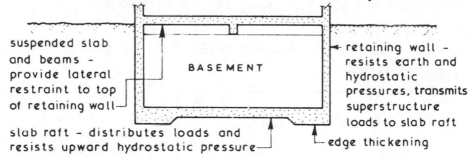

suspended slab and beams - provide lateral restraint to top of retaining wall

BASEMENT

retaining wall - resists earth and hydrostatic pressures, transmits superstructure loads to slab raft

slab raft - distributes loads and resists upward hydrostatic pressure

edge thickening

2. Box and Cellular Raft Basements - similar method to above except that internal walls are used to transmit and spread loads over raft as well as dividing basement into cells.

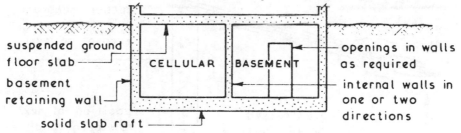

suspended ground floor slab

basement retaining wall

solid slab raft

CELLULAR   BASEMENT

openings in walls as required

internal walls in one or two directions

3. Piled Basements - the main superstructure loads are carried to the basement floor level by columns where they are finally transmitted to the ground via pile caps and bearing piles. This method can be used where low bearing capacity soils are found at basement floor level.

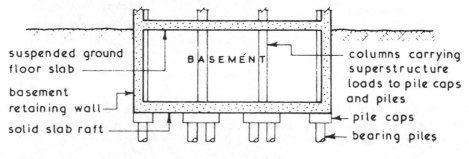

suspended ground floor slab

basement retaining wall

solid slab raft

BASEMENT

columns carrying superstructure loads to pile caps and piles

pile caps

bearing piles

Deep Basement Construction ~ basements can be constructed within a cofferdam or other temporary supported excavation (see Basement Excavations on pages 247 to 249) up to the point when these methods become uneconomic, unacceptable or both due to the amount of necessary temporary support work. Deep basements can be constructed by installing diaphragm walls within a trench and providing permanent support with ground anchors or by using the permanent lateral support given by the internal floor during the excavation period (see next page). Temporary lateral support during the excavation period can be provided by lattice beams spanning between the diaphragm walls (see next page).

Typical Ground Anchor Support Details ~

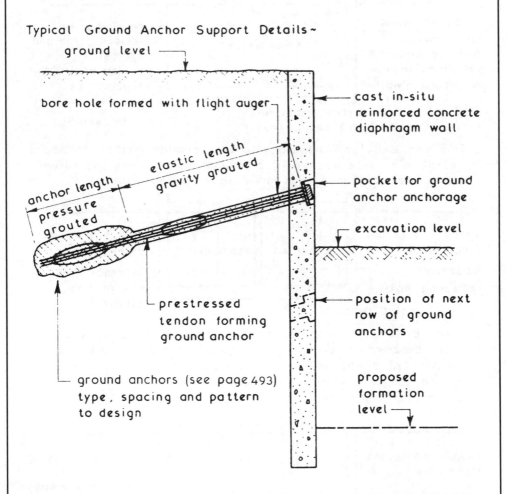

ground level

bore hole formed with flight auger

elastic length
gravity grouted

anchor length

pressure grouted

prestressed tendon forming ground anchor

ground anchors (see page 493) type, spacing and pattern to design

cast in-situ reinforced concrete diaphragm wall

pocket for ground anchor anchorage

excavation level

position of next row of ground anchors

proposed formation level

NB. vertical ground anchors installed through the lowest floor can be used to overcome any tendency to flotation during the construction period

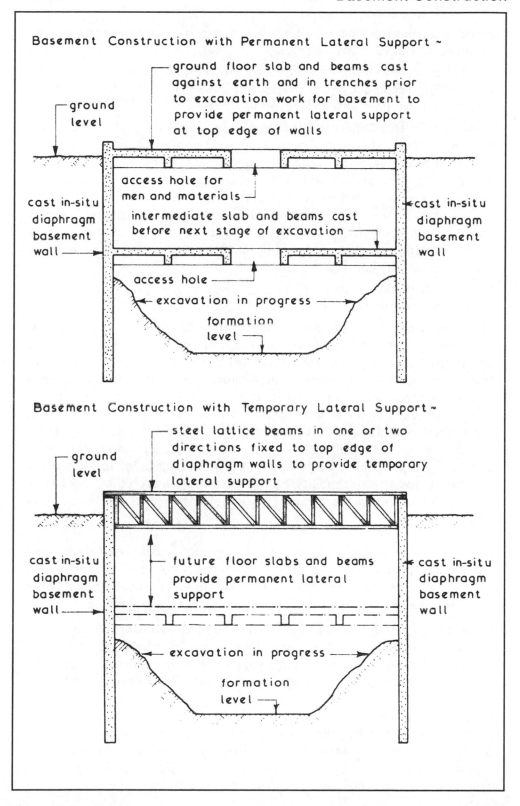

Basement Construction with Permanent Lateral Support ~

ground floor slab and beams cast against earth and in trenches prior to excavation work for basement to provide permanent lateral support at top edge of walls

ground level

cast in-situ diaphragm basement wall

access hole for men and materials

intermediate slab and beams cast before next stage of excavation

cast in-situ diaphragm basement wall

access hole

excavation in progress

formation level

Basement Construction with Temporary Lateral Support ~

steel lattice beams in one or two directions fixed to top edge of diaphragm walls to provide temporary lateral support

ground level

cast in-situ diaphragm basement wall

future floor slabs and beams provide permanent lateral support

cast in-situ diaphragm basement wall

excavation in progress

formation level

Waterproofing Basements ~ basements can be waterproofed by one of three basic methods namely:-

1. Use of dense monolithic concrete walls and floor
2. Tanking techniques (see pages 256 & 257)
3. Drained cavity system (see page 258)

Dense Monolithic Concrete – the main objective is to form a watertight basement using dense high quality reinforced or prestressed concrete by a combination of good materials, good workmanship, attention to design detail and on site construction methods. If strict control of all aspects is employed a sound watertight structure can be produced but it should be noted that such structures are not always water vapourproof. If the latter is desirable some waterproof coating, lining or tanking should be used. The watertightness of dense concrete mixes depends primarily upon two factors:-

1. Water/cement ratio.
2. Degree of compaction.

The hydration of cement during the hardening process produces heat therefore to prevent early stage cracking the temperature changes within the hardening concrete should be kept to a minimum. The greater the cement content the more is the evolution of heat therefore the mix should contain no more cement than is necessary to fulfil design requirements. Concrete with a free water/cement ratio of 0.5 is watertight and although the permeability is three time more at a ratio of 0.6 it is for practical purposes still watertight but above this ratio the concrete becomes progressively less watertight. For lower water/cement ratios the workability of the mix would have to be increased, usually by adding more cement, to enable the concrete to be fully compacted.

Admixtures – if the ingredients of good design, materials and workmanship are present watertight concrete can be produced without the use of admixtures. If admixtures are used they should be carefully chosen and used to obtain a specific objective:-

1. Water-reducing admixtures – used to improve workability
2. Retarding admixtures – slow down rate of hardening
3. Accelerating admixtures – increase rate of hardening – useful for low temperatures – calcium chloride not suitable for reinforced concrete.
4. Water-repelling admixtures – effective only with low water head, will not improve poor quality or porous mixes.
5. Air-entraining admixtures – increases workability – lowers water content.

Joints ~ in general these are formed in basement constructions to provide for movement accommodation (expansion joints) or to create a convenient stopping point in the construction process (construction joints). Joints are lines of weakness which will leak unless carefully designed and constructed therefore they should be simple in concept and easy to construct.

Basement slabs ~ these are usually designed to span in two directions and as a consequence have relatively heavy top and bottom reinforcement. To enable them to fulfil their basic functions they usually have a depth in excess of 250 mm. The joints, preferably of the construction type, should be kept to a minimum and if waterbars are specified they must be placed to ensure that complete compaction of the concrete is achieved.

Typical Basement Slab Joint Details ~

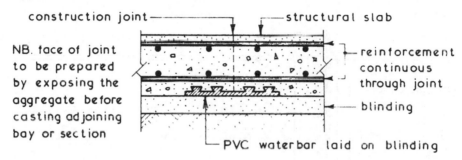

construction joint —————————— structural slab

NB. face of joint to be prepared by exposing the aggregate before casting adjoining bay or section

— reinforcement continuous through joint

— blinding

— PVC waterbar laid on blinding

Basement Walls ~ joints can be horizontal and/or vertical according to design requirements. A suitable waterbar should be incorporated in the joint to prevent the ingress of water. The top surface of a kicker used in conjunction with single lift pouring if adequately prepared by exposing the aggregate should not require a waterbar but if one is specified it should be either placed on the rear face or consist of a centrally placed mild steel strip inserted into the kicker whilst the concrete is still in a plastic state.

Typical Basement Wall Joint Details ~

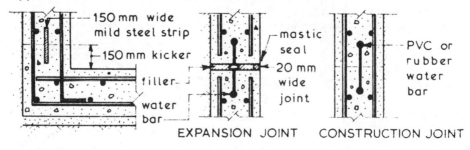

— 150 mm wide mild steel strip

— 150 mm kicker

filler —

water bar —

— mastic seal

20 mm wide joint

— PVC or rubber water bar

EXPANSION JOINT   CONSTRUCTION JOINT

Mastic Asphalt Tanking ~ the objective of tanking is to provide a continuous waterproof membrane which is applied to the base slab and walls with complete continuity between the two applications. The tanking can be applied externally or internally according to the circumstances prevailing on site. Alternatives to mastic asphalt are polythene sheeting: bituminous compounds: epoxy resin compounds and bitumen laminates.

External Mastic Asphalt Tanking ~ this is the preferred method since it not only prevents the ingress of water it also protects the main structure of the basement from aggressive sulphates which may be present in the surrounding soil or ground water.

Typical External Tanking Details ~

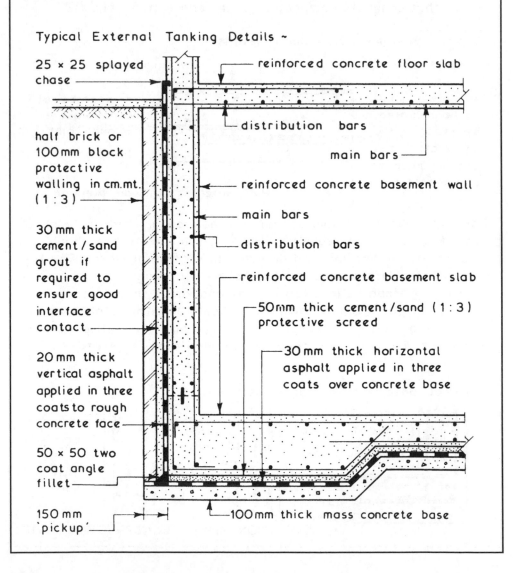

25 × 25 splayed chase

reinforced concrete floor slab

half brick or 100mm block protective walling in cm.mt. (1:3)

distribution bars

main bars

reinforced concrete basement wall

main bars

distribution bars

30 mm thick cement/sand grout if required to ensure good interface contact

reinforced concrete basement slab

50mm thick cement/sand (1:3) protective screed

20 mm thick vertical asphalt applied in three coats to rough concrete face

30 mm thick horizontal asphalt applied in three coats over concrete base

50 × 50 two coat angle fillet

150 mm 'pickup'

100mm thick mass concrete base

Internal Mastic Asphalt Tanking ~ this method should only be adopted if external tanking is not possible since it will not give protection to the main structure and unless adequately loaded may be forced away from the walls and/or floor by hydrostatic pressure. To be effective the horizontal and vertical coats of mastic asphalt must be continuous.

Typical Internal Tanking Details ~

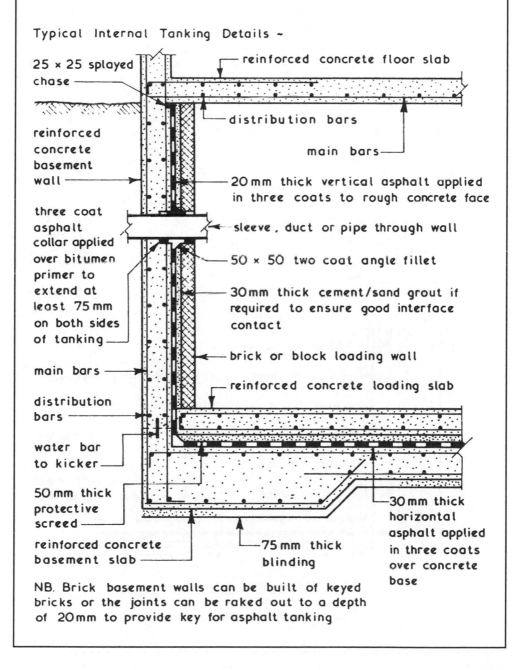

25 × 25 splayed chase

reinforced concrete floor slab

reinforced concrete basement wall

distribution bars

main bars

20 mm thick vertical asphalt applied in three coats to rough concrete face

three coat asphalt collar applied over bitumen primer to extend at least 75 mm on both sides of tanking

sleeve, duct or pipe through wall

50 × 50 two coat angle fillet

30 mm thick cement/sand grout if required to ensure good interface contact

brick or block loading wall

reinforced concrete loading slab

main bars

distribution bars

water bar to kicker

50 mm thick protective screed

reinforced concrete basement slab

75 mm thick blinding

30 mm thick horizontal asphalt applied in three coats over concrete base

NB. Brick basement walls can be built of keyed bricks or the joints can be raked out to a depth of 20 mm to provide key for asphalt tanking

Drained Cavity System ~ this method of waterproofing basements can be used for both new and refurbishment work. The basic concept is very simple in that it accepts that a small amount of water seepage is possible through a monolithic concrete wall and the best method of dealing with such moisture is to collect it and drain it away. This is achieved by building an inner non-load bearing wall to form a cavity which is joined to a floor composed of special triangular tiles laid to falls which enables the moisture to drain away to a sump from which it is either discharged direct or pumped into the surface water drainage system. The inner wall should be relatively vapour tight or alternatively the cavity should be ventilated.

Typical Details ~

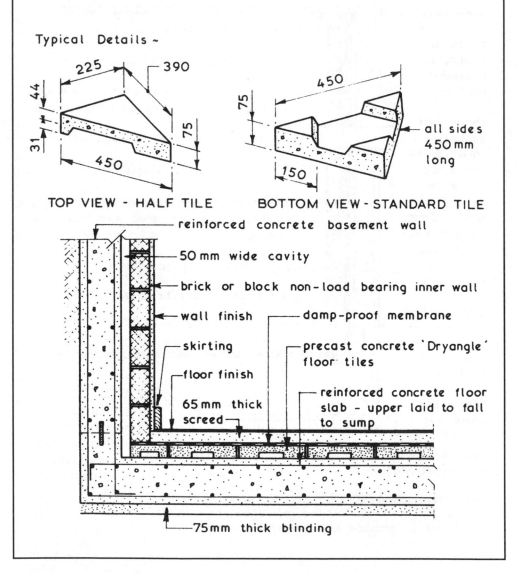

TOP VIEW - HALF TILE   BOTTOM VIEW - STANDARD TILE

reinforced concrete basement wall

50 mm wide cavity

brick or block non-load bearing inner wall

wall finish   damp-proof membrane

skirting   precast concrete 'Dryangle'
                       floor tiles

floor finish

65 mm thick   reinforced concrete floor
screed   slab - upper laid to fall
                       to sump

75 mm thick blinding

Basements benefit considerably from the insulating properties of the surrounding soil. However, that alone is insufficient to satisfy the typical requirements for wall and floor U-values of 0·35 and 0·30 W/m²K, respectively.

Refurbishment of existing basements may include insulation within dry lined walls and under the floor screed or particle board overlay. This should incorporate an integral vapour control layer to minimise risk of condensation.

External insulation of closed cell rigid polystyrene slabs is generally applied to new construction. These slabs combine low thermal conductivity with low water absorption and high compressive strength. The external face of insulation is grooved to encourage moisture run off. It is also filter faced to prevent clogging of the grooves. Backfill is granular.

Typical application -

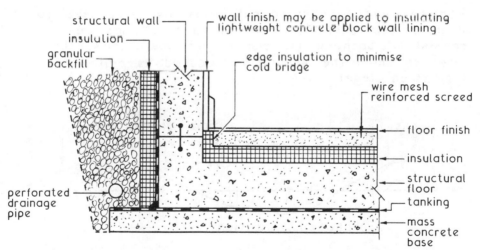

Note: reinforcement in concrete omitted, see details on previous pages.

Tables and calculations to determine U-values for basements are provided in the Building Regulations, Approved Document L and in BS EN ISO 13370: Thermal performance of buildings. Heat transfer via the ground. Calculation methods.

Excavation ~ to hollow out – in building terms to remove earth to form a cavity in the ground.

Types of Excavation ~

Oversite – the removal of top soil (Building Regulations requirement.)

depth varies from site to site but is usually in a 150 to 300mm range. Top soil contains plant life animal life and decaying matter which makes the soil compressible and therefore unsuitable for supporting buildings.

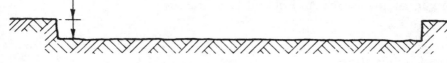

s u b s o i l

Reduce Level – carried out below oversite level to form a level surface on which to build and can consist of both cutting and filling operations. The level to which the ground is reduced is called the formation level.

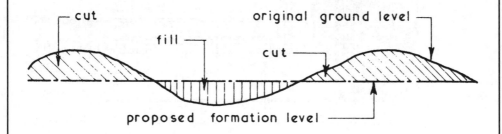

NB. Water in Excavations – this should be removed since it can:~

1. Undermine sides of excavation.
2. Make it impossible to adequately compact bottom of excavation to receive foundations.
3. Cause puddling which can reduce the bearing capacity of the subsoil.

Trench Excavations ~ narrow excavations primarily for strip foundations and buried services – excavation can be carried out by hand or machine.

Typical Examples ~

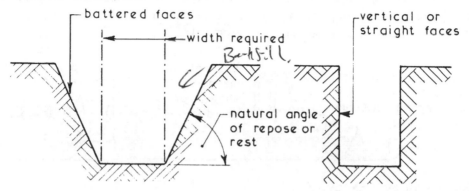

battered faces

width required

Backfill

vertical or straight faces

natural angle of repose or rest

Disadvantage ~ extra cost of over excavating and extra backfilling.

Advantage ~ no temporary support required to sides of excavation.

Disadvantage ~ sides of excavation may require some degree of temporary support.

Advantage ~ minimum amount of soil removed and therefore minimum amount of backfilling.

Pier Holes ~ isolated pits primarily used for foundation pads for columns and piers or for the construction of soakaways.

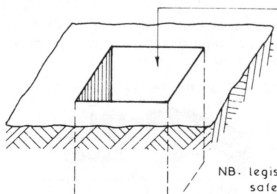

sides of excavation can be battered or straight as described above – deep pier holes may have to be over excavated in plan to provide good access to and good egress from the working area for both men and materials.

NB. legislation affecting safety in excavations is contained in the Construction (Health, Safety and Welfare) Regulations 1996.

Site Clearance and Removal of Top Soil ~

On small sites this could be carried out by manual means using hand held tools such as picks, shovels and wheelbarrows.

On all sites mechanical methods could be used the actual plant employed being dependent on factors such as volume of soil involved, nature of site and time elements.

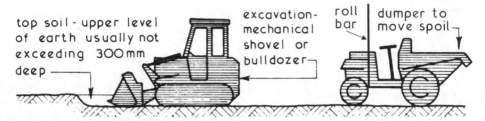

top soil - upper level of earth usually not exceeding 300mm deep

excavation-mechanical shovel or bulldozer

roll bar

dumper to move spoil

Reduced Level Excavations ~
On small sites — hand processes as given above
On all sites mechanical methods could be used dependent on factors given above.

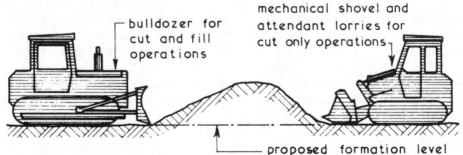

bulldozer for cut and fill operations

mechanical shovel and attendant lorries for cut only operations

proposed formation level

Trench and Pit Excavations ~
On small sites — hand processes as given above but if depth of excavation exceeds 1·200 some method of removing spoil from the excavation will have to be employed.

On all sites mechanical methods could be used dependent on factors given above.

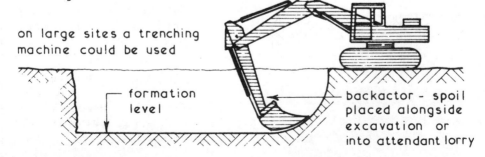

on large sites a trenching machine could be used

formation level

backactor - spoil placed alongside excavation or into attendant lorry

All subsoils have different abilities in remaining stable during excavation works. Most will assume a natural angle of repose or rest unless given temporary support. The presence of ground water apart from creating difficult working conditions can have an adverse effect on the subsoil's natural angle of repose.

Typical Angles of Repose ~
Excavations cut to a natural angle of repose are called battered.

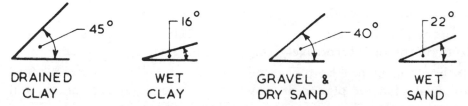

DRAINED CLAY	WET CLAY	GRAVEL & DRY SAND	WET SAND
45°	16°	40°	22°

Factors for Temporary Support of Excavations ~

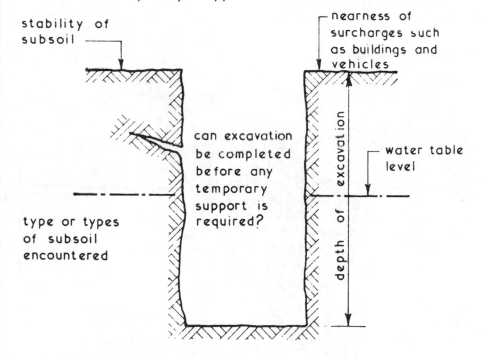

stability of subsoil

nearness of surcharges such as buildings and vehicles

can excavation be completed before any temporary support is required?

water table level

depth of excavation

type or types of subsoil encountered

Time factors such as period during which excavation will remain open and the time of year when work is carried out.

The need for an assessment of risk with regard to the support of excavations and protection of people within, is contained in the Construction (Health, Safety and Welfare) Regulations 1996.

Temporary Support ~ in the context of excavations this is called timbering irrespective of the actual materials used. If the sides of the excavation are completely covered with timbering it is known as close timbering whereas any form of partial covering is called open timbering.

An adequate supply of timber or other suitable material must be available and used to prevent danger to any person employed in an excavation from a fall or dislodgement of materials forming the sides of an excavation.

A suitable barrier or fence must be provided to the sides of all excavations or alternatively they must be securely covered

Materials must not be placed near to the edge of any excavation, nor must plant be placed or moved near to any excavation so that persons employed in the excavation are endangered.

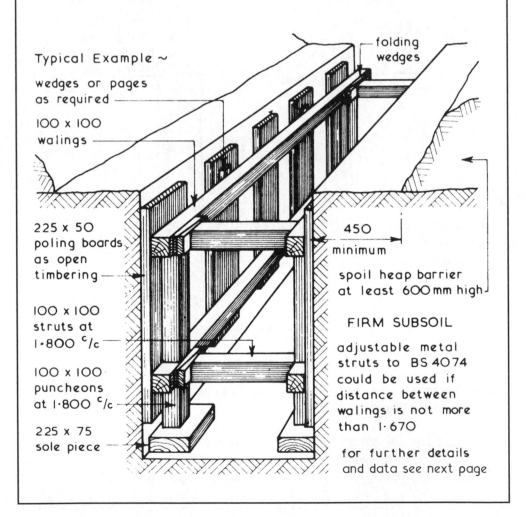

Typical Example ~

wedges or pages as required

100 x 100 walings

folding wedges

225 x 50 poling boards, as open timbering

100 x 100 struts at 1·800 ᶜ/c

100 x 100 puncheons at 1·800 ᶜ/c

225 x 75 sole piece

450 minimum

spoil heap barrier at least 600 mm high

FIRM SUBSOIL

adjustable metal struts to BS 4074 could be used if distance between walings is not more than 1·670

for further details and data see next page

Poling Boards ~ a form of temporary support which is placed in position against the sides of excavation after the excavation work has been carried out. Poling boards are placed at centres according to the stability of the subsoils encountered.

Runners ~ a form of temporary support which is driven into position ahead of the excavation work either to the full depth or by a drive and dig technique where the depth of the runner is always lower than that of the excavation.

Trench Sheeting ~ form of runner made from sheet steel with a trough profile – can be obtained with a lapped joint or an interlocking joint.

Water ~ if present or enters an excavation, a pit or sump should be excavated below the formation level to act as collection point from which the water can be pumped away.

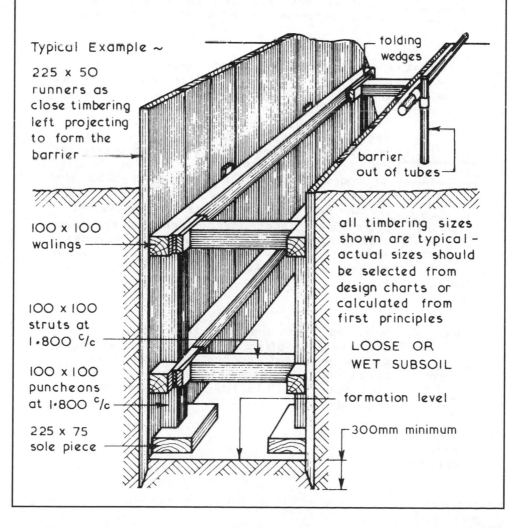

Typical Example ~

225 x 50 runners as close timbering left projecting to form the barrier

folding wedges

barrier out of tubes

100 x 100 walings

all timbering sizes shown are typical - actual sizes should be selected from design charts or calculated from first principles

100 x 100 struts at 1·800 °/c

LOOSE OR WET SUBSOIL

100 x 100 puncheons at 1·800 °/c

formation level

225 x 75 sole piece

300mm minimum

Concrete ~ a mixture of cement + fine aggregate + coarse aggregate + water in controlled proportions and of a suitable quality.

Cement ~ powder produced from clay and chalk or limestone. In general most concrete is made with ordinary or rapid hardening Portland cement, both types being manufactured to the recommendations of BS EN 197-1. Ordinary Portland cement is adequate for most purposes but has a low resistance to attack by acids and sulphates. Rapid hardening Portland cement does not set faster than ordinary Portland cement but it does develop its working strength at a faster rate. For a concrete which must have an acceptable degree of resistance to sulphate attack sulphate resisting Portland cement made to the recommendations of BS 4027 could be specified.

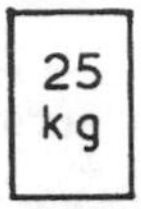

25 kg
BAGS

12t to 50t
SILOS

Aggregates ~ shape, surface texture and grading (distribution of particle sizes) are factors which influence the workability and strength of a concrete mix. Fine aggregates are those materials which pass through a 5mm sieve whereas coarse aggregates are those materials which are retained on a 5mm sieve. Dense aggregates are those with a density of more than 1200kg/m^3 for coarse aggregates and more than 1250kg/m^3 for fine aggregates. These are detailed in BS EN 12620 : Aggregates for concrete. Lightweight aggregates include clinker; foamed or expanded blastfurnace slag and exfoliated and expanded materials such as vermiculite, perlite, clay and sintered pulverized-fuel ash to BS EN 13055-1

coarse aggregate

5mm sieve

fine aggregate

Water ~ must be clean and free from impurities which are likely to affect the quality or strength of the resultant concrete. Pond, river, canal and sea water should not be used and only water which is fit for drinking should be specified.

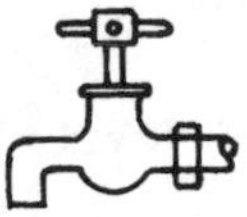

drinking water quality

Cement ~ whichever type of cement is being used it must be properly stored on site to keep it in good condition. The cement must be kept dry since contact with any moisture whether direct or airborne could cause it to set. A rotational use system should be introduced to ensure that the first batch of cement delivered is the first to be used.

Typical Storage Methods ~

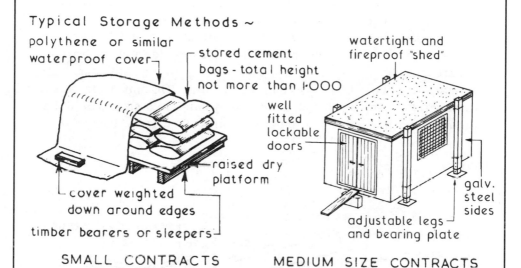

polythene or similar waterproof cover

stored cement bags-total height not more than 1·000

watertight and fireproof "shed"

well fitted lockable doors

raised dry platform

cover weighted down around edges

timber bearers or sleepers

galv. steel sides

adjustable legs and bearing plate

SMALL CONTRACTS   MEDIUM SIZE CONTRACTS

LARGE CONTRACTS — for bagged cement watertight container as above. For bulk delivery loose cement, a cement storage silo.

Aggregates ~ essentials of storage are to keep different aggregate types and/or sizes separate, store on a clean, hard, free draining surface and to keep the stored aggregates clean and free of leaves and rubbish.

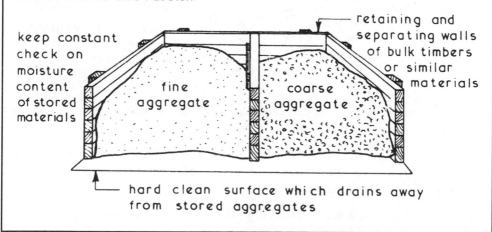

keep constant check on moisture content of stored materials

retaining and separating walls of bulk timbers or similar materials

fine aggregate

coarse aggregate

hard clean surface which drains away from stored aggregates

Concrete Batching ~ a batch is one mixing of concrete and can be carried out by measuring the quantities of materials required by volume or weight. The main aim of both methods is to ensure that all consecutive batches are of the same standard and quality.

Volume Batching ~ concrete mixes are often quoted by ratio such as 1:2:4 (cement : fine aggregate or sand : coarse aggregate). Cement weighing 50 kg has a volume of 0·033 m³ therefore for the above mix 2 × 0·033 (0·066 m³) of sand and 4 × 0·033 (0·132 m³) of coarse aggregate is required. To ensure accurate amounts of materials are used for each batch a gauge box should be employed its size being based on convenient handling. Ideally a batch of concrete should be equated to using 50 kg of cement per batch. Assuming a gauge box 300 mm deep and 300 mm wide with a volume of half the required sand the gauge box size would be – volume = length × width × depth = length × 300 × 300

$$\text{length} = \frac{\text{volume}}{\text{width} \times \text{depth}} = \frac{0 \cdot 033}{0 \cdot 3 \times 0 \cdot 3} = 0 \cdot 366 \text{ m}$$

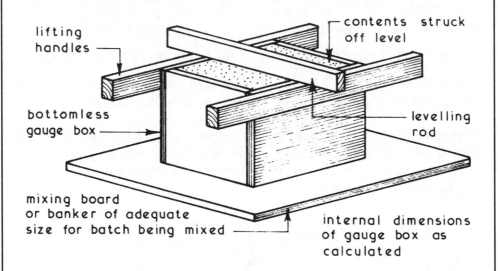

lifting handles

contents struck off level

bottomless gauge box

levelling rod

mixing board or banker of adequate size for batch being mixed

internal dimensions of gauge box as calculated

For the above given mix fill gauge box once with cement, twice with sand and four times with coarse aggregate.

An allowance must be made for the bulking of damp sand which can be as much as 33¹/₃%. General rule of thumb unless using dry sand allow for 25% bulking.

Materials should be well mixed dry before adding water.

Weight Batching ~ this is a more accurate method of measuring materials for concrete than volume batching since it reduces considerably the risk of variation between different batches. The weight of sand is affected very little by its dampness which in turn leads to greater accuracy in proportioning materials. When loading a weighing hopper the materials should be loaded in a specific order –

1. Coarse aggregates – tends to push other materials out and leaves the hopper clean.
2. Cement – this is sandwiched between the other materials since some of the fine cement particles could be blown away if cement is put in last.
3. Sand or fine Aggregates – put in last to stabilise the fine lightweight particles of cement powder.

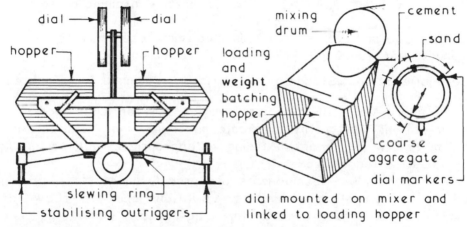

INDEPENDENT WEIGHT BATCHER INTEGRAL WEIGHT BATCHER

Typical Densities ~ cement – 1440 kg/m³ sand – 1600 kg/m³ coarse aggregate – 1440 kg/m³

Water/Cement Ratio ~ water in concrete has two functions –

1. Start the chemical reaction which causes the mixture to set into a solid mass.
2. Give the mix workability so that it can be placed, tamped or vibrated into the required position.

Very little water is required to set concrete (approximately 0·2 w/c ratio) the surplus evaporates leaving minute voids therefore the more water added to the mix to increase its workability the weaker is the resultant concrete. Generally w/c ratios of 0·4 to 0·5 are adequate for most purposes.

Concrete ~ a composite with many variables, represented by numerous gradings which indicate components, quality and manufacturing control.

Grade mixes: C7.5, C10, C15, C20, C25, C30, C35, C40, C45, C50, C55, and C60; F3, F4 and F5; IT2, IT2.5, and IT3.

C = Characteristic compressive
F = Flexural $\left.\right\}$ strengths at 28 days (N/mm^2)
IT = Indirect tensile

NB. If the grade is followed by a 'P', e.g. C30P, this indicates a prescribed mix (see below).

Grades C7.5 and C10 – Unreinforced plain concrete.
Grades C15 and C20 – Plain concrete or if reinforced containing lightweight aggregate.
Grades C25 – Reinforced concrete containing dense aggregate.
Grades C30 and C35 – Post-tensioned reinforced concrete.
Grades C40 to C60 – Pre-tensioned reinforced concrete.

Categories of mix: 1. Standard; 2. Prescribed; 3. Designed; 4. Designated.

1. Standard Mix – BS guidelines provide this for minor works or in situations limited by available material and manufacturing data. Volume or weight batching is appropriate, but no grade over C30 is recognised.
2. Prescribed Mix – components are predetermined (to a recipe) to ensure strength requirements. Variations exist to allow the purchaser to specify particular aggregates, admixtures and colours. All grades permitted.
3. Designed Mix – concrete is specified to an expected performance. Criteria can include characteristic strength, durability and workability, to which a concrete manufacturer will design and supply an appropriate mix. All grades permitted.
4. Designated Mix – selected for specific applications. General (GEN) graded 0–4, 7.5–25 N/mm^2 for foundations, floors and external works. Foundations (FND) graded 2, 3, 4A and 4B, 35 N/mm^2 mainly for sulphate resisting foundations.

Paving (PAV) graded 1 or 2, 35 or 45 N/mm^2 for roads and drives.

Reinforced (RC) graded 30, 35, 40, 45 and 50 N/mm^2 mainly for prestressing.

See also BS EN 206-1: Concrete. Specification, performance, production and conformity, and BS's 8500-1 and -2: Concrete.

Concrete Supply ~ this is usually geared to the demand or the rate at which the mixed concrete can be placed. Fresh concrete should always be used or placed within 30 minutes of mixing to prevent any undue drying out. Under no circumstances should more water be added after the initial mixing.

Small Batches ~ small easily transported mixers with output capacities of up to 100 litres can be used for small and intermittent batches. These mixers are versatile and robust machines which can be used for mixing mortars and plasters as well as concrete.

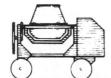

Medium to Large Batches ~ mixers with output capacities from 100 litres to 10 m^3 with either diesel or electric motors. Many models are available with tilting or reversing drum discharge, integral weigh batching and loading hopper and a controlled water supply.

Ready Mixed Concrete ~ used mainly for large concrete batches of up to 6 m^3. This method of concrete supply has the advantages of eliminating the need for site space to accommodate storage of materials, mixing plant and the need to employ adequately trained site staff who can constantly produce reliable and consistent concrete mixes. Ready mixed concrete supply depots also have better facilities and arrangements for producing and supplying mixed concrete in winter or inclement weather conditions. In many situations it is possible to place the ready mixed concrete into the required position direct from the delivery lorry via the delivery chute or by feeding it into a concrete pump. The site must be capable of accepting the 20 tonnes laden weight of a typical ready mixed concrete lorry with a turning circle of about 15·000. The supplier will want full details of mix required and the proposed delivery schedule.

Ref. BS EN 206-1: Concrete. Specification, performance, production and conformity.

Cofferdams ~ these are temporary enclosures installed in soil or water to prevent the ingress of soil and/or water into the working area with the cofferdam. They are usually constructed from interlocking steel sheet piles which are suitably braced or tied back with ground anchors. Alternatively a cofferdam can be installed using any structural material which will fulfil the required function.

Typical Cofferdam Details ~

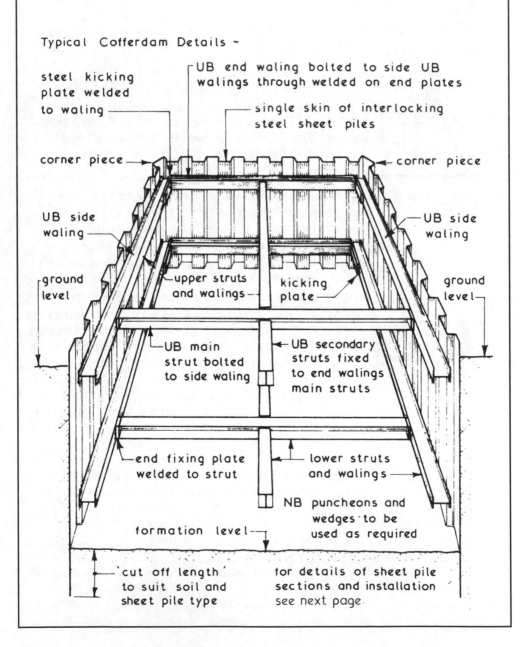

steel kicking plate welded to waling

UB end waling bolted to side UB walings through welded on end plates

single skin of interlocking steel sheet piles

corner piece

corner piece

UB side waling

UB side waling

ground level

upper struts and walings

kicking plate

ground level

UB main strut bolted to side waling

UB secondary struts fixed to end walings main struts

end fixing plate welded to strut

lower struts and walings

NB puncheons and wedges to be used as required

formation level

cut off length to suit soil and sheet pile type

for details of sheet pile sections and installation see next page

Steel Sheet Piling ~ apart from cofferdam work steel sheet can be used as a conventional timbering material in excavations and to form permanent retaining walls. Three common formats of steel sheet piles with interlocking joints are available with a range of section sizes and strengths up to a usual maximum length of 18·000:-

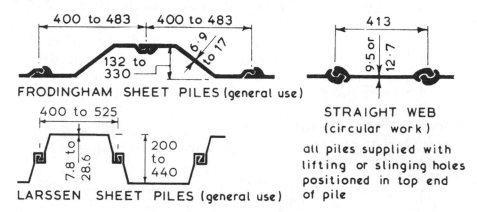

FRODINGHAM SHEET PILES (general use)

LARSSEN SHEET PILES (general use)

STRAIGHT WEB
(circular work)

all piles supplied with lifting or slinging holes positioned in top end of pile

Installing Steel Sheet Piles ~ to ensure that the sheet piles are pitched and installed vertically a driving trestle or guide frame is used. These are usually purpose built to accommodate a panel of 10 to 12 pairs of piles. The piles are lifted into position by a crane and driven by means of percussion piling hammer or alternatively they can be pushed into the ground by hydraulic rams acting against the weight of the power pack which is positioned over the heads of the pitched piles.

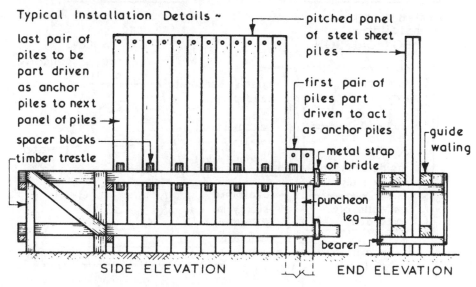

Typical Installation Details ~

last pair of piles to be part driven as anchor piles to next panel of piles →

spacer blocks

timber trestle

pitched panel of steel sheet piles →

first pair of piles part driven to act as anchor piles

metal strap or bridle

guide waling

puncheon

leg →

bearer

SIDE ELEVATION

END ELEVATION

Note: Rot-proof PVC sheet piling is also available.

Caissons ~ these are box-like structures which are similar in concept to cofferdams but they usually form an integral part of the finished structure. They can be economically constructed and installed in water or soil where the depth exceeds 18·000. There are 4 basic types of caisson namely:-

1. Box Caissons
2. Open Caissons
3. Monolithic Caissons
4. Pneumatic Caissons – used in water – see next page.

} usually of precast concrete and used in water being towed or floated into position and sunk – land caissons are of the open type and constructed in-situ.

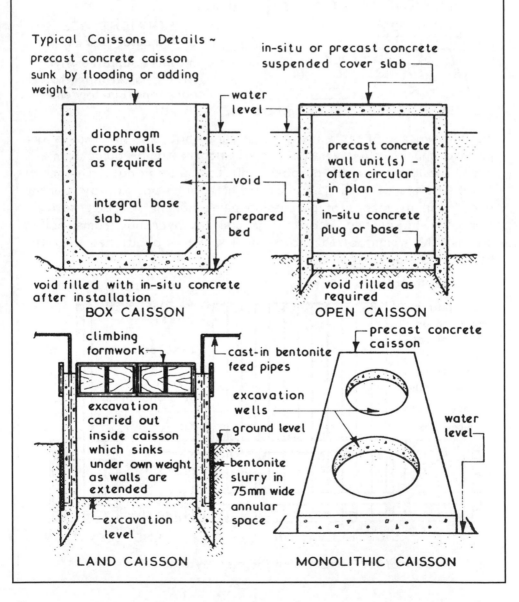

Typical Caissons Details ~

precast concrete caisson sunk by flooding or adding weight

water level

diaphragm cross walls as required

void

integral base slab

prepared bed

void filled with in-situ concrete after installation

BOX CAISSON

in-situ or precast concrete suspended cover slab

precast concrete wall unit(s) – often circular in plan

in-situ concrete plug or base

void filled as required

OPEN CAISSON

climbing formwork

cast-in bentonite feed pipes

excavation carried out inside caisson which sinks under own weight as walls are extended

ground level

bentonite slurry in 75mm wide annular space

excavation level

LAND CAISSON

precast concrete caisson

excavation wells

water level

MONOLITHIC CAISSON

Pneumatic Caissons ~ these are sometimes called compressed air caissons and are similar in concept to open caissons. They can be used in difficult subsoil conditions below water level and have a pressurised lower working chamber to provide a safe dry working area. Pneumatic caissons can be made of concrete whereby they sink under their own weight or they can be constructed from steel with hollow walls which can be filled with water to act as ballast. These caissons are usually designed to form part of the finished structure.

Typical Pneumatic Caisson Details ~

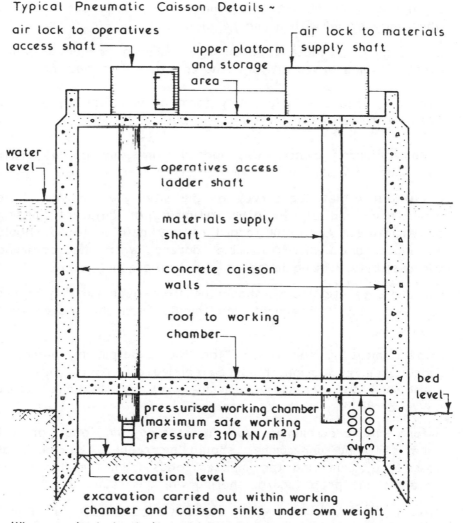

air lock to operatives access shaft

upper platform and storage area

air lock to materials supply shaft

water level

operatives access ladder shaft

materials supply shaft

concrete caisson walls

roof to working chamber

pressurised working chamber (maximum safe working pressure 310 kN/m²)

2·000 - 3·000

bed level

excavation level

excavation carried out within working chamber and caisson sinks under own weight

When required depth is reached a concrete slab or plug is cast over the formation level and chamber sealed with mass concrete

Underpinning ~ the main objective of most underpinning work is to transfer the load carried by a foundation from its existing bearing level to a new level at a lower depth. Underpinning techniques can also be used to replace an existing weak foundation. An underpinning operation may be necessary for one or more of the following reasons:-

1. Uneven Settlement – this could be caused by uneven loading of the building, unequal resistance of the soil action of tree roots or cohesive soil settlement.

2. Increase in Loading – this could be due to the addition of an extra storey or an increase in imposed loadings such as that which may occur with a change of use.

3. Lowering of Adjacent Ground – usually required when constructing a basement adjacent to existing foundations.

General Precautions ~ before any form of underpinning work is commenced the following precautions should be taken:-

1. Notify adjoining owners of proposed works giving full details and temporary shoring or tying.

2. Carry out a detailed survey of the site, the building to be underpinned and of any other adjoining or adjacent building or structures. A careful record of any defects found should be made and where possible agreed with the adjoining owner(s) before being lodged in a safe place.

3. Indicators or 'tell tales' should be fixed over existing cracks so that any subsequent movements can be noted and monitored.

4. If settlement is the reason for the underpinning works a thorough investigation should be carried out to establish the cause and any necessary remedial work put in hand before any underpinning works are started.

5. Before any underpinning work is started the loads on the building to be underpinned should be reduced as much as possible by removing the imposed loads from the floors and installing any props and/or shoring which is required.

6. Any services which are in the vicinity of the proposed underpinning works should be identified, traced, carefully exposed, supported and protected as necessary.

Underpinning to Walls ~ to prevent fracture, damage or settlement of the wall(s) being underpinned the work should always be carried out in short lengths called legs or bays. The length of these bays will depend upon the following factors:-

1. Total length of wall to be underpinned.

2. Wall loading.

3. General state of repair and stability of wall and foundation to be underpinned.

4. Nature of subsoil beneath existing foundation.

5. Estimated spanning ability of existing foundation.

Generally suitable bay lengths are:-

1·000 to 1·500 for mass concrete strip foundations supporting walls of traditional construction.

1·500 to 3·000 for reinforced concrete strip foundations supporting walls of moderate loading.

In all the cases the total sum of the unsupported lengths of wall should not exceed 25% of the total wall length.

The sequence of bays should be arranged so that working in adjoining bays is avoided until one leg of underpinning has been completed, pinned and cured sufficiently to support the wall above.

Typical Underpinning Schedule ~

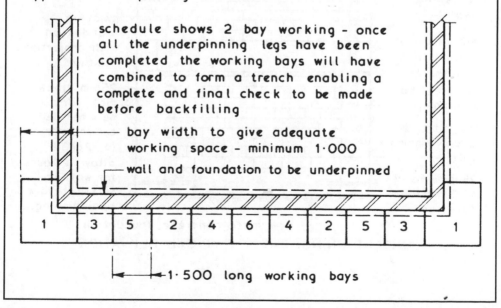

schedule shows 2 bay working - once all the underpinning legs have been completed the working bays will have combined to form a trench enabling a complete and final check to be made before backfilling

bay width to give adequate working space - minimum 1·000

wall and foundation to be underpinned

1 | 3 | 5 | 2 | 4 | 6 | 4 | 2 | 5 | 3 | 1

1·500 long working bays

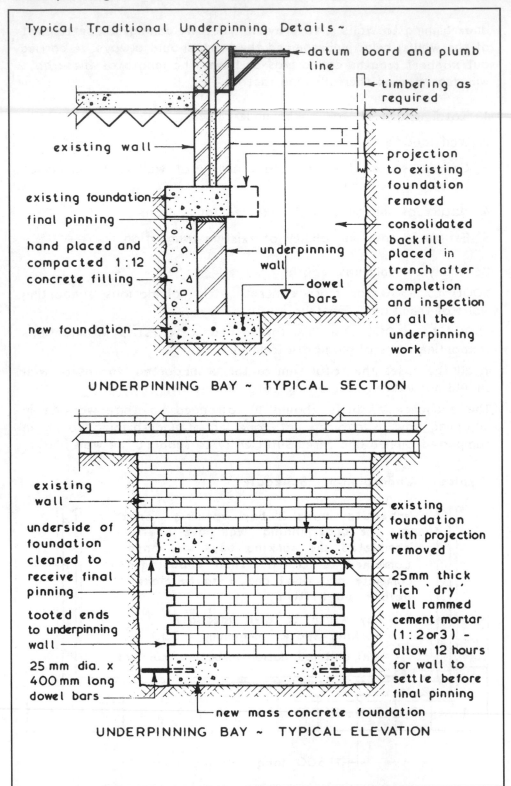

Typical Traditional Underpinning Details ~

datum board and plumb line

timbering as required

existing wall

existing foundation

final pinning

hand placed and compacted 1:12 concrete filling

new foundation

projection to existing foundation removed

consolidated backfill placed in trench after completion and inspection of all the underpinning work

underpinning wall

dowel bars

UNDERPINNING BAY ~ TYPICAL SECTION

existing wall

underside of foundation cleaned to receive final pinning

tooted ends to underpinning wall

25 mm dia. x 400 mm long dowel bars

existing foundation with projection removed

25mm thick rich 'dry' well rammed cement mortar (1:2 or 3) – allow 12 hours for wall to settle before final pinning

new mass concrete foundation

UNDERPINNING BAY ~ TYPICAL ELEVATION

Jack Pile Underpinning ~ this method can be used when the depth of a suitable bearing capacity subsoil is too deep to make traditional underpinning uneconomic. Jack pile underpinning is quiet, vibration free and flexible since the pile depth can be adjusted to suit subsoil conditions encountered. The existing foundations must be in a good condition since they will have to span over the heads of the pile caps which are cast onto the jack pile heads after the hydraulic jacks have been removed.

Typical Details ~

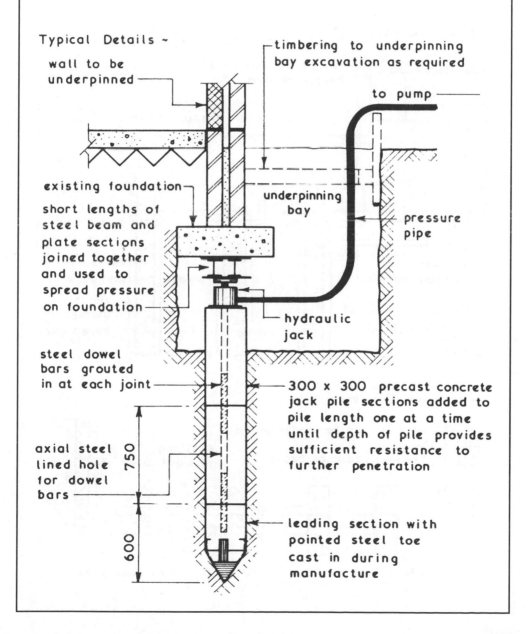

wall to be underpinned

timbering to underpinning bay excavation as required

to pump

existing foundation

underpinning bay

pressure pipe

short lengths of steel beam and plate sections joined together and used to spread pressure on foundation

hydraulic jack

steel dowel bars grouted in at each joint

300 x 300 precast concrete jack pile sections added to pile length one at a time until depth of pile provides sufficient resistance to further penetration

axial steel lined hole for dowel bars

750

leading section with pointed steel toe cast in during manufacture

600

...and Pile Underpinning ~ this method of underpinning can be
...ere the condition of the existing foundation is unsuitable
...ional or jack pile underpinning techniques. The brickwork
...ove the existing foundation must be in a sound condition since
this method relies on the 'arching effect' of the brick bonding to
transmit the wall loads onto the needles and ultimately to the
piles. The piles used with this method are usually small diameter
bored piles – see page 214.

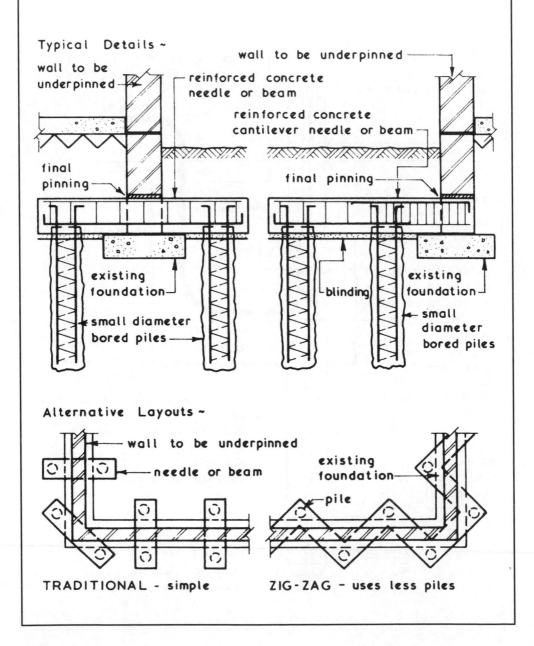

Typical Details ~

wall to be underpinned

wall to be underpinned

reinforced concrete needle or beam

reinforced concrete cantilever needle or beam

final pinning

final pinning

existing foundation

blinding

existing foundation

small diameter bored piles

small diameter bored piles

Alternative Layouts ~

wall to be underpinned

needle or beam

existing foundation

pile

TRADITIONAL - simple

ZIG-ZAG - uses less piles

'Pynford' Stool Method of Underpinning ~ this method can be used where the existing foundations are in a poor condition and it enables the wall to be underpinned in a continuous run without the need for needles or shoring. The reinforced concrete beam formed by this method may well be adequate to spread the load of the existing wall or it may be used in conjunction with other forms of underpinning such as traditional and jack pile.

Typical Details ~

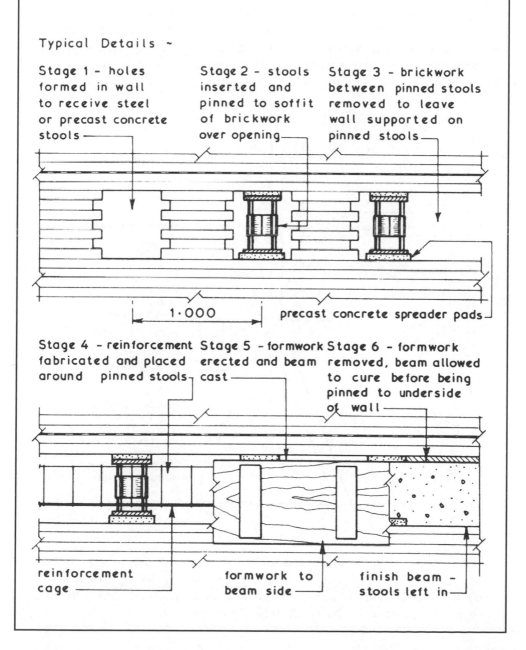

Stage 1 - holes formed in wall to receive steel or precast concrete stools

Stage 2 - stools inserted and pinned to soffit of brickwork over opening

Stage 3 - brickwork between pinned stools removed to leave wall supported on pinned stools

1·000 precast concrete spreader pads

Stage 4 - reinforcement fabricated and placed around pinned stools cast

Stage 5 - formwork erected and beam cast

Stage 6 - formwork removed, beam allowed to cure before being pinned to underside of wall

reinforcement cage

formwork to beam side

finish beam - stools left in

le or Angle Piling ~ this is a much simpler alternative to
al underpinning techniques, applying modern concrete
uipment to achieve cost benefits through time saving. The
ocess is also considerably less disruptive, as large volumes of
excavation are avoided. Where sound bearing strata can be located
within a few metres of the surface, wall stability is achieved
through lined reinforced concrete piles installed in pairs, at
opposing angles. The existing floor, wall and foundation are pre-
drilled with air flushed percussion auger, giving access for a steel
lining to be driven through the low grade/clay subsoil until it
impacts with firm strata. The lining is cut to terminate at the
underside of the foundation and the void steel reinforced prior to
concreting.

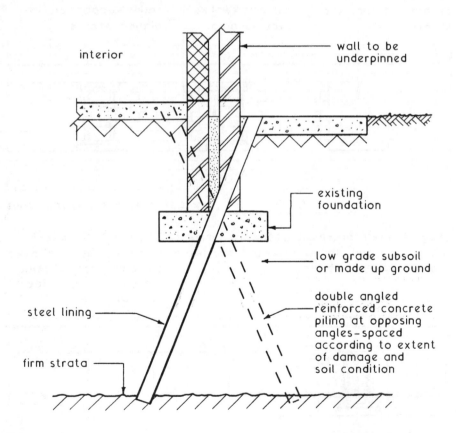

In many situations it is impractical to apply angle piling to both
sides of a wall. Subject to subsoil conditions being adequate, it
may be acceptable to apply remedial treatment from one side only.
The piles will need to be relatively close spaced.

Underpinning Columns ~ columns can be underpinned in the some manner as walls using traditional or jack pile methods after the columns have been relieved of their loadings. The beam loads can usually be transferred from the columns by means of dead shores and the actual load of the column can be transferred by means of a pair of beams acting against a collar attached to the base of the column shaft.

Typical Details ~

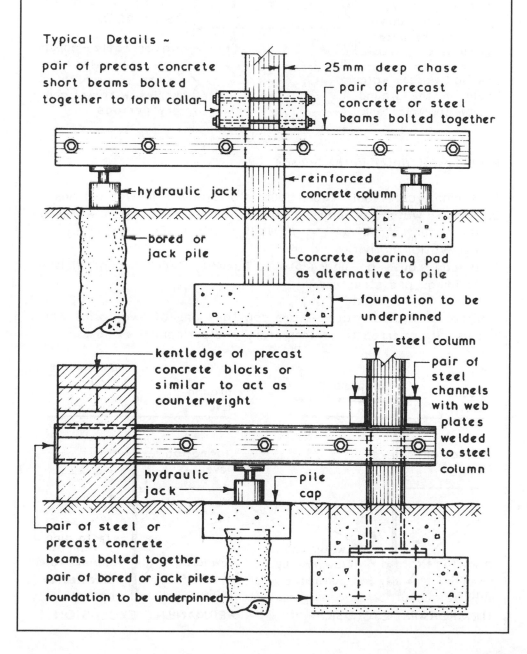

pair of precast concrete short beams bolted together to form collar

25mm deep chase

pair of precast concrete or steel beams bolted together

hydraulic jack

reinforced concrete column

bored or jack pile

concrete bearing pad as alternative to pile

foundation to be underpinned

kentledge of precast concrete blocks or similar to act as counterweight

steel column

pair of steel channels with web plates welded to steel column

hydraulic jack

pile cap

pair of steel or precast concrete beams bolted together

pair of bored or jack piles

foundation to be underpinned

Classification of Water ~ water can be classified by its relative position to or within the ground thus –

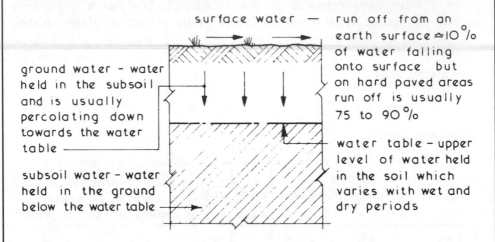

surface water — run off from an earth surface $\simeq 10\%$ of water falling onto surface but on hard paved areas run off is usually 75 to 90 %

ground water - water held in the subsoil and is usually percolating down towards the water table

water table - upper level of water held in the soil which varies with wet and dry periods

subsoil water - water held in the ground below the water table

Problems of Water in the Subsoil ~

1. A high water table could cause flooding during wet periods.
2. Subsoil water can cause problems during excavation works by its natural tendency to flow into the voids created by the excavation activities.
3. It can cause an unacceptable humidity level around finished buildings and structures.

Control of Ground Water ~ this can take one of two forms which are usually referred to as temporary and permanent exclusion –

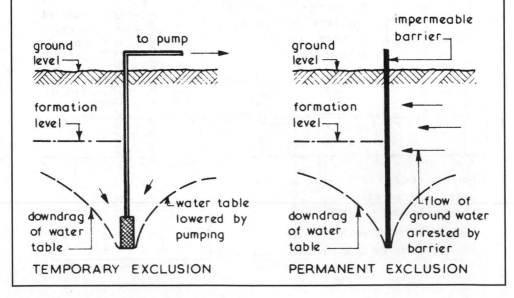

TEMPORARY EXCLUSION

PERMANENT EXCLUSION

Permanent Exclusion ~ this can be defined as the insertion of an impermeable barrier to stop the flow of water within the ground.

Temporary Exclusion ~ this can be defined as the lowering of the water table and within the economic depth range of 1·500 can be achieved by subsoil drainage methods, for deeper treatment a pump or pumps are usually involved.

Simple Sump Pumping ~ suitable for trench work and/or where small volumes of water are involved.

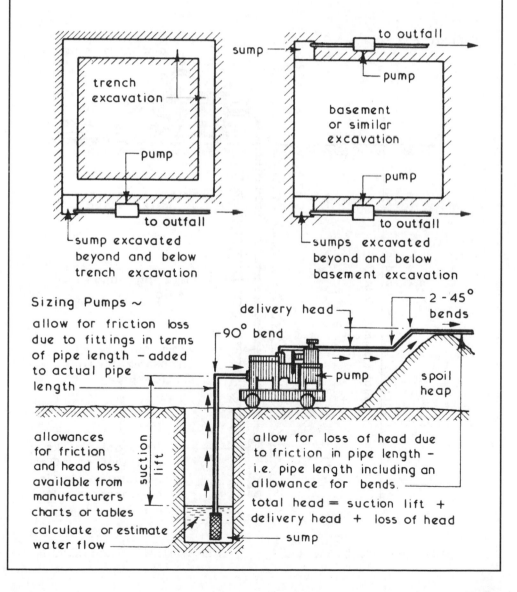

Sizing Pumps ~

allow for friction loss due to fittings in terms of pipe length – added to actual pipe length

allowances for friction and head loss available from manufacturers charts or tables calculate or estimate water flow

allow for loss of head due to friction in pipe length – i.e. pipe length including an allowance for bends.

total head = suction lift + delivery head + loss of head

285

Jetted Sumps ~ this method achieves the same objectives as the simple sump methods of dewatering (previous page) but it will prevent the soil movement associated with this and other open sump methods. A borehole is formed in the subsoil by jetting a metal tube into the ground by means of pressurised water, to a depth within the maximum suction lift of the extract pump. The metal tube is withdrawn to leave a void for placing a disposable wellpoint and plastic suction pipe. The area surrounding the pipe is filled with coarse sand to function as a filtering media.

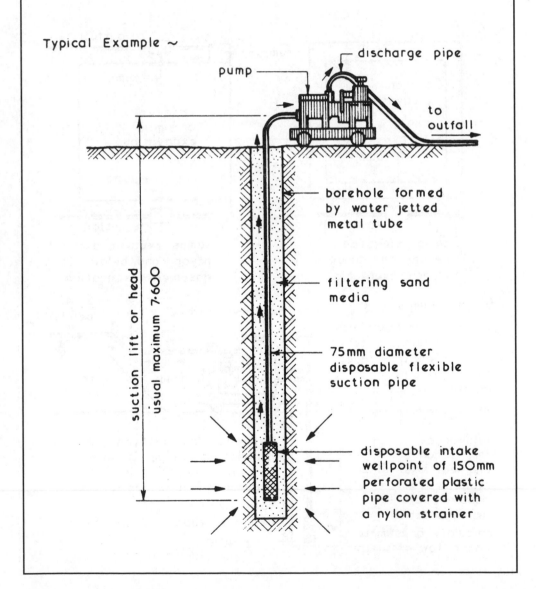

Typical Example ~

discharge pipe

pump

to outfall

suction lift or head usual maximum 7·600

borehole formed by water jetted metal tube

filtering sand media

75mm diameter disposable flexible suction pipe

disposable intake wellpoint of 150mm perforated plastic pipe covered with a nylon strainer

Wellpoint Systems ~ method of lowering the water table to a position below the formation level to give a dry working area. The basic principle is to jet into the subsoil a series of wellpoints which are connected to a common header pipe which is connected to a vacuum pump. Wellpoint systems are suitable for most subsoils and can encircle an excavation or be laid progressively alongside as in the case of a trench excavation. If the proposed formation level is below the suction lift capacity of the pump a multi-stage system can be employed – see next page.

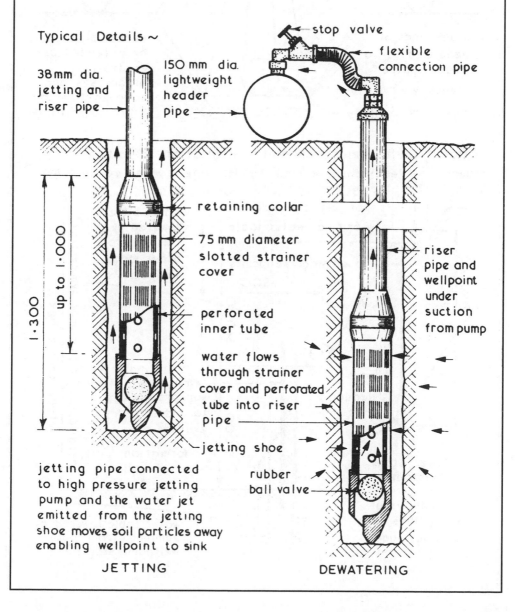

Typical Details ~

38 mm dia. jetting and riser pipe

150 mm dia. lightweight header pipe

stop valve

flexible connection pipe

retaining collar

75 mm diameter slotted strainer cover

perforated inner tube

water flows through strainer cover and perforated tube into riser pipe

jetting shoe

riser pipe and wellpoint under suction from pump

up to 1·000

1·300

jetting pipe connected to high pressure jetting pump and the water jet emitted from the jetting shoe moves soil particles away enabling wellpoint to sink

rubber ball valve

JETTING

DEWATERING

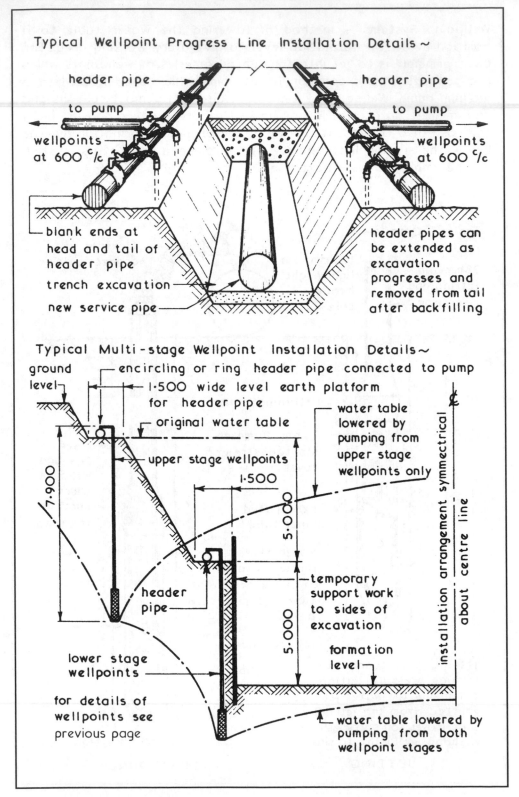

Typical Wellpoint Progress Line Installation Details ~

header pipe

header pipe

to pump

to pump

wellpoints at 600 c/c

wellpoints at 600 c/c

blank ends at head and tail of header pipe

trench excavation

new service pipe

header pipes can be extended as excavation progresses and removed from tail after backfilling

Typical Multi-stage Wellpoint Installation Details ~

ground level

encircling or ring header pipe connected to pump

1·500 wide level earth platform for header pipe

original water table

upper stage wellpoints

water table lowered by pumping from upper stage wellpoints only

1·500

7·900

5·000

5·000

header pipe

temporary support work to sides of excavation

formation level

lower stage wellpoints

for details of wellpoints see previous page

installation arrangement symmetrical about centre line

water table lowered by pumping from both wellpoint stages

Thin Grouted Membranes ~ these are permanent curtain or cut-off non-structural walls or barriers inserted in the ground to enclose the proposed excavation area. They are suitable for silts and sands and can be installed rapidly but they must be adequately supported by earth on both sides. The only limitation is the depth to which the formers can be driven and extracted.

Typical Details ~

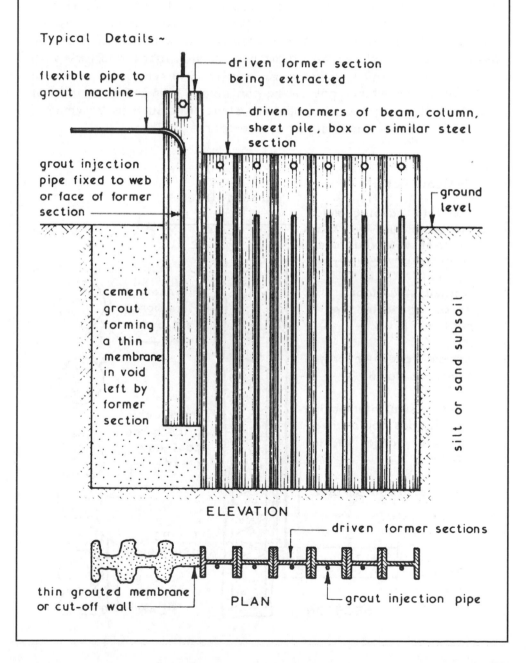

flexible pipe to grout machine

driven former section being extracted

driven formers of beam, column, sheet pile, box or similar steel section

grout injection pipe fixed to web or face of former section

ground level

cement grout forming a thin membrane in void left by former section

silt or sand subsoil

ELEVATION

driven former sections

thin grouted membrane or cut-off wall

PLAN

grout injection pipe

Contiguous or Secant Piling ~ this forms a permanent structural wall of interlocking bored piles. Alternate piles are bored and cast by traditional methods and before the concrete has fully hardened the interlocking piles are bored using a toothed flight auger. This system is suitable for most types of subsoil and has the main advantages of being economical on small and confined sites; capable of being formed close to existing foundations and can be installed with the minimum of vibration and noise. Ensuring a complete interlock of all piles over the entire length may be difficult to achieve in practice therefore the exposed face of the piles is usually covered with a mesh or similar fabric and face with rendering or sprayed concrete. Alternatively a reinforced concrete wall could be cast in front of the contiguous piling. This method of ground water control is suitable for structures such as basements, road underpasses and underground car parks.

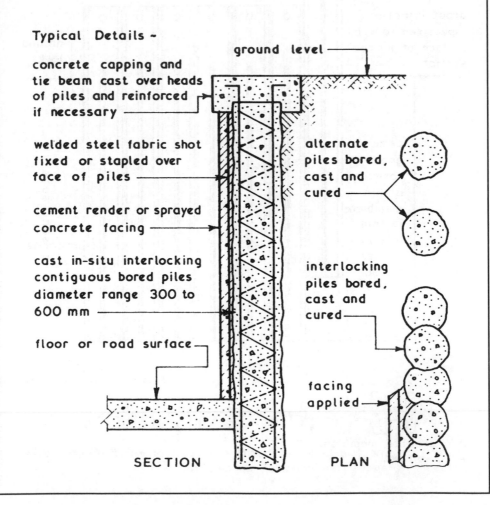

Typical Details ~

concrete capping and tie beam cast over heads of piles and reinforced if necessary

welded steel fabric shot fixed or stapled over face of piles

cement render or sprayed concrete facing

cast in-situ interlocking contiguous bored piles diameter range 300 to 600 mm

floor or road surface

ground level

alternate piles bored, cast and cured

interlocking piles bored, cast and cured

facing applied

SECTION

PLAN

290

Diaphragm Walls ~ these are structural concrete walls which can be cast in-situ (usually by the bentonite slurry method) or constructed using precast concrete components (see next page). They are suitable for most subsoils and their installation generates only a small amount of vibration and noise making them suitable for works close to existing buildings. The high cost of these walls makes them uneconomic unless they can be incorporated into the finished structure. Diaphragm walls are suitable for basements, underground car parks and similar structures.

Typical Cast In-situ Concrete Diaphragm Wall Details ~

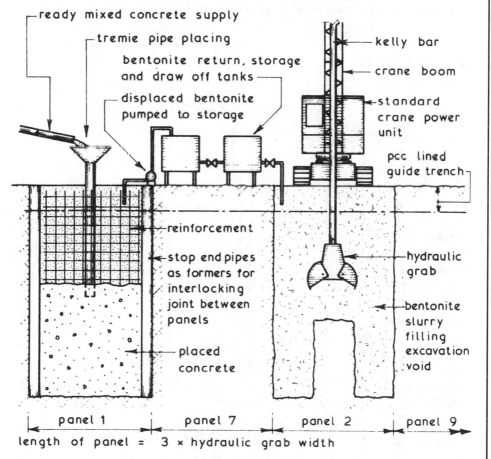

- ready mixed concrete supply
- tremie pipe placing
- bentonite return, storage and draw off tanks
- displaced bentonite pumped to storage
- kelly bar
- crane boom
- standard crane power unit
- pcc lined guide trench
- reinforcement
- stop end pipes as formers for interlocking joint between panels
- placed concrete
- hydraulic grab
- bentonite slurry filling excavation void

| panel 1 | panel 7 | panel 2 | panel 9 |

length of panel = 3 × hydraulic grab width

NB. Bentonite is a controlled mixture of fullers earth and water which produces a mud or slurry which has thixotropic properties and exerts a pressure in excess of earth + hydrostatic pressure present on sides of excavation.

Precast Concrete Diaphragm Walls ~ these walls have the some applications as their in-situ counterparts and have the advantages of factory produced components but lack the design flexibility of cast in-situ walls. The panel or post and panel units are installed in a trench filled with a special mixture of bentonite and cement with a retarder to control the setting time. This mixture ensures that the joints between the wall components are effectively sealed. To provide stability the panels or posts are tied to the retained earth with ground anchors.

Typical Precast Concrete Diaphragm Wall Details ~

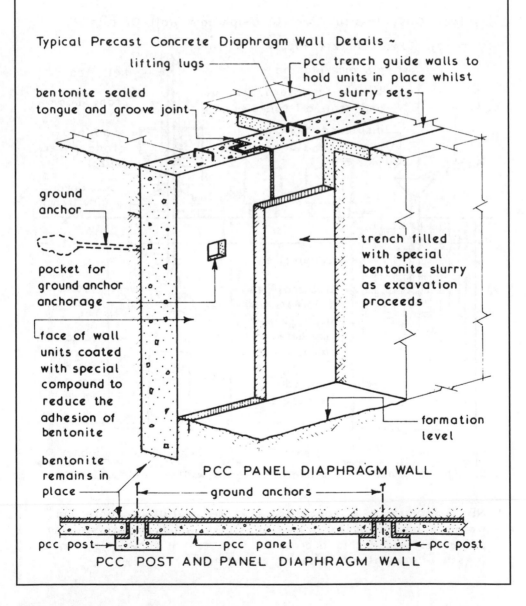

lifting lugs

pcc trench guide walls to hold units in place whilst slurry sets

bentonite sealed tongue and groove joint

ground anchor

pocket for ground anchor anchorage

face of wall units coated with special compound to reduce the adhesion of bentonite

bentonite remains in place

trench filled with special bentonite slurry as excavation proceeds

formation level

PCC PANEL DIAPHRAGM WALL

ground anchors

pcc post

pcc panel

pcc post

PCC POST AND PANEL DIAPHRAGM WALL

Grouting Methods ~ these techniques are used to form a curtain or cut off wall in high permeability soils where pumping methods could be uneconomic. The curtain walls formed by grouting methods are non-structural therefore adequate earth support will be required and in some cases this will be a distance of at least 4·000 from the face of the proposed excavation. Grout mixtures are injected into the soil by pumping the grout at high pressure through special injection pipes inserted in the ground. The pattern and spacing of the injection pipes will depend on the grout type and soil conditions.

Grout Types ~

1. Cement Grouts – mixture of neat cement and water cement sand up to 1 : 4 or PFA (pulverized fuel ash) cement to a 1 : 1 ratio. Suitable for coarse grained soils and fissured and jointed rock strata.

2. Chemical Grouts – one shot (premixed) of two shot (first chemical is injected followed immediately by second chemical resulting in an immediate reaction) methods can be employed to form a permanent gel in the soil to reduce its permeability and at the same time increase the soil's strength. Suitable for medium to coarse sands and gravels.

3. Resin Grouts – these are similar in application to chemical grouts but have a low viscosity and can therefore penetrate into silty fine sands.

Typical Cement Grouting Details ~

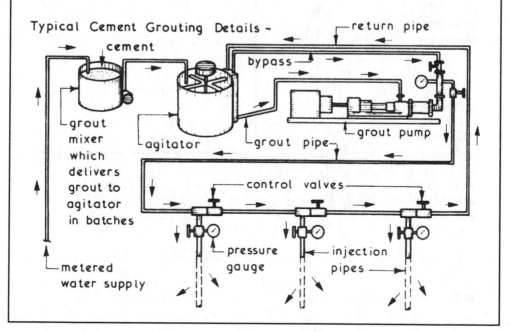

293

Ground Freezing Techniques ~ this method is suitable for all types of saturated soils and rock and for soils with a moisture content in excess of 8% of the voids. The basic principle is to insert into the ground a series of freezing tubes to form an ice wall thus creating an impermeable barrier. The treatment takes time to develop and the initial costs are high, therefore it is only suitable for large contracts of reasonable duration. The freezing tubes can be installed vertically for conventional excavations and horizontally for tunnelling works. The usual circulating brines employed are magnesium chloride and calcium chloride with a temperature of –15° to –25°C which would take 10 to 17 days to form an ice wall 1·000 thick. Liquid nitrogen could be used as the freezing medium to reduce the initial freezing period if the extra cost can be justified.

Typical Ground Freezing Details ~

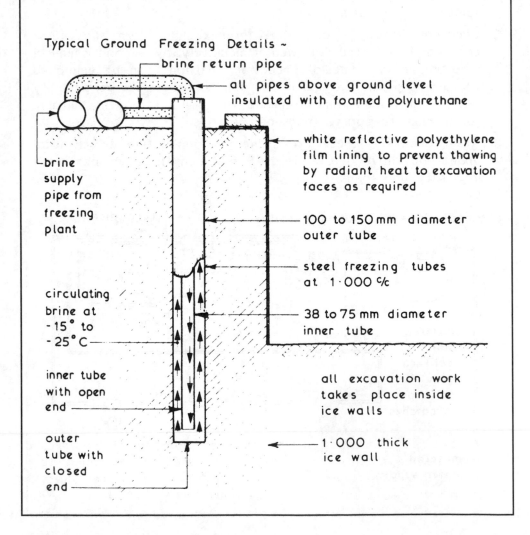

brine return pipe

all pipes above ground level insulated with foamed polyurethane

brine supply pipe from freezing plant

white reflective polyethylene film lining to prevent thawing by radiant heat to excavation faces as required

100 to 150 mm diameter outer tube

steel freezing tubes at 1·000 c/c

circulating brine at –15° to –25°C

38 to 75 mm diameter inner tube

inner tube with open end

all excavation work takes place inside ice walls

outer tube with closed end

1·000 thick ice wall

Soil Investigation ~ before a decision is made as to the type of foundation which should be used on any particular site a soil investigation should be carried out to establish existing ground conditions and soil properties. The methods which can be employed together with other sources of information such as local knowledge, ordnance survey and geological maps, mining records and aerial photography should be familiar to students at this level. If such an investigation reveals a naturally poor subsoil or extensive filling the designer has several options:-

1. Not to Build – unless a new and suitable site can be found building is only possible if the poor ground is localised and the proposed foundations can be designed around these areas with the remainder of the structure bridging over these positions.

2. Remove and Replace – the poor ground can be excavated, removed and replaced by compacted fills. Using this method there is a risk of differential settlement and generally for depths over 4·000 it is uneconomic.

3. Surcharging – this involves preloading the poor ground with a surcharge of aggregate or similar material to speed up settlement and thereby improve the soil's bearing capacity. Generally this method is uneconomic due to the time delay before actual building operations can commence which can vary from a few weeks to two or more years.

4. Vibration – this is a method of strengthening ground by vibrating a granular soil into compacted stone columns either by using the natural coarse granular soil or by replacement – see pages 296 and 297.

5. Dynamic Compaction – this is a method of soil improvement which consists of dropping a heavy weight through a considerable vertical distance to compact the soil and thus improve its bearing capacity and is especially suitable for granular soils – see page 298.

6. Jet Grouting – this method of consolidating ground can be used in all types of subsoil and consists of lowering a monitor probe into a 150 mm diameter prebored guide hole. The probe has two jets the upper of which blasts water, concentrated by compressed air to force any loose material up the guide to ground level. The lower jet fills the void with a cement slurry which sets into a solid mass – see page 299.

Ground Vibration ~ the objective of this method is to strengthen the existing soil by rearranging and compacting coarse granular particles to form stone columns with the ground. This is carried out by means of a large poker vibrator which has an effective compacting radius of 1·500 to 2·700. On large sites the vibrator is inserted on a regular triangulated grid pattern with centres ranging from 1·500 to 3·000. In coarse grained soils extra coarse aggregate is tipped into the insertion positions to make up levels as required whereas in clay and other fine particle soils the vibrator is surged up and down enabling the water jetting action to remove the surrounding soft material thus forming a borehole which is backfilled with a coarse granular material compacted in-situ by the vibrator. The backfill material is usually of 20 to 70 mm size of uniform grading within the chosen range. Ground vibration is not a piling system but a means of strengthening ground to increase the bearing capacity within a range of 200 to 500 kN/m².

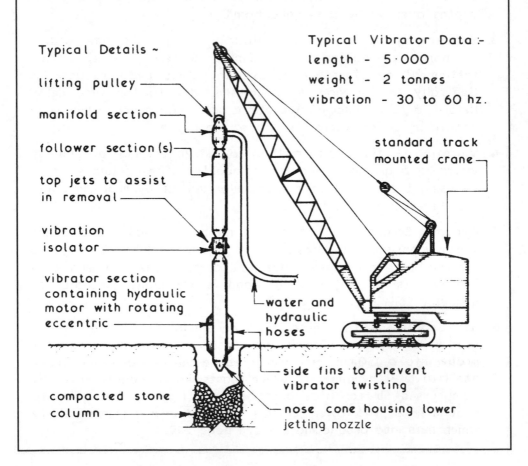

Typical Details ~

lifting pulley

manifold section

follower section (s)

top jets to assist
in removal

vibration
isolator

vibrator section
containing hydraulic
motor with rotating
eccentric

compacted stone
column

Typical Vibrator Data :-
length - 5·000
weight - 2 tonnes
vibration - 30 to 60 hz.

standard track
mounted crane

water and
hydraulic
hoses

side fins to prevent
vibrator twisting

nose cone housing lower
jetting nozzle

Sand Compaction – applied to non-cohesive subsoils where the granular particles are rearranged into a denser condition by poker vibration.

The crane-suspended vibrating poker is water-jetted into the ground using a combination of self weight and water displacement of the finer soil particles to penetrate the ground. Under this pressure, the soil granules compact to increase in density as the poker descends. At the appropriate depth, which may be determined by building load calculations or the practical limit of plant (generally 30 m max.), jetting ceases and fine aggregates or sand are infilled around the poker. The poker is then gradually withdrawn compacting the granular fill in the process. Compaction continues until sand fill reaches ground level. Spacing of compaction boreholes is relatively close to ensure continuity and an integral ground condition.

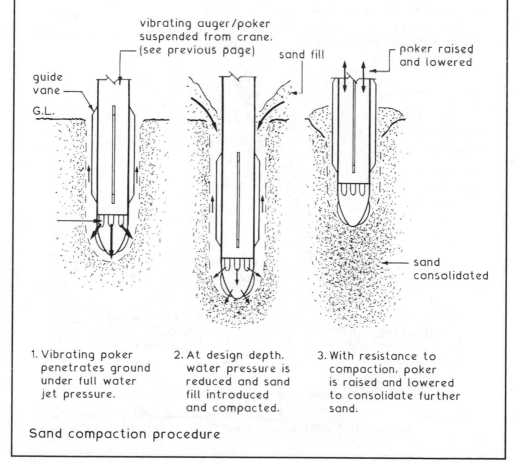

1. Vibrating poker penetrates ground under full water jet pressure.

2. At design depth, water pressure is reduced and sand fill introduced and compacted.

3. With resistance to compaction, poker is raised and lowered to consolidate further sand.

Sand compaction procedure

Dynamic Compaction ~ this method of ground improvement consists of dropping a heavy weight from a considerable height and is particularly effective in granular soils. Where water is present in the subsoil, trenches should be excavated to allow the water to escape and not collect in the craters formed by the dropped weight. The drop pattern, size of weight and height of drop are selected to suit each individual site but generally 3 or 4 drops are made in each position forming a crater up to 2·500 deep and 5·000 in diameter. Vibration through the subsoil can be a problem with dynamic compaction operations therefore the proximity and condition of nearby buildings must be considered together with the depth position and condition of existing services on site.

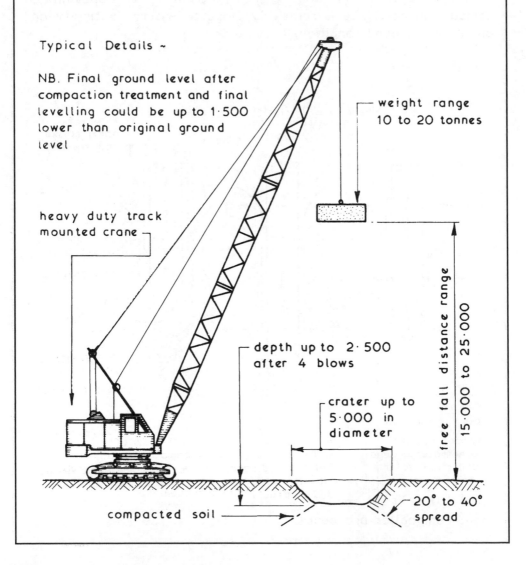

Typical Details ~

NB. Final ground level after compaction treatment and final levelling could be up to 1·500 lower than original ground level

weight range 10 to 20 tonnes

heavy duty track mounted crane

depth up to 2·500 after 4 blows

crater up to 5·000 in diameter

free fall distance range 15·000 to 25·000

compacted soil

20° to 40° spread

Jet Grouting ~ this is a means of consolidating ground by lowering into preformed bore holes a monitor probe. The probe is rotated and the sides of the bore hole are subjected to a jet of pressurised water and air from a single outlet which enlarges and compacts the bore hole sides. At the same time a cement grout is being introduced under pressure to fill the void being created. The water used by the probe and any combined earth is forced up to the surface in the form of a sludge. If the monitor probe is not rotated grouted panels can be formed. The spacing, depth and layout of the bore holes is subject to specialist design.

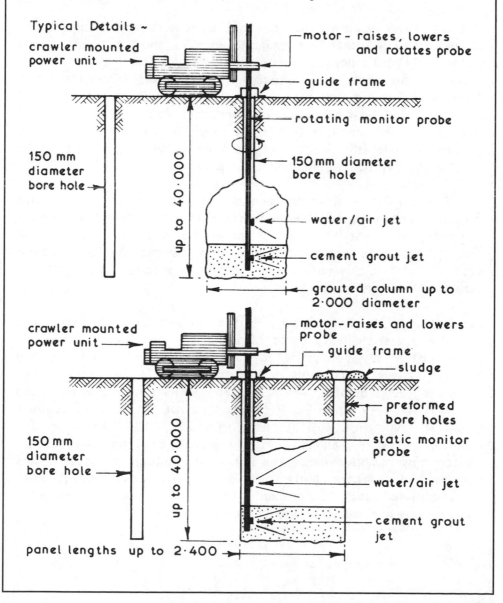

Typical Details ~

crawler mounted power unit

motor - raises, lowers and rotates probe

guide frame

rotating monitor probe

150 mm diameter bore hole

150 mm diameter bore hole

up to 40·000

water/air jet

cement grout jet

grouted column up to 2·000 diameter

crawler mounted power unit

motor - raises and lowers probe

guide frame

sludge

preformed bore holes

150 mm diameter bore hole

static monitor probe

up to 40·000

water/air jet

cement grout jet

panel lengths up to 2·400

Green-Field – land not previously built upon. Usually part of the 'green-belt' surrounding urban areas, designated inappropriate for development in order to preserve the countryside. Limited development for agricultural purposes only may be permitted on 'green-belt' land.

Brown-Field – derelict land formerly a developed site and usually associated with previous construction of industrial buildings. UK government has set an objective to build 60% of the 4 million new homes required by 2016 on these sites.

Site Survey – essential that a geo-technical survey is undertaken to determine whether contaminants are in the soil and ground water. Of particular concern are: acids, salts, heavy metals, cyanides and coal tars, in addition to organic materials which decompose to form the highly explosive gas, methane. Analysis of the soil will determine a 'trigger threshold value', above which it will be declared sensitive to the end user. For example, a domestic garden or children's play area will have a low value relative to land designated for a commercial car park.

Site Preparation – when building on sites previously infilled with uncontaminated material, a reinforced raft type foundation may be adequate for light structures. Larger buildings will justify soil consolidation and compaction processes to improve the bearing capacity. Remedial measures for subsoils containing chemicals or other contaminants are varied.

Legislation – the Environment Protection Act of 1990 attempted to enforce responsibility on local authorities to compile a register of all potentially contaminated land. This proved unrealistic and too costly due to inherent complexities. Since then, requirements under the Environment Act 1995, the Pollution Prevention and Control Act 1999, the PPC Regulations 2000 and the subsequent DCLG Planning Policy Statement (PPS 23, 2004): Planning and Pollution Control (Annex 2: Development of land affected by contamination), have made this more of a planning issue. It has become the responsibility of developers to conduct site investigations and to present details of proposed remedial measures as part of their planning application.

The traditional low-technology method for dealing with contaminated sites has been to excavate the soil and remove it to places licensed for depositing. However, with the increase in building work on brown-field sites, suitable dumps are becoming scarce. Added to this is the reluctance of ground operators to handle large volumes of this type of waste. Also, where excavations exceed depths of about 5 m, it becomes less practical and too expensive. Alternative physical, biological or chemical methods of soil treatment may be considered.

Encapsulation – in-situ enclosure of the contaminated soil. A perimeter trench is taken down to rock or other sound strata and filled with an impervious agent such as Bentonite clay. An impermeable horizontal capping is also required to link with the trenches. A high-specification barrier is necessary where liquid or gas contaminants are present as these can migrate quite easily. A system of monitoring soil condition is essential as the barrier may decay in time. Suitable for all types of contaminant.

Soil washing – involves extraction of the soil, sifting to remove large objects and placing it in a scrubbing unit resembling a huge concrete mixer. Within this unit water and detergents are added for a basic wash process, before pressure spraying to dissolve pollutants and to separate clay from silt. Eliminates fuels, metals and chemicals.

Vapour extraction – used to remove fuels or industrial solvents and other organic deposits. At variable depths, small diameter boreholes are located at frequent intervals. Attached to these are vacuum pipes to draw air through the contaminated soil. The contaminants are collected at a vapour treatment processing plant on the surface, treated and evaporated into the atmosphere. This is a slow process and it may take several months to cleanse a site.

Electrolysis – use of low voltage d.c. in the presence of metals. Electricity flows between an anode and cathode, where metal ions in water accumulate in a sump before pumping to the surface for treatment.

BIOLOGICAL

Phytoremediation – the removal of contaminants by plants which will absorb harmful chemicals from the ground. The plants are subsequently harvested and destroyed. A variant uses fungal degradation of the contaminants.

Bioremediation – stimulating the growth of naturally occurring microbes. Microbes consume petrochemicals and oils, converting them to water and carbon dioxide. Conditions must be right, i.e. a temperature of at least 10°C with an adequate supply of nutrients and oxygen. Untreated soil can be excavated and placed over perforated piping, through which air is pumped to enhance the process prior to the soil being replaced.

CHEMICAL

Oxidation – sub-soil boreholes are used for the pumped distribution of liquid hydrogen peroxide or potassium permanganate. Chemicals and fuel deposits convert to water and carbon dioxide.

Solvent extraction – the sub-soil is excavated and mixed with a solvent to break down oils, grease and chemicals that do not dissolve in water.

THERMAL

Thermal treatment (off site) – an incineration process involving the use of a large heating container/oven. Soil is excavated, dried and crushed prior to heating to 2500°C, where harmful chemicals are removed by evaporation or fusion.

Thermal treatment (in-situ) – steam, hot water or hot air is pressure-injected through the soil. Variations include electric currents and radio waves to heat water in the ground to become steam. Evaporates chemicals.

Ref. Building Regulations, Approved Document, C1: Site preparation and resistance to contaminants. Section 1: Clearance or treatment of unsuitable material. Section 2: Resistance to contaminants.

5 SUPERSTRUCTURE – 1

CHOICE OF MATERIALS

BRICK AND BLOCK WALLS

BRICK BONDING

SPECIAL BRICKS AND APPLICATIONS

CAVITY WALLS

DAMP-PROOF COURSES

GAS RESISTANT MEMBRANES

ARCHES AND OPENINGS

WINDOWS, GLASS AND GLAZING

DOMESTIC AND INDUSTRIAL DOORS

TIMBER FRAME CONSTRUCTION

TIMBER PITCHED AND FLAT ROOFS

TIMBER DECAY AND TREATMENT

GREEN ROOFS

THERMAL INSULATION

U-VALUE CALCULATION

THERMAL BRIDGING

ACCESS FOR THE DISABLED

STAGE 1

Consideration to be given to the following:~

1. Building type and usage.
2. Building owner's requirements and preferences.
3. Local planning restrictions.
4. Legal restrictions and requirements.
5. Site restrictions.
6. Capital resources.
7. Future policy in terms of maintenance and adaptation.

STAGE 2

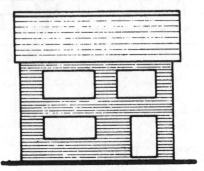

Decide on positions, sizes and shapes of openings.

STAGE 3

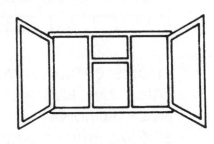

Decide on style, character and materials for openings

STAGE 4

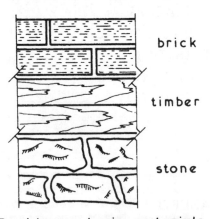

brick

timber

stone

Decide on basic materials for fabric of roof and walls

STAGE 5

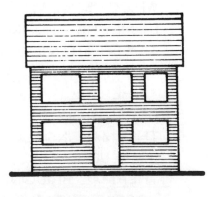

Review all decisions and make changes if required

Bricks ~ these are walling units within a length of 337·5mm, a width of 225mm and a height of 112·5mm. The usual size of bricks in common use is length 215mm, width 102·5mm and height 65mm and like blocks they must be laid in a definite pattern or bond if they are to form a structural wall. Bricks are usually made from clay (BS EN 772-1, BS EN 772-3 and BS EN 772-7) or from sand and lime (BS EN 771-2) and are available in a wide variety of strengths, types, textures, colours and special shaped bricks to BS 4729.

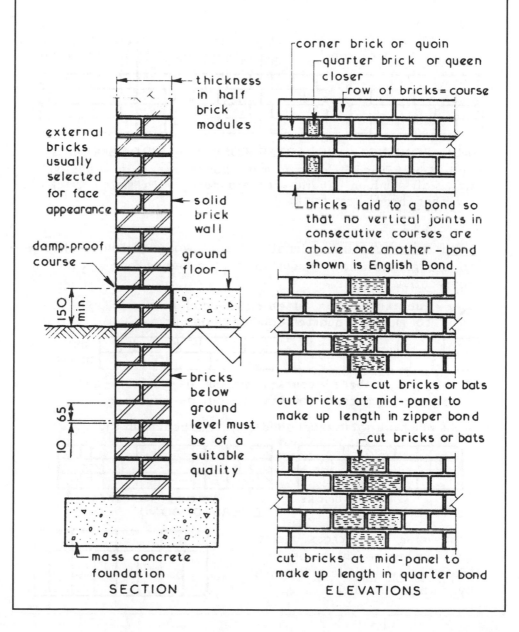

SECTION

ELEVATIONS

Typical Details ~

Bonding ~ an arrangement of bricks in a wall, column or pier laid to a set pattern to maintain an adequate lap.

Purposes of Brick Bonding ~

1. Obtain maximum strength whilst distributing the loads to be carried throughout the wall, column or pier.
2. Ensure lateral stability and resistance to side thrusts.
3. Create an acceptable appearance.

Lap Forms ~

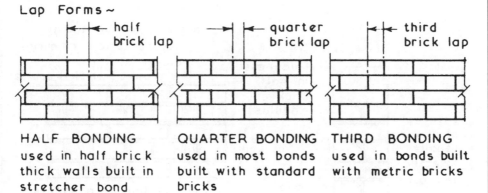

HALF BONDING
used in half brick thick walls built in stretcher bond

QUARTER BONDING
used in most bonds built with standard bricks

THIRD BONDING
used in bonds built with metric bricks

Simple Bonding Rules ~

1. Bond is set out along length of wall working from each end to ensure that no vertical joints are above one another in consecutive courses.

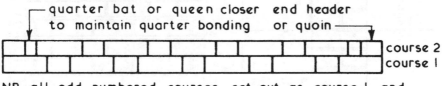

NB all odd numbered courses set out as course 1 and all even numbered courses set out as course 2

2. Walls which are not in exact bond length can be set out thus –

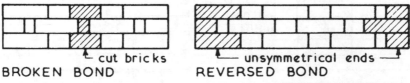

BROKEN BOND   REVERSED BOND

3. Transverse or cross joints continue unbroken across the width of wall unless stopped by a face stretcher.

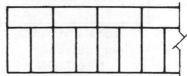

English Bond ~ formed by laying alternate courses of stretchers and headers it is one of the strongest bonds but it will require more facing bricks than other bonds (89 facing bricks per m²)

Typical Example ~

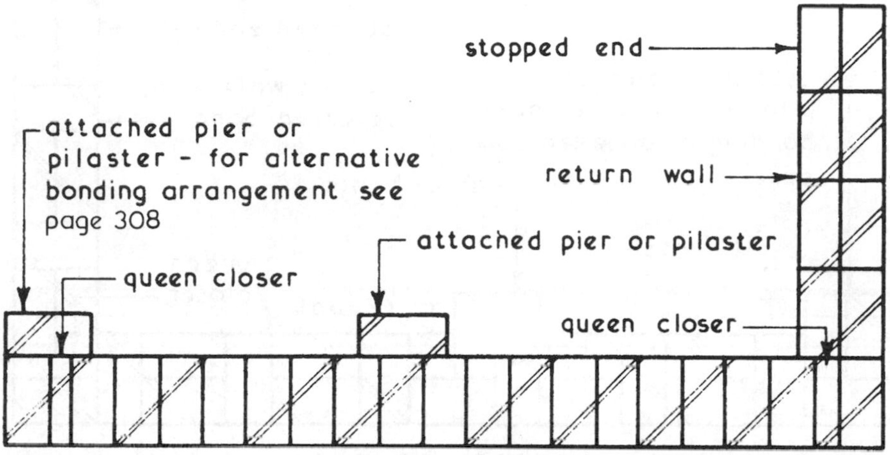

stopped end

attached pier or pilaster - for alternative bonding arrangement see page 308

return wall

attached pier or pilaster

queen closer

queen closer

PLAN ON ODD NUMBERED COURSES

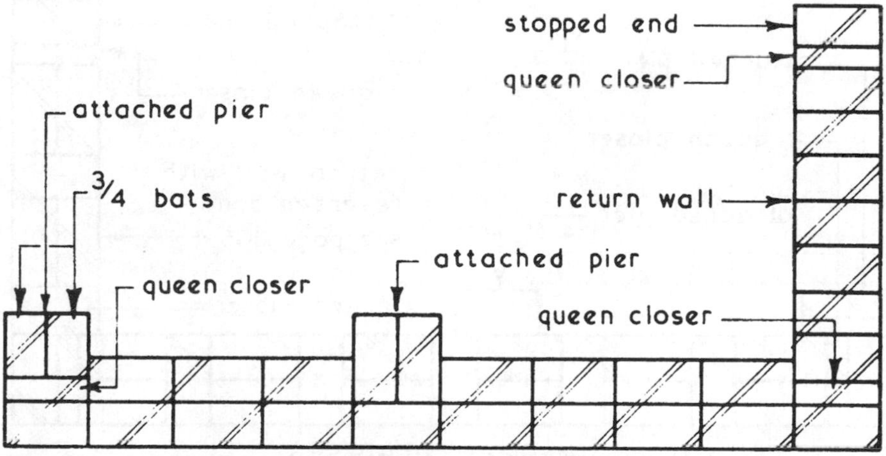

stopped end

queen closer

attached pier

³⁄₄ bats

return wall

attached pier

queen closer

queen closer

PLAN ON EVEN NUMBERED COURSES

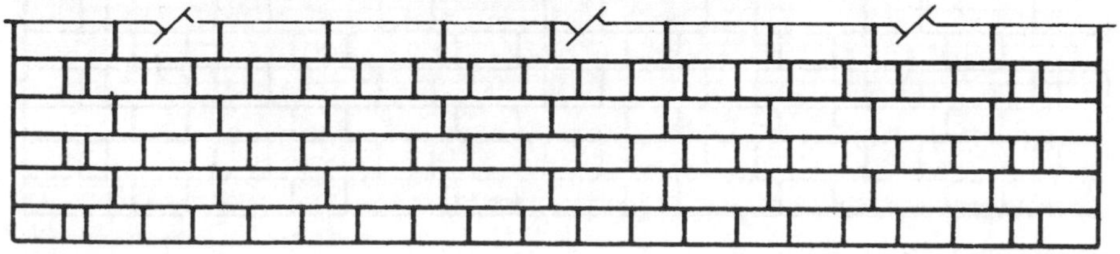

ELEVATION

Flemish Bond ~ formed by laying headers and stretchers alternately in each course. Not as strong as English bond but is considered to be aesthetically superior uses less facing bricks. (79 facing bricks per m²)

Typical Example

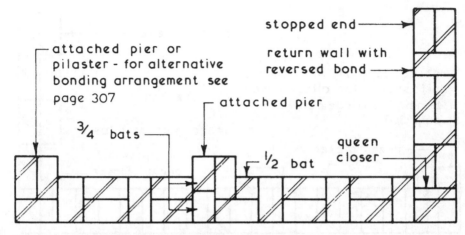

attached pier or pilaster - for alternative bonding arrangement see page 307

stopped end

return wall with reversed bond

attached pier

³/₄ bats

½ bat

queen closer

PLAN ON ODD NUMBERED COURSES

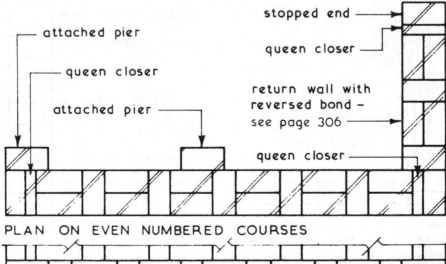

attached pier

queen closer

attached pier

stopped end

queen closer

return wall with reversed bond - see page 306

queen closer

PLAN ON EVEN NUMBERED COURSES

ELEVATION

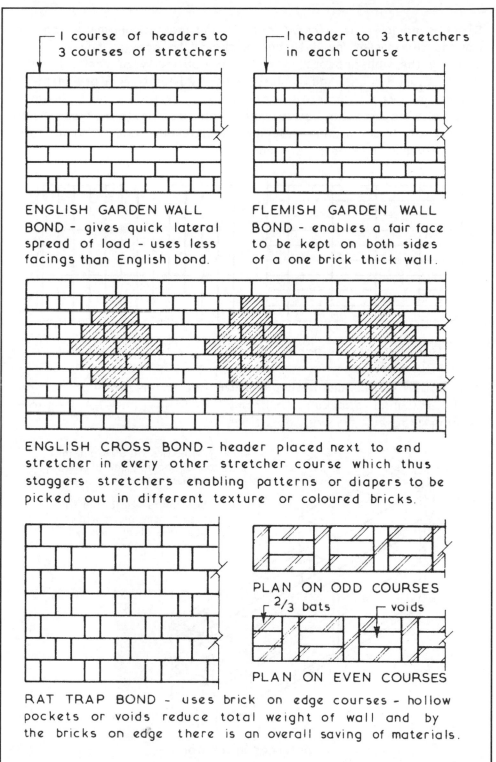

I course of headers to
3 courses of stretchers

I header to 3 stretchers
in each course

ENGLISH GARDEN WALL
BOND - gives quick lateral
spread of load - uses less
facings than English bond.

FLEMISH GARDEN WALL
BOND - enables a fair face
to be kept on both sides
of a one brick thick wall.

ENGLISH CROSS BOND - header placed next to end
stretcher in every other stretcher course which thus
staggers stretchers enabling patterns or diapers to be
picked out in different texture or coloured bricks.

PLAN ON ODD COURSES

$2/3$ bats   voids

PLAN ON EVEN COURSES

RAT TRAP BOND - uses brick on edge courses - hollow
pockets or voids reduce total weight of wall and by
the bricks on edge there is an overall saving of materials.

Stack Bonding – the quickest, easiest and most economical bond to lay, as there is no need to cut bricks or to provide special sizes. Visually the wall appears unbonded as continuity of vertical joints is structurally unsound, unless wire bed-joint reinforcement is placed in every horizontal course, or alternate courses where loading is moderate. In cavity walls, wall ties should be closer than normal at 600mm max. spacing horizontally and 225mm max. spacing vertically and staggered.

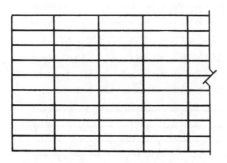

Horizontal stack bond

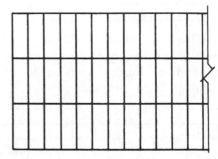

Vertical stack bond

Application – this distinctive uniform pattern is popular as non-structural infill panelling to framed buildings and for non-load bearing exposed brickwork partitions.

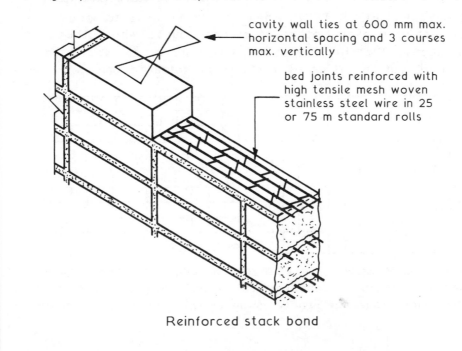

cavity wall ties at 600 mm max. horizontal spacing and 3 courses max. vertically

bed joints reinforced with high tensile mesh woven stainless steel wire in 25 or 75 m standard rolls

Reinforced stack bond

Attached Piers ~ the main function of an attached pier is to give lateral support to the wall of which it forms part from the base to the top of the wall. It also has the subsidiary function of dividing a wall into distinct lengths whereby each length can be considered as a wall. Generally walls must be tied at end to an attached pier, buttressing or return wall.

Typical Examples ~

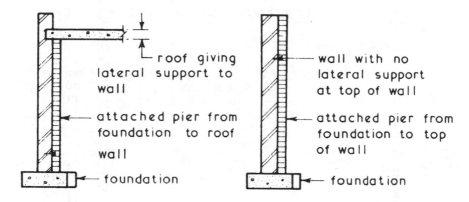

roof giving lateral support to wall

attached pier from foundation to roof wall

foundation

wall with no lateral support at top of wall

attached pier from foundation to top of wall

foundation

Requirements for the external wall of a small single storey non-residential building or annex exceeding 2.5m in length or height and of floor area not exceeding 36 m² ~

- Minimum thickness, 90mm, i.e. 102.5mm brick or 100mm block.
- Built solid of bonded brick or block masonry and bedded in cement mortar.
- Surface mass of masonry, minimum 130kg/m² where floor area exceeds 10 m².
- No lateral loading permitted excepting wind loads.
- Maximum length or width not greater than 9 m.
- Maximum height as shown on page 313.
- Lateral restraint provided by direct bearing of roof and as shown on page 425.
- Maximum of two major openings in one wall of the building. Height maximum 2.1 m, width maximum 5m (if 2 openings, total width maximum 5 m).
- Other small openings permitted, as shown on next page.
- Bonded or connected to piers of minimum size 390 × 190mm at maximum 3m centres for the full wall height as shown above. Pier connections are with pairs of wall ties of 20 × 3mm flat stainless steel type at 300mm vertical spacing.

Attached piers as applied to 1/2 brick (90 mm min.) thick walls ~

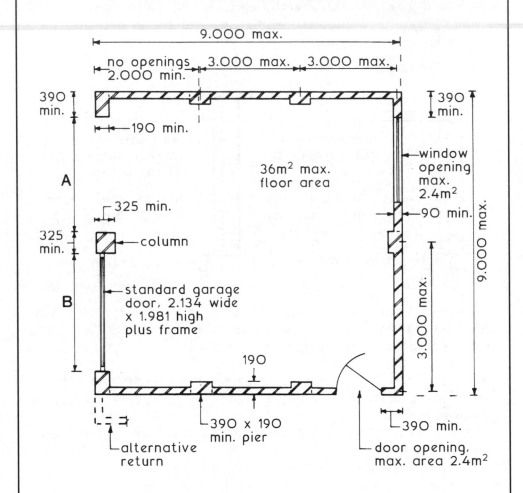

9.000 max.

no openings 2.000 min.

3.000 max.

3.000 max.

390 min.

190 min.

390 min.

A

36m² max. floor area

window opening max. 2.4m²

90 min.

325 min.

325 min.

column

standard garage door, 2.134 wide x 1.981 high plus frame

B

190

390 x 190 min. pier

3.000 max.

9.000 max.

alternative return

door opening, max. area 2.4m²

390 min.

- Major openings A and B are permitted in one wall only. Aggregate width is 5m maximum. Height not greater than 2.1 m. No other openings within 2 m.
- Other walls not containing a major opening can have smaller openings of maximum aggregate area 2.4 m².
- Maximum of only one opening between piers.
- Distance from external corner of a wall to an opening at least 390 mm unless the corner contains a pier.
- The minimum pier dimension of 390 × 190 mm can be varied to 327 × 215 mm to suit brick sizes.

Construction of half-brick and 100mm thick solid concrete block walls (90mm min.) with attached piers, has height limitations to maintain stability. The height of these buildings will vary depending on the roof profile; it should not exceed the lesser value in the following examples ~

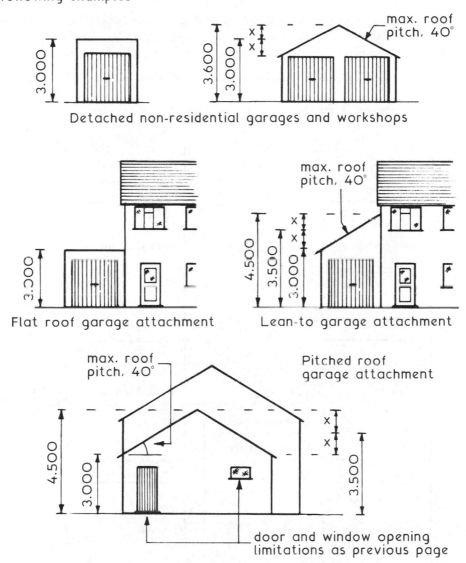

Detached non-residential garages and workshops

Flat roof garage attachment

Lean-to garage attachment

Pitched roof garage attachment

door and window opening limitations as previous page

Note: All dimensions are maximum.

Height is measured from top of foundation to top of wall except where shown at an intermediate position. Where the underside of the floor slab provides an effective lateral restraint, measurements may be taken from here.

The appearance of a building can be significantly influenced by the mortar finishing treatment to masonry. Finishing may be achieved by jointing or pointing.

Jointing – the finish applied to mortar joints as the work proceeds.

Pointing – the process of removing semi-set mortar to a depth of about 20 mm and replacing it with fresh mortar. Pointing may contain a colouring pigment to further enhance the masonry.

Finish profiles, typical examples shown pointed –

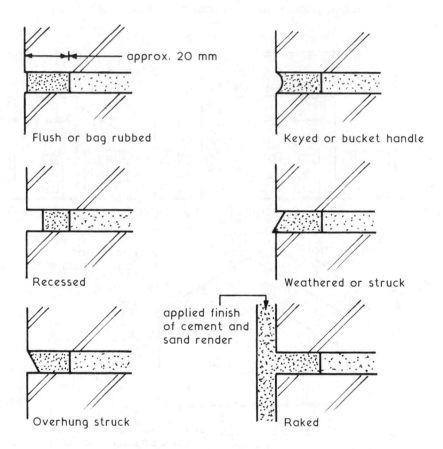

Examples of pointing to masonry

Note: Recessed and overhung finishes should not be used in exposed situations, as rainwater can be detained. This could encourage damage by frost action and growth of lichens.

Specials – these are required for feature work and application to various bonds, as shown on the preceding pages. Bonding is not solely for aesthetic enhancement. In many applications, e.g. English bonded manhole walls, the disposition of bricks is to maximise wall strength and integrity. In a masonry wall the amount of overlap should not be less than one quarter of a brick length. Specials may be machine or hand cut from standard bricks, or they may be purchased as purpose-made. These purpose-made bricks are relatively expensive as they are individually manufactured in hardwood moulds.

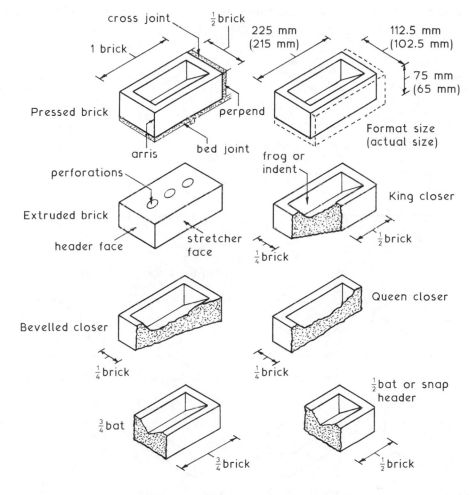

Standard bricks and cut specials

Ref. BS 4729: Clay and calcium silicate bricks of special shapes and sizes. Recommendations.

Brickwork can be repetitive and monotonous, but with a little imagination and skilled application it can be a highly decorative art form. Artistic potential is made possible by the variety of naturally occurring brick colours, textures and finishes, the latter often applied as a sanding to soft clay prior to baking. Furthermore, the range of pointing techniques, mortar colourings, brick shapes and profiles can combine to create countless possibilities for architectural expression.

Bricks are manufactured from baked clay, autoclaved sand/lime or concrete. Clay is ideally suited to hand making special shapes in hardwood moulds. Some popular formats are shown below, but there is no limit to creative possibilities.

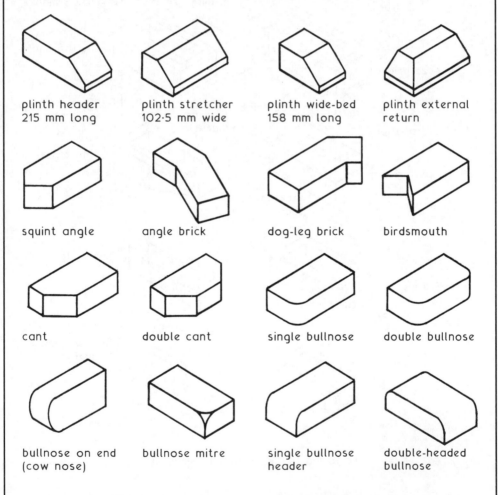

plinth header 215 mm long	plinth stretcher 102·5 mm wide	plinth wide-bed 158 mm long	plinth external return
squint angle	angle brick	dog-leg brick	birdsmouth
cant	double cant	single bullnose	double bullnose
bullnose on end (cow nose)	bullnose mitre	single bullnose header	double-headed bullnose

Purpose-made and special shape bricks

Plinths – used as a projecting feature to enhance external wall appearance at its base. The exposed projection determines that only frost-proof quality bricks are suitable and that recessed or raked out joints which could retain water must be avoided.

Typical external wall base –

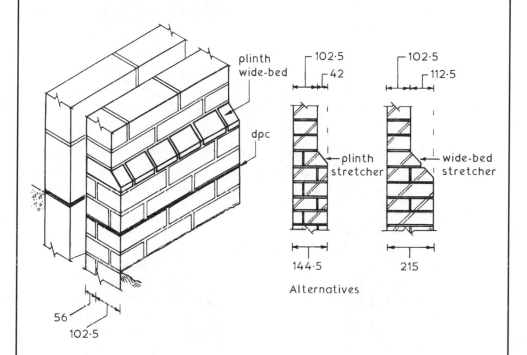

Alternatives

Corbel – a projecting feature at higher levels of a building. This may be created by using plinth bricks laid upside down with header and stretcher formats maintaining bond. For structural integrity, the amount of projection (P) must not exceed one third of the overall wall thickness (T). Some other types of corbel are shown on the next page.

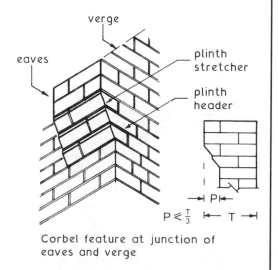

$P \leqslant \frac{T}{3}$

Corbel feature at junction of eaves and verge

317

Corbel – a type of inverted plinth, generally located at the higher levels of a building to create a feature. A typical example is quarter bonded headers as a detail below window openings.

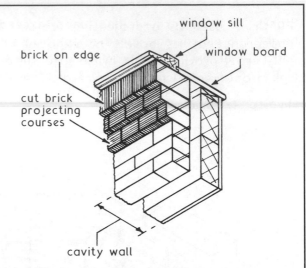

Corbelled sill

Dentil Coursing – a variation on continuous corbelling where alternative headers project. This is sometimes referred to as table corbelling.

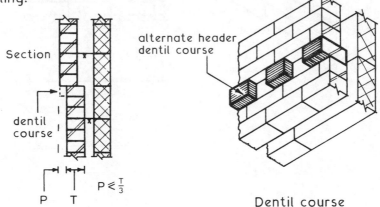

$$P \leqslant \frac{T}{3}$$

Dentil course

Dog Toothing – a variation on a dentil course created by setting the feature bricks at 45°.

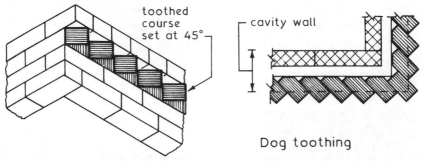

Dog toothing

Note: Cavity insulated as required.

Blocks ~ these are walling units exceeding in length, width or height the dimensions specified for bricks in BS EN 772-16. Precast concrete blocks should comply with the recommendations set out in BS 6073-2 and BS EN 771-3. Blocks suitable for external solid walls are classified as loadbearing and are required to have a minimum average crushing strength of 2.8 N/mm².

Typical Details ~

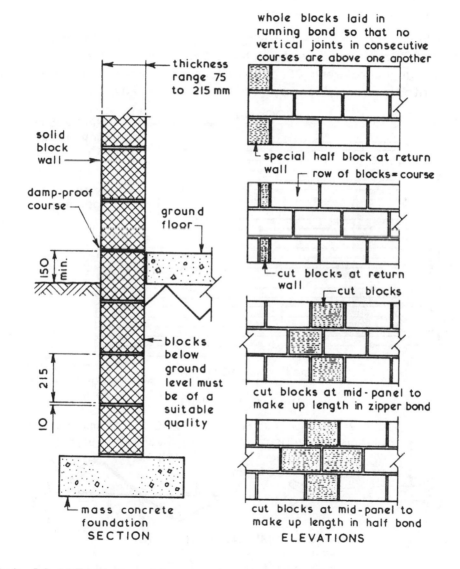

whole blocks laid in running bond so that no vertical joints in consecutive courses are above one another

thickness range 75 to 215 mm

solid block wall

damp-proof course

ground floor

150 min.

215

10

blocks below ground level must be of a suitable quality

mass concrete foundation
SECTION

special half block at return wall

row of blocks = course

cut blocks at return wall

cut blocks

cut blocks at mid-panel to make up length in zipper bond

cut blocks at mid-panel to make up length in half bond
ELEVATIONS

Refs. BS 6073-2: Precast concrete masonry units.
BS EN 772-16: Methods of test for masonry units.
BS EN 771-3: Specification for masonry units.

Cavity Walls ~ these consist of an outer brick or block leaf or skin separated from an inner brick or block leaf or skin by an air space called a cavity. These walls have better thermal insulation and weather resistance properties than a comparable solid brick or block wall and therefore are in general use for the enclosing walls of domestic buildings. The two leaves of a cavity wall are tied together with wall ties located at $2.5/m^2$, or at equivalent spacings shown below and as given in Section 2C of Approved Document A – Building Regulations.

With butterfly type ties the width of the cavity should be between 50 and 75mm. Where vertical twist type ties are used the cavity width can be between 75 and 300mm. Cavities are not normally ventilated and are closed by roof insulation at eaves level.

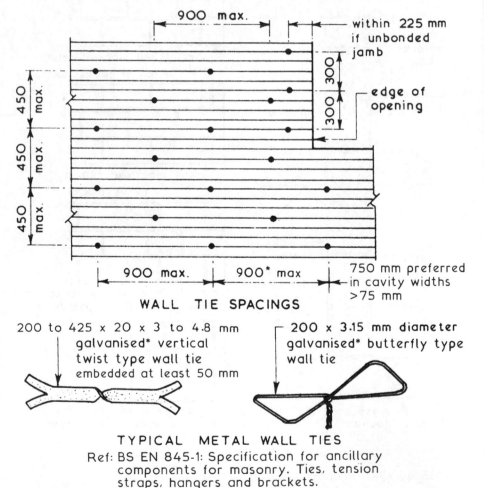

WALL TIE SPACINGS

200 to 425 x 20 x 3 to 4.8 mm galvanised* vertical twist type wall tie embedded at least 50 mm

200 x 3.15 mm diameter galvanised* butterfly type wall tie

TYPICAL METAL WALL TIES
Ref: BS EN 845-1: Specification for ancillary components for masonry. Ties, tension straps, hangers and brackets.

* Note: Stainless steel or non-ferrous ties are now preferred.

Minimum requirements ~

Thickness of each leaf, 90 mm.

Width of cavity, 50 mm.

Wall ties at 2.5/m² (see previous page).

Compressive strength of bricks, 5 N/mm² up to two storeys.*

Compressive strength of blocks, 2.8 N/mm² up to two storeys.*

* For work between the foundation and the surface a 7 N/mm² minimum brick and block strength is normally specified. This is also a requirement where the foundation to underside of the ground floor structure exceeds 1.0 m.

Combined thickness of each leaf + 10 mm whether used as an external wall, a separating wall or a compartment wall, should be not less than 1/16 of the storey height** which contains the wall.

** Generally measured between the undersides of lateral supports, eg. undersides of floor or ceiling joists, or from the underside of upper floor joists to half way up a laterally restrained gable wall. See Approved Document A, Section 2C for variations.

Wall dimensions for minimum combined leaf thicknesses of 90 mm + 90 mm ~

Height	Length
3.5 m max.	12.0 m max.
3.5 m – 9.0 m	9.0 m max.

Wall dimensions for minimum combined leaf thickness of 280 mm, eg. 190 mm + 90 mm for one storey height and a minimum 180 mm combined leaf thickness, ie. 90 mm + 90 mm for the remainder of its height ~

Height	Length
3.5 – 9.0 m	9.0 - 12.0 m
9.0 m – 12.0 m	9.0 m max.

Wall dimensions for minimum combined leaf thickness of 280 mm for two storey heights and a minimum 180 mm combined leaf thickness for the remainder of its height ~

Height	Length
9.0 m – 12.0 m	9.0 m – 12.0 m

Wall length is measured from centre to centre of restraints by buttress walls, piers or chimneys.

For other wall applications, see the reference to calculated brickwork on page 337.

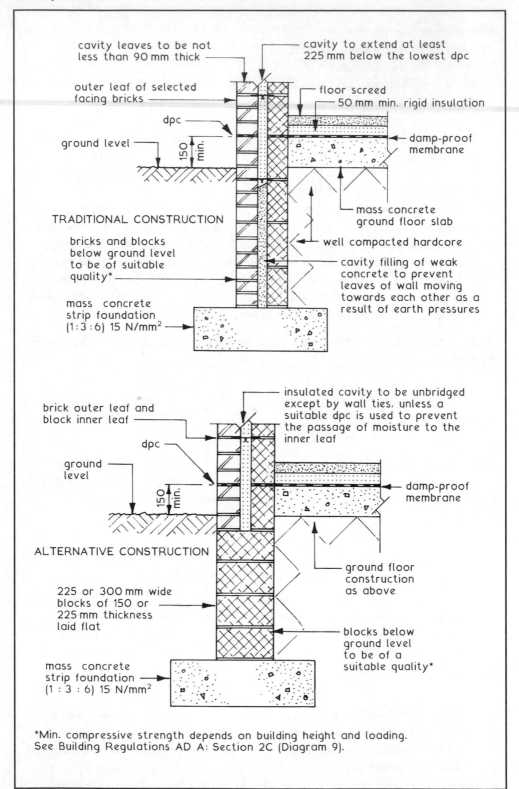

cavity leaves to be not less than 90 mm thick

outer leaf of selected facing bricks

dpc

ground level

150 min.

TRADITIONAL CONSTRUCTION

bricks and blocks below ground level to be of suitable quality*

mass concrete strip foundation (1:3:6) 15 N/mm²

cavity to extend at least 225 mm below the lowest dpc

floor screed

50 mm min. rigid insulation

damp-proof membrane

mass concrete ground floor slab

well compacted hardcore

cavity filling of weak concrete to prevent leaves of wall moving towards each other as a result of earth pressures

brick outer leaf and block inner leaf

dpc

ground level

150 min.

ALTERNATIVE CONSTRUCTION

225 or 300 mm wide blocks of 150 or 225 mm thickness laid flat

mass concrete strip foundation (1 : 3 : 6) 15 N/mm²

insulated cavity to be unbridged except by wall ties, unless a suitable dpc is used to prevent the passage of moisture to the inner leaf

damp-proof membrane

ground floor construction as above

blocks below ground level to be of a suitable quality*

*Min. compressive strength depends on building height and loading. See Building Regulations AD A: Section 2C (Diagram 9).

Parapet ~ a low wall projecting above the level of a roof, bridge or balcony forming a guard or barrier at the edge. Parapets are exposed to the elements on three faces namely front, rear and top and will therefore need careful design and construction if they are to be durable and reliable.

Typical Details ~

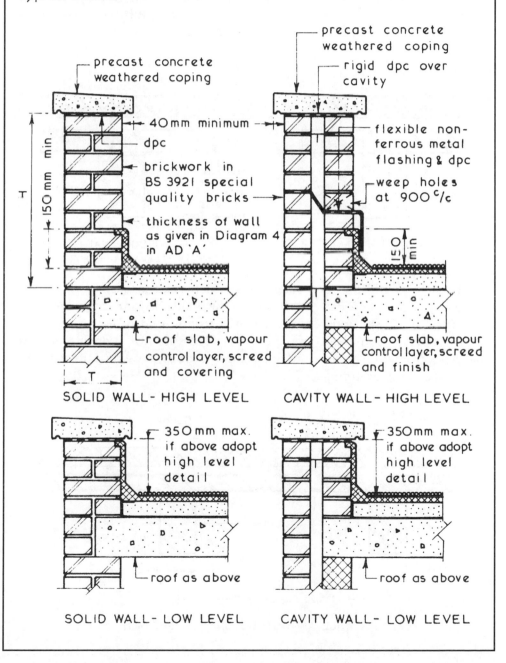

precast concrete weathered coping

precast concrete weathered coping

rigid dpc over cavity

40mm minimum

dpc

brickwork in BS 3921 special quality bricks

flexible non-ferrous metal flashing & dpc

weep holes at 900 c/c

thickness of wall as given in Diagram 4 in AD 'A'

150 mm min.

150 min

roof slab, vapour control layer, screed and covering

roof slab, vapour control layer, screed and finish

SOLID WALL- HIGH LEVEL

CAVITY WALL- HIGH LEVEL

350mm max. if above adopt high level detail

350mm max. if above adopt high level detail

roof as above

roof as above

SOLID WALL- LOW LEVEL

CAVITY WALL- LOW LEVEL

Historically, finned or buttressed walls have been used to provide lateral support to tall single storey masonry structures such as churches and cathedrals. Modern applications are similar in principle and include theatres, gymnasiums, warehouses, etc. Where space permits, they are an economic alternative to masonry cladding of steel or reinforced concrete framed buildings. The fin or pier is preferably brick bonded to the main wall. It may also be connected with horizontally bedded wall ties, sufficient to resist vertical shear stresses between fin and wall.

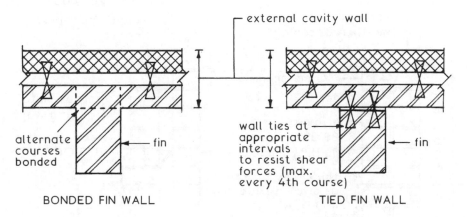

BONDED FIN WALL   TIED FIN WALL

Structurally, the fins are deep piers which reinforce solid or cavity masonry walls. For design purposes the wall may be considered as a series of 'T' sections composed of a flange and a pier. If the wall is of cavity construction, the inner leaf is not considered for bending moment calculations, although it does provide stiffening to the outer leaf or flange.

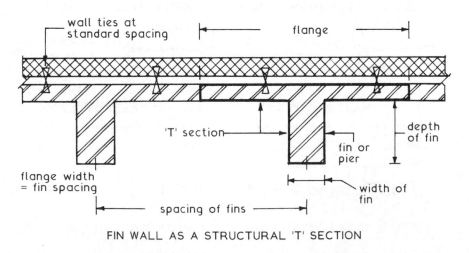

FIN WALL AS A STRUCTURAL 'T' SECTION

Masonry diaphragm walls are an alternative means of constructing tall, single storey buildings such as warehouses, sports centres, churches, assembly halls, etc. They can also be used as retaining and boundary walls with planting potential within the voids. These voids may also be steel reinforced and concrete filled to resist the lateral stresses in high retaining walls.

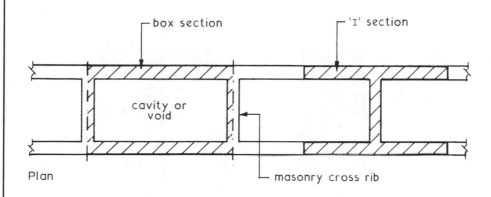

A diaphragm wall is effectively a cavity wall where the two leaves of masonry are bonded together with cross ribs and not wall ties. It is stronger than a conventionally tied cavity wall and for structural purposes may be considered as a series of bonded 'I' sections or box sections. The voids may be useful for housing services, but any access holes in the construction must not disturb the integrity of the wall. The voids may also be filled with insulation to reduce heat energy losses from the building, and to prevent air circulatory heat losses within the voids. Where thermal insulation standards apply, this type of wall will have limitations as the cross ribs will provide a route for cold bridging. U values will increase by about 10% compared with conventional cavity wall construction of the same materials.

Ref. BS 5628-1: Code of practice for use of masonry. Structural use of unreinforced masonry.
   BS 5628-3: Code of practice for use of masonry. Materials and components, design and workmanship.

Function – the primary function of any damp-proof course (dpc) or damp-proof membrane (dpm) is to provide an impermeable barrier to the passage of moisture. The three basic ways in which damp-proof courses are used is to:-

1. Resist moisture penetration from below (rising damp).
2. Resist moisture penetration from above.
3. Resist moisture penetration from horizontal entry.

Typical examples ~

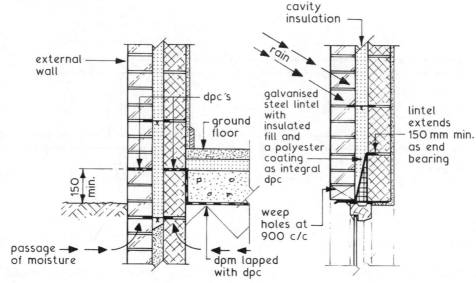

PENETRATION FROM BELOW
(Ground Floor/External Wall)

PENETRATION FROM ABOVE
(Window/Door Head)

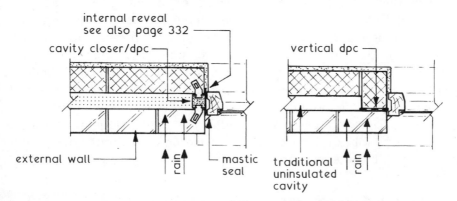

HORIZONTAL ENTRY (Window/Door Jamb)

See also: BS's 743, 8102 and 8215.

Building Regulations, Approved Document C2, Section 5:
A wall may be built with a 'damp-proof course of bituminous material, engineering bricks or slates in cement mortar, or any other material that will prevent the passage of moisture.'

Material			Remarks
Lead	BS EN 12588	Code 4 (1·8 mm)	May corrode in the presence of mortar. Both surfaces to be coated with bituminous paint. Workable for application to cavity trays, etc.
Copper	BS EN 1172	0·25 mm	Can cause staining to adjacent masonry. Resistant to corrosion.
Bitumen in various bases:	BS 6398		Hessian or fibre may decay with age, but this will not affect efficiency.
Hessian		3·8 kg/m²	Tearable if not
Fibre		3·3	protected. Lead bases
Asbestos		3·8	are suited where there
Hessian & lead		4·4	may be a high degree of
Fibre & lead		4·4	movement in the wall. Asbestos is now prohibited.
LDPE (polyethylene)	BS 6515	0·46 mm	No deterioration likely, but may be difficult to bond, hence the profiled surface finish. Not suited under light loads.
Bitumen polymer and pitch polymer		1·10 mm	Absorbs movement well. Joints and angles made with product manufacturer's adhesive tape.
Polypropylene BS 5139 1.5 to 2.0 mm			Preformed dpc for cavity trays, cloaks, direction changes and over lintels.

Note: All the above dpcs to be lapped at least 100 mm at joints and adhesive sealed. Dpcs should be continuous with any dpm in the floor.

Material			Remarks
Mastic asphalt	BS 6925	12 kg/m²	Does not deteriorate. Requires surface treatment with sand or scoring to effect a mortar key.
Engineering bricks	BS EN 771-1 BS EN 772-7	<4·5% absorption	Min. 2 courses laid breaking joint in cement mortar 1:3. No deterioration, but may not blend with adjacent facings.
Slate	BS EN 12326-1	4 mm	Min. 2 courses laid as above. Will not deteriorate, but brittle so may fracture if building settles.

Refs:

BS 743: Specification for materials for damp-proof courses.

BS 5268-3: Code of practice for use of masonry. Materials and components, design and workmanship.

BS 8102: Code of practice for protection of structures against water from the ground.

BS 8215: Code of practice for design and installation of damp-proof courses in masonry construction.

BRE Digest 380: Damp-proof courses.

Note: It was not until the Public Health Act of 1875, that it became mandatory to instal damp-proof courses in new buildings. Structures constructed before that time, and those since, which have suffered dpc failure due to deterioration or incorrect installation, will require remedial treatment. This could involve cutting out the mortar bed joint two brick courses above ground level in stages of about 1m in length. A new dpc can then be inserted with mortar packing, before proceeding to the next length. No two adjacent sections should be worked consecutively. This process is very time consuming and may lead to some structural settlement. Therefore, the measures explained on the following two pages are usually preferred.

Materials – Silicone solutions in organic solvent.
   Aluminium stearate solutions.
   Water soluble silicone formulations (siliconates).

Methods – High pressure injection (0·70 – 0·90 MPa) solvent based.
   Low pressure injection (0·15 – 0·30 MPa) water based.
   Gravity feed, water based.
   Insertion/injection, mortar based.

Pressure injection – 12 mm diameter holes are bored to about two-thirds the depth of masonry, at approximately 150 mm horizontal intervals at the appropriate depth above ground (normally 2–3 brick courses). These holes can incline slightly downwards. With high (low) pressure injection, walls in excess of 120 mm (460 mm) thickness should be drilled from both sides. The chemical solution is injected by pressure pump until it exudes from the masonry. Cavity walls are treated as each leaf being a solid wall.

Gravity feed – 25 mm diameter holes are bored as above. Dilute chemical is transfused from containers which feed tubes inserted in the holes. This process can take from a few hours to several days to effect. An alternative application is insertion of frozen pellets placed in the bore holes. On melting, the solution disperses into the masonry to be replaced with further pellets until the wall is saturated.

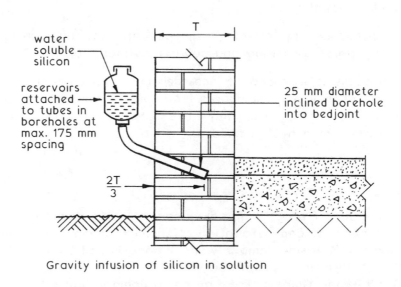

water soluble silicon

reservoirs attached to tubes in boreholes at max. 175 mm spacing

25 mm diameter inclined borehole into bedjoint

T

$\frac{2T}{3}$

Gravity infusion of silicon in solution

Injection mortars – 19mm diameter holes are bored from both sides of a wall, at the appropriate level and no more than 230mm apart horizontally, to a depth equating to three-fifths of the wall thickness. They should be inclined downwards at an angle of 20 to 30°. The drill holes are flushed out with water, before injecting mortar from the base of the hole and outwards. This can be undertaken with a hand operated caulking gun. Special cement mortars contain styrene butadiene resin (SDR) or epoxy resin and must be mixed in accordance with the manufacturer's guidance.

Notes relating to all applications of chemical dpcs:

* Before commencing work, old plasterwork and rendered undercoats are removed to expose the masonry. This should be to a height of at least 300mm above the last detectable (moisture meter reading) signs of rising dampness (1 metre min.).

* If the wall is only accessible from one side and both sides need treatment, a second deeper series of holes may be bored from one side, to penetrate the inaccessible side.

* On completion of work, all boreholes are made good with cement mortar. Where dilute chemicals are used for the dpc, the mortar is rammed the full length of the hole with a piece of timber dowelling.

* The chemicals are effective by bonding to, and lining the masonry pores by curing and solvent evaporation.

* The process is intended to provide an acceptable measure of control over rising dampness. A limited amount of water vapour may still rise, but this should be dispersed by evaporation in a heated building.

Refs.

BS 6576: Code of practice for diagnosis of rising damp in walls of buildings and installation of chemical damp-proof courses.
BRE Digest 245: Rising damp in walls: diagnosis and treatment.
BRE Digest 380: Damp-proof courses.
BRE Good Repair Guide 6: Treating rising damp in houses.

In addition to damp-proof courses failing due to deterioration or damage, they may be bridged as a result of:

* Faults occurring during construction.
* Work undertaken after construction, with disregard for the damp-proof course.

Typical examples ~

Solid walls

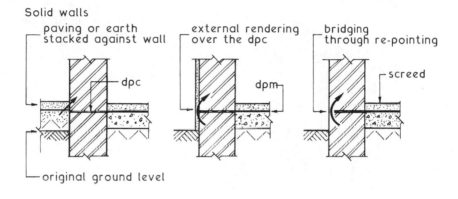

Cavity walls

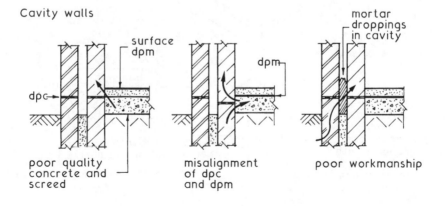

Thermal insulation regulations may require insulating dpcs to prevent cold bridging around window and door openings in cavity wall construction (see pages 452 and 453). By locating a vertical dpc with a bonded insulant at the cavity closure, the dpc prevents penetration of dampness from the outside, and the insulation retains the structural temperature of the internal reveal. This will reduce heat losses by maintaining the temperature above dewpoint, preventing condensation, wall staining and mould growth.

Application –

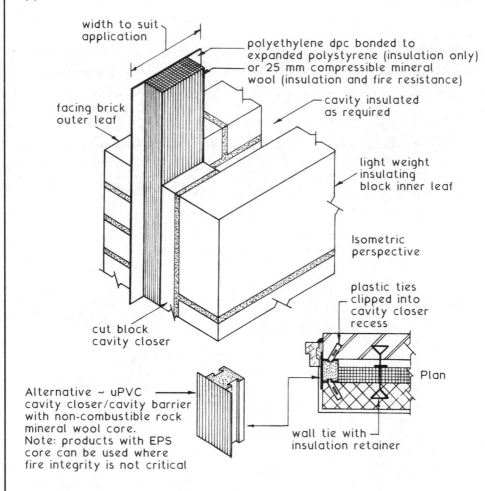

width to suit application

polyethylene dpc bonded to expanded polystyrene (insulation only) or 25 mm compressible mineral wool (insulation and fire resistance)

facing brick outer leaf

cavity insulated as required

light weight insulating block inner leaf

Isometric perspective

plastic ties clipped into cavity closer recess

cut block cavity closer

Plan

Alternative ~ uPVC cavity closer/cavity barrier with non-combustible rock mineral wool core. Note: products with EPS core can be used where fire integrity is not critical

wall tie with insulation retainer

Refs. Building Regulations, Approved Document L: Conservation of fuel and power.
BRE Report – Thermal Insulation: avoiding risks (3rd. ed.).
Building Regulations, Approved Document B3,
Section 6 (Vol. 1): Concealed spaces (cavities).

Penetrating Gases ~ Methane and Radon

Methane – methane is produced by deposited organic material decaying in the ground. It often occurs with carbon dioxide and traces of other gases to form a cocktail known as landfill gas. It has become an acute problem in recent years, as planning restrictions on 'green-field' sites have forced development of derelict and reclaimed 'brown-field' land.

The gas would normally escape to the atmosphere, but under a building it pressurizes until percolating through cracks, cavities and junctions with services. Being odourless, it is not easily detected until contacting a naked flame, then the result is devastating!

Radon ~ a naturally occurring colour/odourless gas produced by radioactive decay of radium. It originates in uranium deposits of granite subsoils as far apart as the south-west and north of England and the Grampian region of Scotland. Concentrations of radon are considerably increased if the building is constructed of granite masonry. The combination of radon gas and the tiny radioactive particles known as radon daughters are inhaled. In some people with several years' exposure, research indicates a high correlation with cancer related illness and death.

Protection of buildings and the occupants from subterranean gases can be achieved by passive or active measures incorporated within the structure.

1. Passive protection consists of a complete airtight seal integrated within the ground floor and walls. A standard LDPE damp proof membrane of 0.3mm thickness should be adequate if carefully sealed at joints, but thicknesses up to 1mm are preferred, combined with foil and/or wire reinforcement.

2. Active protection requires installation of a permanently running extract fan connected to a gas sump below the ground floor. It is an integral part of the building services system and will incur operating and maintenance costs throughout the building's life.

(See next page for construction details)

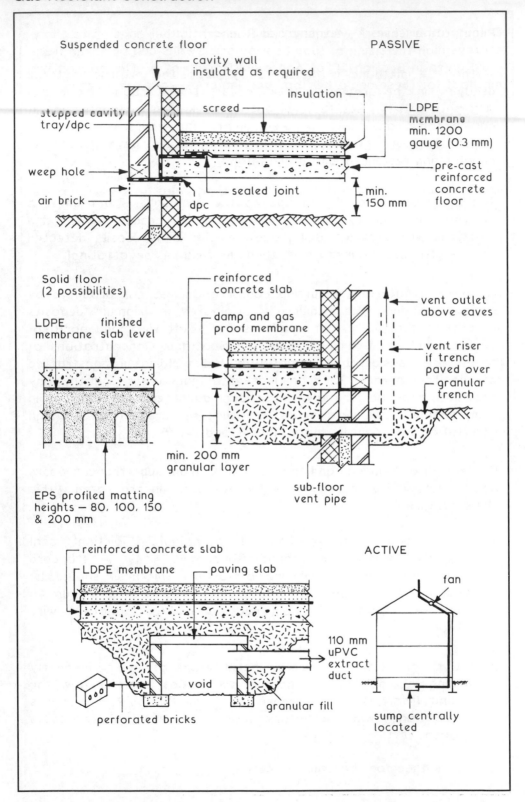

Suspended concrete floor — PASSIVE

cavity wall insulated as required

stepped cavity tray/dpc

insulation

screed

LDPE membrane min. 1200 gauge (0.3 mm)

weep hole

air brick

dpc

sealed joint

pre-cast reinforced concrete floor

min. 150 mm

Solid floor (2 possibilities)

LDPE membrane   finished slab level

reinforced concrete slab

damp and gas proof membrane

vent outlet above eaves

vent riser if trench paved over

granular trench

min. 200 mm granular layer

sub-floor vent pipe

EPS profiled matting heights — 80, 100, 150 & 200 mm

reinforced concrete slab

LDPE membrane   paving slab

ACTIVE

fan

110 mm uPVC extract duct

void

perforated bricks

granular fill

sump centrally located

Calculated Brickwork ~ for small and residential buildings up to three storeys high the sizing of load bearing brick walls can be taken from data given in Section 2C of Approved Document A. The alternative methods for these and other load bearing brick walls are given in:

BS 5628-1: Code of practice for the use of masonry. Structural use of unreinforced masonry, and

BS 8103-2: Structural design of low rise buildings. Code of practice for masonry walls for housing.

The main factors governing the loadbearing capacity of brick walls and columns are:-

1. Thickness of wall.
2. Strength of bricks used.
3. Type of mortar used.
4. Slenderness ratio of wall or column.
5. Eccentricity of applied load.

Thickness of wall ~ this must always be sufficient throughout its entire body to carry the design loads and induced stresses. Other design requirements such as thermal and sound insulation properties must also be taken into account when determining the actual wall thickness to be used.

Effective Thickness ~ this is the assumed thickness of the wall or column used for the purpose of calculating its slenderness ratio – see page 337.

Typical Examples ~

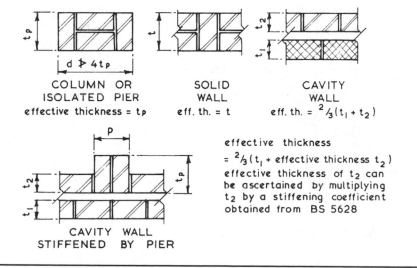

COLUMN OR ISOLATED PIER
effective thickness = t_P

SOLID WALL
eff. th. = t

CAVITY WALL
eff. th. = $\frac{2}{3}(t_1 + t_2)$

CAVITY WALL STIFFENED BY PIER

effective thickness
= $\frac{2}{3}(t_1 +$ effective thickness $t_2)$
effective thickness of t_2 can be ascertained by multiplying t_2 by a stiffening coefficient obtained from BS 5628

Principles of Calculated Brickwork

Strength of Bricks ~ due to the wide variation of the raw materials and methods of manufacture bricks can vary greatly in their compressive strength. The compressive strength of a particular type of brick or batch of bricks is taken as the arithmetic mean of a sample of ten bricks tested in accordance with the appropriate British Standard. A typical range for clay bricks would be from 20 to 170 MN/m² the majority of which would be in the 20 to 90 MN/m² band. Generally calcium silicate bricks have a lower compressive strength than clay bricks with a typical strength range of 10 to 65 MN/m².

Strength of Mortars ~ mortars consist of an aggregate (sand) and a binder which is usually cement; cement plus additives to improve workability; or cement and lime. The factors controlling the strength of any particular mix are the ratio of binder to aggregate plus the water:cement ratio. The strength of any particular mix can be ascertained by taking the arithmetic mean of a series of test cubes or prisms (BS EN 196 and BS EN 1015).

Wall Design Strength ~ the basic stress of any brickwork depends on the crushing strength of the bricks and the type of mortar used to form the wall unit. This relationship can be plotted on a graph using data given in BS 5628 as shown below:-

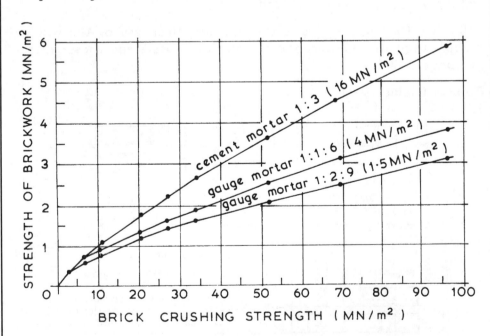

Note: 1 MN/m² equates to 1 N/mm²

336

Slenderness Ratio ~ this is the relationship of the effective height to the effective thickness thus:-

$$\text{Slenderness ratio} = \frac{\text{effective height}}{\text{effective thickness}} = \frac{h}{t} \not> 27 \text{ see BS 5628}$$

Effective Height ~ this is the dimension taken to calculate the slenderness ratio as opposed to the actual height.

Typical Examples – actual height = H   effective height = h

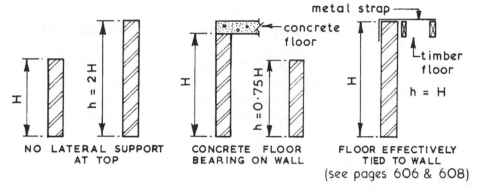

NO LATERAL SUPPORT AT TOP   CONCRETE FLOOR BEARING ON WALL   FLOOR EFFECTIVELY TIED TO WALL

(see pages 606 & 608)

Effective Thickness ~ this is the dimension taken to calculate the slenderness ratio as opposed to the actual thickness.

Typical Examples – actual thickness = T effective thickness = t

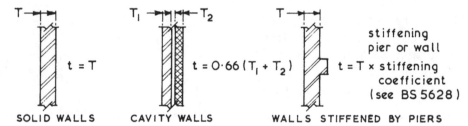

SOLID WALLS   CAVITY WALLS   WALLS STIFFENED BY PIERS

Stress Reduction ~ the permissible stress for a wall is based on the basic stress multiplied by a reduction factor related to the slenderness factor and the eccentricity of the load:-

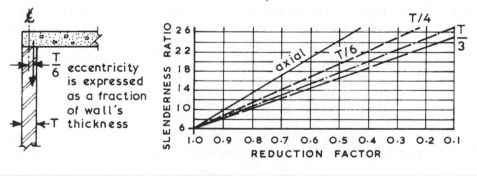

Lime ~ traditional mortars are a combination of lime, sand and water. These mixes are very workable and have sufficient flexibility to accommodate a limited amount of wall movement due to settlement, expansion and contraction. The long term durability of lime mortars is poor as they can break down in the presence of atmospheric contaminants and surface growths. Nevertheless, lime is frequently specified as a supplementary binder with cement, to increase mix workability and to reduce the possibility of joint shrinkage and cracking, a characteristic of stronger cement mortars.

Cement ~ the history of cement type mortar products is extensive. Examples dating back to the Mesopotamians and the Egyptians are not unusual; one of the earliest examples from over 10000 years ago has been found in Galilee, Israel. Modern mortars are made with Portland cement, the name attributed to a bricklayer named Joseph Aspdin. In 1824 he patented his improved hydraulic lime product as Portland cement, as it resembled Portland stone in appearance. It was not until the 1920s that Portland cement, as we now know it, was first produced commercially by mixing a slurry of clay (silica, alumina and iron-oxides) with limestone (calcium carbonate). The mix is burnt in a furnace (calcinated) and the resulting clinker crushed and bagged.

Mortar ~ mixes for masonry should have the following properties:

* Adequate strength
* Workability
* Water retention during laying
* Plasticity during application
* Adhesion or bond
* Durability
* Good appearance ~ texture and colour

Modern mortars are a combination of cement, lime and sand plus water. Liquid plasticisers exist as a substitute for lime, to improve workability and to provide some resistance to frost when used during winter.

Masonry cement ~ these proprietary cements generally contain about 75% Portland cement and about 25% of fine limestone filler with an air entraining plasticiser. Allowance must be made when specifying the mortar constituents to allow for the reduced cement content. These cements are not suitable for concrete.

Refs. BS 6463-101, 102 and 103: Quicklime, hydrated lime and natural calcium carbonate.
BS EN 197-1: Cement. Composition, specifications and conformity criteria for common cements.

338

Ready mixed mortar ~ this is delivered dry for storage in purpose made silos with integral mixers as an alternative to site blending and mixing. This ensures:

* Guaranteed factory quality controlled product
* Convenience
* Mix consistency between batches
* Convenient facility for satisfying variable demand
* Limited wastage
* Optimum use of site space

Mortar and cement strength ~ see also page 336. Test samples are made in prisms of 40×40 mm cross section, 160 mm long. At 28 days samples are broken in half to test for flexural strength. The broken pieces are subject to a compression test across the 40 mm width. An approximate comparison between mortar strength (MN/m^2 or N/mm^2), mortar designations (i to v) and proportional mix ratios is shown in the classification table below. Included is guidance on application.

Proportional mixing of mortar constituents by volume is otherwise known as a prescribed mix or simply a recipe.

Mortar classification ~

Traditional designation	BS EN 998-2 Strength	Proportions by volume cement/lime/sand	cement/sand	Application
i	12	1:0.25:3	1:3	Exposed external
ii	6	1:0.5:4–4.5	1:3–4	General external
iii	4	1:1:5–6	1:5–6	Sheltered internal
iv	2	1:2:8–9	1:7–8	General internal
v	–	1:3:10–12	1:9–10	Internal, grouting

Relevant standards:

BS 5628-3: Code of practice for use of masonry. Materials and components, design and workmanship.
BS EN 196: Methods of testing cement.
BS EN 998-2: Specification for mortar for masonry. Masonry mortar.
PD 6678: Guide to the specification of masonry mortar.
BS EN 1015: Methods of test for mortar for masonry.

Supports Over Openings ~ the primary function of any support over an opening is to carry the loads above the opening and transmit them safely to the abutments, jambs or piers on both sides. A support over an opening is usually required since the opening infilling such as a door or window frame will not have sufficient strength to carry the load through its own members.

Type of Support ~

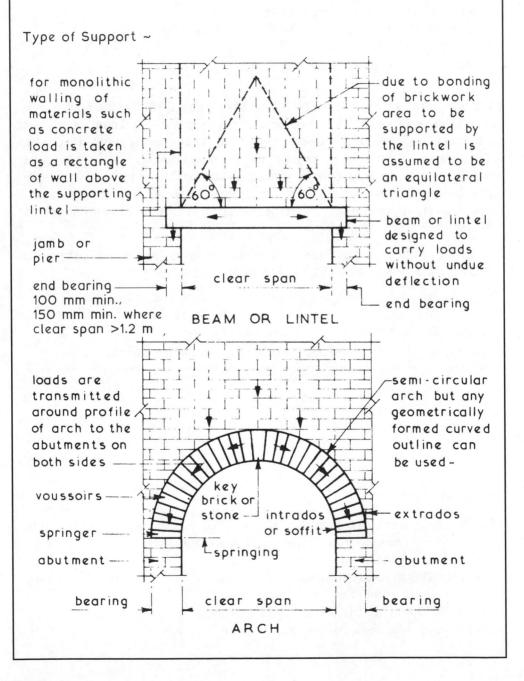

for monolithic walling of materials such as concrete load is taken as a rectangle of wall above the supporting lintel

due to bonding of brickwork area to be supported by the lintel is assumed to be an equilateral triangle

60°   60°

jamb or pier

beam or lintel designed to carry loads without undue deflection

end bearing

end bearing 100 mm min., 150 mm min. where clear span >1.2 m

clear span

BEAM OR LINTEL

loads are transmitted around profile of arch to the abutments on both sides

semi-circular arch but any geometrically formed curved outline can be used –

voussoirs

key brick or stone

intrados or soffit

extrados

springer

springing

abutment

abutment

bearing

clear span

bearing

ARCH

Arch Construction ~ by the arrangement of the bricks or stones in an arch over an opening it will be self supporting once the jointing material has set and gained adequate strength. The arch must therefore be constructed over a temporary support until the arch becomes self supporting. The traditional method is to use a framed timber support called a centre. Permanent arch centres are also available for small spans and simple formats.

Typical Arch Formats ~

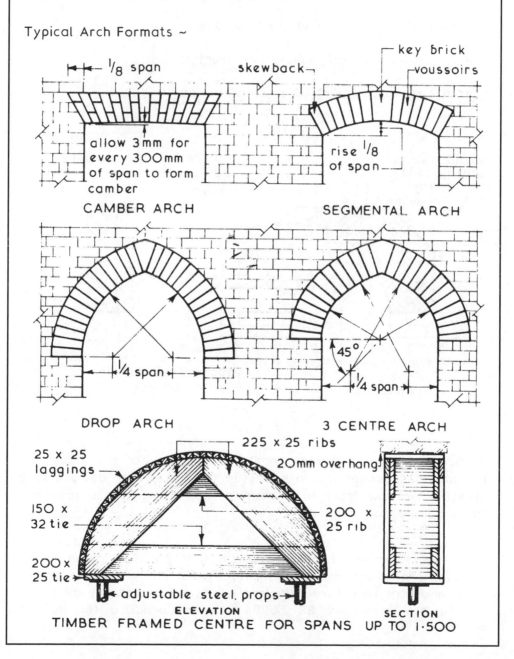

CAMBER ARCH

SEGMENTAL ARCH

DROP ARCH

3 CENTRE ARCH

TIMBER FRAMED CENTRE FOR SPANS UP TO 1·500

The profile of an arch does not lend itself to simple positioning of a damp proof course. At best, it can be located horizontally at upper extrados level. This leaves the depth of the arch and masonry below the dpc vulnerable to dampness. Proprietary galvanised or stainless steel cavity trays resolve this problem by providing:

* Continuity of dpc around the extrados.

* Arch support/centring during construction.

* Arch and wall support after construction.

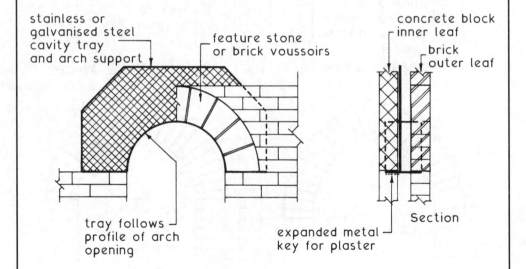

stainless or galvanised steel cavity tray and arch support

feature stone or brick voussoirs

concrete block inner leaf

brick outer leaf

tray follows profile of arch opening

expanded metal key for plaster

Section

Standard profiles are made to the traditional outlines shown on the previous two pages, in spans up to 2 m. Other options may also be available from some manufacturers. Irregular shapes and spans can be made to order.

Note: Arches in semi-circular, segmental or parabolic form up to 2m span can be proportioned empirically. For integrity of structure it is important to ensure sufficient provision of masonry over and around any arch, see BS 5628: Code of practice for use of masonry.

The example in steel shown on the preceding page combines structural support with a damp proof course, without the need for temporary support from a centre. Where traditional centring is retained, a lightweight preformed polypropylene cavity tray/dpc can be used. These factory made plastic trays are produced in various thicknesses of 1.5 to 3mm relative to spans up to about 2 m. Arch centres are made to match the tray profile and with care can be reused several times.

An alternative material is code 4 lead sheet*. Lead is an adaptable material but relatively heavy. Therefore, its suitability is limited to small spans particularly with non-standard profiles.

*BS EN 12588: Lead and lead alloys. Rolled lead sheet for building purposes. Lead sheet is coded numerically from 3 to 8, which closely relates to the traditional specification in lbs./sq. ft.

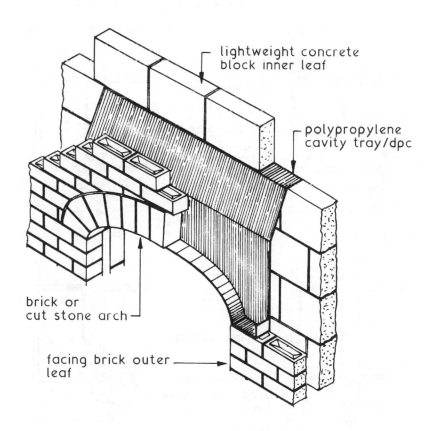

lightweight concrete block inner leaf

polypropylene cavity tray/dpc

brick or cut stone arch

facing brick outer leaf

Ref. BS 5628-3: Code of practice for the use of masonry. Materials and components, design and workmanship.

Openings ~ these consist of a head, jambs and sill and the different methods and treatments which can be used in their formation is very wide but they are all based on the same concepts. Application limited – see pages 452 and 453.

Typical Head Details ~

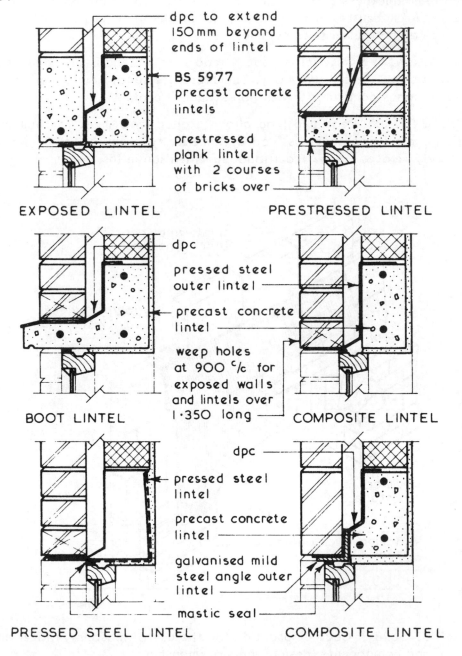

dpc to extend 150 mm beyond ends of lintel

BS 5977 precast concrete lintels

prestressed plank lintel with 2 courses of bricks over

EXPOSED LINTEL   **PRESTRESSED LINTEL**

dpc

pressed steel outer lintel

precast concrete lintel

weep holes at 900 ℃ for exposed walls and lintels over 1·350 long

BOOT LINTEL   **COMPOSITE LINTEL**

dpc

pressed steel lintel

precast concrete lintel

galvanised mild steel angle outer lintel

mastic seal

PRESSED STEEL LINTEL   **COMPOSITE LINTEL**

Jambs ~ these may be bonded as in solid walls or unbonded as in cavity walls. The latter must have some means of preventing the ingress of moisture from the outer leaf to the inner leaf and hence the interior of the building.

Application limited – see pages 452 and 453.

Typical Jamb Details ~

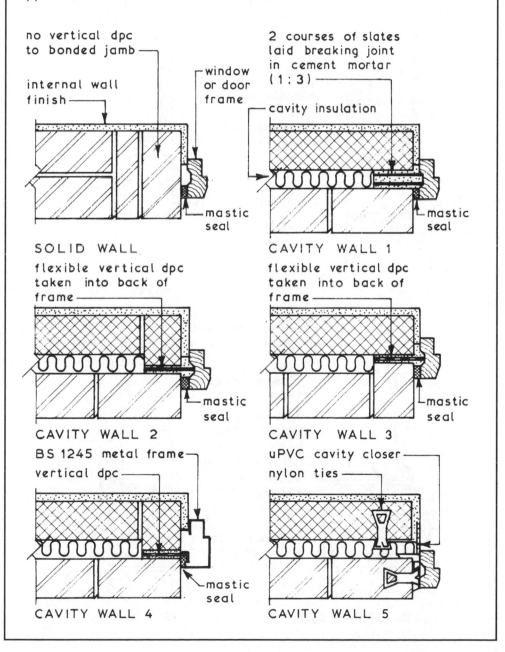

345

Sills ~ the primary function of any sill is to collect the rainwater which has run down the face of the window or door and shed it clear of the wall below.

Application limited – see pages 452 and 453.

Typical Sill Details ~

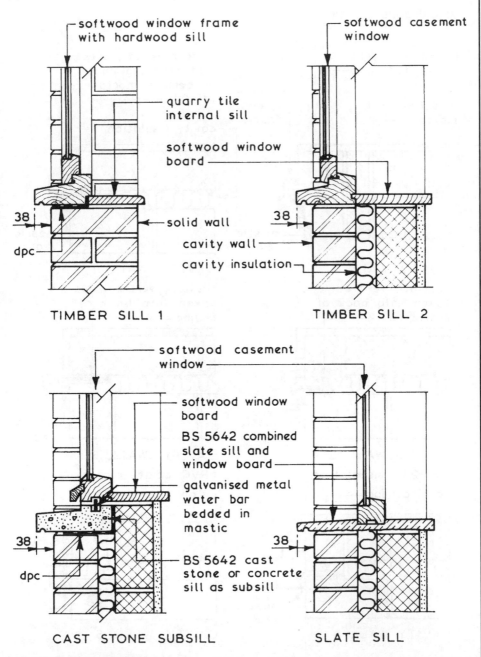

TIMBER SILL 1

TIMBER SILL 2

CAST STONE SUBSILL

SLATE SILL

Traditional Construction – checked rebates or recesses in masonry solid walls were often provided at openings to accommodate door and window frames. This detail was used as a means to complement frame retention and prevent weather intrusion.

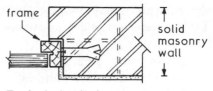

Typical checked masonry (pre 1940s)

Exposure Zones – checked reveal treatment is now required mainly where wind-driven rain will have most impact. This is primarily in the south west and west coast areas of the British Isles, plus some isolated inland parts that will be identified by their respective local authorities.

Typical Checked Opening Details –

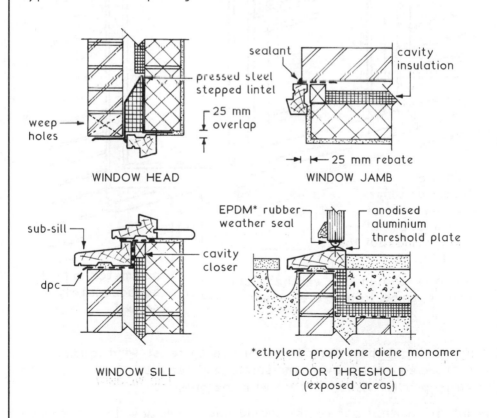

Ref. Building Regulations, Approved Document C2: Resistance to moisture. Driving rain exposure zones 3 and 4.

347

A window must be aesthetically acceptable in the context of building design and surrounding environment

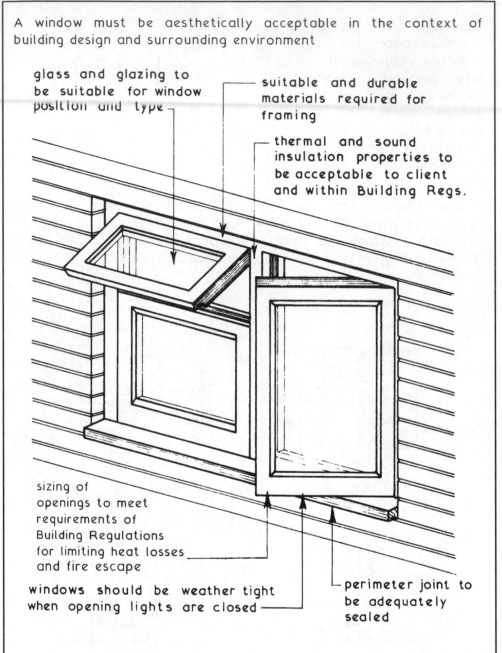

glass and glazing to be suitable for window position and type

suitable and durable materials required for framing

thermal and sound insulation properties to be acceptable to client and within Building Regs.

sizing of openings to meet requirements of Building Regulations for limiting heat losses and fire escape

windows should be weather tight when opening lights are closed

perimeter joint to be adequately sealed

Windows should be selected or designed to resist wind loadings, be easy to clean and provide for safety and security. They should be sited to provide visual contact with the outside.

Habitable upper floor rooms should have a window for emergency escape. Min. opening area, 0.330 m². Min. height and width, 0.450 m. Max height of opening, 1.100 m above floor.

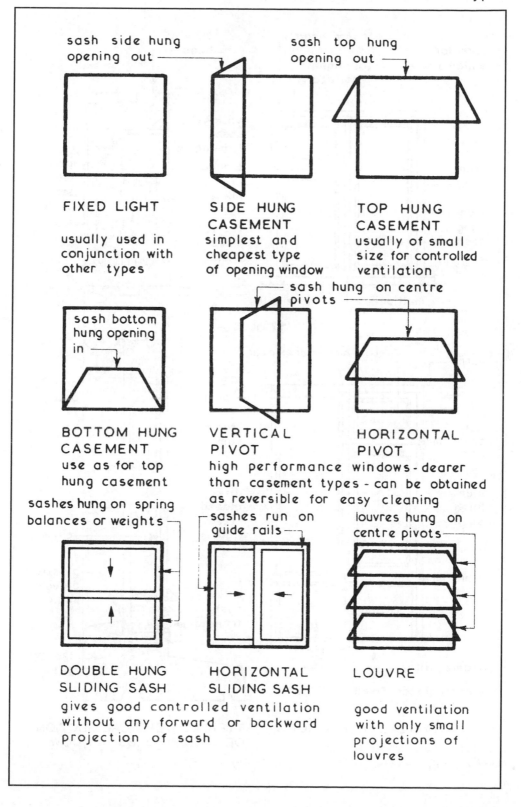

sash side hung
opening out —

sash top hung
opening out —

FIXED LIGHT

usually used in
conjunction with
other types

SIDE HUNG
CASEMENT
simplest and
cheapest type
of opening window

TOP HUNG
CASEMENT
usually of small
size for controlled
ventilation

sash bottom
hung opening
in —

sash hung on centre
pivots —

BOTTOM HUNG
CASEMENT
use as for top
hung casement

VERTICAL
PIVOT

HORIZONTAL
PIVOT

high performance windows - dearer
than casement types - can be obtained
as reversible for easy cleaning

sashes hung on spring
balances or weights —

sashes run on
guide rails —

louvres hung on
centre pivots —

DOUBLE HUNG
SLIDING SASH

HORIZONTAL
SLIDING SASH

LOUVRE

gives good controlled ventilation
without any forward or backward
projection of sash

good ventilation
with only small
projections of
louvres

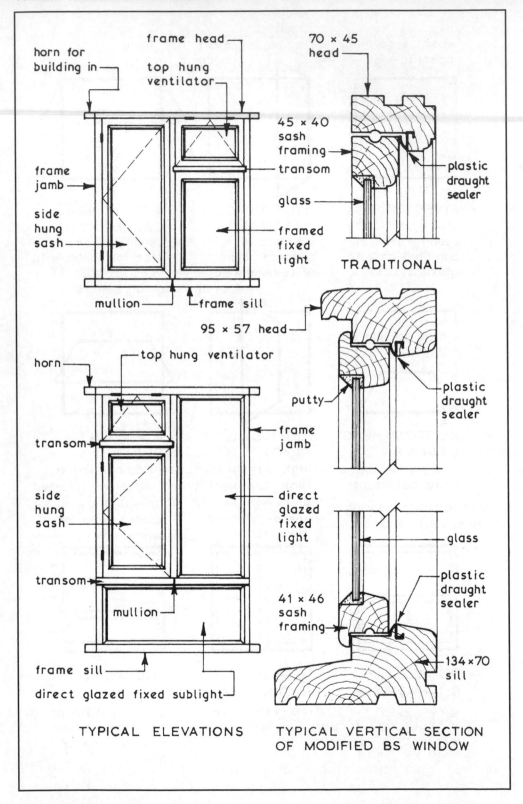

horn for building in

frame head

top hung ventilator

70 × 45 head

45 × 40 sash framing

transom

glass

plastic draught sealer

frame jamb

side hung sash

framed fixed light

mullion

frame sill

TRADITIONAL

95 × 57 head

putty

plastic draught sealer

horn

top hung ventilator

transom

frame jamb

side hung sash

direct glazed fixed light

glass

transom

mullion

plastic draught sealer

frame sill

41 × 46 sash framing

direct glazed fixed sublight

134 × 70 sill

TYPICAL ELEVATIONS

TYPICAL VERTICAL SECTION OF MODIFIED BS WINDOW

The standard range of casement windows used in the UK was derived from the English Joinery Manufacturer's Association (EJMA) designs of some 50 years ago. These became adopted in BS 644: Timber windows. Specification for factory assembled windows of various types. A modified type is shown on the preceding page. Contemporary building standards require higher levels of performance in terms of thermal and sound insulation (Bldg. Regs. Pt. L and E), air permeability, water tightness and wind resistance (BS ENs 1026, 1027 and 12211, respectively). This has been achieved by adapting Scandinavian designs with double and triple glazing to attain U values as low as 1·2 W/m²K and a sound reduction of 50 dB.

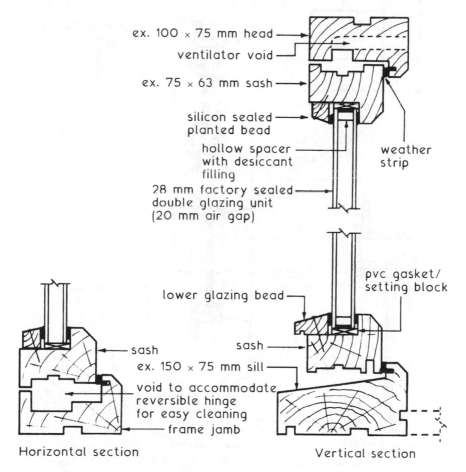

ex. 100 × 75 mm head
ventilator void
ex. 75 × 63 mm sash
silicon sealed planted bead
hollow spacer with desiccant filling
28 mm factory sealed double glazing unit (20 mm air gap)
weather strip
pvc gasket/ setting block
lower glazing bead
sash
sash
ex. 150 × 75 mm sill
void to accommodate reversible hinge for easy cleaning
frame jamb

Horizontal section   Vertical section

Further refs:
BS 6375: Performance of windows.
BS 6375-1: Classification for weather tightness.
BS 6375-2: Operation and strength characteristics.
BS 7950: Specification for enhanced security performance.

Metal Windows ~ these can be obtained in steel (BS 6510) or in aluminium alloy (BS 4873). Steel windows are cheaper in initial cost than aluminium alloy but have higher maintenance costs over their anticipated life, both can be obtained fitted into timber subframes. Generally they give a larger glass area for any given opening size than similar timber windows but they can give rise to condensation on the metal components.

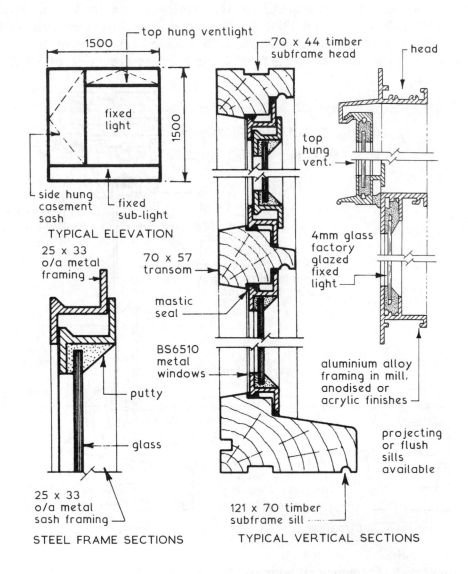

TYPICAL ELEVATION

STEEL FRAME SECTIONS

TYPICAL VERTICAL SECTIONS

Refs.:
BS 4873: Aluminium alloy windows.
BS 6510: Steel-framed windows and glazed doors.

Timber Windows ~ wide range of ironmongery available which can be factory fitted or supplied and fixed on site.

Metal Windows ~ ironmongery usually supplied with and factory fitted to the windows.

Typical Examples ~

malleable iron, curly tail pattern

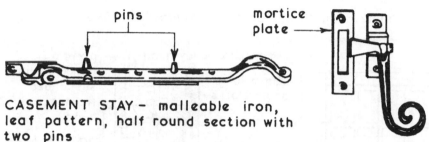

pins

mortice plate

CASEMENT STAY - malleable iron, leaf pattern, half round section with two pins
Sizes: 200; 250 and 300 mm

CASEMENT FASTENER

hot pressed aluminium, plain end pattern

wedge plate

CASEMENT STAY - cast aluminium, plain end pattern with one pin
Sizes: 250 and 300 mm

CASEMENT FASTENER

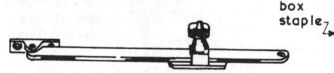

box staple

hot pressed brass

CASEMENT STAY - steel and brass, sliding screw down pattern
Sizes: 250 and 300 mm

VENTLIGHT CATCH
used with bottom hung ventlights

CASEMENT STAY - steel, stayput pattern
Arm Sizes: 100; 140 and 175 mm

malleable iron or brass
Sizes: 150 175 and 200mm

QUADRANT STAY

Sliding Sash Windows ~ these are an alternative format to the conventional side hung casement windows and can be constructed as a vertical or double hung sash window or as a horizontal sliding window in timber, metal, plastic or in any combination of these materials. The performance and design functions of providing daylight, ventilation, vision out, etc., are the same as those given for traditional windows in Windows – Performance Requirements on page 348.

Typical Double Hung Weight Balanced Window Details ~

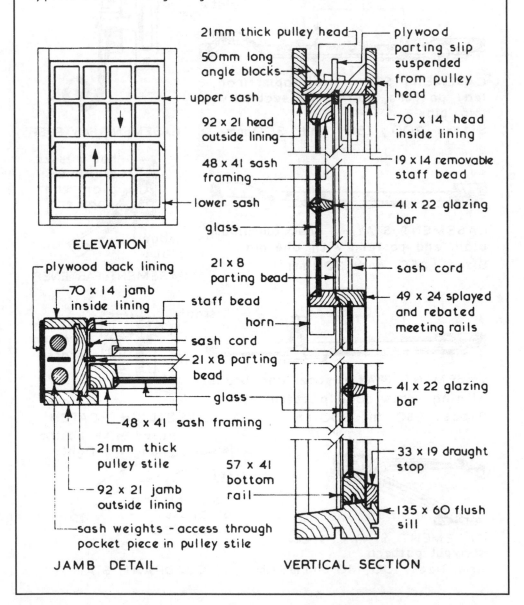

ELEVATION

JAMB DETAIL

VERTICAL SECTION

21mm thick pulley head

50mm long angle blocks

upper sash

92 x 21 head outside lining

48 x 41 sash framing

lower sash

glass

plywood parting slip suspended from pulley head

70 x 14 head inside lining

19 x 14 removable staff bead

41 x 22 glazing bar

21 x 8 parting bead

staff bead

horn

sash cord

21 x 8 parting bead

glass

48 x 41 sash framing

sash cord

49 x 24 splayed and rebated meeting rails

41 x 22 glazing bar

plywood back lining

70 x 14 jamb inside lining

21mm thick pulley stile

92 x 21 jamb outside lining

sash weights – access through pocket piece in pulley stile

57 x 41 bottom rail

33 x 19 draught stop

135 x 60 flush sill

Double Hung Sash Windows ~ these vertical sliding sash windows come in two formats when constructed in timber. The weight balanced format is shown on the preceding page, the alternative spring balanced type is illustrated below. Both formats are usually designed and constructed to the recommendations set out in BS 644.

Typical Double Hung Spring Balanced Window Details ~

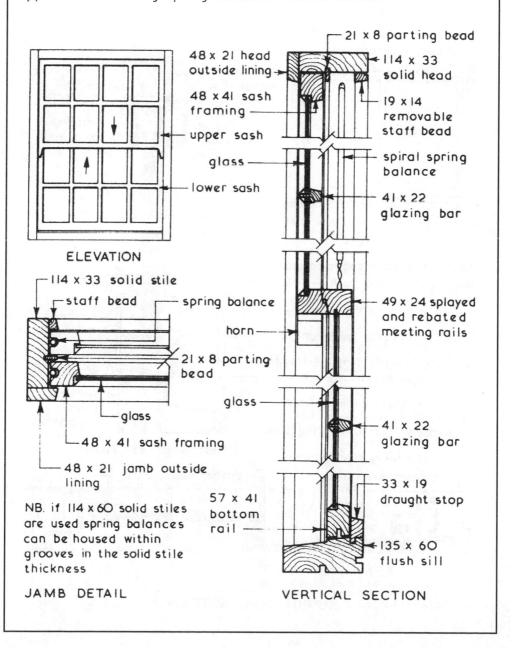

ELEVATION

48 x 21 head outside lining

48 x 41 sash framing

upper sash

glass

lower sash

21 x 8 parting bead

114 x 33 solid head

19 x 14 removable staff bead

spiral spring balance

41 x 22 glazing bar

114 x 33 solid stile

staff bead

spring balance

21 x 8 parting bead

glass

48 x 41 sash framing

48 x 21 jamb outside lining

NB. if 114 x 60 solid stiles are used spring balances can be housed within grooves in the solid stile thickness

JAMB DETAIL

horn

49 x 24 splayed and rebated meeting rails

glass

41 x 22 glazing bar

57 x 41 bottom rail

33 x 19 draught stop

135 x 60 flush sill

VERTICAL SECTION

Horizontally Sliding Sash Windows ~ these are an alternative format to the vertically sliding or double hung sash windows shown on pages 354 & 355 and can be constructed in timber, metal, plastic or combinations of these materials with single or double glazing. A wide range of arrangements are available with two or more sliding sashes which can have a ventlight incorporated in the outer sliding sash.

Typical Horizontally Sliding Sash Window Details ~

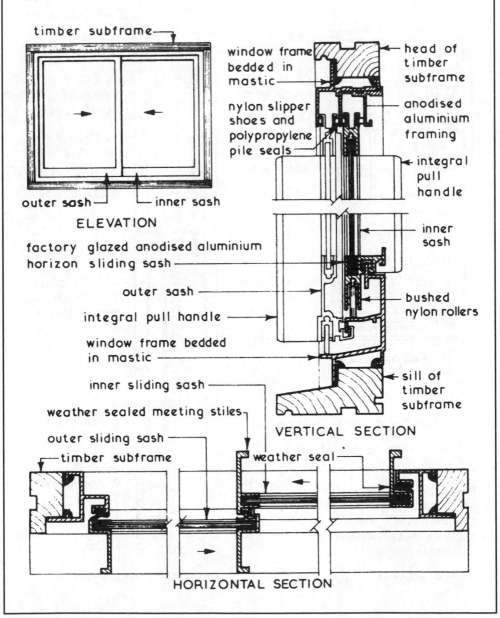

timber subframe

outer sash — — inner sash

ELEVATION

window frame bedded in mastic

nylon slipper shoes and polypropylene pile seals

head of timber subframe

anodised aluminium framing

integral pull handle

inner sash

factory glazed anodised aluminium horizon sliding sash

outer sash

integral pull handle

window frame bedded in mastic

inner sliding sash

weather sealed meeting stiles

outer sliding sash

timber subframe

weather seal

bushed nylon rollers

sill of timber subframe

VERTICAL SECTION

HORIZONTAL SECTION

Pivot Windows ~ like other windows these are available in timber, metal, plastic or in combinations of these materials.

They can be constructed with centre jamb pivots enabling the sash to pivot or rotate in the horizontal plane or alternatively the pivots can be fixed in the head and sill of the frame so that the sash rotates in the vertical plane.

Typical Example ~

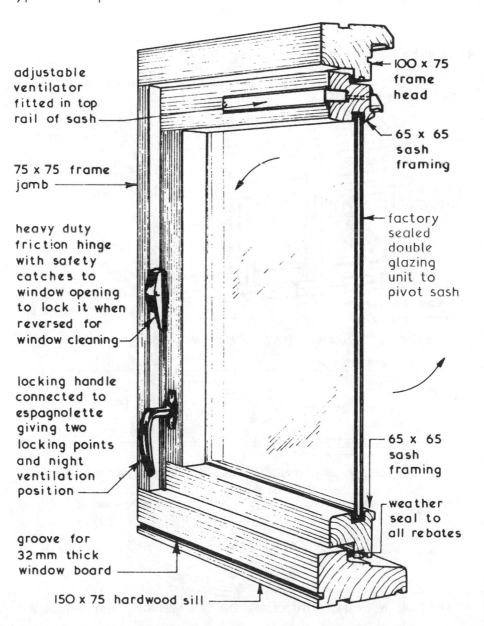

adjustable ventilator fitted in top rail of sash

100 x 75 frame head

65 x 65 sash framing

75 x 75 frame jamb

factory sealed double glazing unit to pivot sash

heavy duty friction hinge with safety catches to window opening to lock it when reversed for window cleaning

locking handle connected to espagnolette giving two locking points and night ventilation position

65 x 65 sash framing

weather seal to all rebates

groove for 32 mm thick window board

150 x 75 hardwood sill

357

Bay Windows ~ these can be defined as any window with side lights which projects in front of the external wall and is supported by a sill height wall. Bay windows not supported by a sill height wall are called oriel windows. They can be of any window type, constructed from any of the usual window materials and are available in three plan formats namely square, splay and circular or segmental. Timber corner posts can be boxed, solid or jointed the latter being the common method.

Typical Examples ~

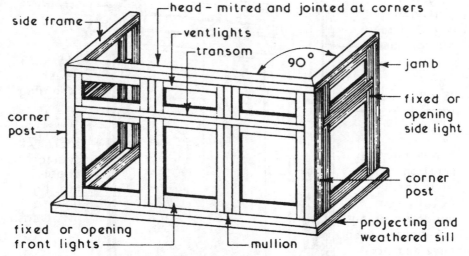

TYPICAL SQUARE BAY WINDOW (665mm projection)

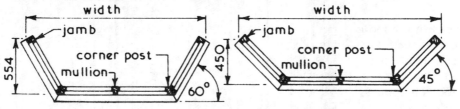

60° SPLAY BAY WINDOW AND 45° SPLAY BAY WINDOW

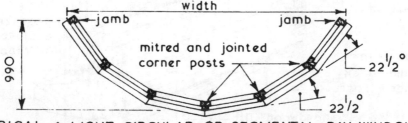

TYPICAL 6 LIGHT CIRCULAR OR SEGMENTAL BAY WINDOW

Schedules ~ the main function of a schedule is to collect together all the necessary information for a particular group of components such as windows, doors and drainage inspection chambers. There is no standard format for schedules but they should be easy to read, accurate and contain all the necessary information for their purpose. Schedules are usually presented in a tabulated format which can be related to and read in conjunction with the working drawings.

Typical Example ~

WINDOW SCHEDULE – Sheet 1 of 1			Drawn By: RC	Date: 14/4/01		Rev.	
Contract Title & Number: Lane End Farm – H 341/80					Drg. Nos. C(31) 450–7		
Number	Type or catalogue ref.	Material	Overall Size w x h	Glass	Ironmongery	Sill External	Sill Internal
2	213 CV	hardwood	1200 x 1350	sealed units as supplied with frames	supplied with casements	2 cos. plain tiles subsill	150×150×15 quarry tiles
4	309 CVC	ditto	1770 x 900	ditto	ditto	ditto	25 mm thick softwood
4	313 CVC	ditto	1770 x 1350	ditto	ditto	sill of frame	ditto

Window manufacturers identify their products with a notation that combines figures with numbers. The objective is to simplify catalogue entries, specification clauses and schedules. For example:

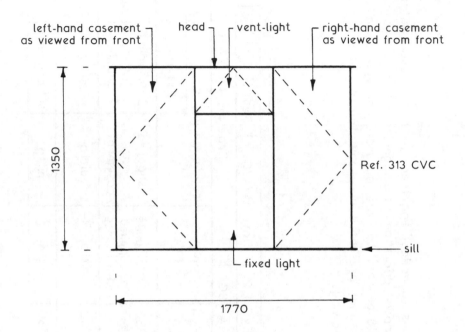

Notation will vary to some extent between the different joinery producers. The example of 313 CVC translates to:

3 = width divided into three units.
13 = first two dimensions of standard height, ie. 1350 mm.
C = casement.
V = ventlight.

Other common notations include:

N = narrow light.
P = plain (picture type window, ie. no transom or mullion).
T = through transom.
S = sub-light, fixed.
VS = vent-light and sub-light.
F = fixed light.
B = bottom casement opening inwards.
RH/LH = right or left-hand as viewed from the outside.

Glass ~ this material is produced by fusing together soda, lime and silica with other minor ingredients such as magnesia and alumina. A number of glass types are available for domestic work and these include:-

Clear Float ~ used where clear undistorted vision is required. Available thicknesses range from 3mm to 25mm.

Clear Sheet ~ suitable for all clear glass areas but because the two faces of the glass are never perfectly flat or parallel some distortion of vision usually occurs. This type of glass is gradually being superseded by the clear float glass. Available thicknesses range from 3mm to 6mm.

Translucent Glass ~ these are patterned glasses most having one patterned surface and one relatively flat surface. The amount of obscurity and diffusion obtained depend on the type and nature of pattern. Available thicknesses range from 4mm to 6mm for patterned glasses and from 5mm to 10mm for rough cast glasses.

Wired Glass ~ obtainable as a clear polished wired glass or as a rough cast wired glass with a nominal thickness of 7mm. Generally used where a degree of fire resistance is required. Georgian wired glass has a 12mm square mesh whereas the hexagonally wired glass has a 20mm mesh.

Choice of Glass ~ the main factors to be considered are:-
1. Resistance to wind loadings. 2. Clear vision required.
3. Privacy. 4. Security. 5. Fire resistance. 6. Aesthetics.

Glazing Terminology ~

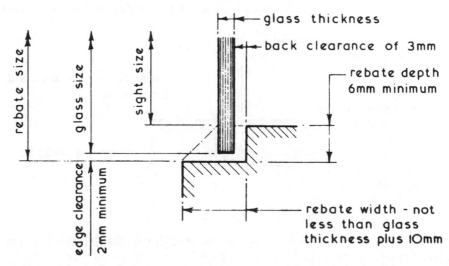

Glazing ~ the act of fixing glass into a frame or surround in domestic work this is usually achieved by locating the glass in a rebate and securing it with putty or beading and should be carried out in accordance with the recommendations contained in BS 6262: Glazing for buildings.

Timber Surrounds ~ linseed oil putty to BS 544 - rebate to be clean, dry and primed before glazing is carried out. Putty should be protected with paint within two weeks of application.

Metal Surrounds ~ metal casement putty if metal surround is to be painted - if surround is not to be painted a non-setting compound should be used.

Typical Glazing Details ~

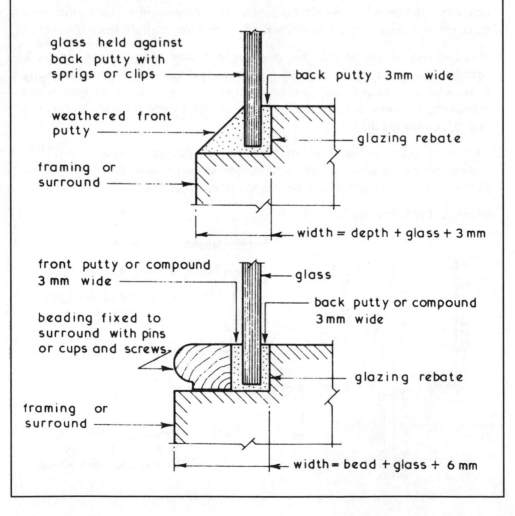

glass held against back putty with sprigs or clips —— back putty 3mm wide

weathered front putty ——→

framing or surround ——→

—— glazing rebate

width = depth + glass + 3 mm

front putty or compound 3 mm wide ——

beading fixed to surround with pins or cups and screws

framing or surround ——→

—— glass

back putty or compound 3 mm wide

—— glazing rebate

width = bead + glass + 6 mm

Double Glazing ~ as its name implies this is where two layers of glass are used instead of the traditional single layer. Double glazing can be used to reduce the rate of heat loss through windows and glazed doors or it can be employed to reduce the sound transmission through windows. In the context of thermal insulation this is achieved by having a small air or argon gas filled space within the range of 6 to 20 mm between the two layers of glass. The sealed double glazing unit will also prevent internal misting by condensation. If metal frames are used these should have a thermal break incorporated in their design. All opening sashes in a double glazing system should be fitted with adequate weather seals to reduce the rate of heat loss through the opening clearance gap.

In the context of sound insulation three factors affect the performance of double glazing. Firstly good installation to ensure airtightness, secondly the weight of glass used and thirdly the size of air space between the layers of glass. The heavier the glass used the better the sound insulation and the air space needs to be within the range of 50 to 300 mm. Absorbent lining to the reveals within the air space will also improve the sound insulation properties of the system.

Typical Examples ~

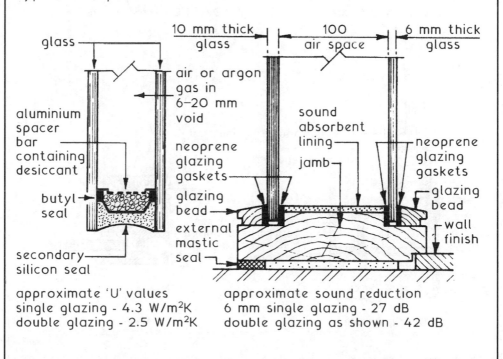

glass

10 mm thick glass   100 air space   6 mm thick glass

air or argon gas in 6–20 mm void

aluminium spacer bar containing desiccant

sound absorbent lining

neoprene glazing gaskets

butyl seal

neoprene glazing gaskets

jamb

glazing bead

glazing bead

secondary silicon seal

external mastic seal

wall finish

approximate 'U' values
single glazing - 4.3 W/m²K
double glazing - 2.5 W/m²K

approximate sound reduction
6 mm single glazing - 27 dB
double glazing as shown - 42 dB

Secondary glazing of existing windows is an acceptable method for reducing heat energy losses at wall openings. Providing the existing windows are in a good state of repair, this is a cost effective, simple method for upgrading windows to current energy efficiency standards. In addition to avoiding the disruption of removing existing windows, further advantages of secondary glazing include, retention of the original window features, reduction in sound transmission and elimination of draughts. Applications are manufactured for all types of window, with sliding or hinged variations. The following details are typical of horizontal sliding sashes -

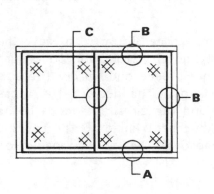

Elevation of frame

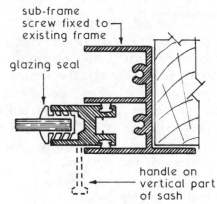

sub-frame
screw fixed to existing frame

glazing seal

handle on vertical part of sash

Detail B - Head and jamb

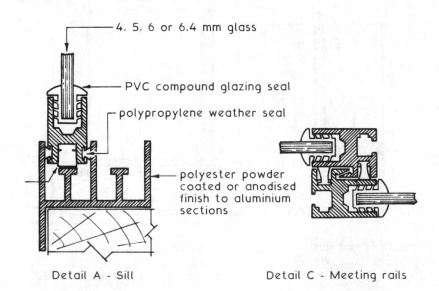

4, 5, 6 or 6.4 mm glass

PVC compound glazing seal

polypropylene weather seal

polyester powder coated or anodised finish to aluminium sections

Detail A - Sill

Detail C - Meeting rails

Low emissivity or "Low E" glass is specially manufactured with a surface coating to significantly improve its thermal performance. The surface coating has a dual function:

1. Allows solar short wave light radiation to penetrate a building.
2. Reflects long wave heat radiation losses back into a building.

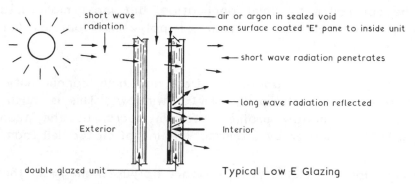

Typical Low E Glazing

Manufacturing processes:

1. Pyrolitic hard coat, applied on-line as the glass is made. Emissivity range, 0.15–0.20, e.g. Pilkington 'K'.
2. A sputtered soft coat applied after glass manufacture. Emissivity range, 0.05–0.10, e.g. Pilkington 'Kappafloat' and 'Suncool High Performance'.

Note: In relative terms, uncoated glass has a normal emissivity of about 0.90.

Indicative U-values for multi-glazed windows of 4 mm glass with a 16 mm void width:

Glazing type	uPVC or wood frame	metal frame
Double, air filled	2.7	3.3
Double, argon filled	2.6	3.2
Double, air filled Low E (0.20)	2.1	2.6
Double, argon filled Low E (0.20)	2.0	2.5
Double, air filled Low E (0.05)	2.0	2.3
Double, argon filled Low E (0.05)	1.7	2.1
Triple, air filled	2.0	2.5
Triple, argon filled	1.9	2.4
Triple, air filled Low E (0.20)	1.6	2.0
Triple, argon filled Low E (0.20)	1.5	1.9
Triple, air filled Low E (0.05)	1.4	1.8
Triple, argon filled Low E (0.05)	1.3	1.7

Notes:

1. A larger void and thicker glass will reduce the U-value, and vice-versa.

2. Data for metal frames assumes a thermal break of 4 mm (see next page).

3. Hollow metal framing units can be filled with a closed cell insulant foam to considerably reduce U-values.

Extruded aluminium profiled sections are designed and manufactured to create lightweight hollow window (and door) framing members.

Finish – untreated aluminium is prone to surface oxidisation. This can be controlled by paint application, but most manufacturers provide a variable colour range of polyester coatings finished gloss, satin or matt.

Thermal insulation – poor insulation and high conductivity are characteristics of solid profile metal windows. This is much less apparent with hollow profile outer members, as they can be considerably enhanced by a thermal infilling of closed cell foam.

Condensation – a high strength 2-part polyurethane resin thermal break between internal and external profiles inhibits cold bridging. This reduces the opportunity for condensation to form on the surface. The indicative U-values given on the preceding page are based on a thermal break of 4mm. If this is increased to 16mm, the values can be reduced by up to 0.2 W/m^2 K.

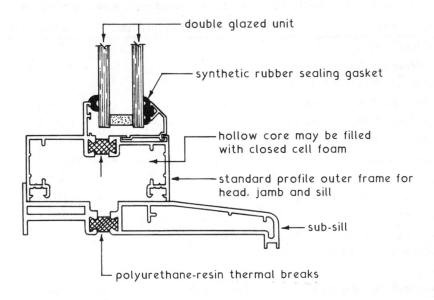

double glazed unit

synthetic rubber sealing gasket

hollow core may be filled with closed cell foam

standard profile outer frame for head, jamb and sill

sub-sill

polyurethane-resin thermal breaks

Hollow Core Aluminium Profiled Window Section

Inert gas fills ~ argon or krypton. Argon is generally used as it is the least expensive and more readily available. Where krypton is used, the air gap need only be half that with argon to achieve a similar effect. Both gases have a higher insulating value than air due to their greater density.

Densities (kg/m^3):
- Air = 1.20
- Argon = 1.66
- Krypton = 3.49

Argon and krypton also have a lower thermal conductivity than air.

Spacers ~ generally hollow aluminium with a desiccant or drying agent fill. The filling absorbs the initial moisture present in between the glass layers. Non-metallic spacers are preferred as aluminium is an effective heat conductor.

Approximate solar gains with ordinary float glass ~

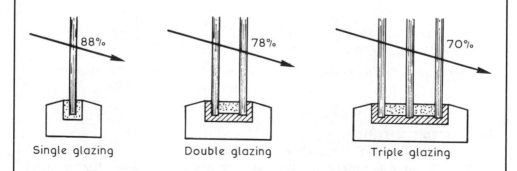

Single glazing   Double glazing   Triple glazing

"Low E" invisible coatings reduce the solar gain by up to one-third. Depending on the glass quality and cleanliness, about 10 to 15% of visible light reduction applies for each pane of glass.

Typical application ~

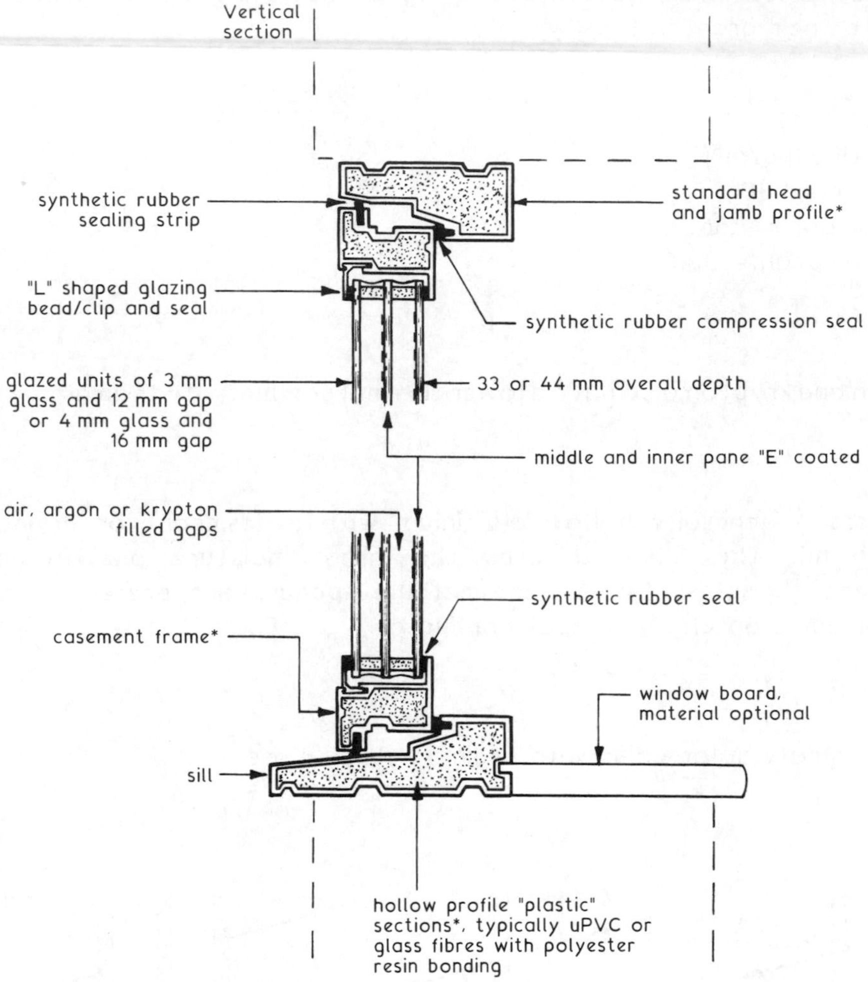

Vertical
section

synthetic rubber
sealing strip

standard head
and jamb profile*

"L" shaped glazing
bead/clip and seal

synthetic rubber compression seal

glazed units of 3 mm
glass and 12 mm gap
or 4 mm glass and
16 mm gap

33 or 44 mm overall depth

middle and inner pane "E" coated

air, argon or krypton
filled gaps

synthetic rubber seal

casement frame*

window board,
material optional

sill

hollow profile "plastic"
sections*, typically uPVC or
glass fibres with polyester
resin bonding

* Hollow profiles manufactured with a closed cell insulant foam/expanded polystyrene core.

Further considerations ~

* U value potential, less than 1.0 W/m²K.
* "Low E" invisible metallic layer on one pane of double glazing gives
  a similar insulating value to standard triple glazing (see page 365).
* Performance enhanced with blinds between wide gap panes.
* High quality ironmongery required due to weight of glazed frames.
* Improved sound insulation, particularly with heavier than air gap fill.

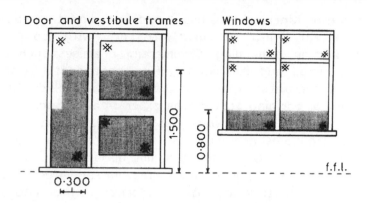

Door and vestibule frames | Windows

In these critical locations, glazing must satisfy one of the following:-

1. Breakage to leave only a small opening with small detachable particles without sharp edges.

2. Disintegrating glass must leave only small detached pieces.

3. Inherent robustness, e.g. polycarbonate composition. Annealed glass acceptable but with the following limitations:-

Thickness of annealed glass (mm)	Max. glazed area. Height (m)	Width(m)
8	1·100	1·100
10	2·250	2·250
12	3·000	4·500
15	no limit	

4. Panes in small areas, <250mm wide and <0.5m² area. e.g. leaded lights (4mm annealed glass) and Georgian pattern (6mm annealed glass).

5. Protective screening as shown:

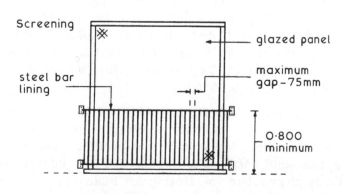

Manifestation or Marking of Glass ~ another aspect of the critical location concept which frequently occurs with contemporary glazed features in a building. Commercial premises such as open plan offices, shops and showrooms often incorporate large walled areas of uninterrupted glass to promote visual depth, whilst dividing space or forming part of the exterior envelope. To prevent collision, glazed doors and walls must have prominent framing or intermediate transoms and mullions. An alternative is to position obvious markings at 1000 and 1500mm above floor level. Glass doors could have large pull/push handles and/or IN and OUT signs in bold lettering. Other areas may be adorned with company logos, stripes, geometric shape, etc.

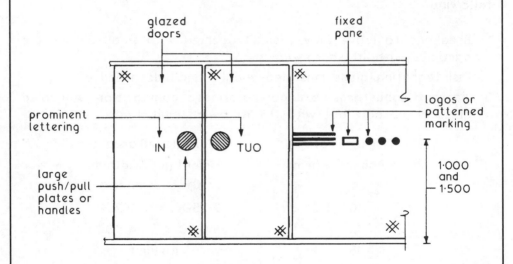

Critical Locations ~ The Building Regulations, Approved Document – N, determines positions where potential personal impact and injury with glazed doors and windows are most critical. In these situations the glazing specification must incorporate a degree of safety such that any breakage would be relatively harmless. Additional measures in British Standard 6206 complement the Building Regulations and provide test requirements and specifications for impact performance for different classes of glazing material. See also BS 6262.

Refs. Building Regulations, A.D. N1: Protection against impact.
A.D. N2: Manifestation of glazing.
BS 6206: Specification for impact performance requirements for flat safety glass and safety plastics for use in buildings.
BS 6262: Code of practice for glazing for buildings.

Glass blocks have been used for some time as internal feature partitioning. They now include a variety of applications in external walls, where they combine the benefits of a walling unit with a natural source of light. They have also been used in paving to allow natural light penetration into basements.

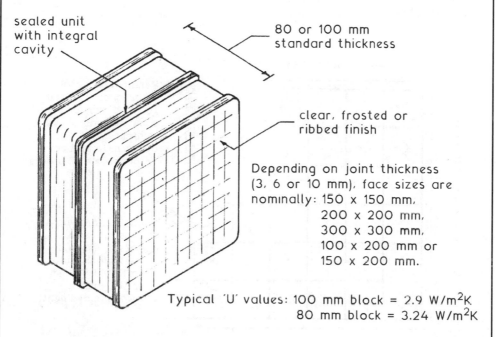

sealed unit with integral cavity

80 or 100 mm standard thickness

clear, frosted or ribbed finish

Depending on joint thickness (3, 6 or 10 mm), face sizes are nominally: 150 x 150 mm, 200 x 200 mm, 300 x 300 mm, 100 x 200 mm or 150 x 200 mm.

Typical 'U' values: 100 mm block = 2.9 W/m^{2}K
80 mm block = 3.24 W/m^{2}K

Fire resistance, BS 476-22 - 1 hour integrity (load bearing capacity and fire containment).
Maximum panel size is 9m^2. Maximum panel dimension is 3 m

Laying – glass blocks can be bonded like conventional brickwork, but for aesthetic reasons are usually laid with continuous vertical and horizontal joints.

Jointing – blocks are bedded in mortar with reinforcement from two, 9 gauge galvanised steel wires in horizontal joints. Every 3rd. course for 150mm units, every 2nd. course for 200mm units and every course for 300mm units. First and last course to be reinforced.

Ref: BS 476-22: Fire tests on building materials and structures. Methods for determination of the fire resistance of non-loadbearing elements of construction.

Mortar – dryer than for bricklaying as the blocks are non-absorbent. The general specification will include: White Portland Cement (BS EN 197-1), High Calcium Lime (BS EN 459-1) and Sand. The sand should be white quartzite or silica type. Fine silver sand is acceptable. An integral waterproofing agent should also be provided. Recommended mix ratios – 1 part cement: 0.5 part lime: 4 parts sand.

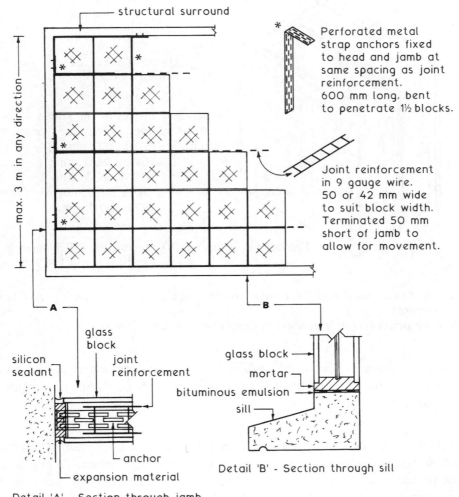

structural surround

max. 3 m in any direction

* Perforated metal strap anchors fixed to head and jamb at same spacing as joint reinforcement. 600 mm long, bent to penetrate 1½ blocks.

Joint reinforcement in 9 gauge wire. 50 or 42 mm wide to suit block width. Terminated 50 mm short of jamb to allow for movement.

A

B

silicon sealant

glass block

joint reinforcement

anchor

expansion material

Detail 'A' - Section through jamb
Note: Same detail for head, except omit reinforcement

glass block

mortar

bituminous emulsion

sill

Detail 'B' - Section through sill

Ref. BS EN 1051-1: Glass in building – glass blocks and glass pavers. Definitions and description.

Doors ~ can be classed as external or internal. External doors are usually thicker and more robust in design than internal doors since they have more functions to fulfil.

Typical Functions ~

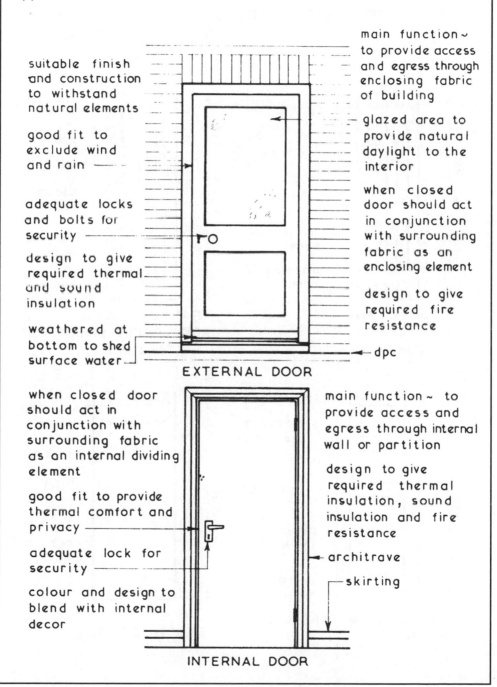

suitable finish and construction to withstand natural elements

good fit to exclude wind and rain

adequate locks and bolts for security

design to give required thermal and sound insulation

weathered at bottom to shed surface water

main function ~ to provide access and egress through enclosing fabric of building

glazed area to provide natural daylight to the interior

when closed door should act in conjunction with surrounding fabric as an enclosing element

design to give required fire resistance

dpc

EXTERNAL DOOR

when closed door should act in conjunction with surrounding fabric as an internal dividing element

good fit to provide thermal comfort and privacy

adequate lock for security

colour and design to blend with internal decor

main function ~ to provide access and egress through internal wall or partition

design to give required thermal insulation, sound insulation and fire resistance

architrave

skirting

INTERNAL DOOR

373

External Doors ~ these are available in a wide variety of types and styles in timber, aluminium alloy or steel. The majority of external doors are however made from timber, the metal doors being mainly confined to fully glazed doors such as 'patio doors'.

Typical Examples of External Doors ~

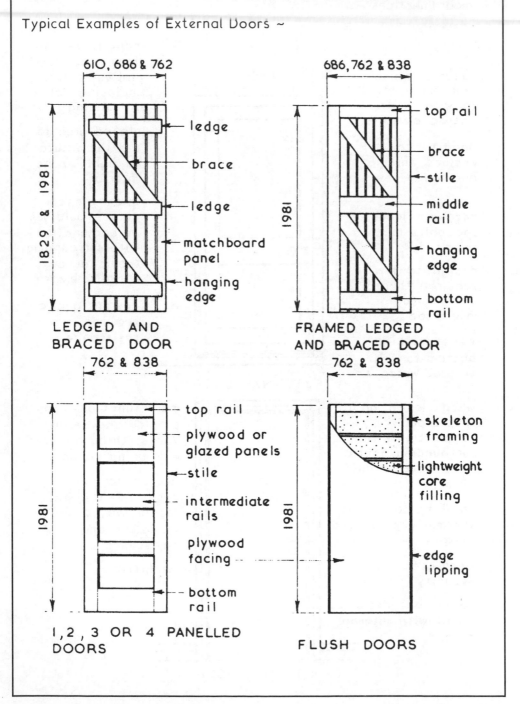

610, 686 & 762

1829 & 1981

- ledge
- brace
- ledge
- matchboard panel
- hanging edge

LEDGED AND BRACED DOOR

686, 762 & 838

1981

- top rail
- brace
- stile
- middle rail
- hanging edge
- bottom rail

FRAMED LEDGED AND BRACED DOOR

762 & 838

1981

- top rail
- plywood or glazed panels
- stile
- intermediate rails
- plywood facing
- bottom rail

1, 2, 3 OR 4 PANELLED DOORS

762 & 838

1981

- skeleton framing
- lightweight core filling
- edge lipping

FLUSH DOORS

Typical examples of purpose made and non-standard external doors ~

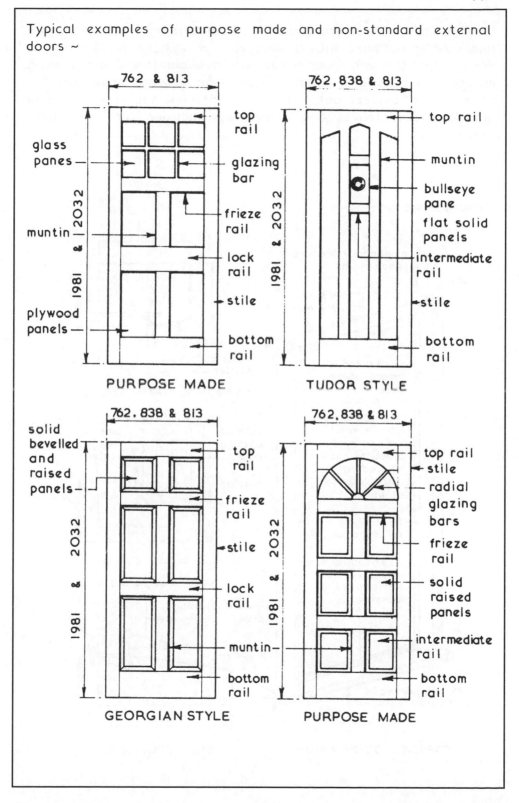

762 & 813

top rail
glass panes
glazing bar
muntin
frieze rail
1981 & 2032
lock rail
stile
plywood panels
bottom rail

PURPOSE MADE

762, 838 & 813

top rail
muntin
bullseye pane
flat solid panels
1981 & 2032
intermediate rail
stile
bottom rail

TUDOR STYLE

762, 838 & 813

solid bevelled and raised panels
top rail
frieze rail
2032
stile
1981 &
lock rail
muntin
bottom rail

GEORGIAN STYLE

762, 838 & 813

top rail
stile
radial glazing bars
frieze rail
2032
solid raised panels
1981 &
intermediate rail
bottom rail

PURPOSE MADE

Door Frames ~ these are available for all standard external doors and can be obtained with a fixed solid or glazed panel above a door height transom. Door frames are available for doors opening inwards or outwards. Most door frames are made to the recommendations set out in BS 4787: Internal and external wood doorsets, door leaves and frames. Specification for dimensional requirements.

Typical Example ~

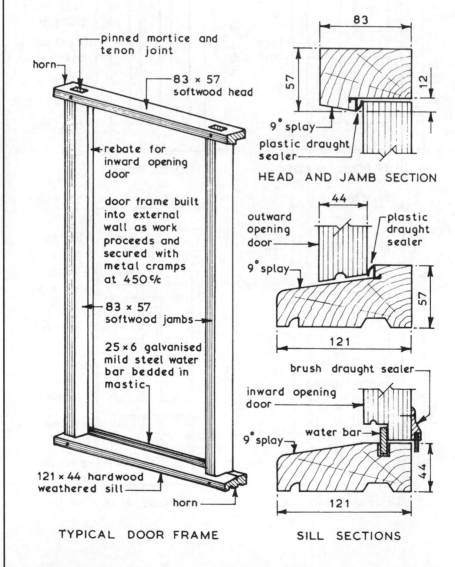

pinned mortice and tenon joint

horn

83 × 57 softwood head

rebate for inward opening door

door frame built into external wall as work proceeds and secured with metal cramps at 450 ℅

83 × 57 softwood jambs

25 × 6 galvanised mild steel water bar bedded in mastic

121 × 44 hardwood weathered sill

horn

TYPICAL DOOR FRAME

83

57

12

9° splay

plastic draught sealer

HEAD AND JAMB SECTION

44

outward opening door

plastic draught sealer

9° splay

57

121

brush draught sealer

inward opening door

9° splay

water bar

44

121

SILL SECTIONS

Door Ironmongery ~ available in a wide variety of materials, styles and finishers but will consist of essentially the same components:- Hinges or Butts — these are used to fix the door to its frame or lining and to enable it to pivot about its hanging edge.

Locks, Latches and Bolts ~ the means of keeping the door in its closed position and providing the required degree of security. The handles and cover plates used in conjunction with locks and latches are collectively called door furniture.

Letter Plates — fitted in external doors to enable letters etc., to be deposited through the door.

Other items include Finger and Kicking Plates which are used to protect the door fabric where there is high usage,

Draught Excluders to seal the clearance gap around the edges of the door and Security Chains to enable the door to be partially opened and thus retain some security.

Typical Examples ~

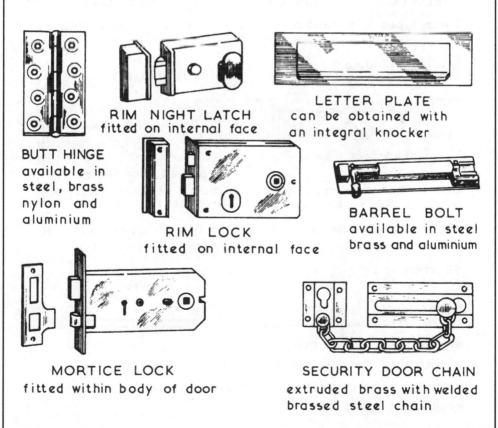

RIM NIGHT LATCH
fitted on internal face

LETTER PLATE
can be obtained with
an integral knocker

BUTT HINGE
available in
steel, brass
nylon and
aluminium

RIM LOCK
fitted on internal face

BARREL BOLT
available in steel
brass and aluminium

MORTICE LOCK
fitted within body of door

SECURITY DOOR CHAIN
extruded brass with welded
brassed steel chain

Industrial Doors ~ these doors are usually classified by their method of operation and construction. There is a very wide range of doors available and the choice should be based on the following considerations:-

1. Movement - vertical or horizontal.
2. Size of opening.
3. Position and purpose of door(s).
4. Frequency of opening and closing door(s).
5. Manual or mechanical operation.
6. Thermal and/or sound insulation requirements.
7. Fire resistance requirements.

Typical Industrial Door Types ~

1. Straight Sliding -

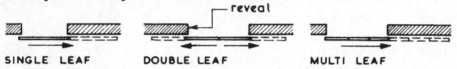

SINGLE LEAF   DOUBLE LEAF   MULTI LEAF

These types can be top hung with a bottom guide roller or hung with bottom rollers and top guides - see page 379

2 Sliding / Folding -

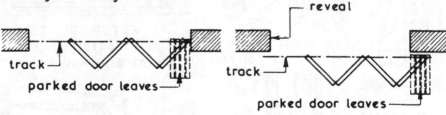

MULTI LEAF END FOLDING
HUNG BETWEEN REVEALS

MULTI LEAF END FOLDING
HUNG BEHIND OPENING

These types can be top hung with a bottom guide roller or hung with bottom rollers and top guides - see page 380

3. Shutters -

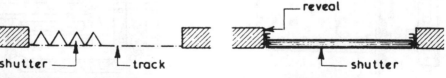

HORIZONTAL FOLDING SHUTTER   ROLLER SHUTTER

Shutters can be installed between, behind or in front of the reveals - see page 381

Straight Sliding Doors ~ these doors are easy to operate, economic to maintain and present no problems for the inclusion of a wicket gate. They do however take up wall space to enable the leaves to be parked in the open position. The floor guide channel associated with top hung doors can become blocked with dirt causing a malfunction of the sliding movement whereas the rollers in bottom track doors can seize up unless regularly lubricated and kept clean. Straight sliding doors are available with either manual or mechanical operation.

Typical Example ~

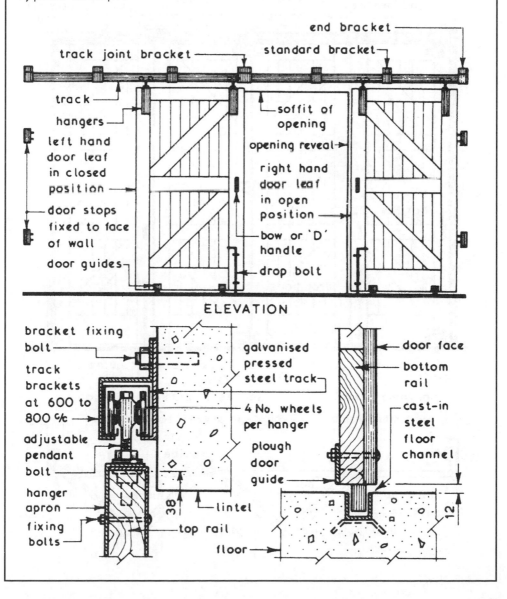

ELEVATION

379

Sliding/Folding Doors ~ these doors are an alternative format to the straight sliding door types and have the same advantages and disadvantages except that the parking space required for the opened door is less than that for straight sliding doors. Sliding/folding are usually manually operated and can be arranged in groups of 2 to 8 leaves.

Typical Example ~

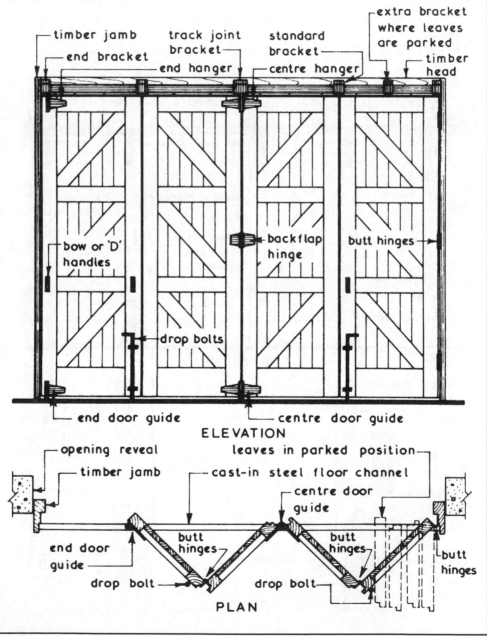

ELEVATION

PLAN

Shutters ~ horizontal folding shutters are similar in operation to sliding/folding doors but are composed of smaller leaves and present the same problems. Roller shutters however do not occupy any wall space but usually have to be fully opened for access. They can be manually operated by means of a pole when the shutters are self coiling, operated by means of an endless chain winding gear or mechanically raised and lowered by an electric motor but in all cases they are slow to open and close. Vision panels cannot be incorporated in the roller shutter but it is possible to include a small wicket gate or door in the design.

Typical Details ~

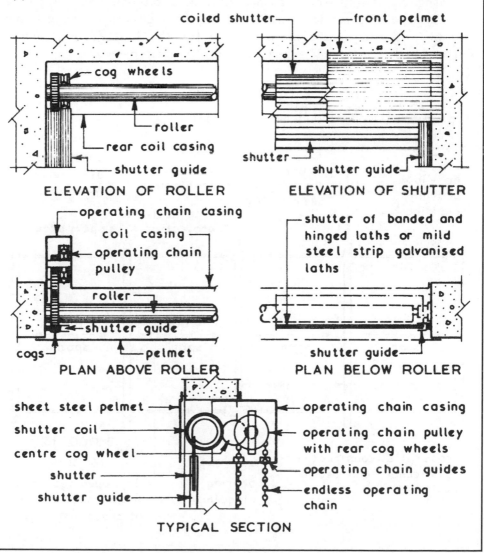

ELEVATION OF ROLLER   ELEVATION OF SHUTTER

PLAN ABOVE ROLLER   PLAN BELOW ROLLER

TYPICAL SECTION

Crosswall Construction ~ this is a form of construction where load bearing walls are placed at right angles to the lateral axis of the building, the front and rear walls being essentially non-load bearing cladding. Crosswall construction is suitable for buildings up to 5 storeys high where the floors are similar and where internal separating or party walls are required such as in blocks of flats or maisonettes. The intermediate floors span longitudinally between the crosswalls providing the necessary lateral restraint and if both walls and floors are of cast in-situ reinforced concrete the series of 'boxes' so formed is sometimes called box frame construction. Great care must be taken in both design and construction to ensure that the junctions between the non-load bearing claddings and the crosswalls are weathertight. If a pitched roof is to be employed with the ridge parallel to the lateral axis an edge beam will be required to provide a seating for the trussed or common rafters and to transmit the roof loads to the crosswalls.

Typical Crosswall Arrangement Details ~

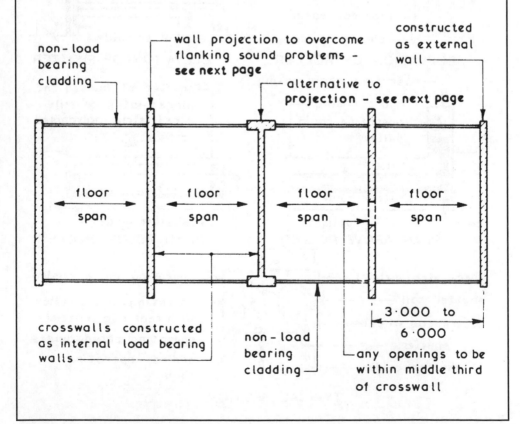

non-load bearing cladding

wall projection to overcome flanking sound problems – **see next page**

constructed as external wall

alternative to **projection - see next page**

floor span   floor span   floor span   floor span

3·000 to 6·000

crosswalls constructed as internal load bearing walls

non-load bearing cladding

any openings to be within middle third of crosswall

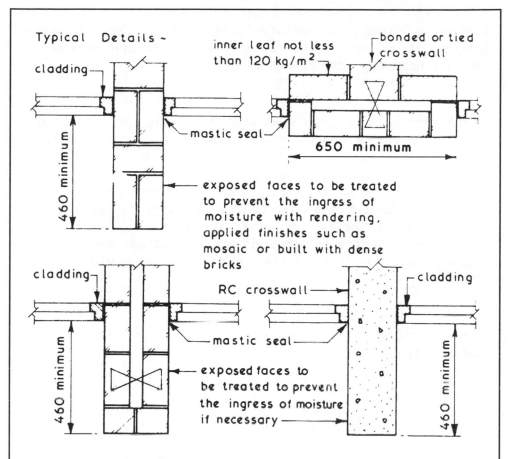

Typical Details ~

cladding

inner leaf not less than 120 kg/m²

bonded or tied crosswall

mastic seal

460 minimum

650 minimum

exposed faces to be treated to prevent the ingress of moisture with rendering, applied finishes such as mosaic or built with dense bricks

cladding

RC crosswall

cladding

mastic seal

460 minimum

460 minimum

exposed faces to be treated to prevent the ingress of moisture if necessary

Advantages of Crosswall Construction:-

1. Load bearing and non-load bearing components can be standardised and in same cases prefabricated giving faster construction times.
2. Fenestration between crosswalls unrestricted structurally.
3. Crosswalls although load bearing need not be weather resistant as is the case with external walls.

Disadvantages of Crosswall Construction:-

1. Limitations of possible plans.
2. Need for adequate lateral ties between crosswalls.
3. Need to weather adequately projecting crosswalls.

Floors:-

An in-situ solid reinforced concrete floor will provide the greatest rigidity, all other form must be adequately tied to walls.

System ~ comprises quality controlled factory produced components of plain reinforced concrete walls and prestressed concrete hollow or solid core plank floors.

Site Assembly ~ components are crane lifted and stacked manually with the floor panel edges bearing on surrounding walls. Temporary support will be necessary until the units are "stitched" together with horizontal and vertical steel reinforcing ties located through reinforcement loops projecting from adjacent panels. In-situ concrete completes the structural connection to provide full transfer of all forces and loads through the joint. Precast concrete stair flights and landings are located and connected to support panels by steel angle bracketing and in-situ concrete joints.

Typical "stitched" joint between precast concrete crosswall components ~

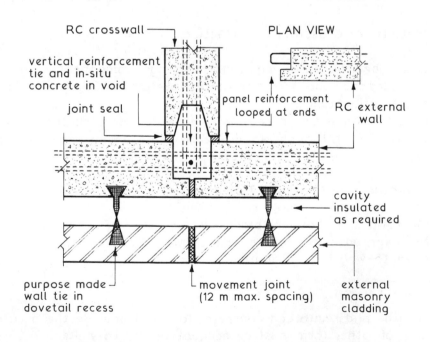

Concept ~ a cost effective simple and fast site assembly system using load-bearing partitions and external walls to transfer vertical loads from floor panels. The floor provides lateral stability by diaphragm action between the walls.

Application ~ precast reinforced concrete crosswall construction systems may be used to construct multi-storey buildings, particularly where the diaphragm floor load distribution is transferred to lift or stair well cores. Typical applications include schools, hotels, hostels apartment blocks and hospitals. External appearance can be enhanced by a variety of cladding possibilities, including the traditional look of face brickwork secured to the structure by in-built ties. Internal finishing may be with paint or plaster, but it is usually dry lined with plasterboard.

Location of "stitched" in-situ reinforced concrete ties ~

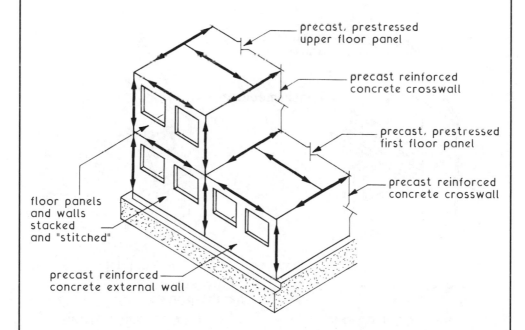

precast, prestressed upper floor panel

precast reinforced concrete crosswall

precast, prestressed first floor panel

precast reinforced concrete crosswall

floor panels and walls stacked and "stitched"

precast reinforced concrete external wall

Fire resistance and sound insulation are achieved by density and quality of concrete. The thermal mass of concrete can be enhanced by applying insulation in between the external precast panel and the masonry or other cladding.

Framing ~ an industry based pre-fabricated house manufacturing process permitting rapid site construction, with considerably fewer site operatives than traditional construction. This technique has a long history of conventional practice in Scandinavia and North America, but has only gained credibility in the UK since the 1960s. Factory-made panels are based on a stud framework of timber, normally ex. 100 × 50 mm, an outer sheathing of plywood, particleboard or similar sheet material, insulation between the framing members and an internal lining of plasterboard. An outer cladding of brickwork weatherproofs the building and provides a traditional appearance.

Assembly techniques are derived from two systems:-

1. Balloon frame

2. Platform frame

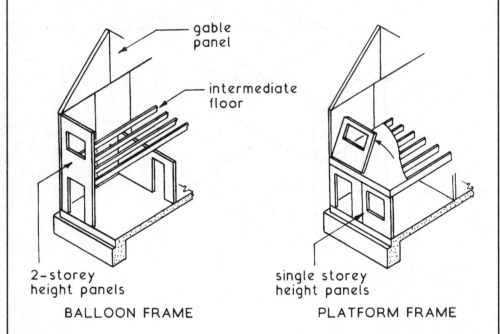

gable panel

intermediate floor

2-storey height panels

BALLOON FRAME

single storey height panels

PLATFORM FRAME

A balloon frame consists of two-storey height panels with an intermediate floor suspended from the framework. In the UK, the platform frame is preferred with intermediate floor support directly on the lower panel. It is also easier to transport, easier to handle on site and has fewer shrinkage and movement problems.

Typical Details ~

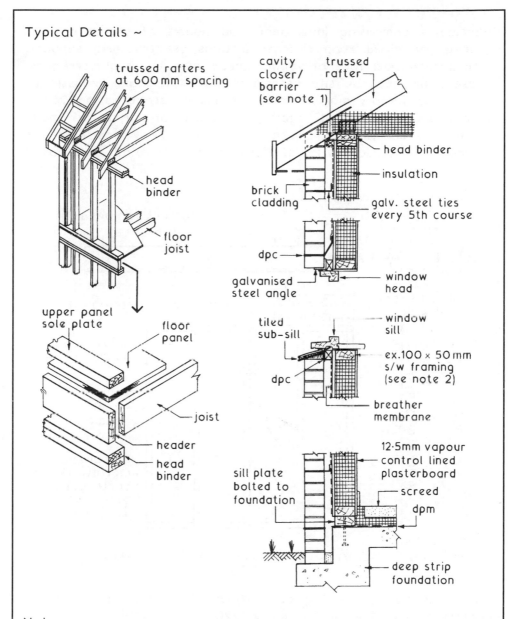

trussed rafters at 600mm spacing

head binder

floor joist

upper panel sole plate

floor panel

joist

header

head binder

cavity closer/ barrier (see note 1)

trussed rafter

head binder

insulation

brick cladding

galv. steel ties every 5th course

dpc

galvanised steel angle

window head

tiled sub-sill

window sill

dpc

ex.100 × 50mm s/w framing (see note 2)

breather membrane

12·5mm vapour control lined plasterboard

sill plate bolted to foundation

screed

dpm

deep strip foundation

Notes:

1. Cavity barriers prevent fire spread. The principal locations are between elements and compartments of construction (see B. Regs. A.D. B3).
2. Thermal bridging through solid framing may be reduced by using rigid EPS insulation and lighter 'I' section members of plywood or OSB.

Framing ~ comprising inner leaf wall panels of standard cold-formed galvanised steel channel sections as structural support, with a lined inner face of vapour check layer under plasterboard. These panels can be site assembled, but it is more realistic to order them factory made. Panels are usually produced in 600mm wide modules and bolted together on site. Roof trusses are made up from steel channel or sigma sections. See page 494 for examples of standard steel sections and BS EN 10162: Cold rolled steel sections.

Standard channel and panel.

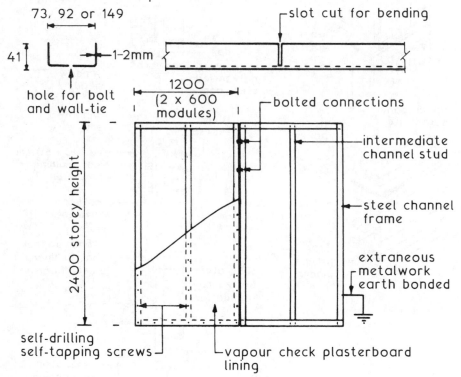

Background/history ~ the concept of steel framing for house construction evolved in the early 1920s, but development of the lightweight concrete "breeze" block soon took preference. Due to a shortage of traditional building materials, a resurgence of interest occurred again during the early post-war building boom of the late 1940s. Thereafter, steel became relatively costly and uncompetitive as a viable alternative to concrete block or timber frame construction techniques. Since the 1990s more efficient factory production processes, use of semi-skilled site labour and availability of economic cold-formed sections have revived an interest in this alternative means of house construction.

Typical Details ~

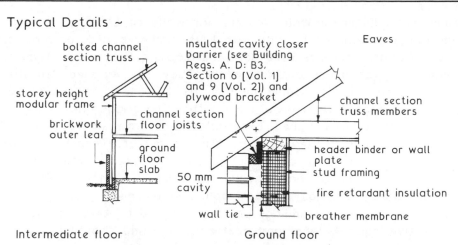

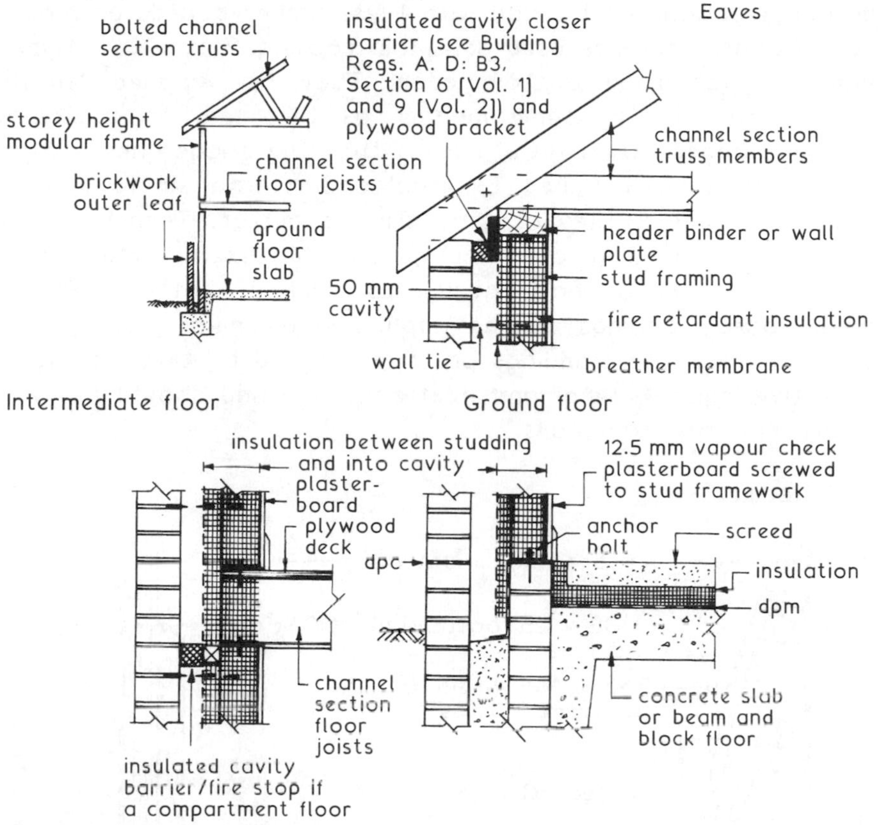

bolted channel section truss ~

storey height modular frame

brickwork outer leaf

channel section floor joists

ground floor slab

insulated cavity closer barrier (see Building Regs. A. D: B3, Section 6 [Vol. 1] and 9 [Vol. 2]) and plywood bracket

Eaves

channel section truss members

header binder or wall plate

stud framing

fire retardant insulation

50 mm cavity

wall tie

breather membrane

Intermediate floor

insulation between studding and into cavity

plaster-board

plywood deck

dpc

channel section floor joists

insulated cavity barrier/fire stop if a compartment floor

Ground floor

12.5 mm vapour check plasterboard screwed to stud framework

anchor bolt

screed

insulation

dpm

concrete slab or beam and block floor

Advantages ~

• Factory made, therefore produced to quality controlled standards and tolerances.
• Relatively simple to assemble on site – bolted connections in pre-formed holes.
• Dimensionally stable, consistent composition, insignificant movement.
• Unaffected by moisture, therefore will not rot.
• Does not burn.
• Inedible by insects.
• Roof spans potentially long relative to weight.

Disadvantages ~

• Possibility of corrosion if galvanised protective layer is damaged.
• Deforms at high temperature, therefore unpredictable in fire.
• Electricity conductor – must be earthed.

Claddings to External Walls ~ external walls of block or timber frame construction can be clad with tiles, timber boards or plastic board sections. The tiles used are plain roofing tiles with either a straight or patterned bottom edge. They are applied to the vertical surface in the same manner as tiles laid on a sloping surface (see pages 401 to 403) except that the gauge can be wider and each tile is twice nailed. External and internal angles can be formed using special tiles or they can be mitred. Timber boards such as matchboarding and shiplap can be fixed vertically to horizontal battens or horizontally to vertical battens. Plastic moulded board claddings can be applied in a similar manner. The battens to which the claddings are fixed should be treated with a preservative against fungi and beetle attack and should be fixed with corrosion resistant nails.

Typical Details ~

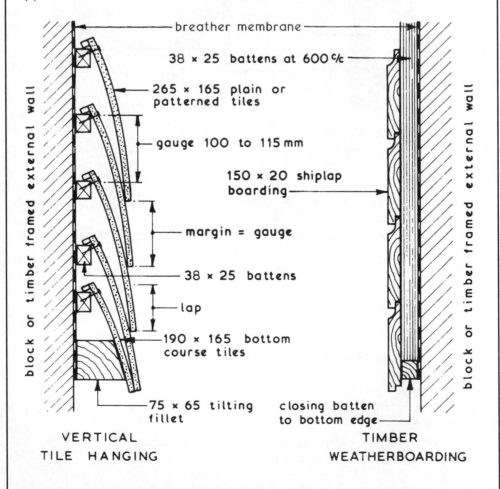

block or timber framed external wall

breather membrane

38 × 25 battens at 600 ℅

265 × 165 plain or patterned tiles

gauge 100 to 115 mm

150 × 20 shiplap boarding

margin = gauge

38 × 25 battens

lap

190 × 165 bottom course tiles

75 × 65 tilting fillet

closing batten to bottom edge

block or timber framed external wall

VERTICAL TILE HANGING

TIMBER WEATHERBOARDING

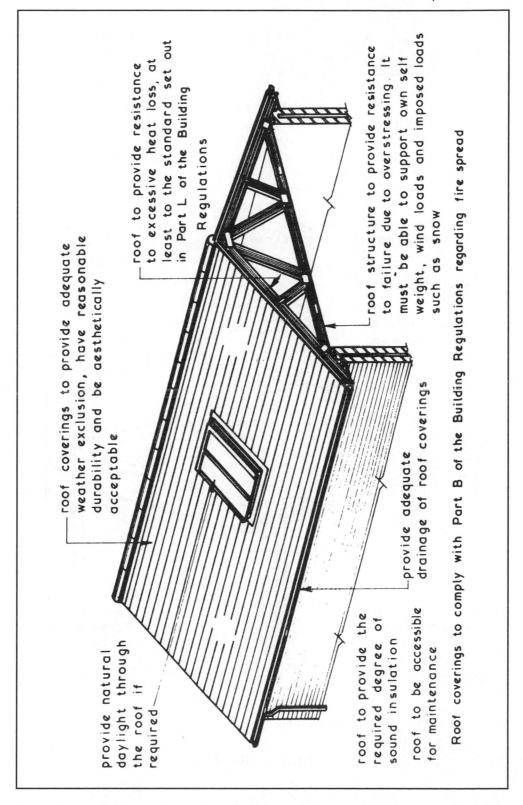

provide natural daylight through the roof if required

roof coverings to provide adequate weather exclusion, have reasonable durability and be aesthetically acceptable

roof to provide resistance to excessive heat loss, at least to the standard set out in Part L of the Building Regulations

roof structure to provide resistance to failure due to overstressing. It must be able to support own self weight, wind loads and imposed loads such as snow

provide adequate drainage of roof coverings

roof to provide the required degree of sound insulation

roof to be accessible for maintenance

Roof coverings to comply with Part B of the Building Regulations regarding fire spread

Roofs ~ these can be classified as either:-

    Flat – pitch from 0° to 10°
    Pitched – pitch over 10°

It is worth noting that for design purposes roof pitches over 70° are classified as walls.

Roofs can be designed in many different forms and in combinations of these forms some of which would not be suitable and/or economic for domestic properties.

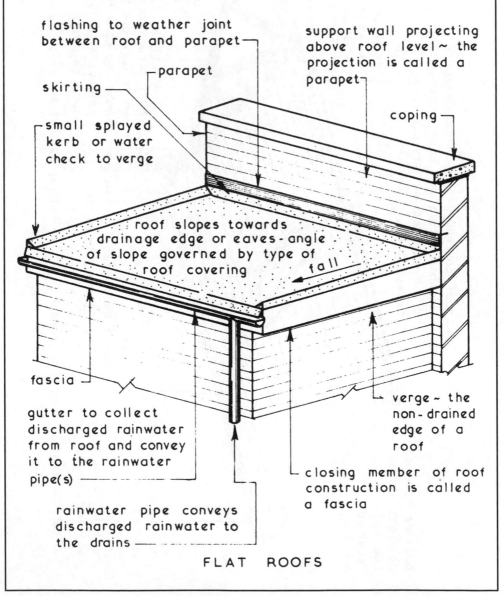

flashing to weather joint between roof and parapet

support wall projecting above roof level~ the projection is called a parapet

parapet

skirting

coping

small splayed kerb or water check to verge

roof slopes towards drainage edge or eaves-angle of slope governed by type of roof covering

fall

fascia

verge ~ the non-drained edge of a roof

gutter to collect discharged rainwater from roof and convey it to the rainwater pipe(s)

closing member of roof construction is called a fascia

rainwater pipe conveys discharged rainwater to the drains

FLAT ROOFS

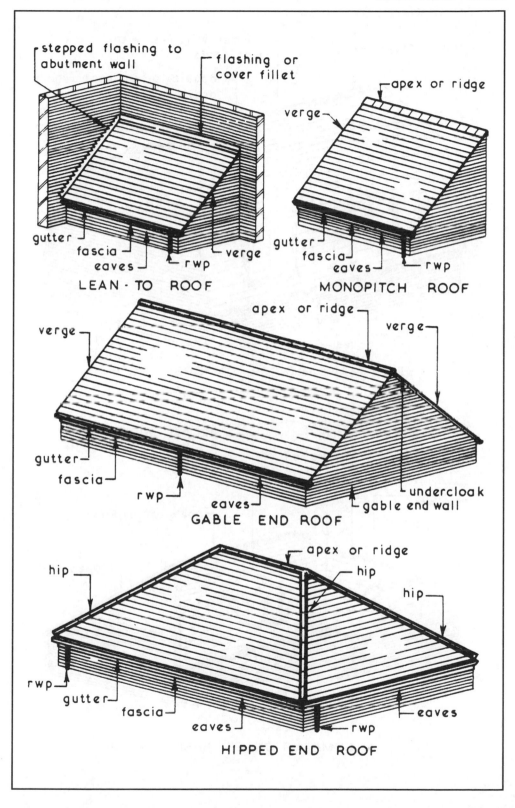

stepped flashing to abutment wall

flashing or cover fillet

gutter
fascia
eaves
rwp
verge

LEAN · TO ROOF

apex or ridge
verge

gutter
fascia
eaves
rwp

MONOPITCH ROOF

verge
apex or ridge
verge

gutter
fascia
rwp
eaves
undercloak
gable end wall

GABLE END ROOF

hip
apex or ridge
hip
hip

rwp
gutter
fascia
eaves
eaves
rwp

HIPPED END ROOF

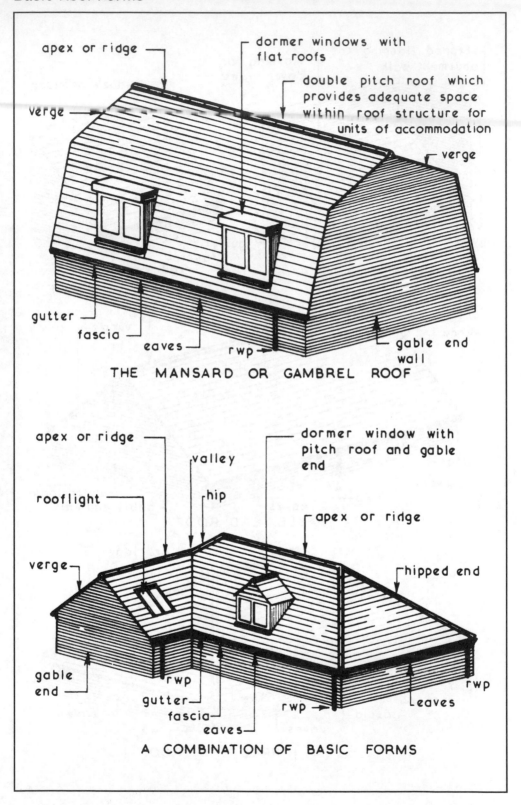

apex or ridge

dormer windows with flat roofs

double pitch roof which provides adequate space within roof structure for units of accommodation

verge

verge

gutter

fascia

eaves

rwp

gable end wall

THE MANSARD OR GAMBREL ROOF

apex or ridge

valley

dormer window with pitch roof and gable end

rooflight

hip

apex or ridge

verge

hipped end

gable end

rwp

gutter

fascia

eaves

rwp

rwp

eaves

A COMBINATION OF BASIC FORMS

Pitched Roofs ~ the primary functions of any domestic roof are to:-

1. Provide an adequate barrier to the penetration of the elements.
2. Maintain the internal environment by providing an adequate resistance to heat loss.

A roof is in a very exposed situation and must therefore be designed and constructed in such a manner as to:-

1. Safely resist all imposed loadings such as snow and wind.
2. Be capable of accommodating thermal and moisture movements.
3. Be durable so as to give a satisfactory performance and reduce maintenance to a minimum.

Component Parts of a Pitched Roof ~

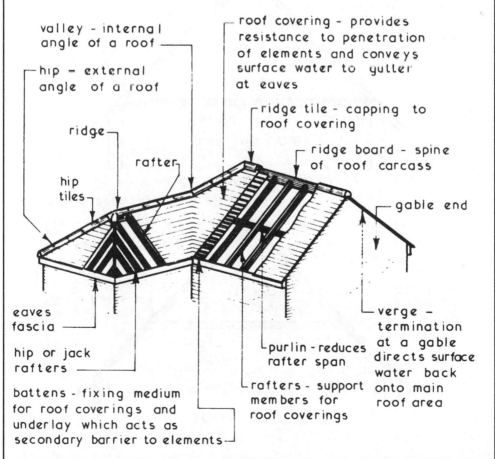

valley - internal angle of a roof

hip - external angle of a roof

ridge

rafter

hip tiles

roof covering - provides resistance to penetration of elements and conveys surface water to gutter at eaves

ridge tile - capping to roof covering

ridge board - spine of roof carcass

gable end

eaves fascia

hip or jack rafters

battens - fixing medium for roof coverings and underlay which acts as secondary barrier to elements

purlin - reduces rafter span

rafters - support members for roof coverings

verge - termination at a gable directs surface water back onto main roof area

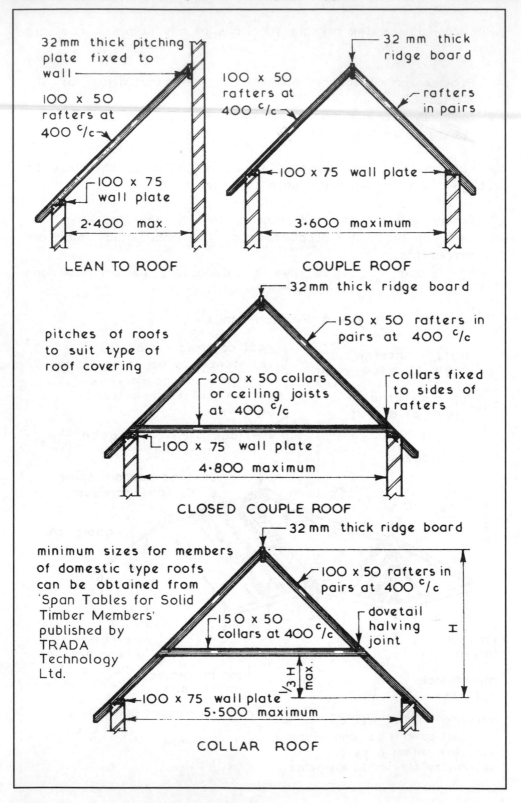

32mm thick pitching plate fixed to wall

100 x 50 rafters at 400 c/c

100 x 75 wall plate

2·400 max.

LEAN TO ROOF

32 mm thick ridge board

100 x 50 rafters at 400 c/c

rafters in pairs

100 x 75 wall plate

3·600 maximum

COUPLE ROOF

32mm thick ridge board

pitches of roofs to suit type of roof covering

150 x 50 rafters in pairs at 400 c/c

200 x 50 collars or ceiling joists at 400 c/c

collars fixed to sides of rafters

100 x 75 wall plate

4·800 maximum

CLOSED COUPLE ROOF

32 mm thick ridge board

minimum sizes for members of domestic type roofs can be obtained from 'Span Tables for Solid Timber Members' published by TRADA Technology Ltd.

100 x 50 rafters in pairs at 400 c/c

150 x 50 collars at 400 c/c

dovetail halving joint

H

1/3 H max.

100 x 75 wall plate

5·500 maximum

COLLAR ROOF

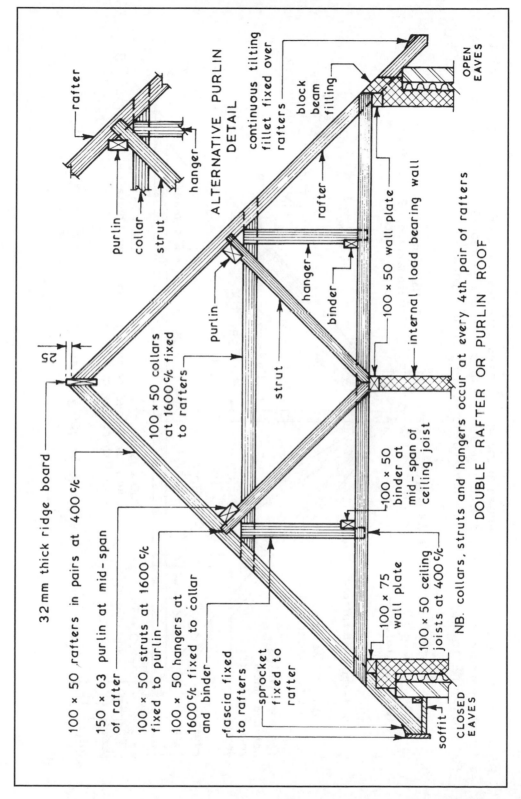

ALTERNATIVE PURLIN DETAIL

rafter

purlin
collar
strut
hanger

continuous tilting fillet fixed over rafters

block
beam
filling

OPEN EAVES

rafter

hanger
binder

100 × 50 wall plate

internal load bearing wall

strut

100 × 50 binder at mid-span of ceiling joist

100 × 75 wall plate

100 × 50 ceiling joists at 400 c/c

CLOSED EAVES

soffit

32 mm thick ridge board

100 × 50 rafters in pairs at 400 c/c

150 × 63 purlin at mid-span of rafter

100 × 50 struts at 1600 c/c fixed to purlin

100 × 50 hangers at 1600 c/c fixed to collar and binder

fascia fixed to rafters

sprocket fixed to rafter

purlin

100 × 50 collars at 1600 c/c fixed to rafters

25

NB. collars, struts and hangers occur at every 4th. pair of rafters

DOUBLE RAFTER OR PURLIN ROOF

397

Roof Trusses ~ these are triangulated plane roof frames designed to give clear spans between the external supporting walls. They are usually prefabricated or partially prefabricated off site and are fixed at 1·800 centres to support purlins which accept loads from the infill rafters.

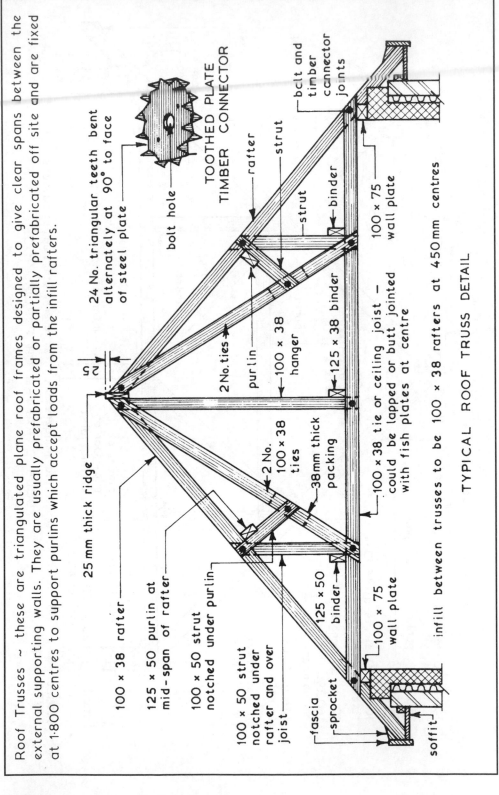

TOOTHED PLATE TIMBER CONNECTOR

24 No. triangular teeth bent alternately at 90° to face of steel plate

bolt hole

rafter

strut

binder

strut

100 × 75 wall plate

100 × 38 tie or ceiling joist — could be lapped or butt jointed with fish plates at centre

bolt and timber connector joints

25 mm thick ridge

2 No. ties

purlin

100 × 38 hanger

125 × 38 binder

2 No. 100 × 38 ties

38 mm thick packing

100 × 38 rafter

125 × 50 purlin at mid-span of rafter

100 × 50 strut notched under purlin

100 × 50 strut notched under rafter and over joist

125 × 50 binder

100 × 75 wall plate

fascia

sprocket

soffit

infill between trusses to be 100 × 38 rafters at 450mm centres

TYPICAL ROOF TRUSS DETAIL

Trussed Rafters ~ these are triangulated plane roof frames designed to give clear spans between the external supporting walls. They are delivered to site as a prefabricated component where they are fixed to the wall plates at 600mm centres. Trussed rafters do not require any ridge board or purlins since they receive their lateral stability by using larger tiling battens (50 × 25mm) than those used on traditional roofs.

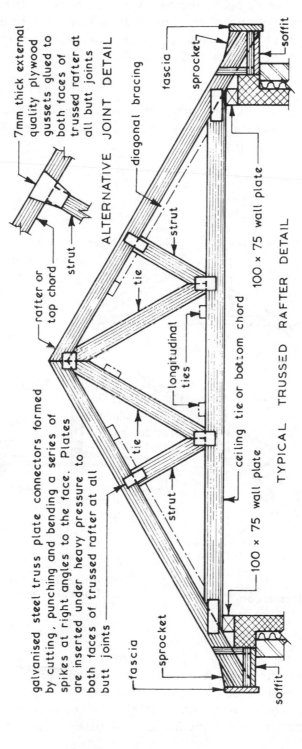

7mm thick external quality plywood gussets glued to both faces of trussed rafter at all butt joints

rafter or top chord

strut

ALTERNATIVE JOINT DETAIL

galvanised steel truss plate connectors formed by cutting, punching and bending a series of spikes at right angles to the face. Plates are inserted under heavy pressure to both faces of trussed rafter at all butt joints

diagonal bracing

fascia

sprocket

soffit

strut

tie

longitudinal ties

ceiling tie or bottom chord

100 × 75 wall plate

strut

tie

100 × 75 wall plate

fascia

sprocket

soffit

TYPICAL TRUSSED RAFTER DETAIL

Longitudinal ties (75 × 38) fixed over ceiling ties and under internal ties near to roof apex and rafter diagonal bracing (75 × 38) fixed under rafters at gable ends from eaves to apex may be required to provide stability bracing – actual requirements specified by manufacturer. Lateral restraint to gable walls at top and bottom chord levels in the form of mild steel straps at 2·000 maximum centres over 2 No. trussed rafters may also be required.

Gambrel roofs are double pitched with a break in the roof slope. The pitch angle above the break is less than 45° relative to the horizontal, whilst the pitch angle below the break is greater. Generally, these angles are 30° and 60°.

Gambrels are useful in providing more attic headroom and frequently incorporate dormers and rooflights. They have a variety of constructional forms.

Typically —

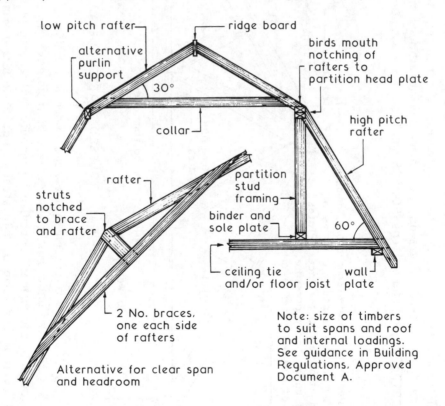

low pitch rafter
ridge board
alternative purlin support
30°
collar
birds mouth notching of rafters to partition head plate
high pitch rafter
rafter
struts notched to brace and rafter
partition stud framing
binder and sole plate
60°
ceiling tie and/or floor joist
wall plate
2 No. braces, one each side of rafters
Alternative for clear span and headroom
Note: size of timbers to suit spans and roof and internal loadings. See guidance in Building Regulations, Approved Document A.

Intermediate support can be provided in various ways as shown above. To create headroom for accommodation in what would otherwise be attic space, a double head plate and partition studding is usual. The collar beam and rafters can conveniently locate on the head plates or prefabricated trusses can span between partitions.

Roof Underlays ~ sometimes called sarking or roofing felt provides the barrier to the entry of snow, wind and rain blown between the tiles or slates. It also prevents the entry of water from capillary action.

Suitable Materials ~

Bitumen fibre based felts — supplied in rolls 1m wide and up to 25m long. Traditionally used in house construction with a cold ventilated roof.

Breather or vapour permeable underlay — typically produced from HDPE fibre or extruded polypropylene fibre, bonded by heat and pressure. Materials permeable to water vapour are preferred as these do not need to be perforated to ventilate the roof space. Also, subject to manufacturer's guidelines, traditional eaves ventilation may not be necessary. Underlay of this type should be installed taut across the rafters with counter battens support to the tile battens. Where counter battens are not used, underlay should sag slightly between rafters to allow rain penetration to flow under tile battens.

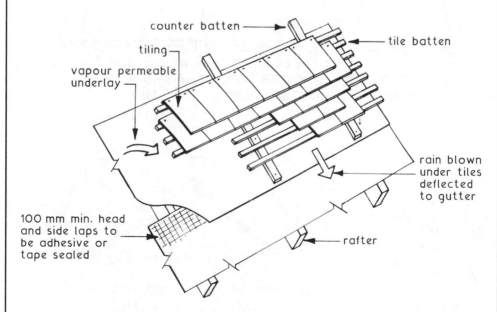

Underlays are fixed initially with galvanised clout nails or st/st staples but are finally secured with the tiling or slating batten fixings

Double Lap Tiles ~ these are the traditional tile covering for pitched roofs and are available made from clay and concrete and are usually called plain tiles. Plain tiles have a slight camber in their length to ensure that the tail of the tile will bed and not ride on the tile below. There is always at least two layers of tiles covering any part of the roof. Each tile has at least two nibs on the underside of its head so that it can be hung on support battens nailed over the rafters. Two nail holes provide the means of fixing the tile to the batten, in practice only every 4th course of tiles is nailed unless the roof exposure is high. Double lap tiles are laid to a bond so that the edge joints between the tiles are in the centre of the tiles immediately below and above the course under consideration.

Typical Plain Tile Details ~

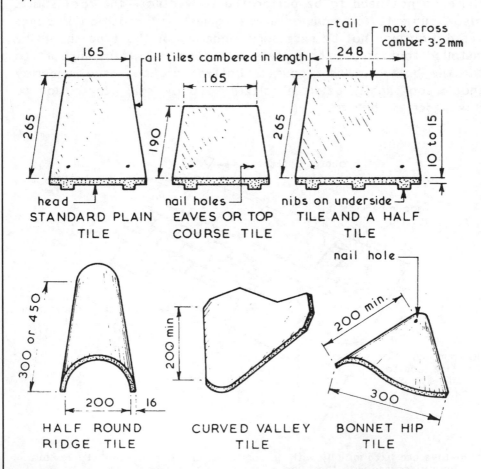

STANDARD PLAIN TILE

EAVES OR TOP COURSE TILE

TILE AND A HALF TILE

HALF ROUND RIDGE TILE

CURVED VALLEY TILE

BONNET HIP TILE

For other types shapes and sections see BS EN 1304: Clay roofing tiles and fittings. Product definitions and specifications.

Typical details ~

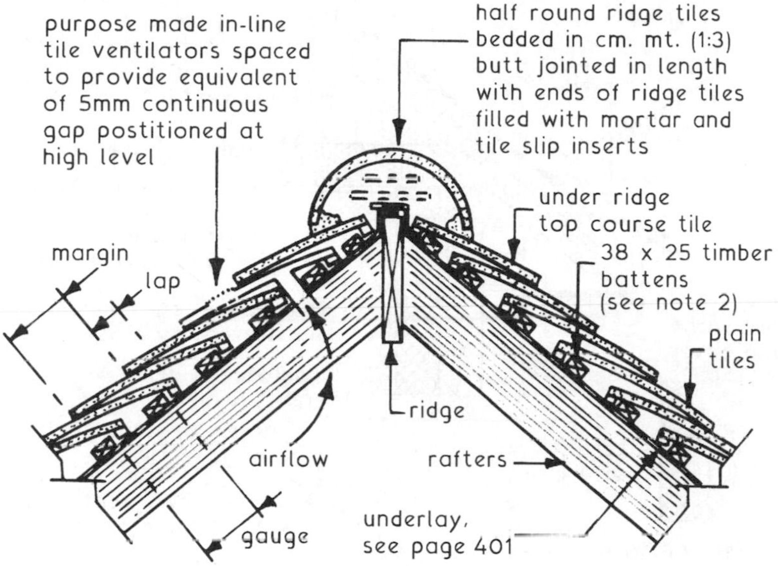

purpose made in-line tile ventilators spaced to provide equivalent of 5mm continuous gap postitioned at high level

half round ridge tiles bedded in cm. mt. (1:3) butt jointed in length with ends of ridge tiles filled with mortar and tile slip inserts

under ridge top course tile
38 x 25 timber battens (see note 2)

plain tiles

margin

lap

ridge

airflow

rafters

gauge

underlay, see page 401

RIDGE DETAIL

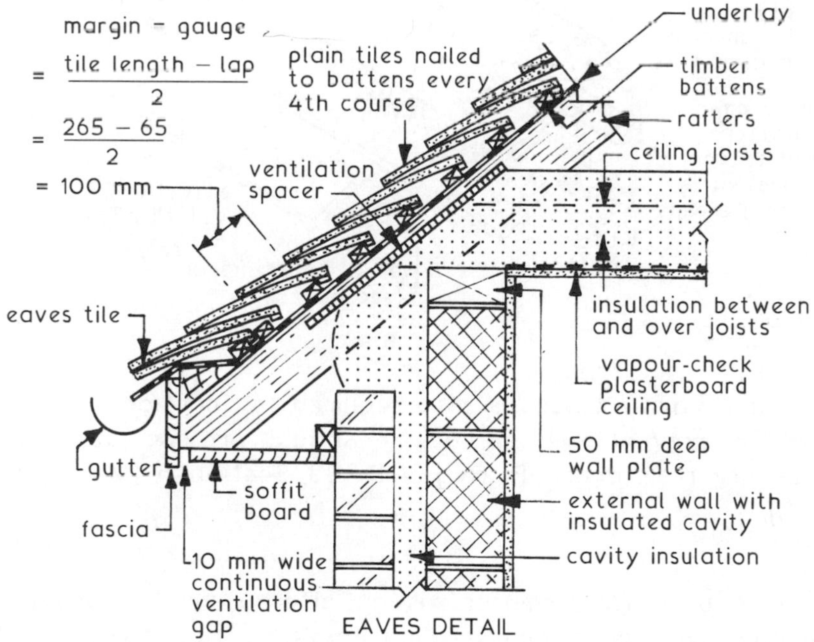

$$\text{margin} = \text{gauge}$$

$$= \frac{\text{tile length} - \text{lap}}{2}$$

$$= \frac{265 - 65}{2}$$

$$= 100 \text{ mm}$$

plain tiles nailed to battens every 4th course

underlay

timber battens

rafters

ceiling joists

ventilation spacer

insulation between and over joists

vapour-check plasterboard ceiling

eaves tile

gutter

soffit board

50 mm deep wall plate

fascia

external wall with insulated cavity

cavity insulation

10 mm wide continuous ventilation gap

EAVES DETAIL

Note 1: Through ventilation is necessary to prevent condensation occurring in the roof space.

Note 2: 50 × 25 where rafter spacing is 600mm.

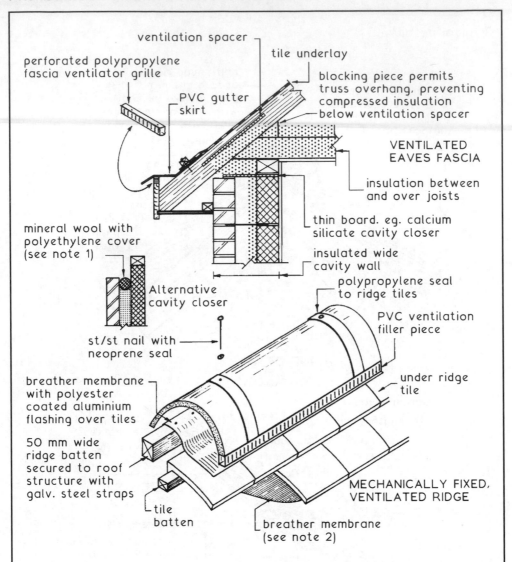

Note 1. If a cavity closer is also required to function as a cavity barrier to prevent fire spread, it should provide at least 30 minutes fire resistance, (B. Reg. A.D. B3 Section 6 [Vol. 1] and 9 [Vol. 2]).

Note 2. A breather membrane is an alternative to conventional bituminous felt as an under-tiling layer. It has the benefit of restricting liquid water penetration whilst allowing water vapour transfer from within the roof space. This permits air circulation without perforating the under-tiling layer.

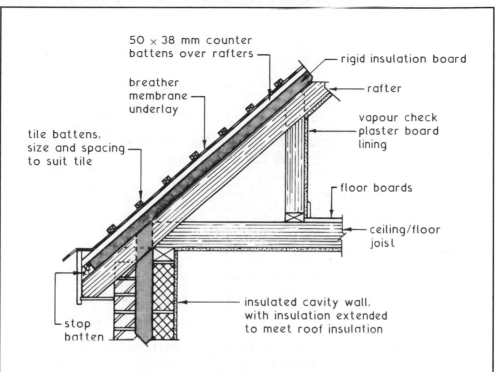

50 × 38 mm counter battens over rafters

rigid insulation board

breather membrane underlay

rafter

tile battens, size and spacing to suit tile

vapour check plaster board lining

floor boards

ceiling/floor joist

insulated cavity wall, with insulation extended to meet roof insulation

stop batten

Where a roof space is used for habitable space, insulation must be provided within the roof slope. Insulation above the rafters (as shown) creates a 'warm roof', eliminating the need for continuous ventilation. Insulation placed between the rafters creates a 'cold roof', requiring a continuous 50 mm ventilation void above the insulation to prevent the possible occurrence of interstitial condensation.

Suitable rigid insulants include; low density polyisocyanurate (PIR) foam, reinforced with long strand glass fibres, both faces bonded to aluminium foil with joints aluminium foil taped on the upper surface; high density mineral wool slabs over rafters with less dense mineral wool between rafters.

An alternative location for the breather membrane is under the counter battens. This is often preferred as the insulation board will provide uniform support for the underlay. Otherwise, extra insulation could be provided between the counter battens, retaining sufficient space for the underlay to sag between rafter positions to permit any rainwater penetration to drain to eaves.

Typical Details ~

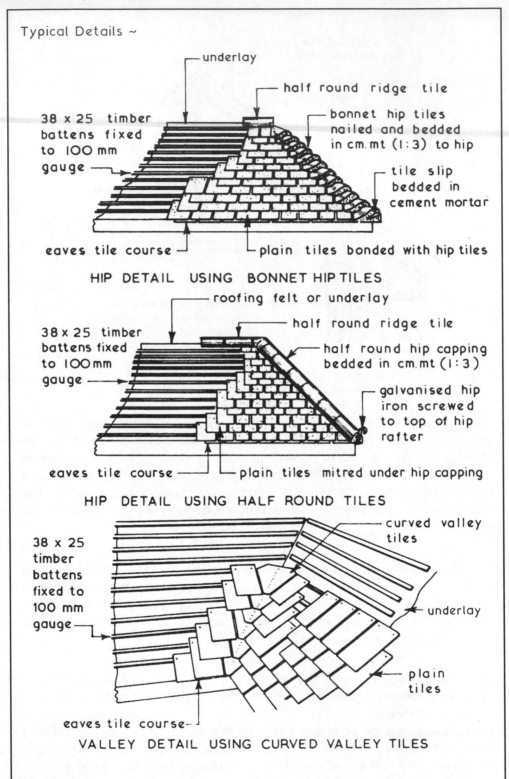

38 x 25 timber battens fixed to 100 mm gauge — underlay

half round ridge tile

bonnet hip tiles nailed and bedded in cm.mt (1:3) to hip

tile slip bedded in cement mortar

eaves tile course — plain tiles bonded with hip tiles

HIP DETAIL USING BONNET HIP TILES

roofing felt or underlay

38 x 25 timber battens fixed to 100 mm gauge — half round ridge tile

half round hip capping bedded in cm.mt (1:3)

galvanised hip iron screwed to top of hip rafter

eaves tile course — plain tiles mitred under hip capping

HIP DETAIL USING HALF ROUND TILES

38 x 25 timber battens fixed to 100 mm gauge — curved valley tiles

underlay

plain tiles

eaves tile course —

VALLEY DETAIL USING CURVED VALLEY TILES

Typical Details ~

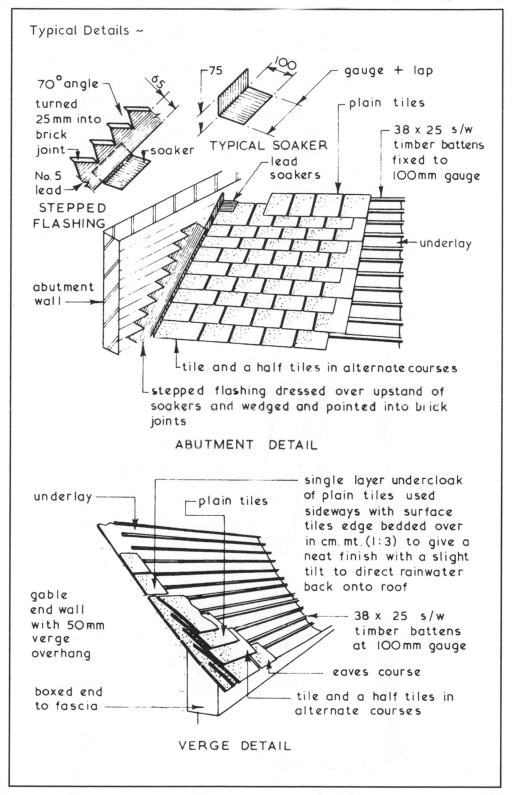

70°angle turned 25mm into brick joint

65

No. 5 lead

STEPPED FLASHING

soaker

75

100

TYPICAL SOAKER

gauge + lap

plain tiles

38 x 25 s/w timber battens fixed to 100mm gauge

lead soakers

underlay

abutment wall

tile and a half tiles in alternate courses

stepped flashing dressed over upstand of soakers and wedged and pointed into brick joints

ABUTMENT DETAIL

underlay

plain tiles

single layer undercloak of plain tiles used sideways with surface tiles edge bedded over in cm. mt. (1:3) to give a neat finish with a slight tilt to direct rainwater back onto roof

gable end wall with 50mm verge overhang

38 x 25 s/w timber battens at 100mm gauge

eaves course

boxed end to fascia

tile and a half tiles in alternate courses

VERGE DETAIL

Single Lap Tiling ~ so called because the single lap of one tile over another provides the weather tightness as opposed to the two layers of tiles used in double lap tiling. Most of the single lap tiles produced in clay and concrete have a tongue and groove joint along their side edges and in some patterns on all four edges which forms a series of interlocking joints and therefore these tiles are called single lap interlocking tiles. Generally there will be an overall reduction in the weight of the roof covering when compared with double lap tiling but the batten size is larger than that used for plain tiles and as a minimum every tile in alternate courses should be twice nailed, although a good specification will require every tile to be twice nailed. The gauge or batten spacing for single lap tiling is found by subtracting the end lap from the length of the tile.

Typical Single Lap Tiles ~

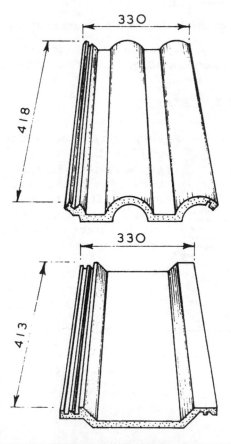

ROLL TYPE TILE

minimum pitch 30°

head lap 75 mm

side lap 30 mm

gauge 343 mm

linear coverage 300 mm

TROUGH TYPE TILE

minimum pitch 15°

head lap 75 mm

side lap 38 mm

gauge 338 mm

linear coverage 292 mm

Typical Details ~

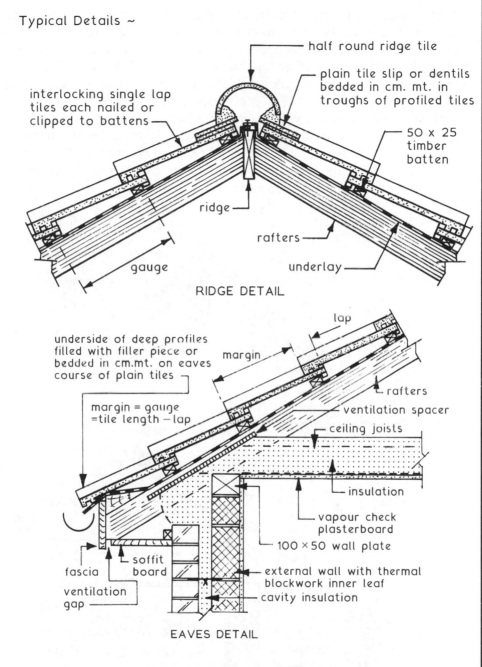

half round ridge tile

plain tile slip or dentils bedded in cm. mt. in troughs of profiled tiles

interlocking single lap tiles each nailed or clipped to battens

50 x 25 timber batten

ridge

rafters

gauge

underlay

RIDGE DETAIL

lap

underside of deep profiles filled with filler piece or bedded in cm.mt. on eaves course of plain tiles

margin

rafters

margin = gauge = tile length − lap

ventilation spacer

ceiling joists

insulation

vapour check plasterboard

100 × 50 wall plate

soffit board

fascia

external wall with thermal blockwork inner leaf

ventilation gap

cavity insulation

EAVES DETAIL

Hips − can be finished with a half round tile as a capping as shown for double lap tiling on page 406.

Valleys − these can be finished by using special valley trough tiles or with a lead lined gutter − see manufacturer's data.

Slates ~ slate is a natural dense material which can be split into thin sheets and cut to form a small unit covering suitable for pitched roofs in excess of 25° pitch. Slates are graded according to thickness and texture, the thinnest being known as 'Bests'. These are of 4mm nominal thickness. Slates are laid to the same double lap principles as plain tiles. Ridges and hips are normally covered with half round or angular tiles whereas valley junctions are usually of mitred slates over soakers. Unlike plain tiles every course is fixed to the battens by head or centre nailing, the latter being used on long slates and on pitches below 35° to overcome the problem of vibration caused by the wind which can break head nailed long slates.

Typical Details ~

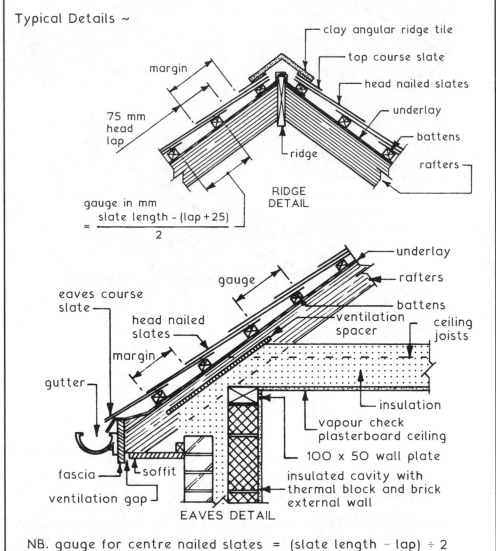

$$\text{gauge in mm} = \frac{\text{slate length} - (\text{lap} + 25)}{2}$$

RIDGE DETAIL

NB. gauge for centre nailed slates = (slate length − lap) ÷ 2

The UK has been supplied with its own slate resources from quarries in Wales, Cornwall and Westermorland. Imported slate is also available from Spain, Argentina and parts of the Far East.

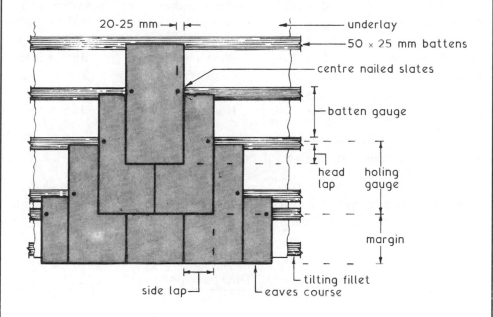

e.g. Countess slate, 510 × 255mm laid to a 30° pitch with 75mm head lap.

Batten gauge = (slate length − lap) ÷ 2
= (510 − 75) ÷ 2 = 218mm.

Holing gauge = batten gauge + head lap + 8 to 15mm,
= 218 + 75 + (8 to 15mm) = 301 to 308mm.

Side lap = 255 ÷ 2 = 127mm.

Margin = batten gauge of 218mm.

Eaves course length = head lap + margin = 293mm.

Traditional slate names and sizes (mm) –

Empress	650 × 400	Wide Viscountess	460 × 255
Princess	610 × 355	Viscountess	460 × 230
Duchess	610 × 305	Wide Ladies	405 × 255
Small Duchess	560 × 305	Broad Ladies	405 × 230
Marchioness	560 × 280	Ladies	405 × 205
Wide Countess	510 × 305	Wide Headers	355 × 305
Countess	510 × 255	Headers	355 × 255
.. ..	510 × 230	Small Ladies	355 × 203
.. ..	460 × 305	Narrow Ladies	355 × 180

Sizes can also be cut to special order.

Generally, the larger the slate, the lower the roof may be pitched. Also, the lower the roof pitch, the greater the head lap.

Slate quality	Thickness (mm)
Best	4
Medium strong	5
Heavy	6
Extra heavy	9

Roof pitch (degrees)	Min. head lap (mm)
20	115
25	85
35	75
45	65

See also:

1. BS EN 12326-1: Slate and stone products for discontinuous roofing and cladding. Product specification.
2. Slate producers' catalogues.
3. BS 5534: Code of practice for slating and tiling.

Roof hip exaples –

Close mitred hip, roof pitch > 30° — hip rafter

jack rafter

underlay in two layers, overlapping at least 300 mm each side of hip

50 × 25 mm batten

code 3 lead soaker under each pair of mitred slates

opposing pairs of mitre cut slates

eaves course

standard slate

Mitred hip with clay or concrete hip tiles

hip rafter — batten

jack rafter

double layer of underlay at hip, at least 300 mm overlap each side of hip

mitred slates

concrete or clay hip tile mortar bedded over mitred slates

hip iron at eaves

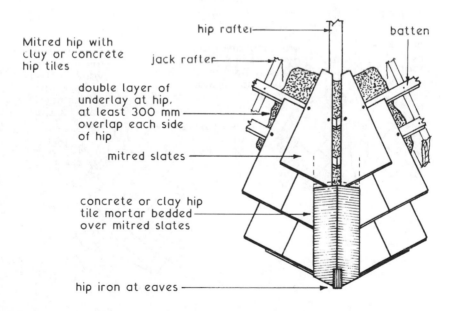

Roof valley examples –

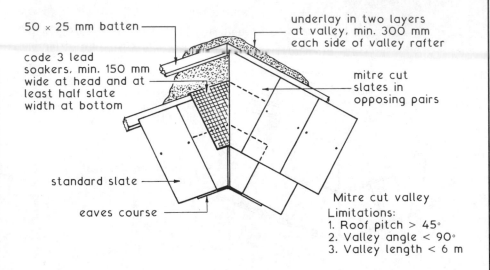

50 × 25 mm batten

code 3 lead soakers, min. 150 mm wide at head and at least half slate width at bottom

underlay in two layers at valley, min. 300 mm each side of valley rafter

mitre cut slates in opposing pairs

standard slate

eaves course

Mitre cut valley
Limitations:
1. Roof pitch > 45°
2. Valley angle < 90°
3. Valley length < 6 m

Alternatives

valley rafter

wide lay boards in valley to support taper cut slates

two supplementary layers of underlay over lay boards to overlap normal underlay

valley slates tapered to a smooth curve

Swept valley

225 mm min. lay board on valley rafter, usually with additional board either side

valley rafter

two layers of underlay at valley

jack rafter

Laced valley

Materials – water reed (Norfolk reed), wheat straw (Spring or Winter), Winter being the most suitable. Wheat for thatch is often known as wheat reed, long straw or Devon reed. Other thatches include rye and oat straws, and sedge. Sedge is harvested every fourth year to provide long growth, making it most suitable as a ridging material.

There are various patterns and styles of thatching, relating to the skill of the thatcher and local traditions.

Typical details –

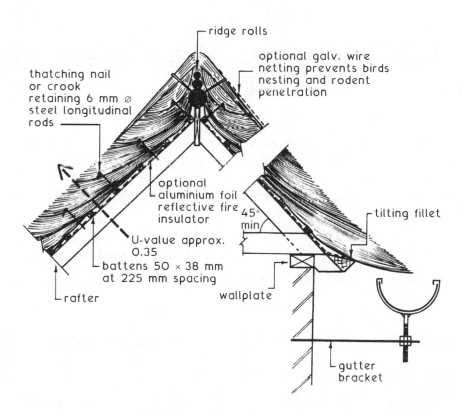

The material composition of thatch with its natural voids and surface irregularities provides excellent insulation when dry and compact. However, when worn with possible accumulation of moss and rainwater, the U-value is less reliable. Thatch is also very vulnerable to fire. Therefore in addition to imposing a premium, insurers may require application of a surface fire retardant and a fire insulant underlay.

Flat Roofs ~ these roofs are very seldom flat with a pitch of 0° but are considered to be flat if the pitch does not exceed 10°. The actual pitch chosen can be governed by the roof covering selected and/or by the required rate of rainwater discharge off the roof. As a general rule the minimum pitch for smooth surfaces such as asphalt should be 1:80 or 0°-43' and for sheet coverings with laps 1:60 or 0°-57'.

Methods of Obtaining Falls ~

1. Joists cut to falls

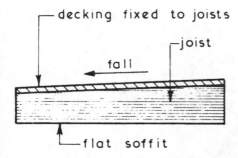

Simple to fix but could be wasteful in terms of timber unless two joists are cut from one piece of timber

2. Joists laid to falls

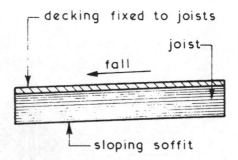

Economic and simple but sloping soffit may not be acceptable but this could be hidden by a flat suspended ceiling

3. Firrings with joist run

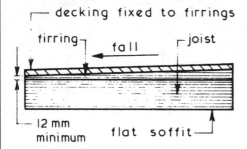

Simple and effective but does not provide a means of natural cross ventilation. Usual method employed.

4. Firrings against joist run

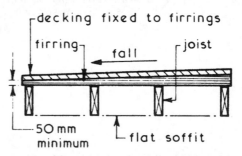

Simple and effective but uses more timber than 3 but does provide a means of natural cross ventilation

Wherever possible joists should span the shortest distance of the roof plan.

Timber Roof Joists ~ the spacing and sizes of joists is related to the loadings and span, actual dimensions for domestic loadings can be taken direct from recommendations in Approved Document A or they can be calculated as shown for timber beam designs. Strutting between joists should be used if the span exceeds 2·400 to restrict joist movements and twisting.

Typical Eaves Details ~

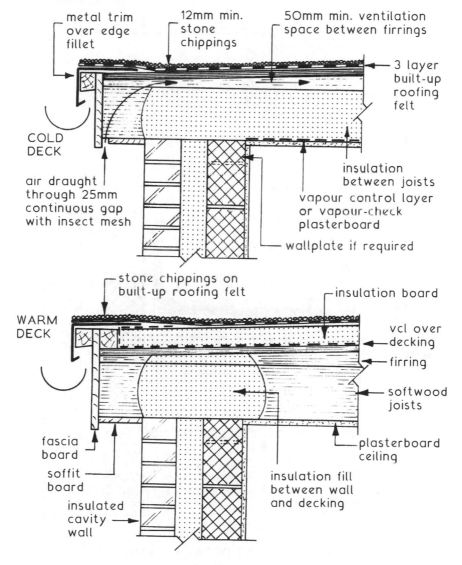

metal trim over edge fillet

12mm min. stone chippings

50mm min. ventilation space between firrings

3 layer built-up roofing felt

COLD DECK

air draught through 25mm continuous gap with insect mesh

insulation between joists

vapour control layer or vapour-check plasterboard

wallplate if required

stone chippings on built-up roofing felt

insulation board

WARM DECK

vcl over decking

firring

softwood joists

fascia board

soffit board

insulated cavity wall

insulation fill between wall and decking

plasterboard ceiling

Ref. BS EN 13707: Flexible sheets for waterproofing. Reinforced bitumen sheets for roof waterproofing. Definitions and characteristics.

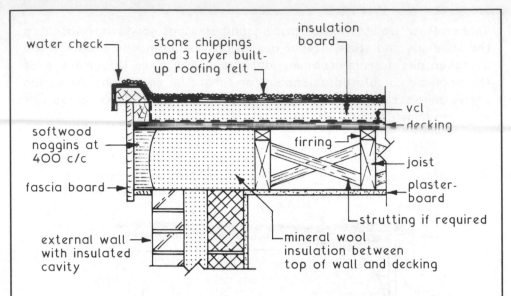

water check

stone chippings
and 3 layer built-
up roofing felt

insulation
board

vcl

decking

softwood
noggins at
400 c/c

firring

joist

fascia board

plaster-
board

strutting if required

external wall
with insulated
cavity

mineral wool
insulation between
top of wall and decking

TYPICAL VERGE DETAILS - WARM DECK

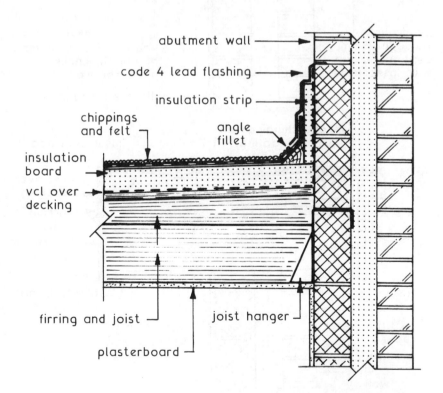

abutment wall

code 4 lead flashing

insulation strip

chippings
and felt

angle
fillet

insulation
board

vcl over
decking

firring and joist

joist hanger

plasterboard

TYPICAL ABUTMENT DETAILS - WARM DECK

Ref. BS 8217: Reinforced bitumen membranes for roofing. Code of
practice.

418

A dormer is the framework for a vertical window constructed from the roof slope. It may be used as a feature, but is more likely as an economical and practical means for accessing light and ventilation to an attic room. Dormers are normally external with the option of a flat or pitched roof. Frame construction is typical of the following illustrations, with connections made by traditional housed and tenoned joints or simpler galvanized steel brackets and hangers.

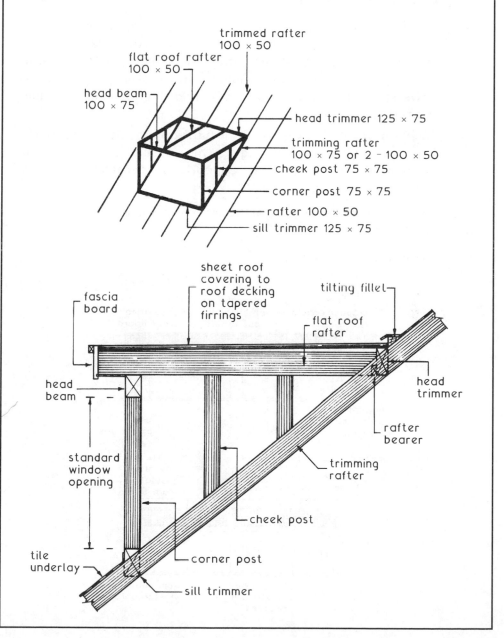

Conservation of Energy ~ this can be achieved in two ways:

1. Cold Deck – insulation is placed on the ceiling lining, between joists. See page 417 for details. A metallized polyester lined plasterboard ceiling functions as a vapour control layer, with a minimum 50 mm air circulation space between insulation and decking. The air space corresponds with eaves vents and both provisions will prevent moisture build-up, condensation and possible decay of timber.

2. (a) Warm Deck – rigid* insulation is placed below the waterproof covering and above the roof decking. The insulation must be sufficient to maintain the vapour control layer and roof members at a temperature above dewpoint, as this type of roof does not require ventilation.

   (b) Inverted Warm Deck – rigid* insulation is positioned above the waterproof covering. The insulation must be unaffected by water and capable of receiving a stone dressing or ceramic pavings.

* Resin bonded mineral fibre roof boards, expanded polystyrene or polyurethane slabs.

Typical Warm Deck Details ~

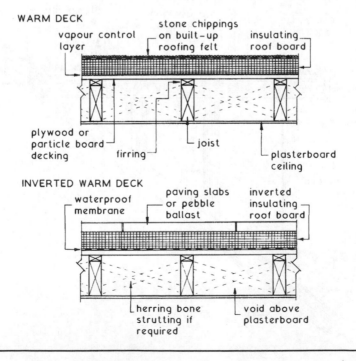

WARM DECK
vapour control layer
stone chippings on built-up roofing felt
insulating roof board
plywood or particle board decking
firring
joist
plasterboard ceiling

INVERTED WARM DECK
waterproof membrane
paving slabs or pebble ballast
inverted insulating roof board
herring bone strutting if required
void above plasterboard

420

Built-up Roofing Felt ~ this consists of three layers of bitumen roofing felt to BS EN 13707, and should be laid to the recommendations of BS 8217. The layers of felt are bonded together with hot bitumen and should have staggered laps of 50mm minimum for side laps and 75mm minimum for end laps – for typical details and references see pages 417 & 418.

Other felt materials which could be used are the two layer polyester based roofing felts which use a non-woven polyester base instead of the woven rag fibre base used in traditional felts.

Mastic Asphalt ~ this consists of two layers of mastic asphalt laid breaking joints and built up to a minimum thickness of 20mm and should be laid to the recommendations of BS 8218. The mastic asphalt is laid over an isolating membrane of black sheathing felt which should be laid loose with 50mm minimum overlaps.

Typical Datails ~

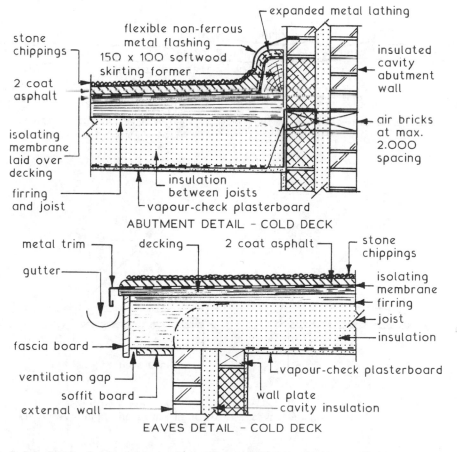

ABUTMENT DETAIL – COLD DECK

EAVES DETAIL – COLD DECK

Ref. BS 8218: Code of practice for mastic asphalt roofing.

Air carries water vapour, the amount increasing proportionally with the air temperature. As the water vapour increases so does the pressure and this causes the vapour to migrate from warmer to cooler parts of a building. As the air temperature reduces, so does its ability to hold water and this manifests as condensation on cold surfaces. Insulation between living areas and roof spaces increases the temperature differential and potential for condensation in the roof void.

Condensation can be prevented by either of the following:

* Providing a vapour control layer on the warm side of any insulation.
* Removing the damp air by ventilating the colder area.

The most convenient form of vapour layer is vapour check plasterboard which has a moisture resistant lining bonded to the back of the board. A typical patented product is a foil or metallised polyester backed plasterboard in 9·5 and 12·5mm standard thicknesses. This is most suitable where there are rooms in roofs and for cold deck flat roofs. Ventilation is appropriate to larger roof spaces.

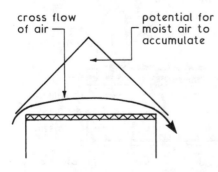

cross flow of air — potential for moist air to accumulate

Partial roof void ventilation through the eaves

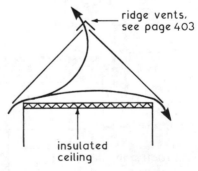

ridge vents, see page 403

insulated ceiling

Total roof void ventilation through eaves and high level vents

Roof ventilation – provision of eaves ventilation alone should allow adequate air circulation in most situations. However, in some climatic conditions and where the air movement is not directly at right angles to the building, moist air can be trapped in the roof apex. Therefore, supplementary ridge ventilation is recommended.

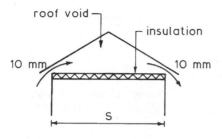

Insulation at ceiling level (1)
S = span < 10 m for
roof pitches 15°-35°

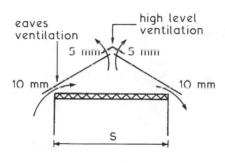

Insulation at ceiling level (2)
S = span > 10 m for
roof pitches 15°-35°
Any span for roof
pitches > 35°

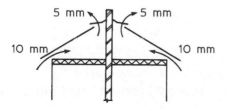

Insulation at ceiling level and
central dividing wall
Roof pitches > 15°
for any span

Note: ventilation dimensions shown relate to a continuous strip (or equivalent) of at least the given gap.

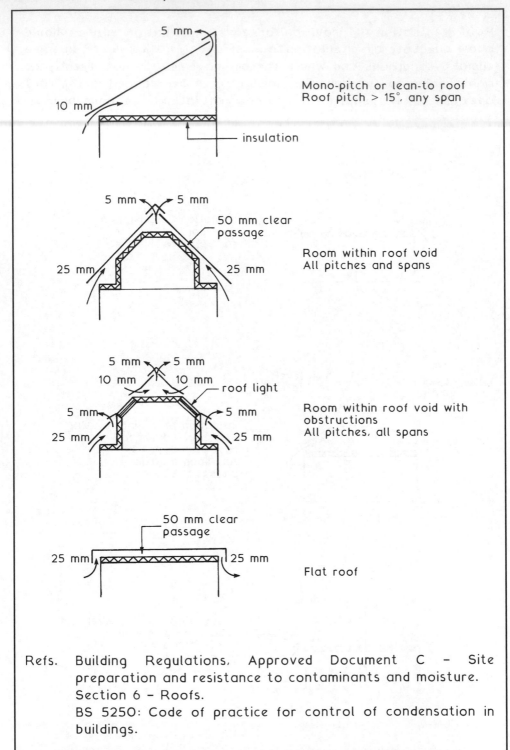

Mono-pitch or lean-to roof
Roof pitch > 15°, any span

insulation

Room within roof void
All pitches and spans

Room within roof void with
obstructions
All pitches, all spans

Flat roof

Refs.   Building Regulations, Approved Document C – Site
preparation and resistance to contaminants and moisture.
Section 6 – Roofs.
BS 5250: Code of practice for control of condensation in
buildings.

BRE report – Thermal Insulation: avoiding risks (3rd. ed.).

Lateral Restraint – stability of gable walls and construction at the eaves, plus integrity of the roof structure during excessive wind forces, requires complementary restraint and continuity through 30 × 5mm <u>cross sectional area galvanised steel straps.</u>

Exceptions may occur if the roof:-

1. exceeds 15° pitch, and
2. is tiled or slated, and
3. has the type of construction known locally to resist gusts, and
4. has ceiling joists and rafters bearing onto support walls at not more than 1·2m centres.

Application ~

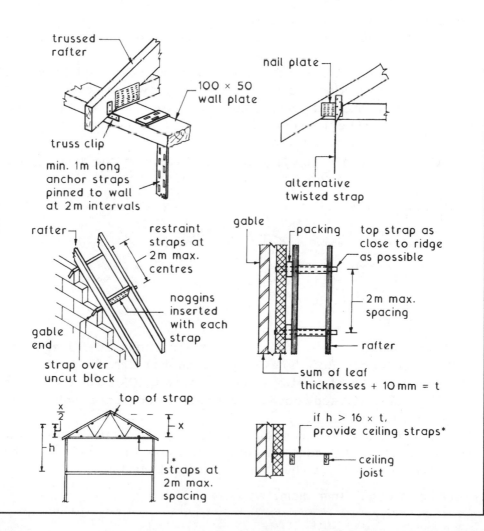

trussed rafter

nail plate

100 × 50 wall plate

truss clip

min. 1m long anchor straps pinned to wall at 2m intervals

alternative twisted strap

rafter

restraint straps at 2m max. centres

gable

packing

top strap as close to ridge as possible

2m max. spacing

noggins inserted with each strap

gable end

rafter

strap over uncut block

sum of leaf thicknesses + 10 mm = t

top of strap

$\frac{x}{2}$

x

h

if h > 16 × t, provide ceiling straps*

ceiling joist

*straps at 2m max. spacing

425

Preservation ~ ref. Building Regulations: Materials and Workmanship. Approved Document to support Regulation 7.

Woodworm infestation of untreated structural timbers is common. However, the smaller woodborers such as the abundant Furniture beetle are controllable. It is the threat of considerable damage potential from the House Longhorn beetle that has forced many local authorities in Surrey and the fringe areas of adjacent counties to seek timber preservation listing in the Building Regulations (see Table 1 in the above reference). Prior to the introduction of pretreated timber (c. 1960s), the House Longhorn beetle was once prolific in housing in the south of England, establishing a reputation for destroying structural roof timbers, particularly in the Camberley area.

House Longhorn beetle data:-

Latin name – Hylotrupes bajulus

Life cycle – Mature beetle lays up to 200 eggs on rough surface of untreated timber.
After 2-3 weeks, larvae emerge and bore into wood, preferring sapwood to denser growth areas. Up to 10 years in the damaging larval stage. In 3 weeks, larvae change to chrysalis to emerge as mature beetles in summer to reproduce.

Timber appearance – powdery deposits (frass) on the surface and the obvious mature beetle flight holes.

Beetle appearence –

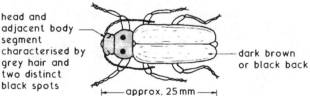

head and adjacent body segment characterised by grey hair and two distinct black spots — dark brown or black back — approx. 25mm

Other woodborers:-

Furniture beetle – dark brown, 6-8mm long, lays 20-50 eggs on soft or hardwoods. Bore holes only 1-2mm diameter.
Lyctus powder post beetle – reddish brown, 10-15mm long, lays 70-200 eggs on sapwood of new hardwood. Bore holes only 1-2mm in diameter.
Death Watch beetle – dark brown, sometimes speckled in lighter shades. Lays 40-80 eggs on hardwood. Known for preferring the oak timbers used in old churches and similar buildings.

Bore holes about 3mm diameter.

Preservation ~ treatment of timber to prevent damage from House Longhorn beetle.

In the areas specified (see previous page), all softwood used in roof structures including ceiling joists and any other softwood fixings should be treated with insecticide prior to installation. Specific chemicals and processes have not been listed in the Building Regulations since the 1976 issue, although the processes detailed then should suffice:-

1. Treatment to BS 4072.*
2. Diffusion with sodium borate (boron salts).
3. Steeping for at least 10 mins in an organic solvent wood preservative.

NB. Steeping or soaking in creosote will be effective, but problems of local staining are likely.

BS 4072 provides guidance on an acceptable blend of copper, chromium and arsenic known commercially as Tanalizing. Application is at specialist timber yards by vacuum/pressure impregnation in large cylindrical containers, but see note below.

Insect treatment adds about 10% to the cost of timber and also enhances its resistance to moisture. Other parts of the structure, e.g. floors and partitions are less exposed to woodworm damage as they are enclosed. Also, there is a suggestion that if these areas received treated timber, the toxic fumes could be harmful to the health of building occupants. Current requirements for through ventilation in roofs has the added benefit of discouraging wood boring insects, as they prefer draught-free damp areas.

Refs. BS 4072: Copper/chromium/arsenic preparations for wood preservation.*
BS 4261: Wood preservation. Vocabulary.
BS 5589: Code of practice for preservation of timber.
BS 8417: Preservation of timber. Recommendations.
BS 5707: Specification for preparations of wood preservatives in organic solvents.

*Note: The EU are processing legislation which will prohibit the use of CCA preservatives for domestic applications and in places where the public may be in contact with it. Ref. CEN/TC 38.

Damp conditions can be the source of many different types of wood-decaying fungi. The principal agencies of decay are –

* Dry rot (Serpula lacrymans or merulius lacrymans), and
* Wet rot (Coniophora cerabella)

Dry rot – this is the most difficult to control as its root system can penetrate damp and porous plaster, brickwork and concrete. It can also remain dormant until damp conditions encourage its growth, even though the original source of dampness is removed.

Appearance – white fungal threads which attract dampness from the air or adjacent materials. The threads develop strands bearing spores or seeds which drift with air movements to settle and germinate on timber having a moisture content exceeding about 25%. Fruiting bodies of a grey or red flat profile may also identify dry rot.

Typical surface appearance of dry rot –

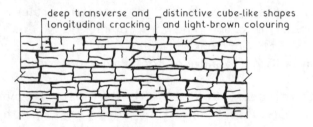

deep transverse and longitudinal cracking | distinctive cube-like shapes and light-brown colouring

Wet rot – this is limited in its development and must have moisture continually present, e.g. a permanent leaking pipe or a faulty dpc. Growth pattern is similar to dry rot, but spores will not germinate in dry timber.

Appearance – fungal threads of black or dark brown colour. Fruiting bodies may be olive-green or dark brown and these are often the first sign of decay.

Typical surface appearance of wet rot –

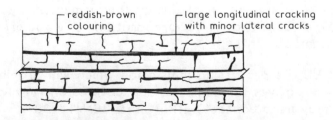

reddish-brown colouring | large longitudinal cracking with minor lateral cracks

Causes –
* Defective construction, e.g. broken roof tiles; no damp-proof course.
* Installation of wet timber during construction, e.g. framing sealed behind plasterboard linings; wet joists under floor decking.
* Lack of ventilation, e.g. blocked air bricks to suspended timber ground floor; condensation in unventilated roof spaces.
* Defective water services, e.g. undetected leaks on internal pipework; blocked or broken rainwater pipes and guttering.

General treatment –
* Remove source of dampness.
* Allow affected area to dry.
* Remove and burn all affected timber and sound timber within 500 mm of fungal attack.
* Remove contaminated plaster and rake out adjacent mortar joints to masonry.

Note: This is normally sufficient treatment where wet rot is identified. However, where dry rot is apparent the following additional treatment is necessary:
* Sterilise surface of concrete and masonry.
  Heat with a blow torch until the surface is too hot to touch. Apply a proprietary fungicide† generously to warm surface. Irrigate badly affected masonry and floors, i.e. provide 12 mm diameter bore holes at about 500 mm spacing and flood or pressure inject with fungicide.

† 20:1 dilution of water and sodium pentachlorophenate, sodium orthophenylphate or mercuric chloride. Product manufacturers' safety in handling and use measures must be observed when applying these chemicals.

Replacement work should ensure that new timbers are pressure impregnated with a preservative. Cement and sand mixes for rendering, plastering and screeds should contain a zinc oxychloride fungicide.

Further reading –
BRE: Timber durability and treatment pack – various Digests, Information Papers and Good Repair Guides.
Remedial timber treatment in buildings – HSE Books.

Ref: Bldg. Regs. Approved Document C, Site preparation and resistance to contaminants and moisture.

Green roof ~ green with reference to the general appearance of plant growths and for being environmentally acceptable. Part of the measures for constructing sustainable and ecologically friendly buildings.

Categories ~
• Extensive ~ a relatively shallow soil base (typically 50 mm) and lightweight construction. Maximum roof pitch is 40° and slopes greater than 20° will require a system of baffles to prevent the soil moving. Plant life is limited by the shallow soil base to grasses, mosses, herbs and sedum (succulents, generally with fleshy leaves producing pink or white flowers).
• Intensive ~ otherwise known as a roof garden. This category has a deeper soil base (typically 400 mm) that will provide for landscaping features, small ponds, occasional shrubs and small trees. A substantial building structure is required for support and it is only feasible to use a flat roof.

Advantages ~
• Absorbs and controls water run-off.
• Integral thermal insulation.
• Integral sound insulation.
• Absorbs air pollutants, dust and CO_2.
• Passive heat storage potential.

Disadvantages ~
• Weight.
• Maintenance.

Construction ~ the following build-up will be necessary to fulfil the objectives and to create stability:
• Vapour control layer above the roof structure.
• Rigid slab insulation.
• Root resilient waterproof under-layer.
• Drainage layer.
• Filter.
• Growing medium (soil).
• Vegetation (grass, etc.)

Examples of both extensive and intensive green roof construction are shown on the next page.

Typical extensive roof build up ~

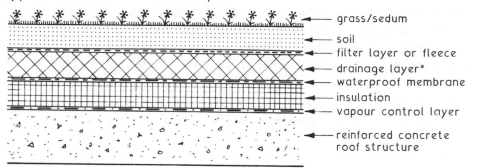

- grass/sedum
- soil
- filter layer or fleece
- drainage layer*
- waterproof membrane
- insulation
- vapour control layer
- reinforced concrete roof structure

* typically, expanded polystyrene with slots

Component	Weight (kg/m²)	Thickness (mm)
vcl	3	3
insulation	3	50
membrane	5	5
drainage layer	3	50
filter	3	3
soil	90	50
turf	40	20
	------------	---------
	147 kg/m²	181 mm

147 kg/m² saturated weight x 9.81 = 1442 N/m² or 1.44 kN/m²

Typical intensive roof build up ~

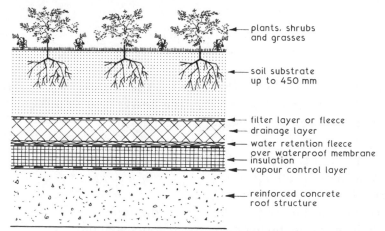

- plants, shrubs and grasses
- soil substrate up to 450 mm
- filter layer or fleece
- drainage layer
- water retention fleece over waterproof membrane
- insulation
- vapour control layer
- reinforced concrete roof structure

Depth to vcl, approximately 560 mm at about 750 kg/m² saturated weight. 750 kg/m² × 9.81 = 7358 N/m² or 7.36 kN/m².

Thermal insulation of external elements of construction is measured in terms of thermal transmittance rate, otherwise known as the U-value. It is the amount of heat energy in watts transmitted through one square metre of construction for every one degree Kelvin between external and internal air temperature, i.e. W/m^2K.

U-values are unlikely to be entirely accurate, due to:

* the varying effects of solar radiation, atmospheric dampness and prevailing winds.

* inconsistencies in construction, even with the best of supervision.

* 'bridging' where different structural components meet, e.g. dense mortar in lightweight blockwork.

Nevertheless, calculation of the U-value for a particular element of construction will provide guidance as to whether the structure is thermally acceptable. The Building Regulations, Approved Document L, Conservation of fuel and power, determines acceptable energy efficiency standards for modern buildings, with the objective of limiting the emission of carbon dioxide and other burnt gases into the atmosphere.

The U-value is calculated by taking the reciprocal of the summed thermal resistances (R) of the component parts of an element of construction:

$$U = \frac{1}{\Sigma R} = W/m^2K$$

R is expressed in m^2K/W. The higher the value, the better a component's insulation. Conversely, the lower the value of U, the better the insulative properties of the structure.

Building Regulations, Approved Document references:
L1A, Work in new dwellings.
L1B, Work in existing dwellings.
L2A, Work in new buildings other than dwellings.
L2B, Work in existing buildings other than dwellings.

Thermal resistances (R) are a combination of the different structural, surface and air space components which make up an element of construction. Typically:

$$U = \frac{1}{R_{so} + R_1 + R_2 + R_a + R_3 + R_4 \, etc \cdots + R_{si}} (m^2K/W)$$

Where: R_{so} = Outside or external surface resistance.

R_1, R_2, etc. = Thermal resistance of structural components.

R_a = Air space resistance, eg. wall cavity.

R_{si} = Internal surface resistance.

The thermal resistance of a structural component (R_1, R_2, etc.) is calculated by dividing its thickness (L) by its thermal conductivity (λ), i.e.

$$R(m^2K/W) = \frac{L(m)}{\lambda(W/mK)}$$

eg. 1. A 102mm brick with a conductivity of 0·84 W/mK has a thermal resistance (R) of: 0·102 ÷ 0·84 = 0·121 m²K/W.

eg. 2.

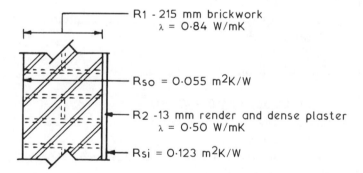

R1 - 215 mm brickwork
λ = 0·84 W/mK

Rso = 0·055 m²K/W

R2 -13 mm render and dense plaster
λ = 0·50 W/mK

Rsi = 0·123 m²K/W

Note: the effect of mortar joints in the brickwork can be ignored, as both components have similar density and insulative properties.

$$U = \frac{1}{R_{so} + R_1 + R_2 + R_{si}}$$

$R_1 = 0.215 \div 0.84 = 0.256$

$R_2 = 0.013 \div 0.50 = 0.026$

$$U = \frac{1}{0.055 + 0.256 + 0.026 + 0.123} = 2.17 W/m^2K$$

Typical values in: m^2K/W

Internal surface resistances (R_{si}):

Walls – 0·123
Floors or ceilings for upward heat flow – 0·104
Floors or ceilings for downward heat flow – 0·148
Roofs (flat or pitched) – 0·104

External surface resistances (R_{so}):

Surface	Exposure		
	Sheltered	Normal	Severe
Wall – high emissivity	0·080	0·055	0·030
Wall – low emissivity	0·110	0·070	0·030
Roof – high emissivity	0·070	0·045	0·020
Roof – low emissivity	0·090	0·050	0·020
Floor – high emissivity	0·070	0·040	0·020

Sheltered – town buildings to 3 storeys.
Normal – town buildings 4 to 8 storeys and most suburban premises.
Severe – > 9 storeys in towns.
> 5 storeys elsewhere and any buildings on exposed coasts and hills.

Air space resistances (R_a):

Pitched or flat roof space – 0·180
Behind vertical tile hanging – 0·120
Cavity wall void – 0·180
Between high and low emissivity surfaces – 0·300
Unventilated/sealed – 0·180

Emissivity relates to the heat transfer across and from surfaces by radiant heat emission and absorption effects. The amount will depend on the surface texture, the quantity and temperature of air movement across it, the surface position or orientation and the temperature of adjacent bodies or materials. High surface emissivity is appropriate for most building materials. An example of low emissivity would be bright aluminium foil on one or both sides of an air space.

Typical values –

Material	Density (kg/m³)	Conductivity (λ) (W/mK)
WALLS:		
Boarding (hardwood)	700	0·18
...... (softwood)	500	0·13
Brick outer leaf	1700	0·84
.... inner leaf	1700	0·62
Calcium silicate board	875	0·17
Ceramic tiles	2300	1·30
Concrete	2400	1·93
....	2200	1·59
....	2000	1·33
....	1800	1·13
.... (lightweight)	1200	0·38
.... (reinforced)	2400	2·50
Concrete block (lightweight)	600	0·18
.... (mediumweight)	1400	0·53
Cement mortar (protected)	1750	0·88
.... (exposed)	1750	0·94
Fibreboard	350	0·08
Gypsum plaster (dense)	1300	0·57
Gypsum plaster (lightweight)	600	0·16
Plasterboard	950	0·16
Tile hanging	1900	0·84
Rendering	1300	0·57
Sandstone	2600	2·30
Wall ties (st/st)	7900	17·00
ROOFS:		
Aerated concrete slab	500	0·16
Asphalt	1900	0·60
Bituminous felt in 3 layers	1700	0·50
Sarking felt	1700	0·50
Stone chippings	1800	0·96
Tiles (clay)	2000	1·00
...... (concrete)	2100	1·50
Wood wool slab	500	0·10

Typical values –

Material	Density (kg/m^3)	Conductivity (λ) (W/mK)
FLOORS:		
Cast concrete	2000	1·33
Hardwood block/strip	700	0·18
Plywood/particle board	650	0·14
Screed	1200	0·41
Softwood board	500	0·13
Steel tray	7800	50·00
INSULATION:		
Expanded polystyrene board	20	0·035
Mineral wool batt/slab	25	0·038
Mineral wool quilt	12	0·042
Phenolic foam board	30	0·025
Polyurethane board	30	0·025
Urea formaldehyde foam	10	0·040
GROUND:		
Clay/silt	1250	1·50
Sand/gravel	1500	2·00
Homogenous rock	3000	3·50

Notes:

1. For purposes of calculating U-values, the effect of mortar in external brickwork is usually ignored as the density and thermal properties of bricks and mortar are similar.

2. Where butterfly wall ties are used at normal spacing in an insulated cavity ≤75mm, no adjustment is required to calculations. If vertical twist ties are used in insulated cavities >75mm, 0·020 W/m^2K should be added to the U-value.

3. Thermal conductivity (λ) is a measure of the rate that heat is conducted through a material under specific conditions (W/mK).

* Tables and charts – Insulation manufacturers' design guides and technical papers (walls, roofs and ground floors).
* Calculation using the Proportional Area Method (walls and roofs).
* Calculation using the Combined Method – BS EN ISO 6946 (walls and roofs).
* Calculation using BS EN ISO 13370 (ground floors and basements).

Tables and charts – these apply where specific U-values are required and standard forms of construction are adopted. The values contain appropriate allowances for variable heat transfer due to different components in the construction, e.g. twisted pattern wall-ties and non-uniformity of insulation with the interruption by ceiling joists. The example below shows the tabulated data for a solid ground floor with embedded insulation of $\lambda = 0.03$ W/mK

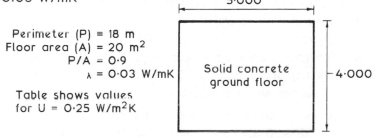

Perimeter (P) = 18 m
Floor area (A) = 20 m^2
P/A = 0.9
λ = 0.03 W/mK

Table shows values
for U = 0.25 W/m^{2}K

Solid concrete ground floor

5.000

4.000

Typical table for floor insulation:

P/A	0.020	0.025	0.030*	0.035	0.040	0.045	W/mK
1.0	61	76	91	107	122	137	mm ins.
0.9*	60	75	90	105	120	135	.. ..
0.8	58	73	88	102	117	132	.. ..
0.7	57	71	85	99	113	128	.. ..
0.6	54	68	82	95	109	122	.. ..
0.5	51	64	77	90	103	115	.. ..

90 mm of insulation required.

Refs. BS EN ISO 6946: Building components and building elements. Thermal resistance and thermal transmittance. Calculation method.

BS EN ISO 13370: Thermal performance of buildings. Heat transfer via the ground. Calculation methods.

Various applications to different ground floor situations are considered in BS EN ISO 13370. The following is an example for a solid concrete slab in direct contact with the ground. The data used is from the previous page.

Floor section

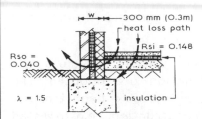

Perimeter = 18 m (exposed)
Floor area = 20 m²
λ for 90 mm insulation = 0.03 W/mK
Characteristic floor dimension = B^1
B^1 = Floor area ÷ (1/2 exp. perimeter)
B^1 = 20 ÷ 9 = 2.222 m

Formula to calculate total equivalent floor thickness for uninsulated and insulated all over floor:

$$dt = w + \lambda \, (R_{si} + R_f + R_{so})$$

where: dt = total equivalent floor thickness (m)
 w = wall thickness (m)
 λ = thermal conductivity of soil (W/mK) [see page 436]
 R_{si} = internal surface resistance (m²K/W) [see page 434]
 R_f = insulation resistance (0.09 ÷ 0.03 = 3 m²K/W)
 R_{so} = external surface resistance (m²K/W) [see page 434]

 Uninsulated: dt = 0.3 + 1.5 (0.148 + 0 + 0.04) = 0.582 m
 Insulated: dt = 0.3 + 1.5 (0.148 + 3 + 0.04) = 5.082 m

Formulae to calculate U-values ~
Uninsulated or poorly insulated floor, $dt < B^1$:

$$U = (2\lambda) \div [(\pi B^1) + dt] \times \ln [(\pi B^1 \div dt) + 1]$$

Well insulated floor, $dt \geq B^1$:

$$U = \lambda \div [(0.457 \times B^1) + dt]$$

 where: U = thermal transmittance coefficient (W/m²/K)
 λ = thermal conductivity of soil (W/mK)
 B^1 = characteristic floor dimension (m)
 dt = total equivalent floor thickness (m)
 ln = natural logarithm

Uninsulated floor ~
U = (2 × 1.5) ÷ [(3.142 × 2.222) + 0.582] × ln [(3.142 × 2.222) ÷ 0.582 + 1]
U = 0.397 × ln 12.996 = 1.02 W/m²K

Insulated floor ~
 U = 1.5 ÷ [(0.457 × 2.222) + 5.082] = 1.5 ÷ 6.097 = 0.246 W/m²K

Compares with the tabulated figure of 0.250 W/m²K on the previous page.

Proportional Area Method (Wall)

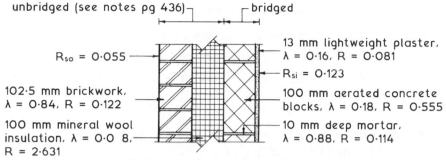

unbridged (see notes pg 436)   bridged

$R_{so} = 0.055$

13 mm lightweight plaster, $\lambda = 0.16$, R = 0.081

$R_{si} = 0.123$

102.5 mm brickwork, $\lambda = 0.84$, R = 0.122

100 mm aerated concrete blocks, $\lambda = 0.18$, R = 0.555

100 mm mineral wool insulation, $\lambda = 0.08$, R = 2.631

10 mm deep mortar, $\lambda = 0.88$, R = 0.114

Fully insulated cavity wall

A standard block with mortar is 450×225 mm = 101250 mm^2

A standard block format of 440×215 mm = 94600 mm^2

The area of mortar per block = 6650 mm^2

Proportional area of mortar = $\dfrac{6650}{101250} \times \dfrac{100}{1} = 6.57\%(0.066)$

Therefore the proportional area of blocks = 93.43% (0.934)

Thermal resistances (R):

Outer leaf + insulation (unbridged)

$\quad R_{so} \qquad = 0.055$
$\quad$ brickwork $= 0.122$
$\quad$ insulation $= \underline{2.631}$
$\qquad\qquad\quad\ 2.808$
$\qquad \times 100\% = 2.808$

Inner leaf (unbridged)

$\quad$ blocks $= 0.555$
$\quad$ plaster $= 0.081$
$\quad R_{si} \qquad = \underline{0.123}$
$\qquad\qquad\quad 0.759$
$\qquad \times 93.43\% = 0.709$

Inner leaf (bridged)

$\quad$ mortar $\qquad = 0.114$
$\quad$ plaster $\qquad = 0.081$
$\quad R_{si} \qquad\quad = \underline{0.123}$
$\qquad\qquad\quad = 0.318$
$\qquad \times\ 6.57\% = 0.021$

$U = \dfrac{1}{\Sigma R} = \dfrac{1}{2.808 + 0.709 + 0.021} = 0.283 \text{W}/\text{m}^2\text{K}$

Combined Method (Wall)

This method considers the upper and lower thermal resistance (R) limits of an element of structure. The average of these is reciprocated to provide the U-value.

Formula for upper and lower resistances = $\dfrac{1}{\Sigma(F_x \div R_x)}$

Where: F_x = Fractional area of a section
R_x = Total thermal resistance of a section

Using the wall example from the previous page:

Upper limit of resistance (R) through section containing blocks – (R_{so}, 0·055) + (brkwk, 0·122) + (ins, 2·631) + (blocks, 0·555) + (plstr, 0·081) + (R_{si}, 0·123) = 3·567 m²K/W

Fractional area of section (F) = 93·43% or 0·934

Upper limit of resistance (R) through section containing mortar – (R_{so} 0·055) + (brkwk, 0·122) + (ins, 2·631) + (mortar, 0·114) + (plstr, 0·081) + (R_{si}, 0·123) = 3·126 m²K/W

Fractional area of section (F) = 6·57% or 0·066

The upper limit of resistance =

$$\frac{1}{\Sigma(0{\cdot}943 \div 3{\cdot}567) + (0{\cdot}066 \div 3{\cdot}126)} = 3{\cdot}533\,m^2K/W$$

Lower limit of resistance (R) is obtained by summating the resistance of all the layers –
(R_{so}, 0·055) + (brkwk, 0·122) + (ins, 2·631) + (bridged layer, 1 ÷ [0·934 ÷ 0·555] + [0·066 ÷ 0·114] = 0·442) + (plstr, 0·081) + (R_{si}, 0·123) = 3·454 m²K/W

Total resistance (R) of wall is the average of upper and lower limits = (3·533 + 3·454) ÷ 2 = 3·493 m²K/W

U-value = $\dfrac{1}{R}$ = $\dfrac{1}{3{\cdot}493}$ = 0·286 W/m²K

Note: Both proportional area and combined method calculations require an addition of 0·020 W/m²K to the calculated U-value. This is for vertical twist type wall ties in the wide cavity. See page 320 and note 2 on page 436.

Proportional Area Method (Roof)

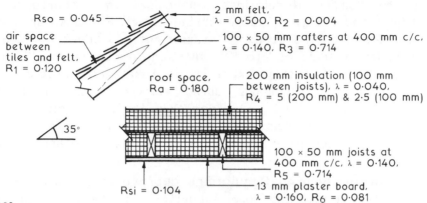

$R_{so} = 0.045$

air space between tiles and felt, $R_1 = 0.120$

2 mm felt, $\lambda = 0.500$, $R_2 = 0.004$

100×50 mm rafters at 400 mm c/c, $\lambda = 0.140$, $R_3 = 0.714$

roof space, $R_a = 0.180$

200 mm insulation (100 mm between joists), $\lambda = 0.040$, $R_4 = 5$ (200 mm) & 2.5 (100 mm)

35°

100×50 mm joists at 400 mm c/c, $\lambda = 0.140$, $R_5 = 0.714$

$R_{si} = 0.104$

13 mm plaster board, $\lambda = 0.160$, $R_6 = 0.081$

Notes:

1. The air space in the loft area is divided between pitched and ceiling components, ie. $R_a = 0.180 \div 2 = 0.090 \, m^2K/W$.
2. The U-value is calculated perpendicular to the insulation, therefore the pitched component resistance is adjusted by multiplying by the cosine of the pitch angle, ie. 0.819.
3. Proportional area of bridging parts (rafters and joists) is $50 \div 400 = 0.125$ or 12.5%.
4. With an air space resistance value (R1) of $0.120 \, m^2K/W$ between tiles and felt, the resistance of the tiling may be ignored.

Thermal resistance (R) of the pitched component:

Raftered part	Non-raftered part
$R_{so} = 0.045$	$R_{so} = 0.045$
$R_1 \ = 0.120$	$R_1 \ = 0.120$
$R_2 \ = 0.004$	$R_2 \ = 0.004$
$R_3 \ = 0.714$	$R_a \ = \underline{0.090}$
$R_a \ = \underline{0.090}$	$\quad 0.259 \times 87.5\%$
$\quad 0.973 \times 12.5\% = 0.122$	$\quad = 0.227$

Total resistance of pitched components =
$$(0.122 + 0.227) \times 0.819 = 0.286 \, m^2K/W$$

Thermal resistance (R) of the ceiling component:

Joisted part	Fully insulated part
$R_{si} = 0.104$	$R_{si} = 0.104$
$R_6 = 0.081$	$R_6 = 0.081$
$R_5 = 0.714$	$R_4 = 5.000$ (200 mm)
$R_4 = 2.500$ (100 mm)	$R_a = \underline{0.090}$
$R_a = \underline{0.090}$	$\quad 5.275 \times 87.5\%$
$\quad 3.489 \times 12.5\% = 0.436$	$\quad = 4.615$

Total resistance of ceiling components = $0.436 + 4.615$
$$= 5.051 \, m^2K/W.$$

$$U = \frac{1}{\Sigma R} = \frac{1}{0.286 + 5.051} = 0.187 \, W/m^2K$$

Standard Assessment Procedure ~ the Approved Document to Part L of the Building Regulations emphasises the importance of quantifying the energy costs of running homes. For this purpose it uses the Government's Standard Assessment Procedure (SAP). SAP has a numerical scale of 1 to 100, although it can exceed 100 if a dwelling is a net energy exporter. It takes into account the effectiveness of a building's fabric relative to insulation and standard of construction. It also appraises the energy efficiency of fuel consuming installations such as ventilation, hot water, heating and lighting. Incidentals like solar gain also feature in the calculations.

As part of the Building Regulations approval procedure, energy rating (SAP) calculations are submitted to the local building control authority. SAP ratings are also required to provide prospective home purchasers or tenants with an indication of the expected fuel costs for hot water and heating. This information is documented and included with the property conveyance. The SAP calculation involves combining data from tables, work sheets and formulae. Guidance is found in Approved Document L, or by application of certified SAP computer software programmes.

As a guide, housing built to 1995 energy standards can be expected to have a SAP rating of around 80. That built to 2002 energy standards will have a SAP expectation of about 90. Current quality construction standards should rate dwellings close to 100.

Ref. Standard Assessment Procedure for Energy Rating of Dwellings. The Stationery Office.

Air Permeability ~ air tightness in the construction of dwellings is an important quality control objective. Compliance is achieved by attention to detail at construction interfaces, e.g. by silicone sealing built-in joists to blockwork inner leafs and door and window frames to masonry surrounds; draft proofing sashes, doors and loft hatches. Dwellings failing to comply with these measures are penalised in SAP calculations.

Compliance with the Building Regulations Part L Robust Details is an acceptable standard of construction. Alternatively, a certificate must be obtained to show pre-completion testing satisfying air permeability of less than 10 m^3/h per m^2 envelope area at 50 Pascals (Pa or N/m^2) pressure.

Ref. Limiting thermal bridging and air leakage: Robust construction details for dwellings and similar buildings. The Stationery Office.

Domestic buildings (England and Wales) ~

Element of construction	Limiting area weighted ave. U-value (W/m²K)	Limiting individual component U-value
Roof	0.25	0.35
Wall	0.35	0.70
Floor	0.25	0.70
Windows, doors, rooflights and roof windows	2.20	3.30

The area weighted average U-value for an element of construction depends on the individual U-values of all components and the area they occupy within that element. E.g. The part of a wall with a meter cupboard built in will have less resistance to thermal transmittance than the rest of the wall (max. U-value at cupboard, 0.45).

Element of construction	U-value targets (W/m²K)
Pitched roof (insulation between rafters)	0.15
Pitched roof (insulation between joists)	0.15
Flat roof	0.15
Wall	0.28
Floor	0.20
Windows, doors, rooflights and roof windows	1.80 (area weighted ave.)

Note: Maximum area of windows, doors, rooflights and roof windows, 25% of the total floor area.

An alternative to the area weighted average U-value for windows, etc., may be a window energy rating of not less than minus 30.

Energy source ~ gas or oil fired central heating boiler with a minimum SEDBUK efficiency rating of 86% (band rating A or B). There are transitional and exceptional circumstances that permit lower band rated boilers. Where this occurs, the construction of the building envelope should compensate with very low U-values.

SEDBUK = Seasonal Efficiency of a Domestic Boiler in the United Kingdom. SEDBUK values are defined in the Government's Standard Assessment Procedure for Energy Rating of Dwellings. There is also a SEDBUK website, www.sedbuk.com.

Note: SEDBUK band A = > 90% efficiency
         band B = 86-90%   ..
         band C = 82-86%   ..
         band D = 78-82%   ..

443

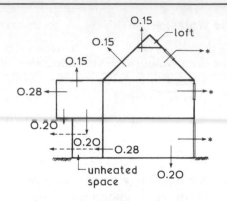

*area weighted average
U-value of rooflights,
roof windows, windows
and doors = 1.80 or a
window energy rating of
not less than –30.
Maximum area 25% of
the total floor area.

Further Quality Procedures (Structure) ~

* Provision of insulation to be continuous. Gaps are unacceptable and if allowed to occur will invalidate the insulation value by thermal bridging.
* Junctions at elements of construction (wall/floor, wall/roof) to receive particular attention with regard to continuity of insulation.
* Openings in walls for windows and doors to be adequately treated with insulating cavity closers.

Further Quality Procedures (Energy Consumption) ~

* Hot water and heating systems to be fully commissioned on completion and controls set with regard for comfort, health and economic use.
* As part of the commissioning process, the sealed heating system should be flushed out and filled with a proprietary additive diluted in accordance with the manufacturer's guidance.

This is necessary to enhance system performance by resisting corrosion, scaling and freezing.

* A certificate confirming system commissioning and water treatment should be available for the dwelling occupant. This document should be accompanied with component manufacturer's operating and maintenance instructions.

Note: Commissioning of heating installations and the issue of certificates is by a qualified "competent person" as recognised by the appropriate body, i.e. CORGI, OFTEC or HETAS.
CORGI ~ Council for Registered Gas Installers.
OFTEC ~ Oil Firing Technical Association for the Petroleum Industry.
HETAS ~ Solid Fuel. Heating Equipment Testing and Approval Scheme.

European Window Energy Rating Scheme (EWERS) ~ an alternative to U-values for measuring the thermal efficiency of windows. U-values form part of the assessment, in addition to factors for solar heat gain and air leakage. In the UK, testing and labelling of window manufacturer's products is promoted by the British Fenestration Rating Council (BFRC). The scheme uses a computer to simulate energy movement over a year through a standard window of 1.480 × 1.230 m containing a central mullion and opening sash to one side.

Data is expressed on a scale from A–G in units of kWh/m²/year.

    A > zero
    B –10 to 0
    C –20 to –10
    D –30 to –20
    E –50 to –30
    F –70 to –50
    G < –70

By formula, rating = (218.6 × g value) – 68.5 (U-value × L value)
Where: g value = factor measuring effectiveness of solar heat block expressed between 0 and 1. For comparison.

    0.48 (no curtains)
    0.43 (curtains open)
    0.17 (curtains closed)

  U value = weighted average transmittance coefficient
  L value = air leakage factor

From the label shown opposite:
Rating = (218·6 × 0·5)
        – 68·5 (1·8 + 0·10)
      = 109·3 – 130·15
      = –20·85 i.e. –21

Typical format of a window energy rating label ~

ABC Joinery Ltd.
Window ref. XYZ 123

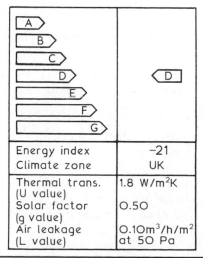

| Energy index | –21 |
| Climate zone | UK |

Thermal trans. (U value)	1.8 W/m²K
Solar factor (g value)	0.50
Air leakage (L value)	0.10m³/h/m² at 50 Pa

445

The Government's Standard Assessment Procedure (SAP) for energy rating dwellings includes a facility to calculate carbon dioxide (CO_2) emissions in kilograms or tonnes per year. The established carbon index method allows for adjustment to dwelling floor area to obtain a carbon factor (CF):

$$CF = CO_2 \div (\text{total floor area} + 45)$$

The carbon index (CI) = $17 \cdot 7 - (9 \log. CF)$

Note: log. = logarithm to the base 10.

e.g. A dwelling of total floor area $125 m^2$, with CO_2 emissions of 2000 kg/yr.

$$CF = 2000 \div (125 + 45) = 11 \cdot 76$$

$$CI = 17 \cdot 7 - (9 \log. 11 \cdot 76) = 8 \cdot 06$$

The carbon index (CI) is expressed on a scale of 0 to 10. The higher the number the better. Every new dwelling should have a CI value of a least 8.

Approved Document L to the Building Regulations includes the Dwelling Carbon Emissions Rate (DER) as another means for assessing carbon discharge. The DER is compared by calculation to a Target Carbon Emissions Rate (TER), based on data for type of lighting, floor area, building shape and choice of fuel.

The DER is derived primarily by appraising the potential CO_2 emission from a dwelling relative to the consumption of fuel (directly or indirectly) in hot water, heating, lighting, cooling (if fitted), fans and pumps.

$$DER \leq TER$$

Buildings account for about half of the UK's carbon emissions. Therefore, there are considerable possibilities for energy savings and reductions in atmospheric pollution.

In new buildings and those subject to alterations, the objective is to optimise the use of fuel and power to minimise emission of carbon dioxide and other burnt fuel gases into the atmosphere. This applies principally to the installation of hot water, heating, lighting, ventilation and air conditioning systems. Pipes, ducting, storage vessels and other energy consuming plant should be insulated to limit heat losses. The fabric or external envelope of a building is constructed with regard to limiting heat losses through the structure and to regulate solar gains.

Approved Document L2 of the Building Regulations is not prescriptive. It sets out a series of objectives relating to achievement of a satisfactory carbon emission standard. A number of other technical references and approvals are cross referenced in the Approved Document and these provide a significant degree of design flexibility in achieving the objectives.

Energy efficiency of buildings other than dwellings is determined by applying a series of procedures modelled on a notional building of the same size and shape as the proposed building. The performance standards used for the notional building are similar to the 2002 edition of Approved Document L2. Therefore the proposed or actual building must be seen to be a significant improvement in terms of reduced carbon emissions by calculation. Improvements can be achieved in a number of ways, including the following:

- Limit the area or number of rooflights, windows and other openings.
- Improve the U-values of the external envelope. The limiting values are shown on the next page.
- Improve the airtightness of the building from the poorest acceptable air permeability of $10\,m^3/hour/m^2$ of external envelope at 50 Pa pressure.
- Improve the heating system efficiency by installing thermostatic controls, zone controls, optimum time controls, etc. Fully insulate pipes and equipment.
- Use of high efficacy lighting fittings, automated controls, low voltage equipment, etc.
- Apply heat recovery systems to ventilation and air conditioning systems. Insulate ducting.
- Install a building energy management system to monitor and regulate use of heating and air conditioning plant.
- Limit overheating of the building with solar controls and appropriate glazing systems.
- Ensure that the quality of construction provides for continuity of insulation in the external envelope.
- Establish a commissioning and plant maintenance procedure. Provide a log-book to document all repairs, replacements and routine inspections.

Buildings Other Than Dwellings (England and Wales) ~

Element of construction	Limiting area weighted ave. U-value (W/m²K)	Limiting individual component U-value
Roof	0.25	0.35
Wall	0.35	0.45
Floor	0.25	0.45
Windows, doors roof-lights and roof windows	2.20	3.30
Curtain wall (full façade)	1.60	2.50
Large and vehicle access doors	1.50	4.00

Notes:
- For display windows separate consideration applies. See Section 5 in A.D., L2A.
- The poorest acceptable thermal transmittance values provide some flexibility for design, allowing a trade off against other thermally beneficial features such as energy recovery systems.
- The minimum U-value standard is set with regard to minimise the risk of condensation.
- The concept of area weighted values is explained on page 443.
- Elements will normally be expected to have much better insulation than the limiting U-values. Suitable objectives or targets could be as shown for domestic buildings.

Further requirements for the building fabric ~

Insulation continuity ~ this requirement is for a fully insulated external envelope with no air gaps in the fabric. Vulnerable places are at junctions between elements of construction, e.g. wall to roof, and around openings such as door and window reveals. Conformity can be shown by producing evidence in the form of a report produced for the local authority building control department by an accredited surveyor. The report must indicate that:

* the approved design specification and construction practice are to an acceptable standard of conformity, OR
* a thermographic survey shows continuity of insulation over the external envelope. This is essential when it is impractical to fully inspect the work in progress.

Air tightness ~ requires that there is no air infiltration through gaps in construction and at the intersection of elements. Permeability of air is tested by using portable fans of capacity to suit the building volume. Smoke capsules in conjunction with air pressurisation will provide a visual indication of air leakage paths.

Thermal Insulation ~ this is required within the roof of all dwellings in the UK. It is necessary to create a comfortable internal environment, to reduce the risk of condensation and to economise in fuel consumption costs.

To satisfy these objectives, insulation may be placed between and over the ceiling joists as shown below to produce a *cold roof* void. Alternatively, the insulation can be located above the rafters as shown on page 404. Insulation above the rafters creates a *warm roof* void and space within the roof structure that may be useful for habitable accommodation.

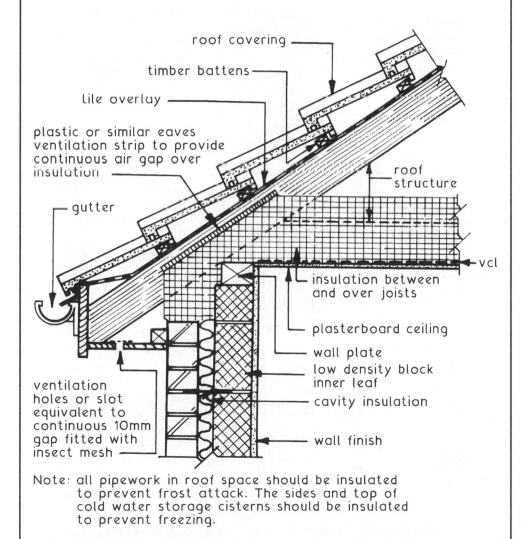

roof covering

timber battens

tile overlay

plastic or similar eaves ventilation strip to provide continuous air gap over insulation

gutter

roof structure

vcl

insulation between and over joists

plasterboard ceiling

wall plate

low density block inner leaf

cavity insulation

wall finish

ventilation holes or slot equivalent to continuous 10mm gap fitted with insect mesh

Note: all pipework in roof space should be insulated to prevent frost attack. The sides and top of cold water storage cisterns should be insulated to prevent freezing.

Thermal insulation to Walls ~ the minimum performance standards for exposed walls set out in Approved Document L to meet the requirements of Part L of the Building Regulations can be achieved in several ways (see pages 439 and 440). The usual methods require careful specification, detail and construction of the wall fabric, insulating material(s) and/or applied finishes.

Typical Examples of existing construction that would require upgrading to satisfy contemporary UK standards ~

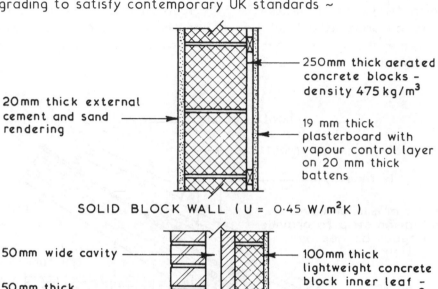

20mm thick external cement and sand rendering

250mm thick aerated concrete blocks – density 475 kg/m³

19 mm thick plasterboard with vapour control layer on 20 mm thick battens

SOLID BLOCK WALL (U = 0·45 W/m²K)

50mm wide cavity

50mm thick mineral wool/fibre cavity batts

102.5mm external brick outer leaf

100mm thick lightweight concrete block inner leaf – density 600 kg/m³

13mm thick lightweight plaster

CAVITY WALL WITH CAVITY INSULATING BATTS (U = 0·39W/m²K)

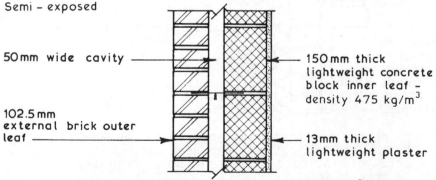

Semi – exposed

50mm wide cavity

102.5mm external brick outer leaf

150mm thick lightweight concrete block inner leaf – density 475 kg/m³

13mm thick lightweight plaster

TRADITIONAL CAVITY WALL (U = 0·58 W/m²K)

Typical examples of contemporary construction practice that achieve a thermal transmittance or U-value below 0.30 W/m²K ~

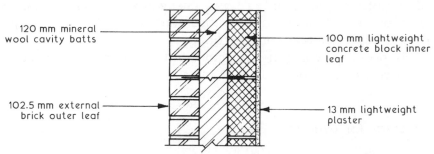

120 mm mineral wool cavity batts

100 mm lightweight concrete block inner leaf

102.5 mm external brick outer leaf

13 mm lightweight plaster

FULL FILL CAVITY WALL, Block density 750 kg/m³   U = 0.25 W/m²K
           Block density 600 kg/m³   U = 0.24 W/m²K
           Block density 475 kg/m³   U = 0.23 W/m²K

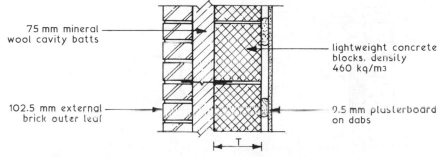

75 mm mineral wool cavity batts

lightweight concrete blocks, density 460 kg/m³

102.5 mm external brick outer leaf

9.5 mm plasterboard on dabs

FULL FILL CAVITY WALL, T = 125 mm   U = 0.28 W/m²K
           T = 150 mm   U = 0.26 W/m²K
           T = 200 mm   U = 0.24 W/m²K

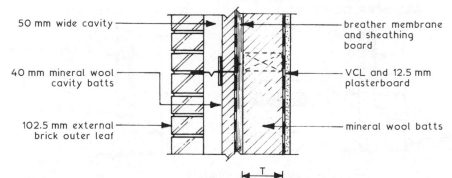

50 mm wide cavity

breather membrane and sheathing board

40 mm mineral wool cavity batts

VCL and 12.5 mm plasterboard

102.5 mm external brick outer leaf

mineral wool batts

TIMBER FRAME PART CAVITY FILL, T = 100 mm   U = 0.26 W/m²K
           T = 120 mm   U = 0.24 W/m²K
           T = 140 mm   U = 0.21 W/m²K

Note: Mineral wool insulating batts have a typical thermal conductivity (λ) value of 0.038 W/mK.

Thermal or Cold Bridging ~ this is heat loss and possible condensation, occurring mainly around window and door openings and at the junction between ground floor and wall. Other opportunities for thermal bridging occur where uniform construction is interrupted by unspecified components, e.g. occasional use of bricks and/or tile slips to make good gaps in thermal block inner leaf construction.

NB. This practice was quite common, but no longer acceptable by current legislative standards in the UK.

Prime areas for concern −

WINDOW SILL

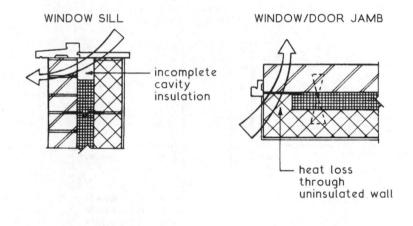

incomplete cavity insulation

WINDOW/DOOR JAMB

heat loss through uninsulated wall

GROUND FLOOR & WALL

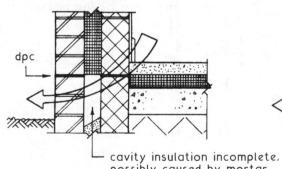

dpc

cavity insulation incomplete, possibly caused by mortar droppings building up and bridging the lower part of the cavity*

WINDOW/DOOR HEAD

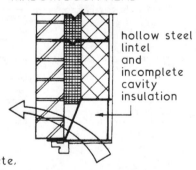

hollow steel lintel and incomplete cavity insulation

*Note: Cavity should extend down at least 225 mm below the level of the lowest dpc (AD, C: Section 5).

As shown on the preceding page, continuity of insulated construction in the external envelope is necessary to prevent thermal bridging. Nevertheless, some discontinuity is unavoidable where the pattern of construction has to change. For example, windows and doors have significantly higher U-values than elsewhere. Heat loss and condensation risk in these situations is regulated by limiting areas, effectively providing a trade off against very low U-values elsewhere.

The following details should be observed around openings and at ground floor ~

WINDOW SILL

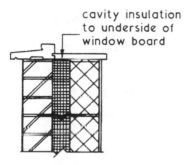

cavity insulation to underside of window board

WINDOW/DOOR JAMB

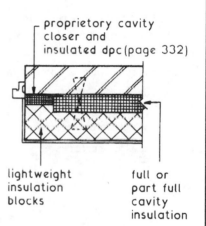

proprietory cavity closer and insulated dpc (page 332)

lightweight insulation blocks

full or part full cavity insulation

GROUND FLOOR & WALL

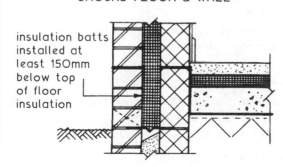

insulation batts installed at least 150mm below top of floor insulation

WINDOW/DOOR HEAD

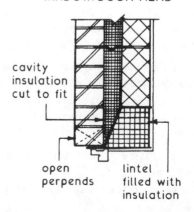

cavity insulation cut to fit

open perpends

lintel filled with insulation

The possibility of a thermal or cold bridge occurring in a specific location can be appraised by calculation. Alternatively, the calculations can be used to determine how much insulation will be required to prevent a cold bridge. The composite lintel of concrete and steel shown below will serve as an example ~

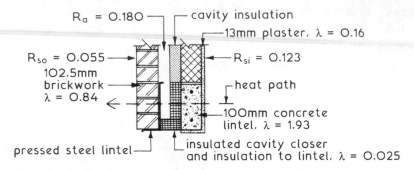

Wall components, less insulation (steel in lintel is insignificant):

102.5 mm brickwork outer leaf,	$\lambda = 0.84$ W/mK
100 mm dense concrete lintel,	$\lambda = 1.93$ ..
13 mm lightweight plaster,	$\lambda = 0.16$ ..

Resistances of above components:

Brickwork,	$0.1025 \div 0.84 = 0.122$ m^{2}K/W
Concrete lintel,	$0.100 \div 1.93 = 0.052$ ..
Lightweight plaster,	$0.013 \div 0.16 = 0.081$ ..

Resistances of surfaces:

Internal $(R_{si}) = 0.123$ ..
Cavity $(R_a) = 0.180$ ..
External $(R_{so}) = 0.055$ ..
Summary of resistances = 0·613 ..

To achieve a U-value of say 0·27 W/m^{2}K,
total resistance required = $1 \div 0.27 = 3.703$ m^{2}K/W

The insulation in the cavity at the lintel position is required to have a resistance of 3·703 − 0·613 = 3·09 m^{2}K/W

Using a urethane insulation with a thermal conductivity (λ) of 0·025 W/mK, 0·025 × 3·09 = 0·077 m or 77 mm minimum thickness.

If the cavity closer has the same thermal conductivity, then:
Summary of resistance = 0·613 − 0·180 (R_a) = 0·433 m^{2}K/W
Total resistance required = 3·703 m^{2}K/W, therefore the cavity closer is required to have a resistance of: 3·703 − 0·433 = 3·270 m^{2}K/W
Min. cavity closer width = 0·025 W/mK × 3·270 m^{2}K/W = 0·082 m or 82 mm.

In practice, the cavity width and the lintel insulation would exceed 82 mm.

Note: data for resistances and λ values taken from pages 434 to 436.

Air Infiltration ~ heating costs will increase if cold air is allowed to penetrate peripheral gaps and breaks in the continuity of construction. Furthermore, heat energy will escape through structural breaks and the following are prime situations for treatment:-

1. Loft hatch
2. Services penetrating the structure
3. Opening components in windows, doors and rooflights
4. Gaps between dry lining and masonry walls

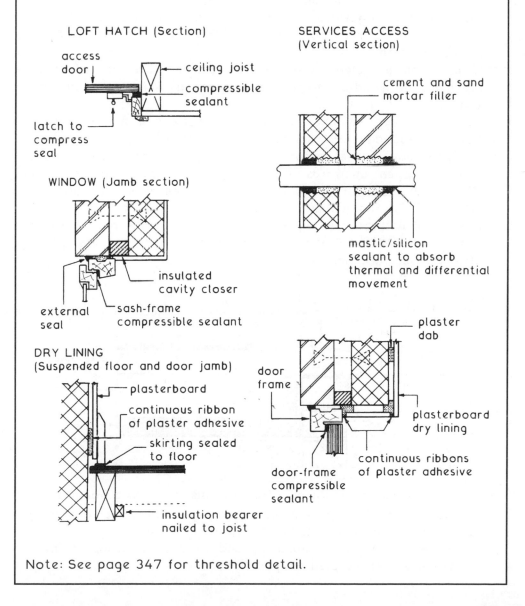

LOFT HATCH (Section)

access door

ceiling joist

compressible sealant

latch to compress seal

WINDOW (Jamb section)

insulated cavity closer

external seal

sash-frame compressible sealant

DRY LINING
(Suspended floor and door jamb)

plasterboard

continuous ribbon of plaster adhesive

skirting sealed to floor

insulation bearer nailed to joist

SERVICES ACCESS
(Vertical section)

cement and sand mortar filler

mastic/silicon sealant to absorb thermal and differential movement

plaster dab

door frame

plasterboard dry lining

continuous ribbons of plaster adhesive

door-frame compressible sealant

Note: See page 347 for threshold detail.

Main features –

* Site entrance or car parking space to building entrance to be firm and level, with a 900mm min. width. A gentle slope is acceptable with a gradient up to 1 in 20 and up to 1 in 40 in cross falls. A slightly steeper ramped access or easy steps should satisfy A.D. Sections 6.14 & 6.15, and 6.16 & 6.17 respectively.
* An accessible threshold for wheelchairs is required at the principal entrance – see illustration.
* Entrance door – minimum clear opening width of 775mm.
* Corridors, passageways and internal doors of adequate width for wheelchair circulation. Minimum 750mm – see also table 1 in A.D. Section 7.
* Stair minimum clear width of 900mm, with provision of handrails both sides. Other requirements as A.D. K for private stairs.
* Accessible light switches, power, telephone and aerial sockets between 450 and 1200mm above floor level.
* WC provision in the entrance storey or first habitable storey. Door to open outwards. Clear wheelchair space of at least 750mm in front of WC and a preferred dimension of 500mm either side of the WC as measured from its centre.
* Special provisions are required for passenger lifts and stairs in blocks of flats, to enable disabled people to access other storeys. See A.D. Section 9 for details.

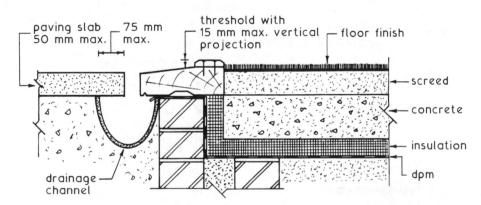

Note: A.D. refers to the Building Regulations, Approved Document.

Refs. Accessible thresholds in new housing – Guidance for house builders and designers. The Stationery Office.
BS 8300: Design of buildings and their approaches to meet the needs of disabled people. Code of practice.

Main features –

* Site entrance, or car parking space to building entrance to be firm and level, ie. maximum gradient 1 in 20 with a minimum car access zone of 1200mm. Ramped and easy stepped approaches are also acceptable.

  * Access to include tactile warnings, ie. profiled (blistered or ribbed) pavings over a width of at least 1200mm, for the benefit of people with impaired vision. Dropped kerbs are required to ease wheelchair use.

  * Special provision for handrails is necessary for those who may have difficulty in negotiating changes in level.

  * Guarding and warning to be provided where projections or obstructions occur, eg. tactile paving could be used around window opening areas.

* Sufficient space for wheelchair manoeuvrability in entrances.

Minimum entrance width of 800mm. Unobstructed space of at least 300mm to the leading (opening) edge of door. Glazed panel in the door to provide visibility from 500 to 1500mm above floor level. Entrance lobby space should be sufficient for a wheelchair user to clear one door before opening another.

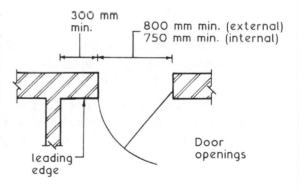

* Internal door openings, minimum width 750mm. Unobstructed space of at least 300mm to the leading edge. Visibility panel as above.

continued......

* Main access and internal fire doors that self-close should have a maximum operating force of 20 Newtons at the leading edge. If this is not possible, a power operated door opening and closing system is required.

* Corridors and passageways, minimum unobstructed width 1200mm. Internal lobbies as described on the previous page for external lobbies.

* Lift dimensions and capacities to suit the building size. Ref. BS EN 81-1 and 2: Lifts and service lifts. Alternative vertical access may be by wheelchair stairlift – BS 5776: *Specification for powered stairlifts,* or a platform lift – BS 6440: *Powered lifting platforms for use by disabled persons.* Code of practice.

* Stair minimum width 1200mm, with step nosings brightly distinguished. Rise maximum 12 risers external, 16 risers internal between landings. Landings to have 1200mm of clear space from any door swings. Step rise, maximum 170mm and uniform throughout. Step going, minimum 250mm (internal), 280mm (external) and uniform throughout. No open risers. Handrail to each side of the stair.

* Number and location of WCs to reflect ease of access for wheelchair users. In no case should a wheelchair user have to travel more than one storey. Provision may be `unisex´ which is generally more suitable, or `integral´ with specific sex conveniences. Particular provision is outlined in Section 5 of the Approved Document.

* Section 4 should be consulted for special provisions for restaurants, bars and hotel bedrooms, and for special provisions for spectator seating in theatres, stadia and conference facilities.

Refs. Building Regulations, Approved Document M: Access to and use of buildings.
Disability Discrimination Act.
BS 5588-8: Fire precautions in the design, construction and use of buildings. Code of practice for means of escape for disabled people.
PD 6523: Information on access to and movement within and around buildings and on certain facilities for disabled people.
BS 8300: Design of buildings and their approaches to meet the needs of disabled people. Code of practice.

6 SUPERSTRUCTURE — 2

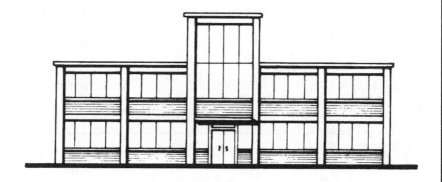

REINFORCED CONCRETE FRAMED STRUCTURES
STRUCTURAL CONCRETE FIRE PROTECTION
FORMWORK
PRECAST CONCRETE FRAMES
PRESTRESSED CONCRETE
STRUCTURAL STEELWORK ASSEMBLY
STRUCTURAL STEELWORK CONNECTIONS
STRUCTURAL FIRE PROTECTION
COMPOSITE TIMBER BEAMS
ROOF SHEET COVERINGS
LONG SPAN ROOFS
SHELL ROOF CONSTRUCTION
ROOFLIGHTS
MEMBRANE ROOFS
RAINSCREEN CLADDING
PANEL WALLS AND CURTAIN WALLING
CONCRETE CLADDINGS
CONCRETE SURFACE FINISHES AND DEFECTS

Simply Supported Slabs ~ these are slabs which rest on a bearing and for design purposes are not considered to be fixed to the support and are therefore, in theory, free to lift. In practice however they are restrained from unacceptable lifting by their own self weight plus any loadings.

Concrete Slabs ~ concrete is a material which is strong in compression and weak in tension and if the member is overloaded its tensile resistance may be exceeded leading to structural failure.

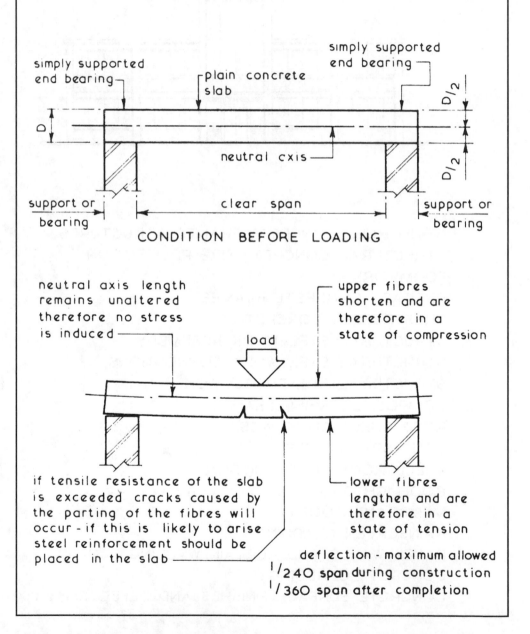

simply supported end bearing⌐ ⌐plain concrete slab

simply supported end bearing ⌐

neutral axis⌐

CONDITION BEFORE LOADING

neutral axis length remains unaltered therefore no stress is induced ⌐

upper fibres shorten and are therefore in a state of compression

load

if tensile resistance of the slab is exceeded cracks caused by the parting of the fibres will occur - if this is likely to arise steel reinforcement should be placed in the slab ⌐

lower fibres lengthen and are therefore in a state of tension

deflection - maximum allowed
$^1/240$ span during construction
$^1/360$ span after completion

460

Reinforcement ~ generally in the form of steel bars which are used to provide the tensile strength which plain concrete lacks. The number, diameter, spacing, shape and type of bars to be used have to be designed; a basic guide is shown on pages 465 and 466. Reinforcement is placed as near to the outside fibres as practicable, a cover of concrete over the reinforcement is required to protect the steel bars from corrosion and to provide a degree of fire resistance. Slabs which are square in plan are considered to be spanning in two directions and therefore main reinforcing bars are used both ways whereas slabs which are rectangular in plan are considered to span across the shortest distance and main bars are used in this direction only with smaller diameter distribution bars placed at right angles forming a mat or grid.

Typical Details ~

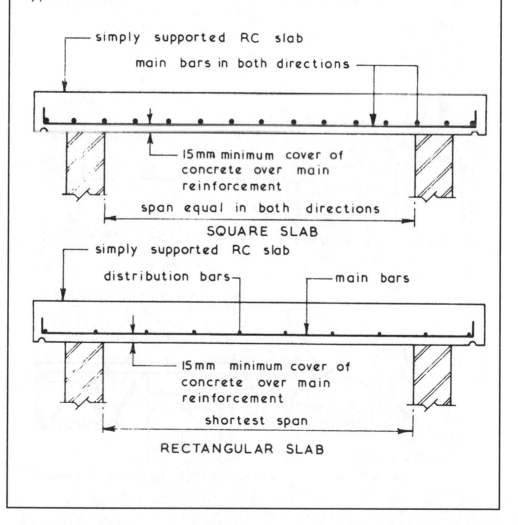

simply supported RC slab

main bars in both directions

15mm minimum cover of concrete over main reinforcement

span equal in both directions

SQUARE SLAB

simply supported RC slab

distribution bars

main bars

15mm minimum cover of concrete over main reinforcement

shortest span

RECTANGULAR SLAB

Construction ~ whatever method of construction is used the construction sequence will follow the same pattern-

1. Assemble and erect formwork.
2. Prepare and place reinforcement.
3. Pour and compact or vibrate concrete.
4. Strike and remove formwork in stages as curing proceeds

Typical Example ~

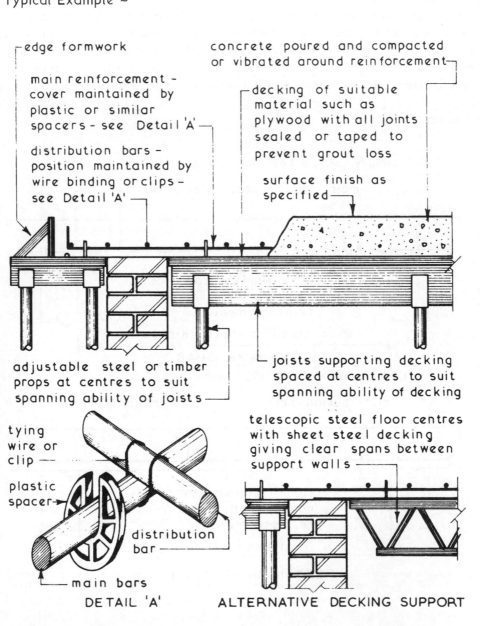

edge formwork

main reinforcement - cover maintained by plastic or similar spacers- see Detail 'A'

distribution bars - position maintained by wire binding or clips - see Detail 'A'

concrete poured and compacted or vibrated around reinforcement

decking of suitable material such as plywood with all joints sealed or taped to prevent grout loss

surface finish as specified

adjustable steel or timber props at centres to suit spanning ability of joists

joists supporting decking spaced at centres to suit spanning ability of decking

tying wire or clip

plastic spacer

distribution bar

main bars

DETAIL 'A'

telescopic steel floor centres with sheet steel decking giving clear spans between support walls

ALTERNATIVE DECKING SUPPORT

462

Profiled galvanised steel decking is a permanent formwork system for construction of composite floor slabs. The steel sheet has surface indentations and deformities to effect a bond with the concrete topping. The concrete will still require reinforcing with steel rods or mesh, even though the metal section will contribute considerably to the tensile strength of the finished slab.

Typical detail –

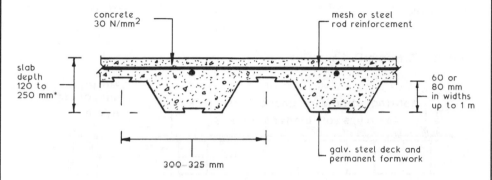

concrete 30 N/mm^2

mesh or steel rod reinforcement

slab depth 120 to 250 mm*

60 or 80 mm in widths up to 1 m

galv. steel deck and permanent formwork

300–325 mm

* For slab depth and span potential, see BS 5950-4: Code of practice for design of composite slabs with profiled steel sheeting.

Where structural support framing is located at the ends of a section and at intermediate points, studs are through-deck welded to provide resistance to shear –

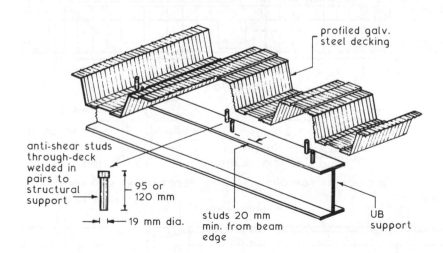

profiled galv. steel decking

anti-shear studs through-deck welded in pairs to structural support

95 or 120 mm

19 mm dia.

studs 20 mm min. from beam edge

UB support

There are considerable savings in concrete volume compared with standard in-situ reinforced concrete floor slabs. This reduction in concrete also reduces structural load on foundations.

Beams ~ these are horizontal load bearing members which are classified as either main beams which transmit floor and secondary beam loads to the columns or secondary beams which transmit floor loads to the main beams.

Concrete being a material which has little tensile strength needs to be reinforced to resist the induced tensile stresses which can be in the form of ordinary tension or diagonal tension (shear). The calculation of the area, diameter, type, position and number of reinforcing bars required is one of the functions of a structural engineer.

Typical RC Beam Details ~

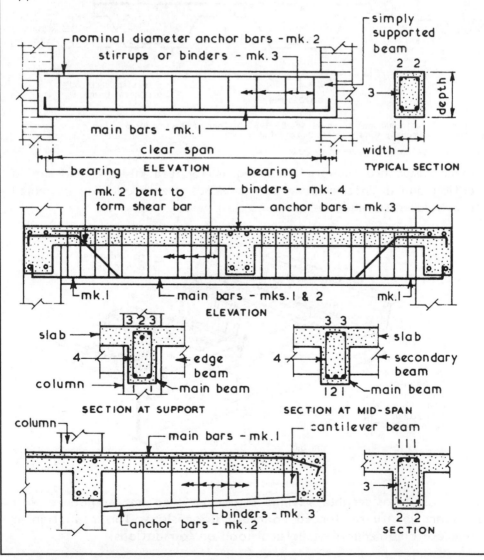

Mild Steel Reinforcement – located in areas where tension occurs in a beam or slab. Concrete specification is normally 25 or 30 N/mm² in this situation.

Simple beam or slab

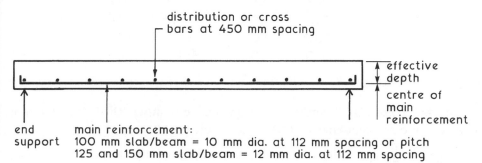

distribution or cross bars at 450 mm spacing

effective depth

centre of main reinforcement

end support

main reinforcement:
100 mm slab/beam = 10 mm dia. at 112 mm spacing or pitch
125 and 150 mm slab/beam = 12 mm dia. at 112 mm spacing

Continuous beam or slab

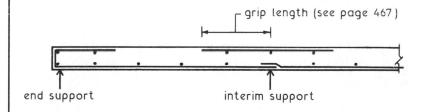

grip length (see page 467)

end support   interim support

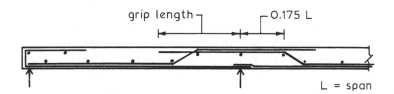

grip length   0.175 L

L = span

Note: Distribution or cross bars function as lateral reinforcement and supplement the units strength in tensile areas. They also provide resistance to cracking in the concrete as the unit contracts during setting and drying.

Pitch of main bars ≤ 3 × effective depth.

Pitch of distribution bars ≤ 5 × effective depth.

Guidance – simply supported slabs are capable of the following loading relative to their thickness:

Thickness (mm)	Self weight (kg/m²)	Imposed load* (kg/m²)	Total load (kg/m²)	Total load (kN/m²)	Span (m)
100	240	500	740	7·26	2·4
125	300	500	800	7·85	3·0
150	360	500	860	8·44	3·6

Note: As a *rule of thumb*, it is easy to remember that for general use (as above), thickness of slab equates to 1/24 span.

* Imposed loading varies with application from 1·5 kN/m² (153 kg/m²) for domestic buildings, to over 10 kN/m² (1020 kg/m²) for heavy industrial storage areas. 500 kg/m² is typical for office filing and storage space. See BS 6399-1: Loading for buildings. Code of practice for dead and imposed loads.

For larger spans – thickness can be increased proportionally to the span, eg. 6 m span will require a 250 mm thickness.

For greater loading – slab thickness is increased proportionally to the square root of the load, eg. for a total load of 1500 kg/m² over a 3 m span:

$$\sqrt{\frac{1500}{800}} \times 125 = 171\cdot2 \quad \text{i.e. 175 mm}$$

Continuous beams and slabs have several supports, therefore they are stronger than simple beams and slabs. The spans given in the above table may be increased by 20% for interior spans and 10% for end spans.

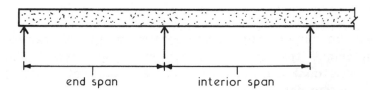

end span   interior span

Deflection limit on reinforced concrete beams and slabs is 1/250 span. BS 8110-1: Structural use of concrete. Code of practice for design and construction. See page 508 for deflection formulae.

Bond Between Concrete and Steel – permissible stress for the bond between concrete and steel can be taken as one tenth of the compressive concrete stress, plus 0·175 N/mm²*. Given the stresses in concrete and steel, it is possible to calculate sufficient grip length.

e.g. concrete working stress of 5 N/mm²
    steel working stress of 125 N/mm²
    sectional area of reinf. bar = $3·142\,r^2$ or $0·7854\,d^2$
    tensile strength of bar = $125 \times 0·7854\,d^2$
    circumference of bar = $3·142\,d$
    area of bar in contact = $3·142 \times d \times L$

Key: r = radius of steel bar
    d = diameter of steel bar
    L = Length of bar in contact

* Conc. bond stress = $(0·10 \times 5\,N/mm^2) + 0·175 = 0·675\,N/mm^2$

  Total bond stress = $3·142\,d \times l \times 0·675\,N/mm^2$

Thus, developing the tensile strength of the bar:

$$125 \times 0.7854\,d^2 = 3·142\,d \times L \times 0·675$$

$$98·175\,d = 2·120\,L$$

$$L = 46\,d$$

As a guide to good practice, a margin of 14 d should be added to L. Therefore the bar bond or grip length in this example is equivalent to 60 times the bar diameter.

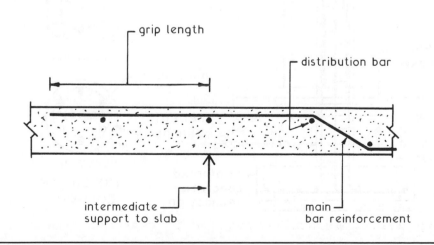

intermediate support to slab

main bar reinforcement

grip length

distribution bar

Columns ~ these are the vertical load bearing members of the structural frame which transmits the beam loads down to the foundations. They are usually constructed in storey heights and therefore the reinforcement must be lapped to provide structural continuity.

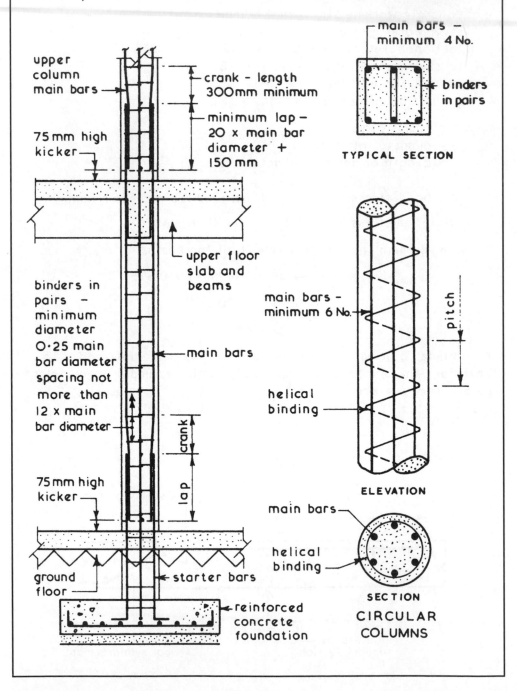

main bars – minimum 4 No.

binders in pairs

TYPICAL SECTION

upper column main bars

crank – length 300mm minimum

minimum lap – 20 × main bar diameter + 150 mm

75 mm high kicker

binders in pairs – minimum diameter 0·25 main bar diameter spacing not more than 12 × main bar diameter

75 mm high kicker

upper floor slab and beams

main bars

crank

lap

main bars – minimum 6 No.

helical binding

pitch

ELEVATION

main bars

helical binding

SECTION

ground floor

starter bars

reinforced concrete foundation

CIRCULAR COLUMNS

With the exception of where bars are spliced ~

BEAMS

The distance between any two parallel bars in the horizontal should be not less than the greater of:

* 25 mm
* the bar diameter where they are equal
* the diameter of the larger bar if they are unequal
* 6 mm greater than the largest size of aggregate in the concrete

The distance between successive layers of bars should be not less than the greater of:

* 15 mm (25 mm if bars > 25 mm dia.)
* the maximum aggregate size

An exception is where the bars transverse each other, e.g. mesh reinforcement.

COLUMNS

Established design guides allow for reinforcement of between 0·8% and 8% of column gross cross sectional area. A lesser figure of 0·6% may be acceptable. A relatively high percentage of steel may save on concrete volume, but consideration must be given to the practicalities of placing and compacting wet concrete. If the design justifies a large proportion of steel, it may be preferable to consider using a concrete clad rolled steel I section.

Transverse reinforcement ~ otherwise known as binders or links. These have the purpose of retaining the main longitudinal reinforcement during construction and restraining each reinforcing bar against buckling. Diameter, take the greater of:

* 6 mm
* 0·25 × main longitudinal reinforcement

Spacing or pitch, not more than the lesser of:

* least lateral column dimension
* 12 × diameter of smallest longitudinal reinforcement
* 300 mm

Helical binding ~ normally, spacing or pitch as above, unless the binding has the additional function of restraining the concrete core from lateral expansion, thereby increasing its load carrying potential. This increased load must be allowed for with a pitch:

* not greater than 75 mm
* not greater than 0.166 × core diameter of the column
* not less than 25 mm
* not less than 3 × diameter of the binding steel

Note: Core diameter is measured across the area of concrete enclosed within the centre line of the binding.

Typical RC Column Details ~

Steel Reinforced Concrete – a modular ratio represents the amount of load that a square unit of steel can safely transmit relative to that of concrete. A figure of 18 is normal, with some variation depending on materials specification and quality.

e.g.

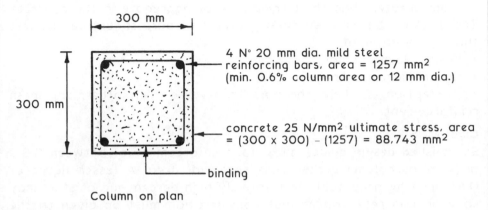

300 mm

300 mm

4 N° 20 mm dia. mild steel reinforcing bars, area = 1257 mm2 (min. 0.6% column area or 12 mm dia.)

concrete 25 N/mm2 ultimate stress, area = (300 x 300) – (1257) = 88,743 mm2

binding

Column on plan

Area of concrete = 88,743 mm²

Equivalent area of steel = 18 × 1257 mm² = 22626 mm²

Equivalent combined area of concrete and steel:

$$\begin{array}{r} 88743 \\ +22626 \\ \hline 111369 \text{ mm}^2 \end{array}$$

Using concrete with a safe or working stress of 5 N/mm², derived from a factor of safety of 5, i.e.

$$\text{Factory of safety} = \frac{\text{Ultimate stress}}{\text{Working stress}} = \frac{25 \text{ N/mm}^2}{5 \text{ N/mm}^2} = 5 \text{ N/mm}^2$$

5 N/mm² × 111369 mm² = 556845 Newtons
kg × 9·81 (gravity) = Newtons

Therefore : $\frac{556845}{9·81}$ = 56763 kg or 56·76 tonnes permissible load

Note: This is the safe load calculation for a reinforced concrete column where the load is axial and bending is minimal or nonexistant, due to a very low slenderness ratio (effective length to least lateral dimension). In reality this is unusual and the next example shows how factors for buckling can be incorporated into the calculation.

Buckling or Bending Effect – the previous example assumed total rigidity and made no allowance for column length and attachments such as floor beams.

The working stress unit for concrete may be taken as 0.8 times the maximum working stress of concrete where the effective length of column (see page 509) is less than 15 times its least lateral dimension. Where this exceeds 15, a further factor for buckling can be obtained from the following:

Effective length ÷ Least lateral dimension	Buckling factor
15	1·0
18	0·9
21	0·8
24	0·7
27	0·6
30	0·5
33	0·4
36	0·3
39	0·2
42	0·1
45	0

Using the example from the previous page, with a column effective length of 9 metres and a modular ratio of 18:

Effective length ÷ Least lateral dimension = 9000 ÷ 300 = 30

From above table the buckling factor = 0·5

Concrete working stress = 5 N/mm²

Equivalent combined area of concrete and steel = 111369 mm²

Therefore: 5 × 0·8 × 0·5 × 111369 = 222738 Newtons

$\dfrac{222738}{9.81}$ = 22705 kg or 22·7 tonnes permissible load

Bar Coding ~ a convenient method for specifying and coordinating the prefabrication of steel reinforcement in the assembly area. It is also useful on site, for checking deliveries and locating materials relative to project requirements. BS EN ISO 3766 provides guidance for a simplified coding system, such that bars can be manufactured and labelled without ambiguity for easy recognition and application on site.

A typical example is the beam shown on page 464, where the lower longitudinal reinforcement (mk·1) could be coded:~

$$2T20\text{-}1\text{-}200B \text{ or, } ①2T\emptyset20\text{-}200\text{-}B\text{-}21$$

    2 = number of bars
    T = deformed high yield steel (460 N/mm², 8–40 mm dia.)
20 or, Ø20 = diameter of bar (mm)
  1 or ① = bar mark or ref. no.
    200 = spacing (mm)
      B = located in bottom of member
      21 = shape code

Other common notation:-

      R = plain round mild steel (250 N/mm², 8–16 mm dia.)
      S = stainless steel
      W = wire reinforcement (4–12 mm dia.)
      T (at the end) = located in top of member
    abr = alternate bars reversed (useful for offsets)

Thus, bar mk.2 = 2R10-2-200T or, ②RØ10-200-T-00
  and mk.3 = 10R8-3-270 or, ③10RØ8-270-54

All but the most obscure reinforcement shapes are illustrated in the British Standard. For the beam referred to on page 464, the standard listing is:-

BS code	Shape	Total bar length on centre line (mm)
OO	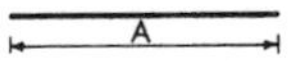 A	A
21	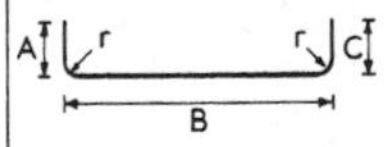	A + B + C − r − 2d (d = bar diameter)
54	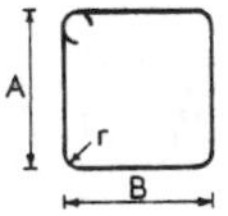	2(A + B) + 12d

Ref. BS EN ISO 3766: Construction drawings. Simplified representation of concrete reinforcement.

Bar Schedule ~ this can be derived from the coding explained on the previous page. Assuming 10 No. beams are required:-

Site ref ....   Schedule ref ....

Prepared by ....   Date ....

Member	Bar mark	Type and size	No. of members	No. of bars in each	Total No.	Bar length (mm)	Shape code	A	B	C	D	E/r
Beam	1	T20	10	2	20	3080	21	200	2700	200		
	2	R10	10	2	20	2700	00	2700				
	3	R8	10	10	100	1336	54	400	220			

Note: r = 2 × d for mild steel
3 × d for high yield steel

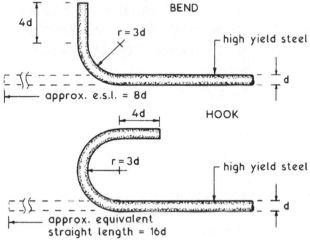

Bar coding ~

1st character	2nd character
O No bends	O Straight bars
1 1 bend	1 90° bends, standard radius, all bends towards same direction
2 2 bends	2 90° bends, non-standard radius, all bends towards same direction
3 3 bends	3 180° bends, non-standard radius, all bends towards same direction
4 4 bends	4 90° bends, standard radius, not all bends towards same direction
5 5 bends	5 Bends <90°, standard radius, all bends towards same direction
6 Arcs of circles	6 Bends <90°, standard radius, not all bends towards same direction
7 Complete helices	7 Arcs or helices

Note: 9 is used for special or non-standard shapes

Material ~ Mild steel or high yield steel. Both contain about 99% iron, the remaining constituents are manganese, carbon, sulphur and phosphorus. The proportion of carbon determines the quality and grade of steel; mild steel has 0·25% carbon, high yield steel 0·40%. High yield steel may also be produced by cold working or deforming mild steel until it is strain hardened. Mild steel has the letter R preceding the bar diameter in mm, e.g. R20 and high yield steel the letter T or Y.

Standard bar diameters ~ 6, 8, 10, 12, 16, 20, 25, 32 and 40 mm.

Grade notation ~

- Mild steel – grade 250 or 250 N/mm² characteristic tensile strength (0·25% carbon, 0·06% sulphur and 0·06% phosphorus).
- High yield steel – grade 460/425 (0·40% carbon, 0·05% sulphur and 0·05% phosphorus).

  460 N/mm² characteristic tensile strength: 6, 8, 10, 12 and 16 mm diameter.

  425 N/mm² characteristic tensile strength: 20, 25, 32 and 40 mm diameter.

Examples of steel reinforcement ~

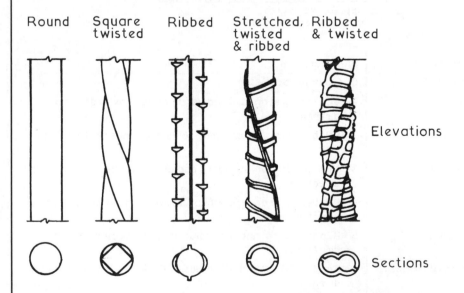

Round Square twisted Ribbed Stretched, twisted & ribbed Ribbed & twisted

Elevations

Sections

Ref. BS 4449: Steel for the reinforcement of concrete, weldable reinforcing steel. Bar, coil and decoiled product. Specification.

Steel reinforcement mesh or fabric is produced in four different formats for different applications:

Format	Type	Typical application
A	Square mesh	Floor slabs
B	Rectangular mesh	Floor slabs
C	Long mesh	Roads and pavements
D	Wrapping mesh	Binding wire with concrete fire protection to structural steelwork

Standard sheet size ~ 4·8 m long × 2·4 m wide.
Standard roll size ~ 48 and 72 m long × 2·4 m wide.

Specification ~ format letter plus a reference number. This number equates to the cross sectional area in mm² of the main bars per metre width of mesh.
E.g. B385 is rectangular mesh with 7 mm dia. main bars, i.e. 10 bars of 7 mm dia. @ 100 mm spacing = 385 mm².

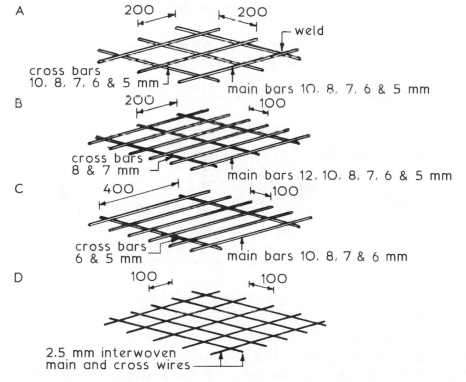

A
200   200   weld
cross bars
10, 8, 7, 6 & 5 mm   main bars 10, 8, 7, 6 & 5 mm

B
200   100
cross bars
8 & 7 mm   main bars 12, 10, 8, 7, 6 & 5 mm

C
400   100
cross bars
6 & 5 mm   main bars 10, 8, 7 & 6 mm

D
100   100
2.5 mm interwoven
main and cross wires

Refs. BS 4483: Steel fabric for the reinforcement of concrete. Specification.

BS 4482: Steel wire for the reinforcement of concrete products. Specification.

475

Cover to reinforcement in columns, beams, foundations, etc. is required for the following reasons:

- To protect the steel against corrosion.
- To provide sufficient bond or adhesion between steel and concrete.
- To ensure sufficient protection of the steel in a fire (see note).

If the cover is insufficient, concrete will spall away from the steel.

Minimum cover ~ never less than the maximum size of aggregate in the concrete, or the largest reinforcement bar size (take greater value). Guidance on minimum cover for particular locations:

Below ground ~

- Foundations, retaining walls, basements, etc., 40 mm, binders 25 mm.
- Marine structures, 65 mm, binders 50 mm.
- Uneven earth and fill 75 mm, blinding 40 mm.

Above ground ~

- Ends of reinforcing bars, not less than 25 mm nor less than 2 × bar diameter.
- Column longitudinal reinforcement 40 mm, binders 20 mm.
- Columns <190 mm min. dimension with bars <12 mm dia., 25 mm.
- Beams 25 mm, binders 20 mm.
- Slabs 20 mm (15 mm where max. aggregate size is <15 mm).

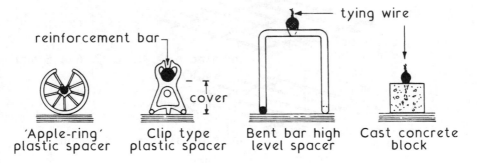

'Apple-ring' plastic spacer   Clip type plastic spacer   Bent bar high level spacer   Cast concrete block

Note: Minimum cover for corrosion protection and bond may not be sufficient for fire protection and severe exposure situations.

For details of fire protection see ~

Building Regulations, Approved Document B: Fire safety
BS 8110-2: Structural use of concrete. Code of practice for special circumstances.

For general applications, including exposure situations, see ~
BS 8110-1: Structural use of concrete. Code of practice for design and construction.

Typical examples using dense concrete of calcareous aggregates (excluding limestone) or siliceous aggregates, eg. flints, quartzites and granites ~

Column fully exposed ~

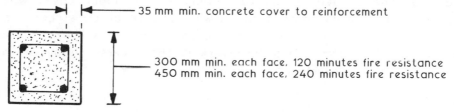

35 mm min. concrete cover to reinforcement

300 mm min. each face, 120 minutes fire resistance
450 mm min. each face, 240 minutes fire resistance

Column, maximum 50% exposed ~

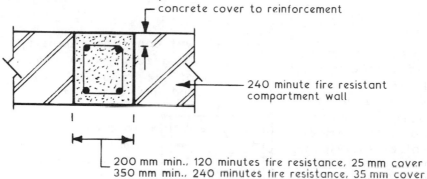

concrete cover to reinforcement

240 minute fire resistant compartment wall

200 mm min., 120 minutes fire resistance, 25 mm cover
350 mm min., 240 minutes fire resistance, 35 mm cover

Column, one face only exposed ~

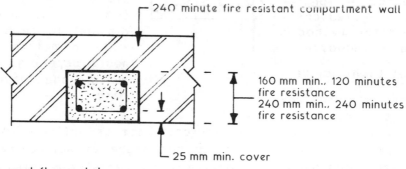

240 minute fire resistant compartment wall

160 mm min., 120 minutes fire resistance
240 mm min., 240 minutes fire resistance

25 mm min. cover

Beam and floor slab ~

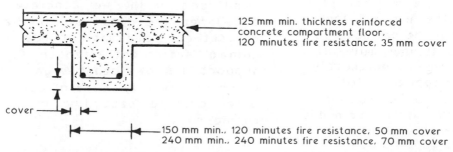

125 mm min. thickness reinforced concrete compartment floor, 120 minutes fire resistance, 35 mm cover

cover

150 mm min., 120 minutes fire resistance, 50 mm cover
240 mm min., 240 minutes fire resistance, 70 mm cover

477

Basic Formwork ~ concrete when first mixed is a fluid and therefore to form any concrete member the wet concrete must be placed in a suitable mould to retain its shape, size and position as it sets. It is possible with some forms of concrete foundations to use the sides of the excavation as the mould but in most cases when casting concrete members a mould will have to be constructed on site. These moulds are usually called formwork. It is important to appreciate that the actual formwork is the reverse shape of the concrete member which is to be cast.

Basic Principles ~

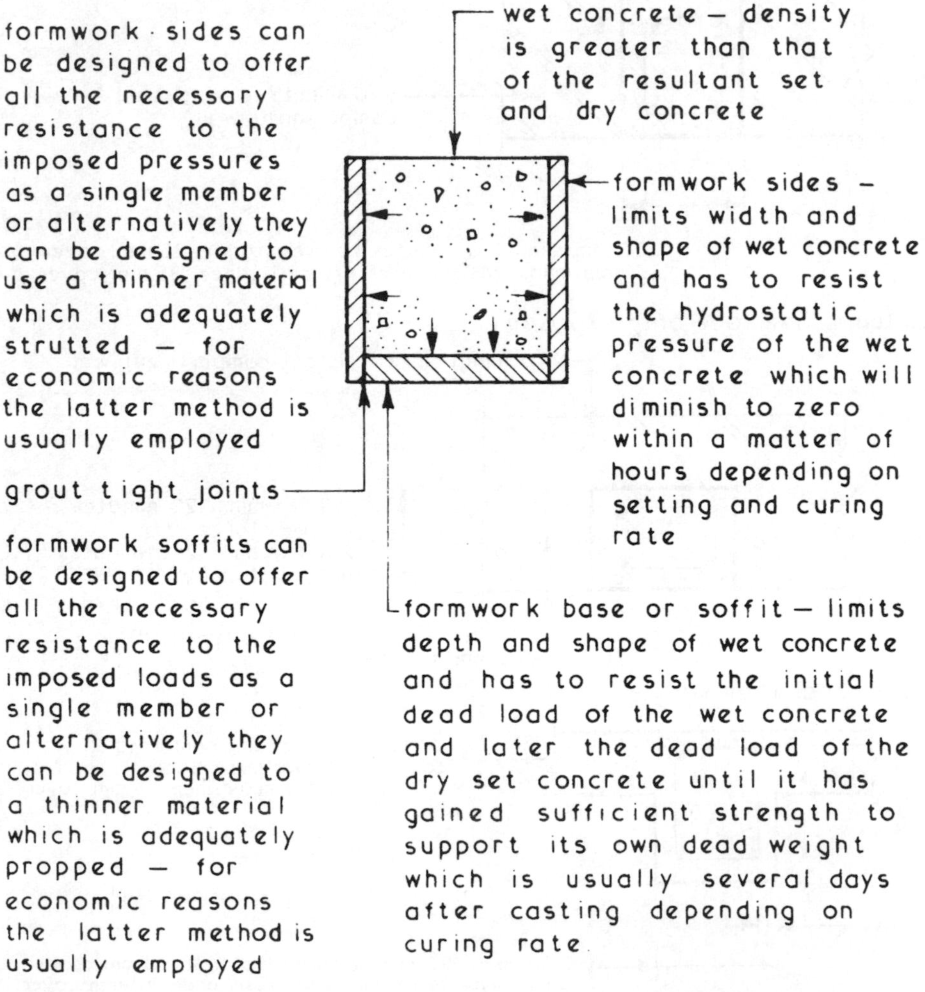

formwork · sides can be designed to offer all the necessary resistance to the imposed pressures as a single member or alternatively they can be designed to use a thinner material which is adequately strutted — for economic reasons the latter method is usually employed

grout tight joints

formwork soffits can be designed to offer all the necessary resistance to the imposed loads as a single member or alternatively they can be designed to a thinner material which is adequately propped — for economic reasons the latter method is usually employed

wet concrete — density is greater than that of the resultant set and dry concrete

formwork sides — limits width and shape of wet concrete and has to resist the hydrostatic pressure of the wet concrete which will diminish to zero within a matter of hours depending on setting and curing rate

formwork base or soffit — limits depth and shape of wet concrete and has to resist the initial dead load of the wet concrete and later the dead load of the dry set concrete until it has gained sufficient strength to support its own dead weight which is usually several days after casting depending on curing rate.

478

Typical Simple Beam Formwork Details ~

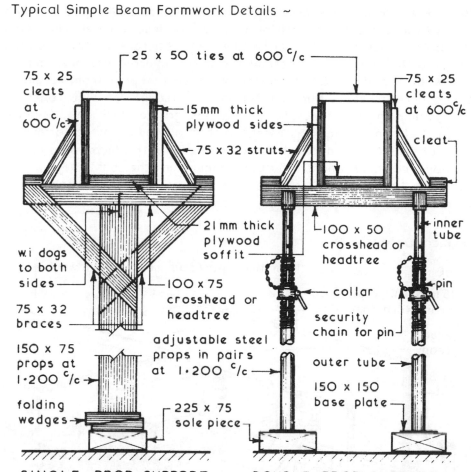

SINGLE PROP SUPPORT

Erecting Formwork

1. Props positioned and levelled through.
2. Soffit placed, levelled and position checked.
3. Side forms placed, their position checked before being fixed.
4. Strutting position and fixed.
5. Final check before casting.

Suitable Formwork Materials~ timber, steel and special plastics.

DOUBLE PROP SUPPORT

Striking or Removing Formwork

1. Side forms as soon as practicable usually within hours of casting this allows drying air movements to take place around the setting concrete.
2. Soffit formwork as soon as practicable usually within days but as a precaution some props are left in position until concrete member is self supporting.

Beam Formwork ~ this is basically a three sided box supported and propped in the correct position and to the desired level. The beam formwork sides have to retain the wet concrete in the required shape and be able to withstand the initial hydrostatic pressure of the wet concrete whereas the formwork soffit apart from retaining the concrete has to support the initial load of the wet concrete and finally the set concrete until it has gained sufficient strength to be self supporting. It is essential that all joints in the formwork are constructed to prevent the escape of grout which could result in honeycombing and/or feather edging in the cast beam. The removal time for the formwork will vary with air temperature, humidity and consequent curing rate.

Typical Details ~

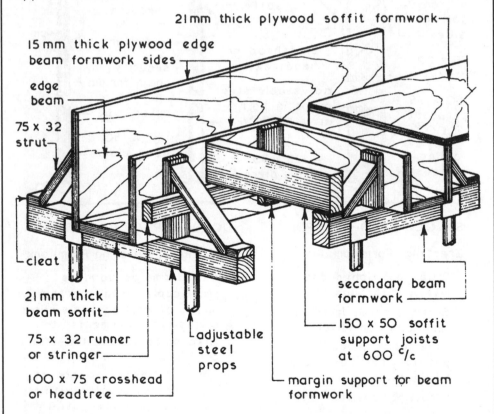

21mm thick plywood soffit formwork

15mm thick plywood edge beam formwork sides

edge beam

75 x 32 strut

cleat

21mm thick beam soffit

75 x 32 runner or stringer

100 x 75 crosshead or headtree

adjustable steel props

secondary beam formwork

150 x 50 soffit support joists at 600 c/c

margin support for beam formwork

Typical Formwork Striking Times ~

Beam Sides – 9 to 12 hours
Beam Soffits – 8 to 14 days (props left under) ⎫ Using OPC-air temp
Beam Props – 15 to 21 days   ⎭ 7 to 16°C

Column Formwork ~ this consists of a vertical mould of the desired shape and size which has to retain the wet concrete and resist the initial hydrostatic pressure caused by the wet concrete. To keep the thickness of the formwork material to a minimum horizontal clamps or yokes are used at equal centres for batch filling and at varying centres for complete filling in one pour. The head of the column formwork can be used to support the incoming beam formwork which gives good top lateral restraint but results in complex formwork. Alternatively the column can be cast to the underside of the beams and at a later stage a collar of formwork can be clamped around the cast column to complete casting and support the incoming beam formwork. Column forms are located at the bottom around a 75 to 100 mm high concrete plinth or kicker which has the dual function of location and preventing grout loss from the bottom of the column formwork.

Typical Details ~

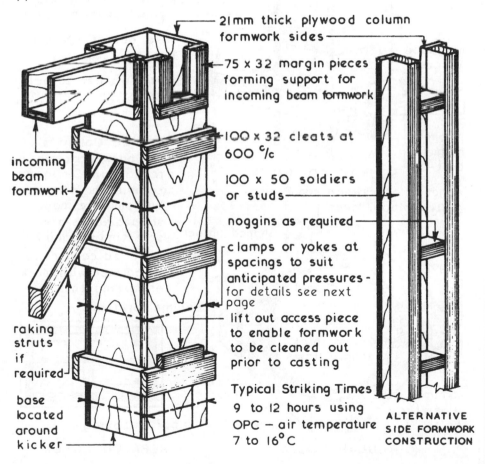

21mm thick plywood column formwork sides

75 x 32 margin pieces forming support for incoming beam formwork

100 x 32 cleats at 600 °/c

100 x 50 soldiers or studs

noggins as required

clamps or yokes at spacings to suit anticipated pressures - for details see next page

lift out access piece to enable formwork to be cleaned out prior to casting

incoming beam formwork

raking struts if required

base located around kicker

Typical Striking Times
9 to 12 hours using OPC – air temperature 7 to 16°C

ALTERNATIVE SIDE FORMWORK CONSTRUCTION

481

Column Yokes ~ these are obtainable as a metal yoke or clamp or they can be purpose made from timber.

Typical Examples ~

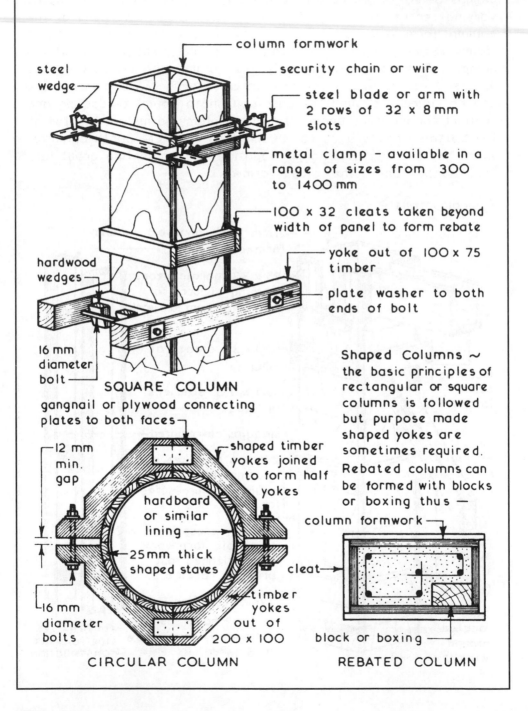

steel wedge

column formwork

security chain or wire

steel blade or arm with 2 rows of 32 x 8mm slots

metal clamp – available in a range of sizes from 300 to 1400 mm

100 x 32 cleats taken beyond width of panel to form rebate

yoke out of 100 x 75 timber

hardwood wedges

plate washer to both ends of bolt

16 mm diameter bolt

SQUARE COLUMN

gangnail or plywood connecting plates to both faces

12 mm min. gap

shaped timber yokes joined to form half yokes

hardboard or similar lining

25mm thick shaped staves

16 mm diameter bolts

timber yokes out of 200 x 100

CIRCULAR COLUMN

Shaped Columns ~ the basic principles of rectangular or square columns is followed but purpose made shaped yokes are sometimes required. Rebated columns can be formed with blocks or boxing thus —

column formwork

cleat

block or boxing

REBATED COLUMN

482

Precast Concrete Frames ~ these frames are suitable for single storey and low rise applications, the former usually in the form of portal frames which are normally studied separately. Precast concrete frames provide the skeleton for the building and can be clad externally and finished internally by all the traditional methods. The frames are usually produced as part of a manufacturer's standard range of designs and are therefore seldom purpose made due mainly to the high cost of the moulds.

Advantages:-

1. Frames are produced under factory controlled conditions resulting in a uniform product of both quality and accuracy.

2. Repetitive casting lowers the cost of individual members.

3. Off site production releases site space for other activities.

4. Frames can be assembled in cold weather and generally by semi-skilled labour.

Disadvantages:-

1. Although a wide choice of frames is available from various manufacturers these systems lack the design flexibility of cast in-situ purpose made frames.

2. Site planning can be limited by manufacturer's delivery and unloading programmes and requirements.

3. Lifting plant of a type and size not normally required by traditional construction methods may be needed.

Typical Site Activities ~

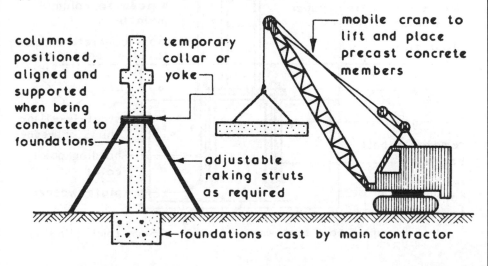

columns positioned, aligned and supported when being connected to foundations—

temporary collar or yoke—

mobile crane to lift and place precast concrete members

←adjustable raking struts as required

←foundations cast by main contractor

Foundation Connections ~ the preferred method of connection is to set the column into a pocket cast into a reinforced concrete pad foundation and is suitable for light to medium loadings. Where heavy column loadings are encountered it may be necessary to use a steel base plate secured to the reinforced concrete pad foundation with holding down bolts.

Typical Details ~

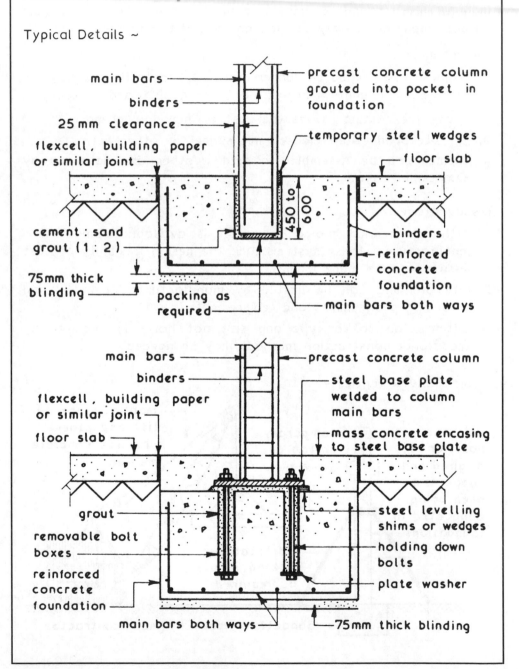

Column to Column Connection ~ precast columns are usually cast in one length and can be up to four storeys in height. They are either reinforced with bar reinforcement or they are prestressed according to the loading conditions. If column to column are required they are usually made at floor levels above the beam to column connections and can range from a simple dowel connection to a complex connection involving in-situ concrete.

Typical Details ~

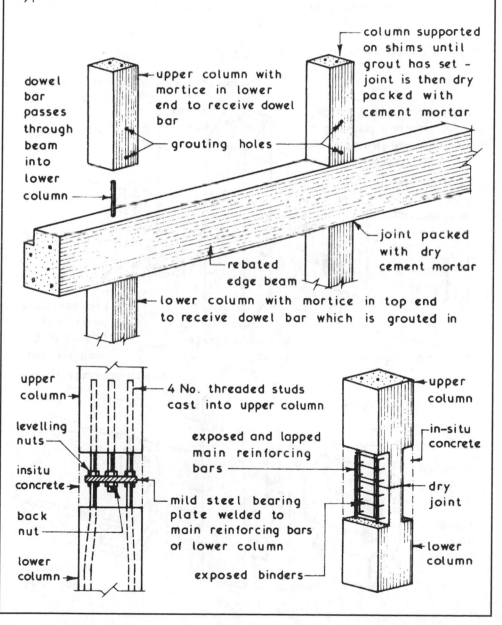

dowel bar passes through beam into lower column

upper column with mortice in lower end to receive dowel bar

grouting holes

column supported on shims until grout has set - joint is then dry packed with cement mortar

joint packed with dry cement mortar

rebated edge beam

lower column with mortice in top end to receive dowel bar which is grouted in

upper column

levelling nuts

insitu concrete

back nut

lower column

4 No. threaded studs cast into upper column

exposed and lapped main reinforcing bars

mild steel bearing plate welded to main reinforcing bars of lower column

exposed binders

upper column

in-situ concrete

dry joint

lower column

Beam to Column Connections ~ as with the column to column connections (see page 485) the main objective is to provide structural continuity at the junction. This is usually achieved by one of two basic methods:-

1. Projecting bearing haunches cast onto the columns with a projecting dowel or stud bolt to provide both location and fixing.
2. Steel to steel fixings which are usually in the form of a corbel or bracket projecting from the column providing a bolted connection to a steel plate cast into the end of the beam.

Typical Details ~

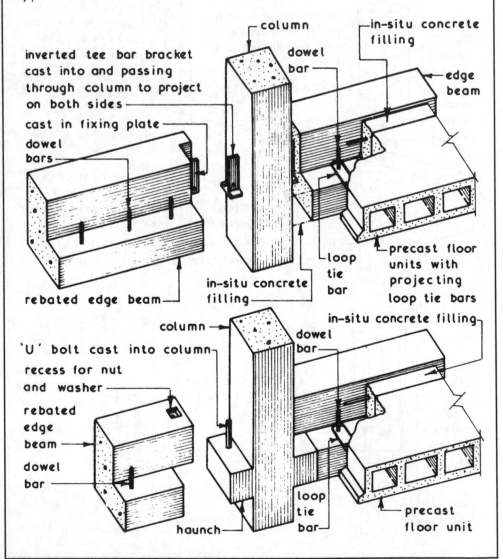

inverted tee bar bracket cast into and passing through column to project on both sides

cast in fixing plate

dowel bars

rebated edge beam

column

dowel bar

in-situ concrete filling

edge beam

in-situ concrete filling

loop tie bar

precast floor units with projecting loop tie bars

'U' bolt cast into column

recess for nut and washer

rebated edge beam

dowel bar

column

dowel bar

in-situ concrete filling

loop tie bar

haunch

precast floor unit

Principles ~ the well known properties of concrete are that it has high compressive strength and low tensile strength. The basic concept of reinforced concrete is to include a designed amount of steel bars in a predetermined pattern to give the concrete a reasonable amount of tensile strength. In prestressed concrete a precompression is induced into the member to make full use of its own inherent compressive strength when loaded. The design aim is to achieve a balance of tensile and compressive forces so that the end result is a concrete member which is resisting only stresses which are compressive. In practice a small amount of tension may be present but providing this does not exceed the tensile strength of the concrete being used tensile failure will not occur.

Comparison of Reinforced and Prestressed Concrete ~

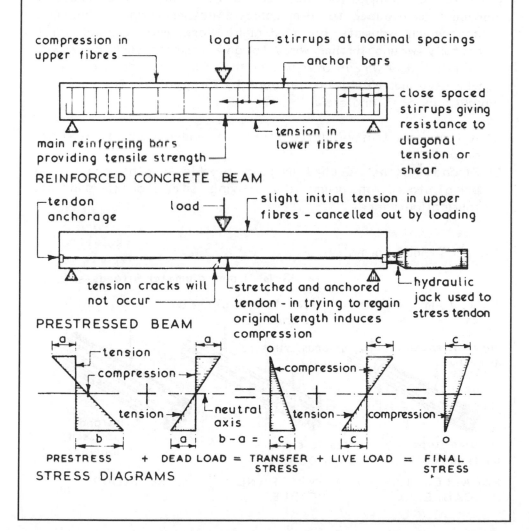

REINFORCED CONCRETE BEAM

PRESTRESSED BEAM

STRESS DIAGRAMS

Materials ~ concrete will shrink whilst curing and it can also suffer sectional losses due to creep when subjected to pressure. The amount of shrinkage and creep likely to occur can be controlled by designing the strength and workability of the concrete, high strength and low workability giving the greatest reduction in both shrinkage and creep. Mild steel will suffer from relaxation losses which is where the stresses in steel under load decrease to a minimum value after a period of time and this can be overcome by increasing the initial stress in the steel. If mild steel is used for prestressing the summation of shrinkage, creep and relaxation losses will cancel out any induced compression, therefore special alloy steels must be used to form tendons for prestressed work.

Tendons ~ these can be of small diameter wires (2 to 7 mm) in a plain round, crimped or indented format, these wires may be individual or grouped to form cables. Another form of tendon is strand which consists of a straight core wire around which is helically wound further wires to give formats such as 7 wire (6 over 1) and 19 wire (9 over 9 over 1) and like wire tendons strand can be used individually or in groups to form cables. The two main advantages of strand are:-

1. A large prestressing force can be provided over a restricted area.
2. Strand can be supplied in long flexible lengths capable of being stored on drums thus saving site storage and site fabrication space.

Typical Tendon Formats ~

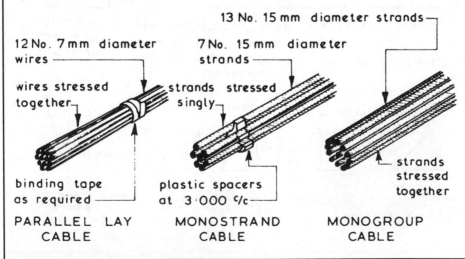

13 No. 15 mm diameter strands

12 No. 7 mm diameter wires

7 No. 15 mm diameter strands

wires stressed together

strands stressed singly

binding tape as required

plastic spacers at 3·000 c/c

strands stressed together

PARALLEL LAY CABLE

MONOSTRAND CABLE

MONOGROUP CABLE

Pre-tensioning ~ this method is used mainly in the factory production of precast concrete components such as lintels, floor units and small beams. Many of these units are formed by the long line method where precision steel moulds up to 120·000 long are used with spacer or dividing plates to form the various lengths required. In pre-tensioning the wires are stressed within the mould before the concrete is placed around them. Steam curing is often used to accelerate this process to achieve a 24 hour characteristic strength of 28 N/mm² with a typical 28 day cube strength of 40 N/mm². Stressing of the wires is carried out by using hydraulic jacks operating from one or both ends of the mould to achieve an initial 10% overstress to counteract expected looses. After curing the wires are released or cut and the bond between the stressed wires and the concrete prevents the tendons from regaining their original length thus maintaining the precompression or prestress.

At the extreme ends of the members the bond between the stressed wires and concrete is not fully developed due to low frictional resistance. This results in a small contraction and swelling at the ends of the wire forming in effect a cone shape anchorage. The distance over which this contraction occurs is called the transfer length and is equal to 80 to 120 times the wire diameter. To achieve a greater total surface contact area it is common practice to use a larger number of small diameter wires rather than a smaller number of large diameter wires giving the same total cross sectional area.

Typical Pre-tensioning Arrangement ~

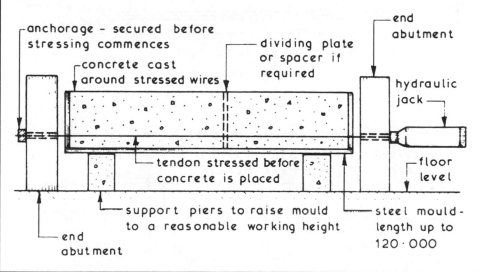

489

Post-tensioning ~ this method is usually employed where stressing is to be carried out on site after casting an in-situ component or where a series of precast concrete units are to be joined together to form the required member. It can also be used where curved tendons are to be used to overcome negative bending moments. In post-tensioning the concrete is cast around ducts or sheathing in which the tendons are to be housed. Stressing is carried out after the concrete has cured by means of hydraulic jacks operating from one or both ends of the member. The anchorages (see next page) which form part of the complete component prevent the stressed tendon from regaining its original length thus maintaining the precompression or prestress. After stressing the annular space in the tendon ducts should be filled with grout to prevent corrosion of the tendons due to any entrapped moisture and to assist in stress distribution. Due to the high local stresses at the anchorage positions it is usual for a reinforcing spiral to be included in the design.

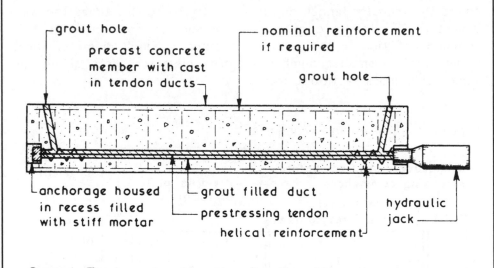

grout hole

precast concrete member with cast in tendon ducts

nominal reinforcement if required

grout hole

anchorage housed in recess filled with stiff mortar

grout filled duct

prestressing tendon

helical reinforcement

hydraulic jack

Curved Tendons for Negative Bending Moments ~

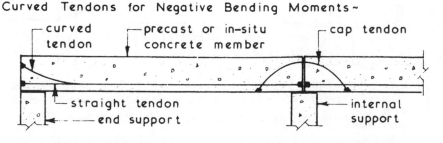

curved tendon

precast or in-situ concrete member

cap tendon

straight tendon

end support

internal support

Typical Post-tensioning Arrangement ~

Anchorages ~ the formats for anchorages used in conjunction with post-tensioned prestressed concrete works depends mainly on whether the tendons are to be stressed individually or as a group, but most systems use a form of split cone wedges or jaws acting against a form of bearing or pressure plate.

Typical Anchorage Details ~

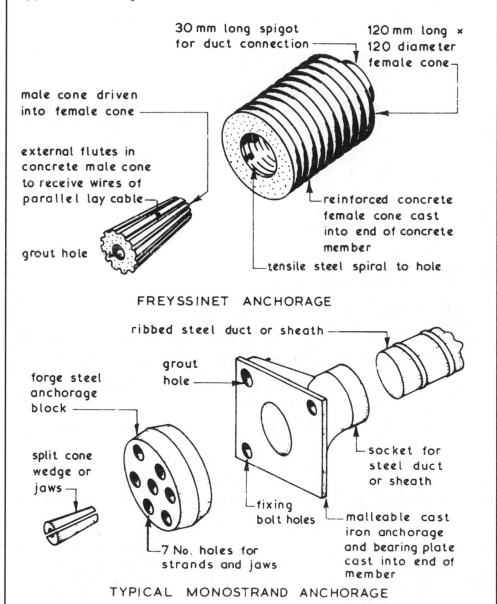

30 mm long spigot for duct connection

120 mm long × 120 diameter female cone

male cone driven into female cone

external flutes in concrete male cone to receive wires of parallel lay cable

grout hole

reinforced concrete female cone cast into end of concrete member

tensile steel spiral to hole

FREYSSINET ANCHORAGE

ribbed steel duct or sheath

grout hole

forge steel anchorage block

split cone wedge or jaws

socket for steel duct or sheath

fixing bolt holes

7 No. holes for strands and jaws

malleable cast iron anchorage and bearing plate cast into end of member

TYPICAL MONOSTRAND ANCHORAGE

Comparison with Reinforced Concrete ~ when comparing prestressed concrete with conventional reinforced concrete the main advantages and disadvantages can be enumerated but in the final analysis each structure and/or component must be decided on its own merit.

Main advantages:-

1. Makes full use of the inherent compressive strength of concrete.
2. Makes full use of the special alloy steels used to form the prestressing tendons.
3. Eliminates tension cracks thus reducing the risk of corrosion of steel components.
4. Reduces shear stresses.
5. For any given span and loading condition a component with a smaller cross section can be used thus giving a reduction in weight.
6. Individual precast concrete units can be joined together to form a composite member.

Main Disadvantages:-

1. High degree of control over materials, design and quality of workmanship is required.
2. Special alloy steels are dearer than most traditional steels used in reinforced concrete.
3. Extra cost of special equipment required to carry out the prestressing activities.
4. Cost of extra safety requirements needed whilst stressing tendons.

As a general comparison between the two structural options under consideration it is usually found that:-

1. Up to 6·000 span traditional reinforced concrete is the most economic method.
2. Spans between 6·000 and 9·000 the two cost options are comparable.
3. Over 9·000 span prestressed concrete is more economical than reinforced concrete.

It should be noted that generally columns and walls do not need prestressing but in tall columns and high retaining walls where the bending stresses are high, prestressing techniques can sometimes be economically applied.

Ground Anchors ~ these are a particular application of post-tensioning prestressing techniques and can be used to form ground tie backs to cofferdams, retaining walls and basement walls. They can also be used as vertical tie downs to basement and similar slabs to prevent flotation during and after construction. Ground anchors can be of a solid bar format (rock anchors) or of a wire or cable format for granular and cohesive soils. A lined or unlined bore hole must be drilled into the soil to the design depth and at the required angle to house the ground anchor. In clay soils the bore hole needs to be underreamed over the anchorage length to provide adequate bond. The tail end of the anchor is pressure grouted to form a bond with the surrounding soil, the remaining length being unbonded so that it can be stressed and anchored at head thus inducing the prestress. The void around the unbonded or elastic length is gravity grouted after completion of the stressing operation.

Typical Ground Anchor Details ~

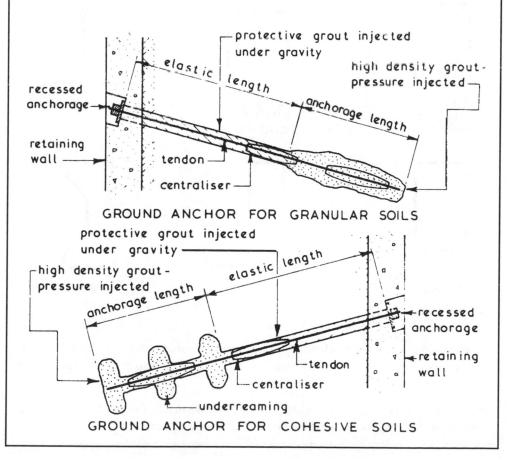

GROUND ANCHOR FOR GRANULAR SOILS

GROUND ANCHOR FOR COHESIVE SOILS

Cold rolled steel sections are a lightweight alternative to the relatively heavy, hot rolled steel sections that have been traditionally used in sub-framing situations, e.g. purlins, joists and sheeting rails. Cold rolled sections are generally only a few millimetres in wall thickness, saving on material and handling costs and building dead load. They are also produced in a wide variety of section profiles, some of which are shown below.

Typical section profiles ~

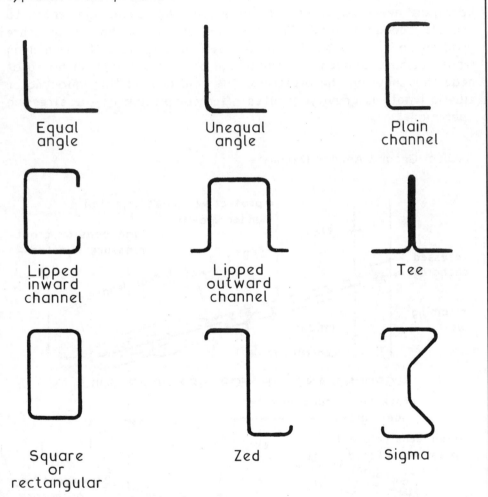

Equal angle	Unequal angle	Plain channel
Lipped inward channel	Lipped outward channel	Tee
Square or rectangular	Zed	Sigma

Dimensions vary considerably and many non-standard profiles are made for particular situations. A range of standard sections are produced to:

BS EN 10162: Cold rolled steel sections. Technical delivery conditions. Dimensional and cross sectional tolerances.

Structural Steelwork ~ the standard sections available for use in structural steelwork are given in BS 4-1 and BS ENs 10056 and 10210. These standards give a wide range of sizes and weights to enable the designer to formulate an economic design.

Typical Standard Steelwork Sections ~

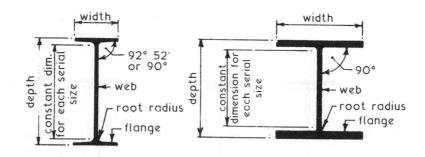

UNIVERSAL BEAMS
127 x 76 x 13 kg/m to
914 x 419 x 388 kg/m

UNIVERSAL COLUMNS
152 x 152 x 23 kg/m to
356 x 406 x 634 kg/m

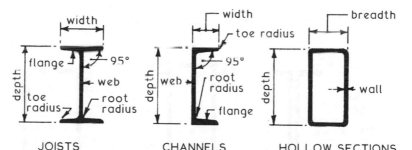

JOISTS
76 x 76 x 13 kg/m to
254 x 203 x 82 kg/m

CHANNELS
100 x 50 x 10 kg/m to
430 x 100 x 64 kg/m

HOLLOW SECTIONS
50 x 30 x 2.89 kg/m to
500 x 300 x 191 kg/m

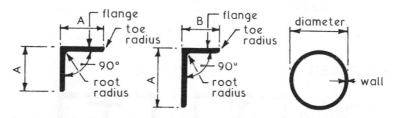

EQUAL ANGLES
25 x 25 x 1.2 kg/m to
200 x 200 x 71.1 kg/m

UNEQUAL ANGLES
40 x 25 x 1.91 kg/m to
200 x 150 x 47.1 kg/m

HOLLOW SECTIONS
21.3 dia. x 1.43 kg/m to
508 dia. x 194 kg/m

NB. Sizes given are serial or nominal, for actual sizes see relevant BS.

Compound Sections – these are produced by welding together standard sections. Various profiles are possible, which can be designed specifically for extreme situations such as very high loads and long spans, where standard sections alone would be insufficient. Some popular combinations of standard sections include:

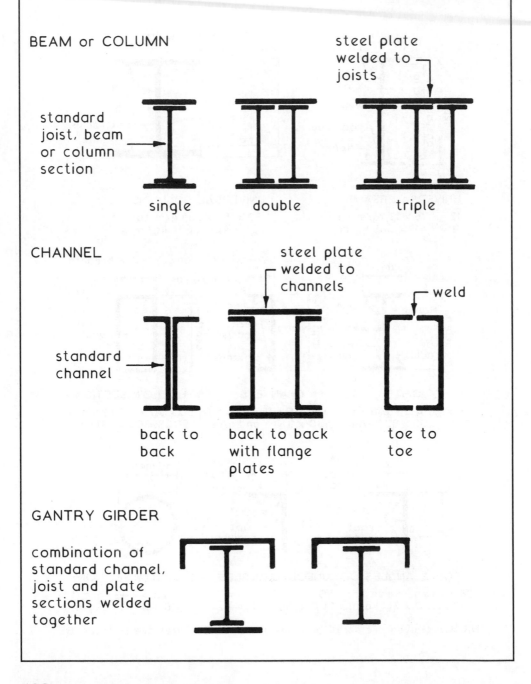

BEAM or COLUMN

steel plate welded to — joists

standard joist, beam or column section →

single   double   triple

CHANNEL

steel plate — welded to channels

standard channel →

— weld

back to back   back to back with flange plates   toe to toe

GANTRY GIRDER

combination of standard channel, joist and plate sections welded together

Open Web Beams – these are particularly suited to long spans with light to moderate loading. The relative increase in depth will help resist deflection and voids in the web will reduce structural dead load.

Perforated Beam – a standard beam section with circular voids cut about the neutral axis.

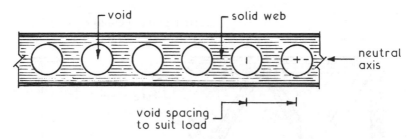

Castellated Beam – a standard beam section web is profile cut into two by oxy-acetylene torch. The projections on each section are welded together to create a new beam 50% deeper than the original.

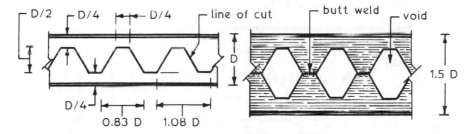

Litzka Beam – a standard beam cut as the castellated beam, but with overall depth increased further by using spacer plates welded to the projections. Minimal increase in weight.

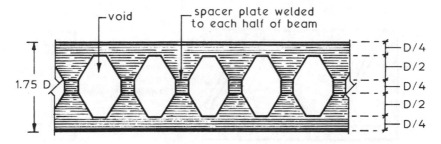

Note: Voids at the end of open web beams should be filled with a welded steel plate, as this is the area of maximum shear stress in a beam.

Lattices – these are an alternative type of open web beam, using standard steel sections to fabricate high depth to weight ratio units capable of spans up to about 15 m. The range of possible components is extensive and some examples are shown below:

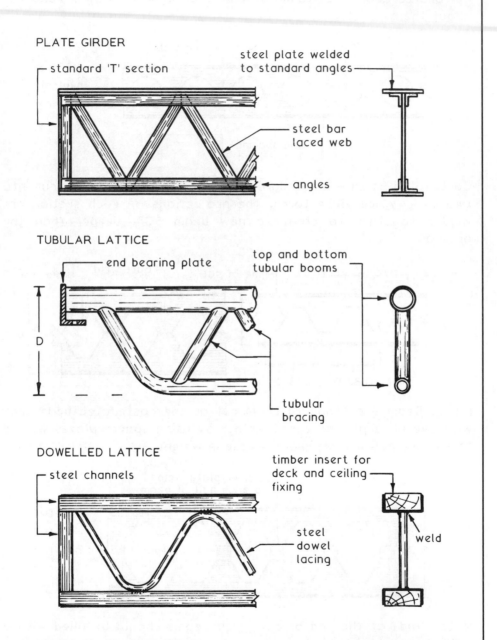

PLATE GIRDER

standard 'T' section

steel plate welded to standard angles

steel bar laced web

angles

TUBULAR LATTICE

end bearing plate

top and bottom tubular booms

D

tubular bracing

DOWELLED LATTICE

steel channels

timber insert for deck and ceiling fixing

steel dowel lacing

weld

Note: span potential for lattice beams is approximately 24 × D

Structural Steelwork Connections ~ these are either workshop or site connections according to where the fabrication takes place. Most site connections are bolted whereas workshop connections are very often carried out by welding. The design of structural steelwork members and their connections is the province of the structural engineer who selects the type and number of bolts or the size and length of weld to be used according to the connection strength to be achieved.

Typical Connection Examples ~

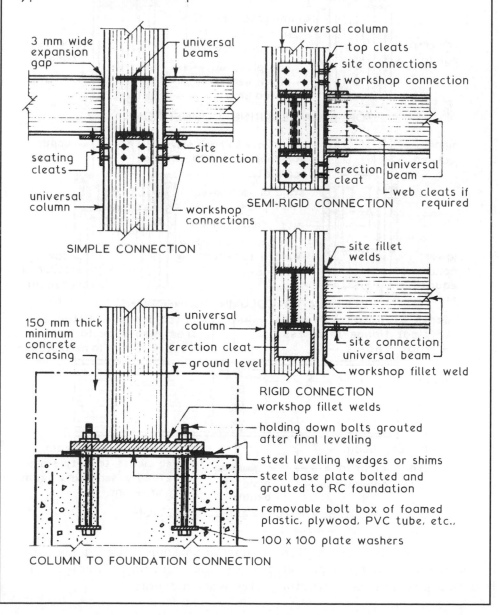

3 mm wide expansion gap

universal beams

seating cleats

site connection

universal column

workshop connections

SIMPLE CONNECTION

universal column

top cleats

site connections

workshop connection

erection cleat

universal beam

web cleats if required

SEMI-RIGID CONNECTION

site fillet welds

universal column

erection cleat

ground level

site connection
universal beam

workshop fillet weld

RIGID CONNECTION

150 mm thick minimum concrete encasing

workshop fillet welds

holding down bolts grouted after final levelling

steel levelling wedges or shims

steel base plate bolted and grouted to RC foundation

removable bolt box of foamed plastic, plywood, PVC tube, etc.,

100 x 100 plate washers

COLUMN TO FOUNDATION CONNECTION

Typical Connection Examples ~

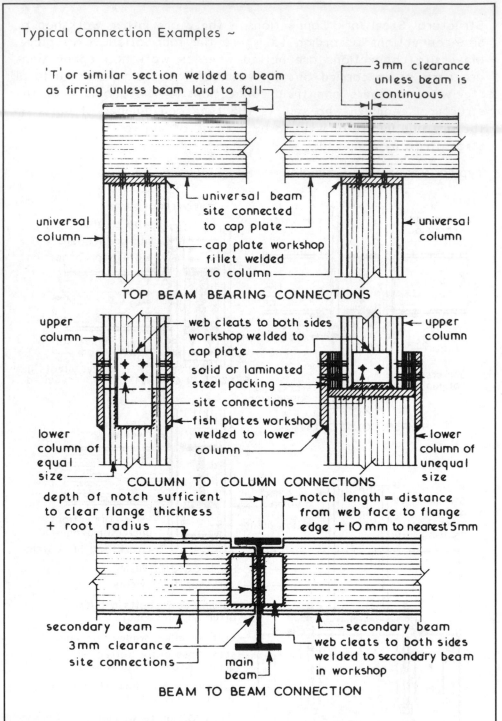

'T' or similar section welded to beam as firring unless beam laid to fall

3mm clearance unless beam is continuous

universal beam site connected to cap plate

cap plate workshop fillet welded to column

universal column

universal column

TOP BEAM BEARING CONNECTIONS

upper column

web cleats to both sides workshop welded to cap plate

solid or laminated steel packing

site connections

fish plates workshop welded to lower column

upper column

lower column of equal size

lower column of unequal size

COLUMN TO COLUMN CONNECTIONS

depth of notch sufficient to clear flange thickness + root radius

notch length = distance from web face to flange edge + 10 mm to nearest 5mm

secondary beam

3mm clearance

site connections

main beam

secondary beam

web cleats to both sides welded to secondary beam in workshop

BEAM TO BEAM CONNECTION

NB. All holes for bolted connections must be made from backmarking the outer surface of the section(s) involved. For actual positions see structural steelwork manuals.

Types ~
Slab or bloom base.
Gusset base.
Steel grillage (see page 209).

The type selected will depend on the load carried by the column and the distribution area of the base plate. The cross sectional area of a UC concentrates the load into a relatively small part of the base plate. Therefore to resist bending and shear, the base must be designed to resist the column loads and transfer them onto the pad foundation below.

SLAB or BLOOM BASE   SLAB BASE WITH ANGLE CLEATS

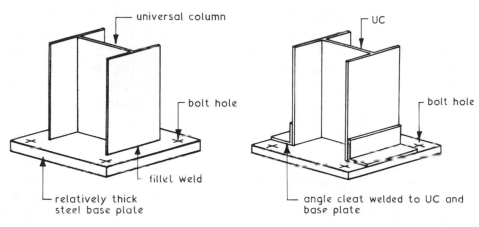

GUSSET BASE   BOLT BOX

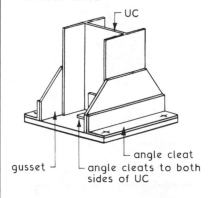

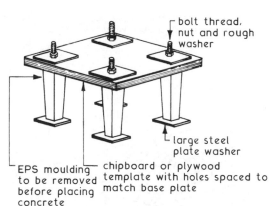

Bolt Box ~ a template used to accurately locate column holding down bolts into wet concrete. EPS or plastic tubes provide space around the bolts when the concrete has set. The bolts can then be moved slightly to aid alignment with the column base.

Welding is used to prefabricate the sub-assembly of steel frame components in the workshop, prior to delivery to site where the convenience of bolted joints will be preferred.

Oxygen and acetylene (oxy-acetylene) gas welding equipment may be used to fuse together light steel sections, but otherwise it is limited to cutting profiles of the type shown on page 497. The electric arc process is preferred as it is more effective and efficient. This technique applies an expendable steel electrode to fuse parts together by high amperage current. The current potential and electrode size can be easily changed to suit the thickness of metal.

Overlapping of sections permits the convenience of fillet welds, but if the overlap is obstructive or continuity and direct transfer of loads is necessary, a butt weld will be specified. To ensure adequate weld penetration with a butt weld, the edges of the parent metal should be ground to produce an edge chamfer. For very large sections, both sides of the material should be chamfered to allow for double welds.

BUTT WELD

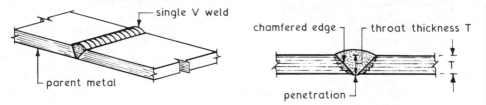

Note: For greater thicknesses of parent metal, both sides are chamfered in preparation for a double weld.

FILLET WELD

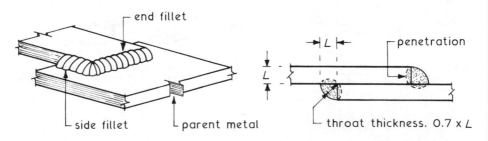

Ref. BS EN 1011-1 and 2: Welding. Recommendations for welding of metallic materials.

Bolts are the preferred method for site assembly of framed building components, although rivets have been popular in the past and will be found when working on existing buildings. Cold driven and 'pop' rivets may be used for factory assembly of light steel frames such as stud walling, but the traditional process of hot riveting structural steel both in the workshop and on site has largely been superseded for safety reasons and the convenience of other practices.

Types of Bolt ~

1. Black Bolts ~ the least expensive and least precise type of bolt, produced by forging with only the bolt and nut threads machined. Clearance between the bolt shank and bolt hole is about 2 mm, a tolerance that provides for ease of assembly. However, this imprecision limits the application of these bolts to direct bearing of components onto support brackets or seating cleats.

2. Bright Bolts ~ also known as turned and fitted bolts. These are machined under the bolt head and along the shank to produce a close fit of 0·5 mm hole clearance. They are specified where accuracy is paramount.

3. High Strength Friction Grip Bolts ~ also known as torque bolts as they are tightened to a predetermined shank tension by a torque controlled wrench. This procedure produces a clamping force that transfers the connection by friction between components and not by shear or bearing on the bolts. These bolts are manufactured from high-yield steel. The number of bolts used to make a connection is less than otherwise required.

Refs.
BS 4190: ISO metric black hexagon bolts, screws and nuts. Specification.

BS 3692: ISO metric precision hexagon bolts, screws and nuts. Specification.

BS 4395 (2 parts): Specification for high strength friction grip bolts and associated nuts and washers for structural engineering.

BS EN 14399 (6 parts): High strength structural bolting assemblies for preloading.

Fire Resistance of Structural Steelwork ~ although steel is a non-combustible material with negligible surface spread of flame properties it does not behave very well under fire conditions. During the initial stages of a fire the steel will actually gain in strength but this reduces to normal at a steel temperature range of 250 to 400°C and continues to decrease until the steel temperature reaches 550°C when it has lost most of its strength. Since the temperature rise during a fire is rapid, most structural steelwork will need protection to give it a specific degree of fire resistance in terms of time. Part B of the Building Regulations sets out the minimum requirements related to building usage and size, BRE Report 128 'Guidelines for the construction of fire resisting structural elements' gives acceptable methods.

Typical Examples for 120 minutes Fire Resistance ~

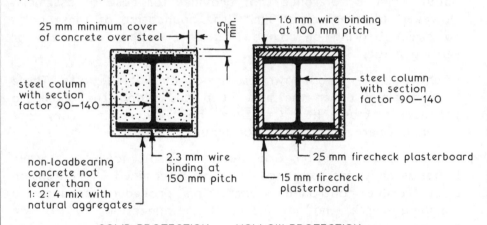

25 mm minimum cover of concrete over steel

25 min.

1.6 mm wire binding at 100 mm pitch

steel column with section factor 90–140

steel column with section factor 90–140

non-loadbearing concrete not leaner than a 1: 2: 4 mix with natural aggregates

2.3 mm wire binding at 150 mm pitch

25 mm firecheck plasterboard

15 mm firecheck plasterboard

SOLID PROTECTION   HOLLOW PROTECTION

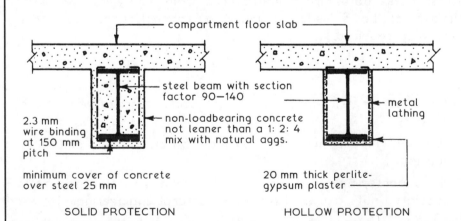

compartment floor slab

steel beam with section factor 90–140

metal lathing

2.3 mm wire binding at 150 mm pitch

non-loadbearing concrete not leaner than a 1: 2: 4 mix with natural aggs.

minimum cover of concrete over steel 25 mm

20 mm thick perlite-gypsum plaster

SOLID PROTECTION   HOLLOW PROTECTION

For section factor calculations see next page.

Section Factors – these are criteria found in tabulated fire protection data such as the Loss Prevention Certification Board's Standards. These factors can be used to establish the minimum thickness or cover of protective material for structural sections. This interpretation is usually preferred by buildings insurance companies, as it often provides a standard in excess of the building regulations. Section factors are categorised: < 90, 90 – 140 and > 140. They can be calculated by the following formula:

Section Factor = Hp/A (m^{-1})

Hp = Perimeter of section exposed to fire (m)

A = Cross sectional area of steel (m^2) [see BS 4-1 or Structural Steel Tables]

Examples:

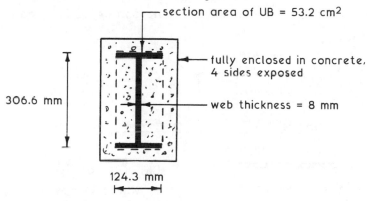

UB serial size, 305 × 127 × 42 kg/m

section area of UB = 53.2 cm^2

fully enclosed in concrete, 4 sides exposed

306.6 mm

web thickness = 8 mm

124.3 mm

Hp = (2 × 124·3) + (2 × 306·6) + 2(124·3 − 8) = 1·0944 m
A = 53·2 cm^2 or 0·00532 m^2
Section Factor, Hp/A = 1·0944/0·00532 = 205

As beam above, but 3 sides only exposed

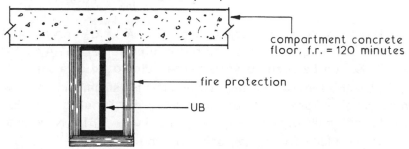

compartment concrete floor, f.r. = 120 minutes

fire protection

UB

Hp = 124·3 + (2 × 306·6) + 2(124·3 − 8) = 0·9701 m
A = 53·2 cm^2 or 0·00532 m^2
Section Factor, Hp/A = 0·9701/0·00532 = 182

References: BS 4-1: Structural steel sections. Specification for hot-rolled sections.

BS 449-2: Specification for the use of structural steel in building. Metric units.

BS 5950-1: Structural use of steelwork in building. Code of practice for design. Rolled and welded sections.

BS EN 1993-1: Eurocode 3. Design of steel structures.

Simple beam design (Bending)

Formula:

$$Z = \frac{M}{f}$$

where: Z = section or elastic modulus (BS 4-1)

M = moment of resistance > or = max. bending moment

f = fibre stress of the material, (normally 165 N/mm² for rolled steel sections)

In simple situations the bending moment can be calculated:-

(a) Point loads

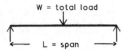

$$BM = \frac{WL}{4}$$

(b) Distributed loads

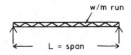

$$BM = \frac{wL^2}{8} \quad or \quad \frac{WL}{8}$$
$$where \quad W = w \times L$$

eg.

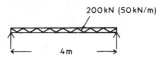

$$BM = \frac{wL^2}{8} = \frac{50 \times 4^2}{8} = 100\,kNm$$

$$Z = \frac{M}{f} = \frac{100 \times 10^6}{165} = 606\,cm^3$$

eg.

From structural design tables, e.g. BS 4-1, a Universal Beam 305 × 127 × 48 kg/m with section modulus (Z) of 612·4 cm³ about the x-x axis, can be seen to satisfy the calculated 606 cm³.

Note: Total load in kN can be established by summating the weight of materials – see BS648: Schedule of Weights of Building Materials, and multiplying by gravity; i.e. kg × 9·81 = Newtons. This must be added to any imposed loading:-

People and furniture = 1·5 kN/m²

Snow on roofs < 30° pitch = 1·5 kN/m²

Snow on roofs > 30° pitch = 0·75 kN/m²

Simple beam design (Shear)

From the previous example, the section profile is:-

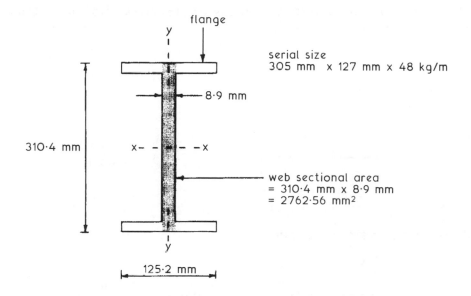

flange

y

serial size
305 mm   x 127 mm x 48 kg/m

8·9 mm

310·4 mm

x— — — — —x

web sectional area
= 310·4 mm x 8·9 mm
= 2762·56 mm²

y

125·2 mm

Maximum shear force normally occurs at the support points, i.e. near the end of the beam. Calculation is made of the average stress value on the web sectional area.

Using the example of 200 kN load distributed over the beam, the maximum shear force at each end support will be 100 kN.

$$\text{Therefore, the average shear stress} = \frac{\text{shear force}}{\text{web sectional area}}$$

$$= \frac{100 \times 10^3}{2762 \cdot 56}$$

$$= 36 \cdot 20 \ N/mm^2$$

Grade S275 steel has an allowable shear stress in the web of 110 N/mm². Therefore the example section of serial size: 305 mm × 127 mm × 48 kg/m with only 36·20 N/mm² calculated average shear stress is more than capable of resisting the applied forces.

Grade S275 steel has a characteristic yield stress of 275 N/mm² in sections up to 40 mm thickness. This grade is adequate for most applications, but the more expensive grade S355 steel is available for higher stress situations.

Ref. BS EN 10025: Hot rolled products of structural steels.

Simple beam design (Deflection)

The deflection due to loading, other than the weight of the structure, should not exceed 1/360 of the span.

The formula to determine the extent of deflection varies, depending on:-

(a) Point loading

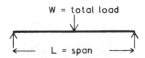

$$\text{Deflection} = \frac{WL^3}{48EI}$$

(b) Uniformly distributed loading

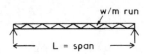

$$\text{Deflection} = \frac{5WL^3}{384EI}$$

where: W = load in kN
L = span in cm
E = Young's modulus of elasticity (typically 21,000 kN/cm² for steel)
I = 2nd moment of area about the x-x axis (see BS 4-1)

Using the example of 200 kN uniformly distributed over a 4 m span:-

$$\text{Deflection} = \frac{5WL^3}{384EI} = \frac{5 \times 200 \times 4^3 \times 100^3}{384 \times 21000 \times 9504} = 0.835\text{cm}$$

Permissible deflection is 1/360 of 4 m = 11.1 mm or 1.11 cm.

Therefore actual deflection of 8.35 mm or 0.835 cm is acceptable. Ref. BS 5950-1: Structural use of steelwork in building. Code of practice for design. Rolled and welded sections.

Simple column design

Steel columns or stanchions have a tendency to buckle or bend under extreme loading. This can be attributed to:

(a) length
(b) cross sectional area
(c) method of end fixing, and
(d) the shape of section.
(b) and (d) are incorporated into a geometric property of section, known as the radius of gyration (r). It can be calculated:-

$$r = \sqrt{\frac{I}{A}}$$

where: I = 2nd moment of area
A = cross sectional area

Note: r, I and A are all listed in steel design tables, eg. BS 4-1.

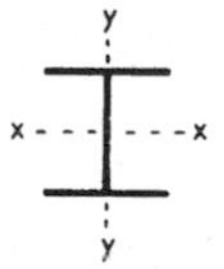

The radius of gyration about the y-y axis is used for calculation, as this is normally the weaker axis.

Effective length of columns

  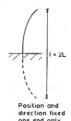

Position and direction fixed is location at specific points by beams or other means of retention. Position fixed only means hinged or pinned. eg. A Universal Column 203 mm × 203 mm × 46 kg/m, 10 m long, position and direction fixed both ends. Determine the maximum axial loading.

Effective length (l) = 0.7 × 10 m = 7 m
(r) from BS 4-1 = 51.1 mm

Slenderness ratio $= \frac{l}{r} = \frac{7 \times 10^3}{51.1} = 137$

Maximum allowable stress for grade S275 steel = 48 N/mm^2 (BS 449-2)

Cross sectional area of stanchion (UC) = 5880 mm^2 (BS 4-1)

The total axial load $= \frac{48 \times 5880}{10^3} = 283$ kN (approx. 28 tonnes)

Portal Frames ~ these can be defined as two dimensional rigid frames which have the basic characteristic of a rigid joint between the column and the beam. The main objective of this form of design is to reduce the bending moment in the beam thus allowing the frame to act as one structural unit. The transfer of stresses from the beam to the column can result in a rotational movement at the foundation which can be overcome by the introduction of a pin or hinge joint. The pin or hinge will allow free rotation to take place at the point of fixity whilst transmitting both load and shear from one member to another. In practice a true 'pivot' is not always required but there must be enough movement to ensure that the rigidity at the point of connection is low enough to overcome the tendency of rotational movement.

Typical Single Storey Portal Frame Formats ~

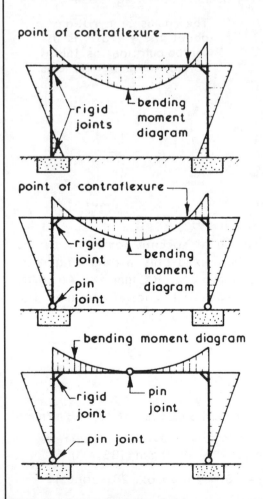

FIXED or RIGID PORTAL FRAME :-

all joints or connections are rigid giving low r bending moments than other formats. Used for small to medium span frames where moments at foundations are not excessive.

TWO PIN PORTAL FRAME :-

pin joints or hinges used at foundation connections to eliminate tendency of base to rotate. Used where high base moments and weak ground are encountered.

THREE PIN PORTAL FRAME :-

pin joints or hinges used at foundation connections and at centre of beam which reduces bending moment in beam but increases deflection. Used as an alternative to a 2 pin frame.

Typical Precast Concrete Portal Frame Details ~

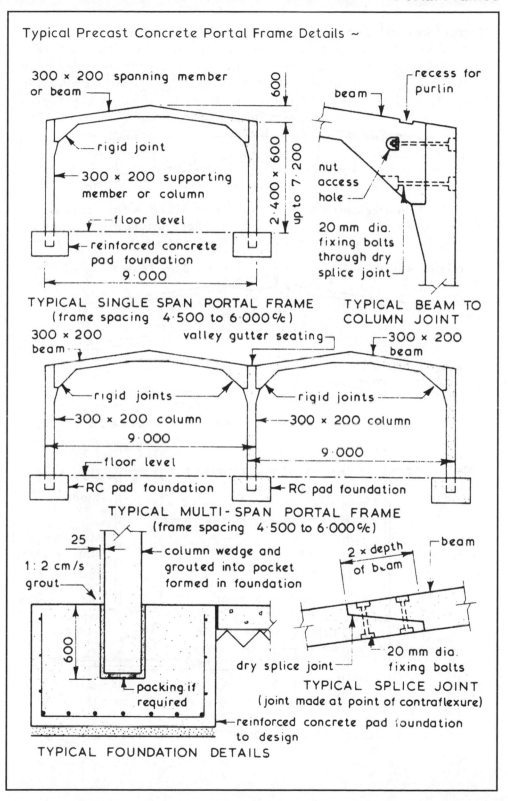

300 × 200 spanning member or beam

600

2·400 × 600 up to 7·200

rigid joint

300 × 200 supporting member or column

floor level

reinforced concrete pad foundation

9·000

TYPICAL SINGLE SPAN PORTAL FRAME
(frame spacing 4·500 to 6·000 ℅)

recess for purlin

beam

nut access hole

20 mm dia. fixing bolts through dry splice joint

TYPICAL BEAM TO COLUMN JOINT

300 × 200 beam

valley gutter seating

300 × 200 beam

rigid joints

300 × 200 column

9·000

rigid joints

300 × 200 column

9·000

floor level

RC pad foundation

RC pad foundation

TYPICAL MULTI-SPAN PORTAL FRAME
(frame spacing 4·500 to 6·000 ℅)

25

1 : 2 cm/s grout

column wedge and grouted into pocket formed in foundation

600

packing if required

reinforced concrete pad foundation to design

TYPICAL FOUNDATION DETAILS

2 × depth of beam

beam

20 mm dia. fixing bolts

dry splice joint

TYPICAL SPLICE JOINT
(joint made at point of contraflexure)

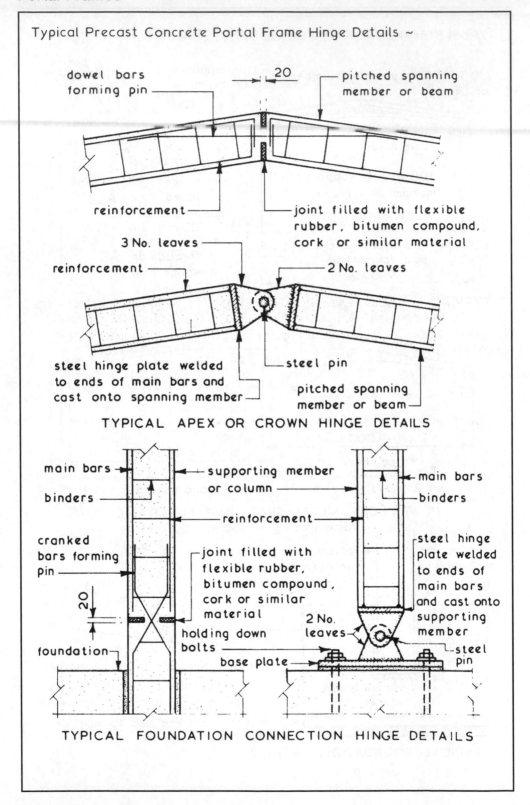

Typical Precast Concrete Portal Frame Hinge Details ~

dowel bars forming pin

20

pitched spanning member or beam

reinforcement

joint filled with flexible rubber, bitumen compound, cork or similar material

3 No. leaves

2 No. leaves

reinforcement

steel hinge plate welded to ends of main bars and cast onto spanning member

steel pin

pitched spanning member or beam

TYPICAL APEX OR CROWN HINGE DETAILS

main bars

supporting member or column

main bars

binders

binders

reinforcement

cranked bars forming pin

joint filled with flexible rubber, bitumen compound, cork or similar material

steel hinge plate welded to ends of main bars and cast onto supporting member

20

2 No. leaves

holding down bolts

foundation

base plate

steel pin

TYPICAL FOUNDATION CONNECTION HINGE DETAILS

Typical Steel Portal Frame Details ~

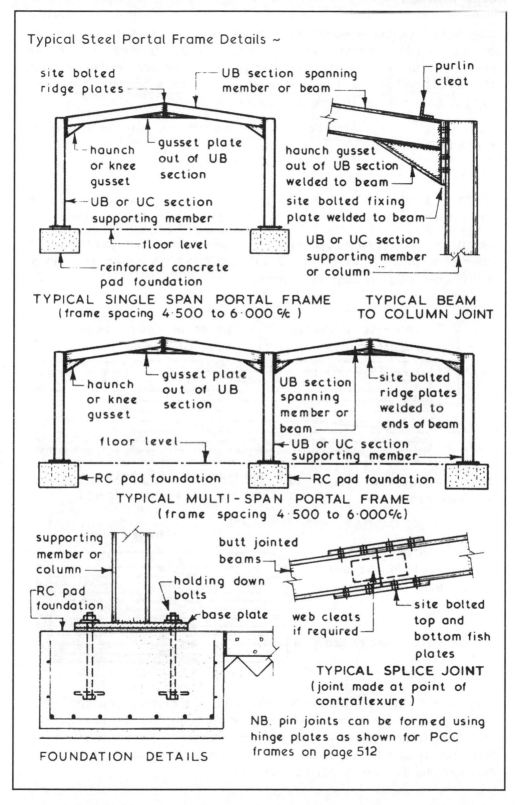

site bolted
ridge plates

UB section spanning
member or beam

purlin
cleat

haunch
or knee
gusset

gusset plate
out of UB
section

haunch gusset
out of UB section
welded to beam

site bolted fixing
plate welded to beam

UB or UC section
supporting member

UB or UC section
supporting member
or column

floor level

reinforced concrete
pad foundation

TYPICAL SINGLE SPAN PORTAL FRAME
(frame spacing 4.500 to 6.000 ℅)

**TYPICAL BEAM
TO COLUMN JOINT**

haunch
or knee
gusset

gusset plate
out of UB
section

UB section
spanning
member or
beam

site bolted
ridge plates
welded to
ends of beam

floor level

UB or UC section
supporting member

RC pad foundation

RC pad foundation

TYPICAL MULTI-SPAN PORTAL FRAME
(frame spacing 4.500 to 6.000℅)

supporting
member or
column

RC pad
foundation

holding down
bolts

base plate

butt jointed
beams

web cleats
if required

site bolted
top and
bottom fish
plates

TYPICAL SPLICE JOINT
(joint made at point of
contraflexure)

NB. pin joints can be formed using
hinge plates as shown for PCC
frames on page 512

FOUNDATION DETAILS

Laminated Timber ~ sometimes called 'Gluelam' and is the process of building up beams, ribs, arches, portal frames and other structural units by gluing together layers of timber boards so that the direction of the grain of each board runs parallel with the longitudinal axis of the member being fabricated.

Laminates ~ these are the layers of board and may be jointed in width and length.

Joints ~

Width — joints in consecutive layers should lap twice the board thickness or one quarter of its width whichever is the greater.

Length — scarf and finger joints can be used. Scarf joints should have a minimum slope of 1 in 12 but this can be steeper (say 1 in 6) in the compression edge of a beam:-

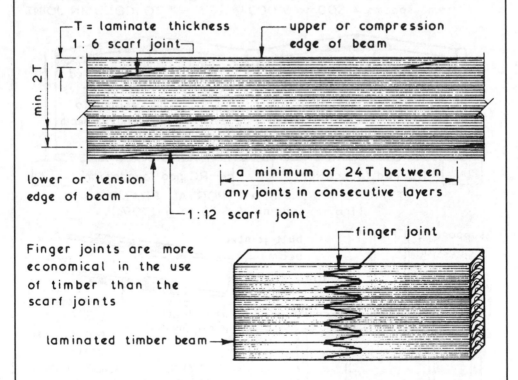

T = laminate thickness

1 : 6 scarf joint

upper or compression edge of beam

min. 2 T

lower or tension edge of beam

1 : 12 scarf joint

a minimum of 24 T between any joints in consecutive layers

finger joint

Finger joints are more economical in the use of timber than the scarf joints

laminated timber beam

Moisture Content ~ timber should have a moisture content equal to that which the member will reach in service and this is known as its equilibrium moisture content; for most buildings this will be between 11 and 15%. Generally at the time of gluing timber should not exceed 15 ± 3% in moisture content.

Vertical Laminations ~ not often used for structural laminated timber members and is unsatisfactory for curved members.

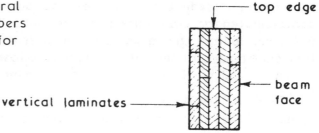

Horizontal Laminations ~ most popular method for all types of laminated timber members. The stress diagrams below show that laminates near the upper edge are subject to a compressive stress whilst those near the lower edge to a tensile stress and those near the neutral axis are subject to shear stress.

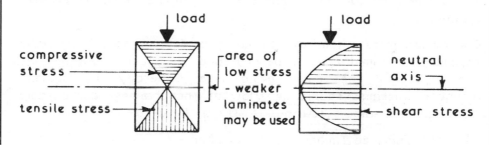

Flat sawn timber shrinks twice as much as quarter sawn timber therefore flat and quarter sawn timbers should not be mixed in the same member since the different shrinkage rates will cause unacceptable stresses to occur on the glue lines.

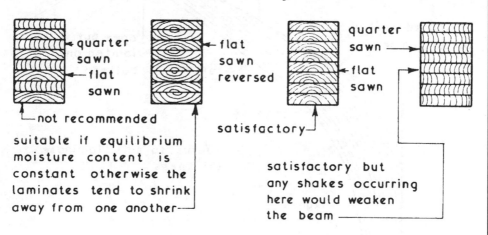

Planing ~ before gluing, laminates should be planed so that the depth of the planer cutter marks are not greater than 0·025 mm.

Gluing ~ this should be carried out within 48 hours of the planing operation to reduce the risk of the planed surfaces becoming contaminated or case hardened (for suitable adhesives see page 517). Just before gluing up the laminates they should be checked for `cupping.' The amount of cupping allowed depends upon the thickness and width of the laminates and has a range of 0·75 mm to 1·5 mm.

Laminate Thickness ~ no laminate should be more than 50 mm thick since seasoning up to this thickness can be carried out economically and there is less chance of any individual laminate having excessive cross grain strength.

Straight Members – laminate thickness is determined by the depth of the member, there must be enough layers to allow the end joints (i.e. scarf or finger joints – see page 514) to be properly staggered.

Curved Members – laminate thickness is determined by the radius to which the laminate is to be bent and the species together with the quality of the timber being used. Generally the maximum laminate thickness should be 1/150 of the sharpest curve radius although with some softwoods 1/100 may be used.

Typical Laminated Timber Curved Member ~

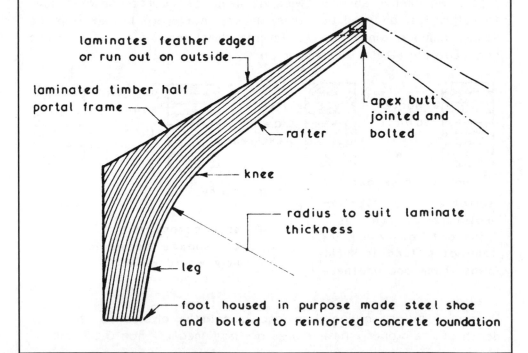

laminates feather edged or run out on outside

laminated timber half portal frame

apex butt jointed and bolted

rafter

knee

radius to suit laminate thickness

leg

foot housed in purpose made steel shoe and bolted to reinforced concrete foundation

Adhesives ~ although timber laminates are carefully machined, the minimum of cupping permitted and efficient cramping methods employed it is not always possible to obtain really tight joints between the laminates. One of the important properties of the adhesive is therefore that it should be gap filling. The maximum permissible gap being 1.25 mm.

There are four adhesives suitable for laminated timber work which have the necessary gap filling property and they are namely:-

1. Casein – the protein in milk, extracted by coagulation and precipitation. It is a cold setting adhesive in the form of a powder which is mixed with water, it has a tendency to stain timber and is only suitable for members used in dry conditions of service.
2. Urea Formaldehyde – this is a cold setting resin glue formulated to MR/GF (moisture resistant/gap filling). Although moisture resistant it is not suitable for prolonged exposure in wet conditions and there is a tendency for the glue to lose its strength in temperatures above 40°C such as when exposed to direct sunlight. The use of this adhesive is usually confined to members used in dry, unexposed conditions of service. This adhesive will set under temperatures down to 10°C.
3. Resorcinol Formaldehyde – this is a cold setting glue formulated to WBP/GF (weather and boilproof/gap filling). It is suitable for members used in external situations but is relatively expensive. This adhesive will set under temperatures down to 15°C and does not lose its strength at high temperatures.
4. Phenol Formaldehyde – this is a similar glue to resorcinol formaldehyde but is a warm setting adhesive requiring a temperature of above 86°C in order to set. Phenol/resorcinol formaldehyde is an alternative, having similar properties to, but less expensive than resorcinol formaldehyde. PRF needs a setting temperature of at least 23°C.

Preservative Treatment – this can be employed if required, provided that the pressure impregnated preservative used is selected with regard to the adhesive being employed. See also page 427.

Ref. BS EN 301: Adhesives, phenolic and aminoplastic, for load-bearing timber structures. Classification and performance requirements.

Composite Beams ~ stock sizes of structural softwood have sectional limitations of about 225 mm and corresponding span potential in the region of 6 m. At this distance, even modest loadings could interpose with the maximum recommended deflection of 0·003 × span.

Fabricated softwood box, lattice and plywood beams are an economic consideration for medium spans. They are produced with adequate depth to resist deflection and with sufficient strength for spans into double figures. The high strength to weight ratio and simple construction provides advantages in many situations otherwise associated with steel or reinforced concrete, e.g. frames, trusses, beams and purlins in gymnasia, workshops, garages, churches, shops, etc. They are also appropriate as purlins in loft conversion.

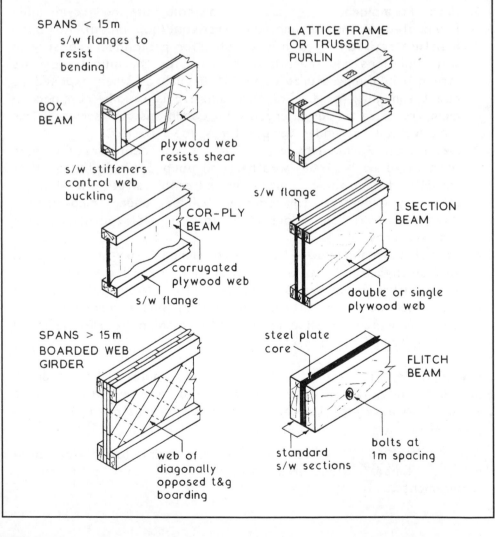

SPANS < 15 m
s/w flanges to resist bending

BOX BEAM

plywood web resists shear

s/w stiffeners control web buckling

LATTICE FRAME OR TRUSSED PURLIN

COR-PLY BEAM

corrugated plywood web

s/w flange

s/w flange

I SECTION BEAM

double or single plywood web

SPANS > 15 m
BOARDED WEB GIRDER

web of diagonally opposed t&g boarding

steel plate core

standard s/w sections

FLITCH BEAM

bolts at 1m spacing

Composite Joist ~ a type of lattice frame, constructed from a pair of parallel and opposing stress graded softwood timber flanges, separated and jointed with a web of V shaped galvanised steel plate connectors. Manufacture is off-site in a factory quality controlled situation. Here, joists can be made in standard or specific lengths to order. Depending on loading, spans to about 8 m are possible at joist spacing up to 600 mm.

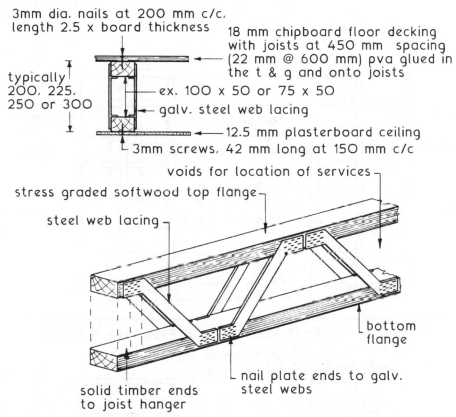

3mm dia. nails at 200 mm c/c, length 2.5 x board thickness

18 mm chipboard floor decking with joists at 450 mm spacing (22 mm @ 600 mm) pva glued in the t & g and onto joists

typically 200, 225, 250 or 300

ex. 100 x 50 or 75 x 50

galv. steel web lacing

12.5 mm plasterboard ceiling

3mm screws, 42 mm long at 150 mm c/c

voids for location of services

stress graded softwood top flange

steel web lacing

bottom flange

nail plate ends to galv. steel webs

solid timber ends to joist hanger

End bearing on inner leaf of cavity wall, silicone sealed to maintain air tightness. Alternatively lateral restraint type joist hanger support or intermedite support from a steel beam.

Advantages over solid timber joists:

Large span to depth ratio.
High strength to weight ratio.
Alternative applications, including roof members, purlins, etc.
Generous space for services without holing or notching.
Minimal movement and shrinkage.
Wide flanges provide large bearing area for decking and ceiling board.

Multi-storey Structures ~ these buildings are usually designed for office, hotel or residential use and contain the means of vertical circulation in the form of stairs and lifts occupying up to 20% of the floor area. These means of circulation can be housed within a core inside the structure and this can be used to provide a degree of restraint to sway due to lateral wind pressures (see next page).

Typical Basic Multi-storey Structure Types ~

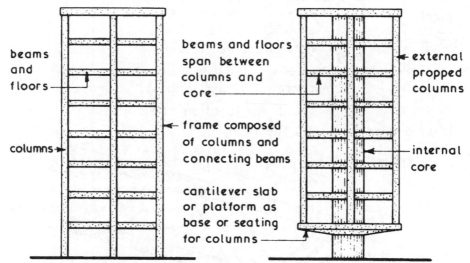

beams and floors
span between
columns and
core

frame composed
of columns and
connecting beams

cantilever slab
or platform as
base or seating
for columns

beams and floors

columns

external propped columns

internal core

TRADITIONAL FRAMED STRUCTURES   PROPPED STRUCTURES

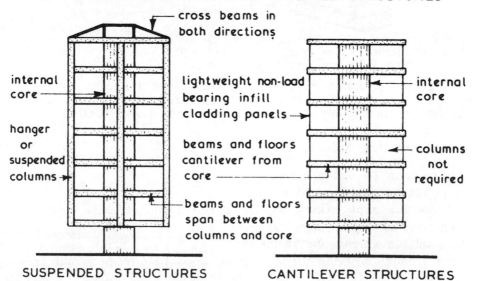

cross beams in
both directions

internal core

hanger or suspended columns

lightweight non-load
bearing infill
cladding panels

beams and floors
cantilever from
core

beams and floors
span between
columns and core

internal core

columns not required

SUSPENDED STRUCTURES   CANTILEVER STRUCTURES

Typical Multi-storey Structures ~ the formats shown below are designed to provide lateral restraint against wind pressures.

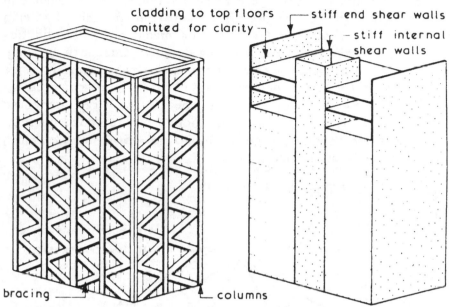

cladding to top floors omitted for clarity

stiff end shear walls

stiff internal shear walls

bracing — columns

BRACED STRUCTURES – bracing used to give stability so that columns can be designed as pure compression members.

SHEAR WALL STRUCTURES – wind pressures transmitted from cladding to shear walls by floors.

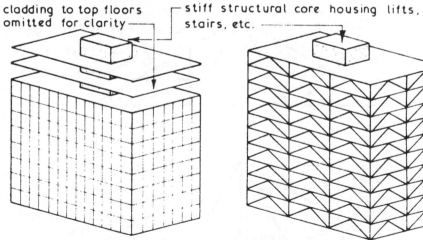

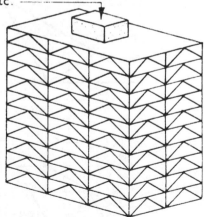

cladding to top floors omitted for clarity

stiff structural core housing lifts, stairs, etc.

CORE STRUCTURES – wind pressures transmitted from cladding to core by floors.

HULL CORE STRUCTURES – rigid and braced framework called the hull acts with core through floors to form a rigid structure.

Steel Roof Trusses ~ these are triangulated plane frames which carry purlins to which the roof coverings can be fixed. Steel is stronger than timber and will not spread fire over its surface and for these reasons it is often preferred to timber for medium and long span roofs. The rafters are restrained from spreading by being connected securely at their feet by a tie member. Struts and ties are provided within the basic triangle to give adequate bracing. Angle sections are usually employed for steel truss members since they are economic and accept both tensile and compressive stresses. The members of a steel roof truss are connected together with bolts or by welding to shaped plates called gussets. Steel trusses are usually placed at 3·000 to 4·500 centres which gives an economic purlin size.

Typical Steel Roof Truss Formats ~

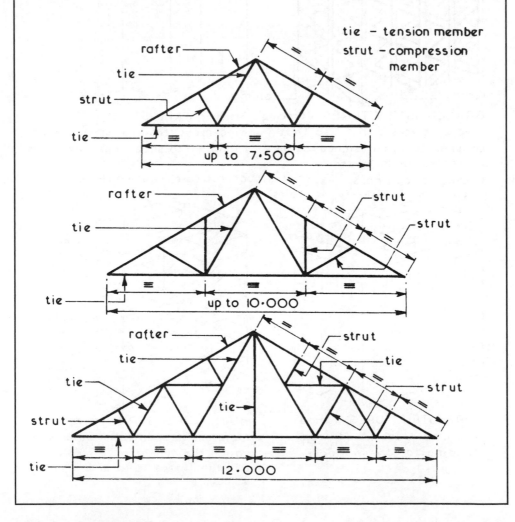

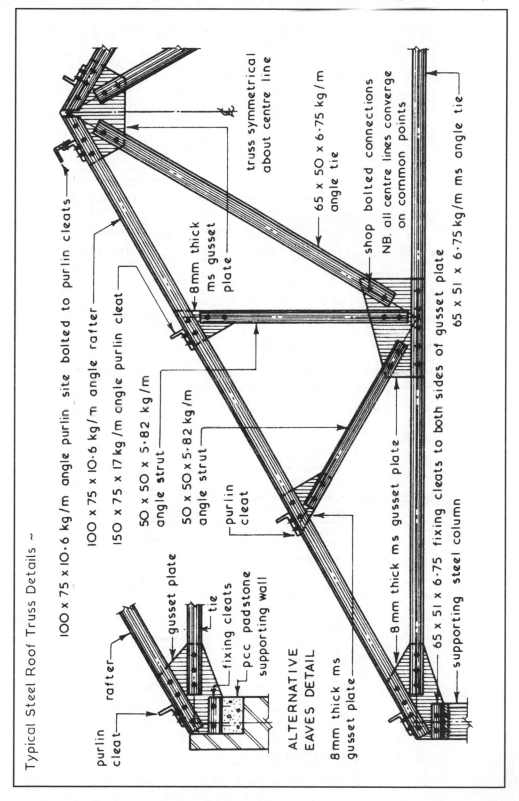

Typical Steel Roof Truss Details ~

100 x 75 x 10·6 kg/m angle purlin site bolted to purlin cleats

100 x 75 x 10·6 kg/m angle rafter

150 x 75 x 17 kg/m angle purlin cleat

50 x 50 x 5·82 kg/m angle strut

50 x 50 x 5·82 kg/m angle strut

purlin cleat

8mm thick ms gusset plate

truss symmetrical about centre line

65 x 50 x 6·75 kg/m angle tie

shop bolted connections

NB. all centre lines converge on common points

65 x 51 x 6·75 kg/m ms angle tie

65 x 51 x 6·75 fixing cleats to both sides of gusset plate

8mm thick ms gusset plate

supporting steel column

65 x 51 x 6·75 fixing cleats to both sides of gusset plate

purlin cleat

rafter

gusset plate

tie

fixing cleats

pcc padstone supporting wall

ALTERNATIVE EAVES DETAIL

8mm thick ms gusset plate

Sheet Coverings ~ the basic functions of sheet coverings used in conjunction with steel roof trusses are to :-

1. Provide resistance to penetration by the elements.
2. Provide restraint to wind and snow loads.
3. Provide a degree of thermal insulation of not less than that set out in Part L of the Building Regulations.
4. Provide resistance to surface spread of flame as set out in Part B of the Building Regulations.
5. Provide any natural daylight required through the roof in accordance with the maximum permitted areas set out in Part L of the Building Regulations.
6. Be of low self weight to give overall design economy.
7. Be durable to keep maintenance needs to a minimum.

Suitable Materials ~

Hot-dip galvanised corrugated steel sheets – BS 3083

Aluminium profiled sheets – BS 4868.

Asbestos free profiled sheets – various manufacturers whose products are usually based on a mixture of Portland cement, mineral fibres and density modifiers – BS EN 494.

Typical Profiles ~

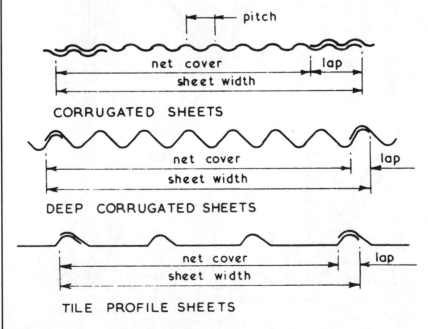

CORRUGATED SHEETS

DEEP CORRUGATED SHEETS

TILE PROFILE SHEETS

Typical Purlin Fixing Details ~

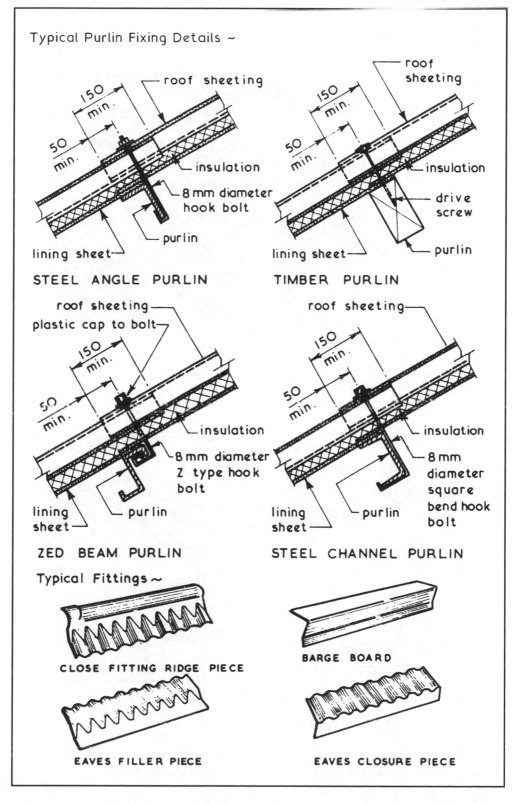

150 min.

50 min.

roof sheeting

insulation

8 mm diameter hook bolt

purlin

lining sheet

STEEL ANGLE PURLIN

roof sheeting

150 min.

50 min.

insulation

drive screw

purlin

lining sheet

TIMBER PURLIN

roof sheeting

plastic cap to bolt

150 min.

50 min.

insulation

8 mm diameter Z type hook bolt

purlin

lining sheet

ZED BEAM PURLIN

roof sheeting

150 min.

50 min.

insulation

8 mm diameter square bend hook bolt

purlin

lining sheet

STEEL CHANNEL PURLIN

Typical Fittings ~

CLOSE FITTING RIDGE PIECE

BARGE BOARD

EAVES FILLER PIECE

EAVES CLOSURE PIECE

525

Typical Details ~

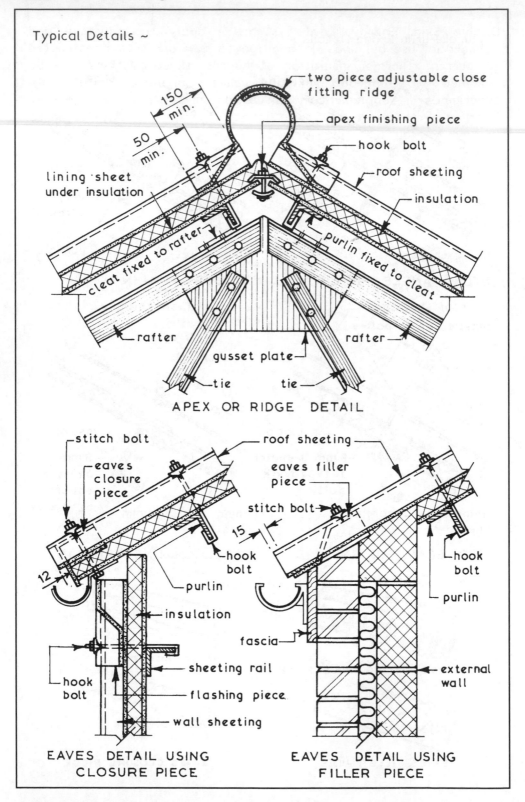

two piece adjustable close fitting ridge

apex finishing piece

hook bolt

roof sheeting

insulation

lining sheet under insulation

150 min.

50 min.

cleat fixed to rafter

purlin fixed to cleat

rafter

gusset plate

rafter

tie

tie

APEX OR RIDGE DETAIL

stitch bolt

roof sheeting

eaves closure piece

eaves filler piece

stitch bolt

hook bolt

15

hook bolt

purlin

12

insulation

purlin

fascia

hook bolt

sheeting rail

external wall

hook bolt

flashing piece

wall sheeting

EAVES DETAIL USING CLOSURE PIECE

EAVES DETAIL USING FILLER PIECE

Double Skin, Energy Roof systems ~ apply to industrial and commercial use buildings. In addition to new projects constructed to current thermal insulation standards, these systems can be specified to upgrade existing sheet profiled roofs with superimposed supplementary insulation and protective decking. Thermal performance with resin bonded mineral wool fibre of up to 250 mm overall depth may provide 'U' values as low as 0·13 W/m²K.

Typical Details ~

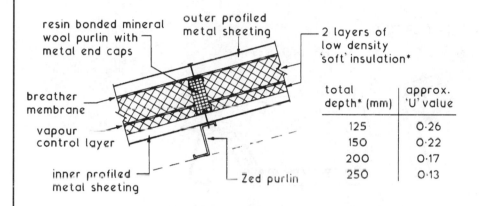

total depth* (mm)	approx. 'U' value
125	0·26
150	0·22
200	0·17
250	0·13

Alternative ~

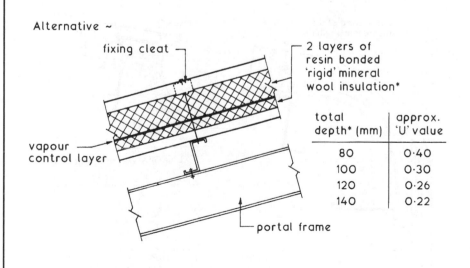

total depth* (mm)	approx. 'U' value
80	0·40
100	0·30
120	0·26
140	0·22

527

Further typical details using profiled galvanised steel or aluminium, colour coated if required ~

RIDGE

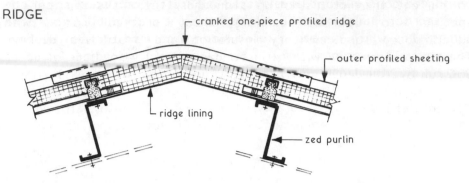

— cranked one-piece profiled ridge

— outer profiled sheeting

— ridge lining

— zed purlin

VALLEY GUTTER

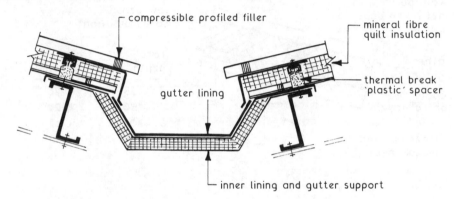

— compressible profiled filler

— mineral fibre quilt insulation

— thermal break 'plastic' spacer

gutter lining

— inner lining and gutter support

EAVES GUTTER

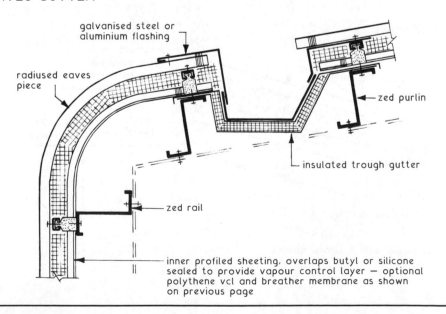

galvanised steel or aluminium flashing

radiused eaves piece

— zed purlin

— insulated trough gutter

— zed rail

— inner profiled sheeting, overlaps butyl or silicone sealed to provide vapour control layer — optional polythene vcl and breather membrane as shown on previous page

Long Span Roofs ~ these can be defined as those exceeding 12·000 in span. They can be fabricated in steel, aluminium alloy, timber, reinforced concrete and prestressed concrete. Long span roofs can be used for buildings such as factories. Large public halls and gymnasiums which require a large floor area free of roof support columns. The primary roof functions of providing weather protection, thermal insulation, sound insulation and restricting spread of fire over the roof surface are common to all roof types but these roofs may also have to provide strength sufficient to carry services lifting equipment and provide for natural daylight to the interior by means of rooflights.

Basic Roof Forms ~

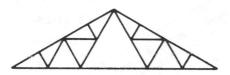

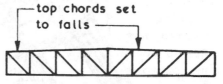
top chords set to falls

Pitched Trusses - spaced at suitable centres to carry purlins to which the roof coverings are fixed. Good rainwater run off - reasonable daylight spread from rooflights - high roof volume due to the triangulated format - on long spans roof volume can be reduced by using a series of short span trusses.

Flat Top Girders - spaced at suitable centres to carry purlins to which the roof coverings are fixed. Low pitch to give acceptable rainwater run off - reasonable daylight spread from rooflights - can be designed for very long spans but depth and hence roof volume increases with span.

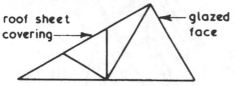
roof sheet covering ──→ ←── glazed face

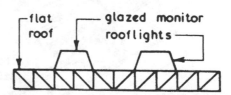
flat roof   glazed monitor rooflights

Northlight - spaced at suitable centres to carry purlins to which roof sheeting is fixed. Good rainwater run off - if correctly orientated solar glare is eliminated - long spans can be covered by a series of short span frames

Monitor - girders or cranked beams at centres to suit low pitch decking used. Good even daylight spread from monitor lights which is not affected by orientation of building.

Pitched Trusses ~ these can be constructed with a symmetrical outline (as shown on pages 522 to 523) or with an asymmetrical outline (Northlight – see detail below). They are usually made from standard steel sections with shop welded or bolted connections, alternatively they can be fabricated using timber members joined together with bolts and timber connectors or formed as a precast concrete portal frame.

Typical Multi-span Northlight Roof Details ~

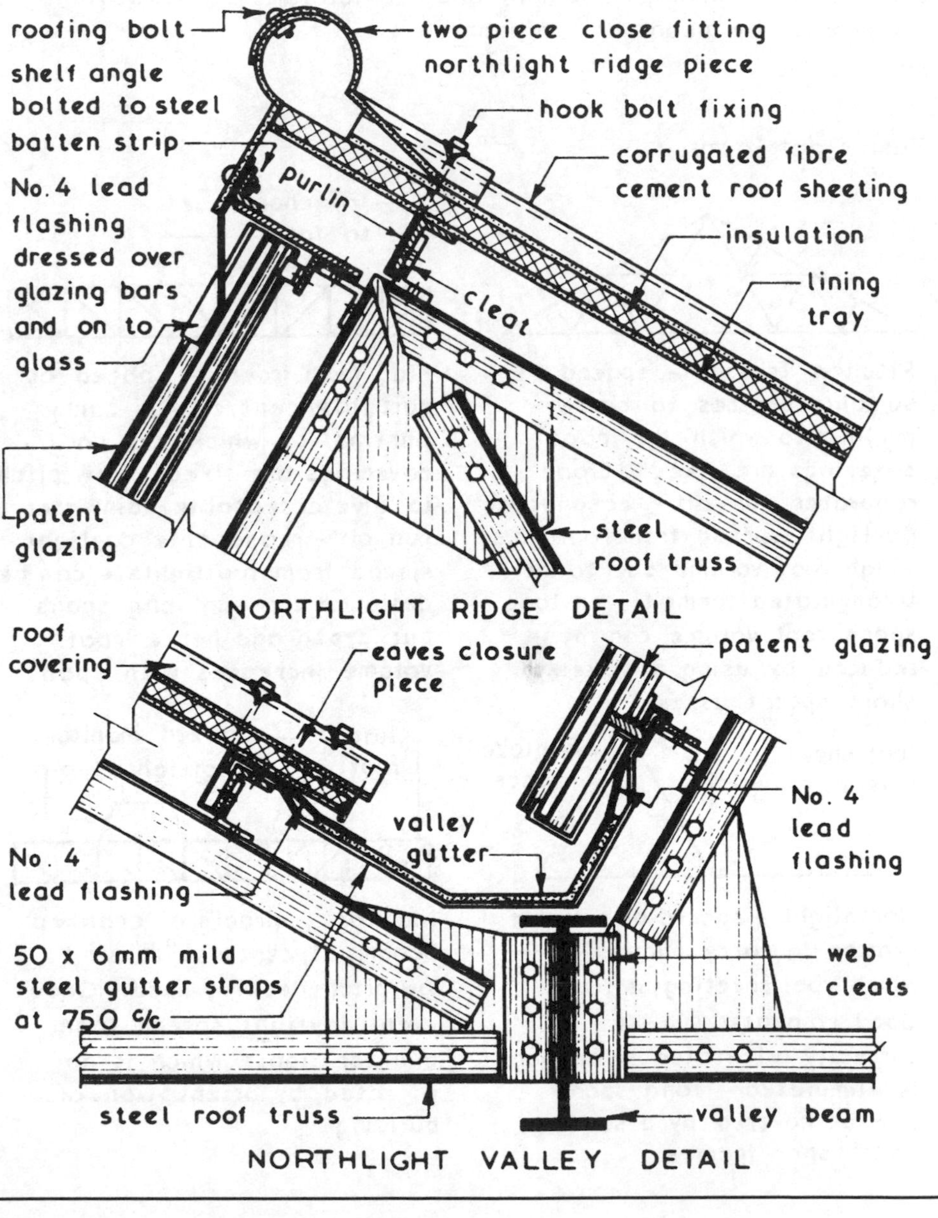

roofing bolt
two piece close fitting northlight ridge piece
shelf angle bolted to steel batten strip
hook bolt fixing
No.4 lead flashing dressed over glazing bars and on to glass
purlin
corrugated fibre cement roof sheeting
insulation
cleat
lining tray
patent glazing
steel roof truss

NORTHLIGHT RIDGE DETAIL

roof covering
eaves closure piece
patent glazing
valley gutter
No. 4 lead flashing
No. 4 lead flashing
50 x 6mm mild steel gutter straps at 750 c/c
web cleats
steel roof truss
valley beam

NORTHLIGHT VALLEY DETAIL

Monitor Roofs ~ these are basically a flat roof with raised glazed portions called monitors which forms a roof having a uniform distribution of daylight with no solar glare problems irrespective of orientation and a roof with easy access for maintenance. These roofs can be constructed with light long span girders supporting the monitor frames, cranked welded beams following the profile of the roof or they can be of a precast concrete portal frame format.

Typical Monitor Roof Details ~

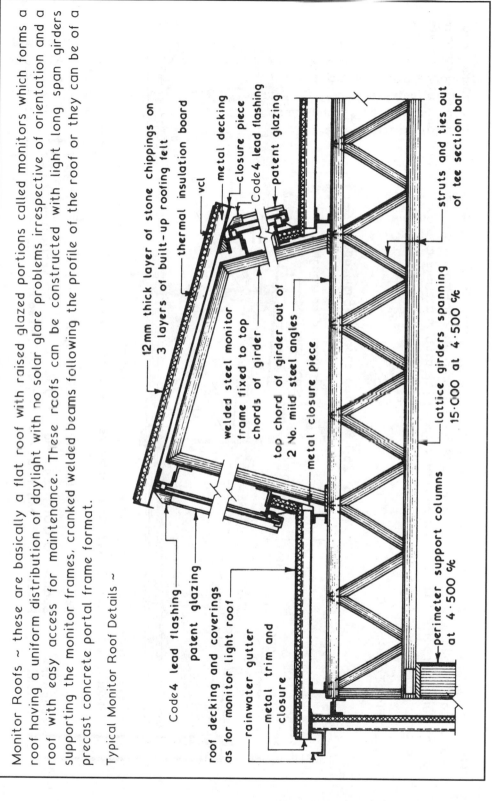

12mm thick layer of stone chippings on 3 layers of built-up roofing felt

thermal insulation board

vcl

metal decking

closure piece

Code 4 lead flashing

patent glazing

welded steel monitor frame fixed to top chords of girder

top chord of girder out of 2 No. mild steel angles

metal closure piece

struts and ties out of tee section bar

lattice girders spanning 15·000 at 4·500 c/c

perimeter support columns at 4·500 c/c

Code 4 lead flashing

patent glazing

roof decking and coverings as for monitor light roof

rainwater gutter

metal trim and closure

Flat Top Girders ~ these are suitable for roof spans ranging from 15·000 to 45·000 and are basically low pitched lattice beams used to carry purlins which support the roof coverings. One of the main advantages of this form of roof is the reduction in roof volume. The usual materials employed in the fabrication of flat top girders are timber and steel.

Typical Flat Top Girder Details ~

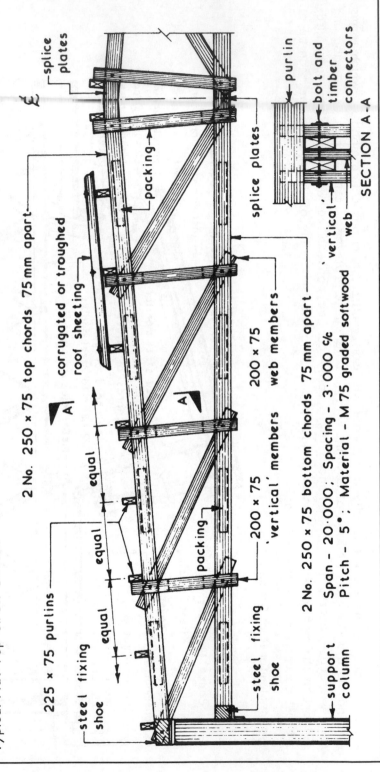

splice plates

C̷L

packing

corrugated or troughed roof sheeting

2 No. 250 × 75 top chords 75 mm apart

A

A

equal

equal

equal

equal

equal

225 × 75 purlins

steel fixing shoe

packing

200 × 75 'vertical' members

200 × 75 web members

splice plates

steel fixing shoe

support column

2 No. 250 × 75 bottom chords 75 mm apart

Span – 20·000; Spacing – 3·000 c/c
Pitch – 5°; Material – M 75 graded softwood

purlin

bolt and timber connectors

'vertical' web

SECTION A-A

Connections ~ nails, screws and bolts have their limitations when used to join structural timber members. The low efficiency of joints made with a rigid bar such as a bolt is caused by the usual low shear strength of timber parallel to the grain and the non-uniform distribution of bearing stress along the shank of the bolt –

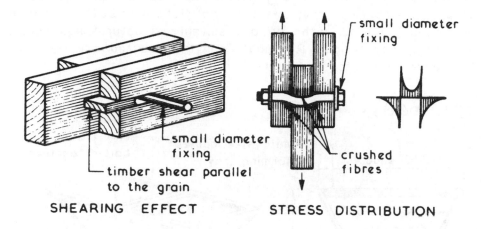

SHEARING EFFECT   STRESS DISTRIBUTION

Timber Connectors ~ these are designed to overcome the problems of structural timber connections outlined above by increasing the effective bearing area of the bolts.

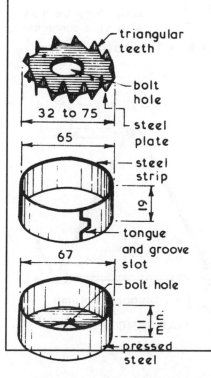

Toothed Plate Connector – provides an efficient joint without special tools or equipment – suitable for all connections especially small sections – bolt holes are drilled 2 mm larger than the bolt diameter, the timbers forming the joint being held together whilst being drilled.

Split Ring Connector – very efficient and develops a high joint strength – suitable for all connections – split ring connectors are inserted into a precut groove formed with a special tool making the connector independent from the bolt.

Shear Plate Connector – counterpart of a split ring connector – housed flush into timber – used for temporary joints.

Space Deck ~ this is a structural roofing system based on a simple repetitive pyramidal unit to give large clear spans of up to 22·000 for single spanning designs and up to 33·000 for two way spanning designs. The steel units are easily transported to site before assembly into beams and the complete space deck at ground level before being hoisted into position on top of the perimeter supports. A roof covering of wood wool slabs with built-up roofing felt could be used, although any suitable structural lightweight decking is appropriate. Rooflights can be mounted directly onto the square top space deck units

Typical Details ~

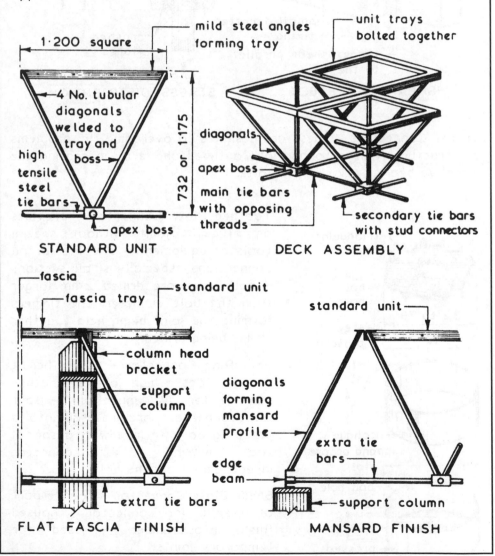

STANDARD UNIT

DECK ASSEMBLY

FLAT FASCIA FINISH

MANSARD FINISH

Space Frames ~ these are roofing systems which consist of a series of connectors which joins together the chords and bracing members of the system. Single or double layer grids are possible, the former usually employed in connection with small domes or curved roofs. Space frames are similar in concept to space decks but they have greater flexibility in design and layout possibilities. Most space frames are fabricated from structural steel tubes or tubes of aluminium alloy although any suitable structural material could be used.

Typical Examples~

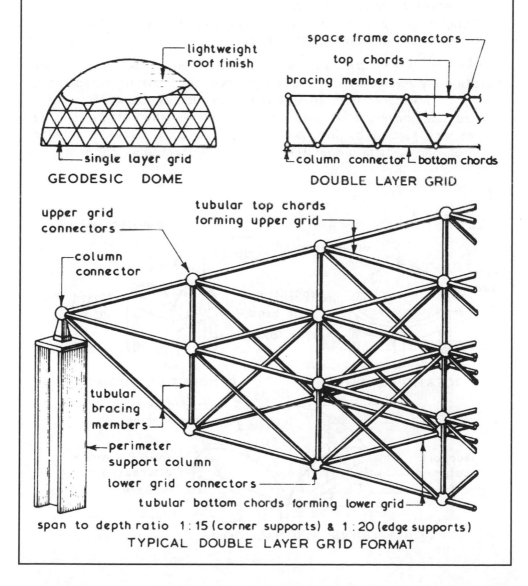

lightweight roof finish

single layer grid

GEODESIC DOME

space frame connectors

top chords

bracing members

column connector — bottom chords

DOUBLE LAYER GRID

upper grid connectors

column connector

tubular top chords forming upper grid

tubular bracing members

perimeter support column

lower grid connectors

tubular bottom chords forming lower grid

span to depth ratio 1:15 (corner supports) & 1:20 (edge supports)
TYPICAL DOUBLE LAYER GRID FORMAT

Shell Roofs ~ these can be defined as a structural curved skin covering a given plan shape and area where the forces in the shell or membrane are compressive and in the restraining edge beams are tensile. The usual materials employed in shell roof construction are in-situ reinforced concrete and timber. Concrete shell roofs are constructed over formwork which in itself is very often a shell roof making this format expensive since the principle of use and reuse of formwork can not normally be applied. The main factors of shell roofs are:-

1. The entire roof is primarily a structural element.
2. Basic strength of any particular shell is inherent in its geometrical shape and form.
3. Comparatively less material is required for shell roofs than other forms of roof construction.

Domes ~ these are double curvature shells which can be rotationally formed by any curved geometrical plane figure rotating about a central vertical axis. Translation domes are formed by a curved line moving over another curved line whereas pendentive domes are formed by inscribing within the base circle a regular polygon and vertical planes through the true hemispherical dome.

Typical Examples ~

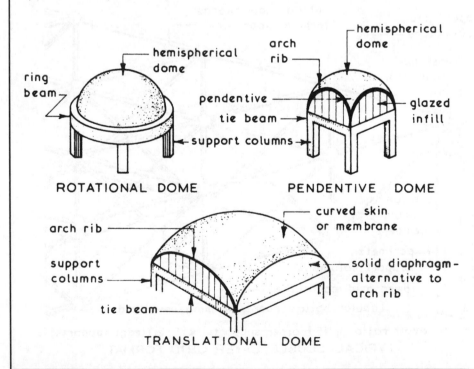

ROTATIONAL DOME

PENDENTIVE DOME

TRANSLATIONAL DOME

Barrel Vaults ~ these are single curvature shells which are essentially a cut cylinder which must be restrained at both ends to overcome the tendency to flatten. A barrel vault acts as a beam whose span is equal to the length of the roof. Long span barrel vaults are those whose span is longer than its width or chord length and conversely short barrel vaults are those whose span is shorter than its width or chord length. In every long span barrel vaults thermal expansion joints will be required at 30·000 centres which will create a series of abutting barrel vault roofs weather sealed together (see next page).

Typical Single Barrel Vault Principles~

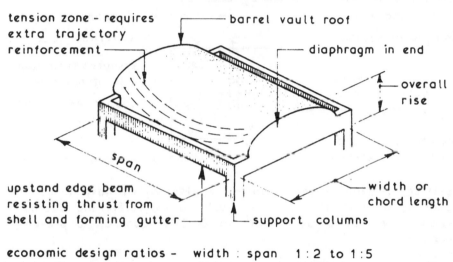

economic design ratios – width : span 1 : 2 to 1 : 5
rise : span 1 : 10 to 1 : 15

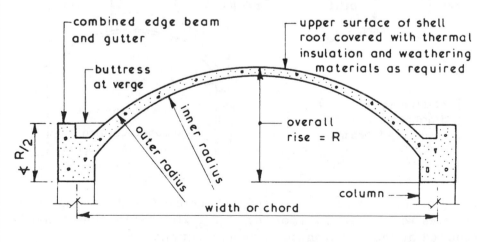

Typical Barrel Vault Expansion Joint Details ~

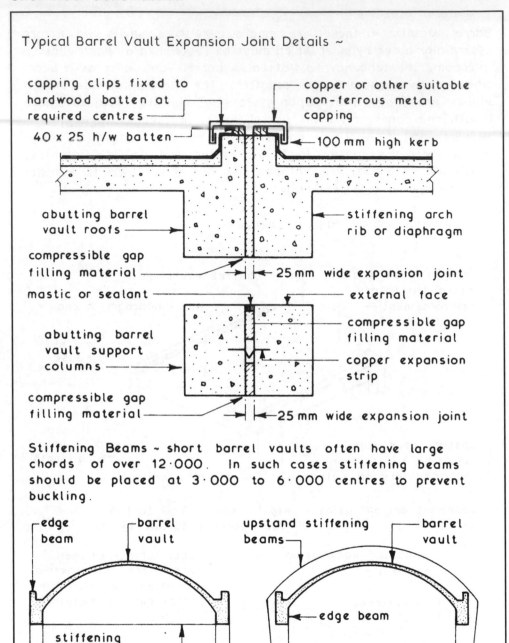

capping clips fixed to hardwood batten at required centres

copper or other suitable non-ferrous metal capping

40 x 25 h/w batten

100 mm high kerb

abutting barrel vault roofs

stiffening arch rib or diaphragm

compressible gap filling material

25 mm wide expansion joint

mastic or sealant

external face

abutting barrel vault support columns

compressible gap filling material

copper expansion strip

compressible gap filling material

25 mm wide expansion joint

Stiffening Beams ~ short barrel vaults often have large chords of over 12·000. In such cases stiffening beams should be placed at 3·000 to 6·000 centres to prevent buckling.

edge beam

barrel vault

upstand stiffening beams

barrel vault

edge beam

stiffening diaphragm or downstand beams

support column

NB. ribs not connected to support columns will set up extra stresses within the shell roof therefore extra reinforcement will be required at the stiffening rib or beam positions.

Other Forms of Barrel Vault ~ by cutting intersecting and placing at different levels the basic barrel vault roof can be formed into a groin or northlight barrel vault roof:-

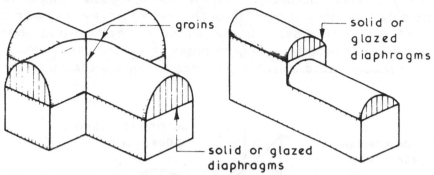

INTERSECTING BARREL VAULTS   STEPPED BARREL VAULTS

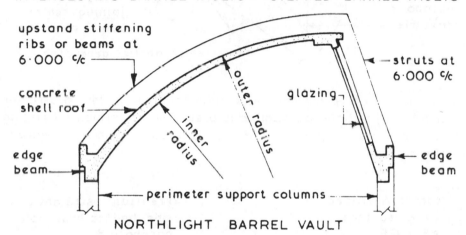

NORTHLIGHT BARREL VAULT

Conoids ~ these are double curvative shell roofs which can be considered as an alternative to barrel vaults. Spans up to 12·000 with chord lengths up to 24·000 are possible. Typical chord to span ratio 2:1.

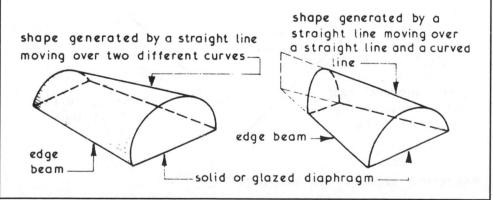

Hyperbolic Paraboloids ~ the true hyperbolic paraboloid shell roof shape is generated by moving a vertical parabola (the generator) over another vertical parabola (the directrix) set at right angles to the moving parabola. This forms a saddle shape where horizontal sections taken through the roof are hyperbolic in format and vertical sections are parabolic. The resultant shape is not very suitable for roofing purposes therefore only part of the saddle shape is used and this is formed by joining the centre points thus:-

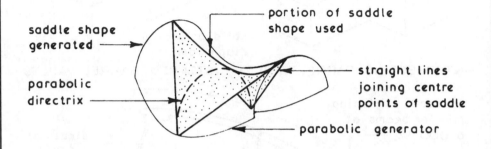

saddle shape generated

portion of saddle shape used

parabolic directrix

straight lines joining centre points of saddle

parabolic generator

To obtain a more practical shape than the true saddle a straight line limited hyperbolic paraboloid is used. This is formed by raising or lowering one or more corners of a square forming a warped parallelogram thus:-

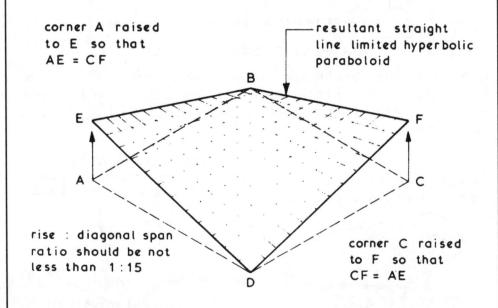

corner A raised to E so that AE = CF

resultant straight line limited hyperbolic paraboloid

rise : diagonal span ratio should be not less than 1:15

corner C raised to F so that CF = AE

For further examples see next page.

Typical Straight Line Limited Hyperbolic Paraboloid Formats ~

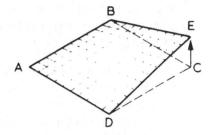

corner C raised to E

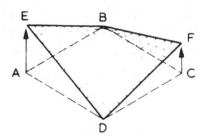

corner A raised to E and
corner C raised to F so that
AE ≠ CF

resultant hyperbolic
paraboloid

original
square

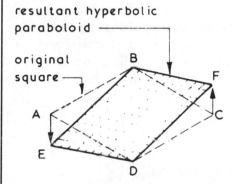

corner A lowered to E and
corner C raised to F so that
AE = CF

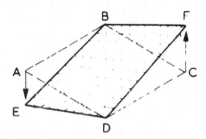

corner A lowered to E and
corner C raised to F so that
AE ≠ CF

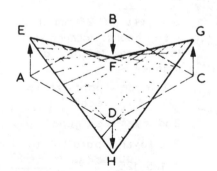

corners A & C raised to E & G
corners B & D lowered to F & H
so that AE = CG & BF = DH

Combination of Hyperbolic
Paraboloid Shell Roofs ~

one corner
raised

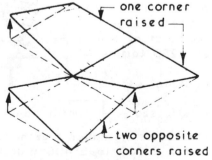

two opposite
corners raised

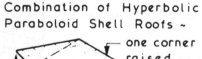

NB. any combination possible

541

Concrete Hyperbolic Paraboloid Shell Roofs ~ these can be constructed in reinforced concrete (characteristic strength 25 or 30 N/mm²) with a minimum shell thickness of 50 mm with diagonal spans up to 35·000. These shells are cast over a timber form in the shape of the required hyperbolic paraboloid format. In practice therefore two roofs are constructed and it is one of the reasons for the popularity of timber versions of this form of shell roof.

Timber Hyperbolic Paraboloid Shell Roofs ~ these are usually constructed using laminated edge beams and layers of t & g boarding to form the shell membrane. For roofs with a plan size of up to 6·000 × 6·000 only 2 layers of boards are required and these are laid parallel to the diagonals with both layers running in opposite directions. Roofs with a plan size of over 6·000 × 6·000 require 3 layers of board as shown below. The weather protective cover can be of any suitable flexible material such as built-up roofing felt, copper and lead. During construction the relatively lightweight roof is tied down to a framework of scaffolding until the anchorages and wall infilling have been completed. This is to overcome any negative and positive wind pressures due to the open sides.

Typical Details ~

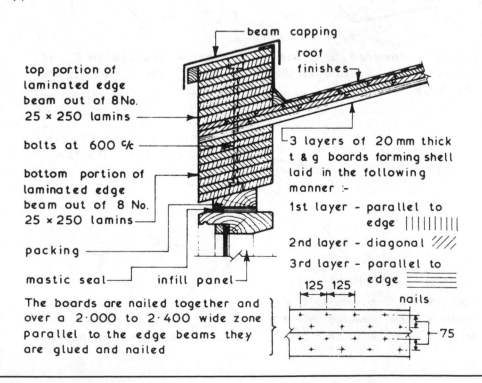

beam capping

roof finishes

top portion of laminated edge beam out of 8 No. 25 × 250 lamins

bolts at 600 ℅

bottom portion of laminated edge beam out of 8 No. 25 × 250 lamins

packing

mastic seal

infill panel

3 layers of 20 mm thick t & g boards forming shell laid in the following manner :-

1st layer - parallel to edge ||||||||||

2nd layer - diagonal ////

3rd layer - parallel to edge ≡

125 125

nails

75

The boards are nailed together and over a 2·000 to 2·400 wide zone parallel to the edge beams they are glued and nailed

Support Considerations ~ in timber hyperbolic paraboloid shell roofs only two supports are required :-

Edge beams are in compression forces P are transmitted to B and D resulting in a vertical force V and a horizontal force H at both positions therefore support columns are required at B and D.

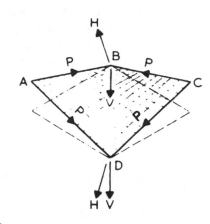

Vertical force V is transmitted directly down the columns to a suitable foundation. The outward or horizontal force H can be accommodated in one of two ways :-

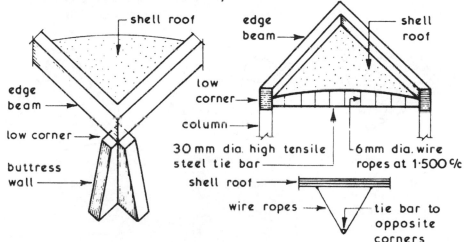

shell roof

edge beam

low corner

column

30 mm dia. high tensile steel tie bar

6 mm dia. wire ropes at 1·500 c/c

shell roof

wire ropes

tie bar to opposite corners

edge beam

low corner

buttress wall

If shell roof is to be supported at high corners the edge beams will be in tension and horizontal force will be inwards. This can be resisted by a diagonal strut between the high corners.

Combination Roof Support Example ~

4 No. roof shells of equal loading joined together

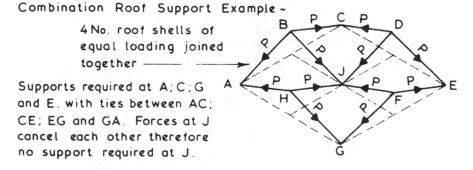

Supports required at A; C; G and E. with ties between AC; CE; EG and GA. Forces at J cancel each other therefore no support required at J.

Membrane Structure Principles ~ a form of tensioned cable structural support system with a covering of stretched fabric. In principle and origin, this compares to a tent with poles as compression members secured to the ground. The fabric membrane is attached to peripheral stressing cables suspended in a catenary between vertical support members.

Form ~ there are limitless three-dimensional possibilities. The following geometric shapes provide a basis for imagination and elegance in design:

- Hyperbolic paraboloid (Hypar)
- Barrel vault
- Conical or double conical

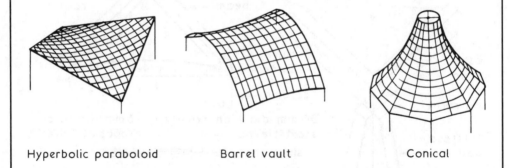

Hyperbolic paraboloid   Barrel vault   Conical

Double conical~

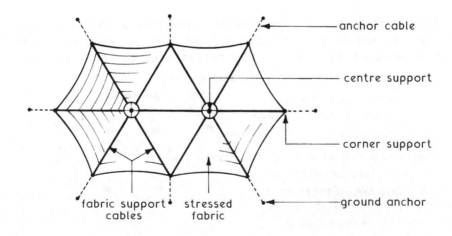

anchor cable

centre support

corner support

ground anchor

fabric support cables   stressed fabric

Simple support structure as viewed from the underside ~

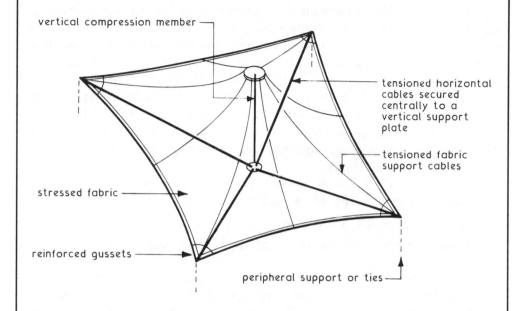

vertical compression member

tensioned horizontal cables secured centrally to a vertical support plate

tensioned fabric support cables

stressed fabric

reinforced gussets

peripheral support or ties

Fabric ~ has the advantages of requiring minimal support, opportunity for architectural expression in colour and geometry and a translucent quality that provides an outside feel inside, whilst combining shaded cover from the sun and shelter from rain. Applications are generally attachments as a feature to entrances and function areas in prominent buildings, notably sports venues, airports and convention centres.

Materials ~ historically, animal hides were the first materials used for tensile fabric structures, but more recently woven fibres of hemp, flax or other natural yarns have evolved as canvas. Contemporary synthetic materials have a plastic coating on a fibrous base. These include polyvinyl chloride (PVC) on polyester fibres, silicone on glass fibres and polytetrafluorethylene (PTFE) on glass fibres. Design life is difficult to estimate, as it will depend very much on type of exposure. Previous use of these materials would indicate that at least 20 years is anticipated, with an excess of 30 years being likely. Jointing can be by fusion welding of plastics, bonding with silicone adhesives and stitching with glass threads.

Rooflights ~ the useful penetration of daylight through the windows in external walls of buildings is from 6·000 to 9·000 depending on the height and size of the window. In buildings with spans over 18·000 side wall daylighting needs to be supplemented by artificial lighting or in the case of top floors or single storey buildings by rooflights. The total maximum area of wall window openings and rooflights for the various purpose groups is set out in the Building Regulations with allowances for increased areas if double or triple glazing is used. In pitched roofs such as northlight and monitor roofs the rooflights are usually in the form of patent glazing (see Long Span Roofs on pages 530 and 531). In flat roof construction natural daylighting can be provided by one or more of the following methods:-

1. Lantern lights – see page 548.

2. Lens lights – see page 548.

3. Dome, pyramid and similar rooflights – see page 549.

Patent Glazing ~ these are systems of steel or aluminium alloy glazing bars which span the distance to be glazed whilst giving continuous edge support to the glass. They can be used in the roof forms noted above as well as in pitched roofs with profiled coverings where the patent glazing bars are fixed above and below the profiled sheets – see page 547.

Typical Patent Glazing Bar Sections ~

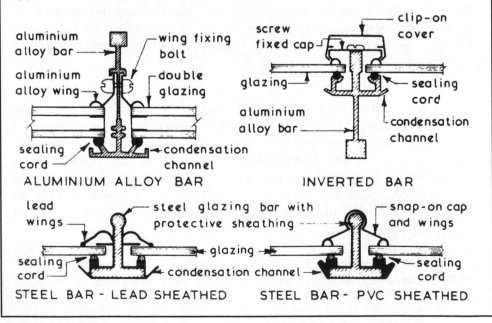

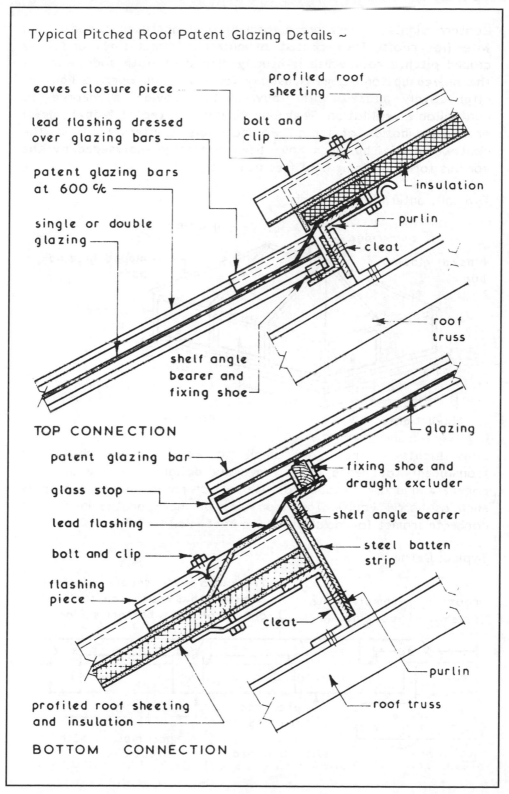

Typical Pitched Roof Patent Glazing Details ~

eaves closure piece

lead flashing dressed over glazing bars

patent glazing bars at 600 %

single or double glazing

profiled roof sheeting

bolt and clip

insulation

purlin

cleat

roof truss

shelf angle bearer and fixing shoe

TOP CONNECTION

glazing

patent glazing bar

glass stop

lead flashing

bolt and clip

flashing piece

fixing shoe and draught excluder

shelf angle bearer

steel batten strip

cleat

purlin

profiled roof sheeting and insulation

roof truss

BOTTOM CONNECTION

Lantern Lights ~ these are a form of rooflight used in conjuction with flat roofs. They consist of glazed vertical sides and fully glazed pitched roof which is usually hipped at both ends. Part of the glazed upstand sides is usually formed as an opening light or alternatively glazed with louvres to provide a degree of controllable ventilation. They can be constructed of timber, metal or a combination of these two materials. Lantern lights in the context of new buildings have been generally superseded by the various forms of dome light (see next page)

Typical Lantern Light Details ~

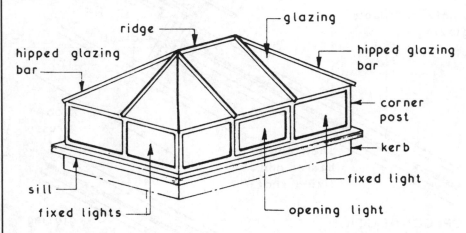

Lens Lights ~ these are small square or round blocks of translucent toughened glass especially designed for casting into concrete and are suitable for use in flat roofs and curved roofs such as barrel vaults. They can also be incorporated in precast concrete frames for inclusion into a cast in-situ roof.

Typical Details ~

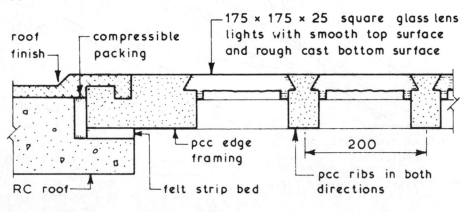

Dome, Pyramid and Similar Rooflights ~ these are used in conjuction with flat roofs and may be framed or unframed. The glazing can be of glass or plastics such as polycarbonate, acrylic, PVC and glass fibre reinforced polyester resin (grp). The whole component is fixed to a kerb and may have a raising piece containing hit and miss ventilators, louvres or flaps for controllable ventilation purposes.

Typical Details ~

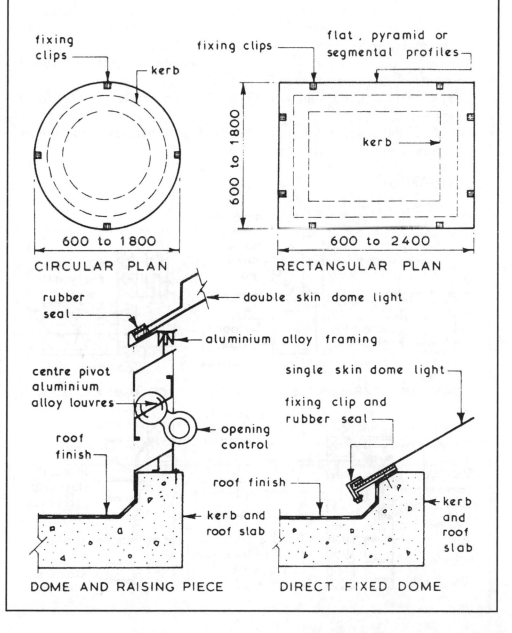

CIRCULAR PLAN

RECTANGULAR PLAN

DOME AND RAISING PIECE

DIRECT FIXED DOME

549

Non-load Bearing Brick Panel Walls ~ these are used in conjunction with framed structures as an infill between the beams and columns. They are constructed in the same manner as ordinary brick walls with the openings being formed by traditional methods.

Basic Requirements ~

1. To be adequately supported by and tied to the structural frame.
2. Have sufficient strength to support own self weight plus any attached finishes and imposed loads such as wind pressures.
3. Provide the necessary resistance to penetration by the natural elements.
4. Provide the required degree of thermal insulation, sound insulation and fire resistance.
5. Have sufficient durability to reduce maintenance costs to a minimum.
6. Provide for movements due to moisture and thermal expansion of the panel and for contraction of the frame.

Typical Details ~

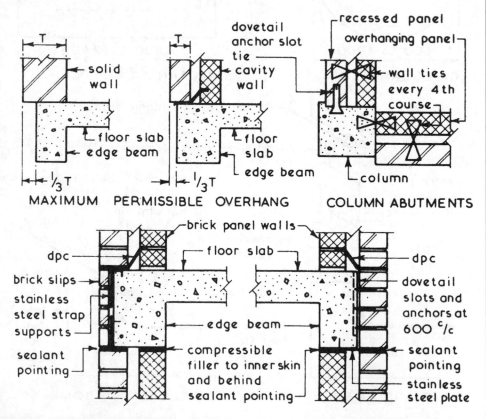

MAXIMUM PERMISSIBLE OVERHANG   COLUMN ABUTMENTS

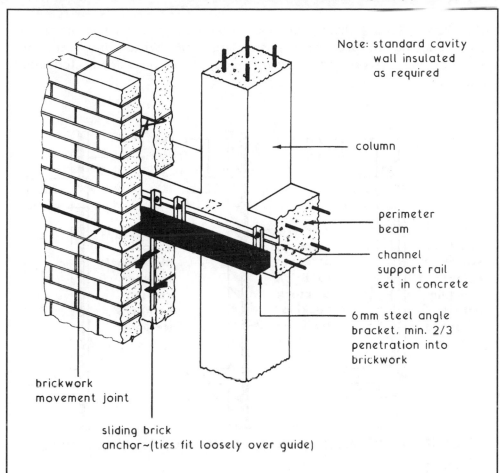

Note: standard cavity
wall insulated
as required

column

perimeter
beam

channel
support rail
set in concrete

6mm steel angle
bracket, min. 2/3
penetration into
brickwork

brickwork
movement joint

sliding brick
anchor~(ties fit loosely over guide)

Application – multi-storey buildings, where a traditional brick façade is required.

Brickwork movement – to allow for climatic changes and differential movement between the cladding and main structure, a 'soft' joint (cellular polyethylene, cellular polyurethane, expanded rubber or sponge rubber with polysulphide or silicon pointing) should be located below the support angle. Vertical movement joints may also be required at a maximum of 12 m spacing.

Lateral restraint – provided by normal wall ties between inner and outer leaf of masonry, plus sliding brick anchors below the support angle.

Infill Panel Walls ~ these can be used between the framing members of a building to provide the cladding and division between the internal and external environments and are distinct from claddings and facing:-

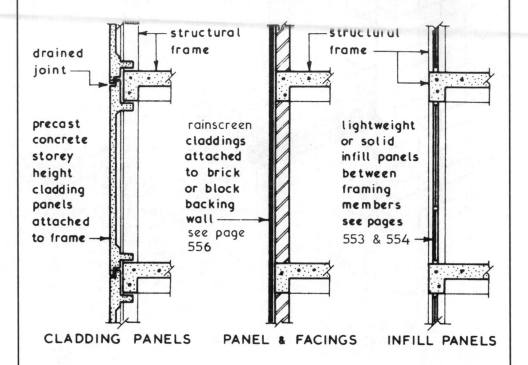

drained joint

precast concrete storey height cladding panels attached to frame ⟶

structural frame

rainscreen claddings attached to brick or block backing wall ⟶ see page 556

structural frame ⟶

lightweight or solid infill panels between framing members see pages 553 & 554 ⟶

CLADDING PANELS   **PANEL & FACINGS**   **INFILL PANELS**

Functional Requirements ~ all forms of infill panel should be designed and constructed to fulfil the following functional requirements:-

1. Self supporting between structural framing members.
2. Provide resistance to the penetration of the elements.
3. Provide resistance to positive and negative wind pressures.
4. Give the required degree of thermal insulation.
5. Give the required degree of sound insulation.
6. Give the required degree of fire resistance.
7. Have sufficient openings to provide the required amount of natural ventilation.
8. Have sufficient glazed area to fulfil the natural daylight and vision out requirements.
9. Be economic in the context of construction and maintenance.
10. Provide for any differential movements between panel and structural frame.

Brick Infill Panels ~ these can be constructed in a solid or cavity format, the latter usually having an inner skin of blockwork to increase the thermal insulation properties of the panel. All the fundamental construction processes and detail of solid and cavity walls (bonding, lintels over openings, wall ties, damp-proof courses etc.,) apply equally to infill panel walls. The infill panel walls can be tied to the columns by means of wall ties cast into the columns at 300 mm centres or located in cast-in dovetail anchor slots. The head of every infill panel should have a compressible joint to allow for any differential movements between the frame and panel.

Typical Details

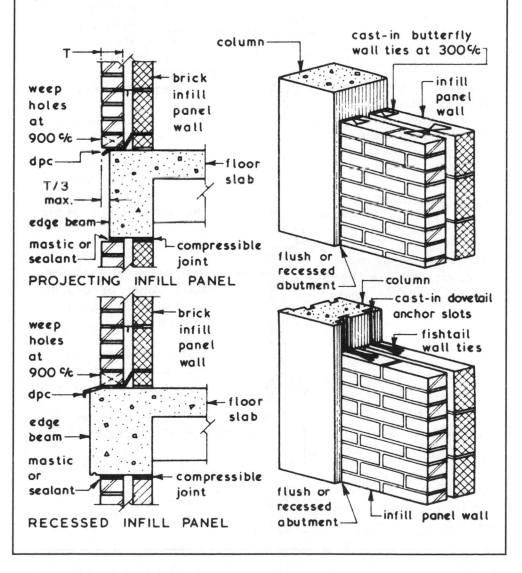

PROJECTING INFILL PANEL

RECESSED INFILL PANEL

Lightweight Infill Panels ~ these can be constructed from a wide variety or combination of materials such as timber, metals and plastics into which single or double glazing can be fitted. If solid panels are to be used below a transom they are usually of a composite or sandwich construction to provide the required sound insulation, thermal insulation and fire resistance properties.

Typical Example ~

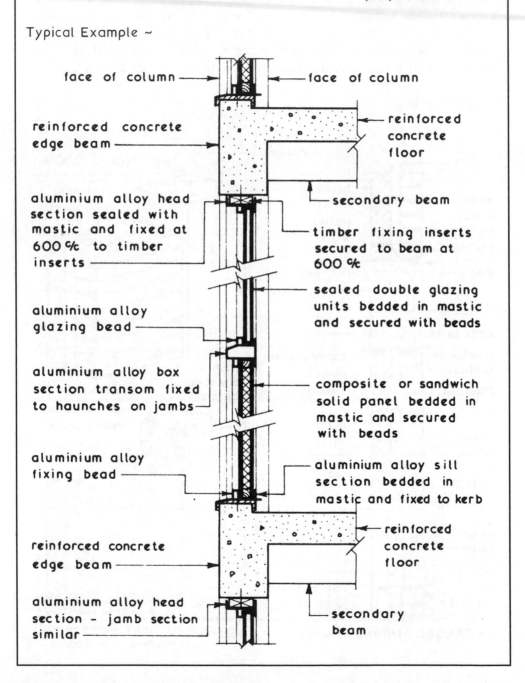

face of column

reinforced concrete edge beam

aluminium alloy head section sealed with mastic and fixed at 600 ℀ to timber inserts

aluminium alloy glazing bead

aluminium alloy box section transom fixed to haunches on jambs

aluminium alloy fixing bead

reinforced concrete edge beam

aluminium alloy head section - jamb section similar

face of column

reinforced concrete floor

secondary beam

timber fixing inserts secured to beam at 600 ℀

sealed double glazing units bedded in mastic and secured with beads

composite or sandwich solid panel bedded in mastic and secured with beads

aluminium alloy sill section bedded in mastic and fixed to kerb

reinforced concrete floor

secondary beam

Lightweight Infill Panels ~ these can be fixed between the structural horizontal and vertical members of the frame or fixed to the face of either the columns or beams to give a grid, horizontal or vertical emphasis to the façade thus –

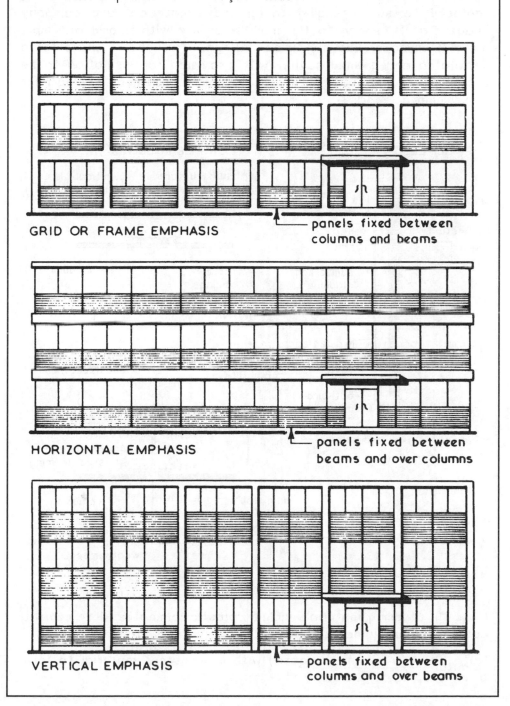

GRID OR FRAME EMPHASIS

panels fixed between columns and beams

HORIZONTAL EMPHASIS

panels fixed between beams and over columns

VERTICAL EMPHASIS

panels fixed between columns and over beams

Overcladding ~ a superficial treatment, applied either as a component of new construction work, or as a façade and insulation enhancement to existing structures. The outer weather resistant decorative panelling is 'loose fit' in concept, which is easily replaced to suit changing tastes, new materials and company image. Panels attach to the main structure with a grid of simple metal framing or vertical timber battens. This allows space for a ventilated and drained cavity, with provision for insulation to be attached to the substructure; a normal requirement in upgrade/refurbishment work.

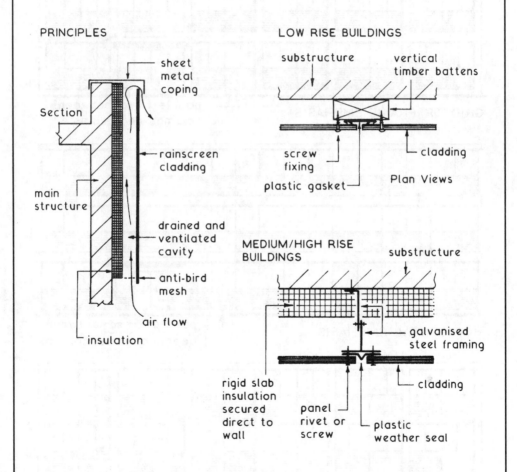

Note (1): Cladding materials include, plastic laminates, fibre cement, ceramics, aluminium, enamelled steel and various stone effects.

Note (2): Anti-bird mesh coated with intumescent material to form a fire stop cavity barrier.

Glazed façades have been associated with hi-tech architecture since the 1970s. The increasing use of this type of cladding is largely due to developments in toughened glass and improved qualities of elastomeric silicone sealants. The properties of the latter must incorporate a resilience to varying atmospheric conditions as well as the facility to absorb structural movement without loss of adhesion.

Systems – two edge and four edge.

The two edge system relies on conventional glazing beads/fixings to the head and sill parts of a frame, with sides silicone bonded to mullions and styles.

The four edge system relies entirely on structural adhesion, using silicone bonding between glazing and support frame – see details.

Structural glazing, as shown on this and the next page, is in principle a type of curtain walling. Due to its unique appearance, it is usual to consider full glazing of the building façade as a separate design and construction concept. BS EN 13830: Curtain walling. Product standard; defines curtain walling as an external vertical building enclosure produced by elements mainly of metal, timber or plastic. Glass as a primary material is excluded.

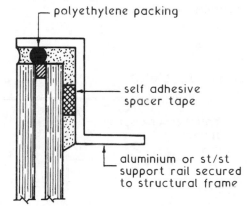

polyethylene packing

self adhesive spacer tape

aluminium or st/st support rail secured to structural frame

Upper edge or head of support frame

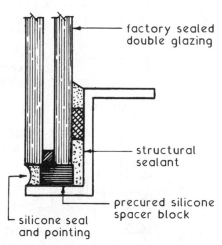

factory sealed double glazing

structural sealant

silicone seal and pointing

precured silicone spacer block

Lower edge of support frame to sill

Note: Sides of frame as head.

Structural glazing is otherwise known as frameless glazing. It is a system of toughened glass cladding without the visual impact of surface fixings and supporting components. Unlike curtain walling, the self-weight of the glass and wind loads are carried by the glass itself and transferred to a subsidiary lightweight support structure behind the glazing.

Assembly principles ~

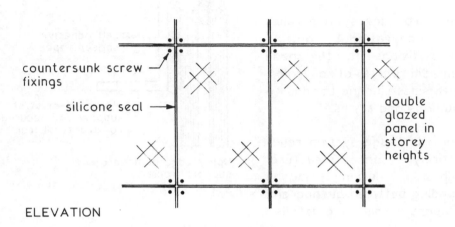

countersunk screw fixings

silicone seal

double glazed panel in storey heights

ELEVATION

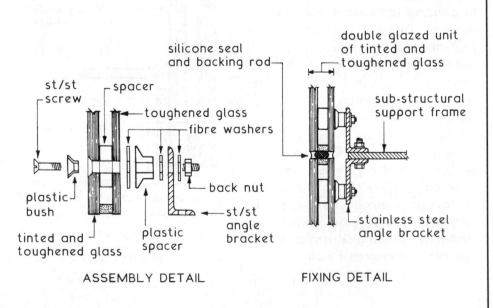

st/st screw

spacer

silicone seal and backing rod

double glazed unit of tinted and toughened glass

toughened glass

fibre washers

sub-structural support frame

plastic bush

back nut

tinted and toughened glass

plastic spacer

st/st angle bracket

stainless steel angle bracket

ASSEMBLY DETAIL

FIXING DETAIL

Curtain Walling ~ this is a form of lightweight non-load bearing external cladding which forms a complete envelope or sheath around the structural frame. In low rise structures the curtain wall framing could be of timber or patent glazing but in the usual high rise context, box or solid members of steel or aluminium alloy are normally employed.

Basic Requirements for Curtain Walls ~

1. Provide the necessary resistance to penetration by the elements.
2. Have sufficient strength to carry own self weight and provide resistance to both positive and negative wind pressures.
3. Provide required degree of fire resistance - glazed areas are classified in the Building Regulations as unprotected areas therefore any required fire resistance must be obtained from the infill or undersill panels and any backing wall or beam.
4. Be easy to assemble, fix and maintain.
5. Provide the required degree of sound and thermal insulation.
6. Provide for thermal and structural movements.

Typical Curtain Walling Arrangement ~

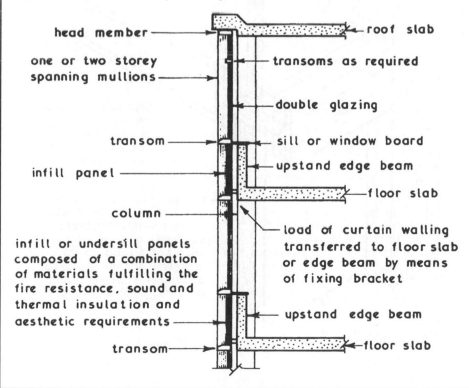

Typical Curtain Walling Details

Principals ~

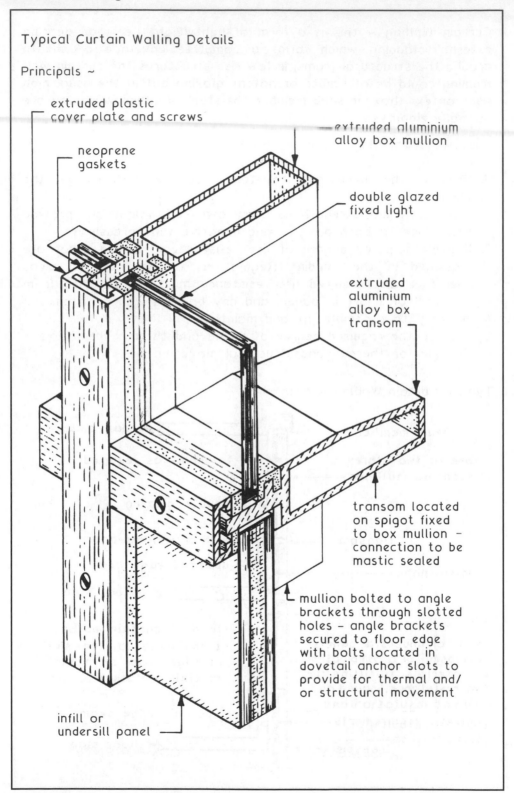

extruded plastic
cover plate and screws

neoprene
gaskets

extruded aluminium
alloy box mullion

double glazed
fixed light

extruded
aluminium
alloy box
transom

transom located
on spigot fixed
to box mullion –
connection to be
mastic sealed

mullion bolted to angle
brackets through slotted
holes – angle brackets
secured to floor edge
with bolts located in
dovetail anchor slots to
provide for thermal and/
or structural movement

infill or
undersill panel

Fixing Curtain Walling to the Structure ~ in curtain walling systems it is the main vertical component or mullion which carries the loads and transfers them to the structural frame at every or alternate floor levels depending on the spanning ability of the mullion. At each fixing point the load must be transferred and an allowance made for thermal expansion and differential movement between the structural frame and curtain walling. The usual method employed is slotted bolt fixings.

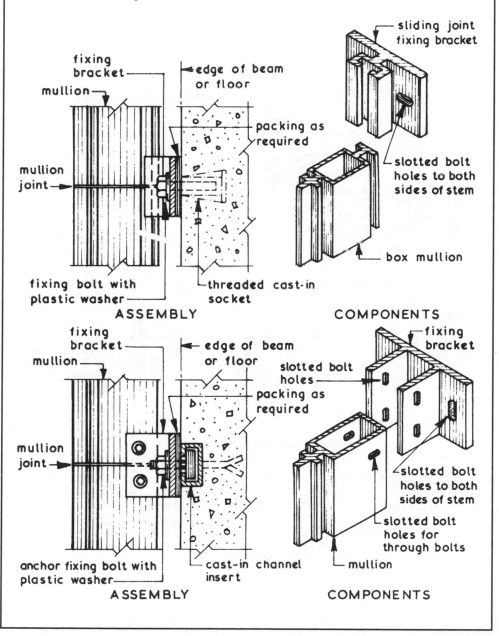

Re-cladding existing framed buildings has become an economical alternative to complete demolition and re-building. This may be justified when a building has a change of use or it is in need of an image upgrade. Current energy conservation measures can also be achieved by the re-dressing of older buildings.

Typical section through an existing structural floor slab with a replacement system attached ~

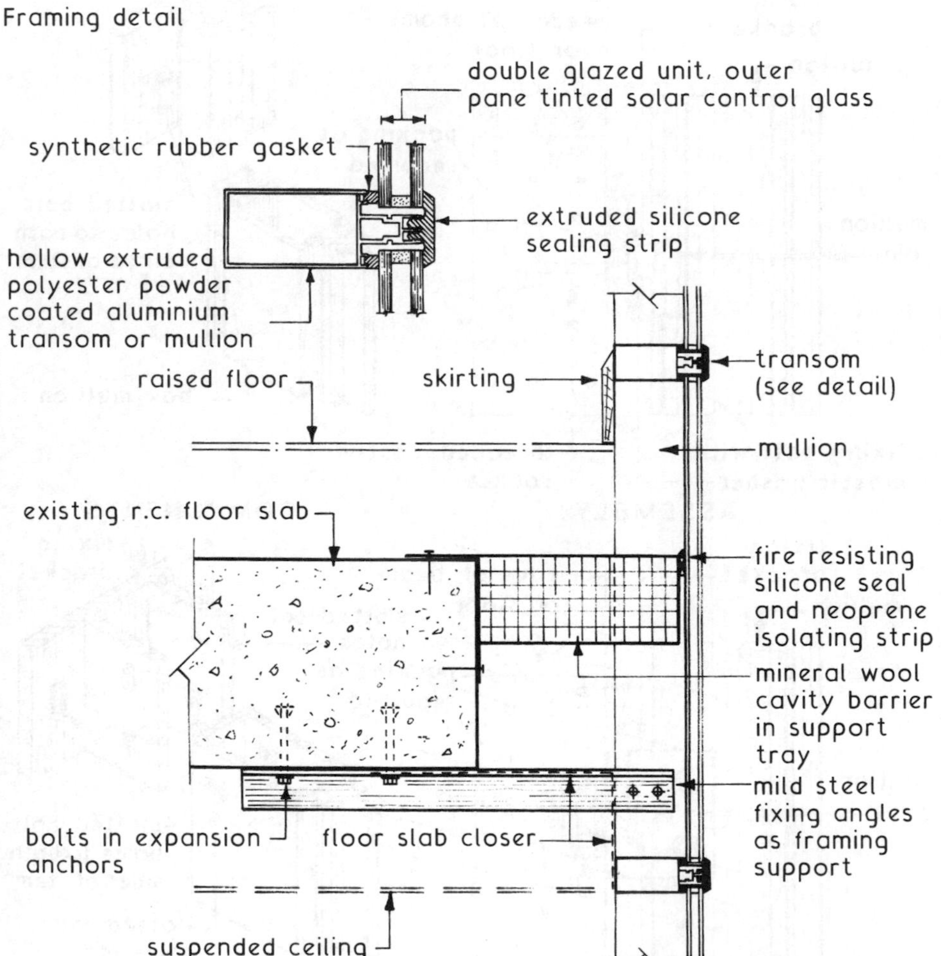

Framing detail

double glazed unit, outer pane tinted solar control glass

synthetic rubber gasket

extruded silicone sealing strip

hollow extruded polyester powder coated aluminium transom or mullion

raised floor

skirting

transom (see detail)

mullion

existing r.c. floor slab

fire resisting silicone seal and neoprene isolating strip

mineral wool cavity barrier in support tray

mild steel fixing angles as framing support

bolts in expansion anchors

floor slab closer

suspended ceiling

VERTICAL SECTION

Load bearing Concrete Panels ~ this form of construction uses storey height load bearing pre-cast reinforced concrete perimeter panels. The width and depth of the panels is governed by the load(s) to be carried, the height and exposure of the building. Panels can be plain or fenestrated providing the latter leaves sufficient concrete to transmit the load(s) around the opening. The cladding panels, being structural, eliminate the need for perimeter columns and beams and provide an internal surface ready to receive insulation, attached services and decorations. In the context of design these structures must be formed in such a manner that should a single member be removed by an internal explosion, wind pressure or similar force, progressive or structural collapse will not occur, the minimum requirements being set out in Part A of the Building Regulations. Load bearing concrete panel construction can be a cost effective method of building.

Typical Details ~

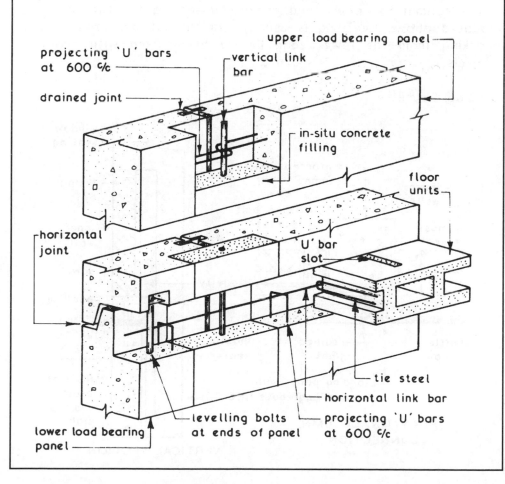

Concrete Cladding Panels ~ these are usually of reinforced precast concrete to an undersill or storey height format, the former being sometimes called apron panels. All precast concrete cladding panels should be designed and installed to fulfil the following functions:-

1. Self supporting between framing members.
2. Provide resistance to penetration by the natural elements.
3. Resist both positive and negative wind pressures.
4. Provide required degree of fire resistance.
5. Provide required degree of thermal insulation by having the insulating material incorporated within the body of the cladding or alternatively allow the cladding to act as the outer leaf of cavity wall panel.
6. Provide required degree of sound insulation.

Undersill or Apron Cladding Panels ~ these are designed to span from column to column and provide a seating for the windows located above. Levelling is usually carried out by wedging and packing from the lower edge before being fixed with grouted dowels.

Typical Details ~

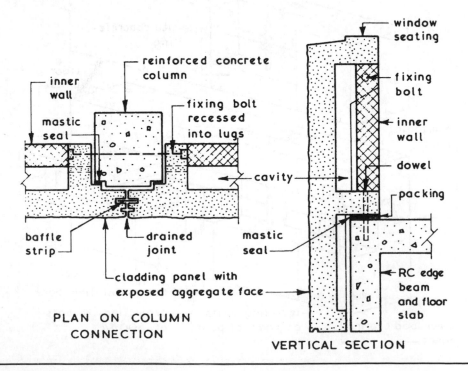

PLAN ON COLUMN
CONNECTION

VERTICAL SECTION

Storey Height Cladding Panels ~ these are designed to span vertically from beam to beam and can be fenestrated if required. Levelling is usually carried out by wedging and packing from floor level before being fixed by bolts or grouted dowels.

Typical Details ~

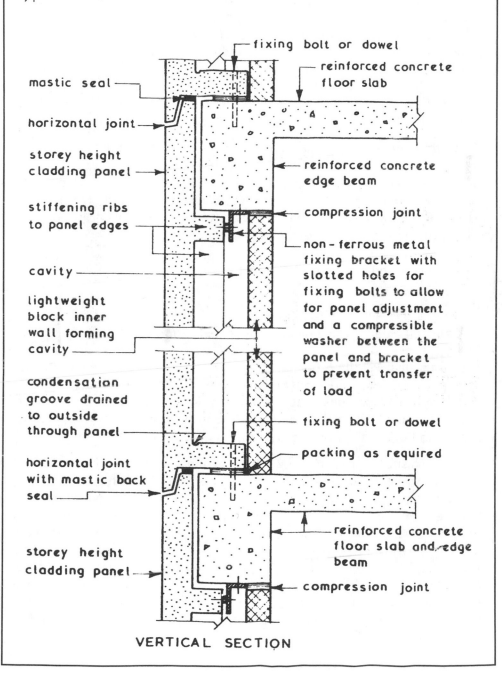

- fixing bolt or dowel
- reinforced concrete floor slab
- mastic seal
- horizontal joint
- storey height cladding panel
- reinforced concrete edge beam
- stiffening ribs to panel edges
- compression joint
- non-ferrous metal fixing bracket with slotted holes for fixing bolts to allow for panel adjustment and a compressible washer between the panel and bracket to prevent transfer of load
- cavity
- lightweight block inner wall forming cavity
- condensation groove drained to outside through panel
- fixing bolt or dowel
- packing as required
- horizontal joint with mastic back seal
- reinforced concrete floor slab and edge beam
- storey height cladding panel
- compression joint

VERTICAL SECTION

565

Concrete Surface Finishes ~ it is not easy to produce a concrete surface with a smooth finish of uniform colour direct from the mould or formwork since the colour of the concrete can be affected by the cement and fine aggregate used. The concrete surface texture can be affected by the aggregate grading, cement content, water content, degree of compaction, pin holes caused by entrapped air and rough patches caused by adhesion to parts of the formworks. Complete control over the above mentioned causes is difficult under ideal factory conditions and almost impossible under normal site conditions. The use of textured and applied finishes has therefore the primary function of improving the appearance of the concrete surface and in some cases it will help to restrict the amount of water which reaches a vertical joint.

Casting ~ concrete components can usually be cast in-situ or precast in moulds. Obtaining a surface finish to concrete cast in-situ is usually carried out against a vertical face, whereas precast concrete components can be cast horizontally and treated on either upper or lower mould face. Apart from a plain surface concrete the other main options are:-

1. Textured and profiled surfaces.
2. Tooled finishes.
3. Cast-on finishes. (see next page)
4. Exposed aggregate finishes. (see next page)

Textured and Profiled Surfaces ~ these can be produced on the upper surface of a horizontal casting by rolling, tamping, brushing and sawing techniques but variations in colour are difficult to avoid. Textured and profiled surfaces can be produced on the lower face of a horizontal casting by using suitable mould linings.

Tooled Finishes ~ the surface of hardened concrete can be tooled by bush hammering, point tooling and grinding. Bush hammering and point tooling can be carried out by using an electric or pneumatic hammer on concrete which is at least three weeks old provided gravel aggregates have not been used since these tend to shatter leaving surface pits. Tooling up to the arris could cause spalling therefore a 10 mm wide edge margin should be left untooled. Grinding the hardened concrete consists of smoothing the surface with a rotary carborundum disc which may have an integral water feed. Grinding is a suitable treatment for concrete containing the softer aggregates such as limestone.

Cast-on Finishes ~ these finishes include split blocks, bricks, stone, tiles and mosaic. Cast on finishes to the upper surface of a horizontal casting are not recommended although such finishes could be bedded onto the fresh concrete. Lower face treatment is by laying the materials with sealed or grouted joints onto the base of mould or alternatively the materials to be cast-on may be located in a sand bed spread over the base of the mould.

Exposed Aggregate Finishes ~ attractive effects can be obtained by removing the skin of hardened cement paste or surface matrix, which forms on the surface of concrete, to expose the aggregate. The methods which can be employed differ with the casting position.

Horizontal Casting - treatment to the upper face can consist of spraying with water and brushing some two hours after casting, trowelling aggregate into the fresh concrete surface or by using the felt-float method. This method consists of trowelling 10 mm of dry mix fine concrete onto the fresh concrete surface and using the felt pad to pick up the cement and fine particles from the surface leaving a clean exposed aggregate finish.

Treatment to the lower face can consist of applying a retarder to the base of the mould so that the partially set surface matrix can be removed by water and/or brushing as soon as the castings are removed from the moulds. When special face aggregates are used the sand bed method could be employed.

Vertical Casting - exposed aggregate finishes to the vertical faces can be obtained by tooling the hardened concrete or they can be cast-on by the aggregate transfer process. This consists of sticking the selected aggregate onto the rough side of pegboard sheets with a mixture of water soluble cellulose compounds and sand fillers. The cream like mixture is spread evenly over the surface of the pegboard to a depth of one third the aggregate size and the aggregate sprinkled or placed evenly over the surface before being lightly tamped into the adhesive. The prepared board is then set aside for 36 hours to set before being used as a liner to the formwork or mould. The liner is used in conjunction with a loose plywood or hardboard baffle placed against the face of the aggregate. The baffle board is removed as the concrete is being placed.

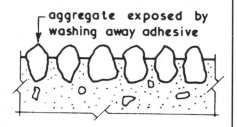

aggregate exposed by washing away adhesive

Concrete – Surface Defects

Discolouration ~ manifests as a patchy surface finish. It is caused where there are differences in hydration or moisture loss during the concrete set, due to concentrations of cement or where aggregates become segregated. Both of these will produce moisture content differences at the surface. Areas with a darker surface indicate the greater loss of moisture, possibly caused by insufficient mixing and/or poorly sealed formwork producing differences in surface absorption.

Crazing ~ surface shrinkage cracks caused by a cement rich surface skin or by too much water in the mix. Out-of-date cement can have the same effect as well as impairing the strength of the concrete.

Lime bloom ~ a chalky surface deposit produced when the calcium present in cement reacts to contamination from moisture in the atmosphere or rainwater during the hydration process. Generally resolved by dry brushing or with a 20:1 water/hydrochloric acid wash.

Scabbing ~ small areas or surface patches of concrete falling away as the formwork is struck. Caused by poor preparation of formwork, ie. insufficient use of mould oil or by formwork having a surface texture that is too rough.

Blow holes ~ otherwise known as surface popping. Possible causes are use of formwork finishes with nil or low absorbency or by insufficient vibration of concrete during placement.

Rust staining ~ if not caused by inadequate concrete cover to reinforcement, this characteristic is quite common where iron rich aggregates or pyrites are used. Rust-brown stains are a feature and there may also be some cracking where the iron reacts with the cement.

Dusting ~ caused by unnaturally rapid hardening of concrete and possibly where out-of-date cement is used. The surface of set concrete is dusty and friable.

7 INTERNAL CONSTRUCTION AND FINISHES

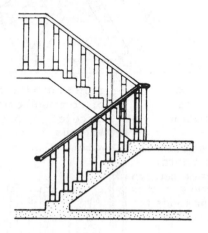

INTERNAL ELEMENTS
INTERNAL WALLS
CONSTRUCTION JOINTS
PARTITIONS
PLASTERS AND PLASTERING
DRY LINING TECHNIQUES
WALL TILING
DOMESTIC FLOORS AND FINISHES
LARGE CAST IN-SITU GROUND FLOORS
CONCRETE FLOOR SCREEDS
TIMBER SUSPENDED FLOORS
TIMBER BEAM DESIGN
REINFORCED CONCRETE SUSPENDED FLOORS
PRECAST CONCRETE FLOORS
RAISED ACCESS FLOORS
SOUND INSULATION
TIMBER, CONCRETE AND METAL STAIRS
INTERNAL DOORS
FIRE RESISTING DOORS
PLASTERBOARD CEILINGS
SUSPENDED CEILINGS
PAINTS AND PAINTING
JOINERY PRODUCTION
COMPOSITE BOARDING
PLASTICS IN BUILDING

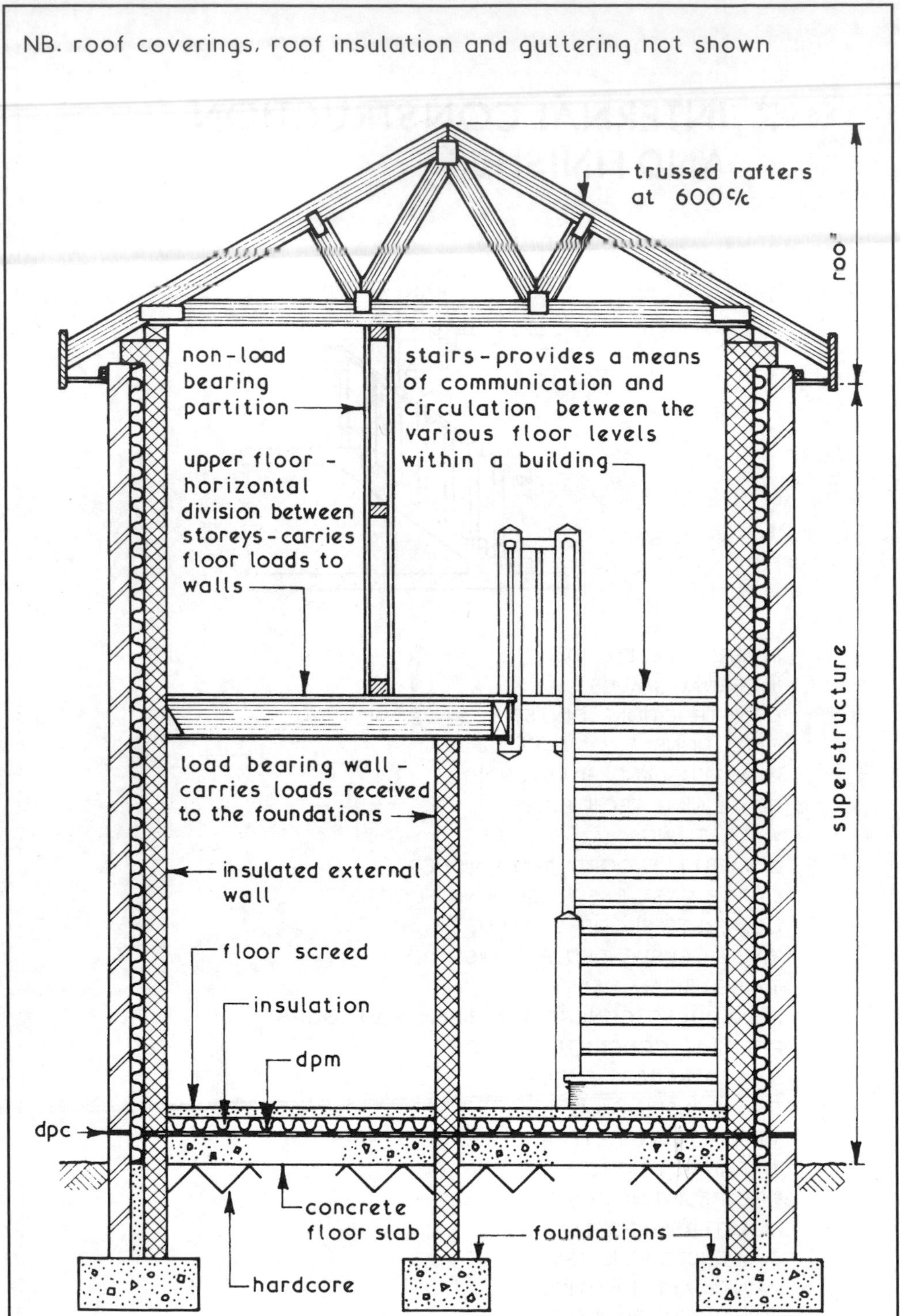

NB. roof coverings, roof insulation and guttering not shown

trussed rafters at 600 c/c

roo"

non-load bearing partition

stairs - provides a means of communication and circulation between the various floor levels within a building

upper floor - horizontal division between storeys - carries floor loads to walls

superstructure

load bearing wall - carries loads received to the foundations

insulated external wall

floor screed

insulation

dpm

dpc

concrete floor slab

foundations

hardcore

NB. all work below dpc level is classed as substructure

Internal Walls ~ their primary function is to act as a vertical divider of floor space and in so doing form a storey height enclosing element.

Other Possible Functions:-

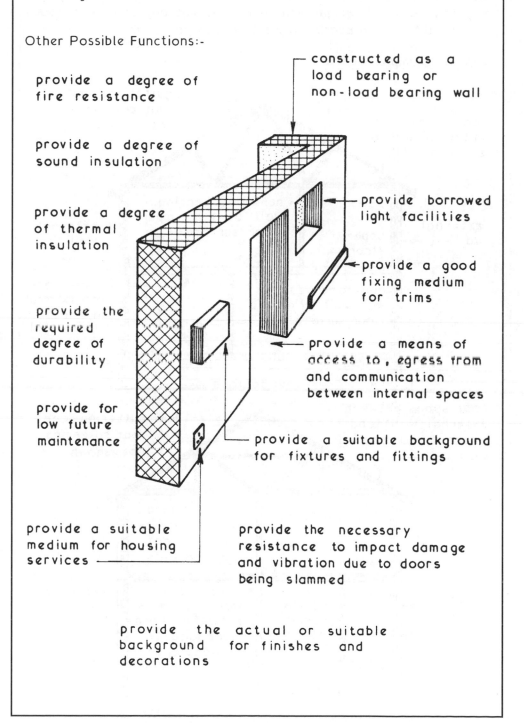

provide a degree of fire resistance

provide a degree of sound insulation

provide a degree of thermal insulation

provide the required degree of durability

provide for low future maintenance

constructed as a load bearing or non-load bearing wall

provide borrowed light facilities

provide a good fixing medium for trims

provide a means of access to, egress from and communication between internal spaces

provide a suitable background for fixtures and fittings

provide a suitable medium for housing services

provide the necessary resistance to impact damage and vibration due to doors being slammed

provide the actual or suitable background for finishes and decorations

Internal Walls ~ there are two basic design concepts for internal walls those which accept and transmit structural loads to the foundations are called Load Bearing Walls and those which support only their own self-weight and do not accept any structural loads are called Non-load Bearing Walls or Partitions.

Typical Examples ~

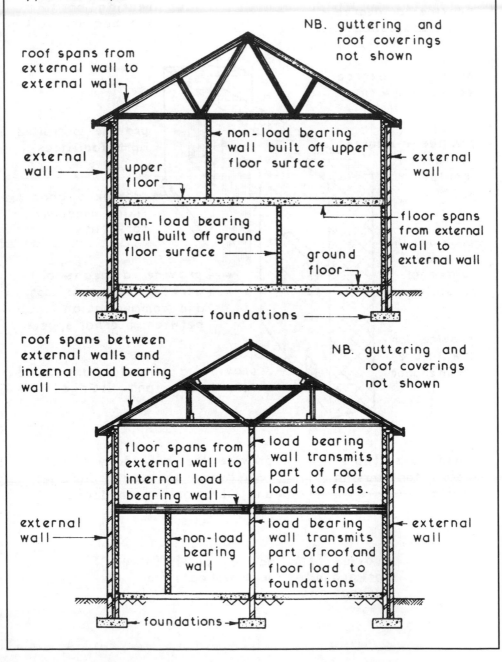

roof spans from external wall to external wall

NB. guttering and roof coverings not shown

external wall

upper floor

non-load bearing wall built off upper floor surface

external wall

non-load bearing wall built off ground floor surface

ground floor

floor spans from external wall to external wall

foundations

roof spans between external walls and internal load bearing wall

NB. guttering and roof coverings not shown

floor spans from external wall to internal load bearing wall

load bearing wall transmits part of roof load to fnds.

external wall

non-load bearing wall

load bearing wall transmits part of roof and floor load to foundations

external wall

foundations

572

Internal Brick Walls ~ these can be load bearing or non-load bearing (see previous page) and for most two storey buildings are built in half brick thickness in stretcher bond.

Typical Details ~

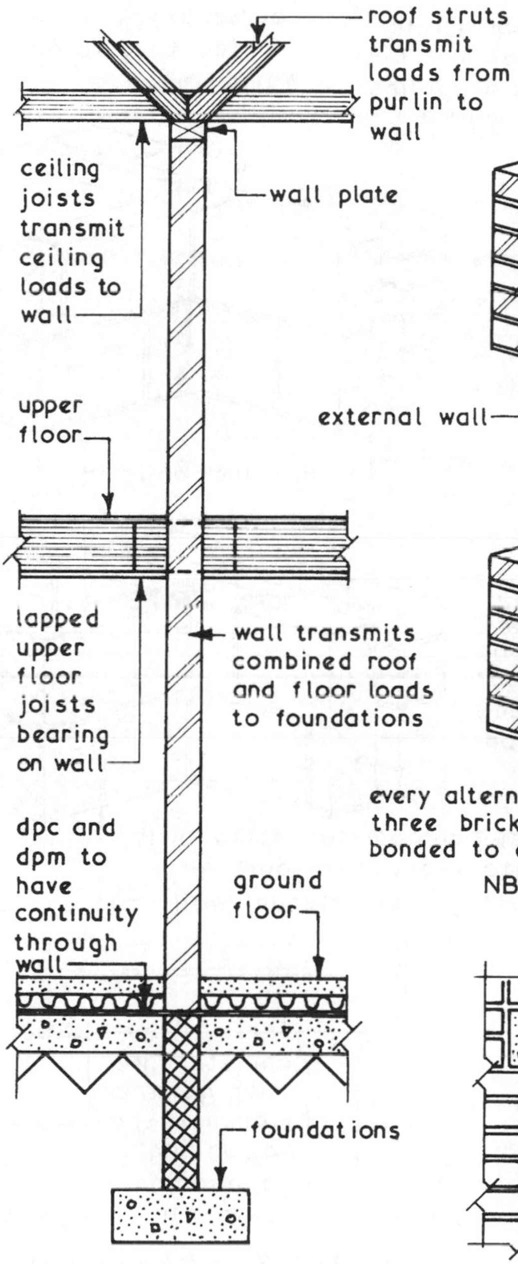

roof struts transmit loads from purlin to wall

wall plate

ceiling joists transmit ceiling loads to wall

every alternate course bonded to external wall

external wall

internal wall

upper floor

lapped upper floor joists bearing on wall

wall transmits combined roof and floor loads to foundations

every alternate group of three brick courses block bonded to external wall

NB. only applicable where cold bridging is not a concern

dpc and dpm to have continuity through wall

ground floor

foundations

concrete lintel- size governed by opening span and loading - end bearing 100 mm minimum

LOAD BEARING WALL

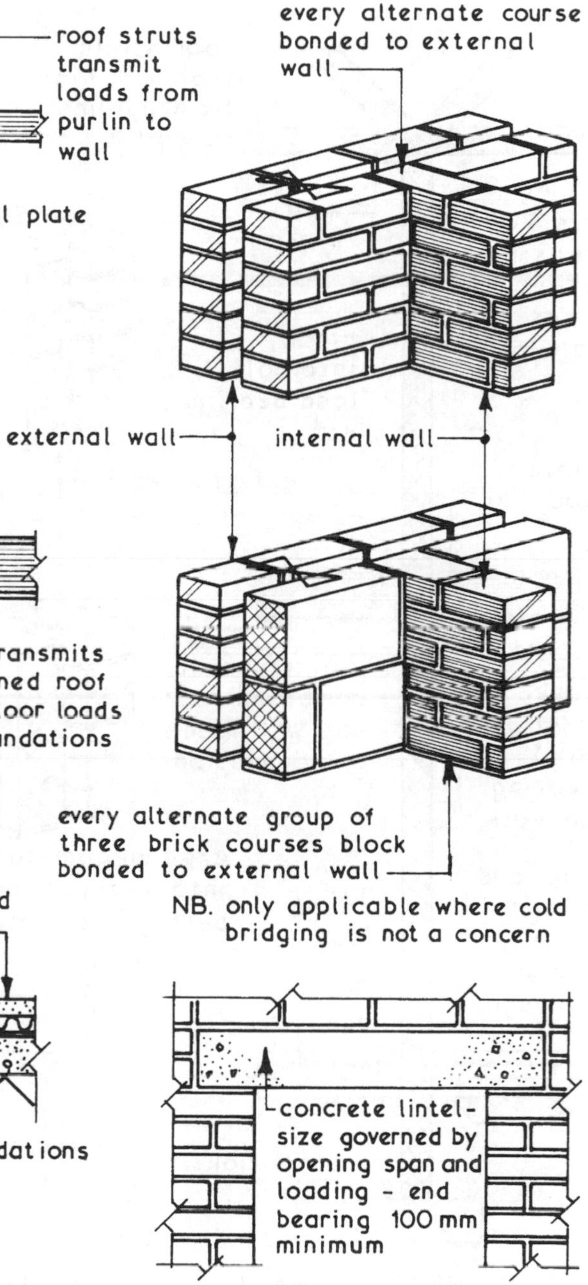

JUNCTIONS AND OPENINGS

Internal Block Walls ~ these can be load bearing or non-load bearing (see page 572) the thickness and type of block to be used will depend upon the loadings it has to carry.

Typical Details ~

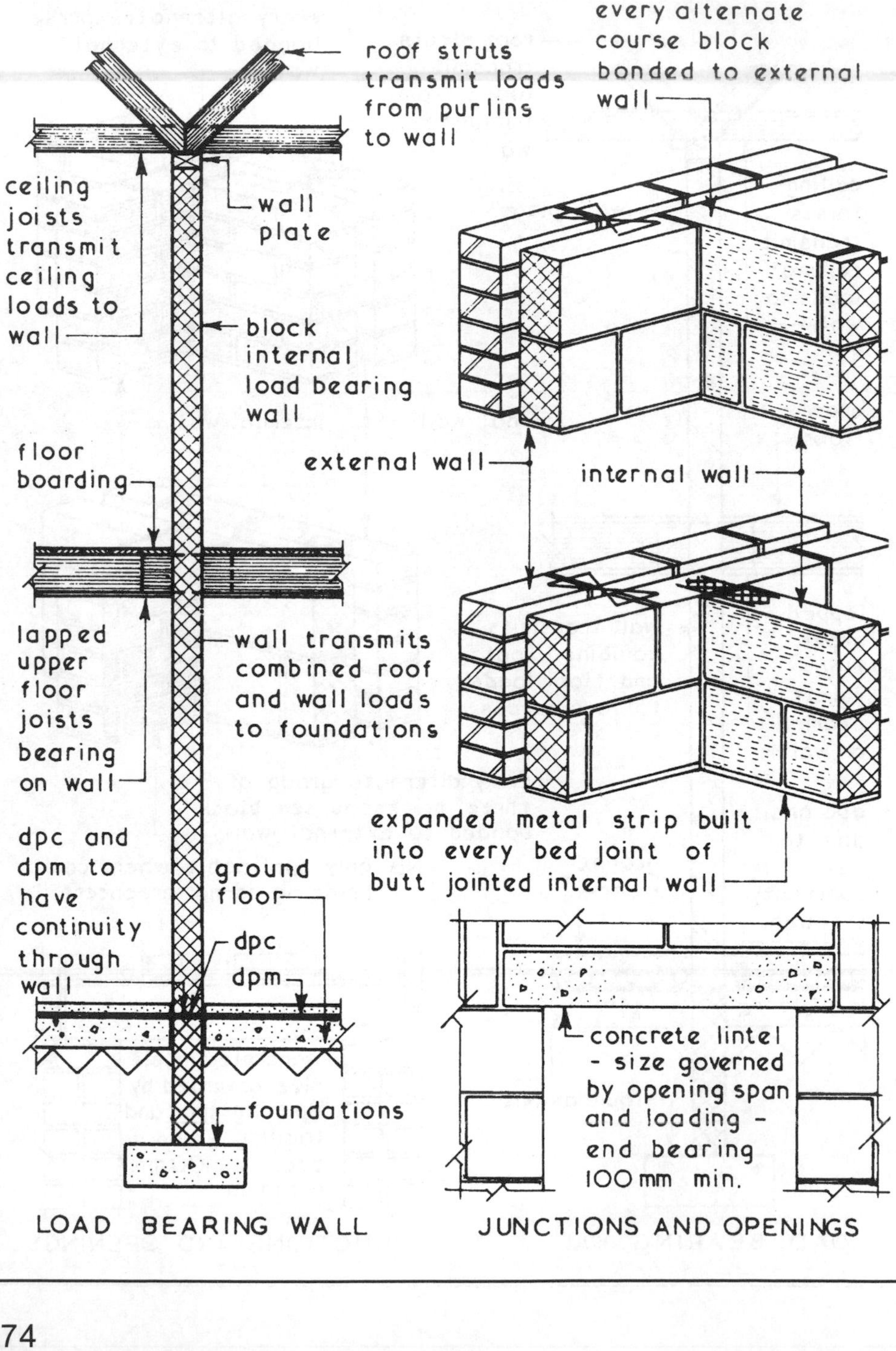

roof struts transmit loads from purlins to wall

every alternate course block bonded to external wall

ceiling joists transmit ceiling loads to wall

wall plate

block internal load bearing wall

external wall

internal wall

floor boarding

lapped upper floor joists bearing on wall

wall transmits combined roof and wall loads to foundations

dpc and dpm to have continuity through wall

ground floor

dpc

dpm

expanded metal strip built into every bed joint of butt jointed internal wall

foundations

concrete lintel - size governed by opening span and loading - end bearing 100 mm min.

LOAD BEARING WALL   JUNCTIONS AND OPENINGS

Internal Walls ~ an alternative to brick and block bonding shown on the preceding two pages is application of wall profiles. These are quick and simple to install, provide adequate lateral stability, sufficient movement flexibility and will overcome the problem of thermal bridging where a brick partition would otherwise bond into a block inner leaf. They are also useful for attaching extension walls at right angles to existing masonry.

Application ~

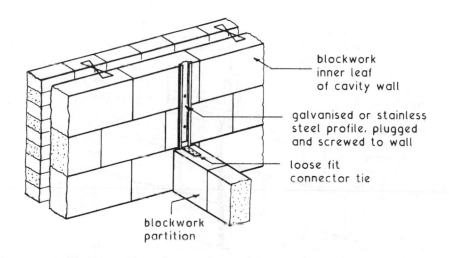

blockwork
inner leaf
of cavity wall

galvanised or stainless
steel profile, plugged
and screwed to wall

loose fit
connector tie

blockwork
partition

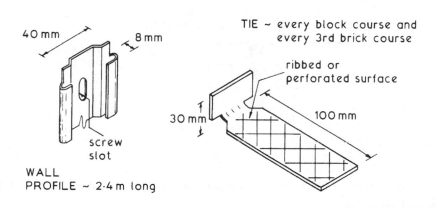

40 mm

8 mm

screw
slot

WALL
PROFILE ~ 2·4 m long

TIE ~ every block course and
every 3rd brick course

ribbed or
perforated surface

30 mm

100 mm

Movement or Construction Joints ~ provide an alternative to ties or mesh reinforcement in masonry bed joints. Even with reinforcement, lightweight concrete block walls are renowned for producing unsightly and possibly unstable shrinkage cracks. Galvanised or stainless steel formers and ties are built in at a maximum of 6m horizontal spacing and within 3m of corners to accommodate initial drying, shrinkage movement and structural settlement. One side of the former is fitted with profiled or perforated ties to bond into bed joints and the other has plastic sleeved ties. The sleeved tie maintains continuity, but restricts bonding to allow for controlled movement.

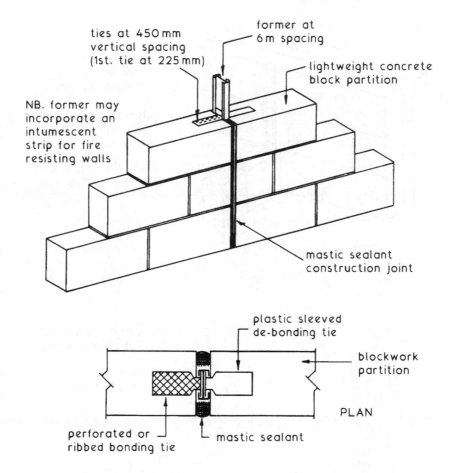

ties at 450mm vertical spacing (1st. tie at 225mm)

former at 6m spacing

lightweight concrete block partition

NB. former may incorporate an intumescent strip for fire resisting walls

mastic sealant construction joint

plastic sleeved de-bonding tie

blockwork partition

PLAN

perforated or ribbed bonding tie

mastic sealant

Note: Movement joints in brickwork should be provided at 12m maximum spacing.
Ref. BS 5628-3: Code of practice for use of masonry. Materials and components, design and workmanship.

Location ~ specifically in positions of high stress.
Reinforcement ~ expanded metal or wire mesh (see page 310).
Mortar Cover ~ 13 mm minimum thickness, 25 mm to external faces.

Openings~

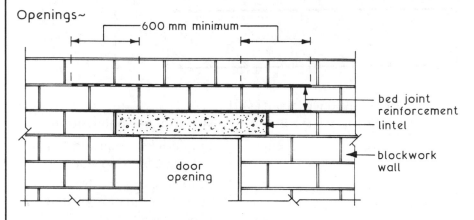

Concentrated Load ~

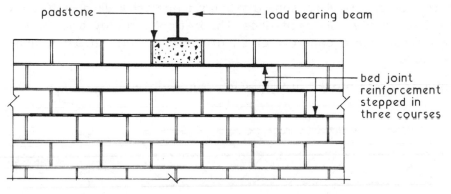

Suspended Floor~

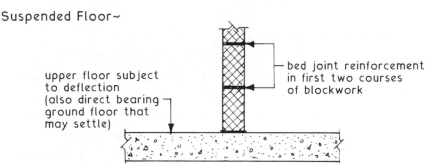

Differential Movement ~ may occur where materials such as steel, brick, timber or dense concrete abut with or bear on lightweight concrete blocks. A smooth separating interface of two layers of non-compressible dpc or similar is suitable in this situation.

Typical examples ~

Solid brickwork

T = thickness

	Fire resistance (minutes)		Material and application
	120	240	
T (mm)	102.5	215	Clay bricks. Load bearing or non-load bearing wall.
T (mm)	102.5	215	Concrete or sand/lime bricks. Load bearing or non-load bearing wall.

Note: For practical reasons a standard one-brick dimension is given for 240 minutes fire resistance. Theoretically a clay brick wall can be 170 mm and a concrete or sand/lime brick wall 200 mm, finishes excluded.

Solid concrete blocks of lightweight aggregate

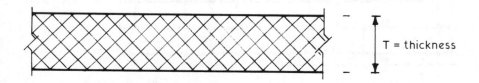

T = thickness

	Fire resistance (minutes)			Material and application
	60	120	240	
T (mm)	100	130	200	Load bearing, 2.8–3.5 N/mm² compressive strength.
T (mm)	90	100	190	Load bearing, 4.0–10 N/mm² compressive strength.
T (mm)	75	100	140	Non-load bearing, 2.8–3.5 N/mm² compressive strength.
T (mm)	75	75	100	Non-load bearing, 4.0–10 N/mm² compressive strength.

Note: Finishes excluded

Party Wall ~ a wall separating different owners buildings, ie. a wall that stands astride the boundary line between property of different ownerships. It may also be solely on one owner's land but used to separate two buildings.

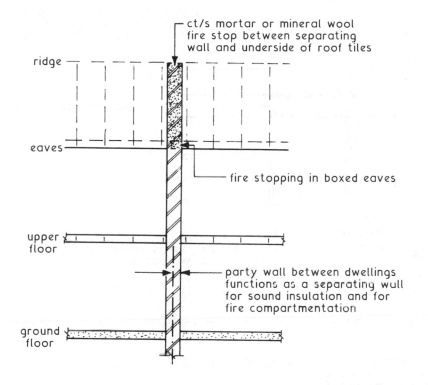

ct/s mortar or mineral wool fire stop between separating wall and underside of roof tiles

ridge

eaves

fire stopping in boxed eaves

upper floor

party wall between dwellings functions as a separating wall for sound insulation and for fire compartmentation

ground floor

Where an internal separating wall forms a junction with an external cavity wall, the cavity must be fire stopped by using a barrier of fire resisting material. Depending on the application, the material specification is of at least 30 minutes fire resistance. Between terraced and semi-detached dwellings the location is usually limited by the separating elements. For other buildings additional fire stopping will be required in constructional cavities such as suspended ceilings, rainscreen cladding and raised floors. The spacing of these cavity barriers is generally not more than 20 m in any direction, subject to some variation as indicated in Volume 2 of Approved Document B.

Refs.
Party Wall Act 1996.
Building Regulations, A.D. B, Volumes 1 and 2: Fire safety.
Building Regulations, A.D. E: Resistance to the passage of sound.

Requirements for fire and sound resisting construction ~

Typical masonry construction~

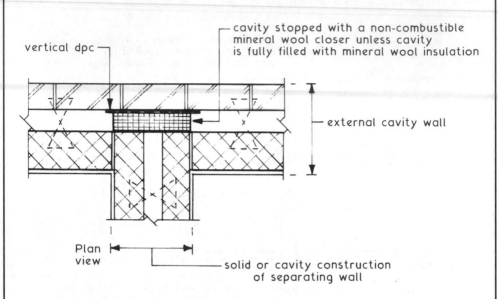

cavity stopped with a non-combustible mineral wool closer unless cavity is fully filled with mineral wool insulation

vertical dpc

external cavity wall

Plan view

solid or cavity construction of separating wall

Typical timber framed construction~

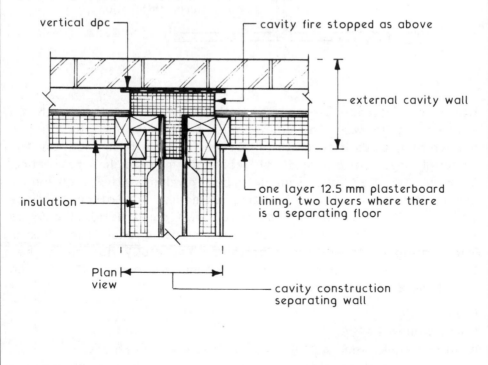

vertical dpc

cavity fire stopped as above

external cavity wall

insulation

one layer 12.5 mm plasterboard lining, two layers where there is a separating floor

Plan view

cavity construction separating wall

Internal Partitions ~ these are vertical dividers which are used to separate the internal space of a building into rooms and circulation areas such as corridors. Partitions which give support to a floor or roof are classified as load bearing whereas those which give no such support are called non-load bearing.

Load Bearing Partitions ~ these walls can be constructed of bricks, blocks or in-situ concrete by traditional methods and have the design advantages of being capable of having good fire resistance and/or high sound insulation. Their main disadvantage is permanence giving rise to an inflexible internal layout.

Non-load Bearing Partitions ~ the wide variety of methods available makes it difficult to classify the form of partition but most can be placed into one of three groups:-

1. Masonry partitions.
2. Stud partitions – see pages 582 & 583.
3. Demountable partitions – see pages 585 & 586.

Masonry Partitions ~ these are usually built with blocks of clay or lightweight concrete which are readily available and easy to construct thus making them popular. These masonry partitions should be adequately tied to the structure or load bearing walls to provide continuity as a sound barrier, provide edge restraint and to reduce the shrinkage cracking which inevitably occurs at abutments. Wherever possible openings for doors should be in the form of storey height frames to provide extra stiffness at these positions.

Typical Details ~

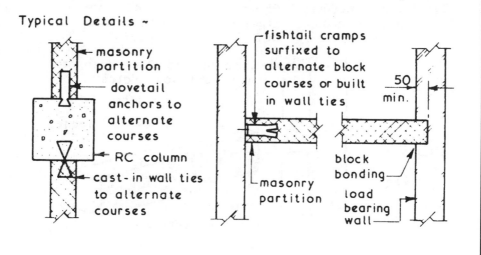

masonry partition

dovetail anchors to alternate courses

RC column

cast-in wall ties to alternate courses

fishtail cramps surfixed to alternate block courses or built in wall ties

50 min.

masonry partition

block bonding

load bearing wall

Timber Stud Partitions ~ these are non-load bearing internal dividing walls which are easy to construct, lightweight, adaptable and can be clad and infilled with various materials to give different finishes and properties. The timber studs should be of prepared or planed material to ensure that the wall is of constant thickness with parallel faces. Stud spacings will be governed by the size and spanning ability of the facing or cladding material.

Typical Details ~

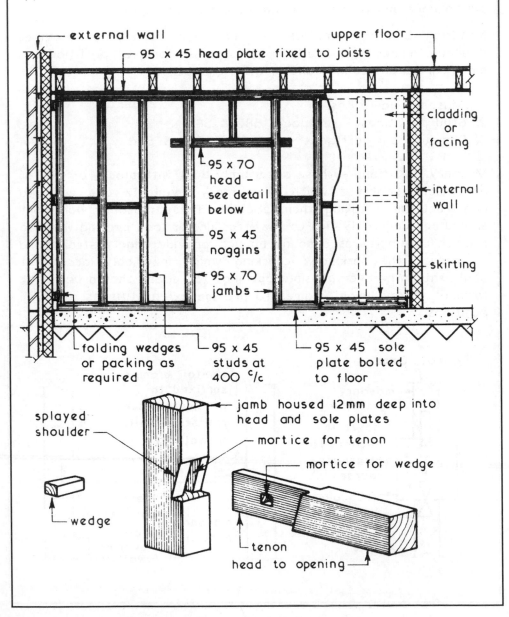

external wall upper floor

95 x 45 head plate fixed to joists

cladding or facing

95 x 70 head - see detail below

internal wall

95 x 45 noggins

95 x 70 jambs

skirting

folding wedges or packing as required

95 x 45 studs at 400 c/c

95 x 45 sole plate bolted to floor

splayed shoulder

jamb housed 12mm deep into head and sole plates

mortice for tenon

mortice for wedge

wedge

tenon

head to opening

Stud Partitions ~ these non-load bearing partitions consist of a framework of vertical studs to which the facing material can be attached. The void between the studs created by the two faces can be infilled to meet specific design needs. The traditional material for stud partitions is timber (see Timber Stud Partitions on previous page) but a similar arrangement can be constructed using metal studs faced on both sides with plasterboard.

Typical Metal Stud Partition Details ~

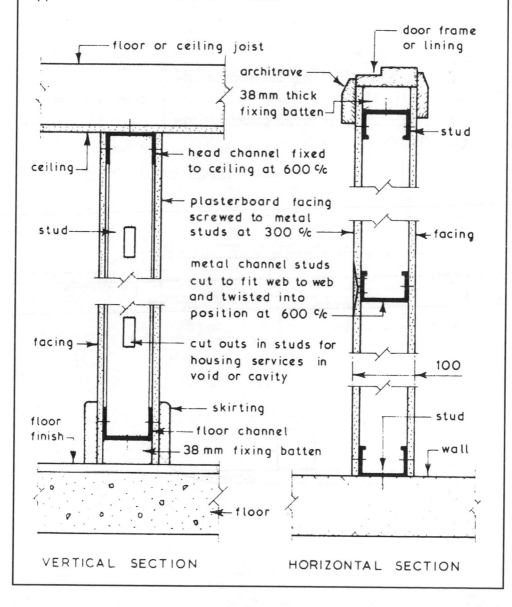

VERTICAL SECTION   HORIZONTAL SECTION

Plasterboard lining to stud framed partition walls satisfies the Building Regulations, Approved Document B – Fire safety, as a material of "limited combustibility" with a Class O rating for surface spread of flame (Class O is better than Classes 1 to 4 as determined by BS 476-7). The plasterboard dry walling should completely protect any combustible timber components such as sole plates. The following shows typical fire resistances as applied to a metal stud frame ~

30 minute fire resistance

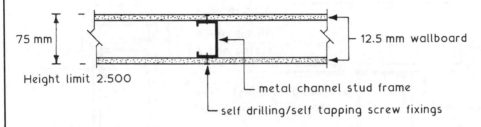

75 mm

Height limit 2.500

12.5 mm wallboard

metal channel stud frame

self drilling/self tapping screw fixings

60(90) minute fire resistance

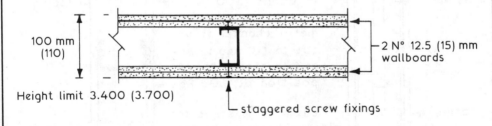

100 mm (110)

Height limit 3.400 (3.700)

2 Nº 12.5 (15) mm wallboards

staggered screw fixings

120 minute fire resistance

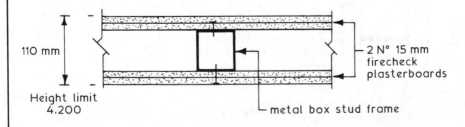

110 mm

Height limit 4.200

2 Nº 15 mm firecheck plasterboards

metal box stud frame

For plasterboard types see page 592.

Partitions ~ these can be defined as vertical internal space dividers and are usually non-loadbearing. They can be permanent, constructed of materials such as bricks or blocks or they can be demountable constructed using lightweight materials and capable of being taken down and moved to a new location incurring little or no damage to the structure or finishes. There is a wide range of demountable partitions available constructed from a variety of materials giving a range that will be suitable for most situations. Many of these partitions have a permanent finish which requires no decoration and only periodic cleaning in the context of planned maintenance.

Typical Example ~

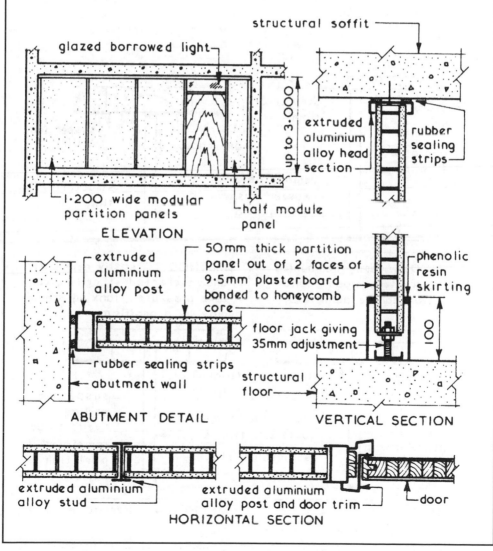

structural soffit

glazed borrowed light

up to 3·000

extruded aluminium alloy head section

rubber sealing strips

1·200 wide modular partition panels

half module panel

ELEVATION

extruded aluminium alloy post

50mm thick partition panel out of 2 faces of 9·5mm plasterboard bonded to honeycomb core

phenolic resin skirting

floor jack giving 35mm adjustment

100

rubber sealing strips

abutment wall

structural floor

ABUTMENT DETAIL

VERTICAL SECTION

extruded aluminium alloy stud

extruded aluminium alloy post and door trim

door

HORIZONTAL SECTION

Demountable Partitions ~ it can be argued that all internal non-load bearing partitions are demountable and therefore the major problem is the amount of demountability required in the context of ease of moving and the possible frequency anticipated. The range of partitions available is very wide including stud partitions, framed panel partitions (see Demountable Partitions on page 585), panel to panel partitions and sliding/folding partitions which are similar in concept to industrial doors (see Industrial Doors on pages 378 and 379) The latter type is often used where movement of the partition is required frequently. The choice is therefore based on the above stated factors taking into account finish and glazing requirements together with any personal preference for a particular system but in all cases the same basic problems will have to be considered:-

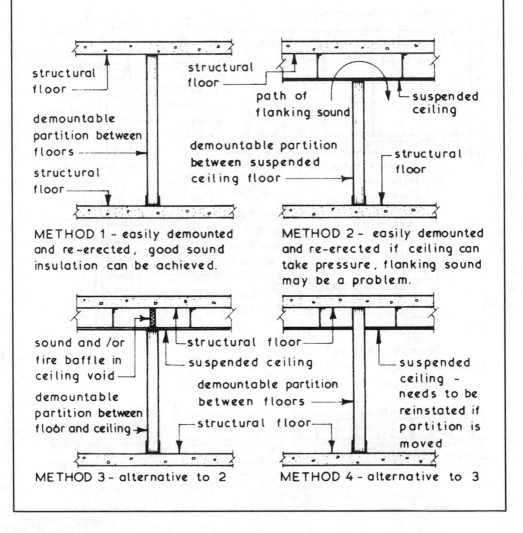

METHOD 1 - easily demounted and re-erected, good sound insulation can be achieved.

METHOD 2 - easily demounted and re-erected if ceiling can take pressure, flanking sound may be a problem.

METHOD 3 - alternative to 2

METHOD 4 - alternative to 3

Plaster ~ this is a wet mixed material applied to internal walls as a finish to fill in any irregularities in the wall surface and to provide a smooth continuous surface suitable for direct decoration. The plaster finish also needs to have a good resistance to impact damage. The material used to fulfil these requirements is gypsum plaster. Gypsum is a crystalline combination of calcium sulphate and water. The raw material is crushed, screened and heated to dehydrate the gypsum and this process together with various additives defines its type as set out in BS EN 13279-1: Gypsum binders and gypsum plasters. Definitions and requirement.

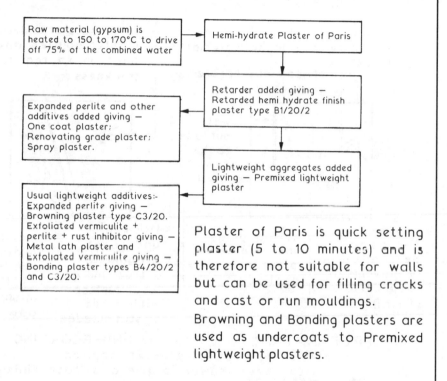

Raw material (gypsum) is heated to 150 to 170°C to drive off 75% of the combined water

Hemi-hydrate Plaster of Paris

Retarder added giving — Retarded hemi hydrate finish plaster type B1/20/2

Expanded perlite and other additives added giving — One coat plaster; Renovating grade plaster; Spray plaster.

Lightweight aggregates added giving — Premixed lightweight plaster

Usual lightweight additives:- Expanded perlite giving — Browning plaster type C3/20. Exfoliated vermiculite + perlite + rust inhibitor giving — Metal lath plaster and Exfoliated vermiculite giving — Bonding plaster types B4/20/2 and C3/20.

Plaster of Paris is quick setting plaster (5 to 10 minutes) and is therefore not suitable for walls but can be used for filling cracks and cast or run mouldings.
Browning and Bonding plasters are used as undercoats to Premixed lightweight plasters.

All plaster should be stored in dry conditions since any absorption of moisture before mixing may shorten the normal setting time of about one and a half hours which can reduce the strength of the set plaster. Gypsum plasters are not suitable for use in temperatures exceeding 43°C and should not be applied to frozen backgrounds.

A good key to the background and between successive coats is essential for successful plastering. Generally brick and block walls provide the key whereas concrete unless cast against rough formwork will need to be treated to provide the key.

Internal Wall Finishes ~ these can be classified as wet or dry. The traditional wet finish is plaster which is mixed and applied to the wall in layers to achieve a smooth and durable finish suitable for decorative treatments such as paint and wallpaper.

Most plasters are supplied in 25kg paper sacks and require only the addition of clean water or sand and clean water according to the type of plaster being used.

Typical Method of Application ~

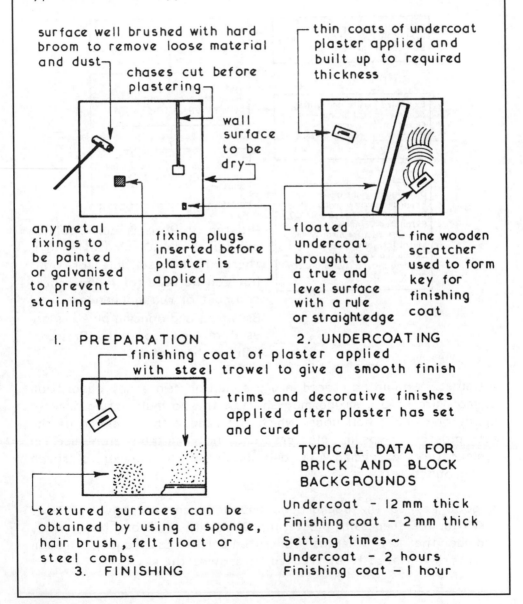

surface well brushed with hard broom to remove loose material and dust

chases cut before plastering

wall surface to be dry

thin coats of undercoat plaster applied and built up to required thickness

any metal fixings to be painted or galvanised to prevent staining

fixing plugs inserted before plaster is applied

floated undercoat brought to a true and level surface with a rule or straightedge

fine wooden scratcher used to form key for finishing coat

1.   PREPARATION

2. UNDERCOATING

finishing coat of plaster applied with steel trowel to give a smooth finish

trims and decorative finishes applied after plaster has set and cured

TYPICAL DATA FOR BRICK AND BLOCK BACKGROUNDS

Undercoat - 12mm thick
Finishing coat - 2mm thick

Setting times ~
Undercoat - 2 hours
Finishing coat - 1 hour

textured surfaces can be obtained by using a sponge, hair brush, felt float or steel combs

3.   FINISHING

Plasterboard ~ a board material made of two sheets of thin mill - board with gypsum plaster between - three edge profiles are available:-

Tapered Edge -

A flush seamless surface is obtained by filling the joint with a special filling plaster, applying a joint tape over the filling and finishing with a thin layer of joint filling plaster the edge of which is feathered out using a slightly damp jointing sponge.

Square Edge - edges are close butted and finished with a cover fillet or the joint is covered with a jute scrim before being plastered.

Bevelled Edge - edges are close butted forming a vee-joint which becomes a feature of the lining.

Typical Details ~

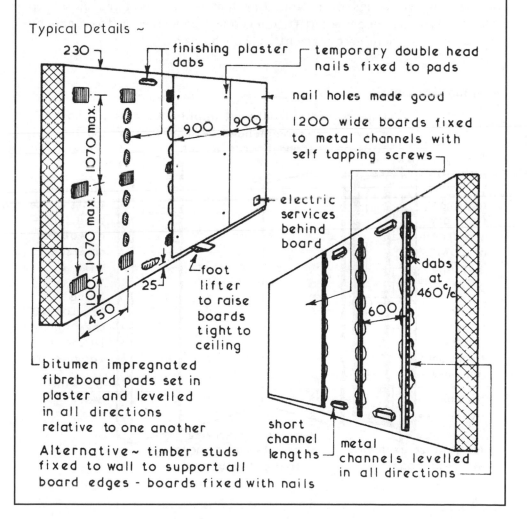

Dry Linings ~ the internal surfaces of walls and partitions are usually covered with a wet finish (plaster or rendering) or with a dry lining such as plasterboard, insulating fibre board, hardboard, timber boards, and plywood, all of which can be supplied with a permanent finish or they can be supplied to accept an applied finish such as paint or wallpaper. The main purpose of any applied covering to an internal wall surface is to provide an acceptable but not necessarily an elegant or expensive wall finish. It is also very difficult and expensive to build a brick or block wall which has a fair face to both sides since this would involve the hand selection of bricks and blocks to ensure a constant thickness together with a high degree of skill to construct a satisfactory wall. The main advantage of dry lining walls is that the drying out period required with wet finishes is eliminated. By careful selection and fixing of some dry lining materials it is possible to improve the thermal insulation properties of a wall. Dry linings can be fixed direct to the backing by means of a recommended adhesive or they can be fixed to a suitable arrangement of wall battens.

Typical Example ~

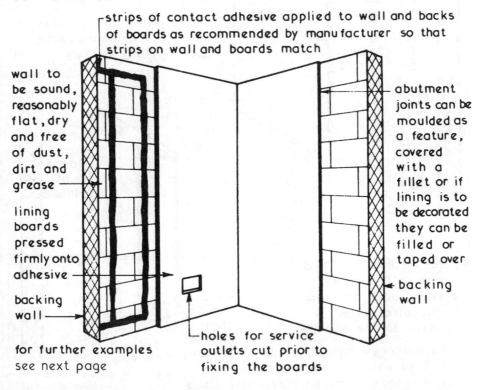

strips of contact adhesive applied to wall and backs of boards as recommended by manufacturer so that strips on wall and boards match

wall to be sound, reasonably flat, dry and free of dust, dirt and grease

lining boards pressed firmly onto adhesive

backing wall

abutment joints can be moulded as a feature, covered with a fillet or if lining is to be decorated they can be filled or taped over

backing wall

holes for service outlets cut prior to fixing the boards

for further examples see next page

Typical Examples ~

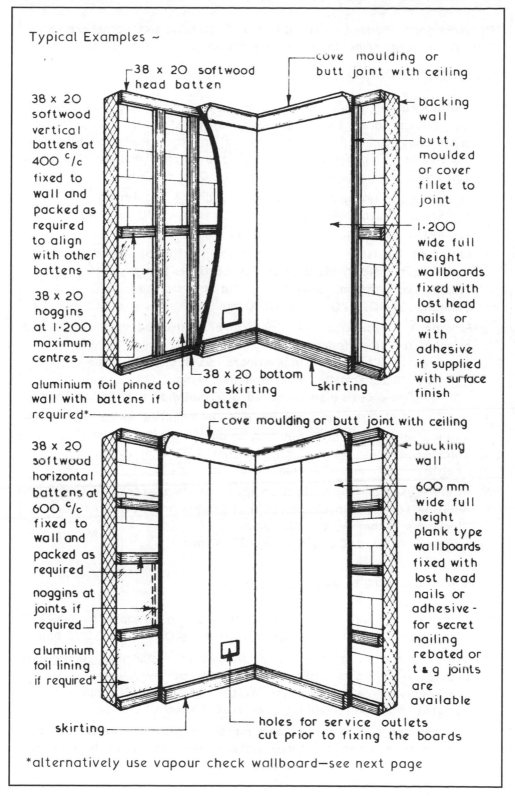

38 x 20 softwood head batten

cove moulding or butt joint with ceiling

38 x 20 softwood vertical battens at 400 ^{c}/c fixed to wall and packed as required to align with other battens

38 x 20 noggins at 1·200 maximum centres

aluminium foil pinned to wall with battens if required*

38 x 20 bottom or skirting batten

skirting

backing wall

butt, moulded or cover fillet to joint

1·200 wide full height wallboards fixed with lost head nails or with adhesive if supplied with surface finish

cove moulding or butt joint with ceiling

38 x 20 softwood horizontal battens at 600 ^{c}/c fixed to wall and packed as required

noggins at joints if required

aluminium foil lining if required*

skirting

holes for service outlets cut prior to fixing the boards

backing wall

600 mm wide full height plank type wallboards fixed with lost head nails or adhesive - for secret nailing rebated or t & g joints are available

*alternatively use vapour check wallboard—see next page

591

Plasterboard Types ~ to BS EN 520: Gypsum plasterboards. Definitions, requirements and test methods.

BS PLASTERBOARDS:~

1. Wallboard – ivory faced for taping, jointing and direct decoration; grey faced for finishing plaster or wall adhesion with dabs. General applications, i.e. internal walls, ceilings and partitions. Thicknesses: 9·5, 12·5 and 15 mm. Widths: 900 and 1200 mm. Lengths: vary between 1800 and 3000 mm.

2. Baseboard – lining ceilings requiring direct plastering.
   Thickness: 9·5 mm. Width: 900 mm. Length: 1220 mm and,
   Thickness: 12·5 mm. Width: 600 mm. Length: 1220 mm.

3. Moisture Resistant – wallboard for bathrooms and kitchens. Pale green colour, ideal base for ceramic tiling.
   Thicknesses: 12·5 mm and 15 mm. Width: 1200 mm.
   Lengths: 2400, 2700 and 3000 mm.

4. Firecheck – wallboard of glass fibre reinforced vermiculite and gypsum for fire cladding.
   Thicknesses: 12·5 and 15 mm. Widths: 900 and 1200 mm.
   Lengths: 1800, 2400, 2700 and 3000 mm.
   A 25 mm thickness is also produced, 600 mm wide × 3000 mm long.

5. Lath – rounded edge wallboard of limited area for easy application to ceilings requiring a direct plaster finish.
   Thicknesses: 9·5 and 12·5 mm. Widths: 400 and 600 mm.
   Lengths: 1200 and 1220 mm.

6. Plank – used as fire protection for structural steel and timber, in addition to sound insulation in wall panels and floating floors.
   Thickness: 19 mm. Width: 600 mm.
   Lengths: 2350, 2400, 2700 and 3000 mm.

NON – STANDARD PLASTERBOARDS:~

1. Contour – only 6 mm in thickness to adapt to curved featurework. Width: 1200 mm. Lengths: 2400 m and 3000 mm.

2. Vapourcheck – a metallized polyester wallboard lining to provide an integral water vapour control layer.
   Thicknesses: 9·5 and 12·5 mm. Widths: 900 and 1200 mm.
   Lengths: vary between 1800 and 3000 mm.

3. Thermalcheck – various expanded or foamed insulants are bonded to wallboard. Approximately 25–50 mm overall thickness in board sizes 1200 × 2400 mm.

Glazed Wall Tiles ~ internal glazed wall tiles are usually made to the various specifications under BS EN 14411: Ceramic tiles. Definitions, classification, characteristics and marking.

Internal Glazed Wall Tiles ~ the body of the tile can be made from ball-clay, china clay, china stone, flint and limestone. The material is usually mixed with water to the desired consistency, shaped and then fired in a tunnel oven at a high temperature (1150°C) for several days to form the unglazed biscuit tile. The glaze pattern and colour can now be imparted onto to the biscuit tile before the final firing process at a temperature slightly lower than that of the first firing (1050°C) for about two days.

Typical Internal Glazed Wall Tiles and Fittings ~

Sizes – Modular   100 × 100 × 5 mm thick and
                         200 × 100 × 6·5 mm thick.

       Non-modular   152 × 152 × 5 to 8 mm thick and
                         108 × 108 × 4 and 6·5 mm thick.

Fittings — wide range available particularly in the non-modular format.

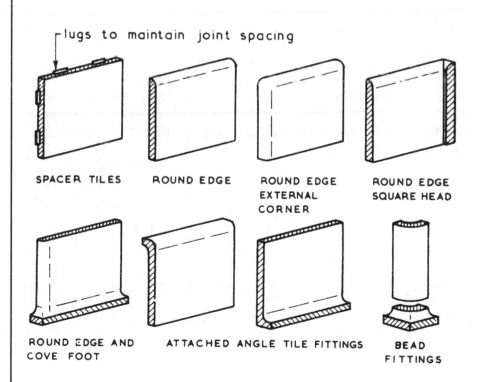

lugs to maintain joint spacing

SPACER TILES   ROUND EDGE   ROUND EDGE   ROUND EDGE
                                           EXTERNAL   SQUARE HEAD
                                           CORNER

ROUND EDGE AND   ATTACHED ANGLE TILE FITTINGS   BEAD
COVE FOOT   FITTINGS

Bedding of Internal Wall Tiles ~ generally glazed internal wall tiles are considered to be inert in the context of moisture and thermal movement, therefore if movement of the applied wall tile finish is to be avoided attention must be given to the background and the method of fixing the tiles.

Backgrounds ~ these are usually of a cement rendered or plastered surface and should be flat, dry, stable, firmly attached to the substrate and sufficiently old enough for any initial shrinkage to have taken place. The flatness of the background should be not more than 3 mm in 2·000 for the thin bedding of tiles and not more than 6 mm in 2·000 for thick bedded tiles.

Fixing Wall Tiles ~ two methods are in general use:-

1. Thin Bedding – lightweight internal glazed wall tiles fixed dry using a recommended adhesive which is applied to the wall in small areas 1 m^2 at a time with a notched trowel, the tile being pressed or tapped into the adhesive.
2. Thick Bedding – cement mortar within the mix range of 1:3 to 1:4 is used as the adhesive either by buttering the backs of the tiles which are then pressed or tapped into position or by rendering the wall surface to a thickness of approximately 10 mm and then applying the lightly buttered tiles (1:2 mix) to the rendered wall surface within two hours. It is usually necessary to soak the wall tiles in water to reduce suction before they are placed in position.

Grouting ~ when the wall tiles have set, the joints can be grouted by rubbing into the joints a grout paste either using a sponge or brush. Most grouting materials are based on cement with inert fillers and are used neat.

Typical Example ~

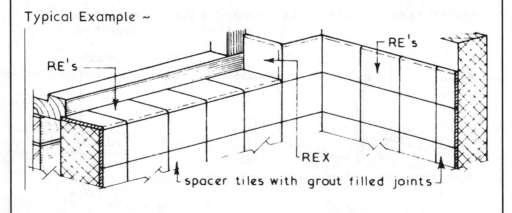

RE's

RE's

RE's

REX

spacer tiles with grout filled joints

Primary Functions ~

1. Provide a level surface with sufficient strength to support the imposed loads of people and furniture.
2. Exclude the passage of water and water vapour to the interior of the building.
3. Provide resistance to unacceptable heat loss through the floor.
4. Provide the correct type of surface to receive the chosen finish.

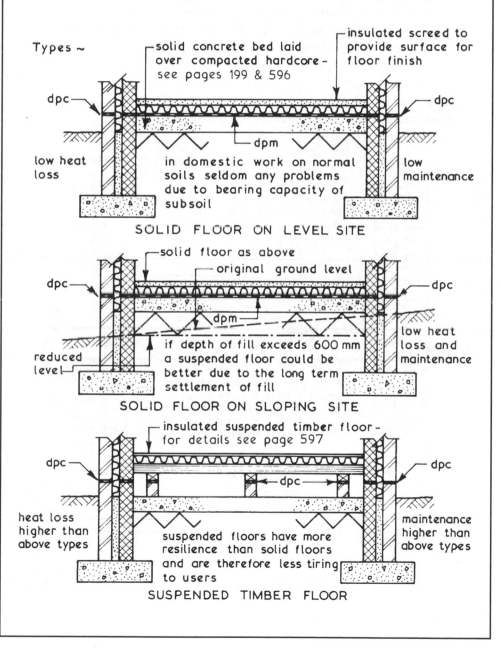

Types ~

solid concrete bed laid over compacted hardcore - see pages 199 & 596

insulated screed to provide surface for floor finish

dpc — — dpc

low heat loss

in domestic work on normal soils seldom any problems due to bearing capacity of subsoil

low maintenance

dpm

SOLID FLOOR ON LEVEL SITE

solid floor as above

original ground level

dpc — — dpc

dpm

reduced level

if depth of fill exceeds 600 mm a suspended floor could be better due to the long term settlement of fill

low heat loss and maintenance

SOLID FLOOR ON SLOPING SITE

insulated suspended timber floor - for details see page 597

dpc — — dpc

dpc

heat loss higher than above types

suspended floors have more resilience than solid floors and are therefore less tiring to users

maintenance higher than above types

SUSPENDED TIMBER FLOOR

This drawing should be read in conjunction with page 199 – Foundation Beds.

A domestic solid ground floor consists of three components:-

1. Hardcore – a suitable filling material to make up the top soil removal and reduced level excavations. It should have a top surface which can be rolled out to ensure that cement grout is not lost from the concrete. It may be necessary to blind the top surface with a layer of sand especially if the damp-proof membrane is to be placed under the concrete bed.
2. Damp-proof Membrane – an impervious layer such as heavy duty polythene sheeting to prevent moisture passing through the floor to the interior of the building.
3. Concrete Bed – the component providing the solid level surface to which screeds and finishes can be applied.

Typical Details ~

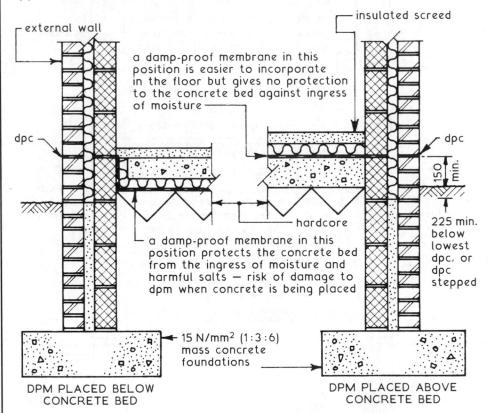

DPM PLACED BELOW
CONCRETE BED

DPM PLACED ABOVE
CONCRETE BED

NB. a compromise to the above methods is to place the dpm in the middle of the concrete bed but this needs two concrete pouring operations.

Suspended Timber Ground Floors ~ these need to have a well ventilated space beneath the floor construction to prevent the moisture content of the timber rising above an unacceptable level (i.e. not more than 20%) which would create the conditions for possible fungal attack.

Typical Details ~

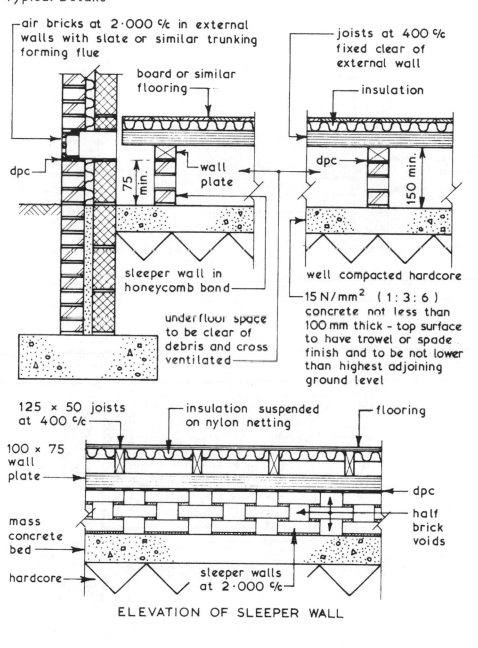

air bricks at 2·000 c/c in external walls with slate or similar trunking forming flue

board or similar flooring

joists at 400 c/c fixed clear of external wall

insulation

dpc

75 min.

wall plate

dpc

150 min.

sleeper wall in honeycomb bond

underfloor space to be clear of debris and cross ventilated

well compacted hardcore

15 N/mm² (1 : 3 : 6) concrete not less than 100 mm thick – top surface to have trowel or spade finish and to be not lower than highest adjoining ground level

125 × 50 joists at 400 c/c

insulation suspended on nylon netting

flooring

100 × 75 wall plate

dpc

mass concrete bed

half brick voids

hardcore

sleeper walls at 2·000 c/c

ELEVATION OF SLEEPER WALL

597

Precast Concrete Floors ~ these have been successfully adapted from commercial building practice (see pages 620 and 621), as an economic alternative construction technique for suspended timber and solid concrete domestic ground (and upper) floors. See also page 334 for special situations.

Typical Details ~

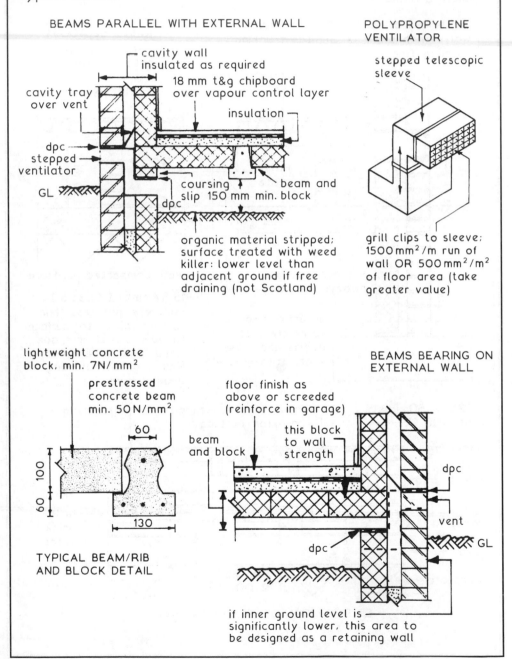

BEAMS PARALLEL WITH EXTERNAL WALL

- cavity wall insulated as required
- 18 mm t&g chipboard over vapour control layer
- insulation
- cavity tray over vent
- dpc
- stepped ventilator
- GL
- coursing
- slip 150 mm min. block
- beam and
- dpc
- organic material stripped; surface treated with weed killer: lower level than adjacent ground if free draining (not Scotland)

POLYPROPYLENE VENTILATOR

- stepped telescopic sleeve
- grill clips to sleeve; 1500 mm^2/m run of wall OR 500 mm^2/m^2 of floor area (take greater value)

- lightweight concrete block, min. 7N/mm^2
- prestressed concrete beam min. 50N/mm^2
- 60
- 100
- 60
- 130

TYPICAL BEAM/RIB AND BLOCK DETAIL

BEAMS BEARING ON EXTERNAL WALL

- floor finish as above or screeded (reinforce in garage)
- beam and block
- this block to wall strength
- dpc
- vent
- dpc
- GL
- dpc
- if inner ground level is significantly lower, this area to be designed as a retaining wall

Precast Reinforced Concrete Beam and Expanded Polystyrene (EPS) Block Floors ~ these have evolved from the principles of beam and block floor systems as shown on the preceding page. The light weight and easy to cut properties of the blocks provide for speed and simplicity in construction. Exceptional thermal insulation is possible, with U-values as low as 0.2 W/m²K.

Typical detail ~

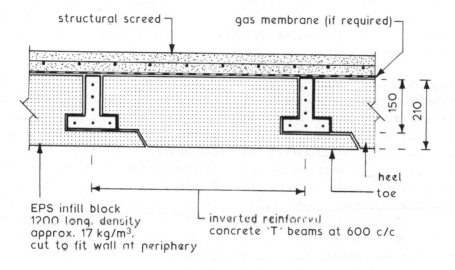

structural screed ⌐

gas membrane (if required) ⌐

150

210

heel

toe

EPS infill block 1200 long, density approx. 17 kg/m³, cut to fit wall at periphery

inverted reinforced concrete 'T' beams at 600 c/c

Cold Bridging ~ this is prevented by the EPS "toe" projecting beneath the underside of the concrete beam.

Structural Floor Finish ~ 50 mm structural concrete (C30grade) screed, reinforced with 10 mm steel Type A square mesh or with polypropylene fibres in the mix. A low-density polyethylene (LDPE) methane/radon gas membrane can be incorporated under the screed if local conditions require it.

Floating Floor Finish ~ subject to the system manufacturer's specification and accreditation, 18 mm flooring grade moisture resistant chipboard can be used over a 1000 gauge polythene vapour control layer. All four tongued and grooved edges of the chipboard are glued for continuity.

Floor Finishes ~ these are usually applied to a structural base but may form part of the floor structure as in the case of floor boards. Most finishes are chosen to fulfil a particular function such as:-

1. Appearance – chosen mainly for their aesthetic appeal or effect but should however have reasonable wearing properties. Examples are carpets; carpet tiles and wood blocks.
2. High Resistance – chosen mainly for their wearing and impact resistance properties and for high usage areas such as kitchens. Examples are quarry tiles and granolithic pavings.
3. Hygiene – chosen to provide an impervious easy to clean surface with reasonable aesthetic appeal. Examples are quarry tiles and polyvinyl chloride (PVC) sheets and tiles.

Carpets and Carpet Tiles – made from animal hair, mineral fibres and man made fibres such as nylon and acrylic. They are also available in mixtures of the above. A wide range of patterns; sizes and colours are available. Carpets and carpet tiles can be laid loose, stuck with a suitable adhesive or in the case of carpets edge fixed using special grip strips.

PVC Tiles – made from a blended mix of thermoplastic binders; fillers and pigments in a wide variety of colours and patterns to the recommendations of BS EN 649: Resilient floor coverings. PVC tiles are usually 305 × 305 × 1·6 mm thick and are stuck to a suitable base with special adhesives as recommended by the manufacturer.

Quarry Tiles ~

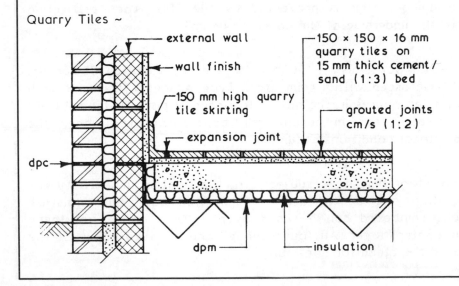

external wall

wall finish

150 mm high quarry tile skirting

expansion joint

dpc

dpm

insulation

150 × 150 × 16 mm quarry tiles on 15 mm thick cement / sand (1:3) bed

grouted joints cm/s (1:2)

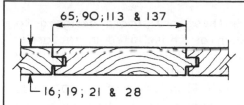

65;90;113 & 137

16;19;21 & 28

Tongue and Groove Boarding ~ prepared from softwoods to the recommendations of BS 1297. Boards are laid at right angles to the joists and are fixed with 2 No. 65 mm long cut floor brads per joists. The ends of board lengths are butt jointed on the centre line of the supporting joist.

Maximum board spans are:-
16 mm thick – 505 mm
19 mm thick – 600 mm
21 mm thick – 635 mm
28 mm thick – 790 mm

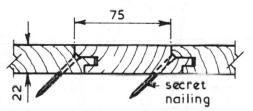

75

22

secret nailing

Timber Strip Flooring ~ strip flooring is usually considered to be boards under 100 mm face width. In good class work hardwoods would be specified the boards being individually laid and secret nailed. Strip flooring can be obtained treated with a spirit-based fungicide. Spacing of supports depends on type of timber used and applied loading. After laying the strip flooring should be finely sanded and treated with a seal or wax. In common with all timber floorings a narrow perimeter gap should be left for moisture movement.

Chipboard ~ sometimes called Particle Board is made from particles of wood bonded with a synthetic resin and/or other organic binders to the recommendations of BS EN 312.

It can be obtained with a rebated or tongue and groove joint in 600 mm wide boards 19 mm thick. The former must be supported on all the longitudinal edges whereas the latter should be supported at all cross joints.

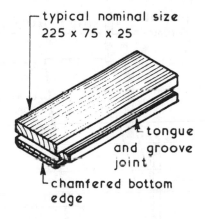

typical nominal size 225 x 75 x 25

tongue and groove joint

chamfered bottom edge

Wood Blocks ~ prepared from hardwoods and softwoods to the recommendations of BS 1187. Wood blocks can be laid to a variety of patterns, also different timbers can be used to create colour and grain effects. Laid blocks should be finely sanded and sealed or polished.

Large Cast-In-situ Ground Floors ~ these are floors designed to carry medium to heavy loadings such as those used in factories, warehouses, shops, garages and similar buildings. Their design and construction is similar to that used for small roads (see pages 118 to 121). Floors of this type are usually laid in alternate 4·500 wide strips running the length of the building or in line with the anticipated traffic flow where applicable. Transverse joints will be required to control the tensile stresses due to the thermal movement and contraction of the slab. The spacing of these joints will be determined by the design and the amount of reinforcement used. Such joints can either be formed by using a crack inducer or by sawing a 20 to 25 mm deep groove into the upper surface of the slab within 20 to 30 hours of casting.

Typical Layout ~

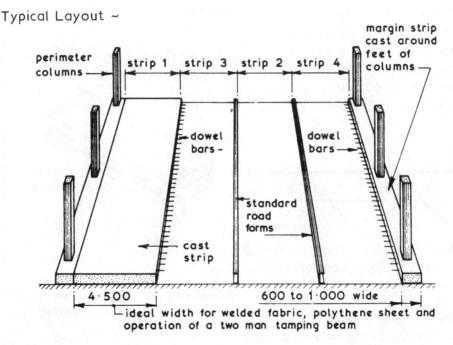

perimeter columns

margin strip cast around feet of columns

strip 1   strip 3   strip 2   strip 4

dowel bars -

dowel bars ➔

standard road forms

cast strip

4·500

600 to 1·000 wide

└ ideal width for welded fabric, polythene sheet and operation of a two man tamping beam

Surface Finishing ~ the surface of the concrete may be finished by power floating or trowelling which is carried out whilst the concrete is still plastic but with sufficient resistance to the weight of machine and operator whose footprint should not leave a depression of more than 3 mm. Power grinding of the surface is an alternative method which is carried out within a few days of the concrete hardening. The wet concrete having been surface finished with a skip float after the initial levelling with a tamping bar has been carried out. Power grinding removes 1 to 2 mm from the surface and is intended to improve surface texture and not to make good deficiencies in levels.

Vacuum Dewatering ~ if the specification calls for a power float surface finish vacuum dewatering could be used to shorten the time delay between tamping the concrete and power floating the surface. This method is suitable for slabs up to 300 mm thick. The vacuum should be applied for approximately 3 minutes for every 25 mm depth of concrete which will allow power floating to take place usually within 20 to 30 minutes of the tamping operation. The applied vacuum forces out the surplus water by compressing the slab and this causes a reduction in slab depth of approximately 2% therefore packing strips should be placed on the side forms before tamping to allow for sufficient surcharge of concrete.

Typical Details~

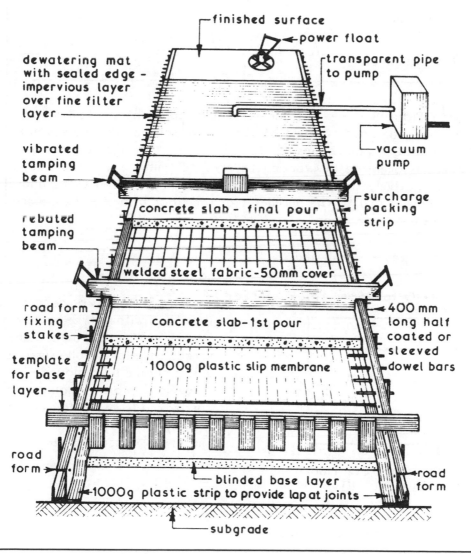

Concrete Floor Screeds ~ these are used to give a concrete floor a finish suitable to receive the floor finish or covering specified. It should be noted that it is not always necessary or desirable to apply a floor screed to receive a floor covering, techniques are available to enable the concrete floor surface to be prepared at the time of casting to receive the coverings at a later stage.

Typical Screed Mixes ~

Screed Thickness	Cement	Dry Fine Aggregate <5 mm	Coarse Aggregate > 5 mm < 10 mm
up to 40 mm	1	3 to 4 1/2	-
40 to 75 mm	1	3 to 4 1/2	-
	1	1 1/2	3

Laying Floor Screeds ~ floor screeds should not be laid in bays since this can cause curling at the edges, screeds can however be laid in 3·000 wide strips to receive thin coverings. Levelling of screeds is achieved by working to levelled timber screeding batten or alternatively a 75 mm wide band of levelled screed with square edges can be laid to the perimeter of the floor prior to the general screed laying operation.

Screed Types ~

Monolithic Screeds –

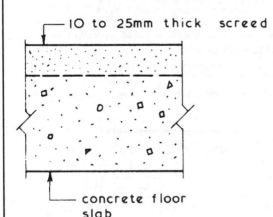

10 to 25mm thick screed

concrete floor slab

screed laid directly on concrete floor slab within three hours of placing concrete – before any screed is placed all surface water should be removed – all screeding work should be carried out from scaffold board runways to avoid walking on the 'green' concrete slab.

Screed Types ⌄

40 mm thick screed

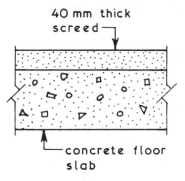

concrete floor slab

50 mm thick screed*

insulation

dpm

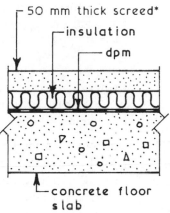

concrete floor slab

65 mm thick screed*

resilient quilt

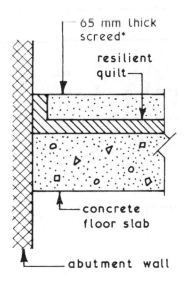

concrete floor slab

abutment wall

Separate Screeds –

screed is laid onto the concrete floor slab after it has cured. The floor surface must be clean and rough enough to ensure an adequate bond unless the floor surface is prepared by applying a suitable bonding agent or by brushing with a cement/water grout of a thick cream like consistency just before laying the screed.

Unbonded Screeds –

screed is laid directly over a damp-proof membrane or over a damp-proof membrane and insulation. A rigid form of floor insulation is required where the concrete floor slab is in contact with the ground. Care must be taken during this operation to ensure that the damp-proof membrane is not damaged.

Floating Screeds –

a resilient quilt of 25 mm thickness is laid with butt joints and turned up at the edges against the abutment walls, the screed being laid directly over the resilient quilt. The main objective of this form of floor screed is to improve the sound insulation properties of the floor.

*preferably wire mesh reinforced

Primary Functions ~

1. Provide a level surface with sufficient strength to support the imposed loads of people and furniture plus the dead loads of flooring and ceiling.
2. Reduce heat loss from lower floor as required.
3. Provide required degree of sound insulation.
4. Provide required degree of fire resistance.

Basic Construction – a timber suspended upper floor consists of a series of beams or joists supported by load bearing walls sized and spaced to carry all the dead and imposed loads.

Joist Sizing – three methods can be used:-

| 1. Building Regs. Approved Document A – Structure. Refs. *BS 6399-1: Code of practice for dead and imposed loads (max. 1·5 kN/m² distributed, 1.4 kN/m² concentrated). *TRADA publication – Span Tables for Solid Timber Members in Dwellings. | 2. Calculation formula:-

$$BM = \frac{fbd^2}{6}$$

where

BM = bending moment

f = fibre stress
b = breadth
d = depth in mm must be assumed | 3. Empirical formula:-

$$D = \frac{\text{span in mm}}{24} + 50$$

where

D = depth of joist in mm

above assumes that joists have a breadth of 50 mm and are at 400c/c spacing |

Support and Restraint ~

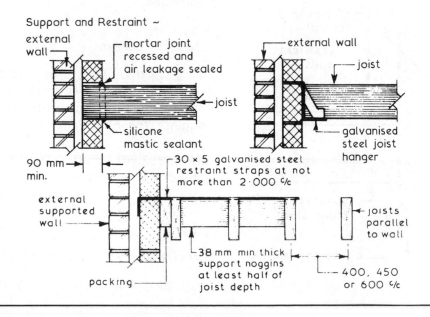

external wall

mortar joint recessed and air leakage sealed

joist

silicone mastic sealant

90 mm min.

external supported wall

packing

30 × 5 galvanised steel restraint straps at not more than 2·000 c/c

38 mm min. thick support noggins at least half of joist depth

external wall

joist

galvanised steel joist hanger

joists parallel to wall

400, 450 or 600 c/c

Strutting ~ used in timber suspended floors to restrict the movements due to twisting and vibration which could damage ceiling finishes. Strutting should be included if the span of the floor joists exceeds 2·5m and is positioned on the centre line of the span. Max. floor span ~ 6m measured centre to centre of bearing (inner leaf centre line in cavity wall).

Typical Details ~

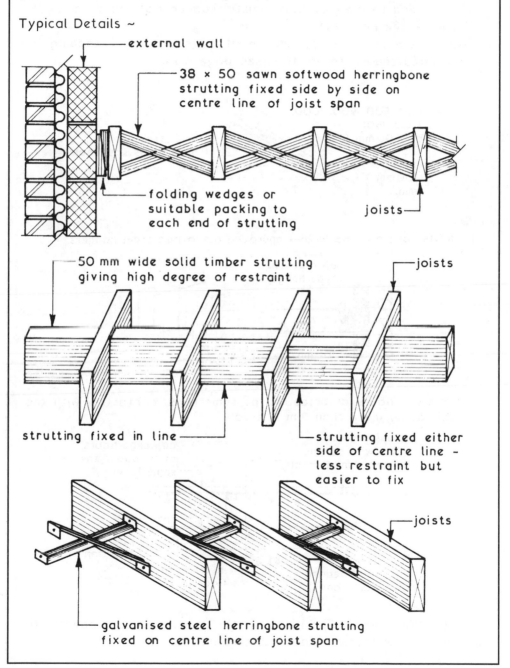

external wall

38 × 50 sawn softwood herringbone strutting fixed side by side on centre line of joist span

folding wedges or suitable packing to each end of strutting

joists

50 mm wide solid timber strutting giving high degree of restraint

joists

strutting fixed in line

strutting fixed either side of centre line – less restraint but easier to fix

joists

galvanised steel herringbone strutting fixed on centre line of joist span

607

Lateral Restraint ~ external, compartment (fire), separating (party) and internal loadbearing walls must have horizontal support from adjacent floors, to restrict movement. Exceptions occur when the wall is less than 3m long.

Methods:

1. 90 mm end bearing of floor joists, spaced not more than 1·2m apart – see page 606.
2. Galvanised steel straps spaced at intervals not exceeding 2m and fixed square to joists – see page 606.

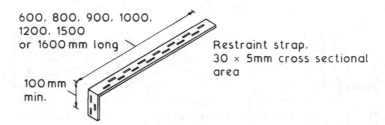

600, 800, 900, 1000, 1200, 1500 or 1600 mm long

100 mm min.

Restraint strap, 30 × 5mm cross sectional area

3. Joists carried by BS 5628–1 approved galvanised steel hangers.

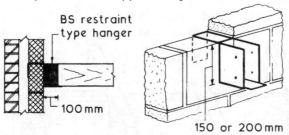

BS restraint type hanger

100 mm

150 or 200 mm

4. Adjacent floors at or about the same level, contacting with the wall at no more than 2 m intervals.

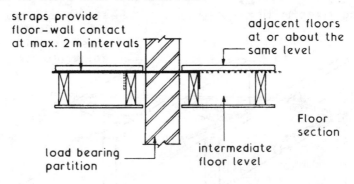

straps provide floor–wall contact at max. 2 m intervals

adjacent floors at or about the same level

Floor section

load bearing partition

intermediate floor level

Ref. BS EN 845-1: Specification for ancillary components for masonry. Ties, tension straps, hangers and brackets.

Wall Stability – at right angles to floor and ceiling joists this is achieved by building the joists into masonry support walls or locating them on approved joist hangers.

Walls parallel to joists are stabilised by lateral restraint straps. Buildings constructed before current stability requirements (see Bldg. Regs. A.D; A – Structure) often show signs of wall bulge due to the effects of eccentric loading and years of thermal movement.

Remedial Measures —

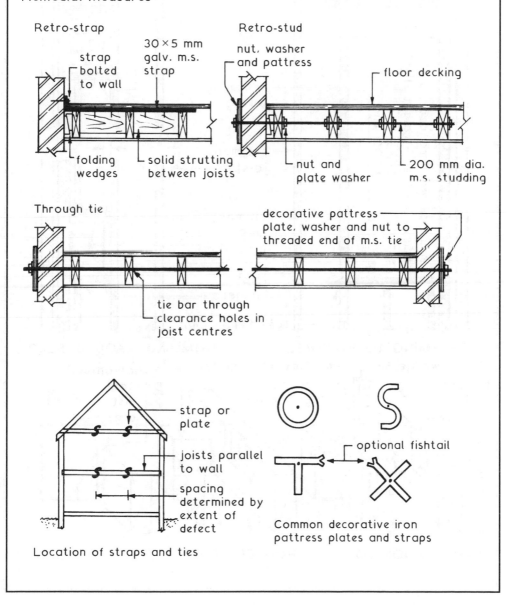

Retro-strap

strap bolted to wall
30×5 mm galv. m.s. strap
folding wedges
solid strutting between joists

Retro-stud

nut, washer and pattress
floor decking
nut and plate washer
200 mm dia. m.s. studding

Through tie

tie bar through clearance holes in joist centres

decorative pattress plate, washer and nut to threaded end of m.s. tie

strap or plate
joists parallel to wall
spacing determined by extent of defect

Location of straps and ties

optional fishtail

Common decorative iron pattress plates and straps

Trimming Members ~ these are the edge members of an opening in a floor and are the same depth as common joists but are usually 25 mm wider.

Typical Details ~

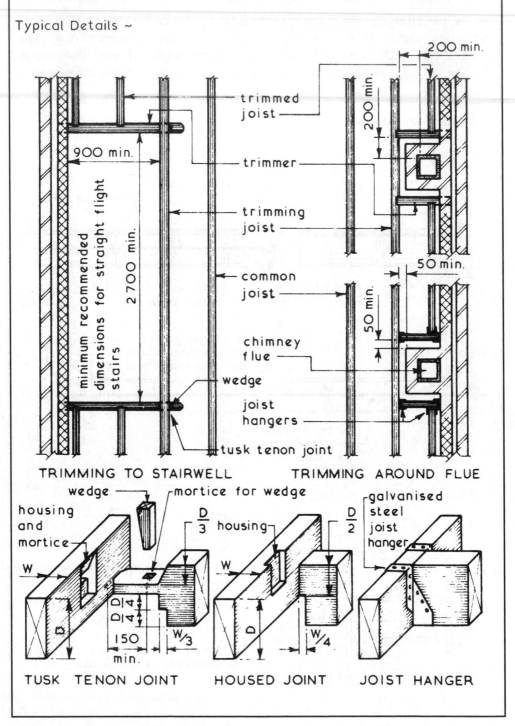

TRIMMING TO STAIRWELL   TRIMMING AROUND FLUE

TUSK TENON JOINT   HOUSED JOINT   JOIST HANGER

Typical spans and loading for floor joists of general structural grade –

Sawn size (mm × mm)	Dead weight of flooring and ceiling, excluding the self weight of the joists (kg/m²)								
	< 25			25–50			50–125		
	Spacing of joists (mm)								
	400	450	600	400	450	600	400	450	600
	Maximum clear span (m)								
38 × 75	1.22	1.09	0.83	1.14	1.03	0.79	0.98	0.89	0.70
38 × 100	1.91	1.78	1.38	1.80	1.64	1.28	1.49	1.36	1.09
38 × 125	2.54	2.45	2.01	2.43	2.30	1.83	2.01	1.85	1.50
38 × 150	3.05	2.93	2.56	2.91	2.76	2.40	2.50	2.35	1.93
38 × 175	3.55	3.40	2.96	3.37	3.19	2.77	2.89	2.73	2.36
38 × 200	4.04	3.85	3.35	3.82	3.61	3.13	3.27	3.09	2.68
38 × 225	4.53	4.29	3.73	4.25	4.02	3.50	3.65	3.44	2.99
50 × 75	1.45	1.37	1.08	1.39	1.30	1.01	1.22	1.11	0.88
50 × 100	2.18	2.06	1.76	2.06	1.95	1.62	1.82	1.67	1.35
50 × 125	2.79	2.68	2.44	2.67	2.56	2.28	2.40	2.24	1.84
50 × 150	3.33	3.21	2.92	3.19	3.07	2.75	2.86	2.70	2.33
50 × 175	3.88	3.73	3.38	3.71	3.57	3.17	3.30	3.12	2.71
50 × 200	4.42	4.25	3.82	4.23	4.07	3.58	3.74	3.53	3.07
50 × 225	4.88	4.74	4.26	4.72	4.57	3.99	4.16	3.94	3.42
63 × 100	2.41	2.29	2.01	2.28	2.17	1.90	2.01	1.91	1.60
63 × 125	3.00	2.89	2.63	2.88	2.77	2.52	2.59	2.49	2.16
63 × 150	3.59	3.46	3.15	3.44	3.31	3.01	3.10	2.98	2.63
63 × 175	4.17	4.02	3.66	4.00	3.85	3.51	3.61	3.47	3.03
63 × 200	4.73	4.58	4.18	4.56	4.39	4.00	4.11	3.95	3.43
63 × 225	5.15	5.01	4.68	4.99	4.85	4.46	4.62	4.40	3.83
75 × 125	3.18	3.06	2.79	3.04	2.93	2.67	2.74	2.64	2.40
75 × 150	3.79	3.66	3.33	3.64	3.50	3.19	3.28	3.16	2.86
75 × 175	4.41	4.25	3.88	4.23	4.07	3.71	3.82	3.68	3.30
75 × 200	4.92	4.79	4.42	4.77	4.64	4.23	4.35	4.19	3.74
75 × 225	5.36	5.22	4.88	5.20	5.06	4.72	4.82	4.69	4.16

Notes:
1. Where a bath is supported, the joists should be duplicated.
2. See pages 35 and 36 for material dead weights.

Joist and Beam Sizing ~ design tables and formulae have limitations, therefore where loading, span and/or conventional joist spacings are exceeded, calculations are required. BS 5268: Structural Use Of Timber and BS EN 338: Structural Timber – Strength Classes, are both useful resource material for detailed information on a variety of timber species. The following example serves to provide guidance on the design process for determining joist size, measurement of deflection, safe bearing and resistance to shear force:-

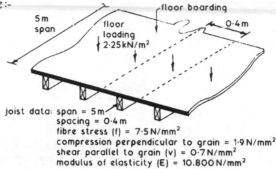

joist data: span = 5 m
spacing = 0.4 m
fibre stress (f) = 7.5 N/mm²
compression perpendicular to grain = 1.9 N/mm²
shear parallel to grain (v) = 0.7 N/mm²
modulus of elasticity (E) = 10.800 N/mm²

Total load (W) per joist = 5 m × 0.4 m × 2.25 kN/m² = 4.5 kN

$$\text{or}: \frac{4.5\ \text{kN}}{5\ \text{m span}} = 0.9\ \text{kN/m}$$

Resistance to bending ~

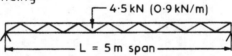

Bending moment formulae are shown on page 506

$$BM = \frac{WL}{8} = \frac{fbd^2}{6}$$

Where: W = total load, 4.5 kN (4500 N)
L = span, 5 m (5000 mm)
f = fibre stress of timber, 7.5 N/mm²
b = breadth of joist, try 50 mm
d = depth of joist, unknown

Transposing:

$$\frac{WL}{8} = \frac{fbd^2}{6}$$

Becomes:

$$d = \sqrt{\frac{6WL}{8fb}} = \sqrt{\frac{6 \times 4500 \times 5000}{8 \times 7.5 \times 50}} = 212\ \text{mm}$$

Nearest commercial size: 50 mm × 225 mm

Joist and Beam Sizing ~ calculating overall dimensions alone is insufficient, checks should also be made to satisfy: resistance to deflection, adequate safe bearing and resistance to shear.

Deflection - should be minimal to prevent damage to plastered ceilings. An allowance of up to 0·003 × span is normally acceptable; for the preceding example this will be:-

0·003 × 5000 mm = 15 mm

The formula for calculating deflection due to a uniformly distributed load (see page 508) is: ~

$$\frac{5WL^3}{384EI} \quad \text{where} \quad I = \frac{bd^3}{12}$$

$$I = \frac{50 \times (225)^3}{12} = 4·75 \times (10)^7$$

So, deflection $= \dfrac{5 \times 4500 \times (5000)^3}{384 \times 10800 \times 4.75 \times (10)^7} = 14·27 \text{ mm}$

NB. This is only just within the calculated allowance of 15 mm, therefore it would be prudent to specify slightly wider or deeper joists to allow for unknown future use.

Safe Bearing ~

$$= \frac{\text{load at the joist end, W/2}}{\text{compression perpendicular to grain} \times \text{breadth}}$$

$$= \frac{4500/2}{1·9 \times 50} = 24 \text{ mm}.$$

therefore full support from masonry (90 mm min.) or joist hangers will be more than adequate.

Shear Strength ~

$$V = \frac{2bdv}{3}$$

where: V = vertical loading at the joist end, W/2
    v = shear strength parallel to the grain, 0.7 N/mm^2
Transposing:-

$$bd = \frac{3V}{2v} = \frac{3 \times 2250}{2 \times 0.7} = 4821 \text{ mm}^2 \text{ minimum}$$

Actual bd = 50 mm × 225 mm = 11,250 mm^2

Resistance to shear is satisfied as actual is well above the minimum.

For fire protection, floors are categorised depending on their height relative to adjacent ground ~

Height of top floor above ground	Fire resistance
Less than 5 m	30 minutes (60 min. for compartment floors)
More than 5 m	60 minutes (30 min. for a three storey dwelling)

Tests for fire resistance relate to load bearing capacity, integrity and insulation as determined by BS 476: Fire tests on building materials and structures. Methods of test for fire propagation for products.

Integrity ~ the ability of an element to resist fire penetration and capacity to bear load in a fire.

Insulation ~ ability to resist heat penetration so that fire is not spread by radiation and conduction.

Typical applications ~

30 MINUTE FIRE RESISTANCE

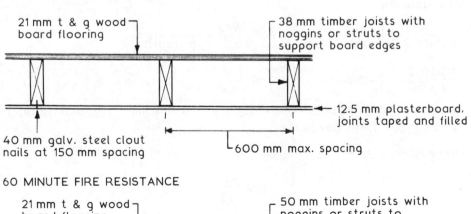

21 mm t & g wood board flooring

38 mm timber joists with noggins or struts to support board edges

12.5 mm plasterboard, joints taped and filled

40 mm galv. steel clout nails at 150 mm spacing

600 mm max. spacing

60 MINUTE FIRE RESISTANCE

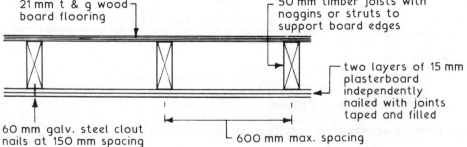

21 mm t & g wood board flooring

50 mm timber joists with noggins or struts to support board edges

two layers of 15 mm plasterboard independently nailed with joints taped and filled

60 mm galv. steel clout nails at 150 mm spacing

600 mm max. spacing

Ref. Building Regulations, AD B Fire safety, Volume 1 – Dwellinghouses.

Reinforced Concrete Suspended Floors ~ a simple reinforced concrete flat slab cast to act as a suspended floor is not usually economical for spans over 5·000. To overcome this problem beams can be incorporated into the design to span in one or two directions. Such beams usually span between columns which transfers their loads to the foundations. The disadvantages of introducing beams are the greater overall depth of the floor construction and the increased complexity of the formwork and reinforcement. To reduce the overall depth of the floor construction flat slabs can be used where the beam is incorporated with the depth of the slab. This method usually results in a deeper slab with complex reinforcement especially at the column positions.

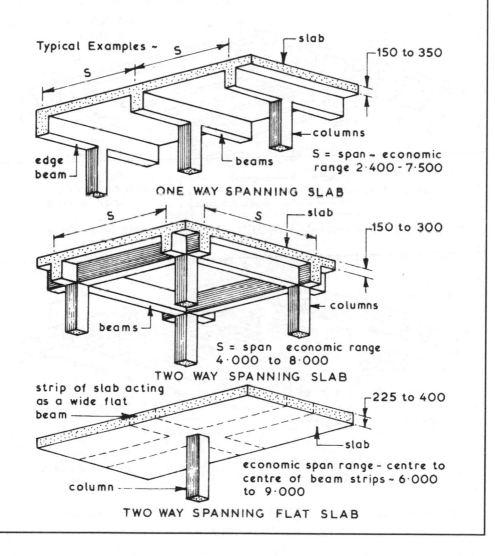

Typical Examples ~

slab
150 to 350

S = span ~ economic range 2·400 - 7·500

edge beam

beams

columns

ONE WAY SPANNING SLAB

slab
150 to 300

S = span economic range 4·000 to 8·000

beams

columns

TWO WAY SPANNING SLAB

strip of slab acting as a wide flat beam

225 to 400

slab

economic span range - centre to centre of beam strips ~ 6·000 to 9·000

column

TWO WAY SPANNING FLAT SLAB

Ribbed Floors ~ to reduce the overall depth of a traditional cast in-situ reinforced concrete beam and slab suspended floor a ribbed floor could be used. The basic concept is to replace the wide spaced deep beams with narrow spaced shallow beams or ribs which will carry only a small amount of slab loading. These floors can be designed as one or two way spanning floors. One way spanning ribbed floors are sometimes called troughed floors whereas the two way spanning ribbed floors are called coffered or waffle floors. Ribbed floors are usually cast against metal, glass fibre or polypropylene preformed moulds which are temporarily supported on plywood decking, joists and props – see page 462.

Typical Examples ~

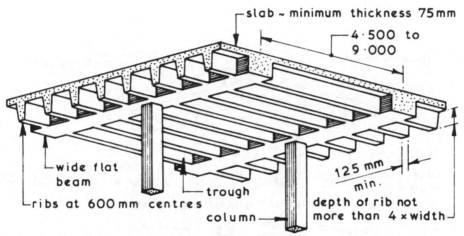

ONE WAY SPANNING RIBBED OR TROUGHED FLOOR

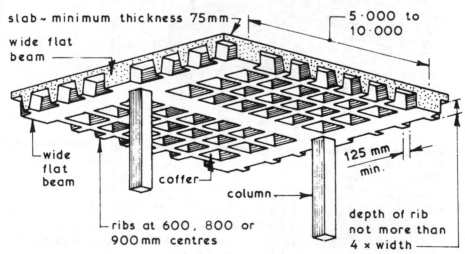

TWO WAY SPANNING COFFERED OR WAFFLE FLOOR

616

Ribbed Floors – these have greater span and load potential per unit weight than flat slab construction. This benefits a considerable reduction in dead load, to provide cost economies in other super-structural elements and foundations. The regular pattern of voids created with waffle moulds produces a honeycombed effect, which may be left exposed in utility buildings such as car parks. Elsewhere such as shopping malls, a suspended ceiling would be appropriate. The trough finish is also suitable in various situations and has the advantage of creating a continuous void for accommodation of service cables and pipes. A suspended ceiling can add to this space where air conditioning ducting is required, also providing several options for finishing effect.

Typical mould profile –

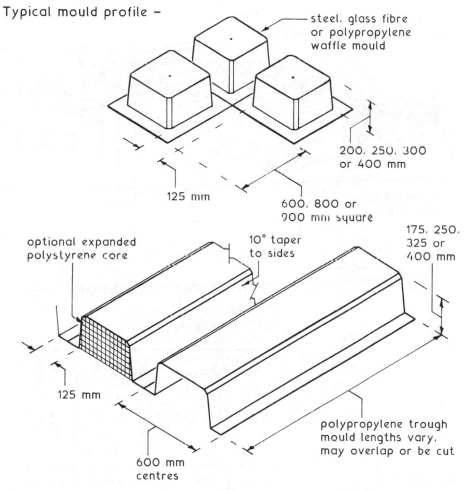

steel, glass fibre or polypropylene waffle mould

200, 250, 300 or 400 mm

125 mm

600, 800 or 900 mm square

optional expanded polystyrene core

10° taper to sides

175, 250, 325 or 400 mm

125 mm

600 mm centres

polypropylene trough mould lengths vary, may overlap or be cut

Note: After removing the temporary support structure, moulds are struck by flexing with a flat tool. A compressed air line is also effective.

Hollow Pot Floors ~ these are in essence a ribbed floor with permanent formwork in the form of hollow clay or concrete pots. The main advantage of this type of cast in-situ floor is that it has a flat soffit which is suitable for the direct application of a plaster finish or an attached dry lining. The voids in the pots can be utilised to house small diameter services within the overall depth of the slab. These floors can be designed as one or two way spanning slabs, the common format being the one way spanning floor.

Typical Example ~

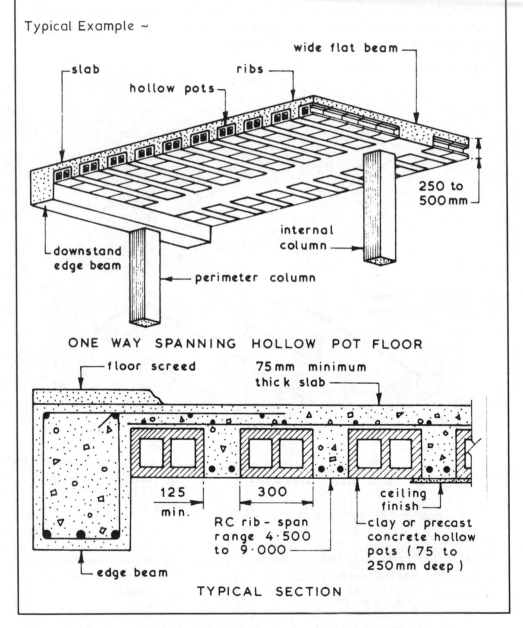

ONE WAY SPANNING HOLLOW POT FLOOR

TYPICAL SECTION

Soffit and Beam Fixings ~ concrete suspended floors can be designed to carry loads other than the direct upper surface loadings. Services can be housed within the voids created by the beams or ribs and suspended or attached ceilings can be supported by the floor. Services which run at right angles to the beams or ribs are usually housed in cast-in holes. There are many types of fixings available for use in conjunction with floor slabs, some are designed to be cast-in whilst others are fitted after the concrete has cured. All fixings must be positioned and installed so that they are not detrimental to the structural integrity of the floor.

Typical Examples ~

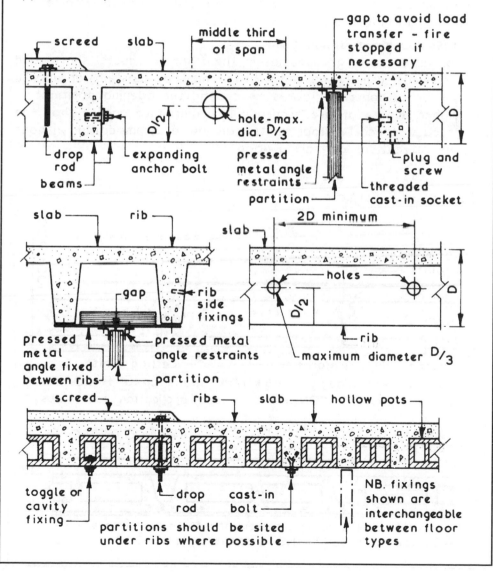

Precast Concrete Floors ~ these are available in several basic formats and provide an alternative form of floor construction to suspended timber floors and in-situ reinforced concrete suspended floors. The main advantages of precast concrete floors are:-

1. Elimination of the need for formwork except for nominal propping which is required with some systems.
2. Curing time of concrete is eliminated therefore the floor is available for use as a working platform at an earlier stage.
3. Superior quality control of product is possible with factory produced components.

The main disadvantages of precast concrete floors when compared with in-situ reinforced concrete floors are:-

1. Less flexible in design terms.
2. Formation of large openings in the floor for ducts, shafts and stairwells usually have to be formed by casting an in-situ reinforced concrete floor strip around the opening position.
3. Higher degree of site accuracy is required to ensure that the precast concrete floor units can be accommodated without any alterations or making good

Typical Basic Formats ~

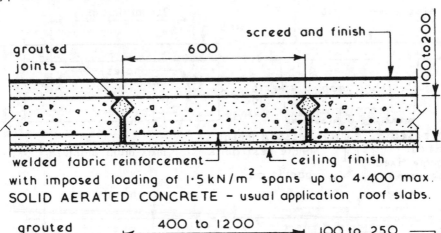

with imposed loading of $1 \cdot 5 \, kN/m^2$ spans up to 4·400 max.
SOLID AERATED CONCRETE – usual application roof slabs.

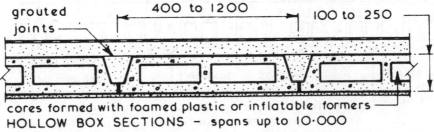

cores formed with foamed plastic or inflatable formers
HOLLOW BOX SECTIONS – spans up to 10·000

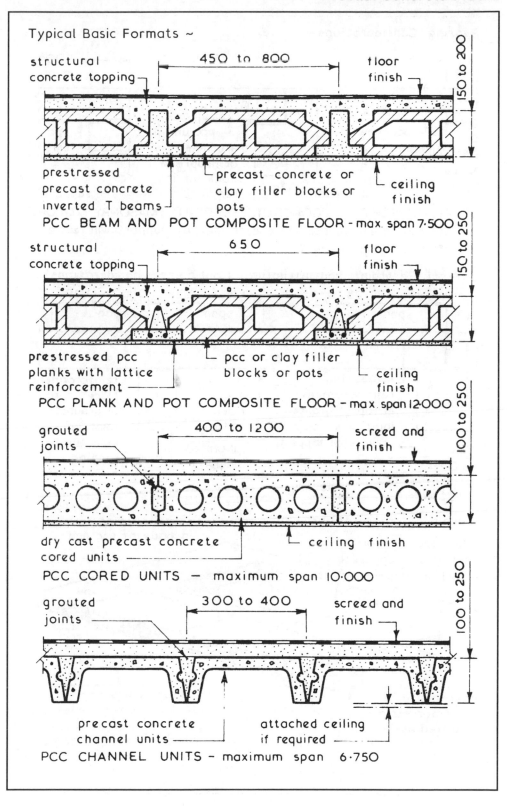

Typical Basic Formats ~

structural concrete topping

450 to 800

floor finish

150 to 200

prestressed precast concrete inverted T beams

precast concrete or clay filler blocks or pots

ceiling finish

PCC BEAM AND POT COMPOSITE FLOOR - max. span 7·500

structural concrete topping

650

floor finish

150 to 250

prestressed pcc planks with lattice reinforcement

pcc or clay filler blocks or pots

ceiling finish

PCC PLANK AND POT COMPOSITE FLOOR - max. span 12·000

grouted joints

400 to 1200

screed and finish

100 to 250

dry cast precast concrete cored units

ceiling finish

PCC CORED UNITS — maximum span 10·000

grouted joints

300 to 400

screed and finish

100 to 250

precast concrete channel units

attached ceiling if required

PCC CHANNEL UNITS - maximum span 6·750

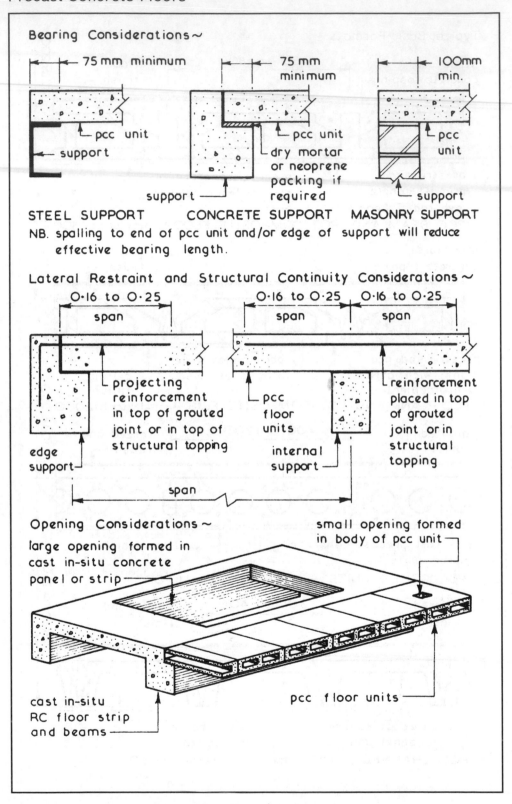

Bearing Considerations~

75 mm minimum

pcc unit

support

STEEL SUPPORT

75 mm minimum

pcc unit

dry mortar or neoprene packing if required

support

CONCRETE SUPPORT

100mm min.

pcc unit

support

MASONRY SUPPORT

NB. spalling to end of pcc unit and/or edge of support will reduce effective bearing length.

Lateral Restraint and Structural Continuity Considerations~

0·16 to 0·25 span

projecting reinforcement in top of grouted joint or in top of structural topping

edge support

0·16 to 0·25 span

pcc floor units

internal support

0·16 to 0·25 span

reinforcement placed in top of grouted joint or in structural topping

span

Opening Considerations~

large opening formed in cast in-situ concrete panel or strip

small opening formed in body of pcc unit

cast in-situ RC floor strip and beams

pcc floor units

Raised Flooring ~ developed in response to the high-tech boom of the 1970s. It has proved expedient in accommodating computer and communications cabling as well as numerous other established services. The system is a combination of adjustable floor pedestals, supporting a variety of decking materials. Pedestal height ranges from as little as 30 mm up to about 600 mm, although greater heights are possible at the expense of structural floor levels. Decking is usually in loose fit squares of 600 mm, but may be sheet plywood or particleboard screwed direct to closer spaced pedestal support plates on to joists bearing on pedestals.

Cavity fire stops are required between decking and structural floor at appropriate intervals (see Building Regulations, A D B, Volume 2, Section 9).

Application ~

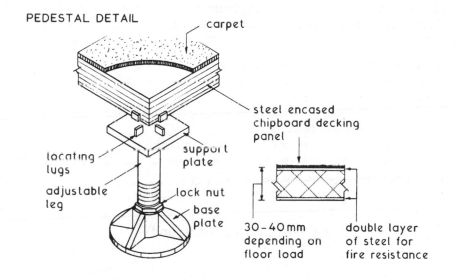

PEDESTAL DETAIL

carpet

steel encased chipboard decking panel

locating lugs

support plate

adjustable leg

lock nut

base plate

30–40 mm depending on floor load

double layer of steel for fire resistance

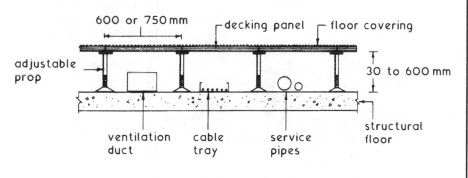

FLOOR SECTION

600 or 750 mm

decking panel

floor covering

adjustable prop

30 to 600 mm

ventilation duct

cable tray

service pipes

structural floor

Sound Insulation ~ sound can be defined as vibrations of air which are registered by the human ear. All sounds are produced by a vibrating object which causes tiny particles of air around it to move in unison. These displaced air particles collide with adjacent air particles setting them in motion and in unison with the vibrating object. This continuous chain reaction creates a sound wave which travels through the air until at some distance the air particle movement is so small that it is inaudible to the human ear. Sounds are defined as either impact or airborne sound, the definition being determined by the source producing the sound. Impact sounds are created when the fabric of structure is vibrated by direct contact whereas airborne sound only sets the structural fabric vibrating in unison when the emitted sound wave reaches the enclosing structural fabric. The vibrations set up by the structural fabric can therefore transmit the sound to adjacent rooms which can cause annoyance, disturbance of sleep and of the ability to hold a normal conservation. The objective of sound insulation is to reduce transmitted sound to an acceptable level, the intensity of which is measured in units of decibels (dB).

The Building Regulations, Approved Document E: Resistance to the passage of sound, establishes sound insulation standards as follows

E1: Between dwellings and between dwellings and other buildings.
E2: Within a dwelling, ie. between rooms, particularly WC and habitable rooms, and bedrooms and other rooms.
E3: Control of reverberation noise in common parts (stairwells and corridors) of buildings containing dwellings, ie. flats.
E4: Specific applications to acoustic conditions in schools.
Note: E1 includes, hotels, hostels, student accommodation, nurses' homes and homes for the elderly, but not hospitals and prisons.

Typical Sources and Transmission of Sound ~

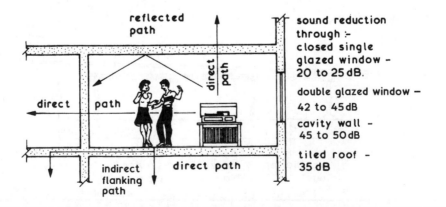

reflected path

direct path

direct   path

direct path

indirect flanking path

sound reduction through :-
closed single glazed window - 20 to 25 dB.

double glazed window - 42 to 45 dB

cavity wall - 45 to 50 dB

tiled roof - 35 dB

Separating Walls ~ types:-

1. Solid masonry
2. Cavity masonry
3. Masonry between isolating panels
4. Timber frame

Type 1 — relies on mass

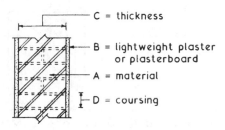

C = thickness

B = lightweight plaster or plasterboard

A = material

D = coursing

Material A	Density of A [Kg/m³]	Finish B	Combined mass A + B (Kg/m²)	Thickness C [mm]	Coursing D [mm]
brickwork	1610	13 mm lwt. pl.	375	215	75
... ..	... ..	12·5 mm pl. brd.	... ..	... ..	... ..
Concrete block	1840	13 mm lwt. pl	415	... ..	110
... ..	1840	12·5 mm pl. brd	... ..	... ..	150
In-situ concrete	2200	Optional	415	190	n/a

Type 2 — relies on mass and isolation

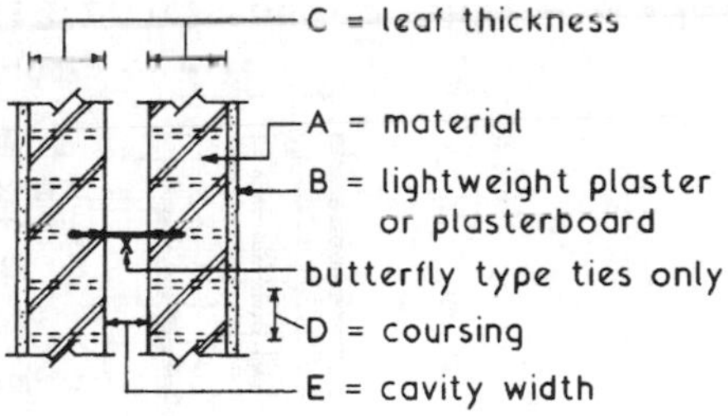

C = leaf thickness

A = material

B = lightweight plaster or plasterboard

butterfly type ties only

D = coursing

E = cavity width

Material A	Density of A [Kg/m³]	Finish B	Mass A + B (Kg/m²)	Thickness C [mm]	Coursing D [mm]	Cavity E [mm]
bkwk.	1970	13 mm lwt. pl.	415	102	75	50
concrete block	1990	..	..	100	225	..
lwt. conc. block	1375	.. or 12.5 mm pl. brd.	300	100	225	75

Type 3 ~ relies on: (a) core material type and mass,
(b) isolation, and
(c) mass of isolated panels.

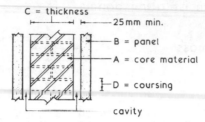

Core material A	Density of A $[kg/m^3]$	Mass A (kg/m^2)	Thickness C (mm)	Coursing D (mm)	Cavity (mm)
brickwork	1290	300	215	75	n/a
concrete block	2200	300	140	110	n/a
lwt. conc. block	1400	150	200	225	n/a
Cavity bkwk. or block	any	any	2 × 100	to suit	50

Panel materials – B

(i) Plasterboard with cellular core plus plaster finish, mass 18 kg/m². All joints taped. Fixed floor and ceiling only.

(ii) 2 No. plasterboard sheets, 12·5 mm each, with joints staggered. Frame support or 30 mm overall thickness.

Type 4 — relies on mass, frame separation and absorption of sound.

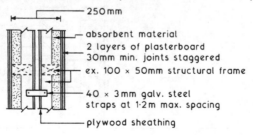

Absorbent material – quilting of unfaced mineral fibre batts with a minimum density of 10 kg/m³, located in the cavity or frames.

Thickness (mm)	Location
25	Suspended in cavity
50	Fixed within one frame
2 × 25	Each quilt fixed within each frame

Separating Floors ~ types:-

1. Concrete with soft covering
2. Concrete with floating layer
3. Timber with floating layer

Type 1. Airborne resistance depends on mass of concrete and ceiling.
Impact resistance depends on softness of covering.

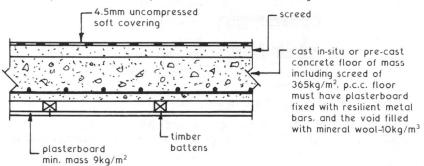

4.5mm uncompressed soft covering

screed

cast in-situ or pre-cast concrete floor of mass including screed of 365kg/m², p.c.c. floor must have plasterboard fixed with resilient metal bars, and the void filled with mineral wool-10kg/m³

plasterboard min. mass 9kg/m²

timber battens

Type 2. Airborne resistance depends mainly on concrete mass and partly on mass of floating layer and ceiling.
Impact resistance depends on resilient layer isolating floating layer from base and isolation of ceiling.

Bases: As type 1. but overall mass minimum 300 kg/m².

Floating layers:

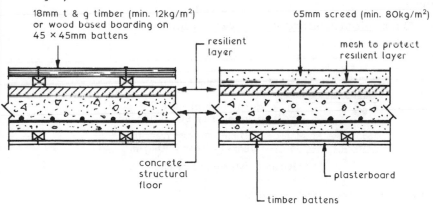

18mm t & g timber (min. 12kg/m²) or wood based boarding on 45 × 45mm battens

resilient layer

65mm screed (min. 80kg/m²)

mesh to protect resilient layer

concrete structural floor

plasterboard

timber battens

Resilient layers:

(a) 25 mm paper faced mineral fibre, density 36 kg/m³.
    Timber floor – paper faced underside.
    Screeded floor – paper faced upper side to prevent screed entering layer.

(b) Screeded floor only:
    13 mm pre-compressed expanded polystyrene (EPS) board, or 5 mm extruded polyethylene foam of density 30–45 kg/m³, laid over a levelling screed for protection.

See BS EN 29052-1: Acoustics. Method for the determination of dynamic stiffness. Materials used under floating floors in dwellings.

Type 3. Airborne resistance varies depending on floor construction, absorbency of materials, extent of pugging and partly on the floating layer. Impact resistance depends mainly on the resilient layer separating floating from structure.

Platform floor ~

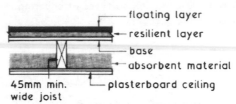

floating layer
resilient layer
base
absorbent material
45mm min. wide joist
plasterboard ceiling

Note: Minimum mass per unit area = 25 kg/m²

Floating layer: 18 mm timber or wood based board, t & g joints glued and spot bonded to a sub-strate of 19 mm plasterboard.

Alternatively, cement bonded particle board in 2 thicknesses – 24 mm total, joints staggered, glued and screwed together.

Resilient layer: 25 mm mineral fibre, density 60–100 kg/m³.

Base: 12 mm timber boarding or wood based board nailed to joists.

Absorbent material: 100 mm unfaced rock fibre, minimum density 10 kg/m³.

Ceiling: 30 mm plasterboard in 2 layers, joints staggered.

Ribbed floor ~

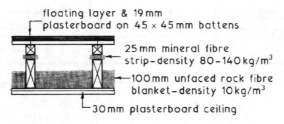

floating layer & 19 mm plasterboard on 45 × 45mm battens
25mm mineral fibre strip-density 80–140 kg/m³
100 mm unfaced rock fibre blanket-density 10 kg/m³
30 mm plasterboard ceiling

Ribbed floor with dry sand pugging ~

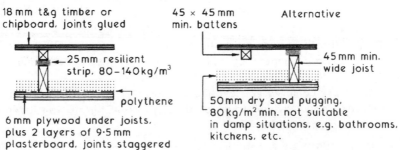

18 mm t&g timber or chipboard, joints glued
45 × 45 mm min. battens
Alternative
25mm resilient strip, 80–140 kg/m³
45mm min. wide joist
polythene
6 mm plywood under joists, plus 2 layers of 9·5 mm plasterboard, joints staggered
50mm dry sand pugging, 80 kg/m² min. not suitable in damp situations, e.g. bathrooms, kitchens, etc.

Primary Functions ~

1 Provide a means of circulation between floor levels.
2. Establish a safe means of travel between floor levels.
3. Provide an easy means of travel between floor levels.
4. Provide a means of conveying fittings and furniture between floor levels.

Constituent Parts ~

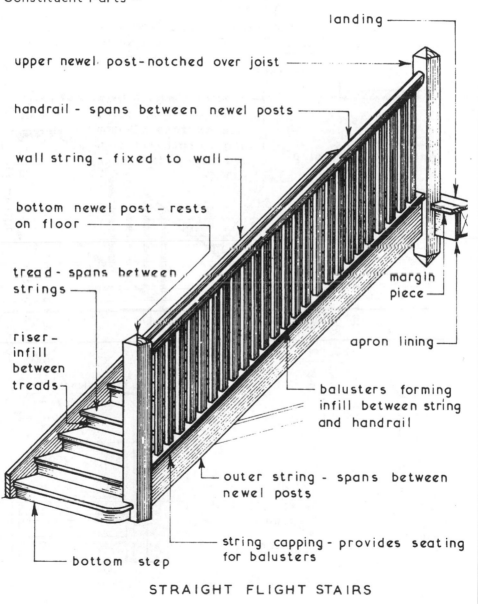

landing

upper newel post-notched over joist

handrail - spans between newel posts

wall string - fixed to wall

bottom newel post - rests on floor

tread - spans between strings

riser - infill between treads

margin piece

apron lining

balusters forming infill between string and handrail

outer string - spans between newel posts

string capping - provides seating for balusters

bottom step

STRAIGHT FLIGHT STAIRS

All dimensions quoted are the minimum required for domestic stairs exclusive to one dwelling as given in Approved Document K unless stated otherwise.

Terminology ~

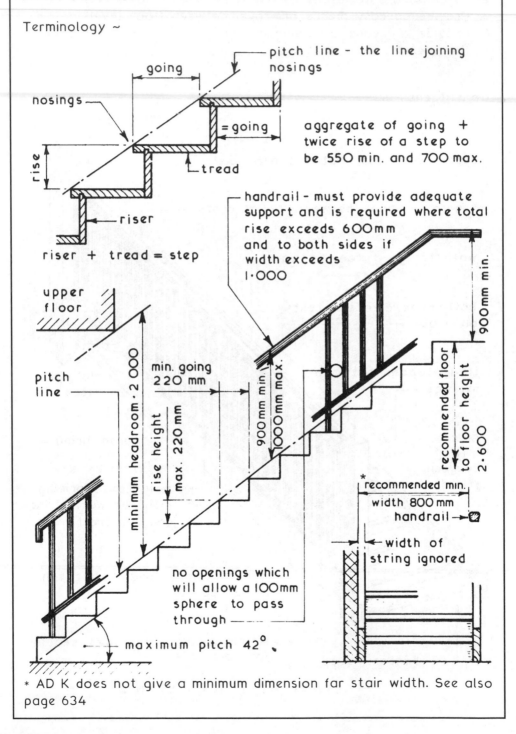

pitch line – the line joining nosings

nosings

going

= going

tread

rise

riser

riser + tread = step

aggregate of going + twice rise of a step to be 550 min. and 700 max.

handrail – must provide adequate support and is required where total rise exceeds 600mm and to both sides if width exceeds 1·000

upper floor

pitch line

minimum headroom · 2 000

min. going 220 mm

rise height max. 220 mm

900mm min.

1000mm max.

recommended floor to floor height 2·600

900mm min.

*recommended min. width 800 mm
handrail

width of string ignored

no openings which will allow a 100mm sphere to pass through

maximum pitch 42°.

* AD K does not give a minimum dimension far stair width. See also page 634

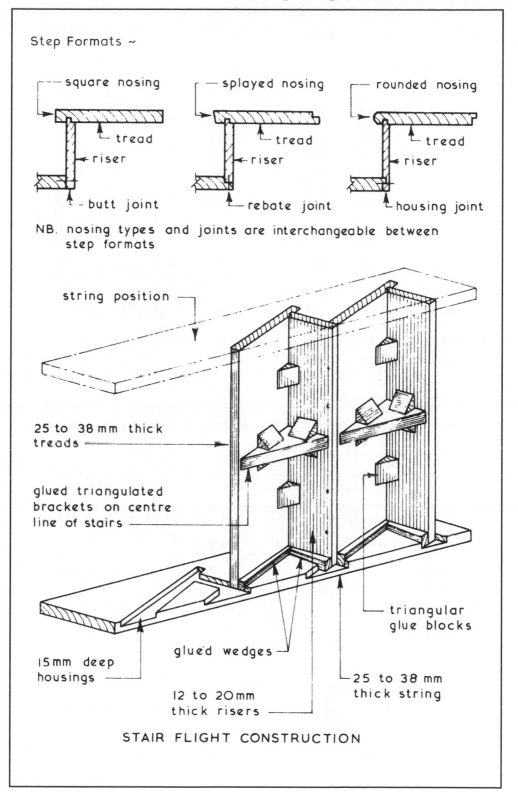

Step Formats ~

square nosing — tread — riser — butt joint

splayed nosing — tread — riser — rebate joint

rounded nosing — tread — riser — housing joint

NB. nosing types and joints are interchangeable between step formats

string position

25 to 38 mm thick treads

glued triangulated brackets on centre line of stairs

triangular glue blocks

15mm deep housings

glued wedges

12 to 20mm thick risers

25 to 38 mm thick string

STAIR FLIGHT CONSTRUCTION

631

Bottom Step Arrangements ~

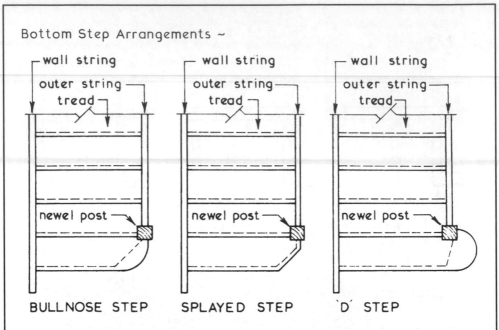

BULLNOSE STEP   SPLAYED STEP   `D´ STEP

Projecting bottom steps are usually included to enable the outer string to be securely jointed to the back face of the newel post and to provide an easy line of travel when ascending or descending at the foot of the stairs.

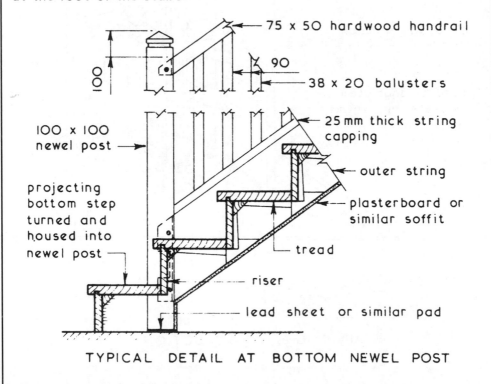

TYPICAL DETAIL AT BOTTOM NEWEL POST

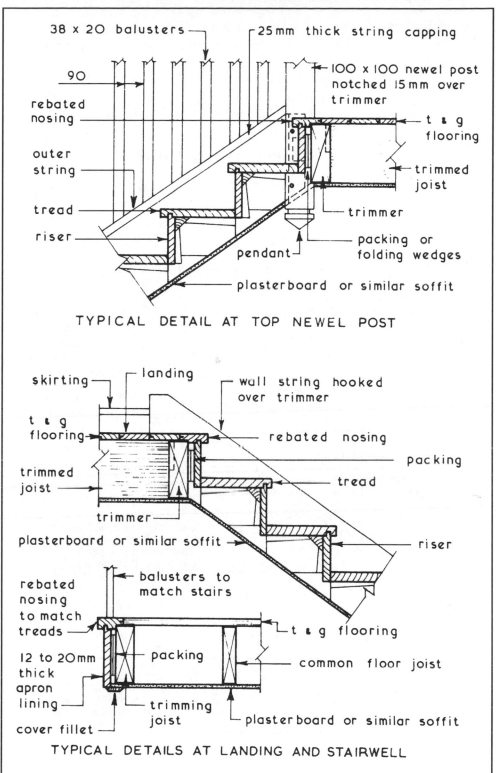

38 x 20 balusters

25mm thick string capping

100 x 100 newel post notched 15mm over trimmer

90

rebated nosing

t & g flooring

outer string

trimmed joist

tread

trimmer

riser

packing or folding wedges

pendant

plasterboard or similar soffit

TYPICAL DETAIL AT TOP NEWEL POST

skirting

landing

wall string hooked over trimmer

t & g flooring

rebated nosing

packing

trimmed joist

tread

trimmer

riser

plasterboard or similar soffit

rebated nosing to match treads

balusters to match stairs

t & g flooring

12 to 20mm thick apron lining

packing

common floor joist

cover fillet

trimming joist

plasterboard or similar soffit

TYPICAL DETAILS AT LANDING AND STAIRWELL

633

Open Riser Timber Stairs ~ these are timber stairs constructed to the same basic principles as standard timber stairs excluding the use of a riser. They have no real advantage over traditional stairs except for the generally accepted aesthetic appeal of elegance. Like the traditional timber stairs they must comply with the minimum requirements set out in Part K of the Building Regulations.

Typical Requirements for Stairs in a Small Residential Building ~

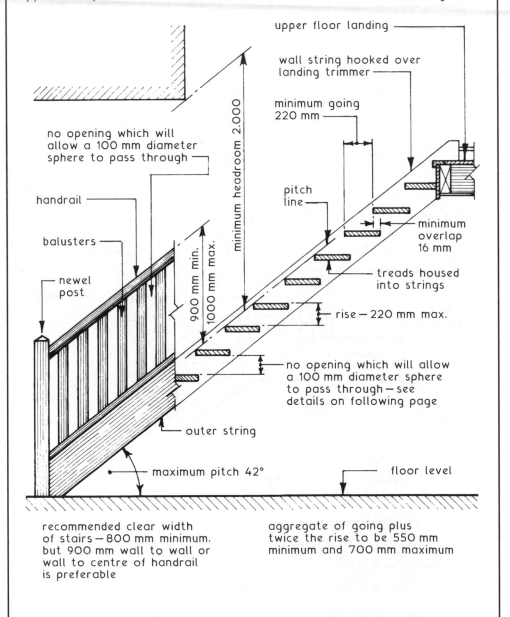

upper floor landing

wall string hooked over landing trimmer

minimum going 220 mm

no opening which will allow a 100 mm diameter sphere to pass through

minimum headroom 2.000

pitch line

minimum overlap 16 mm

handrail

balusters

900 mm min.

1000 mm max.

treads housed into strings

newel post

rise — 220 mm max.

no opening which will allow a 100 mm diameter sphere to pass through — see details on following page

outer string

maximum pitch 42°

floor level

recommended clear width of stairs — 800 mm minimum, but 900 mm wall to wall or wall to centre of handrail is preferable

aggregate of going plus twice the rise to be 550 mm minimum and 700 mm maximum

Design and Construction ~ because of the legal requirement of not having a gap between any two consecutive treads through which a 100 mm diameter sphere can pass and the limitation relating to the going and rise, as shown on the previous page, it is generally not practicable to have a completely riserless stair for residential buildings since by using minimum dimensions a very low pitch of approximately 27½° would result and by choosing an acceptable pitch a very thick tread would have to be used to restrict the gap to 100 mm.

Possible Solutions ~

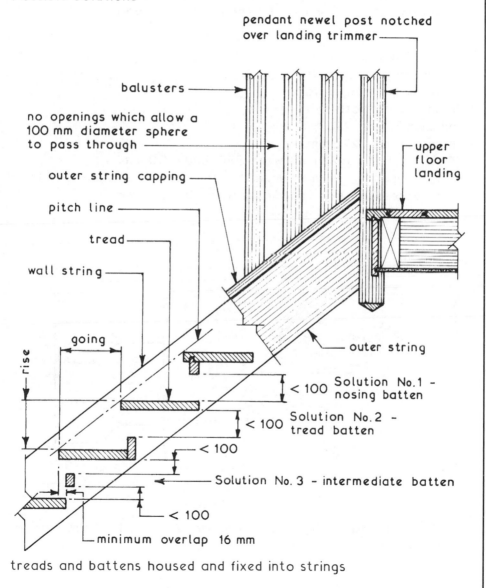

pendant newel post notched over landing trimmer

balusters

no openings which allow a 100 mm diameter sphere to pass through

outer string capping

pitch line

tread

wall string

upper floor landing

outer string

going

rise

< 100   Solution No.1 - nosing batten

Solution No.2 - tread batten   < 100

< 100

Solution No.3 - intermediate batten

< 100

minimum overlap 16 mm

treads and battens housed and fixed into strings

635

Application – a straight flight for access to a domestic loft conversion only. This can provide one habitable room, plus a bathroom or WC. The WC must not be the only WC in the dwelling.

Practical issues – an economic use of space, achieved by a very steep pitch of about 60° and opposing overlapping treads.

Safety – pitch and tread profile differ considerably from other stairs, but they are acceptable to Building Regulations by virtue of "familiarity and regular use" by the building occupants.

Additional features are:

* a non-slip tread surface.
* handrails to both sides.
* minimum going 220 mm.
* maximum rise 220 mm.
* (2 + rise) + (going) between 550 and 700 mm.
* a stair used by children under 5 years old, must have the tread voids barred to leave a gap not greater than 100 mm.

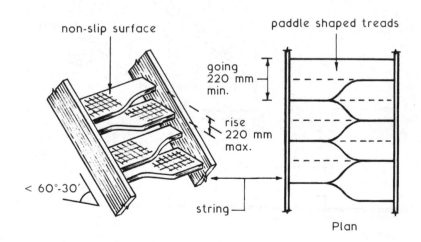

non-slip surface

paddle shaped treads

going 220 mm min.

rise 220 mm max.

< 60°-30'

string

Plan

Ref. Building Regulations, Approved Document K1: Stairs, ladders and ramps: Section 1.29

Timber Stairs ~ these must comply with the minimum requirements set out in Part K of the Building Regulations. Straight flight stairs are simple, easy to construct and install but by the introduction of intermediate landings stairs can be designed to change direction of travel and be more compact in plan than the straight flight stairs.

Landings ~ these are designed and constructed in the same manner as timber upper floors but due to the shorter spans they require smaller joist sections. Landings can be detailed for a 90° change of direction (quarter space landing) or a 180° change of direction (half space landing) and can be introduced at any position between the two floors being served by the stairs.

Typical Layouts ~

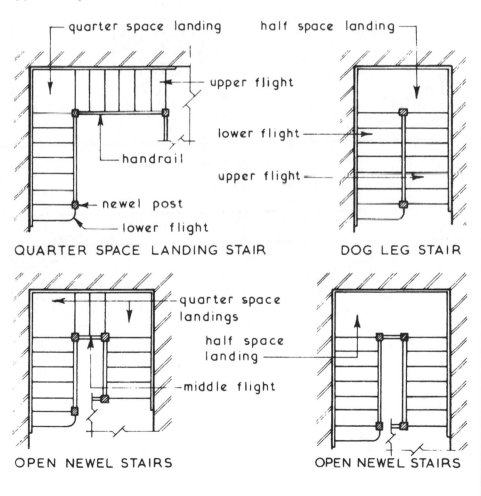

QUARTER SPACE LANDING STAIR   DOG LEG STAIR

OPEN NEWEL STAIRS   OPEN NEWEL STAIRS

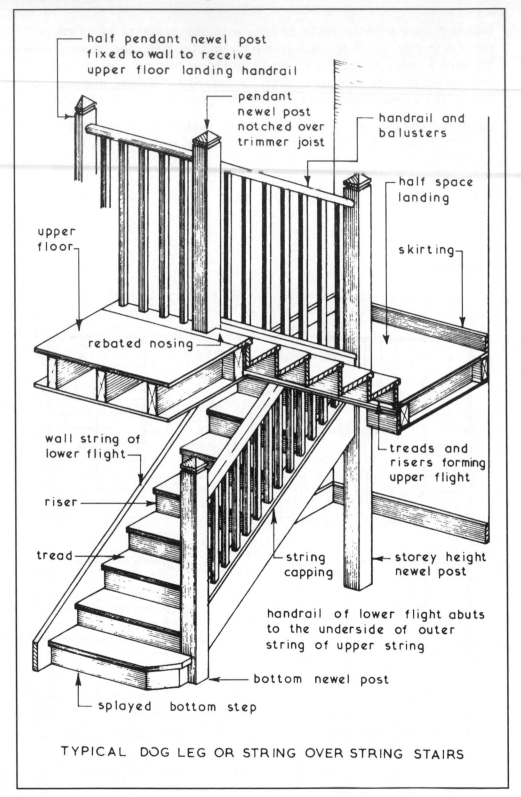

half pendant newel post
fixed to wall to receive
upper floor landing handrail

pendant
newel post
notched over
trimmer joist

handrail and
balusters

half space
landing

upper
floor

skirting

rebated nosing

treads and
risers forming
upper flight

wall string of
lower flight

riser

tread

string
capping

storey height
newel post

handrail of lower flight abuts
to the underside of outer
string of upper string

bottom newel post

splayed bottom step

TYPICAL DOG LEG OR STRING OVER STRING STAIRS

In-situ Reinforced Concrete Stairs ~ a variety of stair types and arrangements are possible each having its own appearance and design characteristics. In all cases these stairs must comply with the minimum requirements set out in Part K of the Building Regulations in accordance with the purpose group of the building in which the stairs are situated.

Typical Examples ~

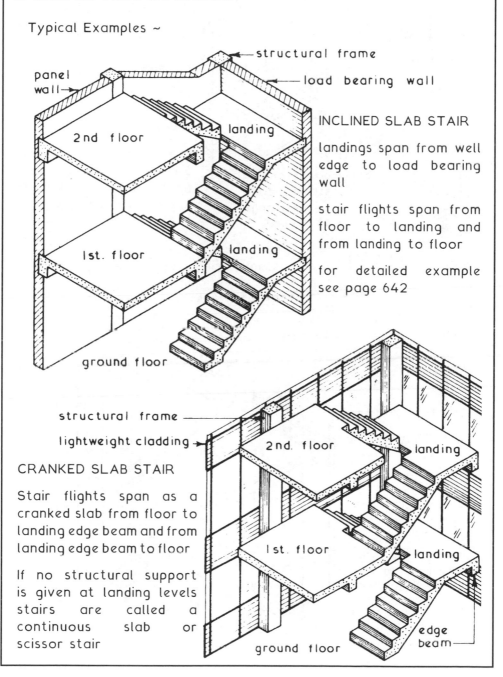

structural frame

panel wall→

load bearing wall

INCLINED SLAB STAIR

2nd floor

landing

landings span from well edge to load bearing wall

stair flights span from floor to landing and from landing to floor

1st. floor

landing

for detailed example see page 642

ground floor

structural frame

lightweight cladding→

2nd. floor

landing

CRANKED SLAB STAIR

Stair flights span as a cranked slab from floor to landing edge beam and from landing edge beam to floor

1st. floor

landing

If no structural support is given at landing levels stairs are called a continuous slab or scissor stair

ground floor

edge beam

Typical Examples ~

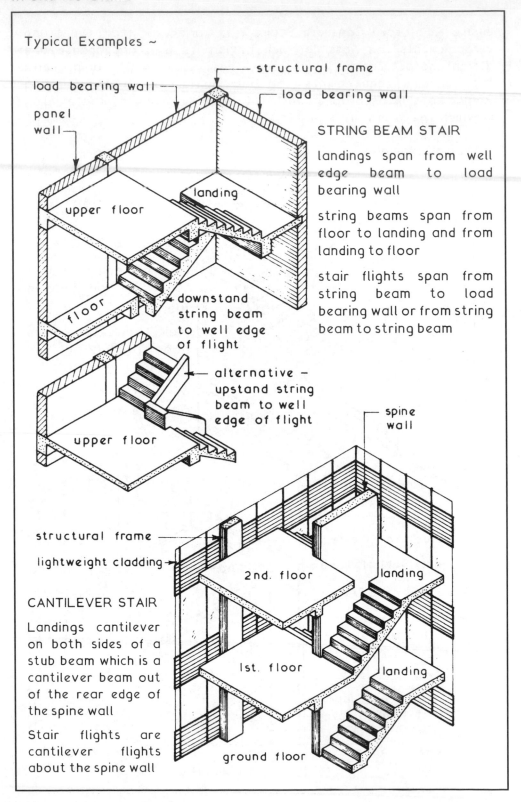

structural frame

load bearing wall

load bearing wall

panel wall

STRING BEAM STAIR

landings span from well edge beam to load bearing wall

string beams span from floor to landing and from landing to floor

stair flights span from string beam to load bearing wall or from string beam to string beam

landing

upper floor

floor

downstand string beam to well edge of flight

alternative – upstand string beam to well edge of flight

upper floor

spine wall

structural frame

lightweight cladding

2nd. floor

landing

CANTILEVER STAIR

Landings cantilever on both sides of a stub beam which is a cantilever beam out of the rear edge of the spine wall

1st. floor

landing

Stair flights are cantilever flights about the spine wall

ground floor

Spiral and Helical Stairs ~ these stairs constructed in in-situ reinforced concrete are considered to be aesthetically pleasing but are expensive to construct. They are therefore mainly confined to prestige buildings usually as accommodation stairs linking floors within the same compartment. Like all other forms of stair they must conform to the requirements of Part K of the Building Regulations and if used as a means of escape in case of fire with the requirements of Part B. Spiral stairs can be defined as those describing a helix around a central column whereas a helical stair has an open well. The open well of a helical stair is usually circular or elliptical in plan and the formwork is built up around a vertical timber core.

Typical Example of a Helical Stair ~

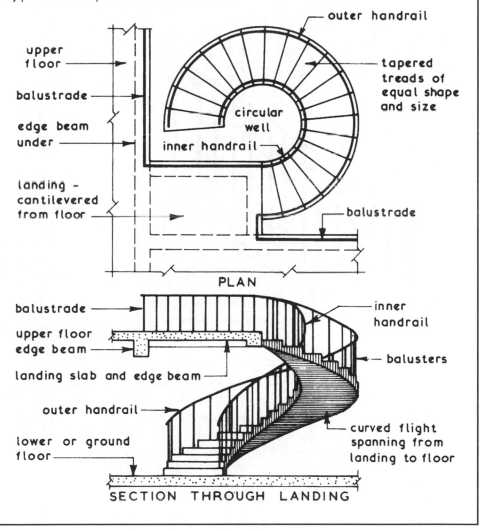

In-situ RC Inclined Slab Stair — Typical Details ~

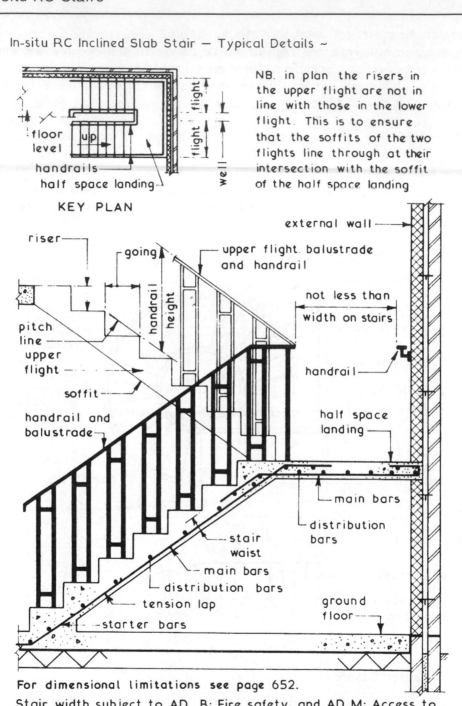

KEY PLAN

floor level

up

handrails

half space landing

flight

flight

well

NB. in plan the risers in the upper flight are not in line with those in the lower flight. This is to ensure that the soffits of the two flights line through at their intersection with the soffit of the half space landing

external wall

riser

going

upper flight. balustrade and handrail

handrail height

not less than width on stairs

pitch line upper flight

handrail

soffit

half space landing

handrail and balustrade

main bars

distribution bars

stair waist

main bars

distribution bars

tension lap

ground floor

starter bars

For dimensional limitations see page 652.

Stair width subject to AD B: Fire safety, and AD M: Access to and use of buildings. Width measured as clear distance between walls or balustrade. Ignore string and handrail if projecting < 100 mm.

In-situ Reinforced Concrete Stair Formwork ~ in specific detail the formwork will vary for the different types of reinforced concrete stair but the basic principles for each format will remain constant.

Typical RC Stair Formwork Details ~ (see page 642 for Key Plan)

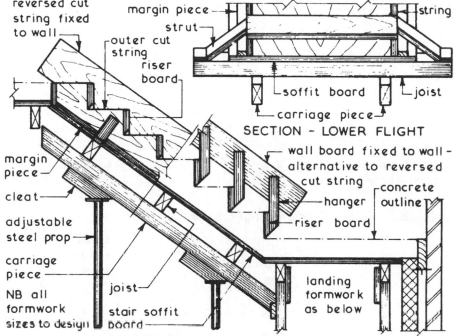

reversed cut string fixed to wall

outer cut string

margin piece
strut
string
riser board
soffit board
joist
carriage piece

SECTION – LOWER FLIGHT

wall board fixed to wall - alternative to reversed cut string

hanger
concrete outline
riser board

margin piece

cleat

adjustable steel prop

carriage piece

joist

NB all formwork sizes to design

stair soffit board

landing formwork as below

TYPICAL FORMWORK TO UPPER FLIGHT

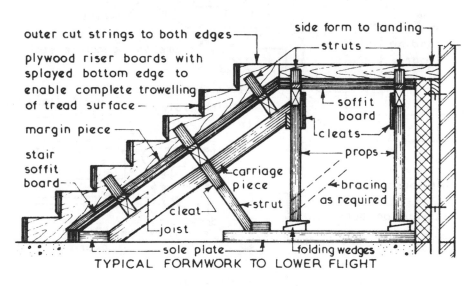

outer cut strings to both edges

side form to landing

struts

plywood riser boards with splayed bottom edge to enable complete trowelling of tread surface

soffit board

cleats

props

margin piece

stair soffit board

carriage piece

bracing as required

strut

cleat

joist

sole plate

folding wedges

TYPICAL FORMWORK TO LOWER FLIGHT

Precast Concrete Stairs ~ these can be produced to most of the formats used for in-situ concrete stairs and like those must comply with the appropriate requirements set out in Part K of the Building Regulations. To be economic the total production run must be sufficient to justify the costs of the moulds and therefore the designers choice may be limited to the stair types which are produced as a manufacturer's standard item.

Precast concrete stairs can have the following advantages:-

1. Good quality control of finished product.
2. Saving in site space since formwork fabrication and storage will not be required.
3. The stairs can be installed at any time after the floors have been completed thus giving full utilisation to the stair shaft as a lifting or hoisting space if required.
4. Hoisting, positioning and fixing can usually be carried out by semi-skilled labour.

Typical Example ~ Straight Flight Stairs

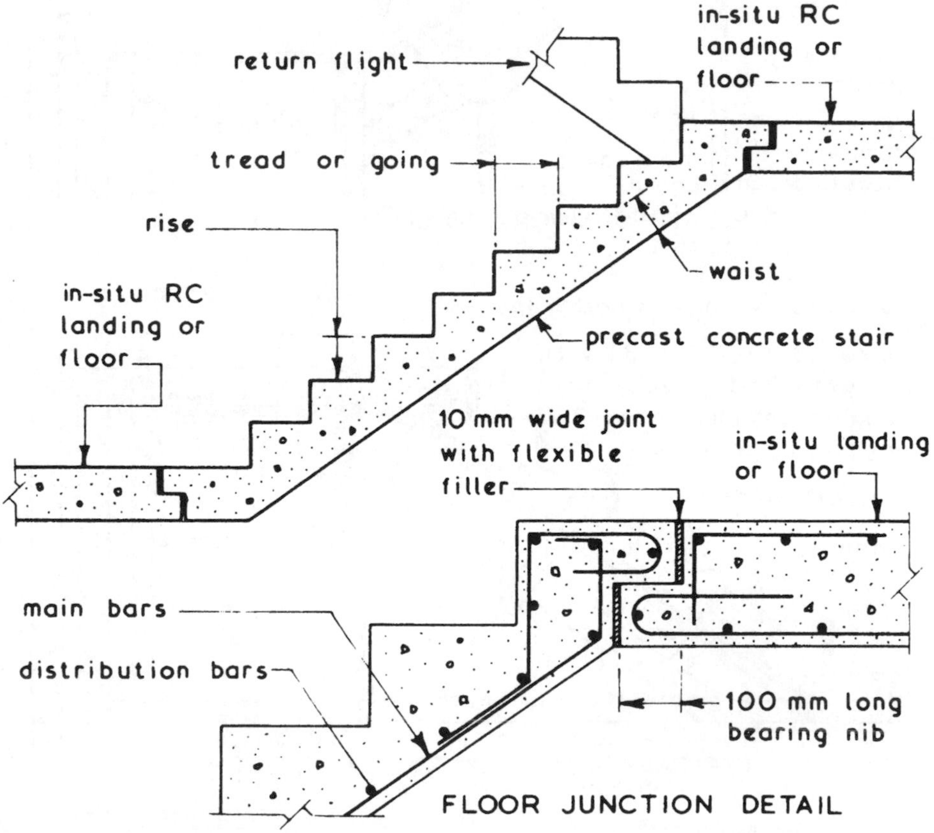

FLOOR JUNCTION DETAIL

Typical Example ~ Cranked Slab Stairs

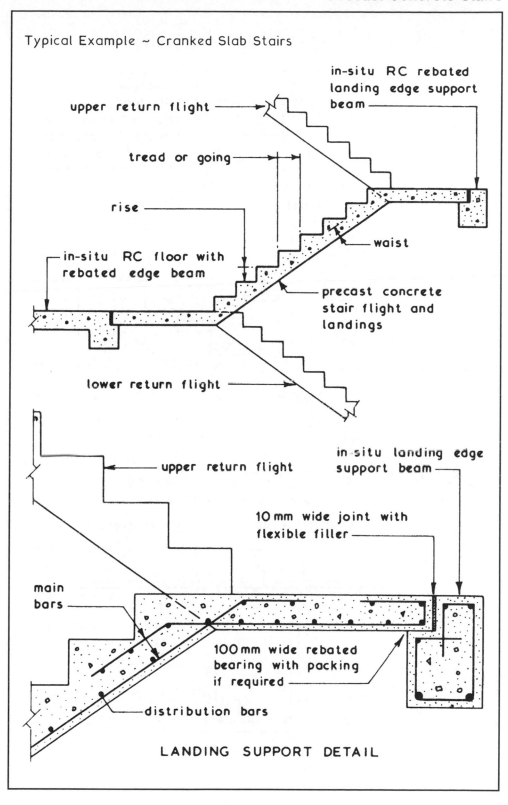

upper return flight

in-situ RC rebated landing edge support beam

tread or going

rise

in-situ RC floor with rebated edge beam

waist

precast concrete stair flight and landings

lower return flight

upper return flight

in-situ landing edge support beam

10 mm wide joint with flexible filler

main bars

100 mm wide rebated bearing with packing if required

distribution bars

LANDING SUPPORT DETAIL

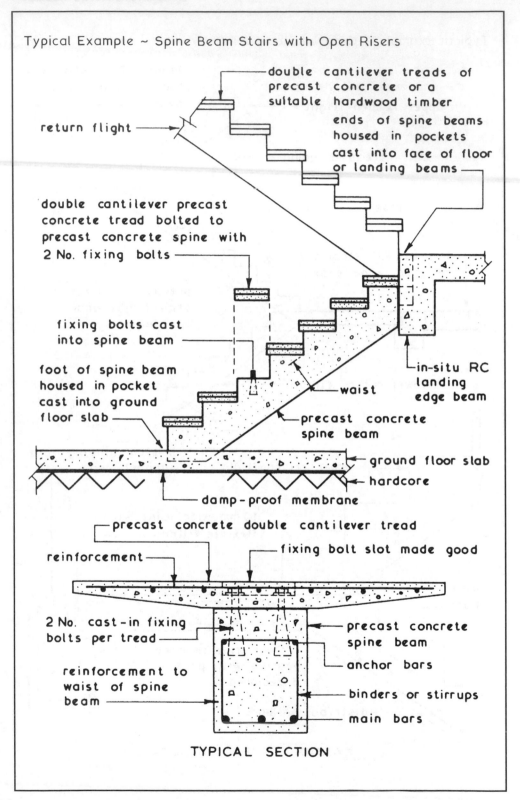

Typical Example ~ Spine Beam Stairs with Open Risers

double cantilever treads of precast concrete or a suitable hardwood timber

ends of spine beams housed in pockets cast into face of floor or landing beams

return flight

double cantilever precast concrete tread bolted to precast concrete spine with 2 No. fixing bolts

fixing bolts cast into spine beam

foot of spine beam housed in pocket cast into ground floor slab

waist

in-situ RC landing edge beam

precast concrete spine beam

ground floor slab

hardcore

damp-proof membrane

precast concrete double cantilever tread

reinforcement

fixing bolt slot made good

2 No. cast-in fixing bolts per tread

precast concrete spine beam

anchor bars

reinforcement to waist of spine beam

binders or stirrups

main bars

TYPICAL SECTION

Precast Concrete Spiral Stairs ~ this form of stair is usually constructed with an open riser format using tapered treads which have a keyhole plan shape. Each tread has a hollow cylinder at the narrow end equal to the rise which is fitted over a central steel column usually filled with in-situ concrete. The outer end of the tread has holes through which the balusters pass to be fixed on the underside of the tread below, a hollow spacer being used to maintain the distance between consecutive treads.

Typical Example ~

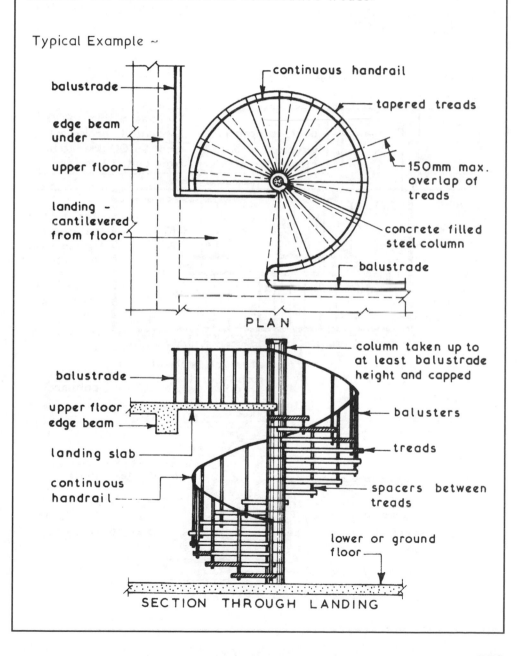

PLAN

SECTION THROUGH LANDING

Metal Stairs ~ these can be produced in cast iron, mild steel or aluminium alloy for use as escape stairs or for internal accommodation stairs. Most escape stairs are fabricated from cast iron or mild steel and must comply with the Building Regulation requirements for stairs in general and fire escape stairs in particular. Most metal stairs are purpose made and therefore tend to cost more than comparable concrete stairs. Their main advantage is the elimination of the need for formwork whilst the main disadvantage is the regular maintenance in the form of painting required for cast iron and mild steel stairs.

Typical Example ~ Straight Flight Steel External Escape Stair

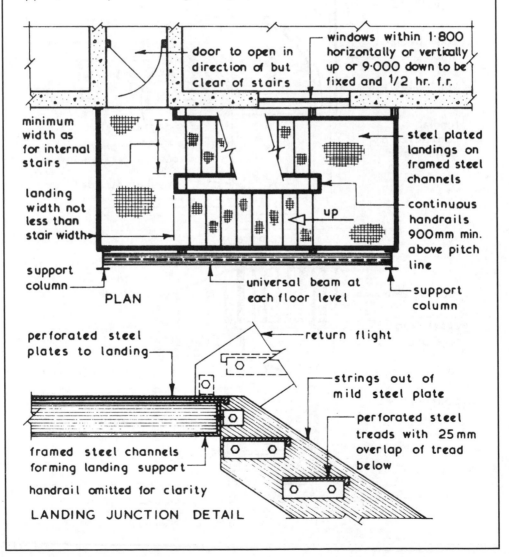

door to open in direction of but clear of stairs

windows within 1·800 horizontally or vertically up or 9·000 down to be fixed and 1/2 hr. f.r.

minimum width as for internal stairs

steel plated landings on framed steel channels

landing width not less than stair width

continuous handrails 900mm min. above pitch line

up

support column

universal beam at each floor level

support column

PLAN

perforated steel plates to landing

return flight

strings out of mild steel plate

perforated steel treads with 25mm overlap of tread below

framed steel channels forming landing support

handrail omitted for clarity

LANDING JUNCTION DETAIL

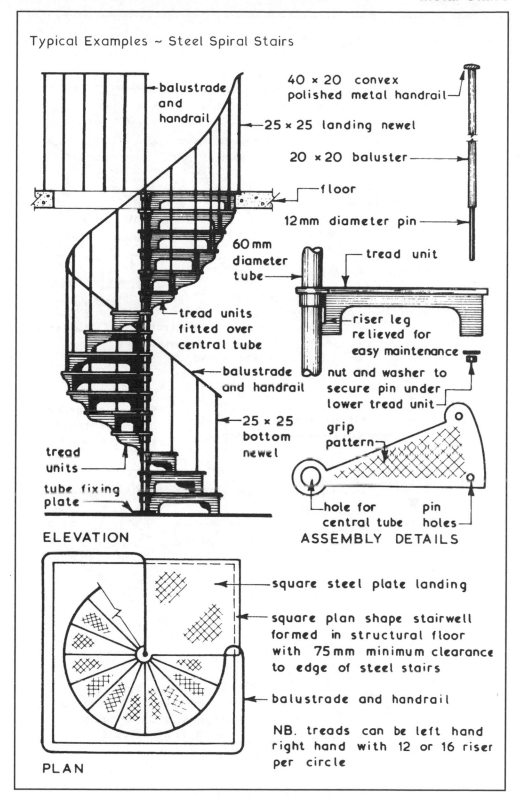

Typical Examples ~ Steel Spiral Stairs

balustrade and handrail

40 × 20 convex polished metal handrail

25 × 25 landing newel

20 × 20 baluster

floor

12 mm diameter pin

60 mm diameter tube

tread unit

tread units fitted over central tube

riser leg relieved for easy maintenance

balustrade and handrail

nut and washer to secure pin under lower tread unit

25 × 25 bottom newel

grip pattern

tread units

tube fixing plate

hole for central tube

pin holes

ELEVATION

ASSEMBLY DETAILS

square steel plate landing

square plan shape stairwell formed in structural floor with 75 mm minimum clearance to edge of steel stairs

balustrade and handrail

NB. treads can be left hand right hand with 12 or 16 riser per circle

PLAN

649

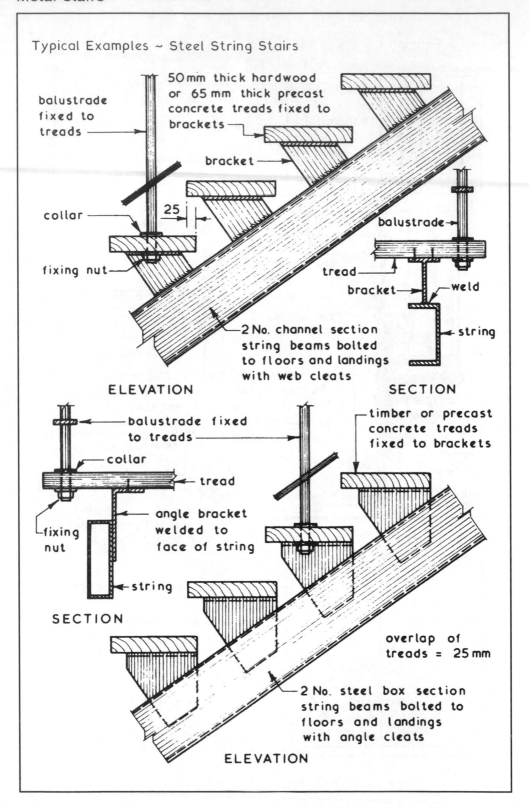

Typical Examples ~ Steel String Stairs

balustrade fixed to treads

50 mm thick hardwood or 65 mm thick precast concrete treads fixed to brackets

bracket

collar

25

fixing nut

balustrade

tread

bracket

weld

string

2 No. channel section string beams bolted to floors and landings with web cleats

ELEVATION

SECTION

balustrade fixed to treads

collar

tread

fixing nut

angle bracket welded to face of string

string

SECTION

timber or precast concrete treads fixed to brackets

overlap of treads = 25 mm

2 No. steel box section string beams bolted to floors and landings with angle cleats

ELEVATION

Balustrades and Handrails ~ these must comply in all respects with the requirements given in Part K of the Building Regulations and in the context of escape stairs are constructed of a non-combustible material with a handrail shaped to give a comfortable hand grip. The handrail may be covered or capped with a combustible material such as timber or plastic. Most balustrades are designed to be fixed after the stairs have been cast or installed by housing the balusters in a preformed pocket or by direct surface fixing.

Typical Details ~

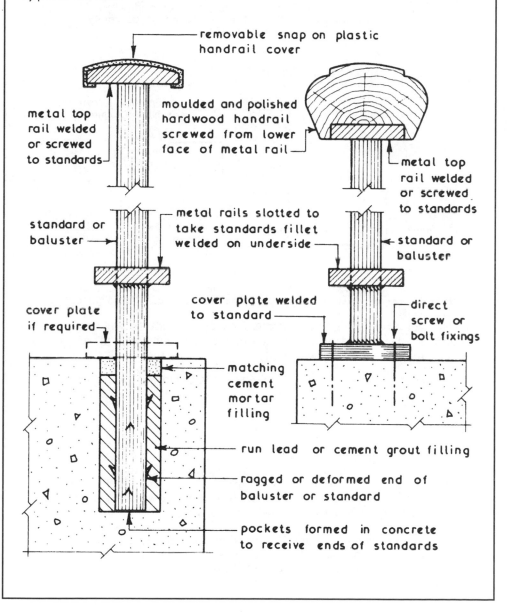

Institutional and Assembly Stairs ~ Serving a place where a substantial number of people will gather.

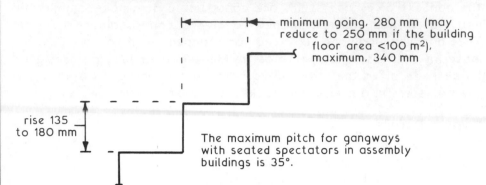

minimum going, 280 mm (may reduce to 250 mm if the building floor area <100 m²), maximum, 340 mm

rise 135 to 180 mm

The maximum pitch for gangways with seated spectators in assembly buildings is 35°.

Other stairs ~ All other buildings.

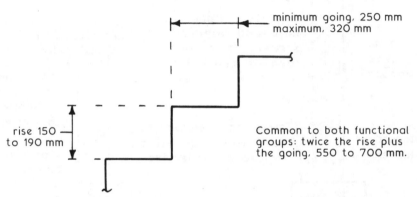

minimum going, 250 mm maximum, 320 mm

rise 150 to 190 mm

Common to both functional groups: twice the rise plus the going, 550 to 700 mm.

Alternative guidance: BS 5395-1: Stairs, ladders and walkways. Code of practice for the design, construction and maintenance of straight stairs and winders.

The rise and going in both situations may be subject to the requirements of Approved Document M: Access to and use of buildings. AD M will take priority and the following will apply:

Going (external steps) ~ 280 mm minimum (300 mm min. preferred).
Going (stairs) ~ 250 mm minimum.
Rise ~ 170 mm maximum.

Other AD M requirements for stairs:

Avoid tapered treads.
Width at least 1200 mm between walls, strings or upstands.
Landing top and bottom, minimum length not less than stair width.
Handrails both sides; circular or oval preferred as shown on next page.
Door openings onto landings to be avoided.
Full risers, no gaps.
Nosings with prominent colour 55 mm wide on tread and riser.
Nosing projections avoided, maximum of 25 mm if necessary.
Maximum 12 risers; exceptionally in certain small premises 16.
Non-slip surface to treads and landing.

Measurement of the going (AD K) ~

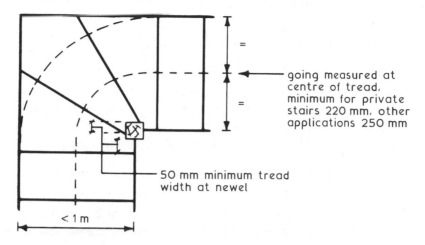

going measured at centre of tread, minimum for private stairs 220 mm, other applications 250 mm

50 mm minimum tread width at newel

< 1 m

For stair widths greater than 1 m, the going is measured at 270 mm from each side of the stair.

Additional requirements:

Going of tapered treads not less than the going of parallel treads in the same stair.
Curved landing lengths measured on the stair centre line.
Twice the rise plus the going, 550 to 700 mm.
Uniform going for consecutive tapered treads.
Other going and rise limitations as shown on the previous page.

Alternative guidance that is acceptable to the requirements of Approved Document K is published in BS 5395-2: Stairs, ladders and walkways. Code of practice for the design of helical and spiral stairs.

Preferred handrail profile (AD M) ~

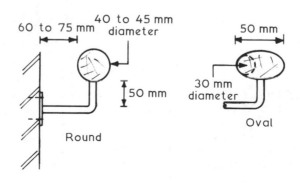

60 to 75 mm
40 to 45 mm diameter
50 mm
Round

50 mm
30 mm diameter
Oval

Functions ~ the main functions of any door are to:

1. Provide a means of access and egress.
2. Maintain continuity of wall function when closed.
3. Provide a degree of privacy and security.

Choice of door type can be determined by:-

1. Position - whether internal or external.
2. Properties required - fire resistant, glazed to provide for borrowed light or vision through, etc.
3. Appearance - flush or panelled, painted or polished, etc.

Door Schedules ~ these can be prepared in the same manner and for the same purpose as that given for windows on page 359.

Internal Doors ~ these are usually lightweight and can be fixed to a lining, if heavy doors are specified these can be hung to frames in a similar manner to external doors. An alternative method is to use door sets which are usually storey height and supplied with prehung doors.

Typical door Lining Details ~

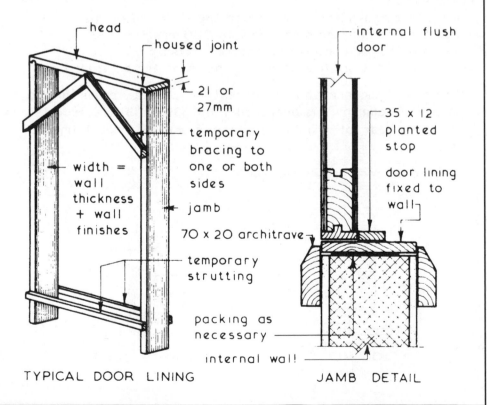

TYPICAL DOOR LINING   JAMB DETAIL

Internal Doors ~ these are similar in construction to the external doors but are usually thinner and therefore lighter in weight.

Typical Examples ~

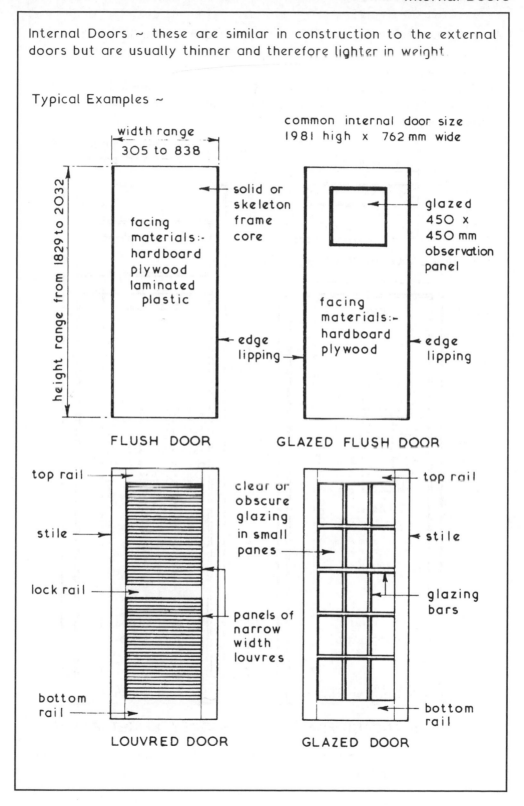

common internal door size
1981 high x 762 mm wide

width range
305 to 838

height range from 1829 to 2032

solid or skeleton frame core

facing materials:-
hardboard
plywood
laminated
plastic

edge lipping

FLUSH DOOR

glazed
450 x
450 mm
observation
panel

facing materials:-
hardboard
plywood

edge lipping

GLAZED FLUSH DOOR

top rail

stile

lock rail

bottom rail

LOUVRED DOOR

clear or obscure glazing in small panes

panels of narrow width louvres

top rail

stile

glazing bars

bottom rail

GLAZED DOOR

Internal Door Frames and linings ~ these are similar in construction to external door frames but usually have planted door stops and do not have a sill. The frames are sized to be built in conjunction with various partition thicknesses and surface finishes. Linings with planted stops ae usually employed for lightweight domestic doors.

Typical Examples ~

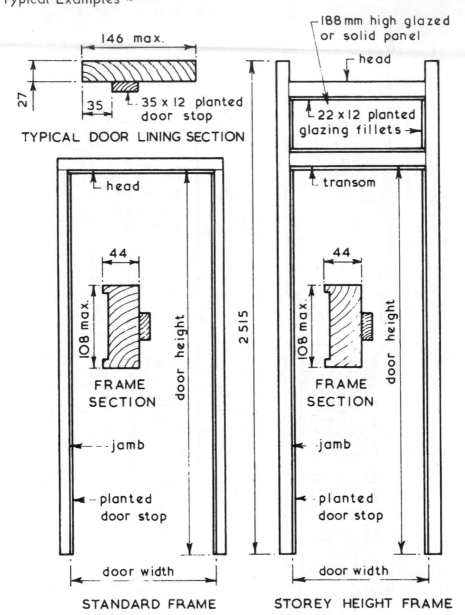

146 max.

27

35

35 x 12 planted door stop

TYPICAL DOOR LINING SECTION

188 mm high glazed or solid panel

head

22 x 12 planted glazing fillets

transom

head

44

108 max.

door height

2 515

FRAME SECTION

jamb

planted door stop

door width

STANDARD FRAME

44

108 max.

door height

FRAME SECTION

jamb

planted door stop

door width

STOREY HEIGHT FRAME

Ref. BS 4787: Internal and external wood doorsets, door leaves and frames. Specification for dimensional requirements.

Doorsets ~ these are factory produced fully assembled prehung doors which are supplied complete with frame, architraves and ironmongery except for door furniture. The doors may be hung to the frames using pin butts for easy door removal. Prehung door sets are available in standard and storey height versions and are suitable for all internal door applications with normal wall and partition thicknesses.

Typical Examples ~

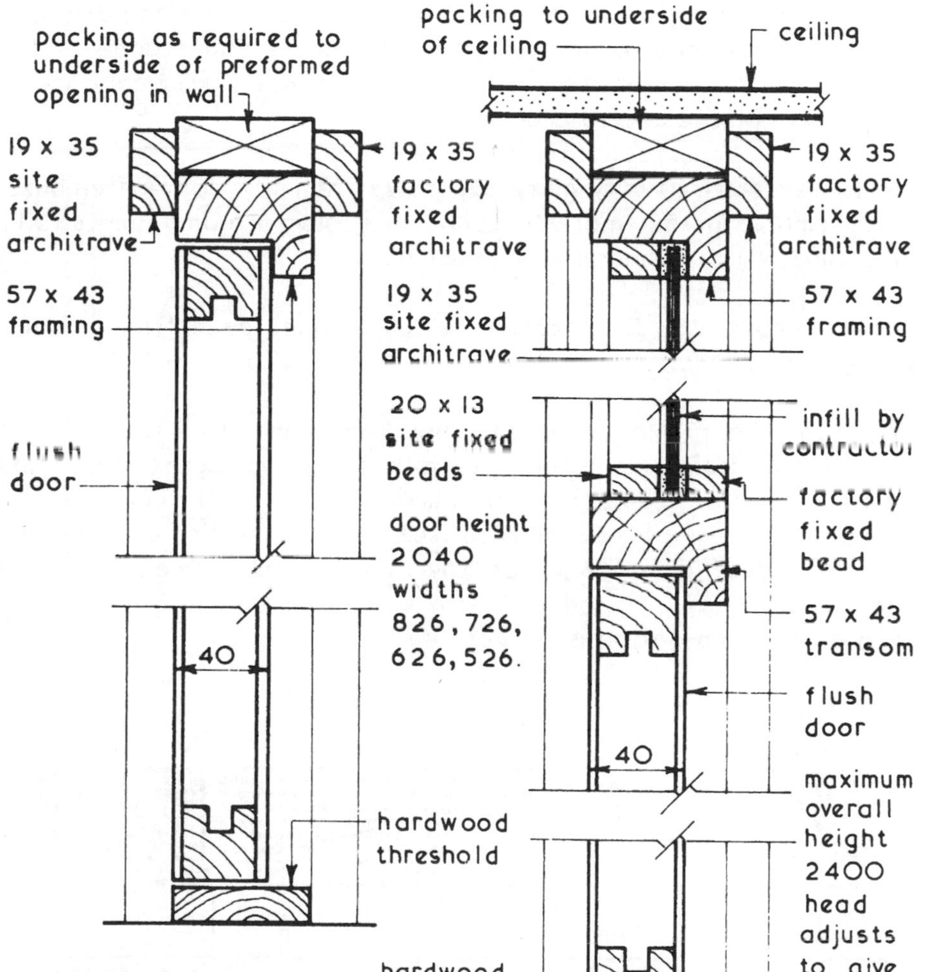

packing as required to underside of preformed opening in wall ⌐

packing to underside of ceiling ──────

⌐ ceiling

19 x 35 site fixed architrave →

← 19 x 35 factory fixed architrave

← 19 x 35 factory fixed architrave

57 x 43 framing ──

19 x 35 site fixed architrave ──────

57 x 43 framing

20 x 13 site fixed beads ──

── infill by contractor

── factory fixed bead

flush door ──→

door height 2040 widths 826, 726, 626, 526.

── 57 x 43 transom

── flush door

40

40

── hardwood threshold

maximum overall height 2400 head adjusts to give overall heights from 2230 to 2380

STANDARD HEIGHT DOORSET

hardwood threshold ──

doorsets fixed to wall or partition with 4 No. wood screws to each jamb

STOREY HEIGHT DOORSET

Fire doorset ~ a "complete unit consisting of a door frame and a door leaf or leaves, supplied with all essential parts from a single source". The difference between a doorset and a fire doorset is the latter is endorsed with a fire certificate for the complete unit. When supplied as a collection of parts for site assembly, this is known as a door kit.

Fire door assembly ~ a "complete assembly as installed, including door frame and one or more leaves, together with its essential hardware [ironmongery] supplied from separate sources". Provided the components to an assembly satisfy the Building Regulations – Approved Document B, fire safety requirements and standards for certification and compatibility, then a fire door assembly is an acceptable alternative to a doorset.

Fire doorsets are usually more expensive than fire door assemblies, but assemblies permit more flexibility in choice of components. Site fixing time will be longer for assemblies.

(Quotes from BS EN 12519: Windows and pedestrian doors. Terminology.)

Fire door ~ a fire door is not just the door leaf. A fire door includes the frame, ironmongery, glazing, intumescent core and smoke seal. To comply with European market requirements, ironmongery should be CE marked (see page 61). A fire door should also be marked accordingly on the top or hinge side. The label type shown below, reproduced with kind permission of the British Woodworking Federation is acceptable.

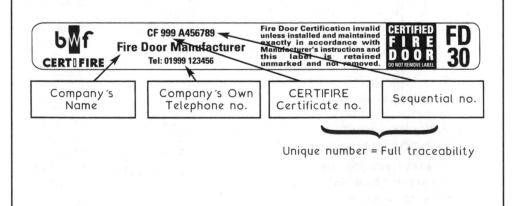

30 Minute Flush Fire Doors ~ these are usually based on the recommendations given in BS 8214. A wide variety of door constructions are available from various manufacturers but they all have to be fitted to a similar frame for testing as a doorset or assembly, including ironmongery.

A door's resistance to fire is measured by:-

1. Insulation – resistance to thermal transmittance, see BS 476–20 & 22: Fire tests on building materials and structures.

2. Integrity – resistance in minutes to the penetration of flame and hot gases under simulated fire conditions.

Typical Details ~

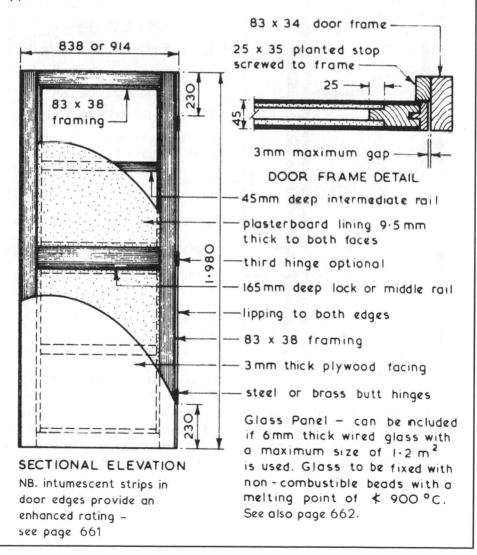

838 or 914

83 x 38 framing

230

1·980

230

SECTIONAL ELEVATION
NB. intumescent strips in door edges provide an enhanced rating – see page 661

83 x 34 door frame

25 x 35 planted stop screwed to frame

25

45

3mm maximum gap

DOOR FRAME DETAIL

45mm deep intermediate rail

plasterboard lining 9·5 mm thick to both faces

third hinge optional

165mm deep lock or middle rail

lipping to both edges

83 x 38 framing

3mm thick plywood facing

steel or brass butt hinges

Glass Panel – can be included if 6mm thick wired glass with a maximum size of $1·2\,m^2$ is used. Glass to be fixed with non-combustible beads with a melting point of $\not< 900\,°C$. See also page 662.

60 Minute Flush Fire Door ~ like the 30 minute flush fire door shown on page 659 these doors are based on the recommendations given in BS 8214 which covers both door and frame. A wide variety of fire resistant door constructions are available from various manufacturers with most classified as having both insulation and integrity ratings of 60 minutes.

Typical Details ~

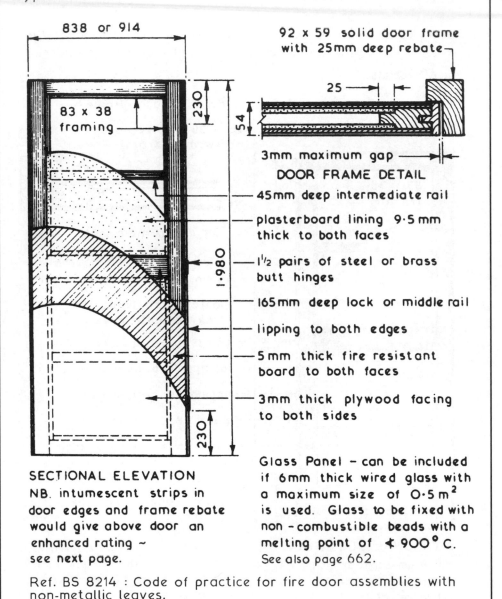

838 or 914

92 x 59 solid door frame with 25mm deep rebate

83 x 38 framing

230

54

25

3mm maximum gap

DOOR FRAME DETAIL

45mm deep intermediate rail

plasterboard lining 9·5 mm thick to both faces

1·980

1½ pairs of steel or brass butt hinges

165mm deep lock or middle rail

lipping to both edges

5 mm thick fire resistant board to both faces

230

3mm thick plywood facing to both sides

SECTIONAL ELEVATION
NB. intumescent strips in door edges and frame rebate would give above door an enhanced rating ~
see next page.

Glass Panel – can be included if 6mm thick wired glass with a maximum size of 0·5 m² is used. Glass to be fixed with non-combustible beads with a melting point of ≮ 900° C.
See also page 662.

Ref. BS 8214 : Code of practice for fire door assemblies with non-metallic leaves.

Fire and Smoke Resistance ~ Doors can be assessed for both integrity and smoke resistance. They are coded accordingly, for example FD30 or FD30s. FD indicates a fire door and 30 the integrity time in minutes. The letter 's' denotes that the door or frame contains a facility to resist the passage of smoke.

Manufacturers produce doors of standard ratings – 30, 60 and 90 minutes, with higher ratings available to order. A colour coded plug inserted in the door edge corresponds to the fire rating. See BS 8214, Table 1 for details.

Intumescent Fire and Smoke Seals ~

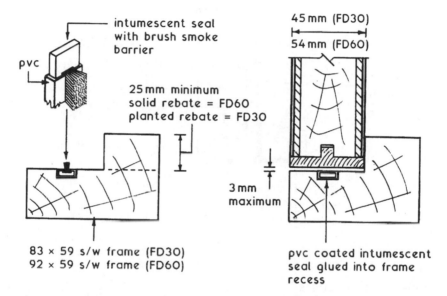

intumescent seal
with brush smoke
barrier

pvc

25 mm minimum
solid rebate = FD60
planted rebate = FD30

45 mm (FD30)
54 mm (FD60)

3 mm
maximum

83 × 59 s/w frame (FD30)
92 × 59 s/w frame (FD60)

pvc coated intumescent
seal glued into frame
recess

The intumescent core may be fitted to the door edge or the frame. In practice, most joinery manufacturers leave a recess in the frame where the seal is secured with rubber based or PVA adhesive. At temperatures of about 150°C, the core expands to create a seal around the door edge. This remains throughout the fire resistance period whilst the door can still be opened for escape and access purposes. The smoke seal will also function as an effective draught seal.

Further references:

BS EN 1634-1: Fire resistance tests for door and shutter assemblies. Fire doors and shutters.

BS EN 13501: Fire classification of construction products and building elements.

Apertures will reduce the potential fire resistance if not appropriately filled. Suitable material should have the same standard of fire performance as the door into which it is fitted.

Fire rated glass types ~

• Embedded Georgian wired glass
• Composite glass containing borosilicates and ceramics
• Tempered and toughened glass
• Glass laminated with reactive fire resisting interlayers

Installation ~ Hardwood beads and intumescent seals. Compatibility of glass type and sealing product is essential, therefore manufacturers details must be consulted.

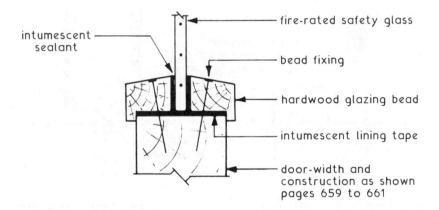

Intumescent products ~

• Sealants and mastics `gun´ applied
• Adhesive glazing strip or tape
• Preformed moulded channel

Note: Calcium silicate preformed channel is also available. Woven ceramic fire-glazing tape/ribbon is produced specifically for use in metal frames.

Building Regulations references:

AD M, Section 2.13, visibility zones/panels between 500 and 1500 mm above floor finish.
AD N, Section 1.6, aperture size (see page 369)

Typical Details ~

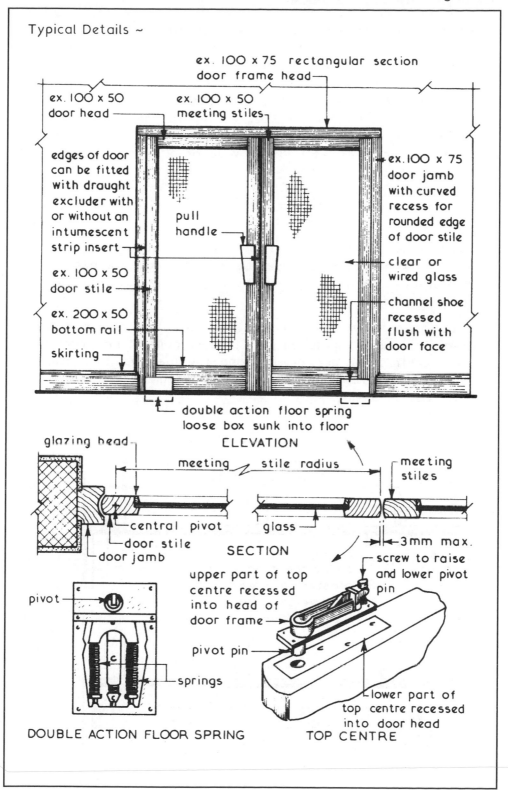

ex. 100 x 75 rectangular section door frame head

ex. 100 x 50 door head

ex. 100 x 50 meeting stiles

edges of door can be fitted with draught excluder with or without an intumescent strip insert

pull handle

ex. 100 x 50 door stile

ex. 200 x 50 bottom rail

skirting

ex. 100 x 75 door jamb with curved recess for rounded edge of door stile

clear or wired glass

channel shoe recessed flush with door face

double action floor spring loose box sunk into floor

ELEVATION

glazing head

meeting stile radius

meeting stiles

central pivot

door stile

door jamb

glass

3mm max.

SECTION

pivot

springs

DOUBLE ACTION FLOOR SPRING

upper part of top centre recessed into head of door frame

pivot pin

screw to raise and lower pivot pin

lower part of top centre recessed into door head

TOP CENTRE

663

Plasterboard ~ this is a rigid board made with a core of gypsum sandwiched between face sheets of strong durable paper. In the context of ceilings two sizes can be considered –

1. Baseboard 2·400 × 1·200 × 9·5mm thick for supports at centres not exceeding 400mm; 2·400 × 1·200 × 12·5mm for supports at centres not exceeding 600mm. Baseboard has square edges and therefore the joints will need reinforcing with jute scrim at least 90mm wide or alternatively a special tape to prevent cracking.

2. Gypsum Lath 1·200 × 406 × 9·5 or 12·5mm thick. Lath has rounded edges which eliminates the need to reinforce the joints.

Baseboard is available with a metallised polyester facing which acts as a vapour control layer to prevent moisture penetrating the insulation and timber, joints should be sealed with an adhesive metallised tape.

The boards are fixed to the underside of the floor or ceiling joists with galvanised or sheradised plasterboard nails at not more than 150mm centres and are laid breaking the joint. Edge treatments consist of jute scrim or plastic mesh reinforcement or a pre-formed plaster cove moulding.

Typical details ~

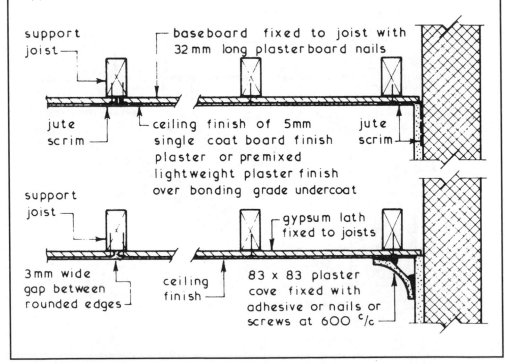

Suspended Ceilings ~ these can be defined as ceilings which are fixed to a framework suspended from main structure thus forming a void between the two components. The basic functional requirements of suspended ceilings are:-

1. They should be easy to construct, repair, maintain and clean.
2. So designed that an adequate means of access is provided to the void space for the maintenance of the suspension system, concealed services and/or light fittings.
3. Provide any required sound and/or thermal insulation.
4. Provide any required acoustic control in terms of absorption and reverberation.
5. Provide if required structural fire protection to structural steel beams supporting a concrete floor and contain fire stop cavity barriers within the void at defined intervals.
6. Conform with the minimum requirements set out in the Building Regulations governing the restriction of spread of flame over surfaces of ceilings and the exemptions permitting the use of certain plastic materials.
7. Flexural design strength in varying humidity and temperature.
8. Resistance to impact.
9. Designed on a planning module, preferably a 300 mm dimensional coordinated system.

Typical Suspended Ceiling Grid Framework Layout ~

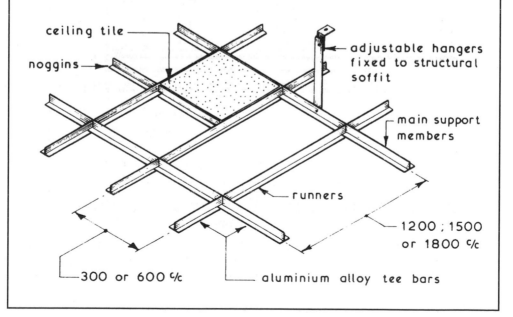

ceiling tile

noggins

adjustable hangers fixed to structural soffit

main support members

runners

1200 ; 1500 or 1800 c/c

300 or 600 c/c

aluminium alloy tee bars

Classification of Suspended Ceiling ~ there is no standard method of classification since some are classified by their function such as illuminated and acoustic suspended ceilings, others are classified by the materials used and classification by method of construction is also very popular. The latter method is simple since most suspended ceiling types can be placed in one of three groups:-

1. Jointless suspended ceilings.
2. Panelled suspended ceilings – see page 667.
3. Decorative and open suspended ceilings – see page 668.

Jointless Suspended Ceilings ~ these forms of suspended ceilings provide a continuous and jointless surface with the internal appearance of a conventional ceiling. They may be selected to fulfil fire resistance requirements or to provide a robust form of suspended ceiling. The two common ways of construction are a plasterboard or expanded metal lathing soffit with hand applied plaster finish or a sprayed applied rendering with a cement base.

Typical Details ~

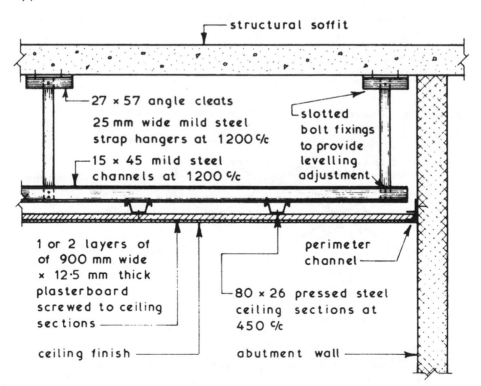

structural soffit

27 × 57 angle cleats

25 mm wide mild steel strap hangers at 1200 c/c

slotted bolt fixings to provide levelling adjustment

15 × 45 mild steel channels at 1200 c/c

1 or 2 layers of of 900 mm wide × 12·5 mm thick plasterboard screwed to ceiling sections

perimeter channel

80 × 26 pressed steel ceiling sections at 450 c/c

ceiling finish

abutment wall

See also: BS EN 13964: Suspended ceilings. Requirements and test methods.

Panelled Suspended Ceilings ~ these are the most popular form of suspended ceiling consisting of a suspended grid framework to which the ceiling covering is attached. The covering can be of a tile, tray, board or strip format in a wide variety of materials with an exposed or concealed supporting framework. Services such as luminaries can usually be incorporated within the system. Generally panelled systems are easy to assemble and install using a water level or laser beam for initial and final levelling. Provision for maintenance access can be easily incorporated into most systems and layouts.

Typical Support Details ~

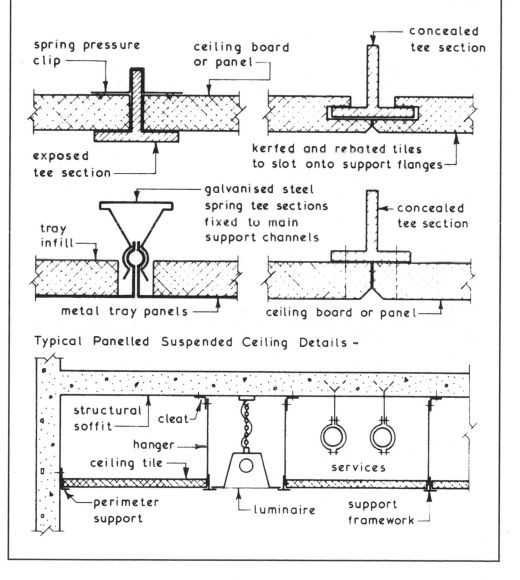

Typical Panelled Suspended Ceiling Details ~

667

Decorative and Open Suspended Ceilings ~ these ceilings usually consist of an openwork grid or suspended shapes onto which the lights fixed at, above or below ceiling level can be trained thus creating a decorative and illuminated effect. Many of these ceilings are purpose designed and built as opposed to the proprietary systems associated with jointless and panelled suspended ceilings.

Typical Examples ~

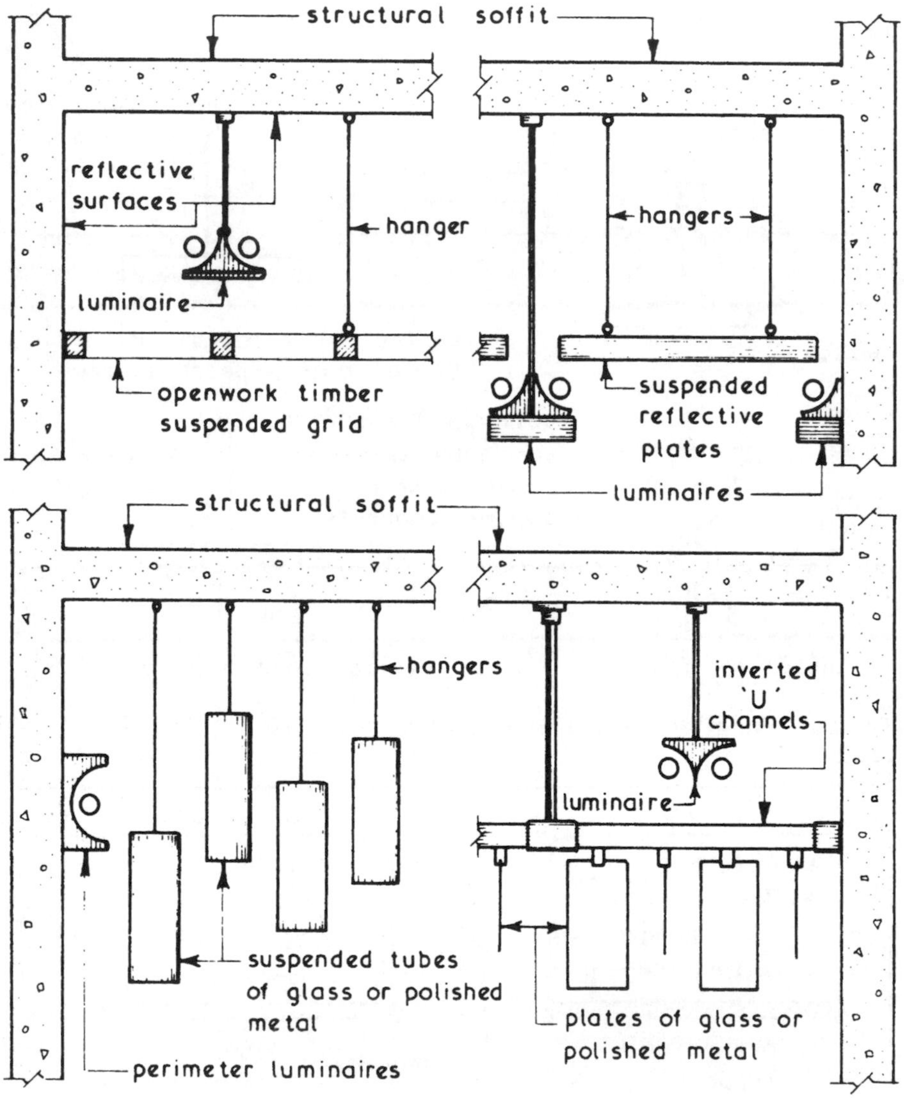

Functions ~ the main functions of paint are to provide:-

1. An economic method of surface protection to building materials and components.
2. An economic method of surface decoration to building materials and components.

Composition ~ the actual composition of any paint can be complex but the basic components are:-

1. Binder ~ this is the liquid vehicle or medium which dries to form the surface film and can be composed of linseed oil, drying oils, synthetic resins and water. The first function of a paint medium is to provide a means of spreading the paint over the surface and at the same time acting as a binder to the pigment.
2. Pigment ~ this provides the body, colour, durability and corrosion protection properties of the paint. White lead pigments are very durable and moisture resistant but are poisonous and their use is generally restricted to priming and undercoating paints. If a paint contains a lead pigment the fact must be stated on the container. The general pigment used in paint is titanium dioxide which is not poisonous and gives good obliteration of the undercoats.
3. Solvents and Thinners ~ these are materials which can be added to a paint to alter its viscosity.

Paint Types – there is a wide range available but for most general uses the following can be considered:-

1. Oil Based paints – these are available in priming, undercoat and finishing grades. The latter can be obtained in a wide range of colours and finishes such as matt, semi-matt, eggshell, satin, gloss and enamel. Polyurethane paints have a good hardness and resistance to water and cleaning. Oil based paints are suitable for most applications if used in conjunction with correct primer and undercoat.
2. Water Based Paints – most of these are called emulsion paints the various finishes available being obtained by adding to the water medium additives such as alkyd resin & polyvinyl acetate (PVA). Finishes include matt, eggshell, semi-gloss and gloss. Emulsion paints are easily applied, quick drying and can be obtained with a washable finish and are suitable for most applications.

Supply ~ paint is usually supplied in metal containers ranging from 250 millilitres to 5 litres capacity to the colour ranges recommended in BS 381C (colours for specific purposes) and BS 4800 (paint colours for building purposes).

Application ~ paint can be applied to almost any surface providing the surface preparation and sequence of paint coats are suitable. The manufacturers specification and/or the recommendations of BS 6150 (painting of buildings) should be followed. Preparation of the surface to receive the paint is of the utmost importance since poor preparation is one of the chief causes of paint failure. The preperation consists basically of removing all dirt, grease, dust and ensuring that the surface will provide an adequate key for the paint which is to be applied. In new work the basic build-up of paint coats consists of:-

1. Priming Coats – these are used on unpainted surfaces to obtain the necessary adhesion and to inhibit corrosion of ferrous metals. New timber should have the knots treated with a solution of shellac or other alcohol based resin called knotting prior to the application of the primer.
2. Undercoats – these are used on top of the primer after any defects have been made good with a suitable stopper or filler. The primary function of an undercoat is to give the opacity and build-up necessary for the application of the finishing coat(s).
3. Finish – applied directly over the undercoating in one or more coats to impart the required colour and finish.

Paint can applied by:-
1. Brush – the correct type, size and quality of brush such as those recommended in BS 2992 (painters and decorators brushes) needs to be selected and used. To achieve a first class finish by means of brush application requires a high degree of skill.
2. Spray – as with brush application a high degree of skill is required to achieve a good finish. Generally compressed air sprays or airless sprays are used for building works.
3. Roller – simple and inexpensive method of quickly and cleanly applying a wide range of paints to flat and textured surfaces. Roller heads vary in size from 50 to 450 mm wide with various covers such as sheepskin, synthetic pile fibres, mohair and foamed polystyrene. All paint applicators must be thoroughly cleaned after use.

Painting ~ the main objectives of applying coats of paint to a surface are preservation, protection and decoration to give a finish which is easy to clean and maintain. To achieve these objectives the surface preparation and paint application must be adequate. The preparation of new and previously painted surfaces should ensure that prior to painting the surface is smooth, clean, dry and stable.

Basic Surface Preparation Techniques ~

Timber – to ensure a good adhesion of the paint film all timber should have a moisture content of less than 18%. The timber surface should be prepared using an abrasive paper to produce a smooth surface brushed and wiped free of dust and any grease removed with a suitable spirit. Careful treatment of knots is essential either by sealing with two coats of knotting or in extreme cases cutting out the knot and replacing with sound timber. The stopping and filling of cracks and fixing holes with putty or an appropriate filler should be carried out after the application of the priming coat. Each coat of paint must be allowed to dry hard and be rubbed down with a fine abrasive paper before applying the next coat. On previously painted surfaces if the paint is in a reasonable condition the surface will only require cleaning and rubbing down before repainting, when the paint is in a poor condition it will be necessary to remove completely the layers of paint and then prepare the surface as described above for new timber.

Building Boards – most of these boards require no special preparation except for the application of a sealer as specified by the manufacturer.

Iron and Steel – good preparation is the key to painting iron and steel successfully and this will include removing all rust, mill scale, oil, grease and wax. This can be achieved by wire brushing, using mechanical means such as shot blasting, flame cleaning and chemical processes and any of these processes are often carried out in the steel fabrication works prior to shop applied priming.

Plaster – the essential requirement of the preparation is to ensure that the plaster surface is perfectly dry, smooth and free of defects before applying any coats of paint especially when using gloss paints. Plaster which contains lime can be alkaline and such surfaces should be treated with an alkali resistant primer when the surface is dry before applying the final coats of paint.

Paint Defects ~ these may be due to poor or incorrect preparation of the surface, poor application of the paint and/or chemical reactions. The general remedy is to remove all the affected paint and carry out the correct preparation of the surface before applying in the correct manner new coats of paint. Most paint defects are visual and therefore an accurate diagnosis of the cause must be established before any remedial treatment is undertaken.

Typical Paint Defects ~

1. Bleeding – staining and disruption of the paint surface by chemical action, usually caused by applying an incorrect paint over another. Remedy is to remove affected paint surface and repaint with correct type of overcoat paint.

2. Blistering – usually caused by poor presentation allowing resin or moisture to be entrapped, the subsequent expansion causing the defect. Remedy is to remove all the coats of paint and ensure that the surface is dry before repainting.

3. Blooming – mistiness usually on high gloss or varnished surfaces due to the presence of moisture during application. It can be avoided by not painting under these conditions. Remedy is to remove affected paint and repaint.

4. Chalking – powdering of the paint surface due to natural ageing or the use of poor quality paint. Remedy is to remove paint if necessary, prepare surface and repaint.

5. Cracking and Crazing – usually due to unequal elasticity of successive coats of paint. Remedy is to remove affected paint and repaint with compatible coats of paint.

6. Flaking and Peeling – can be due to poor adhesion, presence of moisture, painting over unclean areas or poor preparation. Remedy is to remove defective paint, prepare surface and repaint.

7. Grinning – due to poor opacity of paint film allowing paint coat below or background to show through, could be the result of poor application; incorrect thinning or the use of the wrong colour. Remedy is to apply further coats of paint to obtain a satisfactory surface.

8. Saponification – formation of soap from alkali present in or on surface painted. The paint is ultimately destroyed and a brown liquid appears on the surface. Remedy is to remove the paint films and seal the alkaline surface before repainting.

Joinery Production ~ this can vary from the flow production where one product such as flush doors is being made usually with the aid of purpose designed and built machines, to batch production where a limited number of similar items are being made with the aid of conventional woodworking machines. Purpose made joinery is very often largely hand made with a limited use of machines and is considered when special and/or high class joinery components are required.

Woodworking Machines ~ except for the portable electric tools such as drills, routers, jigsaws and sanders most woodworking machines need to be fixed to a solid base and connected to an extractor system to extract and collect the sawdust and chippings produced by the machines.

Saws – basically three formats are available, namely the circular, cross cut and band saws. Circular are general purpose saws and usually have tungsten carbide tipped teeth with feed rates of up to 60·000 per minute. Cross cut saws usually have a long bench to support the timber, the saw being mounted on a radial arm enabling the circular saw to be drawn across the timber to be cut. Band saws consist of an endless thin band or blade with saw teeth and a table on which to support the timber and are generally used for curved work.

Planers – most of these machines are combined planers and thicknessers, the timber being passed over the table surface for planning and the table or bed for thicknessing. The planer has a guide fence which can be tilted for angle planning and usually the rear bed can be lowered for rebating operations. The same rotating cutter block is used for all operations. Planing speeds are dependent upon the operator since it is a hand fed operation whereas thicknessing is mechanically fed with a feed speed range of 6·000 to 20·000 per minute. Maximum planing depth is usually 10 mm per passing.

Morticing Machines – these are used to cut mortices up to 25 mm wide and can be either a chisel or chain morticer. The former consists of a hollow chisel containing a bit or auger whereas the latter has an endless chain cutter.

Tenoning Machines – these machines with their rotary cutter blocks can be set to form tenon and scribe. In most cases they can also be set for trenching, grooving and cross cutting.

Spindle Moulder – this machine has a horizontally rotating cutter block into which standard or purpose made cutters are fixed to reproduce a moulding on timber passed across the cutter.

Purpose Made Joinery ~ joinery items in the form of doors, windows, stairs and cupboard fitments can be purchased as stock items from manufacturers. There is also a need for purpose made joinery to fulfil client/designer/user requirement to suit a specific need, to fit into a non-standard space, as a specific decor requirement or to complement a particular internal environment. These purpose made joinery items can range from the simple to the complex which require high degrees of workshop and site skills.

Typical Purpose Made Counter Details ~

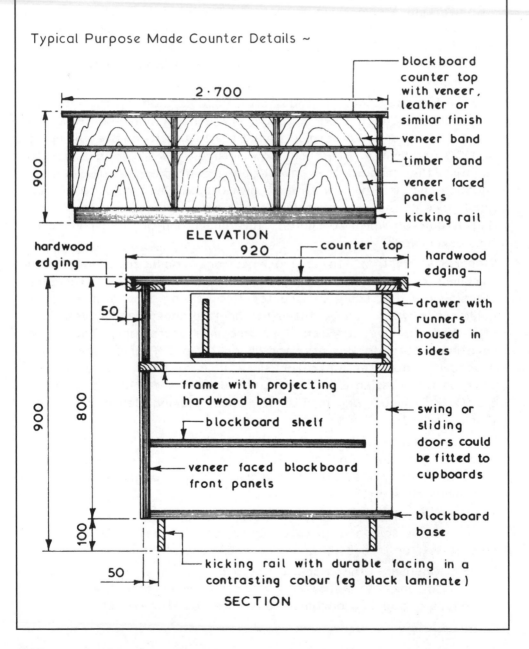

ELEVATION

SECTION

Typical Purpose Made Wall Panelling Details ~

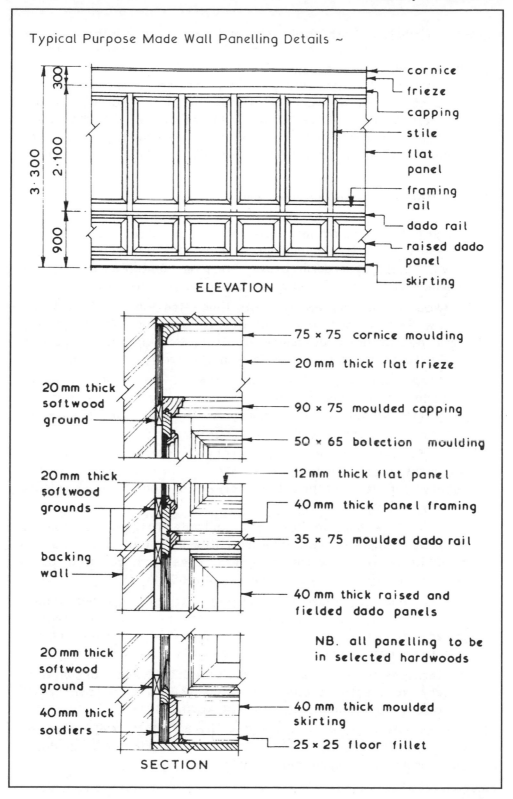

ELEVATION

- cornice
- frieze
- capping
- stile
- flat panel
- framing rail
- dado rail
- raised dado panel
- skirting

3·300
2·100
300
900

20 mm thick softwood ground

20 mm thick softwood grounds

backing wall

20 mm thick softwood ground

40 mm thick soldiers

- 75 × 75 cornice moulding
- 20 mm thick flat frieze
- 90 × 75 moulded capping
- 50 × 65 bolection moulding
- 12 mm thick flat panel
- 40 mm thick panel framing
- 35 × 75 moulded dado rail
- 40 mm thick raised and fielded dado panels

NB. all panelling to be in selected hardwoods

- 40 mm thick moulded skirting
- 25 × 25 floor fillet

SECTION

Joinery Timbers ~ both hardwoods and softwoods can be used for joinery works. Softwoods can be selected for their stability, durability and/or workability if the finish is to be paint but if it is left in its natural colour with a sealing coat the grain texture and appearance should be taken into consideration. Hardwoods are usually left in their natural colour and treated with a protective clear sealer or polish therefore texture, colour and grain pattern are important when selecting hardwoods for high class joinery work.

Typical Softwoods Suitable for Joinery Work ~

1. Douglas Fir – sometimes referred to as Columbian Pine or Oregon Pine. It is available in long lengths and has a straight grain. Colour is reddish brown to pink. Suitable for general and high class joinery. Approximate density 530 kg/m³.

2. Redwood – also known as Scots Pine. Red Pine, Red Deal and Yellow Deal. It is a widely used softwood for general joinery work having good durability a straight grain and is reddish brown to straw in colour. Approximate density 430 kg/m³.

3. European Spruce – similar to redwood but with a lower durability. It is pale yellow to pinkish white in colour and is used mainly for basic framing work and simple internal joinery. Approximate density 650 kg/m³.

4. Sitka Spruce – originates from Alaska, Western Canada and Northwest USA. The long, white strong fibres provide a timber quality for use in board or plywood panels. Approximate density 450 kg/m³.

5. Pitch Pine – durable softwood suitable for general joinery work. It is light red to reddish yellow in colour and tends to have large knots which in some cases can be used as a decorative effect. Approximate density 650 kg/m³.

6. Parana Pine – moderately durable straight grained timber available in a good range of sizes. Suitable for general joinery work especially timber stairs. Light to dark brown in colour with the occasional pink stripe. Approximate density 560 kg/m³.

7. Western Hemlock – durable softwood suitable for interior joinery work such as panelling. Light yellow to reddish brown in colour. Approximate density 500 kg/m³.

8. Western Red Cedar – originates from British Columbia and Western USA. A straight grained timber suitable for flush doors and panel work. Approximate density 380 kg/m³.

Typical Hardwoods Suitable for Joinery Works ~

1. Beech – hard close grained timber with some silver grain in the predominately reddish yellow to light brown colour. Suitable for all internal joinery. Approximately density 700 kg/m³.

2. Iroko – hard durable hardwood with a figured grain and is usually golden brown in colour. Suitable for all forms of good class joinery. Approximate density 660 kg/m³.

3. Mahogany (African) – interlocking grained hardwood with good durability. It has an attractive light brown to deep red colour and is suitable for panelling and all high class joinery work. Approximate density 560 kg/m³.

4. Mahogany (Honduras) – durable hardwood usually straight grained but can have a mottled or swirl pattern. It is light red to pale reddish brown in colour and is suitable for all good class joinery work. Approximate density 530 kg/m³.

5. Mahogany (South American) – a well figured, stable and durable hardwood with a deep red or brown colour which is suitable for all high class joinery particularly where a high polish is required. Approximate density 550 kg/m³.

6. Oak (English) – very durable hardwood with a wide variety of grain patterns. It is usually a light yellow brown to a warm brown in colour and is suitable for all forms of joinery but should not be used in conjunction with ferrous metals due to the risk of staining caused by an interaction of the two materials. (The gallic acid in oak causes corrosion in ferrous metals.) Approximate density 720 kg/m³.

7. Sapele – close texture timber of good durability, dark reddish brown in colour with a varied grain pattern. It is suitable for most internal joinery work especially where a polished finish is required. Approximate density 640 kg/m³.

8. Teak – very strong and durable timber but hard to work. It is light golden brown to dark golden yellow in colour which darkens with age and is suitable for high class joinery work and laboratory fittings. Approximate density 650 kg/m³.

9. Jarrah (Western Australia) – hard, dense, straight grained timber. Dull red colour, suited to floor and stair construction subjected to heavy wear. Approximate density 820 kg/m³.

Composite Boards ~ are factory manufactured, performed sheets with a wide range of properties and applications. The most common size is 2440 × 1220 mm or 2400 × 1200 mm in thicknesses from 3 to 50 mm.

1. Plywood (BS EN 636) – produced in a range of laminated thicknesses from 3 to 25 mm, with the grain of each layer normally at right angles to that adjacent. 3,7,9 or 11 plies make up the overall thickness and inner layers may have lower strength and different dimensions to those in the outer layers. Adhesives vary considerably from natural vegetable and animal glues to synthetics such as urea, melamine, phenol and resorcinol formaldehydes. Quality of laminates and type of adhesive determine application. Surface finishes include plastics, decorative hardwood veneers, metals, rubber and mineral aggregates.

2. Block and Stripboards (BS EN 12871) – range from 12 to 43 mm thickness, made up from a solid core of glued softwood strips with a surface enhancing veneer. Appropriate for dense panelling and doors.

   Battenboard – strips over 30 mm wide (unsuitable for joinery).
   Blockboard – strips up to 25 mm wide.
   Laminboard – strips up to 7 mm wide.

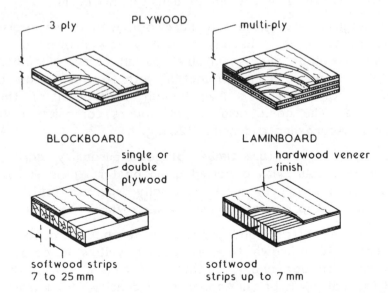

3. Compressed Strawboard (BS 4046) – produced by compacting straw under heat and pressure, and edge binding with paper. Used as panels with direct decoration or as partitioning with framed support. Also, for insulated roof decking with 58 mm slabs spanning 600 mm joist spacing.

4. Particle Board

Chipboard (BS EN 319) – bonded waste wood or chip particles in thicknesses from 6 to 50 mm, popularly used for floors in 18 and 22 mm at 450 and 600 mm maximum joist spacing, respectively. Sheets are produced by heat pressing the particles in thermosetting resins.

Wood Cement Board – approximately 25% wood particles mixed with water and cement, to produce a heavy and dense board often preferred to plasterboard and fibre cement for fire cladding. Often 3 layer boards, from 6 to 40 mm in thickness.

Oriented Strand Board (BS EN 300) – composed of wafer thin strands of wood, approximately 80 mm long × 25 m wide, resin bonded and directionally oriented before superimposed by further layers. Each layer is at right angles to adjacent layers, similar to the structure of plywood. A popular alternative for wall panels, floors and other chipboard and plywood applications, they are produced in a range of thicknesses from 6 to 25 mm.

5. Fibreboards (BS EN 622–4) – basically wood in composition, reduced to a pulp and pressed to achieve 3 categories:

Hardboard density at least 800 kg/m^3 in thicknesses from 3·2 to 8 mm. Provides an excellent base for coatings and laminated finishes.

Mediumboard (low density) 350 to 560 kg/m^3 for pinboards and wall linings in thicknesses of 6·4, 9, and 12·7 mm.

Mediumboard (high density) 560 to 800 kg/m^3 for linings and partitions in thicknesses of 9 and 12 mm.

Softboard, otherwise known as insulating board with density usually below 250 kg/m^3. Thicknesses from 9 to 25 mm, often found impregnated with bitumen in existing flat roofing applications. Ideal as pinboard.

Medium Density Fibreboard, differs from other fibreboards with the addition of resin bonding agent. These boards have a very smooth surface, ideal for painting and are available moulded for a variety of joinery applications. Density exceeds 600 kg/m^3 and common board thicknesses are 9, 12, 18 and 25 mm for internal and external applications.

6. Woodwool (BS EN 13168) – units of 600 mm width are available in 50, 75 and 100 mm thicknesses. They comprise long wood shavings coated with a cement slurry, compressed to leave a high proportion of voids. These voids provide good thermal insulation and sound absorption. The perforated surface is an ideal key for direct plastering and they are frequently specified as permanent formwork.

Plastics ~ the term plastic can be applied to any group of substances based on synthetic or modified natural polymers which during manufacture are moulded by heat and/or pressure into the required form. Plastics can be classified by their overall grouping such as polyvinyl chloride (PVC) or they can be classified as thermoplastic or thermosetting. The former soften on heating whereas the latter are formed into permanent non-softening materials. The range of plastics available give the designer and builder a group of materials which are strong, reasonably durable, easy to fit and maintain and since most are mass produced of relative low cost.

Typical Applications of Plastics in Buildings ~

Application	Plastics Used
Rainwater goods	unplasticised PVC (uPVC or PVC-U).
Soil, waste, water and gas pipes and fittings	uPVC; polyethylene (PE); acrylonitrile butadiene styrene (ABS), polypropylene (PP).
Hot and cold water pipes	chlorinated PVC; ABS; polypropylene; polyethylene; PVC (not for hot water).
Bathroom and kitchen fittings	glass fibre reinforced polyester (GRP); acrylic resins.
Cold water cisterns	polypropylene; polystyrene; polyethylene.
Rooflights and sheets	GRP; acrylic resins; uPVC.
DPC's and membranes, vapour control layers	low density polyethylene (LDPE); PVC film; polypropylene.
Doors and windows	GRP; uPVC.
Electrical conduit and fittings	plasticised PVC; uPVC; phenolic resins.
Thermal insulation	generally cellular plastics such as expanded polystyrene bead and boards; expanded PVC; foamed polyurethane; foamed phenol formaldehyde; foamed urea formaldehyde.
Floor finishes	plasticised PVC tiles and sheets; resin based floor paints; uPVC.
Wall claddings and internal linings	unplasticised PVC; polyvinyl fluoride film laminate; melamine resins; expanded polystyrene tiles & sheets.

8 DOMESTIC SERVICES

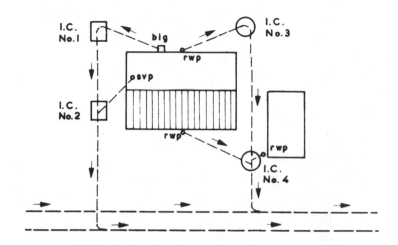

DRAINAGE EFFLUENTS
SUBSOIL DRAINAGE
SURFACE WATER REMOVAL
ROAD DRAINAGE
RAINWATER INSTALLATIONS
DRAINAGE SYSTEMS
DRAINAGE PIPE SIZES AND GRADIENTS
WATER SUPPLY
COLD WATER INSTALLATIONS
HOT WATER INSTALLATIONS
CISTERNS AND CYLINDERS
SANITARY FITTINGS
SINGLE AND VENTILATED STACK SYSTEMS
DOMESTIC HOT WATER HEATING SYSTEMS
ELECTRICAL SUPPLY AND INSTALLATION
GAS SUPPLY AND GAS FIRES
SERVICES FIRE STOPS AND SEALS
OPEN FIREPLACES AND FLUES
COMMUNICATIONS INSTALLATIONS

Effluent ~ can be defined as that which flows out. In building drainage terms there are three main forms of effluent:-

1. Subsoil Water ~ water collected by means of special drains from the earth primarily to lower the water table level in the subsoil. It is considered to be clean and therefore requires no treatment and can be discharged direct into an approved water course.

2. Surface water ~ effluent collected from surfaces such as roofs and paved areas and like subsoil water is considered to be clean and can be discharged direct into an approved water course or soakaway

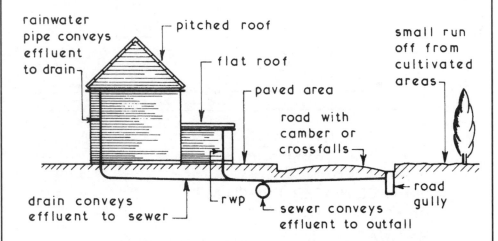

rainwater pipe conveys effluent to drain

pitched roof

flat roof

paved area

road with camber or crossfalls

small run off from cultivated areas

drain conveys effluent to sewer

rwp

sewer conveys effluent to outfall

road gully

3. Foul or Soil Water ~ effluent contaminated by domestic or trade waste and will require treatment to render it clean before it can be discharged into an approved water course.

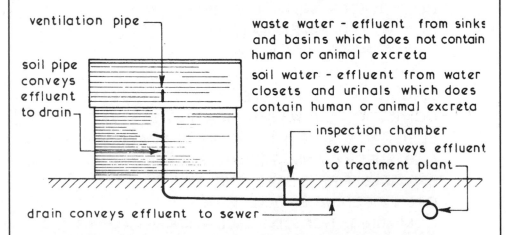

ventilation pipe

soil pipe conveys effluent to drain

waste water - effluent from sinks and basins which does not contain human or animal excreta

soil water - effluent from water closets and urinals which does contain human or animal excreta

inspection chamber

sewer conveys effluent to treatment plant

drain conveys effluent to sewer

Subsoil Drainage ~ Building Regulation C2 requires that subsoil drainage shall be provided if it is needed to avoid:-

a) the passage of ground moisture into the interior of the building or

b) damage to the fabric of the building.

Subsoil drainage can also be used to improve the stability of the ground, lower the humidity of the site and enhance its horticultural properties. Subsoil drains consist of porous or perforated pipes laid dry jointed in a rubble filled trench. Porous pipes allow the subsoil water to pass through the body of the pipe whereas perforated pipes which have a series of holes in the lower half allow the subsoil water to rise into the pipe. This form of ground water control is only economic up to a depth of 1·500, if the water table needs to be lowered to a greater depth other methods of ground water control should be considered (see pages 289 to 293).

The water collected by a subsoil drainage system has to be conveyed to a suitable outfall such as a river, lake or surface water drain or sewer. In all cases permission to discharge the subsoil water will be required from the authority or owner and in the case of streams, rivers and lakes, bank protection at the outfall may be required to prevent erosion (see page 684).

Typical Subsoil Drain Details ~

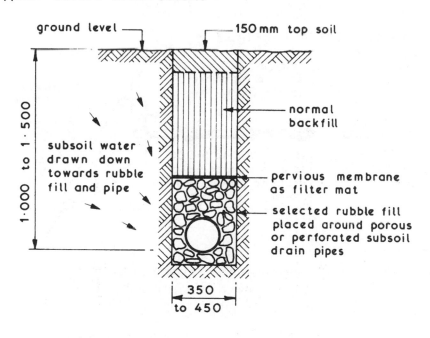

ground level ⎯ ⎯ 150 mm top soil

1·000 to 1·500

subsoil water drawn down towards rubble fill and pipe

normal backfill

pervious membrane as filter mat

selected rubble fill placed around porous or perforated subsoil drain pipes

350 to 450

Subsoil Drainage Systems ~ the lay out of subsoil drains will depend on whether it is necessary to drain the whole site or if it is only the substructure of the building which needs to be protected. The latter is carried out by installing a cut off drain around the substructure to intercept the flow of water and divert it away from the site of the building. Junctions in a subsoil drainage system can be made using standard fittings or by placing the end of the branch drain onto the crown of the main drain.

Typical Examples ~

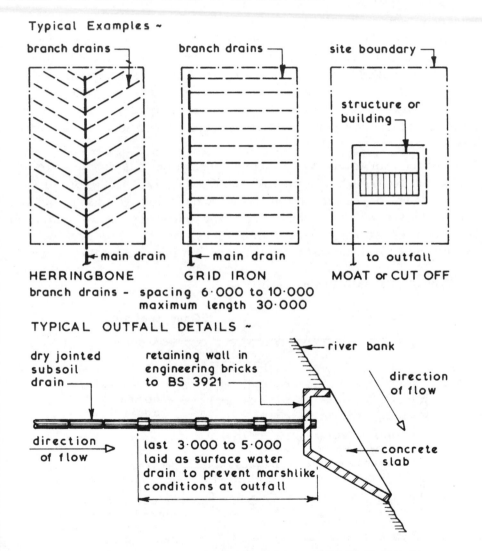

branch drains ─┐   branch drains ─┐   site boundary ─┐

structure or
building ─┐

├─main drain   ├─main drain   to outfall

HERRINGBONE   **GRID IRON**   **MOAT or CUT OFF**

branch drains - spacing 6·000 to 10·000
maximum length 30·000

TYPICAL OUTFALL DETAILS ~

dry jointed
subsoil
drain ──

retaining wall in
engineering bricks
to BS 3921 ──

river bank

direction
of flow

direction
of flow

last 3·000 to 5·000
laid as surface water
drain to prevent marshlike
conditions at outfall

concrete
slab

NB. connections to surface water sewer can be made at inspection chamber or direct to the sewer using a saddle connector – it may be necessary to have a catchpit to trap any silt (see page 688)

General Principles ~ a roof must be designed with a suitable fall towards the surface water collection channel or gutter which in turn is connected to vertical rainwater pipes which convey the collected discharge to the drainage system. The fall of the roof will be determined by the chosen roof covering or the chosen pitch will limit the range of coverings which can be selected.

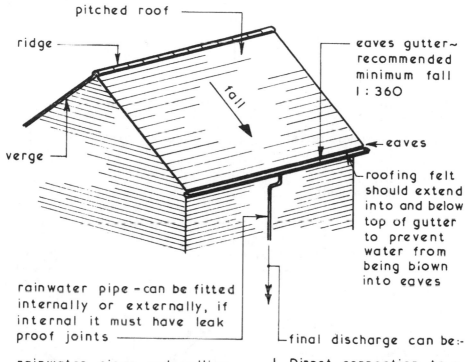

pitched roof

ridge

fall

verge

eaves gutter~ recommended minimum fall 1 : 360

eaves

roofing felt should extend into and below top of gutter to prevent water from being blown into eaves

rainwater pipe -can be fitted internally or externally, if internal it must have leak proof joints

final discharge can be:-

rainwater pipes and gullies must be arranged so as not to cause dampness or damage to any part of the building

Minimum Roof Pitches ~

Slates – depends on width from 25°

Hand made plain tiles – 45°

Machine made plain tiles – 35°

Single lap and interlocking tiles- depends on type from 12½°

Thatch – 45°

Timber shingles – 14°

1. Direct connection to a drain discharging into a soakaway

2. Direct connection to a drain discharging into a surface water sewer

3. Indirect connection to a drain by means of a trapped gully if drain discharges into a combined sewer

See page 691 for details

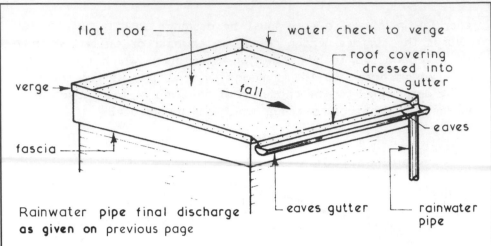

flat roof

water check to verge

roof covering dressed into gutter

verge

fall

eaves

fascia

eaves gutter

rainwater pipe

Rainwater pipe final discharge **as given on** previous page

Minimum Recommended Falls for Various Finishes ~
Aluminium – 1 : 60   Lead – 1:120   Copper – 1:60
Built-up roofing felts – 1 : 60   Mastic asphalt – 1:80

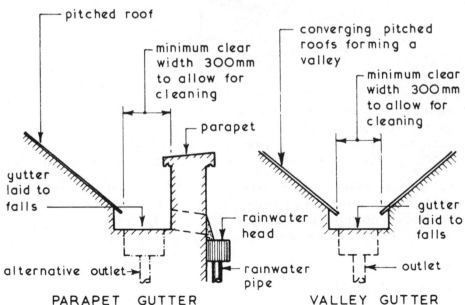

pitched roof

minimum clear width 300mm to allow for cleaning

parapet

converging pitched roofs forming a valley

minimum clear width 300mm to allow for cleaning

gutter laid to falls

rainwater head

gutter laid to falls

alternative outlet

rainwater pipe

outlet

PARAPET GUTTER

gutter formed to discharge into internal rainwater pipes or to external rainwater pipes via outlets through the parapet

VALLEY GUTTER

gutter formed to discharge into internal rainwater pipes or to external rainwater pipes sited at the gable ends

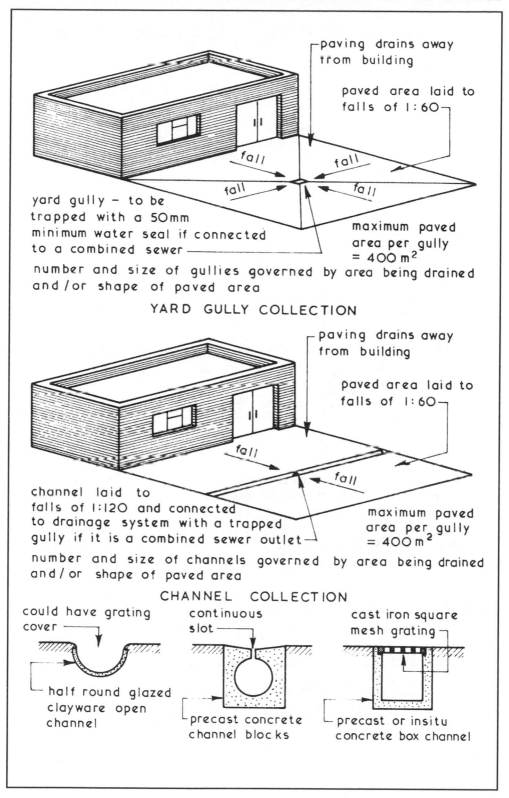

paving drains away from building

paved area laid to falls of 1:60

fall   fall

fall   fall

yard gully – to be trapped with a 50mm minimum water seal if connected to a combined sewer

maximum paved area per gully = 400 m²

number and size of gullies governed by area being drained and / or shape of paved area

YARD GULLY COLLECTION

paving drains away from building

paved area laid to falls of 1:60

fall

fall

channel laid to falls of 1:120 and connected to drainage system with a trapped gully if it is a combined sewer outlet

maximum paved area per gully = 400 m²

number and size of channels governed by area being drained and / or shape of paved area

CHANNEL COLLECTION

could have grating cover

half round glazed clayware open channel

continuous slot

precast concrete channel blocks

cast iron square mesh grating

precast or insitu concrete box channel

687

Highway Drainage ~ the stability of a highway or road relies on two factors –

1. Strength and durability of upper surface
2. Strength and durability of subgrade which is the subsoil on which the highway construction is laid.

The above can be adversely affected by water therefore it may be necessary to install two drainage systems. One system (subsoil drainage) to reduce the flow of subsoil water through the subgrade under the highway construction and a system of surface water drainage.

Typical Highway Subsoil Drainage Methods ~

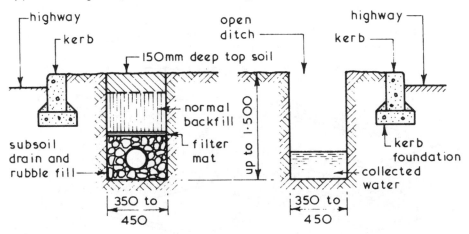

Subsoil Drain – acts as a cut off drain and can be formed using perforated or porous drain pipes. If filled with rubble only it is usually called a French or rubble drain.

Open Ditch – acts as a cut off drain and could also be used to collect surface water discharged from a rural road where there is no raised kerb or surface water drains.

Surface Water Drainage Systems ~

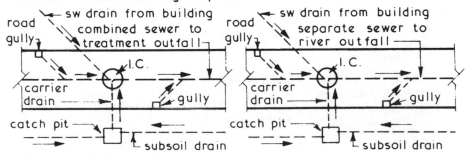

Road Drainage ~ this consists of laying the paved area or road to a suitable crossfall or gradient to direct the run-off of surface water towards the drainage channel or gutter. This is usually bounded by a kerb which helps to convey the water to the road gullies which are connected to a surface water sewer. For drains or sewers under 900mm internal diameter inspection chambers will be required as set out in the Building Regulations. The actual spacing of road gullies is usually determined by the local highway authority based upon the carriageway gradient and the area to be drained into one road gully. Alternatively the following formula could be used:-

$$D = \frac{280\sqrt{S}}{W}$$

where D = gully spacing
S = carriageway gradient (per cent)
W = width of carriageway in metres

∴ If S = 1:60 = 1·66% and W = 4·500

$$D = \frac{280\sqrt{1·66}}{4·500} = \text{say } 80·000$$

Typical Road Gully Detail ~

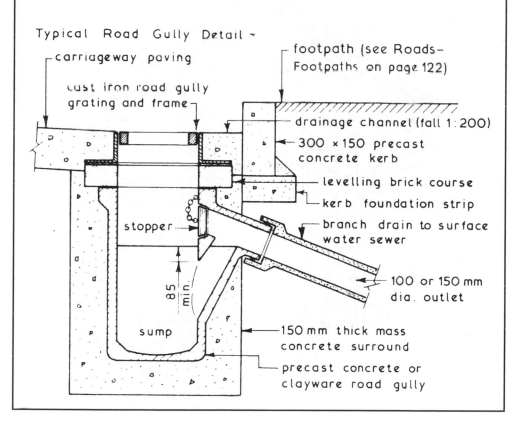

carriageway paving

cast iron road gully grating and frame

footpath (see Roads-Footpaths on page 122)

drainage channel (fall 1:200)

300 × 150 precast concrete kerb

levelling brick course

kerb foundation strip

branch drain to surface water sewer

100 or 150 mm dia. outlet

stopper

85 min.

sump

150 mm thick mass concrete surround

precast concrete or clayware road gully

Materials ~ the traditional material for domestic eaves gutters and rainwater pipes is cast iron but uPVC systems are very often specified today because of their simple installation and low maintenance costs. Other materials which could be considered are aluminium alloy, galvanised steel and stainless steel, but whatever material is chosen it must be of adequate size, strength and durability.

Typical Eaves Details ~

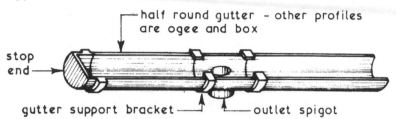

half round gutter - other profiles are ogee and box

stop end

gutter support bracket ──── outlet spigot

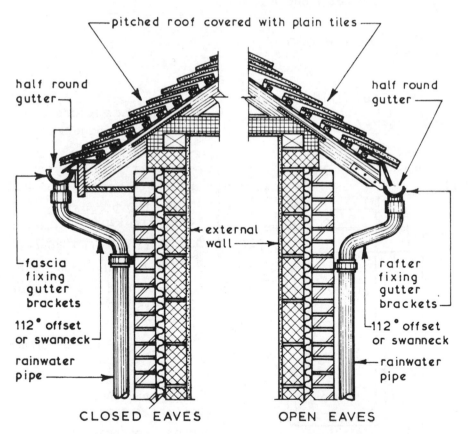

pitched roof covered with plain tiles

half round gutter

half round gutter

external wall

fascia fixing gutter brackets

rafter fixing gutter brackets

112° offset or swanneck

112° offset or swanneck

rainwater pipe

rainwater pipe

CLOSED EAVES

OPEN EAVES

For details of rainwater pipe connection to drainage see next page

690

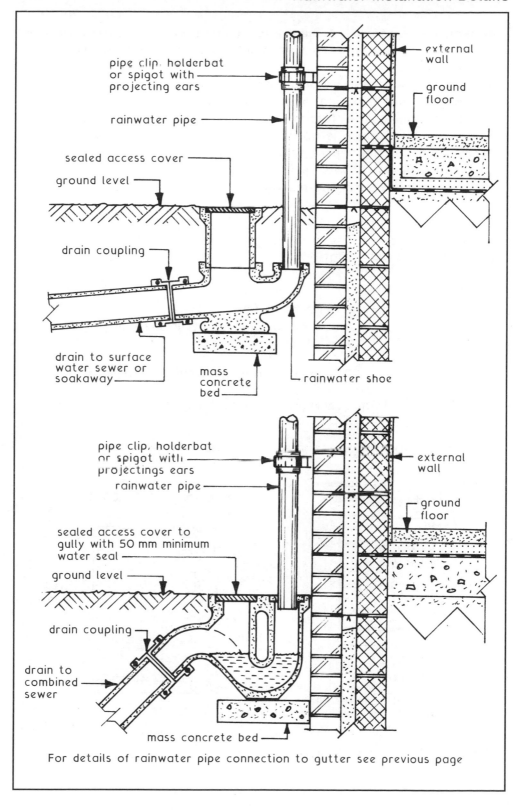

pipe clip, holderbat or spigot with projecting ears

rainwater pipe

sealed access cover

ground level

drain coupling

drain to surface water sewer or soakaway

mass concrete bed

rainwater shoe

external wall

ground floor

pipe clip, holderbat or spigot with projectings ears

rainwater pipe

sealed access cover to gully with 50 mm minimum water seal

ground level

drain coupling

drain to combined sewer

mass concrete bed

external wall

ground floor

For details of rainwater pipe connection to gutter see previous page

691

Soakaways ~ provide a means for collecting and controlling the seapage of rainwater into surrounding granular subsoils. They are not suitable in clay subsoils. Siting is on land at least level and preferably lower than adjacent buildings and no closer than 5m to a building. Concentration of a large volume of water any closer could undermine the foundations. The simplest soakaway is a rubble filled pit, which is normally adequate to serve a dwelling or other small building. Where several buildings share a soakaway, the pit should be lined with precast perforated concrete rings and surrounded in free-draining material.

BRE Digest 365 provides capacity calculations based on percolation tests. The following empirical formula will prove adequate for most situations:-

$$C = \frac{AR}{3}$$

where: C = capacity (m³)
A = area on plan to be drained (m²)
R = rainfall (m/h)

e.g. roof plan area 60 m² and rainfall of 50 mm/h (0·05 m/h)

$$C = \frac{60 \times 0·05}{3} = 1·0 \, m^3 \text{ (below invert of discharge pipe)}$$

FILLED SOAKAWAY

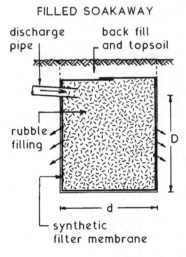

Depth (D) and diameter (d) approximately the same

HOLLOW SOAKAWAY

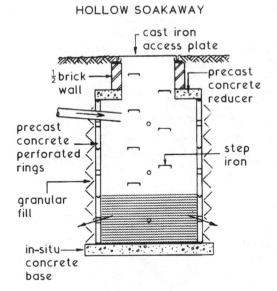

Ref. BRE Digest 365: Soakaways.

Drains ~ these can be defined as a means of conveying surface water or foul water below ground level.

Sewers ~ these have the same functions as drains but collect the discharge from a number of drains and convey it to the final outfall. They can be a private or public sewer depending on who is responsible for the maintenance.

Basic Principles ~ to provide a drainage system which is simple efficient and economic by laying the drains to a gradient which will render them self cleansing and will convey the effluent to a sewer without danger to health or giving nuisance. To provide a drainage system which will comply with the minimum requirements given in Part H of the Building Regulations

Typical Basic Requirements ~

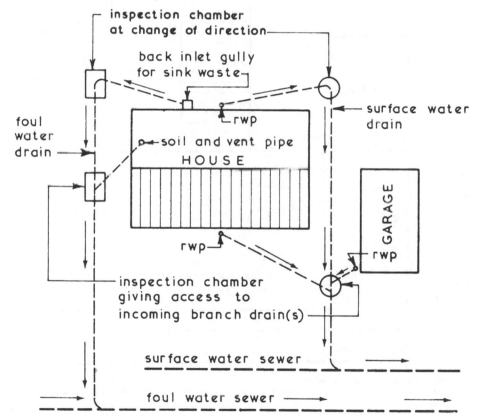

All junctions should be oblique and in direction of flow

There must be an access point at a junction unless each run can be cleared from another access point.

ate System ~ the most common drainage system in use where
urface water discharge is conveyed in separate drains and
s to that of foul water discharges and therefore receives no
treatment before the final outfall.

Typical Example ~

if subsoil is suitable
the rainwater pipes may
be allowed to be connected
direct to soakaways

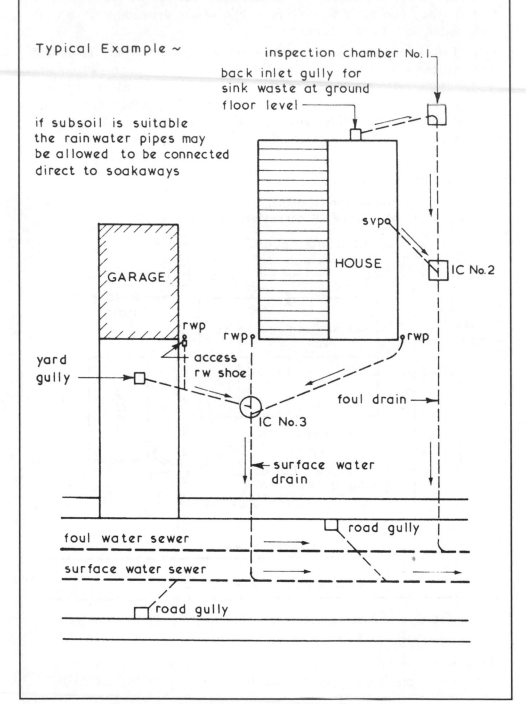

Combined System ~ this is the simplest and least expensive system to design and install but since all forms of discharge are conveyed in the same sewer the whole effluent must be treated unless a sea outfall is used to discharge the untreated effluent.

Typical Example ~

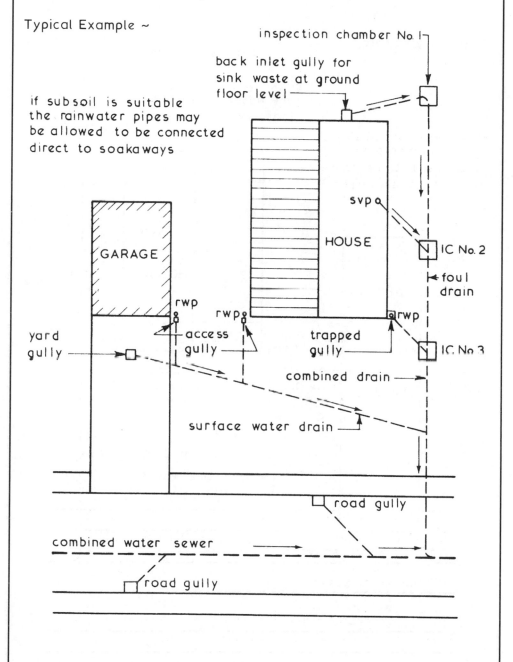

inspection chamber No. 1

back inlet gully for sink waste at ground floor level

if subsoil is suitable the rainwater pipes may be allowed to be connected direct to soakaways

svp

HOUSE

IC No. 2

foul drain

GARAGE

rwp

yard gully

rwp

access gully

trapped gully

rwp

IC No. 3

combined drain

surface water drain

road gully

combined water sewer

road gully

Ref. BS EN 752-1 to -7: Drain and sewer systems outside buildings.

Partially Separate System ~ a compromise system – there are two drains, one to convey only surface water and a combined drain to convey the total foul discharge and a proportion of the surface water.

Typical Example ~

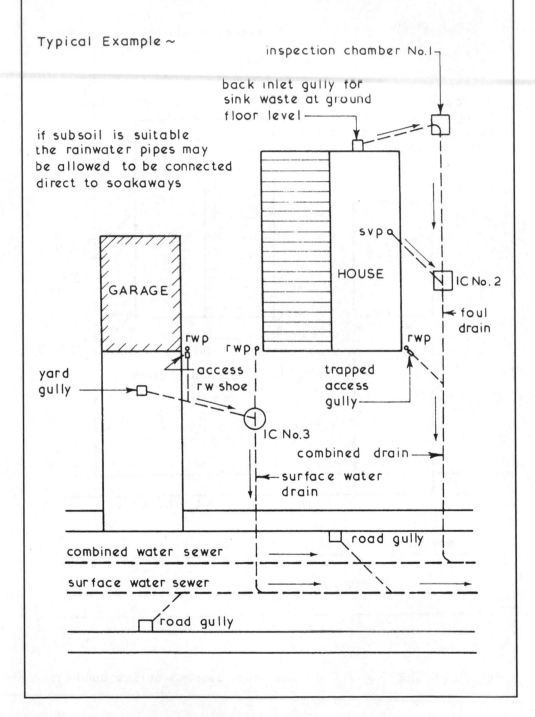

inspection chamber No.1

back inlet gully for sink waste at ground floor level

if subsoil is suitable the rainwater pipes may be allowed to be connected direct to soakaways

svp

HOUSE

IC No. 2

foul drain

GARAGE

rwp

rwp p

rwp

yard gully

access rw shoe

trapped access gully

IC No.3

combined drain

surface water drain

combined water sewer

road gully

surface water sewer

road gully

Inspection Chambers ~ these provide a means of access to drainage systems where the depth to invert level does not exceed 1·000.

Manholes ~ these are also a means of access to the drains and sewers, and are so called if the depth to invert level exceeds 1·000.

These means of access should be positioned in accordance with the requirements of part H of the Building Regulations. In domestic work inspection chambers can be of brick, precast concrete or preformed in plastic for use with patent drainage systems. The size of an inspection chamber depends on the depth to invert level, drain diameter and number of branch drains to be accommodated within the chamber. Ref. BS EN 752: Drain and sewer systems outside buildings.

Typical Details ~

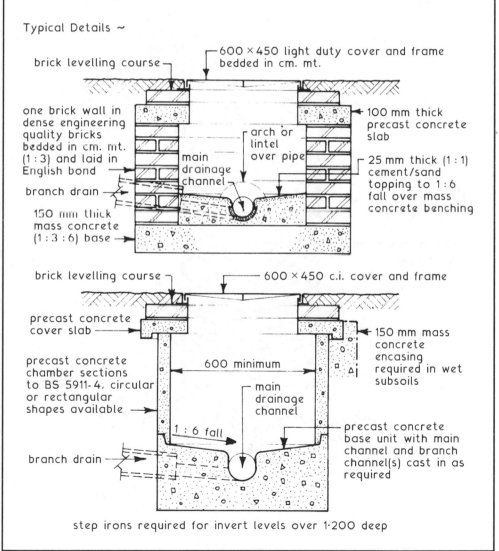

step irons required for invert levels over 1·200 deep

Plastic Inspection Chambers ~ the raising piece can be sawn horizontally with a carpenter's saw to suit depth requirements with the cover and frame fitted at surface level. Bedding may be a 100 mm prepared shingle base or 150 mm wet concrete to ensure a uniform support.

The unit may need weighting to retain it in place in areas of high water table, until backfilled with granular material. Under roads a peripheral concrete collar is applied to the top of the chamber in addition to the 150 mm thickness of concrete surrounding the inspection chamber.

Typical Example ~

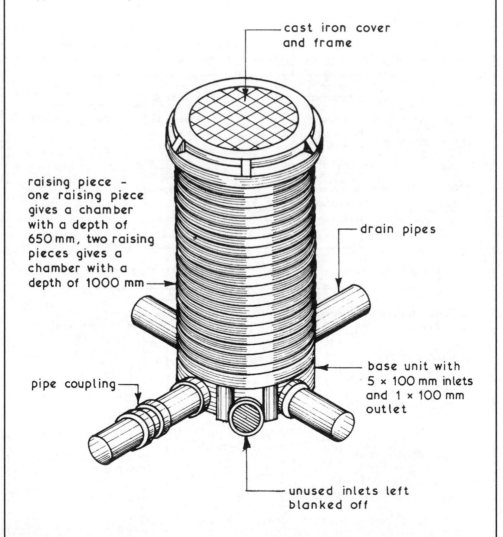

cast iron cover and frame

drain pipes

raising piece - one raising piece gives a chamber with a depth of 650 mm, two raising pieces gives a chamber with a depth of 1000 mm

base unit with 5 × 100 mm inlets and 1 × 100 mm outlet

pipe coupling

unused inlets left blanked off

Means Of Access - provision is required for maintenance and inspection of drainage systems. This should occur at:

* the head (highest part) or close to it
* a change in horizontal direction
* a change in vertical direction (gradient)
* a change in pipe diameter
* a junction, unless the junction can be rodded through from an access point
* long straight runs (see table)

Maximum spacing of drain access points (m)

From: To:	Small access fitting	Large access fitting	Junction	Inspection chamber	Manhole
Drain head	12	12		22	45
Rodding eye	22	22	22	45	45
Small access fitting			12	22	22
Large access fitting			22	45	45
Inspection chamber	22	45	22	45	45
Manhole				45	90

* Small access fitting is 150 mm dia. or 150 mm × 100 mm. Large access fitting is 225 mm × 100 mm.

Rodding Eyes and Shallow Access Chambers - these may be used at the higher parts of drainage systems where the volume of excavation and cost of an inspection chamber or manhole would be unnecessary. SACs have the advantage of providing access in both directions. Covers to all drain openings should be secured to deter unauthorised access.

Ref. Building Regulations, Approved Document H1: Foul Water Drainage.

Excavations ~ drains are laid in trenches which are set out, excavated and supported in a similar manner to foundation trenches except for the base of the trench which is cut to the required gradient or fall.

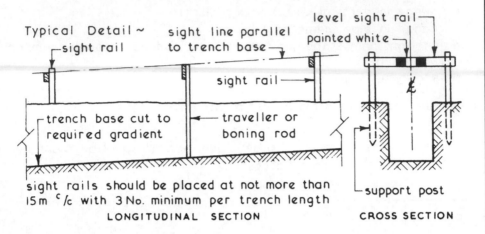

Typical Detail ~ sight line parallel to trench base — sight rail

level sight rail — painted white

sight rail →

trench base cut to required gradient ← traveller or boning rod

support post

sight rails should be placed at not more than 15m ᶜ/c with 3 No. minimum per trench length

LONGITUDINAL SECTION

CROSS SECTION

Joints ~ these must be watertight under all working and movement conditions and this can be achieved by using rigid and flexible joints in conjuntion with the appropriate bedding.

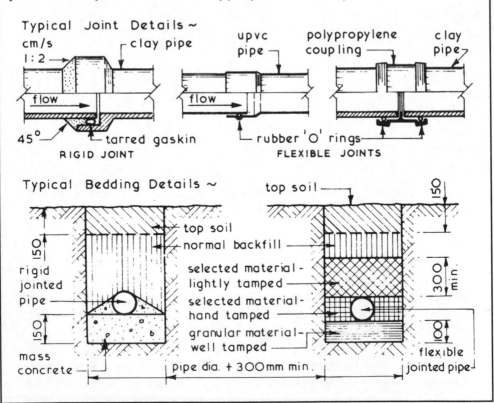

Typical Joint Details ~

cm/s 1:2 → clay pipe

upvc pipe

polypropylene coupling

clay pipe

flow

flow

45° — tarred gaskin

rubber 'O' rings

RIGID JOINT

FLEXIBLE JOINTS

Typical Bedding Details ~

top soil

rigid jointed pipe

top soil

normal backfill →

selected material - lightly tamped

selected material - hand tamped

granular material - well tamped

mass concrete

pipe dia. + 300mm min.

flexible jointed pipe

Watertightness ~ must be ensured to prevent water seapage and erosion of the subsoil. Also, in the interests of public health, foul water should not escape untreated. The Building Regulaions, Approved Document H1: Section 2 specifies either an air or water test to determine soundness of installation.

AIR TEST ~ equipment : manometer and accessories (see page 719) 2 drain stoppers, one with tube attachment

Application ~

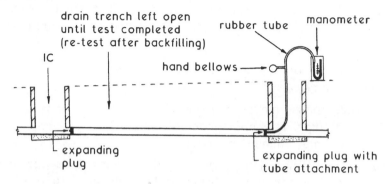

drain trench left open until test completed (re-test after backfilling)

IC

rubber tube

manometer

hand bellows →

expanding plug

expanding plug with tube attachment

Test ~ 100mm water gauge to fall no more than 25mm in 5mins. Or, 50mm w.g. to fall no more than 12mm in 5mins.

WATER TEST ~ equipment : Drain stopper

Test bend

Extension pipe

Application ~

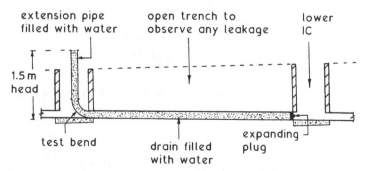

extension pipe filled with water

open trench to observe any leakage

lower IC

1.5 m head

test bend

drain filled with water

expanding plug

Test ~ 1·5m head of water to stand for 2 hours and then topped up. Leakage over the next 30 minutes should be minimal, i.e.

100mm pipe – 0·05 litres per metre, which equates to a drop of 6·4mm/m in the extension pipe, and

150mm pipe – 0·08 litres per metre, which equates to a drop of 4·5mm/m in the extension pipe.

Drainage Pipes ~ sizes for normal domestic foul water applications:-

< 20 dwellings   = 100 mm diameter
20–150 dwellings = 150 mm diameter

Exceptions: 75 mm diameter for waste or rainwater only (no WCs)
            150 mm diameter minimum for a public sewer

Other situations can be assessed by summating the Discharge Units from appliances and converting these to an appropriate diameter stack and drain, see BS EN 12056-2 (stack) and BS EN 752-4 (drain). Gradient will also affect pipe capacity and when combined with discharge calculations, provides the basis for complex hydraulic theories.

The simplest correlation of pipe size and fall, is represented in Maguire's rule:-

4" (100 mm) pipe, minimum gradient 1 in 40
6" (150 mm) pipe, minimum gradient 1 in 60
9" (225 mm) pipe, minimum gradient 1 in 90

The Building Regulations, approved Document H1 provides more scope and relates to foul water drains running at 0·75 proportional depth. See Diagram 9 and Table 6 in Section 2 of the Approved Document.

Other situations outside of design tables and empirical practice can be calculated.

eg. A 150 mm diameter pipe flowing 0·5 proportional depth.

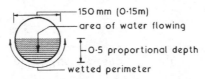

150 mm (0·15m)
area of water flowing
0·5 proportional depth
wetted perimeter

Applying the Chezy formula for gradient calculations:-

$$v = c\sqrt{m \times i}$$

where: v = velocity of flow, (min for self cleansing = 0·8 m/s)

$\qquad$ c = Chezy coefficient (58)

$\qquad$ m = hydraulic mean depth or;
$\qquad\qquad$ $\dfrac{\text{area of water flowing}}{\text{wetted perimeter}}$ for 0·5 p.d. = diam/4

$\qquad$ i = inclination or gradient as a fraction 1/x

Selecting a velocity of 1 m/s as a margin of safety over the minimum:-

$$1 = 58\sqrt{0.15/4 \times i}$$

i = 0·0079 where i = 1/x
So, x = 1/0·0079 = 126, i.e. a minimum gradient of 1 in 126

Water supply ~ an adequate supply of cold water of drinking quality should be provided to every residential building and a drinking water tap installed within the building. The installation should be designed to prevent waste, undue consumption, misuse, contamination of general supply, be protected against corrosion and frost damage and be accessible for maintenance activities. The intake of a cold water supply to a building is owned jointly by the water authority and the consumer who therefore have joint maintenance responsibilities.

Typical Water Supply Arrangement ~

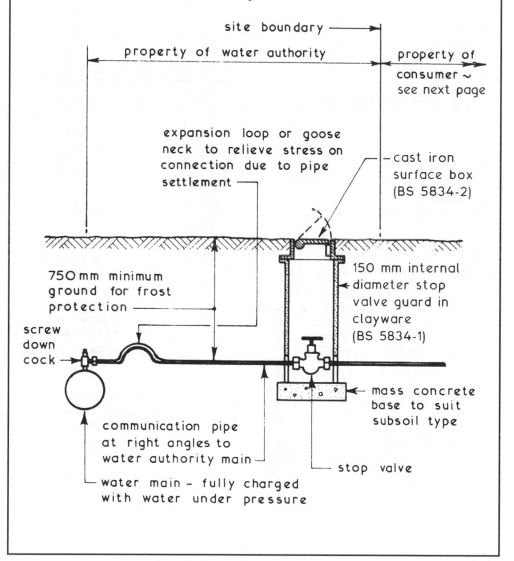

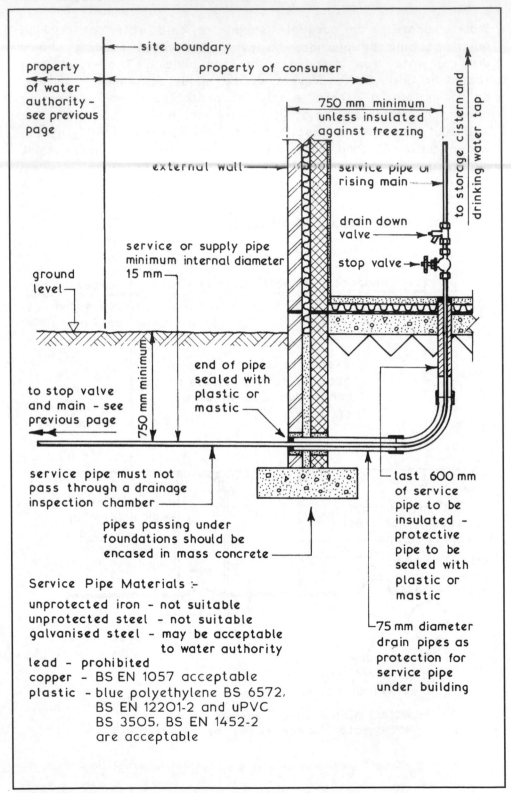

site boundary

property of water authority – see previous page

property of consumer

750 mm minimum unless insulated against freezing

external wall

service pipe or rising main

drain down valve

service or supply pipe minimum internal diameter 15 mm

stop valve

to storage cistern and drinking water tap

ground level

750 mm minimum

end of pipe sealed with plastic or mastic

to stop valve and main – see previous page

service pipe must not pass through a drainage inspection chamber

pipes passing under foundations should be encased in mass concrete

last 600 mm of service pipe to be insulated – protective pipe to be sealed with plastic or mastic

75 mm diameter drain pipes as protection for service pipe under building

Service Pipe Materials :-

unprotected iron – not suitable
unprotected steel – not suitable
galvanised steel – may be acceptable to water authority
lead – prohibited
copper – BS EN 1057 acceptable
plastic – blue polyethylene BS 6572, BS EN 12201-2 and uPVC BS 3505, BS EN 1452-2 are acceptable

General ~ when planning or designing any water installation the basic physical laws must be considered:

1. Water is subject to the force of gravity and will find its own level.
2. To overcome friction within the conveying pipes water which is stored prior to distribution will require to be under pressure and this is normally achieved by storing the water at a level above the level of the outlets. The vertical distance between these levels is usually called the head.
3. Water becomes less dense as its temperature is raised, therefore warm water will always displace colder water whether in a closed or open circuit.

Direct Cold Water Systems ~ the cold water is supplied to the outlets at mains pressure; the only storage requirements is a small capacity cistern to feed the hot water storage tank. These systems are suitable for districts which have high level reservoirs with a good supply and pressure. The main advantage is that drinking water is available from all cold water outlets, disadvantages include lack of reserve in case of supply cut off, risk of back syphonage due to negative mains pressure and a risk of reduced pressure during peak demand periods.

Typical Direct Cold Water System ~

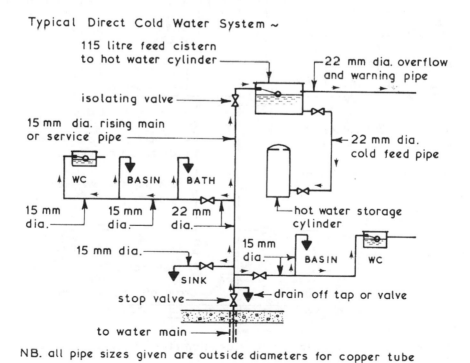

NB. all pipe sizes given are outside diameters for copper tube

Indirect Systems ~ Cold water is supplied to all outlets from a cold water storage cistern except for the cold water supply to the sink(s) where the drinking water tap is connected directly to incoming supply from the main. This system requires more pipework than the direct system but it reduces the risk of back syphonage and provides a reserve of water should the mains supply fail or be cut off. The local water authority will stipulate the system to be used in their area.

Typical Indirect Cold Water System ~

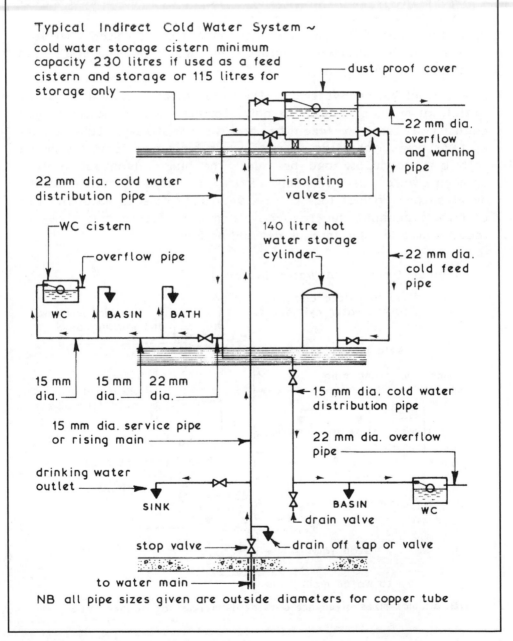

cold water storage cistern minimum capacity 230 litres if used as a feed cistern and storage or 115 litres for storage only

dust proof cover

22 mm dia. overflow and warning pipe

22 mm dia. cold water distribution pipe

isolating valves

140 litre hot water storage cylinder

22 mm dia. cold feed pipe

WC cistern

overflow pipe

WC   BASIN   BATH

15 mm dia.   15 mm dia.   22 mm dia.

15 mm dia. cold water distribution pipe

15 mm dia. service pipe or rising main

22 mm dia. overflow pipe

drinking water outlet

SINK

BASIN   WC

drain valve

stop valve

drain off tap or valve

to water main

NB all pipe sizes given are outside diameters for copper tube

Direct System ~ this is the simplest and least expensive system of hot water installation. The water is heated in the boiler and the hot water rises by convection to the hot water storage tank or cylinder to be replaced by the cooler water from the bottom of the storage vessel. Hot water drawn from storage is replaced with cold water from the cold water storage cistern. Direct systems are suitable for soft water areas and for installations which are not supplying a central heating circuit.

Typical Direct Hot Water System ~

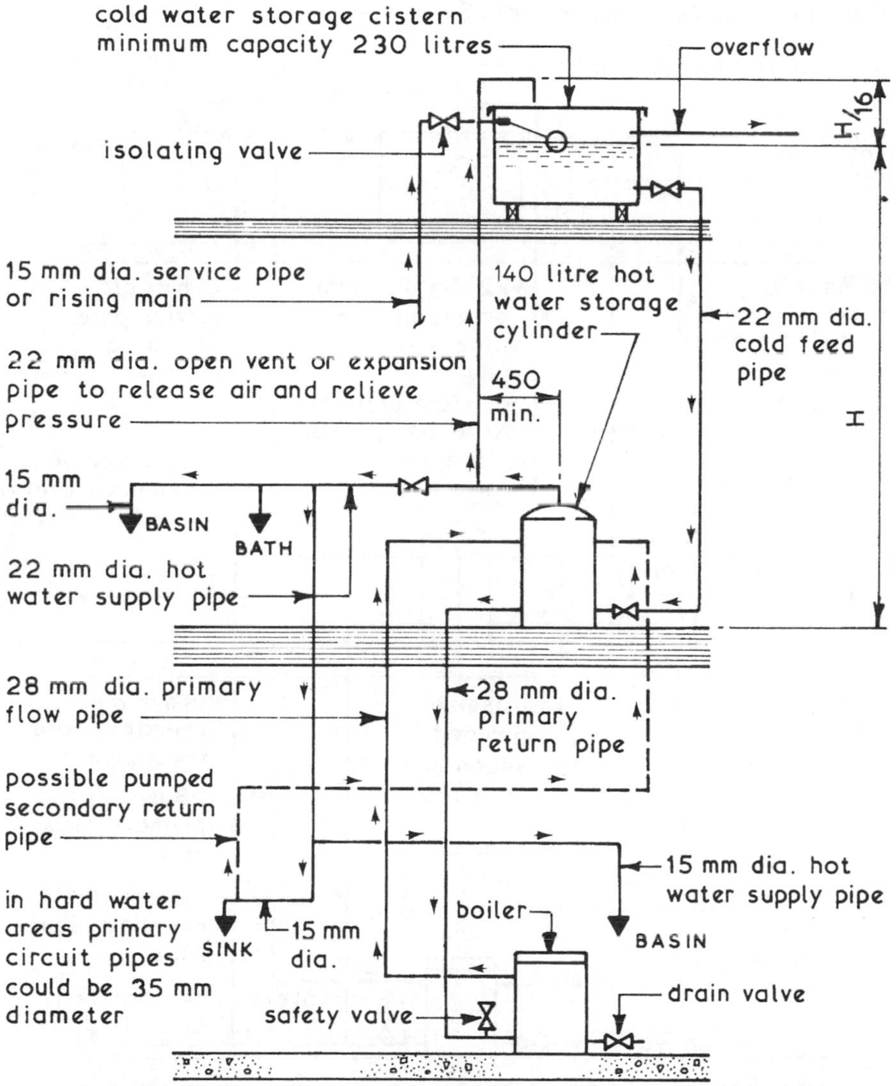

cold water storage cistern
minimum capacity 230 litres ——

overflow

isolating valve ——

$H/16$

15 mm dia. service pipe
or rising main ——

140 litre hot
water storage
cylinder ——

22 mm dia.
cold feed
pipe

22 mm dia. open vent or expansion
pipe to release air and relieve
pressure ——

450
min.

H

15 mm
dia. ——

BASIN   BATH

22 mm dia. hot
water supply pipe ——

28 mm dia. primary
flow pipe ——

28 mm dia.
primary
return pipe

possible pumped
secondary return
pipe ——

15 mm dia. hot
water supply pipe

in hard water
areas primary
circuit pipes
could be 35 mm
diameter

SINK

15 mm
dia.

boiler ——

BASIN

drain valve

safety valve ——

NB all pipe sizes given are outside diameters for copper tube

Indirect System ~ this is a more complex system than the direct system but it does overcome the problem of furring which can occur in direct hot water systems. This method is therefore suitable for hard water areas and in all systems where a central heating circuit is to be part of the hot water installation. Basically the pipe layouts of the two systems are similar but in the indirect system a separate small capacity feed cistern is required to charge and top up the primary circuit. In this system the hot water storage tank or cylinder is in fact a heat exchanger – see page 712.

Typical Indirect Hot Water System ~

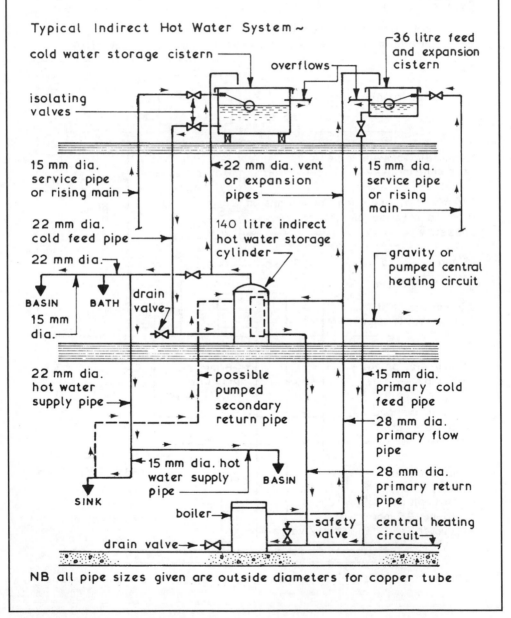

cold water storage cistern

overflows

36 litre feed and expansion cistern

isolating valves

15 mm dia. service pipe or rising main

22 mm dia. vent or expansion pipes

15 mm dia. service pipe or rising main

22 mm dia. cold feed pipe

140 litre indirect hot water storage cylinder

22 mm dia.

BASIN   BATH

drain valve

15 mm dia.

gravity or pumped central heating circuit

22 mm dia. hot water supply pipe

possible pumped secondary return pipe

15 mm dia. primary cold feed pipe

28 mm dia. primary flow pipe

15 mm dia. hot water supply pipe

BASIN

28 mm dia. primary return pipe

SINK

boiler

safety valve

central heating circuit

drain valve

NB all pipe sizes given are outside diameters for copper tube

Mains Fed Indirect System ~ now widely used as an alternative to conventional systems. It eliminates the need for cold water storage and saves considerably on installation time. This system is established in Europe and the USA, but only acceptable in the UK at the local water authority's discretion. It complements electric heating systems, where a boiler is not required. An expansion vessel replaces the standard vent and expansion pipe and may be integrated with the hot water storage cylinder. It contains a neoprene diaphragm to separate water from air, the air providing a 'cushion' for the expansion of hot water. Air loss can be replenished by foot pump as required.

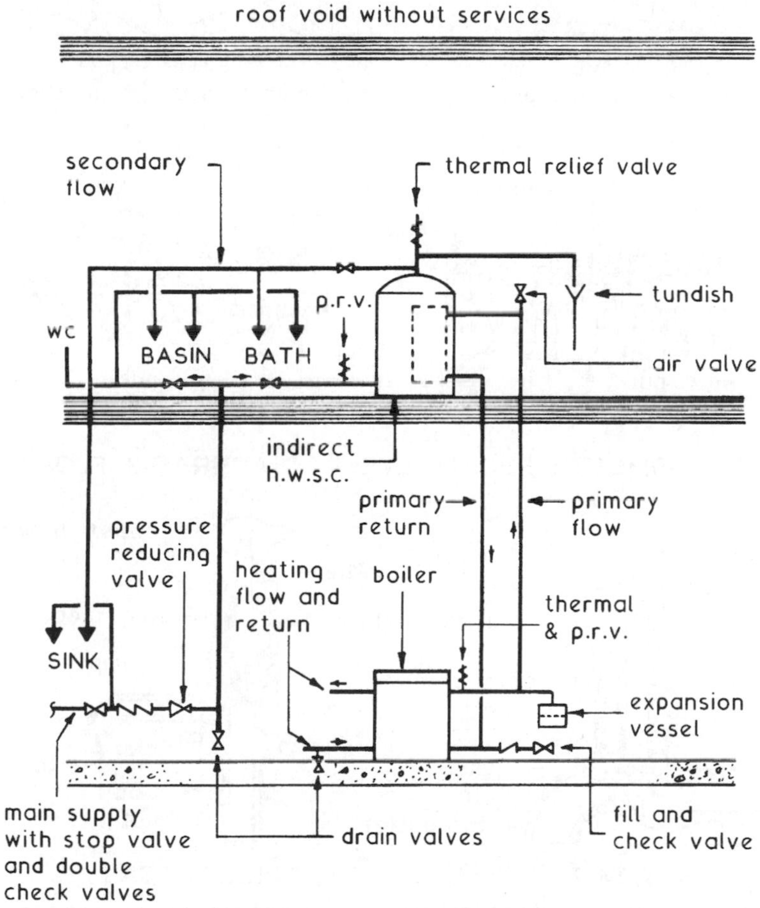

roof void without services

BASIN   BATH
WC

secondary flow
thermal relief valve
p.r.v.
tundish
air valve

indirect h.w.s.c.

primary return
primary flow

pressure reducing valve
heating flow and return
boiler
thermal & p.r.v.

SINK

expansion vessel

main supply with stop valve and double check valves
drain valves
fill and check valve

NB. p.r.v. = pressure relief (safety) valve

Flow Controls ~ these are valves inserted into a water installation to control the water flow along the pipes or to isolate a branch circuit or to control the draw-off of water from the system.

Typical Examples ~

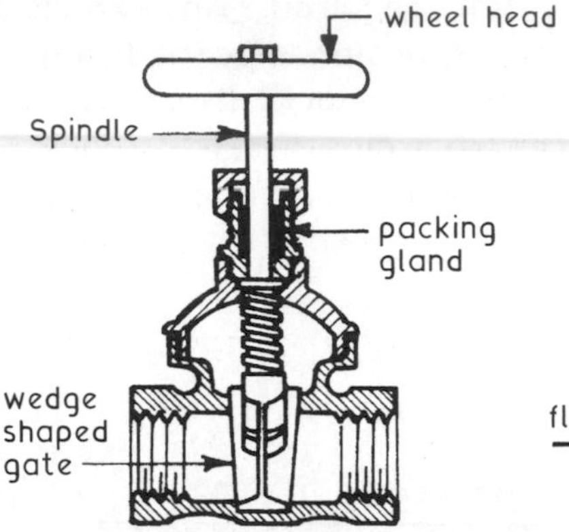

wheel head

Spindle

packing gland

wedge shaped gate

GATE VALVE
low pressure cistern supply

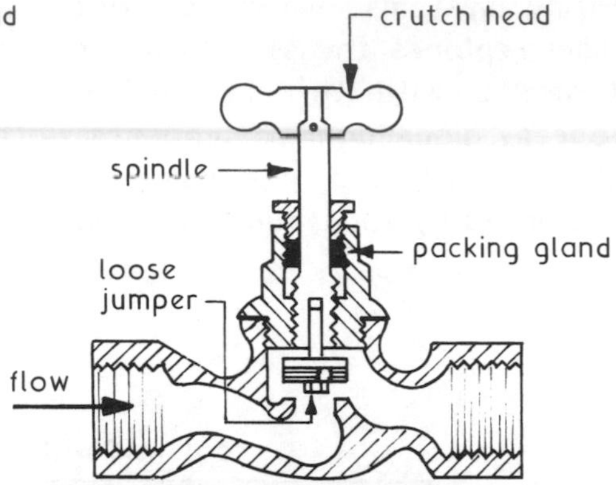

crutch head

spindle

packing gland

loose jumper

flow

STOP VALVE
high pressure mains supply

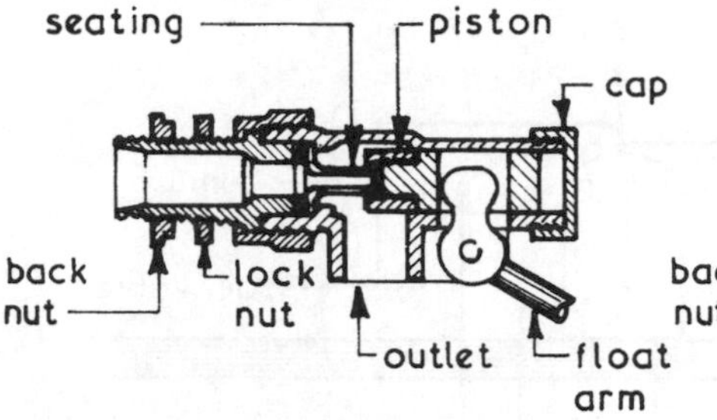

seating — piston

cap

back nut — lock nut

outlet — float arm

PORTSMOUTH FLOATVALVE

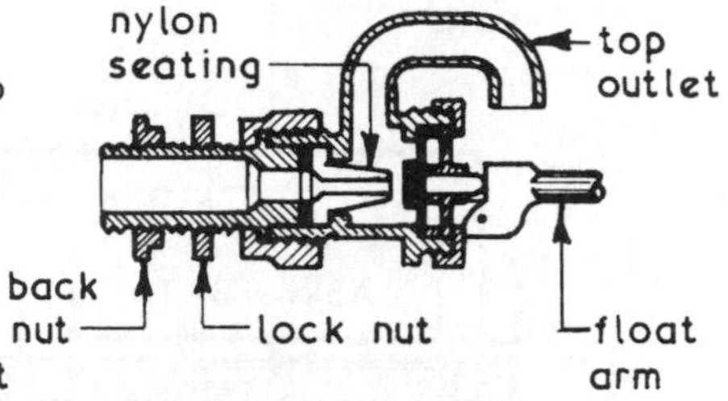

nylon seating

top outlet

back nut — lock nut

float arm

DIAPHRAGM FLOATVALVE

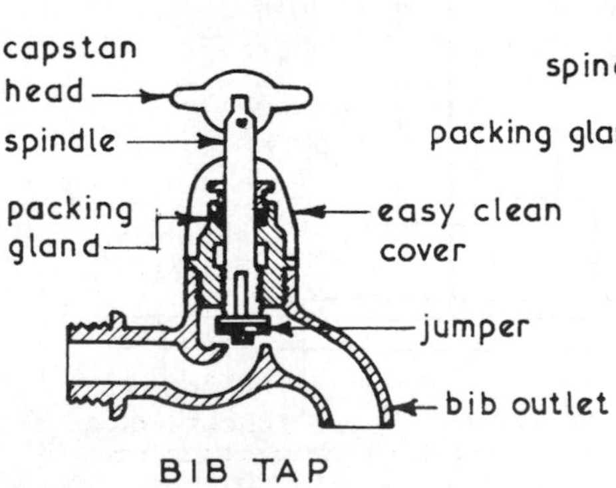

capstan head

spindle

packing gland

easy clean cover

jumper

bib outlet

BIB TAP
horizontal inlet - used over sinks and for hose pipe outlets

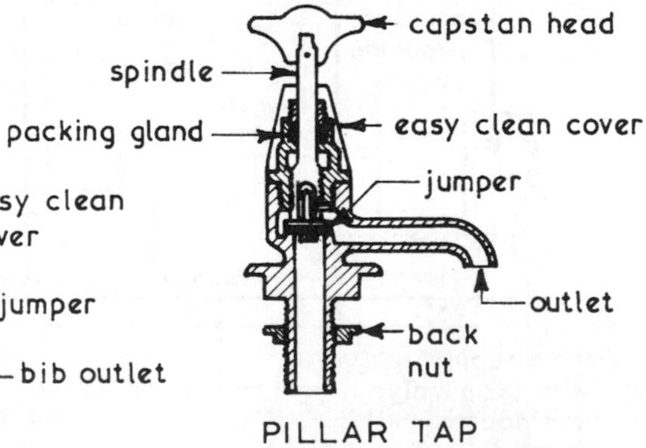

capstan head

spindle

packing gland

easy clean cover

jumper

outlet

back nut

PILLAR TAP
vertical inlet - used in conjunction with fittings

Cisterns ~ these are fixed containers used for storing water at atmospheric pressure. The inflow of water is controlled by a floatvalve which is adjusted to shut off the water supply when it has reached the designed level within the cistern. The capacity of the cistern depends on the draw off demand and whether the cistern feeds both hot and cold water systems. Domestic cold water cisterns should be placed at least 750 mm away from an external wall or roof surface and in such a position that it can be inspected, cleaned and maintained. A minimum clear space of 300 mm is required over the cistern for floatvalve maintenance. An overflow or warning pipe of not less than 22 mm diameter must be fitted to fall away to discharge in a conspicuous position. All draw off pipes must be fitted with a gate valve positioned as near to the cistern as possible.

Cisterns are available in a variety of sizes and materials such as galvanised mild steel (BS 417-2), moulded plastic (BS 4213) and reinforced plastic (BS 4994). If the cistern and its associated pipework are to be housed in a cold area such as a roof they should be insulated against freezing.

Typical Details ~

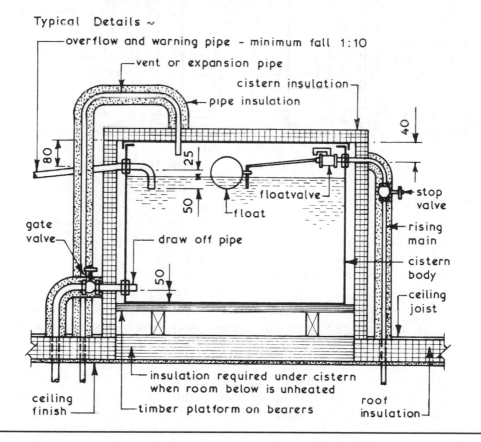

711

Indirect Hot Water Cylinders ~ these cylinders are a form of heat exchanger where the primary circuit of hot water from the boiler flows through a coil or annulus within the storage vessel and transfers the heat to the water stored within. An alternative hot water cylinder for small installations is the single feed or `Primatic´ cylinder which is self venting and relies on two air locks to separate the primary water from the secondary water. This form of cylinder is connected to pipework in the same manner as for a direct system (see page 707) and therefore gives savings in both pipework and fittings. Indirect cylinders usually conform to the recommendations of BS 417-2 (galvanised mild steel) or BS 1566-1 (copper). Primatic or single feed cylinders to BS 1566-2 (copper).

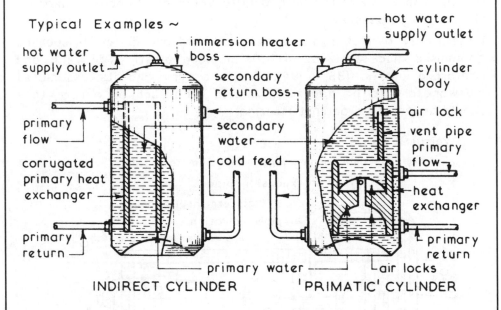

INDIRECT CYLINDER   'PRIMATIC' CYLINDER

Primatic Cylinders ~

1. Cylinder is filled in the normal way and the primary system is filled via the heat exchanger, as the initial filling continues air locks are formed in the upper and lower chambers of the heat exchanger and in the vent pipe.
2. The two air locks in the heat exchanger are permanently maintained and are self-recuperating in operation. These air locks isolate the primary water from the secondary water almost as effectively as a mechanical barrier.
3. The expansion volume of total primary water at a flow temperature of 82°C is approximately 1/25 and is accommodated in the upper expansion chamber by displacing air into the lower chamber; upon contraction reverse occurs.

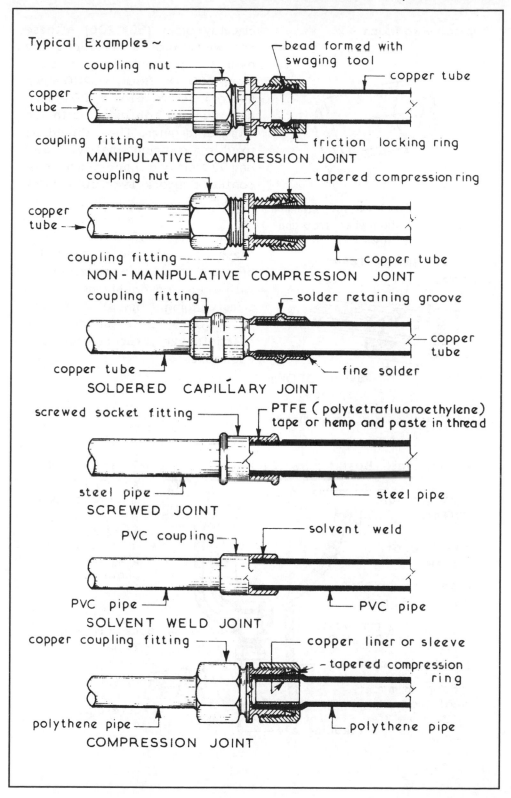

Typical Examples ~

bead formed with swaging tool

coupling nut

copper tube

copper tube

coupling fitting

friction locking ring

MANIPULATIVE COMPRESSION JOINT

coupling nut

tapered compression ring

copper tube

coupling fitting

copper tube

NON-MANIPULATIVE COMPRESSION JOINT

coupling fitting

solder retaining groove

copper tube

copper tube

fine solder

SOLDERED CAPILLARY JOINT

screwed socket fitting

PTFE (polytetrafluoroethylene) tape or hemp and paste in thread

steel pipe

steel pipe

SCREWED JOINT

PVC coupling

solvent weld

PVC pipe

PVC pipe

SOLVENT WELD JOINT

copper coupling fitting

copper liner or sleeve

tapered compression ring

polythene pipe

polythene pipe

COMPRESSION JOINT

Typical Examples ~

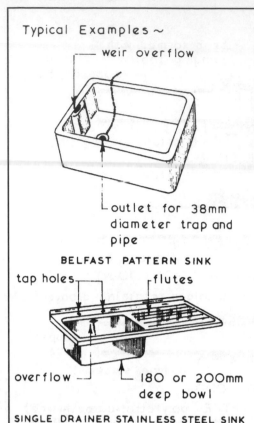

— weir overflow

└outlet for 38mm
diameter trap and
pipe

BELFAST PATTERN SINK

tap holes — flutes

overflow — └ 180 or 200mm
deep bowl

SINGLE DRAINER STAINLESS STEEL SINK

Fireclay Sinks (BS 1206) – these are white glazed sinks and are available in a wide range of sizes from 460 × 380 × 200 deep up to 1220 × 610 × 305 deep and can be obtained with an integral drainer. They should be fixed at a height between 850 and 920 mm and supported by legs, cantilever brackets or dwarf brick walls.

Metal Sinks (BS EN 13310) – these can be made of enamelled pressed steel or stainless steel with single or double drainers in sizes ranging from 1070 × 460 to 1600 × 530 supported on a cantilever brackets or sink cupboards.

Ceramic Wash Basins (BS 1188)

splash back ———

fixing height
790 mm

bowl – depth
165 mm

plan size
457 x 653mm

32 mm waste
to BS EN 274 —

pedestal support ———

38 or 78 mm
seal trap ———

cleaning eye ———

NB sink could
be supported on
cantilever brackets

——— overflow

——— plug and
chain

——— back nut and
washer

——— waste pipe

714

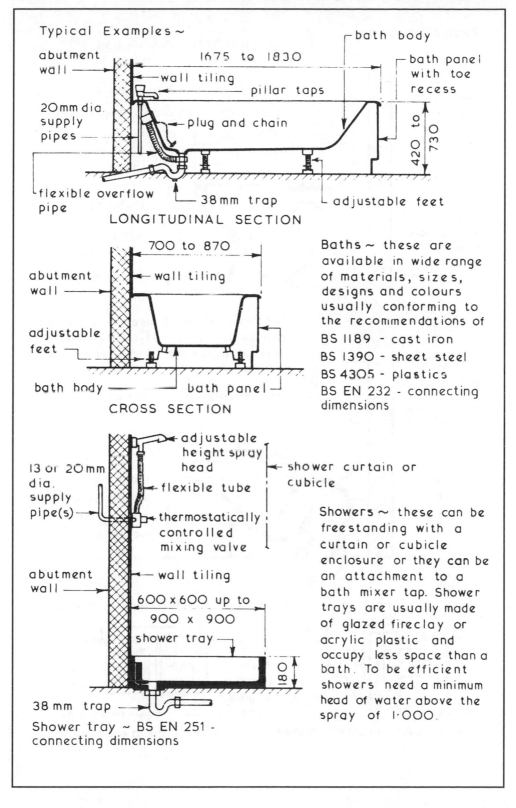

Typical Examples ~

abutment wall

1675 to 1830

bath body

wall tiling

bath panel with toe recess

pillar taps

20mm dia. supply pipes

plug and chain

420 to 730

flexible overflow pipe

38mm trap

adjustable feet

LONGITUDINAL SECTION

700 to 870

abutment wall

wall tiling

adjustable feet

bath body

bath panel

CROSS SECTION

Baths ~ these are available in wide range of materials, sizes, designs and colours usually conforming to the recommendations of

BS 1189 - cast iron

BS 1390 - sheet steel

BS 4305 - plastics

BS EN 232 - connecting dimensions

adjustable height spray head

13 or 20mm dia. supply pipe(s)

flexible tube

thermostatically controlled mixing valve

shower curtain or cubicle

Showers ~ these can be freestanding with a curtain or cubicle enclosure or they can be an attachment to a bath mixer tap. Shower trays are usually made of glazed fireclay or acrylic plastic and occupy less space than a bath. To be efficient showers need a minimum head of water above the spray of 1·000.

abutment wall

wall tiling

600 x 600 up to 900 x 900

shower tray

180

38 mm trap

Shower tray ~ BS EN 251 - connecting dimensions

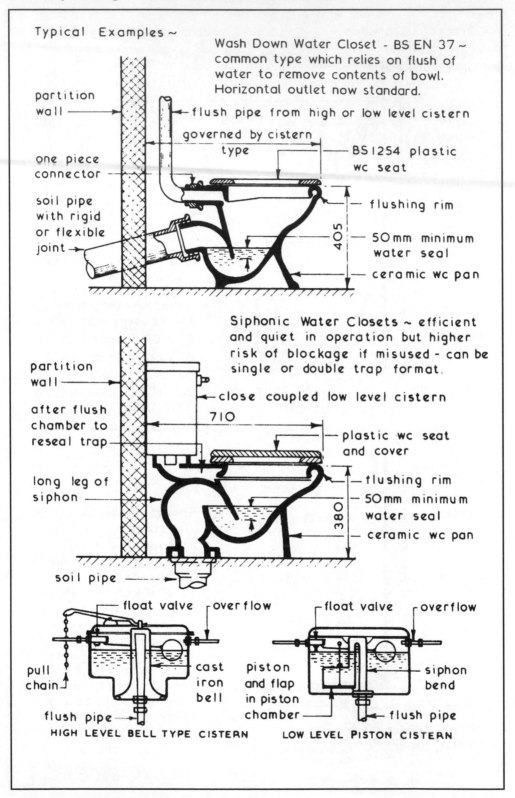

Typical Examples ~

Wash Down Water Closet - BS EN 37 ~ common type which relies on flush of water to remove contents of bowl. Horizontal outlet now standard.

partition wall

flush pipe from high or low level cistern

governed by cistern type

BS 1254 plastic wc seat

one piece connector

flushing rim

405

50 mm minimum water seal

soil pipe with rigid or flexible joint

ceramic wc pan

Siphonic Water Closets ~ efficient and quiet in operation but higher risk of blockage if misused - can be single or double trap format.

partition wall

close coupled low level cistern

after flush chamber to reseal trap

710

plastic wc seat and cover

flushing rim

long leg of siphon

380

50 mm minimum water seal

ceramic wc pan

soil pipe

float valve ~ overflow

float valve ~ overflow

pull chain

cast iron bell

piston and flap in piston chamber

siphon bend

flush pipe

flush pipe

HIGH LEVEL BELL TYPE CISTERN

LOW LEVEL PISTON CISTERN

Single Stack System ~ method developed by the Building Research Establishment to eliminate the need for ventilating pipework to maintain the water seals in traps to sanitary fittings. The slope and distance of the branch connections must be kept within the design limitations given below. This system is only possible when the sanitary appliances are closely grouped around the discharge stack.

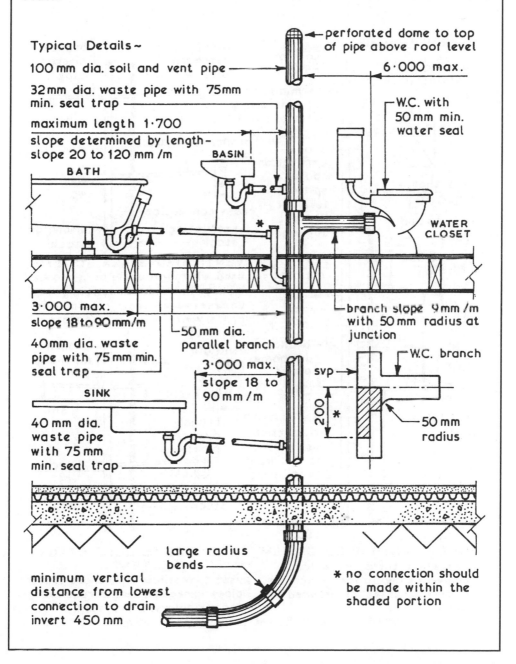

Typical Details ~

100 mm dia. soil and vent pipe

32 mm dia. waste pipe with 75 mm min. seal trap

maximum length 1·700

slope determined by length— slope 20 to 120 mm/m

BATH

BASIN

perforated dome to top of pipe above roof level

6·000 max.

W.C. with 50 mm min. water seal

WATER CLOSET

3·000 max.
slope 18 to 90 mm/m

40 mm dia. waste pipe with 75 mm min. seal trap

SINK

40 mm dia. waste pipe with 75 mm min. seal trap

50 mm dia. parallel branch

3·000 max.
slope 18 to 90 mm/m

branch slope 9 mm/m with 50 mm radius at junction

W.C. branch

svp

200

50 mm radius

large radius bends

minimum vertical distance from lowest connection to drain invert 450 mm

* no connection should be made within the shaded portion

Ventilated Stack Systems ~ where the layout of sanitary appliances is such that they do not conform to the requirements for the single stack system shown on page 717 ventilating pipes will be required to maintain the water seals in the traps. Three methods are available to overcome the problem, namely a fully ventilated system, a ventilated stack system and a modified single stack system which can be applied over any number of storeys.

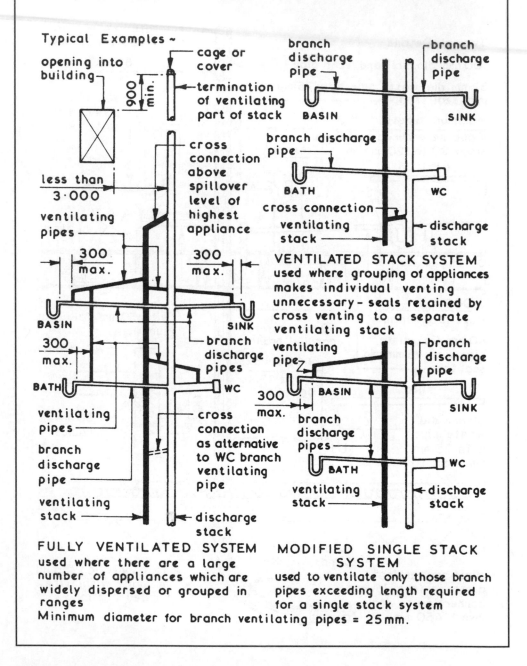

Typical Examples ~

VENTILATED STACK SYSTEM
used where grouping of appliances makes individual venting unnecessary - seals retained by cross venting to a separate ventilating stack

FULLY VENTILATED SYSTEM
used where there are a large number of appliances which are widely dispersed or grouped in ranges

MODIFIED SINGLE STACK SYSTEM
used to ventilate only those branch pipes exceeding length required for a single stack system

Minimum diameter for branch ventilating pipes = 25mm.

Airtightness ~ must be ensured to satisfy public health legislation. The Building Regulations, Approved Document H1: Section 1, provides minimum standards for test procedures. An air or smoke test on the stack must produce a pressure at least equal to 38 mm water gauge for not less than 3 minutes.

Application ~

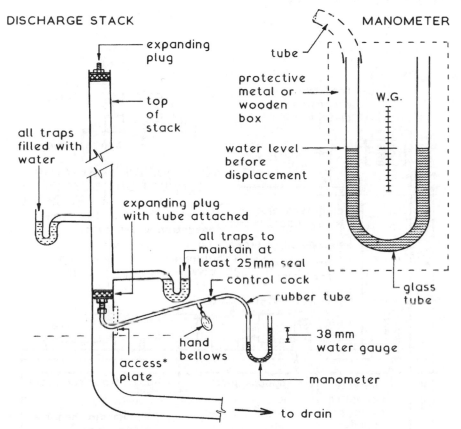

DISCHARGE STACK   MANOMETER

expanding plug

top of stack

all traps filled with water

expanding plug with tube attached

all traps to maintain at least 25 mm seal

control cock

rubber tube

hand bellows

access* plate

to drain

tube

protective metal or wooden box

water level before displacement

W.G.

glass tube

38 mm water gauge

manometer

* if access plate is not provided, top connection to first IC may be plugged and rubber tube inserted through wc pan seal.

NB. Smoke tests are rarely applied now as the equipment is quite bulky and unsuited for use with uPVC pipes. Smoke producing pellets are ideal for leakage detection, but must not come into direct contact with plastic materials.

One Pipe System ~ the hot water is circulated around the system by means of a centrifugal pump. The flow pipe temperature being about 80°C and the return pipe temperature being about 60 to 70°C. The one pipe system is simple in concept and easy to install but has the main disadvantage that the hot water passing through each heat emitter flows onto the next heat emitter or radiator, therefore the average temperature of successive radiators is reduced unless the radiators are carefully balanced or the size of the radiators at the end of the circuit are increased to compensate for the temperature drop.

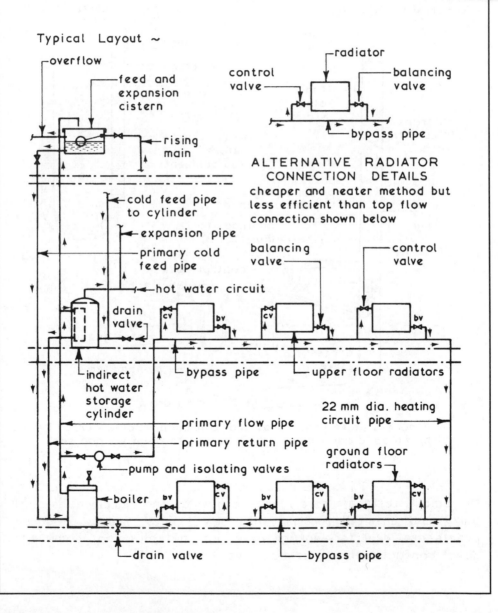

Typical Layout ~

overflow

feed and expansion cistern

rising main

radiator

control valve

balancing valve

bypass pipe

ALTERNATIVE RADIATOR CONNECTION DETAILS
cheaper and neater method but less efficient than top flow connection shown below

cold feed pipe to cylinder

expansion pipe

primary cold feed pipe

hot water circuit

drain valve

balancing valve

control valve

indirect hot water storage cylinder

bypass pipe

upper floor radiators

22 mm dia. heating circuit pipe

primary flow pipe

primary return pipe

ground floor radiators

pump and isolating valves

boiler

drain valve

bypass pipe

Two Pipe System ~ this is a dearer but much more efficient system than the one pipe system shown on the previous page. It is easier to balance since each radiator or heat emitter receives hot water at approximately the same temperature because the hot water leaving the radiator is returned to the boiler via the return pipe without passing through another radiator.

Typical Layout ~

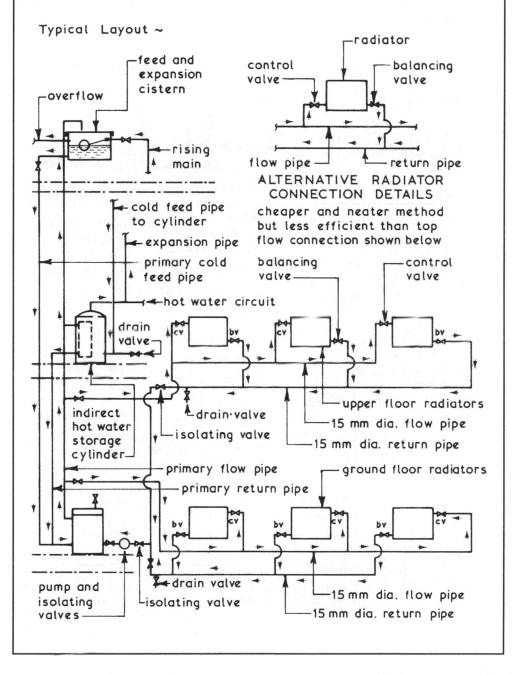

ALTERNATIVE RADIATOR
CONNECTION DETAILS
cheaper and neater method
but less efficient than top
flow connection shown below

Micro Bore System ~ this system uses 6 to 12mm diameter soft copper tubing with an individual flow and return pipe to each heat emitter or radiator from a 22mm diameter manifold. The flexible and unobstrusive pipework makes this system easy to install in awkward situations but it requires a more powerful pump than that used in the traditional small bore systems. The heat emitter or radiator valves can be as used for the one or two pipe small bore systems alternatively a double entry valve can be used.

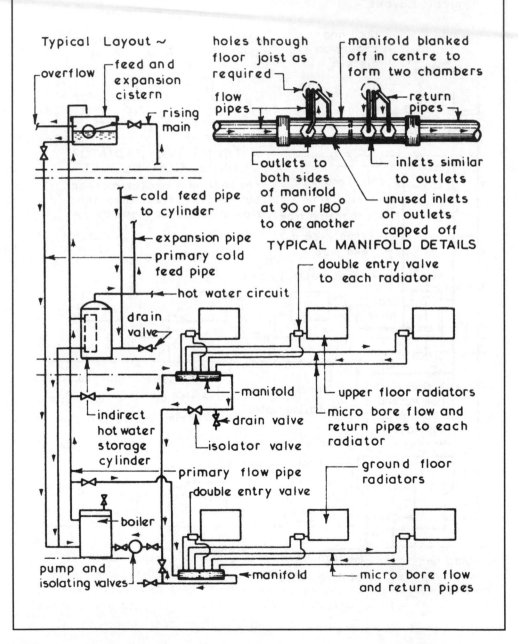

Typical Layout ~

overflow

feed and expansion cistern

rising main

cold feed pipe to cylinder

expansion pipe

primary cold feed pipe

hot water circuit

drain valve

indirect hot water storage cylinder

primary flow pipe

double entry valve

boiler

pump and isolating valves

manifold

drain valve

isolator valve

holes through floor joist as required

manifold blanked off in centre to form two chambers

flow pipes

return pipes

outlets to both sides of manifold at 90 or 180° to one another

inlets similar to outlets

unused inlets or outlets capped off

TYPICAL MANIFOLD DETAILS

double entry valve to each radiator

upper floor radiators

micro bore flow and return pipes to each radiator

ground floor radiators

manifold

micro bore flow and return pipes

Controls ~ the range of controls available to regulate the heat output and timing operations for a domestic hot water heating system is considerable, ranging from thermostatic radiator control valves to programmers and controllers.

Typical Example ~

Boiler – fitted with a thermostat to control the temperature of the hot water leaving the boiler.

Heat Emitters or Radiators – fitted with thermostatically controlled radiator valves to control flow of hot water to the radiators to keep room at desired temperature.

Programmer/Controller – this is basically a time switch which can usually be set for 24 hours, once daily or twice daily time periods and will generally give separate programme control for the hot water supply and central heating systems. The hot water cylinder and room thermostatic switches control the pump and motorised valve action.

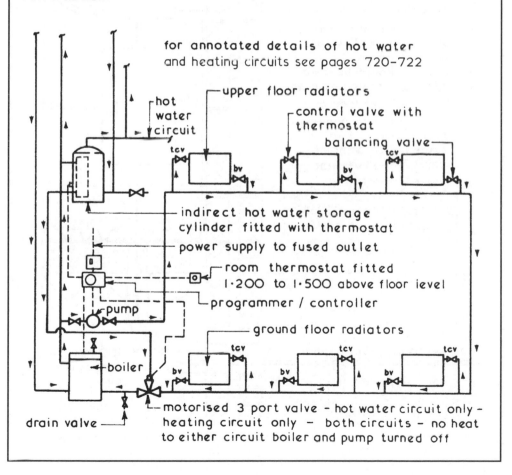

for annotated details of hot water and heating circuits see pages 720–722

upper floor radiators

control valve with thermostat

balancing valve

hot water circuit

indirect hot water storage cylinder fitted with thermostat

power supply to fused outlet

room thermostat fitted 1·200 to 1·500 above floor level

programmer / controller

ground floor radiators

pump

boiler

drain valve

motorised 3 port valve – hot water circuit only – heating circuit only – both circuits – no heat to either circuit boiler and pump turned off

Electrical Supply ~ in England and Wales electricity is generated and supplied by National Power, PowerGen and Nuclear Electric and distributed through regional supply companies, whereas in Scotland it is generated, supplied and distributed by Scottish Power and the Scottish Hydro-Electric Power Company. The electrical supply to a domestic installation is usually 230 volt single phase and is designed with the following basic aims:-

1. Proper earthing to avoid shocks to occupant.
2. Prevention of current leakage.
3. Prevention of outbreak of fire.

Typical Electrical Supply Intake Details ~

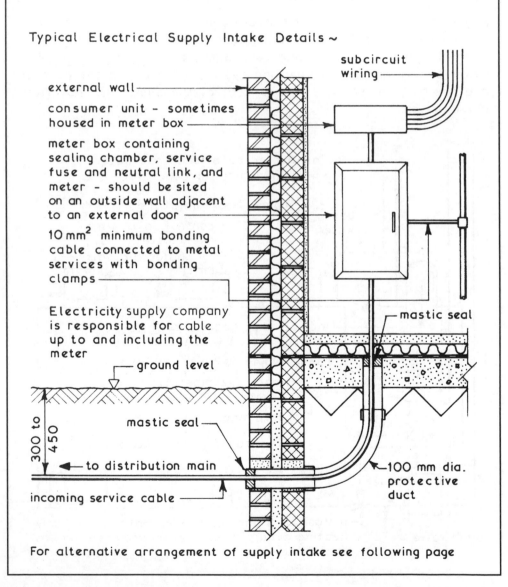

external wall

consumer unit - sometimes housed in meter box

meter box containing sealing chamber, service fuse and neutral link, and meter - should be sited on an outside wall adjacent to an external door

10 mm^2 minimum bonding cable connected to metal services with bonding clamps

Electricity supply company is responsible for cable up to and including the meter

ground level

300 to 450

mastic seal

to distribution main

incoming service cable

subcircuit wiring

mastic seal

100 mm dia. protective duct

For alternative arrangement of supply intake see following page

Electrical Supply Intake ~ although the electrical supply intake can be terminated in a meter box situated within a dwelling, most supply companies prefer to use the external meter box to enable the meter to be read without the need to enter the premises.

Typical Electrical Supply Intake Details ~

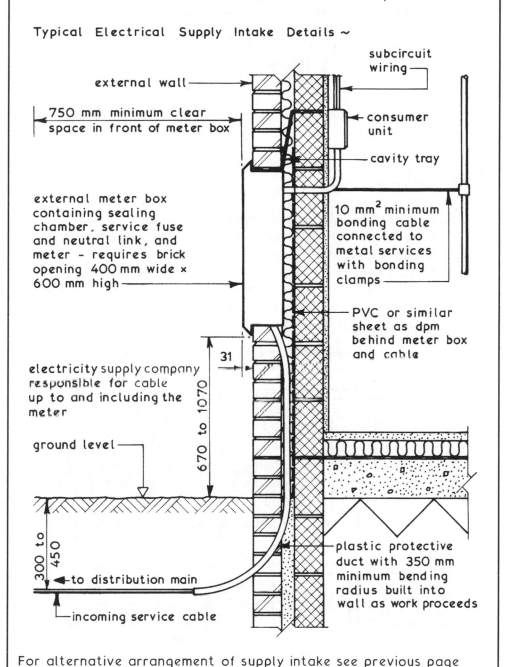

external wall

750 mm minimum clear space in front of meter box

subcircuit wiring

consumer unit

cavity tray

external meter box containing sealing chamber, service fuse and neutral link, and meter - requires brick opening 400 mm wide × 600 mm high

10 mm^2 minimum bonding cable connected to metal services with bonding clamps

PVC or similar sheet as dpm behind meter box and cable

electricity supply company responsible for cable up to and including the meter

31

670 to 1070

ground level

300 to 450

to distribution main

incoming service cable

plastic protective duct with 350 mm minimum bending radius built into wall as work proceeds

For alternative arrangement of supply intake see previous page

725

Entry and Intake of Electrical Service ~ the local electricity supply company is responsible for providing electricity up to and including the meter, but the consumer is responsible for safety and protection of the company's equipment. The supplier will install the service cable up to the meter position where their termination equipment is installed. This equipment may be located internally or fixed externally on a wall, the latter being preferred since it gives easy access for reading the meter – see details on the previous page.

Meter Boxes – generally the supply company's meters and termination equipment are housed in a meter box. These are available in fibreglass and plastic, ranging in size from 450mm wide × 638mm high to 585m wide × 815mm high with an overall depth of 177mm.

Consumer Control Unit – this provides a uniform, compact and effective means of efficiently controlling and distributing electrical energy within a dwelling. The control unit contains a main double pole isolating switch controlling the live phase and neutral conductors, called bus bars. These connect to the fuses or miniature circuit breakers protecting the final subcircuits.

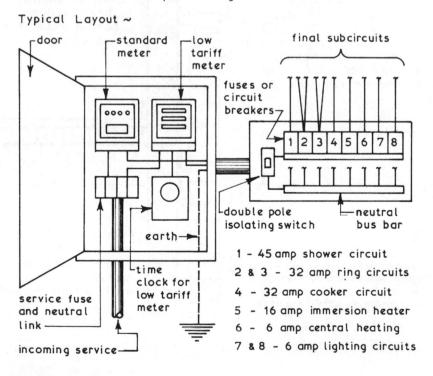

Typical Layout ~

1 – 45 amp shower circuit
2 & 3 – 32 amp ring circuits
4 – 32 amp cooker circuit
5 – 16 amp immersion heater
6 – 6 amp central heating
7 & 8 – 6 amp lighting circuits

Consumer's Power Supply Control Unit – this is conveniently abbreviated to consumer unit. As described on the previous page, it contains a supply isolator switch, live, neutral and earth bars, plus a range of individual circuit over-load safety protection devices. By historical reference this unit is sometimes referred to as a fuse box, but modern variants are far more sophisticated. Over-load protection is provided by miniature circuit breakers attached to the live or phase bar. Additional protection is provided by a split load residual current device (RCD) dedicated specifically to any circuits that could be used as a supply to equipment outdoors, e.g. power sockets on a ground floor ring main.

RCD – a type of electro-magnetic switch or solenoid which disconnects the electricity supply when a surge of current or earth fault occurs. See Part 10 of the Building Services Handbook for more detail.

Typical Split Load Consumer Unit –

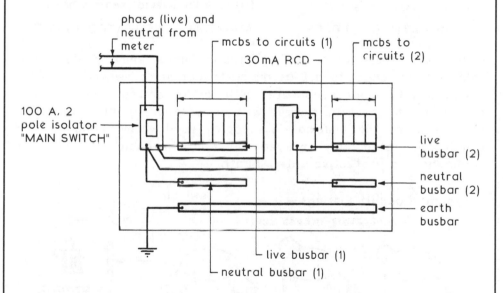

Note that with an overhead supply, the MAIN SWITCH is combined with a 100 mA RCD protecting all circuits.

Note:

Circuits (1) to fixtures, i.e. lights, cooker, immersion heater and smoke alarms.

Circuits (2) to socket outlets that could supply portable equipment outdoors.

Electric Cables ~ these are made up of copper or aluminium wires called conductors surrounded by an insulating material such as PVC or rubber.

Typical Examples ~

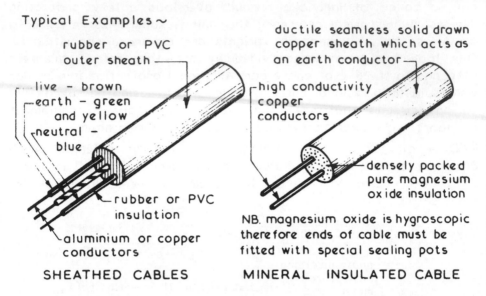

SHEATHED CABLES

MINERAL INSULATED CABLE

NB. magnesium oxide is hygroscopic therefore ends of cable must be fitted with special sealing pots

Conduits ~ these are steel or plastic tubes which protect the cables. Steel conduits act as an earth conductor whereas plastic conduits will require a separate earth conductor drawn in. Conduits enable a system to be rewired without damage or interference of the fabric of the building. The cables used within conduits are usually insulated only, whereas in non-rewireable systems the cables have a protective outer sheath.

Typical Conduit Fittings ~

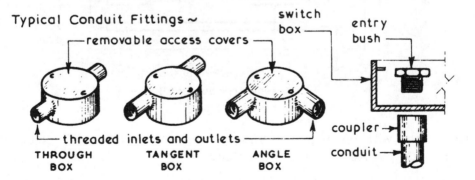

THROUGH BOX

TANGENT BOX

ANGLE BOX

Trunking – alternative to conduit and consists of a preformed cable carrier which is surface mounted and is fitted with a removable or 'snap on' cover which can have the dual function of protection and trim or surface finish.

Wiring systems ~ rewireable systems housed in horizontal conduits can be cast into the structural floor slab or sited within the depth of the floor screed To ensure that such a system is rewireable, draw-in boxes must be incorporated at regular intervals and not more than two right angle boxes to be included between draw-in points. Vertical conduits can be surface mounted or housed in a chase cut in to a wall provided the depth of the chase is not more than one-third of the wall thickness. A horizontal non-rewireable system can be housed within the depth of the timber joists to a suspended floor whereas vertical cables can be surface mounted or housed in a length of conduit as described for rewireable systems.

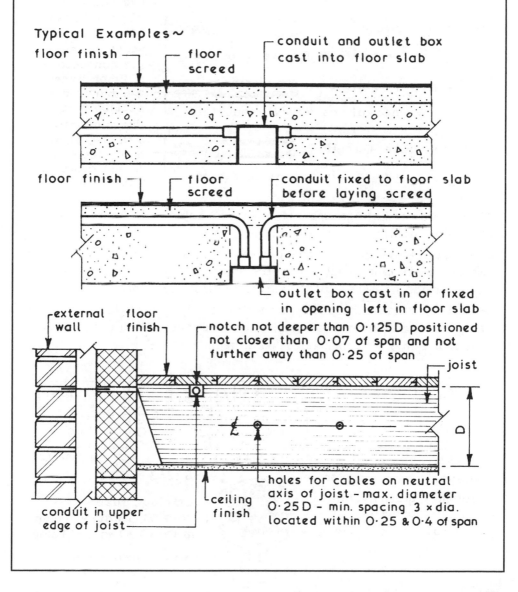

Typical Examples~

floor finish — | — floor screed | — conduit and outlet box cast into floor slab

floor finish — | — floor screed | — conduit fixed to floor slab before laying screed

— outlet box cast in or fixed in opening left in floor slab

external wall | floor finish — | — notch not deeper than 0·125D positioned not closer than 0·07 of span and not further away than 0·25 of span | — joist

conduit in upper edge of joist— | ceiling finish | — holes for cables on neutral axis of joist – max. diameter 0·25D – min. spacing 3 × dia. located within 0·25 & 0·4 of span

Cable Sizing ~ the size of a conductor wire can be calculated taking into account the maximum current the conductor will have to carry (which is limited by the heating effect caused by the resistance to the flow of electricity through the conductor) and the voltage drop which will occur when the current is carried. For domestic electrical installations the following minimum cable specifications are usually suitable –

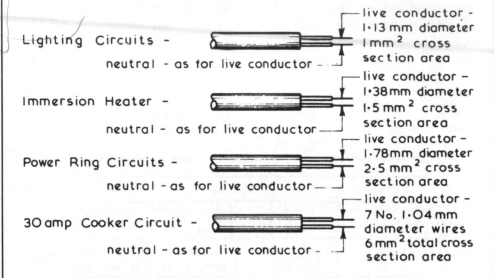

Lighting Circuits –
neutral – as for live conductor
live conductor - 1·13 mm diameter 1 mm² cross section area

Immersion Heater –
neutral – as for live conductor
live conductor – 1·38 mm diameter 1·5 mm² cross section area

Power Ring Circuits –
neutral – as for live conductor
live conductor – 1·78 mm diameter 2·5 mm² cross section area

30 amp Cooker Circuit –
neutral – as for live conductor
live conductor – 7 No. 1·04 mm diameter wires 6 mm² total cross section area

All the above ratings are for one twin cable with or without an earth conductor.

Electrical Accessories ~ for power circuits these include cooker control units and fused connector units for fixed appliances such as immersion heaters, water heaters and refrigerators.

Socket Outlets ~ these may be single or double outlets, switched or unswitched, surface or flush mounted and may be fitted with indicator lights. Recommended fixing heights are –

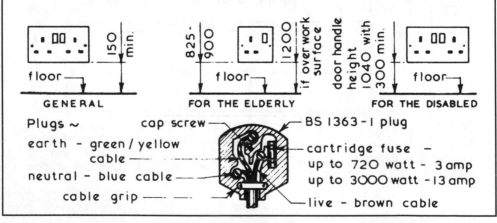

GENERAL — 150 min. — floor

FOR THE ELDERLY — 825 - 900 — 1200 if over work surface — floor

FOR THE DISABLED — door handle height 1040 with 300 min. — floor

Plugs ~
cap screw
earth – green/yellow cable
neutral – blue cable
cable grip

BS 1363-1 plug
cartridge fuse –
up to 720 watt – 3 amp
up to 3000 watt – 13 amp
live – brown cable

Power Circuits ~ in new domestic electrical installations the ring main system is usually employed instead of the older system of having each socket outlet on its own individual fused circuit with unfused round pin plugs. Ring circuits consist of a fuse or miniature circuit breaker protected subcircuit with a 32 amp rating of a live conductor, neutral conductor and an earth looped from socket outlet to socket outlet. Metal conduit systems do not require an earth wire providing the conduit is electrically sound and earthed. The number of socket outlets per ring main is unlimited but a separate circuit must be provided for every 100m² of floor area. To conserve wiring, spur outlets can be used as long as the total number of spur outlets does not exceed the total number of outlets connected to the ring and that there is not more than two outlets per spur.

Typical Ring Main Wiring Diagram ~

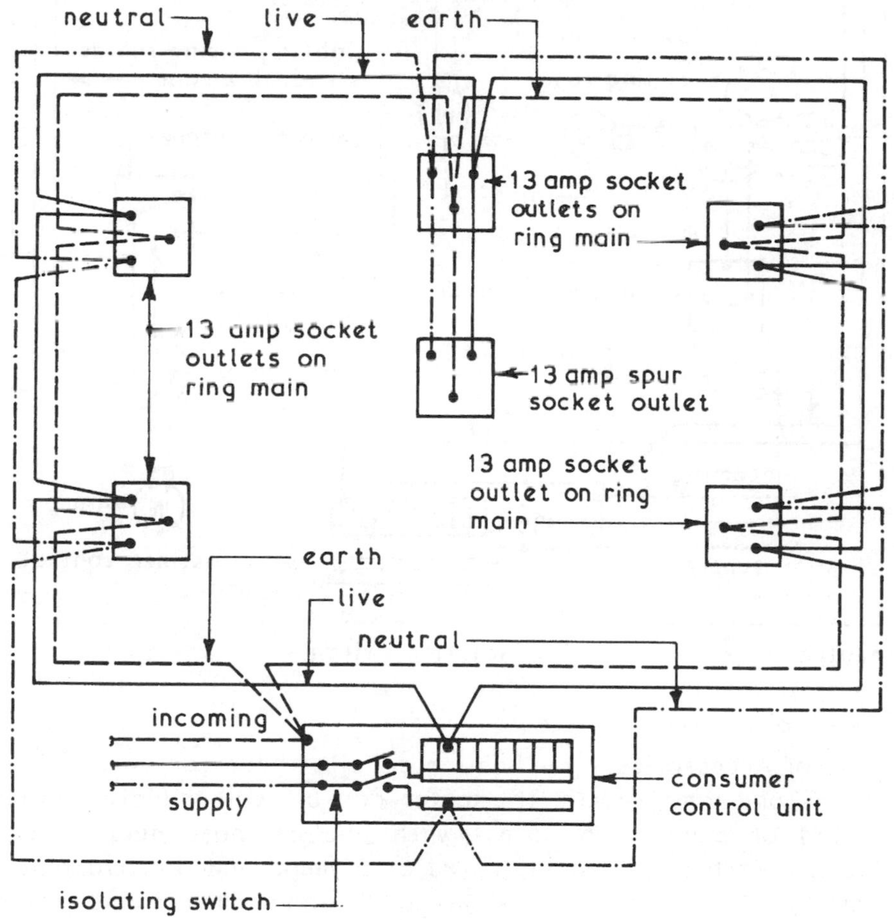

neutral—   live—   earth—

13 amp socket outlets on ring main

13 amp socket outlets on ring main

13 amp spur socket outlet

13 amp socket outlet on ring main

earth

live

neutral

incoming

supply

consumer control unit

isolating switch—

731

Lighting Circuits ~ these are usually wired by the loop-in method using an earthed twin cable with a 6 amp fuse or miniature circuit breaker protection. In calculating the rating of a lighting circuit an allowance of 100 watts per outlet should be used. More than one lighting circuit should be used for each installation so that in the event of a circuit failure some lighting will be in working order.

Typical Lighting Circuit Wiring Diagram ~

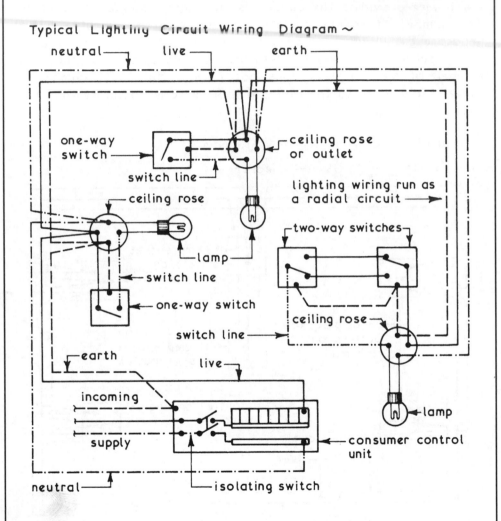

Electrical Accessories ~ for lighting circuits these consist mainly of switches and lampholders, the latter can be wall mounted, ceiling mounted or pendant in format with one or more bulb or tube holders. Switches are usually rated at 5 amps and are available in a variety of types such as double or 2 gang, dimmer and pull or pendant switches. The latter must always be used in bathrooms.

Gas Supply ~ potential consumers of mains gas may apply to their local utilities supplier for connection, e.g. Transco (Lattice Group plc) The cost is normally based on a fee per metre run. However, where the distance is considerable, the gas authority may absorb some of the cost if there is potential for more customers. The supply, appliances and installation must comply with the safety requirements made under the Gas Safety (Installation and Use) Regulations, 1998, and Part J of the Building Regulations.

Typical Gas Supply Arrangement ~

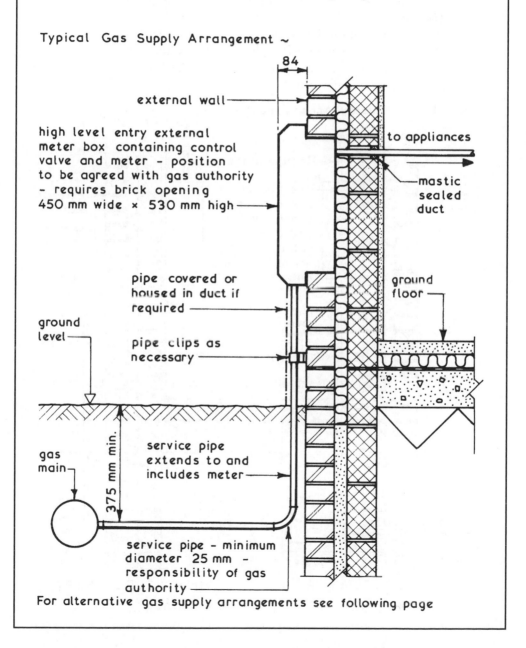

For alternative gas supply arrangements see following page

Gas Service Pipes ~

1. Whenever possible the service pipe should enter the building on the side nearest to the main.
2. A service pipe must not pass under the foundations of a building.
3. No service pipe must be run within a cavity but it may pass through a cavity by the shortest route.
4. Service pipes passing through a wall or solid floor must be enclosed by a sleeve or duct which is and sealed with mastic.
5. No service pipe shall be housed in an unventilated void.
6. Suitable materials for service pipes are copper (BS EN 1057) and steel (BS EN 10255). Polyethylene (BS 7281 or BS EN 1555-2) is normally used underground.

Typical Gas Supply Arrangement ~

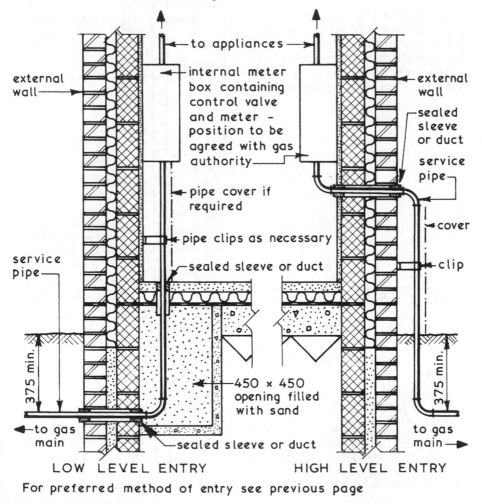

For preferred method of entry see previous page

Gas Fires ~ for domestic use these generally have a low energy rating of less than 7 kW net input and must be installed in accordance with minimum requirements set out in Part J of the Building Regulations. Most gas fires connected to a flue are designed to provide radiant and convected heating whereas the room sealed balanced flue appliances are primarily convector heaters.

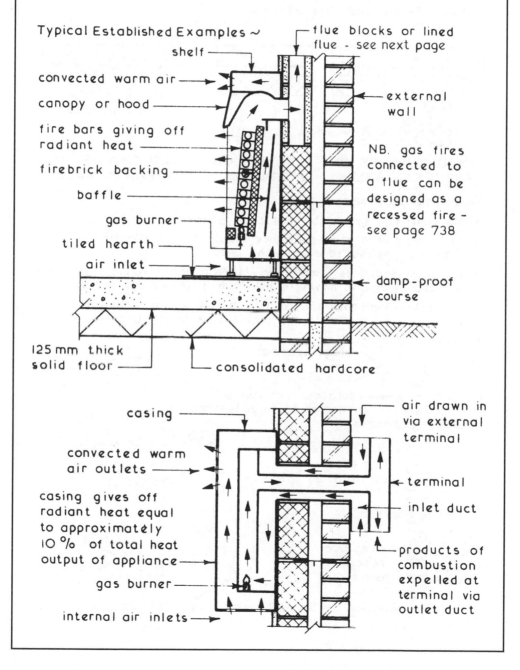

Typical Established Examples ~

flue blocks or lined flue - see next page

shelf

convected warm air

canopy or hood

fire bars giving off radiant heat

firebrick backing

baffle

gas burner

tiled hearth

air inlet

external wall

NB. gas fires connected to a flue can be designed as a recessed fire - see page 738

damp-proof course

125 mm thick solid floor

consolidated hardcore

casing

convected warm air outlets

casing gives off radiant heat equal to approximately 10% of total heat output of appliance

gas burner

internal air inlets

air drawn in via external terminal

terminal

inlet duct

products of combustion expelled at terminal via outlet duct

Gas Fire Flues ~ these can be defined as a passage for the discharge of the products of combustion to the outside air and can be formed by means of a chimney, special flue blocks or by using a flue pipe. In all cases the type and size of the flue as recommended in Approved Document J, BS EN 1806 and BS 5440 will meet the requirements of the Building Regulations.

Typical Single Gas Fire Flues ~

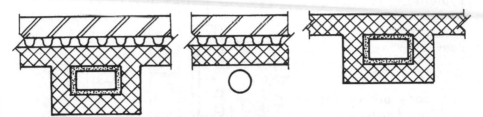

LINED CHIMNEY ON EXTERNAL WALL   FLUE PIPE ON EXTERNAL WALL   LINED CHIMNEY ON INTERNAL WALL

Flue Size Requirements :-

1. No dimension should be less than 63 mm.

2. Flue for a decorative appliance should have a minimum dimension measured across the axis of 175 mm.

3. Flues for gas fires – min. area = 12000 mm^2 if round, 16500 mm^2 if rectangular and having a minimum dimension of 90 mm.

4. Any other appliance should have a flue with a cross-sectional area at least equal to the outlet size of the appliance.

Flue Blocks ~

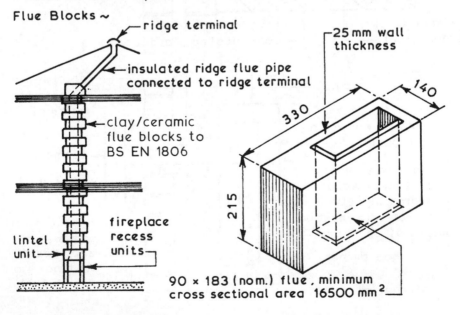

ridge terminal

insulated ridge flue pipe connected to ridge terminal

clay/ceramic flue blocks to BS EN 1806

fireplace recess units

lintel unit

25 mm wall thickness

330

140

215

90 × 183 (nom.) flue, minimum cross sectional area 16500 mm^2

Open Fireplaces ~ for domestic purposes these are a means of providing a heat source by consuming solid fuels with an output rating of under 50 kW. Room-heaters can be defined in a similar manner but these are an enclosed appliance as opposed to the open recessed fireplace.

Components ~ the complete construction required for a domestic open fireplace installation is composed of the hearth, fireplace recess, chimney, flue and terminal.

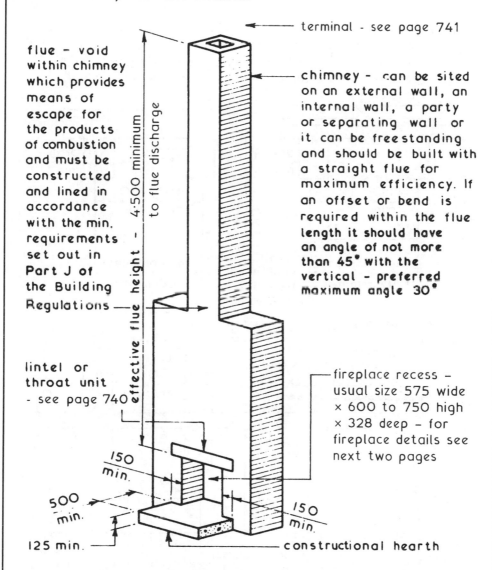

flue - void within chimney which provides means of escape for the products of combustion and must be constructed and lined in accordance with the min. requirements set out in Part J of the Building Regulations

terminal - see page 741

chimney - can be sited on an external wall, an internal wall, a party or separating wall or it can be freestanding and should be built with a straight flue for maximum efficiency. If an offset or bend is required within the flue length it should have an angle of not more than 45° with the vertical - preferred maximum angle 30°

4·500 minimum to flue discharge

effective flue height

lintel or throat unit - see page 740

fireplace recess - usual size 575 wide × 600 to 750 high × 328 deep - for fireplace details see next two pages

150 min.

500 min.

150 min.

125 min.

constructional hearth

See also BS 5854: Code of practice for flues and flue structures in buildings.

Open Fireplace Recesses ~ these must have a constructional hearth and can be constructed of bricks or blocks of concrete or burnt clay or they can be of cast in-situ concrete. All fireplace recesses must have jambs on both sides of the opening and a backing wall of a minimum thickness in accordance with its position and such jambs and backing walls must extend to the full height of the fireplace recess.

Typical Examples ~

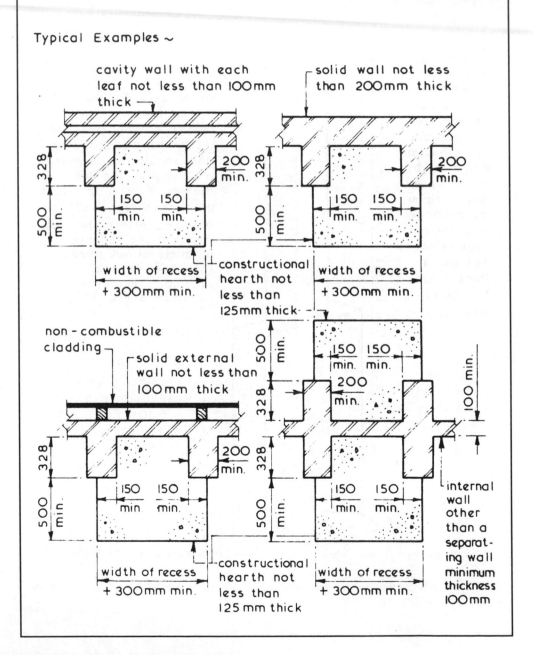

Traditional Fireplace Details ~

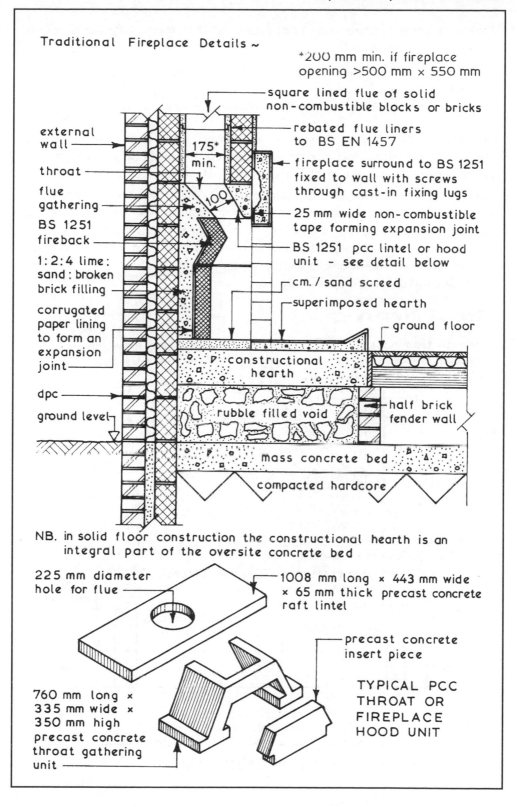

*200 mm min. if fireplace opening >500 mm × 550 mm

square lined flue of solid non-combustible blocks or bricks

rebated flue liners to BS EN 1457

external wall

175* min.

fireplace surround to BS 1251 fixed to wall with screws through cast-in fixing lugs

throat

flue gathering

100

25 mm wide non-combustible tape forming expansion joint

BS 1251 fireback

BS 1251 pcc lintel or hood unit - see detail below

1:2:4 lime: sand: broken brick filling

cm./sand screed

superimposed hearth

corrugated paper lining to form an expansion joint

ground floor

constructional hearth

dpc

ground level

rubble filled void

half brick fender wall

mass concrete bed

compacted hardcore

NB. in solid floor construction the constructional hearth is an integral part of the oversite concrete bed

225 mm diameter hole for flue

1008 mm long × 443 mm wide × 65 mm thick precast concrete raft lintel

precast concrete insert piece

760 mm long × 335 mm wide × 350 mm high precast concrete throat gathering unit

TYPICAL PCC THROAT OR FIREPLACE HOOD UNIT

739

Open Fireplace Chimneys and Flues ~ the main functions of a chimney and flue are to:-

1. Induce an adequate supply of air for the combustion of the fuel being used.
2. Remove the products of combustion.

In fulfilling the above functions a chimney will also encourage a flow of ventilating air promoting constant air changes within the room which will assist in the prevention of condensation

Approved Document J recommends that all flues should be lined with approved materials so that the minimum size of the flue so formed will be 200 mm diameter or a square section of equivalent area. Flues should also be terminated above the roof level as shown, with a significant increase where combustible roof coverings such as thatch or wood shingles are used.

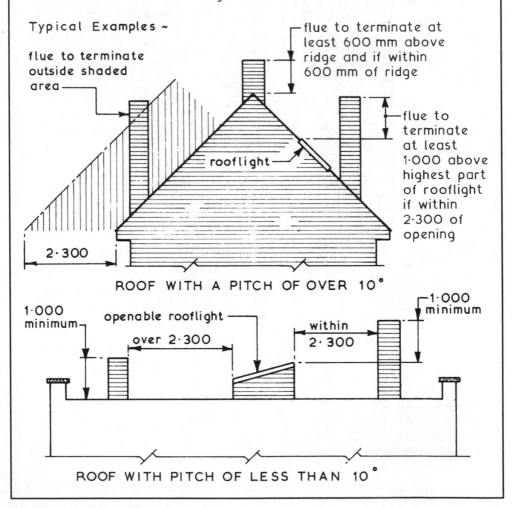

Typical Examples ~

flue to terminate outside shaded area

flue to terminate at least 600 mm above ridge and if within 600 mm of ridge

rooflight

flue to terminate at least 1·000 above highest part of rooflight if within 2·300 of opening

2·300

ROOF WITH A PITCH OF OVER 10°

1·000 minimum

openable rooflight

over 2·300

within 2·300

1·000 minimum

ROOF WITH PITCH OF LESS THAN 10°

Typical Flue Liner and Chimney Pot Details ~

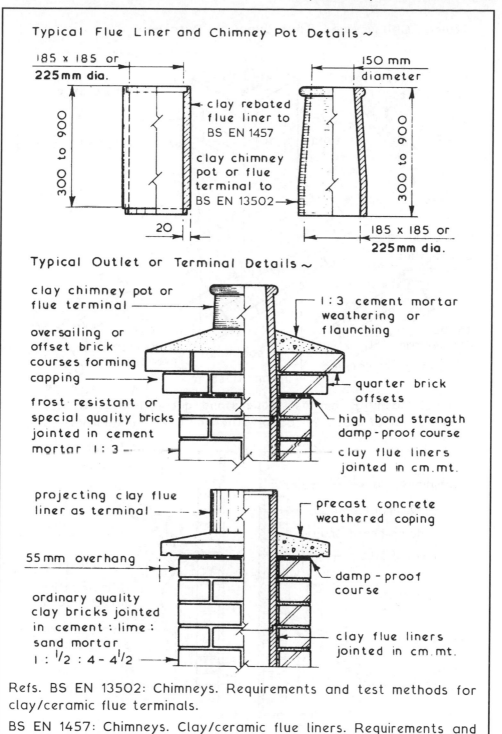

185 x 185 or **225mm dia.**

300 to 900

20

← clay rebated flue liner to BS EN 1457

clay chimney pot or flue terminal to BS EN 13502 →

150 mm diameter

300 to 900

185 x 185 or **225mm dia.**

Typical Outlet or Terminal Details ~

clay chimney pot or flue terminal

oversailing or offset brick courses forming capping

frost resistant or special quality bricks jointed in cement mortar 1:3 –

1:3 cement mortar weathering or flaunching

quarter brick offsets

high bond strength damp-proof course

clay flue liners jointed in cm.mt.

projecting clay flue liner as terminal

55mm overhang

ordinary quality clay bricks jointed in cement : lime : sand mortar 1 : $\frac{1}{2}$: 4 - $4\frac{1}{2}$

precast concrete weathered coping

damp-proof course

clay flue liners jointed in cm.mt.

Refs. BS EN 13502: Chimneys. Requirements and test methods for clay/ceramic flue terminals.

BS EN 1457: Chimneys. Clay/ceramic flue liners. Requirements and test methods.

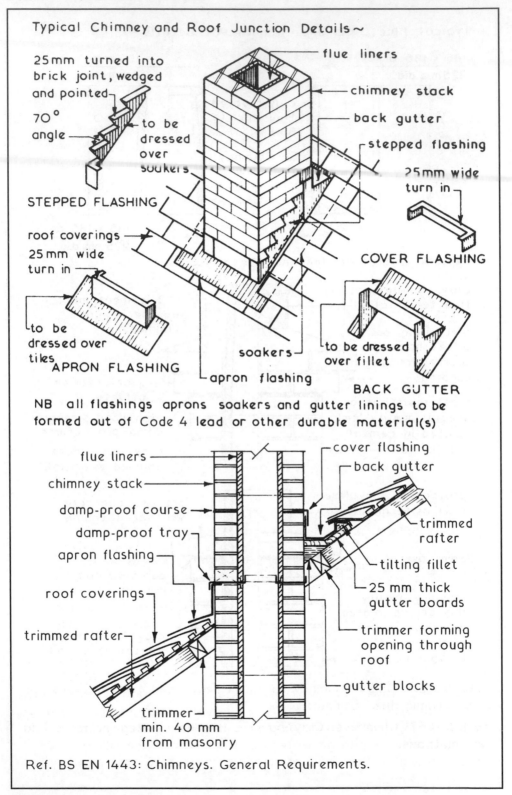

Typical Chimney and Roof Junction Details~

flue liners

chimney stack

back gutter

stepped flashing

25mm turned into brick joint, wedged and pointed

70° angle

to be dressed over soakers

STEPPED FLASHING

25mm wide turn in

COVER FLASHING

roof coverings

25mm wide turn in

to be dressed over tiles
APRON FLASHING

soakers

apron flashing

to be dressed over fillet

BACK GUTTER

NB all flashings aprons soakers and gutter linings to be formed out of Code 4 lead or other durable material(s)

flue liners

chimney stack

damp-proof course

damp-proof tray

apron flashing

roof coverings

trimmed rafter

trimmer min. 40 mm from masonry

cover flashing

back gutter

trimmed rafter

tilting fillet

25 mm thick gutter boards

trimmer forming opening through roof

gutter blocks

Ref. BS EN 1443: Chimneys. General Requirements.

Chimney construction –

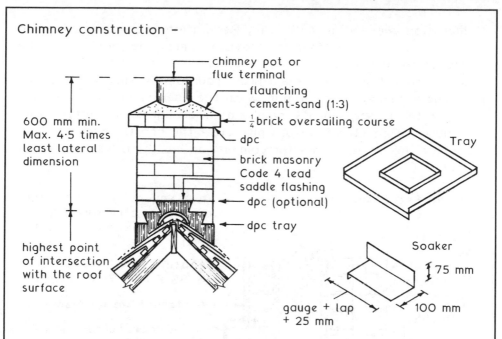

chimney pot or flue terminal

flaunching cement-sand (1:3)

$\frac{1}{4}$ brick oversailing course

dpc

brick masonry
Code 4 lead
saddle flashing

dpc (optional)

dpc tray

600 mm min.
Max. 4·5 times
least lateral
dimension

highest point
of intersection
with the roof
surface

Tray

Soaker

75 mm

100 mm

gauge + lap
+ 25 mm

Typical chimney outlet

Clay bricks – Frost resistant quality. Min. density 1500 kg/m³.

Calcium silicate bricks – Min. compressive strength 20.5 N/mm² (27.5 N/mm² for cappings).

Precast concrete masonry units – Min. compressive strength 15 N/mm².

Mortar – A relatively strong mix of cement and sand 1:3. Cement to be specified as sulphate resisting because of the presence of soluble sulphates in the flue gas condensation.

Chimney pot – The pot should be firmly bedded in at least 3 courses of brickwork to prevent it being dislodged in high winds.

Flashings and dpcs – Essential to prevent water which has permeated the chimney, penetrating into the building. The minimum specification is Code 4 lead (1.80 mm), Code 3 (1.32 mm) for soakers. This should be coated both sides with a solvent-based bituminous paint to prevent the risk of corrosion when in contact with cement. The lower dpc may be in the form of a tray with edges turned up 25 mm, except where it coincides with bedded flashings such as the front apron upper level. Here weep holes in the perpends will encourage water to drain. The inside of the tray is taken through a flue lining joint and turned up 25 mm.

Combustion Air ~ it is a Building Regulation requirement that in the case of open fireplaces provision must be made for the introduction of combustion air in sufficient quantity to ensure the efficient operation of the open fire. Traditionally such air is taken from the volume of the room in which the open fire is situated, this can create air movements resulting in draughts. An alternative method is to construct an ash pit below the hearth level fret and introduce the air necessary for combustion via the ash by means of a duct.

Typical Established Details ~

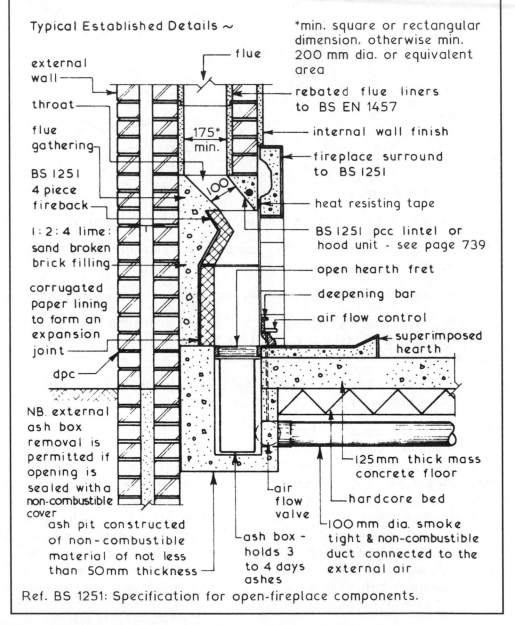

external wall

throat

flue gathering

BS 1251 4 piece fireback

1:2:4 lime: sand broken brick filling

corrugated paper lining to form an expansion joint

dpc

flue

175* min.

100

*min. square or rectangular dimension, otherwise min. 200 mm dia. or equivalent area

rebated flue liners to BS EN 1457

internal wall finish

fireplace surround to BS 1251

heat resisting tape

BS 1251 pcc lintel or hood unit - see page 739

open hearth fret

deepening bar

air flow control

superimposed hearth

125mm thick mass concrete floor

hardcore bed

100 mm dia. smoke tight & non-combustible duct connected to the external air

NB. external ash box removal is permitted if opening is sealed with a non-combustible cover

ash pit constructed of non-combustible material of not less than 50mm thickness

air flow valve

ash box - holds 3 to 4 days ashes

Ref. BS 1251: Specification for open-fireplace components.

Lightweight Pumice Chimney Blocks ~ these are suitable as a flue system for solid fuels, gas and oil. The highly insulative properties provide low condensation risk, easy installation as a supplement to existing or on-going construction and suitability for use with timber frame and thatched dwellings, where fire safety is of paramount importance. Also, the natural resistance of pumice to acid and sulphurous smoke corrosion requires no further treatment or special lining. A range of manufacturer's accessories allow for internal use with lintel support over an open fire or stove, or as an external structure supported on its own foundation. Whether internal or external, the units are not bonded in, but supported on purpose made ties at a maximum of 2 metre intervals.

flue (mm)	plan size (mm)
150 dia.	390 × 390
200 dia.	440 × 440
230 dia.	470 × 470
260 square	500 × 500
260 × 150 oblong	500 × 390

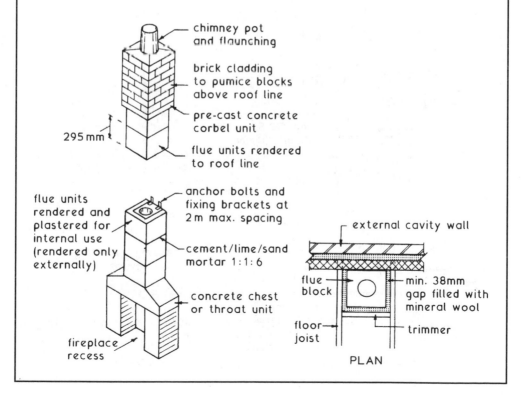

chimney pot and flaunching

brick cladding to pumice blocks above roof line

pre-cast concrete corbel unit

295 mm

flue units rendered to roof line

flue units rendered and plastered for internal use (rendered only externally)

anchor bolts and fixing brackets at 2 m max. spacing

cement/lime/sand mortar 1:1:6

concrete chest or throat unit

fireplace recess

external cavity wall

flue block

min. 38mm gap filled with mineral wool

floor joist

trimmer

PLAN

Fire Protection of Services Openings ~ penetration of compartment walls and floors (zones of restricted fire spread, e.g. flats in one building), by service pipes and conduits is very difficult to avoid. An exception is where purpose built service ducts can be accommodated. The Building Regulations, Approved Document B3:Sections 7 [Vol. 1] and 10 [Vol. 2] determines that where a pipe passes through a compartment interface, it must be provided with a proprietary seal. Seals are collars of intumescent material which expands rapidly when subjected to heat, to form a carbonaceous charring. The expansion is sufficient to compress warm plastic and successfully close a pipe void for up to 4 hours.

In some circumstances fire stopping around the pipe will be acceptable, provided the gap around the pipe and hole through the structure are filled with non-combustible material. Various materials are acceptable, including reinforced mineral fibre, cement and plasters, asbestos rope and intumescent mastics.

Pipes of low heat resistance, such as PVC, lead, aluminium alloys and fibre cement may have a protective sleeve of non-combustible material extending at least 1 m either side of the structure.

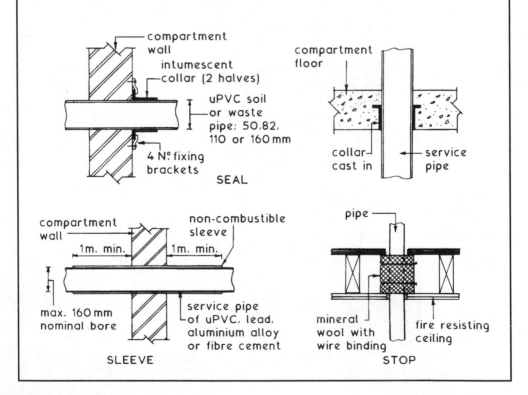

Telephone Installations ~ unlike other services such as water, gas and electricity, telephones cannot be connected to a common mains supply. Each telephone requires a pair of wires connecting it to the telephone exchange. The external supply service and connection to the lead-in socket is carried out by telecommunication engineers. Internal extensions can be installed by the site electrician.

Typical Supply Arrangements ~

underground supply cables must be installed when the building is constructed - they are hidden and therefore have little or no effect on the surrounding environment

overhead supply cables are smaller and cheaper than underground supply cables— convenient but obtrusive

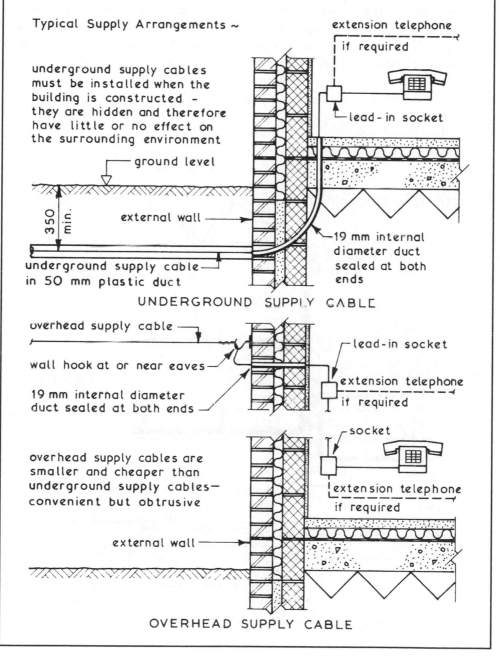

UNDERGROUND SUPPLY CABLE

OVERHEAD SUPPLY CABLE

Electronic Communications Installations

Electronic Installations – in addition to standard electrical and telecommunication supplies into buildings, there is a growing demand for cable TV, security cabling and broadband access to the Internet. Previous construction practice has not foreseen the need to accommodate these services from distribution networks into buildings, and retrospective installation through underground ducting is both costly and disruptive to the structure and surrounding area, particularly when repeated for each different service. Ideally there should be a common facility integral with new construction to permit simple installation of these communication services at any time. A typical installation will provide connection from a common external terminal chamber via underground ducting to a terminal distribution box within the building. Internal distribution is through service voids within the structure or attached trunking.

Electronic communication ducts

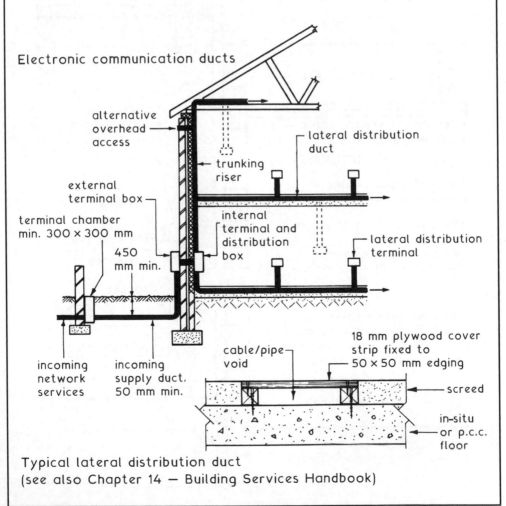

Typical lateral distribution duct
(see also Chapter 14 — Building Services Handbook)

748

INDEX

LIVES *and* VOICES

SOURCES IN EUROPEAN WOMEN'S HISTORY

Lisa DiCaprio

City College of New York, CUNY

Merry E. Wiesner

University of Wisconsin—Milwaukee

HOUGHTON MIFFLIN COMPANY   BOSTON   NEW YORK

Editor-in-Chief: Jean L. Woy
Associate Editor: Leah Strauss
Associate Project Editor: Heather Hubbard
Production/Design Coordinator: Jodi O'Rourke
Manufacturing Manager: Florence Cadran
Senior Marketing Manager: Sandra McGuire

Cover Design: Nina Wishnok, Dynamo Design
Cover Image: Museo Archaeologico Nazionale, Naples/Art Resource NY
Caption: *Idealized portrait of a young girl posing as a poet, perhaps meant to be Sappho. Pompeian wall painting.*

Printed in the U.S.A.

Library of Congress Catalog Card Number: 00-104431

ISBN 0-395-97052-0

5 6 7 8 9 10-CRS-07 06 05

Contents

Preface

Lives and Voices: Sources on Women in European History is the first reader that brings together materials on women's lives from ancient Mesopotamia to contemporary Bosnia. Visual as well as written sources are included which range widely in terms of genre and geography. The reader presents students with issues confronting women of different social classes, educational levels, and family situations, and incorporates political, social, economic, intellectual, religious, and cultural history. It includes both classics in women's history—Sappho, Christine de Pizan, Mary Wollstonecraft, Virginia Woolf—and sources that have never before been published in English, presenting a variety of normative, descriptive, and analytical materials.

Lives and Voices begins by introducing students to some of the central issues in women's history and to the sources available for the study of that history. The introduction also discusses the process of examining and evaluating sources, and suggests some of the special challenges posed by sources in women's history. The chapters are arranged chronologically and each has an introduction that provides an overview of major events and developments which affected women's lives. These introductions allow the reader to be used on its own in European women's history classes or as supplementary reading in more general European history or Western Civilization classes. Within the chapters, sources are arranged into topical sections, so that you may easily assign a group of related sources and adapt the book to your specific needs. Each section includes a brief overview, headnotes for the sources, and discussion questions.

In order to cover the vast amount of material for this reader, we divided the task chronologically; Merry Wiesner wrote the first seven chapters and Lisa DiCaprio the last eight. Each of us read and critiqued the other's chapters, and we also relied on our friends and colleagues in women's history. We thank the following reviewers for their insightful comments on the sourcebook outline: Mary E. Chalmers, Kenyon College; Barbara Corrado Pope, University of Oregon; Marylou Ruud, University of West Florida; Hilda Smith, University of Cincinnati; and Susan Whitney, Carleton University. For suggestions, contributions, and translations, we would particularly like to thank: Joan Afferica, Bonnie Anderson, Polly Beals, Martha Carlin, Linda Clark-Newman, Sandi Cooper, Belinda Davis, Judith DeGroat, Suzanne Desan, Lila Di Caprio, Barbara Einhorn, Barbara Engel, Sagit Fiaritovich Faizov, Patricia Flor, Geraldine Forbes, Laura Frader, Dena Goodman, Dorothy Helly, Marion Kaplan, Susan Karant-Nunn, Cynthia Koepp, Cheryl Koos, Barbara Lesko, Hilary Marland, Jo Ann McNamara, Jeffrey Merrick, Krista O'Donnell, Karen Offen, Sonia Jaffe Robbins, Daniella Sarnoff, Susan Stuard, Molly Tambor, Malgorzata Tarasiewicz, and Lee Whitfield. For her continual encouragement and support, suggestions, meticulous readings and comments on the modern chapters, we would like to thank Renate Bridenthal. For their editorial advice and production assistance, we would like to thank Jean Woy, Leah Strauss, Heather Hubbard, Sharon Donahue, and Jill Dougan.

L. D.
M. E. W.

INTRODUCTION

Over the last thirty years, women's history has developed into a major part of the historical enterprise. Conferences in women's history draw thousands of participants, endowed chairs in women's history may be found at many major universities, specialists in women's history teach and research at hundreds of colleges, where they offer thousands of different graduate and undergraduate courses in women's history. The sections devoted to women's history in libraries have expanded exponentially, and even introductory textbooks include material on women, the family, and gender. Investigations into women's history have led to the development of still newer historical fields, including the study of sexuality, masculinity, and the social construction of gender. The questions asked in women's history are now much more complex than they were thirty years ago, and the answers similarly complex, with scholars more attuned to the differences among women and the variety of their historical experiences.

The proliferation of women's history, in turn, has intensified the search for original sources that will reveal still more about women's pasts. Historians have combed libraries, archives, museums, private collections, attics, drawers, and people's memories for written, oral, and visual sources. They have also read and looked at materials that have been known for centuries in new ways, demonstrating how fresh perspectives can reveal information that was always present but simply never noticed previously. *Lives and Voices* would not have been possible without all of these investigations; many of the materials it contains were not known thirty years ago, or certainly not transcribed from their original form on clay tablets, parchment manuscripts, or hand-written scraps of paper.

Our decision to assemble this collection stems from our own experiences in researching and teaching women's history. We have had the joys—and frustrations—of working in original sources housed in archives, looking at materials that had been long neglected or ignored, and brought questions about women to materials that had been used to write history in which women were not included. We have used some of our own sources in the classroom, and discovered how original sources help students develop their skills of interpretation. We have seen the ways in which original sources such as women's diaries and letters find avid readership beyond the classroom, providing a sense of "you are there" and a unique opportunity to gain an understanding about various historical periods.

1

This sourcebook spans a chronological period from the earliest civilizations to the present and includes material from all aspects of women's lives: social, political, economic, intellectual, and cultural. We have included materials that have become classics in women's history, such as the writings of Sappho, Christine de Pizan, Mary Wollstonecraft, and Virginia Woolf, and sources that have never before been published in English. The documents include visual as well as written sources, and range widely in terms of topic, social class, and geography; both male- and female-authored texts are included to present a range of normative, descriptive, and analytical materials. We hope that our choices will help you both to trace general trends and developments and to see the diversity and variation in women's experiences.

As you progress chronologically in *Lives and Voices,* you will probably notice that sources produced by men predominate in the first half, while those produced by women predominate in the second half. This is not the result of an intentional choice on our part, but of the materials available. For the earliest periods covered in this book, almost no sources survive that were written or produced by women; one recent collection of sources about women in antiquity, for example, includes eight pages of writings by women along with 300 pages by men. Some of this disparity may be the result of differences in the survival rates of male- and female-authored materials; those produced by men were considered more important and valuable and so were recopied and preserved over the course of centuries or millennia. More of the disparity, however, comes from differences in the opportunities afforded men and women based on prevailing ideas about their natures, abilities, and proper roles in society, all of which led men to leave records of their lives while most women did not. Thus the title of this sourcebook, *Lives and Voices,* describes a development as well as the book's contents; in the first half, you will find more sources looking at women's lives from the outside, and in the second, virtually all of the documents consist of women describing their lives in their own voices. Though all of the women whose voices appear in these pages are unusual in that their words were recorded, those from the ancient and medieval periods whose voices may be heard are particularly extraordinary.

STUDYING WOMEN'S HISTORY

History is often presented to students and the general public as facts marching along a timeline, with little information given about how we learned these facts, and none about who decided which facts were important enough to be included on the timeline. In teaching women's history courses, we often illustrate this point by asking students to name events on whatever timeline is appropriate for the course, and then name events during that time period that were especially important to women. The latter request often draws a blank, or occasionally the name of a very famous woman such as Queen Elizabeth I of England, Marie Curie, or Margaret Thatcher. This is changing somewhat as textbooks begin to include more material on women, but students still do not perceive developments that are extremely significant for women, such as the vulcanization of rubber which allowed for flexible baby bottle nipples and freed women from nursing children, legal changes that allowed married women to keep their own wages, or the invention of effective contraceptive techniques like the birth control pill that could be controlled by women themselves, as part of "history."

When women's history began as a field thirty years ago, the first approach taken by scholars interested in this new area was generally to look at the events already on the timeline—the Renaissance, the French Revolution, the Industrial Revolution, the World Wars—and ask "What did these events mean for women?" They thus explored both the role of women in these events and the impact of these events on women. This research has continued, and many of the sources you will read in this book will allow you to investigate what have long been considered major events in European history from the point of view of the women involved. Research into women's roles in major historical developments was accompanied by research into women who were influential in realms that were usually the province of men, those whom the historian Natalie Zemon Davis has termed "women worthies." The names that our students initially put on the timeline are usually those of just such extraordinary women, and many of them—including Elizabeth I, Marie Curie, and Margaret Thatcher—are included in this collection, for their impact on history was indeed significant.

As more research was undertaken, and more women's names and contributions were added to the timeline, scholars and students began to look more closely at the timeline itself. Their research indicated that some of the events on the timeline had little impact on women, and, conversely, that developments which *were* significant for women had never made it onto the timeline of Great Historical Events. Clearly this timeline had not been developed from women's experience and thus was not only incomplete, but faulty. Scholars began to alter the timeline itself by adding and subtracting events, and, in the process, reconceptualizing the story of European history. Certain periods which had always been viewed as high points on the timeline, such as classical Athens or the Italian Renaissance, were now viewed as less golden, and those which had been viewed as low points, such as the so-called "Dark Ages" of the fifth to the tenth centuries, were viewed more positively. Some of the sources you will read in this book have been influential in this rethinking of the timeline, and others will concern issues that have only very recently been added to it, such as the development of new ideas about sexuality and the body.

EXAMINING AND EVALUATING SOURCES

Historical evidence, or what is often referred to as "primary source material," includes anything that provides information about events or cultures in the past. In contrast to archeologists, who rely primarily on physical evidence, and to anthropologists, who frequently use oral testimony, historians tend to focus on written records. In fact, the presence of writing is often used to separate cultures that are "prehistoric" from those that are historic, with the earliest historic cultures those that are featured in the first chapter of this sourcebook, ancient Mesopotamia and Egypt. Written sources include a huge variety of types of materials: law codes, religious texts, letters, newspapers, diaries, court records, speeches, songs, plays, novels, business contracts, philosophical treatises, memoirs, and many more.

You may already have experience analyzing primary sources, but if not, another classroom exercise may help point out some of the issues involved. On a piece of paper, draw your family tree on the front and write several sentences on the back in which you describe yourself, then gather together the papers from the class. What you and your classmates have produced are, of course, written historical sources that can provide evidence about many things. With nothing but these pieces of paper, you can tell, for example, that the culture in which they were produced could make paper and ink and had a system of writing, which implies that there was some sort of training or educational system. This was a culture that prized uniformity, as the pieces of paper are probably roughly the same size and lined, and the writing is probably on the lines.

So far you have treated your sources as archeological artifacts, and for many of the world's cultures this is as far as we can go, for these cultures have either not left written records or their written language has not yet been deciphered. If we assume that you are able to read other students' writing, you will be able to obtain additional insights from your sources. Taken together, the family trees of students in a class will reveal different patterns of family composition. Some may be limited to parents and grandparents while others may include several generations. Similarly, while one student's family tree may correspond to that of the "nuclear family," that of another student may include extended kin. In reading the various self-descriptions written by other students on the opposite sides of their family trees, you will see that, as a general rule, self-image is based on gender and racial identity along with physical and occupational characteristics.

The family tree on the front of the paper is often described as an *objective* source, as it presumes to describe family relations in a purely factual way, while the writing on the back is regarded as a *subjective* source, as it clearly states the opinion of one person. You can readily see some of the problems with subjective sources from the pieces of paper. The self-descriptions might not be completely accurate, as people often do not include negative characteristics when presenting themselves. Even if they are accurate, they are not complete, as no one can be summed up in a few sentences. The family trees, while seemingly more objective, might also be incomplete or inaccurate, as you or your classmates may not know very much about your families or may have been told things that are not true. Though they were supposed to be objective depictions of fact, family trees might actually conceal a relative or a relationship the family is embarrassed about and so, in some ways, they may be just as subjective as the self-descriptions. Doubts about seemingly objective sources such as census data have led some historians to reject the distinction between objective and subjective sources completely. They argue that all written sources were shaped by the point of view of the writer, so that none of them, even numerical records, are completely objective.

Historians may disagree among themselves about the degree of distinction between objective and subjective sources, but they all evaluate their sources in various ways. One of their most important strategies is to search for corroborative sources. If they were to evaluate the accuracy of the family trees and self-descriptions drawn and written in your classroom, they might go to official documents such as birth registries to find out about family members, and interview acquaintances to assess the accuracy of the self-descriptions. Through other sources, they would attempt to learn more about the authors of the family trees and self-descriptions and the cultural milieus in which they were produced. Was the author wealthy or poor? Rural or urban? Why, for whom, and when was the source produced? All of this further research would help provide context for the source, and allow the historian to assess the validity and significance of the information it presents.

As you can see, context is extremely important in understanding and assessing any historical source. For this reason, all of the sources in this book are preceded by headnotes which provide very specific information on the authors' backgrounds, the nature of the sources, and the situation in which they were produced. They are also followed by questions that ask you to analyze the documents on several increasingly complex levels: basic comprehension, the purpose of the document, the internal logic of the document's argument, the information the document provides about the time and place in which it was written. Because the best way to contextualize and understand a source is by reading other sources, each document is grouped with several other related sources, and some of the questions ask you to make specific comparisons among documents in one group. Other questions ask you to use the documents found in several chapters to make broader topical and chronological comparisons.

Though the sources in this book are presented with contextual material, issues of interpretation that emerge are similar to those presented by your family tree and self-description. Each of these sources is untypical in the very fact that it has survived for millennia, centuries, or at least years. The vast majority of the world's historical sources have disappeared or disintegrated, with an even greater amount of information never recorded in the first place. Those that have survived were always written by people with a particular point of view, so that even if they attempted to be accurate, they present only a part of the picture. Until the late eighteenth century, most of the sources that have survived come from people who were at the top of the social and economic heap, and among the tiny group that was literate; they are thus not representative of the experience of most people. Authors belonging to certain classes or strata of society often described events and/or their societies in such a way as to make these descriptions appear to be purely factual but, in fact, they expressed a particular class view. The Athenian general Pericles' extremely idealized view of his own society is one example of this; for centuries this was read as an accurate and complete picture of Athenian justice and democracy, though now we know that people living in the cities of Greece that Athens had conquered had very different views, and that the women and slaves of Athens would probably have told a different story had their opinions been recorded. (For the situation of women in Pericles' Athens, see the sources in Chapter 2.)

An additional problem in interpreting sources is presented by the fact that many of the world's written records are in fact *prescriptive* sources, telling people how they should behave and what they should do, rather than *descriptive* sources, like your family tree and self-description. For many periods of history, especially the earliest ones, prescriptive sources such as law codes, sermons, and advice manuals surpass other extant sources. They can provide us with a great deal of information about how their authors hoped or wished people would behave, but do not provide an accurate picture of people's actual, lived experiences. (Imagine a future historian's false assumptions of American traffic patterns if he or she thought people actually drove the speed limit!) Sometimes distinguishing between prescriptive and descriptive sources is difficult, for very often prescriptive sources masquerade as descriptive ones; writers horrified about the French Revolution, for example, focused only on its most violent incidents rather than presenting the entire picture in order to discourage people in their own countries from similar actions, thus shaping their descriptions in an attempt to shape behavior. (For women's role in the French Revolution, see the sources in Chapter 8.)

All of these concerns also enter into the interpretation of visual sources, of which there a few in almost every chapter. Visual materials include buildings, objects, and artifacts as well as paintings, sculpture, woodcuts, engravings, posters, and, for the nineteenth and twentieth centuries, photographs. Visual sources often reinforce what we discover in written texts; for example, almost all of the art that has survived from the Middle Ages is religious, reinforcing the impression that we get from written works about the importance of religion in people's lives during this period. At times, however, the visual sources contradict the written record. For example, written descriptions and laws about mining during the fifteenth and sixteenth centuries generally mention only male workers, but illustrations of mining show women engaged in various tasks such as washing and hauling ore. As a historian, we are thus confronted with the issue of whether to believe the written or the pictorial record. Like written works, visual evidence can appear to be descriptive when it is actually prescriptive or idealized, showing things the way the artist or photographer wished they would be rather than the way they actually were. Images are an important means of communicating cultural values and teaching people how to behave, especially in historical periods when the majority of the population is illiterate.

As you read or look at each source, use the information provided in the headnote to help you understand the context and the perspective of its author. If it describes or depicts events, was the author an eyewitness, or is the account based on the reports of others? Is what is being described or depicted unique, or can the information it contains be used to make broader generalizations? How does the point of view of the author shape his or her perceptions? How does the source fit with other information you have learned about the topic, and with the other sources we have included in the chapter?

Analyzing historical sources always involves making inferences, what you might want to think of as your best guess about a matter; professional historians do this all the time, and often disagree vehemently with one another about the meaning of historical events. Because the historical record is always incomplete, and all sources are written from a particular perspective, they may even disagree about *what* actually happened as well as *why* it happened. As you discuss the sources in this collection, you may find such disagreements emerging among the members of your class. As long as all points of view can be backed up with evidence from the sources, such debates are welcome.

SOURCES IN WOMEN'S HISTORY

Concerns about accuracy, representativeness, perspective, and context that emerge when evaluating all historical sources are heightened when we look at sources about women. This is particularly true when we seek to disclose the lives of women who are not members of the political, economic, and intellectual elite. For example, information about common people is very often contained in official documents such as court records, which sometimes include testimony in their own words; these are one of the few places prior to the rise of mass literacy where we can hear the voices of people who could not read or write. These official documents, several of which are reproduced in the first seven chapters, are extremely valuable. However, as noted previously, we must also consider their limitations for discerning the nature of women's lives in any given period. Records were written by literate, typically male individuals. Thus, the experiences of ordinary women which are related in these records are viewed through a double filter, both learned and male. Moreover, as married women generally did not control their own property or resources, and single women and widows were often represented by men in legal cases, women appear much less frequently in official documents than men do. There are thus fewer opportunities for cross-checking information or making comparisons than there are for men.

If we turn to women's own writings, we find that, until fairly recently in recorded history, these were typically produced by elite, literate women and therefore represent only a narrow stratum of women in society. The women for whom we have the most sources are generally those who have been involved in activities usually associated with men, such as politics or learning, so their experiences are clearly not typical of most women. Thus, though the works by women in the first seven chapters vary widely—they include poems, speeches, letters, memoirs, sayings, religious writings, books of advice, wills, and autobiographies—most of these are written by women of the upper classes. This one-sidedness changes dramatically with the growth of literacy in the late eighteenth and nineteenth centuries. Various modern developments, including political revolutions, industrialization, feminist movements, and the expansion of education, created unprecedented opportunities for women to shape and record their lives. These new opportunities are reflected in the wide variety of writings by women of various classes reproduced in the last eight chapters of *Lives and Voices*. These include petitions, diaries, letters, articles, pamphlets, speeches, autobiographies, leaflets, and manifestos. The rise of mass culture, the invention of photography, and modernist art movements also expanded the possibilities for women's artistic endeavors. Our samples of women's art from the modern period encompass traditional forms of art such as paintings and drawings as well as photographs, fashion design, suffragist art, and political posters.

Although the absence of sources means that we may never have a female perspective on many issues prior to the modern period, the number of documents in this sourcebook that were unknown thirty years ago leads us to believe that we will eventually know much more than we do now. The expansion in sources that makes this collection in its current shape possible has also led to some difficult decisions on our part. Because we are attempting to present sources on the history of half the European population over more than 4000 years between the covers of one book, we could not include even one document on every development that was significant, nor present all sides on those developments we did include. We have tried, however, to present a balanced range of materials, and hope that the questions these sources leave unanswered will inspire you to investigate further. Finally, for those of you who want to find your own sources, the archives and libraries of Europe—and the rest of the world—are waiting. In the past thirty years, historians have only begun the process of finding new sources about women's history, and there is nothing as exciting as discovering your own women whose stories need to be told.

CHAPTER 1

The Ancient Mediterranean and Western Asia

(2300 B.C.E.–500 B.C.E.)

The history of European culture traditionally begins outside of Europe, in the river valleys of North Africa and western Asia. The cultures of these areas, including Egypt and Mesopotamia, were the first in the world to develop writing systems and thus the first to leave sources that we identify as "historical." The ideas and systems that they developed and described in writing were transmitted northward to Europe over centuries and form one of the bases of later European cultures such as those of the Greeks and the Romans. Thus Europeans look back to these cultures not only because of the age of their records but also because their laws, religious traditions, economic systems, political forms, and technological innovations were important in shaping European civilization.

Women, of course, made up half the population in the earliest historical civilizations, but information about their lives makes up a far smaller share of the historical record. Though only a small part of the population as a whole learned to write, most of those who did were men. Thus the vast majority of our sources from the ancient world were written by men and reflect a male perspective. This is true for most cultures of the world during most periods, but because the total number of sources for the earliest civilizations is so small, the disparity is particularly striking. Much of the material about women is thus actually about male opinions of women, a distinction that earlier scholars sometimes failed to keep in mind.

In some ways, the values and institutions of the earliest Western civilizations were similar. Both Mesopotamia and Egypt regarded the natural world as ruled by various deities, whose favor was essential for continued life and prosperity; that favor was to be maintained by performing rituals and ceremonies honoring local deities. Rulers on earth had connections with the gods, and obedience to rulers was regarded as a religious duty. Both cultures built complex cities and enormous structures, and they developed systems of mathematics. In both, writing served to build up and maintain the power of political and religious authorities, giving permanence to laws, regulations, and sacred texts.

There were also significant differences between Mesopotamia and Egypt, created in part by factors of geography and terrain. Though both civilizations developed in river valleys, in Egypt the Nile's regular and gentle floods brought in new fertile soil and aided agriculture, while in Mesopotamia the Tigris and Euphrates flooded unpredictably and destructively. Egypt was relatively isolated in the Nile Valley by deserts, the Mediterranean, and the cataracts of the Nile, while Mesopotamia had no natural barriers and was frequently invaded from the outside. Egypt was unified very early in its history and remained stable and prosperous, while Mesopotamia saw a battle for resources, frequent warfare, and the rise and fall of many empires—Sumerian, Akkadian, Babylonian, and Assyrian. All of these factors led people in Egypt to be more optimistic in their outlook on life, viewing the gods as benevolent and the afterlife as a continuation of life on earth, while those in Mesopotamia were more pessimistic, viewing the gods as whimsical and harsh.

There were also similarities and differences in terms of women's status, role, and activities in these earliest civilizations. In both Egypt and Mesopotamia, women were generally not supreme rulers or higher-level officials and functionaries, but those who were members of royal or officials' families had power and wealth; they often had religious and sometimes political or administrative duties. More ordinary women worked alongside male family members or independently in a range of occupations. Most people—female and male—were agricultural producers, but sources reveal that in the cities women worked as midwives, beer sellers, healers, market-sellers, weavers, potters, and millers, as well as in other occupations. In the more optimistic and confident Egypt, women inherited land alongside their brothers and retained control of their land even after marriage; they made legal contracts on their own and were frequently out in public for business or pleasure. In the more pessimistic and unstable Mesopotamia, men exerted greater control over the women of their families, and women's rights to work or own property independently were increasingly restricted. Among the Assyrians, respectable women—those who were under the protection of a man as his daughters, wives, or concubines—were required to wear a veil in public; women who were not under male protection, such as slaves and prostitutes, were prohibited from wearing a veil and were judged to be available to all men.

Most of the peoples of Egypt and Mesopotamia were polytheists, honoring a variety of deities, but one group, the Israelites, developed a monotheistic religion in which a single god, Yahweh, demanded certain standards of conduct along with specific rituals. The Israelites apparently migrated from Mesopotamia to Egypt and then back to the area known as Canaan, where for several hundred years (c. 1000–700 B.C.E.) they had an independent kingdom. The body of written religious literature they developed during and after this period, the Hebrew Bible (the Christian Old Testament), was of more lasting significance than their state, and it has formed the core of Israelite religion, termed *Judaism,* ever since. In ancient Judaism, women were excluded from official religious positions, though they emerged as extraordinary leaders during times of turmoil; laws and moral instructions encouraged women to focus on marriage and family life.

WOMEN IN CREATION ACCOUNTS

All cultures develop accounts of how the world began and how people came to inhabit the world. Generally, these accounts involve the actions of creator deities or spirits, who are often conceptualized in human or animal form (or a combination of the two) and given human personality traits such as anger, wisdom, pride, and benevolence. These accounts remained oral in many of the world's cultures, handed down from generation to generation by shamans or other types of religious leaders, and recorded only within the past several centuries. In other cultures, such as those of the ancient Mediterranean and western Asia, creation accounts were written down very early, though they were often modified later as new gods and beliefs were introduced, or as different groups of scribes retold the events from their own perspective.

Many of these creation accounts include discussion of male and female deities or of the creation of men and women; they thus both reflect and shape cultural attitudes about gender roles and differences. Though we do not know exactly who first wrote down the creation accounts from the ancient river civilizations, we do know that most of the people who learned to write in these cultures and most of those with religious authority were men. Thus written creation accounts are told from the point of view of educated men and reflect their values. Women who heard these stories may have developed their own interpretations, but these did not become part of the written record, so we have no way to recover them. It is thus important to remember that creation accounts present images of women but do not reflect the reality of women's lives in ancient societies. They did shape that reality, however, for creation accounts continued to be read or told orally long after they were written down, and they thus became part of the mental world and cultural heritage of most people in a society, not simply the educated elite.

1. Enuma Elish

Enuma Elish, a title meaning "when above" and taken from the opening words of the poem, is the creation account of the Babylonians, one of the peoples who controlled the Tigris and Euphrates valleys. A number of clay tablets with portions of the poem inscribed on them have been found that date to the eighth century B.C.E., though most scholars agree that the poem was probably first composed a thousand years before that, perhaps during the reign of Hammurabi (1728–1686 B.C.E.). It was retold orally every year by the high priest, and portions of it apparently were read to Babylonian women during childbirth. Like the creation accounts of many peoples, *Enuma Elish* involves several generations of gods. The original gods, Apsu and Tiamat, are the parents and grandparents of many gods; the younger gods decide to rebel against them, and Apsu plans to kill them. Ea, one of the younger gods, kills Apsu, and Tiamat vows to take revenge on the younger gods. Marduk, Ea's son (and thus Tiamat's grandson), emerges as the leader of the younger gods and battles with Tiamat.

"Tiamat, our bearer, hates us.
She held a meeting and raged furiously.
All the gods went over to her;
Even those whom ye have created march at her
    side."
They separated themselves(?) and went over to
    the side of Tiamat;
They were angry, they plotted, not resting day
    or night;
They took up the fight, fuming and raging;
They held a meeting and planned the conflict.
Mother Hubur, who fashions all things,
Added (thereto) irresistible weapons, bearing
    monster serpents
Sharp of tooth and not sparing the fang(?).
With poison instead of blood she filled their
    bodies.
Ferocious dragons she clothed with terror,
She crowned them with fear-inspiring glory
    (and) made them like gods,
So that he who would look upon them should
    perish from terror,
So that their bodies might leap forward and
    none turn back their breasts.
She set up the viper, the dragon, and the
    lahamu,

The great lion, the mad dog, and the scorpion-
    man,
Driving storm demons, the dragonfly, and the
    bison,
Bearing unsparing weapons, unafraid of
    battle.
Powerful are her decrees, irresistible are they.
Altogether(?) eleven (kinds of monsters) of this
    sort she brought into being. . . .
(Then) Marduk, the wisest of [the gods, your
    son], came forward.
[His heart prompted (him)] to face Tiamat.
He opened his mouth (and) [said to me]:
"If I am indeed to be [your avenger],
To vanquish Tiamat (and) [to keep you alive],
Convene the assembly and [proclaim my lot
    supreme].
[When ye are joyfully seated together] in the
    Court of Assembly,
[May I] through the utterance of my mouth
    [determine the destinies], instead of [you].
Whatever I create shall remain unaltered,
The command of [my lips] shall not return
    (void), it shall [not be] changed."
Hasten to me (then) and speedily [fix for him]
    your destiny,

[That he may] go to meet your powerful enemy!'"

When Lahha[1] (and) Lahamu[1] heard (this), they cried aloud;

All the Igigi[1] wailed painfully:

"What has happened that she has come to [such a de]cision?

We do not understand Tiamat's ac[tion]!"

They gathered together and departed,

All the great gods who determine [the destinies].

They entered into the presence of Anshar[1] and filled [the Court of Assembly];

They kissed one another [as they came together] in the assembly;

They conversed (and) [sat down] to a banquet.

They ate bread (and) prepared w[ine].

The sweet wine dispelled their fears;

[Their] bod[ies] swelled as they drank the strong drink.

Exceedingly carefree were they, their spirit was exalted;

For Marduk, their avenger, they decreed the destiny[2]. . . .

They erected for him a lordly throne-dais,

And he took his place before his fathers to (receive) sovereignty.

"Thou art (the most) important among the great gods;

Thy destiny is unequaled, thy command is (like that of) Anu.[3]

Marduk, thou art (the most) important among the great gods,

Thy destiny is unequaled, thy command is (like that of) Anu.

From this day onward thy command shall not be changed.

To exalt and to abase—this shall be thy power!

Dependable shall be the utterance of thy mouth, thy command shall not prove vain.

None among the gods shall infringe upon thy prerogative.[4]

Maintenance is the requirement of the sanctuaries of the gods;

And so at (each) place of their shrines shall be appointed a place for thee.

Marduk, thou art our avenger;

To thee we have given kingship over the totality of the whole universe,

So that when thou sittest in the assembly, thy word shall be exalted.

May thy weapons not miss, may they smite thy foes.

O lord, preserve the life of him who puts his trust in thee;

But as for the god who has espoused evil, pour out his life!"[5]

Then they placed a garment in their midst;

To Marduk, their first-born, they said:

"Thy destiny[6], O lord, shall be supreme among the gods.

Command to destroy and to create, (and) they shall be!

By the word of thy mouth, let the garment be destroyed;

Command again, and let the garment be whole!"

He commanded with his mouth, and the garment was destroyed.

He commanded again, and the garment was restored.

When the gods his fathers beheld the power of his word,

They were glad (and) did homage, (saying:) "Marduk is king!"

They bestowed upon him the scepter, the throne, and the royal robe(?);

They gave him an irresistible weapon smiting the enemy, (saying:)

"Go and cut off the life of Tiamat.

---

[1]*Lahha, Lahamu, Igigi,* and *Anshar:* gods and goddesses who were children or grandchildren of Apsu and Tiamat.
[2]I.e., they made him lord of the gods, in conformity with his demands.
[3]The sky-god and at one time the highest god of the pantheon.

[4]*Lit.:* overstep thy boundary.
[5]The expression "to pour out the life of someone" is taken from the pouring-out or shedding of blood, the seat of the element of life.
[6]Thy power and authority.

May the winds carry her blood to out-of-the-way places."

After the gods his fathers had determined the destiny of Bel,[7]

They set him on the road—the way to success and attainment.

He made a bow and decreed (it) as his weapon;

An arrowhead he put (on the arrow and) fastened the bowstring to it.[8]

He took up the club and grasped (it) in his right hand;

The bow and the quiver he hung at his side.

The lightning he set before him;

With a blazing flame he filled his body.

He made a net to inclose Tiamat within (it),

(And) had the four winds take hold that nothing of her might escape;

The south wind, the north wind, the east wind, (and) the west wind,

The gift of his (grand)father, Anu, he caused to draw nigh to the border(s) of the net.

He created the *imhullu:* the evil wind, the cyclone, the hurricane,

The fourfold wind, the sevenfold wind, the whirlwind, the wind incomparable.

He sent forth the winds which he had created, the seven of them;

To trouble Tiamat within, they arose behind him.

The lord raised the rain flood, his mighty weapon.

He mounted (his) irresistible, terrible storm chariot;

He harnessed for it a team of four and yoked (them) to it,

The Destructive, the Pitiless, the Trampler, the Flier.

They were sharp of tooth, bearing poison;

They knew how to destroy, they had learned to overrun;

[. . .] they [smo]te, they were frightful in battle;

To the left [. . .].

He was clad in a terrifying coat of mail;

Terror-inspiring splendor he wore on his head.

The lord took a direct (route) and pursued his way;

Toward the place of raging Tiamat he set his face.

Between his lips he holds {a talisman(?)} of red paste;

An herb to destroy the poison he grasped in his hand.

Then the gods r[un] about him, the gods run about him;

The gods his fathers run about him, the gods run about him.

The lord drew nigh to look into the heart of Tiamat,

(And) to see the plan of Kingu, her spouse.

As he[9] gazes, (Kingu) is confused in his plan;

Destroyed is his will and disordered his action.

As for the gods his helpers, who were marching at his side,

When they saw the valiant hero, their vision became blurred.

Tiamat set up a roar(?) without turning her neck,

Upholding with her li[ps] (her) meanness(?) (and) rebellion.

". . . . . have the gods risen up to thee?

(Or) have they gathered from their [place] to thy place?"

Then the lord [raised] the rain flood, his mighty weapon.

[As for T]iamat, who was furious, thus he answered her: •

"[In arrogance(?)] thou art risen (and) hast highly exalted thyself(?).

[Thou has caused] thy heart to plot the stirring-up of conflict.

[. . .] the sons treat their fathers unjustly;

(And) thou, their bearer, dost hate (them) wi[thout cause(?)].

Thou hast exalted Kingu to be [thy] spouse;

Thine illegal [authority] thou hast set up in place of the authority of Anu.

[Against] Anshar, the king of the gods, thou seekest evil,

And hast proven thy wickedness [against the god]s my fathers.

---

[7]I.e., Marduk.
[8]To the bow.

[9]Marduk.

Let thine army be equipped! let them be girded
with thy weapons!
Come thou forth (alone) and let us, me and
thee, do single combat!"
When Tiamat heard this,
She became like one in a frenzy (and) lost her
reason.
Tiamat cried out loud (and) furiously,
To the (very) roots her two legs shook back and
forth.
She recites an incantation, repeatedly casting
her spell;
As for the gods of battle, they sharpen their
weapons.
Tiamat (and) Marduk, the wisest of the gods,
advanced against one another;
They pressed on to single combat, they ap-
proached for battle.
The lord spread out his net and enmeshed her;
The evil wind, following after, he let loose in
her face.
When Tiamat opened her mouth to devour him,
He drove in the evil wind, in order that (she
should) not (be able) to close her lips.
The raging winds filled her belly;
Her belly became distended, and she opened
wide her mouth.
He shot off an arrow, and it tore her interior;
It cut through her inward parts, it split (her)
heart.
When he had subdued her, he destroyed her
life;
He cast down her carcass (and) stood upon it.
After he had slain Tiamat, the leader,
Her band broke up, her host dispersed.
As for the gods her helpers, who marched at
her side,
They trembled for fear (and) faced about.
They tried to break away to save their lives,
(But) they were completely surrounded, (so
that) it was impossible to flee.
He imprisoned them and broke their weapons,
In the net they lay and in the snare they were;
They hid in the corners (and) were filled with
lamentation;
They bore his wrath, being confined in prison.
As for the eleven (kinds of) creatures which she
had laden with terror-inspiring splendor,

The host of demons that marched impetuously
before her,
He cast (them) into fetters (and) [tied(?)] their
arms [together(?)];
With (all) their resistance, [he tr]ampled
(them) underfoot.
As for Kingu, who had become chief among
them,
He bound him and counted him among the
dead gods.
He took from him the tablet of destinies,
which was not his rightful possession.
He sealed (it) with (his) seal and fastened (it)
on his breast.
After he had vanquished (and) subdued his ene-
mies,
Had overpowered the arrogant foe like a bull(?),
Had fully established Anshar's victory over the
enemy,
Had attained the desire of Nudimmud, the
valiant Marduk
Strengthened his hold upon the captive gods;
And then he returned to Tiamat, whom he had
subdued.
The lord trod upon the hinder part of Tiamat,
And with his unsparing club he split (her)
skull.
He cut the arteries of her blood
And caused the north wind to carry (it) to out-
of-the-way places.
When his fathers saw (this), they were glad and
rejoiced
(And) sent him dues (and) greeting-gifts.
The lord rested, examining her dead body,
To divide the abortion[10] (and) to create inge-
nious things (therewith).
He split her open like a mussel(?) into two
(parts);
Half of her he set in place and formed the sky
(therewith) as a roof.
He fixed the crossbar (and) posted guards;
He commanded them not to let her waters es-
cape.[11]

---

[10]The corpse of Tiamat is here compared to an abortion.
[11]I.e., the waters of Tiamat which were contained in that
half of her body which Marduk used in the construction of
the sky.

He crossed the heavens and examined the regions. . . .
He created stations for the great gods;
The stars their likeness(es), the signs of the zodiac, he set up.
He determined the year, defined the divisions;
For each of the twelve months he set up three constellations.
After he had def[ined] the days of the year [by means] of constellations. . . .
As [Mar]duk hears the words of the gods,

His heart prompts (him) to create ingenious things.
He conveys his idea to Ea,
Imparting the plan [which] he had conceived in his heart:
"Blood will I form and cause bone to be;
Then will I set up *lullu,* Man' shall be his name!
Yes, I will create *lullu:* Man!
(Upon him) shall the services of the gods be imposed that they may be at rest.

2. Genesis

The Hebrew Bible contains a number of different types of literature—history, laws, moral sayings, hymns, prophecy. The first five books of the Bible, called the Torah, relate the story of the Israelite people from creation through the death of Moses, the religious visionary who led them from Egypt to Canaan. Most scholars accept that these books were written down over the period of several centuries beginning in about 900 B.C.E., several centuries after the death of Moses. Like *Enuma Elish,* they continued to be read and told orally as part of religious ceremonies and in less formal settings. In contrast to *Enuma Elish,* Genesis and the other books of the Hebrew Bible continue to be regarded as holy by millions of people today, not only Jews, but also Christians, for whom they are also part of a sacred text.

1 When God began to create[a] heaven and earth—[2]the earth being unformed and void, with darkness over the surface of the deep and a wind from[b] God sweeping over the water—[3] God said, "Let there be light"; and there was light. [4]God saw that the light was good, and God separated the light from the darkness.

[5]God called the light Day, and the darkness He called Night. And there was evening and there was morning, a first day.[c]

[6]God said, "Let there be an expanse in the midst of the water, that it may separate water from water." [7]God made the expanse, and it separated the water which was below the expanse from the water which was above the ex-

---

[a]Others "In the beginning God created."
[b]Others "the spirit of."

---

[c]Others "one day."

panse. And it was so. [8]God called the expanse Sky. And there was evening and there was morning, a second day.

[9]God said, "Let the water below the sky be gathered into one area, that the dry land may appear." And it was so. [10]God called the dry land Earth, and the gathering of waters He called Seas. And God saw that this was good. [11]And God said, "Let the earth sprout vegetation: seed-bearing plants, fruit trees of every kind on earth that bear fruit with the seed in it." And it was so. [12]The earth brought forth vegetation: seed-bearing plants of every kind, and trees of every kind bearing fruit with the seed in it. And God saw that this was good. [13]And there was evening and there was morning, a third day.

[14]God said, "Let there be lights in the expanse of the sky to separate day from night; they shall serve as signs for the set times—the days and the years; [15]and they shall serve as lights in the expanse of the sky to shine upon the earth." And it was so. [16]God made the two great lights, the greater light to dominate the day and the lesser light to dominate the night, and the stars. [17]And God set them in the expanse of the sky to shine upon the earth, [18]to dominate the day and the night, and to separate light from darkness. And God saw that this was good. [19]And there was evening and there was morning, a fourth day.

[20]God said, "Let the waters bring forth swarms of living creatures, and birds that fly above the earth across the expanse of the sky." [21]God created the great sea monsters, and all the living creatures of every kind that creep, which the waters brought forth in swarms, and all the winged birds of every kind. And God saw that this was good. [22]God blessed them, saying, "Be fertile and increase, fill the waters in the seas, and let the birds increase on the earth." [23]And there was evening and there was morning, a fifth day.

[24]God said, "Let the earth bring forth every kind of living creature: cattle, creeping things, and wild beasts of every kind." And it was so. [25]God made wild beasts of every kind and cattle of every kind, and all kinds of creeping things of the earth. And God saw that this was good. [26]And God said, "Let us make man in our image, after our likeness. They shall rule the fish of the sea, the birds of the sky, the cattle, the whole earth, and all the creeping things that creep on earth." [27]And God created man in His image, in the image of God He created him; male and female He created them. [28]God blessed them and God said to them, "Be fertile and increase, fill the earth and master it; and rule the fish of the sea, the birds of the sky, and all the living things that creep on earth."

[29]God said, "See, I give you every seed-bearing plant that is upon all the earth, and every tree that has seed-bearing fruit; they shall be yours for food. [30]And to all the animals on land, to all the birds of the sky, and to everything that creeps on earth, in which there is the breath of life, [I give] all the green plants for food." And it was so. [31]And God saw all that He had made, and found it very good. And there was evening and there was morning, the sixth day.

2 The heaven and the earth were finished, and all their array. [2]On the seventh day God finished the work that He had been doing, and He ceased[a] on the seventh day from all the work that He had done. [3]And God blessed the seventh day and declared it holy, because on it God ceased from all the work of creation that He had done. [4]Such is the story of heaven and earth when they were created.

When the LORD God made earth and heaven—[5]when no shrub of the field was yet on earth and no grasses of the field had yet sprouted, because the LORD God had not sent rain upon the earth and there was no man to till the soil, [6]but a flow would well up from the ground and water the whole surface of the earth—[7]the LORD God formed man[b] from the

---

[a]Or "rested."
[b]Heb. *'adam.*

dust of the earth.[c] He blew into his nostrils the breath of life, and man became a living being.

[8]The LORD God planted a garden in Eden, in the east, and placed there the man whom He had formed. [9]And from the ground the LORD God caused to grow every tree that was pleasing to the sight and good for food, with the tree of life in the middle of the garden, and the tree of knowledge of good and bad.

[10]A river issues from Eden to water the garden, and it then divides and becomes four branches. [11]The name of the first is Pishon, the one that winds through the whole land of Havilah, where the gold is. ([12]The gold of that land is good; bdellium is there, and lapis lazuli.[d]) [13]The name of the second river is Gihon, the one that winds through the whole land of Cush. [14]The name of the third river is Tigris, the one that flows east of Asshur. And the fourth river is the Euphrates.

[15]The LORD God took the man and placed him in the garden of Eden, to till it and tend it. [16]And the LORD God commanded the man, saying, "Of every tree of the garden you are free to eat; [17]but as for the tree of knowledge of good and bad, you must not eat of it; for as soon as you eat of it, you shall die."

[18]The LORD God said, "It is not good for man to be alone; I will make a fitting helper for him." [19] And the LORD God formed out of the earth all the wild beasts and all the birds of the sky, and brought them to the man to see what he would call them; and whatever the man called each living creature, that would be its name. [20]And the man gave names to all the cattle and to the birds of the sky and to all the wild beasts; but for Adam no fitting helper was found. [21]So the LORD God cast a deep sleep upon the man; and, while he slept, He took one of his ribs and closed up the flesh at that spot. [22]And the LORD God fashioned the rib that He had taken from the man into a woman; and He brought her to the man. [23]Then the man said,

"This one at last
Is bone of my bones
And flesh of my flesh.
This one shall be called Woman,[e]
For from man[f] was she taken."

[24]Hence a man leaves his father and mother and clings to his wife, so that they become one flesh. [25]The two of them were naked,[g] the man and his wife, yet they felt no shame.

3 Now the serpent was the shrewdest of all the wild beasts that the LORD God had made. He said to the woman, "Did God really say: You shall not eat of any tree of the garden?" [2]The woman replied to the serpent, "We may eat of the fruit of the other trees of the garden. [3]It is only about fruit of the tree in the middle of the garden that God said: 'You shall not eat of it or touch it, lest you die.'" [4]And the serpent said to the woman, "You are not going to die, [5]but God knows that as soon as you eat of it your eyes will be opened and you will be like [a-]divine beings who know[-a] good and bad." [6]When the woman saw that the tree was good for eating and a delight to the eyes, and that the tree was desirable as a source of wisdom, she took of its fruit and ate. She also gave some to her husband, and he ate. [7]Then the eyes of both of them were opened and they perceived that they were naked; and they sewed together fig leaves and made themselves loincloths.

[8]They heard the sound of the LORD God moving about in the garden at the breezy time of day; and the man and his wife hid from the LORD God among the trees of the garden. [9]The LORD God called out to the man and said to him, "Where are you?" [10]He replied, "I heard the sound of You in the garden, and I was afraid because I was naked, so I hid." [11]Then He asked, "Who told you that you were naked? Did you eat of the tree from which I had forbidden you to eat?" [13]The

---

[c]Heb. *'adamah.*
[d]Others "onyx"; meaning of Heb. *shoham* uncertain.

[e]Heb. *'ishshah.*
[f]Heb. *'ish.*
[g]Heb. *'arummim,* play on *arum* "shrewd" in 3.1.
[a-a]Others "God, who knows."

man said, "The woman You put at my side—she gave me of the tree, and I ate." ¹³And the LORD God said to the woman, "What is this you have done!" the woman replied, "The serpent duped me, and I ate." ¹⁴Then the LORD God said to the serpent,

> "Because you did this,
> More cursed shall you be
> Than all cattle
> And all the wild beasts:
> On your belly shall you crawl
> And dirt shall you eat
> All the days of your life.
> ¹⁵I will put enmity
> Between you and the woman,
> And between your offspring and hers;
> They shall strike at your head,
> And you shall strike at their heel."

¹⁶And to the woman He said,

> "I will make most severe
> Your pangs in childbearing;
> In pain shall you bear children.
> Yet your urge shall be for your husband,
> And he shall rule over you.

¹⁷To Adam He said, "Because you did as your wife said and ate of the tree about which I commanded you, 'You shall not eat of it,'

Cursed be the ground because of you;
By toil shall you eat of it
All the days of your life:
¹⁸Thorns and thistles shall it sprout for you.
But your food shall be the grasses of the field;
¹⁹By the sweat of your brow
Shall you get bread to eat,
Until you return to the ground—
For from it you were taken.
For dust you are,
And to dust you shall return."

²⁰The man named his wife Eve,ᵇ because she was the mother of all the living.ᶜ ²¹And the LORD God made garments of skins for Adam and his wife, and clothed them. ²²And the LORD God said, "Now that the man has become like one of us, knowing good and bad, what if he should stretch out his hand and take also from the tree of life and eat, and live forever!" ²³So the LORD God banished him from the garden of Eden, to till the soil from which he was taken. ²⁴He drove the man out, and stationed east of the garden of Eden the cherubim and the fiery ever-turning sword, to guard the way to the tree of life.

———

ᵇHeb. *hawwah.*
ᶜHeb. *hay.*

QUESTIONS FOR ANALYSIS

1. Marduk was the chief god of the Babylonians, and *Enuma Elish* is often viewed as an allegory for the rise of Babylon over the other peoples of Mesopotamia; just as Marduk defeated his rivals, so did Babylon defeat its. How might *Enuma Elish* also be viewed as an allegory for gender relations in Mesopotamian society?
2. How do the second and third books of Genesis view relationships between women and men? How might the account of events here have been influenced by the fact that the authors of Genesis were male?
3. In *Enuma Elish,* Marduk creates the world out of Tiamat's body, while in Genesis Yahweh creates the world out of chaos. In what ways might Tiamat and chaos be viewed as similar? In what ways are they different?
4. In Genesis, there are two somewhat conflicting versions of creation, one in Genesis 1:1–2:4a and the other in Genesis 2:4b–3:24, with the second one actually the older version. How does the creation of woman differ in the two accounts? What other differences do you see between them?

WOMEN'S RELIGIOUS AND SPIRITUAL ACTIVITIES

Women were not simply figures in creation accounts or goddesses in myths and stories. The lives of real women also included religious and spiritual activities on a regular basis, for human dependence on the gods was an important concept in ancient cultures. In both Egypt and Mesopotamia, elite women served male and female deities as chantresses or priestesses, sometimes gaining a position as the highest religious authority; such women were viewed as communicating or embodying the will of the gods. Because religion was so closely linked with the good of the state, women's positions as priestesses also gave them moral and political authority; they not only oversaw cultic activities but also engaged in diplomacy.

Less exalted women also carried out religious activities on a regular basis, working as temple personnel or maintaining altars in their homes. The honoring of specific deities was associated with being a loyal citizen in the ancient world, akin to patriotism today. In the same way that patriotic citizens of the United States are expected to salute the flag and observe Fourth of July traditions, loyal residents of Egypt, Akkad, or Babylon, male and female, were expected to participate in rituals honoring the local gods.

Along with performing ceremonies and rituals that honored deities, people in both Mesopotamia and Egypt frequently carried out rituals asking for divine protection or assistance in solving daily problems. Many of these have survived as inscriptions, or as paintings on walls or objects such as bowls and amulets. A large number of these rituals concern events that were frequent in the lives of women: family quarrels, difficult pregnancies and deliveries, illnesses in children. Some of the sources indicate that women were not only the intended beneficiaries of such rituals but also their practitioners as shamans, healers, and priestesses. In Egypt, people regarded the dead as well as the gods as influential in human affairs and so sought their favor and help. We have no idea about the exact authorship of most of these sayings and ceremonies, but their frequency on walls and objects suggests they were used by all segments of the population.

3. Hymns to the Goddess Inanna Composed by Enheduanna

Enheduanna was the daughter of King Sargon of Akkad, who created an empire in the Tigris and Euphrates valleys around 2350 B.C.E. Sargon made his daughter the chief priestess of some of the cities he conquered, where she also acted as his representative. Enheduanna is traditionally regarded as the author of a cycle of hymns to Inanna, the goddess of the heavens who was represented as the moon or the evening star. She was connected both with life and fertility, and with death and rebirth. Enheduanna's hymns to Inanna survive in numerous copies, which attests to their popularity in the ancient world.

THE APPEAL TO INANNA

Most precious lady, beloved of An,[1]
Your holy heart is lofty, may it be assuaged on
my behalf!
Beloved bride of Ushumgalanna,
You are the senior queen of the heavenly foun-
dations and zenith.
The Anunna[2] have submitted to you.
From birth on you were the "junior" queen.
How supreme you are over the great gods, the
Anunna!
The Anunna kiss the ground with their lips (in
obeisance) to you.
(But) my own sentence is not concluded, a hos-
tile judgment appears before my eyes as my
judgment.
(My) hands are no longer folded on the ritual
couch,
I may no longer reveal the pronouncements of
Ningal[3] to man.
(Yet) I am the brilliant high priestess of Nanna,[4]
Oh my queen beloved of An, may your heart
take pity on me!

---

[1]*An* or *Anu:* the sky god.
[2]*Anunna:* divine judges of the underworld.
[3]*Ningal:* wife of Nanna.
[4]*Nanna:* the moon-god.

THE EXALTATION OF INANNA

That one has not recited as a "Known! Be it
known!" of Nanna, that one has recited as a
"'Tis Thine!'":
"That you are lofty as Heaven (An)—be it
known!
That you are broad as the earth—be it known!
That you devastate the rebellious land—be it
known!
That you roar at the land—be it known!
That you smite the heads—be it known!
That you devour cadavers like a dog—be it
known!
That your glance is terrible—be it known!
That you lift your terrible glance—be it
known!
That your glance is flashing—be it known!
That you are ill-disposed toward the . . . —be
it known!
That you attain victory—be it known!"
That one has not recited (this) of Nanna, that
one has recited it as a "'Tis Thine!'"—
(That,) oh my lady, has made you great, you
alone are exalted!
Oh my lady beloved of An, I have verily re-
counted your fury!

---

4. Inscription of Queen Hatshepsut from Speos Artemidos and a Carving of Her Head

Queen Hatshepsut (c. 1500 B.C.E.) was one of the few women to rule Egypt in her
own right during its long history, taking over the throne after the death of her husband, Tuthmosis II, from his male heirs. She built and restored a number of temples and other monuments and sent military campaigns and trade expeditions into
foreign territories. Recognizing the importance of a written record of her achievements, she had inscriptions carved detailing her powers and accomplishments, including one at the sacred grotto of Speos Artemidos. Egyptian rulers were
regarded as gods incarnate, a status they achieved through connections with other
gods; this connection was especially important for Hatshepsut, who needed to defend her unusual status as a female ruler. Along with inscriptions, Hatshepsut also

From *Listening to Their Voices: The Rhetorical Activities of Historical Women* edited by Molly Meijer Werthmeier,
pp. 101–102. Copyright © 1997 by University of South Carolina Press. Reprinted by permission

has herself portrayed on many of the monuments she built; some of these portraits were later chiseled off by her male successor attempting to erase her memory. In this carving, Hatshepsut is shown wearing the pharoanic headdress—with a cobra—and the false beard that was commonly worn by male pharoahs.

A.

1. My divine heart is taking thought for the future. The heart of the King of Lower Egypt (herself) has planned for eternity on account of the utterance of the one who opened the Ished tree, Amun,[1] Lord of Millions. I magnified truth which he loved, knowing that he lives on it. It is my bread that I wash down with its dew. I being of one flesh with him. It is to cause that his fame be mighty in this land that he created me.

2. I am the [heir] of Atum,[2] the [beloved] of Kheper, the one who made that which exists, whom Re[3] has predestined when he measured out the lands which are united under my charge, the Black Land and the Red Land being under dread of me. My power causes the foreign lands to bow down, while the uraeus on my forehead pacifies all lands for me. As for (the lands of) Reshwet and Iu[4] they have not been concealed from my majesty. (The land of) Punt[5] [overflows] for me on the fields, its trees bearing fresh myrrh. The paths which were blocked on two sides have been beaten down. My army, which was unequipped, has come into possession of riches since I appeared in glory as king.

3. The temple of the Mistress of Cusae [Hathor],[6] was fallen into ruin, and the earth had swallowed its august sanctuary while children danced on its roof. The serpent-goddess no longer gave terror, and lowly ones were counted *djadja* as perversity, and its festivals no longer appeared. I sanctified it after it had been built anew. It is in order to protect her city that I

fashioned her image from gold, with a barque for a land procession. . . . The face of my majesty gives alertness to the bearers of the god. It is with limestone of Ainu that I built his great temple, its gateways with alabaster of Hatnub. Doors of Asiatic copper, the reliefs thereof of electrum, were consecrated by him who is high of plumes. . . . With a double festival I [extolled] the majesty of this god. . . . The cities thereof are in festival, bearing witness to me, being something completely unknown. Battlements are in the plan; I have set them in order and made them festive while giving chapels to their lords. Every [god] says to himself [about] me. "One who shall spend eternity, One whom Amun has caused to appear as king of eternity on the throne of Horus."[7]

4. Hear you, all patricians and all commoners as many as you are. I have done this by the plan of my heart. I can not sleep as a lowly one. I made to flourish what was ruined. I raised up what was cut up formerly since the Asiatics were in the fold of the Delta at Avaris,[8] with foreigners in their midst overthrowing what had been made. Unmindful of Re they ruled, and he did not act by divine command down to (the time of) my majesty, I having been established on the thrones of Re. I was foretold for an eternity of years as "She will become a conqueror." I have come as the Sole one of Horus flaming against my enemies. I have removed the abomination of the gods, and the earth had brought away their footprints. This was the instruction of the father of my fathers who came at his appointed times, Re. "Never shall occur the destruction of what Amun had com-

---

[1]*Amun:* chief god of Thebes, eventually unified with the sun god Re as Amon-Re.
[2]*Atum:* creator-god of Memphis, city in Egypt.
[3]*Re:* sun god.
[4]*Reshwet* and *Iu:* present-day Sinai.
[5]*Punt:* present-day Somalia.
[6]*Hathor:* goddess of love.

---

[7]*Horus:* divine son of the god Osiris and the goddess Isis.
[8]Here Hatshepsut is talking about the conquest of northern Egypt by the Hyksos, whom she calls "Asiatics."

B. A carving of Queen Hatshepsut. (*Copyright, The British Museum.*)

manded." My command is firm like the mountains and the disk shines and spreads rays over the titulary of my majesty, and my falcon is high over the *serekh* forever and ever.

5. Egyptian Letters to a Deceased Woman from Her Husband and Brother

A message from Merirtifi[1] to Nebitef[2]:
'How are you? Is the west taking care (of you) [as you] desire?
Look, I am your beloved on earth,
(so) fight for me, intercede for my name!
I have not garbled a spell before you, while making your name to live upon earth.[3]
Drive off the illness of my limbs!
May you appear for me as a blessed one before me, that I may see you fighting for me in a dream.
I shall lay down offerings for you when the sun's light has risen,
and I shall establish an altar for you.'
A message from Khuau[4] to his sister:
'[. . .] I have not garbled a spell before you; I have not taken offerings away from you.
Now, I have sought [your benefit(?)]. Fight for me!
Fight for my wife and my children!'

---

[1]*Merirtifi:* the husband.
[2]*Nebitef:* the deceased woman.
[3]I have performed your funerary cult without any error.

[4]*Khuau:* the brother.

From *Voices of Ancient Egypt: An Anthology of Middle Kingdom* by R.B. Parkinson, p. 130. Copyright © 1991 by University of Oklahoma Press. Reprinted by permission.

6. Egyptian Protective Spell for Guarding the Limbs, to Be Recited over a Child When the Sunlight Rises

You rise, Re, you rise!
Have you seen the dead who has come against
    her—
N (born) of N—
to lay a spell on her,
using plans to seize her son from her embrace?
'May you save me, my lord, Re!'

so says N born of N.
'I shall not give you, I shall not give your
    charge
to a male or female robber from the West—
my hand upon you, my seal as your protection!'
so says Re as he rises.
May you break forth! This is a protection.

7. Hittite Ritual Against Domestic Quarrel

These are the words of Mastiggas, the woman from Kizzuwatna: If a father and (his) son, or a husband and his wife, or a brother and (his) sister quarrel, when I reconcile them, I treat them as follows:

She takes black wool and wraps it in mutton fat; *tissatwa* they call it. She presents it to the sacrificer and speaks as follows: "Whatever thou spokest with (thy) mouth (and) tongue—see, here is *tissatwa!* Let it be cut out of your body these days!" She throws the tongues into the hearth.

Afterward the Old Woman takes salt, tissue, *fat* and wax. She makes the wax into tongues and waves [them over the two] sacrificers. She also waves the salt and the [fat and] the tissue over them, [present]s it to them and they [flatten it] with (their) left hands.

The Old Woman speaks as follows: "In whatever curses you indulged, let now the Sungod turn those curses (and) tongues toward the left!" And she throws them into the hearth.

The red wool (and) the blue wool that had been placed upon the bodies of the two sacrificers, the two figures of dough that had been placed before them, and the hands and tongues of

dough that had been placed upon their heads, those the Old Woman removes. She cuts the strings off them, the Old Woman breaks the two hands and the tongues of dough to pieces.

She then waves them over them and speaks as follows: "Let the tongues of these [days] be cut off! Let the words of these days be cut off!" And she throws them [into the hearth].

Afterward the Old Woman takes [a tray] and [places] seven tongues [and seven hands . . .] upon it. She waves it over the two sacrificers and [speaks] as follows: "The day at which ye satisfied your hunger—see here the tongues and the hands of that day. See, the father Sun has now nailed them down." And she puts them into the hearth.

The Old Woman takes water and dough. She sprinkles the water upon them and purified them. Then she waves the dough over them and speaks as follows: "Be ye cleansed of mouth and tongue!" And she puts the dough into the hearth.

They drive up a (white) sheep. The Old Woman presents it to the two sacrificers and speaks as follows: "Here is a substitute for you, a substitute for your persons. Let that tongue and that curse stay in (its) mouth!" They spit into its mouth.

She speaks as follows: "Spit ye out those evil curses!" They dig a hole in the ground, cut the sheep up over it, and then put it into it.

The Old Woman takes water with a cup or an amphora and presents it to the two sacrificers; salt is also put in. The two sacrificers pour the water over their heads, they also rinse their hands (and) their eyes.

Then they pour it into the horn of an ox. The two sacrificers seal it up and the Old Woman speaks as follows:

"On the day when the olden kings return and examine the state of the land,—then, and then only, shall this seal be broken."

8. Akkadian Rituals for Childbirth and Motherhood

A. Akkadian Ritual for a Woman in Labor

The woman in labor is having a difficult labor,
Her labor is difficult, the baby is stuck,
The baby is stuck!
The doorbolt is locked, about to end life,
The door is fastened against the suckling kid, . . .
The woman giving birth is covered with death's dust,
She is covered with the dust of battle, like a chariot,
She is covered with the dust of tuffets, like a plow.
She sprawls in her own blood, like a struggling warrior,
Her eyesight is waning, she cannot see,
Her lips are coated, she cannot open them . . . ,
Her eyesight is dim, . . . she is alarmed, her ears cannot hear!
Her breast is not held in, her headbands are askew,
She is not veiled with . . . , she has no shame:
Stand by me and keep calling out(?), O merciful Marduk,[1]
"Here is the confusion, I'm surrounded, get to me!"[2]

Bring out the one sealed up, created by the gods,
Created by mankind, let him come out and see the light!

B. Akkadian Elegy for a Woman Dead in Childbirth

This poem tells the story of a woman's death in childbirth as if she were narrating it herself. Her pleas and those of her husband fail to move Belet-ili, goddess of birth.

Why are you cast adrift, like a boat in mid-stream,
Your planking stoven, your mooring rope cut?
With shrouded face, you cross the river of the City.

How could I not be cast adrift,
how could my mooring rope not be cut?
The day I carried the fruit, how happy I was,
Happy was I, happy my husband.
The day I went into labor, my face grew overcast,
The day I gave birth, my eyes grew cloudy.

I prayed to Belet-ili with my hands opened out,
'You are mother of those who give birth, save my life!'
Hearing this, Belet-ili shrouded her face,
'You [ ], why do you keep praying to me?'

[My husband, who lov]ed me, uttered a cry,

---

[1] The appeal to Marduk is evidently spoken by the midwife.
[2] This line is evidently spoken by the baby.

From B. R. Foster, *Before the Muses: An Anthology of Akkadian Literature*. Reprinted by permission from CDL Press.

'[   ] me, the wife I adore!'

(gap)

[All . . .] those days I was with my husband,
While I lived with him who was my lover,
Death was creeping stealthily into my bed-
    room,
It forced me from my house,
It cut me off from my lover,
It set my foot toward the land from which I
    shall not return.

C. Akkadian Ritual Against Nurses Harmful to Children

O wetnurse!
Wetnurse, whose breast is (too) sweet,
Wetnurse, whose breast is (too) bitter,
Wetnurse, whose breast is infected,
Wetnurse, who died from an infected breast,
Nursemaid, whose armclasp is relaxed,
Nursemaid, whose armclasp is loose,
Nursemaid, whose armclasp is limp,
Nursemaid, whose armclasp is wrong,
Be conjured by heaven!
Be conjured by the netherworld!

D. Akkadian Ritual for Mothers

This spell was to be recited three times with a piece of bread set by the baby's head. Then the child was to be rubbed all over with the bread and the bread thrown to a dog. After this, the child was supposed to fall silent.

The one who dwelt in darkness, where no light
    shone,
He has come out and has seen sunlight.
Why does he scream till his mother sobs,
Till, in heaven, Antu herself's in tears?

"Who is this, that makes such a racket on
    earth?
"If it's a dog, someone give it some
    food!
"If it's a bird, someone fling a clod at it!
"If it's a human child that's angry,
"Someone cast the spell of Anu and Antu over
    him!
"Let his father lie down, to get the rest of his
    sleep,
"Let his mother, who has her chores to do, get
    her chores done."

QUESTIONS FOR ANALYSIS

1. In the hymns to Inanna and the inscription of Hatshepsut, what attributes do the authors ascribe to female deities? How do they link themselves as royal women to the gods?
2. What religious and military activities does Hatshepsut describe herself as having undertaken, and how does she see these as connected?
3. What religious and ritualistic actions do women take to preserve their own lives and those of their children, or to gain peace within the household? What actions do the husband and brother of Nebitef ask her to take for them now that she has died?
4. How would you compare the relationship to the gods of royal women such as Hatshepsut and Enheduanna, expressed in their writings and portraits, with that of less elite women such as the shaman Mastiggas or the unnamed wives and mothers in the rituals?

MARRIAGE AND FAMILY LIFE

Most women in ancient cultures married or entered into marital-like relationships at some point in their lives. Because illness and injury led to death at an early age for many people, many women married more than once, or they married men who had been married before. Many cultures had specific rules and traditions regarding marriage that specified appropriate marital partners, property arrangements in the case of death or divorce, and the expected treatment of children. Young people in some cultures appear to have engaged in love affairs—or at least wrote poetry about such affairs—but once a couple was married, they were expected to follow certain rules about marital fidelity. These rules were often quite different for women than for men, and the consequences of adultery differed sharply. Divorce was allowed, but it was often difficult, so choosing a good marital partner was an important matter for families as well as the individuals involved.

Households in the river-valley civilizations often included individuals who were not related by blood, such as servants or slaves, as well as children, stepchildren, and perhaps other relatives. People moved in and out of households in ways that are similar to those of contemporary society—such as at marriage—but also in ways that are very different—such as selling themselves or being sold by other family members into slavery or servitude. Girls and women appear to have been sold more often than boys and men, but women were also found among slave-owners. Except in Assyria, women owned and managed property that they had inherited and purchased, including slaves, on their own.

9. Hammurabi's Code

Some of the earliest records that exist in human culture are law codes, designed by rulers to shape their societies. One of the most comprehensive of these from Mesopotamia is that of the Babylonian king Hammurabi (1792–1750 B.C.E.) inscribed on a pillar; like Hatshepsut's inscription, this one also links the ruler with the divine, showing Hammurabi accepting the laws from the Babylonian god of justice, Shamash. Many of the nearly 300 laws deal with marriage and family life and set out what the Babylonian elite regarded as the proper relationship between the sexes and the proper running of a household. It is important to remember that this code, like all laws, sets out an ideal and does not necessarily reflect what really happened in Babylonian families. We have very little information about how or how well these laws were enforced; the social differences set out in them suggest that enforcement probably differed depending on the social status of the parties concerned.

Pritchard, James. *Ancient Near Eastern Texts Relating to the Old Testament*, Third Edition. Copyright © 1969 by Princeton University Press. Reprinted by permission of Princeton University Press.

128: If a seignior[1] acquired a wife, but did not draw up the contracts for her, that woman is no wife.

129: If the wife of a seignior has been caught while lying with another man, they shall bind them and throw them into the water. If the husband of the woman wishes to spare his wife, then the king in turn may spare his subject.

130: If a seignior bound the (betrothed) wife of a(nother) seignior, who had had no intercourse with a male and was still living in her father's house, and he has lain in her bosom and they have caught him, that seignior shall be put to death, while that woman shall go free.

131: If a seignior's wife was accused by her husband, but she was not caught while lying with another man, she shall make affirmation by god and return to her house.

132: If the finger was pointed at the wife of a seignior because of another man, but she has not been caught while lying with the other man, she shall throw herself into the river[2] for the sake of her husband.

133: If a seignior was taken captive, but there was sufficient to live on in his house, his wife [shall not leave her house, but she shall take care of her person by not] entering [the house of another].

133a: If that woman did not take care of her person, but has entered the house of another, they shall prove it against that woman and throw her into the water[3]. . . .

138: If a seignior wishes to divorce his wife who did not bear him children, he shall give her money to the full amount of her marriage-price and he shall also make good to her the dowry which she brought from her father's house and then he may divorce her.

139: If there was no marriage-price, he shall give her one mina of silver as the divorce-settlement.

140: If he is a peasant, he shall give her one-third mina of silver.

141: If a seignior's wife, who was living in the house of the seignior, has made up her mind to leave in order that she may engage in business, thus neglecting her house (and) humiliating her husband, they shall prove it against her; and if her husband has then decided on her divorce, he may divorce her, with nothing to be given her as her divorce-settlement upon her departure. If her husband has not decided on her divorce, her husband may marry another woman, with the former woman living in the house of her husband like a maidservant.

142: If a woman so hated her husband that she has declared, "You may not have me," her record shall be investigated at her city council, and if she was careful and was not at fault, even though her husband has been going out and disparaging her greatly, that woman, without incurring any blame at all, may take her dowry and go off to her father's house.

143: If she was not careful, but was a gadabout, thus neglecting her house (and) humiliating her husband, they shall throw that woman into the water. . . .

153: If a seignior's wife has brought about the death of her husband because of another man, they shall impale that woman on stakes.

154: If a seignior has had intercourse with his daughter, they shall make that seignior leave the city.

155: If a seignior chose a bride for his son and his son had intercourse with her, but later he himself has lain in her bosom and they have caught him, they shall bind that seignior and throw him into the water.

156: If a seignior chose a bride for his son and his son did not have intercourse with her, but he himself has lain in her bosom, he shall pay to her one-half mina of silver and he shall also make good to her whatever she brought from her father's house in order that the man of her choice may marry her. . . .

170: When a seignior's first wife bore him children and his female slave also bore him

---

[1]*Seignior:* a free man of any class.
[2]The woman is to throw herself into the Euphrates and let the god of the river decide her guilt or innocence.
[3]In this case, she is to be drowned.

children, if the father during his lifetime has ever said "My children!" to the children whom the slave bore him, thus having counted them with the children of the first wife, after the father has gone to (his) fate, the children of the first wife and the children of the slave shall share equally in the goods of the paternal estate, with the first-born, the son of the first wife, receiving a preferential share.

171: However, if the father during his lifetime has never said "My children!" to the children whom the slave bore him, after the father has gone to (his) fate, the children of the slave may not share in the goods of the paternal estate along with the children of the first wife; freedom for the slave and her children shall be effected, with the children of the first wife having no claim at all against the children of the slave for service; the first wife shall take her dowry and the marriage-gift which her husband, upon giving (it) to her, wrote down on a tablet for her, and living in the home of her husband, she shall have the usufruct[4] (of it) as long as she lives, without ever selling (it), since her heritage belongs to her children. . . .

_____
[4]*Usufruct:* use.

194: When a seignior gave his son to a nurse and that man has died in the care of the nurse, if the nurse has then made a contract for another son without the knowledge of his father and mother, they shall prove it against her and they shall cut off her breast because she made a contract for another son without the knowledge of his father and mother.

195: If a son has struck his father, they shall cut off his hand.

196: If a seignior has destroyed the eye of a member of the aristocracy, they shall destroy his eye. . . .

209: If a seignior struck a(nother) seignior's daughter and has caused her to have a miscarriage, he shall pay ten shekels of silver for her fetus.

210: If that woman has died, they shall put his daughter to death.

211: If by a blow he has caused a commoner's daughter to have a miscarriage, he shall pay five shekels of silver.

212: If that woman has died, he shall pay one-half mina of silver.

213: If he struck a seignior's female slave and has caused her to a have a miscarriage, he shall pay two shekels of silver.

214: If that female slave has died, he shall pay one-third mina of silver.

10. Israelite Laws from the Books of Numbers and Deuteronomy

Along with the creation accounts in the book of Genesis, the Hebrew Bible includes, in the Torah, several books of laws and admonitions governing many aspects of interpersonal relationships, including marriage. In the same way that Egyptian and Mesopotamian rulers stressed the connections between their rules and actions and the will of the gods, the books of Numbers and Deuteronomy often describe Israelite law as emanating from Yahweh. Marital fidelity in women was a particularly important concern, with the decisions of Yahweh playing a role in determining guilt or innocence as well as establishing the rules themselves.

A. Numbers

5 [11]The LORD spoke to Moses, saying: [12]Speak to the Israelite people and say to them:

If any man's wife has gone astray and broken faith with him [13]in that a man has had carnal relations with her unbeknown to her husband, and she keeps secret the fact that she has defiled herself without being forced, and there is no witness against her—[14]but a fit of jealousy comes over him and he is wrought up about the wife who has defiled herself; or if a fit of jealousy comes over one and he is wrought up about his wife although she has not defiled herself—[15]the man shall bring his wife to the priest. And he shall bring as an offering for her one-tenth of an *ephah* of barley flour. No oil shall be poured upon it and no frankincense shall be laid on it, for it is a meal offering of jealousy, a meal offering of remembrance which recalls wrongdoing.

[16]The priest shall bring her forward and have her stand before the LORD. [17]The priest shall take sacral water in an earthen vessel and, taking some of the earth that is on the floor of the Tabernacle, the priest shall put it into the water. [18]After he has made the woman stand before the LORD, the priest shall bare the woman's head and place upon her hands the meal offering of remembrance, which is a meal offering of jealousy. And in the priest's hands shall be the water of bitterness that induces the spell. [19]The priest shall adjure the woman, saying to her, "If no man has lain with you, if you have not gone astray in defilement while married to your husband, be immune to harm from this water of bitterness that induces the spell. [20]But if you have gone astray while married to your husband and have defiled yourself, if a man other than your husband has had carnal relations with you"—[21]here the priest shall administer the curse of adjuration to the woman, as the priest goes on to say to the woman— "may the LORD make you a curse and an imprecation among your people, as the LORD causes

your thigh to sag and your belly to distend; [22]may this water that induces the spell enter your body, causing the belly to distend and the thigh to sag." And the woman shall say, "Amen, amen!"

[23]The priest shall put these curses down in writing and rub it off into the water of bitterness. [24]He is to make the woman drink the water of bitterness that induces the spell, so that the spell-inducing water may enter into her to bring on bitterness. [25]Then the priest shall take from the woman's hand the meal offering of jealousy, elevate the meal offering before the LORD, and present it on the altar. [26]The priest shall scoop out of the meal offering a token part of it and turn it into smoke on the altar. Last, he shall make the woman drink the water.

[27]Once he has made her drink the water—if she has defiled herself by breaking faith with her husband, the spell-inducing water shall enter into her to bring on bitterness, so that her belly shall distend and her thigh shall sag; and the woman shall become a curse among her people. [28]But if the woman has not defiled herself and is pure, she shall be unharmed and able to retain seed.

[29]This is the ritual in cases of jealousy, when a woman goes astray while married to her husband and defiles herself, [30]or when a fit of jealousy comes over a man and he is wrought up over his wife: the woman shall be made to stand before the LORD and the priest shall carry out all this ritual with her. [31]The man shall be clear of guilt; but that woman shall suffer for her guilt.

B. Deuteronomy

22 [13]A man marries a woman and cohabits with her. Then he takes an aversion to her [14]and makes up charges against her and defames her, saying, "I married this woman; but when I approached her, I found that she was not a virgin." [15]In such a case, the girl's father and mother shall produce the evidence of the girl's virginity before the elders of the town at the gate. [16]And the girl's father shall say to the elders, "I gave this man my daughter to wife, but he has taken an aversion to her; [17]so he has made up charges, saying, 'I did not find your daughter a virgin.' But here is the evidence of my daughter's virginity!" And they shall spread out the cloth before the elders of the town. [18]The elders of that town shall then take the man and flog him, [19]and they shall fine him a hundred [shekels of] silver and give it to the girl's father; for the man has defamed a virgin in Israel. Moreover, she shall remain his wife; he shall never have the right to divorce her.

[20]But if the charge proves true, the girl was found not to have been a virgin, [21]then the girl shall be brought out to the entrance of her father's house, and the men of her town shall stone her to death; for she did a shameful thing in Israel, committing fornication while under her father's authority. Thus you will sweep away evil from your midst.

[22]If a man is found lying with another man's wife, both of them—the man and the woman with whom he lay—shall die. Thus you will sweep away evil from Israel.

[23]In the case of a virgin who is engaged to a man—if a man comes upon her in town and lies with her, [24]you shall take the two of them out to the gate of that town and stone them to death: the girl because she did not cry for help in the town, and the man because he violated another man's wife. Thus you will sweep away evil from your midst. [25]But if the man comes upon the engaged girl in the open country, and the man lies with her by force, only the man who lay with her shall die, [26]but you shall do nothing to the girl. The girl did not incur the death penalty, for this case is like that of a man attacking another and murdering him. [27]He came upon her in the open; though the engaged girl cried for help, there was no one to save her.

[28]If a man comes upon a virgin who is not engaged and he seizes her and lies with her, and they are discovered, [29]the man who lay with her shall pay the girl's father fifty [shekels of] silver, and she shall be his wife. Because he has violated her, he can never have the right to divorce her.

11. Contracts Regarding Marriage, Personal Status, and Property Ownership

In addition to law codes that set out ideals for society, legal documents resulting from real situations have survived from Mesopotamia and Egypt. These documents show women acting in a variety of legal capacities—making or being part of marital contracts, being enslaved and freeing slaves, inheriting and buying land. Though these documents may also be describing a somewhat idealized situation—how children were to act toward their mother, or a slave toward her owner—they reflect real situations and can provide more specific and personal details of women's lives than law codes alone. They do not tell us what the people involved thought of their situation, however, and diaries or similar personal documents describing family life do not exist. These contracts are also weighted toward the elite, as poor people with no property had no reason to worry about who was to inherit or acquire it. They often do refer to slaves, however, who were considered property as well as persons.

HEBREW SLAVE DOCUMENT

Sin-balti, a Hebrew woman, on her own initiative has entered the house of Tehip-tilla as a slave. Now if Sin-balti defaults and goes into the house of another, Tehip-tilla shall pluck out the eyes of Sin-balti and sell her.

(The names of nine persons and the scribe as witnesses, each preceded by the witness-sign.)

BABYLONIAN MARRIAGE CONTRACT

Ama-sukkal, the daughter of Ninurta-mansum, has been taken in marriage by Enlil-izzu, the high priest of Enlil, the son of Lugal-azida; Ama-sukkal has brought 19 shekels of silver to Enlil-izzu, her husband (as dowry).

If Enlil-izzu ever says to Ama-sukkal, his wife, "You are no longer my wife," he shall return the 19 shekels of silver and he shall also weigh out ½ mina as her divorce-settlement. On the other hand, if Ama-sukkal ever says to Enlil-izzu, her husband, "You are no longer my husband," she shall forfeit the 19 shekels of silver and she shall also weigh out ½ mina of silver. In mutual agreement they have sworn together by the king.

(The names of eight men, two women, the scribe, and the notary as witnesses, each preceded by the witness-sign.)

MANUMISSION AND MARRIAGE FROM UGARIT IN SYRIA

As of this day, before witnesses, Gilben, chamberlain of the queen's palace, set free Eliyawe his maidservant, from among the women of the harem, and by pouring oil on her head, made her free, (saying) "Just as I am quit towards her, so is she quit towards me, forever."

Further, Buriyanu, the *namu*,[1] has taken her as his wife, and Buriyanu, her husband has ren-

dered 20 (shekels) of silver into the hands of Gilben. Four witnesses.

(Inscribed on seal:) Should Buriyanu, tomorrow or the following day, refuse to consummate (his marriage) with Eliyawe—[2]

WILL AND TESTAMENT FROM UGARIT IN SYRIA

As of this day, before witnesses, Yarimanu spoke as follows: "Now therefore, whatever I possess (and) that which Bidawe acquired together with me (to wit): my large cattle, my small cattle, my asses, my male slaves, my female slaves, my bronze bowls, bronze kettles, bronze jugs, baskets, the field of Bin-Harasina (bordering) upon the Ra'abani stream—I have bequeathed to Bidawe, my wife.

And now therefore, my two sons—Yatlinu, the elder, and Yanhamu, the younger—whichever of them shall bring a lawsuit against Bidawe, or shall about Bidawe, their mother, shall pay 500 shekels of silver to the king; he shall set his cloak upon the doorbolt,[1] and shall depart into the street. But whichever of them shall have paid respect to Bidawe, his mother—to that one will she bequeath (the possessions).

Five witnesses and the name of the scribe.

JEWISH MARRIAGE CONTRACT

On the 21st of Chisleu, that is the 1st of Mesore, year 6 of King Artaxerxes, Mahseiah b. Yedoniah, a Jew of Elephantine,[1] of the detachment of Haumadata, said to Jezaniah b. Uriah of the said detachment as follows: There is the site of 1 house belonging to me, west of the house belonging to you, which I have given to

---

[2]The penalty is not stated.
[1]He is to be completely disinherited and must even leave his clothing.
[1]Both this and the following document are from Elephantine, a military post on the Nile where Jewish mercenaries were hired to guard the southern border of Egypt. Many documents on papyrus have survived from roughly the fifth century B.C.E.

---

[1]*Namu*: agricultural laborer.

your wife, my daughter Mibtahiah (*Mbthyh*), and in respect of which I have written her a deed. The measurements of the house in question are 8 cubits and a handbreadth by 11, *by the measuring-rod.* Now do I, Mahseiah, say to you, Build and equip that site . . . and dwell thereon with your wife. But you may not sell that house or give it as a present to others; only your children by my daughter Mibtahiah shall have power over it after you two. If tomorrow or some other day you build upon this land, and then my daughter divorces you and leaves you, she shall have no power to take it or give it to others; only your children by Mibtahiah shall have power over it, in return for the work which you shall have done. If, on the other hand, she recovers from you, she [may] take half of the house, and [the] othe[r] half shall be at your disposal in return for the building which you will have done on that house. And again as to that half, your children by Mibtahiah shall have power over it after you. If tomorrow or another day I should institute suit or process against you and say I did not give you this land to build on and did not draw up this deed for you, I shall give you a sum of 10 *karshin* by royal weight, at the rate of 2 *R* to the ten, and no suit or process shall lie. This deed was written by 'Atharshuri b. Nabuzeribni in the fortress of Syene at the dictation of Mahseiah. Witnesses hereto (signatures).

MANUMISSION OF A FEMALE SLAVE AND HER DAUGHTER

On the 20th of Siwan, that is the 7th day of Phamenoth, the year 38 of King Artaxerxes— at that time, Meshullam son of Zakkur, a Jew of the fortress Elephantine, of the detachment of Arpakhu said to the woman Tapmut (as she is called), his slave, who has on her right hand the marking "Of Meshullam," as follows: I have taken kindly thought of you in my lifetime. I hereby declare you released at my death and likewise declare released the daughter Yehoyishma' (as she is called) whom you have borne to me. No son or daughter, close or distant relative, kinsman, or clansman of mine has any right to you or to the daughter Yehoyishma' whom you have borne to me; none has any right to mark you or to *deliver you as a payment of money.* Whoever attempts such action against you or the daughter Yehoyishma' whom you have borne to me must pay you a fine of 50 karsh of silver by the king's weights. You are released, with your daughter Yehoyishma', from the shade for the sun, and no other man is master of you or your daughter Yehoyishma'. You are released for God.

And Tapmut and her daughter Yehoyishma' declared: We shall serve you [as a son or daughter supports his or her father (12) as long as you live; and when you die, we shall support your son Zakkur like a son who supports his father, just as we shall have been doing for you while you were alive. (. . .) If we ever say, "We will not support you as a son supports his father, and your son Zakkur after your death," we shall be liable to you and your son Zakkur for a fine in the amount of 50 karsh of refined silver by the king's weights without suit or process.

Written by Haggai the scribe, at Elephantine, at the dictation of Meshullam son of Zakkur.

QUESTIONS FOR ANALYSIS

1. How would you compare the method of determining guilt or innocence for women accused of adultery in Hammurabi's Code and the Old Testament? The punishment if they were found guilty? The punishment for those who accused them if such accusations proved to be false?
2. How are social differences among women reflected in Hammurabi's Code?
3. What possibilities for divorce exist in Hammurabi's Code and the Old Testament? How are these restricted?

4. What are some of the ways in which women moved from slavery to freedom, and vice versa, in ancient societies? Is women's status in such instances dependent on that of their male relatives?
5. How does the picture of family relationships that emerges from the contracts differ from that which emerges from the law codes? Why might these be different?

✵ CHAPTER 2

The Greek World

(600 B.C.E.–200 C.E.)

Ancient Greece, and more specifically ancient Athens during the
fifth and fourth centuries B.C.E., is often regarded as the originator of many aspects of Western culture that are prized to this
day. In the city-state (or *polis*) of Athens, citizens had the right to participate in direct democracy, debating and voting on important issues
and holding political office. Indeed, citizens were expected to participate in public affairs, with those who did not labeled *idiots,* the Greek
word for those concerned only about private, household matters.
Greek thinkers developed views of the universe based on logic, reason,
and experimentation, rather than on myth or magic, and are thus seen
as the originators of both science and philosophy. Greek art and theater focused on the human as much as the divine, emphasizing the dignity and heroism of the individual, even those caught in tragic moral
dilemmas.

Along with democracy, philosophy, science, drama, and art, classical
Athens is also one of the key sources of Western ideas about the relative value of men and women and their appropriate place in society.
Because citizenship brought political rights, Athenian democracy
made a sharp distinction between citizen and noncitizen. The majority
of people living in Athens were not citizens, but slaves, servants, or
people who had come to Athens from elsewhere. Citizenship was
handed down from father to son, symbolized by a ceremony held on
the tenth day after a child was born in which the father laid his son on
the floor of the house and gave him a name; this ceremony marked a
boy's legal birth. It was thus very important to Athenian citizen men
that their sons be their own; consequently, their wives and daughters
were increasingly secluded in special parts of the house (termed the *gy-
neceum*) and allowed out in public only for religious festivals, funerals,
and perhaps the theater (there is a debate about this among historians).
The wives and daughters of citizens were completely excluded from

public life and thus could never share in the Athenian ideal of active participation in public life; they were, by definition, idiots.

The contrasting pair citizen/noncitizen was only one of many dualities that underlay Athenian conceptions of society and the universe. These dualities also included hot/cold, dry/wet, good/evil, light/dark, right/left, straight/crooked and, of course, male/female; in each case, the female was associated with the quality regarded as less positive. This male/female dualism reflected Athenian society in which women played a smaller and less public role, and it also shaped European thinking from that point on. Athenian philosophers, particularly Aristotle (384–322 B.C.E.) became extremely influential, with their word carrying an authority second only to the Bible; for centuries, in fact, Aristotle was simply referred to as "The Philosopher." Aristotle associated the male/female dichotomy with those of spirit/matter and perfection/imperfection, asserting that the mother provided the material to form the body of a fetus, but the father provided the soul, an idea that was accepted by many scientists and physicians, as well as some later Christian theologians and philosophers. Aristotle described women— and all female animals—as "deformed males," created when the conditions at conception or during pregnancy were less than perfect.

The association of male with perfection and female with imperfection shaped Athenian art as well as philosophy. The perfect human form was the young male, and because the purpose of art was to portray perfection, only young men where shown nude. During the classical period, even Aphrodite, the goddess of beauty, was always shown clothed. Female characters do appear in Athenian theater, which often depicted strong women caught, like men, in tragic situations in which their duties to their family, the gods, or society come into conflict with one another. Such characters were created by male playwrights and portrayed by male actors, however, and the plays may have been performed to an all-male audience. It is difficult to understand how the Greek men who wrote and watched these plays interpreted the actions of these fictional women, and it is even more difficult to know what Athenian women thought about them, as all of our sources from the classical period of Athens come from male authors.

The dichotomy citizen/noncitizen led to restrictions on the activities of citizen women, but it also meant that noncitizen women were generally not included in such restrictions. Female slaves and servants were not secluded; in fact, their work, such as buying and selling goods at the public market or drawing water at neighborhood wells, made the seclusion of citizen women possible. Women from outside Athens were engaged in a number of occupations, the most famous of which was being a *hetaera,* a resident in one of the city-run brothels in which Athenian men spent much of their time. Athenian wives were not expected to entertain their husbands, so the social companions of Athenian men were either *hetaerae* or other men. The Athenian orator Demosthenes put this succinctly: "We have *hetaerae* for our enjoyment, concubines to serve our person, and wives for bearing of legitimate offspring."

Athens was the dominant polis culturally, politically, and intellectually in ancient Greece, and it has left the most records. Another polis, Sparta, was often the strongest militarily, in part because all activity was directed toward military aims. Citizen boys left their homes at age seven in Sparta and lived in military camps until they were thirty, eating and training with boys and men their own age; they married at about eighteen to women of roughly the same age, but saw their wives only when they sneaked out of camp. Military discipline was harsh (thus the origin of the word *spartan*), but severity was viewed as necessary to prepare men both to fight external enemies such as Athens and to control the Spartan slave and unfree farmer population, which vastly outnumbered the citizens.

In this militaristic atmosphere, citizen women were remarkably free. There was an emphasis on child-bearing, but the Spartan leadership viewed maternal health as important for the bearing of healthy, strong children and so encouraged women to participate in athletics and to eat well. With men in military service most of their lives, citizen women owned property and ran the household; they were not physically restricted or secluded. Marriage often began with a trial marriage period to make sure the couple could have children, with divorce and remarriage the normal course if they were unsuccessful. For some Athenian commentators—and almost all information about Sparta comes from Athenians—the system at Sparta was horrible, while for others it was praiseworthy.

For much of the fifth century, Athens and Sparta were at war with each other and with other city-states, which gradually weakened them. In the middle of the fourth century, Greece was conquered by Philip of Macedonia, a kingdom to the north of Greece. Philip's son, Alexander the Great (who had been tutored by Aristotle), went on to conquer much of the eastern Mediterranean and western Asia, including Egypt and Mesopotamia. He died while still a young man, and his empire was broken into several large kingdoms in which Greek-speaking monarchs, military leaders, and officials ruled large non-Greek populations. (These kingdoms and the culture created by Alexander are usually termed *Hellenistic* [Greek-like] to distinguish them from the earlier *Hellenic* [Greek] culture of the era of city-states.)

With the end of the city-states' independence and the development of Hellenistic kingdoms, restrictions on women lifted somewhat. Citizenship no longer brought political power, for this was held by kings, so that women were occasionally made citizens in their own right; as queens they ruled both alongside their husbands and when their sons were young. Some of the philosophical and religious movements that became popular in the Hellenistic world offered women education and opportunities for leadership that had been unavailable in classical Athens. Hellenistic culture was cosmopolitan, with people and ideas moving around the Mediterranean easily, and it has left a much wider variety of sources than earlier Greek culture, particularly those written on papyrus, the paperlike material from Egypt.

THE STATE AND THE HOUSEHOLD
IN CLASSICAL ATHENS

Though in some ways the ancient Athenians regarded the state and the household, or what we might call the public and the private, as a duality, in other ways they viewed them as related. Government was an association of free adult male individuals, but each of these individuals was expected to be the head of a household and to produce (male) children who would inherit his citizenship. Thus the Athenian writers and thinkers who are generally regarded as the founders of Western philosophy all addressed the issue of what went on in households as well as in governmental institutions and of the proper relationships between household members, including the male head and his wife, children, and slaves. This link between the household and the state continued in Western political philosophy and in Western ideas of leadership for millennia. As late as the seventeenth century, families in colonial Massachusetts were described as "little commonwealths," a term normally used for governmental units, such as "the Commonwealth of Massachusetts," and King James I of England described his power as deriving from the power of husbands over their wives and fathers over their children.

12. Plato
The Republic

Plato (427?–347 B.C.E.) was an Athenian philosopher and pupil of Socrates (469–399 B.C.E.), who, like his teacher, was interested in understanding the true nature of virtue and knowledge. Socrates did this by questioning everyone and everything in what has come to be called the Socratic method. He left no writings of his own, but many of Plato's writings are couched in the form of dialogues between Socrates and various other speakers. It is thus sometimes difficult to separate the ideas of Socrates and Plato; on many things they seem to have agreed, especially in the notion that true knowledge comes from contemplating things in their ideal form, rather than in the way they appear in the world around us. In *The Republic,* Plato (through the voice of Socrates speaking to a man named Glaucon) sets out his conception of the ideal state, in which certain people born with a talent for leadership (whom he calls "guardians") are educated to know virtue. They thus rule everyone else, who are trained instead to protect the state (the protectors are called "auxiliaries") or to work at various occupations. Socrates first sets out how male guardians are to be educated and then turns to the issue of female guardians and family relationships among the guardians.

From *The Dialogues of Plato,* Fourth Edition, translated by Benjamin Jowett, Fourth Edition, vol. 2. Copyright © 1965 by Oxford University Press. Reprinted by permission of Oxford University Press.

The drama of the men has been played out, and now properly enough comes the turn of the women, especially in view of your challenge.

For men born and educated like our citizens there can, in my opinion, be no right possession and use of women and children unless they follow the path on which we sent them forth. We proposed, as you know, to treat them as watch-dogs of the herd.

True.

Let us abide by that comparison in our account of their birth and breeding, and let us see whether the result accords with our design.

What do you mean?

What I mean may be put into the form of a question, I said: Are female sheepdogs expected to keep watch together with the males, and to go hunting with them and share in their other activities? or do we entrust to the males the entire and exclusive care of the flocks, while we leave the females at home, because we think that the bearing and suckling their puppies is labour enough for them?

No, he said, they share alike; the only difference between them is that the males are regarded as stronger and the females as weaker.

But can you use different animals for the same purpose, unless they are bred and fed in the same way?

You cannot.

Then if women are to have the same duties as men, they must have the same education?

Yes.

The education which was assigned to the men was music and gymnastic.

Yes.

Then women also must be taught music and gymnastic and military exercises, and they must be treated like the men?

That is the inference, I suppose.

I fully expect, I said, that our proposals, if they are carried out, being unusual, may in many respects appear ridiculous.

No doubt of it.

Yes, and the most ridiculous thing of all will be the sight of women naked in the palaestra, exercising with the men, even when they are no longer young; they certainly will not be a vision of beauty, any more than the enthusiastic old men who in spite of wrinkles and ugliness continue to frequent the gymnasia.

Yes, indeed, he said: according to present notions the proposal would be thought ridiculous.

But then, I said, as we have determined to speak our minds, we must not fear the jests of the wits which will be directed against this sort of innovation; how they will talk of women's attainments both in music and gymnastic, and above all about their wearing armour and riding upon horseback....

Then let us put a speech into the mouths of our opponents. They will say: 'Socrates and Glaucon, no adversary is needed to convict you, for you yourselves, at the first foundation of the State, admitted the principle that everybody was to do the one work suited to his own nature.' And certainly, if I am not mistaken, such an admission was made by us. 'And do not the natures of men and women differ very much indeed?' And we shall reply: Of course they do. Then we shall be asked, 'Whether the tasks assigned to men and to women should not be different, and such as are agreeable to their different natures?' Certainly they should. 'But if so, have you not fallen into a serious inconsistency in saying that men and women, whose natures are so entirely different, ought to perform the same actions?'—What defence will you make for us, my good sir, against these objections?

That is not an easy question to answer when asked suddenly; and I shall and I do beg of you to draw out the case on our side.

These are the objections, Glaucon, and there are many others of a like kind, which I foresaw long ago; they made me afraid and reluctant to take in hand any law about the possession and nurture of women and children.

By Zeus, he said, the problem to be solved is anything but easy.

Why yes, I said, but the fact is that when a man is out of his depth, whether he has fallen into a little swimming-bath or into mid ocean, he has to swim all the same.

Very true.

And must not we swim and try to reach the shore, while hoping that Arion's dolphin or some other miraculous help may save us?

I suppose so, he said.

Well then, let us see if any way of escape can be found. We acknowledged—did we not?—that different natures ought to have different pursuits, and that men's and women's natures are different. And now what are we saying? that different natures ought to have the same pursuits,—this is the inconsistency which is charged upon us.

Precisely.

Verily, Glaucon, I said, glorious is the power of the art of disputation!

Why do you say so?

Because I think that many a man falls into the practice against his will. When he thinks that he is reasoning he is really disputing, just because he does not know how to inquire into a subject by distinguishing its various aspects, but pursues some verbal opposition in the statement which has been made. That is the difference between the spirit of contention and that of fair discussion.

Yes, he replied, that is a fairly common failing, but does it apply at present to us?

Yes, indeed; for there is a danger of our getting unintentionally into verbal contradiction.

In what way?

Why, we valiantly and pugnaciously insist upon the verbal truth that different natures ought to have different pursuits, but we never considered at all what was the meaning of sameness or difference of nature, or with what intention we distinguished them when we assigned different pursuits to different natures and the same to the same natures.

Why, no, he said, that was never considered by us.

I said: Yet it seems that we should be entitled to ask ourselves whether there is not an opposition in nature between bald men and hairy men; and if this is admitted by us, then, if bald men are cobblers, we should forbid the hairy men to be cobblers, and conversely?

That would be a jest, he said.

Yes, I said, a jest; and why? because we were not previously speaking of sameness or difference in *any* sense; we were concerned with one *form* of difference or similarity, namely that which would affect the pursuit in which a man is engaged; we should have argued, for example, that a physician and one who is in mind a physician may be said to have the same nature.

True.

Whereas the physician and the carpenter have different natures?

Certainly.

And if, I said, the male and female sex appear to differ in their fitness for any art or pursuit, we should say that such pursuit or art ought to be assigned to one or the other of them; but if the difference consists only in women bearing and men begetting children, this does not amount to a proof that a woman differs from a man in respect of the sort of education she should receive; and we shall therefore continue to maintain that our guardians and their wives ought to have the same pursuits.

Quite rightly, he said.

Only then shall we ask our opponent to inform us with reference to which of the pursuits or arts of civic life the nature of a woman differs from that of a man?

That will be quite fair.

And perhaps he, like yourself a moment ago, will reply that to give a sufficient answer on the instant is not easy; but that given time for reflection there is no difficulty.

Yes, perhaps.

Suppose then that we invite such an objector to accompany us in the argument, in the hope of showing him that there is no occupation peculiar to women which need be considered in the administration of the State.

By all means.

Let us say to him: Come now, and we will ask you a question:—when you spoke of a nature gifted or not gifted in any respect, did you mean to say that one man will acquire a thing easily, another with difficulty? the first, after brief instruction, is able to discover a great deal more for himself, whereas the other, after much teaching and application, cannot even preserve what he has learnt; or again, did you mean that the one has a body which is a good servant to his mind, while the body of the other is a hindrance to him? Would not these be the sort of differences which distinguish the man gifted by nature from the one who is ungifted?

No one will deny that.

And can you mention any pursuit of mankind in which the male sex has not all these gifts and qualities in a higher degree than the female? Need I waste time in speaking of the art of weaving, and the preparation of pancakes and preserves in which womankind is generally thought to have some skill, and in which for her to be beaten by a man is of all things the most absurd?

You are quite right, he replied, in maintaining that one sex greatly excels the other in almost every field. Although many woman are in many things superior to many men, yet on the whole what you say is true.

And if so, my friend, I said, there is no special faculty of administration in a state which a woman has because she is a woman, or which a man has by virtue of his sex, but the gifts of nature are alike diffused in both; all the pursuits of men can naturally be assigned to women also, but in all of them a woman is weaker than a man.

Very true.

Then are we to impose all our enactments on men and none of them on women?

That will never do.

Because we shall say that a woman too may, or may not, have the gift of healing; and that one is a musician, and another has no music in her nature?

Very true.

And it can hardly be denied that one woman has a turn for gymnastic and military exercises, and another is unwarlike and hates gymnastics?

I think not.

And one woman is a philosopher, and another is an enemy of philosophy; one has spirit, and another is without spirit?

That is also true.

Then one woman will have the temper of a guardian, and another not. For these, as you remember, were the natural gifts for which we looked in the selection of the male guardians.

Yes.

Men and women alike possess the qualities which make a guardian; they differ only in their comparative strength or weakness.

Obviously.

Therefore those women who have such qualities are to be selected as the companions and colleagues of men who also have them and whom they resemble in capacity and in character?

Very true.

But ought not the same natures to be trained in the same pursuits?

They ought.

Then we have come round to the previous point that there is nothing unnatural in assigning music and gymnastic to the guardian women.

Certainly not.

The law which we then enacted was agreeable to nature, and therefore not an impossibility or mere aspiration; it is rather the contrary practice, which prevails at present, that is a violation of nature.

That appears to be true.

We had to consider, first, whether our proposals were possible, and secondly whether they were the most beneficial?

Yes.

And the possibility has been acknowledged?

Yes.

The very great benefit has next to be established?

Quite so.

You will admit that the same education which makes a man a good guardian will make a woman a good guardian; especially if the original nature of both is the same?

Yes.

I should like to ask you a question.

What is it?

Is it your opinion that one man is better than another? Or do you think them all equal?

Not at all.

And in the commonwealth which we were founding do you conceive the guardians who have been brought up on our model system to be more perfect men, or the cobblers whose education has been cobbling?

What a ridiculous question!

You have answered me, I replied: in fact, our guardians are the best of all our citizens?

By far the best.

And will not the guardian women be the best women?

Yes, by far the best.

And can there be anything better for the interests of the State than that the men and women of a State should be as good as possible?

There can be nothing better.

And this is what the arts of music and gymnastic, when present in such manner as we have described, will accomplish?

Certainly.

Then we have made an enactment not only possible but in the highest degree beneficial to the State?

True.

Then let the guardian women strip, for their virtue will be their robe, and let them share in the toils of war and the defence of their country; only in the distribution of labours the lighter are to be assigned to the women, who are the weaker natures, but in other respects their duties are to be the same. And as for the man who laughs at naked women exercising their bodies from the best of motives, in his laughter he is plucking

A fruit of unripe wisdom,

and he himself is ignorant of what he is laughing at, or what he is about;—for that is, and ever will be, the best of sayings, *That the useful is noble and the hurtful is base.*

Very true.

Here, then, is one difficulty in our law about women, which we may say that we have now escaped; the wave has not swallowed us up alive for enacting that the guardians of either sex should have all their pursuits in common; to the utility and also to the possibility of this arrangement the consistency of the argument with itself bears witness.

Yes, that was a mighty wave which you have escaped.

Yes, I said, but you may not think it so impressive when you see the next.

Go on; let me see.

The law, I said, which is the sequel of this and of all that has preceded, is to the following effect,—'that all these women are to be common to all the men of the same class, none living privately together; and, moreover, that their children are to be common, and no parent is to know his own child, nor any child his parent.'

Yes, he said, you will find it much harder to convince anyone either of the possibility or of the usefulness of such a law.

I do not think, I said, that there can be any dispute about the very great utility of having both women and children in common; the possibility is quite another matter, and will, no doubt, be very much disputed.

Both points are sure to be warmly disputed.

You imply that the two questions must be combined, I replied. I hoped that you would admit that the proposal was useful, and so I should escape from at least one of them, and then there would remain only the possibility.

But that attempt to escape is detected, and therefore you will please to give a defence of both.

Well, I said, I submit to my fate. Yet grant me a little favour: let me feast my mind with the dream as day-dreamers are in the habit of feasting themselves when they are walking alone; for before they have discovered any means of effecting their wishes—that is a matter which never troubles them; they would rather not tire themselves by thinking about possibilities—but assuming that what they desire is already granted to them, they proceed with their plan, and delight in detailing what they mean to do when their wish has come true—a recreation which tends to make an idle mind still idler. Now I myself am beginning to lose heart, and I should like, with your permission, to pass over the question of possibility at present. Assuming therefore the possibility of the proposal, I shall now proceed to inquire how the rulers will carry out these arrangements, and I shall demonstrate that our plan, if executed, will be of the greatest possible benefit both to the State and to the guardians. First of all, then, if you have no objection, I will endeavour with your help to consider the advantages of the measure; and hereafter the question of possibility.

I have no objection; proceed.

First, I think that if our rulers and their auxiliaries are to be worthy of the name which they bear, there must be the power of command in the one and willingness to obey in the other; the guardians must themselves obey the laws, and they must also imitate the spirit of them in any details which are entrusted to their care.

That is right, he said.

You, I said, who are their legislator, having selected the men, will now select the women and give them to them;—they must be as far as possible of like natures with them; and they must live in common houses and meet at common meals. None of them will have anything specially his or her own; they will be together, and will be brought up together, and will associate at gymnastic exercises. And so they will be drawn by a necessity of their natures to have intercourse with each other—necessity is not too strong a word, I think?

Yes, he said;—necessity, not geometrical, but another sort of necessity which lovers know, and which is far more convincing and constraining to the mass of mankind.

True, Glaucon, I said; but now we can hardly allow promiscuous unions, or any other kind of disorder; in a city of the blessed, licentiousness is an unholy thing which the rulers will forbid.

Yes, he said, and it ought not to be permitted.

Then clearly the next thing will be to arrange marriages that are sacred in the highest degree; and what is most beneficial will be deemed sacred?

Exactly.

And how can marriages be made most beneficial?—that is a question which I put to you, because I see in your house dogs for hunting, and of the nobler sort of birds not a few. Now, I beseech you, do tell me, have you ever attended to their pairing and breeding?

In what particulars?

Why, in the first place, although they are all of good pedigree, do not some prove to be better than others?

True.

And do you breed from them all indifferently, or do you take care to breed from the best only?

From the best.

From the oldest or the youngest, or only those of ripe age?

From those of ripe age.

And if care was not taken in the breeding, your dogs and birds would greatly deteriorate?

Certainly.

What of horses and of animals in general? Is there any difference?

No, it would be strange if there were.

Good heavens! my dear friend, I said, what consummate skill will our rulers need if the same principle holds of the human species!

Certainly, the same principle holds; but why does this involve any particular skill?

Because, I said, our rulers will often have to practise upon the body corporate with medicines. Now you know that when patients do not require medicines but have only to be put under a regimen, the inferior sort of practitioner is deemed to be good enough; but when medicine has to be given, then the doctor should be more of a man.

That is quite true, he said; but to what are you alluding?

I mean, I replied, that our rulers will find a considerable dose of falsehood and deceit necessary for the good of their subjects: we said before that the use of all these things regarded as medicines might be of advantage.

And we were very right.

And this lawful use of them seems likely to be often needed in the regulations of marriages and births.

How so?

Why, I said, the principle has been already laid down that the best of either sex should be united with the best as often, and the inferior with the inferior as seldom as possible; and that they should rear the offspring of the one sort of union but not of the other, if the flock is to be maintained in first-rate condition. Now these goings-on must be a secret which the rulers only know, in order to keep our herd, as the guardians may be termed, as free as possible from dissension.

Very true.

Had we not better appoint certain festivals at which we will bring together the brides and bridegrooms, and sacrifices will be offered, and suitable hymeneal songs composed by our poets: the number of weddings is a matter which must be left to the discretion of the rulers, whose aim will be to preserve the same total number of guardians, having regard to wars, plagues, and any similar agencies, in order as far as this is possible to prevent the State from becoming either too large or too small.

Certainly, he replied.

We shall have to invent some ingenious kind of lottery, so that the less worthy may, on each occasion of our bringing them together, accuse their own ill luck and not the rulers.

To be sure, he said.

And I think that our braver and better youth, besides their other honours and rewards, might have greater facilities of intercourse with women given them; their bravery will be a reason, and such fathers ought to have as many sons as possible.

True.

And the proper officers, whether male or female or both, for offices are to be held by women as well as by men——

Yes.

The proper officers will take the offspring of the good parents to the pen or fold, and there they will deposit them with certain nurses who dwell in a separate quarter; but the offspring of the inferior, or of the better when they chance to be deformed, will be put away in some mysterious unknown place, as they should be.

Yes, he said, that must be done if the breed of the guardians is to be kept pure.

They will provide for their nurture, and will bring the mothers to the fold when they are full of milk, taking the greatest possible care that no mother recognizes her own child; and other wet-nurses may be engaged if more are required. Care will also be taken that the process of suckling shall not be protracted too long; and the mothers will have no getting up at night or other trouble, but will hand over all this sort of thing to the nurses and attendants.

Maternity, according to you, will be an easy business for these guardian women.

Why, said I, and that is quite proper. Let us, however, proceed with our scheme. We were saying that the parents should be in the prime of life?

Very true.

And what is the prime of life? May it not be defined as a period of about twenty years in a woman's life, and thirty in a man's?

Which years do you mean to include?

A woman, I said, at twenty years of age may begin to bear children to the State, and continue to bear them until forty; a man may begin at five-and-twenty, when he has passed the point at which the pulse of life beats quickest, and continue to beget children until he be fifty-five.

Certainly, he said, both in men and women those years are the prime of physical as well as of intellectual vigour.

Anyone above or below the prescribed ages who presumes to beget children for the commonwealth shall be said to have done an unholy and unrighteous thing; the child of which he is the father, if it steals into life, will have been conceived under auspices very unlike the sacrifices and prayers which at each hymeneal priestesses and priests and the whole city will offer, that the new generation may be better and more useful than their good and useful parents; whereas his child will be the offspring of darkness and strange lust.

Very true, he replied.

And the same law will apply to any one of those within the prescribed age who forms a connexion with any woman in the prime of life without the sanction of the rulers; for we shall say that he is raising up a bastard to the State, unsponsored and unconsecrated.

Very true, he replied.

This applies, however, only to men and women within the specified age: after that we shall probably allow them to range at will, except that a man may not marry his daughter or his daughter's daughter, or his mother or his mother's mother; and women, on the other hand, are prohibited from marrying their sons or fathers, or son's son or father's father, and so on in either direction. And we grant all this, accompanying the permission with strict orders to prevent any embryo which may come into being from seeing the light; and if any force a way to the birth, the parents must understand that the offspring of such a union cannot be maintained, and arrange accordingly.

13. Aristotle
The Politics

Aristotle was born in Macedonia, studied under Plato at the Athenian Academy, and later established his own school. His writings consist mainly of the notes on his lectures made by his students, which were later edited into specific works. These works range widely across many topics, but a few principles underlie all of them. Aristotle asserted that logic was the way to understand things and that true understanding involves an explanation of causality, or why things are the way they are. For him, the most important aspect of causality was the purpose or function of anything, what he termed the "final cause." In *The Politics,* Aristotle asserts that "man is a political animal" and then describes the nature and function of various types of governments; for him, a properly functioning state is based on properly functioning households, so he also sets out how households are to be run.

We shall, I think, in this as in other subjects, get the best view of the matter if we look at the natural growth of things from the beginning. The first point is that those which are incapable of existing without each other must be united as a pair. For example, (a) the union of male and female is essential for reproduction; and this is not a matter of *choice*, but is due to the *natural* urge, which exists in the other animals too and in plants, to propagate one's kind. Equally essential is (b) the combination of the natural ruler and ruled, for the purpose of preservation. For the element that can use its intelligence to look ahead is by nature ruler and by nature master, while that which has the bodily strength to do the actual work is by nature a slave, one of those who are ruled. Thus there is a common interest uniting master and slave....

Thus it was out of the association formed by men with these two, women and slaves, that a household was first formed; and the poet Hesiod was right when he wrote, 'Get first a house and a wife and an ox to draw the plough.' (The ox is the poor man's slave.) This association of persons, established according to nature for the satisfaction of daily needs, is the household....

There are, as we saw, three parts of household-management, one being the rule of a master, which has already been dealt with, next the rule of a father, and a third which arises out of the marriage relationship. This is included because rule is exercised over wife and children—over both of them as free persons, but in other respects differently: over a wife, rule is as by a statesman; over children, as by a king. For the male is more fitted to rule than the female, unless conditions are quite contrary to nature; and the elder and fully grown is more fitted than the younger and undeveloped. It is true that in most cases of rule by statesmen there is an interchange of the role of ruler and ruled, which aims to preserve natural equality and non-differentiation; nevertheless, so long as one is ruling and the other is being ruled, the ruler seeks to mark distinctions in outward dignity, in style of address, and in honours paid. As between male and female this kind of relationship is permanent.

It is clear then that in household-management the people are of greater importance than the inanimate property, and their virtue of more account than that of the property which we call their wealth; and also that the free men are of more account than slaves. About slaves the first question to be asked is whether in addition to their virtue as tools and servants they have another and more valuable one. Can they possess restraint, courage, justice, and every other condition of that kind, or have they in fact nothing but the serviceable quality of their persons?

The question may be answered in either of two ways, but both present a difficulty. If we say that slaves have these virtues, how then will they differ from free men? If we say that they have not, the position is anomalous, since they are human beings and share in reason. Roughly the same question can be put in relation to wife and child. Have not these also virtues? Ought a woman to be 'restrained', 'brave', and 'just', and is a child sometimes 'intemperate', sometimes 'restrained', or not?

All these questions might be regarded as parts of our wider inquiry into the natural ruler and ruled, and in particular whether or not the virtue of the one is the same as the virtue of the other. For if the highest excellence is required of both, why should one rule unqualifiedly, and the other unqualifiedly obey? (A distinction of more or less will not do here; the difference between ruling and obeying is one of kind, and quantitative difference is simply not that at all.) If on the other hand the one is to have virtues, and the other not, we have a surprising state of affairs. For if he that rules is not to be restrained and just, how shall he rule well? And if the ruled lacks these virtues, how shall he be ruled well? For if he is intemperate and feckless, he will per-

form none of his duties. Thus it becomes clear that both ruler and ruled must have a share in virtue, but that there are differences in virtue in each case, as there are also among those who by nature rule. An immediate indication of this is afforded by the soul, where we find natural ruler and natural subject, whose virtues we regard as different—one being that of the rational element, the other of the nonrational. It is therefore clear that the same feature will be found in the other cases too, so that most instances of ruling and being ruled are natural. For rule of free over slave, male over female, man over boy, are all different, because, while parts of the soul are present in each case, the distribution is different. Thus the deliberative faculty in the soul is not present at all in a slave; in a female it is present but ineffective, in a child present but undeveloped.

We should therefore take it that the same conditions inevitably prevail in regard to the moral virtues also, namely that all must participate in them but not all in the same way, but only as may be required by each for his proper function. The ruler then must have moral virtue in its entirety; for his function is in its fullest sense that of a master-craftsman, and reason is a master-craftsman. And the other members must have such amount as is appropriate to each. So it is evident that each of the classes spoken of must have moral virtue, and that restraint is *not* the same in a man as in a woman, nor justice or courage either, as Socrates thought; the one is the courage of a ruler, the other the courage of a servant, and likewise with the other virtues.

If we look at the matter case by case it will become clearer. For those who talk in generalities and say that virtue is 'a good condition of the soul', or that it is 'right conduct' or the like, delude themselves. Better than those who look for definitions in that manner are those who, like Gorgias, *enumerate* the different virtues. For instance, the poet singles out 'silence' as 'bringing credit to a woman'; but that is not so for a man. This is the method of assessment that we should always follow....

Now as it is a lawgiver's duty to start from the very beginning in looking for ways to secure the best possible physique for the young who are reared, he must consider first the union of their parents, and ask what kind of people should come together in marriage, and when. In making regulations about this partnership he should have regard both to the spouses themselves and to their length of life, in order that they may arrive at the right ages together at the same time, and so that the period of the father's ability to beget and that of the mother's to bear children may coincide. A period when one of the two is capable and the other not leads to mutual strife and quarrels. Next, as regards the timing of the children's succession, there should not be too great a gap in age between father and children; for then there is no good that the young can do by showing gratitude to elderly parents, and their fathers are of no help to them. Nor should they be too close in age, for this causes the relationship to be strained: like contemporaries, people in such a position feel less respect, and the nearness in age leads to bickering in household affairs. And further, to go back to the point we started from, one should ensure that the physique of the children that are produced shall be in accordance with the wishes of the legislator.

All these purposes can be fulfilled, or nearly so, if we pay sufficient attention to one thing. Since, generally speaking, the upper limit of age for the begetting of children is for men seventy years and for women fifty, the beginning of their union should be at ages such that they will arrive at this stage of life simultaneously. But the intercourse of a very young couple is not good for child-bearing. In all animals the offspring of early unions are defective, inclined to produce females, and diminutive; so the same kind of results are bound to follow in human beings too. And there is evidence that this is so: in states where early unions are the rule, the people are small in stature and defective. A further objection is that young women have greater difficulty in giving birth and more of them die. (Some say that here we have also

the reason for the oracle given to the people of Troezen: there is no reference to the harvesting of crops, but to the fact that the marrying of girls at too young an age was causing many deaths.) It is also more conducive to restraint that daughters should be no longer young when their fathers bestow them in marriage, because it seems that women who have sexual intercourse at an early age are more likely to be dissolute. On the male side too it is held that if they have intercourse while the seed is just growing, it interferes with their bodily growth; for the seed is subject to a fixed limit of time, after which it ceases to be replenished except on a small scale. Accordingly we conclude that the appropriate age for the union is about the eighteenth year for girls and for men the thirty-seventh. With such timing, their union will take place when they are physically in their prime, and it will bring them down together to the end of procreation at exactly the right moment for both. And the children's succession, if births take place promptly at the expected time, will occur when they are at the beginning of their prime and their parents are past their peak, the father now approaching his seventieth year.

We have spoken now about the time when the union should take place, but not about the seasons of the year best suited for establishing this form of living together. However, the common practice of choosing winter is satisfactory. The spouses too should study for themselves, in good time, the advice of doctors and natural scientists to assist them in bearing children; the former give suitable information about crucial stages in the life of the body, the latter about the winds (they recommend the northerly ones rather than the southerly). To the question of what kind of physique is most advantageous for the offspring that are produced, we must give closer attention in the works on the training of children; for our present purpose the following outline will suffice. The condition of an athlete does not make for the physical fitness needed by a citizen, nor for health and the production of offspring. A condition of much coddling and of unfitness for hard work is equally undesirable. Something between the two is needed, a condition of one inured to hard but not violently hard toil, directed not all in one direction as an athlete's, but towards the various activities typical of men who are free. These requirements are applicable to men and women alike.

Further, it is important that women should look after their bodies during pregnancy. They must not relax unduly, or go on a meagre diet. It is easy for a legislator to ensure this by making it a rule that they shall each day take a walk, the object of which is to worship regularly the gods whose office is to look after childbirth. But while the body should be exercised, the intellect should follow a more relaxed regime, for the unborn infant appears to be influenced by her who is carrying it as plants are by the earth.

With regard to the choice between abandoning an infant or rearing it, let there be a law that no cripple child be reared. But since the ordinance of custom forbids the exposure of infants on account of their numbers, there must be a limit to the production of children. If contrary to these arrangements copulation does take place and a child is conceived, abortion should be procured before the embryo has ac-

quired life and sensation; the presence of life and sensation will be the mark of division between right and wrong here.

Since we have already decided the beginning of the period of life at which male and female should enter on their union, we must also decide upon the length of time during which it is proper that they should render the service of producing children. The offspring of elderly people, like the offspring of the unduly young, are imperfect both in intellect and in body; and those of the aged are feeble. We should therefore be guided by the highest point of intellectual development, and this in most cases is the age mentioned by certain poets who measure life by periods of seven years, that is to say about the fiftieth year of life. Thus anyone who has passed this age by four or five years ought to give up bringing children into the world. But provided it is clearly for the sake of health or other such reason intercourse may continue.

As for extra-marital intercourse, it should, in general, be a disgrace to be detected in intimacy of any kind whatever, so long as one is a husband and so addressed. If anyone is found to be acting thus during the period of his begetting of children, let him be punished by such measure of disgrace as is appropriate to his misdemeanour.

14. Xenophon
Constitution of the Lacedaemonians and *On Household Management*

Xenophon (c. 570–c. 480 B.C.E.) was a disciple of Socrates, historian, and military leader who wrote a large number of treatises on various subjects. Though born in Athens, he fought for Sparta in its wars against Athens and other city-states and was eventually given a rural estate as a reward, where he spent many years writing. His writings include a very positive description of some aspects of Spartan life, which he titled *Constitution of the Lacedaemonians* (*Lacedaemonians* is another name for *Spartans*). This work begins with a discussion of the laws re-

Constitution of the Lacedaemonians by Xenophon, translated by E.C. Marchant. Loeb Classical Library, Xenophon 7 (Cambridge: Harvard University Press, 1914), pp. 137–141.

garding marriage established by the Spartan ruler Lycurgus, whom Xenophon clearly admires. In *On Household Management,* Xenophon portrays a woman living under conditions very different from those in Sparta. This treatise is written in the form of a Socratic dialogue, in which Socrates discusses a conversation he had with a man named Ischomachus, whom everyone has described to him as admirable and a gentleman. Though Xenophon claims Ischomachus was a real person, there is no evidence of his existence, and most analysts regard him as the voice of Xenophon himself.

A. *Constitution of the Lacedaemonians*

It occurred to me one day that Sparta, though among the most thinly populated of states, was evidently the most powerful and most celebrated city in Greece; and I fell to wondering how this could have happened. But when I considered the institutions of the Spartans, I wondered no longer.

Lycurgus, who gave them the laws that they obey, and to which they owe their prosperity, I do regard with wonder; and I think that he reached the utmost limit of wisdom. For it was not by imitating other states, but by devising a system utterly different from that of most others, that he made his country pre-eminently prosperous.

First, to begin at the beginning, I will take the begetting of children. In other states the girls who are destined to become mothers and are brought up in the approved fashion, live on the very plainest fare, with a most meagre allowance of delicacies. Wine is either withheld altogether, or, if allowed them, is diluted with water. The rest of the Greeks expect their girls to imitate the sedentary life that is typical of handicraftsmen—to keep quiet and do wool-work. How, then, is it to be expected that women so brought up will bear fine children?

But Lycurgus thought the labour of slave women sufficient to supply clothing. He believed motherhood to be the most important function of freeborn woman. Therefore, in the first place, he insisted on physical training for the female no less than for the male sex: moreover, he instituted races and trials of strength for women competitors as for men, believing that if both parents are strong they produce more vigorous offspring.

He noticed, too, that, during the time immediately succeeding marriage, it was usual elsewhere for the husband to have unlimited intercourse with his wife. The rule that he adopted was the opposite of this: for he laid it down that the husband should be ashamed to be seen entering his wife's room or leaving it. With this restriction on intercourse the desire of the one for the other must necessarily be increased, and their offspring was bound to be more vigorous than if they were surfeited with one another. In addition to this, he withdrew from men the right to take a wife whenever they chose, and insisted on their marrying in the prime of their manhood, believing that this too promoted the production of fine children. It might happen, however, that an old man had a young wife; and he observed that old men keep a very jealous watch over their young wives. To meet these cases he instituted an entirely different system by requiring the elderly husband to introduce into his house some man whose physical and moral qualities he admired, in order to beget children. On the other hand, in case a man did not want to cohabit with his wife and nevertheless desired children of whom he could be proud, he made it lawful for him to choose a woman who was the mother of a fine family and of high birth, and if he obtained her husband's consent, to make her the mother of his children.

He gave his sanction to many similar arrangements. For the wives want to take charge of two households, and the husbands want to get brothers for their sons, brothers who are members of the family and share in its influence, but claim no part of the money.

Thus his regulations with regard to the begetting of children were in sharp contrast with those of other states. Whether he succeeded in populating Sparta with a race of men remarkable for their size and strength anyone who chooses may judge for himself.

B. *On Household Management*

"Smiling at my question, 'How came you to be called a gentleman?', and apparently well pleased, Ischomachus answered: '...Well now, Socrates, as you ask the question, I certainly do not pass my time indoors; for, you know, my wife is quite capable of looking after the house by herself.'

"'Ah, Ischomachus,' said I, 'that is just what I want to hear from you. Did you yourself train your wife to be of the right sort, or did she know her household duties when you received her from her parents?'

"'Why, what knowledge could she have had, Socrates, when I took her for my wife? She was not yet fifteen years old when she came to me, and up to that time she had lived in leading-strings[1], seeing, hearing and saying as little as possible. If when she came she knew no more than how, when given wool, to turn out a cloak, and had seen only how the spinning is given out to the maids, is not that as much as could be expected? For in control of her appetite, Socrates, she had been excellently trained; and this sort of training is, in my opinion, the most important to man and woman alike.'

"'But in other respects did you train your wife yourself, Ischomachus, so that she should be competent to perform her duties?'

"'Oh no, Socrates; not until I had first offered sacrifice and prayed that I might really teach, and she learn what was best for us both.'

"'Did not your wife join with you in these same sacrifices and prayers?'

"'Oh yes, earnestly promising before heaven to behave as she ought to do; and it was easy to

see that she would not neglect the lessons I taught her.'

"'Pray tell me, Ischomachus, what was the first lesson you taught her, since I would sooner hear this from your lips than an account of the noblest athletic event or horse-race?'

"'Well, Socrates, as soon as I found her docile and sufficiently domesticated to carry on conversation, I questioned her to this effect:

"'"Tell me, dear, have you realised for what reason I took you and your parents gave you to me? For it is obvious to you, I am sure, that we should have had no difficulty in finding someone else to share our beds. But I for myself and your parents for you considered who was the best partner of home and children that we could get. My choice fell on you, and your parents, it appears, chose me as the best they could find. Now if God grants us children, we will then think out how we shall best train them. For one of the blessings in which we shall share is the acquisition of the very best of allies and the very best of support in old age; but at present we share in this our home. For I am paying into the common stock all that I have, and you have put in all that you brought with you. And we are not to reckon up which of us has actually contributed the greater amount, but we should know of a surety that the one who proves the better partner makes the more valuable contribution."

"'My wife's answer was as follows, Socrates: "How can I possibly help you? What power have I? Nay, all depends on you. My duty, as my mother told me, is to be discreet."

"'"Yes, of course, dear," I said, "my father said the same to me. But discretion both in a man and a woman, means acting in such a manner that their possessions shall be in the best condition possible, and that as much as possible shall be added to them by fair and honourable means."

"'"And what do you see that I can possibly do to help in the improvement of our property?" asked my wife.

"'"Why," said I, "of course you must try to do as well as possible what the gods made you capable of doing and the law sanctions."

"'"And pray, what is that?" said she....

---

[1]*Leading-strings:* strings by which mothers and nursemaids held children to help them walk.

"Your duty will be to remain indoors and send out those servants whose work is outside, and superintend those who are to work indoors, and to receive the incomings, and distribute so much of them as must be spent, and watch over so much as is to be kept in store, and take care that the sum laid by for a year be not spent in a month. And when wool is brought to you, you must see that cloaks are made for those that want them. You must see too that the dry corn is in good condition for making food. One of the duties that fall to you, however, will perhaps seem rather thankless: you will have to see that any servant who is ill is cared for."

"'Oh no,' cried my wife," it will be delightful, assuming that those who are well cared for are going to feel grateful and be more loyal than before."

"'Why, my dear,' cried I, delighted with her answer, "what makes the bees so devoted to their leader in the hive, that when she forsakes it, they all follow her, and not one thinks of staying behind? Is it not the result of some such thoughtful acts on her part?"

"'It would surprise me," answered my wife, "if the leader's activities did not concern you more than me. For my care of the goods indoors and my management would look rather ridiculous, I fancy, if you did not see that something is gathered in from outside."

"'And my ingathering would look ridiculous," I countered, "if there were not someone to keep what is gathered in. Don't you see how they who 'draw water in a leaky jar,' as the saying goes, are pitied, because they seem to labour in vain?"

"'Of course,' she said, "for they are indeed in a miserable plight if they do that.''

"'But I assure you, dear, there are other duties peculiar to you that are pleasant to perform. It is delightful to teach spinning to a maid who had no knowledge of it when you received her, and to double her worth to you: to take in hand a girl who is ignorant of housekeeping and service, and after teaching her and making her trustworthy and serviceable to find her worth any amount: to have the power of rewarding the discreet and useful members of your household, and of punishing anyone who turns out to be a rogue. But the pleasantest experience of all is to prove yourself better than I am, to make me your servant; and, so far from having cause to fear that as you grow older you may be less honoured in the household, to feel confident that with advancing years, the better partner you prove to me and the better housewife to our children, the greater will be the honour paid to you in our home. For it is not through outward comeliness that the sum of things good and beautiful is increased in the world, but by the daily practice of the virtues."...

"'When all this was done, Socrates, I told my wife that all these measures were futile, unless she saw to it herself that our arrangement was strictly adhered to in every detail. I explained that in well-ordered cities the citizens are not satisfied with passing good laws: they go further, and choose guardians of the laws, who act as overseers, commending the law-abiding and punishing law-breakers. So I charged my wife to consider herself guardian of the laws to our household. And just as the commander of a garrison inspects his guards, so must she inspect the chattels whenever she thought it well to do so; as the Council scrutinises the cavalry and the horses, so she was to make sure that everything was in good condition: like a queen, she must reward the worthy with praise and honour, so far as in her lay, and not spare rebuke and punishment when they were called for.

"'Moreover, I taught her that she should not be vexed that I assigned heavier duties to her than to the servants in respect of our possessions. Servants, I pointed out, carry, tend and guard their master's property, and only in this sense have a share in it; they have no right to use anything except by the owner's leave; but everything belongs to the master, to use it as he will. Therefore, I explained, he who gains most by the preservation of the goods and loses most by their destruction, is the one who is bound to take most care of them.'

"'Well, now, Ischomachus,' said I, 'was your wife inclined to pay heed to your words?'

"'Why, Socrates,' he cried, 'she just told me that I was mistaken if I supposed that I was laying a hard task on her in telling her that she must take care of our things. It would have been harder, she said, had I required her to neglect her own possessions, than to have the duty of attending to her own peculiar blessings. 'The fact is,' he added, 'just as it naturally comes easier to a good woman to care for her own children than to neglect them, so, I imagine, a good woman finds it pleasanter to look after her own possessions than to neglect them.'" "Now when I heard that his wife had given him this answer, I exclaimed; 'Upon my word, Ischomachus, your wife has a truly masculine mind by your showing!'...

"She did ask whether I could advise her on one point: how she might make herself really beautiful, instead of merely seeming to be so. And this was my advice, Socrates: "Don't sit about for ever like a slave, but try, God helping you, to behave as a mistress: stand before the loom and be ready to instruct those who know less than you, and to learn from those who know more: look after the baking-maid: stand by the housekeeper when she is serving out stores: go round and see whether everything is in its place." For I thought that would give her a walk as well as occupation. I also said it was excellent exercise to mix flour and knead dough; and to shake and fold cloaks and bedclothes; such exercise would give her a better appetite, improve her health, and add natural colour to her cheeks. Besides, when a wife's looks outshine a maid's, and she is fresher and more becomingly dressed, they're a ravishing sight, especially when the wife is also willing to oblige, whereas the girl's services are compulsory. But wives who sit about like fine ladies, expose themselves to comparison with painted and fraudulent hussies. And now, Socrates, you may be sure, my wife's dress and appearance are in accord with my instructions and with my present description.'"

"At this point I said, 'Ischomachus, I think your account of your wife's occupations is sufficient for the present—and very creditable it is to both of you....'"

15. Box with Painting of Domestic Scene, Fifth- to Fourth-Century Athens

Objects meant for household use, such as cups, bowls, jugs, plates, and mirrors, were often decorated with scenes from mythology or daily life in ancient Greece. Respectable women were often shown spinning or carrying out other types of activities done in the home; the fact that they are indoors is indicated by a door in the background (sometimes half-opened to show a bed or couch) or indoor furniture. This object is a *pyxis,* a rounded box that often held cosmetics.

QUESTIONS FOR ANALYSIS

1. Plato has been called the "first feminist" for his suggestion that women should also be trained as guardians. In your opinion, is this title warranted? How would you compare the ideas about the relative capabilities of men and women held by Plato and Aristotle?
2. How do the education and tasks of female guardians in Plato's *Republic* differ from those Xenophon describes for the women of Sparta and for the young wife of Ischomachus? Why might these differ so much in the two works by the same author, Xenophon?
3. What sort of family and sexual relationships does Plato see as essential among guardians for the good of the state? How are these to be accomplished? Does Aristotle differ much from his former teacher in this?

Anxious woman with distaff in front of bridal chamber. (*The Louvre © Photo RMN—Chuzeville #00227308.*)

4. In what ways might the painting on the box serve as an illustration for Xenophon's *On Household Management*? How does the woman portrayed differ from his portrait of the ideal wife?

RELIGIOUS LIFE AND RITUALS

As in ancient Egypt and Mesopotamia, women in ancient Greece participated in a variety of religious activities. The Greeks were polytheistic; each polis had a patron deity or deities, but individuals could choose to honor many other gods or goddesses as well, participating in festivals and rituals. In some of these, such as the Panathenaic procession in honor of Athena, girls and women participated alongside boys and men. Other rituals were for women only. In Athens, girls took part in races in honor of Artemis, and married women spent three days each year on a hilltop for the Thesmophoria festival in honor of Demeter, the goddess of fertility. Though it is uncertain whether women attended the plays held in honor of the god Dionysus, they played important roles in other rituals held in his honor. Because Dionysus was the god of wine, these festivals could often become quite raucous, although it is difficult to know exactly what went on at them because most of our sources come from people who opposed them and who thus emphasized their wild and uncontrolled nature. Women served as priestesses for the cults of Athena and Demeter and brought offerings for the dead at family tombs.

During the Hellenistic period, worship of various deities moved around the Mediterranean just as people and ideas did. Greeks often combined new gods they encountered with ones that they already worshipped, mixing their qualities in a syncretistic blend or developing new ideas about their powers. The Egyptian goddess Isis, who became very popular with women, is a good example of this. She was associated with the Greek goddess Aphrodite, but also came to embody the forces of many other gods and was believed to have power over war, agriculture, marriage, language, the sun, the wind, and justice. Isis had brought her husband Osiris back to life after he had been chopped apart, so she was worshipped as a force of rebirth; she was often depicted holding her son Horus and was honored as the patron of motherhood. The worship of Isis became very popular with Roman as well as Greek women; temples dedicated to her were visited throughout the Roman world until at least the fifth century.

16. Sappho
Hymn to Aphrodite

The only women's writings that survive from classical Greece are a few poems and fragments of poems from women who lived in areas other than Athens, beginning with those of Sappho, who lived on the island of Lesbos in the seventh or sixth century B.C.E. Many stories have been told about Sappho's life, most of which are impossible to verify. Her writings themselves indicate that she lived, as did many Greek women who were wealthy enough to gain an education, in a world where the sexes were separated much of the time and where the most intense emotional relationships—for both men and women—were felt toward members of one's own sex. Sappho's writings—about forty poems or meaningful fragments survive—include poems to Aphrodite and other deities, poems concerning love and friendship, wedding songs, and poems dealing with religious rituals. Her poems often convey a deep sense of loss when a friend or lover has departed or is about to leave; their intensity and use of natural images, such as flowers and the moon, set a pattern for love poetry that has continued to this day.

O immortal Aphrodite of the many-colored
  throne,
child of Zeus, weaver of wiles, I beseech you,
do not overwhelm me in my heart
with anguish and pain, O Mistress,

But come hither, if ever at another time
hearing my cries from afar

you heeded them, and leaving the home of your
  father
came, yoking your golden

Chariot: beautiful, swift sparrows
drew you above the black earth
whirling their wings thick and fast,

From *The Woman and the Lyre: Women Writers in Classical Greece and Rome*, translated by Jane M. Snyder. © 1989 by the Board of Trustees, Southern Illinois University. Reprinted by permission of the Southern Illinois University Press.

from heaven's ether through
   mid-air.

Suddenly they had arrived; but you, O Blessed
   Lady,
with a smile on your immortal face,
asked what I had suffered again and
why I was calling again

And what I was most wanting to happen for me
in my frenzied heart: "Whom again shall I
   persuade
to come back into friendship with you? Who,
O Sappho, does you injustice?

"For if indeed she flees, soon will she
   pursue,
and though she receives not your gifts, she will
   give them,
and if she loves not now, soon she
   will love,
even against her will."

Come to me now also, release me from
harsh cares; accomplish as many things as my
   heart desires
to accomplish; and you yourself
be my fellow soldier.

17. Cup Painting of a *Maenad* (Female Follower of Dionysus), ca. 480 B.C.E.

Depictions of women who worshipped the god Dionysus appear in paintings and poetry from classical Athens. In these, women, termed *maenads* (which means "raving women"), are associated with the wild forces of nature. Here a woman is dressed in an animal-skin cape and a snake headdress; she is carrying

A cup painting of a maenad. (*Staatliche Antikensammlungen und Glyptothek, Munich.*)

a leopard and a thyrsus, a stick topped with ivy leaves. Although it is impossible to know whether Athenian (or other Greek) women actually dressed this way when they worshipped or did so simply in the imagination of the male painter, it does suggest why worship of Dionysus might have been appealing to the secluded women of Athens, or why men might have *assumed* it to be appealing.

18. Pausanius
Description of Greece

Pausanius was a writer who probably came from Asia Minor (present-day Turkey) and traveled widely throughout the eastern Roman Empire in the late second century C.E. He wrote a ten-volume description of different parts of Greece designed as a guide for tourists, describing important religious, cultural, and historical sites. He includes a long discussion of the shrine to Zeus at Olympia in the province of Elis, the most holy place in Greece, at which athletic competitions—the ancient Olympics—were held regularly. These competitions involved only male athletes, but Pausanius also describes competitions for female athletes; like those for men, these games were footraces only.

Every fourth year there is woven for Hera a robe by the Sixteen women, and the same also hold games called Heraea. The games consist of footraces for maidens. These are not all of the same age. The first to run are the youngest; after them come the next in age, and the last to run are the oldest of the maidens. They run in the following way: their hair hangs down, a tunic reaches to a little above the knee, and they bare the right shoulder as far as the breast. These too have the Olympic stadium reserved for their games, but the course of the stadium is shortened for them by about one-sixth of its length. To the winning maidens they give crowns of olive and a portion of the cow sacrificed to Hera. They may also dedicate statues with their names inscribed upon them. Those who administer to the Sixteen are, like the presidents of the games, married women. The games of the maidens too are traced back to ancient times; they say that, out of gratitude to Hera for her

marriage with Pelops, Hippodameia assembled the Sixteen Women, and with them inaugurated the Heraea. They relate too that a victory was won by Chloris, the only surviving daughter of the house of Amphion, though with her they say survived one of her brothers. As to the children of Niobe, what I myself chanced to learn about them I have set forth in my account of Argos. Besides the account already given they tell another story about the Sixteen Women as follows. Damophon, it is said, when tyrant of Pisa did much grievous harm to the Eleans. But when he died, since the people of Pisa refused to participate as a people in their tyrant's sins, and the Eleans too became quite ready to lay aside their grievances, they chose a woman from each of the sixteen cities of Elis still inhabited at that time to settle their differences, this woman to be the oldest, the most noble, and the most esteemed of all the women. The cities from which they chose the women

Description of Greece by Pausanius, translated by H. A. Ormeroid, Loeb Classical Library. (London: William Heinemann, 1918).

were Elis.... The women from these cities made peace between Pisa and Elis. Later on they were entrusted with the management of the Heraean games, and with the weaving of the robe for Hera. The Sixteen Women also arrange two choral dances, one called that of Physcoa and the other that of Hippodameia. This Physcoa they say came from Elis in the Hollow, and the name of the parish where she lived was Orthia. She mated they say with Dionysus, and bore him a son called Naraceus. When he grew up he made war against the neighbouring folk, and rose to great power, setting up moreover a sanctuary of Athena surnamed Narcaea. They say too that Narcaeus and Physcoa were the first to pay worship to Dionysus. So various honours are paid to Physcoa, especially that of the choral dance, named after her and managed by the Sixteen Women. The Eleans still adhere to the other ancient customs, even though some of the cities have been destroyed. For they are now divided into eight tribes, and they choose two women from each. Whatever ritual it is the duty of either the Sixteen Women or the Elean umpires to perform, they do not perform before they have purified themselves with a pig meet for purification and with water. Their purification takes place at the spring Piera. You reach this spring as you go along the flat road from Olympia to Elis.

19. Hellenistic Spell of Attraction, Third Century C.E.

Like the Hittite ritual included in Chapter 1, the Greeks often used professional shamans and magicians to devise spells for various purposes. These shamans would write spells using both Greek words and special "magical" language known only to the gods, and instructing those who purchased them to carry out specific actions to accompany their words. Thousands of such spells survive, many of which are curse tablets—seeking to bring bad luck to a rival or opposing team—or binding spells, attempting to force a person to do something against his or her will. These binding spells include hundreds that seek to make another person love the petitioner; most of these are heterosexual, in which men seek to attract women or women seek to attract men, but a few involve men seeking men or women seeking women. Those from the Hellenistic period often invoke a large number of deities to assist the petitioner, reflecting the mixture of gods that was common in Hellenistic society. Some of them involved actions or speech carried out in public, while others were secret. The spell itself was devised by the professional magician, who may have been a man, but binding spells are one of the few sources we have that provides any evidence about the private life of women who were not members of the elite. This spell, in which a woman named Sophia seeks to attract a woman named Gorgonia, is directed toward Anoubis, the Egyptian jackal- or dog-headed god of the underworld, and mentions a number of Egyptian and Greek deities associated with the underworld.

Fundament of the gloomy darkness, jagged-toothed dog, covered with coiling snakes, turning three heads, traveler in the recesses of the underworld, come, spirit-driver, with the Erinyes,[1] savage with their stinging whips; holy serpents, maenads, frightful maidens, come to my wroth incantations. Before I persuade by force this one and you, render him immediately a fire-breathing daemon. Listen and do everything quickly, in no way opposing me in the performance of this action; for you are the governors of the earth." *Alalachos allēch Harmachimeneus magimeneus athinembēs astazabathos artazabathos ōkoum phlom lonchachinachana thou Azaēl* and *Lykaēl* and *Beliam* and *Belenēa* and *sochsocham somochan sozocham ouzacham bauzacham oueddouch.* By means of this corpse-daemon inflame the heart, the liver,[2] the spirit of Gorgonia, whom Nilogenia bore, with love and affection for Sophia, whom Isara bore. Constrain Gorgonia, whom Nilogenia bore, to cast herself into the bath-house[3] for

the sake of Sophia, whom Isara bore; and you, become a bath-woman. Burn, set on fire, inflame her soul, heart, liver, spirit with love for Sophia, whom Isara bore. Drive Gorgonia, whom Nilogenia bore, drive her, torment her body night and day, force her to rush forth from every place and every house, loving Sophia, whom Isara bore, she surrendered like a slave, giving herself and all her possessions to her, because this is the will and command of the great god, *iartana ouousiō ipsenthanchōchainchoueōch aeēioyō iartana ousiousiou ipsoenpeuthadei annoucheō aeēioyō.* "Blessed lord of the immortals, holding the scepters of Tartaros and of terrible, fearful Styx (?) and of life-robbing Lethe, the hair of Kerberos trembles in fear of you, you crack the loud whips of the Erinyes; the couch of Persephone delights you, when you go to the longed bed, whether you be the immortal Sarapis, whom the universe fears, whether you be Osiris, star of the land of Egypt; your messenger is the all-wise boy; yours is Anoubis, the pious herald of the dead. Come hither, fulfill my wishes, because I summon you by these secret symbols.

---

[1]*Erinyes:* or Furies, Greek goddesses of vengeance, often shown with snake hair and whips.

[2]People in the Hellenistic world saw the liver, along with the heart, as the location of the emotions.

[3]Public baths were common in Hellenistic and Roman society as places where people went for recreation and relaxation as well as cleansing. Here Sophia wants Gorgonia to meet her in a bathhouse and Anoubis to change himself into a bath attendant so he can cast his spell on her more easily.

QUESTIONS FOR ANALYSIS

1. Though Sappho's poem to Aphrodite and the spell of attraction were written 800 years apart, they both involve women invoking deities to help them in human relationships. What do they indicate about women's relationships with other women in the Greek world and with the gods? Do you see any differences between them that might be the result of differences between classical and Hellenistic Greece?

2. How would you compare the body posture and facial expressions of the woman dressed as a *maenad* with those of the women shown spinning? How does this heighten the contrast between respectable and unruly women, at least in the minds of the (probably male) painters of these images?

3. How does Pausanius link the footraces for women held at Olympia in Elis to Greek history and mythology? Why do you think he feels this link is important? How else are these races connected to women's religious practices?

MEDICAL AND ANATOMICAL IDEAS

Along with an interest in human society, many of the Greek philosophers were also interested in the natural world and the physical body. They explored, and attempted to explain, the differences between male and female animals and between men and women; their explanations reflect both their observations and their preconceived ideas about hierarchical polarities. Some of these individuals were physicians, whose interest in the body grew out of a desire to treat disease and maintain health. Medical training was not standardized in the ancient world, nor was there any licensing of doctors; individuals who had gained a reputation through their talents or their writings often attracted followers and developed informal "schools," but these had no set curriculum.

No women appear to have been physicians in classical Athens, though by the Hellenistic period women are described on tomb inscriptions as physicians or midwives or praised for their "healing arts." Some male physicians wrote practical manuals for midwives, noting that the best midwives used theory as well as experience to diagnose and treat their patients; they thus expected that at least some midwives would be able to read. As far as we know, no female physician or midwife wrote a medical treatise in the ancient world; the first in Europe was Tro(c)ta (also called "Trotula") of Salerno in the twelfth century, whose works until very recently were considered by many medical historians to have been actually written by men.

20. Plato
Timaeus

In *Timaeus,* a sequel to *The Republic,* Plato discusses the creation of the world and man, speaking primarily through the voice of the philosopher Timaeus of Locri in Italy, about whom nothing is known outside of this dialogue. At the beginning of the dialogue, Socrates reviews the ideas that he developed in *The Republic* and specifically notes that he included women among the guardians; Timaeus has nothing to say about women until the very end of the dialogue, when he addresses the issue of the reason for sexual differences and the cause of sexual desire.

According to the probable account, all those creatures generated as men who proved themselves cowardly and spent their lives in wrongdoing were transformed, at their second incarnation, into women. And it was for this reason that the gods at that time contrived the love of sexual intercourse by constructing an animate creature of one kind in us men, and of

Timaeus by Plato, translated by R. G. Bury. Loeb Classical Library, Plato, vol. 7. (Cambridge: Harvard University Press, 1914).

another kind in women; and they made these severally in the following fashion. From the passage of egress for the drink,[1] where it receives and joins in discharging the fluid which has come through the lungs beneath the kidneys into the bladder and has been compressed by the air, they bored a hole into the condensed marrow which comes from the head down by the neck and along the spine—which marrow, in our previous account, we termed "seed." And the marrow, inasmuch as it is animate and has been granted an outlet, has endowed the part where its outlet lies with a love for generating by implanting therein a lively desire for emission. Wherefore in men the nature of the genital organs is disobedient and self-willed, like a creature that is deaf to reason, and it attempts to dominate all because of its frenzied lusts. And in women again, owing to the same causes, whenever the matrix or womb, as it is called,—which is an indwelling creature desirous of child-bearing,—remains without fruit long beyond the due season, it is vexed and takes it ill; and by straying all ways through the body and blocking up the passages of the breath and preventing respiration it casts the body into the uttermost distress, and causes, moreover, all kinds of maladies; until the desire and love of the two sexes unite them. Then, culling as it were the fruit from trees, they sow upon the womb, as upon ploughed soil, animalcules[2] that are invisible for smallness and unshapen; and these, again, they mould into shape and nourish to a great size within the body; after which they bring them forth into the light and thus complete the generation of the living creature.

In this fashion, then, women and the whole female sex have come into existence.

---

[1] I.e., the urethra.

[2] *Animalcules:* tiny animals invisible to the naked eye.

21. Hippocrates
On Virgins and *Diseases of Women*

Hippocrates (c. 460–c. 370 B.C.E.) was a Greek physician often termed the "Father of Medicine," who apparently worked and founded a school on the island of Cos. His reputation became so great that many works written by his followers were attributed to him; thus the Hippocratic Corpus, as this body of writings is called, includes works by a number of anonymous individuals from throughout Greek-speaking areas during the fifth and fourth centuries as well as some (probably) by Hippocrates himself. The Hippocratics regarded women as inferior to men because their bodies were generally softer and rounder, which the Hippocratics attributed to excess blood being soaked up in women's flesh. This blood was released once a month as menstrual fluid unless the woman was pregnant, when it instead went to the uterus to nourish the fetus, which was created by the mixing of male and female seed. Too much or too little blood was, therefore, regarded as harmful to women, creating mental as well as physical problems.

Excerpt from "Hippocrates: Diseases of Women 1" by Ann Ellis Hanson, from *Signs* 1 (1975), pp. 575–576.

A. *On Virgins*

As a result of visions, many people choke to death, more women than men, for the nature of women is less courageous and is weaker. And virgins who do not take a husband at the appropriate time for marriage experience these visions more frequently, especially at the time of their first monthly period, although previously they have had no such bad dreams of this sort. For later the blood collects in the womb in preparation to flow out; but when the mouth of the egress is not opened up, and more blood flows into the womb on account of the body's nourishment of it and its growth, then the blood which has no place to flow out, because of its abundance, rushes up to the heart and to the lungs; and when these are filled with blood, the heart becomes sluggish, and then, because of the sluggishness, numb, and then, because of the numbness, insanity takes hold of the woman. Just as when one has been sitting for a long time the blood that has been forced away from the hips and the thighs collects in one's lower legs and feet, it brings numbness, and as a result of the numbness, one's feet are useless for movement, until the blood goes back where it belongs. It returns most quickly when one stands in cold water and wets the tops of one's ankles. This numbness presents no complications, since the blood flows back quickly because the veins in that part of the body are straight, and the legs are not a critical part of the body. But blood flows slowly from the heart and from the phrenes.[1] There the veins are slanted, and it is a critical place for insanity, and suited for madness.

When these places[2] are filled with blood, shivering sets in with fevers. They call these 'erratic fevers'. When this is the state of affairs, the girl goes crazy because of the violent inflammation, and she becomes murderous because of the decay and is afraid and fearful because of the darkness. The girls try to choke themselves because of the pressure on their hearts; their will, distraught

and anguished because of the bad condition of the blood, forces evil on itself. In some cases the girl says dreadful things: [the visions] order her to jump up and throw herself into wells and drown, as if this were good for her and served some useful purpose. When a girl does not have visions, a desire sets in which compels her to love death as if it were a form of good. When this person returns to her right mind, women give to Artemis various offerings, especially the most valuable of women's robes, following the orders of oracles, but they are deceived. The fact is that the disorder is cured when nothing impedes the downward flow of blood. My prescription is that when virgins experience this trouble, they should cohabit with a man as quickly as possible. If they become pregnant, they will be cured. If they don't do this, either they will succumb at the onset of puberty or a little later, unless they catch another disease. Among married women, those who are sterile are more likely to suffer what I have described.

B. *Diseases of Women*

If suffocation occurs suddenly, it will happen especially to women who do not have intercourse and to older women rather than to young ones, for their wombs are lighter. It usually occurs because of the following: when a woman is empty and works harder than in her previous experience, her womb, becoming heated from the hard work, turns because it is empty and light. There is, in fact, empty space for it to turn in because the belly is empty. Now when the womb turns, it hits the liver and they go together and strike against the abdomen—for the womb rushes and goes upward toward the moisture, because it has been unduly heated by hard work, and the liver is, after all, moist. When the womb hits the liver, it produces sudden suffocation as it occupies the breathing passage around the belly.

Sometimes, at the same time the womb begins to go toward the liver, phlegm flows down from the head to the abdomen (that is, when the woman is experiencing the suffocation) and sometimes, simultaneously with the flow of

---

[1]*Phrenes:* the "mind," which Hippocrates located near the lungs, not in the brain.
[2]The heart and the phrenes.

phlegm, the womb goes away from the liver to its normal place and the suffocation ceases. The womb goes back, then, when it has taken on moisture and has become heavy. The womb makes a gurgling sound whenever it goes back to its own position. When, in fact, the womb does go back, occasionally the stomach is more moist after these circumstances than it was previously, because now the head releases phlegm to the body cavity. When the womb is near the liver and the abdomen and when it is suffocating, the woman turns up the whites of her eyes and becomes chilled; some women become livid. She grinds her teeth and saliva flows out of her mouth. These women resemble those who suffer from Herakles' disease.[1] If the womb lingers near the liver and the abdomen, the woman dies of the suffocation.

Sometimes, if a woman is empty and she overworks, her womb turns and falls toward the neck of her bladder and produces strangury—but no other malady seizes her. When such a woman is treated, she speedily becomes healthy; sometimes recovery is even spontaneous.

In some women the womb falls toward the lower back or toward the hips because of hard work or lack of food, and produces pain.

_____

[1]*Herakles' disease:* epilepsy.

22. Galen
On the Usefulness of the Parts of the Body

Galen (c. 130–c. 200 C.E.) was a native of Pergamum in Asia Minor (present-day Turkey) and became the best-known physician and medical writer in the ancient world; he eventually served as physician to the Roman Emperor Marcus Aurelius. In contrast to the Hippocratics and Aristotle, who emphasized the differences between men and women, Galen emphasized their similarities in terms of anatomy in what some historians have since come to call the "one-sex" model of the human body. Like the Hippocratics and Aristotle, Galen regarded heat as the main agent creating sex differences and saw illness as the result of an imbalance in four fluids (termed "humors") in the body—blood, phlegm, black bile, and yellow bile. Treatment thus consisted primarily of trying to get the humors back into balance, which was generally attempted by "letting" blood from the veins either by incision or the use of leeches. The theory of bodily humors lasted in Europe until the seventeenth century—and even later in some areas—with bloodletting frequently prescribed for pregnant women or those with menstrual difficulties.

The female is less perfect than the male for one, principal reason—because she is colder; for if among animals the warm one is the more active, a colder animal would be less perfect than a warmer....

All the parts, then, that men have, women have too, the difference between them lying in only one thing, which must be kept in mind throughout the discussion, namely, that in women the parts are within [the body], whereas in men they are outside, in the region called the perineum. Consider first whichever ones you please, turn outward the woman's, turn inward, so to speak, and fold double the man's, and you will find them the same in both in every respect. Then think first, please, of the man's turned in and extending inward between the rectum and the bladder. If this should happen, the scrotum would necessarily take the place of the uteri, with the testes lying outside, next to it on either side; the penis of the male would become the neck of the cavity that had been formed; and the skin at the end of the penis, now called the prepuce, would become the female pudendum [the vagina] itself. Think too, please, of the converse, the uterus turned outward and projecting. Would not the testes [the ovaries] then necessarily be inside it? Would it not contain them like a scrotum? Would not the neck [the cervix], hitherto concealed inside the perineum but now pendent, be made into the male member? And would not the female pudendum, being a skinlike growth upon this neck, be changed into the part called the prepuce? It is also clear that in consequence the position of the arteries, veins, and spermatic vessels [the ductus deferentes and Fallopian tubes] would be changed too. In fact, you could not find a single male part left over that had not simply changed its position; for the parts that are inside in woman are outside in man.

Now just as mankind is the most perfect of all animals, so within mankind the man is more perfect than the woman, and the reason for his perfection is his excess of heat, for heat is Nature's primary instrument. Hence in those animals that have less of it, her[1] workmanship is necessarily more imperfect, and so it is no wonder that the female is less perfect than the male by as much as she is colder than he. In fact, just as the mole has imperfect eyes, though certainly not so imperfect as they are in those animals that do not have any trace of them at all, so too the woman is less perfect than the man in respect to the generative parts. For the parts were formed within her when

she was still a fetus, but could not because of the defect in the heat emerge and project on the outside, and this, though making the animal itself that was being formed less perfect than one that is complete in all respects, provided no small advantage (χρεία) for the race; for there needs must be a female. Indeed, you ought not to think that our Creator would purposely make half the whole race imperfect and, as it were, mutilated, unless there was to be some great advantage in such a mutilation.

Let me tell what this is. The fetus needs abundant material both when it is first constituted and for the entire period of growth that follows. Hence it is obliged to do one of two things: it must either snatch nutriment away from the mother herself or take nutriment that is left over. Snatching it away would be to injure the generant, and taking left over nutriment would be impossible if the female were perfectly warm; for if she were, she would easily disperse and evaporate it. Accordingly, it was better for the female to be made enough colder so that she cannot disperse all the nutriment which she concocts and elaborates. Now an animal that is too cold cannot concoct, and on the other hand, one that is perfectly warm, though strong enough to concoct, is also strong enough to disperse. Hence one that falls not very far short of being perfectly warm is able both to concoct, being no longer cold, and to leave some [nutriment] over, since it is not exceedingly warm. This is the reason (χρεία) why the female was made cold, and the immediate consequence of this is the imperfection of the parts, which cannot emerge on the outside on account of the defect in the heat, another very great advantage for the continuance of the race. For, remaining within, that which would have become the scrotum if it had emerged on the outside was made into the substance of the uteri, an instrument fitted to receive and retain the semen and to nourish and perfect the fetus.

Forthwith, of course, the female must have smaller, less perfect testes, and the semen generated in them must be scantier, colder, and wetter (for these things too follow of necessity from the deficient heat). Certainly such semen

---

[1] *Her* in this case refers to Nature.

would be incapable of generating an animal, and, since it too has not been made in vain, I shall explain in the course of my discussion what its use is. The testes of the male are as much larger as he is the warmer animal. The semen generated in them, having received the peak of concoction, becomes the efficient principle of the animal. Thus, from one principle devised by the Creator in his wisdom, that principle in accordance with which the female has been made less perfect than the male, have stemmed all these things useful for the generation of the animal: that the parts of the female cannot escape to the outside; that she accumulates an excess of useful nutriment and has imperfect semen and a hollow instrument to receive the perfect semen; that since everything in the male is the opposite [of what it is in the female], the male member has been elongated to be most suitable for coitus and the excretion of semen; and that his semen itself has been made thick, abundant, and warm.

QUESTIONS FOR ANALYSIS

1. According to Timaeus, how did women come into the world? What does the "animate creature" with a "love for sexual intercourse" cause in men? In women? How does the language used to describe these two sets of consequences differ?

2. How would you compare Plato's (or at least Plato's as expressed by Timaeus) and Hippocrates' opinions about the results of a lack of sexual intercourse in women? How do these differ, according to Hippocrates, depending on the age or social status of the woman?

3. To what does Galen attribute sex differences, and how does he value these differences? How are his explanations and judgments about value different from those of Plato?

4. As you have seen, both Plato and Aristotle wrote about physical and social reasons for what we would call gender differences. In their view, which was most important in determining the gender hierarchy, politics or biology? How are politics and biology related on this issue?

CHAPTER 3

The Roman World

(Fifth Century B.C.E.–Sixth Century C.E.)

At the same time that Athenians were establishing a democracy of male citizens, Romans—Latin-speaking people who lived in central Italy—overthrew the Etruscan kings who ruled them and established a republic. The republic was governed by a senate made up of male members of the land-owning patrician class, who also held all public and religious offices. The vast majority of free Romans were plebs, who, almost from the beginning of the republic, sought to gain greater political voice; by about 300 B.C.E. plebs had gained the right to hold many offices. During the republican period, Rome gradually conquered all of Italy and eventually much of southern Europe, the Near East, and northern Africa.

Most of the sources from the early Roman Republic have been lost, and historians writing during the later imperial period frequently idealized early Roman citizens. They included women as well as men in their stories about heroism and virtue. The standard account of the founding of the Republic tells the story of Lucretia, a faithful Roman wife, who was raped by the son of Tarquin, the king of the Etruscans. Though her husband was willing to forgive her, Lucretia stabbed herself rather than live in dishonor, which led her husband and the other Roman men to vow never to give in to a king again. We have no way of assessing the truth of this story, but the frequency with which it was repeated indicates a clear link in the minds of later writers between female virtue and the good of the state. (The story of Lucretia was a common theme in Roman art as well as literature and reemerged as the subject of many paintings in the Renaissance.)

This link was made even more clear with the vestal virgins, women from prominent families chosen while still girls to act as priestesses in the temple of Vesta, the goddess of the hearth. Many people had altars to Vesta in their homes, and the public temple in her honor had a sacred fire which the vestal virgins tended. This fire was viewed as a symbol of Rome's success and was never to go out; vestals who neglected it, particularly those who broke their vow of chastity, were entombed alive.

The vestals were the only women from well-to-do families in republican Rome who were not expected to marry, and large amounts of Roman law concerned itself with marriage and family life. The Roman family was supposed to be patriarchal, with fathers holding strong control over their female and male children, and husbands control over their wives. Women's economic dependence was symbolized by the dowry, an amount of property, goods, cash, and often slaves that was given to a woman by her father when she married; the dowry was under her husband's control during the marriage, but remained her property. It, or items of equivalent value, was to be returned to her if the marriage ended, providing her with some financial security.

Both men and women married early at roughly the same age, a marital pattern both supported by and supporting the Roman ideal that husbands and wives should share interests, property, and activities; spousal loyalty was often used as a metaphor for political loyalty in a way that would have seemed very strange to Greek thinkers like Aristotle. If the marriage was less than ideal, however, by the late Republic, divorce was possible at the instigation of either the husband or wife. Marriage was an obligation for those with property, but not a right open to all; marriages between individuals of widely different social statuses were prohibited, though an upper-class man could take a lower-class woman as his concubine, which carried some legal advantages for the woman. (The reverse was not possible; a free woman who lived with a slave man could be made a slave herself.) Slaves could also not marry, even to each other; at most, they could enter a relationship called a *contubernium* which was not legally binding.

During the late republican and imperial period, upper-class men were often away from home on military or government duties in Rome's rapidly expanding territories, leaving women responsible for running estates and managing family businesses and property. Some women owned businesses such as brickyards through which they became quite wealthy. Though they could not hold political office, women could act as legal witnesses, an important right in a culture like Rome in which law was viewed as the cement that held society together. About one-quarter of the Roman population were slaves, many of whom were captured in war. Male slaves ranged from highly educated tutors, physicians, and artists to those who worked in the mines and galley ships. The experience of female slaves was less variable, for most of them were engaged in domestic chores or agricultural tasks; because they were property, they were also sexually available to their masters.

During the first century B.C.E., the Roman Republic fell apart as power passed to warring generals; gradually, one of these, Caesar Augustus, took over sole authority and slowly took all power into his own hands. His successors took on the title of emperor and increasingly relied on members of their families, including their mothers and wives, for assistance and advice. Because there was no clear line of succession, women often maneuvered to get their male family members chosen as the next emperor; their actions may have included murdering rivals, although it is hard to know for certain as our information about them comes primarily from later male historians who were very hostile to women's power of any type. No Roman woman was named emperor in her own right, but several were given special titles, had towns named after them, and had coins issued in their honor.

Augustus interfered in family life more than the Senate had done, attempting to increase the marriage and birth rate with a variety of laws. Bachelors were penalized (a practice the Italian dictator Mussolini returned to in the twentieth century) and women who had more than three children—four if the women were freed slaves—were given the right to be legal persons, no longer under male guardianship. Adultery was made a crime punishable by the state, and marriages across class lines continued to be prohibited. The Right of Three Children did give many women greater opportunities, and sources indicate that they owned and transferred all forms of property, including land and slaves, and ran businesses on their own.

Women's increasing independence also extended to the religious sphere. Traditional Roman religion, like that of Mesopotamia and Egypt, was a civic or state religion akin to patriotism, in which honoring the gods was viewed as essential to the health and well-being of the state. By the first century B.C.E., the expansion of Roman holdings brought Romans into contact with many different religions and philosophies, and people began to develop more individualistic belief systems and patterns of worship. In the first century C.E., Christianity, a movement centered on the teaching of a Jewish reformer, Jesus of Nazareth, began among Jews in the Roman province of Palestine but soon spread among many different groups throughout the Roman Empire. Christianity differed from other classical religions in that it required its adherents to stop participating in ceremonies that honored other gods; many Romans viewed this as provoking the wrath of the gods who protected Rome, and sporadic persecutions of Christians began under emperors such as Nero and Domitian.

Women took an active role in the spread of Christianity, preaching, acting as missionaries, and being martyred alongside men. Early Christians expected Jesus, who had been executed under Roman authority, to return to earth again very soon and so taught that one should concentrate on this Second Coming. Because of this, marriage and normal family life should be abandoned, and Christians should depend on their new spiritual family of co-believers. Early Christians often met in people's homes and called each other brother and sister, a metaphorical use of family terms that was new in the Roman Empire. This, too, made Christians seem dangerous to many Romans, especially when becoming Christian actually led some young people to avoid marriage, viewed by Romans as the foundation of society and the proper patriarchal order.

WOMEN'S LEGAL POSITION

Roman law is often looked at as one of the great achievements of both the republic and the empire; it remains the basis of the law codes of many European countries and of the state of Louisiana in the United States. Roman law began during the republican period as a set of rules governing the private lives of citizens and, as such, had much to say about marital and family relations. It includes statutes and decrees—first those of the senate and then of the emperors—the opinions of learned jurists, and the rulings of lesser officials and judges. Legal interpreters termed *praetors* and judges termed *judices* made decisions based on explicit statutes and on their own notions of what would be fair and equitable, which gave them a great deal of flexibility. *Praetors* generally followed the laws set by their predecessors, announcing publicly at the beginning of their terms of office that they would do this, but they also added to the body of law as new issues arose. Thus Roman law was adaptable to new conditions, with jurists in the empire regarding their work as building on that of earlier centuries rather than negating it.

23. The Twelve Tables

The first written codification of Roman law, the Twelve Tables, was apparently made in the middle of the fifth century B.C.E. and posted publicly on bronze tables, giving at least those Romans who could read direct access to it. Each of the Twelve Tables concerns one topic and is made up of a number of laws; the code as a whole is quite short. Some of its provisions were later ignored, but the Twelve Tables were included in all subsequent codifications of Roman law, including that of Justinian.

1. A notably deformed child shall be killed immediately.

2a. To a father...shall be given over a son the power of life and death.

2b. If a father thrice surrenders a son for sale[1] the son shall be free from the father.

3. To repudiate his wife her husband shall order her...to have her own property for herself, shall take the keys, shall expel her.

4. A child born within ten months of the father's death shall enter into the inheritance...[2]

1. ...Women, even though they are of full age, because of their levity of mind shall be under guardianship...except vestal virgins, who...shall be free from guardianship...[3]

2. The conveyable possessions of a woman who is under guardianship of male agnates[4]

---

[1] I.e., sold him into slavery.

[2] I.e., will be considered the child and legal heir of his or her father.
[3] Thus the vestal virgins are the only women not under the legal control of a man.
[4] *Agnates:* relatives in the male line.

From *Ancient Roman Statues* as translated by Allan Chester Johnson et al. Copyright © 1961 by University of Texas Press.

shall not be acquired by prescriptive right unless they are transferred by the woman herself with the authorization of her guardian....

5. ...If any woman is unwilling to be subjected in this manner to her husband's marital control she shall absent herself for three successive nights in every year and by this means

shall interrupt his prescriptive right of each year.[5]

---

[5]Through this they were not divorced, but the husband lost his complete legal control over his wife.

24. Gaius
Institutes

Gaius was a Roman jurist and teacher of law who lived in the second century C.E. His work, officially titled "The Four Commentaries of Gaius on the Institutes of the Civil Law," was a textbook designed for law students; the four parts discuss laws relating to persons, things, estates, and actions.

(56) Roman citizens are understood to have contracted marriage according to the Civil Law and to have the children begotten by them in their power if they marry Roman citizens, or even Latins or foreigners whom they have the right to marry; for the result of legal marriage is that the children follow the condition of the father and not only are Roman citizens by birth, but also become subject to paternal authority.

(57) Therefore, certain veterans are usually granted permission by the Imperial Constitutions to contract civil marriage with those Latin or foreign women whom they first marry after their discharge, and the children born of such unions become Roman citizens by birth, and are subject to the authority of their fathers.

(57a) Marriage, however, cannot take place with persons of servile condition.... The issue of a Latin man and a foreign woman, as well as that of a foreign man and a Latin woman, follows the condition of the mother....

(82) The result of this is that the child of a female slave and a freeman is, by the Law of Nations, born a slave; and, on the other hand,

the child of a free woman and a male slave is free by birth....

(108) Now let us consider those persons who are in our hand, which right is also peculiar to Roman citizens.

(109) Both males and females are under the authority of another, but females alone are placed in the hands.[1]

(110) Formerly this ceremony was performed in three different ways, namely, by use, by confarreation, and by coemption.

(111) A woman came into the hand of her husband by use when she had lived with him continuously for a year after marriage; for the reason that she was obtained by usucaption, as it were, through possession for the term of a year, and passed into the family of her husband where she

---

[1]A woman could be placed "in the hands" of her husband by the three ways Gaius discusses here. In doing this, her father no longer had any power over her. Marriage could also take place "without the hand" (*sine manu*), in which a father retained legal control over his married daughter and could order her to divorce. Marriage *sine manu* was the most common form by the time of Gaius.

From *Corpus Juris Civilis: The Civil Law*, Vol. 1, translated by S. P. Scott. Published by AMS Press.

occupied the position of a daughter. Hence it is provided by the Law of the Twelve Tables that if a woman was unwilling to be placed in the hand of her husband in this way, she should every year absent herself for three nights, and in this manner interrupt the use during the said year; but all of this law has been partly repealed by legal enactments, and partly abolished by disuse.

(112) Women are placed in the hand of their husbands by confarreation, through a kind of sacrifice made to Jupiter Farreus, in which a cake is employed, from whence the ceremony obtains its name; and in addition to this, for the purpose of performing the ceremony, many other things are done and take place, accompanied with certain solemn words, in the presence of ten witnesses. This law is still in force in our time, for the principal flamens, that is to say, those of Jupiter, Mars, and Quirinus, as well as the chief of the sacred rites, are exclusively selected from persons born of marriages celebrated by confarreation. Nor can these persons themselves serve as priests without marriage by confarreation.

(113) In marriage by coemption, women become subject to their husbands by mancipation, that is to say by a kind of fictitious sale; for the man purchases the woman who comes into his hand in the presence of not less than five witnesses, who must be Roman citizens over the age of puberty....

(144) Parents are permitted to appoint testamentary guardians for their children who are subject to their authority, who are under the age of puberty, and of the male sex; and for those of the female sex, no matter what their age may be, and even if they are married; for the ancients required women, even if they were of full age, to remain under guardianship on account of the levity of their disposition....

(190) There does not seem to be any good reason, however, why women of full age should be under guardianship, for the common opinion that because of their levity of disposition they are easily deceived, and it is only just that they should be subject to the authority of guardians, seems to be rather apparent than real; for women of full age transact their own affairs, but in certain cases, as a mere form, the guardian interposes his authority, and he is often compelled to give it by the Prætor, though he may be unwilling to do so....

(194) Moreover, a freeborn woman is released from guardianship if she is the mother of three children, and a freedwoman[2] if she is the mother of four, and is under the legal guardianship of her patron.

------

[2] A freedwoman was a slave who had been freed.

25. Justinian
Corpus Juris Civilis

During the fourth century, the Roman emperor Constantine split the Empire into two parts in order to rule it more effectively. During the fifth century, the western half of the Roman Empire collapsed, but the eastern half, usually termed the Byzantine Empire, survived with an emperor until the mid-fifteenth century. One of the most powerful Byzantine emperors was Justinian (reigned 527–565), who ordered the compilation of all imperial laws and legal opinions into one comprehensive code. The result was an enormous code of civil law, officially termed the *Corpus Juris Civilis,* which became the basis of Byzantine legal procedure for

Excerpted from *Corpus Juris Civilis: The Civil Law,* Vol. 1, translated by S. P. Scott. Published by AMS Press.

nearly a millennium and the core of legal education in western Europe when law schools were established in the twelfth century. Justinian's code consists of three parts: the *Codex,* which was actual imperial legislation; the *Digest,* the opinions of various jurists from throughout the history of Rome; and the *Institutes,* an officially prescribed course for first-year law students modeled on the much shorter work of Gaius. Each of these parts has many sections that refer to women's legal situation. The following come from the *Digest,* where they are identified by the name of the jurist who wrote the opinion. These opinions were not the same as laws and the jurists sometimes disagreed with one another, but because judicial decisions were based on precedent (as they still are), they carried great weight.

CONCERNING BETROTHALS

1. *Florentinus, Institutes, Book III.*
A betrothal is the mention and promise of a marriage to be celebrated hereafter. . . .

12. *Ulpianus, On Betrothals.*
A girl who evidently does not resist the will of her father is understood to give her consent. A daughter is only permitted to refuse to consent to her father's wishes, where he selects someone for her husband who is unworthy on account of his habits or who is of infamous character.

13. *Paulus, On the Edict, Book V.*
Where a son under paternal control refuses his consent, a betrothal cannot take place, so far as he is concerned.

14. *Modestinus, Differences, Book IV.*
In contracting a betrothal, there is no limit to the age of the parties, as is the case in marriage. Wherefore, a betrothal can be made at a very early age, provided what is being done is understood by both persons, that is to say, where they are not under seven years of age.

CONCERNING THE CEREMONY OF MARRIAGE

1. *Modestinus, Rules, Book I.*
Marriage is the union of a man and a woman, forming an association during their entire lives, and involving the common enjoyment of divine and human privileges.

2. *Paulus, On the Edict, Book XXXV.*
Marriage cannot take place unless all the parties consent, that is to say those who are united, as well as those under whose authority they are. . . .

10. *Paulus, On the Edict, Book XXXV.*
It is doubtful what course to pursue where the father is absent, and it is not known where he is, or even whether he is still alive. If three years should elapse from the time when the father's whereabouts or whether he was living began to be unknown, his children of both sexes will not be prevented from legally contracting marriage. . . .

24. *Modestinus, Rules, Book I.*
Where a man lives with a free woman, it is not considered concubinage but genuine matrimony, if she does not acquire gain by means of her body . . .

26. *The Same, Opinions, Book V.*
Modestinus says that women accused of adultery cannot marry during the lifetime of their husbands, even before they have been convicted.

41. *Marcellus, Digest, Book XXXVI.*
It is understood that disgrace attaches to those women who live unchastely, and earn money by prostitution, even if they do not do so openly.
(1) If a woman should live in concubinage with someone besides her patron, I say that she does not possess the virtue of the mother of a family.

42. *Modestinus, On the Rite of Marriage.*

In unions of the sexes, it should always be considered not only what is legal, but also what is decent.

(1) If the daughter, granddaughter, or great-granddaughter of a senator should marry a freedman, or a man who practices the profession of an actor, or whose father or mother did so, the marriage will be void.

43. *Ulpianus, On the Lex Julia et Papia, Book I.*

We hold that a woman openly practices prostitution, not only where she does so in a house of ill-fame, but also if she is accustomed to do this in taverns, or in other places where she manifests no regard for her modesty.

(1) We understand the word "openly" to mean indiscriminately, that is to say, without choice, and not if she commits adultery or fornication, but where she sustains the role of a prostitute.

(2) Moreover, where a woman, having accepted money, has intercourse with only one or two persons, she is not considered to have openly prostituted herself.

(3) Octavenus, however, says very properly that where a woman publicly prostitutes herself without doing so for money, she should be classed as a harlot.

(4) The law brands with infamy not only a woman who practices prostitution, but also one who has formerly done so, even though she has ceased to act in this manner; for the disgrace is not removed even if the practice is subsequently discontinued.

(5) A woman is not to be excused who leads a vicious life under the pretext of poverty.

(6) The occupation of a pander is not less disgraceful than the practice of prostitution.

(7) We designate those women as procuresses who prostitute other women for money.

(8) We understand the term "procuress" to mean a woman who lives this kind of a life on account of another.

(9) Where one woman conducts a tavern, and keeps others in it who prostitute themselves, as many are accustomed to do under the pretext of employing women for the service of the house; it must be said that they are included in the class of procuresses.

(10) The Senate decreed that it was not proper for a senator to marry or keep a woman who had been convicted of a criminal offence, the accusation for which could be made by any of the people; unless he was prohibited by law from bringing such an accusation in court.

(11) Where a woman has been publicly convicted of having made a false accusation, or prevarication,[1] she is not held to have been convicted of a criminal offence.

(12) Where a woman is caught in adultery, she is considered to have been convicted of a criminal offence. . . .

47. *Paulus, On the Lex Julia et Papia, Book II.*

The daughter of a senator who has lived in prostitution, or has exercised the calling of an actress[2] or has been convicted of a criminal offence, can marry a freedman with impunity; for she who has been guilty of such depravity is no longer worthy of honor. . . .

51. *Licinius Rufinus, Rules, Book I.*

When a female slave has been manumitted for the purpose of matrimony, she cannot marry anyone else than the party by whom she was set free, unless her patron renounces the right of marriage with her.

CONCERNING THE LAW OF DOWRY

1. *Paulus, On Sabinus, Book XIV.*

The right to a dowry is perpetual, and, in accordance with the desire of the party who bestows it, the contract is made with the understanding that the dowry will always remain in the hands of the husband.

---

[1]Prevarication was revealing confidential information with the intent to harm.

[2]Actors and actresses were considered dishonorable because most of them were slaves or freedmen and because some of them also worked as prostitutes.

2. *The Same, On the Edict, Book LX.*

It is to the interest of the State that women should have their dowries preserved, in order that they can marry again.

3. *Ulpianus, On the Edict, Book LXIII.*

The term dowry does not apply to marriages which are void, for there cannot be a dowry without marriage. Therefore, where the name of marriage does not exist, there is no dowry.

7. *Ulpianus, On Sabinus, Book XXXI.*

Equity demands that the profits of a dowry shall belong to the husband, for, as he sustains the burdens of matrimony, it is but just that he should receive the profits. . . .

42. *Gaius, On the Provincial Edict, Book XI.*

Where property which can be weighed, counted, or measured, is given by the way of dowry, this is done at the risk of the husband, because it is given to enable him to sell it at his pleasure; and when the marriage is dissolved, he must return articles of the same kind and quality, or his heir must do so. . . .

73. *Paulus, Sentences, Book II.*

A person who is dumb, deaf, or blind, is liable on account of a dowry, because each of them can contract a marriage.

(1) While marriage exists, the dowry can be returned to the wife for the following reasons, provided she does not squander it, namely: in order that she may support herself and her children, or may purchase a suitable estate, or may provide sustenance for her father banished to some island, or may relieve her brother or sister who is in want.

CONCERNING DONATIONS BETWEEN HUSBAND AND WIFE

1. *Ulpianus, On Sabinus, Book XXXII.*

In accordance with the custom adopted by us, gifts between husband and wife are not valid. This rule has been adopted to prevent married persons from despoiling themselves through mutual affection, by setting no limits to their generosity, but being too profuse toward one another through the facility afforded them to do so.

2. *Paulus, On Sabinus, Book VII.*

Another reason is that married persons might otherwise not have so great desire to educate their children. Sextus Cæcilius also added still another, namely, because marriage would often be dissolved where the husband had property and could give it, but did not do so; and therefore the result would be that marriage would become purchasable.

3. *Ulpianus, On Sabinus, Book XXXII.*

This reason is also derived from a Rescript of the Emperor Antoninus, for it says: "Our ancestors forbade donations between husband and wife, being of the opinion that true affection was based upon their mutual inclination, and also taking into consideration the reputation of the parties who were united in matrimony, lest their agreement might seem to be brought about for a price, and to prevent the better one of the two from becoming poor, and the worse one from becoming more wealthy."

CONCERNING DIVORCES AND REPUDIATIONS

1. *Paulus, On the Edict, Book XXXV.*

Marriage is dissolved by divorce, death, captivity, or by any other kind of servitude which may happen to be imposed upon either of the parties.

3. *Paulus, On the Edict, Book XXXV.*

It is not a true or actual divorce unless the purpose is to establish a perpetual separation. Therefore, whatever is done or said in the heat of anger is not valid, unless the determination becomes apparent by the parties persevering in their intention, and hence where repudiation takes place in the heat of anger and the wife returns in a short time, she is not held to have been divorced.

IN WHAT WAY THE DOWRY CAN BE RECOVERED AFTER THE MARRIAGE HAS BEEN DISSOLVED

1. *Pomponius, On Sabinus, Book XV.*

The cause of the dowry always and everywhere take precedence, for it is to the public interest for dowries to be preserved to wives, as it is absolutely necessary that women should be endowed for the procreation of progeny, and to furnish the state with freeborn citizens.

2. *Ulpianus, On Sabinus, Book XXXV.*

Where marriage is dissolved, the dowry should be delivered to the woman. . . .

24. *Ulpianus, On the Edict, Book XXXIII.*

If, during the existence of the marriage, the wife desires to institute proceedings on account of the impending insolvency of her husband, what time must we fix for her to claim the dowry? It is settled that it can be demanded from the time when it is perfectly apparent that the pecuniary resources of the husband are not sufficient for the delivery of the dowry.[1]

30. *Julianus, Digest, Book XVI.*

A woman who is married a second time is not prevented from instituting proceedings against her first husband for the recovery of her dowry.

CONCERNING ACCUSATIONS AND INSCRIPTIONS[1]

1. *Pomponius, On Sabinus, Book I.*

A woman is not permitted to accuse anyone in a criminal case unless she does so on account of the death of her parents or children, her patron or patroness, and their son, daughter, grandson, or granddaughter. . . .

CONCERNING THE JULIAN LAW FOR THE PUNISHMENT OF ADULTERY

2. *Ulpianus, On Adultery, Book VIII.*

(2) The crime of pandering is included in the Julian Law on Adultery, as a penalty has been prescribed against a husband who profits pecuniarily by the adultery of his wife; as well as against one who retains his wife after she has been taken in adultery.

(3) Moreover, he who permits his wife to commit this offence, holds his marriage in contempt; and where anyone who does not become indignant on account of such pollution, the penalty for adultery is not inflicted. . . .

8. *Papinianus, On Adultery, Book II.*

Anyone who knowingly lends his house to enable debauchery or adultery to be committed there with a matron who is not his wife, or with a male, or who pecuniarily profits by the adultery of his wife, no matter what may be his status, is punished as an adulterer. . . .

10. *Papinianus, On Adultery, Book II.*

A matron means not only a married woman, but also a widow.

(1) Women who lend their houses, or have received any compensation for debauchery which they have committed, are also liable under this Section of the law.

(2) A woman who gratuitously acts as a bawd for the purpose of avoiding the penalty for adultery, or hires her services to appear in the theatre, can be accused and convicted of adultery under the Decree of the Senate. . . .

13. *Ulpianus, On Adultery, Book II.*

(5) The judge who has jurisdiction of adultery must have before his eyes, and investigate whether the husband, living modestly, has afforded his wife the opportunity of having good morals; for it would be considered extremely unjust for the husband to require chastity for his wife, which he himself does not practice. This, indeed, may condemn the husband, but cannot afford a set-off for mutual crime when committed by both parties. . . .

---

[1]In other words, if her husband was going bankrupt, a woman could sue him for return of her dowry.

[1]An inscription was a formal denunciation of a crime against a person.

(8) Where a girl, less than twelve years old, brought into the house of her husband, commits adultery, and afterwards remains with him until she has passed that age, and begins to be his wife; she cannot be accused of adultery by her husband, for the reason that she committed it before reaching the marriageable age. . . .

20. *Papinianus, On Adultery, Book I.*

The right is granted to the father to kill a man who commits adultery with his daughter while she is under his control. Therefore no other relative can legally do this, nor can a son under paternal control, who is a father, do so with impunity.

24. *Macer, Public Prosecutions, Book I.*

A husband is also permitted to kill a man who commits adultery with his wife, but not everyone without distinction, as the father is; for it is provided by this law that the husband can kill the adulterer if he surprises him in his own house, but not if he surprises him in the house of his father-in-law; nor if he was formerly a panderer; or had exercised the profession of a mountebank, by dancing or singing on the stage; or had been convicted in a criminal prosecution and not been restored to his civil rights.

26. *Ulpianus, Disputations, Book III.*

A woman cannot be accused of adultery during marriage by anyone who, in addition to the husband, is permitted to bring the accusation; for a stranger should not annoy a wife who is approved by her husband, and disturb a quiet marriage, unless he has previously accused the husband of being a panderer.

(1) When, however, the charge has been abandoned by the husband, it is proper for it to be prosecuted by another.

33. *Marcianus, Public Prosecutions, Book I.*

Where anyone alleges that adultery has been committed by his slave, with a woman whom he had for his wife, the Divine Pius stated in a Rescript[1] that he must accuse the woman before subjecting his slave to torture to her prejudice.

34. *Modestinus, Rules, Book I.*

He is guilty of fornication who keeps a free woman for the purpose of cohabiting with her, but not with the intention of marrying her, excepting, of course, a concubine.

(1) Adultery is committed with a married woman; fornication with a widow, a virgin, or a boy.

------

[1]A rescript was an imperial enactment, here by the Emperor Antoninus Pius; Roman emperors were regarded as gods after death, so he is termed "Divine Pius."

26. Flavius Josephus
Against Apion

As the Roman Republic and then Empire expanded, territory in which many Jews were living was taken over. Jews began to migrate throughout Roman territory; these migrations, often called the *diaspora,* increased after Jerusalem was destroyed by the Romans after an unsuccessful Jewish revolt against Roman overlordship in 66–73 C.E. Some Romans were very attracted to Jewish ideas and ways of living, but others were hostile to them, leading Jewish leaders to write explanations and defenses of their religion. Flavius Josephus was a historian of Judaism, who in about 96 C.E. wrote one of these explanations, answering an attack by the anti-Jewish writer Apion and presenting Jewish ideas and laws.

THE FAMILY

What are our marriage laws? The Law recognizes no connection of the sexes but the natural connection between a man and his wife, and that only for the procreation of children; it abhors sodomy, and death is the punishment for that crime. The law commands us also, when we marry, not to have regard to dowry, nor to take a woman by violence, nor to persuade her by deceit and guile, but to demand her in marriage of him who has power to give her away and is fit to do so because of his nearness of kin. For the legislator says that a woman is inferior to her husband in all things. Let her, therefore, be obedient to him; not that he should ill-treat her, but that she may be directed; for God has given the authority to the husband. A husband is to lie only with his wife, and to seduce another man's wife is a wicked thing, which, if any one ventures upon, death is inevitably his punishment, as it is also his who forces a virgin betrothed to another man.

The Law, moreover, enjoins us to bring up all our offspring, and forbids women to cause abortion of what is begotten, or to destroy the foetus in any other way, for she will be an infanticide who thus destroys life and diminishes the human race. If any one, therefore, commits adultery or seduction, he cannot be considered pure. Why, even after the regular union of man and wife, the Law enjoins that they shall both wash themselves [   ]

Moreover the Law does not permit us to feast at the births of our children, and so make excuses for drinking to excess, but it ordains that the very beginning of life should be sober. It also commands us to bring our children up in learning, and to make them conversant with the laws and acquainted with the acts of their forefathers, that they may imitate them, and, being grounded in them, may neither transgress them nor have any excuse for ignorance of them.

The Law also ordains, that parents should be honored next after God himself, and orders the son who does not requite them for the benefits he has received from them, but comes short on any occasion, to be handed over to justice and stoned. It also says that young men should pay due respect to every elder, because God is the eldest of all beings.

QUESTIONS FOR ANALYSIS

1. The Roman laws presented here stretch over more than a millennium. What changes over time do you see in laws regarding marriage, divorce, adultery, and guardianship for women?
2. How would you compare Roman and Jewish laws regarding marriage and adultery? What role do the wishes of the father and the spouses play in marriage in these two systems? How would you compare the expectations for parent-child relations in these two systems?
3. What type of actions make a woman dishonorable in the opinions of the Roman jurists in the *Corpus Juris Civilis?* How does being dishonorable shape a woman's chances of marriage?
4. What is the definition of adultery in Roman and Jewish law? In Roman law, who can punish adultery and what are the limitations on their actions?

IDEALIZED WIVES AND MOTHERS

Laws are very clearly prescriptive sources; that is, they indicate how women and men are supposed to act and suggest what the consequences of not acting in the prescribed manner will be. The Romans left a large number of sources that seem at first glance to be descriptive—relating the actions of real individuals— but that actually provide models for the way people were supposed to live. Romans loved to tell stories with a moral and recounted their history and the life stories of individuals as stories with a moral. Sometimes this was implicit, but in many works it was explicit, with the author addressing his readers directly with comments about what they should learn from the life he has retold. (We discuss Roman authors as "he" because, other than a few love poems by the author Sulpicia, preserved in the work of a male poet, and a few fragments of religious poetry by several women, no works by Roman women have survived.)

Roman biographies and descriptions of men generally center on their public roles as thinkers, statesmen, and generals, while those of women focus on their roles as wives and mothers. For some women, those roles were private ones, and they were praised for qualities that were highly gender-specific, such as chastity. For others, particularly women of the upper classes, being an ideal wife and mother might require public bravery, intellectual achievements, or firm leadership. The lives of such women were judged according to the same standards as those of men, and they were regarded as possessing great *virtue,* a word that means "manliness" in Latin.

27. Descriptions of Cornelia, Mother of the Gracchi

Tiberius and Caius Gracchus were two brothers who sought to bring about political and economic reforms during the period of the Punic Wars in the mid-second century B.C.E. The wars led to a great concentration of wealth in the hands of the already wealthy, and the Gracchi—as the brothers are called—attempted to have laws passed that would redistribute some of the land to smaller landholders who had been dispossessed. Both Gracchi were killed in riots, and their reforms were revoked. The mother of the Gracchi, Cornelia, herself the daughter and wife of generals, raised the boys on her own after her husband's death and was praised by countless Roman authors for centuries afterward as the perfect Roman matron, training sons to do their duty. The first excerpt is the most commonly told story of Cornelia, recounted by many authors, including Valerius Maximus (c. 49 B.C.E.–c. 30 C.E.). Plutarch (c. 46 C.E.–c. 120 C.E.), the author of the second and third excerpts, was the biographer of many famous Greeks and Romans; he opens and closes his works on the Gracchi brothers with stories of Cornelia.

Plutarch's Lives, translated by Bernadette Perrin. Loeb Classical Library, Plutarch, vol. 10. (Cambridge: Harvard University Press, 1914).

Lefkowitz, Mary R. and Maureen B. Fant, eds. *Women's Life in Greece and Rome: A Source Book in Translation.* © 1992. Reprinted by permission from the Johns Hopkins University Press.

A. Valerius Maximus

A Campanian matron who was staying with Cornelia, mother of the Gracchi, was showing off her jewels, the most beautiful of that period. Cornelia managed to prolong the conversation until her children got home from school. Then she said, 'These are *my* jewels.'

B. Plutarch, *Life of Tiberius Gracchus*

. . . Tiberius and Caius . . . were sons of Tiberius Gracchus, who, although he had been censor at Rome, twice consul, and had celebrated two triumphs, derived his more illustrious dignity from his virtue. Therefore, after the death of the Scipio who conquered Hannibal, although Tiberius had not been his friend, but actually at variance with him, he was judged worthy to take Scipio's daughter Cornelia in marriage. We are told, moreover, that he once caught a pair of serpents on his bed, and that the sooth-sayers, after considering the prodigy, forbade him to kill both serpents or to let both go, but to decide the fate of one or the other of them, declaring also that the male serpent, if killed, would bring death to Tiberius, and the female, to Cornelia. Tiberius, accordingly, who loved his wife, and thought that since she was still young and he was older it was more fitting that he should die, killed the male serpent, but let the female go. A short time afterwards, as the story goes, he died, leaving Cornelia with twelve children by him.

Cornelia took charge of the children and of the estate, and showed herself so discreet, so good a mother, and so magnanimous, that Tiberius was thought to have made no bad decision when he elected to die instead of such a woman. For when Ptolemy[1] the king offered to

share his crown with her and sought her hand in marriage, she refused him, and remained a widow. In this state she lost most of her children, but three survived; one daughter, who married Scipio the Younger, and two sons, Tiberius and Caius, whose lives I now write. These sons Cornelia reared with such scrupulous care that although confessedly no other Romans were so well endowed by nature, they were thought to owe their virtues more to education than to nature.

C. Plutarch, *Life of Caius Gracchus*

. . . Cornelia is reported to have borne all her misfortunes in a noble and magnanimous spirit, and to have said of the sacred places where her sons had been slain that they were tombs worthy of the dead which occupied them. She resided on the promontory called Misenum, and made no change in her customary way of living. She had many friends, and kept a good table that she might show hospitality, for she always had Greeks and other literary men about her, and all the reigning kings interchanged gifts with her. She was indeed very agreeable to her visitors and associates when she discoursed to them about the life and habits of her father Africanus, but most admirable when she spoke of her sons without grief or tears, and narrated their achievements and their fate to all enquirers as if she were speaking of men of the early days of Rome. Some were therefore led to think that old age or the greatness of her sorrows had impaired her mind and made her insensible to her misfortunes, whereas, really, such persons themselves were insensible how much help in the banishment of grief mankind derives from a noble nature and from honourable birth and rearing, as well as of the fact that while Fortune often prevails over virtue when it endeavours to ward off evils, she cannot rob virtue of the power to endure those evils with calm assurance.

---

[1]Ptolemy VI, king of Egypt, 181–146 B.C.E.

28. Funerary Inscriptions and Tomb Sculpture, Second Century B.C.E.–Second Century C.E.

Roman tombstones and funerary inscriptions carved in stone provide us with a great deal of information about such issues as life expectancy, family life and sentiments, and the activities and aspirations of people who were not among the elite, for the inscriptions of quite ordinary people, including slaves, have survived. Like contemporary funeral eulogies, they of course reveal only those things the survivors wished us to know and so present idealized depictions of the deceased. In addition to the words of the inscription, sculptures and reliefs on tombs and grave markers give us an idea of what qualities of the deceased their survivors regarded as particularly admirable. Roman tomb sculpture is one of the few sources from the ancient world that highlights men's roles as husbands, rather than as public figures. Funerary inscriptions for men sometimes include mention of the fact that the deceased was a husband and father, but they pay more attention to his occupation and public accomplishments. Funerary inscriptions for women have a more domestic tone.

A. Here lies Amymone wife of Marcus best and most beautiful, worker in wool, pious, chaste, thrifty, faithful, a stayer-at-home.

B. Sacred to the gods of the dead. To Urbana my sweetest, chastest, and rarest wife. Surely no one more distinguished ever existed. She deserved honour also for this reason, that she lived every day of her life with me with the greatest kindness and the greatest simplicity, both in her conjugal love and the industry typical of her character. I added this so that those who read may understand how much we loved one another. Paternus set this up in honour of his deserving wife.

C. Friend, I have not much to say; stop and read it. This tomb, which is not fair, is for a fair woman. Her parents gave her the name Claudia. She loved her husband in her heart. She bore two sons, one of whom she left on earth, the other beneath it. She was present to talk with, and she walked with grace. She kept the house and worked in wool. That is all. You may go.

D. Of Graxia Alexandria, distinguished for her virtue and fidelity. She nursed her children with her own breasts. Her husband Pudens the emperor's freedman [dedicated this monument] as a reward to her. She lived 24 years, 3 months, 16 days.

E. ... In this tomb lies Amemone, a bar-maid known [beyond the boundaries] of her own country, [on account of whom] many people used to frequent Tibur.[1] [Now the supreme] god has taken [fragile life] from her, and a kindly light receives her spirit [in the aether]. I, ... nus, [put up this inscription] to my holy wife. [It is right that her name] remain forever.

---

[1]*Tibur:* a city near Rome where this inscription was found.

F.

G.

Tomb sculpture from first century C.E. Rome. *Above* (F.), the wife is kissing her husband's hand; *right* (G.), the spouses are clasping right hands in a gesture that demonstrated the legitimacy of their marriage. Above, (*The British Museum.*) Right, (*Nimatallah/Art Resource, NY, S0019788.*)

29. Turia Inscription

Though most funeral and other inscriptions that recount the details of a person's life are very short, a few are quite lengthy. Most of these are, not surprisingly, about the deeds of famous men. One of the few long inscriptions that has survived about a woman dates from the first century B.C.E. Neither she nor her husband—the speaker in the inscription—has been identified with certainty, though she is traditionally called "Turia" after a woman praised by Valerius Maximus in *Memorable Deeds and Sayings.*

(*heading*) . . . of my wife.

Left-hand column (line 1) . . . through the honesty of your character . . .

(2) . . . you remained . . .

(3) You became an orphan suddenly before the day of our wedding, when both your parents were murdered together in the solitude of the countryside. It was mainly due to your efforts that the death of your parents was not left unavenged. For I had left for Macedonia, and your sister's husband Cluvius had gone to the Province of Africa.

(7) So strenuously did you perform your filial duty by your insistent demands and your pursuit of justice that we could not have done more if we had been present. But these merits you had in common with that most virtuous lady your sister.

(10) While you were engaged in these things, having secured the punishment of the guilty, you immediately left your own house in order to guard your modesty and you came to my mother's house, where you awaited my return. (13) Then pressure was brought to bear

Rome, first century BC.

on you and your sister to accept the view that your father's will, by which you and I were heirs, had been invalidated by his having contracted a [fictitious purchase] with his wife. If that was the case, then you together with all your father's property would necessarily come under the guardianship of those who pursued the matter; your sister would be left without any share at all of that inheritance, since she had been transferred to the power of Cluvius. How you reacted to this, with what presence of mind you offered resistance, I know full well, although I was absent.

(18) You defended our common cause by asserting the truth, namely, that the will had not in fact been broken, so that we should both keep the property, instead of your getting all of it alone. It was your firm decision that you would defend your father's written word; you would do this anyhow, you declared, by sharing your inheritance with your sister, if you were unable to uphold the validity of the will. And you maintained that you would not come under the state of legal guardianship, since there was no such right against you in law, for there was no proof that your father belonged to any gens that could by law compel you to do this. For even assuming that your father's will had become void, those who prosecuted had no such right since they did not belong to the same gens.[1]

(25) They gave way before your firm resolution and did not pursue the matter any further. Thus you on your own brought to a successful conclusion the defence you took up of your duty to your father, your devotion to your sister, and your faithfulness towards me.

(27) Marriages as long as ours are rare, marriages that are ended by death and not broken by divorce. For we were fortunate enough to see our marriage last without disharmony for fully 40 years. I wish that our long union had come to its final end through something that had befallen me instead of you; it would have been

more just if I as the older partner had to yield to fate through such an event.

(30) Why should I mention your domestic virtues: your loyalty, obedience, affability, reasonableness, industry in working wool, religion without superstition, sobriety of attire, modesty of appearance? Why dwell on your love for your relatives, your devotion to your family? You have shown the same attention to my mother as you did to your own parents, and have taken care to secure an equally peaceful life for her as you did for your own people, and you have innumerable other merits in common with all married women who care for their good name. It is your very own virtues that I am asserting, and very few women have encountered comparable circumstances to make them endure such sufferings and perform such deeds. Providentially Fate has made such hard tests rare for women.

We have preserved all the property you inherited from your parents under common custody, for you were not concerned to make your own what you had given to me without any restriction. We divided our duties in such a way that I had the guardianship of your property and you had the care of mine. Concerning this side of our relationship I pass over much, in case I should take a share myself in what is properly yours. May it be enough for me to have said this much to indicate how you felt and thought.

(42) Your generosity you have manifested to many friends and particularly to your beloved relatives. On this point someone might mention with praise other women, but the only equal you have had has been your sister. For you brought up your female relations who deserved such kindness in your own houses with us. You also prepared marriage-portions for them so that they could obtain marriages worthy of your family. The dowries you had decided upon Cluvius and I by common accord took upon ourselves to pay, and since we approved of your generosity we did not wish that you should let your own patrimony suffer diminution but substituted our own money

---

[1] *gens:* clan; extended family.

and gave our own estates as dowries. I have mentioned this not from a wish to commend ourselves but to make clear that it was a point of honour for us to execute with our means what you had conceived in a spirit of generous family affection.

(52) A number of other benefits of yours I have preferred to mention . . . (*several lines missing*)

Right-hand column (2a) You provided abundantly for my needs during my flight and gave me the means for a dignified manner of living, when you took all the gold and jewellery from your own body and sent it to me and over and over again enriched me in my absence with servants, money and provisions, showing great ingenuity in deceiving the guards posted by our adversaries.

(6a) You begged for my life when I was abroad—it was your courage that urged you to this step—and because of your entreaties I was shielded by the clemency of those against whom you marshalled you words. But whatever you said was always said with undaunted courage.

(9a) Meanwhile when a troop of men collected by Milo, whose house I had acquired through purchase when he was in exile, tried to profit by the opportunities provided by the civil war and break into our house to plunder, you beat them back successfully and were able to defend our home. (*About 12 lines missing*)

(0) . . . exist . . . that I was brought back to my country by him (Caesar Augustus), for if you had not, by taking care for my safety, provided what he could save, he would have promised his support in vain. Thus I owe my life no less to your devotion than to Caesar.

(4) Why should I now hold up to view our intimate and secret plans and private conversations: how I was saved by your good advice when I was roused by startling reports to meet sudden and imminent dangers; how you did not allow me imprudently to tempt providence by an overbold step but prepared a safe hiding-place for me, when I had given up my ambitious designs, choosing as partners in your

plans to save me your sister and her husband Cluvius, all of you taking the same risk? There would be no end, if I tried to go into all this. It is enough for me and for you that I was hidden and my life was saved.

(11) But I must say that the bitterest thing that happened to me in my life befell me though what happened to you. When thanks to the kindness and judgment of the absent Caesar Augustus I had been restored to my country as a citizen, Marcus Lepidus, his colleague, who was present, was confronted with your request concerning my recall, and you lay prostrate at his feet, and you were not only not raised up but were dragged away and carried off brutally like a slave. But although your body was full of bruises, your spirit was unbroken and you kept reminding him of Caesar's edict with its expression of pleasure at my reinstatement, and although you had to listen to insulting words and suffer cruel wounds, you pronounced the words of the edict in a loud voice, so that it should be known who was the cause of my deadly perils. This matter was soon to prove harmful for him.

(19) What could have been more effective than the virtue you displayed? You managed to give Caesar an opportunity to display his clemency and not only to preserve my life but also to brand Lepidus' insolent cruelty by your admirable endurance.

(22) But why go on? Let me cut my speech short. My words should and can be brief, lest by dwelling on your great deeds I treat them unworthily. In gratitude for your great services towards me let me display before the eyes of all men my public acknowledgment that you saved my life.

(25) When peace had been restored throughout the world and the lawful political order re-established, we began to enjoy quiet and happy times. It is true that we did wish to have children, who had for a long time been denied to us by an envious fate. If it had pleased Fortune to continue to be favourable to us as she was wont to be, what would have been lacking for either of us? But Fortune took a different course, and

our hopes were sinking. The courses you considered and the steps you attempted to take because of this would perhaps be remarkable and praiseworthy in some other women, but in you they are nothing to wonder at when compared to your other great qualities and I will not go into them.

(31) When you despaired of your ability to bear children and grieved over my childlessness, you became anxious lest by retaining you in marriage I might lose all hope of having children and be distressed for that reason. So you proposed a divorce outright and offered to yield our house free to another woman's fertility. Your intention was in fact that you yourself, relying on our well-known conformity of sentiment, would search out and provide for me a wife who was worthy and suitable for me, and you declared that you would regard future children as joint and as though your own, and that you would not effect a separation of our property which had hitherto been held in common, but that it would still be under my control and, if I wished so, under your administration: nothing would be kept apart by you, nothing separate, and you would thereafter take upon yourself the duties and the loyalty of a sister and a mother-in-law.

(40) I must admit that I flared up so that I almost lost control of myself; so horrified was I by what you tried to do that I found it difficult to retrieve my composure. To think that separation should be considered between us before fate had so ordained, to think that you had been able to conceive in your mind the idea that you might cease to be my wife while I was still alive, although you had been utterly faithful to me when I was exiled and practically dead!

(44) What desire, what need to have children could I have had that was so great that I should have broken faith for that reason and changed certainty for uncertainty? But no more about this! You remained with me as my wife, for I could not have given in to you without disgrace for me and unhappiness for both of us.

(48) But on your part, what could have been more worthy of commemoration and praise

than your efforts in devotion to my interests: when I could not have children from yourself, you wanted me to have them through your good offices, and since you despaired of bearing children, to provide me with offspring by my marriage to another woman.

(51) Would that the life-span of each of us had allowed our marriage to continue until I, as the older partner, had been borne to the grave—that would have been more just—and you had performed for me the last rites, and that I had died leaving you still alive and that I had had you as a daughter to myself in place of my childlessness.

(54) Fate decreed that you should precede me. You bequeathed me sorrow through my longing for you and left me a miserable man without children to comfort me. I on my part will, however, bend my way of thinking and feeling to your judgments and be guided by your admonitions.

(56) But all your opinions and instructions should give precedence to the praise you have won so that this praise will be a consolation for me and I will not feel too much the loss of what I have consecrated to immortality to be remembered for ever.

(58) What you have achieved in your life will not be lost to me. The thought of your fame gives me strength of mind and from your actions I draw instruction so that I shall be able to resist Fortune. Fortune did not rob me of everything since it permitted your memory to be glorified by praise. But along with you I have lost the tranquillity of my existence. When I recall how you used to foresee and ward off the dangers that threatened me, I break down under my calamity and cannot hold steadfastly by my promise.

(63) Natural sorrow wrests away my power of self-control and I am overwhelmed by sorrow. I am tormented by two emotions: grief and fear—and I do not stand firm against either. When I go back in thought to my previous misfortunes and when I envisage what the future may have in store for me, fixing my eyes on your glory does not give me strength to bear

my sorrow with patience. Rather I seem to be destined to long mourning.

(67) The conclusions of my speech will be that you deserved everything but that it did not fall to my lot to give you everything as I ought; your last wishes I have regarded as law; what-

ever it will be in my power to do in addition, I shall do.

(69) I pray that your Di Manes[2] will grant you rest and protection.

———

[2]*Di Manes:* Souls of the dead.

QUESTIONS FOR ANALYSIS

1. What qualities mark the ideal wife and mother in Roman eyes; that is, what words are used to describe her? How does she relate to her parents, her husband, and her children?
2. A woman's duties as wife or mother sometime require her to become involved in actions beyond the household. How do the husband in the Turia inscription and Plutarch in *Life of Tiberius Gracchus* and *Life of Caius Gracchus* describe their female subjects' more public activities? How do these authors make these activities fit with their depictions of idealized wives and mothers?
3. The wife in the Turia inscription was apparently unable to have children. What do her actions regarding this situation indicate about the importance of children for Romans? How do they fit with the stories about Cornelia?
4. What do the portrayals of spouses on tombstones suggest about Roman marriages? How do these reinforce or conflict with the verbal portraits?

CHRISTIANITY

In some of its teachings on gender, Christianity broke with Roman traditions. The words and actions of Jesus of Nazareth were very favorable toward women, and a number of women are mentioned as his followers in the books of the New Testament, the key religious text for Christians. Women emerged as leaders in some early Christian groups, regarded by their followers as having gained special spiritual insight directly from God. Some women enthusiastically avoided marriage, embracing the ideal of virginity and either singly or in communities declaring themselves "virgins in the service of Christ."

Not all Christian teachings about gender were radical, however. Many of Jesus' early followers, particularly the Apostle Paul whose letters make up a major part of the New Testament, had ambivalent ideas about women's proper role in the church and began in the first century C.E. to place restrictions on female believers. Women were gradually excluded from holding official positions within Christianity. As we have seen, both Jewish and classical Mediterranean culture viewed female subordination as natural and proper, so that in limiting the activities of female believers Christianity was following well-established patterns, in the same way that it later patterned its official hierarchy after that of the Roman Empire.

Christian rejection of marriage and family could also have negative consequences for women. Many early Christian writers—often called the "Church Fathers"—began to disparage marriage and sexuality not because they were preparing for the Second Coming, but because they were influenced by a strain of Greek philosophical thought usually termed *dualism,* which made sharp distinctions between the spiritual and the material, the soul and the body. In dualistic thinking, the purpose of life was to allow the soul to escape the domination of the body. Sexual activity was bad both because it demonstrated the power of the body and because it created children, thereby imprisoning more souls in bodies. Some Christian writers, such as Clement of Alexandria, opposed dualistic thinking, while others, such as Augustine (354–430 C.E.), accepted it and clearly regarded virginity as the preferred state of existence. The writings of male authors—which form the vast majority of Christian writings—often view women as an obstacle to the spiritual fulfillment offered by virginity and thus have a strong streak of misogyny.

30. New Testament, Letters of Paul and Ascribed to Paul (1 Corinthians, Galatians, and 1 Timothy)

Paul was a Jewish convert to Christianity who was well educated in Greek and Roman ways of thinking. Though he never met Jesus, he became the most important missionary in the early church, transforming its mission from one directed primarily at Jews to one that included Gentiles (non-Jews). Paul traveled and wrote extensively, and many of his letters to various Christian groups throughout the Roman Empire survived. These letters later became part of Christian Scripture in the New Testament; his reputation was so great that works probably written by others were also attributed to him. The Epistles of 1 and 2 Timothy and Titus (the "Pastorals") are now considered by almost all biblical scholars not to be Paul's words, though they were considered Pauline for most of Christian history. The majority of modern scholars view Ephesians, Colossians, and 2 Thessalonians also as deutero-Pauline—that is, as written by someone other than Paul.

A. 1 Corinthians 7
Now concerning the matters about which you wrote. It is well for a man not to touch a woman. ² But because of the temptation to immorality, each man should have his own wife and each woman her own husband. ³The husband should give to his wife her conjugal rights, and likewise the wife to her husband. ⁴For the wife does not rule over her own body, but the husband does; likewise the husband does not rule over his own body, but the wife does. ⁵Do not refuse one another except perhaps by agreement for a season, that you may devote yourselves to prayer; but then come together again, lest Satan tempt you through lack of self-control. ⁶I say this by way of concession, not of command. ⁷I wish that all were as I myself am. But each has his own special gift from God, one of one kind and one of another.

⁸To the unmarried and the widows I say that it is well for them to remain single as I do. ⁹But if they cannot exercise self-control, they should marry. For it is better to marry than to be aflame with passion.

¹⁰To the married I give charge, not I but the Lord, that the wife should not separate from her

husband [11](but if she does, let her remain single or else be reconciled to her husband)—and that the husband should not divorce his wife.

[12]To the rest I say, not the Lord, that if any brother has a wife who is an unbeliever, and she consents to live with him, he should not divorce her. [13]If any woman has a husband who is an unbeliever, and he consents to live with her, she should not divorce him. [14]For the unbelieving husband is consecrated through his wife, and the unbelieving wife is consecrated through her husband. Otherwise, your children would be unclean, but as it is they are holy. [15]But if the unbelieving partner desires to separate, let it be so; in such a case the brother or sister is not bound. For God has called us to peace. [16]Wife, how do you know whether you will save your husband? Husband, how do you know whether you will save your wife?

[17]Only, let every one lead the life which the Lord has assigned to him, and in which God has called him. This is my rule in all the churches. [18]Was any one at the time of his call already circumcised? Let him not seek to remove the marks of circumcision. Was any one at the time of his call uncircumcised? Let him not seek circumcision. [19]For neither circumcision counts for anything nor uncircumcision, but keeping the commandments of God. [20]Every one should remain in the state in which he was called. [21]Were you a slave when called? Never mind. But if you can gain your freedom, avail yourself of the opportunity. [22]For he who was called in the Lord as a slave is a freedman of the Lord. Likewise he who was free when called is a slave of Christ. [23]You were bought with a price; do not become slaves of men. [24]So, brethren, in whatever state each was called, there let him remain with God.

[25]Now concerning the unmarried, I have no command of the Lord, but I give my opinion as one who by the Lord's mercy is trustworthy. [26]I think that in view of the present distress it is well for a person to remain as he is. [27]Are you bound to a wife? Do not seek to be free. Are you free from a wife? Do not seek marriage. [28]But if you marry, you do not sin, and if a girl marries she does not sin. Yet those who marry will have worldly troubles, and I would spare you that. [29]I mean, brethren, the appointed time has grown very short; from now on, let those who have wives live as though they had none, [30]and those who mourn as though they were not mourning, and those who rejoice as though they were not rejoicing, and those who buy as though they had no goods, [31]and those who deal with the world as though they had no dealings with it. For the form of this world is passing away.

[32]I want you to be free from anxieties. The unmarried man is anxious about the affairs of the Lord, how to please the Lord; [33]but the married man is anxious about worldly affairs, how to please his wife, [34]and his interests are divided. And the unmarried woman or girl is anxious about the affairs of the Lord, how to be holy in body and spirit; but the married woman is anxious about worldly affairs, how to please her husband. [35]I say this for your own benefit, not to lay any restraint upon you, but to promote good order and to secure your undivided devotion to the Lord.

[36]If any one thinks that he is not behaving properly toward his betrothed if his passions are strong, and it has to be, let him do as he wishes: let them marry—it is no sin. [37]But whoever is firmly established in his heart, being under no necessity but having his desire under control, and has determined this in his heart, to keep her as his betrothed, he will do well. [38]So that he who marries his betrothed does well; and he who refrains from marriage will do better.

[39]A wife is bound to her husband as long as he lives. If the husband dies, she is free to be married to whom she wishes, only in the Lord. [40]But in my judgment she is happier if she remains as she is. And I think that I have the Spirit of God.

B. 1 Corinthians 11

Be imitators of me, as I am of Christ.

[2]I commend you because you remember me in everything and maintain the traditions even as I have delivered them to you. [3]But I want you to understand that the head of every man is Christ, the head of a woman is her husband, and the head of Christ is God. [4]Any man who

prays or prophesies with his head covered dishonors his head, [5]but any woman who prays or prophesies with her head unveiled dishonors her head—it is the same as if her head were shaven. [6]For if a woman will not veil herself, then she should cut off her hair; but if it is disgraceful for a woman to be shorn or shaven, let her wear a veil. [7]For a man ought not to cover his head, since he is the image and glory of God; but woman is the glory of man. [8](For man was not made from woman, but woman from man. [9]Neither was man created for woman, but woman for man.) [10]That is why a woman ought to have a veil on her head, because of the angels. [11](Nevertheless, in the Lord woman is not independent of man nor man of woman; [12]for as woman was made from man, so man is now born of woman. And all things are from God.) [13]Judge for yourselves; is it proper for a woman to pray to God with her head uncovered? [14]Does not nature itself teach you that for a man to wear long hair is degrading to him, [15]but if a woman has long hair, it is her pride? For her hair is given to her for a covering. [16]If any one is disposed to be contentious, we recognize no other practice, nor do the churches of God.

C. Galatians 4

[27]For as many of you as were baptized into Christ have put on Christ. [28]There is neither Jew nor Greek, there is neither slave nor free, there is neither male nor female; for you are all one in Christ Jesus.

D. 1 Timothy 2

[8]I desire then that in every place the men should pray, lifting holy hands without anger or quarreling; [9]also that women should adorn themselves modestly and sensibly in seemly apparel, not with braided hair or gold or pearls or costly attire [10]but by good deeds, as befits women who profess religion. [11]Let a woman learn in silence with all submissiveness. [12]I permit no woman to teach or to have authority over men; she is to keep silent. [13]For Adam was formed first, then Eve; [14]and Adam was not deceived, but the woman was deceived and became a transgressor. [15]Yet woman will be saved through bearing children, if she continues in faith and love and holiness, with modesty.

31. Gospel of Mary Magadalene

Along with the letters of various missionaries and accounts of Jesus' life circulating among early Christians that eventually became part of the New Testament, there were other writings that told slightly different versions of events and that were not included. Some of these come from a movement termed *Christian gnosticism,* whose adherents regarded the public teachings of Jesus—told in the gospels which became part of the New Testament—as only a part of the story. In their eyes, Jesus also had imparted secret, special knowledge to a small group of his followers, which would enable those who truly grasped this secret message to overcome evil and understand the world. (There were also gnostics who were not Christian, for whom the secret knowledge was communicated in some other way or by some other person.) Gnosticism was a diffuse body of complex beliefs, and gnostics often disagreed with one another. Some gnostics were dualists who rejected sex and marriage—and thus were often very misogynist—while other gnostics appear to have been more open to female leadership.

Until recently, scholars had only fragments of certain gnostic texts and discussions of gnostic ideas in the writings of authors who were hostile to them, but this changed in 1945, when Egyptian peasants near the town of Nag Hammadi accidentally found a jar filled with gnostic books. These were copies made, probably in 350–400 C.E., of texts first written down as much as three centuries earlier; they were gradually transcribed and translated. They turned out to be fifty-two texts, some of them labeled *gospels,* a Greek word meaning "good news" and given to texts that describe the life of Jesus. The following is from the Gospel of Mary Magadalene, a woman who also appears in the New Testament gospels as a follower of Jesus (see, for example, Mark 15 and 16) and about whom large numbers of legends have grown up.

When the blessed one[1] had said this, he greeted them all, saying, "Peace be with you. Receive my peace to yourselves. Beware that no one lead you astray, saying, 'Lo here!' or 'Lo there!' For the Son of Man is within you. Follow after him! Those who seek him will find him. Go then and preach the gospel of the kingdom. Do not lay down any rules beyond what I appointed for you, and do not give a law like the lawgiver lest you be constrained by it." When he had said this, he departed.

But they were grieved. They wept greatly, saying, "How shall we go to the gentiles and preach the gospel of the kingdom of the Son of Man? If they did not spare him, how will they spare us?" Then Mary stood up, greeted them all, and said to her brethren, "Do not weep and do not grieve nor be irresolute, for his grace will be entirely with you and will protect you. But rather let us praise his greatness, for he has prepared us and made us into men." When Mary said this, she turned their hearts to the Good, and they began to discuss the words of the [Savior].

Peter said to Mary, "Sister, we know that the Savior loved you more than the rest of women. Tell us the words of the Savior which you remember—which you know (but) we do not, nor have we heard them." Mary answered and said, "What is hidden from you I will proclaim to you." And she began to speak to them these words: "I," she said, "I saw the Lord in a vision and I said to him, 'Lord, I saw you today in a vision.' He answered and said to me, 'Blessed are you, that you did not waver at the sight of me. For where the mind is, there is the treasure.' I said to him, 'Lord, now does he who sees the vision see it (through) the soul (or) through the spirit?' The Savior answered and said, 'He does not see through the soul nor through the spirit, but the mind which [is] between the two—that is [what] sees the vision and it is [. . .].'

[The next several pages are lost, and then Mary goes on to describe the ascent of the soul.] "[. . .] it. And desire that, 'I did not see you descending, but now I see you ascending. Why do you lie, since you belong to me?' The soul answered and said, 'I saw you. You did not see me nor recognize me. I served you as a garment, and you did not know me.' When it had said this, it went away rejoicing greatly.

"Again it came to the third power, which is called ignorance. [It (the power)] questioned the soul saying, 'Where are you going? In wickedness are you bound. But you are bound; do not judge!' And the soul said, 'why do you judge me although I have not judged? I was bound though I have not bound. I was not recognized. But I have recognized that the All is being dissolved, both the earthly (things) and the heavenly.'

When the soul had overcome the third power, it went upwards and saw the fourth power, (which) took seven forms. The first form is darkness, the second desire, the third ignorance, the fourth is the excitement of death, the

---

[1]*The blessed one:* Jesus.

fifth is the kingdom of the flesh, the sixth is the foolish wisdom of flesh, the seventh is the wrathful wisdom. These are the seven [powers] of wrath. They ask the soul, 'Whence do you come, slayer of men, or where are you going, conqueror of space?' The soul answered and said, 'What binds me has been slain, and what surrounds me has been overcome, and my desire has been ended, and ignorance has died. In a [world] I was released from a world, [and] in a type from a heavenly type, and (from) the fetter of oblivion which is transient. From this time on will I attain to the rest of the time, of the season, of the aeon, in silence.'"

When Mary had said this, she fell silent, since it was to this point that the Savior had spoken with her. But Andrew[1] answered and said to the brethren, "Say what you (wish to) say about what she has said. I at least do not believe that the Savior said this. For certainly these teachings are strange ideas." Peter answered and spoke concerning these same things. He questioned them about the Savior: "Did he really speak with a woman without our knowledge (and) not openly? Are we to turn about and all listen to her? Did he prefer her to us?"

Then Mary wept and said to Peter,[1] "My brother Peter, what do you think? Do you think that I thought this up myself in my heart, or that I am lying about the Savior?" Levi[2] answered and said to Peter, "Peter, you have always been hot-tempered. Now I see you contending against the woman like the adversaries. But if the Savior made her worthy, who are you indeed to reject her? Surely the Savior knows her very well. That is why he loved her more than us. Rather let us be ashamed and put on the perfect man and acquire him for ourselves as he commanded us, and preach the gospel, not laying down any other rule or other law beyond what the Savior said." When [. . .] and they began to go forth [to] proclaim and to preach.

---

[1]Andrew and Peter were two of Jesus' disciples.

[2]Levi was another follower of Jesus.

32. Martyrdom of Perpetua

Beginning in the middle of the first century, some Roman authorities reacted to the spread of Christianity by ordering Christians to be arrested and, if they would not recant their beliefs, to be executed, often in gruesome ways. There are many accounts of these martyrdoms, most of which are wholly or partly fiction, for the stories of martyrs became important teaching tools for missionaries and others spreading Christianity. The *Martyrdom of Perpetua* is accepted by most scholars as more authentic than most accounts; it is unusual in that it is partly a firsthand account written by Vibia Perpetua, a Roman woman martyred in Carthage during the persecutions of the Emperor Septimus Serverus in 202–203. The whole text consists of an introduction written by someone else, Perpetua's personal narrative, the narrative of a man who was martyred with her, and a closing section describing the martyrdom itself; linguistic scholars have demonstrated that each of these was probably written by a different person. Perpetua's account is thus the first surviving piece of Christian literature by a woman and one of the very few works by a woman from the entire Roman era.

2. Arrested were some young catechumens[1]; Revocatus and Felicitas (both servants), Saturninus, Secundulus, and Vibia Perpetua, a young married woman about twenty years old, of good family and upbringing. She had a father, mother, two brothers (one was a catechumen like herself), and an infant son at the breast. The following account of her martyrdom is her own, a record in her own words of her perceptions of the event.

3. While I was still with the police authorities (she said) my father out of love for me tried to dissuade me from my resolution. "Father," I said, "do you see here, for example, this vase, or pitcher, or whatever it is?" "I see it," he said. "Can it be named anything else than what it really is?" I asked, and he said, "No." "So I also cannot be called anything else than what I am, a Christian." Enraged by my words, my father came at me as though to tear out my eyes. He only annoyed me, but he left, overpowered by his diabolical arguments.

For a few days my father stayed away. I thanked the Lord and felt relieved because of my father's absence. At this time we were baptized and the Spirit instructed me not to request anything from the baptismal waters except endurance of physical suffering.

A few days later we were imprisoned. I was terrified because never before had I experienced such darkness. What a terrible day! Because of crowded conditions and rough treatment by the soldiers the heat was unbearable. My condition was aggravated by my anxiety for my baby. Then Tertius and Pomponius, those kind deacons who were taking care of our needs, paid for us to be moved for a few hours to a better part of the prison where we might refresh ourselves. Leaving the dungeon we all went about our own business. I nursed my child, who was already weak from hunger. In my anxiety for the infant I spoke to my mother about him, tried to console my brother, and asked that they care for

my son. I suffered intensely because I sensed their agony on my account. These were the trials I had to endure for many days. Then I was granted the privilege of having my son remain with me in prison. Being relieved of my anxiety and concern for the infant, I immediately regained my strength. Suddenly the prison became my palace, and I loved being there rather than any other place.

4. Then my brother said to me, "Dear sister, you already have such a great reputation that you could ask for a vision indicating whether you will be condemned or freed." Since I knew that I could speak with the Lord, whose great favors I had already experienced, I confidently promised to do so. I said I would tell my brother about it the next day. Then I made my request and this is what I saw.

There was a bronze ladder of extraordinary height reaching up to heaven, but it was so narrow that only one person could ascend at a time. Every conceivable kind of iron weapon was attached to the sides of the ladder: swords, lances, hooks, and daggers. If anyone climbed up carelessly or without looking upwards, he/she would be mangled as the flesh adhered to the weapons. Crouching directly beneath the ladder was a monstrous dragon who threatened those climbing up and tried to frighten them from ascent.

Saturus went up first. Because of his concern for us he had given himself up voluntarily after we had been arrested. He had been our source of strength but was not with us at the time of the arrest. When he reached the top of the ladder he turned to me and said, "Perpetua, I'm waiting for you, but be careful not to be bitten by the dragon." I told him that in the name of Jesus Christ the dragon could not harm me. At this the dragon slowly lowered its head as though afraid of me. Using its head as the first step, I began my ascent."

At the summit I saw an immense garden, in the center of which sat a tall, grey-haired man dressed like a shepherd, milking sheep. Standing around him were several thousand white-robed people. As he raised his head he noticed

---

[1]*Catechumen:* a person receiving instruction in the Christian faith before baptism.

me and said, "Welcome, my child." Then he beckoned me to approach and gave me a small morsel of the cheese he was making. I accepted it with cupped hands and ate it. When all those surrounding us said "Amen," I awoke, still tasting the sweet cheese. I immediately told my brother about the vision, and we both realized that we were to experience the sufferings of martyrdom. From then on we gave up having any hope in this world.

5. A few days later there was a rumor that our case was to be heard. My father, completely exhausted from his anxiety, came from the city to see me, with the intention of weakening my faith. "Daughter", he said, "have pity on my grey head. Have pity on your father if I have the honor to be called father by you, if with these hands I have brought you to the prime of your life, and if I have always favored you above your brothers, do not abandon me to the reproach of men. Consider your brothers; consider your mother and your aunt; consider your son who cannot live without you. Give up your stubbornness before you destroy all of us. None of us will be able to speak freely if anything happens to you."

These were the things my father said out of love, kissing my hands and throwing himself at my feet. With tears he called me not daughter, but woman. I was very upset because of my father's condition. He was the only member of my family who would find no reason for joy in my suffering. I tried to comfort him saying, "Whatever God wants at this tribunal will happen, for remember that our power comes not from ourselves but from God." But utterly dejected, my father left me.

6. One day as we were eating we were suddenly rushed off for a hearing. We arrived at the forum and the news spread quickly throughout the area near the forum, and a huge crowd gathered. We went up to the prisoners' platform. All the others confessed when they were questioned. When my turn came my father appeared with my son. Dragging me from the step, he begged: "Have pity on your son!"

Hilarion, the governor, who assumed power after the death of the proconsul Minucius Timinianus, said, "Have pity on your father's grey head; have pity on your infant son; offer sacrifice for the emperors' welfare". But I answered, "I will not." Hilarion asked, "Are you a Christian?" And I answered, "I am a Christian." And when my father persisted in his attempts to dissuade me, Hilarion ordered him thrown out, and he was beaten with a rod. My father's injury hurt me as much as if I myself had been beaten, and I grieved because of his pathetic old age. Then the sentence was passed; all of us were condemned to the beasts. We were overjoyed as we went back to the prison cell. Since I was still nursing my child who was ordinarily in the cell with me, I quickly sent the deacon Pomponius to my father's house to ask for the baby, but my father refused to give him up. Then God saw to it that my child no longer needed my nursing, nor were my breasts inflamed. After that I was no longer tortured by anxiety about my child or by pain in my breasts.

7. A few days later while all of us were praying, in the middle of a prayer I suddenly called out the name "Dinocrates." I was astonished since I hadn't thought about him till then. When I recalled what had happened to him I was very disturbed and decided right then that I had not only the right, but the obligation, to pray for him. So I began to pray repeatedly and to make moaning sounds to the Lord in his behalf. During that same night I had this vision: I saw Dinocrates walking away from one of many very dark places. He seemed very hot and thirsty, his face grimy and colorless. The wound on his face was just as it had been when he died. This Dinocrates was my blood-brother who at the age of seven died very tragically from a cancerous disease which so disfigured his face that his death was repulsive to everyone. It was for him that I now prayed. But neither of us could reach the other because of the great distance between. In the place where Dinocrates stood was a pool filled with water, and the rim of the pool was so high that it extended far

above the boy's height. Dinocrates stood on his toes as if to drink the water but in spite of the fact that the pool was full, he could not drink because the rim was so high!

I realized that my brother was in trouble, but I was confident that I could help him with his problem. I prayed for him every day until we were transferred to the arena prison where we were to fight wild animals on the birthday of Geta Caesar. And I prayed day and night for him, moaning and weeping so that my petition would be granted.

8. On the day that we were kept in chains, I had the following vision: I saw the same place as before, but Dinocrates was clean, well-dressed, looking refreshed. In place of the wound there was a scar, and the fountain which I had seen previously now had its rim lowered to the boy's waist. On the rim, over which water was flowing constantly, there was a golden bowl filled with water. Dinocrates walked up to it and began to drink; the bowl never emptied. And when he was no longer thirsty, he gladly went to play as children do. Then I awoke, knowing that he had been relieved of his suffering.

9. A few days passed. Pudens, the official in charge of the prison (the official who had gradually come to admire us for our persistence), admitted many prisoners to our cell so that we might mutually encourage each other. As the day of the games drew near, my father, overwhelmed with grief, came again to see me. He began to pluck out his beard and throw it on the ground. Falling on his face before me, he cursed his old age, repeating such things as would move all creation. And I grieved because of his old age.

10. The day before the battle in the arena, in a vision I saw Pomponius the deacon coming to the prison door and knocking very loudly. I went to open the gate for him. He was dressed in a loosely fitting white robe, wearing richly decorated sandals. He said to me, "Perpetua, come. We're waiting for you!" He took my hand and we began to walk over extremely rocky and winding paths. When we finally arrived short of breath, at the arena, he led me to

the center saying, "Don't be frightened! I'll be here to help you." He left me and I stared out over a huge crowd which watched me with apprehension. Because I knew that I had to fight with the beasts, I wondered why they hadn't yet been turned loose in the arena. Coming towards me was some type of Egyptian, horrible to look at, accompanied by fighters who were to help defeat me. Some handsome young men came forward to help and encourage me. I was stripped of my clothing, and suddenly I was a man. My assistants began to rub me with oil as was the custom before a contest, while the Egyptian was on the opposite side rolling in the sand. Then a certain man appeared, so tall that he towered above the amphitheatre. He wore a loose purple robe with two parallel stripes across the chest; his sandals were richly decorated with gold and silver. He carried a rod like that of an athletic trainer, and a green branch on which were golden apples. He motioned for silence and said, "If this Egyptian wins, he will kill her with the sword; but if she wins, she will receive this branch." Then he withdrew.

We both stepped forward and began to fight with our fists. My opponent kept trying to grab my feet but I repeatedly kicked his face with my heels. I felt myself being lifted up into the air and began to strike at him as one who was no longer earth-bound. But when I saw that we were wasting time, I put my two hands together, linked my fingers, and put his head between them. As he fell on his face I stepped on his head. Then the people began to shout and my assistants started singing victory songs. I walked up to the trainer and accepted the branch. He kissed me and said, "Peace be with you, my daughter." And I triumphantly headed towards the Sanavivarian Gate.[2] Then I woke up realizing that I would be contending not with wild animals but with the devil himself. I knew, however, that I would win. I have recorded the events which occurred up to the day before the final contest. Let anyone who

---

[2]*Sanavivarian Gate:* or Gate of Life, the gate through which the victors left the arena.

wishes to record the events of the contest itself, do so." . . .

15. As for Felicitas, she too was touched by God's grace in the following manner. She was pregnant when arrested, and was now in her eighth month. As the day of the contest approached she became very distressed that her martyrdom might be delayed, since the law forbade the execution of a pregnant woman. Then she would later have to shed her holy and innocent blood among common criminals. Her friends in martyrdom were equally sad at the thought of abandoning such a good friend to travel alone on the same road to hope.

And so, two days before the contest, united in grief they prayed to the Lord. Immediately after the prayers her labor pains began. Because of the additional pain natural for an eight-month delivery, she suffered greatly during the birth, and one of the prison guards taunted her; "If you're complaining now, what will you do when you'll be thrown to the wild beasts? You didn't think of them when you refused to sacrifice." She answered, "Now it is I who suffer, but then another shall be in me to bear the pain for me, since I am now suffering for him." And she gave birth to a girl whom one of her sisters reared as her own daughter.

16. Since the Holy Spirit has permitted, and by permitting has willed, that the events of the contest be recorded, we have no choice but to carry out the injunction (rather, the sacred trust) of Perpetua, in spite of the fact that it will be an inferior addition to the magnificent events already described. We are adding an instance of Perpetua's perseverance and lively spirit. At one time the prisoners were being treated with unusual severity by the commanding officer because certain deceitful men had intimated to him that the prisoners might escape by some magic spells. Perpetua openly challenged him; "Why don't you at least allow us to freshen up, the most noble of the condemned, since we belong to Caesar and are about to fight on his birthday? Or isn't it to your credit that we should appear in good condition on that day?" The officer grimaced and

blushed, then ordered that they be treated more humanely and that her brothers and others be allowed to visit and dine with them. By this time the prison warden was himself a believer.

17. On the day before the public games, as they were eating the last meal commonly called the free meal, they tried as much as possible to make it instead an *agape*.[3] In the same spirit they were exhorting the people, warning them to remember the judgment of God, asking them to be witnesses to the prisoners' joy in suffering, and ridiculing the curiosity of the crowd. Saturus[4] told them, "Won't tomorrow's view be enough for you? Why are you so eager to see something you hate? Friends today, enemies tomorrow! Take a good look so you'll recognize us on that day." Then they all left the prison amazed, and many of them began to believe.

18. The day of their victory dawned, and with joyful countenances they marched from the prison to the arena as though on their way to heaven. If there was any trembling it was from joy, not fear. Perpetua followed with quick step as a true spouse of Christ, the darling of God, her brightly flashing eyes quelling the gaze of the crowd. Felicitas too, joyful because she had safely survived child-birth and was now able to participate in the contest with the wild animals, passed from one shedding of blood to another; from midwife to gladiator, about to be purified after child-birth by a second baptism. As they were led through the gate they were ordered to put on different clothes; the men, those priests of Saturn, the women, those of the priestesses of Ceres. But that noble woman stubbornly resisted even to the end. She said, "We've come this far voluntarily in order to protect our rights, and we've pledged our lives not to recapitulate on any such matter as this. We made this agreement with you." Injustice bowed to justice and the guard conceded

---

[3]*Agape:* or "love-meal," a meal shared by early Christian communities in honor of the love of Jesus Christ.
[4]*Saturus:* probably the instructor of the catechumens listed at the beginning.

that they could enter the arena in their ordinary dress. Perpetua was singing victory psalms as if already crushing the head of the Egyptian. Revocatus, Saturninus and Saturus were warning the spectators, and as they came within sight of Hilarion they informed him by nods and gestures: "You condemn us; God condemns you." This so infuriated the crowds that they demanded the scourging of these men in front of the line of gladiators. But the ones so punished rejoiced in that they had obtained yet another share in the Lord's suffering.

19. Whoever said, "Ask and you shall receive," granted to these petitioners the particular death that each one chose. For whenever the martyrs were discussing among themselves their choice of death, Saturus used to say that he wished to be thrown in with all the animals so that he might wear a more glorious crown. Accordingly, at the outset of the show he was matched against a leopard but then called back; then he was mauled by a bear on the exhibition platform. Now Saturus detested nothing as much as a bear and he had already decided to die by one bite from the leopard. Consequently, when he was tied to a wild boar the professional gladiator who had tied the two together was pierced instead and died shortly after the games ended, while Saturus was merely dragged about. And when he was tied up on the bridge in front of the bear, the bear refused to come out of his den; and so a second time Saturus was called back unharmed.

20. For the young women the devil had readied a mad cow, an animal not usually used at these games, but selected so that the women's sex would be matched with that of the animal. After being stripped and enmeshed in nets, the women were led into the arena. How horrified the people were as they saw that one was a young girl and the other, her breasts dripping with milk, had just recently given birth to a child. Consequently both were recalled and dressed in loosely fitting gowns.

Perpetua was tossed first and fell on her back. She sat up, and being more concerned with her sense of modesty than with her pain,

covered her thighs with her gown which had been torn down one side. Then finding her hair-clip which had fallen out, she pinned back her loose hair thinking it not proper for a martyr to suffer with dishevelled hair; it might seem that she was mourning in her hour of triumph. Then she stood up. Noticing that Felicitas was badly bruised, she went to her, reached out her hands and helped her to her feet. As they stood there the cruelty of the crowds seemed to be appeased and they were sent to the Sanavivarian Gate. There Perpetua was taken care of by a certain catechumen, Rusticus, who stayed near her. She seemed to be waking from a deep sleep (so completely had she been entranced and imbued with the Spirit). She began to look around her and to everyone's astonishment asked, "When are we going to be led out to that cow, or whatever it is." She would not believe that it had already happened until she saw the various markings of the tossing on her body and clothing. Then calling for her brother she said to him and to the catechumen, "Remain strong in your faith and love one another. Do not let our excruciating sufferings become a stumbling block for you."

21. Meanwhile, at another gate Saturus was similarly encouraging the soldier, Pudens. "Up to the present," he said, "I've not been harmed by any of the animals, just as I've foretold and predicted. So that you will now believe completely, watch as I go back to die from a single leopard bite." And so at the end of that contest, Saturus was bitten once by the leopard that had been set loose, and bled so profusely from that one wound that as he was coming back the crowd shouted in witness to his second baptism: "Salvation by being cleansed; Salvation by being cleansed;"[5] And that man was truly saved who was cleansed in this way.

Then Saturus said to Pudens the soldier, "Goodbye, and remember my faith. Let these happenings be a source of strength for you,

---

[5]This statement was a customary greeting Romans used when going to the public baths. The crowd here is using it ironically.

rather than a cause for anxiety." Then asking Pudens for a ring from his finger, he dipped it into the wound and returned it to Pudens as a legacy, a pledge and remembrance of his death. And as he collapsed he was thrown with the rest to that place reserved for the usual throat-slitting. And when the crowd demanded that the prisoners be brought out into the open so that they might feast their eyes on death by the sword, they voluntarily arose and moved where the crowd wanted them. Before doing so they kissed each other so that their martyrdom would be completely perfected by the rite of the kiss of peace.

The others, without making any movement or sound, were killed by the sword. Saturus in particular, since he had been the first to climb the ladder and was to be Perpetua's encouragement, was the first to die. But Perpetua, in order to feel some of the pain, groaning as she was struck between the ribs, took the gladiator's trembling hand guided it to her throat. Perhaps it was that so great a woman, feared as she was by the unclean spirit, could not have been slain had she not herself willed it.

O brave and fortunate martyrs, truly called and chosen to give honor to our Lord Jesus Christ! And anyone who is elaborating upon, or who reverences or worships that honor, should read these more recent examples, along with the ancient, as sources of encouragement for the Christian community. In this way, there will be new examples of courage witnessing to the fact that even in our day the same Holy Spirit is still efficaciously present, along with the all powerful God the Father and Jesus Christ our Lord, to whom there will always be glory and endless power. Amen.

33. Clement of Alexandria
Stromateis

Clement of Alexandria (died c. 215) was a Greek theologian and writer who was active in Alexandria. He was the first to attempt to bring together Christian and Platonic teachings, a task that would occupy many later thinkers as well. Only a few of his texts survive, including *Stromateis* ("Miscellanies"), in which he attacked those who disparaged marriage.

Marriage is a union between a man and a woman; it is the primary union; it is a legal transaction; it exists for the procreation of legitimate children. The comic dramatist Menander says,

I give you my own daughter
for the sowing of true children.

We ask the question whether it is right to marry. This is one of those questions named after their relation to an end. Who is to marry? In what situation? With what woman? What about her situation? It is not right for everyone to marry; it is not right at all times. There is a time when it is appropriate; there is a person for whom it is appropriate; there is an age up to which it is appropriate. It is not right for every man to marry every woman, on every occasion, in absolutely all circumstances. It depends on the circumstances of the man, the character of the woman, the right time, the prospect of children, the total compatibility of the woman and the absence of any violence or compulsion to drive her to look after the man who loves her. . . .

So there is every reason to marry—for patriotic reasons, for the succession of children, for the fulfillment of the universe (insofar as it is our business). The poets regret a marriage which is "half-fulfilled" and childless, and bless the marriage which is "abundant in growth." Physical ailments demonstrate the necessity of marriage particularly well. A wife's care and her patient attention seem to surpass all the earnest devotion of other family and friends; she likes to excel all others in sympathy and present concern; she really and truly is, in the words of Scripture, a necessary "helper."

Menander, the writer of comic drama, runs down marriage, but he also sets its advantages in the other scale in answering the man who says,

(S) I am not well disposed to the matter.
(B) Yes, you take it in left-handed style.

He goes on thus:

You are seeing in it the problems, the things
you take exception to. You are not concentrat-
    ing on the blessings.

And so on. Marriage helps those of advanced years too. It provides a wife to take care of them, and nurtures children from her to be supports of old age. Children

Are the glory of the dead.
Like corks, they support the net
and preserve the flaxen cord from the depths,

as the tragic dramatist Sophocles puts it. The legislators do not permit the highest offices of state to pass to unmarried men. The legislator of the Spartans[1] went further than laying a penalty on the unmarried state; he extended it to im-

proper marriages, late marriages, and celibacy. Our good Plato enunciated that an unmarried man should hand over the equivalent of a wife's upkeep into the public treasury, paying the appropriate costs to the authorities. If they do not marry or produce children, they will be playing their part in reducing the population and undermining the cities and the world they compose.

Such behavior is irreverent. It undermines generation, which is a gift of God. It is a sign of weakness and unmanliness to try to escape from a partnership in life with wife and children. A state which it is wrong to reject must be totally right to procure. So with all the rest. In fact, they say the loss of children is one of the gravest evils. It follows that the acquisition of children is a good thing. If so, so is marriage. The poet says,

Without a father there could be no children,
without a mother, not even conception of
    children.

Marriage makes a man a father, a husband makes a woman a mother. . . .

Marriage must be kept pure, like a sacred object to be preserved from all stain. We are with our Lord when we wake from sleep; we go to sleep with gratitude, praying

alike when you go to bed, and when the holy
    light appears.

We witness to our Lord throughout our lives, with reverence held within our soul and applying self-control to our body as well. God really does approve of seemliness escorted from tongue to action. Coarse language is the road to a loss of a sense of shame, and shameful behavior results from the combination. Scripture advises us to marry. It tells us never to divorce the conjugal yoke. The law says expressly, "You shall not divorce your wife except by reason of unchastity," [Matthew 5:32] and regards remarriage while the other of the divorced pair is alive as adultery.

---

[1]Lycurgus.

A wife is clearly exempt from all suspicion if she refrains from prettifying or adorning herself beyond the proper limit, if she applies herself consistently to prayers and intercessions, if she avoids going out all the time, if she shuts herself off as far as possible from seeing those she ought not to be gazing at, if she puts the care of the home as a more valuable pursuit than gossiping. Scripture says, "The man who marries a divorcee is committing adultery," for "if a man divorces his wife, he is committing adultery against her"; in other words, he is forcing her into adultery [Mark 10:11–12]. Guilt in this does not attach merely to the man who divorces her; it attaches to the man who takes her on, since he provides the starting point for the woman's sin. If he were not to accept her, she would return to her husband. . . .

What about those who use religious language for irreligious practices involving abstinence against creation and the holy creator, the one and only almighty God, and teach that we ought not to accept marriage and childbearing or introduce yet more wretches in their turn into the world to provide fodder for death? This is what we must say to them; first, in the words of the apostle John: "Now many antichrists have come, from which we know that it is the last hour. They went out from us, but were not of our company: if they had been, they would have stayed with us." [1 John 2:18–19] Next we must turn their statements on the grounds that they destroy the sense of their citations. Here is an example: When Salome asked, "How long will death maintain its power?" the Lord said, "As long as you women bear children." He is not speaking of life as evil and the creation as rotten. He is giving instruction about the normal course of nature. Death is always following on the heels of birth.

The design of the Law is to divert us from extravagance and all forms of disorderly behavior; this is its object, to draw us from unrighteousness to righteousness, making us responsible in marriage, engendering children, and living well. The Lord "comes to fulfill, not to destroy the Law." [Matthew 5:17] Fulfillment does not mean that it was defective. The prophecies which followed the Law were accomplished through his presence, since the qualities of an upright way of life were announced to people of righteous behavior before the coming of the Law by the Word. The majority know nothing of self-discipline. They live by the body, not by the spirit. Without the spirit the body is earth and dust. The Lord condemns adultery in thought. Well? Is it not possible to practice self-discipline within marriage without trying to pull apart "that which God has joined"? That is the sort of thing taught by the dissolvers of the marriage bond. Through them the name of Christian comes into bad repute. These people say that sexual intercourse is polluted. Yet they owe their existence to sexual intercourse! Must they not be polluted? Personally, I think that the seed coming from consecrated people is sacred too.

So it is not just our spirit which ought to be consecrated. It is our character, our life, our body. What is the sense of the Apostle Paul's words that the wife is consecrated by her husband, and the husband by his wife? What was it that the Lord said to those who questioned him about divorce, asking whether it was permissible to get rid of one's wife on the authority of Moses? He said, "Moses wrote this with an eye to your hardheartedness. But have you not read what God said to the first-formed male: 'You two shall come into one single flesh'? So, anyone who disposes of his wife except by reason of sexual immorality is making an adulteress of her." [Matthew 19:3–9] . . . It follows that celibacy is not particularly praiseworthy unless it arises through love of God.

34. Jerome
Against Jovinian

Jerome (347–420?) was a Christian scholar and writer best known as the author of the official translation of the Bible into Latin, called the Vulgate. Jerome was a difficult and quarrelsome individual who was often involved in scholarly controversies; he spent his later life in Jerusalem because of conflicts with various leaders in Rome. In both Rome and Jerusalem he was financially supported by several wealthy Roman women who left their families to pursue lives of Christian devotion. These women, usually widows or unmarried young women, chose not to marry (or not to remarry if they were widows) and developed ideas about the value of virginity, which they shared with their male associates, like Jerome. Unfortunately, they did not leave writings of their own; consequently, the most influential advocates of virginity for subsequent Christians were male authors. This piece was written by Jerome in answer to another Christian author, Jovinian, who had suggested that marriage and virginity were of equal spiritual worth; it was frequently quoted by later authors throughout the Middle Ages and Renaissance.

The battle[1] must be fought with the whole army of the enemy, and the disorderly rabble, fighting more like brigands than soldiers, must be repulsed by the skill and method of regular warfare. In the front rank I will set the Apostle Paul, and, since he is the bravest of generals, will arm him with his own weapons, that is to say, his own statements. . . .

[Jerome then repeats Pauls' statements from 1 Corinthians 7.] Let us turn back to the chief point of the evidence: "It is good," he says, "for a man not to touch a woman." If it is good not to touch a woman, it is bad to touch one: for there is no opposite to goodness but badness. But if it be bad and the evil is pardoned, the reason for the concession is to prevent worse evil. But surely a thing which is only allowed because there may be something worse has only a slight degree of goodness. He would never have added "let each man have his own wife," unless he had previously

used the words "but, because of fornications." Do away with fornication, and he will not say "let each man have his own wife." Just as though one were to lay it down: "It is good to feed on wheaten bread, and to eat the finest wheat flour," and yet to prevent a person pressed by hunger from devouring cow-dung, I may allow him to eat barley. Does it follow that the wheat will not have its peculiar purity, because such an one prefers barley to excrement? That is naturally good which does not admit of comparison with what is bad, and is not eclipsed because something else is preferred. At the same time we must notice the Apostle's prudence. He did not say, it is good not to have a wife: but, it is good not to touch a woman: as though there were danger even in the touch: as though he who touched her, would not escape from her who "hunteth for the precious life," who causeth the young man's understanding to fly away. "Can a man take fire in his bosom, and his clothes not be burned? Or can one walk upon hot coals, and his feet not be scorched?" As then he who touches fire is instantly burned, so by the mere touch the peculiar

---

[1]Jerome is referring here to the battle against Jovinian and those who agree with him about the equal spiritual worth of marriage and virginity.

nature of man and woman is perceived, and the difference of sex is understood. . . .

[Jovian argued that Solomon accomplished a great deal though he was married, and Jerome answers.] . . . when our opponent adduced Solomon, who, although he had many wives, nevertheless built the temple, I briefly replied that it was my intention to run over the remaining points. Now that he may not cry out that both Solomon and others under the law, prophets and holy men, have been dishonoured by us, let us show what this very man[2] with his many wives and concubines thought of marriage. For no one can know better than he who suffered through them, what a wife or woman is. Well then, he says in the Proverbs: "The foolish and bold woman comes to want bread." What bread? Surely that bread which cometh down from heaven: and he immediately adds "The earth-born perish in her house, rush into the depths of hell." Who are the earth-born that perish in her house? They of course who follow the first Adam, who is of the earth, and not the second, who is from heaven. And again in another place: "Like a worm in wood, so a wicked woman destroyeth her husband." But if you assert that this was spoken of bad wives, I shall briefly answer: What necessity rests upon me to run the risk of the wife I marry proving good or bad? "It is better," he says, "to dwell in a desert land, than with a contentious and passionate woman in a wide house." How seldom we find a wife without these faults, he knows who is married.

[Jerome also argued against second marriages, citing the authority of the Greek philosopher Theophrastus, who lived in the fourth century B.C.E. and whose treatise on marriage is unknown except in Jerome's citing of it.]

. . . But what am I to do when the women of our time press me with apostolic authority, and before the first husband is buried, repeat from morning to night the precepts which allow a second marriage? Seeing they despise the fidelity which Christian purity dictates, let them at least learn chastity from the heathen.

A book *On Marriage,* worth its weight in gold, passes under the name of Theophrastus. In it the author asks whether a wise man marries. And after laying down the conditions—that the wife must be fair, of good character, and honest parentage, the husband in good health and of ample means, and after saying that under these circumstances a wise man sometimes enters the state of matrimony, he immediately proceeds thus: "But all these conditions are seldom satisfied in marriage. A wise man therefore must not take a wife. For in the first place his study of philosophy will be hindered, and it is impossible for anyone to attend to his books and his wife. Matrons want many things, costly dresses, gold, jewels, great outlay, maid-servants, all kinds of furniture, litters and gilded coaches. Then come curtain-lectures the live-long night: she complains that one lady goes out better dressed than she: that another is looked up to by all: 'I am a poor despised nobody at the ladies' assemblies.' 'Why did you ogle that creature next door?' 'Why were you talking to the maid?' 'What did you bring from the market?' 'I am not allowed to have a single friend, or companion.' She suspects that her husband's love goes the same way as her hate. There may be in some neighbouring city the wisest of teachers; but if we have a wife we can neither leave her behind, nor take the burden with us. To support a poor wife, is hard: to put up with a rich one, is torture. Notice, too, that in the case of a wife you cannot pick and choose: you must take her as you find her. If she has a bad temper, or is a fool, if she has a blemish, or is proud, or has bad breath, whatever her fault may be—all this we learn after marriage. Horses, asses, cattle, even slaves of the smallest worth, clothes, kettles, wooden seats, cups, and earthenware pitchers, are first tried and then bought: a wife is the only thing that is not shown before she is married, for fear she may not give satisfaction. Our gaze must always be directed to her face, and we must always praise her beauty: if you look at another woman, she thinks that she is out of favour. She must be called my lady, her birth-day must be

---

[2] I.e., Solomon.

kept, we must swear by her health and wish that she may survive us, respect must be paid to the nurse, to the nurse-maid, to the father's slave, to the foster-child, to the handsome hanger-on, to the curled darling who manages her affairs, and to the eunuch who ministers to the safe indulgence of her lust: names which are only a cloak for adultery. Upon whomsoever she sets her heart, they must have her love though they want her not. If you give her the management of the whole house, you must yourself be her slave. If you reserve something for yourself, she will not think you are loyal to her; but she will turn to strife and hatred, and unless you quickly take care, she will have the poison ready. If you introduce old women, and sooth-sayers, and prophets, and vendors of jewels and silken clothing, you imperil her chastity; if you shut the door upon them, she is injured and fancies you suspect her. But what is the good of even a careful guardian, when an unchaste wife cannot be watched, and a chaste one ought not to be? For necessity is but a faithless keeper of chastity, and she alone really deserves to be called pure, who is free to sin if she chooses. If a woman be fair, she soon finds lovers; if she be ugly, it is easy to be wanton. It is difficult to guard what many long for. It is annoying to have what no one thinks worth possessing. But the misery of having an ugly wife is less than that of watching a comely one. Nothing is safe, for which a whole people sighs and longs. One man entices with his figure, another with his brains, another with his wit, another with his open hand. Somehow, or sometime, the fortress is captured which is attacked on all sides. Men marry, indeed, so as to get a manager for the house, to solace weariness, to banish solitude; but a faithful slave is a far better manager, more submissive to the master, more observant of his ways, than a wife who thinks she proves herself mistress if she acts in opposition to her husband, that is, if she does what pleases her, not what she is commanded. But friends, and servants who are under the obligation of benefits received, are better able to wait upon us in sickness than a wife who makes us responsible for her tears (she will sell you enough to make a deluge for the hope of a legacy), boasts of her anxiety, but drives her sick husband to the distraction of despair. But if she herself is poorly, we must fall sick with her and never leave her bedside. Or if she be a good and agreeable wife (how rare a bird she is!), we have to share her groans in childbirth, and suffer torture when she is in danger. A wise man can never be alone. He has with him the good men of all time, and turns his mind freely wherever he chooses. What is inaccessible to him in person he can embrace in thought. And, if men are scarce, he converses with God. He is never less alone than when alone.

35. Sayings of the Ascetic Desert Mother Sarah

Some of the women and men who opted for a life of ascetic virginity chose to go to the Egyptian desert, where they could escape the temptations of urban life and get away from the increasing wealth of the Church itself. Many of their sayings and anecdotes about their lives became part of Christian tradition, though it is difficult to assess their authenticity. Most of the sayings are attributed to men—the desert fathers—though a few are attributed to women. The following are attributed to a woman living in the fourth or fifth century, about whom nothing else is known.

From *Sayings of the Desert Fathers* by Benedicta Ward. Copyright © 1975 by A.R. Mowbray.

1. It was related of Amma[1] Sarah that for thirteen years she waged warfare against the demon of fornication. She never prayed that the warfare should cease but she said, "O God, give me strength."

2. Once the same spirit of fornication attacked her more intently, reminding her of the vanities of the world. But she gave herself up to the fear of God and to asceticism and went up onto her little terrace to pray. Then the spirit of fornication appeared corporally to her and said, "Sarah, you have overcome me." But she said, "It is not I who have overcome you, but my master, Christ."

3. It was said concerning her that for sixty years she lived beside a river and never lifted her eyes to look at it.

4. Another time, two old men, great anchorites,[2] came to the district of Pelusia to visit her. When they arrived one said to the other, "Let us humiliate this old woman." So they said to her, "Be careful not to become conceited thinking of yourself: 'Look how anchorites are coming to see me, a mere woman.'" But Amma Sarah said to them, "According to nature I am a woman, but not according to my thoughts."

5. Amma Sarah said, "If I prayed God that all men should approve of my conduct, I should find myself a penitent at the door of each one, but I shall rather pray that my heart may be pure towards all."

6. She also said, "I put out my foot to ascend the ladder, and I place death before my eyes before going up it."

7. She also said, "It is good to give alms for men's sake. Even if it is only done to please men, through it one can begin to seek to please God."

8. Some monks of Scetis came one day to visit Amma Sarah. She offered them a small basket of fruit. They left the good fruit and ate the bad. So she said to them, "You are true monks of Scetis."

9. She also said to the brothers, "It is I who am a man, you who are women."

---

[1] *Amma:* Mother.
[2] *Anchorite:* hermit who lived in a small cell.

QUESTIONS FOR ANALYSIS

1. How would you compare the opinions about women and marriage expressed by Paul in 1 Corinthians with the opinions in the two books now judged to be by someone other than Paul (1 Timothy, Ephesians)?
2. Clement of Alexandria and Jerome have widely different opinions of the value of marriage. What arguments do they make for and against it. How much do their arguments consider women's perspective on the issue?
3. In Perpetua's vision, she becomes a man, and Mother Sarah describes herself also as a man. How do these gender transformations fit with Roman and Christian ideas about female and male natures?
4. How would you compare the ideas about the importance and function of marriage and family life in the writings of the Roman and Jewish authors with those of the Christian ones? Are some Christian authors more similar to the Roman and Jewish authors than others? Why do you think this might be?
5. How would you compare pagan and Christian ideas about the ideal woman? In what ways did earlier Roman ideas influence Christian ideals?

CHAPTER 4

The Early and High Middle Ages

(Fifth–Fourteenth Centuries)

The disintegration of the Roman Empire in western Europe in the fifth century traditionally marks the end of the classical period and the beginning of the Middle Ages. Dividing European history into three periods—ancient, medieval, and modern—was begun by thinkers of the Italian Renaissance in the fourteenth and fifteenth centuries, who saw themselves as the first "modern" people. They admired the ancient Romans and Greeks and thought less of those who had lived in the intervening thousand years since the end of the Roman Empire; in their eyes, this long period was simply a "middle" one, a trough between two high points.

Recently, some historians have begun to question this three-part periodization, noting that both the fifth and the fifteenth centuries were ones of continuity along with change. They point out that for most people in Europe, the gradual decline of the Roman Empire and the establishment of smaller Germanic kingdoms made little difference. Trade and town life certainly declined from the levels they had achieved in the Roman Empire, which led to a decrease in cultural and intellectual achievements, as these were largely urban phenomena. Most people had not been town dwellers in touch with culture and education, however, and they continued—as they had for centuries—to live in small villages and support themselves by farming. They kept on paying taxes to their social superiors, though these were now local nobles instead of officials sent from Rome. Christianity gradually spread throughout Europe, continuing to develop institutions and ideas that gave it economic and political power as well as spiritual authority.

Historians of women have been prominent among those stressing continuities between the ancient and medieval worlds, noting that the valuations ascribed to certain historical periods—as "golden" or "dark" ages—are generally drawn from male experience. Cultures that may appear "golden" for elite men, such as classical Athens or Renaissance Florence, may not have been very positive for women. Conversely, those that have been labeled "dark" actually may have been times when women had greater opportunities.

The period usually called the Early Middle ages (450–1050) is one of the times that may have been more favorable for women. Until recently, this period was often called the "Dark Ages" even in textbooks, as historians continued to accept Renaissance notions about its lack of education, culture, literature, and art. As we have learned more about women's experiences during this period, however, it is clear that this pejorative label does not tell the whole story. Though European societies continued to restrict women's rights and legal status as they had in the classical period, the decentralization of government in the Early Middle Ages allowed some queens and noble women to rule huge territories in their sons' or husbands'—and occasionally their own—names. Most Christian authorities taught that women were morally and spiritually inferior, but they also encouraged women to convert and then act as domestic missionaries converting their husbands and families. Elite women established religious communities where women, and occasionally men, devoted themselves to prayer and other religious activities, living apart from the world (though supported by donations from their families and pious laypeople and the work of the peasants who lived on land owned by the community). Women continued to have fewer rights to property than men, but they were not as disenfranchised as they had been in classical Athens; widows and heiresses often controlled their own property and even owned and controlled churches and other ecclesiastical property.

At a lower social level, with the slow decline of slavery and the Church's encouragement of marriage, the basic unit of agricultural production became the peasant household, with households grouped in villages—what are often termed *manors*—under the control of an upper-class landlord or the Church. These households were centered on a married couple, and women's productive and reproductive activities were essential. Women's worth as child-bearers was recognized in laws, derived from earlier Germanic customs, which punished the murder or injury of a woman of child-bearing years more harshly than that of a man of similar age and social status. Women's role as workers is indicated in records from landlords listing the labor and tax obligations of women as well as men. Their contributions did not give women of any social group a status equal to that of their male relatives, but the leadership and creative energies shown by women during this period have led some historians to view it as a sort of "golden age" for women.

This positive assessment of women's situation in the Early Middle ages comes not only by comparing it to the earlier classical period, but also by comparing it to the later High Middle Ages (1050–1300). During the eleventh century, agricultural improvements and warmer weather created a slight increase in the food supply, which caused a slow increase in population and freed more people to work at things other than agriculture. This led to the growth of cities and an increase in the amount and complexity of trade and commerce, with wealth joining inherited social status as an important means of differentiating people. This process is often described as the "rise of the middle class," in which merchants and to some degree artisans became more important economically, politically, and socially. New types of educational institutions, such as universities, were established in these cities, where scholars attempted to fuse classical and Christian teachings, or, more broadly stated, reason and faith, in a movement called "scholasticism."

Women as well as men moved into the growing cities, where the basic unit of production was the household, just as it was in rural areas. Some of them became involved in trade and in crafts such as weaving, leatherwork, brewing, and baking. In all of these occupations, however, women tended to be clustered in the lower-paying, lower-status end of the job spectrum. They were rarely long-distance merchants, for family responsibilities and ideas about what was respectable for women made it difficult for them to travel, and they rarely had access to as much capital as male merchants. They worked in craft production but were only occasionally allowed to join craft guilds or open shops on their own, instead working in the shops of their fathers or husbands. Women were not allowed to enroll in the new universities, which meant they could not become physicians or lawyers because university training became a requirement in these professions. Reforms within the Church restricted women's opportunities to control church property, and efforts to enforce clerical celibacy on male priests disrupted priestly families and encouraged the more open expression of misogynist sentiment. Ideas about gender—that is, about the proper place and role of women and men—became more rigid and polarized.

MARRIAGE AND THE HOUSEHOLD ECONOMY

Because marital couples and the households they formed were the basic social and economic units, marriage was regarded as an important matter in the Early and High Middle Ages by families, kin groups, rulers, and church authorities, as well as the individuals entering the marriage. Almost all law codes, secular and church, include articles dealing with the formation and breakup of marriages, how husbands and wives were to relate to one another, and how unrelated men and women were to behave toward one another so that village harmony was maintained. Such laws give us an idea about the theoretical status of women in the minds of male authorities and the legal context within which women and men operated. They do not tell us how, or whether, such laws were enforced, and we have very few records from actual legal proceedings for the period before the Black Death.

Learning about how women and men actually lived and worked together requires looking beyond law codes and pulling small bits of evidence from widely scattered sources. Though most manors left no records during this period, some landlords did keep track of the obligations of their tenants, and a few of these have survived. Paintings and manuscript illuminations may be somewhat idealized, but their artists probably showed a gender division of labor that fit with contemporary experience. For some parts of Europe, legal sources such as court records or coroners' reports occasionally give eyewitness accounts of events that led to tragedy or dispute or that involved criminal conduct. Like court records for more modern times, these provide information both on actual events and on community standards, for the only conduct that shows up in court is that which deviated from the norm.

36. Laws of the Salian Franks

The Early Middle Ages have left relatively few written sources of any type, and especially few regarding women's lives. Originally, even law in Germanic society was oral, though by the fifth century Germanic kings in western Europe supported written codifications of what had been oral customary law. These codes usually bore the name of the tribe—such as the Lombard Law, the Burgundian Law, and the Salian Law (the law of the Salian Franks). Though the initial codifications claimed to be simply the recording of long-standing customs, in reality the laws often modified customs that no longer fit the needs of Germanic peoples as they became more settled and adapted some aspects of the more sophisticated Roman culture. Later kings were also not hesitant to make new laws when situations demanded it and to state explicitly that this is what they were doing. Thus Germanic codes gradually evolved from being records of tribal customs based on moral sanctions and notions of a common tradition into collections of royal statutes based on the political authority of kings. Germanic society was one in which murder, injuries, and insults to honor had resulted in feuds between individuals and families; but by the time the law codes were written down, a system of monetary compensatory payments—usually termed *wergeld*—was being devised as a substitute. These compensatory payments were set according to the severity of the loss or injury and according to the social status of the perpetrator and the victim. The following is a selection from the original Frankish Code, the *Pactus Legis Salicae,* issued by King Clovis in about 510 and emended and revised by many of his successors.

XIII: CONCERNING THE ABDUCTION OF FREEMEN OR FREE WOMEN

1. If three men take a free girl from her house or workroom, the three shall be liable to pay thirty solidi. . . .

7. If it was a servant of the king or a half-freeman who abducted the free girl, he shall make composition with his life [i.e., he shall be given as a slave to the family of the girl].

8. But if the free girl voluntarily followed one of these she shall lose her freedom.

9. The freeman who takes another man's female slave in marriage shall remain in servitude with her.

10. He who associates another man's half-free woman (*litam*) with himself in marriage shall be liable to pay thirty solidi.

11. He who joins to himself in profane marriage the daughter of his sister or brother, or a cousin of further degree [i.e., daughter of a

Excerpted from *Laws of the Salian Franks,* translated by Katherine Fischer Drew. Copyright 1991 by the University of Pennsylvania Press. Reprinted by permission of the publisher.

niece or nephew or of a grandniece or grand-nephew], or the wife of his brother, or of his mother's brother, shall be subjected to this punishment: the couple shall be separated from such a union and, if they had children, these will not be legitimate heirs but will be marked with disgrace.

12. He who takes a woman betrothed to someone else and joins her to himself in marriage shall be liable to pay sixty-two and one-half solidi [to her family or guardian].

13. He shall be liable to pay fifteen solidi to the man to whom she was betrothed.

14. He who attacks on the road a betrothed girl with her bridal party being led to her husband and forcefully has intercourse with her shall be liable to pay two hundred solidi [to her family or guardian or to her betrothed husband?].

XV: CONCERNING HOMICIDE OR THE MAN WHO TAKES ANOTHER MAN'S WIFE WHILE HER HUSBAND STILL LIVES

1. He who kills a freeman or takes another man's wife while her husband lives, and it is proved against him, shall be liable to pay two hundred solidi.

2. He who rapes a free girl and it is proved against him shall be liable to pay sixty-two and one-half solidi.

3. He who secretly has intercourse with a free girl with the consent of both and it is proved against him shall be liable to pay forty-five solidi.

XX: CONCERNING THE MAN WHO TOUCHES THE HAND OR ARM OR FINGER OF A FREE WOMAN

1. The freeman who touches the hand or arm or finger of a free woman or of any other woman, and it is proved against him shall be liable to pay fifteen solidi.

2. If he touches her arm [below the elbow], he shall be liable to pay thirty solidi.

3. But if he places his hand above her elbow and it is proved against him, he shall be liable to pay thirty-five solidi.

4. He who touches a woman's breast or cuts it so that the blood flows shall be liable to pay forty-five solidi.

XXIV: ON KILLING CHILDREN AND WOMEN

1. He who kills a free boy less than twelve years old up to the end of his twelfth year, and it is proved against him, shall be liable to pay six hundred solidi.

2. He who cuts the hair of a long-haired free boy without the consent of his relatives, and it is proved against him, shall be liable to pay forty-five solidi.

3. If he cuts the hair of a free girl without the consent of her relatives, and it is proved against him, he shall be liable to pay forty-five solidi.

4. He who kills a long-haired boy, and it is proved against him, shall be liable to pay six hundred solidi.

5. He who strikes a pregnant free woman, and it is proved against him, shall be liable to pay seven hundred solidi.

6. He who kills an infant in its mother's womb or within nine days of birth before it has a name, and it is proved against him, shall be liable to pay one hundred solidi.

7. If a boy under twelve years old commits some offense, a fine (*fredus*) will not be required of him.

8. He who kills a free woman after she has begun to bear children, if it is proved against him, shall be liable to pay six hundred solidi.

9. He who kills a woman after she is no longer able to bear children, if it is proved against him, shall be liable to pay two hundred solidi.

XXV: ON HAVING INTERCOURSE WITH SLAVE GIRLS OR BOYS

1. The freeman who has intercourse with someone else's slave girl, and it is proved

against him, shall be liable to pay fifteen solidi to the slave girl's lord.

2. The man who has intercourse with a slave girl belonging to the king and it is proved against him, shall be liable to pay thirty solidi.

3. The freeman who publicly joins himself with (i.e., marries) another man's slave girl, shall remain with her in servitude.

4. And likewise the free woman who takes someone else's slave in marriage shall remain in servitude.

5. If a slave has intercourse with the slave girl of another lord and the girl dies as a result of this crime, the slave himself shall pay six solidi to the girl's lord or he shall be castrated; the slave's lord shall pay the value of the girl to the lord.

6. If the slave girl has not died, the slave shall receive three hundred lashes or, to spare his back, he shall pay three solidi to the girl's lord.

7. If a slave joins another man's slave girl to himself in marriage without the consent of her lord, he shall be lashed or clear himself by paying three solidi to the girl's lord.

C: CONCERNING THE WIDOW WHO WANTS TO GO TO ANOTHER HUSBAND

1. If a widow after the death of her husband wants to go to another husband, first he who wishes to receive her must give a betrothal fine (*reipus*) for her according to law. And if the woman had children by her former husband, she ought to consult her children's relatives. If she had received twenty-five solidi as a dos [from her previous husband], she should give three solidi as a fee for release of her mundium (*achasius*) to the closest relatives of the dead husband...and [the remainder of] the dos which the earlier husband had given her will be claimed and defended by his children, after the mother's death, without any share going to him [the second husband]. The mother may not presume to sell or to give away any of this dos. But if the woman does not have children by the former husband and

wishes to enter another marriage with her dos, she must give the *achasius,* as said above. And afterwards let her cover a bench and prepare the bed with coverlet; and with nine witnesses let her summon the relatives of the dead husband and say: "You are my witnesses that I have given the *achasius* in order to have peace with his [my former husband's] relatives, and I leave here the covered bed and the worthy bedspread, the prepared bench and an armchair that I brought with me from my father's house." Then she may give herself to another husband with two parts of her dos.

3. But if she does not do this, she loses the two parts of her dos and in addition she shall be liable to pay sixty-two and one-half solidi to the fisc.

CIV: CONCERNING THE WOMAN WHO HAS HER HAIR CUT OR PULLED

1. If anyone pulls a woman's hair so that her hood (*obbonis*) falls to the ground, he shall be liable to pay fifteen solidi.

2. But if he undoes her headband so that her hair falls to her shoulders, he shall be liable to pay thirty solidi.

3. If a slave strikes a free woman or pulls her hair, he shall lose his hand or pay five solidi.

4. He who strikes a pregnant free woman in the stomach or in the kidney with fist or foot, and she does not lose her fetus but she is weighed down almost to death on account of this, shall be liable to pay two hundred solidi.

5. If the fetus emerges dead but she herself lives, he [who struck her] shall be liable to pay six hundred solidi.

6. But if the woman dies because of this, he shall be liable to pay nine hundred solidi.

7. If the woman who died had been placed under the protection of the king for any reason, he [who struck her] shall be liable to pay twelve hundred solidi.

8. If the child which was aborted was a girl, he shall pay twenty-four hundred solidi.

37. Gratian
Canon Law on Marriage

Gratian (fl. 1140) was an Italian monk and legal scholar who synthesized what were often contradictory decisions and rulings in church law (termed *canon law*) into one comprehensive system in a lengthy treatise titled the *Decretum.* Not much is known about his life, but he was clearly educated in Roman law as well as Christian teachings, for principles that were central to Roman law appear throughout his work. Because church courts, rather than secular courts, had jurisdiction over marriage, canon law was central to medieval understandings of marriage, betrothal, remarriage, adultery, and separation. The *Decretum,* including its rules about marriage, served as the basis for Christian canon law in western Europe from that point on, and for Roman Catholic law until today; the most recent major revision of canon law occurred in 1983.

CASE 27, QUESTION II

1. [According to John Chrysostom:] "Coitus does not make a marriage; consent does; and therefore the separation of the body does not dissolve it, but the separation of the will. Therefore he who forsakes his wife, and does not take another, is still a married man. For even if he is now separated in his body, yet he is still joined in his will. When therefore he takes another woman, then he forsakes fully. Therefore he who forsakes is not the adulterer, but he who takes another woman."

2. ... When therefore there is consent, which alone makes a marriage, between those persons, it is clear that they have been married.

20. A woman who has been sent to a monastery without the consent of her husband is not prohibited from returning to live with him....

21. He who has become a monk without his wife's consent must return to her....

22. A man may not make a monastic vow without his wife's consent....

If any married man wishes to join a monastery, he is not to be accepted, unless he has first been released by his wife, and she makes a vow of chastity. For if she, through incontinence, marries another man while he is still living, without a doubt she will be an adulteress....

23. A husband and a wife may not turn to the religious life without the knowledge of their bishop. According to the synod of Pope Eugenius, "If a husband and wife have agreed between themselves to turn their life to religion, by no means let this be done without the knowledge of their bishop, so that they may be individually and properly examined by him. For if the wife is not willing, or indeed the husband, the marriage is not dissolved, even for such a reason."

24. A husband is not permitted to be celibate without his wife's consent....

26. A wife is not permitted to take a vow of celibacy, unless her husband chooses the same way of life.

27. A girl who is betrothed is not prohibited from choosing the monastery. [According to Pope Eusebius:] "Parents may not give a betrothed girl to another man; but she herself may choose a monastery."

29. If a woman proves that her husband had never known her carnally, there may be a separation....

35. Marriage is initiated in the betrothal....

36. The union of the couple completes the marriage. For according to St. Ambrose, "In all marriage the union is understood to be spiritual, and it is confirmed and completed by the bodily union of the couple."

50. One man may not take [in marriage] a girl betrothed to another man....

51. If a man plights his troth to any woman, he is not permitted to marry another....

CASE 30, QUESTION II

Betrothals may not be contracted before the age of seven. For only the consent is contracted, which cannot happen unless it is understood by each party what is being done between them. Therefore it is shown that betrothals cannot be contracted between children, whose weakness of age does not admit consent. This same is attested by Pope Nicholas: before the time of consent a marriage cannot be contracted. He says, "Where there is no consent from either party, there is no marriage. Therefore those who give girls to boys while they are still in the cradle, and vice versa, achieve nothing, even if the father and mother are willing and do this, unless both of the children consent after they have reached the age of understanding."

CASE 30, QUESTION V

1. Clandestine marriages should not be made....

2. It is not permitted to perform a marriage in secret....

6. No one shall marry a wife without a public ceremony....

CASE 31, QUESTION I

10. So that the peril of fornication may be avoided, a second marriage is allowed....

11. Neither a second nor a third nor successive marriages shall be condemned....

CASE 31, QUESTION II

1. A girl whose own agreement has never been shown is not required by the oath of her father to marry....

3. Those who are to be of one body ought also to be of one spirit, and therefore no woman who is unwilling ought ever to be joined to anyone....

CASE 32, QUESTION I

Many authorities and arguments show that an immoral woman should not be taken to wife. For she who is found guilty of adultery is not supposed to be kept in marital fellowship except after the completion of penance. John Chrysostom said this:

1. He who does not wish to forsake his adulterous wife is a protector of vice. "Just as the man who forsakes a chaste woman is cruel and unjust, so he who keeps an immoral woman is foolish and unfair. For he who conceals the crime of his wife is a protector of vice."

2. A man may forsake his wife because of her fornication, but he may not marry another....

4. Let him who does not wish to forsake his adulterous wife do penance for two years, if she pays her conjugal debt[1] to him....

5. After penance for adultery the man may receive his wife....But it is another thing to marry an immoral woman, or to keep an adulterous wife, whom you dress in your own habits, your chastity and your modesty....

14. It is no sin to marry an immoral woman....

CASE 32, QUESTION II

1. Childbirth is the sole purpose of marriage for women....

3. Immoderate conjugal union is not an evil of marriage, but a venial sin, because of the good of marriage....

7. Those who obtain drugs of sterility are fornicators, not spouses....

CASE 32, QUESTION V

19. A man is not permitted to forsake his wife except because of fornication....

21. Let a man who forsakes his wife for a cause short of fornication be deprived of communion....

23. Adultery in either sex is punished in the same way....

---

[1] Pay the conjugal debt means to have sexual relations with one's spouse.

CASE 32, QUESTION VII

1. The bond of marriage cannot be dissolved by fornication....

2. Once a marriage has been proved to have begun, it cannot be dissolved for any reason....

3. If a man leaves his wife, or a woman her husband, on account of fornication, he or she is not permitted to marry another....

5. Let a man forsaken by his wife, or a woman forsaken by her husband, be brought to penitence, unless they wish either to live continently or be reconciled to each other....

6. He who dares to wed a woman forsaken by her husband is a fornicator....

8. A faithful wife who leaves her adulterous husband may not marry another....

10. He who forsakes his wife on account of fornication and marries another is proved a fornicator....

14. Unnatural acts are more filthy and disgraceful than fornication or adultery....

25. A marriage cannot be dissolved because of a bodily infirmity or wound....

26. A madman or a madwoman cannot contract marriage....

27. No man may forsake an infertile wife and marry another for her fertility....

CASE 33, QUESTION II

5. Marriage is completely forbidden to men who have killed their wives....

6. No man may kill his adulterous wife....

38. Peasants' Manorial Obligations

The economy of the Early Middle Ages was one in which cash played a relatively minor role; most people paid their debts, taxes, and rents either in labor or in kind—that is, in goods. The Church owned much of the land in many parts of Europe, and clerical landholders, who could usually read and write, were more likely than their secular neighbors to keep records of the obligations of their tenants. The following is from a land register done in 893 for the abbey of Prüm in central Germany; it shows some of the activities through which peasant women were able to pay the rents and taxes they owed.

Widrad has possession of a full hide.[1] As swine-tax he must pay a boar worth twenty pennies, a pound of yarn, three hens, eighteen eggs. In addition, annually he must carry out the transport of wine, and must bring five wagonloads of his manure. He gives five skins for tanning, one cord of wood, six feet wide, twelve feet long, for twelve wagonloads. He bakes bread and brews beer. Every household that possesses a hide pays the abbey fifty planks of wood and watches the pigs in the forest for one week, whenever it is their turn. They cultivate three morgen[2] of land three days per week for the whole year. They carry five modius[3] of grain to the abbey from Holler [a town near the abbey] and serve their turn on the watch. At harvest time, every smallholder[4] protects [the grain] from six square rods and [the hay] from three square rods of meadow. He must clean a part of the landlord's garden. His wife must sew leggings.

---

[1] *Hide:* a measure of land, usually between 80 and 120 acres.

[2] *Morgen:* a measure of land, usually 0.6 to 0.9 acre.
[3] *Modius:* a measure of volume.
[4] *Smallholder:* a holder of less than a hide of land.

From *Urkundenbuch zur Geschichte der jetzt die Preussischen Regierungsbezirke Coblenz und Trier bildenden mittelrheinischen Territorien* (Koblenz, 1860), Vol. 1, Nr. 135, edited by H. Beyer et al. Translated by Merry E. Wiesner.

Reimbald has three days fieldwork and a farm. For that he must serve as a messenger for two days of every week, perform compulsory service for one week in every two, and give a hen. The wives of the pensioners must give two days of service every week; one of these days they will receive their meals, and one of these days they will not. In addition each must prepare one-half pound flax or finish one-half piece of shirt material. Aeduin's wife has a farm and land. She does service and finishes one-half piece of shirt material. Nantcher has a farm and a meadow. His wife has a farm and pays four pennies. . . .

Every woman who finishes linen shirt material, harvests the flax, roasts it and prepares it. . . .

Every woman who finished shirt material, shears sheep and washes the wool. . . .

. . . His wife gives ten modius of wine. She gathers one quart of blackberries and one basket of mustard. She picks one field of green onions and one and one half fields of leeks. She must harvest and prepare flax, wash sheep and shear them. At the hay-, grain-, and grape-harvest every man and woman should serve daily with their team [of oxen]. . . . The woman gives five young hens and 25 eggs. . . .

There are five women there, who work one day every week during the harvest. . . . Every one of the hay-renters gives twelve pennies. Women, however, either pay linen shirt material, or twelve pennies, or they must work two days per week at the grain- and hay-harvest and give six pennies. There are also two positions as farmhands; each of them serves with his wife at the hay harvest three days per week. . . .

39. Manuscript Illuminations of Women's Work, ca. 1340

Written records from the Middle Ages tend to focus on religious subjects and the social elite; the same is true of most of the visual records, including sculpture and painting. One type of visual record that does occasionally depict the lives of ordinary people is manuscript illumination, the small paintings that adorned the borders of handwritten books. In this era before the invention of the printing press, people who wanted a specific book had to arrange for an existing examplar to be copied; if the purchaser were quite wealthy, he or she would also pay an illuminator to decorate it. The purchaser might specify the illuminations wanted, or leave this up to the illuminator. The following comes from one of the most famous illuminated manuscripts in the world, the *Luttrell Psalter,* commissioned by the English nobleman Sir Geoffrey Luttrell in early fourteenth-century England. Psalters were personal prayerbooks and varied from very simple to highly ornate. The *Luttrell Psalter* has over 200 pages with elaborate decorations, many of which show rural life on estates like those owned by Sir Geoffrey.

A. A harvesting scene from the Luttrell Psalter. *(By permission of the British Library).*

40. Coroners' Rolls

In the late twelfth century, the office of the coroner was established in England. Each county was to elect several upper-class men to act as coroners who would investigate any accidental or suspicious death. The person who found the body was to call the coroner as quickly as possible; he then held an inquest as to the circumstances of the death and made a written report about the incident. Kings were interested in keeping track of murders and suicides because these were criminal actions; they were interested in accidental deaths because after the twelfth century the item that killed the person was sold, with the proceeds going to the royal treasury. (This item was called the *deodand,* the gift to God, because earlier it had been sold to pay for prayers for the dead person's soul.) These reports were written on parchment and stitched together in long rolls, as were many other records from medieval courts in England. The majority of the coroners' rolls that have survived are from the thirteenth and fourteenth centuries, and they often reveal details about ordinary people's daily lives that are available through no other source. Unfortunately, other countries in Europe did not establish an office of coroner until much later, so that there are no similar sources outside of England for this early period. These extracts are from the county of Bedfordshire and the city of London.

Coroners' Rolls. Published in the Bedfordshire Historical Record Society, Vol. XVI. (Streatley, England, 1961) and in *Calendar of Coroners' Rolls of the City of London, A.D., 1300–1378* by Reginald R. Sharpe, editor. (London: Richard Clay and Sons, 1913).

A. Bedfordshire 13. Soon after nones[1] on 22 July 1267 Emma, Christine de Furnevall's washerwoman, tried to draw water from a leaden vat full of boiling water with a bowl in Cadbury and by misadventure fell into it. Richard the Brewer of Christine's house was present, tried to drag her from the vat, lost his foot-hold and fell in. Gregory de Canmori arrived, saw them lying in the vat, raised the hue and called his servant Richard, who dragged them both out. Emma died about vespers[2] on 24 July, having had the rites of the church.

35. About nones on 2 Oct. 1270 Amice daughter of Robert Belamy of Staploe and Sibyl Bonchevaler were carrying a tub full of grout between them in the brewhouse of Lady Juliana de Beauchamp (*de Bello Campo*) in the hamlet of Staploe in Eaton Socon, intending to empty it into a boiling leaden vat, when Amice slipped and fell into the vat and the tub upon her. Sibyl immediately jumped towards her, dragged her from the vat and shouted; the household came and found her scalded almost to death. A chaplain came and Amice had the rites of the church and died by misadventure about prime[3] the next day.

58. After nones on 24 May 1270 Emma daughter of Richard Toky of Southill went to 'Houleden' in Southill to gather wood. Walter Garglof of Stanford came, carrying a bow and a small sheaf of arrows, took hold of Emma and tried to throw her to the ground and deflower her, but she immediately shouted and her father came. Walter immediately shot an arrow at him, striking him on the right side of the forehead and giving him a mortal wound. He struck him again with another arrow under the right side and so into the stomach. Seman of Southill immediately came and asked him why he wished to kill Richard, and Walter immediately shot an arrow at him, striking him in the back, so that his life

was despaired of. Walter then immediately fled. Later Emma, Richard's wife, came and found her husband wounded to the point of death and shouted. The neighbours came and took him to his house. He had the rites of the church, made his will and died at twilight on the same day.

County Court of 15 Jan. 1274
165. On 1 Jan. Emma of Hatch (*dil Hacche*) came from Beeston, where she had been begging bread from door to door, and towards vespers she returned towards Beeston to seek lodging. She came to a piece of cultivated land called 'Pokebrokforlong' in Northill, was seized by the cold and died by misadventure. Ranulf le Cras came towards Northill, found her dead.

County Court of 14 May 1274
187. About prime on ?19 April Joan, a poor child aged 5, went through Riseley to beg for bread, came to a bridge called 'Fordebrugge' and, as she tried to cross it, fell into the water and drowned by misadventure. Her mother Alice daughter of Bicke first found her drowned.

228. Biggleswade Hundred. About prime on 28 Feb., while William Sagar of Sutton was at the plough, his wife Emma took a bundle of straw inside the court-yard of his house in Sutton, intending to go to heat an oven. She came to a part of the court-yard which was near their dwelling-house and near a well on the north of the house, and by misadventure fell into the well and drowned. Maud daughter of Ellis Batte of Sutton was sitting in William's house guarding Emma's child Rose, who was lying in a cradle, heard the noise made by Emma as she sank, immediately went outside, found Emma drowned and continuously raised the hue and cry.[4] The township came and followed the hue according to the custom of the realm.

---

[1]*Nones:* noon.
[2]*Vespers:* roughly, nightfall.
[3]*Prime:* roughly, dawn.

[4]People who saw a crime or found something suspicious were expected to yell loudly and get the community involved in searching for the perpetrator; this is called "raising the hue and cry."

255. At twilight on 4 Sept. 1300 Nicholas le Swon of Bedford came to his house there, when his wife Isabel was at Robert Asplon's house giving milk to Robert's son, and asked his daughter where her mother was. She said: at Robert Asplon's house; whereupon he immediately went after her because she stayed there too much. As he left his house he met his wife and told her to come home to sleep, saying that he wanted to go to his bed. While Isabel was making his bed, Nicholas drew his sword and struck her in the back so that she immediately died. He immediately fled. His chattels were 3 bushels of corn worth 15d., 2 bushels of oats worth 4d., 8 lbs. Of wool worth 2s., wood worth 4d., 2 pigs worth 3d. and a chest worth 4d., for which Bedford will answer. His daughter Joan, the first finder, raised the hue.

B. London On Wednesday after the Feast of Translation of St. Thomas [3 July] a° 16 Edward II [A.D. 1322], information given to the aforesaid Coroner and Sheriffs that many poor people lay dead of a death other than their rightful death within Ludgate around the gate of the Preaching Friars in the Ward of Farndone. On hearing this, the aforesaid Coroner and Sheriffs proceeded thither, and having summoned good men of that Ward and of the three nearest Wards, viz.: Castle Baynard, Bredstrete and Aldresgate, they diligently enquired how it happened. The jurors[1] say that when at daybreak of that day a great multitude of poor people were assembled at the gate of the Friars Preachers seeking alms. Robert Fynel, Simon, Robert and William his sons and 22 other male persons, names unknown, Matilda, daughter of Robert le Carpenter, Beatrix Cole, Johanna "le Peyntures," Alice la Norice and 22 other women, names unknown, whilst entering the gate were fatally crushed owing to the numbers, and immediately died thereof and of no other felony. They suspect no one of their death

except the misadventure and crushing. As regards who were present or who first saw the corpses, they are unable to say owing to the crowd and it being night-time. The corpses so crushed were viewed on which no other hurt, wound, or bruise appeared.

On Monday before the Feast of St. Michael [29 Sept.] the year aforesaid [A.D. 1322], it happened that a certain Lucy Faukes lay dead of a death other than her rightful death in a certain shop which Richard le Sherman held of John Priour, senior, in the parish of St. Olave in the Ward of Alegate. On hearing this, the aforesaid Coroner and Sheriffs proceeded thither, and having summoned good men of that Ward and of the three nearest Wards, viz.: Portsokne, Tower, and Langebourne, they diligently enquired how it happened. The jurors say that on Sunday before the Feast of St. Matthew [21 Sept.] a° 16 [Edward II., A.D. 1322], about the hour of curfew, the aforesaid Lucy came to the said shop in order to pass the night there with the said Richard le Sherman and Cristina his wife, as she oftentimes was accustomed, and because the said Lucy was clad in good clothes, the said Richard and Cristina began to quarrel with her in order to obtain a reason for killing her for her clothes. At length the said Robert took up a staff called "Balstaf," and with the force and assistance of the said Cristina, struck her on the top of the head, and mortally broke and crushed the whole of her head, so that she forthwith died; that the said Richard and Cristina stript the said Lucy of her aforesaid clothes, and immediately fled, but whither they went or who received them, they (the jurors) know not. Being asked who were present when this happened, they say No one except the said Richard, Cristina and Lucy, nor do they suspect anyone of the death except the said Richard and Cristina. Being asked of the goods and chattels of the said Richard and Cristina, the jurors say they had nothing except what they took away with them. Being asked who found the dead body of the aforesaid Lucy, they say a certain Giles le Portour who raised the cry so that the country came. Precept to the

---

[1]*Jurors:* here, the men who had been summoned to give information about the events.

Sheriffs to attach the said Richard and Cristina when found in their bailiwick.[2]

On Wednesday the eve of All Saints [1 Nov.] aº 18 Edward II. [A.D. 1324], a certain Elena Gubbe lay drowned in the water of the Thames under the wharf of John le White in the parish of St. Martin in the Ward of Vintry. On hearing this, the aforesaid Coroner and Sheriffs proceeded thither, and having summoned good men of that Ward and of the three nearest Wards, viz.: Queenhithe, Cordewanerstret and Douegate, they diligently enquired how it happened. The jurors say that when on Monday after the Feast of St. Luke [18 Oct.] at the hour of curfew the said Elena went to the Thames with two earthenware pitchers (*duobus picher' terre*) for water, and had come to the wharf called "La Lauenderebrigge"[3] and filled her vessels, by accident she fell into the water and was drowned, nobody being present; that she remained under water until the aforesaid Wednesday when Ralph Gubbe, her father, found her submerged and raised the cry so that the country came. The stair of the wharf from which the said Elena fell, valued by the jury at 4*d.,* for which Benedict de Suffolk the Sheriff will answer.[4]

Wednesday after the Feast of Nativ. B.V. [8 Sept.] the same year, [1336] information given to aforesaid Coroner and Sheriffs that Alice, wife of John Ryvet, lay dead of a death other than her rightful death in a shop which he held under John Spray in the parish of St. Botulph without Aldresgate in the Ward of Aldresgate. Thereupon they proceeded thither, and having summoned good men of that Ward and of the three nearest Wards, viz.: Crepulgate, Farndone Within and Castle Baynard, they diligently enquired how it happened. The jurorsy . . . say that on the preceding Monday, the

Feast of the Nativity aforesaid, the said John Ryvet and Alice his wife were alarmed at midnight by a fire which had been caused by the fall of a lighted candle as they were going to sleep, and hurriedly left the burning shop; that the said John, blaming the said Alice for causing the disaster, violently pushed her back into the shop and fled, but whither, they (the jurors) know not; the said Alice was thus injured by the fire, and again leaving the shop lingered until the following Tuesday, when she had her ecclesiastical rights and died of her burns. The corpse viewed, &c. Precept to the Sheriffs to attach the said John when found in their bailiwick and four nearest neighbours.

Monday the morrow of St. Valentine [14 Feb.] the same year, information given to the aforesaid Coroner and Sheriffs that Alice Warde of York lay dead of a death other than her rightful death in the rent of John de Blackwell in the lane called "Faitoreslane" in the parish of St. Andrew de Holbourne in the Ward of Farndone Without. Thereupon, they proceeded thither and having summoned good men of that Ward and of the Ward of Farnedone Within, they diligently enquired how it happened. The jurors . . . say that on the preceding Sunday, at dusk, Geoffrey le Perler, a groom of the mistery of Lormerie came to the rent where the above Alice was living, intending to find Emma de Brakkele, a harlot,[5] and to lie with her, but failing to find her, a quarrel arose between the said Geoffrey and Alice; and that thereupon the said Geoffrey secretly drew his knife called "trenchour," and therewith struck the said Alice on the side under the right arm, inflicting a mortal wound 5 inches deep, so that she immediately died. The felon fled but whither &c., the jurors know not. Chattels none. The corpse viewed, &c. Precept to the Sheriffs, &c.

---

[2]In other words, the sheriffs of nearby areas were instructed to arrest the suspects if they were found.
[3]The wharf had this name as it was where women who laundered clothing and household items came to do their work.
[4]The stair in this case is the *deodand,* sold for 4*d.* to the sheriff, who had to pay the royal treasury this sum.

---

[5]*Harlot:* prostitute.

QUESTIONS FOR ANALYSIS

1. How did the punishments set in the Salian Law for injuring or killing a person differ depending on the sex and social status of the victim? The social status of the perpetrator? How were these punishments shaped by ideas about women's honor?
2. According to Gratian, what were the defining characteristics of marriage? How was extramarital sex on the part of husband or wife to be handled? Is divorce with remarriage possible?
3. What types of work did women perform—and occasionally die while performing—in the medieval economy? How did the gender division of labor act as an encouragement of marriage?
4. How would you compare actual incidents of personal violence in the coroners' reports with the theoretical treatment of violence in the Salian Law? The spousal and family relationships in the coroners' reports with those envisioned in the Salian Law and Gratian's decretals?

EXCEPTIONAL WOMEN

In many ways, all women from the Middle Ages whose names we know are exceptional in that they made it into the historical record. (This might also be said for men of the Middle Ages, of course, though there are many more sources about men than women.) The women whose lives we know the most about were generally exceptional for one of three reasons: they were members of the nobility and thus of elite social status, they gained fame through special religious devotion or leadership in the Church, or they broke laws or regulations in a way that was serious enough to leave a record of the case. Some women actually fit in more than one of these categories. Wealthy noblewomen, for example, often established religious houses for men and women, later serving as the abbess. (As abbesses, they supervised both nuns and male monks who lived in a separate part of the house and did work expected of men; these arrangements have at times been called "double monasteries," but they were not formally set up as such.) Women who gained public reputations through pious acts or religious visions frequently came into conflict with Church authorities who doubted their motives or the authenticity of their mystical experiences. Our sources regarding exceptional women are extremely varied, and comparisons among them must be made very carefully; we must also consider their uniqueness when using information about their lives to make generalizations about women's more widespread status and role.

41. Dhuoda
Handbook for William

Dhuoda was a Frankish noblewoman who in the middle of the ninth century wrote a guide for her adolescent son, then being held as a hostage at the court of the Frankish King Charles the Bald in order to make sure that his father

(Dhuoda's husband) remained loyal to the king. It is the only major text written by a woman to survive from the Carolingian period (roughly 750–900) and provides evidence that at least some elite women were able to read and write Latin. It is full of quotations from the Bible, St. Augustine and other Church fathers, and the leading intellectuals of Dhuoda's own day, most prominently Alcuin of York, the director of Charlemagne's palace school. The introduction and the postscript give us some glimpses of Dhuoda's understanding of herself, and the body of the text instructs her son in those matters she finds fundamental to the proper life, both in terms of beliefs and actions.

Here begins the text. The little book before you branches out in three directions. Read it through and, by the end, you will understand what I mean. I would like it to be called three things at once, as befits its contents—rule, model, and handbook. These terms all mirror each other. The rule comes from me, the model is for you, and the handbook is as much from me as for you—composed by me, received by you....

From the beginning of this book to the end, both in form and in content, in the meter and rhythm of the poetry as well as in the prose passages here—know that everything, through it all, in it all, is intended entirely for you, for the health of your soul and body. I wish that you eagerly take this work in your own hand and fulfill its precepts, after my hand has addressed it to you. I wish you to hold it, turn its pages and read it, so that you may fulfill it in worthy action. For this little model-book, called a handbook, is a lesson from me and a task for you. As someone said, *I have planted, Apollo watered, but God gave the increase.*[1] What further can I say, my son, except that—thinking on your past good deeds—I have in this work *fought the good fight, I have kept the faith, I have finished my course?* And how is what I say of worth unless in him who said, *It is consummated?* For whatever I have accomplished in this volume, from its beginning on, according to the Hebrew speech and to Greek letters and to the Latin language, I have completed in him who is called God.[2]

In the name of Holy Trinity. In the name of Holy Trinity, here begins the handbook of Dhuoda, which she sent to her son William.

I am well aware that most women rejoice that they are with their children in this world, but I, Dhuoda, am far away from you, my son William. For this reason I am anxious and filled with longing to do something for you. So I send you this little work written down in my name,[3] that you may read it for your education, as a kind of mirror. And I rejoice that, even if I am apart from you in body, the little book before you may remind you, when you read it, of what you should do on my behalf....

Here begins the prologue. Things that are obvious to many people often escape me. Those who are like me lack understanding and have dim insight, but I am even less capable than they. Yet always there is he at my side who *opened the mouths of the dumb, and made the tongues of infants eloquent.* I, Dhuoda, despite my weakness of mind, unworthy as I am among worthy women—I am still your mother, my son William, and it is to you that I now address the

---

[1] All texts in italics are biblical quotations.

[2] Dhuoda's reference to Hebrew is unclear. She knew a little Greek, but no Hebrew.

[3] *Written down in my name* refers to the scribe who did the actual writing. This was common practice for members of the elite, even those who could write themselves.

Reprinted from *Handbook for William: A Carolingian Woman's Counsel for Her Son* by Dhuoda, translated and with an introduction by Carol Neel, by permission of the University of Nebraska Press. © 1993 by the University of Nebraska Press.

words of my handbook. From time to time children are fascinated by dice more than all the other games that they enjoy. And sometimes women are absorbed in examining their faces in mirrors, in order then to cover their blemishes and be more beautiful, for the worldly intention of pleasing their husbands. I hope that you may bring the same care, burdened though you may be by the world's pressures, to reading this little book addressed to you by me. For my sake, attend to it—according to my jest—as children do to their dice or women to their mirrors.

Even if you eventually have many more books, read this little work of mine often. May you, with God's help, be able to understand it to your own profit. You will find in it all you may wish to know in compact form. You will find in it a mirror in which you can without hesitation contemplate the health of your soul, so that you may be pleasing not only in this world, but to him who formed you out of dust. What is essential, my son William, is that you show yourself to be such a man on both levels that you are both effective in this world and pleasing to God in every way.

My great concern, my son William, is to offer you helpful words. My burning, watchful heart especially desires that you may have in this little volume what I have longed to be written down for you, about how you were born through God's grace. I shall best begin there.

Preface.   In the eleventh year of the imperial rule of our lord Louis, who then reigned by Christ's favor—on the twenty-ninth of June 824—I was given in marriage at the palace of Aachen to my lord Bernard, your father, to be his legitimate wife. It was still in that reign, in its thirteenth year on the twenty-ninth of November, that with God's help, as I believe, you were born into this world, my firstborn and much-desired son.

Afterward, as the wretchedness of this world grew and worsened, in the midst of the many struggles and disruptions in the kingdom, that emperor followed the path common to all men. For in the twenty-eighth year of his reign, he

paid the debt of his earthly existence before his time. In the year after his death, your brother was born on the twenty-second of March in the city of Uzès. This child, born after you, was the second to come forth from my body by God's mercy. He was still tiny and had not yet received the grace of baptism when Bernard, my lord and the father of you both, had the baby brought to him in Aquitaine in the company of Elefantus, bishop of Uzès, and others of his retainers.[4]

Now I have been away from you for a long time, for my lord constrains me to remain in this city. Nonetheless I applaud his success. But, moved by longing for both of you, I have undertaken to have this little book—a work on the scale of my small understanding—copied down and sent to you. Although I am besieged by many troubles, may this one thing be God's will, if it please him—that I might see you again with my own eyes. I would think it certain that I would, if God were to grant me some virtue. But since salvation is far from me, sinful woman that I am, I only wish it, and my heart grows weak in this desire.

As for you, I have heard that your father, Bernard, has given you as a hostage to the lord king Charles. I hope that you acquit yourself of this worthy duty with perfect good will. Meanwhile, as Scripture says, *Seek ye therefore the kingdom of God . . . and all these things shall be added unto you,* that is all that is necessary for the enjoyment of your soul and your body.

So the preface comes to an end.

1. On Loving God.   God must be loved and praised—not only by powers on high, but also by every human creature who walks upon the earth and reaches toward heaven. I beseech you, my son, since you are among these, always to try your best to find the way to climb to its secure height along with those others who are worthy and who are able to love God. Then, along with them, you will be able to reach his kingdom without end.

----

[4]Dhuoda reveals nothing further about the fate of this son.

I ask of you, and I humbly suggest to your noble youth—just as if I were with you in person—and to those to whom you may offer this little book for perusal, that they not condemn me or hold it against me that I am so rash as to take upon myself so lofty and perilous a task as to speak to you about God. Indeed, knowing my human frailty, I never cease to chastise myself, *whereas I am* wretched, *dust and ashes.* And what shall I say? If the patriarchs and prophets and the other saints, from the first-made man up until now, have been unable to understand entirely the accounts of holy mysteries, how much less should I be able to—I who am but weak, born of a lowly people? And if, as Scripture says, *the heaven of heavens cannot contain thee* on account of your greatness, Lord, what can I, unlearned as I am, say about you? . . .

Still, although I am as weak as a shadow, I must bring to your awareness, my son William, what you can understand of God above. For I neither can nor have the strength to nor should set forth for you a complete discourse. Instead, I now begin my partial attempt at such a task, putting together those things that are most important to understand . . .

1. On the reverence you should show your father throughout your life.   Now I must do my best to guide you in how you should fear, love, and be faithful to your lord and father, Bernard, in all things, both when you are with him and when you are apart from him. . . .

4. Direction on your comportment toward your lord. You have Charles[5] as your lord; you have him as lord because, as I believe, God and your father, Bernard, have chosen him for you to serve at the beginning of your career, in the flower of your youth. Remember that he comes from a great and noble lineage on both sides of his family. Serve him not only so that you please him in obvious ways, but also as one clearheaded in matters of both body and soul. Be steadfastly and completely loyal to him in all things. . . . I urge

you to keep this loyalty as long as you live, in your body and in your mind. For the advancement that it brings you will be of great value both to you and to those who in turn serve you. May the madness of treachery never, not once, make you offer an angry insult. May it never give rise in your heart to the idea of being disloyal to your lord. There is harsh and shameful talk about men who act in this fashion. I do not think that such will befall you or those who fight alongside you because such an attitude has never shown itself among your ancestors. It has not been seen among them, it is not seen now, and it will not be seen in the future.

Be truthful to your lord, my son William, child of their lineage. Be vigilant, energetic, and offer him ready assistance as I have said here. In every matter of importance to royal power take care to show yourself a man of good judgment—in your own thoughts and in public—to the extent that God gives you strength. Read the sayings and the lives of the holy Fathers who have gone before us. You will there discover how you may serve your lord and be faithful to him in all things. When you understand this, devote yourself to the faithful execution of your lord's commands. Look around as well and observe those who fight for him loyally and constantly. Learn from them how you may serve him. Then, informed by their example, with the help and support of God, you will easily reach the celestial goal I have mentioned above. And may your heavenly Lord God be generous and benevolent toward you. May he keep you safe, be your kind leader and your protector. May he deign to assist you in all your actions and be your constant defender. *As it shall be the will of God in heaven so be it done.* Amen. . . .

Still I offer you this as a guiding principle—that you detest and flee the wicked, the immoral, the sluggish, and the proud and that in all you do you avoid those who are abhorrent in spirit. Why? Because they cast out strings, like mousetraps, in order to deceive. They never cease to prepare a road to scandal and offense, so that they themselves fall down recklessly and make others like them tumble with them. This has happened

---

[5]Charles the Bald (ruled 843–877), king of the West Franks, a territory roughly corresponding to modern France.

in the past, and I urge you that you avoid it now and in the future. May God forbid that your fate be linked in any way with theirs.

Seek out, hold close, and observe faithfully the examples of great men in the past, present, and future—men who are known to be pleasing in their faith to both God and this world, men who have persevered. For that is the meaning of the Scripture that commands us to hold the written names of the twelve patriarchs in our hands and to wear them on our foreheads, and to keep our eyes looking backward and forward; this Scripture concerns virtue. While the men it describes were in this present world, they were always headed toward heaven, growing and flourishing toward God. Wise in faith and spirit, happy in their path, they undertook and accomplished worthy ends through thought and deed. Then they left behind for us an example, so that in seeking it, we may do as they did.

6. ... If, because of the persuasion of the devil, fornication or some other spur of the flesh should drive your heart to frenzy, set chastity against it, and remember the continence of the blessed patriarch Joseph, and of Daniel, and of those others who faithfully maintained purity in spirit and body in respect to their lords and their neighbors. They were therefore found worthy to be saved and honored, gathered up by the Lord full of praise among the number of his saints. For as the Apostle says, *fornicators and adulterers God will judge.* And the Psalmist says, for behold they who fornicate away from thee shall perish. And the Apostle says likewise, *Every sin that man doth, is without the body; but he that committeth fornication, sinneth against his own body,* and other comments of this sort.

Therefore, my son, flee fornication and keep your mind away from any prostitute. It is written, *Go not after thy lusts, but turn away from thy own will.* Do not *give to thy soul* to fly away after her evil desires. Surely, if you attend to one or another of these ills and if you consent to them, they will make you fall onto the sword and into the hands of your enemies. They will say with the Prophet, *Bow down, that we may go over you.* May this not happen to you. But if those evils

come and sting your mind through an angel of Satan sent against you, fight them, pray, and say with the Psalmist, *Deliver not up to the beasts of the earth my soul, I beseech you, and forget not the soul of thy poor servant; give me not the haughtiness of my eyes; let not the lusts of the flesh take hold of me, and give me not over to a shameless and foolish mind.*

The haughtiness of my eyes has, I think, not only an outer, corporeal sense but also an inner sense. For if it did not have inner meaning, this saying would be empty. *I have made a covenant with my eyes, that I would not so much as think upon a virgin,* and many similar sayings in many places. You will find consolation in such great accounts so that you may, in petitioning God, escape the thrill of such embraces and the temptation of such turmoil. And although it is in the head that the eyes of the flesh are turned to desire, the struggle against such evils is fought within. For it is written of eyes turning in passionate desire that they perform their wrongful outrages carnally. *For death is come up through our windows,* and again, *whosoever shall look on a woman to lust after her* does so carnally.

As for those who adhere to continence and who crush beneath their heels the desire of the flesh, you find it written, *The light of the body is thy eye.* And *if thy eye be single, thy whole body shall be lightsome.* He who said this, *Turn away my eyes that they may not behold vanity,* and many things like it, wished that chastity be inviolable. For as learned men say, "chastity is the angelic life," and it makes whoever participates in it a citizen of heaven. "O," someone says, "how short, short indeed is that moment of fornication by which future life is lost! And how great is the strength and the enduring splendor of chastity, which makes a mortal man like a fellow citizen of the angels."

For learned authors do not refuse sacred marriage to the union of the flesh, but they try to root out from among us libidinous and wrongful fornication. For Enoch was chaste, and so were Noah, Abraham, Isaac, Jacob, Joseph, and Moses, and all the others who struggled to keep their hearts pure in Christ in the union of marriage. And what more shall I say?

So my son, whether you keep your body in virginity, a resplendent gift, or in the chastity of the union of marriage, you will be free from the origin of this sin. Your mind will be *secure... like a continual feast,* and will rest throughout all you do in all of the eight beatitudes. And there will be fulfilled in you in the company of other good men, as it is written, the worthy praise offered for many; take courage, *blessed are the clean of heart, for they shall see God....*

Love the poor, and gather them to you. Do your duty to them at all times in the spirit of mildness and gentleness, lest you forget fraternal compassion for those who are beneath you. Always let your nobility be clothed in a suppliant heart, in the poverty of the spirit. Then you will be able to listen untroubled and to share in the kingdom with those of whom it is written, *Blessed are the poor in spirit: for theirs is the kingdom of heaven.*

Love justice, so that you may be recognized as just in legal matters. *For the Lord is just, and hath loved justice.* He loves it always. *His countenance beholds righteousness.* Another man loved it greatly long ago, and he directed that it be loved when he said, *Love justice, you that are the judges of the earth.* Yet another said, *if in very deed you speak justice, judge right things.* It is written, *for with what judgment you judge, you shall be judged.*

Therefore, my son William, avoid iniquity, love fairness, follow justice, and fear to hear the saying of the Psalmist: *he that loveth iniquity hateth his own soul.* That Lord who is true and pure has given you a soul that is true, pure, and immortal, though in a fragile body. May it not befall you, then, to prepare evil snares for that soul, for the sake of desire for transient things, by doing or saying or consenting to acts of injustice or pitilessness. For many are tormented for the wrongdoings of others.

3. A postscript on public life. Here the words of this little book conclude. I have dictated them with an eager mind and have had them copied down for your benefit, as a model for you.

For I wish and urge that, when with God's help you have grown to manhood, you may arrange your household well, in appropriate order. As is written of another man who lived in this fashion, a man *like the most tender little worm of the wood,* perform all the duties of your public life with loyalty, in a well-ordered fashion.

As for whether I survive to that time when I may see this with my own eyes, I am uncertain—uncertain in my own merits, uncertain in my strength, battered as I am among the waves in my frail toil. Although such is what I am, all things are possible for the Almighty. It is not in man's power to do his own will, rather whatever men accomplish is according to God's will. In the words of scripture, *it is not of him that willeth, nor of him that runneth, but of God that showeth mercy.* Now, trusting in him, I say nothing else but *as it shall be the will of God in heaven so it be done.* Amen.

4. Returning to myself, I grieve. The sweetness of my great love for you and my desire for your beauty have made me all but forget my own situation. I wish now, *the doors being shut,* to return to my own self. But because I am not worthy to be numbered among those who are mentioned above, I still ask that you—among the innumerable people who may do so—pray without ceasing for the remedy of my soul on account of your special feeling for me, which can be measured.

You know how much, because of my continual illnesses and other circumstances, I have suffered all these things and others like them in my fragile body—according to the saying of a certain man, *in perils from my own nation, in perils from the Gentiles*—because of my pitiful merits. With God's help and because of your father, Bernard, I have at last confidently escaped these dangers, but my mind still turns back to that rescue. In the past I have often been lax in the praise of God, and instead of doing what I should in the seven hours of the divine office, I have been slothful seven times seven ways. That is why, with a humble heart and with all my strength, I pray that I may take my pleasure in continually beseeching God for my sins and my transgressions. May he deign to raise even me into heaven, shattered and heavy though I am.

And since you see me as I live in the world, strive with watchful heart—not only in vigils and prayer but also in alms to the poor—that I may be found worthy, once I am liberated from the flesh and from the bonds of my sins, to be freely received by the good Lord who judges us.

Your frequent prayer and that of others is necessary to me now. It will be more and more so in time to come if, as I believe, my moment is upon me. In my great fear and grief about what the future may bring me, my mind casts about in every direction. And I am unsure how, on the basis of my merits, I may be able to be set free in the end. Why? Because I have sinned in thought and in speech. Ill words themselves lead to evil deeds. Nevertheless I will not despair of the mercy of God. I do not despair now and I will never despair. I leave no other such as you to survive me, noble boy, to struggle on my behalf as you do and as many may do for me because of you, so that I may finally come to salvation.

I acknowledge that, to defend the interests of my lord and master Bernard, and so that my service to him might not weaken in the March[6] and elsewhere—so that he not abandon you and me, as some men do—I know that I have gone

greatly into debt. To respond to great necessities, I have frequently borrowed great sums, not only from Christians but also from Jews. To the extent that I have been able, I have repaid them. To the extent that I can in the future, I will always do so. But if there is still something to pay after I die, I ask and I beg you to take care in seeking out my creditors. When you find them, make sure that everything is paid off either from my own resources, if any remain, or from your assets—what you have now or what you eventually acquire through just means, with God's help.

What more shall I say? As for your little brother, I have above directed you time and again concerning what you should do for him. What I ask now is that he too, if he reaches the age of manhood, deign to pray for me. I direct both of you, as if you were together here before me, to have the offering of the sacrifice and the presentation of the host made often on my behalf.

Then, when my redeemer commands that I depart this world, he will see fit to prepare refreshment for me. And if this transpires through your prayers and the worthy prayers of others, he who is called God will bring me into heaven in the company of his saints.

This handbook ends here. Amen. Thanks be to God.

---

[6]The Spanish March, a region where Dhuoda's husband had a position as a count.

42. Elisabeth of Schönau's Vision of the Assumption of the Virgin Mary

The most exceptional woman in all of Christian history was the Virgin Mary, the mother of Jesus, who was widely venerated and worshipped in medieval Europe by both women and men. Churches and cathedrals were built in her honor (often named "Our Lady"—"Notre dame"), statues and paintings depicted her, and special prayers and hymns were written for her. She was widely reported to perform miracles and appeared to people in visions. Women (and men) who saw visions were also regarded as exceptional in medieval society, though mystical experiences were viewed as an acceptable way to gain religious insights. A few women became well known through their mystical visions, which they shared with others orally or in writing. Elisabeth of Schönau (c. 1128–1164) was one of these, a Benedictine nun from the Rhineland area of Germany and a friend of the famous abbess and mystic Hildegard of Bingen. Elisabeth's ecstatic visions included some of the Virgin Mary, which were frequently translated into other languages and retold; the details they revealed became part of the popular traditions surrounding Mary and occasionally part of official Catholic theology. The vision here concerns the Assumption, or bodily resurrection of Mary shortly after her death, which had been debated since early Christianity. Those favoring Mary's Assumption argued that because she—alone of all people—was sinless, she did not have to suffer death; those arguing against it pointed out the lack of biblical basis for this doctrine. Despite these debates, the Assumption of Mary was widely accepted and celebrated throughout Europe, and in 1950 Pope Pius XII declared it "divinely revealed dogma."

In the year the angel announced the *Book of the Paths of God* to me [c. 1155], on the day on which the church celebrates the octave of Our Lady's Assumption, during the holy sacrifice of the mass I fell into an ecstasy and that comforter, my Lady of heaven, appeared to me in her usual manner. Then, following the advice of one of our superiors, I asked her: "My benevolent Lady, may it please you to make known to us whether you were assumed into heaven only in spirit or also in the flesh?" I mentioned this because as they say, there is uncertainty about it in the books the [Church] Fathers wrote. She said to me: "What you ask you may not know until the future when it shall be revealed through you." For the space of a whole year I dared ask nothing more about this either from the angel who was my spiritual companion or from her when she revealed herself to me. But the brother who had urged me to ask the question put many requests to me to ask her for a revelation in answer to it. After a year had passed, as the feast of her Assumption again approached, I languished sick for many days. But as I was lying in my small bed during the holy sacrifice of the mass I fell, with great effort, into an ecstasy. And I saw in a very distant place a sepulcher surrounded by many lights with what appeared to be the shape of a woman inside it, and a great band of angels encircled it. After a bit, she was raised up from the sepulcher and along with that throng surrounding her, she was lifted on high. And when I had seen this, behold, coming to meet her from the

heavenly heights was a man glorious beyond estimate, carrying in his right hand the sign of the cross on which there was a banner, whom I knew to be the Lord Savior, and infinite thousands of angels accompanied him. Eagerly greeting her they carried her to the pinnacle of heaven with great harmonious song. When I had observed this, after a while my Lady approached me through the door of light in which I usually saw her, and standing there, she revealed to me her glory. In that same hour an angel of the Lord attended me, who came to tell me the tenth discourse of the aforementioned book, and I asked him, "Lord, what does this great vision I have seen mean?" He answered, "Through this vision, you were shown how Our Lady was assumed into heaven in body as well as spirit." Later, on the eighth day [after the feast] I asked the angel (who had also come to me to state the conclusion for the book) how many days after her death did her physical resurrection occur. Again, he kindly assured me, saying, "Since she passed from this life on the day on which her Assumption is now celebrated [August 15], she rose from the dead forty days after that on September 23. The holy fathers, who ordered the church to celebrate the feast of the Assumption, were uncertain about her physical assumption, so they set the feast on the day of her death which they called her Assumption even though they really did believe that she had also been bodily assumed." Afterwards, when I hesitated to make this revelation known in writing, fearing I would be considered the

"Elizabeth of Schönau's Vision of the Assumption" as translated and edited by John Shinner in *Medieval Popular Religion, 1000–1500: A Reader,* pp. 117–119. Copyright © 1997 by Broadview Press. All Rights Reserved. Reprinted with permission of Broadview Press.

inventor of novelties, when two years had passed and it was again the feast of her Assumption, my Lady appeared to me. I asked her: "Lady, may we not make public the news you revealed to me about your resurrection?" But she said, "This is not to be made known to the people, for the world is wicked, and those who would hear it are ensnared in it and do not know how to untangle themselves." Again I said: "Do you wish us to erase completely what we have written about this revelation?" She answered: "These things were not revealed to you for you to erase them and consign them to oblivion, but for my glory to be multiplied among those of you who especially love me. Let them be revealed by you to those in my service; they will be open to those who open their heart to me. Through this let them show me special praise and they will receive a special reward from me. For there are many people who will hear this news with great rejoicing and veneration." Because of these words, we celebrated that feast in our community on the day mentioned above the best way we could, offering devout praises to our venerable Lady....

It was while we were celebrating the feast of the Lord's Annunciation [March 25] when my Lady again showed me her glorious face that I was bold enough to ask her how old she was when, at the angel's news, she conceived the Word of God in her virgin womb. She deigned to give me this answer to my question: "I had lived fifteen years," she said, "from the time of my birth to the feast of the Lord's Annunciation.

43. Rule of St. Clare

Like her model Francis of Assisi, Clare of Assisi (1193–1253) rejected the wealth of her Italian urban family and wanted to live in poverty, devoting herself to charitable works. She refused to marry and became a follower of Francis, who established a place for her to live in the church of San Damiano. She was joined by other women, and they attempted to establish a rule for life in their community that would closely follow Francis's ideals of absolute poverty. These ideals were opposed even before Francis's death in 1226 by many leaders of the church, who saw them as creating hostility to church wealth and property ownership. Clare's rule for the women in her community—the first monastic rule written by a woman—thus took decades to gain papal approval, which actually came two days before her death. Like other monastic rules, this set out both the purposes and structure of life for women in Clare's community, and later for other houses in the order, which became known as the Poor Clares.

CHAPTER I: IN THE NAME OF THE LORD BEGINS THE FORM OF LIFE OF THE POOR SISTERS

1. The form of life of the Order of the Poor Sisters which the Blessed Francis established, is this: 2. to observe the holy Gospel of our Lord Jesus Christ, by living in obedience, without anything of one's own, and in chastity.

3. Clare, the unworthy handmaid of Christ and the little plant of the most blessed Father Francis, promises obedience and reverence to the Lord Pope Innocent and to his canonically elected successors, and to the Roman Church. 4. And, just as at the beginning of her conversion, together with her sisters she promised obedience to the Blessed Francis, so now she promises his successors to observe the same [obedience] invio-

From *Francis and Clare,* translation and introduction by Regis J. Armstrong, O.F.M. CAP. © 1982 by Paulist Press. Used by permission of Paulist Press.

lably. 5. And the other sisters shall always be obliged to obey the successors of the blessed Francis and [to obey] Sister Clare and the other canonically elected Abbesses who shall succeed her.

CHAPTER II: THOSE WHO WISH TO ACCEPT THIS LIFE AND HOW THEY ARE TO BE RECEIVED

1. If, by divine inspiration, anyone should come to us with the desire to embrace this life, the Abbess is required to seek the consent of all the sisters; and if the majority shall have agreed, having had the permission of our Lord Cardinal Protector, she can receive her. 2. And if she judges [the candidate] acceptable, let [the Abbess] carefully examine her, or have her examined, concerning the Catholic faith and the sacraments of the Church. 3. And if she believes all these things and is willing to profess them faithfully and to observe them steadfastly to the end; and if she has no husband, or if she has [a husband] who has already entered religious life with the authority of the Bishop of the diocese and has already made a vow of continence; and if there is no impediment to the observance of this life, such as advanced age or some mental or physical weakness, let the tenor of our life be clearly explained to her.

4. And if she is suitable, let the words of the holy Gospel be addressed to her: that she should *go and sell* all that she has and take care to distribute the proceeds *to the poor* (cf. Mt 19:21). If she cannot do this, her good will suffices. 5. And let the Abbess and her sisters take care not to be concerned about her temporal affairs, so that she may freely dispose of her possessions as the Lord may inspire here. If, however, some counsel is required, let them send her to some prudent and God-fearing men, according to whose advice her goods may be distributed to the poor.

6. Afterward, once her hair has been cut off round her head and her secular dress set aside, she is to be allowed three tunics and a mantle. 7. Thereafter, she may not go outside the monastery except for some useful, reasonable, evident, and approved purpose. 8. When the year of probation is ended, let her be received into obedience, promising to observe always our life and form of poverty.

9. During the period of probation no one is to receive the veil. 10. The sisters may also have small cloaks for convenience and propriety in serving and working. 11. Indeed, the Abbess should provide them with clothing prudently, according to the needs of each person and place, and seasons and cold climates, as it shall seem expedient to her by necessity.

12. Young girls who are received into the monastery before the age established by law should have their hair cut round [their heads]; and, laying aside their secular dress, should be clothed in religious garb as the Abbess has seen [fit]. 13. When, however, they reach the age required by law, in the same way as the others, they may make their profession. 14. The Abbess shall carefully provide a Mistress from among the more prudent sisters of the monastery both for these and the other novices. She shall form them diligently in a holy manner of living and proper behavior according to the form of our profession.

15. In the examination and reception of the sisters who serve outside the monastery, the same form as above is to be observed. 16. These sisters may wear shoes. 17. No one is to live with us in the monastery unless she has been received according to the form of our profession.

18. And for the love of the most holy and beloved Child Who *was wrapped in* the poorest of *swaddling clothes and laid in a manger* (cf. Lk 2:7–12), and of His most holy Mother, I admonish, entreat, and exhort my sisters that they always wear the poorest of garments.

CHAPTER III: THE DIVINE OFFICE AND FASTING, CONFESSION AND COMMUNION

1. The Sisters who can read shall celebrate the Divine Office according to the custom of the Friars Minor,[1] for this they may have breviaries, but they are to read it without singing. 2. And those

---

[1]*Friars Minor:* Franciscans, members of the order founded by Francis of Assisi.

who, for some reasonable cause, sometimes are not able to read and pray the Hours, may, like the other sisters, say the Our Father's.

3. Those who do not know how to read shall say twenty-four Our Father's for Matins; five for Lauds; for each of the hours of Prime, Terce, Sext, and None, seven; for Vespers, however, twelve; for Compline, seven. 4. For the dead, let them also say seven Our Father's with the *Requiem aeternam* in Vespers; for Matins, twelve: 5. because the sisters who can read are obliged to recite the Office of the Dead. 6. However, when a sister of our monastery shall have departed this life, they are to say fifty Our Father's.

7. The sisters are to fast at all times. 8. On Christmas, however, no matter on what day it happens to fall, they may eat twice. 9. The younger sisters, those who are weak, and those who are serving outside the monastery may be dispensed mercifully as the Abbess sees fit. 10. But in a time of evident necessity the sisters are not bound to corporal fasting.

11. At least twelve times a year they shall go to confession, with the permission of the Abbess. 12. And they shall take care not to introduce other talk unless it pertains to confession and the salvation of souls. 13. They should receive Communion seven times [a year], namely, on Christmas, and Thursday of Holy Week, Easter, Pentecost, the Assumption of the Blessed Virgin, the Feast of Saint Francis, and the Feast of All Saints. 14. [In order] to give Communion to the sisters who are in good health or to those who are ill, the Chaplain may celebrate inside [the enclosure].

CHAPTER IV: THE ELECTION AND OFFICE OF THE ABBESS; THE CHAPTER. THOSE WHO HOLD OFFICE AND THE DISCREETS

1. In the election of the Abbess the sisters are bound to observe the canonical form.[1]

---

[1] The election of an abbess was the only time women voted, with a few exceptions, until the nineteenth century.

2. However, they should arrange with haste to have present the Minister General or the Minister Provincial of the Order of Friars Minor. Through the Word of God he will dispose them to perfect harmony and to the common good in the choice they are to make. 3. And no one is to be elected who is not professed. And if a nonprofessed should be elected or otherwise given them, she is not to be obeyed unless she first professes our form of poverty.

4. At her death the election of another Abbess is to take place. 5. Likewise, if at any time it should appear to the entire body of the sisters that she is not competent for their service and common welfare, the sisters are bound to elect another as Abbess and mother as soon as possible according to the form given above.

6. The one who is elected should reflect upon the kind of burden she has undertaken, and to Whom she is *to render an account* (Mt 12:36) of the flock committed to her. 7. She should strive as well to preside over the others more by her virtues and holy behavior than by her office, so that, moved by her example, the sisters might obey her more out of love than out of fear. 8. She should avoid particular friendships, lest by loving some more than others she cause scandal among all. 9. She should console those who are afflicted, and be, likewise, the last refuge for those who are disturbed; for, if they fail to find in her the means of health, the sickness of despair might overcome the weak.

10. She should preserve the common life in everything; especially regarding all in the church, dormitory, refectory, infirmary, and in clothing. Her vicar is bound to do likewise.

11. At least once a week the Abbess is required to call her sisters together in Chapter. 12. There both she and her sisters must confess their common and public offenses and negligences humbly. 13. There, too, she should consult with all her sisters on whatever concerns the welfare and good of the monastery; for the Lord often reveals what is best to the lesser [among us].

14. No heavy debt is to be incurred except with the common consent of the sisters and by reason of an evident need. This should be done

through a procurator. 15. The Abbess and her sisters, however, should be careful that nothing is deposited in the monastery for safekeeping; often such practices give rise to troubles and scandals.

16. To preserve the unity of mutual love and peace, all who hold offices in the monastery should be chosen by the common consent of all the sisters. 17. And in the same way at least eight sisters are to be elected from among the more prudent, whose counsel the Abbess is always bound to heed in those things which our form of life requires. 18. Moreover, if it seems useful and expedient, the sisters can and must sometimes depose the officials and discreets, and elect others in their place.

CHAPTER V: SILENCE, THE PARLOR, AND THE GRILLE

1. The sisters are to keep silence from the hour of Compline until Terce,[1] except those who are serving outside the monastery. 2. They should also keep silence continually in the church, in the dormitory, and, only while they are eating, in the refectory. 3. In the infirmary, however, they may speak discreetly at all times for the recreation and service of those who are sick. 4. However, they may briefly and quietly communicate what is really necessary always and everywhere.

5. The sisters may not speak in the parlor or at the grille without the permission of the Abbess or her Vicar. 6. And those who have permission should not dare to speak in the parlor unless they are in the presence and hearing of two sisters. 7. Moreover, they should not presume to go to the grille unless there are at least three sisters present [who have been] appointed by the Abbess or her Vicar from the eight discreets who were elected by all the sisters as the council of the Abbess. 8. The Abbess and her vicar are themselves bound to observe this custom in speaking. 9. [The sisters should speak] very rarely at the grille and, by all means, never at the door.

_____
[1]*Compline:* the last hour of the day; *Terce:* the pre-dawn hours.

10. At the grille a curtain is to be hung inside which is not to be removed except when the Word of God is being preached, or when a sister is speaking to someone. 11. The grille should also have a wooden door which is well provided with two distinct iron locks, bolts, and bars, so that, especially at night, it can be locked by two keys, one of which the Abbess is to keep and the other the sacristan; it is to be locked always except when the Divine Office is being celebrated and for reasons given above. 12. Under no circumstances whatever is any sister to speak to any one at the grille before sunrise or after sunset. 13. Moreover, in the parlor there is always to be a curtain on the inside, which is never to be removed....

CHAPTER VII: THE MANNER OF WORKING

1. The sisters to whom the Lord has given the grace of working are to work faithfully and devotedly, [beginning] after the Hour of Terce, at work which pertains to a virtuous life and to the common good. 2. They must do this in such a way that, while they banish idleness, the enemy of the soul, they do not extinguish the Spirit of holy prayer and devotion to which all other things of our earthly existence must contribute.

3. And the Abbess or her vicar is bound to assign at the Chapter, in the presence of all, the manual work each is to perform. 4. The same is to be done if alms have been sent by anyone for the needs of the sisters, so that the donors may be remembered by all in prayer together. 5. And all such things are to be distributed for the common good by the Abbess or her vicar with the advice of the discreets.

CHAPTER IX: THE PENANCE TO BE IMPOSED ON THE SISTERS WHO SIN

1. If any sister, at the instigation of the enemy, shall have sinned mortally against the form of our profession, and if, after having been

admonished two or three times by the Abbess or other sisters, she will not amend, she shall eat bread and water on the floor before all the sisters in the refectory for as many days as she has been obstinate; and if it seems advisable to the Abbess she shall undergo even greater punishment. 2. Meanwhile, as long as she remains obstinate, let her pray that the Lord will enlighten her heart to do penance. 3. The Abbess and her sisters, however, must beware not to become angry or disturbed on account of anyone's sin: for anger and disturbance prevent charity in oneself and in others.

4. If it should happen—God forbid—that through [some] word or gesture an occasion of trouble or scandal should ever arise between sister and sister, let she who was the cause of the trouble, at once, before offering the gift of her prayer to the Lord, not only prostrate herself humbly at the feet of the other and ask pardon, but also beg her earnestly to intercede for her to the Lord that He might forgive her. 5. The other sister, mindful of that word of the Lord: *If you do not forgive from the heart, neither will your* heavenly *Father forgive you* (Mt 6:15; 18:35), should generously pardon her sister every wrong she has done her.

CHAPTER X: THE ADMONITION AND CORRECTION OF THE SISTERS

4. Indeed, I admonish and exhort in the Lord Jesus Christ that the sisters be on their guard against all pride, vainglory, envy, greed, worldly care and anxiety, detraction and murmuring, dissension and division. 5. Let them be ever zealous to preserve among themselves the unity of mutual love, which is the bond of perfection.

6. And those who do not know how to read should not be eager to learn. 7. Rather, let them devote themselves to what they must desire to have above all else: the Spirit of the Lord and His holy manner of working, to pray always to Him with a pure heart, and to have humility, patience in difficulty and weakness, and

to love those who persecute, blame, and accuse us; for the Lord says: *Blessed are they who suffer persecution for justice's sake, for theirs is the kingdom of heaven* (Mt 5:10). But *he who shall have persevered to the end will be saved* (Mt 10:22).

CHAPTER XI: THE CUSTODY OF THE ENCLOSURE

1. The portress is to be mature in her manners and prudent, and of suitable age. During the day she should remain in an open cell without a door. 2. A suitable companion should be assigned to her who may, whenever necessary, take her place in all things.

3. The door is to be well secured by two different iron locks, with bars and bolts, 4. so that, especially at night, it may be locked with two keys, one of which the portress is to have, the other the Abbess. 5. And during the day the door must not be left unguarded on any account, but should be firmly locked with one key.

6. They should take utmost care to make sure that the door is never left open, except when this can hardly be avoided gracefully. 7. And by no means shall it be opened to anyone who wishes to enter, except to those who have been granted permission by the Supreme Pontiff or by our Lord Cardinal. 8. The sisters shall not allow anyone to enter the monastery before sunrise or to remain within after sunset, unless an evident, reasonable, and unavoidable cause demands otherwise.

9. If a bishop has permission to offer mass within the enclosure, either for the blessing of an Abbess or for the consecration of one of the sisters as a nun or for any other reason, he should be satisfied with as few and virtuous companions and assistants as possible.

10. Whenever it is necessary for other men to enter the monastery to do some work, the Abbess shall carefully post a suitable person at the door who is to open it only to those assigned for the work, and to no one else. 11. At such times all the sisters should be extremely careful not to be seen by those who enter.

44. University of Paris Case Against a Female Doctor Jacoba Felicie, 1322

In the Early Middle Ages, medical practitioners were trained through informal systems of apprenticeship. Medical training became more formalized and professionalized in the High Middle Ages, with guilds of barber-surgeons and apothecaries developing in many cities, and faculties of medicine established as parts of the new universities. These faculties began to issue licenses to individuals who had successfully completed a formal course of study and to pressure city and state governments to prohibit those without licenses to treat patients, particularly for internal ailments. (University-trained physicians rarely handled external ailments such as skin diseases or wounds; these were treated by barber-surgeons.) The following is a case from the University of Paris against a woman who continued to treat patients despite official prohibitions.

Witnesses were brought before us . . . in the inquisition made at the instance of the masters in medicine at Paris against Jacoba Felicie and others practising the art of medicine and surgery in Paris and the suburbs without the knowledge and authority of the said masters, to the end that they be punished, and that this practice be forbidden them. . . .

The dean and the regent masters of the faculty of medicine at Paris intend to prove against the accused, Mistress Jacoba Felicie, (1) that the said Jacoba visited, in Paris and in the suburbs, many sick persons afflicted with grave illness, inspecting their urine[1] both in common and individually, and touching, feeling, and holding their pulses, body, and limbs. Also (2) that after this inspection of urine and this touching, she said and has said to those sick persons: "I shall make you well, God willing, if you will have faith in me," making an agreement concerning the cure with them, and receiving money for it. Also (3) that after an agreement had been made between the said defendant and the sick persons or their friends, concerning curing them of their internal illness, or from a wound or ulcer appearing on the outside of the bodies of the said sick persons, the said defendant visited and visits the sick persons very often, inspecting their urine carefully and continually, after the manner of physicians and doctors, and feeling their pulse, and their body, and touching and holding their limbs. Also (4) that after these actions, she gave and gives to the sick persons syrups to drink, comforting, laxative, digestive, both liquid and nonliquid, and aromatic, and other drinks, which they take and drink and have drunk very often in the presence of the said defendant, she prescribing and giving them. Also (5) that she has exercised and exercises, in the aforesaid matters, this function of practising medicine in Paris and in the suburbs, that she has practised and practises from day to day although she has not been approved in any official *studium* at Paris or elsewhere. . . .

Jean Faber, living near the Tower in Paris . . . when he was asked if he knew the parties, said that he knew some of the masters by sight, others not, but that he knew the said Jacoba, because she has done well by him, as he said. Asked what he knows of those matters contained in the articles, he replied that he was suffering from a certain sickness in his head and ears at a time of great heat, that is, before the

---

[1] Inspecting urine was the most common means of diagnosing illness at the time.

feast of the nativity of St. John [June 24], and that the said Jacoba had visited him, and had shown such great care for him that he was cured from his illness by the potations which she gave him, and by the aid of God. When he was asked what potions she had given to him, he answered that Jacoba had administered potions to him, of which the first was green, and the second and third more colourless, but how they were made he did not know. Asked if the said Jacoba has been wont to visit the sick, he said that she has, as he has heard many say. When he was asked if he had made a contract with her about curing him, he said that he had not. After he had been made well he paid her as he wished....

The lord Odo de Cormessiaco, a brother of the hospital of Paris, a witness, when he was asked what he knew of the matters contained in the articles, etc., answered this by law on his oath, that is, that when, around the feast of the nativity of St. John, he had been seized by a severe illness, to such an extent that his own limbs could not support him, Master Jean de Turre had visited him, and many other masters in medicine, Masters Martin and Herman and many others. And he had had himself taken to the house of the said Jacoba, and was there for a while, and then afterwards this Jacoba visited him both at the baths and in the aforesaid hospital. And Master Jean, who lives with this Jacoba, gave him a purgative, and they prepared many baths and bandages for him, and anointed him very often. The said Jacoba and Jean worked over him with such great care that he was completely restored to health. They also gave him herbs, that is, camomile leaves, melilot, and very many others, which he did not recall. Also, on the advice of the said Jacoba, the said Jean made a certain charcoal fire of the length and breadth of the witness, and upon this fire he placed many herbs, and afterwards he had him lie down on these herbs, and lie there until this made him sweat exceedingly. Afterwards they wrapped him in linen cloth and put him in his own bed, and cared for him with such diligence that by means of God's help and the said care, he was cured. When he was asked if Ja-

coba had made a contract with him concerning visiting and healing him, he said she had not. He paid as he wished when he got well, and he believed better than otherwise....

Jeanne, wife of Denis called Bilbaut, living in the Rue de la Ferronnerie in Paris... answered on oath that around the feast of St. Christopher [July 25], just passed, she had been seized with a fever, and very many physicians had visited her in the said illness, that is, a certain brother de Cordelis, Master Herman, Manfred, and very many others. And she was so weighed down by the said illness that on a certain Tuesday around the said feast, she was not able to speak, and the said physicians gave her up for dead. And so it would have been, if the said Jacoba had not come to her at her request. When she had come, she inspected her urine and felt her pulse, and afterwards gave her a certain clear liquid to drink, and gave her also a syrup, so that she would go to the toilet. And Jacoba so laboured over her that by the grace of God she was cured of the said illness....

These are the arguments which Jacoba said and set forth in her trial....

The said Jacoba said that if the statute, decree, admonition, prohibition, and excommunication which the said dean and masters are trying to use against her, Jacoba, had ever been made, this had been only once, on account of and against ignorant women and inexperienced fools, who, untrained in the medical art and totally ignorant of its precepts, usurped the office of practising it. From their number the said Jacoba is excepted, being expert in the art of medicine and instructed in the precepts of said art. For these reasons, the statute, decree, admonition, prohibition, and excommunication aforesaid are not binding and cannot be binding on her, since when the cause ceases, the effect ceases....

Also the said statute and decree, etc., had been made on account of and against ignorant women and foolish usurpers who were then exercising the office of practice in Paris, and who are either dead or so ancient and decrepit that they are not able to exercise the said office, as it appears from the tenor of the said statute and decree, etc., which were made a hundred and two years ago, at which time the said Jacoba was not, nor was she for sixty years afterwards, in the nature of things; indeed, she is young, thirty or thereabouts, as it appears from her aspect....

Also it is better and more becoming that a woman clever and expert in the art should visit a sick woman, and should see and look into the secrets of nature and her private parts, than a man, to whom it is not permitted to see and investigate the aforesaid, nor to feel the hands, breasts, belly and feet, etc., of women.[2] Indeed a man ought to avoid and to shun the secrets of women and their intimate associations as much as he can. And a woman would allow herself to die before she would reveal the secrets of her illness to a man, because of the virtue of the female sex and because of the shame which she would endure by revealing them. And from these causes many women and also men have

perished in their illnesses, not wishing to have doctors, lest they see their secret parts....

Also, supposing without prejudice that it is bad that a woman should visit, care for, and investigate, as has been said, it is, however, less bad that a woman wise, discreet, and expert in the aforesaid matters has done and does these things, since the sick persons of both sexes, who have not dared to reveal the aforesaid secrets to a man, would not have wished to die. Thus it is that the laws say that lesser evils should be permitted, so that greater ones may be avoided. For this reason, since the said Jacoba is expert in the art of medicine, it is, therefore, better that she be permitted to make visits, in order to exercise the office of practice, than that the sick should die, especially since she has cured and healed all those....

Also, it has been ascertained and thus proved, that some sick persons of both sexes, seized by many severe illnesses and enduring the care of very many expert masters in the art of medicine, have not been able to recover at all from their illnesses, although the masters applied as much care and diligence to these as they were able. And the said Jacoba, called afterwards, has cured these sick persons in a short time, by an art which is suitable for accomplishing this.

---

[2] Male doctors commonly did not touch female patients. Indeed, many physicians never touched any patients at all; they just looked at their eyes and urine.

QUESTIONS FOR ANALYSIS

1. Why does Dhuoda write the book for her son? What advice does she give him about his public and private life? What does she say about her own situation? How might this work have been different had it been written by a male noble for his son? By a noblewoman for her daughter?
2. Elisabeth of Schönau reports that she waited several years after her first vision of Mary to make her revelations known. What reason does she give for this? Can you think of other reasons she might have waited?
3. What does Clare of Assisi see as the purpose of her religious community, and how is life to be structured to fulfill this purpose? How does she balance the power of the abbess and the power of the community as a whole? What role do enclosure, silence, work, and learning play in convent life?
4. What reasons does Jacoba Felicie give in arguing that the statute prohibiting women from practicing medicine should not apply to her? How does women's honor enter into her arguments?
5. How would you compare the sense of self exhibited by the four exceptional women presented in this section? In what other ways are they similar and different?

IDEAS ABOUT WOMEN

Very often discussions of women's lives in any period begin with information on ideas about women, as these formed the intellectual structures within which both men and women operated. For this chapter, we have chosen to place sources regarding ideas about women at the end, for two reasons. One of these is simply chronological; there are no extended discussions of women's nature and proper role that have survived from the Early Middle Ages. The other is both chronological and substantive. The discussions of women's place in creation, the reasons for differences among women and between women and men, and the role of sex and marriage in God's plan for the world which became more vigorous in the High Middle Ages actually marked the beginning of an extended debate about women that would continue for many centuries. They thus shaped women's lives more in subsequent periods than in the period covered in this chapter.

45. Hildegard of Bingen
Causes and Cures

Hildegard of Bingen (1098–1179) was the abbess of two monastic houses for women, a visionary, and a prolific author. She entered a Benedictine monastery when she was eight, and spent most of her life in various women's religious communities, two of which she founded herself. When she was over fifty, she left her community to preach to audiences of clergy and laity, and she was the only woman of her time whose opinions on religious matters were considered authoritative by the Church. She was the first woman encouraged by the pope to write works of theology, which she did along with poetry, plays, and scientific works. She was also a talented artist and composer of chants, liturgy, and other types of music. (Many of her musical compositions have been recorded recently by various artists, and are available on compact disk and on several Web sites; this current fame plus her theological status means that she could easily have been included in the "Exceptional Women" section of this chapter.) The following comes from one of her medical works, *Causes and Cures,* which includes discussion of illnesses and their treatment, but also conveys her theories about the creation of the universe and the causes and consequences of differences between men and women and among women.

31. *Adam's fall.*[1] God has created the human such that all living beings are subject to him in order to serve him [Gen. 1:26]. But when the human transgressed God's command he was transformed both in body and mind. For the pureness of his blood was turned into something different so that he emits the foam of semen instead of pureness. Had the human stayed in Paradise, he would have remained in an immutable and perfect state. But after his

---

[1]Here and elsewhere, Hildegard contrasts the situation at Creation with that after Adam and Eve disobeyed god in teh garden of Eden, commonly called the Fall.

transgression he was turned into something different and bitter....

When he consented to evil and relinquished good he was made similar to the earth that produces good and useful as well as bad and useless plants and that has good and bad moisture and sap within itself. With the taste for evil the blood of Adam's children was changed into the poison of semen from which the humans' offspring are propogated. Therefore their flesh is ulcerous and perforated. Those ulcers and perforations cause some kind of storm and a vaporous moisture in human beings. From this develop and coagulate *flegmata*[2] that affect the human body with various infirmities. All of this arose from the first evil that a human first performed. Because had Adam remained in Paradise he would have the sweetest health and the best dwelling, just as the strongest balsam gives off the best fragrance. But now, by contrast, human beings have poison in them, *flegma* and various infirmities....

Why Eve fell first. Had Adam transgressed prior to Eve, this transgression would have been so severe and incorrigible that the human would have fallen into such a grave state of incorrigibility that he would not have wanted to be saved nor could he have been saved. Because Eve transgressed first, [her sin] could be more easily eradicated since she was weaker than the male. Adam's and Eve's flesh and skin were stronger and harder than that of humans now, because Adam was formed from earth and Eve from him. Yet after they had procreated children their flesh became weaker and weaker, and thus it shall be until the Last Day.

The creation of Adam. God made the human being from clay [Gen. 2:7]. But the man was transformed from clay to flesh and therefore he is the particular cause of and the ruler over created things. He works the earth so that it brings forth fruit. There is a strength in his

bones, blood vessels and flesh. His head is whole and his skin is thick. He has his [reproductive] strength within him and produces semen as the sun brings forth light. Woman, however, was not transformed because taken from flesh she remained flesh. Therefore a more artful work was given to her to be done with her hands. She is airy, as it were. She carries the child in her womb and gives birth to it. She has a divided head and thin skin, so that the child she is carrying in her womb can get air....

Diversity in conception. When a man who has intercourse with a woman has an emission of strong semen and feels proper affectionate love for the woman and when the woman at that hour feels proper love for the man, then a male is conceived because it was ordained by God. It cannot be otherwise than that a male is conceived, since Adam too was formed from clay which is stronger matter than flesh. This male will be intelligent and virtuous because he was conceived from strong semen and with mutual affectionate love. If, however, the woman's love for the man is lacking, so that the man alone at that hour feels proper affectionate love for the woman and the woman does not feel the same for the man, and if the man's semen is strong, then still a male is conceived because the man's affectionate love is predominant. Yet this male child will be weak and not virtuous because the woman's love for the man was lacking. If the man's semen is thin, yet he feels affectionate love for the woman and she feels the same love for him, then a virtuous female is procreated. If, on the other hand, the man feels affectionate love for the woman and the woman does not feel the same for the man, or if the woman feels affectionate love for the man and the man does not feel the same for the woman, and if, further, the man's semen is thin at that hour, then a female is born due to the semen's weakness. But if the man's semen is strong, yet the man feels no affectionate love for the woman and the woman does not feel any for the man, then a male is procreated because the semen was stong, but he will be bitter on account of his parents' bitterness. And if a man's semen is thin and if at that hour neither feels affectionate love for

---

[2]*flegma* (pl.: flegmata) is a fluid in the body

the other, a female of bitter disposition is born. The warmth of women who are obese by nature outweighs the man's semen, such that the child's face often resembles the mother's. But women who are thin by nature often generate a child whose face resembles the father's....

The creation of Adam and the formation of Eve. When God created Adam, Adam felt great love in the deep sleep that God imparted to him. And God made a form for the love of man, and so woman is the love of man. As soon as woman was formed, God gave man the power of creation, that he might procreate children through his love which is woman. For when Adam looked at Eve he was wholly filled with wisdom, because he saw in her the mother through whom he would procreate children. And when Eve looked at Adam, she looked at him as though she were looking into heaven and as the soul, longing for heavenly things, strives upward, because her hope was placed in man. Therefore the same love and no other will be and must be in man and in woman.

But, man's love with its blazing heat, compared with woman's love is like the fire of blazing mountains that is difficult to extinguish compared with a wood fire that is easily extinguished. Woman's love compared with man's love is like sweet warmth proceeding from the sun which brings forth fruit compared with an ardent wood fire, because woman sweetly brings forth fruit in her offspring. In Adam's transgression, both the great love that he felt when Eve proceeded from him and the sweetness of the sleep which he then slept were turned into a contrary mode of sweetness. Therefore, because man still feels and has this strong sweetness in himself, as a deer longs for water [Ps. 42:1] so he rushes toward woman and woman toward him, like a threshing floor that is shaken by many strokes and brought to heat when grain is threshed on it....

Conception. When a woman has intercourse with a man, a warm pleasurable feeling in her brain announces the sensation of this pleasure in intercourse and the outpouring of semen. After

the semen has fallen into its place, this extremely strong warmth in the brain will attract it and hold it. Soon the woman's loins, too, contract, and all the members of her body that were prepared to open at menstruation close at once very tightly like a strong man enclosing something in his hand. Then menstrual blood intermingles with semen, makes it sanguineous and turns it into flesh. When it has become flesh, this same blood draws a vessel around it, like a little worm preparing its dwelling out of itself. And so the blood prepares this vessel day after day until a human being is formed in it and until this human being receives the breath of life. Then this vessel grows with the human being and is so firmly set that it cannot move from its place until this human being leaves it.

Eve. The first mother of humankind was created similar to ether. For as ether contains all the stars so she, pure and uncorrupted, carried humankind within herself when she was told: "Be fruitful and multiply" [Gen. 1:28]. And this comes to pass with much pain....

Sanguine women.[3] Some women are plump by nature. They have soft and delightful flesh, slender blood vessels, and good untainted blood. Because their blood vessels are slender they contain less blood and their flesh grows that much better and is that much more permeated with blood. They have a clear and light facial coloring, are lovable in the embrace of love and meticulous in arts. Their mental disposition is such that they are capable of self-control. They suffer only a moderate effusion of blood from the rivulets of menstruation and their uterus is strongly developed to bear children. Thus they are also fertile and able to receive man's semen. Still, they do

---

[3]These four categories of women derive from ideas about four fluids (also called "humors") in the body that date back to the ancient Greeks. Europeans from the Greeks to the seventeenth century saw individuals as dominated by one of these humors (in the same way that we might describe someone according to psychologized personality traits, such as "Type A" or "anal retentive"), and saw illness as caused by an imbalance in these humors. Hildegard is the first to interpret these in terms of sexual behavior and to discuss them extensively in regard to women.

not bear many children. If they remain without a husband so that they do not bear offspring, they will possibly suffer physical pain, but if they have a husband they are healthy. If, at menstruation, drops of blood are locked up in them before the natural time, so that they do not flow out, then these women will occasionally be melancholic, or suffer a pain in the side, or a worm will grow in their flesh, or lymph nodes, called scrofulae, will burst, or a rather mild form of leprosy will develop.

Phlegmatic women. There are other women whose flesh does not grow much because they have thick blood vessels and rather healthy white blood containing, however, a small amount of poison which gives it its light color. They have a severe expression and a darkish coloring. They are industrious and useful and possess a somewhat virile mind. At menstruation their rivulets of blood flow moderately, neither too little nor too much. Because they have thick blood vessels they are extremely fertile with offspring and conceive easily, since also their uteri and all their viscera are strongly developed. They attract men and cause men to pursue them, and therefore men love them. If they wish to abstain from men, they are able to abstain from intercourse with them with only some, though not too much, debilitating effect. Yet if they have avoided intercourse with men, they will become morose and disagreeable in their demeanor. But if they have been together with men so that they do not wish to abstain from intercourse with them, they will be unrestrained and excessive in their lust, as has been observed by men. Because they are somewhat virile they will occasionally, due to the greenness within them, grow a little fluff on the chin. If, at menstruation, the rivulet of blood is constricted in them before the natural time, they will occasionally incur an unsoundness of the head, madness in other words. Or they will suffer from the spleen or from dropsy, or a swelling will develop as always in tumors, or they will develop wildly growing flesh on a limb, like a gall on a tree or on an apple.

Choleric women. There are other women who have soft flesh but big bones, average blood vessels and dense red blood. Their facial coloring is pallid, they are intelligent and kind. People show them respect and are afraid of them. At menstruation they suffer severe blood loss. Their uteri are strongly developed and they are fertile. Men like their disposition but stay out of their way because these women draw men's attention but do not attract them. In a marital union they are chaste and faithful as wives, and together with their husbands they are physically healthy. If they are deprived of husbands they will suffer physically and be debilitated by that, because they do not know in which man to put their womanly faith and also because they do not have a husband. If the flow at menstruation ceases sooner than it should, they will easily become paralyzed and their humors will dissolve, so that their humors become weak and these women will either feel pain in the liver or will easily incur a black tumor as from a dracunculus[4] or their breasts will swell with cancer.

Melancholic women. But there are other women who have haggard flesh, thick blood vessels, average-sized bones, and blood that is more bluish than sanguineous. Their faces are a blend of greyish and black color. These women are also windy, and wavering in their thoughts and wearisome when they waste away as a result of annoyance. They are not very resilient, so that at times they are weary from melancholia. At menstruation they suffer severe blood loss, and they are sterile because they have a weak and fragile uterus. Therefore they can neither receive nor retain nor warm man's semen. Consequently they are healthier, stronger, and happier without husbands than with them, because they will become weak if they have been with husbands. But men avoid them and shy away from them because they do not talk pleasantly to men and because men love them only a little. If, at some time, these women feel carnal pleasure, it will, however, pass quickly in them. But if they have robust and sanguine husbands, occasionally some

---

[4]*from a dracunculus: dragunculi,* the Guinea worm (*Dracunculus medinensis*) of North Africa and Asia, which causes eruptions on the skin of its hosts. Hildegard will have learned of it via the ancient medical authors.

of these women can bear at least one child when they reach a sound age like fifty. But if they have had different husbands whose nature is weak they will not conceive from them, but remain sterile. If their menstruation ceases sooner than is right for the nature of women, they will at times suffer gout and swelling of the legs, or they will incur an unsoundness of the head, brought on by melancholia. Or they will suffer back or kidney pain or a rapid swelling of the body because waste matter and foulness, from which menstruation should have purged their bodies, remain enclosed in them. If they do not receive any help in their infirmity, so that they are not freed from it by the help of God or by medicine, they will die very soon.

46. Thomas Aquinas
Summa Theologica

Thomas Aquinas (1225–1274) was a professor of theology and philosophy at the University of Paris and the most important scholastic philosopher of the Middle Ages. In 1879, Pope Leo XIII declared his ideas to be the official philosophy of the Catholic Church. Aquinas was a follower of Aristotle and so believed in the importance of logic and reason; one of his aims was to demonstrate that reason and faith are not contradictory, but complement one another. Aquinas wrote many works, the most significant of which was the *Summa Theologica,* a systematic discussion of Christian theology based on philosophical principles. Like many works of the scholastics, the *Summa* is organized into series of questions and answers, in which Aquinas states a question, sets out several negative answers to the question (termed *objections*), and then answers these objections. In this section, Aquinas adapts dichotomous notions of male and female nature derived from Aristotle to Christian teachings.

PART I, QUESTION XCII
THE PRODUCTION OF THE WOMAN (IN FOUR ARTICLES)

We must next consider the production of the woman. Under this head there are four points of inquiry: (1) Whether the woman should have been made in that first production of things? (2) Whether the woman should have been made from man? (3) Whether of man's rib? (4) Whether the woman was made immediately by God?

First Article

Whether the woman should have been made in the first production of things?

We proceed thus to the First Article:—

Objection 1. It would seem that the woman should not have been made in the first production of things. For the Philosopher[1] says, that the female is a misbegotten male. But nothing misbegotten or defective should have been in the first production of things. Therefore woman should not have been made at that first production.

Obj. 2: Further, subjection and limitation were a result of sin, for to the woman was it said after sin (Genesis iii. 16): Thou shalt be

------

[1] *The Philosopher:* Aristotle.

Summa Theologica by Thomas Aquinas, translated by Fathers of the English Dominican Provinces, Second Edition, Vol. 4. (London: Barns and Oats and Wahborune, Ltd., 1922).

under the man's power; and Gregory[2] says that, Where there is no sin, there is no inequality. But woman is naturally of less strength and dignity than man; for the agent is always more honourable than the patient, as Augustine[3] says. Therefore woman should not have been made in the first production of things before sin.

Obj. 3. Further, occasions of sin should be cut off. But God foresaw that the woman would be an occasion of sin to man. Therefore He should not have made woman.

On the contrary, It is written (Genesis ii. 18): It is not good for man to be alone; let us make him a helper like to himself.

I answer that, It was necessary for woman to be made, as the Scripture says, as a helper to man; not, indeed, as a helpmate in other works, as some say, since man can be more efficiently helped by another man in other works; but as a helper in the work of generation.

This can be made clear if we observe the mode of generation carried out in various living things. Some living things do not possess in themselves the power of generation, but are generated by some other specific agent, such as some plants and animals by the influence of the heavenly bodies, from some fitting matter and not from seed: others possess the active and passive generative power together; as we see in plants which are generated from seed; for the noblest vital function in plants is generation. Wherefore we observe that in these the active power of generation invariably accompanies the passive power. Among perfect animals the active power of generation belongs to the male sex, and the passive power to the female. And as among animals there is a vital operation nobler than generation, to which their life is principally directed; therefore, the male sex is not found in continual union with the female in perfect animals, but only at the time of coition;

so that we may consider that by this means the male and female are one, as in plants they are always united; although in some cases one of them preponderates, and in some the other. But man is yet further ordered to a still nobler vital action, and that is intellectual operation. Therefore there was greater reason for the distinction of these two forces in man; so that the female should be produced separately from the male; although they are carnally united for generation. Therefore directly after the formation of woman, it was said: And they shall be two in one flesh (Genesis ii. 24).

Reply Obj. 1. As regards the individual nature, woman is defective and misbegotten, for the active force in the male seed tends to the production of a perfect likeness in the masculine sex; while the production of woman comes from defect in the active force or from some material indisposition, or even from some external influence; such as that of a south wind, which is moist, as the Philosopher observes. On the other hand, as regards human nature in general, woman is not misbegotten, but is included in nature's intention as directed to the work of generation. Now the general intention of nature depends on God, Who is the universal Author of nature. Therefore, in producing nature, God formed not only the male but also the female.

Reply Obj. 2. Subjection is twofold. One is servile, by virtue of which a superior makes use of a subject for his own benefit; and this kind of subjection began after sin. There is another kind of subjection, which is called economic or civil, whereby the superior makes use of his subjects for their own benefit and good; and this kind of subjection existed even before sin. For good order would have been wanting in the human family if some were not governed by others wiser than themselves. So by such a kind of subjection woman is naturally subject to man, because in man the discretion of reason predominates. Nor is inequality among men excluded by the state of innocence, as we shall prove.

Reply Obj. 3. If God had deprived the world of all those things which proved an occasion of sin, the universe would have been imperfect.

---

[2]Gregory (c. 540–604), usually called "the Great," was the pope from 590 to 604 and a strong advocate of clerical celibacy and papal supremacy.
[3]Augustine (354–430) was an extremely influential Christian bishop and theologian.

Nor was it fitting for the common good to be destroyed in order that individual evil might be avoided; especially as God is so powerful that He can direct any evil to a good end.

Second Article

Whether woman should have been made from man?

We proceed thus to the Second Article:—

Objection 1. It would seem that woman should not have been made from man. For sex belongs both to man and animals. But in the other animals the female was not made from the male. Therefore neither should it have been so with man.

Obj. 2. Further, things of the same species are of the same matter. But male and female are of the same species. Therefore, as man was made of the slime of the earth, so woman should have been made of the same, and not from man.

Obj. 3. Further, woman was made to be a helpmate to man in the work of generation. But close relationship makes a person unfit for that office; hence near relations are debarred from intermarriage, as is written (Leviticus xviii. 6). Therefore woman should not have been made from man.

On the contrary, It is written: He created of him, that is, out of man, a helpmate like to himself, that is, woman.

I answer that, When all things were first formed, it was more suitable for the woman to be made from the man than (for the female to be from the male) in other animals. First, in order thus to give the first man a certain dignity consisting in this, that as God is the principle of the whole universe, so the first man, in likeness to God, was the principle of the whole human race. Wherefore Paul says that God made the whole human race from one (Acts xvii. 26). Secondly, that man might love woman all the more, and cleave to her more closely, knowing her to be fashioned from himself. Hence it is written (Genesis ii. 23, 24): She was taken out of man, wherefore a man shall leave father and mother, and shall cleave to his wife. This was most neces-

sary as regards the human race, in which the male and female live together for life; which is not the case with other animals. Thirdly, because, as the Philosopher says, the human male and female are united, not only for generation, as with other animals, but also for the purpose of domestic life, in which each has his or her particular duty, and in which the man is the head of the woman. Wherefore it was suitable for the woman to be made out of man, as out of her principle. Fourthly, there is a sacramental reason for this. For by this is signified that the Church takes her origin from Christ.[4] Wherefore the Apostle[5] says (Ephesians v. 32): This is a great sacrament; but I speak in Christ and in the Church.

Reply Obj. 1 is clear from the foregoing.

Reply Obj. 2. Matter is that from which something is made. Now created nature has a determinate principle; and since it is determined to one thing, it has also a determinate mode of proceeding. Wherefore from determinate matter it produces something in a determinate species. On the other hand, the Divine Power, being infinite, can produce things of the same species out of any matter, such as a man from the slime of the earth, and a woman from a man.

Reply Obj. 3. A certain affinity arises from natural generation, and this is an impediment to matrimony. Woman, however, was not produced from man by natural generation, but by the Divine Power alone. Wherefore Eve is not called the daughter of Adam; and so this argument does not prove.

Third Article

Whether the woman was fittingly made from the rib of man?

We proceed thus to the Third Article:—

Objection 1. It would seem that the woman should not have been formed from the rib of

---

[4]Here and elsewhere Aquinas compares the relationship between husband and wife with that between Christ and the church.

[5]*The Apostle:* Paul.

man. For the rib was much smaller than the woman's body. Now from a smaller thing a larger thing can be made only—either by addition (and then the woman ought to have been described as made out of that which was added, rather than out of the rib itself);—or by rarefaction, because, as Augustine says: A body cannot increase in bulk except by rarefaction. But the woman's body is not more rarefied than man's—at least, not in the proportion of a rib to Eve's body. Therefore Eve was not formed from a rib of Adam.

Obj. 2. Further, in those things which were first created there was nothing superfluous. Therefore a rib of Adam belonged to the integrity of his body. So, if a rib was removed, his body remained imperfect; which is unreasonable to suppose.

Obj. 3. Further, a rib cannot be removed from man without pain. But there was no pain before sin. Therefore it was not right for a rib to be taken from the man, that Eve might be made from it. On the contrary, It is written (Genesis ii. 22): God built the rib, which He took from Adam, into a woman.

I answer that, It was right for the woman to be made from a rib of man. First, to signify the social union of man and woman, for the woman should neither use authority over man, and so she was not made from his head; nor was it right for her to be subject to man's contempt as his slave, and so she was not made from his feet. Secondly, for the sacramental signification; for from the side of Christ sleeping on the Cross the Sacraments flowed—namely, blood and water—on which the Church was established. . . .

QUESTIONS FOR ANALYSIS

1. What differences does Hildegard of Bingen see in the role of men and women in procreation? What effects did the Fall have on human sexuality? In Hildegard's opinion, what role does love play in procreation and in creating differences among women?
2. Aquinas states his own ideas most forcefully in his answers to the objections. In these, how does he describe the qualities inherent in men and women? Did these exist in the original creation or only after the Fall?
3. How would you compare Hildegard and Aquinas's ideas about the role of women in procreation and conception? About the sources and nature of gender inequality?
4. Aquinas is generally regarded as the most important Christian philosopher and theologian of the Middle Ages; many would judge him second in influence only to Augustine in all of Christian history. By contrast, Hildegard's works have been translated and reprinted only very recently and are familiar to only a small group of people. Given this difference, how would you expect ideas about gender differences to develop in the centuries after they wrote? How might they have developed differently had the prominence of these two authors been reversed?

CHAPTER 5

The Late Middle Ages and the Renaissance

(Fourteenth–Sixteenth Centuries)

European culture during the High Middle Ages, with growing cities, increasing trade, expanding population, and new institutions such as guilds and universities, is generally described in very glowing terms, such as *vital, vibrant,* and *vigorous.* By contrast, the Late Middle Ages—the fourteenth and fifteenth centuries—is usually described in negative terms, such as *in decline, in crisis,* and even *calamitous.* These harsh assessments stem from a number of factors. Beginning around 1300, the European climate became colder and wetter, leading to crop failures and famine, sometimes localized and sometimes widespread. Famine led to population decline as deaths outnumbered births and people became less resistant to disease. This gradual decrease in population was speeded up dramatically in 1347 with the introduction of the bubonic plague in Europe, which killed between one-quarter and one-third of the population of many parts of Europe in the following three years. Though subsequent outbreaks were never as deadly as the first one, the plague stayed in Europe for centuries, joining other diseases to make a recovery of pre-plague population levels slow in coming.

The fourteenth century also saw a number of revolts in the country-side and in towns, as peasants sought to lessen the labor obligations imposed on them and lower-class urban residents sought to gain a political voice in cities ruled by wealthy merchants. England and France were at war for much of the period in the Hundred Years' War (1337–1453), and the Christian Church was split into factions as first two and later three men each claimed to be pope simultaneously.

Gender made little difference in terms of the impact of these various calamities. Bubonic plague is not gender-specific, nor is major famine. (There is some evidence that in many cultures when food supplies are

140

low, men and boys receive more food than women or girls; when there is no food at all, however, both sexes starve equally.) Women participated in revolts, though they were not usually leaders, and they suffered in the reprisals when these revolts failed. They fought only rarely in the Hundred Years' War—Joan of Arc is an exception, not a model for other women—but as armies confiscated crops and animals and burned villages, women also lost their homes and means of existence. No woman, of course, held high office within the Church, but as the various claimants to the papal throne each appointed local officials and declared the other pope and his supporters to be the Antichrist, women along with men had doubts about whether the Church hierarchy was concerned with their salvation. Both sexes turned to more interior and personal forms of religion or to groups that declared the existing Church to be flawed and misguided.

At the same time that Europe was suffering from plague, war, and famine, an intellectual movement began in northern Italian cities that looked back to classical antiquity for models of human behavior and paid more attention to worldly issues than religious ones. This movement, termed *humanism,* was initially concerned primarily with reforming education to put more emphasis on Greek and Latin literature and less on theology. Humanists opened academies, where they taught practical skills in writing and public speaking through classical examples, seeing their own age as one of a rebirth of classical culture, a Renaissance. They created a new ideal for men, the "Renaissance man" who excelled at everything he did and was proud of his accomplishments as an individual. The ideal "Renaissance man" was an active participant in urban culture and civic life, not a contemplative scholar in his study. These ideas came to be shared by many artists, who developed artistic styles based on those of antiquity, signed their works, and gradually came to be considered creative geniuses.

The extent to which women shared in Renaissance culture has been hotly debated since the publication, more than twenty years ago, of an essay by the historian Joan Kelly in which she asked, "Did women have a Renaissance?" Kelly answered her own question with "no," and both she and many subsequent historians have pointed to the disparity between male and female experience in the very things that most people think of when they hear the word *Renaissance.* Like universities, humanist academies were for male students only, with the few women who received an advanced education doing so at home with private tutors. The talents most highly praised in men, such as ability in public speaking or writing, were suspect in women, for as one fifteenth-century Italian humanist commented, "An eloquent woman is never chaste." Women were generally not accepted into programs for training artists or musicians, and they were discouraged from publishing any written work that was not religious. Thus creativity, along with reason, became another quality regarded as limited to men, or to women who had somehow transcended their sex. The arts themselves became gendered: the "high arts"—painting, sculpture, and architecture—were usually done by men; and the "minor arts" (sometimes further diminished and called "decorative arts" or "crafts")—textiles, porcelain manufacture and decoration, flower painting, collage, and needlework—involved women. Women's skill in these "arts" was judged to be the result of industriousness, copying a male example, or nimble fingers, not genius. The ideal Renaissance woman looked very much like her medieval counterpart—chaste, demure, pious, and obedient—rather than like a female counterpart to a "Renaissance man."

Some historians have viewed Kelly's answer to her own question as too sweeping and note that the experience of women during this period—like that of men—varied so much as to make generalizations about increasing or decreasing status meaningless. They note that we now often use the term *Renaissance* to mean not simply an intellectual and artistic movement, but a time period (roughly 1350 to 1600), so that issues other than humanism and art must be taken into account when we are assessing women's experience in the Renaissance. They point out that a few women did gain a humanist education and artistic training—often through their fathers—and that Renaissance culture touched very few men in any case. The secularism and individualism generally seen as hallmarks of the Renaissance were not accepted by most people—men as well as women—for whom religion and family remained important. Processes of social and economic change, such as the growth of cities, increasing commercialism and wage labor, and regional specialization, ultimately affected the lives of far more people than humanism, and the impact on women of these processes is more mixed. The debate that Kelly's question engendered continues, for it has led to very fruitful discussions of just what we mean by the term *Renaissance* and just who we mean when we talk about *women*.

RELIGION AND FAMILY

Though the Christian Church in western Europe was plagued by many problems during the Late Middle Ages and the Renaissance, religion remained very important to most people, including most women. The number of women in monasteries outnumbered that of men, and women experimented with different ways of creating lives of religious devotion without becoming nuns. Some of these women joined together in small communities of like-minded women; such women were often called "Beguines" and were both praised and attacked for their piety and independence. Other women sought to increase their level of devotion while continuing to live in a family by engaging in activities regarded as spiritually rewarding, such as giving to the poor or making donations to the Church. Still others joined groups that regarded official Church teachings and practices as deviating from the true message of Christianity; such groups, which the official Church judged to be "heresies," attracted large numbers of women and men in France and England and provoked sharp responses, including executions and military attacks.

Women's religious life in this period was linked in a variety of ways with their family life. For some women, familial and religious duties and loyalties conflicted with one another, as parents or husbands opposed their devotional practices or choices. For others, these two supported one another, and women viewed their obligations to and dependence on God and their family as connected. This was true for Jewish as well as Christian women in late medieval Europe, for Jewish women's religious activities were largely centered on the household rather than the temple or school that was the center of Jewish men's religious life.

47. *The Autobiography of Leonor López de Córdoba, c. 1412*

Leonor López de Córdoba (1361–1420) was the daughter of a Spanish knight caught up in a war between the king of Castile, Pedro I, and his half-brother Enrique de Trastámara, who killed Pedro and placed himself on the throne. Her father's loyalty to King Pedro resulted in a long imprisonment and other problems for the whole family, which Doña Leonor relates in a narrative account. This work is the first in Spain by a known woman author and the first Spanish autobiography by an author of either sex.

In the name of God the Father and of the Son and of the Holy Ghost, three persons, and one single God true in trinity, to which glory be given, to the Father and to the Son and to the Holy Ghost, thus as it was in the beginning, so it is now, and for ever and ever, amen. In the name of which above mentioned Lord and of the Holy Virgin Mary his mother, and lady and advocate of sinners, and to the honor and exaltation of all the angels and saints in the Court of Heaven amen.

Therefore, know all who see this document, how I, Doña Leonor López de Córdoba, daughter of my Lord Grand Master Don Martín López de Córdoba and Doña Sancha Carrillo, to whom God grant glory and heaven, swear by this sign † which I worship, that all that is written here is true for I saw it and it happened to me, and I write it to the honor and glory of my Lord Jesus Christ, and of the Holy Virgin Mary his mother who bore him so that all creatures who suffered might be certain that I believe in her mercy,

that if they commend themselves from the heart to the Holy Virgin Mary she will console and succor them as she consoled me. And so that whoever might hear it know the tale of my deeds and miracles that the Holy Virgin Mary showed me, it is my intention that it be left as a record. I ordered it written as you see before you. I am the daughter of the aforesaid Grand Master of Calatrava, in the time of Lord King Don Pedro, who bestowed the honor of giving him the Commandery of Alcantara, which is in the city of Seville. And then he made him Grand Master of Alcantara, and at last of Calatrava, and that Grand master my father was a descendant of the House of Aguillar and nephew of Don Juan Manuel, son of a niece of his, a daughter with a brother. And he rose to very high rank, as can be found in the *Chronicles of Spain.*

As I have said, I am the daughter of Doña Sancha Carrillo, niece and maid of the most illustriously remembered Lord King Don Alfonso (whom God grant Holy paradise), father of the aforementioned Lord King Don Pedro. My mother died very young, and so my father married me at the age of seven years to Ruy Gutiérrez de Henestrosa, son of Juan Ferrandez de Henestrosa, High Chamberlain of Lord King Don Pedro and his High Chancellor of the Secret Seal, and High Steward of Queen Doña Blanca his wife, who married Doña María de Haro, Lady of Haro and los Cameros. My husband inherited many goods from his father and many offices. And his men on horseback numbered three hundred, and forty skeins of pearls as fat as chickpeas, and five hundred Moors, men and women, and two thousand marks of silver in tableware, and the jewels and gems of his household you could not write down on two sheets of paper. And his father and his mother left all this to him because they had no other son or heir. To me, my father gave twenty thousand gold coins upon marriage. And my husband and I resided in Carmona with the daughters of Lord King Don Pedro, and my brothers-in-law, husbands of my sisters, and a brother of mine whose name was Don Lope López de Córdoba Carrillo. My brothers-in-law were named Fernán Rodríguez de Aza, Lord of Aza and Villalobos, and the other Ruy García de Aza, the other Lope Rodríguez de Aza, who were sons of Alvaro Rodríguez de Aza, and of Doña Constanza de Villalobos.

And thus it was, that when Lord King Don Pedro was surrounded in the Castle of Montiel by his brother Lord King Don Enrique, my father went down to Andalusia to bring people to help him, and on the road back discovered that Don Pedro was dead at the hands of his brother. Seeing this misfortune he took the road for Carmona where the lady princesses, daughters of Lord King Don Pedro were, and very close relatives of my husband and mine by my mother. And Lord King Don Enrique seeing himself King of Castile came to Seville and put a blockade around Carmona, and as it was such a strong town, it was blockaded for many months. My father having chanced to leave there, and the people of Real del Rey learning

that he had left that town and that there would no longer be such good protection there, twelve knights offered to climb into the town but were captured after scaling the wall. And then my father, advised of that deed, came and ordered their heads cut off for their effrontery. And Lord King Don Enrique, seeing that he could not enter there to get satisfaction for this act by force of arms, sent the High Constable of Castile to negotiate with my father. My father discussed two terms, the first that the lady princesses had to be set free in England with their treasures before he gave the town over to the king. And so it was done, for which reason he sent some illustrious gentlemen, kinsmen of his, born in Córdoba, to accompany them and the other people he deemed necessary. The other term was that he and his children and defenders, and those who had been present in that town by his order be pardoned by the King, and that they and their estates be considered loyal. Thus it was granted him, signed by the above mentioned High Constable in the name of the King. And by this agreement he gave the town over to the Constable in the King's name. And from there he, his family, and the rest of his people went to kiss the hand of the King. But the Lord King Don Enrique ordered them taken prisoner and put in the Arsenal of Seville. And the previously mentioned Constable, seeing that Lord King Don Enrique had not kept his word, which he had given in his name to the aforementioned Grand Master, left his court and never again returned. The Lord King ordered that they cut off my father's head in the Plaza de San Francisco in Seville, and that his property be confiscated, and that of his son-in-law, defenders, and servants. Going there to be beheaded, he met Mosén Beltrán de Clequin, the French knight trusted by King Don Pedro whom the King set free during the siege of Montiel; and not keeping his promise, Clequin handed him over to King Don Enrique instead so that he could kill him. And when he encountered the Grand Master he said to him: "Lord Grand Master did I not tell you that your exploits would end this way?" And my father answered him: "It is better to die loyal, as I have done, than to live as you have lived, a traitor."

The rest of us were kept prisoner for nine years until Lord King Enrique died. Our husbands had sixty pounds of iron each on their feet, and my brother Don Lope López had a chain on top of the irons, in which there were seventy links. He was a child of thirteen years, the most beautiful creature there was in the world. And they singled out my husband to be put in the hunger tank, where they held him for six or seven days without food or drink because he was a cousin of the lady princesses, daughters of Lord King Don Pedro. At this juncture a plague came, and my two brothers, my brother-in-law, and thirteen knights of the house of my father all died. And Sancho Míñez de Villendra, his high Chamberlain, said to me and to my brothers and sisters: "My lord's children: Pray God that I live, for if I do, you will never die poor." But it pleased God that he died three days later without speaking. And they took them all out to the ironsmith's like slaves to remove their irons. After they were dead my sad little brother Don Lope López asked the jailer who held us to tell Gonzalo Ruiz Bolante to do us a great kindness and a great honor for the love of God: "Sir jailer be so kind as to strike these irons from me before my soul departs, and do not let them take me out to the ironsmith's." He answered him as if he were speaking to a slave: "If it were up to me I would do it," and at that moment his soul departed in my hands. He was but one year older than I, and they took him out on a plank to the ironsmith's like a slave, and they buried him with my brothers and with my sisters, and with my brothers-in-law in San Francisco of Seville. My five brothers-in-law each wore chains of gold on their necks, for they had put on those necklaces at Santa María de Guadalupe and promised not to take them off, until all five might remove them at Santa María. But for their sins one died in Seville, another in Lisbon, another in England, and so they scattered, having willed they should be buried with their gold necklaces. But the friars greedily took his

necklace off him after he was buried. And no one from my lord the Grand Master's house remained in the Seville Arsenal but my husband and me.

Then the very eminent and very honorable, most illustriously and saintedly remembered Lord King Don Enrique died, and he ordered in his will that they let us out of prison and return to us all that was ours. And I stayed at the home of my lady aunt María García Carrillo, and my husband went to reclaim his property and those who held it esteemed him little, because he had no rank nor means to claim it, and you well know how rights depend on the station you have on which to base a claim. And thus was my husband lost, and he wandered seven years through the world, a wretched man, and never did he find a relative or friend who did him a good turn or had pity on him. And at the end of seven years, they told my husband, who was in Badajoz with his uncle Lope Fernández de Pidilla in the War of Portugal, that I was doing very well in the house of my lady and aunt Doña María García Carrillo, that my relatives had done me much kindness. He rode on his mule, which was worth very little money, and what he wore was not worth thirty *maravedís*. And he came through the doorway of my lady and aunt.

As I had known that my husband was wandering lost through the world, I consulted with my lady and aunt, sister of my lady and mother, whose name was Doña Teresa Fernández Carrillo (she was in the Order of Guadalajara, which my great grandparents founded and endowed with the money for forty rich females of their lineage to come into that order).[1] I asked her to request that I please be admitted into that order, since, for my sins, my husband and I were undone. And she and the entire order were happy to do it, because my lady mother had been raised in those monasteries, and King Don Pedro had taken her

---

[1]Wealthy families often endowed convents or other types of religious houses as honorable places for female family members who did not marry. The level of religious devotion in such houses varied widely.

from there. And he had given her to my father to marry, because she was the sister of Gonzalo Díaz Carrillo, and of Diego Carrillo, sons of Don Juan Fernández Carrillo, and of Doña Sancha de Roxas. These uncles of mine were afraid of the aforementioned Lord Don Pedro, for he had killed and exiled many of this family, and he had brought down my grandfather's house and given what he had to another; so these uncles of mine went off to serve King Don Enrique (when he was count) because of this affront. And I was born in Calatayud in the house of the Lord King. His daughters, the lady princesses, were my godmothers, and they brought me with them to the castle in Segovia with my lady mother who died there, and I was of such an age that I never knew her.

After my husband came, as I have said, he went to the house of my lady aunt, which was in Córdoba next to San Hipólito, and she took me in with my husband there in some houses adjacent to hers. And seeing that we had so little peace, for thirty days I said a prayer to the Holy Virgin Mary of Bethlehem, praying every night on my knees three hundred Hail Marys, that she might put it into my lady aunt's mind to consent and open a doorway into her dwellings. And two days before I finished the prayer, I asked my aunt if she would allow me to open that passageway so that we would not have to walk through the street, among all the knights who were in Córdoba, to come and eat at her table. And her grace responded that she would be happy to do so, and I was greatly consoled. When on the following day I tried to open the passageway, maids of hers had turned her against me, so that she would not do it, and I was so disconsolate I lost my patience, and the one who had most set my lady aunt against me died in my hands, swallowing her tongue. And the next day, Saturday, only one day remaining of my thirty days of prayer, I dreamed that in passing by San Hipólito with the dawn bells ringing, I saw on the courtyard walls a very large and very high arch, and that I entered there and picked flowers from the mountainside, and I saw the vast Heavens. At this point I

woke up and placed my hope in the Holy Virgin Mary that she would give me a house.

Then there was a raid on the Jewish quarter, and I took an orphan child who was there and had him baptized so that he might be instructed in the faith.[2] And one day, walking back with my lady aunt from mass at San Hipólito, I saw the clerics of San Hipólito dividing up those courtyards where I dreamt the great arch was, and I begged my lady aunt Doña Mencía Carrillo that she be so kind as to buy that place for me, since I had been in her company for seventeen years. She bought them for me and gave them with the condition that a chaplaincy be laid upon the stipulated houses for the soul of Lord King Don Alfonso, who built that church in the name of San Hipólito, because he was born on that day. These chaplains have another six or seven chaplaincies belonging to Don Gonzalo Fernández, husband of the said lady my aunt, and to their sons, the Marshall and Don Alfonso Fernández, Lord of Aguilar. This favor done, I raised my eyes to God and to the Virgin Mary, giving her thanks. And then there came a servant of the Grand Master, my lord and father, who lives with Martín Fernández, Castellan of los Donceles, who was there hearing mass; and I sent him a request with that servant that as a kinsman he give thanks to my lady aunt for the favor she had done for me. And he was very happy to do so, and did graciously, saying to her that he received this favor as if it had been done for him personally.

The title having been given me, I cut open a door on the very site where I had seen the arch the Virgin showed me. And it disturbed the abbots that they should hand the lot over to me because I was of grand lineage and my sons would be great, and they were abbots and had no need of great knights near them. But I took it for a good sign, and told them I hoped in God it would be thus, and I came to an agreement with them so that I placed the door where I wanted it. I believe

---

[2]Jewish children who had been orphaned—in this case by an attack on the Jewish section of town—were sometimes made the wards or servants of Christian families and forcibly baptized.

that for the charitable act I performed in raising that orphan in the faith of Jesus Christ, God helped me in giving me the beginning of a house. Before this time, I had gone barefoot in the wind and rain for thirty days to morning prayer to [the shrine of] María el Amortecida, which is in the order of San Pablo de Córdoba, and I prayed to her sixty-three times this prayer which is followed by sixty-six Hail Marys, in homage to the sixty-six years that she lived with bitterness in this world, that she might give me a house; and she gave me a house, and because of her mercy, houses better than I deserved. And the prayer begins: "Holy Mother Mary, great pain did you feel because of your son—you saw him tormented with his great suffering, and your heart came close to death. After his agony he gave you comfort, so intercede with my lady, for you know my pain." At this time, it pleased God that with the help of my lady aunt and of the labor of my hands I built in that courtyard two palaces and a garden and another two or three houses for the servants.

Then there came a very cruel pestilence, and my lady did not want to leave the city. I begged her for mercy to flee with my little children so that they would not die, and this did not please her, but she gave me permission, and I departed from Córdoba, and I went to Santaella with my children. The orphan I brought up lived in Santaella, and he gave me lodging in his house, and all the residents of the town were very happy with my going there, and received me very warmly because they had been servants of my lord and father. And thus they gave me the best house that there was in the place, which belonged to Fernando Alonso Mediabarba. My lady aunt arrived unexpectedly with her daughters, and I removed myself to a small apartment. And her daughters, my cousins, were never favorably disposed toward me because of the kindness their mother did me, and from then on I suffered so much bitterness that it cannot all be written down. And a pestilence came, and my lady departed with her people for Aguilar, and she took me with her, although her daughters thought that was doing too much, because she loved me

greatly and had a high opinion of me. And I sent the orphan whom I had raised to Ecija.

The night we arrived in Aguilar the young man came in from Ecija with two tumors on his throat and three dark blotches on his face and a very high fever. Don Alfonso Fernandez, my cousin, was there, and his wife and all his household, and although all of them were my nieces and my friends, they came to me when they found out that my servant had come in that state. They said to me: "Your servant Alonso has come with pestilence,[3] and if Don Alfonso Fernandez sees him, he will wreak havoc being in the presence of such an illness." You who hear this story can well understand the pain that came to my heart for I was angered and bitter. And think that such great suffering had entered the house on my account, I had a servant of the lord my father, the Grand Master, called, whose name was Miguel de Santaella, and I begged him to take that young man to his house, and the poor man became afraid and said: "My lady, how shall I take him sick with the pestilence, for it may kill me?" And I said to him: "Son, God shall not will it so." And he took him out of shame. And because of my sins, thirteen people, who kept vigil over him during the night, all died. And I offered a prayer that I had heard a nun say before a crucifix. It seems she was very devoted to Jesus Christ, and it is said that after she heard morning prayer, she came before the crucifix and prayed on her knees seven thousand times: "Merciful son of the Virgin, take pity." And one night the nun heard that the crucifix answered her and said: "You called me merciful and merciful I shall be." I placed great faith in these words and prayed this prayer every night, entreating God that he should want to free me and my children, and if any of them had to be taken away, it should be the older one for he was in great pain. And it was God's will that one night I could not find anyone to watch over that suffering young man, because all those who had watched over him up to then had died. And that son of mine whose name was

---

[3]Probably the bubonic plague.

Juan Fernández de Henestrosa, after his grandfather, and who was twelve years and four months of age, came to me and said: "My lady, is there no one who will watch over Alonso tonight?" And I said to him: "You watch over him for the love of God." And he answered me: "My lady, now that the others have died, do you want it to kill me?" And I said to him: "For the charitable act I am performing, God will take pity on me." And my son, so as not to disobey me, went to watch over him, and because of my sins, that night he came down with the plague and the next day I buried him. And the sick one survived, but all those stated above died. And Doña Teresa, wife of Don Alfonso Fernández my cousin, became very angry that my son was dying for that reason in her house, and with death in his mouth, she ordered him to be taken out. And I was so wrought with anguish that I could not speak for the shame that those noble people made me bear. And my sad little son said: "Tell my lady Doña Teresa that she not have me cast out, for my soul will soon depart for Heaven." And that night he died, and he was buried in Santa María la Coronada, which is in the town. But Doña Teresa had designs against me, and I did not know why. And she had ordered

that he not be buried within the town, and thus, when they took him to be buried I went with him. And when I was going down the street with my son, the people, offended for me, came out shouting: "Come out good people and you will see the most unfortunate, forsaken and condemned woman in the world," with cries that rent the Heavens. Since the residents of that place were all liege and subject to my lord father, and although they knew it troubled their masters, they made great display of the grief they shared with me, as if I were their lady.

That night, as I came back from burying my son, they told me that I should go to Córdoba, and I approached my lady aunt to see if she would order me to do it. But she said to me: "Lady niece, I cannot fail to do so, as I have promised my daughter-in-law and my daughters who are of one mind; since they have pressed me to remove you from my presence, I have granted it to them. I do not know what vexation you have caused my daughter-in-law, Doña Teresa, that she feels such ill will toward you." And I said to her with many tears: "My lady, may God not save me if I deserved this." And thus I came to my houses in Córdoba.

48. The Book of Margery Kempe

Margery Kempe (c. 1373–c. 1438) was the daughter of a prosperous urban merchant in the city of Lynn in England. She married at about twenty and had fourteen children; after the birth of the first, she went through a period of depression and began a series of intense religious experiences, during which she saw visions and heard voices. She related these ecstatic experiences to her friends and neighbors—often loudly and with tears while in church or other public places—and was inspired by them to go on a pilgrimage to Jerusalem. She was suspected of heretical opinions and investigated by religious authorities. She decided to tell her story in writing, using the services of a priest as she herself was illiterate. This text, which was lost except for a small segment until the mid-twentieth century, is the first autobiography in English; in it, Margery refers to herself in the third person, as "this creature."

Reprinted by permission from The Council of the Early English Text Society.

On a night, as this creature lay in her bed with her husband, she heard a sound of melody so sweet and delectable, that she thought she had been in Paradise, and therewith she started out of her bed and said:—

'Alas, that ever I did sin! It is full merry in Heaven.'

This melody was so sweet that it surpassed all melody that ever might be heard in this world, without any comparison, and caused her, when she heard any mirth or melody afterwards, to have full plenteous and abundant tears of high devotion, with great sobbings and sighings after the bliss of Heaven, not dreading the shames and the spites of this wretched world. Ever after this inspiration, she had in her mind the mirth and the melody that was in Heaven, so much, that she could not well restrain herself from speaking thereof, for wherever she was in any company she would say oftentimes:—'It is full merry in Heaven.'

And they that knew her behaviour beforetime, and now heard her speaking so much of the bliss of heaven, said to her:—

'Why speak ye so of the mirth that is in Heaven? Ye know it not, and ye have not been there, any more than we.' And were wroth with her, for she would not hear nor speak of worldly things as they did, and as she did beforetime.

And after this time she had never desired to commune fleshly with her husband, for the debt of matrimony was so abominable to her that she would rather, she thought, have eaten or drunk the ooze and the muck in the gutter than consent to any fleshly communing, save only for obedience.

So she said to her husband:—'I may not deny you my body, but the love of my heart and my affections are withdrawn from all earthly creatures, and set only in God.'

He would have his will and she obeyed, with great weeping and sorrowing that she might not live chaste. And oftentimes this creature counselled her husband to live chaste, and said that they often, she knew well, had displeased God by their inordinate love, and the great delectation they each had in using the other,

and now it was good that they should, by the common will and consent of them both, punish and chastise themselves wilfully by abstaining from the lust of their bodies. Her husband said it was good to do so, but he might not yet. He would when God willed. And so he used her as he had done before. He would not spare her. And ever she prayed to God that she might live chaste; and three or four years after, when it pleased Our Lord, he made a vow of chastity, as shall be written afterwards, by leave of Jesus.

And also, after this creature heard this heavenly melody, she did great bodily penance. She was shriven[1] sometimes twice or thrice on a day, and specially of that sin she so long had (hid), concealed and covered, as is written in the beginning of the book.[2]

She gave herself up to great fasting and great watching; she rose at two or three of the clock, and went to church, and was there at her prayers unto the time of noon and also all the afternoon. Then she was slandered and reproved by many people, because she kept so strict a life. She got a hair-cloth from a kiln, such as men dry malt on, and laid it in her kirtle as secretly and privily as she might, so that her husband should not espy it.[3] Nor did he, and she lay by him every night in his bed and wore the hair-cloth every day, and bore children in the time.

Then she had three years of great labour with temptations which she bore as meekly as she could, thanking Our Lord for all His gifts, and was as merry when she was reproved, scorned and japed for Our Lord's love, and much more merry than she was beforetime in the worship of the world. For she knew right well she had sinned greatly against God and was worthy of more shame and sorrow than any man could

---

[1]*Shriven:* received forgiveness and penance for a sin after confessing it to a priest.
[2]It is never clear exactly what this sin is; after the birth of her first child, at the beginning of the book, dread about her lack of forgiveness drives Kempe into postpartum depression.
[3]Hair-cloth was very scratchy; Kempe wore it under her clothing as a form of self-mortification and penance.

cause her, and despite of the world was the right way Heavenwards, since Christ Himself had chosen that way. All His apostles, martyrs, confessors and virgins, and all that ever came to Heaven, passed by the way of tribulation, and she, desiring nothing so much as Heaven, then was glad in her conscience when she believed that she was entering the way that would lead her to the place she most desired.

And this creature had contrition and great compunction with plenteous tears and many boisterous sobbings for her sins and for her unkindness against her Maker. She repented from her childhood for unkindness, as Our Lord would put it in her mind, full many a time. Then, beholding her own wickedness, she could but sorrow and weep and ever pray for mercy and forgiveness. Her weeping was so plenteous and continuing, that many people thought she could weep and leave off, as she liked. And therefore many men said she was a false hypocrite, and wept before the world for succour and worldly goods. Then full many forsook her that loved her before while she was in the world, and would not know her. And ever, she thanked God for all, desiring nothing but mercy and forgiveness of sin....

At the time that this creature had revelations, Our Lord said to her:—'Daughter, thou art with child.'

She said to Him:—'Ah! Lord, what shall I do for the keeping of my child?'

'Our Lord said:—'Dread thee not. I shall arrange for a keeper.'

'Lord, I am not worthy to hear Thee speak, and thus to commune with my husband. Nevertheless, it is to me great pain and great disease.'

'Therefore it is no sin to thee, daughter, for it is rather to thee reward and merit, and thou shalt have never the less grace, for I will that thou bring Me forth more fruit.'

Then said the creature:—'Lord Jesus, this manner of living belongeth to Thy holy maidens.'

'Yea, daughter, trow thou right well that I love wives also, and specially those wives who would live chaste if they might have their will and do their business to please Me as thou dost; for, though the state of maidenhood be more perfect and more holy than the state of widowhood, and the state of widowhood more perfect than the state of wedlock, yet, daughter, I love thee as well as any maiden in the world. No man may hinder Me in loving whom I will, and as much as I will, for love, daughter, quencheth all sin. And therefore ask of Me the gifts of love. There is no gift so holy as is the gift of love, nor anything to be desired so much as love, for love may purchase what it can desire. And therefore, daughter, thou mayest no better please God than continually to think on His love.... As this creature lay in contemplation for weeping, in her spirit she said to Our Lord Jesus Christ:—

'Ah! Lord, maidens dance now merrily in Heaven. Shall not I do so? For, because I am no maiden, lack of maidenhood is to me now great sorrow; methinketh I would I had been slain when I was taken from the font-stone,[4] so that I should never have displeased Thee, and then shouldst Thou, blessed Lord, have had my maidenhood without end. Ah! dear God, I have not loved Thee all the days of my life and that sore rueth[5] me; I have run away from Thee, and Thou hast run after me; I would fall into despair, and Thou wouldst not suffer me.' ...

'Daughter, when thou art in Heaven, thou shalt be able to ask what thou wilt, and I shall grant thee all thy desire. I have told thee beforetime that thou art a singular lover, and therefore thou shalt have a singular love in Heaven, a singular reward, and a singular worship. And, forasmuch as thou art a maiden in thy soul, I shall take thee by the one hand in Heaven, and My Mother by the other hand, and so shalt thou dance in Heaven with other holy maidens and virgins, for I may call thee dearly bought, and Mine own dearworthy darling.

---

[4]*font-stone:* baptismal font.
[5]*rueth:* troubles; bothers.

49. Ordinances Regarding Monasteries for Women, Florence 1435

Like the majority of women in medieval Europe, Doña Leonor and Margery Kempe married and had children, practicing their religion from within a family. A significant minority—in some places perhaps as many as 15 percent—of women did not marry, however. Some of these women lived on their own or with their families, supporting themselves through various types of occupations, while others entered convents or other religious institutions. Medieval and Renaissance secular governments joined the Church in attempting to regulate and oversee convent life. Both church and state officials—who were all male except for a handful of female rulers—regarded marriage as the norm for women and thought of women in convents as "brides of Christ." This notion of nuns as married to Christ—which was accepted by many nuns themselves—shaped attitudes toward those who broke their vows. The following is from the 1435 deliberations of the Officials of the Curfew and the Convents, a government body in Florence.

In the name of Christ our Savior, amen. The trumpet of the Lord and the voice of the Highest shall call out on the day of judgment: "You who are worthy, come, O blessed of my Father; and you who are unworthy, O accursed ones, go into the eternal fire...." For above there will be the angry judge, and below will be chaos; on the right will be those accused of sin, and on the left, an infinite number of demons. Without and within the earth burns.... All of this was described by Augustine,[1] the doctor of the church.

Furthermore, natural law has ordained that the human species should multiply and that man and woman be joined together by holy matrimony, and that this [institution] is and should be of such gravity and dignity that it should be respected by everyone. Nothing is more pleasing to God than the preservation of matrimony; nothing is more displeasing to him

than its violation. And if a mortal being is angered by this violation, how much more outraged ought to be the Creator, the Father of every living creature. There are many nuns, brides of God and dedicated to Him, who are enclosed in convents in order to serve him with their virginity, but who through carnal desire have failed in their reverence to Him. As a result of this, divine providence is perturbed and has afflicted the world with the evils of wars, disorders, epidemics, and other calamities and troubles. In order to avoid these, the severity of the Florentine people has decreed that no one is permitted to enter any convent, and that a heavy penalty be meted out to delinquents. Not only carnal relations with nuns but also access to the convents are prohibited, so that these nuns, whose sex is weak and fragile... will be preserved in security and honor and the convents flourish in liberty.... Thus, these nuns will not be transformed, by the audacity of evil men, from virtue into dishonor, from chastity into luxury, and from modesty into shame....

---

[1] Augustine (354–430 C.E.) was a bishop and leading thinker in Christianity.

From *The Society of Renaissance Florence: A Documentary Study* edited by Gene Brucker. Copyright © 1971 by Harper. Reprinted by permission from the author.

50. Anonymous Poems Opposing Entrance to a Convent, France and Spain

During the Renaissance, dowries appear to have gone up in many parts of Europe, particularly for wealthy families, and families with several daughters could not afford to properly dower all of them. Without a high enough dowry, a young woman could not attract an appropriate husband, a situation that continued in Europe for many centuries, as readers of Jane Austen's novels, or viewers of their screen adaptations, know well. Convents also demanded a fee to enter, but this was lower than the dowry expected of upper-class girls, and many girls were sent to convents. In Florence in the mid-sixteenth century, perhaps as many as one-quarter of all upper-class young women were in convents. Girls and women sent to convents sometimes objected, and the situation was widespread enough to be reflected in popular songs and poetry. The following are anonymous ballads from France and Spain, taken from songbooks which began to be compiled in the fifteenth century. We cannot know the sex of their authors, but they express a familiar situation.

A. Motet

I am merry,
Pretty, pleasing,
A young maiden;
Not quite fifteen;
Little breasts swelling
In keeping with the season;
I should indeed be learning
Of love, and discovering
Its charming
Face.
But I am put in prison.
God's curse on
Him who put me in this place!
Evil and villainy
And sin did he
To send to a nunnery
Such a maid as I.

I'faith, he did too great a wrong;
In religion I live in misery,
O God, for I am too young.
I feel beneath my belt the sweet pain:
May God curse him who made me nun.

—*Muriel Kittel* (trans.)

B. Now That I'm Young

Now that I'm young
  I want my fun,
I can't serve God
  being a nun.

Now that I'm young
  and come of age,
why be a nun
  in a convent caged?
I can't serve God

Anonymous, "Motet," c. twelfth century. English translation by Muriel Kittel. Reprinted, by permission of The Feminist Press at The City University of New York, from *The Defiant Muse: French Feminist Poems from the Middle Ages to the Present, A Bilingual Anthology,* edited by Domna C. Stanton. Translation copyright © 1986 by Muriel Kittel. Compilation copyright © 1986 by Domna C. Stanton.

"Now That I'm Young" translated by William M. Davis.

Anonymous (a nun from Alcalá), "My Parents, As If Enemies" (Mis padres, como enemigos), c. late seventeenth century. English translation by Kate Flores. Reprinted, by permission of The Feminist Press at The City University of New York, from *The Defiant Muse: Hispanic Feminist Poems from the Middle Ages to the Present, A Bilingual Anthology,* edited by Angel Flores and Kate Flores. Translation copyright © 1986 by Kate Flores. Compilation copyright © 1986 by Angel Flores and Kate Flores.

being a nun!

Now that I'm young
  I want my fun,
I can't serve God
  being a nun.

—*William M. Davis* (trans.)

C. My Parents, as If Enemies

My parents, as if enemies

of the life they gave me,
alive have buried me here
between wickets and iron bars. . . .

Where all that I can feel
is that a pleasing mate,
even if imaginary,
is more pleasing than a convent grate.

—*Kate Flores* (trans.)

51. The Will of Anne Latimer, 1402

Along with letters and the very rare autobiography, wills were another place where women could express their religious devotion and ties to family and household members. Wills were generally drawn up by a professional notary, and some of the language is formulaic. In some parts of Europe, testators—persons making a will—were restricted in how they could divide up their goods, but in England testators were largely free to do what they wanted. This will comes from a woman of the gentry class who had a moderate amount of wealth. She apparently had no children and was less concerned with arranging for priests to say masses for her soul than were many of her contemporaries.

In the name of God, amen. On 13 July 1402, I Anne Latimer, thanking God in his mercy for having such mind as he vouchsafes, and desiring that God's will be fulfilled in me and in the use of all the goods that I have been given to keep, make my testament with that in mind in this form. First I place my soul in the hands of God, praying to him humbly that by his grace he will take as poor a present as my wretched soul to his mercy. I will that my body be buried at Braybrooke beside my lord and husband Thomas Latimer, if God wills. I bequeath to the repair of the chancel and the parsonage of the church at Braybrooke 40s. 40s also to make the bridge which my lord[1] began. I also bequeath £20 to be doled out to needy poor men known by the discretion of the overseers and executors of my will. To Roger my brother, 40s. To Alison Bretoun, five

marks. To Kalyn Okham, 20s. To Anneys, 20s. To Magote Deye, 20s. To Thomas Fetplace, 26s 8d. To William, my brother's man, 3s 4d. To William Leycestrechyre, 10s. I also bequeath 40s to be divided among the rest of my servants by the discretion of the executors and overseers of this testament. The rest of my goods I wish to be sold and doled out to needy poor men according to the law of God, by the advice and discretion of the overseers and executors of this will. To execute, ordain and accomplish this will truly, I principally desire and pray Master Philip, abbot of Leicester, and Sir Lewis Clifford, and Robert parson of Braybrooke to be overseers, so that all these things may be fulfilled according to the law of God. For my executors of this will I ask Sir Robert Lethelade parson of 'Kynmerton', Thomas Wakeleyn, Sir Henry Slayer parson of Warden, and John Pulton. May these things be done in the name and worship of God, amen. In

---

[1] *My lord:* in this case, her husband.

From *Women for the English Nobility and Gentry, 1066–1509,* by Jennifer Ward (editor), 1995, Manchester University Press, Manchester, U.K. Reprinted by permission.

witness of this will I affix my seal. Witnessed by: Sir Robert priest of Braybrooke, Thomas Fet- place and Alison Bretoun. Written on the year and day aforesaid.

52. Inquisition Trial of Béatrice de Planissoles, 1320

During the fourteenth century, some heretical movements gained large numbers of adherents, and the Church set up special inquisitorial courts to investigate those suspected of heretical beliefs. One of the largest was the movement termed the *Cathars* or *Albigensians* in southern France and northern Spain, whose adherents denounced the church for its wealth and worldliness. Albigensians were dualists, regarding the material world as evil and the spiritual world as good. Women along with men were attracted to these ideas and were questioned by inquisitors. This source is from the inquisition held by Jacques Fournier, the bishop of Pamiers in France who later became Pope Benedict XII. The woman he is questioning, Béatrice de Planissoles, was a member of the minor nobility; she was condemned to death for her beliefs, but the sentence was later lessened to wearing a double-cross, as a sign of heretical beliefs, on her clothing.

Twenty-six years ago during the month of August (I do not recall the day), I was the wife of the late knight Bérenger de Roquefort, castelain of Montaillou. The late Raimond Roussel, of Rades, was the intendant and the stewart of our household which we held at the castle of Montaillou. He often asked me to leave with him and to go to Lombardy with the good Christians[1] who are there, telling me that the Lord had said that man must quit his father, mother, wife, husband, son and daughter and follow him, and that he would give him the kingdom of heaven. When I asked him, "How could I quit my husband and my sons?" he replied that the Lord had ordered it and that it was better to leave a husband and sons whose eyes rot than to abandon him who lives for eternity and who gives the kingdom of heaven.

When I asked him "How is it possible that God created so many men and women if many of them are not saved?" he answered that only the good Christians will be saved and no others, neither religious nor priests, nor anyone except

these good Christians. Because, he said, just as it is impossible for a camel to pass through the eye of a needle, it is impossible for those who are rich to be saved. This is why the kings and princes, prelates and religious, and all those who have wealth, cannot be saved, but only the good Christians. They remained in Lombardy, because they did not dare live here where the wolves and the dogs would persecute them. The wolves and the dogs were the bishops and the Dominicans[2] who persecute the good Christians and chase them from this country.

He said that he had listened to some of these good Christians. They were such that once one had heard them speak one could not do without them and if I heard them one time, I would be one of theirs for ever.

When I asked how we could flee together and go to the good Christians, because, when my husband found out, he would follow us and kill us, Raimond answered that when my husband would take a long trip and be far from our country, we could leave and go to the good Christians. I asked

---

[1]*Good Christians:* in this case, Cathars.

[2]Dominicans were a religious order whose members often served as inquisitors.

"Inquisition Trial of Béatrice de Planissoles" from *Readings in Medieval History* translated and edited by Patrick Geary. Copyright © 1989 by Broadview Press. Reprinted with permission.

him how we would live when we were there. He answered that they would take care of us and give us enough with which to live. "But," I told him, "I am pregnant. What could I do with the child that I am carrying when I leave with you for the good Christians?" "If you give birth to it in their presence, it will be an angel. With the help of God they will make a king and a holy thing of him because he will be born without sin, not having frequented the people of this world, and they would be able to educate him perfectly in their sect, since he would know no other."...

About 21 years ago, a year after the death of my husband, I wanted to go to the church of Montaillou to confess during Lent. When I was there, I went to Pierre Clergue, the rector, who was hearing confessions behind the altar of Saint Mary. As soon as I had knelt before him, he embraced me, saying that there was no woman in the world that he loved as much as me. In my surprise, I left without having confessed.

Later, around Easter, he visited me several times, and he asked me to give myself to him. I said one day that he so bothered me in my home that I would rather give myself to four men than to a single priest because I had heard it said that a woman who gave herself to a priest could not see the face of God. To which he answered that I was an ignorant fool because the sin is the same for a woman to know her husband or another man, and the same whether the man were husband or priest. It was even a greater sin with a husband he said, because the wife did not think that she had sinned with her husband but realized it with other men. The sin was therefore greater in the first case.

I asked him how he, who was a priest, could speak like that, since the church said that marriage had been instituted by God, and that it was the first sacrament instituted by God between Adam and Eve, as a result of which it was not a sin when spouses know each other. He answered, "If it was God who instituted marriage between Adam and Eve and if he created them, why didn't he protect them from sin?" I understood then that he was saying that God did not create Adam and Eve and that he had not instituted marriage between them. He

added that the Church taught many things which were contrary to truths. Ecclesiastics said these things because without them it would inspire neither respect nor fear. Because, except for the Gospels and the Lord's Prayer, all of the other texts of Scripture were only "affitilhas," a word in the vernacular which designates what one adds on one's own to what one has heard.

I answered that in this case ecclesiastics were throwing the people into error.

(August 8, 1320, in the Chamber of the bishop before the bishop and Gaillard de Pomiès)

Speaking of marriage, he told me that many of the rules concerning it do not come from divine will who did not forbid people to marry their sisters or other persons related by blood, since at the beginning brothers knew their sister. But when several brothers had one or two pretty sisters, each wanted to have her or them. The result was many murders among them and this is why the Church had forbidden brothers to know their sisters or blood relatives carnally. But for God the sin is the same whether it is an outside woman, a sister, or another relative, because the sin is as great with one woman as with another, except that it is a greater sin between a husband and wife, because they do not confess it and they unite themselves without shame.

He added that the marriage was complete and consummated as soon as a person had promised his faith to the other. What is done at the church between spouses, such as the nuptial benediction, was only a secular ceremony which had no value and had been instituted by the Church only for secular splendor.

He further told me that a man and a woman could freely commit any sort of sin as long as they lived in this world and act entirely according to their pleasure. It was sufficient that at their death they be received into the sect or the faith of the good Christians to be saved and absolved of all the sins committed during this life. Because, he said, Christ said to his apostles to leave father, mother, spouse, and children, and all that they possessed, to follow him, in order to have the kingdom of heaven. Peter answered Christ, "If we, who have left everything and followed you, will have the kingdom of heaven, what will be the share of those who are ill and cannot follow you?" The Lord answered Peter that his "friends" would come and impose their hands on the heads of the ill. The ill would be healed and healed, they would follow him and have the kingdom of heaven.

The rector said that these "friends of God" were the good Christians, whom others call heretics. The imposition of the hands that they give to the dying saves them and absolves them of all their sins.

To prove that it was better for the world if brothers married sisters, he told me, "You see that we are four brothers. I am a priest and do not want a wife. If my brothers Guillaume and Bernard had married Esclarmonde and Guillemette, our sisters, our house would not have been ruined by the need of giving them a dowry. It would have remained whole. With a wife who would have been brought into the house for Raimond, our brother, we would have had enough spouses and our house would have been richer. It would therefore have been better if the brother married the sister or the sister the brother, because when she leaves her paternal house with great wealth in order to marry an outsider, the house is ruined."

And with these opinions and many others he influenced me to the point that in the octave of Saints Peter and Paul I gave myself to him one night in my home. This was often repeated and he kept me like this for one and one half years, coming two or three times each week to spend the night in my house near the chateau of Montaillou. . . .

Pierre Clergue, the rector, told me that this world which the devil had made is decaying and will totally destroy itself, and before this happens, God will assemble his friends and draw them to himself so that they do not see so great a catastrophe as that of the destruction of the world.

When I went down from the region of Alion to contract a marriage with my second husband, Othon de Lagleize of Dalou, this rector told me that he was very displeased that I was going down into that low country because no one would dare speak further with me about the good Christians or come see me to save my soul. I was going to live with wolves and dogs, of whom, he said, none would be saved. He called all the Catholics who are not good Christians wolves and dogs.

He nevertheless said that if one day my heart told me to be received into the sect of good Christians, I should let him know early because he would do all that he could so that the good Christians would receive me into their sect and save my soul. I answered him that I did not want to be received into that sect but that I wanted to be saved in the faith in which I was, quoting my sister Gentille, who had first said this.

These heretical conversations continued between us for around two years and this priest taught me all of this.

—Did you believe and do you still believe these heresies that this rector of the church of Montaillou, Pierre Clergue, told you and in which he instructed you?—The last year, when I left the region of Alion from Easter until the following August, I completely and fully believed these errors to the point that I would not have hesitated to undergo any suffering to defend them. I believed that they were truth taught by the priest whom, because he was a priest, I believed what he said. But when I was at Crampagna with my second husband and I heard the preaching of the Dominicans and Franciscans[3] and I visited with faithful Christians, I abandoned these errors and heresies and I confessed at the penitential court of a Franciscan of the convent of Limoux in the Church of Notre Dame de Marseille where I had gone to see my sister Gentille, who lived at Limoux and who was the wife of the late Paga de Post. I made this confession fifteen years ago and I remained around five years without confessing heretical opinions that I had heard and believed although I confessed my other sins during these five years.

At the time when I believed these heresies, I did not see (nor did I see before or after) any heretics that I knew to be heretics, although I believed that they were good men because they suffered martyrdom for God and this priest had taught me that it was only in their sect that one could be saved.

---

[3]Franciscans were also a religious order.

I greatly regret having ever heard these here-tical opinions and even more for having be-lieved these heresies, and I am ready to accept the penance that my lord the bishop may wish to impose on me for them.

QUESTIONS FOR ANALYSIS

1. To whom is Doña Leonor's religious devotion especially addressed? Margery Kempe's? Anne Latimer's?
2. When Doña Leonor is describing her family in the opening of her autobiography, who does she mention most prominently? Are these individuals the same as those she interacts with throughout her life? Why might they be different?
3. Along with family members, who else appears to be significant to Doña Leonor and Anne Latimer? What evidence suggests this?
4. Why might Margery Kempe's ideas and actions have provoked the reaction they did from her neighbors? Why does her status as a wife and mother trouble her?
5. Given the reasons why women entered convents, why might the restrictions placed on convent life by the Florentine government have seemed necessary? Why might the families of upper-class girls in particular—the type who would write poetry—be especially eager to place them in convents?
6. What distinctive opinions about marriage and the family do the Cathars, to whom Béatrice de Planissoles speaks, hold? How do these differ from those of the official Church, Doña Leonor, and Margery Kempe?

THE HOUSEHOLD ECONOMY

Despite the growing importance of money in the economy, the household con-tinued to be the basic unit of economic organization in the Late Middle Ages. Wages were usually paid to individuals, but this often masked an entire family's labor. In mining areas, for example, men were paid for cleaned and sorted ore; adult men did the actual mining underground, while their wives and children did the cleaning and sorting above ground. Married women's property and wages were considered part of their husband's holdings, though unmarried women and widows could exert some control over their own property.

Toward the end of the Middle Ages, the European economy grew increasingly specialized by region, with vast grain-growing areas in the east supplying growing cities in Italy and the Low Countries with basic foodstuffs and the production of such items as wine and silk developing in certain areas. The merchants who handled this international trade, and who became fabulously wealthy through it, were almost all male, although women occasionally invested in their ventures or sold the prod-ucts of international trade, such as pepper and oranges, at local markets. Women were often employed in the production of items for trade; for example, growing grain in Russia, unwinding silk cocoons in Italy, and drying fish in Denmark.

Like trade, organizations that structured production in many European cities also offered more opportunities for men than for women. Craft guilds set the rules by which most items were manufactured, including training requirements, quality and price levels, hours of operation, and size of shops. Each shop was headed by a master craftsman, with journeymen and apprentices working for him and hoping to become masters themselves someday. There were a few all-women's guilds in large cities like Paris and Cologne, but in general most guild members were men. The master's wife, daughters, and female domestic servants assisted in the shop and fed and clothed the apprentices and journeymen, but they played no formal role in the guild.

53. Duties of Rural Workers

Agricultural production in the Later Middle Ages continued to be carried out by peasant households in most parts of Europe, with a gender division of labor typical of that in grain-growing societies. Men were responsible for clearing new land, plowing, and the care of large animals; women were responsible for the care of small animals, spinning, and food preparation. These gender-specific tasks were often maintained when agriculture was gradually commercialized and wage labor was added to subsistence agriculture. Larger landowners began to hire workers at busy times of the year, such as planting and harvest, as early as the fourteenth century and to hire servants throughout the year for general labor or specific tasks. Women were hired as day-laborers (usually at wages about half those of men) and as permanent employees. The following is an extract from the ordinances regulating work at a large country estate at Erfurt, Germany, in 1490.

THE CHEESEMOTHER

The cheesemother shall make sure that the cows and the calves are fed and watered, given straw, and had their manure removed at the right times. She is to make sure that the cows are well milked at the right time, and assist in this, and bring the milk into the cellar and pour it into the trough. She should skim the cream from the milk at the right time, make it into butter, salt the butter, take it to the kitchen master and put it into tubs [to be sold].

She and the dairymaid shall make the cheese, and put it on shelves in the cheese-room. She is to turn it over and press it, and take it to the kitchen-clerk.[1] When it is time, she and the dairymaid are

to take the towels, washcloths, breadcloths, pillow- and cushion-covers from the kitchen clerk after he has counted them and wash them vigorously, press them, and give them back to the kitchen clerk. Only the kitchen clerk and the cheesemother are to have keys to the cheese room.

If people want to bathe, she and the dairymaid are to make the suds, heat the bathing-room and wash the seats and floor, the footstools and the cushions.

The cheesemother is to feed the geese, and if they are to be eaten, pluck them and give the feathers to the kitchen clerk.

During harvest time, if she and the dairymaid have time they are to help with the harvest.

If she takes oats from the baker, she is to give these only to the geese and not to the calves.

She is to pay attention so that the oxen are taken in at the right times, the calves are

---

[1]The kitchen clerk kept records of everything that went into and out of the kitchen.

From *Frauen im Mittelalter, Bd. I: Frauenarbeit im Mittlealter* by Peter Ketsch, pp. 96–97. © 1963 by Schwann. Translated by Merry Wiesner.

weaned and eaten before Lent, and that during Lent the milk is made into butter and cheese.[2]

She and the dairymaid are to sleep by each other in the same room, each in her own bed.

When the baker bakes, she and the dairymaid are to help him.

THE DAIRYMAID

The dairymaid is to obey the cheesemother, follow her instructions, and, as it says above, help her.

She is to give food and drink to the cows, calves, old and young pigs, and clean their manure. The shepherds are to help her.

In winter she is to give food to the fatted pigs, and when it is time she should drive them into the water and wash them. The shepherds are to help her.

In winter she is to take straw to the cows in the cowstalls, feed the young calves and milk cows in the shed, pick up the chaff from the floor and soften it in hot water for the cows and the pigs. The shepherds are to help her.

If baths are wanted, and the servant normally responsible for this is not free to do this, she and the shepherds should carry wood and draw water, fill the bath-basin, and warm the bathing-room.

---

[2]Lent was the forty days before Easter during which Christians did not eat meat or animal products. Male calves—other than those few allowed to grow into bulls—were thus killed and eaten before Lent, and milk produced by cows during Lent was made into cheese and butter to be eaten at Easter feasts or post-Easter meals.

54. Weavers' Ordinances Regarding Women's Work in England

The earliest guild regulations in Europe, from the thirteenth and fourteenth centuries, were very short and made little explicit mention of women. Widows appear to have been able to run shops after the death of their husbands and perhaps even take on new apprentices; masters' wives and daughters worked at whatever tasks were needed. This informal participation began to change in some parts of Europe with the economic dislocation of the Late Middle Ages, and ordinances such as the following became common in many crafts in many cities. These two are from two towns in England, the first from Shrewsbury in 1448 and the second from Bristol in 1461.

[1.] Also that no woman shall occupy the craft of weaving after the death of her husband except for one quarter of the year, within which time it shall be lawful to her to work out her stuff that remains with her unworked, so that she be ruled and governed by the wardens and stewards of the said craft during the said terms as for the good rule of the said craft....

[2.] Also, it is agreed, ordained, and assented by William Canynges, mayor of the town of Bristol,

Thomas Kempson, sheriff of the same, and all the Common Council of the said town of Bristol held in the Guildhall there the 24th day of September in the first year of the reign of King Edward the Fourth after the Conquest, that for as much as various persons of the weavers' craft of the said town of Bristol direct, employ, and engage their wives, daughters, and maids some to weave on their own looms and some to engage them to work with other persons of the

From *Women of the English Nobility and Gentry, 1066–1509*, by Jennifer Ward (editor), 1995, Manchester University Press, Manchester, U.K. Reprinted by permission.

said craft, whereby many and various of the king's subjects, men liable to do the king service in his wars and in defence of this his land, and sufficiently skilled in the said craft, go vagrant and unemployed, and may not have work for their livelihood, therefore that no person of the said craft of weavers within the said town of Bristol from this day forward set, put, or engage his wife, daughter, or maid to any such occupation of weaving at the loom with himself or with any other person of the said craft within

the said town of Bristol, and that upon pain of losing...6s 8d to be levied, half the use of the Chamber of Bristol aforesaid and half to the contribution of the said craft, always provided and except that this act does not pertain to the wife of any man of the said craft now living at the making of this act, but that they may employ their wives during the natural life of the said women in the manner and form as they have done before the making of this said act etc.

55. Household Slaves in Florence

Rural slavery slowly died out in western Europe during the Middle Ages, but the growth of cities provided a new market for household slaves, most of them women. Girls and young women were brought into western European cities from the Balkans, North Africa, and the Black Sea area to serve in wealthy urban households. This practice was especially common in the cities of northern Italy where the Renaissance began; Florence officially legalized slavery and the importation and sale of slaves during the fourteenth century, and most well-to-do households had at least one female slave (*ancilla*). Domestic slaves had a range of household duties, sometimes including sexual ones, though they might also later be freed, given a dowry, and allowed to marry. In theory, slaves were not supposed to be Christians, though this restriction was often ignored, as the second source below indicates. The first source is a letter from a Florentine merchant to an associate in Venice written in 1392; the second is a ruling of one of the city's courts in 1399.

[1.]...If you haven't written to Spalato [in Dalmatia] to Bartolomeo or to others to send you the two slave girls about whom I wrote you in other letters, I beg you to write him...so that he will send you the slaves and the documents of purchase. You can transport them from Venice and send them to me.... These slave girls should be between twelve and fifteen years old, and if there aren't any available at that age, but a little older or younger, don't neglect to send them. I would prefer to have them younger than twelve instead of older than fifteen, as long as they are not under ten. I don't care if they are pretty or ugly, as long as they are healthy and able to do hard work....

[2.]...We condemn...Romeo di Lapo of Florence, a vagabond with no fixed residence...a man of base condition, life, and reputation, a thief and a vendor of women and Christians.... Romeo went to the suburb of Narente in the city of Ragusa [in Dalmatia] and there...with bland and deceptive words he cajoled Ciaola of Albania and Mazia Scosse of Bosnia and her daughter Caterina, all baptized Christians and free, saying to them: "Come with me to Italy and there I will find good husbands for you, and you will remain free." Knowing that they were free and Christian women, Romeo with his associates took Ciaola, Mazia, and Caterina to the city of Lesina in Dalmatia and there he pursued his evil intention of

From *The Society of Renaissance Florence: A Documentary Study* edited by Gene Brucker. Copyright © 1971 by Harper. Reprinted by permission from the author.

depriving them of their liberty. When the women were absent, he had a document drawn up by a notary who (so he said) was named Ser Domenico di Cobutio of Viterbo, which stated that Romeo had bought these women, who were heretics and unbaptized, for the price of 57 gold florins.... Then he took Ciaola, Mazia, and Cate-rina to Florence and sold the woman called Ciaola to Ser Stefano di Rainieri del Forese, of the parish of S. Trinita of Florence, for 49 florins, asserting that she was a heretic and unbaptized. [Romeo escaped the authorities; he was sentenced to death *in absentia* and the three women were freed.]

56. Sumptuary Laws and Their Enforcement

Households were not only producers in the Middle Ages and the Renaissance; like modern households, they were also consumers and buying units. City governments, similar to those of contemporary governments, enacted laws that protected consumers. These laws, for example, set fines for merchants who shortchanged their customers and forbid the adulteration of food and beverages with unhealthy substances. The governments also passed laws, which have no modern equivalent, limiting people's ability to consume. Such laws, termed *sumptuary laws,* set upper limits on expenditures for weddings, baptisms, and funerals and outlawed certain styles of dress and amounts of jewelry. Very often such laws were set according to social class, and they frequently focused on women's clothing. The first two sources give some of the reasoning behind sumptuary laws on the part of city officials, and the third presents a few actual cases of their enforcement; all three are from Florence in the late fourteenth and early fifteenth centuries.

[1.] Considering the Commune's need for revenue to pay current expenses...they have enacted...the following:

First, all women and girls, whether married or not, whether betrothed or not, of whatever age, rank, and condition...who wear—or who wear in future—any gold, silver, pearls, precious stones, bells, ribbons of gold or silver, or cloth of silk brocade on their bodies or heads... for the ornamentation of their bodies...will be required to pay each year...the sum of 50 florins...to the treasurer of the gabelle on contracts....[The exceptions to this prohibition are] that every married woman may wear on her hand or hands as many as two rings....And every married woman or girl who is betrothed may wear...a silver belt which does not exceed fourteen ounces in weight....

[2.]...After diligent examination and mature deliberation, the lord priors have seen, heard, and considered certain regulations issued in the current year in the month of August by Francesco di Andrea da Quarata and other officials who are called "the officials to restrain female ornaments and dress." They are aware of the authority granted to the lord priors and the officials responsible for ornaments by a provision passed in July of the current year.... They realize the great desire of these officials to restrain the barbarous and irrepressible bestiality of women, who, not considering the fragility of their nature, but rather with that reprobate and diabolical nature, they force their men, with their honeyed poison, to submit to them. But it is not in accordance with nature for women to be burdened by so many expensive ornaments, and on account of these unbearable expenses, men are avoiding matrimony.... But women were created to replenish this free city, and to live chastely in matrimony, and not to spend gold and silver on clothing and jewelry. For did not God himself say: "Increase and multiply and replenish the earth"? The lord priors, having diligently considered the regulations recorded below...and in order that the city be reformed with good customs, and so that the bestial audacity of these women be restrained, approve and confirm the following regulations....

[3.] [June 10, 1378] We prosecute Nicolosa, daughter of Niccolò Soderini, of the parish of S. Frediano, aged ten years. Nicolosa was discovered wearing a dress made of two pieces of silk, with tassels and bound with various pieces of black leather, in violation of the Communal statutes. [She confessed through her procurator and paid a fine of 14 lire.]

[November 12, 1378] We prosecute Flora, a servant of Messer Jacopo Sacchetti.... Flora was discovered wearing a knitted gown and also a cap of samite[1] and slippers on her feet in violation of the Communal statutes. [She confessed and was fined 22½ lire.]

[June 27, 1397] We prosecute Monna Agnella, wife of Giovanni di Messer Michele de'Medici, parish of S. Lorenzo.... Agnella was found wearing a prohibited gown, one part of sky blue cloth and another part of velvet with sleeves wider than one yard in circumference, in violation of the Communal statutes. [She confessed and paid a fine of 28 lire.]

----

[1]*Samite:* fancy silk fabric with gold or silver threads.

57. Women's Response to Sumptuary Legislation

It is clear from cases in Florence and many other cities that women broke sumptuary laws all the time, wearing forbidden clothing and jewelry or buying expensive food for family celebrations. We cannot tell anything about their motivations from their actions alone, nor even whether it was they or their male family members who decided they should flaunt the law. There are occasional literary glimpses of women's reactions, as in this song written by an anonymous female troubadour in southern France in the thirteenth century, taken from a slightly later manuscript. All of the female troubadours whose names and social status can be known were noblewomen, who could afford the type of clothing that sumptuary laws proscribed.

In heavy grief, in heavy dismay,
and in dreadful pain, I weep and sigh.
When I gaze at myself my heart all but cracks,
and I nearly go blind when I look at my clothes
(rich and noble,
trimmed with fine gold,
worked with silver)
or look at my crown.
May the Pope in Rome
send him to the fire
who untrims our clothes.

I will not observe this custom,
this law they've just made,
for Iacme the King wasn't there,
nor was the Pope; let the order be lifted;
they've harmed and dishonored
our rich clothing.
May the law's author suffer
to see every woman resolve
not to wear veil or wimple
but garlands of flowers
in the summer for love.

Whenever our lord the King may come
(from him comes all merit)
let pity move him to hear our outcry
against the offense brought on by his stewards,
who have torn from our clothing
its chains
and its buttons.
See that our persons
are no longer shamed:
pray, have them restored
to us, high, honored King.

Let us, lord goldsmiths and jewelers,
and ladies and girls who are of their trade,
ask the Pope in a message
to excommunicate council and councilmen,
and the friars minor,
who are greatly to blame for this,
the preachers
and penitentials
who show their ill will in it,
and other regulars
accustomed to preach it.

Go, my sirventes, to the good King of Aragon
and to the Pope; let them undo the law,
for—as God grant me grace—our ignoble hus-
    bands
have done a vile deed.

. . .

that way I'll be happier.
The girdle I used to fasten
dismays me. Alas!
I dare not wear it.

I grieve for my white blouse
embroidered with silk—
jonquil, vermilion and black mixed to-
    gether,
white, blue, gold and silver.
Alas! I dare not put it on.
My heart feels like breaking,

. . .

and it's no wonder.
Lords, make me a coarse cloak;
I prefer to wear that
when my clothes have no trimmings.

QUESTIONS FOR ANALYSIS

1. What restrictions were placed on women working as weavers? What justifications do officials in Bristol give for these restrictions?
2. In what ways were the lives of slave girls and women such as those in Florence different from those of rural workers such as the cheesemother and dairymaid at Erfurt? In what ways might they have been similar?
3. What reasons do the authorities of Florence give for passing sumptuary laws? What do these indicate about their concepts of women's nature?
4. Who does the author of the poem against sumptuary laws blame for passing them? Do the process and restrictions s/he discusses suggest that the sumptuary laws in France were similar to those in Florence?

WOMEN'S NATURE AND EDUCATION
IN THE RENAISSANCE

As we have seen, debates about women's nature were part of European culture from its beginning, with Aristotle, Plato, Plutarch, the apostle Paul, Jerome, Thomas Aquinas, Hildegard of Bingen, and many others debating what made women good or bad and what made them different from men. From the classical period through the Middle Ages, criticism of women—in scholarly treatises and in vernacular plays, poems, and songs—vastly outweighed defenses of women. During the fourteenth century, a few authors began to answer this misogynist tradition, beginning a debate about women (often described in its French variant, the *querelle des femmes*) that would last for centuries. Supporters of women often couched their defenses as long lists of virtuous and brave women from mythology, ancient history, Christian tradition, and contemporary politics, though their message was somewhat ambiguous, as the highest praise they gave a woman was that she was like a man. Most of this debate was carried out by male authors, some of whom argued both sides of the question to show off their knowledge and debating skills, so it is hard to gauge their true sentiments. The attacks on women continued to be reprinted more often and argued more forcefully, however, and to find visual expression in woodcuts, engravings, and paintings on objects of daily use such as plates and chests.

Defenders of women sometimes conceded that women were inferior to men but attributed this to gaps in their education; part of the Renaissance debate about women was a discussion about the merits and limits of women's education. No one suggested that universities or humanist academies should be opened to women, but some writers, both male and female, argued that women with enough property to support themselves should be able to hire a tutor or study on their own. Even those writers who favored women's education often limited the range of subjects, however, and were ambivalent about the purpose of educating women when public careers were closed to them.

58. Christine de Pizan
The Book of the City of Ladies

Christine de Pizan (ca. 1364–1430) was the daughter and wife of highly educated men who held positions at the court of the king of France. She was widowed at twenty-five with young children and an elderly mother to support. Christine had received an excellent education herself and decided to support her family through writing, an unusual choice for anyone in this era before the printing press and unheard of for a woman. She began to write prose works and poetry, sending them to wealthy individuals in hopes of receiving their support. Her works were well received, and Christine gained commissions to write specific works, including biographies, histories, and books of military tactics. She became the first woman in Europe to make her living as a writer and eventually turned her attention to the debate about women in a number of works. The most important of these was *The Book of the City of Ladies,* written in 1405, in which Christine directly ponders the issue of why misogynist ideas are so widely held and whether the attacks on women might be true. The book was first translated into English in 1521 but did not become well known in any language until the eighteenth century.

Selections, as attached, from *The Selected Writings of Christine de Pizan,* translated by Renate Blumenfield-Kosinski and Kevin Brownlee.

Here begins the Book of the City of Ladies, the First Chapter of Which Tells Why and for What Reason This Book was Made

These are my habits and the way I spend my life: studying literature. One day, doing just that, I was sitting in my study surrounded by several volumes on a variety of topics; at that moment my mind was occupied with the weighty opinions of various authors that I had studied for a long time. I looked up from my book, wondering whether I should leave alone for now these subtle problems and enjoy myself by looking at some cheerful poetry. And as I was looking around for some small book of that kind, I came across a strange book that did not belong to me but which had been given to me for safekeeping. When I opened it I saw on that title page that it was by Mathéolus.[1] Then I smiled, for although I had never seen it, I had heard people say that it, like other books, spoke well and with reverence of women, and so I thought that I would take a look at it to enjoy myself. But I had looked at it only for a moment when my good mother called me to supper—for it was that time already—and I put down the book with the intention of looking at it the next day.

The next morning, again sitting in my study as usual, I did not forget that I had wanted to look at this book by Mathéolus; so I began reading it and made a little progress. But since the subject matter did not seem very agreeable to people who do not enjoy slander, and since it did not contribute anything to the building up of virtue or good manners, in view of its dishonest themes and subject, I read a little bit here and there and looked at the end, and then put it down in order to devote myself to the study of higher and more useful things. But the sight of this book, although it was of no authority, made me think along new lines which made me wonder about the reasons why so many different men, learned and nonlearned, have been and are so

ready to say and write in their treatises so many evil and reproachful things about women and their behavior. And not just one or two, and not just this Mathéolus, who has no particular reputation and writes in a mocking manner, but more generally it seems that in all treatises philosophers, poets, and orators, whose names it would take too long to enumerate, all speak with the same mouth and all arrive at the same conclusion: that women's ways are inclined to and full of all possible vices.

Thinking these things over very deeply, I began to examine myself and my behavior as a natural woman, and likewise I thought about other women that I see frequently, princesses, great ladies as well as a great many ladies of the middle and lower classes who were gracious enough to tell me their private and hidden thoughts. From all this I hoped that I could judge in my conscience, without prejudice, whether these things, to which so many notable men bear witness, could be true. But according to everything I could know about this problem, however I looked at it and peeled away its various layers, it was clear to me that these judgments did not square with the natural behavior and ways of women. Nevertheless, I argued strongly against women, saying that it would be unlikely that so many famous men, such solemn scholars of such vast understanding, so clear-sighted in all things as these men seemed to be, could have lied in so many places that I could hardly find a book on morals, no matter who was its author, in which, even before finishing it, I would not find some chapters or certain sections speaking ill of women. This reason alone, in short, made me conclude that, although my intellect in its simplicity and ignorance did not recognize the great defects in myself and in other women, it must nonetheless be so. And thus I relied more on the judgment of others than on what I myself felt and knew.

I was so deeply and for such a long time transfixed by this thought that I seemed to be in a trance; and a large number of authors on this subject passed through my mind, one after the other, just like a gushing fountain. And eventually I

---

[1]Christine is apparently reading *The Lamentations of Mathéolus,* a minor author who gathers together many misogynist texts.

concluded that God made a vile thing when He formed woman, wondering how such a worthy artisan could have stooped to making such an abominable piece of work which is the vessel, as they say, as well as the hiding-place and shelter of every evil and vice. As I was thinking this, a great unhappiness and sadness rose up in my heart, and I despised myself and the entire feminine sex, just as if it were a monstrosity in nature. And in my grief I spoke the following words:

"Oh, God, how can this be? For lest I be mistaken in my faith, I am not allowed to doubt that Your infinite wisdom and perfect goodness created anything that is not good. Did not You Yourself form woman in a very singular way, and did You not give her all those inclinations which it pleased You she should have? And how could it be that You should have erred in anything? And nevertheless, here are all the great accusations against them, all judged, determined, and concluded. I cannot understand this hostility. And if this is so, dear Lord God, that it is true that all these abominations abound in the feminine sex—as so many testify—and if You Yourself say that the testimony of more than one witness should be believed, why should I doubt that this is true? Alas, God, why did You not let me be born into this world as a man, so that I would be inclined to serve you better and so that I would not err in anything and be of such perfection as man is said to be? But since Your kindness does not extend to me, forgive my negligence in Your service, good Lord God, and let it not displease You: for the servant who receives fewer gifts from his lord is less obliged to be of service to him."

In my grief I spoke these words and many more to God for a very long time, and in my folly I behaved as if I should be most unhappy because God had me exist in this world in a female body.

Here Christine Tells How Three Ladies Appeared to Her, and How the One Who Was in Front Addressed Her First and Comforted Her in Her Unhappiness

Lost in these painful thoughts, my head bowed in shame, my eyes full of tears, my hand supporting my cheek and my elbow on the pommel of my chair's armrest, I suddenly saw a ray of light descending onto my lap as if it were the sun. And as I was sitting in a dark place where the sun could not shine at this hour, I was startled as if awakened from sleep. And as I lifted my head to see where this light was coming from, I saw standing before me three crowned ladies of great nobility. The light coming from their bright faces illuminated me and the whole room. Now, no one would ask whether I was surprised, given that my doors were closed, and nevertheless they had come here. Wondering whether some phantom had come to tempt me, in my fright I made the sign of the cross on my forehead.

Then the first of the three began to address me as follows: "Dear daughter, do not be afraid, for we have not come to bother or to trouble you but rather to comfort you, having taken pity on your distress, and to move you out of the ignorance that blinds your own intelligence so that you reject what you know for certain and believe what you do not know, see, and recognize except through a variety of strange opinions. You resemble the fool in that funny story who was dressed in a woman's dress while he slept in a mill. When he woke up those who made fun of him told him that he was a woman and he believed their lies more readily than the certainty of his own being.

"Fair daughter, what has happened to your good sense? Have you forgotten that when fine gold is tested in the furnace, it does not change or vary in strength, but rather gets purer the more it is hammered and handled in various ways? Do you not know that the best things are those that are most debated and argued about? If you just look at the highest things, which are ideas and celestial things, try to see whether the greatest philosophers, those whom you use to argue against your own sex, have ever determined what is false and contrary to the truth and whether they have not contradicted and blamed each other. You have seen this yourself in the *Metaphysics,* where Aristotle argues against some opinions and speaks of Plato and others in this way. And note, moreover, how Saint Augustine and other doctors of the

Church have criticized certain places in Aristotle even though he is called the prince of philosophers and was a supreme master of both natural and moral philosophy.

"And it seems that you think that all the words of the philosophers are articles of faith and that they cannot be wrong. And as for the poets of whom you speak, don't you know that they have spoken of many things in fables, and that many times they mean the opposite of what their texts seem to say? And one can approach them through the grammatical figure of *antiphrasis,* which means, as you know, that if someone says this is bad, it actually means it is good and vice versa. I therefore advise you to profit from their texts and that you interpret the passages where they speak ill of women that way, no matter what their intention was. And perhaps this man who called himself Mathéolus understood things in his own book in this way: for there are many things in it which, if taken literally, would be pure heresy. And as for the accusations against the holy estate of marriage ordained by God, put forth not only by him and others but even by the *Romance of the Rose*[2] to which people give greater credence because of its author's great authority, it is clearly proved by experience that the contrary of the evil that they say exists in this estate through the fault of women is true. For where has there ever been a husband who would permit his wife to dominate him in such a way that she could have the right to abuse and insult him, as those who speak of women claim? I believe that, no matter what you have seen written, you will never with your own eyes see such a husband; these lies are painted too badly. I tell you in conclusion, dear friend, that simplicity has brought you to your current opinion. Come back to yourself, recover your good sense and do not bother yourself anymore with these absurdities. For you should know that all evil things that are said about women in such a

general way only hurt those who say them, and not women themselves.". . .

"There is another greater and more special reason for our coming which you learn from what we tell you: . . . the simple good ladies, following the examples of suffering that God commands, have gladly suffered the great offenses that, by speaking and by writing, have wrongfully and sinfully been done to them by those who claim to have the right to do so from God. But now it is time that their just cause be taken from Pharaoh's hands, and this is why we three ladies, that you see here, moved by pity, have come to you to announce a certain edifice built like the wall of a city, with strong stones and well constructed, which you are predestined and made to build with our help and counsel, and where no one will live except all ladies of renown and worthy of praise: for to those who are without virtue the walls of our city will be closed.". . .

[The lady who is speaking explains that she is Reason, and that her companions are Righteousness and Justice. Each of these allegorical figures helps Christine build her "city of ladies," and provides her with many examples of strong and virtuous women. The rest of the book is primarily a dialogue between Christine and the three Figures. Speaking to Reason, Christine says:]

"Lady, I remember well that you told me earlier, on the subject of why so many men have blamed and continue to blame the behavior of women, that the longer gold is in the furnace the purer it gets: which means that the more wrongfully they are blamed, the greater is the merit of their glory. But I get you to tell me why and for what reason so many different authors have spoken against women in their books, since I already know from you that they are wrong: is it that Nature makes them do it or do they do it out of hatred, and where does all this come from?"

Then she replied: "Daughter, to give you a way of entering more deeply, I will remove this first basketful of soil. You should know that all this does not come from Nature but is in opposition to her; for in this world there is no greater and stronger bond than that of the great love that Nature, by the will of God, forged

---

[2]The *Romance of the Rose* was a long poem written by two different authors. It begins as an allegory of a lover's quest and ends as a sharp critique of love and women.

between man and woman. But the causes that have moved and still move men to blame women are diverse and varied, and the same goes for the authors in their books, as you have found. For some have done this with good intentions: that is, to get men that were led astray away from frequenting vicious and dissolute women, with whom they may be besotted or to keep them from getting besotted in the first place, and so that men avoid a lewd and lascivious life. They have blamed all women in general because they believe that they should all be abominated."

"Lady," I said then, "forgive me if I interrupt your words: have they done well, then, because they were motivated by good intentions? For the intention, so people say, judges the man." "This is badly put, dear daughter," she said, "for one should never excuse gross ignorance. If someone killed you with good intentions moved by crazy thoughts, would this then be well done? Rather, those that act like this, whoever they may be, misinterpreted the law: for to harm and wrong one party in order to help another is not justice, and nor is to blame all feminine behavior in opposition to the truth, as I will demonstrate by this hypothesis: Let us suppose that they did it in order to get fools away from foolish behavior. It would be as if I blamed fire, which is, after all, very good and necessary, just because some people burned themselves, or water because people drown in it. And the same could be said of all good things that can be used well or badly. Nonetheless, one should not blame them just because fools abuse them; you yourself have touched upon this point quite well elsewhere in your texts.[3] But those who have spoken abundantly on the subject, whatever their intentions might be, have cast their net rather widely just to achieve their goal. Just like someone who has a long and wide robe cut from a large piece of cloth that costs him nothing and that no one refuses him: he takes and unsurps the rights of others.

---

[3]A reference to Christine's *The God of Love's Letter*, lines 341–47.

"But, as you have said earlier, if these writers had looked for ways and means to get men away from folly and to keep them away by untiringly blaming those women who show themselves to be vicious and dissolute—which is exactly, to tell the straight truth, what an evil, dissolute, and perverse woman does, a woman who is like a monster in nature, a counterfeit far removed from her true natural condition, which must be simple, quiet, and honest—if they had done this, then I would agree that they would have built a good and beautiful piece of work. But to blame all of them, when there are so many excellent women, I can assure you that this did not come from me and that whoever does this and whoever follows this approach is making a great mistake. So now throw away these dirty, black, and knobbed stones, for they will never be part of the beautiful edifice of your city.

"Other men have blamed women for other reasons: some have invented blame because of their own vices and others have been motivated by the defects in their own bodies, others through pure envy, and some others by the sheer pleasure they experience from slander. Others, in order to show that they have read many texts, base themselves on what they have found in books and repeat others and cite authorities.

"Those who have invented blame because of their own vices are men who wasted their youth in dissolution and had a great many love affairs with different women, used deception in many instances, have grown old in their sins without repentance, and now regret their past follies and the dissolute life they led in their time. But nature, which does not allow the fulfillment of the heart's desire without sufficient power of the appetite, has grown cold in them. They are mournful when they see that the life that they used to call good times is over for them, and it seems to them that the young, who are now what they used to be, are enjoying the good times. They do not know how to make their sadness go away except by blaming women, believing that in this way they will make them displeasing to others. And one sees commonly such old men speaking obscenely and dishonestly, just as you can see with Mathéolus, who admits himself that he was an old man with plenty of will but no potency. Through him you can prove that what I tell you is true, and you can firmly believe that the same holds true for many others.

"But these corrupt old men who are like incurable leprosy, are not the good, valiant men of ancient times whom I made perfect in virtue and wisdom—for not all old men have such corrupt desires, it would too bad if such were the case—and in whose mouths are, according to their hearts, good, exemplary, honest, and discreet words. And these men hate misdeeds and slander and neither blame nor defame men or women, they hate vices and blame them in general without indicting or charging anyone in particular, they counsel the avoidance of evil and the pursuit of virtue and the straight path.

"Those men who are motivated by the defect of their own bodies are impotent and have deformed limbs but sharp and malicious minds, and they cannot avenge the pain of their impotence except by blaming those women who bring joy to many: and thus they hope to spoil for others the pleasure that they themselves cannot enjoy.

"Those who blame women out of jealousy are those wicked men who have seen and perceived many women of greater intelligence and nobler conduct than they themselves possess, and thus they are full of sorrow and disdain; and for these reasons their great jealousy has made them blame all women, hoping to suppress and diminish their glory and praise, just like I do not know which man who in his text entitled *On Philosophy* makes a great effort to prove that women should not be honored by men, and he says that those men who make so much of women pervert the name of his book: that is to say that out of 'philosophy' they make 'philofolly.'[4] But I assure and swear to you that he himself, through the deduction—filled with lies—of the case he makes there, transforms the content of his book into a true philofolly.

"As for those who are slanderers by nature, it is no wonder that they blame women since they blame everyone. Nevertheless, I assure you that any man who willingly slanders women, does so because of a vile heart, for he acts against reason and nature. Against reason, in so far that he is most ungrateful and ignorant of the great good that woman has brought him, so often and continuously catering to his needs; it is so great a good that he could never pay her back. And it is against nature, in so far as there is no mute beast anywhere, nor is there a bird who does not by nature dearly love its companion: and that is the female! And thus it is quite unnatural when a reasonable man does the opposite.

"And just as there has never been a work so worthy and made by such a good master that some people did not, and still do not want, to counterfeit it, there are many who want to try their hand at writing poetry.[5] And it seems to them that they cannot go wrong, since others have stated in their books what they want to say—or rather misstated, as I well know. Some want to embark on expressing themselves by making poems of water without salt, such as they are, or ballads without feeling, speaking of the behavior of women or princes or other people, while they themselves cannot recognize or correct their own miserable behavior and inclinations. But the simple people, who are as ignorant as they are, say that these poems are the best in the world."...

Against Those Who Say That It Is Not Good That Women Should Pursue Learning

After hearing these things, I, Christine, spoke as follows: "Lady, I see well that many good things have come about through women, and if any evil resulted from evil women, it still seems to me that the good things brought about by good women, and also by the wise women and those learned in

---

[4]*Sophia* means "wisdom"; *philosophy* is the "love of wisdom." Christine invents the counterpart "love of folly."

[5]The word in French is *dicter*. Eustache Deschamps (or Eustache Morel), to whom Christine wrote a letter in 1404 (see p. 109), had written an *Art de dictier* [Art of poetry] in 1393.

literature and the sciences I mentioned above, are more numerous. For this reason I am extremely amazed by the opinion of some men who say that they do not want their daughters, wives, or female relatives to study the sciences and that their morals would worsen through this."

Answer: "In this you can clearly see that not all the opinions of men are based on reason and that these men are wrong. For it should not be assumed that knowing the moral sciences, which teach the virtues, would worsen morals, rather, there is no doubt that they improve and ennoble them. How is it possible to believe that a person who follows good teaching and doctrine could be the worse for it? Such a thing cannot be uttered or supported. I do not say that it would be good for a man or a woman to study the science of sorcery, or those areas that are forbidden—for the Holy Church did not remove them from common usage for nothing—but that women should get worse by knowing good things is not believable."

The End of the Book: Christine Speaks to the Ladies

My most revered ladies, God be praised, for now our City is all finished and completed, where all of you who love virtue, glory, and praise may be lodged, ladies from past times just as much as from the present and future, for it has been constructed and founded for every honorable lady. And my dearest ladies, it is natural for the human heart to be joyful when it finds itself to have gained victory in any enterprise and the enemies are confounded. Thus you have reason to rejoice virtuously in God and in good comportment when you see this new city perfected, which can be not only the refuge for all of you, that is, virtuous women, but also the defense and protection against your enemies and assailants, if you keep them up well. For you can see that the material from which it is made is entirely of virtue, indeed it is so resplendent that you can all mirror yourselves in it and especially in the upper structures built in this last part, as well as in the other parts that might apply to you.

And, my dear ladies, do not misuse this new heritage, like the arrogant people who become proud when their prosperity increases and their riches multiply, but rather live by the example of your Queen, the sovereign Virgin, who, after the great honor of her being the mother of the Son of God was announced to her, humbled herself all the more by calling herself the handmaiden of God. Thus, my ladies, as it is true that the greater the virtues are in human beings the more humble and kind they make them, may this city be the reason for you to have good morals and to be virtuous and humble.

And you, married ladies, do not resent being subject to your husbands: for sometimes it is not the best thing for a human being to be free. And the angel of God testified to this to Esdras: Those, he said, who used their free will fell into sin and despised Our Lord and oppressed the just, and for this they were destroyed. And those women who have peaceful, good, and discreet husbands that love them greatly, should praise God for this favor, which is not a small thing, for a greater good in the world could not be given to them. And they should be diligent in serving, loving, and cherishing them with all their heart, as it is fitting, keeping their peace and praying to God that he maintain and safeguard it for them. And those that have husbands in between good and bad should also praise God that they do not have worse ones, and should try to moderate their perverse behavior and pacify them, according to their condition. And those that have husbands who are evil, cruel, and savage should make an effort to endure them so that they can try and oppose their evil ways and lead them back, if they can, to a reasonable and good life. And if husbands are so obstinate that the wives cannot succeed, at least they will acquire great merit for their souls through the virtue of patience. And everyone will bless them and be on their side.

So, my ladies, be humble and patient; and God's grace will increase in you, and praise will be given to you as well as the kingdom of heaven. For St. Gregory says that patience is the doorway to Paradise and the way to Jesus Christ. And may none of you be obstinate or hardened by holding frivolous opinions that have no basis in reason, or by jealousies, or by a disturbed mind, or by haughty speech, or by outrageous actions. For these are things that corrupt the mind and make a person crazy. Such conduct is improper and unseemly for women.

And you, maidens in the state of virginity, be pure, simple, and peaceful, without vagueness, for the snares of evil men are set for you. Your gaze should be lowered, few words should be in your mouths, respect should govern all of your actions. And be armed with the virtuous strength against the ruses of the deceivers and avoid their company.

And, widowed ladies, may there be modesty in your dress, behavior and speech, piety in your actions and way of life; prudence in your conduct; patience, which is so much needed; strength and resistance in tribulations and important affairs; humility in your hearts, countenance, and speech; and charity in your works.

And, to make matters brief, all women, whether of the upper, middle, or lower classes, be well informed in all things and take care in mounting a defense against the enemies of your honor and chastity. My ladies, see how these men accuse you of so many vices from all sides. Make liars of them all by showing your virtue and prove by your good actions that those who reprimand you are lying, so that you can say with the psalmist: "the wickedness of the evil will fall on their heads." Chase away the deceiving flatterers who use various tricks in their intrigues to try and get that which you should supremely guard, that is, your honors and the beauty of your reputation. Oh, ladies, flee, flee the foolish love with which they beseech you. Flee it, for God's sake, flee: for nothing good can come of it for you. Rather, you can be certain that, with all its deceptive attractions, it will always end badly for you. And do not believe the contrary: for it cannot be otherwise. Remember, dear ladies, how these men call you weak, light-minded, and quickly persuaded; and how, nevertheless, they make a great effort to seek out all sorts of strange and deceptive tricks to catch you, just as one does in trapping animals. Flee, flee, my ladies,, and shun the kind of companions under whose smiles are hidden grievous poisons that kill people. And so may it please you, my most honored ladies, to increase and multiply our City through the adherence to virtues and the rejection of vices, and to rejoice and act well. And may I, your servant, be commended to you by praying to God, who by His grace has granted me to live in this world and to persevere in His holy service, and at the end may He be merciful toward my great sins and grant me the joy that lasts forever, which by His grace He may grant you as well. Amen.

Here ends the third and last part of the Book of the City of Ladies.

59. Visual Depictions of Gender Inversion

One of the standard themes in the debate about women was the possibility of gender role reversal or gender ambiguity. This was often portrayed visually in woodcuts and engravings that accompanied literary works or that were produced independently. Such images became increasingly widespread after the development of the printing press with movable metal type in the mid-fifteenth century. The first image here (A.) is an engraving by an anonymous fifteenth-century artist known as the Housebook Master, depicting the legend of Phyllis and Aristotle, according to which the great philosopher was supposed to have been so infatuated by a much younger woman that he let her ride him around the yard. The second (B.) is a woodcut by Hans Schäuffelein, produced in about 1530 to accompany a now-lost poem of the German poet Hans Sachs, *Ho, Ho*

Diaper Washer. "Diaper-washer" (*Windelwasher* in German) was a derogatory term for a henpecked husband.

A. Aristotle and Phyllis. (© *Rijksmuseum-Stichting Amsterdam.*)

B. Der Windelwascher by Hans Leonhard Schäufelein. (*Kunstsammlungen der Veste Coburg.*)

60. Leonardo Bruni
Letter to Lady Baptista Malatesta, ca. 1405

Leonardo Bruni (1370–1444) was a Florentine humanist who wrote a long history of the city of Florence and later served as its chancellor. In his writings for men and in his own career, Bruni favored the active life of public service over the quiet life of scholarly contemplation. He was a strong advocate of the new type of humanistic education centering on the classics, and he considered the issue of women's education. The following is from his treatise on women's education, written in ca. 1405 in the form of a letter to Lady Baptista Malatesta, the daughter of the duke of Urbino (a small state near Florence) and thus a member of the highest Italian nobility.

There are certain subjects in which, whilst a modest proficiency is on all accounts to be desired, a minute knowledge and excessive devotion seem to be a vain display. For instance, subtleties of Arithmetic and Geometry are not worthy to absorb a cultivated mind, and the same must be said of Astrology. You will be surprised to find me suggesting (though with

"Letter to Lady Baptista Malatesta" by Leonardo Bruni. Found in *Vittorino de Feltre and Other Huminist Educators* edited by W.H. Woodward. (London: Cambridge University Press, 1897).

much more hesitation) that the great and complex art of Rhetoric should be placed in the same category. My chief reason is the obvious one, that I have in view the cultivation most fitting to a woman. To her neither the intricacies of debate nor the oratorical artifices of action and delivery are of the least practical use, if indeed they are not positively unbecoming. Rhetoric in all its forms—public discussion, forensic argument, logical fence, and the like—lies absolutely outside the province of women.

What Disciplines then are properly open to her? In the first place she has before her, as a subject peculiarly her own, the whole field of religion and morals. The literature of the Church will thus claim her earnest study. Such a writer, for instance, as St. Augustine affords her the fullest scope for reverent yet learned inquiry. Her devotional instinct may lead her to value the help and consolation of holy men now living; but in this case let her not for an instant yield to the impulse to look into their writings, which, compared with those of Augustine, are utterly destitute of sound and melodious style, and seem to me to have no attraction whatever.

Moreover, the cultivated Christian lady has no need in the study of this weighty subject to confine herself to ecclesiastical writers. Morals, indeed, have been treated of by the noblest intellects of Greece and Rome. What they have left to us upon Continence, Temperance, Modesty, Justice, Courage, Greatness of Soul, demands your sincere respect....

But we must not forget that true distinction is to be gained by a wide and varied range of such studies as conduce to the profitable enjoyment of life, in which, however, we must observe due proportion in the attention and time we devote to them.

First amongst such studies I place History: a subject which must not on any account be neglected by one who aspires to true cultivation. For it is our duty to understand the origins of our own history and its development; and the achievements of Peoples and of Kings.

For the careful study of the past enlarges our foresight in contemporary affairs and affords to citizens and to monarchs lessons of incitement or warning in the ordering of public policy. From History, also, we draw our store of examples of moral precepts....

The great Orators of antiquity must by all means be included. Nowhere do we find the virtues more warmly extolled, the vices so fiercely decried. From them we may learn, also, how to express consolation, encouragement, dissuasion or advice....

I come now to Poetry and the Poets—a subject with which every educated lady must shew herself thoroughly familiar. For we cannot point to any great mind of the past for whom the Poets had not a powerful attraction.... Hence my view that familiarity with the great poets of antiquity is essential to any claim to true education. For in their writings we find deep speculations upon Nature, and upon the Causes and Origins of things, which must carry weight with us both from their antiquity and from their authorship. Besides these, many important truths upon matters of daily life are suggested or illustrated. All this is expressed with such grace and dignity as demands our admiration.

But I am ready to admit that there are two types of poet: the aristocracy, so to call them, of their craft, and the vulgar, and that the latter may be put aside in ordering a woman's reading. A comic dramatist may season his wit too highly: a satirist describe too bluntly the moral corruption which he scourges: let her pass them by....

But my last word must be this....All sources of profitable learning will in due proportion claim your study. None have more urgent claim than the subjects and authors which treat of Religion and of our duties in the world; and it is because they assist and illustrate these supreme studies that I press upon your attention the works of the most approved poets, historians and orators of the past.

61. Moderata Fonte (Modesta Pozzo) "Women's Worth"

Modesta Pozzo (1555–1592) was a well-to-do Venetian woman whose male relatives and husband were open to her obtaining an education and writing. Early in her literary career, she took the pen name Moderata Fonte, perhaps to hide her identity but also to highlight her talents (Modesta Pozzo, her given name, means "modest well" in Italian, while Moderata Fonte means "moderate fountain"). The poem reprinted here comes from a chivalric romance in verse, *Il Florido,* published in Venice in 1581, which tells the story of identical twin princesses, one given the normal women's education and the other kidnapped and raised as a knight. Later Fonte also wrote a longer prose work on the same subject, *The Worth of Women,* which was apparently finished the night before her death in childbirth at thirty-seven.

WOMEN'S WORTH

Women in every age by Nature were
   With sound judgment and brave hearts endowed,
   And no less fit to demonstrate with zeal
Men's wisdom, care, and worth, were born,
   And why then, if they bear a common stamp,
   If their substances be not different,
   If they have one food, one speech alike,
   Should they then differ in good sense and courage?

There have been always and still are (whene'er
   A woman decides to put her mind to it)
Successful women warriors, more than one,
   Who wrest both prize and rank from many men;
   So too in letters and in every
Enterprise man undertakes or speaks of,
   Women have been and are full well so fruitful
That they have no cause to envy men.

And, though worthy in itself, their number
   In positions of high fame is not large;
   This is because to splendid and heroic acts
   They have not addressed their hearts in more respects.
Gold left hidden deep within the mines
Is no less gold for being buried there,
   And when extracted and well worked upon
Is just as rich and fine as other gold.

If, when a daughter to a father is born,
   He were to set her with his son to equal tasks,
   She would not in lofty enterprise or light
Inferior or unequal to her brother be;
   Or were he to place her in armed squadrons
With him, or allow her to learn some liberal art;
   But because she is reared to other things,
Small esteem her education brings.

QUESTIONS FOR ANALYSIS

1. According to Lady Reason, why do men produce misogynist attacks on women? Do the two images, both made by male artists, fit with her argument?
2. What does Lady Reason, and later Christine de Pizan, describe as the chief virtues of women? How do these compare with Moderata Fonte's description of women's strengths?
3. Both of the women in the two images carry objects with which they can beat the men. Why might the artists have included these in their depictions of male humiliation and female power?
4. How would you compare Leonardo Bruni's, Lady Reason's, and Moderata Fonte's justifications for educating women?
5. What reasons do Christine de Pizan and Moderata Fonte give for women's lesser achievements?

CHAPTER 6

Early Modern Social and Economic Developments
(1500–1750)

The end of the Middle Ages and beginning of the early modern era is traditionally set at around 1500, a date marked by the rough conjuncture of several events—Columbus's first trip to America and the Protestant Reformation—and some long-term changes, such as the emergence of the nation-state and capitalist production. Though all of these would eventually have an impact on women's lives, some historians argue that continuity outweighs change and that the whole period from the fourteenth through the eighteenth century should be looked at as a whole. A few go even further and assert—as we discussed in Chapter 4 about the ancient/medieval break—that the medieval/modern break should be challenged for all of European history and not only the part that focuses on women.

The arguments of those who stress continuity rely primarily on social and economic evidence, for social and economic structures are generally slower to change than political ones. We may speak, for example, of a "Commercial Revolution" or an "Industrial Revolution" just as we speak of an "American Revolution" or a "French Revolution," but the former take centuries while the latter take years. The "rise of modern capitalism" and the "rise of the modern family," both viewed as aspects of the development of modernity, have taken even longer. As the modern period itself grows ever longer, it has been subdivided into "early modern" (roughly 1500–1750 or 1789) and "modern" (roughly 1750 or 1789 to the present), with some scholars also using "late modern" or "post-modern" to describe our own times. Some of this debate has led to a questioning of just what the term *modern* means and whether it means the same thing for men and women, Europeans and non-Europeans, and so on; the debate has pointed out that chronological divisions are intellectual constructs and not part of the natural world, a point also made in debates about the new millennium.

In many ways, then, the aspects of life discussed in this chapter did not differ dramatically from those in earlier chapters. During the early modern period, most Europeans continued to live in small villages and make their living by farming, though the gradual increase in the size of cities that had begun in the High Middle Ages continued. Most people continued to marry and live in a nuclear family, and most work was carried out by family units. Child mortality remained high, as did the number of women who died in childbirth.

Social and economic changes during this period were thus subtle, rather than dramatic. Political leaders had long regarded the household as the basic unit of society and saw stable, orderly households as essential to the proper functioning of a city or state. Beginning in the fifteenth century, however, they were increasingly activist in attempting to promote this ideal and to punish or exclude those who deviated from it. Sumptuary laws, such as those we saw in Chapter 5, were increasingly justified with moral as well as economic reasons, forbidding necklines that were too low on women or doublets that were too short on men along with fancy fabrics and jewelry. Public brothels, which had been a common part of European city life in the Middle Ages, were often closed, or prostitutes were required to wear distinctive clothing that clearly distinguished them from respectable women. This drive for order and morality became more intense after the Protestant and Catholic Reformations, with religious leaders arguing that these were marks of divine favor and that disorder and deviance would bring God's wrath. All sides in the religious controversy supported moves to increase social discipline, with Church and state authorities usually cooperating to regulate people's sexual and moral lives. Sexual behavior that disrupted marriage patterns and family life, such as adultery and premarital intercourse, was prosecuted more stringently, particularly if this behavior resulted in a pregnancy. Harsh laws were passed against homosexuality—usually termed *sodomy* or *buggery*—prescribing death by burning, although these were enforced erratically. State authorities cooperated when families sought to discipline their wayward members, allowing disinheritance or sterner measures for moral infractions, and they intervened in family life if they thought control was lacking. Laws were passed against anyone who was "masterless" or perceived as such. These included servants between positions, itinerant actors and musicians, wandering journeymen, demobilized soldiers, and vagrants.

This emphasis on order and morality also shaped working life. Guild leaders increasingly justified restrictions on women's work by noting that not only did women take jobs away from men but that women could not control what went on in a shop if they were in charge. As commercial capitalism brought wealth to some urban families, they hired more servants to make life easier and to care for the increasing amount of consumer goods that were becoming a mark of bourgeois status. These servants were predominantly women, who were often suspected of theft and immorality and sternly ordered to obey their employer. Firmly hierarchical relationships were also to structure the growing opportunities for wage labor available to women in some areas, such as those in cloth production, and employers were instructed not to pay women a salary that would allow them to become too independent.

MARRIAGE AND FAMILY LIFE

Early modern political and religious authorities all regarded the marital couple and the male-headed household as the basis of society, a "little commonwealth" on which the larger commonwealths of city and state were based. They thus passed laws and ordinances designed to set ideals for family life and punish those who deviated from them, and they established various types of legal bodies to investigate, try, and punish malefactors. In all of western Europe before the Protestant Reformation and in Catholic areas after the Reformation, Church courts handled many types of cases involving marriage and the family; in Protestant areas after the Reformation these came largely under the jurisdiction of secular courts. Court records thus provide us with invaluable information about how laws regarding family life were enforced. They do not provide evidence about day-to-day life in families that did not deviate from the norm, however, nor do they provide much information about people's sentiments regarding their spouses, unless these were expressed in criminal activity. For this we have to turn to personal sources such as letters and diaries, or to literary evidence, which may sometimes reflect literary conventions along with the author's feelings.

62. Marriage in Law Codes

During the late fifteenth and sixteenth centuries, under the influence of lawyers trained in Roman law by university law faculties, many cities and states developed more comprehensive civil and criminal law codes. These codes set out an ideal of marital and family relations and sought to bring about this idea through regulations regarding inheritance, property ownership, and personal family relations. Cities and states borrowed heavily from earlier codes when they were writing their own, and the codes grew longer and more elaborate: The following is a relatively early code from the Austrian territory of Salzburg in 1526.

From *Die Salzburger Landesordnung von 1526* by Franz V. Spechtler and Rudolf Uminsky, Göppinger Arbeiten zur Germanistik, Nr. 305, pp. 119, 154, 197. © 1981 by Kümmerle. Translated by Merry Wiesner.

It is to be accepted that both spouses have married themselves together from the time of the consummation of their marriage, body to body and goods to goods....

The husband shall not spend away the dowry[1] or other goods of his wife unnecessarily with gambling or other useless frivolous pastimes, wasting and squandering it. Whoever does this is guilty of sending his wife into poverty. His wife, in order to secure her legacy and other goods she has brought to the marriage, may get an order requiring him to pledge and hold in trust for her some of his property. In the same way he is to act in a suitable manner in other things that pertain to their living together and act appropriately toward her. If there is no cause or she is not guilty of anything, he is not to hit her too hard, push her, throw her or carry out any other abuse. For her part, the wife should obey her husband in modesty and honorable fear, and be true and obedient to him. She should not provoke him with word or deed to disagreement or displeasure, through which he might be moved to strike her or punish her in unseemly ways. Also, without his knowledge and agreement she is not to do business [with any household goods] except those which she has brought to the marriage; if she does it will not be legally binding....

The first and foremost heirs are the children who inherit from their parents. If a father and mother leave behind legitimate children out of their bodies, one or more sons or daughters, then these children inherit all paternal and maternal goods, landed property and movables, equally with each other....

Women who do not have husbands, whether they are young or old, shall have a guardian and advisor in all matters of consequence and property, such as the selling of or other legal matters regarding landed property. Otherwise these transactions are not binding. In matters which do not involve court actions and in other matters of little account they are not to be burdened with guardians against their will....

---

[1]*Dowry:* an amount of goods, cash, and/or property that was brought by a wife to her husband on marriage; the husband could do with the dowry whatever he wished within the course of the marriage, though legally it continued to belong to his wife.

63. Family Portrait

While most art in the Middle Ages had religious themes, individual and group portraits along with other art with secular subjects became increasingly popular during the Renaissance and early modern period. Well-to-do families paid artists to memorialize them and then hung their paintings on the walls of their comfortable homes. Like law codes, family portraits provide evidence of idealized family and marital relationships. This is a painting attributed to the artist Paulus Moreelse; the two children painted in front of the table are most likely those who have died.

"Portrait of a Family Saying Grace Before a Meal, with a Servant Stoking the Fire and a Landscape Seen Through an Open Door," attributed to Paulus Moreelse. (*Christie's Images Inc. #OMP310589090+01.*)

64. Vittoria Colonna
"Epistle to Ferrante Francesco d'Avalos, Her Husband, after the Battle of Ravenna"

Official sources such as law codes provide evidence for the ideal marital relationship in the minds of male lawmakers. Evidence for women's ideas about marriage, either in ideal form or reality, is more difficult to obtain. Most writing by women during the early modern period was religious in purpose and subject matter and rarely dealt with personal issues. Occasionally, women did express their feelings, however, or at least those feelings they were willing to share with the world. The following is a poem written by the Roman poet Vittoria Colonna (1492–1547), the daughter and wife of two military leaders active in Italy's many wars during this period. Both her father and husband were taken prisoner in the battle of Ravenna in 1512, though released shortly afterward, and her husband later died from war wounds, leaving Colonna a widow for more than twenty years. She became friends with many prominent artists and learned men, including Michelangelo, and in the 1530s her poetry began to appear in print. The following was one of the first of her poems to appear in print; she may have written it much earlier, as it relates to events that happened more than two decades before it was published.

From *Women Writers of the Renaissance and Reformation* edited by Katharina M. Wilson, pp. 34–35. Copyright © 1987 by University of Georgia Press. Reprinted by permission.

My most noble lord, I write you this
to recount to you how sadly—and amid so
    many
uncertain desires and harsh torments—I live.
I did not expect pain and sorrow from you.
. . . .

I did not believe a marquis and a Fabrizio,[1]
one a husband, the other a father, would be
the cruel, pitiless beginning of my suffering.
Love of my father and love of you,
like two famished and furious snakes,
have always lived, gnawing, in my heart.
I believed the fates had more kindness.
. . . .

But now in this perilous assault,
in this horrible, pitiless battle
that has so hardened my mind and heart,
your great valor has shown you an equal
to Hector and Achilles.[2] But what good is
this to me, sorrowful, abandoned?
My mind has always been uncertain:

those seeing me sad have thought
me hurt by absence and jealousy.
But I, alas, have always had in mind
your daring courage, your audacious soul,
with which wicked fortune ill accords.
Others called for war, I always for peace,
saying it is enough for me if my marquis
remains quietly at home with me.
Your uncertain enterprises do not hurt you;
but we who wait, mournfully grieving,
are wounded by doubt and by fear.
You men, driven by rage, considering nothing
but your honor, commonly go off, shouting,
with great fury, to confront danger.
We remain, with fear in our heart and
grief on our brow for you; sister longs for
    brother, wife for husband, mother for son.
. . . .

You live happily and know no sorrow;
thinking only of your newly acquired fame,
you carelessly keep me hungry for your love.
But I, with anger and sadness in my face,
lie in your bed, abandoned and alone,
feeling hope intermingled with pain,
and with your rejoicing I temper my grief.

---

[1] The title and family name of her father and husband
[2] *Hector* and *Achilles:* two heroes of the Trojan War.

65. Anna Bijns
"Happy the Woman Without a Man"

The religious controversies of the sixteenth century led to hundreds of pamphlets and books arguing for or against Protestantism; the vast majority of these were written by men, but occasionally a woman published her religious convictions as well. Anna Bijns (1493–1575) was a middle-class woman from Antwerp in the Netherlands who ran a school for a time and wrote many strongly pro-Catholic poems in Dutch. Several volumes of these, in which she calls Martin Luther a "wolf in sheep's clothing" and his followers "children of darkness and stupidity," were published in her lifetime by Catholics in Antwerp. Toward the end of her life, a third book of her poetry, containing secular along with religious poems, was published as a fund-raising attempt by a Catholic prior. Included among these is the following, which was accompanied by a companion poem "Happy the Man Without a Woman." The strict opposition to marriage it expresses parallels Bijns' harshness toward Lutherans in her religious poetry, and it may also reflect her own feelings. We know that she did not marry and that her younger sister married quickly and unhappily.

How good to be a woman, how much better to
   be a man!
Maidens and wenches, remember the lesson
   you're about to hear.
Don't hurtle yourself into marriage far too soon.
The saying goes: "Where's your spouse?
   Where's your honor?"
But one who earns her board and clothes
Shouldn't scurry to suffer a man's rod.
So much for my advice, because I suspect—
Nay, see it sadly proven day by day—
'T happens all the time!
However rich in goods a girl might be,
Her marriage ring will shackle her for life.
If however she stays single
With purity and spotlessness foremost,
Then she is lord as well as lady. Fantastic, not?
Though wedlock I do not decry:
Unyoked is best! Happy the woman without a
   man.

Fine girls turning into loathly hags—
'Tis true! Poor sluts! Poor tramps! Cruel mar-
   riage!
Which makes me deaf to wedding bells.
Huh! First they marry the guy, luckless dears,
Thinking their love just too hot to cool.
Well, they're sorry and sad within a single year.
Wedlock's burden is far too heavy.
They know best whom it harnessed.
So often is a wife distressed, afraid.
When after troubles hither and thither he goes
In search of dice and liquor, night and day,
She'll curse herself for that initial "yes."
So, beware ere you begin.
Just listen, don't get yourself into it.
Unyoked is best! Happy the woman without a
   man.

A man oft comes home all drunk and pissed
Just when his wife had worked her fingers to
   the bone

(So many chores to keep a decent house!),
But if she wants to get in a word or two,
She gets to taste his fist—no more.
And that besotted keg she is supposed to obey?
Why, yelling and scolding is all she gets,
Such are his ways—and hapless his victim.
And if the nymphs of Venus he chooses to fre-
   quent,
What hearty welcome will await him home.
Maidens, young ladies: learn from another's
   doom,
Ere you, too, end up in fetters and chains.
Please don't argue with me on this,
No matter who contradicts, I stick to it:
Unyoked is best! Happy the woman without a
   man.

A single lady has a single income,
But likewise, isn't bothered by another's whims.
And I think: that freedom is worth a lot.
Who'll scoff at her, regardless what she does,
And though every penny she makes herself,
Just think of how much less she spends!
An independent lady is an extraordinary prize—
All right, of a man's boon she is deprived,
But she's lord and lady of her very own
   hearth.
To do one's business and no explaining sure is
   lots of fun!
Go to bed when she list, rise when she list, all
   as she will,
And no one to comment! Grab tight your inde-
   pendence then.
Freedom is such a blessed thing.
To all girls: though the right Guy might come
   along:
Unyoked is best! Happy the woman without a
   man.

Prince,
Regardless of the fortune a woman might bring,

Many men consider her a slave, that's all.
Don't let a honeyed tongue catch you off guard,
Refrain from gulping it all down. Let them rave,
For, I guess, decent men resemble white ravens.
Abandon the airy castles they will build for you.
Once their tongue has limed a bird:
Bye bye love—and love just flies away.
To women marriage comes to mean betrayal

And the condemnation to a very awful fate.
All her own is spent, her lord impossible to
  bear.
It's *peine forte et dure* instead of fun and games.
Oft it was the money, and not the man
Which goaded so many into their fate.
Unyoked is best! Happy the woman without a
  man.

66. Reflections on Widowhood

Many women in early modern Europe, like those of earlier (and later) eras, spent part of their lives as widows. The death of a spouse brought a more significant change in a woman's life than it did in a man's, a situation reflected in the fact that the word for *widower* in most European languages derives from the word for *widow,* rather than the more common reverse pattern. Widows' situation varied widely; widowhood generally brought a decline in economic status—similar to that experienced by divorced women today—and the households of widows were usually among the poorest in any city. On the other hand, widows had a wider range of action than married women, and, if they owned property, could handle it as they wished. This freedom was disturbing to many authorities, who, in contrast to the opinion of Jerome that we read in Chapter 3, advocated quick remarriage to bring widows under male control again. Despite this suspicion of widows—a suspicion arising from the fact that widows were sexually experienced as well as possibly financially independent—widows remarried at a lower rate than widowers. The following are extracts from letters written by two well-educated eighteenth-century German women about their situation as widows.

A. Isabella von Wallenrodt

I laugh away the prophecy that I would quickly remarry, for I had firmly decided to fulfill the wish of my husband and not make a second choice. It was easy to strengthen my resolve in this, for I found the independence which I enjoyed unbelievably sweet. I repeat, that I did every duty to my husband willingly and avoided none of them, but it was also natural that I began to feel that this was a burden that I carried, and who would not finally be worn out from this?...Though I had the truest and most just husband, who got along with me very well, and appeared to let me have my will in all things, I was still a slave; I prized him, whose chains I wore, I loved him, but still I felt the burden. So I firmly decided not to lay this yoke on myself again....I want to enjoy my freedom completely. It seems very agreeable to me to get up, go to sleep, stay at home, or go out, etc., when I want to, without having the expression of my will get in the way of someone else.

B. Sophie Rosine Richter

[To her brother-in-law] Please don't take offense that I am so free and write to you and in-

From *Frauenleben im 18. Jahrhundert*, pp. 309–311. © 1992 by C.H. Beck. Translated by Merry Wiesner.

quire about your health. I would not have burdened you with this letter, if I had not been in
such great need. It is so expensive to live here
and I have had to pay off [so many debts] for
my husband; the legal processes also cost
a lot of money and I have five children to
raise. Now next Monday I have to move out
of my lodgings, and because I owe rent for the
last quarter-year, I must also pay 15 gulden or

the people [who own the place], who are so
rude and lacking in sympathy, will not hand
over any of my things. Therefore I send this
obliging request to you, that you would be so
kind and loan me 15 gulden, with my garden-
house as security [for the loan]. You would save
me from great misery and God would reward
you for this. I hope that you would not refuse
the request of a widow and a relative.

QUESTIONS FOR ANALYSIS

1. The phrase *traditional family values* is frequently used by contemporary commentators and politicians to describe their ideal of family life. Based on the sources reproduced here, what were the "family values" of early modern society? What
   similarities and differences emerge in the impact of these on men and women?
2. The poems by Vittoria Colonna and Anna Bijns both discuss disadvantages of marriage for women, but quite different ones. Compare them.
3. What aspects of the relationships between husband and wife set out in the Salzburg
   law code does Bijns highlight in her poem? What aspects are highlighted by the
   artist of the family portrait?
4. Bijns and Isabella von Wallenrodt both note advantages to being single or widowed, while Sophie Rosine Richter's letter points out some disadvantages. Why
   might their opinions be so different?

CONTROL OF WOMEN'S TONGUES AND BODIES

The emphasis on order and discipline in the early modern period often led to
stricter control of both sexual behavior and speech, two activities that, as we saw
in Chapter 5, were linked in people's opinions of women. Gossip, slander, scolding, and aggressive speech were criminalized, as were flirting (described as "lewd
and lascivious carriage"), giving one's attention to several suitors ("wanton dal-
liance"), and courting without parental approval. Laws regarding more serious
sexual acts, such as adultery and premarital sexual activity, were stiffened. In some
cases, such laws increased parental control of children and husbandly control of
wives; for example, in many parts of Europe, parental approval became obligatory
for marriage, with those marrying against the wishes of their parents subject to arrest as well as disinheritance. In other cases, political authorities interfered with
patriarchal control in families, ordering men to set aside money for their wives and
children if the authorities thought they were spendthrifts or drunkards, or punishing men whose physical coercion went beyond acceptable limits. (In England,
men who beat their wives with a stick wider than their thumb were subject to arrest; this is the origin of the phrase *rule of thumb*.)

Laws and punishments were not the only means of accomplishing social control and discipline, for such external measures could go only so far. People also heard these ideals preached in sermons, learned about them in the growing number of primary schools, and read about them in moral literature and conduct books. They thus developed internal measures of control—what we usually call "self-control"—and grew more guarded in their speech and actions. Some historians view this as a positive process that made Europeans more "civilized," while others see it as negative, repressive, or neurotic. A few argue that this development of internal controls never extended to the majority of the European population, while others stress that stability and order had been important to the villages where most Europeans lived all along, and so measures to assure them were not really new or different. All of these points of view find support in the sources, for people continued to violate laws and standards, but also helped to impose such standards on their neighbors and family members. Because most women's best chances for well-being lay in a good marriage ("good" in both economic and emotional terms), members of the community rarely supported women who deviated from the norm and instead reported suspicious words and actions to the authorities.

67. Regulation of City Brothels

During the Late Middle Ages, most cities in Europe established licensed or official brothels and often regulated the hours, prices, and clothing of prostitutes. Some cities issued very specific and detailed ordinances regarding life in the brothel, forbidding men to carry weapons, forbidding prostitutes to have special favorites, and forbidding the brothel manager to sell women to other houses. The brothel was to be closed on Sundays and holidays, and a minimum age—usually fourteen—was set for its residents. In many cities, particularly those of northern Europe, further restrictions were added beginning in the late fifteenth century as civic leaders became more concerned about enforcing morality and excluding those who deviated from the norm of male-headed households. These restrictions culminated in the sixteenth century with the closing of brothels in many cities of northern Europe, which made all prostitution illegal, but, of course, did not end it. (The timing of these restrictions was attributed by earlier historians to fears about the spread of syphilis, a disease that first appeared in Europe during the 1490s. It is now clear, however, that worries about order and decorum played a much greater role than concerns about disease.) Italian cities generally chose to regulate prostitution more strictly but chose not to suppress it completely. The following come from the records of the city government of Florence in 1415 and Nuremberg in 1562.

"Regulation of Prostitution: Florence: Florentine State Archives, Provvisioni, 105, fols. 248r 248v" From *The Society of Renaissance Florence: A Documentary Study* edited by Gene Brucker. Copyright © 1971 by Harper. Reprinted by permission from the author. Nuremberg: Bavarian State Archives in Nuremberg, Ratsbücher 31, fols. 316, 350; 36, fol. 15; Ratschlagbücher, 36, fol. 150–153. Translation by Merry E. Wiesner.

A. Florence

Desiring to eliminate a worse evil by means of a lesser one, the lord priors... [and their colleges] have decreed that... the priors... [and their colleges] may authorize the establishment of two public brothels in the city of Florence, in addition to the one which already exists: one in the quarter of S. Spirito and the other in the quarter of S. Croce. [They are to be located] in suitable places or in places where the exercise of such scandalous activity can best be concealed, for the honor of the city and of those who live in the neighborhood in which these prostitutes must stay to hire their bodies for lucre, as other prostitutes stay in the other brothel. For establishing these places... in a proper manner and for their construction, furnishing, and improvement, they may spend up to 1,000 florins....

B. Nuremberg

[January 5, 1562] The high honorable [city] council asks for learned opinions about whether it should close the city brothel (*Frauenhaus*), or if it were closed, whether other dangers and still more evil would be the result.

[January 19, 1562] The learned counsellors, pastors and theologians discussed whether closing the house would lead the journeymen and foreign artisans to turn instead to their masters' and landlords' wives, daughters, and maids. The pastors and theologians urged the city not to break God's word just because of foreigners, and one argued that the brothel caused journeymen to have impure thoughts about women. If they were never introduced to sex, they would not bother other woman. The argument that a man performed a good deed when he married a woman from the brothel, which he could no longer do if the brothel was closed, is to be rejected, as closing the brothel would also pull the women out of the devil's grip. One jurist added his opinion that because there were only ten or twelve women in the brothel, they couldn't possibly be taking care of all the journeymen, so closing the house would not make that much difference. The council then asked for an exact report from Augsburg about the numbers of illegitimate children before and after it closed its brothel [in 1532] to see if it had increased or decreased.

[March 18, 1562] On the recommendations written and read by the high honorable theologians and jurists, why the men of the council are authorized and obliged to close the common brothel, it has been decided by the whole council to follow the same recommendation and from this hour on forbid all activity in that house, to post a guard in the house and let no man enter it any more. Also to send for the brothel manager and say to him he is to send all women that he has out of the house in two days and never take them in again. From this time on he is to act so blamelessly and unsuspiciously that the council has no cause to punish him. When this has been completed, the preachers should be told to admonish the young people to guard themselves from such depravity and to keep their children and servants from it and to lead such an irreproachable life that the council has no cause to punish anyone for this vice.

[May 18, 1577] The high honorable city council asks for learned opinions, because adultery, prostitution, immorality and rape have gotten so out of hand here in the city and the countryside.

68. Local Control of Sex and Scolding

Along with national and municipal laws, people in Europe lived under a variety of local, village, and church regulations and ordinances, and could be charged before local or church courts for many different infractions. In the late fifteenth century, these bodies became increasingly intent on regulating interpersonal relations to promote order and investigating those suspected of moral infractions. Church officials, in particular, did not always wait until cases were brought to court, but went around to villages checking on rumors and reports. These investigations, termed *visitations,* were designed to gain information about those who broke church laws, which in the case of women often involved sexual activities. During the course of the Protestant Reformation, visitations were used by authorities to investigate the religious opinions along with the suspicious activities of both clergy and laypeople. Not surprisingly, when authorities wished to argue that the existing church was in need of serious reform, the visitation reports stress low moral standards and lack of religious devotion. The intent of their authors thus has to be kept in mind, but they still provide a window on daily life that emerges from few other sources.

BOROUGH ORDINANCE, HEREFORD, 1486

Also, concerning scolds, it was agreed that through such women many ills in the city arose, viz. quarrelling, beating, defamation, disturbing the peace of the night, discord frequently stirred between neighbours, as well as opposing the bailiffs, officers, and others and abusing them in their own person, and often raising hue and cry and breaking the peace of the lord king, and to the disturbance of the city's tranquility. Consequently, whenever scolds shall be taken and convicted, they shall have their judgement of the cuckingstool[1] without making any fine. And they shall stand there with bare feet and the hair of the head hanging loose for the whole time that they may be seen by all travelling on the road, according to the will of the lord king's bailiff, and not that of the bailiff of any other fee whatever. And afterwards when judgement has been made they shall be taken to the lord king's gaol and shall stay there until they make fine at the will of the bailiff whoever's tenants they be. And if they refuse to be punished by such a judgement, they shall be thrown out of the city....

CHURCH COURT, LINCOLN

[1517]...Alice Ridyng, unmarried, the daughter of John Ridyng of Eton in the diocese of Lincoln appeared in person and confessed that she conceived a boy child by one Thomas Denys, then chaplain to Master Geoffrey Wren, and gave birth to him at her father's home at Eton one Sunday last month and immediately after giving birth, that is within four hours of the birth, killed the child by putting her hand in the baby's mouth and so suffocated him. After she had killed the child she buried it in a dung heap in her father's orchard. At the time of the delivery she had no midwife and nobody was ever told as such that she was pregnant, but some women of Windsor and Eton had suspected and said that she was pregnant, but Alice always denied this saying that something else was wrong with her belly. On the Tuesday after the delivery of the child, however, the women and honest wives of Windsor and Eton took her and inspected her belly and her breasts by which they knew for certain that she had given birth. She then confessed everything to them and showed them the place where she had put the dead child. She said further that neither her father nor her mother ever knew that she was pregnant since she always denied it until she was taken by the wives as described. Examined further she said by virtue of the oath she had taken on the gospels that she had never been known carnally by anyone other than the said Thomas and that nobody else urged or agreed to the child's death. She also said that the child had been conceived on the feast of the Purification of

---

[1]*Cuckingstool:* a chair set up in a public place in which scolds were subject to ridicule and sometimes dunked in a pond.

From *Women in England, c. 1275–1525* by P.J.P. Goldberg (trans), 1995, Manchester University Press, Manchester, U.K. Reprinted by permission.

the Blessed Virgin Mary last at the time of high mass in the house of Master Geoffrey Wren at Spytell where Master Geoffrey was then infirmarer.

[1519]...John Asteley [rector of Shepshed] confessed that he had made Agnes Walles, unmarried, pregnant and that she had given birth to a girl child at Hauxley before Christmas. He had supported the child there from his tithes. He also confessed that he had made pregnant Margaret Swynerton, unmarried, and had had three children by her. Margaret is now dead. He also had another child by one Joan Chadwyk, now married, then single. Joan lives at Dunstable. John does not know where Agnes lives now. Because he confessed these things... the vicar general ordered that from henceforth no other woman should serve in his home and that he should live continently.

[1520] Agnes Plumrige of Fingerst, the wife of John Plumrige, appeared and confessed that she had been known and made pregnant by her husband before their marriage. The bishop ordered her to go round her neighbours on the feast of the Purification of Mary openly carrying a candle costing four pence.

VISITATIONS, DIOCESE OF LINCOLN, 1518–1519

Alconbury....Edward Kyng has in his house a woman who is the object of suspicion. Henry Man and Christopher Walton's wife are suspected of adultery.

St Ives. Agnes Haward, wife of John Haward was made pregnant before the solemnisation of the marriage between herself and her husband by John Wiks of Ramsey.

Bringhurst. Sir Robert Worthington made pregnant his servant Alice. The same vicar previously begot two children by one Elizabeth his servant to his great shame. The mother gave birth in his vicarage.

Ashby-de-la-Zouche....One Amy Barton of the same, unmarried, became pregnant there.

Amersham....John Denton keeps company with Alice Fulmer and has done so for the past

year, and he is said to have betrothed her last year, but does not want to have it solemnised.

Surfleet....A baby boy was turned away at the door of Thomas Leeke, but his mother claimed that Thomas Leeke was the father of this baby. Thomas, however, denied this and so the baby was taken away to different places, ill treated, and died.

Baumber....They say also that one Robert Burwhite betrothed one Alice, formerly a servant with Richard Sawcoen of the same parish, and she lives suspectly in the same Richard's house. They refuse to solemnise marriage at the church door.

Grainthorpe. Thomas Dam slandered Isabel Gremmesby so that William Pennyngton will not have her for his wife.

Hartford....The wife of the same John Kareles arouses discord between neighbours and parishioners there and resorted to the vicar with tales and flattery.

Church of St Leonard, Leicester....Also, Margaret Stanton is a common slanderer. Also Elizabeth Whythrop, Margaret Frost, Margaret Watkyn, Joan Walker, and Agnes Slytpeny, and the same Agnes receives suspect persons.

Thornton. The parishioners say all is well. The rector there has in his home a certain woman, Joan Thakham. The parishioners, however, suspect nothing ill of her, except that the woman is too proud and speaks haughtily to the parishioners there. And the said Joan Thakham is a common whore and lives in the rectory and keeps a common tavern there. Also the same Joan is a common scold. The rector lives with her incontinently.

Surfleet....Item, the wives of John Austyn, Robert Carter, and William Hebburn keep public taverns which many use at time of divine service and, for that reason, absent themselves from divine service, especially John Robert, Thomas Barret, Edward Laborer, John Robynson, James Tailour, Richard Dalley, Robert Bacheler, and Thomas Tofte.

Hemingby....Robert Grene [and] Richard Wyrth committed fornication with Alice Waytt. Thomas Norcotes and Alice Waytt defamed their neighbours. William Maltby does not live with his wife. Alice Waytt forged a false will of someone who had died.

Haugh. The vicar keeps a girl in his house. The same vicar and the girl labour in the fields.

Willoughton.... Margaret Fraunces is a common scold.... One Margaret Spynk is a common scold. The same Margaret Spynk and Margaret Fraunces are fornicators.... Margaret Fraunces is lodged in the home of Richard Frauncis and he is a promoter of the same Margaret Fraunces' wickedness.

Sandy.... The wife of John Clark always has players at dice in her house at time of divine service to the bad example of everyone else.

Sutton. John Alen and Agnes his wife receive servants and labourers to drink in their house at time of divine service.

VISITATIONS, DIOCESE OF LINCOLN, 1530

Harpenden. John Hunt lives incontinently with Joan Willys his servant.... Afterwards they appeared at the priory of St Giles and confessed that they had contracted marriage together.

Also that we charge you that you gave the said woman advice and persuaded her to take and drink certain drinks to destroy the child that she is with.[1] The bishop instructed him to appear the next day at the priory of St Giles and

he confessed that he had known her carnally. The bishop directed them that on a Sunday that the vicar general would assign they were to do public penance in a penitential manner before the cross at Harpenden and that they should have their marriage solemnised as soon as they conveniently could. He ruled that under pain of excommunication they should not live incontinently until their marriage was solemnised.

Gaddesden. Thomas Clarke confesses that he had known Agnes Dryvar carnally around Michaelmas and made her pregnant, but he denies that he had contracted marriage with her. The bishop ordered that as soon as he were able to see that the child was supported. The bishop directed the said Agnes to go before the procession on the next Sunday after the [feast of] the Purification in a penitential manner bearing a burning candle in her hand and give the candle into the priest's hand at the time of the offertory.

[St Peter-le-Bailey, Oxford] Widow Colls receives pregnant women in her home in which they are cared for. Widow Gybscott is a common defamer of her neighbours

[Stokenchurch] Joan Schower it is reported was made pregnant by one William Hewes. Midwives examined her to see whether she was or not and having examined her they concluded that she was not pregnant. The aforesaid Joan, however, told the midwives that she had been known carnally by the said William and had become pregnant, but she took a potion by which means she obtained an abortion. She previously had two children for which offence she had not yet been punished at all.

---

[1]This follows immediately from the previous entry and represents a charge made by the bishop to John Hunt in respect of Joan Willys.

69. Laws Restricting Single Women

In many parts of Europe beginning in the Late Middle Ages, women waited until they were in their mid-twenties to marry, in contrast to most cultures in the premodern world in which women married soon after puberty. Young women thus spent years working for their families or an employer before marriage, and some women did not marry at all. Their living arrangements varied, but beginning in the sixteenth century city and state governments attempted to prohibit unmarried women from living on their own, lumping them in with other "masterless" groups as suspect. In England, city officials were given the authority to force any unmarried woman between the ages of twelve and forty into becoming a servant in a male-headed household, and in France and Germany cities forbade unmarried women to immigrate or to work at occupations that allowed them to be independent, such as selling food at the public market. The following 1665 ordinance from Strasbourg, a city on the border between France and Germany, indicates how far cities were willing to go to control the activities of young women.

Numerous complaints have been made that some widows living here have two, three, or more daughters living with them at their expense. These girls go into service during the winter but during the summer return to their mothers, partly because they want to wear more expensive clothes than servants are normally allowed to and partly because they want to have more freedom to walk around, to saunter back and forth whenever they want to. It is our experience that this causes nothing but shame, immodesty, wantonness, and immorality, so that a watchful eye should be kept on this, and if it is discovered, the parents as well as the daughters should be punished with a fine, a jail sentence, or even banishment from the city in order to serve as an example to others.

70. Laws Requiring Declaration of Pregnancy

Unmarried women were often in situations where sexual contacts were difficult to avoid, working unchaperoned in fields with men or as domestic servants. Some of them did become pregnant, for which the usual solution was a quick marriage; in many parts of Europe, sex between persons intending to marry was not regarded as very serious, though it was officially prohibited. If the woman did not or could not marry—usually because the father was married, was of high social status, or had vanished—the consequences varied. In rural areas with labor shortages, unwed motherhood did not bring long-lasting stigma and women with children often later made respectable marriages; in other situations, especially those of domestic servants, the results could be disastrous. Women thus sometimes attempted to induce an abortion through strenuous labor or herbs, or tried to hide the birth, taking the infant to one of the new foundling homes or killing it. Medieval jurists rarely prosecuted infanticide, as they recognized they had no clear way of telling between a stillbirth, death by natural causes, and infanticide; coroners' methods did not improve in the sixteenth century, but governments became increasingly concerned about infanticide and called for the death penalty. More women were executed for infanticide than any other crime except witchcraft in early modern Europe, and some governments went even further. The following French royal edict was issued in 1556; similar statutes requiring a public declaration of all out-of-wedlock pregnancies were passed in England in 1624 and Scotland in 1690.

"Restrictions on Single Women" Strasbourg, Archives Municipales, Statuten, vol. 33, no. 61 (1665). Translated by Merry E. Wiesner.

Reprinted by permission of the publishers from *A History of the Family, Volume II: The Impact of Modernity,* ed. André Burguiere, Christiane Klapisch-Zuber, Martine Segalen and Françoise Zonabend, translated by Sarah Hanbury-Tenison, Cambridge, Mass: The Belknap Press of Harvard University Press. Copyright © 1996 by the President and Fellows of Harvard College.

Having been duly aprised of a great and execrable crime occurring frequently in our kingdom, which is that several women, having conceived children by dishonest or other means and prompted by ill will and bad advice, disguise, conceal and hide their pregnancies, without revealing or declaring anything; and when the time of their parturition and deliverance of their fruit occurs, they give birth secretly, then suffocate, strike or otherwise suppress the child without having imparted the holy sacrament of baptism to it; having done this, they throw it into secret and disgusting places or bury it in profane ground, in these ways depriving it of the customary burial of Christians.... We have, in order to obviate this, said, commanded and ordered...that every woman on finding herself duly accused and convicted of having hidden, covered and concealed both her pregnancy and her childbearing, without having declared one or the other... .be held and reputed to have murdered her child and, in reparation, be punished by death and the last agony.

71. Request for a *Lettre de Cachet* Ordering a Wife's Imprisonment

A *lettre de cachet* was a warrant of arrest signed by the king of France and closed with a seal (*cachet*), usually ordering the imprisonment of a person without trial until further notice. In some well-known cases, these *lettres de cachet* were used against political enemies of the crown and became one of the grievances leading to the French Revolution. More commonly, they were used against ordinary criminals and religious radicals, or by families wishing to prevent young people from marrying against family wishes or otherwise harming the family's reputation and honor. In some cases, they were also used by husbands seeking to control their wives, as in the following request for a second *lettre de cachet,* from the archives of the Bastille in Paris in 1758.

To my Lord the Lieutenant General of Police

My Lord,

Michael Pierre Corneille, journeyman cabinetmaker living on the main street of the Faubourg Saint-Antoine [working-class district on the Right Bank], has the honor to represent very humbly to Your Lordship that the bad conduct and remarkable disorder of Anne Doisteau, his wife, whose inclination was disposed not only to prostitution but even to committing acts most contrary to honesty, obliged the petitioner, who, although in poverty, makes a duty of living with honor, to file complaints about them with your court, upon which complaints and after they had been carefully verified, it pleased you to grant him a royal order by virtue of which his said wife was arrested on 30 November 1756 and taken to the Salpêtrière [hospital/prison for women], where she was detained until the month of August 1757. Because the petitioner was in no condition to pay any board, it pleased Your Lordship to have her released. Upon her promises and the greatest protestations of living with her husband in the future with the correctness of a good wife, he, touched with compassion at seeing her reduced to such a sad condition, received her into his house, expecting that his mercy and the punishment that she had just undergone would have changed her, but very far from that, she spent

From Arlette Farge and Michel Foucault, eds., *Le Désordre des Familles: Lettres de cachet des Archives de la Bastille,* pp. 74–75. © 1982 by Gallimand, Paris. Translated by Jeffrey Merrick.

only three months in harmony, that is to say until she saw that she was cured of the ulcers that the bad air [in the Salpêtrière] had caused her. Then she began her libertine life again and even went farther in unmasking herself and in leading a continuous life of debauchery and, what is still worse, devotes herself to theft and threatens her husband with having him murdered.

The petitioner and his wife's family, fearing for their honor, one and all living in the strictest honesty for the longest time, take the liberty of having recourse to Your Lordship's goodness and justice in order to protect him from this danger by having a woman so inclined to vices imprisoned. They will not stop making wishes and saying prayers for the preservation of your precious health.

Mr. P. Corneille, Marguerite Desoteau, Laure Perlot (cousin), Marie Dorineau (cousin)

I certify that Madam Corneille, working for me, wronged me.

Dernier (dealer in used goods), Patipas, Mason, Pubi

Working for me, whom she wronged (Pubi's wife).

72. Trial of Two Women for Same-Sex Relations

Statutes against sodomy in early modern Europe varied in their inclusion of women; they were explicitly included in the Holy Roman Empire, but they were not mentioned in the English statutes. There is no record of female homosexual subcultures developing in cities the way male homosexual subcultures did in early modern Amsterdam, Florence, Paris, and London, and there is debate among historians about how to interpret intense female friendships that emerge in letters and poetry. Actual trials of women for same-sex relations were very rare, and the cases that did come to trial generally involved women who wore men's clothing or used a dildo or other device to effect penetration (sex, in the minds of male authorities, always involved penetration). In a very few instances, women actually married other women, thus fully usurping the male social as well as sexual role. The following is one of these rare cases, from the German city of Halberstadt in 1721. The case was heard first by the municipal court of Halberstadt in Saxony, then sent to the law faculty in Duisberg in the Rhineland for review; both of these cities were under Prussian rule at the time, and the ultimate decision rested with the ruler of Prussia.

According to her mother's statement, the defendant, A. [*sic*] Marg. Lincken, approximately 27 years of age, was illegitimately born, after Mr. Lincken's death, to another man, at Gehowen. She was baptized there and then was sent to the orphanage in Halle. There she was instructed in the Christian religion until her fourteenth year.

The mother still resides in Halle and freely admitted and confessed that the defendant, after leaving the orphanage, had remained for some time in Halle where she learned the trade of button-making and of printing cotton. Subsequently, she had stayed with her friends in Calbe. In order to lead a life of chastity, she had

"A Lesbian Execution in Germany, 1721: The Trial of Records" from *Journal of Homosexuality,* 6 (1980/1981); 28, 30–38, 40. Translated by permission of the author.

disguised herself in men's clothes. Dressed in this way she went to see her mother in Halle but took leave of her because she wanted to lead a holy life. She left for Sora with a troop of Inspirants [religious radicals] she encountered in the Strohhof outside Halle and then went to Sechsstädten and Nuremberg. There she was baptized by the so-called prophetess Eva Lang in the river outside the town in the name of Jehova Almajo Almejo in the presence of a multitude of Inspirants from far and wide and took on the name of Anastasius Lagrantinus Rosenstengl....

Following this she had gone back to Halle. After a few months, she joined the Hanoverian troops as a musketeer in Colonel Stallmeister's regiment, serving for 3 years under the name of Anastasius Lagrantinus Beuerlein or, according to page 89 of the affidavit from Hanover, as Caspar Beuerlein. In 1708, she deserted at Brabant; however, she was apprehended near Antwerp and condemned to death by hanging. But when she disclosed her sex and a letter in her behalf arrived from Professor Francken at Halle [a near-by university], she was reprieved.

After some time she joined the Volunteer Company of Royal Prussian troops in Soest, under General Horn, using the name of Augustus or Caspar Beuerlein. Her captain's name was Becker. When she had served approximately one year, Professor Francken wrote to the garrison priest and informed him that she was a female, whereupon she had been sent away with identification papers. Thereupon she returned to Halle, put on her woman's clothes, and stayed there one summer. Then she went to Wittenberg in men's clothing and joined the royal Polish troops as a musketeer with the name of Peter (or Lagrantinus) Wannich. When she participated in a campaign and was captured by the French on a march near Brussels, she ran away again. Having served for one year with the Hessian troops in Rheinfels in Major Briden's Volunteers, there was a fight for which she was to have run the gauntlet, but again she escaped. Serving amongst these troops she had called herself Cornelius Hubsch and had changed her name often so that, in case of desertion, interrogation would be more difficult.

During her war service she had fluctuated between Catholicism and Lutheranism, taking both Catholic and Lutheran communion. After she had run away from the Hessian troops, she went back to Halle and made flannel for the university shoemaker; also she did spinning and printing and frequently kept 8, 9, and more spinning maids busy. She continued in this way of life for three or four years, sometimes wearing female, sometimes male clothing. Once she had been arrested by the soldiers in Halle, but, because of Professor Francken and his disclosure of her femaleness, she had been let go again. At this occasion she had been inspected at the Rathhaus to see whether she was a man or a woman.

After this inspection she once again changed her way of life, put on men's clothing, and, in 1717, went to work for a French stockingmaker, where she had become acquainted with the codefendant, Catharina Margaretha Mühlhahn. She became engaged to her and went to the former pastor of St. Paul's, Lic. Chauden, to inform him of her engagement to the Mühlhahn woman. On this occasion she had passed herself off as a dyer of fine colors and cloth maker and as a son of Cornelius Joseph Rosenstengl, former mine superintendent in Gültenberg near Prague. She asked for the usual reading of the banns and the wedding actually occurred shortly before Michaelmas in the year of 1717 in the church of St. Paul. At the first reading of the banns someone had called out that she had a wife and children in Halle; however, the defendant showed a letter from her mother, and also procured two witnesses to prove that such was not the case. After this, the reading of the banns and the wedding had continued.

After the wedding they lived together as an alleged married couple and kept the same table and the same bed. She had made a penis of stuffed leather with two stuffed testicles made from pig's bladder attached to it and had tied it to her pubes with a leather strap. When she went to bed with her alleged wife she put this leather object into the other's body and in this way had actually accomplished intercourse. When she had gone to bed with the bride for the first time she is said to have told her that she wanted to have intercourse 24 times, but she had done so only three or four times. Making love never lasted more than a quarter of an hour because the defendant was unable to perform any more. At these times, she petted and fondled somewhat longer. The defendant also added that during intercourse, whenever she was at the height of her passion, she felt tingling in her veins, arms, and legs.

She said that while she was a soldier she had hired many a woman whom she excited with the leather object. At times she ran for miles after a beautiful woman and spent all her earnings on her. Often when a woman touched her, even slightly, she became so full of passion that she did not know what to do.

Once she put the leather object, which she had first used down below, into the codefendant's mouth. Whether the latter had complained about this to her mother, the defendant did not know. The codefendant complained that her genitals had become swollen and that her pelvic bones hurt so much that she had been unable to walk; but this was a lie, since the defendant had at first used a thin leather instrument, and only when the Mühlhahn woman's vagina had stretched had she acquired a thicker instrument. The Mühlhahn woman had frequently held the leather instrument in her hands and had stuck it into her vagina, which she would not have done if it had not felt pleasurable to her. The mother-in-law had taken her daughter away from the defendant and had insisted on a divorce; however, the defendant complained to the priests at St. Paul's so that the mother had to restore her daughter to the defendant....

...The mother had charged the defendant with being a woman and not a man, and when they got into a fight about it, the mother, together with a woman named Peterson, attacked her, took her sword, ripped open her pants, examined her, and discovered that she was indeed not a man but a woman. They also tore the leather instrument from her body, [as well as] the leather-covered horn through which she urinated and [which she had] kept fastened against her nude body. When the defendant had, never-

theless, insisted that she was a man, they had spread open her vulva and found not the slightest sign of anything masculine. They beat her up, and the mother-in-law submitted to the courts the leather instrument as well as the horn; the defendant was promptly interrogated.

The defendant had made the leather instrument herself while she was with the Hanoverian soldiers, and, using her ingenuity, she had used it with several girls when she was a soldier. Since she had to act like all the other soldiers, she caressed many a widow as well, who touched the leather penis and played with it and yet had not realized what it was; however, she had never inserted it into any of the widows' bodies.

The mother-in-law and the bride had known before the wedding that she was a woman, and she mentioned several incidents in this regard. Once in Münster, when the codefendant tore the leather object from the defendant's body and was therefore fully aware that she was no man, she nevertheless later let herself be tickled with it and they lived together even more intimately. Later, in Münster, they had been married a second time....

Concerning the codefendant Catharina Margaretha Mühlhahn, age 22, she confessed likewise and declared at the summary interrogation that she had had the banns read and entered marriage in the year 1717, 14 days before Michalmas, with the so-called Rosenstengel in St. Paul's Church at Halberstadt. Also, at night she had shared the bridal bed with him. However, when the time came for actual intercourse, her supposed husband had been unable to insert his sexual organ into hers but had tortured and tormented her for about a week so that she endured great pains and her sexual organ had become very swollen. After a week it finally worked, but he had never been able to get in more than half a finger's length. Her bones and her vagina had always pained her a great deal since she had to suffer him in this manner morning, evening, and midnight, too. Often he had tormented her like this for a whole hour. Whenever night approached she had started shuddering. She had been a very naive maiden and had not realized that she had been deceived in this way. That is why the defendant had become so bold and made her handle his supposed male organ which had become quite warm in her body so that she had not noticed that it was made of leather. The defendant had always slept in pants and she had not been allowed to reach into them. It's true she had noticed that whenever he pissed he always wet his shoes and this made her suspicious so that she said to him: "Other men can piss quite a ways, but you always piss on your shoes." But he had called her a beast and scum and had threatened to beat her. In the year 1718, shortly before Gallen [St. Gall's Day], in Münster, she had been lying on the bed with her husband. Since she had been sick he had unwittingly taken off his pants. When she noticed that he had fallen asleep, she had inspected him closely since his shirt had slipped up, and she had found he had the leather sausage, which her mother had later handed over to the court, tied to his body, but that otherwise he was fashioned exactly like herself. This scared and amazed her terribly. Her supposed husband awakened, and, since she had torn off the leather male member with the bag, he looked for it in their bed under the pretext of having lost his cuff links. When she reproached him with his deception, he had begged her not to get him into trouble; what good would there be in her getting another's blood on her hands? Henceforth he would live with her as brother and sister and even offered to take her back home, but she had been afraid of him and had worried he might kill her on the way, and so she had stayed with him in Münster until Shrovetide. Neither was he to bother her any more with the object; she had thrown it into the river by their house....

When the defendant and the codefendant were confronted with each other, the defendant stated that once, when she put the leather thing into her mouth, she, Catherine Linck, had been stark naked and the codefendant, Margaretha Mühlhahn, had fondled her breasts and therefore should have known and felt that the object was made of leather or a fake. Here, however, the codefendant contradicted her and said that the Linck woman had not been naked but that

she had been wrapped in a shirt; that although the codefendant had felt her breasts, the Linck woman had commented that many men had such breasts.

When questioned about her daughter's nature, the defendant's mother made a deposition that when her daughter was young she had not noticed anything masculine about her, but that she had not been perfectly female either, since in her youth the vagina had practically no opening and that because of this she might not have been capable of intercourse. Of course the witness had not examined the defendant since she had become an adult.

On the other hand, the testimony of the city physician, Dr. Bornemann, and the surgeon, Dr. Röper, instructs us that they inspected the defendant very carefully and that they found in her nothing hermaphroditic, much less masculine, but found her to be fashioned like a woman; and judging from the size of the breasts and the rather large womb and the largeness of the vulva, which was examined by a midwife, it could be concluded that her female member had not altogether been left alone, but that, during her extensive vagabonding, it undoubtedly had been disgracefully misused....

The Halberstadt Municipal Government respectfully sent this decision with a report on reasons for compliance or modification. They would like to suggest respectfully that since the outrages perpetrated by the Linck woman were hideous and nasty, and it can furthermore not be denied that she engaged in repeated baptism for which she should be punished, the jurists are also of the opinion that a woman who commits sodomy with another woman with such an instrument could therefore be given poena ordinaria [that is capital punishment].

According to Carpzow[1] and several others, the fact alone that she had practiced oral intercourse in the Mühlhahn woman's mouth with the leather instrument is enough to warrant capital punishment; however, since in this particular case and with these instruments

---

[1]*Carpzow:* probably Benedict Carpzow (1595–1666), a well-known jurist.

semen cannot be sucked off or ejaculated, this reasoning may not be applicable. It further becomes evident from article 116 of the criminal code for capital offenses [the code of Charles V] that in cases of crimes of sodomy, when a woman is involved with another woman, the penalty is burning alive. It has been the practice to kill the evildoers before such burning in order to prevent despair; however, a few jurists have been of the opinion that according to Saxon law it is beyond doubt that, for people who commit sodomy, only the sword comes into consideration, in order that there may be a differentiation in penalties for sodomy between humans and sodomy between humans and animals. The court would like to leave it most respectfully to the discretion of your Royal Majesty whether to defer to the judgment of the Duisberg Judicial Faculty or whether to choose the latter, milder way and graciously prescribe the sword alone for the defendant; in any case they see no reason why the defendant should not first be killed with the halter rather than with the sword, and then burned, particularly since she is a female.

Concerning the codefendant Mühlhahn, she should expect a serious punishment since even after she discovered the deceit and the fact that the Linck woman, her alleged husband, was not a man, she nevertheless continued to commit these sexual acts....

Because of those circumstances, the Duisberg sentence, as far as it prescribes the death penalty for Cath. Marg. Linck, alias Anast. Lagrantinus Rosenstengel, is to be confirmed. The manner of execution is to be performed according to the suggestions of the Halberstadt government, and the accused, because of her many committed and confessed serious crimes, is to be put to death by the sword, in accordance with the gravity of her crime....

Considering the codefendant, C. M. Mühlhahn, in order that the means by which the truth is obtained should not be harsher than the actual sentence, and since the death penalty is not applicable in the case of this simple-minded person who let herself be seduced into depravity, the Duisberg sentence should be changed so that instead of the second degree [torture] she should be punished *extraordinarie* and should be condemned to three years in the penitentiary or spinning room and afterwards should be banished from the country. Furthermore, both are responsible for paying all court costs incurred in this case, two-thirds of which are to be paid by the Linck woman....

QUESTIONS FOR ANALYSIS

1. Italian and German treatment of prostitutes is often contrasted, as prostitution remained largely legal in Italy but was criminalized in Germany. What differences and similarities do you see in the discussion of prostitution by Florentine and Nuremberg civic authorities?
2. What examples of the scrutiny of women's bodies do you find in the laws, church court records, and visitation reports? Who orders these, and who carries them out?
3. What role do personal and family shame and honor play in the legal treatment of women? How do their punishments for harsh words and illicit sexual relations compare with those of the men involved?
4. How would you compare the opinion of unmarried women that emerges from the Strasbourg laws and French edict with that in Anna Bijns' poem? What might be the reasons for the differences you find?
5. What actions of Catherine Linck do authorities find particularly disturbing? How does the handling of this very rare type of case reflect more general ideas of ideal behavior in women?

WORK

In the earlier chapters of this book, women's work has generally been discussed within the context of the household economy, rather than as a separate topic, for most work occurred within a family setting. For many women in early modern Europe this situation continued: rural households continued to be the main agricultural producers, marital couples ran shops together in cities, and even the most highly developed merchant companies were essentially family businesses with some additional employees. All of these enterprises were increasingly linked together by international trading networks, however, with monetary transactions becoming more common and more complex. This process is usually described as the growth of commercial capitalism, and the changes it brought were different for women than for men. Women rarely controlled enough financial resources to become major traders, and ideas about women's honor often prevented them from engaging in occupations that required them to travel. As we saw in Chapter 5, universities and academies were closed to women, an exclusion that grew increasingly significant as formal credentials became more important in occupations such as medicine and law.

Capitalism also brought a change in the definition of work, from something one did to support one's family to something one did for money; this meant that women's work, which was often unpaid, came to be regarded as not really "work," but as "housework" or at most "helping out." Such value judgments were used as a further rationalization to pay women low wages, with the argument made that they were just helping out or working for "pin money" (money to provide small luxuries), even if the woman did have a family to support. This gender division between "real" work (that done by men) and "women's work" also shaped the division between "skilled" and "unskilled" labor developing in this period, with jobs traditionally done by men regarded as requiring more skill than those done by women, even if the actual skill level was the same. Glass cutting and goldsmithing, for example, were both viewed as men's occupations—and thus well-paid—while lace making was a women's occupation, whose practitioners were never adequately rewarded for their skill, dexterity, and concentration.

These value judgments regarding men's and women's work, along with other factors, shaped subsequent economic developments, including the growth of large-scale production. During the fifteenth century, the production of some types of goods, particularly cloth, came to be organized on a larger scale than individual guild shops, with an investor (almost always male) hiring many households to produce for him and owning both the raw materials and finished product (and sometimes the tools). This new organization of production is usually termed *pre-industrial* or *proto-industrial capitalism,* and it came to provide wage work for both women and men in many parts of Europe. Wages were highly skewed in terms of gender—just as they had been in earlier agricultural labor—which meant that in many places investors chose to hire women and children. Both international trade and larger-scale production provided a steadily increasing amount of consumer goods in Europe, thus increasing the amount of time women spent on shopping and housework.

73. Artemisia Gentileschi
Letters to Her Patron Don Antonio Ruffo

The judgment that men's work was more skilled than women's affected the visual arts as well as other types of production in the early modern period. Certain branches of art—in particular, painting, sculpture, and architecture—were judged "major" arts, while others—such as needlework, goldsmithing, porcelain manufacture, and textiles—were judged "minor" or "decorative" arts, or even "crafts." True art increasingly came to be seen as the product of individual genius, not hard work, and linked with characteristics judged to be masculine, such as power, forcefulness, and singularity of purpose. A hierarchy also developed within painting itself during the early modern period, with large history paintings viewed as the most important, followed by portraits, landscapes, miniatures, still lifes, and finally flower paintings. This hierarchy was clearly gendered, as women were not allowed to study the male nude, essential if one wanted to paint history paintings, and over half the known flower painters were women. A few women, however, did not accept the view that only men could paint professionally; they received training privately—often from their fathers—and sought out patrons just like their male counterparts. The following are letters to a patron from one of the most prolific early modern women painters, Artemisia Gentileschi (c. 1597–1651), who used her skills at painting women to produce many mythological and biblical scenes, including numerous depictions of the Old Testament heroine Judith, who cut off the head of the tyrant Holofernes during his war with the Israelites.

A. Letter to Don Antonio Ruffo
Naples, January 30, 1649

Most Illustrious Sir and My Master,

By God's will, Your Most Illustrious Lordship has received the painting and I think that by now you must have seen it. I fear that before you saw the painting you must have thought that I was arrogant and presumptuous. But I hope to God that after seeing it you will agree that I was not totally wrong. In fact, were it not for Your Most Illustrious Lordship, of whom I am so affectionate a servant, I would not have given it for one hundred and sixty, because in every other place where I have been, I was paid one hundred scudi per figure. And this was in Florence as well as in Venice, and in Rome and even in Naples when there was more money. Whether this is due to merit or luck, Your Most Illustrious Lordship, a discriminating nobleman with all of the virtues of the world, will judge what I am.

I sympathize greatly with Your Lordship, because a woman's name causes doubt until her work is seen. Please forgive me, for God's sake, if I gave you reason to think me greedy. As for the rest I will not trouble you any longer. I will only say that on other occasions I will serve you with greater perfection, and if Your Lordship likes my work, I will send you also my portrait so that you can keep it in your gallery as all the other Princes do.

And thus I end this letter and I most humbly bow to Your Most Illustrious Lordship with the assurance that as long as I live I will be ready for any orders from you. To end, I kiss your hands.

Your Most Illustrious Lordship's most humble servant Artemisia Gentileschi

B. Letter to Don Antonio Ruffo
Naples, November 13, 1649

My Most Illustrious Sir,

I prefer not to discuss our business in this letter in case that gentleman [the bearer]

will read it. With regard to your request that I reduce the price of the paintings, I will tell Your Most Illustrious Lordship that I can take a little from the amount that I asked, but the price must not be less than four hundred ducats, and you must send me a deposit as all other gentlemen do. However, I can tell you for certain that the higher the price, the harder I will strive to make a painting that will please Your Most Illustrious Lordship and that will conform to my taste and yours. With regard to the painting which I have already finished for Your Most Illustrious Lordship, I cannot give it to you for less than I asked, as I have already overextended myself to give the lowest price. I swear, as your servant, that I would not have given it even to my father for the price that I gave you. Don Antonio, my Lord, I beg you, for God's sake, not to reduce the price because I am sure that when you see it, you will say that I was not presumptuous. Your nephew, the Duke, thinks that I must have great affection for you to charge you such a price. I only wish to remind you that there are eight [figures], two dogs and landscape and water. Your Most Illustrious Lordship will understand that the expense for models is staggering.

I am going to say no more except what I have in my mind, that I think Your Most Illustrious Lordship will not suffer any loss with me and that you will find the spirit of Caesar in the soul of a woman.

And thus I most humbly bow to you.

Your Most Illustrious Lordship's most humble servant Artemisia Gentileschi.

C. Letter to Don Antonio Ruffo
Naples, November 13, 1649

My Most Illustrious Sir,

I received a letter of 26th October which I greatly appreciated, particularly noting how my Master always concerns himself with favoring me despite my unworthiness. In it, you tell me about that gentleman who wishes to have some paintings by me, that he would like a Galatea and a Judgment of Paris, and that Galatea should be different from the one that Your Most Illustrious Lordship owns. There

was no need for you to suggest this to me, since, by the grace of God and of the Most Holy Virgin, it would occur to a woman with my kind of talent to vary the subjects in my paintings; never has anyone found in my pictures any repetition of invention, not even of one hand.

As for the fact that this gentleman wishes to know the price before the work is done, believe me, as I am your servant, that I do it most unwillingly since it is very important to me not to err and thus burden my conscience, which I value more than all the gold in the world. I know that by erring I will offend my Lord God and I thus fear that God will not bestow his grace on me. Therefore, I never quote a price for my works until they are done. However, since Your Most Illustrious Lordship would like me to do it, I will do what you command. Tell this gentleman that I want five hundred ducats for both; he can show them to the whole world and, should he find anyone who does not think that the paintings are worth two hundred scudi more, I do not want him to pay me the agreed price. I assure Your Most Illustrious Lordship that these are paintings with nude figures requiring very expensive female models, which is a big headache. When I find good ones they fleece me and at other times, one must suffer their trivialities with the patience of Job.

As for my doing a drawing and sending it, I have made a solemn vow never to send my drawings because people swindled me. In particular I just today found out that, in order to spend less, the Bishop of St. Gata, for whom I did a drawing of souls in Purgatory, commissioned another painter to do the painting using my work. If I were a man I cannot imagine it would turn out this way, because when the concept [*inventione*] has been realized and defined with lights and darks, and established by means of planes, the rest is a trifle. I think that this gentleman is wrong to ask for drawings since he can see the design and the composition of the Galatea.

I don't know what else to say except that I kiss Your Most Illustrious Lordship's hands and most humbly bow to you, praying for the greatest happiness from Heaven.

Your Most Illustrious Lordship's most humble servant Artemisia Gentileschi

I advise Your Most Illustrious Lordship that when I ask a price I don't do as they do in Naples, where they ask thirty and then give it to you for four. I am Roman and thus I want to act in the Roman manner.

74. Glickl bas Judah Leib
Trading Activities in *Memoirs*

Though most merchants in early modern Europe were men, women did invest in trading ventures, and a few conducted them on their own. One of the best known of these is Glickl bas Judah Leib (1646 or 1647–1724, traditionally known as "Glückel of Hameln"), a Jewish woman born in Hamburg. Glickl married while in her teens and had eight children. She also assisted her husband in his growing trade in gold, pearls, jewels, and money. When she was in her early forties, her husband died accidentally, and Glickl continued his business, traveling to the fairs where many commercial transactions were made. (Jewish women were freer to engage in business beyond their own town than Christian women and often knew people in faraway cities, as they relied on such networks to find marriage partners for their children. Glickl was still accompanied on her trips by one of her older sons, as respectable women—Jewish or Christian—did not travel alone.) She also began to write her memoirs to help her get over her grief, composing these in Yiddish, the everyday language of Jews in central Europe. Her memoirs contain much about her family and business life, which were intertwined, as well as many stories drawn from history and tradition through which she sought to understand or explain the events of her life. The text survived in two family copies to the nineteenth century, when it was first published.

[The memoirs open with Glickl's description of her childhood.]

My mother had already learned the trade of making gold and silver lace, and God in His mercy saw to it that she received orders from the Hamburg merchants. At first Jacob Ree, of blessed memory, went surety for her; but when the merchants found that she knew her business and was prompt in her deliveries, they trusted her without surety.[1] Next she taught the trade to a number of young girls and engaged them to work by her side, so that finally she was able to provide a living for her mother and clean, decent clothes for herself. Little enough, however, remained over, and often my dear mother had nothing but a crust of bread the livelong day. She never complained, but put her faith in God who had never forsaken her....

[After the death of her first husband, both Glickl and her son Loeb engage in business, she in Hamburg and he in Berlin.]

At that time I had a manufactory for Hamburger stockings, many thousands' worth of

---

[1]*Surety* meant that someone else would guarantee to pay back her loans if she could not.

"Memoirs" by Glickl bas Judah Leib, as translated by Marvin Lowental.

which I turned out for my own account. And my unlucky son writes me to send him a thousand thalers and more of stockings, and I did so.

Then I meet at the Brunswick Fair certain Amsterdam merchants who hold my son's notes for about 800 Reichsthalers. My son Loeb writes me I can safely take up the notes—he will forward me the money to Hamburg. As I always stood by my children, I said to myself, I shall not put him to shame by protesting the notes, and I proudly paid them.

When I returned from the Brunswick Fair I expected to find bills of exchange from my son Loeb. But nothing awaited me, and when I wrote to him, he sent me all kinds of answers none of which pleased me. What was I to do? I needs must content myself.

Two weeks later, a good friend came to me and said, "I cannot keep it from you, I must tell you that your son Loeb's business mislikes me, for he is heavily plunged in debt. He owes his brother-in-law Model 4000 Reichsthalers, and Model sits in his store, that is to say, he watches after things. But he is a child and cannot attend to his business. He is out gulping food and drink at all hours and everyone is lord and master of the store. Your son Loeb is too nice and good, and easily led by the nose. Added to that, the Berliners are bleeding him with their interest. Moreover, he has two wolves at his flanks, one is that Wolf Mirels—son of the Hamburg rabbi Solomon Mirels—and the other is Wolf the brother-in-law of the learned Benjamin Mirels.[2] Every day this second Wolf goes to the store and makes off with what he pleases. Finally, your son does business with Polish Jews, so much so, I know, that he has already rid himself of more than 4000 thalers."

Such and more of the like my good friend told me, and my soul nearly died within me, and I fainted on the spot.

When my friend saw my shock, he tried to console me and said he believed that with some one to stand by him my son could still be saved.

I told all I had heard to my sons Nathan and Mordecai. They shrank with fright and said he owed them several thousands. God knows what it meant for me—my son Loeb owed me alone more than 3000 Reichsthalers—but I had little minded it were not his brothers so deeply immersed. But what could we do in our distress? We dare speak of it to no one.

We agreed that I should accompany my son Mordecai to the Leipzig Fair and see how matters stood. When we reached Leipzig we found my son Loeb already on hand, as was his wont, and laden with goods.

I now began to talk with him. "They are saying," I said, "thus and so of you. Bethink yourself of God and of your good and honest father, that you bring us not to shame." He answered, "You need not worry over me. But recently—it was not a month ago—my father-in-law had visiting him his brother-in-law Wolf of Prague, and we reckoned up my accounts and he found me, praise God, in excellent shape." Whereat I said to him, "Show me your balance sheet." He replied, "I haven't it with me, but do me the favour to come to Berlin and I will show you everything, to your content." "In any case," I concluded, "buy not a jot more of goods."

But my back was no sooner turned than Reb Isaac and Reb Simon, son of Rabbi Mann of Hamburg, sold him on credit more than 1400 thalers of goods. When I learned of it, I went to them and begged them in Heaven's name withdraw the sale, for my son needs must give over the merchandise trade, else it be his ruin. But it was all to no purpose, and they forced my son to take the wares.

After the fair, I accompanied my son Mordecai, Hirschel Ries and the other Berliners to Berlin.

Once I was in his house, my son Loeb said to me, "I fancy my one mistake is to have tied up too much money in goods." Whereupon I told him, "You owe me more than 3000 Reichsthalers—for my part I am satisfied to take it in goods at the price they cost you." "Mother

---

[2] Glickl's play on words—that the two men named Wolf are like wolves to her son—works in both Yiddish and English.

dear," he said, "if you are willing to do that, it will ease me of my difficulties, and no one need lose a penny through me."

The next day I went with my son to his store, and truly, he was badly overladen with goods. He gave me 3000 Reichsthalers of merchandise at the price it cost him. And you can imagine the face I made. But regardless of everything, I only sought to help my children.

We had the goods packed in bales to send on to Hamburg. Then I noticed the two bales of goods my son had bought in Leipzig from Reb Isaac and Reb Simon the Hamburg merchants, and I said to my son, "Send back those two bundles of goods, and I shall see to it they are accepted, even if I pay for them from my own pocket. And now," I continued, "that you have repaid your debt to me, what of my sons Nathan and Mordecai?" He had on hand bills of exchange and Polish paper amounting to over 12,000 Reichsthalers, and he gave them to my son Mordecai by way of payment.

After sitting the whole day in his store, we went home together; and you would be right in thinking I did not enjoy my supper....

At that time I was busied in the merchandise trade, selling every month to the amount of five or six hundred Reichsthalers. Further, I went twice a year to the Brunswick Fair and each time made my several thousands profit, so in all, had I been left in peace, I would have soon repaired the loss I suffered through my son.

My business prospered, I procured me wares from Holland, I bought nicely in Hamburg as well, and disposed of the goods in a store of my own. I never spared myself, summer and winter I was out on my travels, and I ran about the city the livelong day.

What is more, I maintained a lively trade in seed pearls. I bought them from all the Jews, selected and assorted them, and then resold them in towns where I knew they were in good demand.

My credit grew by leaps and bounds. If I had wanted 20,000 Reichsthalers *banko* during a session of the Bourse, it would have been mine.

"Yet all this availeth me nothing." I saw my son Loeb, a virtuous young man, pious and skilled in Talmud, going to pieces before my eyes.

One day I said to him, "Alas, I see nothing ahead of you. As for me, I have a big business, more indeed than I can manage. Come then, work for me in my business and I will give you two per cent of all the sales."

My son Loeb accepted the proposal with great joy. Moreover, he set to work diligently, and he could soon have been on his feet had not his natural bent led him to his ruin. He became, through my customers, well known among the merchants, who placed great confidence in him. Nearly all my business lay in his hands.

[Under her guidance, Loeb's business skills improve and he died solvent. This was not true for Glickl's second husband, whom she married a decade after the death of her first. He went bankrupt, and the couple had to be helped by their children; after his death, Glickl lived with one of her daughters.]

75. Catharina Schrader
Midwife's Activities from
Memoirs of the Women

Catharina Schrader (1656–1746) was a professional midwife in Frisia (part of the Netherlands) who kept notebooks of all her cases between the years 1693 and 1745. These number over 3,000—in her most active years, Schrader attended over 130 deliveries per year—making them the most complete record of any midwife's activities that exist from this period and among the most complete records of any medical practitioner, male or female. When she was in her eighties, Schrader decided to pull together her more complicated cases into a single book, which she titled *Memoirs of the Women,* dedicating it to the women she had delivered. During the early modern period, childbirth was a female matter, handled by neighbors, friends, and, where they were available, by professional midwives who were self-taught or trained by other midwives. A few men—usually called "man-midwives"—were beginning to venture into the field, but generally doctors or surgeons were called only in the most serious cases, and their own training rarely included obstetrics. From her notebooks, it is clear that Schrader became known as someone who could handle difficult cases—she was often called when things were going badly—and she also handled other types of surgical and gynecological matters along with childbirth. As men moved into the field, they attempted to portray female midwives as superstitious, bungling, and inept, but Schrader's memoirs, along with guides written by other prominent midwives such as Jane Sharp in England, Louise Bourgeois in France, and Justina Siegemund in Germany, suggest otherwise.

Thereupon in my eighty-forth year of old age in my empty hours I sat and thought over what miracles The Lord had performed through my hands to unfortunate, distressed women in childbirth. So I decided to take up the pen in order to refresh once more my memory, to glorify and make great God Almighty for his great miracles bestowed on me. Not me, but You oh Lord be the honour, the glory till eternity. And also in order to alert my descendants so that they can still become educated. And I have pulled together the rare occurrences from my notes.

In my thirty-eight years living in Hallum in Friesland I saw my good, learned and highly esteemed, and by God and the people loved husband, go to his God to the great sadness of me and the inhabitants, leaving six small children in my thirty-eight years of age. But then it pleased God to choose me for this important work: by force almost through good doctors and the townspeople because I was at first

struggling against this, because it was such a weighty affair. Also I thought that it was for me and my friends below my dignity; but finally I had myself won over. This was also The Lord's wish.

1. 1693 on 9 January fetched to Jan Wobes's wife, Pittie, in Hallum. A very heavy labour. Came with his face upwards. A dangerous birth for the child and very difficult for me. The afterbirth had to be pulled loose. But everything well. (The notebook mentions that Schrader was called at five o'clock in the morning to the labour. With God's "grace and help," Schrader delivered a boy.)[1]

3. 1693 on Shrove Tuesday (26 February) in the evening I was fetched for the very first jour-

---

[1]Sometimes the notebook of all cases contains additional information; this is included here in parentheses.

Excerpted from *Memoirs: Mother and Child Were Saved: The Memoirs (1693–1740) of the Frisian Midwife Catherina Schrader.* Translated and annotated by Hilary Marland, Nieuwe Nederlandse Bijdragen tot de Geschiedenis der Geneeskunde en der Natuurwetenschappen, No. 22 (Amsterdam: Rodopi, 1984). Reprinted by permission.

ney in my life to Wijns to a widow whose husband was called Chlas Jansen, in terrible weather, stormy wind, hard frost. The three of us travelled by sleigh over the ice. The wind blew so hard that one could not stand. Pieces of ice got stuck in my legs, so that blood dripped into my hose. And came at last by sleigh to Wijns, three hours going, we were almost dead. The people carried me into the house and forced my mouth open; and poured brandy into my mouth. There was a good fire. I thawed out a little. First I demanded a bowl with snow and rubbed my hands and feet with it until life came into them. Otherwise I would have been ruined for life. After I recovered again, I went to help the woman. And also her dead husband's brothers had taken everything away from her and had said that she would not give birth; therefore the life of this child was of great consequence.[2] The woman had a very heavy labour, like her previous labours had also been; she had had two midwives from Leeuwarden [in her previous labours]. I prayed to [The] Lord, and he answered me and delivered the woman of a good, big daughter to the great delight of her and me. This introduction was oppressive for the first time. The Lord be thanked. All well. And the woman got all her belongings back....

20. 1694 on 27 January I was fetched to the wife of Derck Jans, Antie, after another [midwife] had been with her two days and nights. Everything was in a terrible state. The child was deeply embedded, with the feet round the neck [and] trapped behind the pubic bone, the cord round the legs and round the neck. Must be choked. Was stuck two hours in the birth canal. Had to loosen it with enormous difficulty. I had almost given up, but The Lord brought solution. The mother does

well. (The notebook mentions that she was fetched on Sunday morning. The child was dead.)...

161. 1697 on 30 June fetched to Oostrum to wife of Gerrben Teyepkes, farmer. There had already been another midwife there for two days. Could only help her with the first [child]. A dead child. But The Lord be thanked, I delivered her of the afterbirth within one hour. They both lay strangely. Turned them. The middle one was dead. The afterbirth was stuck, so that [there were] three children; one living, two dead. So that there were three children. They were big; the parents small, delicate people. The woman fresh and healthy. (The notebook states that the first child was dead, the second alive, and the third dead. Two boys and a girl, who was alive. There were three afterbirths. Schrader managed the deliveries in an hour. The smallest and last child presented with its bottom.)...

743. 1702 on 4 May [I] was fetched to Rinsumageest to the former sweetheart of the town clerk, Veenema, who had promised to be hers in marriage, but who had left her on the advice of friends. Was four days in labour. Could not be helped. Then I was fetched and delivered her quickly through the help of my God. Yet a heavy birth, because of the heartache caused to her.[3] (The notebook mentions that she delivered a son. Another midwife had been there for three days. According to the baptismal register of Rinsumageest, the child was a daughter.)

796. 1702 on 12 October delivered two sons to the knitter, Swaantie. The first came well, the second with his stomach [first]. Had difficulty with turning [it]. Still all was well for mother and children. The Lord be praised and thanked....

---

[2] Because the woman's husband was dead and she had no children yet, his brothers stood to inherit much of his property and goods and had already taken it.

[3] I.e., because this was a child born out of wedlock.

1672. 1710 on 5 February with Jan Gorrt-zacke's daughter, Hinke, whose husband, Wattse, was a corn merchant, who was visiting her mother. And delivered her quickly of a son. Lived but half an hour. But, The Lord works mysteriously, I [was] terrified. Found that between the stomach and the belly [there] was an opening as big as a gold guilder, all round it grew a horny border. Out of this hung the intestines with the bowels. Had grown outside the body. One saw there the heart, liver, lungs clear and sharp, without decay. One could touch wholely under the breast. It was worthy to be seen by an artist, but she did not want it to be shown. I inquired [of] the woman if she had also had a fright or mishap. She declared that she was unaware of anything, but that [when] it had been the killing time they had slaughtered a pig. They had hung it on the meat hook, and the butcher had cut out the intestines and the bowels.[4]

1795. 1711 on 10 February I was fetched to Nijkerk to Wattse Jennema, whose wife was called Alltie Jouwkes. She wanted me to attend her, but didn't call for me. And fetched a midwife from Morra, who tortured her for three days. She turned it over to the man-midwife, doctor Van den Berrg. He said, he must cut off the child's arms and legs. He took her for dead. And he said, the child was already dead. Then I was fetched in secret. When I came there her husband and friends were weeping a great deal. I examined the case, suspected that I had a chance to deliver [her]. The woman was very worn out. I laid her in a warm bed, gave her a cup of caudle[5], also gave her something in it;

sent the neighbours home, so that they would let her rest a bit. An hour after her strength awakened again somewhat. And I had the neighbours fetched again. And after I had positioned the woman in labour, [I] heard that the doctor came then to sit by my side. I pulled the child to the birth canal and in half of a quarter of an hour I got a living daughter. And I said to the doctor, here is your dead child, to his shame. He expected to earn a hundred guilders there.[6] The friends and neighbours were very surprised. The mother and the child were in a very good state. (The notebook maintains that altogether it took Schrader three hours to deliver the child.)...

Now I have had more than a hundred bad and heavy complicated births. There was much writing involved. Of these mentioned all were dangerous. And yet there is not one of them, whether or not they had the life and bodily health left to them, that The Lord alone had not ordained; otherwise it would have been impossible many times. Him alone the praise. I have seen over this, and to my wonder, [the memoirs are] there to be used as a guide, or after my death, so someone still may get use or learning from it, to the advantage of my fellowmen. I have [written] this in my eighty-fifth year of old age, 1740 on 18 September. And it shall now be my last light. And I have during the time of my sinful life had a heavy time. And about over four thousand children helped into the world, these including 64 twins and three triplets.

---

[4]Like most early modern people, Schrader believes strongly in the power of the maternal imagination, that what a woman saw or dreamed during pregnancy could affect the child. Pregnant women were advised not to watch executions or butchering or to look at injured or handicapped people.

[5]*Caudle*: a warm, thin soup made of grain, sugar, wine, and spices given to invalids and women in childbirth to strengthen them. After a birth, the women who had helped drank caudle together in celebration of the delivery.
[6]A hundred guilders was a huge sum of money. Elite families might pay an experienced midwife fifty guilders for an extremely complicated delivery. For most of her cases, Schrader was paid between one and six guilders; a few were as high as forty guilders and in some she received no fee at all.

76. Arthur Young
Report on Rural Industry from *Six-Months Tour Through the North of England*

During the seventeenth and eighteenth centuries, rulers, government officials, and well-educated private individuals became increasingly concerned with developing methods of improving the economic prosperity of their countries. They began to travel around their own and foreign countries observing agricultural and manufacturing techniques, measuring productivity, and suggesting ways of improvement. One of the most prolific of these commentators was Arthur Young (1741–1820), who became widely known as an agricultural expert, with a reputation that extended from Russia to America. (Young corresponded with George Washington, who was very concerned about the state of American farming.) On his travels, Young often found examples of rural proto-industry as well as farming, which he appears to have viewed as one means of helping to alleviate rural poverty, one of his great concerns. Young's analysis of English rural life has been criticized as overly biased toward his own interests in agricultural improvements, but his reports on wages and prices have not been disputed.

... From thence to *Boynton,* the seat of Sir *George Strickland.* Sir *George* was so obliging as to shew me his woollen manufactory; a noble undertaking, which deserves the greatest praise. In this country, the poor have no other employment than what results from a most imperfect agriculture; consequently three-fourths of the women and children were idle. It was this induced Sir *George* to found a building large enough to contain on one side a row of looms of different sorts, and on the other a large space for women and children to spin. The undertaking was once carried so far as to employ 150 hands, who made very sufficient earnings for their maintenance; but the decay of the woollen exportation reduced them so much, that now those employed are, I believe, under a dozen....

The town of *Bedford* is noted for nothing but its lace manufactory, which employs above 500 women and girls. They make it of various sorts up to 25 *s.* a yard; women that are very good

hands, earn 1 *s.* a day, but in common only 8 *d.* 9 *d.* and 10 *d.*[1] Girls from eight to fifteen, earn 6 *d.* 8 *d.* 9 *d.* a day. This manufacture is of infinite use to the town, employing advantageously those who otherwise would have no employment at all.

Sheffield contains about 30,000 inhabitants, the chief of which are employed in the manufacture of hard-ware: The great branches are the plating-work, the cutlery, the lead works, and the silk mill.

In the plated work some hundreds of hands are employed; the men's pay extends from 9 *s.* a week to 60 *l.* a year: In works of curiosity, it must be supposed that dexterous hands are paid very great wages. Girls earn 4 *s.* 6 *d.* and 5 *s.* a week; some even to 9 *s.* No men are employed

---

[1]*d.* is the abbreviation for pence, and *s.* for shillings; one shilling equals twelve pence. Twenty shillings equals one pound, for which the abbreviation is *l.*

Excerpted from *A Six Months Tour Through the North of England* by Arthur Young. © 1770. Reprinted in 1967 by Augustus M. Kelly, New York.

that earn less than 9 *s.* Their day's work, including the hours of cessation, is thirteen....

Here is likewise a silk mill, a copy from the famous one at *Derby,* which employs 152 hands, chiefly women and children; the women earn 5 or 6 *s.* a week by the pound; girls at first are paid but 1 *s.* or 1 *s.* 2 *d.* a week, but rise gradually higher, till they arrive at the same wages as the women. It would be preposterous to attempt a description of this immense mechanism; but it is highly worthy of observation, that all the motions of this complicated system are set at work by one water-wheel, which communicates motion to others, and they to many different ones, until many thousand wheels and powers are set at work from the original simple one. They use *Bengal, China, Turkey, Piedmont,* and *American* raw silk; the *Italian* costs them 35 *s.* a pound, but the *American* only 20 *s.* It is a good silk, though not equal to the *Piedmont.* This mill works up 150 *lb.* of raw silk a week all the year round, or 7800 *per annum.* The erection, of the whole building, with all the mechanism it contains, cost about 7000 *l.* ...

Upon the whole, the manufacturers of *Sheffield* make immense earnings: There are men who are employed in more laborious works, that do not earn above 6 or 7 *s.* a week, but their number is very small; in general they get from 9 *s.* to 20 *s.* a week; and the women and children are all employed in various branches, and earn very good wages, much more than by spinning wool in any part of the kingdom.

Leeds cloth market is well known, and has been often described. They make chiefly broad cloths from 1 *s.* 8 *d.* a yard to 12 *s.* but mostly of 4 *s.* 6 *d.* and 5 *s.* Good hands at this branch, would earn about 10 *s.* 6 *d.* a week the year round, if they were fully employed; but as it is, cannot make above 8 *s.* This difference of 2 *s.* 6 *d.* is a melancholy consideration. A boy of 13 or 14, about 4 *s.* a week, some women earn by weaving as much as the men. The men, at what they call offal work, which is the inferior branches, such as picking, rinting, &c. are paid 1 *d.* an hour. Besides broad cloths, there are some shaloons, and many stuffs made at *Leeds,* particularly *Scotch* camblets, grograms, burdies, some callimancoes, &c.[2] The weavers earn from 5 *s.* to 12 *s.* a week; upon an average 7 *s.* Boys of 13 or 14, 5 *s.* a week. But they are all thrown out in bad weather; men in general at an average the year round, about 6 *s.* or 6 *s.* 6 *d.* a week. They never want work at weaving. Dressers earn from 1 *s.* to 3 *s.* a day, but are much thrown out by want of work. The women by weaving stuffs, earn 3 *s.* 6 *d.* or 4 *s.* a week. Wool-combers, 6 *s.* to 12 *s.* a week. The spinning trade is constant, women earn about 2 *s.* 6 *d.* or 3 *s.* a week. Girls of 13 or 14, earn 1 *s.* 8 *d.* a week. A boy of 8 or 9 at ditto $2\frac{1}{2}$ *d.* a day; of six years old, 1 *d.* a day. The business of this town flourished greatly during the war,[3] but sunk much at the peace, and continued very languid till within these two years, when it began to rise again.

PROVISIONS, &c.

Much oat bread eat, 10 or 11 ounces for 1 *d.*

Butter, — 8 *d. per* lb. 18 or 19 ounces.

Cheese,	—	4	Pork —	4 *d.*
Mutton,	—	4	Bacon, —	7
Beef,	—	4	Veal, —	$2\frac{1}{2}$

Milk, a pint in summer $\frac{1}{2}$, in winter $1\frac{1}{2}$ *d.* and 1 *d.*

Manufacturer's house rent, 40 *s.*

Their firing,[4] 20 *s.*

From *Asgarth* returning by *Crakehill,* I took the road once more to *Richmond;* and from thence to *Darlington,* in the county of *Durham.* At that town is a considerable manufacture of *Huckerback* cloths, in which the workmen earn from 10 *d.* to 2 *s.* 6 *d.* a day, and women and children proportionably. One master manufacturer employs above 50 looms, and asserts, that he could easily set many more at work, and employ numerous women and children, if the idle part of the poor of the town could be persuaded to turn industri-

---

[2]These are all different types of cloth.
[3]The Seven Years War (1756–1763).
[4]*Firing:* heating.

ous; but numbers of hands, capable of working, remain in total indolence; and that in general, there need never be an unemployed person in *Darlington*.... The poor women and children in total idleness. They do not drink tea, but smoke tobacco unconscionably....

Kendal is a well built and well paved town, pleasantly situated, in the midst of the beautiful country just described. It is famous for several manufactories; the chief of which is that of knit stockings, employing near five thousand hands by computation. They reckon one hundred and twenty wool-combers, each employing five spinners, and each spinner four or five knitters; if four, the amount is two thousand four hundred; this is the full work, supposing them all to be industrious; but the number is probably much greater. They make five hundred and fifty dozen a week the year round, or twenty-eight thousand six hundred dozen annually: The price *per* pair is from 22 *d.* to 6 *s.* but in general from 22 *d.* to 4 *s.* some boys at 10 *d.* If we suppose the average 3 *s.* or 36 *s.* a dozen, the amount is 51,480 *l.*

The wool they use is chiefly *Leicestershire, Warwickshire,* and *Durham:* They generally mix *Leicestershire* and *Durham* together. The price 8 *d.* 9 *d.* and 10 *d. per lb.* They send all the manufactures to *London* by land carriage, which is said to be the longest, for broad wheel waggons, of any stage in *England*. The earnings of the manufacturers in this branch are as follow:

		s.	d.
The combers, *per* week,	—	10	6
The spinners, women,	—	3	0
Ditto, children of ten or twelve years,	— — —	2	0
The knitters,	— —	2	6

All the work-people may have constant employment if they please.

During the late war business was exceedingly brisk, very dull after the peace, but now as good as ever known.

The making of cottons is likewise a considerable manufacture in this town. They are called *Kendal* cottons, chiefly for exportation, or

sailors jackets, about 10 *d.* or 1 *s.* a yard, made of *Westmoreland* wool, which is very coarse, selling only at 3 *d.* or 4 *d. per lb*. This branch employs three or four hundred hands, particularly shearmen, weavers, and spinners.

	s.	d.
The shearmen earn *per* week,	10	6
The weavers, (chiefly women,)	4	3
The spinners, — —	3	3

All have constant employment. During the war this manufacture was more brisk than ever, very dull after the peace, and has continued but indifferent ever since.

Their third branch of manufacture is the linsey woolsey, made chiefly for home consumption, of *Westmoreland, Lancashire,* and *Cumberland* wool; the hands are chiefly weavers and spinners. The first earn 9 *s.* or 10 *s.* a week; the second (women) 4 *s.* 6 *d.* or 5 *s.*

The farmers and labourers spin their own wool, and bring the yarn to market every week: There are about five hundred weavers employed, and from a thousand to thirteen hundred spinners in town and country. The business during the war was better than it has been since, but is now better than after the peace.

Their fourth manufacture is the tannery, which employs near a hundred hands, who earn from 7 *s.* to 7 *s.* 6 *d.* a week. They tan many hides from *Ireland.*

They have likewise a small manufactory of cards, for carding cloth. Another also of silk: They receive the waste silk from *London,* boil it in soap, which they call scowering, then it is combed by women (there are about thirty or forty of them,) and spun, which article employs about an hundred hands; after this it is doubled and dressed, and sent back again to *London.* This branch is upon the increase.

PROVISIONS, &c.

Bread—oatmeal baked in thin hard cakes, called clap-bread, costs 1 *d. per lb.*

Cheese, $3\frac{1}{2}d.$

Butter, $6\frac{1}{2}d.$ to 16 oz.

Mutton, 2 *d.* to $2\frac{1}{2}$ *d.*

Beef, $2\frac{1}{2}d.$ to 3 *d.*

Veal, $2\frac{1}{2}d.$

Pork, $4\frac{1}{2}d.$

Bacon, $6\frac{1}{2}d.$

Milk, $\frac{1}{2}$ *d.* a pint.

Potatoes, 10 *d.* four gallons.

Poors house-rent, 30 *s.*

————firing, 45 *s.* to 50 *s* ...

At *Warrington* the manufactures of sail-cloth and sacking are very considerable. The first is spun by women and girls, who earn about 2 *d.* a day. It is then bleached, which is done by men, who earn 10 *s.* a week; after bleaching it is wound by women, whose earnings are 2 *s.* 6 *d.* a week; next it is warped by men, who earn 7 *s.* a week; and then starched, the earnings, 10 *s.* 6 *d.* a week. The last operation is the weaving, in which the men earn 9 *s.* the women 5 *s.* and boys 3 *s.* 6 *d.* a week.

The spinners in the sacking branch earn 6 *s.* a week, women; then it is wound on bobbins by women and children, whose earnings are 4 *d.* a day; then the starchers take it, they earn 6 *s.* a week; after which it is wove by men, at 9 *s.* a week. The sail-cloth employs about three hundred weavers, and the sacking an hundred and fifty; and they reckon twenty spinners and two or three other hands to every weaver.

During the war the sail-cloth branch was very brisk, grew a little faint upon the peace, but is now, and has been for some time, pretty well recovered, though not to be so good as in the war. The sacking manufacture was also better in the war; but is always brisk.

The spinners never stand still for want of work; they always have it if they please; but weavers sometimes are idle for want of yarn, which, considering the number of poor within reach, (the spinners of the sacking live chiefly in *Cheshire,*) is melancholy to think of.

Here is likewise a small pin-manufactory, which employs two or three hundred children, who earn from 1 *s.* to 2 *s.* a week.

Another of shoes for exportation, that employs four or five hundred hands, (men,) who earn 9 *s.* a week.

QUESTIONS FOR ANALYSIS

1. The Renaissance notion of the artist as an individual creative genius stood in sharp contrast to ideals for women that emphasized modesty and silence. How does Artemisia Gentileschi balance these two in the letters to her patron? In what other ways is her situation shaped by her gender?
2. What kinds of activities do Glickl and her mother undertake and invest in? How is Glickl's business affected by her son's actions?
3. Why does Catharina Schrader become a midwife, and why does she decide to write up her cases? How would you compare her attitude toward her work and her sense of self with that of Glickl and Gentileschi?
4. According to Young's report, what types of work did proto-industry provide for women in England? How did their wages compare with those of men and with prices for basic commodities such as food and rent?
5. Based on these four sources, what factors most shaped the working life of women in early modern Europe?

CHAPTER 7

Early Modern Religious and Intellectual Developments

(1500–1800)

The social and economic developments of the early modern period traced in the last chapter were closely related with religious and intellectual changes that brought new ideas about women's proper place and role. During the sixteenth century, about half of Europe broke with the Roman Catholic Church, and separate Protestant churches were established in many countries and territories, mostly in northern and central Europe. Though they differed on many points of doctrine, Protestant reformers generally downplayed the importance of good works in achieving salvation and reduced the distinction—at least in theory—between clergy and laypeople (the phrase often used to express this is "the priesthood of all believers"). They rejected the notion that clergy should be celibate and saw no value in the monastic life; thus monasteries and convents were closed, and priests who staffed parish churches generally married and had children. In contrast to Catholicism, there was no special religious vocation open to women in Protestantism, who were told to express their devotion through marriage and motherhood. In fact, family life was proclaimed as the ideal for all men and women; spouses were increasingly urged to view one another as companions, and children to honor and obey their parents. Such exhortations, expressed in thousands of sermons and published moral guides and stories, were buttressed, as we saw in the last chapter, by laws that prohibited married women from carrying out economic transactions without their husband's permission and laws that allowed families to imprison disobedient wives and children.

In Catholic areas—largely southern and western Europe—virginity continued to be valued over marriage, though some Catholic authors did write guides for married life similar to those written by Protestants. Catholic response to the Protestant Reformation led to calls for the reform of abuses and problems, which in turn resulted in a stricter enforcement of cloister for the female religious. Catholic response also led to various moves that would win people back to loyalty to Rome or gain converts in Europe's new colonial empires, but these exciting new roles of defender of Catholicism within Europe and missionary to lands outside of Europe were to be for men only.

Religious differences led to open warfare at various points in early modern Europe, and they also led rulers to crack down on those who held different religious opinions within their own territories. This drive for religious uniformity played a role in the upsurge of trials and executions for witchcraft, called the "Great Witch Craze," of the sixteenth and seventeenth centuries. Europeans—like people in most of the pre-modern world—believed firmly in the ability of witches to do harm, and during the Late Middle Ages also came to believe that witches made a pact with the Devil and were doing his bidding. Witches were regarded as enemies of God and the community, and their extermination was seen as a way that political and religious authorities could prove their piety and religious commitment. Protestants and Catholics were united on this, though the pattern and extent of actual witch persecutions varied significantly across Europe. The percentage of accused witches who were women also varied, but in most parts of Europe women made up the vast majority of both the accused and the executed.

Discussions of why women were most likely to be witches generally centered on their disorderly nature; lack of reason; strong sexual drive; vulnerability to the Devil's wiles; and lack of political, physical, or economic power (which would lead them to seek demonic assistance). These qualities were also emphasized in more general discussions about women, particularly discussions about whether women should rule over men or should have any political rights. The debate about female rule—gynecocracy—was not simply a theoretical one, for dynastic accidents left many women as the actual monarchs of countries and territories or the effective monarchs during the period when their sons were young. This debate involved questioning what we would term the social construction of gender: Could a woman's social rank (for instance, as a queen) ever make up for the limitations of her sex? Marital status also became involved in this discussion, as writers argued whether a woman could be a ruler in public life while she was legally dependent on her husband in private life. (Elizabeth I avoided this conundrum by never marrying.)

Women's dependent position within marriage became one reason given for excluding women from political rights once broader discussions of rights began in the seventeenth century. Most political leaders and writers—including those who stressed men's rights vis-à-vis their monarch—asserted that women were represented by their husbands in everything beyond the household. Marriage was not the only grounds given for denying women political rights, however, for rights became increasingly associated with reason, a quality that women—since Aristotle—were widely regarded as lacking. This long-standing association of reason and masculinity was called into question by some authors in the seventeenth and eighteenth centuries, however, as well-educated Europeans debated issues of equality and difference involving gender as well as race and social class: Was a well-ordered society one that emphasized the fundamental equality of men and women, or the (perhaps complementary) fundamental differences between them? What were the sources of this equality or these differences? What were their implications? How was the "Woman Question"—similar in some ways to the debate about women that had already gone on for centuries—reframed in a world in which change was increasingly perceived as good and individual capacity was seen to play a greater role in the determination of individual lives and choices than inherited status and wealth? These debates were increasingly carried out by women as well as men, in literary salons which they ran in their own homes or in the pages of books, periodicals, pamphlets, and newspaper articles which they wrote and read.

THE PROTESTANT AND CATHOLIC REFORMATIONS

Though the Protestant Reformation is one of the key events traditionally regarded as separating the medieval from the modern era, there were also strong continuities in religious ideas and practices across many centuries. Protestants broke with the institutional structure and denounced many of the theological ideas of the Catholic Church, but they did not break sharply with medieval theologians in their ideas about women. Except for a very few religious radicals, Protestant reformers—and their Catholic counterparts—viewed women as created by God and spiritually equal, but in every other respect subordinate and inferior to men. Protestants rejected vows of celibacy and praised marriage, but their arguments in favor of marriage often retained the suspicion of sexuality and women's sexual attractiveness that had been present in many ancient and medieval Christian thinkers. They also retained the same three purposes of marriage that pre-Reformation writers did—the procreation of children, the avoidance of sin, and mutual companionship. Both Protestants and Catholics reinforced their message more firmly through oral and written instruction in religious beliefs than had the pre-Reformation Church—a process known as confessionalization—and both engaged in stricter disciplining of wayward members in their attempts to build a godly society.

Neither Protestant nor Catholic reformers encouraged women to play larger public roles in terms of religion, but some women became reformers themselves, often in the face of great male hostility. They preached and published religious works, left their husbands to join a different religious denomination, changed the religion of territories they governed, and, if they were Catholic, formed groups that were active in social service outside convent walls. Thus, like Christianity itself, the Reformation both expanded and diminished women's opportunities.

77. Argula von Grumbach
Letter to the University of Ingolstadt

Argula von Grumbach (1492?–1564?) was a German noblewoman who came into contact with Protestant ideas through books and personal contact and took the message about the priesthood of all believers to heart. In 1523 she published a letter defending Arsacius Seehofer, a student at the University of Ingolstadt who had been arrested for Protestant leanings and forced to recant. Her letter sparked a furor among Ingolstadt theologians and pastors, who in letters and sermons called her a "silly bag," "female devil," "heretical bitch," "shameless whore," and "wretched and pathetic daughter of Eve." None of them answered her letters, and her husband was immediately dismissed from his position and ordered to force her to stop writing. She did publish a few more letters and apparently remained in contact with Protestant reformers all her life, but after 1524 she wrote nothing more for publication.

The account of a Christian woman of the Bavarian nobility whose open letter, with arguments based on divine Scripture, criticises the University of Ingolstadt for compelling a young follower of the gospel to contradict the word of God.

Preface

Brothers: it is time to rouse ourselves from sleep. For our salvation is closer than we think. Therefore, my Christian reader, and you, too, you blind, raging, deluded Pharisees—you who have always resisted the Holy Spirit—if you refuse to believe the words of Christ, at least believe the works which he achieves through them. Put off your great cloak of pride, greed and fleshly lust.

See now, and understand, that, in these last days, Christ our savior is enticing and strengthening us by his divine and saving word in such diverse, gracious and wondrous ways (as hap-

pened at the beginning of his Church); not only through those learned in Scripture, but also through the great constancy, pain, martyrdom and death of many others, young and old, men and women; while bringing their persecutors to such humiliating and total confusion.

Lest, like Pharaoh, Exodus 4, your hearts become callous and hard; lest you remain untouched; 'for if the children are dumb', Luke 19, 'the very stones will cry out'. And, Joel 2, 'after this time I will pour out my spirit upon all flesh, and your sons and your daughters will prophesy; they will speak words of wisdom, your manservants and maidservants, too; and I will work wonders in heaven and on earth, before the great and awesome day of God comes to pass'.

Many are now aware of this saying, and now it is quite evident in the person of the woman mentioned above,[1] since it can be seen from

---

[1] von Grumbach refers to herself in the third person in the preface and in the first person in the letter itself.

her open letter, which is reproduced here, that she criticises the biblical scholars at the University of Ingolstadt for their persecution of the holy Gospel (as Judith, chapter eight, the false priests), and exhorts and instructs them, citing a host of 'insuperable' divine writings. (This is scarcely credible, something very rare for the female sex, and completely unheard of in our times.) And what's more, in the same letter she offers to appear before the same biblical scholars and to be interrogated by them. It can be seen from this that her writing comes from the spirit of God and not from the instruction of others. Moreover, just as holy Esther faced death and destruction in order to save the people, Esther 4, she, too, refuses to let herself be deterred from this Christian initiative of hers by the gruesome punishments imposed in recent times on so many advocates of the divine word. Like the holy Susanna, (Daniel 13) she would prefer to fall into the hands of men for what she does than to sin against God by keeping silent about the truth. And so we should pray to God that these incredibly arrogant and powerful enemies of Christ may be swept aside and vanquished, saying, like Judith 9: 'O Lord, what honour will be done to your name, if the hands of a woman overcome them'. And we should rejoice and sing with holy Zachariah: 'Blessed be the Lord God of Israel, who has visited and redeemed his people.'

Now follows the Christian letter of the woman we have mentioned, whose name will be found at the end.

The Lord says, John 12, 'I am the light that has come into the world, that none who believe in me should abide in darkness.' It is my heartfelt wish that this light should dwell in all of us and shine upon all callous and blinded hearts. Amen. I find there is a text in Matthew 10 which runs: 'Whoever confesses me before another I too will confess before my heavenly Father.' And Luke 9: 'Whoever is ashamed of me and of my words, I too will be ashamed of when I come in my majesty', etc. Words like these, coming from the very mouth of God, are always

before my eyes. For they exclude neither woman nor man.

And this is why I am compelled as a Christian to write to you. For Ezekiel 33 says: 'If you see your brother sin, reprove him, or I will require his blood at your hands.' In Matthew 12, the Lord says: 'All sins will be forgiven; but the sin against the Holy Spirit will never be forgiven, neither here nor in eternity.' And in John 6 the Lord says: 'My words are spirit and life . . . .'.

How in God's name can you and your university expect to prevail, when you deploy such foolish violence against the word of God; when you force someone to hold the holy Gospel in their hands for the very purpose of denying it, as you did in the case of Arsacius Seehofer? When you confront him with an oath and declaration such as this, and use imprisonment and even the threat of the stake to force him to deny Christ and his word?

Yes, when I reflect on this my heart and all my limbs tremble. What do Luther or Melanchthon[2] teach you but the word of God? You condemn them without having refuted them. Did Christ teach you so, or his apostles, prophets, or evangelists? Show me where this is written! You lofty experts, nowhere in the Bible do I find that Christ, or his apostles, or his prophets put people in prison, burnt or murdered them, or sent them into exile. . . . Don't you know that the Lord says in Matthew 10? 'Have no fear of him who can take your body but then his power is at an end. But fear him who has power to despatch soul and body into the depths of hell.'

One knows very well the importance of one's duty to obey the authorities. But where the word of God is concerned neither Pope, Emperor nor princes—as Acts 4 and 5 make so clear—have any jurisdiction. For my part, I have to confess, in the name of God and by my soul's salvation, that if I were to deny Luther and Melanchthon's writing I would be denying God and his word, which may God forfend for ever. Amen. . . .

---

[2]*Martin Luther* (1483–1546) and *Philip Melanchthon* (1497–1560) were two early Protestant reformers.

I had to listen for ages to your Decretal preacher crying out in the Church of Our Lady: *Ketzer/ketzer,* 'Heretic, heretic!' Poor Latin, that[3] I could say as much myself, no doubt; and I have never been to university. But if they are to prove their case they'll have to do better than that. I always meant to write to him, to ask him to show me which heretical articles the loyal worker for the gospel, Martin Luther, is supposed to have taught.

However I suppressed my inclinations; heavy of heart, I did nothing. Because Paul says in 1 Timothy 2: 'The women should keep silence, and should not speak in church.' But now that I cannot see any man who is up to it, who is either willing or able to speak, I am constrained by the saying: 'Whoever confesses me', as I said above. And I claim for myself Isaiah 3: 'I will send children to be their princes; and women, or those who are womanish, shall rule over them.' And Isaiah 29: 'Those who err will know knowledge in their spirit, and those who mutter will teach the law.' And Ezekiel 20: 'I raise up my hand against them to scatter them. They never followed my judgements, they rejected my commandments, and their eyes were on the idols of their fathers. Therefore I gave them commandments, but no good ones; and judgements by which they could never live.' And Psalm 8: 'You have ordained praise out of the mouth of children and infants at the breast, on account of your enemies.' And Luke 10: 'Jesus rejoiced in the Spirit, and said: "Father, I give you thanks, that you have hidden these things from the wise, and revealed them to the little ones".' Jeremiah 3: 'They will all know God, from the least to the greatest.' John 6, and Isaiah 54: 'They will all be taught of God.' Paul in 1 Corinthians 12: 'No one can say "Jesus", without the spirit of God.' Just as the Lord says of the confession of Peter in Matthew 16: 'Flesh and blood has not revealed this to you, but my heavenly Father'.

Do you hear this? That it is God who gives us understanding, not any human being? As Paul, too, says in 1 Corinthians 2: 'Your faith should not be in human wisdom . . . '. You, with your papal laws, will not be able to coerce us, not by a long chalk. We have witness enough from Scripture that they have no right to make laws without God's command, as Jeremiah 23 says. Where, however, it is based in the Bible, the book which contains all God's commands, we will be happy and pleased to accept it. But where it is not, it has no validity for us at all. Or only in so far as it is my duty to spare my weak and foolish brother, until he, too, has been instructed.

Jurisprudence cannot harm me; for it avails nothing here; I can detect no divine theology in it. Therefore I have no fears for myself, as long as you wish to instruct me by writing, and not by violence, prison or the stake. Joel 2: 'Turn again; return to the Lord. For he is kind and merciful.' The Lord laments in the words of Jeremiah 2: 'They have forsaken me, the well of living water, and have dug out broken cisterns which cannot hold any water.'

With Paul, 1 Corinthians 2, I say: 'I am not ashamed of the gospel which is the power of God to salvation to those who believe.' The Lord says, in Matthew 10: 'Should you be called forward do not worry about what you will say. It is not you who speak. In that same hour you will be given what you have to say. And the spirit of your Father will speak through you.'

I have no Latin; but you have German, being born and brought up in this tongue. What I have written to you is no woman's chit-chat, but the word of God; and (I write) as a member of the Christian Church, against which the gates of Hell cannot prevail. Against the Roman, however, they do prevail. Just look at that church! How is it to prevail against the gates of Hell? God gives us his grace, that we all may be saved, and may (God) rule us according to his will. Now may his grace carry the day. Amen.

My signature.

Argula von Grumbach,
von Stauff by birth.

---

[3]von Grumbach is criticizing a Catholic preacher for taking his themes from canon law (the Decretals) rather than the Bible, and using German rather than Latin to call Luther a heretic (*Ketzer* in German).

78. Ducal Order
"How to Proceed Against Anabaptist Women," 1584

Once a few individuals such as Martin Luther and Ulrich Zwingli broke with the Catholic Church, many other people developed their own ideas about what was proper Christian doctrine and practice. They often attempted to return to what they viewed as the practices of the first generations of Christians, rejecting private ownership of property and infant baptism in favor of the baptism of adults who could understand and testify to their own beliefs. Such groups were often termed *Anabaptists* (re-baptizers) or radicals by their opponents and were harshly persecuted by both more conservative Protestants and Catholics. Their punishments included execution, imprisonment, whipping, and banishment, which often penalized the entire family. This was appropriate, in the eyes of authorities, in the case of men, whose actions were supposed to determine the fate of the family, but less appropriate in the case of women whose husbands were not themselves Anabaptists. Church and state authorities debated whether "innocent" husbands might be given the right to divorce their Anabaptist wives but generally decided against this and favored policies like that given below.

For a few years it has often occurred and come to pass that women whose husbands are not Anabaptists but of the correct religion are misled and drawn into error by others. They do not let themselves be led back to proper, circumspect behavior and well-directed energy by the commissioners and magistrates or by the chancellery, and they are unaffected by the reminders and fervent petitions of their husbands. Instead they persist in their error, and for this reason they have previously been dealt with by banishment from the land, as is the case with men by virtue of ordinance. But this has caused many great burdens to small untrained, even nursing children. For this reason, and also because of the earnest, plaintive requests of the husbands, such cases have been dealt with leniently, and such wives have been handed over to their husbands to be chained in the house. Officials have earnestly been instructed to permit no access to them and to take care that such erring women not presume to lead their families or others astray.

This procedure should also be followed in the future in similar cases, and the pastors in such places should be commanded to go to them—even though by their conception of the struggle they do not want to permit this but to turn them away—and deal with them gently and moderately in order that no possible opportunity might be lost for our dear God to accomplish something by His grace and Holy Spirit. Previously, however, we have learned that when it was thought that the women were chained and secured, they had been freed secretly with the help of their husbands or their servants. And even though sometimes the deputy commissioners or others came to the place to inquire, it seems to have occurred that the freed woman stood on the chain and would not budge until the visitors had left. Now since they have used deception, the command must be all the more serious from now on, and a considerable punishment of fine or imprisonment must be enjoined and inflicted upon the men given such a serious, high trust. For

Ducal Order "How to Proceed Against Anabaptist Women," 1584.

themselves there will be no mercy, no release or leniency. And when punishment is finally imposed on such a one, his servants or those who at their own instigation, on their own attempt and will, opened the lock or chain, the magistrates and their subordinates are commanded to betake themselves to these same houses unexpectedly and without warning in about a month or six weeks.

There they are to gather their own information and not let themselves be deceived or turned away easily. For when such erroneous and headstrong women, who have such ill-timed zeal and now and then could well scream, are freed, they may sometimes wander secretly at night to other places and among people, especially other women, where they may cause harm.

79. Luisa de Carvajal
Letter to Father Joseph Creswell, S.J., 1608

Luisa de Carvajal y Mendoza (1566–1614) was a wealthy noble Spanish woman whose parents died when she was very young. She was raised by an uncle who obliged her to engage in extreme penance and self-mortification—including frequent whippings—and she developed a strong sense of religious calling. She put herself under the protection of the Jesuits, vowed to become a martyr, established a house for religious women in Madrid, and in 1605 went to England to minister to Catholics there. (At that time, Catholicism was illegal in England; many Catholics were jailed and Catholic priests executed.) Though Catholic leaders generally opposed such public activities on the part of women, the situation in England was viewed as a special case. She was arrested and jailed for a few days in 1608, which she recounts in the letter below, and again in 1613; she died shortly after her second release from prison. The following is an excerpt from one of her letters to Joseph Creswell, the director of the English Jesuits in Spain and Portugal.

1. I receive much mercy and consolation from Your Grace's letters, and I hope that Your Grace has received consolation with my last correspondence, seeing the great constancy of the holy martyrs Garves and Fludder.[1] And of me, I can tell Your Grace that I have walked between the cross and holy water, as they say there, because I have been in prison, and since it was in the public jail, it would be useless for me to keep silent about it.

2. The reason was because, arriving one day at a store in Cheapside,[2] from outside, as is my custom, leaning on the door sill, the situation offered itself to ask one of the young attendants if he was Catholic, and he responded, "No, God forbid!" And I replied, "May God not permit that you not be, which is what matters for you." At this the mistress and master of the shop came over, and another youth and neighboring merchants, and a great chat about religion

---

[1]*Garves* and *Fludder:* Catholic martyrs.

[2]*Cheapside:* at the time, a part of London.

ensued. They asked a lot about the mass, about priests, about confession, but what we spent the most time on (over two hours) was whether the Roman religion was the only true one, and whether the Pope is the head of the Church, and whether St. Peter's keys have been left to them [the Popes] forever in succession.

3. Some listened with pleasure, others with fury, and so much that I sensed some danger, at least of being arrested. But I thought nothing of it, in exchange for setting that light before their eyes in the best way I could. And in these simple matters of faith there are known methods [of convincing] which are very handy for anyone, and with which one can wage war on error. And although they might not take it very well at first, in the end those truths remain in their memories, to be meditated upon and open to holy inspirations, and God's cause for their salvation or condemnation justifies this a great deal. And there are very many who never manage to find out even where the priests are, and among the lay Catholics, not many want to run that risk [of contact with priests] without a guaranteed benefit. And the merchants of Cheapside exceed the rest of the city in malice, error, and hatred for the Pope, as well as in the quantity of its residents and money. And some of this can be observed in the fact that, my having spoken on several occasions with others about exactly the same things, they always took it affably.

4. The mistress of the shop tried to stir everyone to anger, as did another infernal young man who was there, younger in age but with greater malice. The woman said it was a shame that they were tolerating me and that, without a doubt, I was some Roman [Catholic] priest dressed like a woman so as to better persuade people of my religion. Our Lord saw fit that I speak the best English I've spoken since I've been in England, and they thought I was Scottish because of the way I spoke. . . .

6. And hearing behind me that someone calling Mr. Garves a traitor, and my Ann, [insisting he was] a martyr, were disputing, I prohibited her from continuing, fearing that she might say something impolitic.[3] And I asked him to tell

me why Garves had died. He said only because he was a Roman Catholic. "And for no other reason?" I replied. He said yes. "Well then," I said to him, "don't be shocked that he is called a martyr." And he seemed to take it well.

7. With this I returned to my house and I left them like lions against me. And two weeks later they managed to spot me, for it was necessary for me to go out, which I do but few times without a very specific need to buy necessary things or to go see the felicitous confessors of Christ in the prisons, or something similar, and never to visit anyone (for my natural condition so inclines me and my poor health and strength require it). And in the end they surrounded me, looking at me like basilisks,[4] and with a sheriff they brought, they said I had to go to Sir Thomas Bennet's house, the justice of the peace, not far from there. And although they had no warrant, I didn't resist, so they wouldn't grab me by the arm or raise a ruckus right in the middle of the street. And it wasn't a bad moment for my soul either. And all three of us went along agreeably, I mean Ann and Faith, my companions, and myself (for the other two [companions] had stayed at home). And our servant, who is an old and virtuous man of honor and long-standing Catholic faith, went with us.

8. We found the judge seated beneath a little roof on his patio, where he probably does his business, and he had us there, examining witnesses and questioning people from six in the evening or a bit later until around nine, when it started to get dark. The witnesses swore on their Bible what truths they said, with a few lies, but more or less within the limits of what I've already touched upon, without inventing anything else. And they talked so much nonsense sometimes that they made me recall that line, *"Et testimonia convenientia non erant."*[5] And there were two or three of them stirring up the people of the nearby streets against me, saying I

---

[3]Ann and Faith are her two English Catholic companions.
[4]*Basilisks:* small, snakelike creatures mentioned in Pliny's *Natural History;* believed to kill with their sight and their breath.
[5]"But their witness agreed not together", a line from the Bible describing accusations against Christ.

was a priest in a woman's clothes who was walking around persuading people of my faith, and since it was something so unheard of, I believe in half an hour there were more than two hundred people, so they were saying, at the judge's door, with the street full of a great, confusing noise. And among them they were already saying that there were three priests, with their long black gowns, which is our garb. The judge got up to calm them down a few times, because they were trying hard to get in. And he told me that if he were to send me to the jail then, the people would have at me. I told him I thought he had more charity than that.

9. He asked my homeland, name, address, and the reason I was in England, and in truth I cut it short saying my name was Luisa de Carvajal, and I was Spanish, and I lived close to don Pedro's house, where I went to hear mass, and that I had come to follow the example of many saints of the holy church who voluntarily exiled themselves into foreign lands, being unprotected and poor. And although it was all gibberish to the pitiful old man, that was the best answer, without a doubt.

10. He laughed as if he were crazy and asked me if it was the case that I affirmed the Pope to be the head of the Church and his religion the only true one. I said yes. He asked me if I wanted to remain in said opinions. And I responded that yes I did, and that I was prepared to die for them. . . .

13. The judge's daughters kept coming in and out, as well as his wife; it must have been so they could see us. In the end they took us to the jail. . . . He told the jailer to treat us well, but that must not have been able to happen

that night, and so they put us in the highest part of a narrow little attic, with a lit candle and the door locked with a key the jailer took with him, without our being able to get so much as a drop of water or beer or even a bite of bread. And what with this and not being very well, and without being able to get into a bed, I slept little indeed, but with notable consolation, and this diminished when I considered what little the whole business amounted to.

14. I had asked them if, for money, they would put me close to the jailer's wife and female servants, even though it were less comfortable than that lodging, and in the morning they put us in one of her rooms, around ten o'clock. And although dingy and without air, it was reasonable, and the women were all courteous and affable. . . .

15. We were in there for four days, from Saturday until Wednesday at ten at night, when the Council sent orders for them to set me free, with the judge having sent them my papers. . . .

16. While in jail I spoke about religion much more than I had out of it, with all the jailers and officials and their family and friends whom, with my permission, they brought to speak with me. And they listened nicely. And I didn't want to let the chance slip by, remembering the Holy Apostle who says that the word of God is not tied down. . . . If some day one [letter] should arrive saying that these people have sent me to heaven, it would be a fortunate end to my pilgrimage, and then my relations and friends could really be happy. May the will of God be done in everything, amen, for that alone brought me here, and that I hope will guide all my behavior until it puts me in divine obeisance.

80. William Whately
A Bride Bush, 1617

Protestant preachers frequently conveyed their ideas about marriage and gender relations in sermons held at weddings, which might later be published in an expanded version and given to newlyweds as advice literature. William Whately (1583–1639) was a prominent English Protestant clergyman who published several such books, of which *A Bride Bush* was the first. The subtitle to this work, "a direction for married persons Plainely describing the duties common to both, and peculiar to each of them," accurately describes its contents. Whately's views on the importance of marital fidelity—typical of Protestant thinkers—led him to support divorce for adultery and desertion in the first edition of this work; this position got him into trouble with English religious authorities and he retracted his statements in subsequent editions.

Marriage is honourable amongst all men: but whoremongers and adulterers God will judge.

Heb. 13:4

27. Now proceed we to the woman's duty, and giving the men leave to chew the cud awhile, request the women to listen with more diligence than before. The whole duty of the wife is referred to two heads. The first is to acknowledge her inferiority, the next to carry herself as inferior. First then, the wife's judgement must be convinced that she is not her husband's equal, yea that her husband is her better by far: else there can be no contentment, either in her heart or in her house. If she stand upon terms of equality, much more of being better than he is, the very root of good carriage is withered, and the fountain thereof is dried up. Out of place, out of peace. And woe to these miserable aspiring shoulders, that content not themselves to take their room next below the head. If ever thou purpose to be a good wife, and to live comfortably, set down this with thy self. *Mine husband is my superior, my better:* he hath authority and rule over me: nature hath given it him, having framed our bodies to tenderness, men's to more hardness. God hath given it to him, saying to our first mother Eve, *thy desire shall be subject to him, and he shall rule over thee.* His will is the tie and tedder even of my desires and wishes, I will not strive against GOD and nature. Though my sin hath made my place tedious, yet I will confess the truth, *mine husband is my superior, my better.* If the wife do not learn this lesson perfectly, if she have it not without book, even at her fingers' ends as

we speak, if her very heart condescend not to it, there will be wrangling, repining, striving, vying to be equal with him, or above him. And thus their life will be but a battle, and a trying of masteries. A woeful living.

28. Secondly, the wife being resolved that her place is the lower, must carry herself as an inferior. It little boots to carry his authority in word, if she frame not submission indeed. Now she shall testify her inferiority in a Christian manner by practising those two virtues of reverence and obedience, which are appropriate to the place of inferiors.

29. And first for reverence, the wife owes as much of that to her husband as the children or servants do to her, yea, as they do to him: only it is allowed that it be sweetened with more love and more familiarity. The wife should not think so erroneously of her place as if she were not bound equally with the children and servants to reverence her husband: all inferiors owe reverence alike. The difference is only this: she may be more familiar, not more rude than they, as being more dear, not less subject to him.

30. Also this reverence of hers must be both inward and outward. First her heart must she keep inwardly in a dutiful respect of him, and she must regard him as God's deputy, not looking to his person but his place; nor thinking so much who and what an one he is, as whose officer. This the apostle doth very strictly enjoin, saying: *let the wife see that she fear the husband* [Eph. 5]. As if he had said: of all things let her most carefully labour not to fail in this duty: for if she do, her whole life besides will be rude and unbeseeming. And you must know that the apostle means here

"A Bride Bush" by William Whately, ca. 1617.

not a slavish but a loving fear, such as may stand with the nearest union of hearts as between Christ and his church. And this fear is when, in consideration of his place, she doth abhor it as the greatest evil, next to the breach of God's commandment, to displease and offend her husband. Men stand in right awe of God, when they loath it as the greatest of all evils to break his commandment and grieve his spirit: and the wife fears her husband in good manner when she doth shun it as the next evil, to displease, grieve and disobey her husband who is next to God above her in the family. Such regard her heart must have of her head, that it keep hand and tongue and all from disorder. I know this is not customable, nay it is scarce thought seemly amongst many women; nay they care as little for their husbands as they for them; yea they despise them; yea they have inverted this precept, and cause their husbands to fear them. This impudence, this unwomanhood tracks the way to the harlot's house, and gives all wise men to know that such have, or would, or soon will, cast off the care of honesty, as of loyalty.

31. And as the heart principally, so next the outward behaviour must be regarded in three special things. First in speeches and gestures unto him. These must carry the stamp of fear upon them, and not be cutted,[1] sharp, sullen, passionate, tetchy; but meek, quiet, submissive, which may show that she consider who herself is, and to whom she speaks. The wife's tongue towards her husband must be neither keen, nor loose; her countenance neither swelling nor deriding; her behaviour not flinging, nor puffing, not discontented; but favouring of all lowliness and quietness of affection. Look what kinds of words or behaviour thou wouldst dislike from thy servant or child, those must thou not give to thine husband: for thou art equally commanded to be subject . . . how subject women are to disreverent behaviour, and withal how loathsome, how unwomanly they be. Yet for all these warnings we have some women that can chase and scold with their husbands, and rail upon them, and revile them, and shake them together with such terms and carriage, as were unsufferable towards a servant. Stains of womankind, blemishes of their sex, monsters in nature, botches of human society, rude, graceless, impudent, next to harlots, if not the same with them. Let such words leave a blister behind them, and let the canker eat out these tongues. . . .

34. Obedience follows: as concerning which duty a plain text avers it to the full, saying: *let the wife be subject to her husband in all things, in the Lord.* What need we further proof? Why is she his wife if she will not obey? And how can she require obedience of the children and servants, if she will not yield to the husband? Doth not she exact it in his name, and as his deputy? But the thing will not be so much questioned, as the measure: not whether she must obey, but how far. Wherefore we must extend it as far as the apostle, to a generality of things, to all things, so it be in the Lord. In whatsoever thing obeying of him doth not disobey God, she must obey: and if not in all things, it were as good in nothing. It is a thankless service if not general. To yield alone in things that please herself, is not to obey him but her own affections. The trial of obedience is when it crosseth her desires. To do that which he bids when she would have done without his bidding, what praise is it? But this declares conscionable submission: when she chooseth to do what herself would not, because her husband wills it. And seeing she requireth the like largeness of duty in his name from the servants, herself shall be judge against herself, if she give not what she looks to receive. But it sufficeth not that her obedience reach to all things that are lawful, unless it be also willing, ready, without brawling, contending, thwarting, sourness. A good work may be marred in the manner of doing. . . . Then it is laudable, commendable, a note of a virtuous woman, a dutiful wife when she submits herself with quietness, cheerfully, even as a well-broken horse turns at the least turning, stands at the least check of the rider's bridle, readily going and standing as

---

[1]*Cutted:* short; curt.

he wishes that sits upon his back: if you will have your obedience worth anything, make no tumult about it outwardly, allow none within.

35. And for the less principal duties of husband and wife concerning their ordinary society, thus much. I come now to such as concern the marriage bed, which are as needful to be known as the former, because offences in that kind are more capital and dangerous, though not so public. Their matrimonial meetings must have three properties. First, it must be cheerful: they must lovingly, willingly, and familiarly communicate themselves unto themselves, which is the best means to continue and nourish their mutual natural love, and by which the true and proper ends of matrimony shall be attained in best manner: for the husband is not his own but the wife's, and the wife the husband's. Secondly, their meeting must be sanctified. *Paul* saith meat, drink and marriage are good and sanctified by prayer. Men and women must not come together as brute creatures and unreasonable beasts through the heat of desire; but must see their maker in that his ordinance, and crave his blessing solemnly as at meals (the apostle speaks of both alike), that marriage may indeed be blessed unto them. To sanctify the marriage bed and use it reverently with prayer and thanksgiving will make it moderate and keep them from growing weary of each other (as in many it falls out), and cause that lust shall be assuaged, which else shall be increased by these meetings. Propagation and chastity, the two chief ends of marriage, are best attained by prayer and thanksgiving in the use thereof, without which they will hardly come, or not with comfort. Neither is it more than needs, to see God in that which so nearly toucheth ourselves, as the hope of posterity;

him, as the increase of his kingdom. Let Christians therefore know the fruit of prayer, even in all things. Thirdly, their nuptial meetings must be seasonable, and at lawful times. There is a season when God and Nature sejoins[2] man and wife in this respect. The woman is made to be fruitful; and therefore also moist and cold of constitution. Hence it is that their natural heat serves not to turn all their sustenance into their own nourishment; but a quantity redounding is set apart in a convenient place to cherish and nourish the conception, when they shall conceive. Now this redundant humour (called their flowers or terms) hath (if no conception be) its monthly issue or evacuation (and in some oftener) unless there be extraordinary stoppings and obstructions, lasting for six or seven days in the most. Sometimes also this issue, through weakness and infirmity of nature, doth continue many more days. Always after childbirth there is a larger and longer emptying because of the former retention, which continueth commonly for four, five or six weeks, and some longer. Now in all these three times and occasions, it is simply unlawful for a man to company with his own wife. The Lord tells us so. *Leviticus* 15 vers 19–25; also ch. 18 ver. 19; also ch. 20 ver. 18. Of which places it is needful that married persons should take note: to which I send them. Neither let women think themselves disgraced, because I have laid this matter open in plain but modest speeches. Where God threatens death to the offender, can the minister be faithful if he do not plainly declare the offence? This fault is by GOD condemned to the punishment of death, *Leviticus* 20:18.

---

[2]*Sejoins:* separates.

QUESTIONS FOR ANALYSIS

1. How does Argula von Grumbach justify her speaking out in defense of Seehofer? Who does she see as her models?

2. How did the ducal authorities attempt to balance their concerns about women spreading Anabaptism with the authority of husbands over their wives?

3. How does Luisa de Carvajal describe her efforts at converting others to Catholicism? Her reaction to arrest? Given her actions and the language she uses, how do you read her description of herself as weak and sickly, and her avowed hesitancy in telling her story to Creswell?

4. What does Whately see as a wife's chief obligations to her husband? What metaphors does he use to describe their relationship? In his opinion, what role do sexual relations play in a proper marriage?

5. Both Protestants and Catholics taught that faith and salvation were individual matters, a message that Argula von Grumbach, female Anabaptists, and Luisa de Carvajal clearly internalized. How did this fit with the requirements for wifely obedience that Whately (and many others) prescribed?

WITCHCRAFT

Explanations for the "Great Witch Craze"—during which somewhere between 100,000 and 200,000 people, the vast majority of them women, were tried and between 50,000 and 100,000 executed—have pointed to a number of factors: economic uncertainties, legal changes that allowed for secret interrogation and torture, rising numbers of unmarried women who were regarded with suspicion, desires for social control on the part of church and state authorities, misogyny, new demonological concepts, and social changes that led to greater suspicions between neighbors. The balance between these, and perhaps other factors, differed in various parts of Europe, and the timing and nature of witch persecutions thus also differed. In some areas, witch trials were few and involved isolated individuals, while in others they grew into large-scale panics; in some, the number of men among the accused was significant, while in others the accused were all women; in some, Christian demonology was central to accusations, while in others, particularly in Scandinavia and eastern Europe, it was not important; in some, almost all accused witches were executed, while in others they were more likely to be given a lesser punishment or have the whole case dismissed. (The latter was most common for cases that came before the Roman or Spanish Inquisitions; despite their reputations, the Inquisitions actually regarded witchcraft as much less threatening than did other types of courts.)

There are a few similarities despite these many differences. Most trials took place on a small scale, involving only one or a handful of accused; most of the people originally accused were women who did not fit the model of acceptable feminine behavior—they were scolds, quarrelsome, looked or behaved oddly, or were not under the control of a man. In the minority of cases that grew to mass panics, other types of women—and men—were more likely to be swept up in a wave of accusations. Women themselves often brought charges, accusing other women of actions that were the inverse of those expected of good women—harming children rather than nurturing them, killing animals rather than caring for them, spoiling food rather than preparing it.

81. Heinrich Krämer and Jacob Sprenger
Malleus Maleficarum, 1486

During the Late Middle Ages, Christian philosophers and theologians developed a complex learned demonology, in which the most important aspect of witchcraft became a pact with the Devil. This demonology was described at great length by two German Dominican inquisitors, Heinrich Krämer (also called Institorius, the Latin form of his name) and Jacob Sprenger, in the *Malleus Maleficarum* ("The Hammer of [Female] Witches"), which also served as a guide for witch-hunters, advising them how to recognize and question witches. The *Malleus* was translated from Latin into the vernacular and reprinted many times; it taught judges and lawyers over large parts of northern Europe what to expect of witches and what questions to ask of them. Similarities in the answers they received helped convince people that witchcraft was an international conspiracy, and Europeans carried their demonology with them when they established colonies outside of Europe.

As for the first question, why a greater number of witches is found in the fragile feminine sex than among men; it is indeed a fact that it were idle to contradict, since it is accredited by actual experience, apart from the verbal testimony of credible witnesses. And without in any way detracting from a sex in which God has always taken great glory that His might should be spread abroad, let us say that various men have assigned various reasons for this fact, which nevertheless agree in principle. Wherefore it is good, for the admonition of women, to speak of this matter; and it has often been proved by experience that they are eager to hear of it, so long as it is set forth with discretion.

For some learned men propound this reason; that there are three things in nature, the Tongue, an Ecclesiastic, and a Woman, which know no moderation in goodness or vice. . . . Others again have propounded other reasons why there are more superstitious women found than men. And the first is, that they are more credulous; and since the chief aim of the devil is to corrupt faith, therefore he rather attacks them. See *Ecclesiasticus* xix: He that is quick to believe is light-minded, and shall be diminished. The second reason is, that women are naturally more impressionable, and more ready to receive the influence of a disembodied spirit; and that when they use this quality well they are very good, but when they use it ill they are very evil.

The third reason is that they have slippery tongues, and are unable to conceal from their fellow-women those things which by evil arts they know; and, since they are weak, they find an easy and secret manner of vindicating themselves by witchcraft. See *Ecclesiasticus* as quoted above: I had rather dwell with a lion and a dragon than to keep house with a wicked woman. All wickedness is but little to the wickedness of a woman. And to this may be added that, as they are very impressionable, they act accordingly.

There are also others who bring forward yet other reasons, of which preachers should be very careful how they make use. For it is true that in the Old Testament the Scriptures have much that is evil to say about women, and this because of the first temptress, Eve, and her imitators; yet afterwards in the New Testament we find a change of name, as from Eva to Ave[1] (as S. Jerome

---

[1] *Ave* ("Hail") is the first word the angel bringing God's message to the Virgin Mary says in the Latin translation of the Bible. Because it is the reverse of Eve ("Eva" in Latin), many commentators used this as the beginning of their contrast between good women like Mary and evil women like Eve.

Malleus Maleficarum by Heinrich Krämer and Jacob Sprenger, 1486. Translated by Montague Summers (London: 1928).

says), and the whole sin of Eve taken away by the benediction of Mary. Therefore preachers should always say as much praise of them as possible.

But because in these times this perfidy is more often found in women than in men, as we learn by actual experience, if anyone is curious as to the reason, we may add to what has already been said the following: that since they are feebler both in mind and body, it is not surprising that they should come more under the spell of witchcraft.

For as regards intellect, or the understanding of spiritual things, they seem to be of a different nature from men; a fact which is vouched for by the logic of the authorities, backed by various examples from the Scriptures. Terence says: Women are intellectually like children. And Lactantius (*Institutiones,* III): No woman understood philosophy except Temeste. And *Proverbs* xi, as it were describing a woman, says: As a jewel of gold in a swine's snout, so is a fair woman which is without discretion.

But the natural reason is that she is more carnal than a man, as is clear from her many carnal abominations. And it should be noted that there was a defect in the formation of the first woman, since she was formed from a bent rib, that is, a rib of the breast, which is bent as it were in a contrary direction to a man. And since through this defect she is an imperfect animal, she always deceives. For Cato says: When a woman weeps she weaves snares. And again: When a woman weeps, she labours to deceive a man. And that is shown by Samson's wife, who coaxed him to tell her the riddle he had propounded to the Philistines, and told them the answer, and so deceived him. And it is clear in the case of the first woman that she had little faith; for when the serpent asked why they did not eat of every tree in Paradise, she answered: Of every tree, etc.—lest perchance we die. Thereby she showed that she doubted, and had little faith in the word of God. And all this is indicated by the etymology of the word; for *Femina* comes from *Fe* and *Minus,* since she is ever weaker to hold and preserve the faith.[2]

---

[2]This is a false derivation. *Femina* ("women") actually comes from the Greek word meaning "to produce," not from lacking (*minus*) faith (*fe*).

And this as regards faith is of her very nature; although both by grace and nature faith never failed in the Blessed Virgin, even at the time of Christ's Passion, when it failed in all men.

Therefore a wicked woman is by her nature quicker to waver in her faith, and consequently quicker to abjure the faith, which is the root of witchcraft.

And as to her other mental quality, that is, her natural will; when she hates someone whom she formerly loved, then she seethes with anger and impatience in her whole soul, just as the tides of the sea are always heaving and boiling. . . .

And indeed, just as through the first defect in their intelligence they are more prone to abjure the faith; so through their second defect of inordinate affections and passions they search for, brood over, and inflict various vengeances, either by witchcraft, or by some other means. Wherefore it is no wonder that so great a number of witches exist in this sex.

Women also have weak memories; and it is a natural vice in them not to be disciplined, but to follow their own impulses without any sense of what is due; this is her whole study. . . .

If we inquire, we find that nearly all the kingdoms of the world have been overthrown by women. Troy, which was a prosperous kingdom, was, for the rape of one woman, Helen, destroyed, and many thousands of Greeks slain. The kingdom of the Jews suffered much misfortune and destruction through the accursed Jezebel, and her daughter Athaliah, queen of Judah, who caused her son's sons to be killed, that on their death she might reign herself; yet each of them was slain. The kingdom of the Romans endured much evil through Cleopatra, Queen of Egypt, that worst of women. And so with others. Therefore it is no wonder if the world now suffers through the malice of women.

And now let us examine the carnal desires of the body itself, whence has arisen unconscionable harm to human life. . . .

Let us consider another property of hers, the voice. For as she is a liar by nature, so in her speech she stings while she delights us. Wherefore her voice is like the song of the Sirens, who with their sweet melody entice the passers-by and kill them. . . .

Let us consider also her gait, posture, and habit, in which is vanity of vanities. There is no man in the world who studies so hard to please the good God as even an ordinary woman studies by her vanities to please men.

To conclude. All witchcraft comes from carnal lust, which is in women insatiable. See *Proverbs* xxx: There are three things that are never satisfied, yea, a fourth thing which says not, It is enough; that is, the mouth of the womb. Wherefore for the sake of fulfilling their lusts they consort even with devils. More such reasons could be brought forward, but to the understanding it is sufficiently clear that it is no matter for wonder that there are more women than men found infected with the heresy of witchcraft. And in consequence of this, it is better called the heresy of witches than of wizards, since the name is taken from the more powerful party. And blessed be the Highest Who has so far preserved the male sex from so great a crime: for since He was willing to be born and to suffer for us, therefore He has granted to men this privilege. . . .

The method by which they profess their sacrilege through an open pact of fidelity to devils varies according to the several practices to which different witches are addicted. And to understand this it first must be noted that there are, as was shown in the First Part of this treatise, three kinds of witches; namely, those who injure but cannot cure; those who cure but, through some strange pact with the devil, cannot injure; and those who both injure and cure. And among those who injure, one class in particular stands out, which can perform every sort of witchcraft and spell, comprehending all that all the others individually can do. Wherefore, if we describe the method of profession in their case, it will suffice also for all the other kinds. And this class is made up of those who, against every instinct of human or animal nature, are in the habit of eating and devouring the children of their own species.

And this is the most powerful class of witches, who practise innumerable other harms also. For they raise hailstorms and hurtful tempests and lightnings; cause sterility in men and animals; offer to devils, or otherwise kill, the children whom they do not devour. But these are only the children who have not been re-born by baptism at the font, for they cannot devour those who have been baptized, nor any without God's permission. They can also, before the eyes of their parents, and when no one is in sight, throw into the water children walking by the water side; they make horses go mad under their riders; they can transport themselves from place to place through the air, either in body or in imagination; they can affect Judges and Magistrates so that they cannot hurt them; they can cause themselves and others to keep silence under torture; they can bring about a great trembling in the hands and horror in the minds of those who would arrest them; they can show to others occult things and certain future events, by the information of devils, though this may sometimes have a natural cause; they can turn the minds of men to inordinate love or hatred; they can at times strike whom they will with lightning, and even kill some men and animals; they can make of no effect the generative desires, and even the power of copulation, cause abortion, kill infants in the mother's womb by a mere exterior touch; they can at times bewitch men and animals with a mere look, without touching them, and cause death; they dedicate their own children to devils; and in short, as has been said, they can cause all the plagues which other witches can only cause in part, that is, when the Justice of God permits such things to be. All these things this most powerful of all classes of witches can do, but they cannot undo them.

But it is common to all of them to practise carnal copulation with devils. . . .

A Succubus devil draws the semen from a wicked man; and if he is that man's own particular devil, and does not wish to make himself an Incubus to a witch, he passes that semen on to the devil deputed to a woman or witch; and this last, under some constellation that favours his purpose that the man or woman so born should be strong in the practice of witchcraft, becomes the Incubus to the witch. . . .

Now the method of profession is twofold. One is a solemn ceremony, like a solemn vow. the other is private, and can be made to the devil at any hour alone. The first method is when witches meet together in conclave on a set day, and the devil appears to them in the assumed body of a man, and urges them to keep faith with him, promising them worldly prosperity and length of life; and they recommend a novice to his acceptance. And the devil asks whether she will abjure the Faith, and forsake the holy Christian religion and the worship of the Anomalous Woman (for so they call the Most Blessed Virgin Mary), and never venerate the Sacraments; and if he finds the novice or disciple willing, then the devil stretches out his hand, and so does the novice, and she swears with upraised hand to keep that covenant. And when this is done, the devil at once adds that this is not enough; and when the disciple asks what more must be done, the devil demands the following oath of homage to himself: that she give herself to him, body and soul, for ever, and do her utmost to bring others of both sexes into his power. He adds, finally, that she is to make certain unguents from the bones and limbs of children, especially those who have been baptized; by all which means she will be able to fulfil all her wishes with his help.

82. Witch Pamphlets: Strasbourg 1583 and Scotland 1591

Hysteria about witchcraft was fueled by the publication of pamphlets—often called "witch news" or "witch newspapers"—detailing the alleged actions of witches and their punishment. In these pamphlets, popular ideas of witches as those who cause harm mix with more learned ideas about witches making pacts with the Devil. Such pamphlets were usually printed cheaply, with paper covers and poor-quality paper, so they were affordable to most literate people; there was a steady market for them throughout Europe until the late seventeenth century.

A. Strasbourg, 1583

Truthful and Believable Report about 134 Fiends Who Were Burned to Death because of Witchcraft.

. . . On the 29th of August Duke William of Darmstadt had ten women burned, and also a 17-year old boy and a 13-year old girl. They confessed to over 331 points; these should be publicized and read to young Christian people. The Christian reader should pay attention to these short writings, since the injuries are even larger than are described here. Here is enumerated how the monsters [i.e., witches] have been burned one after the other, and what useless things they have done.

Completely believable reports have come in from Mümpelgarten, that on a mountain a pile of all types of fruit and other strange things has been found, which the witches made themselves; 134 were near it, and, as is their custom, they held a dance on the 21st of August. Among them was a female cow-herd who was supposed to upset the pile, but she cried heartily. She was asked by her associates why she was crying, and answered, "I pity the poor children who will be killed and the innocent blood that will be spilled during the coming

"Witch Pamphlets: Strasbourg 1583 and 1591."

storm." Nevertheless she knocked over the pile until it was only half as tall [as it had been before], and a horrible, terrible thunderstorm followed immediately, that, God be merciful, brought much irreparable harm, as one can see with one's own eyes. For this 44 women and three men were arrested, and burned on the 24th of October in Mümpelgarten.

Also, on the same hill, the apothecary's wife from Mümpelgarten held a wedding, and gave her own daughter to the devil as his wife. One also found more than 1500 gulden worth of silver objects, and three tables with all kinds of foods, but no bread or salt.[1] All of these things were brought into the city, and the silver pieces given to a goldsmith to look at. He knew the coats of arms and marks on them, and for that reason so many were taken in and burnt.

In the same way 42 monsters were assembled together on the 9th of September between Rottenburg and Tübingen and held a dance. They made such a wretched thunderstorm that the trees were ripped out of the ground along with their roots, doing hundreds of gulden worth of damage to the vineyards and grainfields and other crops. Basically everything was destroyed.

Further, at Reyte, a half mile from Waldtkrich in Breisgau, on the 19th of October 38 women were burned to death; among these were four midwives who had done shocking things shortly before. The first, a farmer's wife from the Black Forest, killed 19 children and three expectant mothers. The second, a smith's wife, killed nine children and four expectant mothers. The third, a miller's wife, killed nine children and four mothers. The fourth was a tailor's wife and killed four children and eight mothers. The other twelve women were for the most part rich and well-off women, the wives of inn-keepers, farmers, and millers. They did such uncountable damage to people and animals that it cannot be described here.

There are three men in the hospital in Freiburg in Breisgau who have been robbed of their reason and sense [by the witches].

In the same way, at Duercken in Alsace on the 28th of October 36 women were burned to death. On the ninth of August near Adlaw in Alsace they held a dance on a mountain, and made such a thunderstorm that brought irreparable damage to houses, vineyards, and other fields, which is to be pitied. Again, on the 29th of October six women were imprisoned in Tuercke, along with a [male] hay-cutter from Colmar who was their witchmaster [*Hexenmeister*]; they will all soon be offered to the fire.

May God will that such terrible things come to an end, and that the accursed Satan no longer watch over and be the doom of the weaker vessels of the female sex! May God will that all people be enlightened by the Holy Spirit to withstand the arch-enemy of the human race and know to guard themselves from his thousand tricks, so that at the Resurrection they will receive a gracious judgment!

B. Scotland, 1591 Within the towne of Trenent, in the kingdome of Scotland, there dwelleth one David Seaton, who, being deputie bailiffe in the said towne, had a maid called Geillis Duncane, who used secretlie to absent and lie forth of hir maister's house every other night: This Geillis Duncane tooke in hand to helpe all such as were troubled or grieved with anie kinde of sicknes or infirmitie, and in short space did perfourme many matters most miraculous; which things, for asmuche as she began to do them upon a sodaine, having never done the like before, made her maister and others to be in great admiration, and wondered thereat: by means whereof, the saide David Seaton had his maide in great suspition that shee did not those things by naturall and lawful waies, but rather supposed it to bee done by some extraordinarie and unlawfull meanes. Whereupon, her maister began to grow verie inquisitive, and examined hir which way and by what means shee was able to performe matters of so great impor-

---

[1]Food at witches' sabbats was often described as lacking salt, or being in other ways without flavor.

tance; whereat shee gave him no aunswere: nevertheless, her maister, to the intent that hee might the better trie and finde out the truth of the same, did with the help of others torment her with the torture of the pilliwinkes[1] upon her fingers, which is a grievous torture; and binding or wrinching her head with a cord or roape, which is a most cruell torment also; yet would she not confess anie thing; whereuppon, they suspecting that she had beene marked by the Devill (as commonly witches are), made diligent search about her, and found the enemies mark to be in her fore crag, or fore part of her throate; which being found, she confessed that al her doings was done by the wicked allurements and entisements of the Devil, and that she did them by witchcraft. After this her confession, she was committed to prison, where shee continued a season, where immediately shee accused these persons following to bee notorious witches, and caused them forthwith to be apprehended, one after another, viz. Agnes Sampson the eldest witche of them all, dwelling in Haddington; Agnes Tompson of Edenbrough; Doctor Fian alias John Cuningham, master of the schoole at Saltpans in Lowthian, of whose life and strange acts you

shal heare more largely in the end of this discourse. These were by the saide Geillis Duncane accused, as also George Motts wife, dwelling in Lowthian; Robert Grierson, skipper; and Jannet Blandilands; with the porter's wife of Seaton: the smith at the Brigge Hallis, with innumerable others in those parts, and dwelling in those bounds aforesaid; of whom some are alreadie executed, the rest remaine in prison to receive the doome of judgment at the Kinges Majesties will and pleasure.

The saide Geillis Duncane also caused Ewphame Mecalrean to bee apprehended, who conspired and performed the death of her godfather, and who used her art upon a gentleman, being one of the Lordes and Justices of the Session, for bearing good will to her daughter. Shee also caused to be apprehended one Barbara Naper, for bewitching to death Archibalde lait Earle of Angus, who languished to death by witchcraft, and yet the same was not suspected; but that hee died of so straunge a disease as the Phisition knewe not how to cure or remedie the same. But of all other the said witches, these two last before recited, were reputed for as civill honest women as anie that dwelled within the cittie of Edenbrough, before they were apprehended. Many other besides were taken dwelling in Lieth, who are detayned in prison untill his Majesties further will and pleasure be knowne. . . .

---

[1] An instrument of torture similar to the thumbscrews later in use.

83. Trial of Suzanne Gaudry for Witchcraft, 1652

Some of the best-known witch trials involved dozens or even hundreds of people like the trials reported in the Strasbourg witch pamphlets and in Salem, Massachusetts. Most trials were much smaller and involved only one or two accused. In some of these, demonological ideas were not very important and the focus remained on the witch's supposed evil deeds (*maleficia* in Latin). In others, however, demonological ideas were there from the start, particularly as these became more widely known in the seventeenth century with the repeated publication and translation of witch-hunting guides like the *Malleus*. The following comes from the records of the local secular court at Rieux in France;

the accused, an illiterate older woman, is asked about others who participated in demonic rituals, but this trial apparently did not lead to a mass panic.

At Ronchain, 28 May, 1652. . . . Interrogation of Suzanne Gaudry, prisoner at the court of Rieux. Questioned about her age, her place of origin, her mother and father.

—Said that she is named Suzanne Gaudry, daughter of Jean Gaudry and Marguerite Gerné, both natives of Rieux, but that she is from Esgavans, near Odenarde, where her family had taken refuge because of the wars, that she was born the day that they made bonfires for the Peace between France and Spain, without being able otherwise to say her age.

Asked why she had been taken here.

—Answers that it is for the salvation of her soul.

—Says that she was frightened of being taken prisoner for the crime of witchcraft.

Asked for how long she has been in the service of the devil.

—Says that about twenty-five or twenty-six years ago she was his lover, that he called himself Petit-Grignon, that he would wear black breeches, that he gave her the name Magin, that she gave him a pin with which he gave her his mark on the left shoulder, that he had a little flat hat; said also that he had his way with her two or three times only.

Asked how many times she has been at the nocturnal dance.

—Answers that she had been there about a dozen times, having first of all renounced God, Lent and baptism; that the site of the dance was at the little marsh of Rieux, understanding that there were diverse dances. The first time, she did not recognize any-one there, because she was half blind. The other times, she saw and recognized there Noelle and Pasquette Gerné, Noelle the wife of Nochin Quinchou and the other of Paul Doris, the widow Marie Nourette, not having recognized others because the young people went with the young people and the old people with the old. And that when the dance was large, the table also was accordingly large.

Questioned what was on the table.

—Says that there was neither salt nor napkin, that she does not know what there was because she never ate there. That her lover took her there and back.

Asked if her lover had never given her some powder.

—Answers that he offered her some, but that she never wanted to take any, saying to her that it was to do with what she wanted, that this powder was gray, that her lover told her she would ruin someone but good, and that he would help her, especially that she would ruin Elisabeth Dehan, which she at no time wanted to do, although her lover was pressing her to do it, because this Elisabeth had battered his crops with a club.

Interrogated on how and in what way they danced.

—Says that they dance in an ordinary way, that there was a guitarist and some whistlers who appeared to be men she did not know; which lasted about an hour, and then everyone collapsed from exhaustion.

Inquired what happened after the dance.

—Says that they formed a circle, that there was a king with a long black beard dressed in black, with a red hat, who made everyone do his bidding, and that after the dance he made a . . . [the word is missing in the text], and then everyone disappeared. . . .

Interrogated on how long it has been since she has seen Grignon, her lover.

—Says that it has been three or four days.

Questioned if she has abused the Holy Communion.

"Trial of Suzanne Gaudry for Witchcraft," 1612.

—Says no, never, and that she has always swallowed it. Then says that her lover asked her for it several times, but that she did not want to give it to him.

After several admonitions were sent to her, she has signed this

Mark

X

Suzanne Gaudry[1]

Second Interrogation, May 29, 1652, in the presence of the afore-mentioned.

This prisoner, being brought back into the chamber, was informed about the facts and the charges and asked if what she declared and confessed yesterday is true.

—Answers that if it is in order to put her in prison it is not true; then after having remained silent said that it is true.

Asked what is her lover's name and what name has he given himself.

—Said that his name is Grinniou and that he calls himself Magnin.

Asked where he found her the first time and what he did to her.

—Answers that it was in her lodgings, that he had a hide, little black breeches, and a little flat hat; that he asked her for a pin, which she gave to him, with which he made his mark on her left shoulder. Said also that at the time she took him oil in a bottle and that she had thoughts of love.

Asked how long she has been in subjugation to the devil.

—Says that it has been about twenty-five or twenty-six years, that her lover also then made her renounce God, Lent, and baptism, that he has known her carnally three or four times, and that he has given her satisfaction. And on the subject of his having asked her if she wasn't afraid of having a baby, says that she did not have that thought.

Asked how many times she found herself at the nocturnal dance and carol and who she recognized there.

—Answers that she was there eleven or twelve times, that she went there on foot with her lover, where the third time she saw and recognized Pasquette and Noelle Gerné, and Marie Homitte, to whom she never spoke, for the reason that they did not speak to each other. And that the sabbat took place at the little meadow. . . .

Interrogated on how long it is since she saw her lover, and if she also did not see Marie Hourie and her daughter Marie at the dance.

—Said that it has been a long time, to wit, just about two years, and that she saw neither Marie Hourie nor her daughter there; then later said, after having asked for some time to think about it, that it has been a good fifteen days or three weeks [since she saw him], having renounced all the devils of hell and the one who misled her.

Asked what occurred at the dance and afterwards.

—Says that right after the dance they put themselves in order and approached the chief figure, who had a long black beard, dressed also in black, with a red hat, at which point they were given some powder, to do with it what they wanted; but that she did not want to take any.

Charged with having taken some and with having used it evilly.

—Says, after having insisted that she did not want to take any, that she took some, and that her lover advised her to do evil with it; but that she did not want to do it.

Asked if, not obeying his orders, she was beaten or threatened by him, and what did she do with this powder.

—Answers that never was she beaten; she invoked the name of the Virgin [and answered] that she threw away the powder that she had, not having wanted to do any evil with it.

Pressed to say what she did with this powder. Did she not fear her lover too much to have thrown it away?

—Says, after having been pressed on this question, that she made the herbs in her garden die at the end of the summer, five to six years ago,

---

[1]Gaudry made an X because she did not know how to write.

by means of the powder, which she threw there because she did not know what to do with it.

Asked if the devil did not advise her to steal from Elisabeth Dehan and to do harm to her.

—Said that he advised her to steal from her and promised that he would help her; but urged her not to do harm to her; and that is because she [Elisabeth Dehan] had cut the wood in her [Suzanne Gaudry's] fence and stirred up the seeds of her garden, saying that her lover told her that she would avenge herself by beating her.

Charged once more with having performed some malefice with this powder, pressed to tell the truth.

—Answers that she never made any person or beast die; then later said that she made Philippe Cornié's red horse die, about two or three years ago, by means of the powder, which she placed where he had to pass, in the street close to her home.

Asked why she did that and if she had had any difficulty with him.

—Says that she had had some difficulty with his wife, because her cow had eaten the leeks.

Interrogated on how and in what way they dance in the carol.

—Says that they dance in a circle, holding each others' hands, and each one with her lover at her side, at which she says that they do not speak to each other, or if they speak that she did not hear it, because of her being hard-of-hearing. At which there was a guitarist and a piper, whom she did not know; then later says that it is the devils who play.

After having been admonished to think of her conscience, was returned to prison after having signed this

Mark

X

Suzanne Gaudry

Deliberation of the Court of Mons—June 3, 1652

The under-signed advocates of the Court of Mons have seen these interrogations and answers. They say that the aforementioned Suzanne Gaudry confesses that she is a witch, that she has

given herself to the devil, that she has renounced God, Lent, and baptism, that she has been marked on the shoulder, that she has cohabited with him and that she has been to the dances, confessing only to have cast a spell upon and caused to die a beast of Philippe Cornié, but there is no evidence for this, excepting a prior statement. For this reason, before going further, it will be necessary to become acquainted with, to examine and to probe the mark, and to hear Philippe Cornié, on the death of the horse and on when and in what way he died. . . .

Deliberation of the Court of Mons—June 13, 1652

[The Court] has reviewed the current criminal trial of Suzanne Gaudry, and with it the trial of Antoinette Lescouffre, also a prisoner of the same office.

It appeared [to the Court] that the office should have the places probed where the prisoners say that they have received the mark of the devil[2], and after that, they must be interrogated and examined seriously on their confessions and denials, this having to be done, in order to regulate all this definitively. . . .

Deliberation of the Court of Mons, June 22, 1652

The trials of Antoinette Lescouffre and Suzanne Gaudry having been described to the undersigned, advocates of the Court of Mons, and [the Court] having been told orally that the peasants taking them to prison had persuaded them to confess in order to avoid imprisonment, and that they would be let go, by virtue of which it could appear that the confessions were not so spontaneous:

They are of the opinion that the office, in its duty, would do well, following the two preced-

---

[2]"Devils' marks" were places that supposedly felt no pain and did not bleed. They were probed with pins or needles to see if this was indeed the case, a procedure also called "pricking a witch." Suspected witches were often completely shaved, head and body, in this search for marks.

ing resolutions, to have the places of the marks that they have taught us about probed, and if it is found that these are ordinary marks of the devil, one can proceed to their examination; then next to the first confessions, and if they deny [these], one can proceed to the torture, given that they issue from bewitched relatives, that at all times they have been suspect, that they fled to avoid the crime [that is to say, prosecution for the crime of witchcraft], and that by their confessions they have confirmed [their guilt], notwithstanding that they have wanted to revoke [their confessions] and vacillate. . . .

Third Interrogation, June 27, in the presence of the afore-mentioned.

This prisoner being led into the chamber, she was examined to know if things were not as she had said and confessed at the beginning of her imprisonment.

—Answers no, and that what she has said was done so by force.

Asked if she did not say to Jean Gradé that she would tell his uncle, the mayor, that he had better be careful . . . and that he was a Frank.

—Said that that is not true.

Pressed to say the truth, that otherwise she would be subjected to torture, having pointed out to her that her aunt was burned for this same subject.

—Answers that she is not a witch.

Interrogated as to how long she has been in subjection to the devil, and pressed that she was to renounce the devil and the one who misled her.

—Says that she is not a witch, that she has nothing to do with the devil, thus that she did not want to renounce the devil, saying that he has not misled her, and upon inquisition of having confessed to being present at the carol, she insisted that although she had said that, it is not true, and that she is not a witch.

Charged with having confessed to having made a horse die by means of a powder that the devil had given her.

—Answers that she said it, but because she found herself during the inquisition pressed to

say that she must have done some evil deed; and after several admonitions to tell the truth:

She was placed in the hands of the officer of the *haultes oeuvres* [the officer in charge of torture], throwing herself on her knees, struggling to cry, uttering several exclamations, without being able, nevertheless, to shed a tear. Saying at every moment that she is not a witch.

The Torture

On this same day, being at the place of torture.

This prisoner, before being strapped down, was admonished to maintain herself in her first confessions and to renounce her lover.

—Said that she denies everything she has said, and that she has no lover.

Feeling herself being strapped down, says that she is not a witch, while struggling to cry.

Asked why she fled outside the village of Rieux.

—Says that she cannot say it, that God and the Virgin Mary forbid her to; that she is not a witch. And upon being asked why she confessed to being one, said that she was forced to say it.

Told that she was not forced, that on the contrary she declared herself to be a witch without any threat.

—Says that she confessed it and that she is not a witch, and being a little stretched [on the rack] screams ceaselessly that she is not a witch, invoking the name of Jesus and of Our Lady of Grace, not wanting to say any other thing.

Asked if she did not confess that she had been a witch for twenty-six years.

—Says that she said it, that she retracts it, crying Jésus-Maria, that she is not a witch.

Asked if she did not make Philippe Cornié's horse die, as she confessed.

—Answers no, crying Jésus-Maria, that she is not a witch.

The mark having been probed by the officer, in the presence of Doctor Bouchain, it was adjudged by the aforesaid doctor and officer truly to be the mark of the devil.

Being more tightly stretched upon the torture-rack, urged to maintain her confessions.

—Said that it was true that she is a witch and that she would maintain what she had said.

Asked how long she has been in subjugation to the devil.

—Answers that it was twenty years ago that the devil appeared to her, being in her lodgings in the form of a man dressed in a little cow-hide and black breeches.

Interrogated as to what her lover was called.

—Says that she said Petit-Grignon, then, being taken down [from the rack] says upon interrogation that she is not a witch and that she can say nothing.

Asked if her lover has had carnal copulation with her, and how many times.

—To that she did not answer anything; then, making believe that she was ill, not another word could be drawn from her.

As soon as she began to confess, she asked who was alongside of her, touching her, yet none of those present could see anyone there. And it was noticed that as soon as that was said, she no longer wanted to confess anything.

Which is why she was returned to prison.

Verdict

July 9, 1652

In the light of the interrogations, answers and investigations made into the charge against Suzanne Gaudry, coupled with her confessions, from which it would appear that she has always been ill-reputed for being stained with the crime of witchcraft, and seeing that she took flight and sought refuge in this city of Valenciennes, out of fear of being apprehended by the law for this matter; seeing how her close family were also stained with the same crime, and the perpetrators executed; seeing by her own confessions that she is said to have made a pact with the devil, received the mark from him, which in the report of *sieur* Michel de Roux was judged by the medical doctor of Ronchain and the officer of *haultes oeuvres* by Cambrai, after having proved it, to be not a natural mark but a mark of the devil, to which they have sworn with an oath; and that following this, she had renounced God, Lent, and baptism and had let herself be known carnally by him, in which she received satisfaction. Also, seeing that she is said to have been a part of nocturnal carols and dances. Which are crimes of divine lése-majesty:

For expiation of which the advice of the under-signed is that the office of Rieux can legitimately condemn the aforesaid Suzanne Gaudry to death, tying her to a gallows, and strangling her to death, then burning her body and burying it there in the environs of the woods.

At Valenciennes, the 9th of July, 1652. To each [member of the Court] 4 *livres*, 16 *sous*. . . . And for the trip of the aforementioned Roux, including an escort of one soldier, 30 *livres*.

84. Woodcuts of Witches' Activities, from F.-M. Guazzo, *Compendium Maleficarum*, 1610

Books and pamphlets about witchcraft and witch-hunting were often illustrated with woodcuts and engravings depicting witches' evil deeds and demonic rituals. The artists frequently copied one another so that visual materials, like verbal descriptions, helped spread stereotypes. The following come from an Italian guide to witches; the dress of the female and male witches is quite elegant, very different from what women such as Suzanne Gaudry could have afforded.

A. Kissing the Devil's posterior. (*By permission of the Houghton Library, Harvard University.*)

B. Causing illness. (*By permission of the Houghton Library, Harvard University.*)

QUESTIONS FOR ANALYSIS

1. According to the authors of the *Malleus Maleficarum,* why are women more likely to be witches? How do gender and sexual issues figure in their discussions of acts done by witches?
2. How do popular ideas of witches as individuals who do harm blend with Christian notions of witches as devil worshippers in the witch newspapers, trial records, and illustrations? Are there issues on which they lend particular support to each other, or others on which they seem to conflict?
3. Suzanne Gaudry tells her story slightly differently at each interrogation, and she fully retracts her confession at one point. Why does this happen? Why do you think this fails to make her interrogators skeptical about the charges against her?
4. The number and names of accomplices are mentioned prominently in the witch newspapers and trial accounts, and several of the illustrations show witches in groups. Why might this have been viewed as important in reports about witchcraft? What patterns do you see in the type of individuals who are accused or mentioned together?

WOMEN'S RULE AND RIGHTS

As we saw in Chapter 5, the "debate about women" in the Late Middle Ages and Renaissance largely focused on women's education and moral character. These concerns continued in the early modern period, but issues involving politics and power were added to the discussion as well. During the sixteenth century, the controversy centered on women's rule, a burning issue because of the number of women ruling in their own right or as advisors to child kings—Isabella in Castile, Mary and Elizabeth Tudor in England, Catherine de' Medici and Anne of Austria in France, Mary Stuart in Scotland, and many noblewomen in the smaller states of the Holy Roman Empire and Italy. Those opposing women's rule generally asserted that nothing could ever overcome the disabilities of being born female, while those supporting women's rule argued that a few women could "overcome their sex" through noble birth, strength of character, and excellent training.

During the seventeenth century, discussions of political rights for men and actual revolts against monarchies led to occasional debates about whether women should be included in new notions of the "body politic" that were developing. If subjects had the right to overthrow the authority of a tyrannous monarch and become citizens, didn't women have the right to overthrow the authority of a tyrannous husband? These discussions emerged most pointedly in England, where the monarch was indeed overthrown in the English Civil War of 1640–1660. During this period, women occasionally acted as if they had a political voice rather than simply discussing this as a possibility. Such actions appalled men of all political persuasions; even those men who supported extending political rights to all men, were convinced that women's secondary status was "natural."

85. John Knox
First Blast of the Trumpet Against the Monstrous Regiment of Women, 1158

John Knox (1514?–1572) was a Scottish Protestant reformer and founder of the Scottish Presbyterian Church. He went into exile on the continent when the Catholic Mary Tudor became queen in England, and from there he wrote many letters and pamphlets urging people to oppose "ungodly" (meaning Catholic) monarchs. Among these was the long treatise excerpted here directed against Mary Tudor and Mary of Guise (the regent of Scotland), which Knox described as only the first of three blasts he would eventually direct against women's rule. The publication of Knox's treatise coincided with Mary Tudor's death and Elizabeth's accession to the throne, which made his position as both a Protestant and opponent of female rule rather tricky. Other authors quickly wrote defenses of female rule, hoping these would help them win favor in Elizabeth's eyes, but Knox never retracted his words. He also never wrote the other two blasts, but Elizabeth was still so angry that she never allowed him to enter England.

The first blast of the trumpet to awake women degenerate

To promote a woman to bear rule, superiority, dominion or empire above any realm, nation, or city is repugnant to nature, contumely[1] to God, a thing most contrarious to his revealed will and approved ordinance, and finally it is the subversion of good order, and all equity and justice.

In the probation[2] of this proposition, I will not be so curious as to gather whatsoever may amplify, set forth, or decore[3] the same, but I am purposed, even as I have spoken my conscience in most plain and few words, so to stand content with a simple proof of every member, bringing in for my witness God's ordinance in nature, his

---

[1]*Contumely:* insulting.

[2]*Probation:* proof.
[3]*Decore:* embellish.

First Blast of the Trumpet Against the Monstrous Regiment of Women by John Knox, 1158.

plain will revealed in his word, and the minds of such as be most ancient amongst godly writers.

And first, where I affirm the empire of a woman to be a thing repugnant to nature, I mean not only that God by the order of his creation hath spoiled women of authority and dominion, but also that man hath seen, proved and pronounced just causes why that it so should be. Man, I say, in many other cases blind, doth in this behalf see very clearly. For the causes be so manifest that they cannot be hid. For who can deny that it repugneth to nature, that the blind shall be appointed to lead and conduct such as do see? That the weak, the sick and the impotent persons shall nourish and keep the whole and strong, and finally, that the foolish, mad and frenetic shall govern the discreet, and give counsel to such as be sober of mind? And such be all women compared unto man, in bearing of authority. For their sight in civil regiment is but a blindness: their strength weakness: their counsel foolishness: and judgement frenzy, if it be rightly considered.

I except such as God by singular privilege and for certain causes known only to himself hath exempted from the common rank of women, and speak of women as nature and experience do this day declare them. Nature, I say, doth paint them forth to be weak, frail, impatient, feeble and foolish: and experience hath declared them to be unconstant, variable, cruel, and lacking the spirit of counsel and regiment. And these notable faults have men in all ages espied in that kind, for the which not only they have removed women from rule and authority, but also some have thought that men subject to the counsel or empire of their wives were unworthy of all public office. For thus writeth Aristotle in the second of his *Politics: what difference shall we put,* saith he, *whether that women bear authority, or the husbands that obey the empire of their wives be appointed to be magistrates? For what ensueth the one, must needs follow the other, to wit: injustice, confusion, disorder.* The same author further reasoneth, that the policy or regiment of the Lacedomians[4] (who other ways amongst the Grecians were most excellent) was not worthy or reputed to be accompted amongst the number of commonwealths that were well governed, because the magistrates and rulers of the same were so much given to please and obey their wives. What would this writer (I pray you) have said to that realm or nation where a woman sitteth crowned in parliament amongst the midst of men? Oh, fearful and terrible are thy judgements (oh, Lord) which thus hast abased man from his iniquity! . . . Women are removed from all civil and public office, so that they neither may be judges, neither may they occupy the place of the magistrate, neither yet may they be speakers for others. . . . The law in the same place doth further declare that a natural shamefastness ought to be in womankind, which most certainly she loseth whensoever she taketh upon her the office and estate of man.

———

[4]*Lacedomians:* Spartans.

86. Elizabeth I
Tilbury Speech, 1588

Though one could never view the life of Queen Elizabeth I of England (1533–1603) as that of a typical early modern European woman, it provides an unusually well-documented example of the way in which gender shaped experience, even for someone who was arguably the most powerful woman in the world. Her father was Henry VIII, who divorced his first wife, Catherine of

"Tilbury Speech" by Queen Elizabeth I.

Aragon, because she had not had a son; to do so, however, required Henry to remove England from the power of the papacy and create his own Church of England. Henry's disappointment at Elizabeth's birth—he had her mother, Anne Boleyn, executed so he could try for a son with yet another wife—did not prevent her from attaining a fine Renaissance education, and she learned to speak and write well in several languages. The death of her younger brother, Edward VI, and older sister, Mary I, made Elizabeth queen at age twenty-five. Though she had many suitors, Elizabeth never married, and in both her actions and words skillfully used her unusual status as a virgin queen (immortalized in "Virginia," the name given originally by the English to all of North America not held by the Spanish or French) and as a person who combined masculine and feminine qualities. She wrote letters, prayers, poems, and translations, and she gave many speeches, which were written down for the official record or survive in listeners' notes. The following is a speech she apparently gave after the English defeat of the Spanish Armada in 1588, when reviewing her troops at Tilbury.

"My loving people: we[1] have been persuaded by some that are careful of our safety to take heed how we commit ourself to armed multitudes for fear of treachery, but I assure you I do not desire to live to distrust my faithful and loving people. Let tyrants fear. I have always so behaved myself that, under God, I have placed my chiefest strength and safeguard in the loyal hearts and goodwill of my subjects. And therefore I am come amongst you, as you see, at this time, not for my recreation and disport, but being resolved in the midst and heat of the battle to live or die amongst you all, to lay down for my God, and for my kingdom, and for my people, my honor and my blood, even in the dust. I know I have the body but of a weak and feeble woman, but I have the heart and stomach of a king—and of a king of England too—and think foul scorn that Parma,[2] or Spain, or any prince of Europe should dare to invade the borders of my realm. To which, rather than any dishonor shall grow by me, I myself will take up arms, I myself will be your general, judge, and rewarder of every one of your virtues in the field. I know already for your forwardness you have deserved rewards and crowns, and we do assure you, in the word of a prince, they shall be duly paid you."

---

[1]Monarchs customarily used the plural to refer to themselves.

[2]*Parma* refers to the duke of Parma, one of the leaders of the Spanish Armada.

87. T.E.
Law's Resolution of Women's Rights, 1632

This book, with the descriptive subtitle "a methodical collection of such statutes and customs as do properly concern women," was a compilation of existing laws and customs regarding women, published anonymously in London. The preface

The lavves, resolutions of womens' rights: or, The lavves provision for women. A methodicall collection of such statutes and customes as doe properly concerne women (London; I. More, 1632.)

to the reader is signed "T.E." and there has been much speculation, but no consensus, about who the author was. The book may originally have been put together by another author and circulated in manuscript before it was published, as T.E. comments that he has "added many reasons, opinions, cases and resolutions of cases to the author's store."

Because . . . women . . . have nothing to do in constituting laws or consenting to them, in interpreting of laws or in hearing them interpreted at lectures, leets[1] or charges, and yet they stand strictly tied to men's establishments, little or nothing excused by ignorance, methinks it were pity and impiety any longer to hold from them such customs, laws and statutes as are in a manner proper or principally belonging unto them. . . . I will in this treaty with as little tediousness as I can, handle that part of the English law which containeth the immunities, advantages, interests, and duties of women, not regarding so much to satisfy the deep learned or searchers for subtility, as womenkind, to whom I am a thankful debtor by nature.

The punishment of Adam's sin

Return a little to Genesis, in the 3. chapter whereof is declared our first parents' transgression in eating the forbidden fruit: for which Adam, Eve, the serpent first, and lastly the earth itself is cursed: and besides the participation of Adam's punishment, which was subjection to mortality, exiled from the garden of Eden, enjoined to labour; Eve, because she had helped to seduce her husband, hath inflicted on her an especial bane: *in sorrow shalt thou bring forth thy children, thy desires shall be subject to thy husband, and he shall rule over thee.*

See here the reason of that which I touched before, that women have no voice in parliament. They make no laws, consent to none, they abrogate none. All of them are understood either married or to be married, and their desires are subject to their husband: I know no remedy, though some women can shift it[2] well enough. The common law here shaketh hand[3] with divinity. . . .

The baron[4] may beat his wife

If a man beat an outlaw, a traitor, a pagan, his villein,[5] or his wife, it is dispunishable, because by the law common these persons can have no action: God send gentlewomen better sport or better company. . . .

That which the husband hath is his own

But the prerogative of the husband is best discerned in his dominion over all external things . . . whatsoever the husband had before coverture,[6] either in goods or land, it is absolutely his own, the wife hath therein no seisin[7] at all. If anything when he is married be given him, he taketh it by himself distinctly to himself. . . .

The very goods which a man giveth to his wife, are still his own: her chain, her bracelets, her apparel, are all the goodman's goods.

That which the wife hath is the husband's

For thus it is, if before marriage the woman were possessed of horses, neat,[8] sheep, corn, wool, money, plate, and jewels, all manner of moveable substance is presently by conjunction the husband's to sell, keep, or bequeath if he die.

---

[1]*Leets:* local courts.

[2]*Shift it:* get along.
[3]*Shaketh hand:* agrees.
[4]*Baron:* here, husband.
[5]*Villein:* serf.
[6]*Coverture:* here, marriage.
[7]*Seisin:* possession.
[8]*Neat:* cattle.

And though he bequeath them not, yet are they the husband's executor's, and not the wife's which brought them to her husband.

Appeal of rape

Now let us consider a little how these laws ought to be put in practice, if any virgin, widow, or single woman be ravished, she herself may sue an appeal of rape, prosecute the felon to death, and the king's pardon (it seemeth) cannot help him.

If a feme covert[9] be ravished, she cannot have an appeal without her husband, as appears 8 Hen. 4, fol. 21.[10] But if a feme covert be ravished and consent to the ravisher, the husband alone may have an appeal, and this by statute 6 Rich. 2, cap. 6.[10] The husband that this statute speaketh of, which may sue the appeal, must be a lawful husband in right and possession, for *ne unques accouple* [failure to consummate] in loyal matrimony is a good plea against him.

---

[9]*Feme covert:* married women.

[10]These are references to royal statutes, the first from the reign of Henry VIII and the second from that of Edward VI.

88. *A True Copy of the Petition of the Gentlewomen and Tradesmen's Wives in and about the City of London, 1642*

During the English Civil War, many groups used petitions to Parliament as a way to get their political and religious grievances aired and push for changes. Most of these groups were male, but occasionally groups of women used this tactic, petitioning Parliament to improve trade laws, end imprisonment for debt, release prisoners, and end martial law in times of peace. The following is a petition from a group of London women who wanted Parliament to take sterner moves against Catholics. We do not know who wrote it; many contemporaries assumed women's petitions must have been written by men, but the women who brought them defended them orally so this may, indeed, have a female author. This petition was apparently received respectfully, though many were not.

Delivered, to the Honorable, the Knights, Citizens, and Burgesses, of the House of Commons in Parliament, the 4th of February, 1642 Together, with their several reasons why their sex ought thus to petition, as well as the men; and the manner how both their petition and reasons was delivered.

. . . And whereas we, whose hearts have joined cheerfully with all those petitions which have been exhibited unto you in the behalf of the purity of religion, and the liberty of our husbands'

persons and estates, recounting ourselves to have an interest in the common privileges with them, do with the same confidence assure ourselves to find the same gracious acceptance with you, for easing of those grievances, which in regard of our frail condition, do more nearly concern us, and do deeply terrify our souls: our domestical dangers with which this kingdom is so much distressed, especially growing on us from those treacherous and wicked attempts already are such as we find ourselves to have as deep a share as any other.

A True Copy of the Petition of the Gentlewomen and Tradesmen's Wives (London, 1642).

We cannot but tremble at the very thoughts of the horrid and hideous facts which modesty forbids us now to name. . . . We wish we had no cause to speak of those insolencies, and savage usage and unheard-of rapes, exercised upon our sex in Ireland,[1] and have we not just cause to fear they will prove the forerunners of our ruin, except Almighty God by the wisdom and care of this Parliament be pleased to succor us, our husbands and children, which are as dear and tender unto us as the lives and blood of our hearts, to see them murdered and mangled and cut in pieces before our eyes, to see our children dashed against the stones, and the mothers' milk mingled with the infants' blood, running down the streets, to see our houses on flaming fire over our heads: oh how dreadful would this be? . . .

We humbly signify that our present fears are, that unless the blood-thirsty faction of the Papists and Prelates[2] be hindered in their designs, ourselves here in England as well as they in Ireland, shall be exposed to the misery which is more intolerable than that which is already past, as namely to the rage not of men alone, but of devils incarnate (as we may so say), besides the thralldom of our souls and consciences in matters concerning God, which of all things are most dear unto us.

Now the remembrance of all these fearful accidents aforementioned do strongly move us from the example of the woman of Tekoa (II Samuel 14.2–20) to fall submissively at the feet of his Majesty, our dread sovereign, and cry Help, oh King, help oh ye the noble Worthies now sitting in Parliament: And we humbly beseech you, that you will be a means to his Majesty and the House of Peers,[3] that they will be pleased to take our heartbreaking grievances into timely consideration, and to add strength and encouragement to your noble endeavors, and further that you would move his Majesty with our humble requests, that he would be graciously pleased according to the example of the good King Asa, to purge both the court and kingdom of that great idolatrous service of the Mass, which is tolerated in the Queen's court,[4] this sin (as we conceive) is able to draw down a greater curse upon the whole kingdom than all your noble and pious endeavors can prevent, which was the cause that the good and pious King Asa would not suffer idolatry in his own mother, whose example if it shall please his Majesty's gracious goodness to follow, in putting down Popery and idolatry both in great and small, in court and in the kingdom throughout, to subdue the Papists and their abettors, and by taking away the power of the Prelates, whose government by long and woeful experience we have found to be against the liberty of our conscience and the freedom of the Gospel, and the sincere profession and practice thereof, then shall our fears be removed, and we may expect that God will pour down his blessings in abundance both upon his Majesty, and upon this Honorable Assembly, and upon the whole land.

For which your new petitioners shall pray affectionately.

The reasons follow.

It may be thought strange and unbeseeming our sex to show ourselves by way of petition to this Honorable Assembly: but the matter being rightly considered, of the right and interest we have in the common and public cause of the church, it will, as we conceive (under correction) be found a duty commanded and required.

First, because Christ hath purchased us at as dear a rate as he hath done men, and therefore requireth the like obedience for the same mercy as of men.

---

[1]The Irish, who were Catholic, had just rebelled against English overlordship and had killed English Protestant settlers.

[2]*Papists and prelates:* Catholics and bishops.

[3]*House of Peers:* House of Lords, the upper house of the English Parliament.

[4]The Queen, Henrietta Maria, was the daughter of the king of France and a Catholic. Though the Catholic Mass was officially prohibited in England, it was held at her court, and both foreign and English Catholics attended there.

Secondly, because in the free enjoying of Christ in his own laws, and a flourishing estate of the church and commonwealth, consisteth the happiness of women as well as men.

Thirdly, because women are sharers in the common calamities that accompany both church and commonwealth, when oppression is exercised over the church or kingdom wherein they live; and an unlimited power have been given to Prelates to exercise authority over the consciences of women, as well as men, witness Newgate, Smithfield,[5] and other places of persecution, wherein women as well as men have felt the smart of their fury.

Neither are we left without example in scripture, for when the state of the church, in the time of King Ahasuerus, was by the bloody enemies thereof sought to be utterly destroyed, we find that Esther the Queen and her maids fasted and prayed, and that Esther petitioned to the King in the behalf of the church: and though she enterprised this duty with the hazard of her own life, being contrary to the law to appear before the King before she were sent for, yet her love to the church carried her through all difficulties, to the performance of that duty.

On which grounds we are emboldened to present our humble petition unto this Honorable Assembly, not weighing the reproaches which may and are by many cast upon us, who (not well weighing the premises) scoff and deride our good intent. We do it not out of any self-conceit, or pride of heart, as seeking to equal ourselves with men, either in authority or wisdom: But according to our places to discharge that duty we owe to God, and the cause of the church, as far as lieth in us, following herein the example of the men which have gone in this duty before us.

---

[5]Newgate and Smithfield were places where Protestants had been imprisoned or executed.

QUESTIONS FOR ANALYSIS

1. How would you compare Knox's attitude toward women's rule with Elizabeth I's? Which takes precedence for them, gender or rank?

2. What are the consequences of viewing, as T.E. notes, "all women as either married or to be married" in terms of women's rights to their property and to their persons in cases of domestic violence and rape? Why might granting married women the right to own their own property have been viewed as so threatening? (England did not allow this until 1870.)

3. Knox and T.E. both agree that women's secondary status is permanent. What similarities and differences do you find in the reasons they give for that secondary status?

4. What do the women petitioners want the House of Commons to do? What justifications do they give for their actions? How do these compare with those given by Argula von Grumbach for publishing her pamphlet?

5. How would you compare Knox's view of women's nature with that of the authors of the *Malleus Maleficarum*? What other parallels do you see in the literature about witchcraft and that about women's rule in a country or a household?

THE ENLIGHTENMENT

In the growing cities of early modern Europe, new places for social and intellectual interaction among both men and women developed outside of the formal institutions of universities and academies. Reading clubs, voluntary associations, and weekly gatherings known as salons brought people together to discuss books and poetry, the latest plays, and, by the eighteenth century, political developments. At such meetings, like-minded individuals increasingly criticized what they viewed as unjust and irrational traditions and called for judging everything according to the "enlightened" standards of reason.

Among the many topics open for debate in the Enlightenment was that perennial favorite—"The Woman Question." Did women's secondary status result from innate inferiority or social traditions? If the latter, were these unjust and thus in need of change, or proper and in need of reinforcing? Were the problems in modern society the result of women having too much independence or too little? Was marriage an institution ordained by God or a form of despotism worse than absolute monarchy? Enlightenment considerations of these questions, which emerged in novels, letters, newspapers, and treatises written and read by women and men, created a foundation for later calls for both greater equality and "separate spheres" for the two sexes.

89. Antoine-Léonard Thomas
Essay on the Character, Morals and Mind of Women Across the Centuries, Paris 1772

Antoine-Léonard Thomas (1732–1785) was a writer, teacher, and government official who was a frequent recipient of prestigious prizes for his works. In 1772 he published a long essay surveying the role of women throughout history, building on the lists of noble and praiseworthy women that had been part of defenses of women since Christine de Pizan. Like many Enlightenment writers, he compared his own culture with those of earlier times and other parts of the world, seeing progress but also worrying that the emphasis on sociability, conversation, and literature had gone too far. The end of the book, from which this excerpt is taken, highlights both the problems Thomas sees in contemporary society and the way changes in gender relations would solve them.

In the last years of Louis XIV's reign some terrible sort of seriousness and sadness had spread through the court and part of the nation. At heart, penchants were the same, but they were more repressed. A new court and new ideas changed everything. A bolder sensuality became

From *Qu-est-ce qu-une femme?* edited by Elisabeth Badinter, pp. 150–161. © 1989 by P.O.L. Reprinted by permission. Translated by Dena Goodman and Katherine Ann Jensen.

fashionable. Audacity and impetuosity colored all desires, and a part of the veil which covered gallantry was torn away. The decency which had been respected as a duty, was no longer even maintained as a pleasure. Shame was dispensed with on both sides. Fickleness matched excess, and a corruption took shape that was at once profound and frivolous, that, blushing at nothing, laughed instead at everything.

The upsetting of fortunes precipitated this change. Extreme misery and extreme luxury followed from it; and we know what influence these have had. Rarely does a people receive such a quick jolt in regard to property without a prompt alteration in manners and morals. . . .

At the same time, and by the same general trend which carries everything along, the taste for the society of women grew. Seduction was made easier, the opportunities for it increased everywhere. Men lived together less; less timid women became accustomed to throwing off a constraint which honors them. The two sexes were denatured; one placed too much value on being agreeable, the other on independence.

Thus the weight of time, the desire to please, necessarily spread the spirit of society farther and farther; and the time had to come when this sociability pushed to excess, in mixing everything up, succeeded in spoiling everything; and this is perhaps where we are today.

Among a people in whom the spirit of society is carried so far, domestic life is no longer known. Thus all the sentiments of Nature that are born in retreat, and which grow in silence, are necessarily weakened. Women are less often wives and mothers. . . .

Put all of this together and a disturbing frivolity in the two sexes must arise, along with a serious and busy vanity. But what must above all characterize morals, is the fury of appearance, the art of putting everything on the surface, a great importance given to small duties, and a great value to small successes. One must speak gravely of petty things from last night and from the next morning. In the end the soul and the mind must be engaged in a cold activity, which spreads them over a thousand objects without interesting them in any, and which gives them movement without direction. . . .

As the general mass of enlightenment is greater, and as it is communicated through greater movement, women, without taking the least trouble, are necessarily better instructed; but faithful to their plan, they do not seek enlightenment, except as an ornament of the mind. In learning they wish to please rather than to know, and to amuse rather than to learn.

Moreover, in a state of society where there is a rapid movement and an eternal succession of works and ideas, women, occupied with following this panorama which flees and changes around them constantly, will better understand the idea of the moment in each genre, than eternal ones, and know the dominant ideas, rather than those which are forming. They will thus know the language of the arts better than their principles, and have more specific ideas than systems of thought.

It seems to me that in the sixteenth century women learned through enthusiasm for learning itself. There was in them a profound taste which derived from the spirit of the age and that was nourished even in solitude. In this century, it is less a real taste than a flirtation of the mind; and as with all objects, a luxury, [although] more represented than based on wealth. . . .

It would perhaps be interesting to examine now what must result among us, from all this mixture of movement and ideas, of frivolity and wit, of philosophy in the head and of liberty in morals. It would be interesting to compare the present character of women with that which they have had in all ages; with their timid reserve, and their sweet modesty in England; their mixture of devotion and sensuality in Italy; their ardent imagination and their jealous sensitivity in Spain; their deep retreat in China, and the barriers that, for four thousand years have separated them in this empire from the gaze of men; finally, with the character and the morals that must result for them from their enclosure in almost all of Asia, where, existing for one person alone, not being able to cultivate either their character, or their reason, and destined to have only senses, they are forced by the bizarreness of their state, to join modesty with sensuality, flirtation with retreat: but to make this parallel, it is enough to point it out.

I will observe only that in this century there is less praise of women than ever before. The sad dignity of funeral panegyrics is now reserved almost exclusively for women who have occupied or were destined to occupy thrones. The philosophe orators only celebrate that which has been useful to humanity as a whole, or to nations. The poets seem to have lost that delicate gallantry which for a long time characterized them. They sing more of pleasures than of love, and are more sensual than sensitive. This general taste for women which is neither love, nor passion, nor even gallantry, but the effect of a cold and artificial habit, no longer arouses anywhere either the imagination or the mind. Among social circles, in the eternal mixing of the sexes, one learns to praise less, because one learns to be more severe. Egoism, judge and rival, sometimes indulgent out of pride, but almost always cruel out of jealousy, has never been more vigilant, in looking for faults and planting the seeds of ridicule. Praise is produced out of enthusiasm; and never during any century has there been less of it, although perhaps it is affected more. Enthusiasm is born of an ardent soul, which creates objects rather than seeing them. Today we see too much: and thanks to enlightenment, we see everything coldly. Vice itself ranks among our pretensions. The less we esteem women, the more we appear to know them. Each one prides himself on not believing in their virtues; and he who would like to be a fop and can not succeed by speaking ill of them, prides himself on a satire that, as the height of ridiculousness, he has no right to make. Such is, with regard to women themselves, the influence of this general spirit of society which is their work, and which they do not cease to vaunt. They are like those Asian rulers who are never honored more than when they are seen less: by communicating too much to their subjects, they encourage them to revolt.

However, despite our morals and our eternal satires, despite our fury to be esteemed without merit, and our even greater fury never to find anything worthy of esteem, there are in our cen-

tury, and in this capital itself, women who would do honor to another century than ours. Several bring together a strong soul and a truly cultivated reason, and bring forth by their virtues, feelings of courage and honor. There are some who could think with Montesquieu,[1] and with whom Fénelon[2] would love to be moved. There are some who, in opulence, and, surrounded by this luxury that today practically forces avarice to join with pomp and renders souls small, vain, and cruel, take from their property each year a portion for the unfortunate; they are familiar with shelters for the poor, and will learn to be sensitive in shedding their tears there. There are tender wives, who, young and beautiful, pride themselves on their duties, and in the sweetest of attachments offer a ravishing spectacle of innocence and love. Finally, there are mothers who dare to be mothers. In several houses beauty can be seen taking charge of the most tender cares of nature, and by turns pressing in her arms or to a breast the son whom she nourishes with her own milk, while in silence the husband divides his tender regards between son and mother.

Oh! if these examples could restore among us nature and morals! If we could learn how much superior are virtues to pleasures for happiness itself; how much a simple and calm life where nothing is affected, where one exists only for oneself, and not for the gaze of others, where one enjoys by turn friendship, nature, and oneself, is preferable to this anxious and turbulent life, in which one is forever running after a feeling that one does not find! Ah! then, women would regain their empire. It is then that beauty, embellished with morals, would command men, happy to be subjected, and great in their weakness. Then an honorable and pure sensuality seasoning every moment, would make of life an enchanted dream. Then troubles, not being poisoned by remorse, troubles softened by love and shared in friendship, would be a touching sadness rather than a tor-

---

[1]The Baron de Montesquieu (1689–1755) was a French jurist and political philosopher.
[2]François Fénelon (1651–1715) was a French theologian and writer.

ment. In this state society would doubtless be less active, but the interior of families would be sweeter. There would be less ostentation and more pleasure; less movement and more happiness. One would talk less to please, and one would please onself more. The days would flow on pure and tranquil: and if in the evening one did not have the sad satisfaction of having during the course of the day played at having the most tender interest in thirty random people, one would have at least lived with the one whom one loves; one would have added for the morrow a new charm to the feelings of yesterday. Must such a sweet image be only an illusion? And in this burning and vain society, is there no refuge for simplicity and happiness?

90. Louise d'Epinay's Letter to Abbé Ferdinando Galiani

Thomas's essay sparked responses from several writers, among them the writer Louise d'Epinay (1726–1783), whose home in Paris was a place where Enlightenment philosophers and authors gathered to discuss literary and political issues. She shared her thoughts on it in a letter to Abbé Ferdinando Galiani (1728–1787), a brilliant Neopolitan scholar and writer who had been a regular guest at her home but who had returned to Naples. Galiani answered her letter, but focused on another subject and did not address her concerns.

You haven't written me at all this week, my dear Abbé. I am not well: therefore, I don't have much to tell you. So, I am resolved to read in front of the fire Monsieur Thomas' book: *On the Character, the Morals and the Spirit of Women.* This work just appeared a few days ago; and, if it gives rise to certain ideas, I will share them with you. I will tell you, as usual, everything that comes into my head, provided that my views remain between you and me.

Well then! I've read it and to anyone but you I would be careful about saying what I think, or of taking so definite a tone in the world; but I confess that this [book] seems to me nothing but pompous chattering, very eloquent, a little pedantic, and very monotonous. One finds there some little dressed-up sentences, the sort of sentences that, heard in a small circle, cause people to say of their author, that day and the next: "He has the wit of an angel! He is charming! he is charming!" But when I find them in a work that has the pretention to be serious, I have real difficulty being satisfied with them. This one doesn't add up to anything. One does not know, after one has read it, what the author thinks, and if his opinion about women is anything but received opinion. He writes with great erudition the history of famous women in all fields. He discusses a bit dryly what they owe to nature, to the institution of society and to education; and then, in showing them as they are, he attributes endlessly to nature that which we obviously owe to education or to institutions, etc.

And then so many commonplaces! "Are they more sensitive? More devoted in friendship than men? Are they more this? Are they more that?" "Montaigne,"[1] he says, "decides the question clearly against women, perhaps like that judge who so feared to be partial that on princi-

---

[1] Michel de Montaigne (1533–1592) was an essayist who wrote, among other topics, on friendship.

From *Qu-est-ce qu-une femme?* edited by Elisabeth Badinter, pp. 150–161. © 1989 by P.O.L. Reprinted by permission. Translated by Dena Goodman and Katherine Ann Jensen.

ple he always lost cases in which his friends were involved." And then, in another place: "Nature," he says, "makes them like flowers in order to shine softly in the garden from which they rise. One ought thus perhaps to desire a man for a friend for the great occasions, and for everyday happiness, one ought to wish for the friendship of a woman." How small, common, and unphilosophical these details are!

He claims that they are not able to transact business with as much continuity and constancy as men, nor with as much courage in their resolutions. This is, I think, a very false vision; there are a thousand examples of the contrary; there are even some very recent and rather remarkable ones. Moreover, constancy and courage in the pursuit of an object could be, it seems to me, calculated out of idleness, and this would be a strong argument in our favor. I don't have the time to work out this idea to the extent that I would like. But fortunately this is not necessary with you, and you will figure out the rest. "We have seen," says monsieur Thomas, "in [times of] great danger, examples of great courage among women; but this is always when a great passion or an idea that moves them strongly carries them beyond themselves," etc. But is courage anything else among men? Opinion or ambition is what moves them strongly. Were you to attach, in the institution and the education of women, the same prejudice to valor, you will find as many courageous women as men, since cowards are found among them, despite opinion, and the number of courageous women is as great as the number of cowardly men. Of the sum total of physical ailments spread over the face of the earth, women's share is more than two-thirds. It is quite constant also that they suffer them with infinitely more constancy and courage than do men. There is in this neither prejudice nor vanity for support: [woman's] physical constitution has, moreover, become weaker than man's as a result of education. One can thus conclude that courage is a gift of nature among women, just as it is among men, and, to carry this view farther, that it is of the essence of humanity in general to struggle against pain, difficulties, obstacles, etc.

One could, to even greater advantage, make the same calculation regarding moral troubles.

In speaking of the minority of Louis XIV, he says: "All women of this era had this sort of restless agitation produced by partisan spirit: a spirit less far from their character than one would think." That's true, Monsieur Thomas. But, since you would like to be scientific, here was a case for examining whether this restless disposition, which they have by nature, is particular to them and is not found equally among men; whether men, deprived as they are, of serious occupations, excluded from business and strangers to all great causes, would not display this same restless disposition, which is, in your eyes, extinguished by the nourishment given them by the role they play in society. The proof of this is that [this restless disposition] is noticed nowhere as much as among monks and in religious houses. Your work is not at all philosophic, you examine nothing on a grand scale, and once again I do not find any point. . . .

He finishes his work by expressing a wish for a return to morality and to virtue. So be it, certainly! These last four pages are the most agreeable of his book because of the picture he paints of woman as she ought to be; but he sees it as a chimera.

It is well established that men and women have the same nature and the same constitution. The proof lies in that female savages are as robust, as agile as male savages: thus the weakness of our constitution and of our organs belongs definitely to our education, and is a consequence of the condition to which we have been assigned in society. Men and women, being of the same nature and the same constitution, are susceptible to the same faults, the same virtues, and the same vices. The virtues that have been ascribed to women in general, are almost all virtues against nature, which only produce small artificial virtues and very real vices. It would no doubt take several generations to get us back to how nature made us. We could perhaps reach that point; but men would thereby lose too much. They are quite lucky that we are no worse than we are, after all that they have done to denature us through their lovely institutions, etc. This is so obvious that it is no more worth the trouble of saying than all that monsieur Thomas has said.

It was difficult to say anything new on this subject, and, in general, as monsieur Grimm[2] was saying the other day, there are no longer any new subjects or new ideas: all that we need are new heads to get us to imagine things from different points of view. But where are we to find them? I know two of them, nonetheless:

the abbé Galiani and the marquis de Crois-mare.[3] The marquis is to society what you are to philosophy and administration.

Goodby, my abbé! I do not know if women are constant, brave, etc.; but I know at least that they are as chatty as philosophers. You will agree in reading this letter, but I hope nevertheless that you will not disdain to respond and to give me your views on this delicate question.

---

[2]Friedrich-Melchior Grimm (1723–1807) was a German diplomat and at one point ambassador to France; he was good friends with many Enlightenment philosophers and was Louise d'Epinay's lover for three decades.

[3]The Marquis de Croismore (1718–1781) was a friend of many Enlightenment philosophes who spent much time in the salons and drawing rooms of Paris.

91. Jean-Jacques Rousseau
Emile, 1762

Jean-Jacques Rousseau (1712–1778) was born in Geneva, Switzerland, associated with Enlightenment writers and philosophers in Paris and elsewhere in France (he was financially supported for a while by Louise d'Epinay), and wrote several extremely influential books ranging from works of fiction to political theory and autobiography. Like Antoine-Léonard Thomas, he challenged the view that civilization meant progress and sought to establish society on a "natural" basis by championing the virtue of women and the freedom and equality of men. Somewhat contradictorily, he viewed proper education as a way to bring out people's positive natural qualities. His wildly popular novel *Emile* describes at great length the training its hero receives at the hands of his tutor, and in the final chapter discusses the upbringing of Sophie, the girl who had been trained to be Emile's perfect companion. Rousseau was never successful at finding his own Sophie, but lived with his illiterate mistress; she bore him five children, all of whom he sent to foundling homes in Paris.

Sophie ought to be a woman as Emile is a man—that is to say, she ought to have everything which suits the constitution of her species and her sex in order to fill her place in the physical and moral order. Let us begin, then, by examining the similarities and the differences of her sex and ours.

In everything not connected with sex, woman is man. She has the same organs, the same needs, the same faculties. The machine is constructed in the same way; its parts are the same; the one functions as does the other; the form is similar; and in whatever respect one considers them, the difference between them is only one of more or less.

In everything connected with sex, woman and man are in every respect related and in every respect different. The difficulty of comparing them comes from the difficulty of determining what in their constitutions is due to sex and what is not. On the basis of comparative anatomy and even just by inspection, one finds general differences between them that do not appear connected with sex. They are, nevertheless, connected with sex, but by relations which we are not in a position to perceive. We do not know the extent of these relations. The only thing we know with certainty is that everything man and woman have in common belongs to the species, and that everything which distinguishes them belongs to the sex. From this double perspective, we find them related in so many ways and opposed in so many other ways that it is perhaps one of the marvels of nature to have been able to construct two such similar beings who are constituted so differently.

These relations and these differences must have a moral influence. This conclusion is evident to the senses; it is in agreement with our experience; and it shows how vain are the disputes as to whether one of the two sexes is superior or whether they are equal—as though each, in fulfilling nature's ends according to its own particular purpose, were thereby less perfect than if it resembled the other more! In what they have in common, they are equal. Where they differ, they are not comparable. A perfect woman and a perfect man ought not to resemble each other in mind any more than in looks, and perfection is not susceptible of more or less.

In the union of the sexes each contributes equally to the common aim, but not in the same way. From this diversity arises the first assignable difference in the moral relations of the two sexes. One ought to be active and strong, the other passive and weak. One must necessarily will and be able; it suffices that the other put up little resistance.

Once this principle is established, it follows that woman is made specially to please man. If man ought to please her in turn, it is due to a less direct necessity. His merit is in his power; he pleases by the sole fact of his strength. This is not the law of love, I agree. But it is that of nature, prior to love itself.

If woman is made to please and to be subjugated, she ought to make herself agreeable to man instead of arousing him. Her own violence is in her charms. It is by these that she ought to constrain him to find his strength and make use of it. The surest art for animating that strength is to make it necessary by resistance. Then *amour-propre* unites with desire, and the one triumphs in the victory that the other has made him win. From this there arises attack and defense, the audacity of one sex and the timidity of the other, and finally the modesty and the shame with which nature armed the weak in order to enslave the strong. . . .

Do you want, then, to inspire young girls with the love of good morals? Without constantly saying to them "Be pure," give them a great interest in being pure. Make them feel all the value of purity, and you will make them love it. It does not suffice to place this interest in the distant future. Show it to them in the present moment, in the relationships of their own age, in the character of their lovers. Depict for them the good man, the man of merit; teach them to recognize him, to love him, and to love him for themselves; prove to them that this man alone can make the women to whom he is attached—wives or beloveds—happy. Lead them to virtue by means of reason. Make them feel that the empire of their sex and all its advantages depend not only on the good conduct and the morals of women but also on those of men, that they have little hold over vile and base souls, and that a man will serve his mistress no better than he serves virtue. You can then be sure that in depicting to them the morals of our own days, you will inspire in them a sincere disgust. In showing them fashionable people, you will make them despise them; you will only be keeping them at a distance from their maxims and giving them an aversion for their sentiments and a disdain for their vain gallantry. You will cause a nobler ambition to be born in them—that of reigning over great and strong souls, the ambition of the women of Sparta, which was to command men. A bold, brazen, scheming woman who knows how to attract her lovers only by coquetry and to keep them only by favors makes them obey her like valets in servile and common things; however, in important and weighty things she is without authority over them. But the woman who is at once decent, lovable, and self-controlled, who forces those about her to respect her, who has reserve and modesty, who, in a word, sustains love by means of esteem, sends her lovers with a nod to the end of the world, to combat, to glory, to death, to anything she pleases. This seems to me to be a noble empire, and one well worth the price of its purchase. . . .

It also makes a great difference for the good order of the marriage whether the man makes an alliance above or below himself. The former case is entirely contrary to reason; the latter is more conformable to it. Since the family is connected with society only by its head, the position of the head determines that of the entire family. When he makes an alliance in a lower rank, he does not descend, he raises up his wife. On the other hand, by taking a woman above him, he lowers her without raising himself. Thus, in the first case there is good without bad, and in the second bad without good. Moreover, it is part of the order of nature that the woman obey the man. Therefore, when he takes her from a lower rank, the natural and the civil order agree, and everything goes well. The contrary is the case when the man allies himself with a woman above him and thereby faces the alternative of curbing either his rights or his gratitude and of being either ungrateful or despised. Then the woman, pretending to authority, acts as a tyrant toward the head of the house, and the master becomes a slave and finds himself the most ridiculous and most miserable of creatures. Such are those unfortunate favorites whom the Asian kings honor and torment by marrying them to their daughters, and who are said to dare to approach only from the foot of the bed in order to sleep with their wives.

I expect that many readers, remembering that I ascribe to woman a natural talent for governing man, will accuse me of a contradiction here. They will, however, be mistaken.

There is quite a difference between arrogating to oneself the right to command and governing him who commands. Woman's empire is an empire of gentleness, skill, and obligingness; her orders are caresses, her threats are tears. She ought to reign in the home as a minister does in a state—by getting herself commanded to do what she wants to do. In this sense, the best households are invariably those where the woman has the most authority. But when she fails to recognize the voice of the head of the house, when she wants to usurp his rights and be in command herself, the result of this disorder is never anything but misery, scandal, and dishonor.

There remains the choice between one's equals and one's inferiors; and I believe that some restriction must be placed upon the latter, for it is difficult to find among the dregs of the people a wife capable of making a gentleman happy. It is not that they are more vicious in the lowest rank than in the highest, but that they have few ideas of what is beautiful and decent, and that the injustice of the older estates makes the lowest see justice in its very vices.

By nature man hardly thinks. To think is an art he learns like all the others and with even more difficulty. In regard to relations between the two sexes, I know of only two classes which are separated by a real distinction—one composed of people who think, the other of people who do not think; and this difference comes almost entirely from education. A man from the first of these two classes ought not to make an alliance in the other, for the greatest charm of society is lacking to him when, despite having a wife, he is reduced to thinking alone. People who literally spend their whole lives working in order to live have no idea other than that of their work or their self-interest, and their whole mind seems to be at the end of their arms. This ignorance harms neither probity nor morals. Often it even serves them. Often one compromises in regard to one's duties by dint of reflecting on them and ends up replacing real things with abstract talk. Conscience is the most enlightened of philosophers. One does not need to know Cicero's *Offices* to be a good man, and the most decent woman in the world perhaps has the least knowledge of what decency is. But it is no less true that only a cultivated mind makes association agreeable, and it is a sad thing for a father of a family who enjoys himself in his home to be forced to close himself up and not be able to make himself understood by anyone.

Besides, how will a woman who has no habit of reflecting raise her children? How will she discern what suits them? How will she incline them toward virtues she does not know, toward merit of which she has no idea? She will know only how to flatter or threaten them, to make them insolent or fearful. She will make mannered monkeys or giddy rascals of them, never good minds or lovable children.

Therefore, it is not suitable for a man with education to take a wife who has none, or, consequently, to take a wife from a rank in which she could not have an education. But I would still like a simple and coarsely raised girl a hundred times better than a learned and brilliant one who would come to establish in my house a tribunal of literature over which she would preside. A brilliant wife is a plague to her husband, her children, her friends, her valets, everyone. From the sublime elevation of her fair genius she disdains all her woman's duties and always begins by making herself into a man after the fashion of Mademoiselle de l'Enclos.[1] Outside her home she is always ridiculous and very justly criticized; this is the inevitable result as soon as one leaves one's station and is not fit for the station one wants to adopt. All these women of great talent never impress anyone but fools. It is always known who the artist or the friend is who holds the pen or the brush when they work. It is known who the discreet man of letters is who secretly dictates their oracles to them. All this charla-

tanry is unworthy of a decent woman. Even if she had some true talents, her pretensions would debase them. Her dignity consists in her being ignored. Her glory is in her husband's esteem. Her pleasures are in the happiness of her family. Readers, I leave it to you. Answer in good faith. What gives you a better opinion of a woman on entering her room, what makes you approach her with more respect—to see her occupied with the labors of her sex and the cares of her household, encompassed by her children's things, or to find her at her dressing table writing verses, surrounded by all sorts of pamphlets and letters written on tinted paper? Every literary maiden would remain a maiden for her whole life if there were only sensible men in this world:

Quaeris cur nolim te ducere, Galla? diserta es.[2]

After these considerations comes that of looks. It is the first consideration which strikes one and the last to which one ought to pay attention, but still it should count for something. Great beauty appears to me to be avoided rather than sought in marriage. Beauty promptly wears out in possession. After six weeks it is nothing more for the possessor, but its dangers last as long as it does. Unless a beautiful woman is an angel, her husband is the unhappiest of men; and even if she were an angel, how will she prevent his being ceaselessly surrounded by enemies? If extreme ugliness were not disgusting, I would prefer it to extreme beauty; for in a short time both are nothing for the husband, and thus beauty becomes a drawback and ugliness an advantage. But ugliness which produces disgust is the greatest of misfortunes. This sentiment, far from fading away, increases constantly and turns into hatred. Such a marriage is a hell. It would be better to be dead than to be thus united.

---

[1] Ninonde Lenclos (or l'Enclos) was a well-known seventeenth-century French courtesan, and friend to prominent writers and political leaders.

[2] This is a quotation from the Roman author Martial, who was speaking to a woman named Galla: "You ask Galla, why I do not want to marry you? You are eloquent."

92. Mary Wollstonecraft
Vindication of the Rights of Woman, 1792

Rousseau's ideas about natural virtues and gender complementarity did not remain limited to philosophical treatises but were widely spread in children's books, such as *The History of Little Goody Two-Shoes* and *The History of Sandford and Merton*—in which poor children with a simple education invariably turn out to be more heroic and virtuous than their sophisticated social betters—and in novels—in which poor but moral young women maintain their sexual purity and eventually marry someone better than the rake who has been trying to seduce them. They also provoked strong counterreactions, the best known of which became Mary Wollstonecraft's *Vindication of the Rights of Woman,* written in response to both Rousseau and to the ideas of the early French Revolution. Wollstonecraft (1759–1797) came from a middle-class English family and worked as a governess, teacher, and journalist; she died shortly after giving birth to a daughter, who later gained fame as Mary Shelley, the author of *Frankenstein.*

My own sex, I hope, will excuse me, if I treat them like rational creatures, instead of flattering their *fascinating* graces, and viewing them as if they were in a state of perpetual childhood, unable to stand alone. I earnestly wish to point out in what true dignity and human happiness consists. I wish to persuade women to endeavour to acquire strength, both of mind and body, and to convince them that the soft phrases, susceptibility of heart, delicacy of sentiment, and refinement of taste, are almost synonymous with epithets of weakness, and that those beings who are only the objects of pity, and that kind of love which has been termed its sister, will soon become objects of contempt.

Dismissing, then, those pretty feminine phrases, which the men condescendingly use to soften our slavish dependence, and despising that weak elegancy of mind, exquisite sensibility, and sweet docility of manners, supposed to be the sexual characteristics of the weaker vessel, I wish to show that elegance is inferior to virtue, that the first object of laudable ambition is to obtain a character as a human being, regardless of the distinction of sex, and that secondary views should be brought to this simple touchstone. . . .

Contending for the rights of woman, my main argument is built on this simple principle, that if she be not prepared by education to become the companion of man, she will stop the progress of knowledge and virtue; for truth must be common to all, or it will be inefficacious with respect to its influence on general practice. And how can woman be expected to co-operate unless she knows why she ought to be virtuous? unless freedom strengthens her reason till she comprehends her duty, and see in what manner it is connected with her real good. If children are to be educated to understand the true principle of patriotism, their mother must be a patriot; and the love of mankind, from which an orderly train of virtues spring, can only be produced by considering the moral and civil interest of mankind; but the education and situation of woman at present shuts her out from such investigations. . . .

Consequently, the most perfect education, in my opinion, is such an exercise of the understanding as is best calculated to strengthen the body and form the heart. Or, in other words, to enable the individual to attain such habits of virtue as will render it independent. In fact, it is a farce to call any being virtuous whose virtues do not result from the exercise of its own reason. This was Rousseau's opinion respecting men; I extend it to women, and confidently assert that they have been drawn out of their sphere by false refinement, and not by an endeavour to acquire masculine qualities. . . .

Women are therefore to be considered either as moral beings, or so weak that they must be entirely subjected to the superior faculties of men.

Let us examine this question. Rousseau declares that a woman should never for a moment feel herself independent, that she should be governed by fear to exercise her *natural* cunning, and made a coquettish slave in order to render her a more alluring object of desire, a *sweeter* companion to man, whenever he chose to relax himself. He carries the arguments, which he pretends to draw from the indications of nature, still further, and insinuates that truth and fortitude, the corner-stones of all human virtue, should be cultivated with certain restrictions, because, with respect to the female character, obedience is the grand lesson which ought to be impressed with unrelenting rigour.

What nonsense! When will a great man arise with sufficient strength of mind to puff away the fumes which pride and sensuality have thus spread over the subject? If women are by nature inferior to men, their virtues must be the same in quality, if not in degree, or virtue is a relative idea; consequently their conduct should be founded on the same principles, and have the same aim.

Connected with man as daughters, wives, and mothers, their moral character may be estimated by their manner of fulfilling those simple duties; but the end, the grand end, of their exertions should be to unfold their own faculties, and acquire the dignity of conscious virtue. They may try to render their road pleasant; but ought never to forget, in common with man, that life yields not the felicity which can satisfy an immortal soul. I do not mean to insinuate that either sex should be so lost in abstract reflections or distant views as to forget the affections and duties that lie before them, and are, in truth, the means appointed to produce the fruit of life; on the contrary, I would

warmly recommend them, even while I assert that they afford most satisfaction when they are considered in their true sober light.

Probably the prevailing opinion that woman was created for man, may have taken its rise from Moses' poetical story; yet as very few, it is presumed, who have bestowed any serious thought on the subject ever supposed that Eve was, literally speaking, one of Adam's ribs, the deduction must be allowed to fall to the ground, or only be so far admitted as it proves that man, from the remotest antiquity, found it convenient to exert his strength to subjugate his companion, and his invention to show that she ought to have her neck bent under the yoke, because the whole creation was only created for his convenience or pleasure. . . .

I wish to sum up what I have said in a few words, for I here throw down my gauntlet, and deny the existence of sexual virtues, not excepting modesty. For man and woman, truth, if I understand the meaning of the word, must be the same; yet the fanciful female character, so prettily drawn by poets and novelists, demanding the sacrifice of truth and sincerity, virtue becomes a relative idea, having no other foundation than utility, and of that utility men pretend arbitrarily to judge, shaping it to their own convenience.

Women, I allow, may have different duties to fulfil; but they are *human* duties, and the principles that should regulate the discharge of them, I sturdily maintain, must be the same.

To become respectable, the exercise of their understanding is necessary, there is no other foundation for independence of character; I mean explicitly to say that they must only bow to the authority of reason, instead of being the *modest* slaves of opinion. . . .

As the care of children in their infancy is one of the grand duties annexed to the female character by nature, this duty would afford many forcible arguments for strengthening the female understanding, if it were properly considered.

The formation of the mind must be begun very early, and the temper, in particular, requires the most judicious attention—an attention which woman cannot pay who only love their children because they are their children, and seek no further for the foundation of their duty, than in the feelings of the moment. It is this want of reason in their affections which makes women so often run into extremes, and either be the most fond or most careless and unnatural mothers.

To be a good mother, a woman must have sense, and that independence of mind which few women possess who are taught to depend entirely on their husbands. Meek wives are, in general, foolish mothers; wanting their children to love them best, and take their part, in secret, against the father, who is held up as a scarecrow. When chastisement is necessary, though they have offended the mother, the father must inflict the punishment; he must be the judge in all disputes; but I shall more fully discuss this subject when I treat of private education. I now only mean to insist, that unless the understanding of woman be enlarged, and her character rendered more firm, by being allowed to govern her own conduct, she will never have sufficient sense or command of temper to manage her children properly. . . .

To render mankind more virtuous, and happier of course, both sexes must act from the same principle; but how can that be expected when only one is allowed to see the reasonableness of it? To render also the social compact truly equitable, and in order to spread those enlightening principles, which alone can ameliorate the fate of man, women must be allowed to found their virtue on knowledge, which is scarcely possible unless they be educated by the same pursuits as men. For they are now made so inferior by ignorance and low desires, as not to deserve to be ranked with them; or, by the serpentine wrigglings of cunning, they mount the tree of knowledge, and only acquire sufficient to lead men astray.

It is plain from the history of all nations, that women cannot be confined to merely domestic pursuits, for they will not fulfil family duties, unless their minds take a wider range, and whilst they are kept in ignorance they become in the same proportion the slaves of pleasure as they are the slaves of man. Nor can they be shut out of great enterprises, though the narrowness of their minds often make them mar, what they are unable to comprehend.

. . . [The] weakness that makes woman depend on man for a subsistence, produces a kind of cattish affection, which leads a wife to purr about her husband as she would about any man who fed and caressed her.

Men are, however, often gratified by this kind of fondness, which is confined in a beastly manner to themselves; but should they ever become more virtuous, they will wish to converse at their fireside with a friend after they cease to play with a mistress.

Besides, understanding is necessary to give variety and interest to sensual enjoyments, for low indeed in the intellectual scale is the mind that can continue to love when neither virtue nor sense give a human appearance to an animal appetite. But sense will always preponderate; and if women be not, in general, brought more on a level with men, some superior women like the Greek courtesans, will assemble the men of abilities around them, and draw from their families many citizens, who would have stayed at home had their wives had more sense, or the graces which result from the exercise of the understanding, and fancy, the legitimate parents of taste. A woman of talents, if she be not absolutely ugly, will always obtain great power—raised by the weakness of her sex; and in proportion as men acquire virtue and delicacy, by the exertion of reason, they will look for both in women, but they can only acquire them in the same way that men do. . . .

Asserting the rights which women in common with men ought to contend for, I have not attempted to extenuate their faults; but to prove them to be the natural consequence of their education and station in society. If so, it is reasonable to suppose that they will change their character, and correct their vices and follies, when they are allowed to be free in a physical, moral, and civil sense.

Let woman share the rights, and she will emulate the virtues of man; for she must grow more perfect when emancipated, or justify the authority that chains such a weak being to her duty. If the latter, it will be expedient to open a fresh trade with Russia for whips: a present which a father should always make to his son-in-law on his wedding day, that a husband may keep his whole family in order by the same means; and without any violation of justice reign, wielding this sceptre, sole master of his house, because he is the only thing in it who has reason:—the divine, indefeasible earthly sovereignty breathed into man by the Master of the universe. Allowing this position, women have not any inherent rights to claim; and, by the same rule, their duties vanish, for rights and duties are inseparable.

Be just then, O ye men of understanding; and mark not more severely what women do amiss than the vicious tricks of the horse or the ass for whom ye provide provender—and allow her the privileges of ignorance, to whom ye deny the rights of reason, or ye will be worse than Egyptian task-masters, expecting virtue where Nature has not given understanding.

QUESTIONS FOR ANALYSIS

1. According to Thomas, what was wrong with gender relations in modern society? What does he hold up as ideal female behavior, and what does he anticipate would be the effects of women behaving as he suggests?

2. What does d'Epinay see as most open to criticism in Thomas's work? To what does she attribute the differences between men and women? Why is she so angry, when Thomas seeks only to praise and glorify women?

3. By contrast, what does Rousseau see as the source of differences between men and women? In his opinion, how should these shape a woman's education and a man's choice of a wife?

4. What does Wollstonecraft view as the chief weaknesses in Rousseau's ideas about gender differences? What does she see as the reasons for women's secondary status? What would be the benefits of greater gender egalitarianism?

5. Most sections of this chapter include materials by authors who supported and authors who opposed women's inferiority and dependence. How do the authors' justifications for their ideas change over the three centuries covered in this chapter? Pushing these comparisons back somewhat further, how would you compare Mary Wollstonecraft's arguments in favor of women with those of Christine de Pizan from Chapter 5?

The French Revolution, Industrialization, and Midcentury Working-Class Movements

T he French Revolution ushered in the era of modern, republican politics and transformed the political and geographical contours of Europe. Although denied full political rights, such as the right to vote, women capaciously exercised the new political rights accorded all citizens which included the rights of petition and assembly. They also gained such important civil rights as the right to divorce and an equal share of inheritance. Following the declaration of war in April 1792, women participated in the first homefront mobilization of a modern republic. The war dramatically reoriented the direction of the French Republic as it drained resources intended for societal reform, including a new, secular system of health care and welfare. At the same time, the exigencies of war strengthened the popular movement of the *sans-culottes* which demanded that the National Convention assume ever more radical economic, social, and political measures to defend the Republic. The Thermidorian coup against Robespierre and the Jacobins in the National Convention on 9 Thermidor Year II (July 27, 1794) ended the radical phase of the Revolution. The last of the popular protests took place in the spring of 1795 and failed to reverse the government's decision to eliminate price controls. Despite its short-lived nature (1789–1799), the French Revolution provided an essential reference point for the radical social and feminist movements that emerged in the nineteenth century.

In England, the Industrial Revolution served as the economic counterpart to the French Revolution. "All that is solid melts into air," wrote Marx of the "creative destruction" intrinsic to capitalism. The growth of mechanization and the factory system dramatically altered family as well as economic relations. At first, whole families were hired in the new factories, but manufacturers soon favored the cheap labor of women and children. Government commissions as well as a variety of social critics carried out investigations into the living and working conditions of the working class. The horrors of industrialization, seen as especially harmful to women, led to demands for the prohibition of married women from factories and the "male breadwinner wage" by which a male worker could support his entire family. Such a wage never fully materialized, however, and married women were often relegated to the sweated system of labor in which they worked at home for subcontractors. This form of work actually expanded along with the factory system, as it provided a flexible pool of superexploited labor which could be expanded and contracted to meet the vagaries of industry. Women also continued to work in such traditional areas as domestic service.

Industrialization generated new forms of working-class organization and political movements. In the midnineteenth century, various forms of political formations existed in Europe, ranging from a parliamentary monarchy with a limited franchise in England to absolutism in czarist Russia. Germany did not even exist as a nation, but rather consisted of hundreds of principalities in which a common language was spoken. From Chartism in England to the Revolutions of 1848 in Paris and Berlin, workers' movements at this time shared these common aims: to extend or secure male suffrage and/or to create new forms of government in which they could best advance their specific class aims. Through their participation in these movements, women challenged prevailing notions of political democracy and advanced new visions of social and economic organization.

THE FRENCH REVOLUTION

The French Revolution created unprecedented opportunities for women even though it did not accord them full political rights. The Constitution of 1791 divided the population into active and passive citizens. About a third of all men without sufficient income and all women were denied active citizenship; specifically, the right to select and vote on candidates for electoral office. While income qualifications were eliminated for men in 1793, women did not obtain the vote until the post–World War II period.[1] But they did gain important legal rights

---

[1] Women in France first voted in municipal elections in 1945 and in national elections in 1946. On the campaign for female suffrage, see Source 128 in Chapter 10.

and benefited socially and economically from revolutionary family, health care, and social welfare legislation. Women took advantage of new political channels to participate actively in the popular movement, including the popular protests that compelled the National Convention to institute the Terror whose economic aspects included price controls on basic necessities and the death penalty for hoarders. Overall, the possibilities for women's organizing followed the trajectory of the popular *sans-culotte* movement which was suppressed in 1795.

93. A *Cahier* from Women of the Third Estate

In anticipation of the convening of the Estates-General in 1789, and at the request of the monarchy, all three estates (the clergy, the nobility, and the Third Estate) drew up *cahiers de doléances,* or notebooks of grievances. Typically, the *cahiers* combined traditional and new demands, as in this *cahier* of women merchant flower sellers in Paris.

The liberty accorded to all citizens to expose before the nation's representatives the abuses which assail them from all sides is without doubt a sure sign of impending reform.

In this confidence, the *marchandes bouquetières*[1] formerly comprising the *communauté des maîtresses bouquetières et marchandes chapelières en fleurs*[2] of the City and Faubourgs of Paris venture to address themselves to you, our Lords (the Estates General). It is not for ordinary abuses they ask correction. It is their trade, their whole existence, which they lost by the doubtless involuntary error of one of His Majesty's former ministers, which they reclaim at this time.

Even before forming a *communauté* and constituting themselves as a *corps de jurande,*[3] the *marchandes bouquetières* already had statutes which they enforced among themselves under the authority and jurisdiction of the *prévôt*[4] of Paris or of his lieutenant-general of police.

This kind of *jurande,* as imperfect as it was, undoubtedly prevented abuses. However, it was soon realized that these precautions would not be adequate for maintaining order unless the responsibility for internal discipline was placed in the hands of the *bouquetières.* Consequently, they were established as a *communauté.*

In 1735, thanks to the kindheartedness of Louis XV, the *maîtresses bouquetières* obtained verification of their *communauté* and new regulations which were ordered executed by *lettres-patentes*[5] of November 26, 1736, [and] registered in Parlement on December 18, 1737.

The *maîtresses bouquetières* who had undergone a three-year apprenticeship and who had paid substantial fees for the *maîtrise*[6] were peacefully

[1]*Marchandes bouquetières:* merchant flower sellers, the authors of this *cahier.*

[2]*Communauté des maîtresses bouquetières et marchandes chapelières en fleurs:* a corporation or guild of women flower sellers and sellers of flower decorated hats. The corporations were an essential aspect of the Old Regime organization of production and retailing.

[3]*Corps de jurande:* usually referred to simply as the *jurande;* the official, internal governing body of a corporation.

[4]*Prévôt:* the functional equivalent of a mayor.

[5]*Lettres-patentes:* royal orders that conveyed legislation to the Parlement for registration; in this case, they refer to the royal orders for the establishment of a guild.

[6]*Maîtrise:* the status of master in a corporation which an apprentice aspired to attain after several years of formal training and several more years as a probationary journeyman.

plying their trade when they found themselves suddenly deprived of it by the suppression of their *communauté.*

The ministers' oppressive silence, the menaces with which they were armed—doubtless without their knowledge—and the repellent coldness of their clerks stifled their [the *bouquetières'*] cries at that time, but today, when a better order is in the making, they hope for everything from the Prince's justice and from that of the Estates.

In the eyes of anyone else except the enlightened representatives of the nation, the petitioners' reclamation perhaps would not seem worth the sacrifice of some of the precious time being allocated to the examination of the major interests which will occupy them. But they [the *bouquetières*] need not fear being rebuffed. They know, these worthy representatives of the French people, that they have a greater responsibility towards the most indigent class in particular. The more unfortunate a man is, the more sacred are his rights, especially where his trade constitutes all the wealth he has. Its protection cannot fail to be of concern to the nation's deputies and to hold their attention.

Besides, the petitioners' reclamation is bound up with an important issue which is submitted to the Estates for a decision: that of knowing whether it would be useful or not to allow every individual an indeterminate liberty to engage in all kinds of commerce.

Let the petitioners be allowed to venture some reflections here which may perhaps pave the way for a decision on this capital issue.

In the regulation of commerce and trade, two things should be taken into consideration. The entire *corps*[7] should be organized so that each individual who commits himself to a trade can earn a living for himself and his children. However, access to trades, and above all to those which are designated in particular for the class with meager means, should not be made

so difficult that the industry of the poor is discouraged and emulation stifled.

This combination, on which the general happiness and the prosperity of commerce rest, necessarily would be destroyed by indeterminate liberty.

Too great a facility allowed under the edicts of 1776 and 1777[8] has made all the professions only too acutely aware of the inconveniences of this liberty. A large number of merchants does not produce—far from it—the salutary effect which it would appear must have been expected from competition. Because the number of consumers does not increase in proportion to the number of artisans, the latter necessarily hurt one another; they can obtain the buyer's preference only by lowering the price of their goods. Distress and the need to sell at a low price force them to turn out shoddy [goods] or to feign ruin, which brings mortal blows to commerce.

The more restricted a trade is, [and] the fewer resources it holds out, the more important it is to reduce the number of competitors. Unfortunately, since the suppression of their *communauté,* that is what the petitioners have found out. Their trade, although limited, offered them, before this suppression, a means for making a living and raising their children. Today, when everybody can sell flowers and arrange bouquets, their modest gains are divided up to the point that they can no longer make enough to live on. The lure of this gain, restricted though it may be, and even more, a strong propensity for laziness, nonetheless determined a crowd of young people of the feminine sex to enter the petitioners' trade; and because their trade cannot feed them, they look to libertinage and the most shameful debauchery for the means they are lacking. The petitioners' cause is also that of morals.

---

[7]*Corps:* refers to the corporate or guild system.

[8]These royal edicts, which first abolished the guilds and then recreated some guilds in new configurations, met with concerted opposition. Guild members accustomed to the old guild structures generally felt less protected under the new system. Some guilds, however, like the *communauté des maîtresses bouquetières et marchandes chapelières en fleurs,* were not recreated at all, as this *cahier* points out.

The flower trade in itself does not suffer any the less from this anarchy. There is no more policing of the market. All the unprincipled girls whom no law, no decency can restrain, throw themselves on the goods which the flower growers bring, pillage them or crush them, [and] arbitrarily set the price. And only too often it happens that because of this, the flower growers lose the precious fruit of their hard work. It's because of this [situation] that brawls arise during which these hawkers are backed up by soldiers and by disreputable characters. It is because of this that you get this abuse which set in when liberty was accorded in the trade of the *marchandes bouquetières:* girls who are not merchants have thought up the idea of fastening orange blossoms with pins or artificial flowers which they attach to the aforementioned branches [or] they attach several carnations together and put them on cards and sell them as a single one.

All these abuses, which can only be ended with the reestablishment of the *communauté,* are contrary to order and discouraging for the growers, who devote themselves to the cultivation of flowers. The inconveniences produced by these disorders have been felt in all times, and it is because of them that in 1731 the new statutes for the *communauté des bouquetières maîtresses* were confirmed.

Therefore, the petitioners dare expect from the equity of the Estates the reestablishment of their corporation. Besides, this is a justice that is due them since they paid the King considerable sums for enjoying the advantages of their trade, advantages they are deprived of by too much competition and the disorders it brings with it.

The petitioners ask that a police force be provided to stop large numbers of people who, claiming free trade, hang around the market every night (and especially the nights before holidays for patron saints), waiting for the flower growers and intending to abuse their good faith. They lay waste to the markets, either arbitrarily or by leave of the authorities, before the usual hour for selling the above-mentioned flowers. These same people have strayed from honesty to the point of going off into the countryside, where they lay waste to the flower beds and orange groves belonging to seigneurs and private homes.

The former *journalières bouquetières* are reduced to the most extreme poverty by the different kinds of people who have taken to selling flowers. Since liberty was declared in this branch of trade, the petitioners have been pained to see mothers of families without work, [women] whom they used to support with wages of thirty *sols* a day, plus food; four *livres,*[9] ten *sols* for two and one-half days' work during the minor holidays; and nine *livres,* also for two and one-half days, during the principal holidays—which allowed them to raise their little families.

Liberty, which all orders of the State reclaim, cannot constitute an obstacle to the petitioners' request. Liberty is an enemy of license, and citizens are free so long as they obey only the laws which they impose upon themselves.

The petitioners entrust their just reclamation to the deputies of the Third Estate above all. They [the deputies of the Third Estate] more than the others [that is, more than the deputies of the First and Second Estates] are their [the petitioners'] representatives, their friends, their brothers, and it is up to them to plead the cause of indigent people.

Based on these considerations, let an irrevocable order be promulgated, with interdictions, from His Majesty to the effect that under no pretext whatsoever shall a *marchande bouquetière* or other person be permitted to buy or sell flowers before 6 A.M. between [the religious holidays of] Pâques and Saint-Martin, and also before 6 A.M. between Saint-Martin and Pâques.

The petitioners will not cease addressing vows to the heavens for the protection and prosperity of the nation's representatives.

Signed: the above-mentioned *marchandes,* represented by Madame Marlé, syndic[10] of the corporation.

---

[9]*Sols* and *livres* (pounds): the prevailing units of currency. There were 20 *sols* to one *livre.*

[10]*syndic:* a warden and elected leader of the corporation or guild.

94. Olympe de Gouges "Declaration of the Rights of Woman"

In her 1791 "Declaration of the Rights of Woman and the Female Citizen," dedicated to Marie Antoinette, Olympe de Gouges offered a feminist version of the "Declaration of the Rights of Man and Citizen" which included a model "Social Contract Between Man and Woman." Olympe de Gouges openly opposed the execution of the King and Queen and was guillotined for her royalist views in November 1793.

The Rights of Woman

Man, are you capable of being just? It is a woman who poses the question; you will not deprive her of that right at least. Tell me, what gives you sovereign empire to oppress my sex? Your strength? Your talents? Observe the Creator in his wisdom; survey in all her grandeur that nature with whom you seem to want to be in harmony, and give me, if you dare, an example of this tyrannical empire. Go back to animals, consult the elements, study plants, finally glance at all the modifications of organic matter, and surrender to the evidence when I offer you the means; search, probe, and distinguish, if you can, the sexes in the administration of nature. Everywhere you will find them mingled; everywhere they cooperate in harmonious togetherness in this immortal masterpiece.

Man alone has raised his exceptional circumstances to a principle. Bizarre, blind, bloated with science and degenerated—in a century of enlightenment and wisdom—into the crassest ignorance, he wants to command as a despot a sex which is in full possession of its intellectual faculties; he pretends to enjoy the Revolution and to claim his rights to equality in order to say nothing more about it.

Declaration of the Rights of Woman and the Female Citizen

For the National Assembly to decree in its last sessions, or in those of the next legislature:

Preamble

Mothers, daughters, sisters [and] representatives of the nation demand to be constituted into a national assembly. Believing that ignorance, omission, or scorn for the rights of woman are the only causes of public misfortunes and of the corruption of governments, [the women] have resolved to set forth in a solemn declaration the natural, inalienable, and sacred rights of woman in order that this declaration, constantly exposed before all the members of the society, will ceaselessly remind them of their rights and duties; in order that the authoritative acts of women and the authoritative acts of men may be at any moment compared with and respectful of the purpose of all political institutions; and in order that citizens' demands, henceforth based on simple and incontestable principles, will always support the constitution, good morals, and the happiness of all.

Consequently, the sex that is as superior in beauty as it is in courage during the sufferings of maternity recognizes and declares in the presence and under the auspices of the Supreme Being, the following Rights of Woman and of Female Citizens.

Article I

Woman is born free and lives equal to man in her rights. Social distinctions can be based only on the common utility.

Article II

The purpose of any political association is the conservation of the natural and imprescriptible rights of woman and man; these rights are liberty, property, security, and especially resistance to oppression.

Article III

The principle of all sovereignty rests essentially with the nation, which is nothing but the union of woman and man; no body and no individual can exercise any authority which does not come expressly from it [the nation].

Article IV

Liberty and justice consist of restoring all that belongs to others; thus, the only limits on the exercise of the natural rights of woman are perpetual male tyranny; these limits are to be reformed by the laws of nature and reason.

Article V

Laws of nature and reason proscribe all acts harmful to society; everything which is not prohibited by these wise and divine laws cannot be prevented, and no one can be constrained to do what they do not command.

Article VI

The law must be the expression of the general will; all female and male citizens must contribute either personally or through their representatives to its formation; it must be the same for all: male and female citizens, being equal in the eyes of the law, must be equally admitted to all honors, positions, and public employment according to their capacity and without other distinctions besides those of their virtues and talents.

Article VII

No woman is an exception; she is accused, arrested, and detained in cases determined by law. Women, like men, obey this rigorous law.

Article VIII

The law must establish only those penalties that are strictly and obviously necessary, and no one can be punished except by virtue of a law established and promulgated to the crime and legally applicable to women.

Article IX

Once any woman is declared guilty, complete rigor is [to be] exercised by the law.

Article X

No one is to be disquieted for his very basic opinions; woman has the right to mount the scaffold; she must equally have the right to mount the rostrum, provided that her demonstrations do not disturb the legally established public order.

Article XI

The free communication of thoughts and opinions is one of the most precious rights of woman, since that liberty assures the recognition of children by their fathers. Any female citizen thus may say freely, I am the mother of a child which belongs to you, without being forced by a barbarous prejudice to hide the truth; [an exception may be made] to respond to the abuse of this liberty in cases determined by the law.

Article XII

The guarantee of the rights of woman and the female citizen implies a major benefit; this guarantee must be instituted for the advantage of all, and not for the particular benefit of those to whom it is entrusted.

Article XIII

For the support of the public force and the expenses of administration, the contributions of woman and man are equal; she shares all the duties [*corvées*] and all the painful tasks; therefore, she must have the same share in the distribution of positions, employment, offices, honors, and jobs [*industrie*].

Article XIV

Female and male citizens have the right to verify, either by themselves or through their representatives, the necessity of the public contribution. This can only apply to women if they

The French Revolution   263

are granted an equal share, not only of wealth, but also of public administration, and in the determination of the proportion, the base, the collection, and the duration of the tax.

Article XV

The collectivity of women, joined for tax purposes to the aggregate of men, has the right to demand an accounting of his administration from any public agent.

Article XVI

No society has a constitution without the guarantee of rights and the separation of powers; the constitution is null if the majority of individuals comprising the nation have not cooperated in drafting it.

Article XVII

Property belongs to both sexes whether united or separate; for each it is an inviolable and sacred right; no one can be deprived of it, since it is the true patrimony of nature, unless the legally determined public need obviously dictates it, and then only with a just and prior indemnity....

Form for a Social Contract Between Man and Woman

We, _____ and _____, moved by our own will, unite ourselves for the duration of our lives, and for the duration of our mutual inclinations, under the following conditions: We intend and wish to make our wealth communal, meanwhile reserving to ourselves the right to divide it in favor of our children and of those toward whom we might have a particular inclination, mutually recognizing that our property belongs directly to our children, from whatever bed they come, and that all of them without distinction have the right to bear the name of the fathers and mothers who have acknowledged them, and we are charged to subscribe to the law which punishes the renunciation of one's own blood. We likewise obligate ourselves, in case of separation, to divide our wealth and to set aside in advance the portion the law indicates for our children, and in the event of a perfect union, the one who dies will divest himself of half his property in his children's favor, and if one dies childless, the survivor will inherit by right, unless the dying person has disposed of half the common property in favor of one whom he judged deserving.

That is approximately the formula for the marriage act I propose for execution. Upon reading this strange document, I see rising up against me the hypocrites, the prudes, the clergy, and the whole infernal sequence. But how it [my proposal] offers to the wise the moral means of achieving the perfection of a happy government! I am going to give in a few words the physical proof of it. The rich, childless Epicurean finds it very good to go to his poor neighbor to augment his family. When there is a law authorizing a poor man's wife to have a rich one adopt their children, the bonds of society will be strengthened and morals will be purer. This law will perhaps save the community's wealth and hold back the disorder which drives so many victims to the almshouses of shame, to a low station, and into degenerate human principles where nature has groaned for so long. May the detractors of wise philosophy then cease to cry out against primitive morals, or may they lose their point in the source of their citations.[1]

Moreover, I would like a law which would assist widows and young girls deceived by the false promises of a man to whom they were attached; I would like, I say, this law to force an inconstant man to hold to his obligations or at least [to pay] an indemnity equal to his wealth. Again, I would like this law to be rigorous against women, at least those who have the effrontery to have recourse to a law which they themselves had violated by their misconduct, if proof of that were given. At the same time, as I showed in *Le Bonheur primitif de l'homme,* in 1788, that prostitutes

---

[1] Abraham had some very legitimate children by Agar, the servant of his wife.

should be placed in designated quarters.[2] It is not prostitutes who contribute the most to the depravity of morals, it is the women of society. In regenerating the latter, the former are changed.

[2]See Olympe de Gouges, *Le Bonheur primitif de l'homme, ou les Rêveries patriotiques* (Amsterdam and Paris, 1789). She wrote numerous plays and pamphlets on a variety of topics, including the abolition of slavery.—Ed.

This link of fraternal union will first bring disorder, but in consequence it will produce at the end a perfect harmony.

I offer a foolproof way to elevate the soul of women; it is to join them to all the activities of man; if man persists in finding this way impractical, let him share his fortune with woman, not at his caprice, but by the wisdom of laws. Prejudice falls, morals are purified, and nature regains all her rights.

95. Women and Revolutionary Divorce Legislation

Revolutionary family legislation aimed to eliminate the patriarchal legal structure of the Old Regime family (which replicated the relationship of the king to his subjects) and to create an egalitarian family as befit the new Republican government. The new laws eliminated primogeniture, the inheritance of most property by the first-born son, and permitted younger daughters as well as sons to inherit. Legislation also eliminated the legal distinction between "legitimate" and "illegitimage" children, all of whom were now declared children of the Republic. The divorce law of September 1792 gave both women and men the right to initiate divorce and remained in force until the Napoleonic Code of 1804,[1] which made divorce much more difficult to obtain, especially for women. Women's concerns and support for the new divorce legislation are illustrated in this letter addressed to the president of the Republic in September, the same month as the legislation, by *Femme* Berlin, a woman who was separated from her husband for several years, unable to gain access to her dowry because of his legal obstructions, and the sole provider for her three children.

Monsieur, the President of the French Republic:

The National Assembly took up the issue of divorce; several articles have already been decreed, but others were not yet legislated when

the Convention replaced the Assembly.[1] I believe it is urgent, as did the Assembly, to com-

[1]The National Convention first convened in Paris on September 20, 1792.

[1]For an excerpt from the Napoleonic Code, see Source 98 in this chapter.

From Archives Nationales, D III 361. Contributed by Suzanne Desan and translated by Lisa DiCaprio. Used by permission.

plete this project. The fate of children has not yet been legislated.[2]

Allow a woman who, for the more than nine years since she was forced to flee her husband's house (not counting the seven years that she lived with him), has been a prey to all of his clever tricks, allow me to plead with you to complete this divorce legislation.

Separated by a decree by the former Parlement in 1786, I could not obtain from my litigious and quarrelsome husband, who is also of bad faith, the restitution of a considerable dowry whose liquidation he has obstructed with all of the ability of a specialist of deceit.

I can only ask for a divorce after you have decreed on the fate of children, for I have three who are as dear to me as they are of little interest to their father, and on me alone falls the responsibility for their education and financial support, but this is not my main complaint. I would have presented myself with them to the Assembly, if I had been able to overcome the timidity that is natural for women, for we are either the object of honest men's compassion or of libertines' jokes. I would have been able to request from you the rest and peace that was taken away from me by a despot who only desired my fortune, all the more so since his own was reduced to nothing. But I loved him and he loved my possessions, an additional reason for not reconciling with him. Also, since the day after our marriage, he dropped the mask that he wore to deceive me and I have been his victim for the past sixteen years.

I have related these details, Monsieur President, because I am convinced that the Convention intends to recognize that a woman is someone and to remind certain men (what they have forgotten) that a woman is not their slave, but rather their companion, and that she must not be the victim of their moods and bad treatment.

I dare say that by legalizing divorce you will regenerate morals, which will affect future generations even more strongly than our own, for a man knowing that his wife can ask for a divorce, will not treat her like a slave, and will not think of himself as a *grand seigneur* [great lord]. Likewise, if a woman has bad morals, she will think that a man can divorce her and this certainty will prevent promiscuity among those who retain such tendencies.

I have been told, Monsieur President, that you are a good husband and father, you are consequently endowed with a sensitive heart, which gives me hope that you will consider my request, especially as a great number of unfortunate women are impatiently waiting, like me, for you to decide on their fate, so that those who have spent their springtime in tears, can hope that their three remaining seasons will be less stormy than the first.

I have expressed myself with the frankness of a *citoyenne*[3] that you have freed and I have cast aside the base flatteries which one used to address important individuals in the Old Regime. I will only say to you that the Convention is engraved on the hearts of good patriots as the liberator of France and from the Convention we expect a solid peace and happiness.

> Berlin
> Rue Geoffroy L'Asnier No. 36

Paris, 27 September
Year I of the Republic

P.S. I am only including my address so that you will not regard me as an imposter, but I wish that in reading my petition you will not disclose my identity for I do not like publicity.

---

[2] Writing in the same month as divorce was being legislated, *Femme* Berlin was apparently unaware that the most recently decreed aspects of the law did, in fact, concern child custody and support. The law stipulated that in cases of divorce by mutual consent or incompatibility, children under the age of seven were to remain with their mother while those above this age were to live with the parent of the same sex. In contested divorces, the family assembly was to determine the children's custody. In divorces granted to those who had already separated under the Old Regime, as in *Femme* Berlin's situation, children of all ages generally were to remain with their current parent but custody was sometimes determined by the family assembly. In all instances, however, the parent without custody was to contribute financial support to the children to the best of his or her ability. *Femme* Berlin's letter was referred to the Committee on Legislation.

[3] *Citoyenne*, a female citizen.

96. Petition of the Society of Revolutionary Republican Women

With few exceptions, the popular societies denied women the right to participate as active members, but Jacobin women formed hundreds of their own clubs throughout France. These clubs made a significant contribution to the homefront war mobilization. In May 1793, lower-middle-class women in Paris formed the Society of Revolutionary Republican Women which combined a militant defense of the Republic with demands specific to women. The Society closely allied with groups to the left of the Jacobins, such as the Cordeliers Club, as illustrated in this petition presented in a joint deputation to a Jacobin Society meeting. All women's clubs, including the Society, were closed down by the government in October 1793, the month of Marie Antoinette's execution. The Jacobin repression of the Left opposition, including the Cordeliers, soon followed.

Session of Sunday, May 19, 1793.

...A deputation from the Cordeliers Club and the *citoyennes*[1] of the Revolutionary Society of Women is admitted. The *orator* announces a petition drawn up by the members of these two societies joined together and reads this petition, the substance of which is as follows:

"Representatives of the people, the country is in the most imminent danger; if you want to save it, the most energetic measures must be taken...." (*Noise.*)

"I demand," the orator cries out, "the fullest attention."

Calm is restored.

He continues: If not, the people will save themselves. You are not unaware that the conspirators are waiting only the departure of the volunteers, who are going to fight our enemies in the Vendée, to immolate the patriots and everything they cherish most. To prevent the execution of these horrible projects, hasten to decree that suspect men will be placed under arrest immediately, that revolutionary tribunals will be set up in all the Departments and in the Sections of Paris.

For a long while the Brissots, the Gaudets, the Vergniauds, the Gensonnés, the Buzots, the Barbarouxes, etc., have been pointed out as being the general staff of the counterrevolutionary army. Why do you hesitate to issue charges against them? Criminals are not sacred anywhere.

Legislators, you cannot refuse the French people this great act of justice. That would be to declare yourselves their accomplices; that would be to prove that several among you fear the light which the trial investigation of these suspect members would cause to flash.

We ask that you establish in every city revolutionary armies composed of *sans-culottes*, proportional in size to the population; that the army of Paris be increased to forty thousand men, paid at the expense of the rich at a rate of forty *sous* a day. We ask that in all public places workshops be set up where iron be converted into all kinds of weapons.

Legislators, strike out at the speculators, the hoarders, and the egotistical merchants. A horrible plot exists to cause the people to die of hunger by setting an enormous price on goods. At the head of this plot is the mercantile aristocracy of an insolent caste, which wants to assimilate itself to royalty and to hoard all riches by forcing up the price of goods of prime necessity

---

[1]*Citoyennes*, female citizens.

in order to satisfy its cupidity. Exterminate all these scoundrels; the Fatherland will be rich enough if it is left with the *sans-culottes* and their virtues. Legislators! Come to the aid of all unfortunate people. This is the call of nature; this is the vow of true patriots. Our heart is torn by the spectacle of public misery. Our intention is to raise men up again; we do not want a single unfortunate person in the Republic. Purify the Executive Council; expel a Gohier, a Garat, a Le Brun, etc.; renew the directory of the postal service and all corrupted administrations, etc.

A large number of people, the orator cries out, must bear this address to the Convention. What! Patriots are still sleeping and are busy with insignificant discussions while perfidious journals openly provoke the people! We will see whether our enemies will dare show themselves opposed to measures on which the happiness of a republic depends.

The President. The Society hears with the keenest satisfaction the accents of the most ardent patriotism; it will second your efforts with all its courage, for it has the same principles, and it has evinced the same opinion. Whatever the means and the efforts of our enemies, liberty will not perish because there will remain forever in the heart of Frenchmen this sentiment that insurrection is the ultimate reason of the people. (Applauded.)

97. Petition of Indigent Women Workers

In 1790, the National Assembly established a Committee on Mendicity to eliminate the causes of begging in France. Headed by the liberal noble La Rochefoucauld-Liancourt, the committee aimed to create a new centralized, secular system of social welfare. Liancourt declared the right of subsistence[1] to be a right of citizenship and work the preferred method of government assistance. Subsequently, the National Assembly established outdoor public workshops for men and two spinning workshops for women and children. While the men's workshops were closed down in 1791, those created for women remained in existence until 1795. At any given time, they employed some 2,500 workers to clean and card cotton and spin thread from cotton, hemp, and flax. During the war, the workshops were converted into centers of war production to provide thread for military uniforms and sailcloth. The women workers utilized the new political channels created by the Revolution to organize for the improvement of working conditions and wages. Their initial methods of protest consisted mainly of letters and petitions to the administrators of the workshops, the mayor of Paris, and officials of the welfare committees that existed in each of the forty-eight sections of Paris. Later, during the Jacobin period of the Revolution, the women expanded their activities to include protests to the National Convention, the Committee of Public Safety, and the popular societies. Finally, as indicated in this 1795 petition, the women opposed the suppression decree of the Thermidorian government[2] which denounced the workshops as centers of workers' organizing and stipulated that raw materials should be provided to women at home rather than in the spinning workshops.

---

[1] The right of individuals to be assured of sufficient means to obtain the necessities of life.

[2] The new government derived its name from the fact that Robespierre and the Jacobins in the National Convention were overthrown on 9 Thermidor Year II (July 27, 1794).

From Archives Nationales, F/15/3567. Translated by Lisa DiCaprio.

The Spinners of the *Atelier du Nord*[1]
To the *Citoyens*[2] of the Commission of Agriculture and the Arts[3]
19 Pluviôse [February 7, 1795][4]

Citoyens:

We are writing on behalf of the workers in the *atelier de filature du Nord* to appeal to your humanity to suspend your decree which stipulates that all workers must work at home and imposes requirements that they cannot fulfill.

We are convinced that when we have brought the specific details of our misfortune to your attention, that you will be persuaded to assist the mothers of six children, the wives of the defenders of the *patrie*,[5] [and] the infirm *citoyennes*[6] who have no means of subsistence other than earning a living in the workshops.

In order to convince you, *citoyens,* that it is impossible to carry out your decree, it will suffice for us to describe our unfortunate situation.

Your decree requires, we are told, that we must all work at home. How could you think, *citoyens,* that women compelled by indigence to work in the workshops would have lodgings that could contain a spinning wheel and all that is required for this work? No, *citoyens,* we all live in nooks which can scarcely contain a bed in poor condition or a bundle of straw; and almost all of us have children, therefore the impossibility of working at home is all the more evident. In addition, the majority of spinners of cotton do not know how to card, who then will card their cotton?[7] Another point, which is no less important, we are required to provide guarantors or a deposit in *assignats*[8] to cover the cost of the cotton. In the first place, most of us have lived in the past month or two in furnished rooms. It is certain that no landlord will guarantee a sum of 40 *francs* for women who do not possess a *sol.* We [point out] in addition that several of us have pawned our belongings in order to obtain this sum and still cannot attain the full amount.

Prior to appealing to you, *citoyens,* we presented all of these grievances to our director, who did not want to listen to them, and who refused to let us read your decree. Unfortunately for us, he is egotistical, inhumane, and vexatious, and long before our protests, he seemed to rejoice in our misfortune, and seemed angry that a poor female worker could earn enough to live on.

Citoyens, if we have complained to you about the suffering we have endured during the extreme cold of this winter, we would also like to thank you for the orders which you issued to pay us during those times when we lack work.[9]

Today, we are therefore appealing to your humanity and we are convinced that if we had a director who was more aware of the responsibil-

---

[1]The two workshops were called the *atelier du Midy* and the *atelier du Nord* based on their respective locations in central Paris and a northern *faubourg* (suburb) of the city.

[2]*Citoyen* refers to a male citizen while *citoyenne* refers to a female citizen. Revolutionary protocol eliminated Old Regime forms of salutation based on deference. In this petition, therefore, workers and administrators alike are referred to as citizens.

[3]At this time, the Commission of Agriculture and the Arts had assumed administrative responsibility for the spinning workshops.

[4]Only the day and month appear on the petition, but it must date from 1795 as this was the year of the suppression decree which is the topic of the petition.

[5]Defenders of the *patrie*: soldiers in the war declared by the French government in April, 1792. In this context, the defense of the *patrie* (fatherland) means the defense of the Revolution.

[6]In both workshops, cotton was cleaned by hundreds of women who were "infirm" in the sense that they were too old or disabled to be employed in private industry even if such work was available.

[7]Carding was a separate skill that made cotton suitable for spinning.

[8]*Assignats:* paper currency.

[9]The women lacked work because of shortages of raw materials that occurred frequently during the war.

ities of his position than is ours [and] who did not give false accounts which dispose you against us, that we would merit more of the consideration that our *état* [status], justice, and humanity demand.

We count on your sense of justice and you may be assured that we will always comply with your decrees because we are convinced that you do not wish us to suffer.

Femme Miroir, *Femme* Lesa, *Femme* Roucel, *Femme* Degodet, *Femme* Vigoreu, *Femme* Levasseur.[10]

---

[10]Only one of the spinners may have known how to write, as the names of all six signatories appear in the same rough handwriting. The petition itself is written in a secretarial hand which suggests that the spinners obtained assistance from an *écrivain public*, or public letter writer, a common practice at this time.

QUESTIONS FOR ANALYSIS

1. What are the grievances of the merchant flower sellers as expressed in their *cahier de doléance?* What solutions do they propose? How does the *cahier* combine old and new demands?

2. What rights does Olympe de Gouges enumerate for women in her "Declaration of the Rights of Woman and the Female Citizen"? On what basis does she claim these rights? How does her model "Social Contract Between Man and Woman" envision the transformation of marriage? What new laws does Olympe de Gouges propose to assist women?

3. What is the purpose of *Femme* Berlin's letter to the president of the French Republic on the subject of divorce? What is her appraisal of the new divorce law? How does she believe it will affect marriage in the future? Compare *Femme* Berlin's views on marriage and divorce with those of Olympe de Gouges.

4. How does the joint petition of the Society of Revolutionary Republican Women and the Cordeliers Club assess the political situation in France? What are the main points of the petition's critique of the policies of the National Convention? What measures does it call on the Convention to implement?

5. On what basis does the petition of the spinners call for the government workshops to remain in operation? What does the petition reveal about their living conditions? Compare the spinners' attitudes toward the director and administrators of the workshops. What is the sense of entitlement expressed by the spinners? How does their petition compare with the *cahier* of the women flower sellers in Source 93 of this chapter?

6. What do the documents in this section reveal about women's roles in the French Revolution and the new channels for their activism that were created by the revolutionary process?

THE NAPOLEONIC CODE AND THE UTOPIAN SOCIALIST CRITIQUE OF MARRIAGE AND THE FAMILY

The Napoleonic Code of 1804 restored the patriarchal family in France by denying women property rights and depriving them of the right to initiate divorce, which was now restricted to men. The Code became the focus of women utopian socialists who demanded a return to French Revolution legislation. In the process, they also critiqued women's unequal economic and political status in society. In addition to utilizing their right to present petitions to the National Assembly, women organized their own newspapers and journals to promote socialist feminist issues. French feminism thus emerged as a movement with a definite set of political beliefs with the French Revolution providing an important precedent for women's nineteenth-century militancy.

98. The Napoleonic Code

The Napoleonic Code (1804) established a uniform code of civil law throughout France. Elaborated over a period of three years, the Code comprises three books (two of which are on property), and contains 2,281 articles which regulate property ownership and acquisition, citizenship, and family relations. Favoring propertied men, the Code restored patriarchal authority in the family by overturning various aspects of revolutionary legislation on marriage and the family, including women's right to initiate divorce, and severely diminishing women's civil status. Women outside of France were also adversely affected by the Napoleonic Code which came into effect in European territories under French rule during the Napoleonic period. Reproduced here are the articles which comprise Chapter VI "Of the Respective Rights and Duties of Married Persons."

Book I: Of Persons

Title I: Of the Enjoyment and Deprivation of Civil Rights

Chapter VI: Of the Respective Rights and Duties of Married Persons

-212-

Married persons owe to each other fidelity, succor, assistance.

-213-

The husband owes protection to his wife, the wife obedience to her husband.

-214-

The wife is obliged to live with her husband, and to follow him to every place where he may judge it convenient to reside: the husband is obliged to receive her, and to furnish her with every necessity for the wants of life, according to his means and station.

-215-

The wife cannot plead in her own name, without the authority of her husband, even though she should be a public trader, or non-communicant, or separate in property.

-216-

The authority of the husband is not necessary when the wife is prosecuted in a criminal manner, or relating to police.

-217-

A wife, although non-communicant or separate in property, cannot give, pledge, or acquire by free or chargeable title, without the concurrence of her husband in the act, or his consent in writing.

-218-

If the husband refuses to authorize his wife to plead in her own name, the judge may give her authority.

-219-

If the husband refuses to authorize his wife to pass an act, the wife may cause her husband to be cited directly before the court of the first instance, of the circle of their common domicil[e], which may give or refuse its authority, after the husband shall have been heard, or duly summoned before the chamber of council.

-220-

The wife, if she is a public trader, may, without the authority of her husband, bind herself for that which concerns her trade; and in the said case she binds also her husband, if there be a community between them.

She is not reputed a public trader, if she merely retails goods in her husband's trade, but only when she carries on a separate business.

-221-

When the husband is subjected to a condemnation, carrying with it an afflictive or infamous punishment, although it may have been pronounced merely for contumacy, the wife, though of age, cannot, during the continuance of such punishment, plead in her own name or contract, until after authority given by the judge, who may in such case give his authority, without hearing or summoning the husband.

-222-

If the husband is interdicted or absent, the judge, on cognizance of the cause, may authorize his wife either to plead in her own name or to contract.

-223-

Every general authority, though stipulated by the contract of marriage, is invalid, except as respects the administration of the property of the wife.

-224-

If the husband is a minor, the authority of the judge is necessary for his wife, either to appear in court, or to contract.

-225-

A nullity, founded on defect of authority, can be opposed by the wife, the husband, or by their heirs.

-226-

The wife may make a will without the authority of her husband.

99. Suzanne Voilquin
Tribune des femmes

Suzanne Voilquin (1801–1877) was a writer, an activist in the utopian socialist Saint-Simonian movement in Paris in the 1830s, and a participant in the Revolution of 1848. As she recounts in her autobiography, *Souvenirs d'une fille du peuple* (1866; *Recollections of a Daughter of the People*), Voilquin's interest in social justice and politics was inspired by her father's participation in the French Revolution while her mother, a devout Catholic, imparted an interest in Christianity, an integral aspect of Saint-Simonian theories. Voilquin joined the Saint-Simonian movement in 1830 and subsequently became involved in a new women's journal founded by two working-class Saint-Simonian women. Published between 1832 and 1834, the journal assumed various names, including *La Femme libre (The Free Woman)*, *Apostolat des femmes (Apostolate of Women)*, and *Tribune des femmes (Women's Tribune)*. Voilquin co-edited all but the first issues and contributed extensively to the journal, writing on a broad range of topics including her own unhappy marriage which made her a fervent advocate of women's right to divorce. Reproduced here are two of her articles from the *Tribune des femmes*. In the first, "The Two Mothers," Voilquin examines the class differences among women, while in the second, "The Justice of Men," she critiques the provisions of the Napoleonic Code concerning relations between husbands and wives and parents and children. During the Revolution of 1848, Voilquin served on the daily women's newspaper, *La Voix des femmes (Women's Voice)*, along with other women who had participated in the utopian socialist movements of the 1830s.

A. "The Two Mothers"

Since hours spent in observation are not always lost hours, and since, moreover, Paris is so beautiful at the beginning of each year, so adorned with all those pretty *nothings* that make so little profit for the ingenious worker who invents them, that the desire takes me to try out the profession of an idler today, and walk without *purpose*.... Let us stop in front of this store. This place will no doubt provide me with some useful observations. But you, poor woman huddled in that corner, what are you doing with your three poor little children by that door? You are not there out of curiosity. Oh no, no, in your lusterless eye, your pale face, and the rags that cover you, I see the reason. It is need, the cruel need to exist, that keeps you there. Let us see if I cannot perceive on all these rich faces that pass before me a compassionate movement that would indicate a soul. But it is not in evidence. This cold and disdainful expression is definitely theirs. Oh, rich people! since your souls are too narrow to understand association, then at least practice philanthropy. That is today's virtue. You, poor woman, do not lose heart. Here is an elegant young lady getting out of her carriage. She is also rich, that is true, but a woman's heart guesses so many things that, without having suffered, she may understand your grief. How brilliant her attire, what extravagant horses and carriage! Oh! with what servility her servants care for her! I believe I notice that this is not affection which gives nor the heart which grants, but rather wealth which is paying cupidity its wage. And would the honest merchant not feel himself obliged to bow reverentially to a woman who alights from a carriage to go into his shop! Certainly, in seeing his hurried manner, one can be assured that

he is far from wishing to leave Fortune waiting at the door. [....] Fortune! the sole divinity of the day. It seems, when we observe how she surrounds her favorites with attention, that they epitomize in their persons all the merit and happiness in life, and that to attain the favor of this capricious goddess, one must sacrifice everything, everything including one's human dignity. But why are you waiting, poor woman; don't you see by the servile but impertinent tone of her valets that their mistress's selfish and unfeeling heart is about to spurn your lament. What expressive looks you give this brilliant carriage and the tattered clothes that barely cover you; and then, raising your eyes to heaven, you seem to question the justice in the mystery of this inequality! Nevertheless, there is still hope. Like you, this woman is a mother. She must sympathize with your misery. Oh! oh! good merchant, your beaming air shows that your excessive civilities are not for naught. What trifles are brought from your shop to the wealthy carriage! Poor woman, hasten to present your mother's request. The elegant young lady is herself ready to get in with her children. Now insinuate yourself amongst them. Quickly.... Alas! only an outburst of anger answers your touching prayer, and she informs the servants and the merchant himself that they must hasten to push away the troublesome solicitor.... And I, watching this carriage roll off and flee with a wringing heart, need to think of eternal justice in order to breath more freely. *To each according to their work.*[1]

---

[1]A Saint-Simonian slogan, indicating their position against inherited wealth.

Yes, rich woman, your life will go on like that of the poor woman! You, mother, have not deigned to help end the sorrows of a mother. May God help you achieve progress and soften your soul by making you suffer through your children all the anguish of hunger, cold, and poverty that you have not prevented. Let me remind you, rich woman: *To each according to their work!*

B. "The Justice of Men"

Some time ago my interest in a case concerning a member of the popular class drew me to the criminal court. [....] By giving you the details of another case that was also being tried that day, I hope you will feel as I do how little we are protected by men, even *those* of the law.

A young woman first appeared before the court. She moved me by her pallor, her moral suffering, and a poverty she did not deserve. Before these cold, severe men's faces, this poor young woman recounted simply but truthfully how her husband, after promising her support and protection before all of *society* (*represented* in the person of *Monsieur the Mayor*), then sold everything they owned, so that the only material resources left to her were their common debts. And afterward he spent all this money with an immoral, degraded girl. Poor little *legitimate* wife!... You who religiously bent your head under the yoke of their laws. What consolation do these men bring to your broken heart when one of them destroyed your existence by abandonment, contempt, and poverty? [...] In spite of the all too just grievances this woman had against her husband, she appeared pained at the obligation to enter into all these details. It was not against him that she complained, but against the unfortunate girl who, not content to bring disorder into her marriage, had brutally hit her one day, so as not to endure from the poor abandoned wife a few all too just and deserved reproaches. There were proofs, medical certificates and another court had already sentenced her to six months of imprisonment and a fine. But vice of this magnitude is

audacious; this girl had appealed, and the court had appointed her a counsel. Oh! How can I explain to you the indignation I felt when I heard this lawyer playing on the emotions of the court and audience with his despicable argument: "You see, gentlemen, the plaintiff's jealousy prompted her to these insults, and my client retorted a little vigorously. Gentlemen, you will overturn the sentence for lack of cause, for, you see, this is *only a woman's quarrel.*" (The accused was not completely pardoned, but her penalty was cut in half.) If a woman had been presiding beside the judge, such an important question as the relations between the sexes would not have been resolved so lightly as a *woman's quarrel!* What indecency for a lawyer to express himself that way! Where then is morality? Where then is the protection that one owes to the weak and unfortunate, if they do not find it in the sanctuary of justice? In this society with no *bonding,* there is only hurt for us, whether we follow or discard your laws. [...]

To prove to you, dear readers, that this [...] is not exaggerated [...], I will go back to the root cause, that is to say I will examine civilization's masterpiece, the code of laws that men impose upon us.

Whoever says "code of laws" speaks of the social regulations made in *everyone's* interest and approved and consented to by *everyone;* but who in truth are *we?* Humanity is not composed only of men. Legislators of all ages, tell me, if we are half of *everyone,* have you ever at any time admitted women among yourselves to uphold the rights of their sex? And if we have never had representatives to *discuss* and *prevent* the oppressive laws that you have drawn up against us, explain by what right you would have us remain forever submissive to these laws? Men! be therefore no longer surprised by the disorder that reigns in your society. It is an energetic protest against what you have done *alone.*

Frail woman that I am, I feel the strength and need today to *protest* boldly against what is arbitrary and depraved in your social system, as it is summed up in several articles of the law that I attack as the forced consequence of a bad principle. For example, how can we in the nine-teenth century listen with composure to a dele-gate under your authority who says in all seriousness to us: "The wife must obey her hus-band" (Art. 213). I have already asked *why* in another article, and my gauntlet remained on the ground. No one picked it up. Only *La Revue des Deux Mondes* joked about it, but *joking* does not *prove* anything. (Art. 214): "The wife is *obligated*" (emphasis in the text) "to live with the husband, and to follow him wherever he judges it appropriate to live." Does not the spirit of his laws establish our slavery, so that we cannot be ourselves?[...]

The spirit of your law is even more malevo-lent for the *mother* than for the *wife*. Against the wife there is the arbitrary despotism from which she escapes through a constant silent struggle. But it is the mother's heart that you wound and break by the mistrust and injustice which are palpable in the following articles. (Art. 373): "The *father* alone exercises authority over his children during the marriage." (Art. 374): "The child cannot leave the *paternal* home without the father's approval." What then is the mother in the family? *Everything;* her influ-ence is immense. And her rights? *None.* Oh, justice of men!...(Art. 389): "During the mar-riage, the father is the administrator of the per-sonal belongings of his children in their minority." More than once while reading this article I asked myself: But does the father alone then have innate knowledge? Does infallibility then find refuge in the little conjugal fortress? Probably so in the mind of the legislator. For if woman in her maternal love finds the strength to control her husband's acts, the *law of man* is there ready to tell her: *back!* usurper, this is a right that you have appropriated and do not possess. (Art. 390): "After the death of one of the spouses, the guardianship of minors belongs to the survivor." This article appears to estab-lish too much equality between spouses; what follows serves as a corrective. (Art. 391): "Nev-ertheless the father will be able to appoint a special advisor to the surviving mother and guardian, without whose advice she will not be able to execute any action over her ward." Since it is generally known that maternal love is the strongest and deepest of all feelings, why this mistrust? [...] Who more than the mother is in a position to supervise her children's happi-ness? Who more than she knows with certainty that they are definitely a part *of her*, definitely *her own*? If the law wants to prevent presumed misconduct by one of the spouses, then why does the mother not have the same right to consideration? Why not allow her tranquillity after death, by letting her consolidate *her* chil-dren's future. Cannot everything that can be prejudged about the subsequent conduct of women once widowed be applied in the same case to men? They do not have as we do a feel-ing for the intimate family; and besides, if a new love were to bring the woman new chil-dren, would they not *all* be drawn from the same *life* source that is common to them all? Does she not unite them all in her mind and maternal heart? And if, forced by this law of in-heritance (which is impious since it is against nature)[1] to divide in an unequal manner the property among *all,* she at least does not disin-herit them from any of her care or love. It is not the same for the man. The children from a first bed are almost always driven away from the fa-ther's house by the maternal selfishness of the second wife, and meanwhile the father *alone* has the right in dying to appoint a guardian-coun-sel. Oh, justice of men!...

Young mothers, chase such dark thoughts away from your mind, so that the love of your children may prepare you for years of happiness if marriage is only a long disappointment for you. Find refuge in your children's future, so that your imagination may embellish their ex-istence with those things of which you have been deprived; for these children are assuredly yours, young mothers. Find joy in your daugh-ter; watch how her charms unfold in adoles-cence, how her face blushes and grows pensive

---

[1]An attack on inheritance was central to Saint-Simonian-ism.

when she feels her heart pound for the first time. Oh yes, tender mothers, become her confidante, for you know well the needs of a woman's heart. Prepare her from her earliest years for her first love, for it is often a woman's whole destiny. You know that the sentence your society hurls against you will still be the irrevocable sentence against us for a long time. Therefore, kind mother, do not mistakenly think that your child's heart *alone* will be consulted. (Art. 148): "To form a marriage the father's consent is sufficient." (Art. 150): "If the father and mother are dead and there is a disagreement between the grandfather and grandmother, the grandfather's consent is sufficient."

Oh, justice of men! truly the time is near when you will be declared impious. Soon the mother will no longer be martyred in her spirit or her flesh. *God* entrusts the *certitude* of the family to the mother *alone.* In the young girl's bosom lies the *living bond* that links forever the generations to come with those that pass away. A mystery that realizes itself in *God's* bosom under the great name of *humanity*!

QUESTIONS FOR ANALYSIS

1. How does Chapter VI of the Napoleonic Code define the relations between husbands and wives? What are the legal and property rights of married women? What is the overall purpose of this chapter of the Code?
2. What is Suzanne Voilquin's critique of society in "The Two Mothers"? Why has she chosen an encounter between two mothers for this purpose? How does Voilquin scrutinize the Napoleonic Code in "The Justice of Men"? How does she fuse socialist and feminist concerns in these two articles?

WOMEN AND INDUSTRIALIZATION

The rise of the factory system and the creation of whole new factory towns, such as Manchester, England, dramatically transformed the living and working conditions of the working class, which now became an object of systematic inquiry. By investigating and publicizing these conditions, social critics aimed to reform or eliminate the horrors of capitalist industrialization. Their proposed solutions varied widely, from the abolition of capitalism, as advocated by Marx and Engels in *The Communist Manifesto,* to legislation that would exclude women, especially married women, and children from work in factories and mines. While the factory system became increasingly dominant, urban women continued to engage in traditional forms of work such as domestic service and street vending. Women's piecework labor at home persisted and even expanded as a flexible auxiliary to factory production while prostitution[1] grew as women's wages were not sufficient to support themselves or their families.

---

[1]On the subject of prostitution, see Josephine Butler, Source 123 in Chapter 10.

100. Julie-Victoire Daubié
Women Workers in France

Julie-Victoire Daubié (1824–1874) was the first woman in France to receive the baccalaureat. In her activism and writings, she focused on the interrelated themes of women's education, work, and suffrage. In *Du progrès dans l'enseignement primaire. Justice et Liberté!* (1862; *On the Progress of Primary Education. Justice and Liberty!*), Daubié advocated for women's equality in education both as students and as women in the teaching profession. She regarded educational opportunities as the key to improving women's economic status and to securing their independence from men. Women's public schooling at this time was limited to primary education and not until 1880 did the government legislate secondary education for women. Daubié also argued for women's right to vote as essential for improving their social and economic status, although she believed that the franchise should be limited to men and women who were literate. Women's work, education, and moral condition are the subjects of Daubié's three-volume study, *La femme pauvre aux xix siècle* (1866–1870; *The Poor Woman in the Nineteenth Century*). Here, she uses the term *femme pauvre* in an expansive way to mean not only indigent women, but all women who had to work for a living. Unmarried herself, Daubié was especially concerned with the plight of single women who, along with widows, comprised nearly half of all adult women in midcentury France. In the first volume of *La femme pauvre,* she examines women's factory, domestic, and professional work. In these excerpts from the section "Quels moyens des subsistence ont les femmes?" ("How Do Women Earn a Living?"), Daubié discusses how industrialization transformed women's lives at work and at home with particular reference to women workers in the silk industry in Lyon.

If we turn to industry in the region of Lyon, we see that the single *arrondissement* of Saint-Étienne employs more than thirty thousand workers of both sexes: one third of the 72,000 weaving looms of Lyon and two-thirds of those in the suburbs are occupied by women suspended fourteen hours a day from a strap so that they may simultaneously maneuver the looms with their feet and hands. According to Blanqui's study,[1] in 1848 they earned less than 300 francs per year. Moreover, the majority of Lyon's women workers are paid by piecework. Their low wages are due to commercial crises and to the status of working women in our society. Conditions are better for women with sufficient skill to weave cloth of high quality, but those who prepare the looms and work thirteen hours a day earn very little.

The city of Lyon, this second capital of France, embodies for the Midi, as Paris does for the Nord, all the sorrows of the worker's existence. Lyon's commerce is based on silk, of which five-sixths is sold abroad; therefore, it is especially vulnerable to industrial fluctuations which sometimes diminish production by 30 to 50 million francs per year. I do not know if free

---

[1] *Des classes ouvrières en France.*

Translated by Lisa DiCaprio.

trade[2] has ameliorated the situation of the workers of Lyon, but that of the female workers will certainly worsen if motherhood and childhood do not, at last, gain the protection that they deserve. Certain employers, for whom women workers are an easy prey, seduce and abandon them. One has even seen factory owners who, after hurling insults at women workers and debasing the value of their work, boast about these crimes with impunity.[3]

The supervisors of the workshops who are not seducers in their own right, remain indifferent and do not condescend to protect the girls for whom they alone are guardians.[4]

This neglect of women, excluded from professional schools and overwhelmed by the responsibilities and sorrows of motherhood, has led large numbers of women workers to vice or to suicide. To cite only a few examples, one woman threw herself out of a window because, in her profound destitution, she could not pay her rent; another poisoned herself, as she despaired from not finding work; a third asphyxiated herself after an illness of fifteen days which exhausted all of her financial resources.[5]

One orphan with a festering finger infection at the end of her finger which prevented her from earning a living, became completely destitute. She attempted to enter a hospital, but was refused admission. Returning home in total despair, she killed herself by drinking a glass of vinegar mixed with pepper.

A poor girl worked day and night to support an aged mother, infirm and mentally unsound; but her own health weakened, her wages became insufficient, and she lacked work. Suc-

cumbing under the weight of her heavy burdens, she killed herself, saying: "Since my life is useless, at least my death will enable my mother to enter a charitable institution."[6]

It is noteworthy that almost all suicides carried out by women are motivated by destitution or social immorality. I will provide only two additional examples, citing again M. Brierre de Boismont who has scrupulously studied this sad situation in Paris. At the moment of killing herself, he relates, one of the women wrote: "I have taken a thousand steps to find work; I have only found hearts of stone, or the debauched from whom I did not want to hear insulting propositions."

A young girl of great beauty, at the point of ending her life, left a note in which she announced that, after having exhausted her resources, she had pawned all of her possessions at the Mont-de-Piété. "It was only up to me; I could have had a storehouse of riches, but I would rather die with integrity than live in sin."

Before seeking the rational remedies to such serious wrongs, it remains for us to examine the conditions of the workers who, numbering more than 300,000 in our manufactures, earn an average wage of one franc per day.

Many of the manufacturers require workers to travel long distances and separate mothers from their families for fifteen hours each day. Different economists have observed that it is the women workers who work harder than slaves. The boiling water of the basins is also painful to the fingers of the spinners who pull thread from the silk cocoons. The putrid emanations from the chrysalis cause various illnesses, called illnesses of the silkworm or of the basin, which lead to long periods of unemployment.

From the carding and preparation of cotton, the women often contract a terrible pulmonary tuberculosis which is called cotton consumption in the creative idiom of the workshop. One

---

[2]Here and later when she speaks of a treaty of commerce, Daubié is referring to the Cobden-Chevalier Treaty signed between England and France in 1860. This free trade treaty lowered the high tariffs that had protected French manufacture from international competition, especially from British factories.—Ed.

[3]E. Buret, *De la misère des classes laborieuses en France et an Angleterre.*

[4]Louis Reynaud, *Études sur le régime des manufactures,* Paris, 1859.

[5]*Presse* du 2 octobre 1856, *Siècle* du 17 avril 1857.

[6]Brierre de Boismont, *De la folie-suicide.*

might believe that the constitutions of women, who are generally employed in these deadly forms of work, were especially suited to resisting their pernicious effects if the statistics of medical science and the reports of hygiene and public health councils did not show the contrary, that for a given number of workers, women are more likely than men to contract this lung disease. A similar observation has been made concerning the manufacture of white lead, the processes requiring the use of mercury and arsenic, and the making of phosphate matches which cause necrosis of the jawbone, designated as a chemical illness. The industry, however, employs the weak and the strong without distinction for this work. The department of the Seine alone employs 1,500 men, women, and children in the match factories.

In the workshops for the printing of calico cloth, male workers have the best-paying jobs which require skill. The women workers employed at Scottish finishing or stiffening of fabrics spend their twelve hour workdays in temperatures which range from 26 to 40 degrees and suffer greatly from the frequent shifts of temperature from hot to cold. These are manufactures where the women work in all seasons, for twelve hours a day, with their feet in water.

This, then, explains the mortality rates which strike the children of working women. According to doctor Villermé,[7] the children of the factory managers and other employers typically reach the age of twenty-nine, while those of women workers in spinning, deprived of their mother's affection and milk, often succumb before they are two years old. The death rate is generally two and three times higher among children of workers of all kinds as compared to those of the well-to-do families.[8] This appalling level of mortality is also attributed to the very young mother's custom of providing their infants with the milk of cows or goats, and of making them fall asleep with the assistance of drugs.[9] When required to work to earn their daily bread, they silence their excessive crying with this slow poison.[10]

The children of these workers who do survive, already weakened by the deprivation of maternal care, are so sickly that in our manufacturing towns two-thirds of them are unfit for military service.

We have seen that the causes of the injustices which reduce the wages of the woman worker are to be found in her ignorance and in the absence of legislation that protects children.

Lacking the knowledge required for the intelligent exercise of their occupations, these women often do not know how to read or write. In our manufactures and provincial industries, the proportion of women workers who are illiterate is much higher than that of male workers. In addition, there is an excess of work for the woman in mechanized industry who does not know how to darn stockings, mend clothing, keep accounts of expenses, calculate savings, prepare meals; who has lost, along with the titles of housewife and mother, the knowledge of the thousand productive tasks which, at every turn, contribute to the prosperity of the individual household and the national wealth.

---

[7]Daubié is referring here to the studies of factory conditions in the cotton, wool, and silk industries carried out by Louis René Villermé, such as the *Tableau de l'état physique et moral des ouvriers employés dans les manufactures de coton, de laine, et de soie* (Paris, 1840).—Ed.

[8]According to information provided to me by Dr. Devilliers at the Academy of Medicine, children of weavers suffer a mortality rate of 35 percent in their first year, as compared to the rate of 10 percent or 5 percent experienced by children of agricultural workers and the well-off.

[9]Most likely, this was laudanum, a form of liquid opium, which was also commonly given by British women workers to their children—Ed.

[10]M. Jean Dollfus, [a factory owner] known for his philanthropy, has saved thirteen out of every hundred children from death by allowing their mothers to remain at home for six weeks after childbirth and paying them the daily equivalent of their wages at the factory.

It is easy to convince oneself that this decline of the woman worker is also related to the decline of motherhood. In speaking of the isolated, individual, single worker, I have shown some of the effects of our social immorality which is more prevalent in manufacture than elsewhere.

The conditions of modern society, the extension of the factory system has appreciably altered women's fate; her appearance on the battlefield of industry has brought an anxious, I will say almost maternal, concern from the legislators for the safeguarding of the moral dignity of man, the principle of the family, and even of civilization itself. By an inexplicable aberration, the day when manufacture tore away the women from the household, French law allowed it to develop at full speed with an unbridled freedom and made the daughter of the people the target of greed freed from all social obligations.

The reports of all the observers and writers are unanimous on this sad question. . . .

Men who are distressed by the dissolution of family bonds have made fruitless attempts, in the manufacturing towns like Lyon and Lille, to urge the [male] weavers to transport their looms to the countryside; they obstinately refuse to do so, even when they have the certainty of receiving the same amount of payment due to the reduction of their expenses. It is worth taking the trouble to figure out the motives for their preference for the city. In the countryside, they say, one is often compelled to enter into formal marriages in which the husband is responsible for his wife and child, while the moral laxity of the city allows for more freedom. . . .

In effect, debauchery gives privileges to men which are rightfully revolting to morality. Investigations have shown that in prosperous industries which offer a daily wage of seven francs [to the male workers] one finds ten free unions for every one legal marriage, and that earnings are consumed by personal and harmful expenditures. This explains why our workers are expelled from European workshops when they also attempt to convert other people to our anarchic principles and lack of moral doctrine.

This license, in corrupting the man, devastates motherhood and childhood: young girls who are weakened at age ten and mothers at fifteen, having neither maternal feelings nor obligations, abandon their infants or deliver their young girls themselves to the promiscuity of the factory owners and male employees.

One speaks a great deal of the vices of the people, but it is easy to convince oneself that these vices are borrowed from the ruling classes; as long as a single industrialist can exploit a single worker in the name of his egoism, and a single woman worker in the name of his passions, the social order will be badly constituted. Our opulent bourgeoisie, unacquainted with industry, offers no more morality than certain leading manufacturers and one knows the habits of our celebrated dandies whose irresponsible immorality in bringing immense ravages to the popular ranks troubles the economic order at the same time as the moral order.

Freedom of trade, then, requires a preliminary uniformity of the European code on a host of industrial questions, such as the rights of the factory owner concerning holidays, child labor in the manufactures, the rights of illegitimate children to financial assistance from their fathers, the obligations of the owners towards disabled workers, etc. So far, we have only known how to rupture the ties which bind the economic order to the civil and moral order, and one could attribute the suffering caused by the treaty of commerce at least, in part, to our lack of principles.

When the law prevents the industrialist from committing injustice, it will look after those who accomplish good; it will invite him to combat drunkenness, to promote the economy with savings banks and family values; to cultivate intelligence and reason with schools, courses, lectures, and libraries, etc.

From the investigations and the reports which will certify the progress achieved, this individual [the industrialist], however humble he may be, with his life in his hands, will demand the recognition due to him for his morally uplifting activities. These immediate means of eliminating our social antagonisms supposes, above all, moral solidarity and association.

QUESTIONS FOR ANALYSIS

1. What is Julie-Victoire Daubie's analysis of the living and working conditions of women in the silk industry in Lyon? How has industrialization transformed motherhood and family relations?
2. What sources does Daubié draw on for her study?
3. What reforms and changes in societal attitudes does she propose?

WOMEN AND MIDCENTURY MOVEMENTS FOR DEMOCRATIC AND ECONOMIC RIGHTS

In midcentury Europe, workers' movements combined the struggle for democratic forms of government with demands for social and economic rights for workers. In England, the Chartist movement demanded male working-class suffrage as a means of influencing legislation beneficial to workers; in the Revolution of 1848 in France, the most radical workers called for the creation of a Social Republic, and in Germany in 1848 and 1849, workers participated in the revolutionary movement for national unification and the creation of a constitutional form of government. Women's participation in these movements included demands for their economic and political rights, such as the right to work and vote.

101. Chartist Women Demand Equal Political Rights

The Chartist movement was inspired by the Reform Bill of 1832 which expanded suffrage for men with property but excluded working men and all women. Deriving its name from the Magna Carta, or Great Charter, the six-point People's Charter of 1838 called for universal male suffrage, annual elections, secret ballots, equal electoral districts, the elimination of property qualifications for office, and payment for members of Parliament. Women and men in the manufacturing districts were also mobilized by the Chartist movement's opposition to the Poor Law of 1834. Embodying Malthusian ideas, the law required whole families to enter workhouses on a sex-segregated basis in order to receive assistance. Chartist women participated in demonstrations, meetings, and strikes, and organized female associations which advocated passage of the Charter. This 1839 "Address of the Female Political Union of Newcastle-Upon-Tyne" was published in the *Northern Star,* a Chartist newspaper.

From the *Northern Star,* February 1839.

"ADDRESS OF THE FEMALE POLITICAL UNION OF NEWCASTLE-UPON-TYNE TO THEIR FELLOW COUNTRYWOMEN"

Well ye know
What woman is, for none of woman born
Can chose but drain the bitter dregs of
   woe
Which ever to the oppressed from the oppressors
   flow.

             SHELLEY

FELLOW-COUNTRYWOMEN,—We call upon you to join us and help our fathers, husbands, and brothers, to free themselves and us from political, physical, and mental bondage, and urge the following reasons as an answer to our enemies and an inducement to our friends.

We have been told that the province of woman is her home, and that the field of politics should be left to men; this we deny; the nature of things renders it impossible, and the conduct of those who give the advice is at variance with the principles they assert. Is it not true that the interests of our fathers, husbands, and brothers, ought to be ours? If they are oppressed and impoverished, do we not share those evils with them? If so, ought we not to resent the infliction of those wrongs upon them? We have read the records of the past, and our hearts have responded to the historian's praise of those women, who struggled against tyranny and urged their countrymen to be free or die.

Acting from those feelings when told of the oppression exercised upon the enslaved negroes in our colonies, we raised our voices in denunciation of their tyrants, and never rested until the dealers in human blood were compelled to abandon their hell-born traffic; but we have learned by bitter experience that slavery is not confined to colour or clime, and that even in England cruel oppression reigns—and we are compelled by our love of God and hatred of wrong to join our countrywomen in their demand for liberty and justice.

We have seen that because the husband's earnings could not support his family, the wife has been compelled to leave her home neglected and, with her infant children, work at a soul and body degrading toil. We have seen the father dragged from his home by a ruffian press-gang, compelled to fight against those that never injured him, paid only 34/- per month, while he ought to have had £6; his wife and children left to starve or subsist on the scanty fare doled out by hired charity. We have seen the poor robbed of their inheritance and a law enacted to treat poverty as a crime, to deny misery consolation, to take from the unfortunate their freedom, to drive the poor from their homes and their fatherland, to separate those whom God has joined together, and tear the children from their parents' care,—this law was passed by men and supported by men, who avow the doctrine that the poor have no right to live, and that an all wise and beneficent Creator has left the wants of his children unprovided for.

For years we have struggled to maintain our homes in comfort, such as our hearts told us should greet our husbands after their fatiguing labours. Year after year has passed away, and even now our wishes have no prospect of being realised, our husbands are over wrought, our houses half furnished, our families ill-fed, and our children uneducated—the fear of want hangs over our heads; the scorn of the rich is pointed towards us; the brand of slavery is on our kindred, and we feel the degradation. We are a despised caste; our oppressors are not content with despising our feelings, but demand the control of our thoughts and wants!—want's bitter bondage binds us to their feet, we are oppressed because we are poor—the joys of life, the gladness of plenty, and the sympathies of nature, are not for us; the solace of our homes, the endearments of our children, and the sympathies of our kindred are denied us—and even in the grave our ashes are laid with disrespect.

We have searched and found that the cause of these evils is the Government of the country being in the hands of a few of the upper and middle classes, while the working men who form the millions, the strength and wealth of the country, are left without the pale of the Constitution, their wishes never consulted, and their interests sacrificed by the ruling factions, who have created useless officers and enormous salaries for their own aggrandisement—burthened the country with a debt of eighteen hundred millions sterling, and an enormous taxation of fifty-four millions sterling annually, which ought not to be more than eight millions; for these evils there is no remedy but the just measure of allowing every citizen of the United Kingdom, the right of voting in the election of the members of Parliaments, who have to make the laws that he has to be governed by, and grant the taxes he has to pay; or, in other words, to pass the people's Charter into a law and emancipate the white slaves of England. This is what the working men of England, Ireland, and Scotland are struggling for, and we have branded ourselves together in union to assist them; and we call on all our fellow countrywomen to join us.

We tell the wealthy, the high and mighty ones of the land, our kindred shall be free. We tell their lordly dames we love our husbands as well as they love theirs, that our homes shall be no longer destitute of comfort, that in sickness, want, and old age, we will not be separated from them, that our children are near and dear to us and shall not be torn from us.

We harbour no evil wishes against any one, and ask for nought but justice; therefore, we call on all persons to assist us in this good work, but especially those shopkeepers which the Reform Bill[1] enfranchised. We call on them to remember it was the unrepresented working men that procured them their rights, and that they ought now to fulfil the pledge they gave to assist them to get theirs—they ought to remember that our pennies make their pounds, and that we cannot in justice spend the hard earnings of our husbands with those that are opposed to their rights and interests.

Fellow-Countrywomen, in conclusion, we entreat you to join us to help the cause of freedom, justice, honesty, and truth, to drive poverty and ignorance from our land, and establish happy homes, true religion, righteous government, and good laws.

_____

[1]The Reform Bill of 1832.

102. Flora Tristan
The Workers' Union

Flora Tristan (1801–1844) was one of the most prolific utopian socialists of the early nineteenth century. She became involved in feminist activities and in 1827 began to attend meetings of the journal *Gazette des femmes* which was especially concerned with the issue of divorce.[1] In *The Workers' Union* (1843), excerpted here, Tristan called for a Universal Workers' Union transcending craft and gender divisions to press for the "right to work" and the right to organize. She also outlined a blueprint for the creation of workers' palaces, each to comprise two

_____

[1]While the Napoleonic Code had permitted men to initiate divorce proceedings, the Catholic-dominated legislature of the Restoration government abolished the right to divorce altogether in 1816 and it did not become legal again until 1884.

thousand to three thousand individuals, which would provide education for children, medical care for the sick and injured, and housing for workers no longer able to work, including the elderly and disabled. To promote *The Workers' Union* and bring her message directly to the workers, Tristan undertook a tour of France in 1844 which included some seventeen cities. She also maintained a detailed record of responses she received to *The Workers' Union* and the conditions of workers in the cities she visited. During the tour, Tristan contracted an illness and died in Bordeaux. In a ceremony dedicating her tomb there, some 1,500 workers from various trades marched behind the tricolor flag of the Republic, now veiled with black tissue, and on which was written, "Association, Right to Work." Tristan's writings include *Pérégrinations d'une paria* (1833–1834), a diary of her travels in South America; *Méphis* (1838), a novel; and *Promenades dans Londres* (1840), her observations of London, including descriptions of working women's conditions which preceded Friedrich Engels' *The Condition of the Working Class in England in 1844* (1845). Tristan's account of her tour, originally intended for publication in 1845, was finally published in 1973 as *Le Tour de France: Journal inédit 1843–1844.*

"PLAN FOR THE UNIVERSAL UNIONIZATION OF WORKING MEN AND WOMEN"

I am going to provide a quick glimpse of the steps to be followed if one wishes promptly to consolidate the Workers' Union on a solid footing....

A. How the Workers Must Proceed to Establish the Workers' Union

1. In their respective trade, union, or welfare groups,[1] the workers must begin by forming one or several committees (according to the number of members) composed of seven members (five men and two women),[2] chosen from the most capable.

2. These committees may not receive any contributions. Their function will be temporarily limited to inscribing in a great register book the sex, age, names, addresses, and occupations of all those who want to become members of the Workers' Union, and the amount each pledges to contribute.

3. That the person in question is a working man or woman must be verified in order to have a name put in the book.[3] And by working man or woman, we mean any individual who *works with his or her hands* in any fashion. Thus, servants, porters, messengers, laborers, and all the so-called odd-jobbers will be considered workers. Only soldiers and sailors will have to be exempted. Here are the reasons for this exception: (*a*) the State comes to the aid of soldiers and sailors through the disabled fund;

---

[1] The societies in Paris and the suburbs number 236, with 15,840 subscribers and about three million francs in funds (*De la Condition des ouvriers de Paris de 1789 jusqu'en 1841,* p. 254).

[2] If I am not allowing for an equal number of men and women on the committees, it is because it has been observed that today's working women are much less educated and intellectually less developed than the male workers. But of course this inequality will only be transitory.

[3] The Workers' Union, proceeding in the name of *universal unity,* must not make any distinction between nationalities or male and female workers belonging to whichever nation on earth. Therefore, for anyone called a foreigner, the Union benefits will be absolutely the same as for the French.

The Workers' Union will have to set up branch committees in all the main towns in England, Germany, and Italy—in a word, in all the European capitols—so that men and women workers of all the European nations can be listed in the Workers' Union register as members. The same steps will have to be taken for these committees as for the French ones....

(*b*) and the soldiers are knowledgeable only in destructive work and the sailors in sea work, so neither would be usefully employed in the Workers' Union palace.

4. Yet, since the soldiers and sailors belong to the working class and by this reason have the right to adhere to the Workers' Union, they will be inscribed separately as brothers. They may contribute toward their children's admission to the palaces. In a third book, all those who want to cooperate in the prosperity of the working class will be entered as sympathizers.

5. In no case may the professional beggar put his name on the list. But the workers who have signed up at the welfare office and who receive aid because their work is not enough to support their families may not be excluded. Misfortune is respectable; only idleness debases and degrades and must be pitilessly rejected.

6. In view of the Union, and this is of the utmost importance, the workers must make it their duty and mission to use all their influence to get their mothers, wives, sisters, daughters, and girl friends to join in with them. They themselves must urge them and escort them to the committee so that they can enter their names in the Union's great-book. That is a proper mission for the workers.

7. As soon as the working men and women are represented by the committees they will have elected, these committes will then elect a central committee from among themselves for all of France. Its headquarters will be in Paris or Lyons (in the city with the most workers). This committee will be composed of fifty members (forty men and ten women) nominated from the most capable....

9. Once the central committee is elected, the Workers' Union will be established....

C. The Intellectual Point of View

22. ....At this time, it is very important for the working class to know exactly what to expect with regard to the sympathy or animosity the other social classes might have toward it.

23. Here is the outline for these sorts of appeals as I conceive of them. It is up to the central committee to modify them, as they see fit.

24. APPEAL TO THE KING OF FRANCE, as the appointed national ruler[4]

Sire,

....In 1830, the nation's representatives, considering it a time of peace, liberty, equality and work, and no longer seeing a need for a military leader, pronounced the fall of the King of France. And in the midst of the Chamber of Deputies they elected a king for the French.[5]

Sire, in accepting the title of King of the French, you contracted the sacred obligation of defending the interests of all the French. Sire, in the name of the mandate you received from the French people, the Workers' Union has come to call your Majesty's attention to the fact that the sufferings of the most populous and useful class have been hidden from you. The Workers' Union asks for no privilege; it simply requests the recognition of a right it has been denied and without which its life is in jeopardy; it asks for the RIGHT TO WORK.

Sire, as head of State, you can propose a bill. You can ask the Chambers to pass a law granting the RIGHT TO WORK for all men and women....

As head of State, you can provide a shining sign of sympathy and gratitude to the Workers' Union. Sire, you own several magnificent estates located on French soil; you could immortalize your name by offering one of your beautiful properties to the Workers' Union as a mark of your sympathy and gratitude to the most populous and useful class so it can build its first palace....

Sire, by acting thus, you will provide a grand and healthy example which all future heads of State will be *forced to imitate.* This act of gen-

---

[4]The dictionary defines *King* (from the Latin *rex, regis,* derived from *regere,* to rule, to govern) as the one who exercises sovereign power in a kingdom and *leader* as the one who is at the head, who commands, directs, leads, etc.

[5]Louis-Philippe, elected King of the French on August 9, 1830.

erosity will proclaim that the kings' primary duty is to be concerned with the defense of the interests of the most populous and useful class.

25. TO THE CATHOLIC CLERGY

Catholic priests,

The Workers' Union has come to request your aid and support....

We know that the term *Catholic Church* means *universal association;* that the word *communion* means *universal brotherhood;* we know that the Catholic Church is based on the principle of UNITY and has as its goal the fusion of all peoples so as to consolidate the world through a great religious, social body. Catholic priests, it is up to you to realize the great notions of UNITY posed by Christ and his apostles. Think about it; you cannot do this work unless you become priests to the most populous and useful class. The Workers' Union is pursuing absolutely the same goal as the Catholic Church. The Workers' Union wants peace, brotherhood, and equality among all— HUMAN UNITY. Catholic priests, if you are truly men of peace and real Catholics, your place is among the people. You must march with them and at their head.

You priests, who have huge churches where the townspeople and peasants gather, you who can speak from your pulpits to both the rich and poor, preach justice to the rich and unity to the poor.

However, understand that the proletarians do not ask the ten million owners for alms. No, they are calling for the right to work; so once assured of always being able to make a living, they will no longer be debased and degraded by the alms the wealthy scornfully throw to them....

27. TO THE MANUFACTURERS

Sirs and Employers,

By making us work, you and your families live like English bankers. You amass more or less huge fortunes. In working for you, we can scarcely survive and feed our poor families. This is a legal issue. Thus, take note that we are not blaming or accusing you; we are simply observing what is. Today at last the workers are aware of the cause of their pains, and in their wish to put a stop to them, they have UNITED.

The Workers' Union has judged that it has to make an appeal to the generosity of the employers. It thought that the gentlemen owning the factories, deeply conscious of the gratitude they owe the working class, would be pleased to show a mark of their sympathy. The Workers' Union, motivated by purely fraternal feelings and completely peaceful intentions, has reason to be able to count on your support. Thus it confidently has come to ask you for your real patronage and active cooperation....

28. TO THE FINANCIERS, OWNERS, AND BOURGEOIS

This would be the same letter as above in substance with a few variations in the form.

29. Finally, the central committee ought to make a last appeal, the one I would count on the most, to women.

30. APPEAL TO WOMEN OF ALL STATIONS, AGES, OPINIONS, AND COUNTRIES

Women,

You, whose souls, hearts, minds, and senses are so impressionable that, without realizing it, you shed a tear for all suffering,... you, women, will you remain silent and always hidden, when the most populous and useful class, your proletarian brothers and sisters, working, suffering, weeping, and moaning, come and beg you to help them leave their misery and ignorance?

Women, the Workers' Union has looked your way. It has understood that it cannot have more devoted, intelligent, and powerful allies. Women, the Workers' Union has a right to your gratitude. It was the first to recognize *in theory* women's rights. Today its cause and yours are becoming one and the same. Rich women, educated, intelligent, enjoying the power afforded by instruction, merit, status, and wealth, who can influence your men, children, servants, and workers, lend your powerful protection to the men who have only numbers and rights to make them strong. In turn, these men with bare arms will lend you their support. You are oppressed by law and prejudice. Unite with the oppressed, and this legitimate, sacred alliance will enable us to struggle legally and loyally against oppressive laws and prejudices.

Women, what is your mission in society? None. Well, do you want a worthy way to spend your life? Devote it to the victory of the most sacred of causes: the Workers' Union.

Women, who feel that holy fire called faith, love, devotion, intelligence, and action, you must become the preachers for the Workers' Union.

Women writers, poets, artists, write to instruct the people and use the union as the text for your songs.

Rich women, get rid of all those cosmetic frivolities absorbing enormous sums and learn to use your wealth more effectively and magnificently. Donate to the Workers' Union.

Women of the people, join the Workers' Union. Enlist your daughters and sons to sign up in the union book.

Women of all of France and the whole earth, place your glory in proudly and publicly becoming the defenders of the union.

Oh, women, our sisters, do not remain deaf to our appeal! Come to us, we need your help, assistance, and protection.

Women, in the name of your suffering and ours, we ask for your cooperation in our great work.

31. The central committee might also make an appeal to artists. They are usually very generous. They could give their cooperation in constructing the first palace and decorating it with their paintings and sculpture. The dramatic artists and musicians could give shows and concerts to benefit the Workers' Union, and the proceeds would be used to buy blocks of marble, canvases, paint, and everything necessary for the artists to carry out their work....

35. I repeat, the central committee would be committing a *huge mistake* if it neglected to attract the sympathy of all social classes to the Workers' Union....

E. Building the Palaces

38. We have reached an era in which the social state is progressively moving toward a complete transformation. The construction of the Workers' Union palaces does not have to be solid enough to last centuries. The essential thing is that the palaces be built so as to offer simultaneously: (*a*) healthfulness in terms of space, daylight, sunniness, ventilation, and heating, (*b*) convenience in terms of ease and rapidity of communication among the different parts of the buildings, (*c*) interior practical distribution of rooms for the elderly, employees, and children, (*d*) outside: workshops, schools, exercise rooms, and finally a farm to meet agricultural needs....

F. Conditions for Admission to the Palaces for the Elderly, the Injured, and Children

...45. First, individuals will be admitted to the Union palace by department in proportion to the number of subscribers. To avoid preferences, special passes, and unfairness, straws could be drawn from a hat.

46. For instance, 600, 1,000, 1,500, or 2,000 persons will be admitted. Then as the resources increase, new palaces will be built. At this rate, in thirty years all working men and women will be sure of having their children raised in Union palaces and of finding a bed for their old age.

47. As a general rule, half those admitted will be children (entrance age will be six years) and the other half elderly or disabled.

48. I neither want nor can set a rule for admission; these regulations will change as the Union's resources grow. I believe, however, that preference ought to be given to orphans, sons of widows, or those whose parents are disabled or very old, and finally, for any worker's family with more than five children; the sixth, seventh, eighth, and beyond would enter automatically. As for the disabled, widows and widowers would have preference; but that is only a minor indication.

G. Labor Organization in the Palaces

49. The Workers' Union palaces will offer in every respect the most suitable milieu to try out one or several experiments in labor organization. Men, women, and children will all work.... But

until there is agreement on the system to follow for labor organization, the central committee will institute a labor board in each palace....

H. Moral, Intellectual, and Vocational Instruction to Be Given to the Children

54. In my opinion, there cannot be a healthy, true morality except when it logically follows from the belief in a good and just God, wisely, providentially, and carefully creating and guiding his creation.... By every means possible, the child would have to be led to understand that our globe is a large humanitarian body, whose different nations represent its internal organs, members, and main arteries and whose individuals represent the other arteries, veins, nerves, muscles, and even the tiniest fibers. All the parts of this great body are as closely connected to each other as the various parts of the human body, all helping each other and receiving life from the same source. A nerve, a muscle, a vessel, or a fiber cannot suffer without the whole body's feeling it....

57. By separating love from intelligence, a mortal blow was delivered to Jesus's religion. Catholicism said, "Believe and do not analyze." What is the result? Those with natures more *intelligent* than loving, the scholars and philosophers, finding no suitable nourishment for their minds in the Catholic religion, abjured the Church, heaping much disdain, distrust, and insult upon it. From disdain they moved to anger and indignation, and hitting twice as hard, they demolished the grand edifice stone by stone. On the other hand, those who are more *loving* than intelligent, seduced by the attractive power of ecstasy, ruined and lost themselves in the emptiness. For loving God outside humanity is scornful and insulting, an outrage to God in His manifestation.

58. Thus the teachers ought to have as their fundamental law the simultaneous development of the loving and intellectual faculties of every child.

59. If one wants to obtain this double result, a very powerful element must be introduced into the method—the *why*. The Jacotot method[6] lies partly in posing the question, *"Why?"* Consequently, I would like to see it more widely accepted. Applying the *why* to solving great moral, social, and philosophical questions in the daily education given to children of the working class would be the way to make human intelligence take gigantic strides....

67. As for vocational training, each child would choose the trade he feels the most suited for. Besides all the other work he would have to do, upon leaving the palace he will have to be a competent worker in at least two trades.

68. In order for him to become interested in work, as of the age of ten the child will be eligible to share in the profits produced by the work in the establishment. This amount will increase every year until his departure at the age of eighteen. Half will be given to him as a trousseau made in the establishment and the other half in money....

I. The Inevitable Results of This Education

71. The results the Workers' Union ought to have are immeasurable. This union is a bridge erected between a dying civilization and the harmonious social order foreseen by superior minds. First of all, it will bring about the rehabilitation of manual labor diminished by thousands of years of slavery. And this is a capital point. As soon as it is no longer dishonorable to work with one's hands, when work is even an honorable deed,[7] the rich and the poor alike will work. For idleness is both a torture for mankind and the cause of its ills. All will work, and for this reason alone, prosperity will rule for everyone. Then, there will be no more

---

[6]Jean-Joseph Jacotot described his "universal" theory of education in *Enseignement universel* (1822). Tristan also drew on the educational theories of such utopian socialists as Charles Fourier and Robert Owen.—Ed.

[7]I am totally of Fourier's opinion that a means must be found to make work *attractive*; but I think that before reaching this ultimate goal, work must first cease being considered *dishonorable*.

poverty; and poverty ceasing, ignorance will too. Who causes the evil we suffer from today? Isn't it that thousand-headed monster, *selfishness?* But selfishness is not the primary cause; poverty and ignorance are what produce selfishness.

73. Only when all men and women work with their hands and are dignified by it, will this great, desirable productivity take place. And this is the only way to eradicate the vices fostered by selfishness, and consequently to civilize men.

74. The second, but not lesser, result necessarily brought about by the Workers' Union will be to establish de facto real equality among all men. In fact, as soon as the day comes when working-class children are carefully raised and trained to develop their intellects, faculties, and physical strength—in a word, all that is good and beautiful in human nature—and as soon as there is no distinction between rich and

poor children in their education, talent, and good manners, I ask: where could there be inequality? Nowhere, absolutely nowhere. Then only one inequality will be recognized, but that one must be experienced and accepted, for God is the One who established it. To one, he gives genius, love, intelligence, wit, strength, and beauty; to the other, he denies all these gifts and makes him stupid, dull-minded, weak-bodied, and ill-shapen. That is natural inequality before which man's pride must humble itself; that inequality indiscriminately touches the sons of kings as well as the sons of the poor.

75. I stop here, wanting to leave my readers the sweet joy of counting for themselves the important and magnificent results the Workers' Union will doubtless obtain. In this institution the country will find elements of order, prosperity, wealth, morality, and happiness, such as they can be desired.

103. Women's Requests for Work During the Revolution of 1848

In response to demands by workers, the Provisional Government of France, established in February 1848, proclaimed "the right to work" in a February 25 decree and four days later created the Government Committee for the Workers, popularly called the Luxembourg Commission as it met in the Luxembourg Palace. The commission was headed by Louis Blanc, one of only two workers represented in the government. Workshops subsequently established by the government for the unemployed resembled those created during the French Revolution rather than the cooperative or "social workshops" elaborated in Blanc's *Organisation du travail* (1840; *Organization of Work*) and advocated by the most radical workers as the only means of guaranteeing "the right to work." Moreover, the government initially provided work only for men and not until later were workshops established for unemployed women. In these letters, women address their requests for work and grievances about mechanization to government officials.

Sources A–C from Archive de la Seine, VD6 619 n7 and Source D from Archives Nationales, F/12/4898. Contributed and translated by Judith DeGroat. Used by permission.

A. MONSIEUR [MAYOR][1]

Pardon the liberty which I take to impose on you, but your well-known generosity serves as my excuse.

For several days, posters have announced that in each town hall, work will be given to women or they will be provided with assistance. Not knowing anything about the formalities required to apply, I address myself to you, monsieur, who are known for your generosity, your humanity for the unfortunate, hoping that you will take pity on my sad situation.

Without work for three months, my husband ill with a chest ailment, an infirm mother 72 years old for whom we are responsible, our belongings, already held for two quarters of the year at the *mont-de-piété* [a pawnshop], sold because we were unable to renew our claim.

There it is, monsieur, the position in which I find myself. I hope that you will take pity on us. I am not asking for charity. I ask, I implore, that I will be given work which can at least prevent us from dying of starvation....

I anxiously await your response which is my only hope.

Respectfully,

Jeanne Pierre Goursault
65 Rue du Four St. Germain

B. TO THE *CITOYEN* MAYOR OF THE ELEVENTH ARRONDISSEMENT

I come with confidence to tell you of my sad situation. Since my youth I have been constantly in service. Until two years ago, I had for twenty years given devoted care to an aged lady. I left that respectable house when her family placed her in an institution. Since then, I have had only temporary employment which does not pay well. Today, I am without resources, having had the misfortune to lose my savings in a fraud.

I hope, that in your kindness, *Citoyen* Mayor, you would wish to help me find the means to earn my living by helping me find a position as a cook or some other post which would save me from the extreme hardship that I find myself in. Although I am older, 45 years of age, I am strong and of good health and am able to offer good service.

I am confident, *Citoyen* Mayor, that you will give attention to my case. I thank you in advance and assure you of my sincere gratitude.

Respectfully,

Elénore Déchambre
24 Rue Sevandoni

C. *CITOYENS*[1]

I am the daughter of proletarian parents, married to a born republican[2] as well as one in my own heart....

Previously, I had my own establishment; my courage, my perseverance failed against the unjust organization of society and work. I have faith in what our representatives have promised us. If I did not follow my husband and my parents into combat to achieve my liberty, it is because I knew that God was with them, God whom I begged for the success of our holy cause, at the same time as, with my feeble hands, I prepared the weapons needed by our warriors and assisted in dressing their wounds. They were as strong in defending us as we

---

[1]At this time, each *arrondissement* of Paris had its own elected mayor and town hall, the *mairie,* which continue to exist today. This letter is addressed to the mayor of the eleventh *arrondissement* which housed a mix of homeworkers in the garment and luxury trades as well as workshops of various types.

[1]The letter is addressed to the administration of the eleventh *arrondissement.*

[2]She means that her husband is a republican by instinct or "birth" because of his social milieu.

became in demonstrating that our sex is as useful as theirs, in this and other ways, in serving our beautiful fatherland.

Actually, owing to a crisis that, I hope, will pass, I am without work, like our other *citoyennes.* I took your proclamation seriously. I hasten to put my strength at the service of the republic in whatever ways I may be useful to the most beautiful, the most pure of republics. In my youth, I was in a private boarding school; soon I returned to the working class to support my family which had made sacrifices for me. My establishment was in fine linens and notions. I know equally well all of the kinds of work of our sex: upholstery, sewing, etc. I can keep accounts and supervise work; finally, see and judge for yourself. I have confidence in God and in the republic he has given us.

There is no need for me to tell you that, lacking work I lack resources.

Please, accept my best wishes,

Julia Jacquier
50 Rue de la Harpe

D. MONSIEURS, MEMBERS OF THE PROVISIONAL GOVERNMENT

The women workers exercising the *état*[1] of rabbit fur cutters for the hat industry, two or three thousand in number, all diligent workers and mothers of families, have the honor of showing that this *état,* which is only practiced by women, gives them the means to live, to feed their children, and to give an *état* to their daughters. It enabled them to earn, according to their ability, 10 or 12 *francs* per week, while today they are reduced to being able to earn 50 or 75 *centimes*[2] a day, which makes it impossible for them to support themselves or their families. This state of things, monsieurs, arises from machines which have been adopted by the richest owners... which reduces the numbers of workers, aggravates their sad situation, and takes away their bread. These cutters using machines still harbor a great prejudice against the other owners who prefer to support those who work by hand, but they cannot compete with the machines and are forced to abandon their businesses and their workers. In addition, for fifteen years, ten thousand foreign workers[3] have arrived to bring misery to an *état,* which once flourished but is now almost extinct.

For these reasons, the petitioners join together to beg the provisional government, from which all justice arises, to look with compassion on their unfortunate fate and to restore their livelihood by stopping these machines which cause the loss and ruin of the workers. It is their most ardent wish and they have complete confidence that your humanity and justice will not fail them and will protect them against the egotism of the wealthy cutters.

Their gratitude will match the profound respect with which they have the honor, Monsieurs, to be your most humble servants.

Paris, March 13, 1848.

[The letter is signed by nine women]

---

[1] *État:* in its narrowest sense, refers to a trade or profession practiced by an individual; also refers to one's station and status in society generally.

[2] In French currency, one hundred *centimes* comprise a *franc.*

[3] In French, *étrangère,* or foreigner, can refer either to an individual from outside of France or someone who previously lived outside of the city, in this case Paris.

104. Louise Otto
Women's Right to Work

Louise Otto (1819–1895) was a founder of the women's movement that emerged in midnineteenth-century Germany. A writer and an editor, her first novel, *Schloss und Fabrik* (1846; *Castle and Factory*) depicted the class tensions caused by industrialization and captured the economic unrest of the 1840s. During the Revolution of 1848–1849, Otto gained public attention with a letter on women's work addressed to the Minister of Interior of Saxony and a recently convened Saxon Commission on Workers. The letter, reproduced here, was published in several journals and newspapers. Otto also founded and edited the *Frauen-Zeitung* (*Women's Newspaper*) which appeared from 1849 to 1852 and aimed to represent the interests of women workers. In 1865, she established and served as president of the General German Women's Organization, the first national women's organization in Germany. In the next year, Otto wrote *Das Recht der Frauen auf Erwerb* (1866; *The Right of Women to Employment*) and initiated the organization's journal *Neue Bahnen* (*New Paths*) which she co-edited for nearly thirty years. In 1894, Otto organized the Alliance of German Women's Organizations which brought together various middle-class women's organizations formed since 1865.

Gentlemen:[1]

My taking the liberty of addressing you with no signature other than the simple one of "a girl" can only be excused by the unlimited confidence I place in the Ministry of the Interior, by the importance I attach to the Workers' Committee, and by the interest I have long taken in the fate of the working classes.

Gentlemen, do not misunderstand me: I do not write this Address *despite* the fact that I am a weak woman, but *because* I am one. Indeed I take it as my holiest *duty* to lend my voice to the cause of those who do not have the courage to represent themselves. You will not be able to accuse me of arrogance, because the history of all times, and especially of today, has taught us that those who forget to think about their rights will also be forgotten. This is why I will exort you about my poor sisters, about poor working women.

Gentlemen, when you deliberate about the great problem of our time—*the organization of labor*—you should not forget that it is not enough to organize work for *men*, you must also organize it for *women*.

Everyone knows that in the working classes women as well as men must labor for their daily bread. I will not dwell on how but since women are admitted only to a few kinds of work, competition within these has so depressed wages that when you look at the whole picture, the fate of working women is much more miserable than that of working men. You all know that this is so and if you do not yet know it, form committees which must corroborate it for you. Now, one can say, that when men are better paid in the future than now, they can better provide for their wives, who can devote themselves to the care of their children, instead of working for others. For one thing, I fear that the fate of the working classes cannot be improved to this degree right away

---

[1]The letter was entitled "A Girl's Address to the Most Honorable Minister, Mr. Oberlander, and the Committee on Workers Convened by Him, and to All Workers."

Contributed by Bonnie Anderson and translated by Bonnie Anderson and Renate Bridenthal. Used by permission.

and, after all, there would still remain the multitudes of widows and orphans, and full-grown girls in general, even excepting wives and mothers. Furthermore, it would mean that half of mankind pronounces for minors and children and makes complete and total dependents of the others. It means this, to put it bluntly, to on courage immorality and crime. A girl who can barely earn a living will direct all her energies to getting a husband, through whom she will be relieved of these concerns. If she is already corrupted, she will give herself to the first man who comes along so that he will marry her—if not for herself, then for her children. Or, if she has not sunk so low, she will marry the first comer, no matter whether she loves him and suits him or not. In any case, the number of unhappy, immoral, heedless marriages, of unhappy children, and of the unhappiest proletarian families is considerably increased precisely because *the lot of the unmarried working women is such a sad one.* I have not yet called attention here to the worst fate of the female proletariat—it is prostitution. I blush to use this word before you—but I blush even more over the social circumstances of a state which is able to give to thousands of its poor daughters no other bread than the poison of a hideous occupation, based on the depravity of men!

Gentlemen—in the name of morality, in the name of the fatherland, in the name of humanity, I charge you: *Do not forget women in the organization of labor!*

You, most honorable Minister, will not forget them, since your heart has room for *all* the sufferings of the people. You thought about the poor hungry lacemakers, about the general state of distress, even then, when you prophesied: that if things continue in this way, there will be only a hundred rich people and millions of poor. In the Chamber of Deputies, your word faded away without a trace and only outside did it fall into the grateful hearts of the poor and their friends. You will also take the fate of the poor working women into your (and so the best) hands and will not be angry at me for raising my weak voice on behalf of a part of the people who do not yet dare to plead for their interests themselves.

And you, gentlemen, who are appointed to the investigation and regulation of workers' conditions, think also about the weaker sex, which since it cannot help itself has a sacred right to claim this help from you, the stronger sex. Also do not forget the women who work in factories, the women day laborers, the knitters, the seamstresses, etc. Ask them about their wages, the pressure under which they languish and you will discover how necessary your help is here.

And also for you, gentlemen, and for you also, *the entire great multitude of working men,* have I written this address. As the stronger sex, you also have the obligation to help the weaker! Are these not your wives, sisters, mothers, and daughters, whose interests must be watched over as much as your own. Instead of that, it was possible for the male factory workers in Berlin, desirous of bettering their lot, to demand that all women be fired from the factory!—This is an abuse of the right of the stronger! Workers! I am convinced that the majority of you are filled with a different spirit! No, do not yet concede that misery will force your daughters *to sell* their only possession, their *honor,* to lecherous rich men, since their labor is disdained! Do not allow this shame to accompany poverty anymore! Think not only how you can get a livelihood for yourselves, but also for your wives and daughters!

I am certain that my poor sisters share my feelings, but their days pass in such need and dullness that they do not dare to express publicly their pleas and wishes the way men do. So I have dared to do this alone for them through the only means available to me, to at least attempt an effect on the general good—through the press. May I have succeeded in directing your attention to the condition of working women and the need for its improvement!

Louise Otto
Leipziger Arbeiter-Zeitung
May 20, 1848

105. Jeanne Deroin and Pauline Roland
Prison Letter, 1851

Jeanne Deroin (1805–1894) and Pauline Roland (1805–1852) participated in the Saint-Simonian utopian socialist movement of the 1830s and were leading activists in the Revolution of 1848. In 1849, Deroin ran unsuccessfully as the first woman candidate to the new Legislative Assembly on a platform calling for equal citizenship for men and women. In this same year, she also founded a group of workers' associations and, with Roland, helped to create and served on the central commission of an expanded Union of the Fraternal Associations of Workers. The government considered these associations subversive and in May 1850 Deroin and Roland were arrested and detained in prison until November when they were tried, convicted, and sentenced to six months in prison. Midcentury international connections among French, American, and British feminists are revealed in letters Deroin and Roland wrote from Saint Lazare prison in Paris to the American Women's Suffrage Convention and, as reproduced here, a letter to the Sheffield Female Political Rights Association formed by Chartist women. In 1852, Deroin published the first issue of the *Almanach des femmes.* Fearful of political repression in the wake of Louis Napoleon's coup d'état of December 1851, she went into exile in London where she published the journal for the next two years. Roland was arrested in February 1852 and deported to Algeria. She received an unsolicited pardon five months later but died on her way back to France from the harsh conditions she had experienced in the Algerian prison.

To the Political Rights Union of the Women of Sheffield[1]

Dear Sisters,—Your appeal has resounded in our prison, and filled our souls with inexpressible joy. The women of France would establish a hospitable tribune to welcome the complaints of the oppressed and of the suffering, and claim in the name of humanity the social rights of women as well as men.

This hospitable tribunal will claim the right of true liberty and the complete unfolding of all our faculties, one half of which are in women.—She must therefore be emancipated. Without this no social work can be accomplished.

The darkness of reaction has obscured the sun of 1848. Why? Because the storm, in overthrowing the throne and the scaffold—in breaking the chain of the black slave, had forgotten to break also the chain of the woman,—this pariah of humanity—for after, as before the revolution, she is nothing, and can do nothing of herself; she is not reckoned as a member of society; she is without a name or a country. Her name? It is the name of her master, of the father, or the husband. Her country? Whether she be born on the banks of the Ganges, the Thames, or of the Seine, it is the country of her master; for she ever bears the law imposed on her by man.

"There are no more slaves!" said our brothers in 1848. All will have the right of electing deputies, &c. On hearing this appeal, Woman

---

[1]The letter was read to members of the organization at their weekly meeting on June 11, 1851, by Abiah Higginbotham, the corresponding secretary and a signatory of the 1851 petition to the House of Lords demanding the vote for women. It was subsequently published in the August 9 issue of the *Northern Star,* a Chartist newspaper. On women and Chartism, see Source 101 in this chapter.

Contributed by Bonnie Anderson.

arises to exercise her right, but the barrier of privilege interposes and says,—*"You must wait."*

And soon, indeed, on the fatal days of June, 1848, liberty glides away from her pedestal in the blood of the victims of the reaction. Based on the right of the strongest, she falls, overthrown by the right of the strongest.

The Constituent Assembly keeps silence on the right of one half of humanity. There is no mention made of the rights of woman in a constitution framed in the name of Liberty, Fraternity, and Equality.

It is in the name of these principles that Woman claims the right to be a member of the Legislative Assembly who are to frame the laws to govern the society of which she is a member.... But, while the elected half of the people, of the men alone, call out for brute force to stifle liberty, and forge restrictive laws to establish order by compression, the woman, guided by fraternity, and foreseeing incessant conflict—inspired by the hope of putting an end to it—comes to make an appeal to the labourers to establish Liberty and Equality on fraternal solidarity. Woman gave to this work of affranchisement a character eminently pacific; male workers did not disown the right of Woman, as the companion of their labour.

The delegates of 104 associations assembled without distinction of sex. The union of the association[2] had for its object the organisation of labour. Here was laid the foundation of a society, indeed, based on liberty, &c.—in the name of the law, framed by men who are now shut up in the walls of a prison. But the Rights of Women have been acknowledged by the labour[ers] and they have consecrated this right by the election of those women, who, after having accomplished the mission of enfranchisement, partake at the present hour their captivity.

It is from their prison, that they address to you, of Sheffield, the relation of facts, which comprise in themselves a high instruction. It is by labour, and by enlisting themselves in the ranks of the labourers that women will acquire civil and political equality, on which depends the happiness of the world. As to moral equality, has she not obtained it by the power of her moral feeling?

Sisters of Sheffield, your Sisters of France unite with you for claiming the Rights of Woman, both civil and political; they have the sound conviction that it is only by the power of association, by the union of the labours of the two sexes to organize labour, that we can acquire, completely and specifically the civil and political equality of Woman, and of all the members of the labouring classes. It is in this confidence that from the depths of the gaol, which for a time incarcerates their bodies, without being able to imprison their hearts, that we repeat to you the cry of faith, love, hope! and we send you our most fraternal salutations.

Jeanne Deroin and Pauline Roland
Paris, St. Lazare, May 31, 1851

---

[2]Delegates from the 104 workers' associations met in Paris on October 5, 1849, and adopted a proposal to form the Union of the Fraternal Associations of Workers.

QUESTIONS FOR ANALYSIS

1. On what basis does the "Address of the Female Political Union of Newcastle-upon-Tyne" claim women's right to participation in politics? How does it appeal to upper- and middle-class women?

2. How does Flora Tristan propose to create a Workers' Union? Discuss the main points of her appeals to the king, the Catholic clergy, the manufacturers, and women. What do they reveal about her views on societal transformation? What is the purpose of the workers' palaces? How will they complement the Workers' Union?

3. What were the life circumstances of the women who requested work during the Revolution of 1848? How did they regard various governmental officials and the members of the Provisional Government?

4. How does Louise Otto portray the conditions of women workers in Germany? What is the purpose of her letter to the Minister of the Interior and its publication in the *Leipziger Arbeiter-Zeitung,* a workers' newspaper?

5. How do Jeanne Deroin and Pauline Roland assess their participation in the Revolution of 1848 and the present political situation in France? What does their letter to the Political Rights Union of the Women of Sheffield reveal about women's international connections?

6. Compare the requests, demands, and proposals of women workers in this section with those made by women in the Old Regime and the French Revolution. To what factors do you attribute the transformation of working women's demands and methods of organizing? What were the main points of commonality and difference in women's participation in political movements in midcentury Europe?

Women and the Development of Bourgeois Society, Politics, and Culture

*I*n the nineteenth century, European bourgeois society, politics, and culture developed in relation to nationalism, colonialism, and the growth of a mass, consumer society which accompanied industrialization.

Women's political demands and movements in the latter part of the century took place in the context of wars for national unification and the democratic extension of the franchise. The unification of Italy in 1860 and Germany in 1871 provided working- and middle-class women with a new, national basis on which to demand political, legal, and economic rights. In France, the collapse of the Second Empire during the Franco-Prussian War of 1870 led to the declaration of the Third Republic and the short-lived Paris Commune. The struggle for a Social Republic was now renewed within a framework of national defense. Although the parliamentary system in England was the most developed in Europe, a large percentage of the male population as well as all women were denied the vote. In the 1860s, as men organized to demand a lowering of property qualifications, women began the struggle for female suffrage which they would not achieve until after World War I. In Russia, by contrast, men and women alike were denied political rights and women carried out revolutionary activities against the autocracy as participants in various populist movements that emerged in the 1860s and 1870s.

Nationalism also provided a new impetus for colonialism. By the end of the nineteenth century, virtually all of Africa was divided up among the colonial powers. Colonialism assumed an essential role in European industrialization as colonial territories provided cheap raw materials, captive markets for European industrial goods, and control of commercial sea lanes. The possession of colonies also provided an important basis for national prestige. England's empire was the largest as it encompassed vast territories in Asia and Africa. The subcontinent of India assumed an especially prominent role in British colonialism and Queen Victoria was given the title of "Empress of India" in 1877. Thousands of "surplus" women emigrated to the colonies in search of work and new career opportunities. Typically, colonial women adhered to the racist beliefs that provided the rationale for the "civilizing" mission. However, European women also worked with colonized women to challenge certain patriarchal traditions and colonial laws, and some British women even argued for India's right to a limited form of self-rule, if not outright independence.

The emergence of a new consumer society was the third major factor that defined nineteenth-century bourgeois culture. The accumulation of bourgeois wealth increased consumer demand for traditional forms of art as well as new forms of artistic representation, fashion, advertising, and leisure activities. Female writers and readers were crucial to the expansion of the market for novels and mass-circulating periodicals, while female artists contributed to the creation of bourgeois culture with artistic activities ranging from traditional oil paintings to the new art of photography. The subject matter of female artists also spanned a wide spectrum, from domestic European scenes to drawings of women in harems in the Middle East.

NATIONAL AND DEMOCRATIC STRUGGLES

European governments in the last third of the nineteenth century ran the gamut from autocracy in Russia, where political rights did not exist at all, to the parliamentary system in England, where some men and all women were still denied the vote. These documents from Italy, England, France, and Russia illustrate how national and democratic struggles presented women with new challenges and opportunities.

106. Anna Maria Mozzoni
Women and the New Civil Code in Italy

Anna Maria Mozzoni (1837–1920) was an Italian women's rights activist. For more than sixty years, she wrote about and participated in struggles concerning women's legal rights, suffrage, education, employment, divorce, prostitution, and the political influence of the Catholic church. Mozzoni first became politically active in the 1860s around the issue of women's legal status in the newly unified Italy. The Kingdom of Italy was formed under the leadership of the Piedmont monarchy and contrasted sharply with the democratic republic envisioned by Mazzini and Garibaldi. The proposal for a new, uniform civil code intended to eliminate local customs and laws was modeled on the conservative Piedmontese Code of 1937, which in turn was inspired by the Napoleonic Code.[1] During the 1865 Italian Senate debate on the code, Mozzoni wrote "La donna in faccia al progetto del nuovo Codice Civile Italiano" ("Women and the Project for the New Italian Civil Code"), the first part of which is excerpted here.[2]

In the 1860s and 1870s, she promoted her ideas on women's emancipation as a member of Mazzini-oriented groups and journals. One of the most important advocates for women's suffrage in Italy, Mozzoni edited the newspaper *La donna* (*Woman*) which linked women's emancipationist and suffrage groups throughout Italy. This was especially important given the absence of a national, feminist organization in Italy. In 1870, Mozzoni translated John Stuart Mill's *On the Subjection of Women* (1869) into Italian. She participated in the first International Congress on Women's Rights held in Paris in 1878 and, four years later, established the League for the Promotion of Women's Interests to inform women about their legal rights and responsibilities within society. In the latter part of the 1880s, Mozzoni played an important role in preparations leading up to the formation in 1892 of the Italian Socialist Party. While working with the party, she always insisted on the specificity of women's interests in politics. In 1919, just one year before her death, Mozzoni participated in a House of Deputies debate on female suffrage.

WOMEN AND THE PROJECT FOR THE NEW ITALIAN CIVIL CODE

Can legislation not but take into consideration those principles acknowledged by philosophy?

Can judicial laws be anything other than conventional, yet still follow the laws of nature?

Can civil laws be content with the safeguarding, more or less, of property and persons, without referring to the principle on which the human community is based and the goal toward which it is moving?

Once a right is acknowledged and established, can the law prevent its explication and suppress its application?

---

[1]On the Napoleonic Code, see Source 98 in Chapter 8.

[2]Mozzoni wrote this article a year after "La donna e i suoi rapporti" ("Women and Their Social Relations"), her first critique of the new Civil Code proposed for Italy.

Translated by Lila Di Caprio. Used by permission.

These are the questions which I posed to myself in discussing the conditions of women in relation to the law in my essay which has just been published, "La donna e i suoi rapporti sociali" ("Women and Their Social Relations").

These are complicated questions for which I was seeking a solution based on a logic founded in law, and the difficulty of this solution was in my finding imperfections, contradictions and barbarisms in our existing laws.

But I would have aided my sex but little, and would have poorly served the cause which I support, had I limited myself to calling the attention of legislative bodies and logical-minded and honest citizens to the miserable conditions women are subject to under the dominion of laws now in effect (imperfections which more or less all Italians admit to, since reforms have been proposed), limited myself, I mean, to so little, putting my trust, in spite of reason and experience, in a victory which numerous interests, widespread prejudices and centuries-old habits make extremely difficult, if not impossible.

The affirmation of the rights of women, in principle, is accepted by the spirit of the masses and this principle has already been embodied in the customs of all civilized peoples. If men represent the family in business, women represent it in society. Besides, there is no husband who still seriously believes in his legal dominion over his wife, not a son who does not acknowledge maternal rights and does not respect his mother's wishes, wherefore the law should do nothing more than sanction this reality. And what is more, it should do this because not a single paragraph, no matter how forcefully it is conceived, could destroy this reality, as it should first destroy reason, blood ties, sentiments and the rules of nature.

But there is more; since the law, by placing the destiny of families in the hands of men and entrusting it to their abilities, does not and cannot always give them abilities, therefore, it is not rare that the family is guided by the woman who administers and manages it not only "de facto" but also by right, as those who

are capable must supplant the inept and those with sight, guide the blind. Therefore, the law must impotently witness its own abolition and bow to necessity.

The law is partially responsible for placing itself in this state of infirmity and impotence each time it denies the principles of natural law, which is not the law of any given place or people or time, but is the law of all places, all peoples, and all times, and the insufficiency of the law is evident in its eternal battle with custom.

Faced with these facts, which strongly support my position, I will not reformulate theories on laws, which have met with public approval and which seem to satisfy common sense. It is not my task to theorise here, but to pose these doctrines in relation to the conditions which the project for the new Civil Code will create for Italian women. And my task will be so much the easier, in so much as I will not need to draw conjectures or look for interpretations regarding the spirit and the intention of the law in succinct and arid paragraphs; but my efforts can be reduced to following the Most Honourable Minister Pisanelli's report, and to considering the development of his doctrines and their more or less exact application as proposed by him.

The Minister makes it his duty to motivate, discuss and demonstrate all of his proposals, to anticipate objections which may be directed at him and to prepare his rebuttal.

He appeals first to reason, then to sentiment, sometimes to the application and more often to the tradition of Roman Law, to Napoleonic laws, to one or the other of the codices which are currently in force on Italian soil. The Minister's eclecticism is obvious, a bee could not do better!

Although I am by nature adverse to eclecticism, whose results must necessarily be hybrids, and being rather a lover of lucid principles which result in applications which are logical, spontaneous, sure and imprinted with all their original characteristics, I see the necessity of resigning myself to the facts, since this is not new legislation, but rather a reform, a modification which is intended; and I now pin down the developments in the new civil laws with regards to the condition of women.

I cannot but applaud the minister, and agree with him when, considering marriage as a social institution, in that it is fundamental to the family, or rather the nest of humanity, he calls upon it to depend on the State and to receive legal sanction from it. This situation, of preparing the State to emancipate itself from a dominant religion, which carries with it the implicit suppression of tolerated sects and the privilege for itself of being the dominant religion, perfectly obeys the principles of liberty of conscience, so highly acclaimed by philosophy....[1]

The project begins by assigning guardianship[2] to mothers; and in her absence, by logical corollary, to her ascendants; then takes into consideration that nepotism has always played a great role in social drama, and that for the woman who has no family of her own, it is naturally constituted by her nieces and nephews. And that is fine. But all of a sudden, the project pauses as on the brink of a steep slope and realises that guardianship has a public function and as such is not suitable for women and puts a stop to concessions. It is perhaps in vain that one points out that motherhood is a guardianship and consequently as private or public as it may be, there is no function more suitable for women than this. The project does not respond, but clings obstinately to not granting women any public functions. We therefore throw the gauntlet that it is indeed public.

What, in practical terms, is public in guardianship, in as much as it takes place behind domestic walls; that if contact with a magistrate and a court for minors is all it takes to make it public, then we would have to say that everything we do is public, in as much as

there is not a single citizen in daily life who is not exposed to such a possibility, whether male or female. This problem is so elastic that it is impossible to pose it in lucid detail.

What we would have wished from the minister is that instead of excluding women from public functions for its own sake,[3] that he would have demonstrated and proven through the above documents what makes it incompatible for women to play a public role. When society employs the arms of a woman for physical labor and in factories and is content without concerning itself with whether her muscles are the strongest, I see no reason why it cannot also employ her mind, which is not as empty as it contends.

No matter how he twists and turns this syllogism, our Minister cannot but, in the end, reach this conclusion which has not yet become custom, and so I will answer him in the words of Viennet:

"L'usage est un vieux sot qui governe le monde."[4]

The second reason for exclusion is household duties.[5] It is decisively a disgrace for manly criteria not to be able to divest itself of vagueness, uncertainty, nebulousness, and abstractness to establish the rules of civil procedure in truth, determination, the practical and the concrete. One would say that the philosopher, like the poet, has to avoid certain realities where his flights of fancy would become mired.

What are household duties?

It is the material and daily catering to material and daily needs.

And what are the material and daily needs of man? Clothing and food. Now let us analyse its value.

---

[1]Minister's report. Civil marriage.

[2]Guardianship refers not only to the physical custody of children, but also to the legal right to make various decisions affecting their upbringing and future.—Ed.

[3]Minister's report. "The principle of equality, on which the project is based, does not seem to go so far as to admit legally that women can play a public role."

[4]"Custom is an old fool that governs the world."—Ed.

[5]Minister's report. "On the other hand, the household duties which are specifically the domain of women, the natural reserve with which all her efforts are devoted to the benefit of the family, must be highly respected by legislators."

All men cover their bodies and nourish themselves, but not all do this in the same manner. The poor man satisfies these needs in a few minutes with just a few things. The means he uses to secure them takes up his whole day. Men and women in this class are equal. Household duties do not by any means keep the woman from being employed out of the house in some factory all day. The woman who considers herself occupied exclusively in family matters can already consider herself well-off....

107. Barbara Leigh Smith Bodichon
Female Suffrage in England

Barbara Leigh Smith Bodichon (1827–1891) was a leading proponent in England of women's right to educational and career opportunities, the Married Women's Property Bill, and female suffrage. Soon after attending Ladies College in Bedford Square, Bodichon established an experimental school for girls and boys. She later cofounded and served as an important benefactress of Girton College for women at Cambridge University. Bodichon's interest in legal discrimination faced by women led to her 1854 pamphlet, *A Brief Summary, in Plain Language, of the Most Important Laws Concerning Women.* Two years later, she formed a committee to petition for a Married Women's Property Act. It was defeated when first presented in 1857 and finally passed in 1870, but not until 1882 did married women achieve full property rights. To provide a forum for women's rights issues, Bodichon cofounded the *English Woman's Journal* in 1858 with Bessie Rayner Parkes who became its editor. Published in Langham Place, London, the journal served as a focal point for women activists who became known as the Langham Place Group. In 1859, some members of the group formed the Association (later Society) for Promoting the Employment of Women. In the 1860s, the *English Woman's Journal* took up the issue of women's suffrage and Bodichon, Parkes, and Emily Davies worked on John Stuart Mill's successful 1865 campaign for Parliament in which he advocated female suffrage. Bodichon subsequently organized a women's suffrage committee to prepare and collect signatures on the petition that Mill presented to Parliament during the 1867 Reform Act debates. The petition called for the extension of the vote to women on the same terms as men. By lowering property qualifications, the 1867 law nearly doubled the male electorate but still denied the vote to all women. The following document is excerpted from Bodichon's 1869 article, "Reasons for and against the Enfranchisement of Women." Women won the right to vote in municipal elections in 1870, but it was not until 1918 that women in England and Ireland over the age of thirty could vote in national elections.[1]

---

[1]On the suffrage movement in the twentieth century, see Emmeline Pankhurst, Source 129 in Chapter 10.

1869 by Barbara Leigh Smith Bodichon.

That a respectable, orderly, independent body in the State should have no voice, and no influence recognised by the law, in the election of the representatives of the people, while they are otherwise acknowledged as responsible citizens, are eligible for many public offices, and required to pay all taxes, is an anomaly which seems to require some explanation. Many people are unable to conceive that women can care about voting. That some women do care, has been proved by the petitions presented to Parliament. I shall try to show why some care—and why those who do not, ought to be made to care.

There are now a very considerable number of open minded, unprejudiced people, who see no particular reason why women should not have votes, if they want them; but, they ask, what would be the good of it? What is there that women want which male legislators are not willing to give? And here let me say at the outset, that the advocates of this measure are very far from accusing men of deliberate unfairness to women. It is not as a means of extorting justice from unwilling legislators that the franchise is claimed for women. In so far as the claim is made with any special reference to class interests at all, it is simply on the general ground that under a representative government, any class which is not represented is likely to be neglected. Proverbially, what is out of sight is out of mind; and the theory that women, as such, are bound to keep out of sight, finds its most emphatic expression in the denial of the right to vote....

...And among all the reasons for giving women votes, the one which appears to me the strongest, is that of the influence it might be expected to have in increasing public spirit.... And I know no better means, at this present time, of counteracting the tendency to prefer narrow private ends to the public good, than this of giving to all women, duly qualified, a direct and conscious participation in political affairs. Give some women votes, and it will tend to make all women think seriously of the concerns of the nation at large, and their interest having once been fairly roused, they will

take pains, by reading and by consultation with persons better informed than themselves, to form sound opinions. As it is, women of the middle class occupy themselves but little with anything beyond their own family circle. They do not consider it any concern of theirs, if poor men and women are ill-nursed in workhouse infirmaries, and poor children ill-taught in workhouse schools. If the roads are bad, the drains neglected, the water poisoned, they think it is all very wrong, but it does not occur to them that it is their duty to get it put right. These farmer-women and business-women have honest, sensible minds and much practical experience, but they do not bring their good sense to bear upon public affairs, because they think it is men's business, not theirs, to look after such things. It is this belief—so narrowing and deadening in its influence—that the exercise of the franchise would tend to dissipate. The mere fact of being called upon to enforce an opinion by a vote, would have an immediate effect on awakening a healthy sense of responsibility. There is no reason why these women should not take an active interest in all the social questions—education, public health, prison, discipline, the poor laws, and the rest—which occupy Parliament, and they would be much more likely to do so, if they felt that they had importance in the eyes of members of Parliament, and could claim a hearing for their opinions.

Besides these women of business, there are ladies of property, whose more active participation in public affairs would be beneficial both to themselves and the community generally. The want of stimulus to energetic action is much felt by women of the higher classes. It is agreed that they ought not to be idle, but what they ought to do is not so clear. Reading, music and drawing, needlework, and charity are their usual employments. Reading, without a purpose, does not come to much. Music and drawing, and needlework, are most commonly regarded as amusements intended to fill up time. We have left, as the serious duty of independent and unmarried women, the care of the poor in all its branches, including visiting the sick and the aged, and ministering to their wants, looking after the schools, and in every possible way giving help wherever help is needed. Now education, the relief of the destitute, and the health of the people, are among the most important and difficult matters which occupy the minds of statesmen, and if it is admitted that women of leisure and culture are bound to contribute their part towards the solution of these great questions, it is evident that every means of making their co-operation enlightened and vigorous should be sought for. They have special opportunities of observing the operation of many of the laws. They know, for example, for they see before their eyes, the practical working of the law of settlement—of the laws relating to the dwellings of the poor—and many others, and the experience which peculiarly qualifies them to form a judgment on these matters ought not to be thrown away. We all know that we have already a goodly body of rich, influential working-women, whose opinions on the social and political questions of the day are well worth listening to. In almost every parish there are, happily for England, such women. Now everything should be done to give these valuable members of the community a solid social standing....

...Now, let us calmly consider all the arguments we have heard against giving the franchise to women.

Among these, the first and foremost is— women do not want votes. Certainly that is a capital reason why women should not have votes thrust upon them, and no one proposes compulsory registration. There are many men who do not care to use their votes, and there is no law compelling them either to register themselves or to vote. The statement, however, that women do not wish to vote, is a mere assertion, and may be met by a counter-assertion. Some women do want votes, which the petitions signed, and now in course of signature, go very largely to prove. Some women manifestly do; others, let it be admitted, do not. It is impossible to say positively which side has the

majority, unless we could poll all the women in question; or, in other words, without resorting to the very measure which is under discussion. Make registration possible, and we shall see how many care to avail themselves of the privilege.

But, it is said, women have other duties. The function of women is different than that of men, and their function is not politics. It is very true that women have other duties—many and various. But so have men. No citizen lives for his citizen duties only. He is a professional man, a tradesman, a family man, a club man, a thousand things as well as a voter. Of course these occupations sometimes interfere with a man's duties as a citizen, and when he cannot vote, he cannot. So with women; when they cannot vote, they cannot.

The proposition we are discussing, practically concerns only single women and widows who have 40s. freeholds, or other county qualifications, and for boroughs, all those who occupy, as owners or tenants, houses of the value of £10 a year. Among these there are surely a great number whose time is not fully occupied, not even so much as that of men. Their duties in sickrooms and in caring for children, leave them a sufficient margin of leisure for reading newspapers, and studying the *pros* and *cons* of political and social questions. No one can mean seriously to affirm that widows and unmarried women would find the mere act of voting once in several years arduous. One day, say once in three years, might surely be spared from domestic duties. If it is urged that it is not the time spent in voting that is in question, but the thought and the attention which are necessary for forming political opinions, I reply that women of the class we are speaking of, have, as a rule, more time for thought than men, their duties being of a less engrossing character, and they ought to bestow a considerable amount of thought and attention on the questions which occupy the Legislature. Social matters occupy every day a larger space in the deliberations of Parliament, and on many of these questions women are led to think and to judge in the fulfilment of those duties which, as a matter of course, devolve upon them in the ordinary business of English life. And however important the duties of home may be, we must bear in mind that a woman's duties do not end there. She is a daughter, a sister, the mistress of a household; she ought to be, in the broadest sense of the word, a neighbour, both to her equals and to the poor. These are her obvious and undeniable duties, and within the limits of her admitted functions; I should think it desirable to add to them—duties to her parish and to the State. A woman who is valuable in all the relations of life, a woman of a large nature, will be more perfect in her domestic capacity, and not less....

...We do not want to compel women to act; we only wish to see them free to exercise or not, according as they themselves desire, political and other functions.

The argument that 'women are ignorant of politics,' would have great force if it could be shown that the mass of the existing voters are thoroughly well-informed on political subjects, or even much better informed than the persons to whom it is proposed to give votes. Granted that women are ignorant of politics, so are many male ten-pound householders. Their ideas are not always clear on political questions, and would probably be even more confused if they had not votes. No mass of human beings will or can undertake the task of forming opinions on matters over which they have no control, and on which they have no practical decision to make. It would by most persons be considered a waste of time....

The fear entertained by some persons that family dissension would result from encouraging women to form political opinions, might be urged with equal force against their having any opinions on any subject at all. Differences on religious subjects are still more apt to rouse the passions and create disunion than political differences. As for opinions causing disunion, let it be remembered that what is a possible cause of disunion is also a possible cause of deeply-founded union. The more rational women become, the more real union there will be in families, for nothing separates so much as unreasonableness and frivolity....

An assertion often made, that women would lose the good influence which they now exert indirectly on public affairs if they had votes, seems to require proof. First of all, it is necessary to prove that women have this indirect influence,—then that it is good,—then that the indirect good influence would be lost if they had direct influence,—then that the indirect influence which they would lose is better than the direct influence they would gain. From my own observation I should say, that the women who have gained by their wisdom and earnestness a good indirect influence, would not lose that influence if they had votes. And I see no necessary connexion between goodness and indirectness. On the contrary, I believe that the great thing women want is to be more direct and straightforward in thought, word, and deed. I think the educational advantage of citizenship to women would be so great, that I feel inclined to run the risk of sacrificing the subtle indirect influence, to a wholesome feeling of responsibility, which would, I think, make women give their opinions less rashly and more conscientiously than at present on political subjects.

A gentleman who thinks much about details, affirms that 'polling-booths are not fit places for women.' If this is so, one can only say that the sooner they are made fit the better. That in a State which professes to be civilised, a solemn public duty can only be discharged in the midst of drunkenness and riot, is scandalous and not to be endured. It is no doubt true, that in many places polling is now carried on in a turbulent and disorderly manner. Where that is unhappily the case, women clearly must stay away. Englishwomen can surely be trusted not to force their way to the polling-booth when it would be manifestly unfit....

Nor is it needful to discuss the extreme logical consequences which may be obtained by pressing to an undue length the arguments used in favour of permitting women to exercise the suffrage. The question under consideration is, not whether women ought logically to be members of Parliament, but whether, under existing circumstances, it is for the good of the State that women, who perform most of the duties, and enjoy nearly all the rights of citizenship, should be by special enactment disabled from exercising the additional privilege of taking part in the election of the representatives of the people. It is a question of expediency, to be discussed calmly, without passion or prejudice....

108. The Social and Economic Measures of the Paris Commune

The Paris Commune was formed on March 18, 1871, in defiance of the decision of the national government to surrender to Prussia during the Franco-Prussian War. The political views of female Communards, like those of their male counterparts, reflected the heterogeneous nature of the European workers' movement at this time. They ranged from anarchists like Louise Michel to Elizabeth Dmitrieff, a representative of the First International, who created the Women's Union for the Defense of Paris and for Aid to the Wounded. In existence for less than two months, the Commune nonetheless implemented several social and economic measures that benefited women. It established a Commission of Labor and Exchange to which Communards could address proposals for work. The Commune agreed to pay a minimum wage to workers in military production and to give workers' cooperatives a preference in contracts for military supplies. These decisions were especially important for women workers who were the main producers of clothing for the National Guard which played a crucial role in the defense of Paris. The Commune also created a Commission for the Organization of Education, an all-female Commission for Girls' Education, and supported initiatives to provide a secular education to all children under sixteen. Only one child in three received an education at this time and half of the schools in Paris were under the control of the Church.

The Paris Commune posed a tremendous threat to the European powers. Prussia and Russia, in particular, pressured the Versailles government led by Adolphe Thiers to take decisive action. The French military attacked on May 21, 1871, and crushed the Commune following a bloody week of fighting during which an estimated 20,000 to 25,000 were killed and 40,000 were taken prisoner. The Communards made their last stand at the Père Lachaise cemetery. Several buildings were burned during the fighting, including the Hôtel-de-Ville which conservatives blamed on female incendiaries or *petroleuses*. The following four documents illustrate the range of working women's demands for work and educational opportunities during the Paris Commune.

A. The Organization of Women's Work

The Revolution of 18 March was spontaneously accomplished by the people in circumstances unique in history. It is a major victory for the rights of the people in their relentless battle against tyranny, a battle that was first waged by the slave, was continued by the serf and will be gloriously brought to an end by the proletarian through the Revolution of social equality.

The new movement was so unexpected and so radical that it was beyond the understanding of professional politicians, who merely saw it as an insignificant, aimless revolt.

Others have tried to belittle the spirit of the Revolution by reducing it to a mere demand for 'municipal rights', for some kind of administrative autonomy.

But the people are not taken in by the illusions perpetrated by governments, nor by so-called parliamentary representation; in proclaiming the Commune they are not demanding certain municipal prerogatives but communal autonomy in its greatest sense.

To the people the Commune does not merely signify administrative autonomy; above all it represents a sovereign authority, a legislative authority. It stands for the entire and absolute right of the community to create its own laws and political structure as a means to achieving the aims of the Revolution. These aims are the emancipation of labour, the end of monopolies and privileges, the abolition of the bureaucracy and of the feudalism of industrialists, speculators and capitalists, and finally the creation of an economic order in which the reconciliation of interests and a fair system of exchange will replace the conflicts and disorders begotten by the old social order of inaction and *laissez-faire.*

For the people the Commune is the new order of equality, solidarity and liberty, the crowning of the communal Revolution that Paris is proud to have initiated....

Today it is the duty of the Commune to the workers who created it to take all necessary steps to achieve constructive results.... Action must be taken and it must be taken fast. However, we must not resort to expedients or makeshift solutions that may sometimes be appropriate in abnormal situations but which only create formidable problems in the long run, such as those resulting from the closure of the National Workshops in 1848.[1]...The Commune must abandon the mistaken ideas of old, it must gather inspiration from the very difficulties of the situation and apply methods that will survive the circumstances that first led to their use.

We will achieve this through the creation of special workshops for women and trading centres where finished products may be sold.

Each arrondissement would open premises where the raw materials would be taken in and distributed to individual women workers or to groups according to their skills. Other buildings would receive the finished products for their sale and storage.

The necessary organization for the application of this scheme would be under the control of a committee of women appointed in each municipal district.

The Commune's Commission of Labour and Exchange could organize the distribution of raw materials to the arrondissements from a vast central building.

Finally the Finance Delegate would make a weekly credit available to the municipalities so that work for women can be organized immediately....

Proposal for the organization of women's work from a printer member of the Commission of Labour and Exchange.

B. Demand for Free Producers' Co-operatives

We consider that the only way to reorganize labour so that the worker enjoys the product of

---

[1]The closure of the National Workshops in June 1848 provoked a workers' uprising in Paris. See Source 103 in Chapter 8 for women's requests for work during the Revolution of 1848.

his work is by forming free producers' co-operatives which would run the various industries and share the profits.

These co-operatives would deliver Labour from capitalist exploitation and thus enable the workers to control their own affairs. They would also facilitate urgently needed reforms in techniques of production and in the social relations of workers, as follows:

a The diversification of work within each trade to counter the harmful effects on body and mind of continually repeating the same manual operation;

b A reduction of working hours to prevent physical exhaustion leading to loss of mental faculties;

c The abolition of all competition between men and women workers since their interests are absolutely identical and their solidarity is essential to the success of the final and universal strike of Labour against Capital;

And therefore:

1 Equal pay for equal hours worked;

2 A federation of the various sections of the trades on a local and international level to facilitate the sale and exchange of products by centralizing the international interests of the producers.

The general development of these producers' co-operatives calls for:

1 Propaganda and organization among the working masses; every co-operative member shall therefore be expected to join the International Working Men's Association;

2 Financial aid from the State for the setting up of these co-operatives in the form of a social loan repayable in yearly instalments at 5 per cent interest.

We also believe that in the social order of the past women's work has been particularly subject to exploitation and therefore urgently needs to be reorganized.

Given the present situation, with poverty increasing at a terrifying rate because of the unjustified cessation of all work, it is to be feared that the women of Paris, having had their revolutionary moment, will relapse under the pressure of continuous hardship to the passive and more or less reactionary role that the social order of the past had cut out for them. This would endanger the revolutionary and international interests of the peoples of the world and consequently the Commune.

For these reasons the Central Committee of the Women's Union asks the Commune's Commission of Labour and Exchange to entrust it with the reorganization and allocation of women's work in Paris and to begin by placing it in charge of military supplies. Since this work is naturally not sufficient to employ the majority of women the federated producers' co-operatives should be given the necessary funds to take over those factories and workshops abandoned by the bourgeois where work is mainly carried out by women....

Address from the Central Committee of the Women's Union for the Defence of Paris and for Aid to the Wounded to the Commission of Labour and Exchange.

C. An Industrial School for Girls

Citoyen Editor of *Le Vengeur,*

I submitted the following proposal to the Hôtel de Ville yesterday, and hope you will see fit to print it.

A proposal for the setting up of an industrial school for training in women's occupations

The aim of the industrial school is to revise and complete the scientific education of girls while affording at the same time sound vocational training.

To fulfil this aim groups of working women and groups of teachers or sufficiently educated women, more suited to intellectual than practical work, would be elected. The latter would undergo examination by a competent panel.

Together these groups would form the teaching staff of the industrial school.

The exchange of knowledge among women of different types of intelligence working side by side would provide a most favourable setting for a progressive education entirely free of prejudice.

The pupils would attend the industrial school from the age of twelve; practical work would alternate with the study of scientific theories and the industrial arts.

The State or the municipality, depending on their resources, would be able to assist the industrial schools, fix and guarantee their rent, pay the teachers and collect the manufactured goods from the workshop. As soon as they are skilled enough to produce, the pupils will be able to receive some remuneration.

The Industrial School would be a great improvement on the needlework school at present managed by nuns; it would be a truly professional school.

V. Manière,
Headmistress of the Provisional
Industrial School,
38 Rue Turenne.

Proposal for an industrial school for girls, 2 April 1871: *Le Vengeur,* 3 April 1871.

D. Day Nurseries

Education starts from the very first day of life; it is therefore important to decide how much and what kind of education is suited to the very young child, while recognizing, however, that the main aim at this stage of life is physical development.

If we accept the saying *mens sana in corpore sano,*[1] then it is obvious that the development of the child's mind is affected by its state of health. A healthy mind cannot dwell in a sick body.

In present-day urban society, mothers pay nurses to look after their new-born children. The rich women do this often for reasons of

vanity, the tradeswomen because the cost of months of wet-nursing is offset by their gainful employment, and the working-class women because it is impossible for them to do a heavy day's work and also provide the continuous care that an infant requires.

In the country children are usually raised by their mothers; the examination of the facts, that is, positive science, has established that this method of upbringing is the best and that we must return to natural practices. This confirms the principle that J.-J. Rousseau regarded as universal: that the so-called progress of civilization has brought about the degeneration of man.

Among all animals in the wild state the female suckles her young, and in these animals the degeneration of the races is unknown. Rickets have been produced experimentally by weaning animals prematurely, but they are never present in the natural state.

Given these facts, and since in an ideal society the product we must seek to perfect above all others is the child, it follows that the mother ought not to engage, during gestation and lactation, in any occupation that might be harmful to the health of her child or the quality of her milk.

To achieve this result economic reforms are necessary. Either the breadwinner's wages must suffice to maintain his family or the State must intervene, for our aim should be to arrest the process of physical and moral degeneration of the French people and to eliminate the consequences of involuntary poverty.

Until such a time as society is rebuilt on new political and social foundations, we must accept it as it is and apply such reforms as we can where the radical cure, in other words the Revolution, has yet to come about.

What are these reforms?

1 Up to now Public Assistance has paid an allowance to all unmarried mothers as an incentive to look after their children, and it has also given aid to needy women. But these payments were not sufficient to enable the mothers to

---

[1] A healthy mind in a healthy body.

nurse their children and thus build generations of sturdy youngsters.

2 Public Assistance has maintained institutions for children abandoned by their parents. These institutions are very costly to maintain; out of every hundred children admitted only three reach the age of twenty. These modern *oubliettes,* which will be regretted by no one except perhaps the staff whose livelihood they provide, should be closed down, and the capital they consume should be used for providing a substantial allowance for nursing mothers.

To those who fear that the payment of an allowance to unmarried mothers might be an incentive to promiscuity we can only reply that poverty and inadequate wages already have such dire effects that it is impossible for the result to be worse. In any case, if a mother nurses and brings up her child this is more likely to have a moralizing effect on her than if she abandons it to a poor-house. The latter solution leaves her free almost the day after the child's birth, without relieving her in any way from the poverty that will cause her inevitable relapse.

Until such time as mothers can be relieved of all outside work during nursing through the social reforms we advocate, the day nursery can be of considerable value to the mother and the child and therefore to society. It is a temporary remedy that preserves the family ties practically intact; it enables the child to be fed almost entirely by the mother, while she in turn has time to engage in some form of work outside the home. However, the existing day nurseries must be modified to be genuinely beneficial to society. We consequently advocate the following changes:

1 Since the nursing of the infant by its mother, according to natural laws and scientific observation, is the only means of obtaining strong and healthy subjects, social reforms should be made to enable all mothers to breast-feed their children.

2 Day nurseries should be maintained on a temporary basis as the least defective means of promoting breast-feeding and fostering the natural bond between mother and child. The following reforms must be made in their organization:

The Premises

These should be scattered throughout the working-class areas, near the large factories. They are to comprise four rooms and a garden. The neighbouring houses should be low, or else the nursery should be situated above street level so as to receive plenty of air and light; proper ventilation and cleanliness are essential.

Each nursery should provide for a hundred children, both toddlers and infants. There should be one room for infants, one for toddlers, a dining-room and a play-room. The premises should contain a kitchen.

In the infants' room the cots should stand one and a half feet apart and be draped with white curtains; they are to be made of iron and the mattresses of material that is easy to dry.

In the toddlers' room the cots are to be used for resting and the floor is to be covered with carpets on which the children may romp around.

The dining-room should contain a semi-circular table and benches of the right height for children. The inner area of the semi-circle allows for circulation while serving. Around the table there is to be a gallery with a double handrail on which the children may practise walking; this replaces the leading-strings and walking carts that deform the children's shoulders by lifting them too high. The children are to take their meals together, and, so that the wait between mouthfuls does not provoke screams, the entire staff should be present at mealtime.

The play-room is to contain everything to keep children amused; boredom is their greatest affliction. There should be a walking-ramp with double handrail around the table; all sorts of toys should be available, such as carts, an organ, an aviary full of birds; paintings or sculptures should be displayed, showing animals or trees, that is, real objects and not religious fabrications.

The garden is to be used as the season and the regulations permit.

Staffing and Regulations

No minister or representative of a religion is to be accepted on the staff. Each member of staff should be expected at any time to perform the most humble tasks.

A staff of ten is necessary for a hundred children: a matron, four women for the infants, three for the toddlers, one for the kitchen and one for the laundry. All these functions are to be rotated from week to week among those capable of undertaking them. The nurses in charge of the infants and toddlers should change duties every day, since the same daily routine would soon pall and make them dull and disgruntled. It is important, that the children be looked after as far as possible by young and cheerful women. Dress should not be drab and black should be banished from the nursery.

The children are to be looked after during the night only in cases of absolute necessity.

The nursery regulations should be posted up in each room.

A physician and a chemist are to be designated by the civil authority on the recommendation of the staff of each nursery.

To prevent the spread of infectious diseases, children should not be admitted without examination by the doctor.

Proposal for the establishment of day nurseries for the children of women workers: *Journal Officiel de la République Française,* 15 and 17 May 1871.

109. Vera Figner
Revolutionary Activities in Russia

Vera Figner (1852–1942) was a member of the People's Will group which assassinated Tsar Alexander II on March 1, 1881. Born into a gentry family, she arrived in Zurich in 1872 to obtain a medical education which was denied to women in Russia. The Zurich students, numbering a hundred Russian women, followed the feminist and populist movements in Russia with intense interest and formed various revolutionary circles. Although they were near completion, Figner terminated her studies in 1876 to return to Russia and participate in the populist movement. She soon joined the secret Land and Freedom organization which called for a redistribution of land among the peasants. In 1879, however, it divided into the Black Partition and the People's Will. While still committed to the economic program of Land and Freedom, the Black Partition now advocated a policy of terrorist attacks or "economic terror" against landlords, factory owners, and other local agents of economic exploitation. The People's Will, by contrast, focused on political issues and sought to eliminate absolutism by assassinating political officials who represented centralized authority.

Following numerous arrests in the immediate aftermath of the tsar's assassination, Figner alone remained of the original People's Will leadership. Police infiltration foiled her attempt to revitalize the group and she was arrested in 1883 along with several other active members. During the September 1884 trial of the

"Fourteen," Figner was the only People's Will member permitted to give a speech. She explained the events that led to her revolutionary activities as well as the aims and actions of the organization. Her speech is excerpted here as it appears in the chapter, "The Court Is in Session," in her *Memoirs of a Revolutionist* (1927). Figner received the death penalty which was commuted to life imprisonment in Schlüsselburg Fortress, some fifty miles from St. Petersburg. Released in 1904 after serving twenty years, she was one of the People's Will revolutionaries to survive the prison. Figner initially carried out work on behalf of Russian political prisoners and then went into exile in Europe from 1907 to 1917. She returned to Petrograd after the February Revolution of 1917 which overthrew the tsar. Figner did not become a Bolshevik but instead carried out charitable work and supported the Society of Political Exiles. Figner was living in Moscow when German troops approached the city in the fall of 1941, but refused the evacuation offered to certain individuals and groups and was buried with honor when she died in 1942.

There came at last the most memorable day of my life, the most profoundly moving moment in any trial, when the president turning to the accused, says in a peculiarly solemn voice, "Defendant, the last word is yours!" . . .

Under the circumstances created by the investigation, I was the central figure of the trial, the person of most importance in the case under consideration. The previous trials (dating from 1879 to 1884), . . . during which my name had been frequently mentioned, had created for me, the last of them all to be arrested, an exceptional position. This position demanded that, as the last member of the Executive Committee, and as a representative of The Will of the People, I should speak at the trial.

But I was in no mood for making speeches. I was crushed by the general situation in our native land. There was no doubt that the conflict and the protest were ended; a long, dark period of reaction had come upon us, all the more difficult to endure morally, because we had not expected it, but had hoped for an entirely new form of social life and government. The warfare had been waged by methods of unheard-of cruelty, but we had paid for such methods with our lives, and had believed and hoped. But the common people had remained silent, and had not understood. The advanced elements had re-

mained silent, though they had understood. The wheel of history had been against us; we had anticipated by twenty-five years the course of events, the general political development of the city people and the peasantry—and we were left alone. The carefully selected and organised forces, small in number, but audacious in spirit, had been swept from life's arena, suppressed and annihilated. . . .

And at the very time when my body was shaken and weakened by the conditions of my preliminary imprisonment in the Fortress, when my spirit was broken and devastated by all that I had lived through, the moment arrived in which I was inexorably bound to fulfil my duty to my dead comrades and to our shattered party, to confess my faith, to declare before the court the spiritual impulses which had governed our activities, and to point out the social and political ideal to which we had aspired.

The presiding judge had spoken; my name was called. There was an unnatural silence, and the eyes of those present, strangers and comrades alike, turned to me, and they were all listening, though as yet I had not uttered a word.

I was nervous and timid: what if in the midst of my carefully thought-out speech that mental darkness should suddenly descend upon me, which in those decisive days frequently overwhelmed me, without causing me to lose consciousness?

And in the midst of the stillness, vibrant with the general attention, I spoke my last words in a voice wherein sounded my repressed emotion.

"The court has been examining my revolutionary activities since the year 1879. The public prosecutor has expressed astonishment in his speech of indictment, both with respect to the character and to the extent of those activities. But these crimes, like all others, have their own history. They are logically and closely bound up with my whole previous life. During the period of my preliminary imprisonment, I have often debated whether my life could have followed a different course, or could have ended in any other spot than this Criminal Court. And every time I have replied to myself, No!

"I began my life under very happy surroundings. I had no lack of guides in the formation of my character; it was not necessary to keep me in leading strings. My family was intelligent and affectionate, so that I never experienced the disharmony which often exists between the older and younger generations. I had no knowledge of material want, and no anxiety concerning daily necessities or self-support. When, at the age of seventeen, I left the Institute, the thought was borne in upon me for the first time that not every one lived under such happy conditions as I. The vague idea that I belonged to the cultured minority, aroused in me the thought of the obligations which my position imposed upon me with respect to the remaining uneducated masses, who lived from day to day, submerged in manual toil, and deprived of all those things which are usually called the blessings of civilisation. This visualisation of the contrast between my position and the position of those who surrounded me, aroused in me the first thought of the necessity of creating for myself a purpose in life which should tend to benefit those others.

"Russian journalism of that period, and the feminist movement which was in full swing at the beginning of the seventies, gave a ready answer to the questions which arose in my mind, and indicated the medical profession as being a form of activity which would satisfy my philanthropic aspirations.

"The Women's Academy in St. Petersburg had already been opened, but from its very beginning it was characterised by the weakness for which it has been distinguished up to the present time, in its constant struggle between life and death; and since I had firmly made up my mind, and did not wish to be forced to abandon the course which I had undertaken, I decided to go abroad.

"And so, having considerably recast my life, I departed for Zurich, and entered the University. Life abroad presented a sharp contrast to Russian life. I saw there things which were entirely new to me. I had not been prepared for them by what I had previously seen and known; I had not been prepared to make a correct evaluation of everything which came into my life. I accepted the idea of socialism at first almost instinctively. It seemed to me that it was nothing more than a broader conception of that altruistic thought which had earlier awakened in my mind. The teaching which promised equality, fraternity, and universal happiness, could not help but dazzle me. My horizon became broader; in place of my native village and its inhabitants, there appeared before me a picture of the common people, of humanity. Moreover, I had gone abroad at the time when the events which had taken place in Paris, and the revolution which was progressing in Spain, were evoking a mighty echo from the entire labouring world of the west. At the same time I became acquainted with the doctrines and the organisation of the International.[1] Not till later did I begin to realise that much of what I saw there was only the brighter side of the picture. Moreover, I did not regard the working-class movement with which I had become acquainted, as a product of western-European life,

---

[1]The first International Working Men's Association founded in 1864.

but I thought that the same doctrine applied to all times, and to every locality.

"Attracted by socialist ideas while abroad, I joined the first revolutionary circle in the work of which my sister Lydia was engaged. Its plan of organisation was very weak; each member might take up revolutionary work in any form he chose, and at any time suited to his convenience. This work consisted of spreading the ideas of socialism, in the optimistic hope that the common people of Russia, already socialists because of their poverty and their social position, could be converted to socialism by a mere word. What we termed at that time the social revolution was rather in the nature of a peaceful social reorganisation; that is, we thought that the minority who opposed socialism, on seeing the impossibility of carrying on the strife, would be forced to yield to the majority who had become conscious of their own interests; and so there was no mention of bloodshed.

"I remained abroad for almost four years. I had always been more or less conservative, in the sense that I did not make speedy decisions, but having once made them, I withdrew from them only with great difficulty. Even when in the spring of 1874, almost the entire circle left for Russia, I remained abroad to continue my medical studies.

"My sister and the other members of the circle ended their careers most miserably. Two or three months' work as labourers in factories, secured for them two- and three-year terms of preliminary detention, after which a trial condemned some of them to penal servitude, others, to lifelong exile in Siberia. While they were in prison, the summons came to me: they asked me to return to Russia to support the cause of the circle. Inasmuch as I had already received a sufficiently thorough medical education, so that the conferring of the title of doctor of medicine and surgery upon me would satisfy only vanity, I cut short my course and returned to Russia.

"There, from the very first, I found a critical and difficult situation. The movement 'To the People' had already suffered defeat. Nevertheless, I found a fairly large group of persons who

seemed congenial, whom I trusted, and with whom I became intimate. Together with them I participated in the working out of that programme which is known as the Programme of the Populists [*Narodniki*].

"I went to live in the country. The programme of the Populists, as the court knows, had aims which the law could not sanction, for its problem was to effect the transfer of all the land into the hands of the peasant communes. But before this could be accomplished, the rôle which the revolutionists living among the people must play, consisted in what is called in all countries, cultural activity. So it was that I too went to live in the country with designs of a purely revolutionary nature, and yet I do not think that my manner towards the peasants, or my actions in general, would have aroused persecution in any other country save Russia; elsewhere I might even have been considered a useful member of society.

"I became an assistant surgeon in the Zemstvo.[2]

"A whole league was formed against me in a very short time, at the head of which stood the marshal of the nobility, and the district police captain, while in the rear were the village constable, the country clerk and others. Rumours of every kind were spread about me. . . .

"Public and secret inquiries were made: the police captain came; several of the peasants were arrested; my name figured in the cross-questioning; two complaints were made to the governor, and it was only through the efforts made by the president of the executive board of the County Zemstvo that I was left in peace. Police espionage rose up around me; people began to be afraid of me. The peasants came to my house by stealthy and circuitous routes.

"These obstacles naturally led me to the question: what could I accomplish under such conditions?

---

[2] The Zemstvo (Rural Boards) was established in 1864 as a self-governing institution for the management of the material and cultural needs of the village. Eventually it was curtailed like other reforms of Alexander II.

"I shall speak frankly. When I settled in the village, I was at an age when I could no longer make gross mistakes through any lack of tact, at an age when people become more tolerant, more attentive to the opinions of others. I wanted to study the ground, to learn what the peasant himself thought, what he wished. I saw that there were no acts of mine which could incriminate me, that I was being persecuted only for my spirit, for my private views. They did not think that it was possible for a person of some culture to settle in the village without some horrible purpose.

"And so I was deprived of the possibility of even physical contact with the people, and was unable not only to accomplish anything, but even to hold the most simple, everyday relations with them.

"Then I began to ponder: had I not made some mistake from which I could escape by moving to another locality and repeating my attempt? It was hard for me to give up my plans. I had studied medicine for four years and had grown accustomed to the thought that I was going to work among the peasants.

"On considering this question, and hearing the stories that others had to tell, I became convinced that it was not a question of my own personality, or the conditions of a given locality, but of conditions in general, namely, the absence of political freedom in Russia.

"I had already received more than one invitation from the organisation Land and Freedom, to become one of its members, and to work among the intelligentsia. But as I always clung fast to a decision once made, I did not accept these invitations, and stayed in the village as long as there was any possibility of my doing so. Thus, not vacillation but bitter necessity forced me to give up my original views and to set out on another course.

"At that time individual opinions had begun to arise to the effect that the political element was to play an important rôle in the problems of the revolutionary party. Two opposing divisions grew up in the society Land and Freedom, and pulled in opposite directions. When I had come to the end of my attempts in the country, I notified the organisation that I now considered myself free.

"At that time two courses were open to me: I could either take a step backwards, go abroad and become a physician—no longer for the peasants, to be sure, but for wealthy people—which I did not wish to do, or I could choose the other course, which I preferred: employ my energy and strength in breaking down that obstacle which had thwarted my desires. After entering Land and Freedom, I was invited to attend the conference at Voronezh, where there took place no immediate split in the party, but where the position of each member was more or less clearly defined. Some said that we must carry on our work on the old basis, that is, live in the village and organise a popular insurrection in some definite locality; others believed that it was necessary to live in the city and direct our efforts against the imperial authority.

"From Voronezh I went to St. Petersburg, where shortly afterwards Land and Freedom broke up, and I received and accepted an invitation to become a member of the Executive Committee of The Will of the People. My previous experience had led me to the conviction that the only course by which the existing order of things might be changed was a course of violence. Peaceful methods had been forbidden me; we had of course no free press, so that it was impossible to think of propagating ideas by means of the printed word. If any organ of society had pointed out to me another course than violence, I might have chosen it, at least, I would have tried it. But I had seen no protest either from the Zemstvo, or from the courts, or from any institutions whatsoever; neither had literature exerted any influence to change the life which we were leading, and so I concluded that the only escape from the position in which we found ourselves, lay in militant resistance.

"Having once taken this position, I maintained my course to the end. I had always required logical and harmonious agreement of word and action from others, and so, of course, from myself; and it seemed to me that if I admitted theoretically that only through violence could we accomplish anything, I was in duty bound to take active part in whatever programme of violence might be undertaken by the organisation which I had joined. Many things forced me to take this attitude. I could not with a quiet conscience urge others to take part in acts of violence if I myself did not do so; only personal participation could give me the right to approach other people with various proposals. The organisation really preferred to use me for other purposes, for propaganda among the intelligentsia, but I desired and demanded a different rôle. . . .

"The most essential part of the programme in accordance with which I worked, and which had the greatest significance for me, was the annihilation of the autocratic form of government. I really ascribe no practical importance to the question whether our programme advocates a republic or a constitutional monarchy.

We may dream of a republic, but only that form of government will be realised for which society proves itself ready—and so this question has no special meaning for me. I consider it most important, most essential, that such conditions should be established as will allow the individual to develop his abilities to the fullest extent, and to devote them wholeheartedly to the good of society. And it seems to me that under our present order, such conditions do not exist."

When I had finished, the president asked gently, "Have you said all that you wish to say?"

"Yes," I replied.

And no earthly power could have urged me to speak further, so great was my agitation and weariness.

The sympathetic glances, handshakes and congratulations of my comrades and defenders at the end of my speech, and in the following intermission, convinced me that my address had produced an impression.

QUESTIONS FOR ANALYSIS

1. In Anna Maria Mozzoni's view, what are the general, philosophical principles that must be embodied in the new Italian Civil Code? What is her critique of the proposed code and how does she counter the argument that women do not have the same civil rights as men due to their "private" role in the family?

2. On what basis did Barbara Leigh Smith Bodichon advocate women's right to vote? In her view, how will women's suffrage transform women's lives as well as the nation as a whole? Does the property qualification (which she does not challenge) affect Bodichon's overall perspective?

3. How did female Communards envision the general aims of the Commune and the reorganization of women's work and education? Compare their proposals with Flora Tristan's concept of a Workers' Union in Source 102 in Chapter 8. What concepts of motherhood, child rearing, and the role of the state inform the proposal for the establishment of day nurseries?

4. How does Vera Figner analyze the political situation in Russia at the time of her arrest? How did her political consciousness evolve? What kind of political activities did Figner become involved in and how does she justify the violent methods of the People's Will? What was the overall purpose of her speech to the court?

PHILANTHROPY, CHARITY, AND SOCIAL REFORM

The impoverished living and working conditions of the working class inspired middle- and upper-class women to become involved in a wide variety of issues: prisons, child protection, prostitution, education, working-class housing, the protection of motherhood, and the workhouses. Many women subsequently became involved in reform efforts to alleviate the sources of the social problems faced by the working class, including campaigns to pass new laws to protect women and children. Middle- and upper-class women often imposed their own values on the working class, but their charitable and reform activities made them increasingly aware of the numerous constraints they faced within a patriarchal society. They began to demand new career and educational opportunities for themselves, reform of marriage and property laws, and the transformation of fashions that inhibited women's physical movements. The pressing need for these legislative reforms also provided an important impetus for the women's suffrage movement.

110. Pauline Kergomard
The French Association for Child Rescue

Pauline Kergomard (1838–1925) was an activist and school reformer who advanced the cause of children's education and protection in France. She served on the Conseil Supérior de l'Instruction Publique, the highest educational advisory council in France, and was an inspectress general of *écoles maternelles,* or nursery schools. Kergomard avidly promoted these schools as a means to assist mothers in balancing employment with the care of children. *Écoles maternelles* were fully integrated into the elementary school system in 1886 and remain an essential aspect of French education. As an inspector, Kergomard came into contact with thousands of children every year and personally encountered victims of child abuse. She shared her discoveries with other inspectors and organized street inspections to find children who were abused or abandoned. These experiences inspired Kergomard to cofound the Union Française pour le sauvetage de l'enfance (French Association for Child Rescue) with Caroline Barrau, the director of the Oeuvre des Libérées de Saint-Lazare (Society for the Released Female Prisoners of Saint-Lazare). Initiated with notices placed in French newspapers calling for adherents, the French Association for Child Rescue soon developed a formal organizational structure with Kergomard as vice president and expanded beyond its original supporters to comprise some eight hundred members. The organization's program, reproduced here, was presented by Kergomard in a speech she gave as a delegate to an International Congress of Women's Charitable Organizations and Institutions held in conjunction with the 1889 International Exposition in Paris. A law passed on July 24 of this year gave the French government the right to intervene in child abuse cases.

Pauline Kergomard, 1890.

The purpose of the *French Association* is to seek out, bring to those in authority, or gather in under its wing children who are physically abused or morally endangered, under the conditions set forth in the first article of its statutes.

Individuals under sixteen years of age are considered *children*.

Abused (children) are:

1. Children who are the subjects of habitual and excessive physical mistreatment;
2. Children who, as a result of criminal negligence by their parents, are habitually deprived of proper care;
3. Children habitually involved in mendacity, delinquency, or dissipation;
4. Children employed in dangerous occupations;
5. Children who are physically abandoned.

Children who are morally endangered are:

1. Children whose parents live in a notoriously uproarious and scandalous state;
2. Children whose parents are habitually in a state of drunkenness;
3. Children whose parents live by mendacity;
4. Children whose parents have been convicted of crimes;
5. Children whose parents have been convicted of theft, habitually encouraging the delinquency of minors, of committing an offense against public decency or an immoral act.

The *Association* establishes local Committees wherever possible.

Each local Committee is invited to establish a shelter for its own use. Local Committees shall keep the central Committee informed of all their undertakings.

Every active member of the *Association* promises to take an interest in every mistreated or morally endangered child he shall discover or encounter or who is brought to his attention.

First, he shall take down the first and last names of the child and the child's address.

If possible, he shall go that same day to the indicated address and conduct a brief preliminary investigation, the results of which he shall convey immediately to the local Committee, or if one is not available, to the central Committee.

If he is unable to do this, he shall send the name and address to the Committee the same day.

In either case, the Committee shall proceed without delay to conduct a meticulous investigation.

In the event that the investigation establishes that the child falls into one of the above categories, the Committee's representative shall suggest that the person exercising parental custody entrust the child to the Committee until the child's future shall be resolved.

If this person consents, the child shall be taken immediately to the temporary shelter.

Next, the Committee shall determine as rapidly as possible the conditions of the [child's] placement, in conformity with the first article of its statutes, and demand that the person having custody of the child agree to this placement. To this end it shall have that person sign a release identical to that now in use by the Public Assistance Administration for the parents of children taken in under the program for the morally abandoned.

If consent is refused, or if the child does not fall into one of the categories detailed above, the Committee shall prepare a report summarizing the results of the investigation it has conducted. In every case a copy of this report shall be directed to the mayor of the city (or in Paris, the district) where the parents live.

If circumstances appear to warrant it, another copy shall be sent to the prefect or subprefect and, in Paris, to the director of Public Assistance.

Finally, a copy shall be sent to the public prosecutor of the Republic in every case in which acts revealed by the investigation appear to fall under the jurisdiction of the law. A summary of pending action shall be sent to each member of the *Association*.

The *Association* is functioning, as I have said. Certainly, it is far from realizing our ideal of perfection. But as each day brings its progress we hope, bit by bit, to make it a model society, completely liberal, that is, it is to be above all sectarian antagonisms.

While we wait for the law to arm us solidly against the undeserving parents from whom we have already wrenched some victims; while we wait for it to permit us to take all the children who roam the streets to our temporary shelter; while we wait for it to permit us to search the hovels where there is torture and depravity, we have set ourselves the task of:

1. Snatching children from the horrible hornets' nest of a police record by saving them from first convictions;
2. preventing them from being confined or released from confinement too early, which would fatally return them to the streets and bring them back before the courts;
3. improving their conditions at the police station or detention center while they await trial.

To this end we have established our headquarters at the Palais de Justice and, thanks to the humanitarian feelings of almost all the magistrates . . . we make ourselves known to the accused on their arrival, we negotiate with certain parents to have them sign over custody to us, we send children to farmers in the country (of whom we now have one hundred twenty for this purpose), we release to the Public Assistance Administration those who fall under their jurisdiction; as for the others, M. Rollet [the Association's representative] pleads that they be kept in custody until their twentieth year with conditional freedom, that is to say that the penal authorities should send them to us after a period of confinement so that they can then be placed in the country like the others.

Then, too, we have improved the conditions under which children are detained at the Police Station and Detention Center, girls being in greater need of this than boys. Not long ago the girls held in the Detention Center—some were four years old, some sixteen—were all placed together; now there is a separation [by ages], quite an elementary one, alas! But at least there are now two divisions.

Not long ago they slept three to a cell—without surveillance. Today there is a dormitory where observation is easy.

Not long ago they had nothing to keep them busy and spent their days telling each other their sad pasts. Today they sew—the Association has furnished them with material and sent a sewing teacher—today they read, because the Association has assembled a library.

Not long ago they stayed in the Detention Center in the same sordid clothing and unwashed state as the day they were arrested. Today there are baths and changes of clothes.

Oh! there is a great deal still to be done, but the Association is happy with the results it has achieved. . . .

111. Mary Carpenter
Prison Reform

Mary Carpenter (1807–1877) was active in several areas of social reform: children's education, juvenile delinquency, prisons, and women's education. She wrote extensively and spoke to audiences in Great Britain, North America, and

From *Reformatory Prison Discipline, as Developed by the Rt. Hon. Sir Walter Crofton, in the Irish Conflict Prisons* by Mary Carpenter. © 1872 by Longman, Longman, Green, Longman.

India. Her writings on prisons include *Juvenile Delinquents* (1853), *On Convicts* (1864), and *Reformatory Prison Discipline* (1872). Like many Quakers who pioneered prison reform in the United States and Europe, Carpenter believed that prisons must rehabilitate as well as punish and that rehabilitation would require a thorough transformation of the existing prison system. She especially admired the reforms carried out in the Irish Convict Prisons under the direction of Sir Walter Crofton and advocated their adoption in England. The publication of *Reformatory Prison Discipline,* in which she summarizes the Crofton system, was intended to coincide with an international conference on prisons to be held in London. The present selection is from that book's Chapter 4, "Female Convicts."

All who have had any practical acquaintance with the management of convicted women, are fully aware that it is one of the most difficult problems to be satisfactorily solved.

The organization of women, both mental and physical, is much more delicate and sensitive than that of men, and also is subject to peculiar conditions;—it follows from this that when morally diseased and in an abnormal state, their reformation, and restoration to a healthy condition, is far more difficult than that of the other sex. The structure of society, besides, precludes the adoption of such a system for women as has been found to work admirably for men. This is well known to all who have undertaken the care of females, young or old, and in whatever condition of life, who are mentally, morally, or physically diseased.

The evidence submitted to the Royal Commission on Prison Discipline in 1863, by the Directors of the English Convict Prisons, and other official gentlemen, painfully confirm these statements. The Director of the Female Prisons, frankly avowed his inability to check the evils existing in them, which had been made known to the public by some volumes which excited at the time considerable notice.

The women of this degraded portion of society will be generally found to differ in many respects from those belonging to a higher sphere. Their intellectual powers are low, and from having been left uncultivated, are in a state of torpidity from which it is very difficult to rouse them. This peculiarly low intellectual condition in females of the lowest social grade is accompanied by a very strong development of the passions and of the lower nature. Extreme excitability, violent and even frantic outbursts of passion, a duplicity and disregard of truth hardly conceivable in the better classes of society, render all attempts to improve them peculiarly difficult. And if, added to all this, what is holiest and best in woman has been perverted and diseased by unlawful intercourse with the other sex, as is very frequently the case, there is engendered in her a hardness of heart, a corruption of the whole nature, which would seem to make absolute reformation almost impossible. . . .

In order to have any prospect of success in the reformation of women in this very degraded and, we may say, abnormal condition, for their characteristics differ essentially from those of the labouring, middle and upper classes, there must exist, in the first place, firm steady controul, against which it is evidently hopeless to rebel, combined with a strict and vigilant discipline, administered with the most impartial justice. In the next place, to provide abundance of active useful work is absolutely necessary. The restless excitable nature of these women requires a vent in something; they should have full employment, of a kind which will exercise their muscles and fully occupy their minds, so as to calm their spirits and satisfy them with the feeling of having accomplished something. These two primary conditions having been arranged satisfactorily, considerable attention must at the same time be paid to the culture of the intellectual powers. These, we have already stated, are more deadened, or perverted to a bad use in women than in men. There is far greater difficulty in stimulating to mental exertion girls who have passed their childhood in neglect, than boys. The effort of learning to read is to such often positively painful, and without the greatest skill, kindness, and firmness combined on the part of the teacher, the young person succumbs to the difficulty. The effort once made and a triumph achieved, an important step in reformation is attained, for stores of interesting information are now open which will fill the mind, instead of the pernicious thoughts which formerly harboured there. Intellectual effort, which would be very easy and pleasant to a child of six years old, is extremely difficult and unpleasant to a girl of sixteen; still more so to a woman of thirty or upwards;—a mastery over it once gained, not only an intellectual but a moral power is acquired, both of which facilitate the work of reformation. Another essential part of the work of reforming such women as have been described, is the healthy development of their affections. These are peculiarly strong in the female sex, and may be made the means of calling out the highest virtues, the most genuine self-devotion; when perverted, they may be, and are frequently, made an instrument of much evil; but in a woman they can never be utterly lost. It will then be essential to the success of any system which has as its object the reformation of women, that scope should be given to the affectional part of the woman's nature, and that this should be enlisted on the side of virtue.

That all these conditions should be fulfilled in a Convict Prison does certainly appear very difficult; yet, if they are essential to success, no labour, no expense, should be deemed too great to develop a system which should embody them all, and do the work required,—reform female convicts. The expense which a bad woman is to the public, who comes forth from a lengthened confinement in a Government gaol unreformed, is far greater than any possible cost which might have been incurred in reforming her; the evil she has done within the prison to those around her is very great, and extends the

poisonous influence to a widely-extending circle, when the women she has corrupted go out into the world; on her own discharge she emerges from her seclusion only to plunge into greater excesses than before, and, to perpetuate and intensify the pollution of the moral atmosphere from which she had been temporarily withdrawn.

Keeping in view the foregoing remarks, we shall now proceed to give an account of the system successfully pursued in the Irish Female Convict Prisons, under the direction of Sir Walter Crofton.

The condition of the Female Convict Prisons in Ireland was even worse than that of those for males, when the Directors first undertook the charge. The female convicts who had been transported to Western Australia had been so bad that the colony absolutely refused to receive any others. The Directors say in their first report:—

"Our proportion of female criminals is very large, and it is much to be deplored that such is the case, considering the influence for good or evil that women must exercise on the rising generation. This large proportion may, in a great measure, be ascribed to the circumstances of the country, and want of industrial employment. A prison is now erecting at Mountjoy for the reception of 600 female convicts; which will, we trust, enable us, from its construction, to carry out such penal and reformatory treatment as will induce habits of reflection and amendment, and will also relieve the County Gaols from the great inconvenience to which they are subjected through the reception of Government prisoners. Pending its erection, however, we are endeavouring to ameliorate, if possible, the condition of those confined in Grangegorman and Cork Prisons, which, unfortunately, can only hold a portion of our convicts. Towards attaining this object, education adapted to the wants of that class, and engendering habits of industry, are the great adjuncts to the religious influence inculcated by their chaplains. With regard to education, the Female Prison Schools, in common with the others, will be placed under the inspection of the National Board of Education. Heretofore instruction has been limited to those under twenty-seven or twenty-eight years: we have given directions that there should be no limit as to age provided there is a disposition to acquire information.

"Respecting industrial training, we have desired that all the convicts should, in turn, receive instruction in cooking, laundry, sewing, knitting, cleaning, &c., instead of confining a certain number to a particular occupation; although this plan tends to the work not being so well performed, we prefer it on account of the advantages gained by the individuals receiving general instruction.

"It has been a custom to admit convicts into the prison with their children sometimes at the age of five or six years; we cannot consider such places, with their necessary associations, advantageous for education of the young, and recommend its discontinuance, excepting in cases of children under two years of age."

In their second report they show that immediate good results have followed the adoption of their plans. They say:—

"With regard to female convicts, we have devoted much attention to carry out the plans proposed in our last year's report concerning them, and have observed a manifest improvement in their general demeanour and conduct. This we attribute in some measure to the efforts made by our teachers to open their minds by education, and to engender habits of self-controul. Many, instead of sullenly brooding over their past life, now look forward with hope to the future. Even women advanced in life, who have spent most of their career in prison, and who at first would not attend school, and seemed incapable of understanding the advantages of education, are now amongst the most assiduous in their classes. A difference in their conduct is already apparent; they are more orderly and obedient to the rules, and make efforts to exercise that self-command, the want of which has so often led them into crime. We trust that under the new arrange-

ments in the prisons, and a system of refuges and patronage on discharge, which we are now advocating, many convicts formerly considered irreclaimable, will finish their career as good members of society...."

The minds of the Directors were awakened to the importance of devising some plan for the gradual introduction to liberty of the female convicts, while at the same time they should be brought into personal contact of ladies unconnected with the prisons, who would devote to them their voluntary benevolent effort. They continue:—

"Great difficulties present themselves in the final disposal of female convicts. A man can obtain employment in various ways in out-door service, not requiring, in all cases, special reference to character, and at work which is not open to females in this country. A woman, immediately on discharge from prison, is totally deprived of any honest means of obtaining a livelihood.... Persons of her own class will object to associate in labour with her, even if employers were willing to give her work; and the well-conducted portion of the community object to receive with their families, or domestic servants, persons so circumstanced, without a stronger guarantee and proof of their real and permanent reformation, than would be afforded by a prison character."...

To give such confidence to the public in the reformation of these unhappy women, as to make families willing to receive them into their domestic circle, it was necessary that the female convicts should not only have gone through some such intermediate stage as the men, but that they should have had some kind of trial of the sincerity of their reformation without the restraint of the prison walls, or the guardianship of government officials. The plan proposed by the Directors admirably combined these objects. "For this reason," they continue, "instead of increasing the existing Government Prison Establishments—a plan attended with much expense, delay, and difficulty—we proposed, in December last, to the Irish Government, that convicts whose conduct had been exemplary should be drafted into existing private charitable institutions willing to receive them, where the disposition of each inmate would be studied, and the certificate of character founded on that study, together with recommendations, which would then be considered sufficiently satisfactory to obtain her employment; the prisoners, in all such institutions, should be under the general supervision and inspection of the Convict Directors. In order to carry out this plan, a certain number of exemplary convicts should be selected from the Government Prisons, at periods varying according to circumstances, previous to the time when in the usual course they would become eligible for discharge, and be sent to such private establishments, and not released therefrom under at least three months; and not then unless immediate and proper employment should offer, excepting, however, cases where prisoners become regularly entitled to their discharges, from having completed their sentence, and special cases to be determined on by the Directors and sanctioned by the Executive. Should, however, a prisoner misconduct herself, she would be liable to recommittal to the Convict Depôt, to undergo her original sentence. It is obviously most desirable to enlist public sympathy and interest in any scheme for the employment of discharged female prisoners; this object we consider will be best attained in the manner proposed."

Here we have the first sketch of a plan which has succeeded admirably.

Two Refuges were at once established. One was a large convent of the Sisters of Mercy at Golden Bridge, near Dublin, the other was a Protestant refuge in Heytesbury Street, established purposely by some benevolent ladies.

It required some moral courage, or rather a strong faith and a devoted love in these ladies, unaided by means of punishment, or of physical restraint, to undertake the custody and care of women who had sprung from "a class so depraved, and hitherto deemed so incorrigible," continues the Report, "as to be absolutely rejected by the colonists of Western Australia, a colony whose vitality at the present moment depends on an increase of the female sex."

These Refuges form a valuable link to society, or they are accessible to the public, whose cöoperation is so important. Many visitors from England who, in 1861, attended the Social Science Association in Dublin, closely inspected them, and received every desired information as to their working. All were struck with the changed look and manner of the women from what had been noticed in the earlier stages. There was nothing to remind one that they had even been in prison; and they were ready to converse with visitors with full assurance of sympathy respecting their future prospects....

It is a sufficient proof of the efficiency of this system, if worked in accordance with its intention and principles, that of 510 female convicts who were licensed during the seven years from 1856 to 1862 inclusive, only 21 had their licenses revoked for misconduct in the refuges, and 5 after; in all 26, viz., 5 per cent. Only 4 of the whole number were reconvicted, viz., 0.8 per cent, or less than one.

In Ireland, the public has fully cöoperated in this work undertaken by the managers of the refuges, in restoring these women to society. Increased experience only confirms the truth of the principle on which they are founded. The ladies who take an interest in these refuges have full opportunity of judging of the competency of the women, and the sincerity of their reformation; they are, therefore, in a position to recommend them, and the public place confidence in their recommendation. The women also find themselves still, on their actual discharge, under the friendly surveillance of those who have already proved their true interest in them, by their earnest efforts for their reformation.

QUESTIONS FOR ANALYSIS

1. How does the program of the French Association for Child Rescue define child abuse? What is the organizational structure of the Association and how did it rescue children from abusive situations? What concepts of childhood, the family, and the role of the state informed the Association's purpose and activities?

2. How did Mary Carpenter compare female and male convicts and the prospects for their rehabilitation? What were the main features of the Crofton system that was implemented in the Irish Female Convict Prisons? What were the roles of women and religious orders in this system?

3. Compare and contrast the underlying philosophical assumptions of the French Association for Child Rescue and the Crofton system of prison rehabilitation.

WOMEN AND COLONIALISM

Women from middle- and working-class backgrounds frequently emigrated to the colonies because of financial hardship. In England, in particular, emigration was promoted as a solution to the issue of "redundant women" whose plight was revealed by an 1851 census which predicted that marriageable women would always outnumber men by nearly a half million. Various agencies recruited women to become governesses, nurses, servants, and wives in the colonies. European women's organizations not only facilitated female emigration, but also often demanded an active role for women in the "domestication of empire." For the most part, women in the colonies did not question the racist basis of colonialism or the many privileges and benefits that they derived from it. Some women, however, did challenge patriarchal norms and colonial practices harmful to women and a few also advocated for the right of colonial nations to a limited form of self-government.

112. Flora Annie Steel
The Complete Indian Housekeeper and Cook

Flora Annie Steel (1847–1929) moved to India in 1867 to accompany her husband, a civil engineer, whom she had recently married, and lived in India for the duration of his twenty-two year tour of duty. In the Punjab, Steel initiated various cultural and educational activities, became a member of the Provincial Educational Board, and was appointed an inspectress of girls' schools. Shortly after she accepted the position as inspectress, Steel's husband was transferred to another region, but she remained to complete her appointment. When reunited with her husband, Steel became the manager of a large household staffed with several Indian servants.

From *The Complete Indian Housekeeper & Cook,* by F.A. Steel and G. Gardiner. © 1902 by William Heinemann.

These experiences inspired *The Complete Housekeeper and Cook* which she co-authored with Grace Gardiner in 1888. The book was dedicated "to the English girls to whom fate may assign the task of being house-mothers in our Eastern empire."[1] But it was also intended for experienced housekeepers unfamiliar with the specific requirements of household management in India "who have found themselves almost as much at sea as their more ignorant sisters." *The Complete Indian Housekeeper and Cook* discusses the duties of mistresses and servants; horses and stable management, cows and dairy, gardening, child rearing, a table of wages and weights, hints on clothing, advice to the cook, and over a hundred pages of recipes which include "native dishes." The following excerpts appear in the first chapter, "The Duties of the Mistress."[2] Chapters on the duties of servants were published as pamphlets in Urdu and Hindi: "[I]t is believed that they will be found of great use, as, even when the servants cannot read, they can get some one to read to them." The recipes also appeared in a separate cookbook in Urdu. *The Complete Indian Housekeeper and Cook* was very popular and went through several editions, each carefully revised to provide accurate prices for various goods and services. When she returned to England in 1889, Steel pursued her writing career in earnest and wrote short stories and novels, including *On The Face of the Waters* (1896), which featured British and Indian characters, and *The Garden of Fidelity,* her unfinished autobiography published posthumously in 1929.

Housekeeping in India, when once the first strangeness has worn off, is a far easier task in many ways than it is in England, though it none the less requires time, and, in this present transitional period, an almost phenomenal patience....

And, first it must be distinctly understood that it is not necessary, or in the least degree desirable, that an educated woman should waste the best years of her life in scolding and petty supervision. Life holds higher duties, and it is indubitable that friction and over-zeal is a sure sign of a bad housekeeper. But there is an appreciable difference between the careworn Martha vexed with many things, and the absolute indifference displayed by many Indian mistresses, who put up with a degree of slovenliness and dirt which would disgrace a den in St. Giles[1] on the principle that it is no use attempting to teach the natives.

They never go into their kitchens, for the simple reason that their appetite for breakfast might be marred by seeing the *khitmutgâr*[2] using his toes as an efficient toast-rack (*fact*); or their desire for dinner weakened by seeing the soup strained through a greasy *pugri*.[3] ...

Easy, however, as the actual housekeeping is in India, the personal attention of the mistress is quite as much needed here as at home. The Indian servant, it is true, learns more readily, and is guiltless of the sniffiness with which Mary Jane receives suggestions, but a few days of absence or neglect on the part of the mistress, results in the servants falling into their old habits with the inherited conser-

---

[1]*St. Giles;* a slum in London.

[2]*Khitmutgâr:* the male servant who waited on the table.

[3]*Pugri:* a turban.

---

[1]This quotation and others cited here appear in the Preface to the first edition which was then reproduced in future editions.

[2]These excerpts from Chapter One are from the 1902 edition.

vatism of dirt. This is, of course, disheartening, but it has to be faced as a necessary condition of life, until a few generations of training shall have started the Indian servant on a new inheritance of habit. It must never be forgotten that at present those mistresses who aim at anything beyond keeping a good table are in the minority, and that pioneering is always arduous work.

The first duty of a mistress is, of course, to be able to give intelligible orders to her servants; therefore it is necessary she should learn to speak Hindustani. No sane Englishwomen would dream of living, say, for twenty years, in Germany, Italy, or France, without making the *attempt,* at any rate, to learn the language....

The next duty is obviously to insist on her orders being carried out. And here we come to the burning question, "How is this to be done?" Certainly, there is at present very little to which we can appeal in the average Indian servant, but then, until it is implanted by training, there is very little sense of duty in a child; yet in some well-regulated nurseries obedience is a foregone conclusion. The secret lies in making rules, and *keeping to them.* The Indian servant is a child in everything save age, and should be treated as a child; that is to say, kindly, but with the greatest firmness. The laws of the household should be those of the Medes and Persians, and first faults should never go unpunished. By overlooking a first offence, we lose the only opportunity we have of preventing it becoming a habit.

But it will be asked, How are we to punish our servants when we have no hold either on their minds or bodies?—when cutting their pay is illegal, and few, if any, have any real sense of shame.

The answer is obvious. Make a hold.

In their own experience the authors have found a system of rewards and punishments perfectly easy of attainment. One of them has for years adopted the plan of engaging her servants at so much a month—the lowest rate at which such servant is obtainable—and so much

extra as *buksheesh,*[4] conditional on good service. For instance, a *khitmutgâr* is engaged permanently on Rs. 9 a month, but the additional rupee which makes the wage up to that usually demanded by good servants is a fluctuating assessment! From it small fines are levied, beginning with one pice for forgetfulness, and running up, through degrees of culpability, to one rupee for lying. The money thus returned to imperial coffers may very well be spent on giving small rewards; so that each servant knows that by good service he can get back his own fines. That plan has never been objected to, and such a thing as a servant giving up his place has never been known in the author's experience. On the contrary, the household quite enters into the spirit of the idea, infinitely preferring it to volcanic eruptions of fault-finding.

To show what absolute children Indian servants are, the same author has for years adopted castor oil as an ultimatum in all obstinate cases, on the ground that there must be some cause for inability to learn or to remember. This is considered a great joke, and exposes the offender to much ridicule from his fellow-servants; so much so, that the words, *"Mem Sahib*[5] *tum ko zuroor kâster ile pila dena hoga" (The Mem Sahib will have to give you castor oil),* is often heard in the mouths of the upper servants when new-comers give trouble....

To turn to the minor duties of a mistress, it may be remarked that she is primarily responsible for the decency and health of all persons living in her service or compound. With this object, she should insist upon her servants living in their quarters, and not in the bazaar;

---

[4]*Buksheesh:* cash (in this case, *rupees*) given to ensure a desired result, as in a tip, and not part of a regular wage. Here, Steel advocates reducing the servants' wages and allowing them to make up the lost portion only on the basis of good service.

[5]*Mem Sahib:* a term of respect with which servants addressed the European woman who was the mistress of the household.

but this, on the other hand, is no reason why they should turn your domain into a caravanserai for their relations to the third and fourth generation. As a rule, it is well to draw a very sharp line in this respect, and if it be possible to draw it on the other side of the mothers-in-law, so much the better for peace and quietness.

Of course, if the rule that all servants shall live in quarters be enforced, it becomes the mistress's duty to see that they are decently housed, and have proper sanitary conveniences. The bearer should have strict orders to report any illness of any kind amongst the servants or their belongings; indeed, it is advisable for the mistress to inquire every day on this point, and as often as possible—once or twice a week at least—she should go a regular inspection round the compound, not forgetting the stables, fowl-houses, &c.

With regard to the kitchen, every mistress worthy the name will insist on having a building suitable for this use, and will not put up with a dog-kennel. On this point the authors cannot refrain from expressing their regret, that where the power exists of forcing landlords into keeping their houses in repair, and supplying sanitary arrangements, as in cantonments, this power has not been exercised in regard to the most important thing of all; that is, to the procuring of kitchens, where the refuse and offal of ages cannot percolate through the mud floors, and where the drain water does not most effectually apply sewage to a large surrounding area. With existing arrangements many and many an attack of typhoid might be traced to children playing near the kitchen and pantry drain, and as in large stations the compounds narrow from lessening room, the evil will become greater.

In regard to actual housekeeping, the authors emphatically deny the common assertion that it must necessarily run on different lines to what it does in England. Economy, prudence, efficiency are the same all over the world, and because butcher meat is cheap, that is no excuse for its being wasted. Some modification, of course, there must be, *but as little as possible.* It is, for instance, most desirable that the mistress should keep a regular storeroom, containing not merely an assortment of tinned foods, as is usually the case, but rice, sugar, flour, potatoes, &c.; everything, in short, which, under the common custom, comes into the *khânsâmâh's*[6] daily account, and helps more than larger items to swell the monthly bills. For it is *absolutely impossible* for him to give a true account of consumption of these things daily....

A good mistress in India will try to set a good example to her servants in routine, method, and tidiness. Half-an-hour after breakfast should be sufficient for the whole arrangements for the day; but that half-hour should be given as punctually as possible. An untidy mistress invariably has *untidy,* a weak one, *idle* servants. It should never be forgotten that—though it is true in both hemispheres that if you want a thing done you should do it yourself—still, having to do it is a distinct confession of failure in your original intention. Anxious housewives are too apt to accept defeat in this way; the result being that the lives of educated women are wasted in doing the work of lazy servants.

The authors' advice is therefore—

"Never do work which an ordinarily good servant ought to be able to do. If the one you have will not or cannot do it, get another who can."...

Finally, when all is said and done, the whole duty of an Indian mistress towards her servants is neither more or less than it is in England. Here, as there, a little reasonable human sympathy is the best oil for the household machine. Here, as there, the end and object is not merely personal comfort, but the formation of a home—that unit of civilisation where father and children, master and servant, employer and employed, can learn their several duties. When all is said and done also, herein lies the natural outlet for most of the talent peculiar to women.

---

[6]*Khânsâmâh:* a servant, the chief table steward of the household.

It is the fashion nowadays to undervalue the art of making a home; to deem it simplicity and easiness itself. But this is a mistake, for the proper administration of even a small household needs both brain and heart....

It is astonishing how few women know how to keep accounts; yet this is the first step towards economy, and a little method and care in this point saves infinite friction....

Make up your mind as to the fair price. Do not take an article at an exorbitant one simply because you want it. Some housekeepers in India nowadays cheerfully give the equivalent of one and sixpence per pound for bad beef, and by so doing are ruining the poor. Accept past deceit or fraud as *past;* tell the offender quietly that you do not intend to pay such a price again, and in cases of extravagance, give the order that in future the ingredients are to be brought to you before being put into use. If meat is bad, or dear, resort for a week to doing your own marketing, but do not sit for half-an-hour and squabble over it with your *khânsâmâh.* Make it a rule that all food which is to be used that day shall be personally inspected by the mistress, otherwise you may pay good prices for bad things. It must be remembered that half the faults of native servants arise from want of thought and method, and that mere fault-finding will never mend matters. A mistress must know not only where the fault lies, but how to mend it. So, in keeping accounts, a mistress *must* take the lead, and, knowing the proper prices of the different articles, and the amount which ought to be consumed, set aside all objections with a high hand.

Having thus gone generally into the duties of the mistress, we may detail what in our opinion should be the daily routine.

The great object is to secure three things—smooth working, quick ordering, and subsequent peace and leisure to the mistress. It is as well, therefore, with a view to the preservation of temper, to eat your breakfast in peace before venturing into the pantry and cookroom; it is besides a mistake to be constantly on the worry.

Inspection parade should begin, then, immediately after breakfast, or as near ten o'clock as circumstances will allow. The cook should be waiting—in clean raiment—with a pile of plates, and his viands for the day spread out on a table. With everything *en evidence,* it will not take five minutes to decide on what is best, while a very constant occurrence at Indian tables—the serving up of stale, sour, and unwholesome food—will be avoided. It is perhaps *not* pleasant to go into such details, but a good mistress will remember the breadwinner who requires blood-forming nourishment, and the children whose constitutions are being built up day by day, sickly or healthy, according to the food given them; and bear in mind the fact that, in India especially, half the comfort of life depends on clean, wholesome, digestible food.

Luncheon and dinner ordered, the mistress should proceed to the storeroom, when both the bearer and the *khitmutgâr* should be in attendance. Another five minutes will suffice to give out everything required for the day's consumption, the accounts, writing of orders, &c., will follow, and then the mistress (with a sinking heart) may begin the daily inspection of pantry, scullery, and kitchen....

We do not wish to advocate an unholy haughtiness; but an Indian household can no more be governed peacefully, without dignity and prestige, than an Indian Empire. For instance, if the mistress wishes to teach the cook a new dish, let her give the order for everything, down to charcoal, to be ready at a given time, and the cook in attendance; and let her do nothing herself that the servants can do, if only for this reason, that the only way of teaching is to *see* things done, not to let others see *you* do them.

Another duty which must never be omitted, so long as copper vessels are used for cooking, is their weekly inspection to see that the tinning is entire. No more fruitful source of danger and disease can be imagined than a dirty copper saucepan; and unless proper supervision is given, we advise every mistress who has any desire to avoid serious risk, both to her own family and her *guest,* to use nothing but steel or enamelled utensils. In many jails also, nowadays, glazed earthenware utensils may be had, which will stand a charcoal fire, and are invaluable for milk, fruit, and soups. In fact, a much larger variety of articles can now be had of native manufacture than was possible ten years ago. It need hardly be said that, wherever practicable, it is clearly the Indian housekeeper's duty to encourage native industry, and at the same time, by never offering an unfair price or submitting to one, striving to prevent the evils of our civilisation going hand in hand with the good.

113. Annie Besant
"The Education of Indian Girls"

Annie Besant (1847–1933) was active in the women's rights and socialist movements in England in the 1870s and 1880s prior to departing for India in 1893 where she was to live for the rest of her life. In England, Besant advocated birth control, suffrage, and improved employment opportunities and protection for working women, including unionization. She joined the Fabian Society in 1885 and served on its Executive Committee for six years, but increasingly felt that socialism was bereft of spiritual inspiration. Besant became involved with Madame Blavatsky's Theosophical Society, which espoused "universal brotherhood" and the study of comparative religions, and subsequently left the Fabians. Following the death of Madame Blavatsky, who had designated Besant as her successor, she headed the Theosophical Society's Esoteric Section and, in 1893, moved to Madras to assist in the management of the Society's headquarters. Several years later, Besant was elected President of the entire Theosophical Society, a position she held from 1907 to 1933. To India, Besant brought a commitment to women's rights and a critical view of British imperialism. She studied Sanskrit and various religious traditions, often citing indigenous Indian customs and Vedic texts to offer precedents for women's equality. In 1913, Besant gave a series of lectures in Madras on such topics as education, child marriage, the status of Indian industries, and the caste system. While her "Mass Education" lecture examined education for the masses of Indian boys and girls, "The Education of Indian Girls," excerpted here, focused on the education of upperclass Indian girls. The Theosophical Publishing House published the lectures as *Wake Up, India. A Plea for Social Reform* (1913).

Besant consistently argued that India should be granted the status of a "self-governing nation" that "owed allegiance to the British Crown" but only after it was "uplifted" through internal reforms which included improvements in women's status, abolition of the caste system, and the guarantee of religious tolerance. She made the case for Indian Home Rule with a restricted franchise in various public forums and writings including a pamphlet, *India: A Nation, A*

Plea for Indian Self-Government (1915). Besant organized the Women's College in Benares in 1916 and, in the following year, became the first president of the Women's Indian Association which she founded with Margaret Cousins and Dorothy Jinarajadasa. In 1917, British authorities in Madras arrested Besant for refusing to cease publication of articles advocating Home Rule in her newspaper, *New India*. Released after three months under house arrest, she was elected president of the Indian National Congress, but failed to win reelection in 1920 as moderates within the Congress began to support the more radical views on independence advocated by Mohandas Gandhi.

Now to come to the important question as to the kind of education that has to be given if you decide it ought to be given. I have been trying to urge upon you the reasons why you should do something, because the *wish* to raise women is the thing which is most wanted. Suppose you are with me on that, let us see what sort of education shall be given to Indian girls. Clearly it must be one which will not denationalise them. It is bad enough to denationalise your men. It would be a thousand times worse to denationalise your girls, for that would be the death-knell of India. The first thing you will have to recognise is that you cannot give the same education to all the women of the cultivated classes. You must give some of them education, as our chairman suggested, which will enable the great lack of teachers in girls' schools to be supplied. That of course means University education. But that, I think, will never be the education of the great number, the great mass, of educated women in India. But there are always some who will find their happiness along lines of public work, and I cannot but think that a very large number of the childless widows who exist in the community might there have a career open to them which would make them at once happy in themselves and useful to their country. The more Widows' Homes multiply in which this higher education is given to fit those women who have no real homes, as it were, of their own, for teachers, for skilled nurses, for doctors, the better will it be for the homes that they will later bless with their presence. There must be more

women than men in order that the work of the nation may be carried on. In the West those women—surplus women as they are very often called—do not marry. Either they remain unmarried in the outer world, or become sisters in some religious order—women of the most noble and most self-sacrificing character, whom to know is to revere and to love. Among the Little Sisters of the Poor, among the Sisters of Mary, and under many other names—names do not matter—you have thousands of good women who are verily mothers of the motherless, the nurses of the sick, the helpers of the miserable. Every nation has some of these. Now in India you have avoided the problem so far by your joint family system, for under that there were no homeless women, and there were none needing help in the house who could not be looked after by the widows in the family. That was their natural work within the circle of the joint family system. Women who had no children of their own nursed the sick children, perhaps of the sister, of the aunt, of the brother.... The childless widow took the place which the nun, or the sister, or the skilled nurse took in the small English home. But as your system is changing and as your young men are no longer willing to live in the joint family, and as they are taking up the western system of separate independent homes, so that each family shall be an isolated unit, there comes the absolute necessity, with the breaking up of the older system, to adapt yourself to it by training widows properly, so that they may perform their ancient duties but perform them in a way suited to modern conditions. That is the secret, of course, of national growth. You cannot bring back the old system which the people have outgrown, and which you are no longer willing to follow. But you can take what you had of good in the old and adapt it to the conditions of the new; and so I submit that you ought to have Widows' Homes everywhere, where this higher University education should be given, and where technical education for skilled nursing as well as for medical women should be provided. That is a clear line so far marked out for the great widow population.

But the enormous majority, of course, will be wives in their own homes, and their education is the next to which we turn. Widows for teachers, for skilled nurses, for doctors, for those special lines of work for which higher education is wanted. In England higher education is wanted largely, because women are going into professions and competing with men. The day is far off here, when you will have your wives competing with you in every walk of life, for the place of the wife is clearly in the home where the children claim her care. How then will you educate the wife and the mother? Religion is clearly a part of her education; religion and morals must be a part of the education. Morals, I think, come almost instinctively to her. May I say one word about religion? Too much of the religion of the Indian ladies to-day is devotion without knowledge, zeal without understanding. I do not want in a public lecture to go into details on a subject so sacred. I will only ask the men at once whether it be not true that when they go back into the home they go into an entirely different atmosphere in matters of religious thought from their own, and whether they do not find there observances which are childish rather than elevating; whether it would not be well to add to the women's devotion a knowledge of religious topics, which will make her worship intelligent as well as loving; for remember that your young boys are brought up at their mothers' knees; if the boy worships as the mother now worships, he will later think all religion is childish, and

only fit for women and children. How many young men of sixteen, seventeen and eighteen reject their mothers' faith because it is unintelligent, and her observances because they are regarded as superstitious; so I say educate your women in religion, not to diminish their devotion but to render it more intelligent, so that they may prevent the boy from growing into a sceptic, and that he may learn at his mother's knee a religion of which he need never be ashamed.

Outside religion and morals, literary education. What should that be? I submit it should include first a thorough literary knowledge of the vernacular, the vernacular of the family to which the girl belongs. That is fundamental, so that the great vernacular literature may be studied by her to the enrichment of her life. Then she should learn, according to her religion, the elements at least of Samskrt [*sic*] or Arabic, according as she is Hindū or Mussalmān; for so you open to her the great treasures of sacred literature, which cannot be rivalled by modern writings. These are fundamentals. I plead next for knowledge of English, and I will tell you quite frankly why. English is the language here of politics, of social matters, of discussions among the men. You cannot discuss these things in the vernaculars conveniently, because the vernaculars are so different. ...Are you not going to open to her the treasures you have found through knowledge of English, and enable her to sympathise with you in all those great problems which have to be solved, and for which knowledge of English is clearly a necessity under the conditions of the present time? So I plead that an English-educated man should also have an English-educated wife, English-educated in this sense only, that the English language shall form part of the school curriculum. I leave you to think over and decide the question, for the question should be decided by Indians and not by foreigners, however sympathetic the foreigners may be.

Outside religious, moral and literary education, I will ask you to give her a simple scientific education, so that she shall know the laws of sanitation; that she shall know the laws of hygiene as her grandmother knew them; that she shall know the value of foodstuffs, so that she may know how to build up the bodies of her children, that she may distinguish between the effects of different kinds of food.... Then she should learn First Aid, and what to do in an accident. You know how men and women stand around at present, not knowing what to do when a little accident happens. First Aid is wanted constantly. You may even save a life by knowing the mere elements as to what to do at once. Then she should also learn simple medicines, and how to deal with children's trifling ailments. That is the scientific side that I suggest for women's education. I do not know whether cookery is by itself supposed to be scientific, but it ought to be, if it is to be carried on for the good of the household.

To all that I add some little knowledge of art. It may be music. Music when I came to Madras was thought of as a rather disreputable thing for a girl to learn.... So many girls are learning music, and that largely for the benefit of the boys; because if they have music at home, they won't run off after dancing girls and undesirable associations. Teach the sisters to sing, if only that they may keep their brothers safe. You will not only have done a good thing for the home, but you will have given pleasant amusement to the girls, who will feel happy in their art, and spend the time well, instead of only eating sweetmeats and talking gossip. The girl may learn embroidery, drawing or painting; for a girl ought to have some form of amusement to fill the idle hours which occur in life....

The only other point in girls' education that I would add to this, is physical education. I have said religious, moral, literary, scientific and artistic. They are all very big names. Lastly I come to physical. Do not forget the bodies of the girls. Let them have plenty of exercise which shall develop and strengthen them. In the old days when they did all the housework, that kept them healthy. Now that they are too

fine to do it, and sit upon English-made sofas, they are apt to become very anæmic and very weak. With early marriage on the top of that, you cannot wonder at the enormous death-rate of women under twenty-five years of age....

These are the lines which I venture to suggest to you as those which are wanted for the education at present of an Indian girl. By using them all you will prepare her for the life of a wife and mother, to be a helpmate for man, to sympathise with him, to train her children well. Not one of these tends to 'unsex' them, or to make them less feminine than they would otherwise be. I am putting them down as a necessary minimum. I am suggesting them now to you as a student of the subject, who has had a good deal of experience in it, as plans—plans that some of us are now trying to introduce, bit by bit, as one school after another for girls comes into our hands. I submit that these questions ought to be discussed, so that you may form your own opinions on the best form of girls' education, and then carry them out.

All I would say, in conclusion, is that along some such lines of education you must guide your girl-children for the sake of India.... Indian women have fallen between two stools; English education of men and the Pandit education[1] they used to have in earlier days, when every family had a Pandit to read to the women in the evening, and so they had literary education, if they had not the outer education of the world. Both have been taken from the girls, and yet they remain with brilliant intelligence. I would back Indian women against Indian men for brilliance of intelligence, for willingness to

---

[1]Pandits were Brahmins learned in Sanskrit and Indian philosophy. In traditional Indian schools, open only to males belonging to the three upper castes, a group of students would gather around a pandit. In some cases, a pandit would be brought into an Indian household to teach the girl. This was not a typical practice, however, as suggested by Besant who tended to exaggerate the extent of women's private education in traditional Indian society. When the first Indian schools for girls were organized, there was a shortage of educated women and pandits served as the instructors.

study, and above all for that power of self-sacrifice which appears as public spirit, when it is carried out into public life.

What you want most in this country is that practical spirit of self-sacrifice, that public spirit which looks on the interests of the country as greater than the interests of the individual. You can learn this from women. They sacrifice themselves every day and every night for the interests of the home; they realise the subordination of the one to the benefit of the larger self of the family. Learn that from your women and then you will become great, and India will become great; for if you carry into public life the self-sacrifice of women, then the redemption of India will be secured. But you will do it best, if you will go with them into the world hand in hand, men and women together. The perfect man is made up, remember, of the man, the wife and the child, and not the man alone. If such men become citizens of India, then her day is not far off.

114. Women and German Colonialism in Southern Africa

Germany first took formal control of colonial territories in Africa and Asia in 1884–1885, much later than the empires of England, France, Portugal, and Spain. By the 1890s the German government was eager to increase European settlement and economic activity in its most important colony, Southwest Africa (Namibia). The German Colonial Society, an association of influential men interested in promoting imperialist causes, agreed to subsidize the travel of German women to the colony as brides and servants. The Women's League of the German Colonial Society was formed in 1907. It selected applicants for emigration subsidies and arranged for their transportation. The League's biweekly magazine, *Kolonie und Heimat (Colony and Home),* published news and stories about German colonialism, including reports from women colonists. Several schools were also established in Germany to prepare middle-class women for the new demands of colonial farm life. The most important of these was the Colonial Women's School which opened in 1908 in Bad Weilbach in Wiesbaden. Less rigorous in its curriculum than the academic men's German Colonial School, it provided training in practical home economics, gardening, and animal care as well as an orientation to geography and ethnography. Many supporters of women's colonization believed that emigration would help some of the many unmarried, or "surplus," German women to find husbands. Some also thought that the presence of female colonists would discourage sexual unions between male settlers and African subjects, whose children they viewed as racially inferior to Europeans. During the decades of German colonization, a period that included terrible German brutalities against the territory's African population, imperialists and feminists publicly debated what roles German women should play in the empire and how to select the best potential women settlers to transport to the colonies.

Contributed and translated by Krista O'Donnell. Used by permission.

A. Minna Cauer, "The Women's Colonization Question," *The Women's Movement: Review for the Interests of Women* (April 1, 1898)

Alerted by a reader's inquiry, feminist Minna Cauer's newsletter, *The Women's Movement,*[1] published a call for more information on a rumored plan to settle German women in Southwest Africa. The promoters of women's colonization, including the missionaries of the Protestant Africa Society[2] responded to her call for information. Cauer published her reaction to the scheme, demanding greater involvement and oversight by her organization. The German governor of Southwest Africa, Theodor Leutwein,[3] successfully demanded that Cauer's journal publish his rebuttal, which follows her remarks.

The editors have received the following notice from [the Protestant Africa Society] regarding the "Women's Question in Southwest Africa." . . .

"We desire the assistance and participation of women in this field, but we have not yet been successful in enlisting the various Protestant women's associations, such as the Women's Society for Nursing in the Colonies.[4] The Protestant Africa Society (a men's organization) is working with the knowledge of the government in order to offer individual reputable girls the possibility of emigrating to Southwest Africa. They provide the girls with funding for transportation (the cost is approximately 500 marks per person) and shelter with a missionary family, since these are the only completely reliable families over there."[5]

So much for our inquiries. Here lies before us a field of activity for women, which, however, can only be initiated if women occupy the same position as men in the new colonies, which truly have no need to reckon with traditions and antiquated prejudices. The churches, schools, and communities there, which are only making their smallest beginnings, must immediately count women as members who are born equal to men and with equal rights. We surely should not condemn the introduction of individual girls for the purpose of their foreseeable marriages, for that is the dubious aim of their settlement (as the proceedings of the Protestant Africa Society reveal). Women can only exercise a moral influence over there, if they are given a place and a voice in all activities and enterprises from the very beginning. The women who believe in the principles of the women's movement and pursue its goals would certainly be ready to assist in such an important endeavor, if only they could be convinced that it was not solely a response to the marriage question, but rather that women's contributions to solving the cultural problems of the colonies would be valued from the beginning.

---

[1]Minna Cauer (1841–1922), a leading liberal bourgeois feminist and head of the Berlin Women's Welfare Association, was active in promoting women's suffrage, professional education, and employment. She edited *The Women's Movement: Review for the Interests of Women,* the newsletter of the Women's Welfare Association, and numerous regional middle-class feminist organizations. The newsletter discussed women's colonization in several issues between 1898 and 1899.

[2]The Protestant Africa Society was a private Lutheran missionary society which aimed to spread Christianity throughout Africa. In 1896, one of its missionaries in Southwest Africa proposed new measures to settle German women in the region in order to provide eligible brides for male colonists and deter them from cohabiting with African women.

[3]Theodor Leutwein (1849–1921) served as commander of the colonial army and governor of Southwest Africa from 1893–1905, where he did much to promote German economic development and settlement in the region. His appointment ended during the brutal German suppression of the Herero rebellion. The Herero, a Bantu population distinguished for its cattle-herding culture, was the largest of several ethnic groups in German Southwest Africa, but German atrocities decimated their numbers.

---

[4]The German Women's Society for Nursing in the Colonies was a charitable association of women that organized travel subsidies and employment for nurses to serve the German empire's settler population, bureaucrats, and armed forces. Conservative, nationalist, and militarist, the association overtly rejected feminist views.

[5]Typical urban maids, who generally earned 20–50 marks per month, with room and board, would find the fare to Southwest Africa far beyond their means.

This, first of all, requires an organization from which these virtuous and self-sacrificing women would go forth, possessed with the strength, courage, and resolve to assume so great a work—women of equal rights assisting in carrying forth the civilization and human development of the colonies in every area. Unfortunately, experience, until now, has demonstrated that savagery, special interests, and outdated viewpoints have led to the shocking descent to barbarism there. Can we, as things now stand, permit women to emigrate? This is a genuine and important problem. Women may engage it by promising or witholding their cooperation in given cases.

B. Governor Theodor Leutwein, "A Public Letter to Minna Cauer," *The Women's Movement* (May 1, 1898)

To Minna Cauer, Berlin,
My most worthy, honored lady

A copy of number 7 of your newspaper, *The Women's Movement,* was sent to me here, containing an article on the women's colonization question. This deliberate contact forced me to take a position on the article, and I will loyally provide one to you with the hope that you will be equally loyal in making room for the piece in your paper.

The article demanded that women in the colonies occupy the same position as that of men. In the church, school and community women must immediately count as equal members. Regarding these statements, I must reply that we have carried on our affairs in church, school, and community quite well alone and hope to do the same in the future. What we cannot achieve alone, however, is to create a German household, where the German housewife lives and strives for her husband and her children according to what until now has been a good custom. This household establishes German family life in the colonies and offers the assurance that German children will grow into good German citizens and wives. The women who wish to do this may

come to us. They will be received with open arms and constant esteem. Others may stay far away.

You state in your article that women can *only* exercise a moral influence when, "They are given a place and a voice in all public activities and enterprises." This, we do not understand at all out here. Women can achieve this completely in their work in the circle of their families as well. As it further states in the article, "Unfortunately, experience, until now, has demonstrated that savagery, special interests, and outdated viewpoints have led to a shocking descent to barbarism there. Can we, as things now stand, permit women to emigrate?" This could only have been written by someone who has never been here, since we have never noticed such things.

The so-called "Women's Movement" may have a large core of justification in the old Fatherland, with its surplus of women. But women will only harm themselves if they bring their movement over to the colonies, where the powerful minority of women are sought after and valued as wives and only as wives. There is the danger that the men there will respond by becoming marriage-shy and do just what we hope to prevent, namely tie themselves to native women, who do not raise these types of demands.

The Imperial Governor
Theodor Leutwein
April 6, 1898

C. Frieda Dahl, "Letter from a Potential Emigrant" (1906)

Ordinary women, excited by newspaper reports that described the German government's support for subsidizing female emigration to German Southwest Africa, wrote letters to German officials requesting travel funds. The German emperor, his colonial administration, the Colonial Society, and Southwest African officials, received hundreds of letters like the following from women seeking free passage and employment in the colony.

Dear Imperial Governor of Windhoek, Southwest Africa:

Inspired by a newspaper notice that women or girls who wish to enter into servant contracts in Southwest Africa can travel cost-free from Germany, I introduced myself at the German Colonial Society headquarters at Schilling Street [Berlin]. There, I requested more information and sought perhaps to fulfill my long-held wish to take a position in the German colonies. I was told to direct my request to you, and I have not hesitated to follow this instruction.

I was born in 1876 in the village of Kroepetin in Mecklenburg. I am 1.65 meters tall, healthy and strong. Experienced in all aspects of housework, I have a practical nature, learn easily, and do not shy away from any kind of work. I am especially adept in sewing, which I carry out to perfection, as well as all kinds of handiwork, fine ironing, hairdressing, cooking, etc. I would like to work as a housekeeper for a fine German employer or master. I would like to be of service and hope that my many abilities and skills will be better utilized there than here in Germany.

Unfortunately, I must inform you that I am married, but therein lies absolutely no impediment to my leaving, since my husband abandoned me years ago because he was not in a position to support me. I have lived for a long time on my own two feet. After my education in Rostock, I served for several years in the home of the Master builder Vogel in Rostock, and for Mayor Fink in Plan as a housekeeper's assistant. After the break-up with my husband, I supported myself as a seamstress, but gave it up three years ago to take up my current position, and I moved last summer with Mr. Vogel's family from Rostock to Berlin.

My long-cherished wish is to take a job overseas, and I gave my notice on April 5 with this intention. My most humble plea to Your Honor is that you will have the kindness to inform me of the steps that I must take to find a suitable (well-paid) position in your territory. Is your Honor acquainted with an appropriate employer? If so, I am quite ready to immediately accept a firm offer which includes all details about pay and duties. Regardless of employer, please let me know what equipment is necessary and anything else I need to know, as well as if I may receive an advance on my salary to outfit myself.

If there are any newspapers there where a person can find such employment by advertising, I humbly ask you to forward them my request, and I will gladly assume the cost.

In the hope that you will excuse my many demands and not take them amiss, since I want so badly to go there, and await your answer with great anticipation, I enclose my photograph and references. Hoping not to have made a useless inquiry, I sign in the greatest respect, most humbly,

Frieda Dahl, née Küchenmeister, Charlottenburg, Berlin (1906).

D. Hedwig Pohl,[1] "A Thank-You Letter from a Nurserymaid in Southwest Africa," *Colony and Home* (August 1, 1909)

Very honored Baroness von Liliencron,

I send you from Africa, as I promised, a small description of our life and affairs here. It is already ten weeks since I left our Fatherland. I have seen so much that is new and beautiful in this short time. Stood on Belgian soil, seen Britons in their homeland, and traded fruit and blankets with

---

[1]Hedwig Pohl received a travel subsidy to emigrate from the Women's League of the German Colonial Society and wrote to thank the Baroness von Liliencron, the head of the organization, and to describe her experiences on a farm in Southwest Africa. Her letter was published in *Colony and Home* as a sample of letters sent to the Women's League by women whom it had helped to find employment in the colonies.

Spaniards on Tenerife. And then Africa! A couple of hours before we arrived in Swakopmund, our new home came into sight in bright tones....

I won't hide it. I was soon homesick. Now that I am better accustomed, I find it pleasing to be here in Windhoek. I live with the children in a little house, five minutes from the hotel. The four older children go to school from 7:00–11:00 and from 3:00–5:00 in the afternoon in the Catholic mission school. Then, I am alone with my little darling, Bertha, whose funny faces and babbling cheer me up. It is pretty in Klein-Windhoek, two kilometers from here. Mr. and Mrs. S., the children and I went there one Sunday. There, everything is always green: grapevines, spinach, herbs, and much more thrive there. We picked carnations in the churchyard and made a pretty bouquet to take home. There are many birds here. We saw one tree along our journey with about 50 nests in it. A pair of hens has nested under my roof, and a hen once laid an egg in the yard. The children saw it and they came to me laughing and jubilant about this small miracle. Our children have already made friends with the blacks and use many of their expressions. So, they are always *steeve* (hungry), and then eat *moy* (food). They click with their tongues, just like the natives do. Many of the *Kaffirs*[2] here speak and understand German well. Once I asked a little one, "Do you go to school?"

"Yes, Miss."

"What do you learn there?"

"About our dear Lord."

"Anything else?"

"About Master Jesus."

As I gave him a piece of bread, he sprang up in delight, and shared it with the others, who stood a ways off.

Behind our house, we have a beautiful view. There, over on a mountain, stands the maternity hospital, a pretty building. It lies in the West, and when the sun goes down it is so lovely. One sees the "mowing sheep," as our children call them, return home and hurry to drink, oxen and horses herded together.

We go to bed early and at about 9:00 p.m. everything is peaceful.

So, one day turns into another and the favorite days are always when the post arrives from Germany: Hurrah, a letter from home. Now it is cold in dear Germany and in Hamburg the citizenry complain about the rain. Here, it gets hotter every day. Hopefully, we will stay healthy, then we can gladly suffer every hardship.

Now, I send you many greetings to the dear German Fatherland, and also to you, dear Baroness von Liliencron. I remain, with all respect, your humble Hedwig Pohl, nursery maid with Mr. S. in Windhoek.

E. Anonymous, "Letters from Our Readers," *Swakopmund Newspaper* (April 3, 1912)

After the colonial troops in Southwest Africa murdered or drove out the majority of the African population during the armed African uprisings of 1904–1907, colonists had difficulty hiring enough servants and farm workers. Native Ordinances were then decreed which prohibited free movement, mandated employment, and established uniform wages and working conditions. The indigenous population resisted these measures whenever possible. Colonial officials, who were alarmed at the low African population figures, forbade the physical punishment of African women and excused many African wives from paid labor because they wanted these women to concentrate on being mothers. German colonial wives, who generally managed household labor, reacted negatively to these restrictions on their power over indigenous women.

The local housewives complain a great deal about the shortage of colored female workers, although there are many available in this place. Few are willing to work, however, since by inquiry most "have a husband, no pass, or herd

---

[2]*Kaffir:* a demeaning term for African natives.

cattle."[1] This last claim seldom matches the truth.

The police administration may perhaps bring about a change, as they are tightening their control by conducting raids in order to seek out the lazy ones. Also, the full power of the law must be applied to native women in all cases. Declarations such as those stated above must be given absolutely no consideration.

The native women are becoming and are already much too bold and believe that they do not need to work. The majority believe that they need only to move from one group of huts to the next and to visit each other in turn [to escape forced labor].

With increased oversight from employers and the authorities, additional numbers of native women will be discovered to be without work who can be employed in households, so that the great demand can be met.

The police will surely be thanked if they announce when extra hands are available.

[signed] One for Many

F. Clara Brockmann, "The Necessity of the Immigration of German Women" (1910)

The German writer Clara Brockmann emigrated independently to Southwest Africa, which was highly unusual. She was a frequent contributor to the *Colony and Home* magazine and a strong supporter of the Women's League of the German Colonial Society. Excerpted here is the opening essay of her first book, *The German Woman in Southwest Africa* (1910), which outlines a racial argument for increasing women's colonization in Southwest Africa.

The immigration of the German woman to our colony is much discussed and desired. The purposes stand clear before us: hindering miscegenation,[1] which results in the spiritual and economic ruin of settlers; establishing a profitable farm, which cannot grow into a thriving enterprise without a housewife's assistance; and, above all, the civilizing power of German domesticity and customs which the presence of a woman fosters. Moreover, there is the caring for the younger generation, a factor that should not be undervalued because it defines the cultural level of an expanding population....

In the North, and especially the middle of the territory, the shortage of white women is now felt somewhat less. However, it is otherwise in the Southern region of the colony. Here, many a native woman occupies the position of a white wife, exercising complete authority over the home. In the North I have also seen Herero[2] women who were the companions of German farmers on some individual homesteads. And their bastard[3] children play in the house and yard, tragic and unwanted offspring who will bear their native blood until the end of time. Neither a religious nor a civil marriage is permitted any longer and the courts have even gone so far as to declare all existing unions with native women invalid. This decision was reached concerning the case of an old Herero woman who demanded a divorce from her white husband for extreme abuse. The court's verdict has been much debated, since [annulling

---

[1]Married African women could escape compulsory labor by claiming that they were needed at home to care for their husbands and children or to tend cattle herds while male relatives worked. Since many natives who lived outside of German settlements did not have passes or identity cards, African men and women who wanted to abandon unpleasant jobs could discard their papers and run away. Many of these escapees moved frequently and changed their names. Without any means of permanent identification, they could not be returned to their former employers.

[1]Miscegenation: Interracial unions.

[2]African women of the Herero ethnic group.

[3]"Bastard" has multiple meanings here. It indicates a person who is "hybrid," or mixed-race. It also suggests a child who is illegitimate, since interracial unions had been declared illegitimate in the colonial courts. The term primarily describes the members of the Rehoboth Baster population, a centuries-old fusion group of Dutch, English, and Khoi-San ancestry and culture, who frequently intermarried with colonists in southern Africa.

longstanding interracial couple's marriages, declaring their off-spring illegitimate, and stripping them of German citizenship] resulted in a considerable hardship. Even the governor, who has been misquoted in the German press, declared his personal conviction that he could not support this decision. But this decisive measure also has brought good, for in the future, native women increasingly will be driven from these illicit unions.

It is an old and sad fact of experience that the black woman has never made the slightest advance to a higher cultural level and that her husband instead descends closer to the level of his colored living companion and, not uncommonly, sinks completely to her level. The bastard children comprise a dubious element, which hinders the development of the colony. Two natures fight within them. The constructive abilities and intelligence, as well as a pronounced sense of good and evil—the legacy of their white fathers—wrestles against the complete moral inferiority of the native race.

The decrease of such mixed marriages is to be greeted as a positive development. How long will it last, since before no settler would ever think to make a native woman his housewife and the mother of his children? The time is past when former officers and the sons of admirals give their names to a black.

QUESTIONS FOR ANALYSIS

1. According to *The Complete Indian Housekeeper and Cook,* what kinds of challenges will British women encounter in managing a household in India? How does the guidebook define the duties of the mistress of the household and what suggestions does it offer for ensuring their proper fulfillment? What colonial attitudes and prejudices are revealed in these suggestions?

2. In Annie Besant's view, how has the transformation of the family in Indian society affected the educational needs of girls and women? What subjects does she propose for the education of Indian girls and why? How does Besant link educational issues to the struggle for Indian independence?

3. How did Minna Cauer envision women's roles in German colonial settlements in Africa and how were her views challenged by Governor Leutwein? What were the life circumstances, ambitions, and actual experiences of women who emigrated to the colonies? How were racial arguments used to promote women's emigration to the colonies?

4. What do the writings in this section reveal about the diversity of women's colonial experiences? How do they enhance our understanding of the nature of British and German colonialism?

WOMEN ARTISTS AND BOURGEOIS CULTURE

Although women often faced discrimination in training and in the exhibition of their work, they played an important role in the creation of bourgeois culture as amateurs and professionals. In addition to traditional art forms such as painting and sculpting, women contributed to the popularization of new mass art forms such as photography.

115. Margaretta Higford Burr
"Interior of a Hareem, Cairo," 1846

An amateur English watercolor artist, Margaretta Higford Burr drew sketches of her travels in Europe and the Middle East over a period of twenty-five years. She was in Cairo in 1844 and from there journeyed with her husband to Syria, Jerusalem, and Constantinople. In 1846, Burr published a portfolio of her sketches, *Sketches from the Holy Lands,* which included "Interior of a Hareem, Cairo." In the autumn of 1848, she traveled up the Nile River and then overland to a point about two hundred miles below Khartoum, Sudan. Burr painted mainly for her own enjoyment and that of her friends, but she also exhibited her paintings at various times. In 1859, her paintings were shown by the Society of Female Artists which promoted amateur as well as professional artists and thus became an especially important venue for the drawings and paintings of women travelers. Burr also contributed to amateur exhibits and sold her pictures for such charitable causes as Lady De Grey's Home for Needlewomen. Burr's painting of a harem contrasts with the more exotic paintings of Ingres, who never traveled to Turkey or the Middle East but rather relied on vivid descriptions of harems and bathhouses provided by Lady Mary Wortley Montague, the wife of the British ambassador to Turkey, whose letters were published in France in 1805.

A. *Interior of a Hareem, Cairo,* lithograph by Margaretta Burr from *Sketches from the Holy Lands,* 1846. (*Victoria and Albert Picture Library*).

116. Berthe Morisot
"Woman and Child on a Balcony," 1872

Berthe Morisot (1841–1895) was a leading Impressionist painter. The Impressionists challenged the basic precepts of Academy art and focused on scenes of daily life that could be painted without permanent studios, state sponsorship, and large canvases. Reflecting the democratization and commercialization of bourgeois culture in the Third Republic, this shift provided important, new opportunities for women painters. Female Impressionists, however, could not move as freely in public spaces as their male counterparts and so generally focused on scenes of bourgeois domesticity. Mothers and daughters were a favorite subject of Morisot's paintings. She often depicted her sisters, Edma and Yves, with their daughters. Julie Manet, her own daughter, also inspired many portraits. Morisot contributed to the first Impressionist exhibition in 1874—and to all but one of the subsequent exhibitions—as well as to avant-garde exhibitions. The first exhibition devoted entirely to Morisot's work was held in 1892. Following her death, Degas, Monet, Renoir, and Mallarmé organized a showing of nearly half of the some 850 paintings in oil, pastel, and watercolor which Morisot produced in her thirty-year career. In this 1872 watercolor, "Woman and Child on a Balcony," Morisot's sister Yves and her daughter are looking across the Seine from a balcony in the Morisot family garden.

On the Balcony, 1874 by Berth Morisot. Water-color, 20.6 × 17.3 cm. (The Art Institute of Chicago. Gift of Mrs. Charles Netcher in memory of Charles Netcher II.)

117. Julia Margaret Cameron
"Mrs. Herbert Duckworth" April 1867

Julia Margaret Cameron (1815–1879) was born in Calcutta, India, the daughter of an East India Company official. By 1848, when Cameron's family moved to England, French daguerrotype portraits were popular among the upper and middle classes. Family albums came into existence in the 1850s as did photographic exhibitions. By the 1860s, portrait photography was highly commercialized and millions of *cartes-de-visites*, albumin prints on cardboard of family members or famous individuals, were sold in Great Britain alone. Cameron criticized such photographs as too mechanical and not sufficiently elevating of the spirit. Her own photographs, taken mainly between 1864 and 1875, included portraits of Charles Darwin; Robert Browning; Alfred Tennyson; Alice Liddell,

Mrs. Herbert Duckworth. (*The Beaumont and Nancy Newhall Estate, Courtesy of Scheinbaum & Russek Ltd., Santa Fe, New Mexico.*)

the muse for Lewis Carroll; and Julia Jackson, her niece and the mother of Virginia Woolf. Cameron also photographed models with whom she recreated painted portraits and staged scenes inspired by biblical, classical, and literary sources. She exhibited in Paris, Berlin, and London and her photographs were privately owned by such individuals as Queen Victoria, George Eliot, and Victor Hugo. Virginia Woolf assembled a collection of Cameron's photographs for publication by Hogarth Press which she had founded with Leonard Woolf. *Victorian Photographs of Famous Men and Fair Women* appeared in 1926. In her introductory essay, "Julia Margaret Cameron," Woolf noted that photography provided Cameron with

an outlet for the energies which she had dissipated in poetry and fiction and doing up houses and concocting curries and entertaining her friends. Now she became a photographer. All her sensibility was expressed, and, what was more to the purpose, controlled in the newborn art. The coalhouse was turned into a dark room; the fowl-house was turned into a glass-house. Boatmen were turned into King Arthur; village girls into Queen Guenevere . . . the parlour-maid sat for her portrait and the guest had to answer the bell. "I worked fruitlessly but not hopelessly," Mrs. Cameron wrote of this time. Indeed, she was indefatigable.[1]

Virginia Woolf's mother was one of Cameron's favorite subjects. This photograph was taken in 1867, the year she married Herbert Duckworth. He died three years later and in 1878 she married Sir Leslie Stephen, Virginia Woolf's father.

QUESTIONS FOR ANALYSIS

1. How does Margaretta Higford Burr's painting "Interior of a Hareem, Cairo" portray the women of the harem? How does her painting compare with those of such male artists as Ingres? What might account for their different approaches?
2. Discuss the composition of Berthe Morisot's "Woman and Child on a Balcony" and how it depicts the relations between a mother and her daughter. What sentiments does this painting evoke and how does it enhance our understanding of nineteenth-century notions of bourgeois domesticity?
3. Consider the pose and lighting in Julia Margaret Cameron's 1867 photograph of Mrs. Herbert Duckworth (Julia Stephens). What does it convey about her character and appearance? Compare the photograph with one of Virginia Woolf, her daughter. How does this photograph differ from a painted portrait and what does it reveal about the possibilities and limitations of the new medium of photography? Why are photographs considered a form of art and not simply an "objective" reflection of reality?
4. Compare and contrast the artistic mediums and subjects chosen by Margaretta Higford Burr, Berthe Morisot, and Julia Margaret Cameron. What do they reveal about the range of women's artistic expression?

---

[1]Virginia Woolf, "Julia Margaret Cameron," in *A Bloomsbury Group Reader,* ed. S. P. Rosenbaum (Oxford: Basil Blackwell Ltd., 1993), p. 87.

CHAPTER 10

The Development
of Bourgeois and
Working-Class Feminism

The growth of bourgeois society, industrialization, and national unification in Italy and Germany inspired new forms of liberal, bourgeois, and socialist feminism. The development of national bureaucracies and social services led to the expansion as well as creation of new forms of white-collar and professional work. Typically involved in "respectable" philanthropic activities, middle-class women now began to demand more educational and career opportunities in education, social welfare, nursing, and the civil service. Nineteenth-century feminist writings on marriage and sexual norms also covered a wide range of issues relevant to working-class as well as middle-class women, such as the inequality of marriage and divorce laws, prostitution, women's fashion, and homosexuality.

Following the defeat of the Paris Commune in France in 1871 and the achievement of national unification in Italy in 1870 and in Germany in 1871, working-class struggles assumed the form of mass, socialist working-class parties. Socialist women insisted that these parties take up women's issues, oppose demands for the exclusion of women from industry, and make special efforts to reach out to working-class women. Middle- and working-class women participated in the suffrage movement that began in England in the 1860s and continued throughout Europe into the twentieth century. Women first obtained the vote in Finland in 1906 and then in Norway in 1913. The outbreak of World War I divided suffrage movements and most terminated their activities for the war's duration.

NEW CAREERS IN MEDICINE AND SCIENCE

The development of bourgeois society and national unification expanded careers for middle-class women in nursing, medicine, teaching, the postal and telegraph services, social welfare, the civil service, and science. The following writings by Florence Nightingale, Sophia Jex-Blake, and Marie Curie illustrate their pioneering roles in expanding careers for women in medicine and science.

118. Florence Nightingale
The Crimean War

Immortalized as the "Lady with the Lamp," Florence Nightingale (1820–1910) played a key role in the modernization of nursing as a profession. She studied nursing in Prussia and France and had served for eighteen months as superintendent of the Hospital for Invalid Gentlewomen in Harley Street at the outbreak of the Crimean War (1854–1856). Nightingale was arranging with King's College Hospital to create a training establishment for nurses when she read letters in *The Times* describing the abysmal medical care received by British soldiers in the war. Sidney Herbert, Secretary at War, accepted Nightingale's offer to serve as a nurse in the war with enthusiasm, as his earlier attempts to allow female nurses to work in military hospitals had been thwarted by the military and medical establishments as well as the absence of experienced nursing personnel. On November 4, 1854, Nightingale arrived with thirty-eight nurses at the Scutari Barrack Hospital near Constantinople and became chief of the Nursing Department. The Crimean War provided her with an unprecedented opportunity to transform nursing practices and hospital administration. New military technologies and military incompetence were responsible for the extraordinarily high numbers of military casualties in the Crimea. Nightingale implemented various reforms—in particular, strict enforcement of sanitary conditions—and reduced the casualty fatality rate from 45 to 5 percent within a year of her arrival. She became famous when British newspapers published pictures of her treating wounded soldiers. Nightingale wrote over three hundred letters detailing her work and proposals for reform, including this November 14, 1854 letter to Dr. William Bowman of King's College Hospital. Returning to London in 1856, Nightingale researched the conditions prevailing in British military hospitals which she summarized in *Notes on Matters Affecting the Health, Efficiency, and Hospital Administration of the British Army* (1858). Nightingale also prompted the formation of a Royal Commission on the Health of the Army to which she provided information. Her *Notes on Nursing: What It Is, and What It Is Not* (1859) became the basis for the Nightingale Training School for Nurses established in June, 1860 at St. Thomas's Hospital. Soon afterwards, Nightingale turned her attention to the conditions faced by British soldiers in India. She received the Order of Merit in 1907.

Written on November 14, 1854.

To Dr. William Bowman

"I came out Ma'am, prepared to submit to every thing—to be put upon in every way—but there are some things, Ma'am, one can't submit to—There is caps, Ma'am, that suits one face, and some that suits anothers, and if I'd known, Ma'am, about the caps, great as was my desire to come out to nurse at Scutari, I wouldn't have come, Ma'am." Speech of Mrs. Lawfield,[1] Nov. 5th

> Barrack Hospital, Scutari,
> Asiatic Side
> 14th November 1854

Dear Sir,

Time must be at a discount with the man who can adjust the balance of such an important question as the above—and I, for one, have none, as you will easily suppose when I tell you that on Thursday last we had 1715 sick and wounded in this Hospital, (among whom 120 Cholera Patients), and 650 severely wounded in the other building, called the General Hospital,[2] of which we also have charge—when a message came to me to prepare for 570 wounded on our side of the Hosp*l.* who were arriving from the dreadful affair of the 5th of Nov*ber.* at Balaclava, where some 1763 wounded, & 442 killed, besides 96 Officers wounded & 38 killed.[3] I always expected to end my days as Hospital Matron, but I never expected to be Barrack Mistress. We had but half an hour's notice before they began landing the wounded. Between one and nine o'clock, we had the mattresses stuffed, sewn up, and laid down, alas! only upon matting on the floors, the men washed and put to bed, & all their wounds dressed—I wish I had time or I would write you a letter dear to a surgeon's heart, I am as good as a "Medical Times".

But oh! you gentlemen of England, who sit at home in all the well-earned satisfaction of your successful cases, can have little idea from reading the newspapers, of the horror & misery (in a military Hosp*l.*) of operating upon these dying and exhausted men—a London Hosp*l.* is a garden of flowers to it—we have had such a sea in the Bosphorus, and the Turks, the very men for whom we are fighting, carry our wounded so cruelly, that they arrive in a state of agony—one amputated stump died two hours after we received him—one compound fracture just as we were getting him into bed, in all 24 cases on the day of landing—the dysentery cases have died at the rate of one in two—then the day of operations which follows—I have no doubt that Providence is quite right and that the Kingdom of Hell is the best beginning for the Kingdom of Heaven, but that this is the Kingdom of Hell no one can doubt. We are very lucky in our Medical Heads—two of them are brutes, and four of them are angels—for this is a work which makes *either* angels or devils of men, and of women too. As for the Assistants, they are all cubs, and will, while a man is breathing his last breath under the knife, lament the 'arrogance of being called up from their dinners by such a fresh influx of wounded'. But wicked cubs grow up into good old bears, tho' I don't know how—for certain it is, the old bears are good. We have now four miles of beds—and not eighteen inches apart. We have our quarters in one Tower of the Barrack—and all this fresh influx has been laid down between us and the Main Guard in two corridors with a line of beds down each side, just room for one man to step between, and four wards.

Yet, in the midst of this appalling horror (we are steeped up to our necks in blood)—there is

---

[1]Mrs. Rebecca Lawfield was one of six nurses from St. John's House. She later became one of Nightingale's most valued assistants.

[2]Sidney Herbert placed Florence Nightingale in charge of two military hospitals in Scutari, which is located on the Bosphorus just opposite Constantinople. "This hospital" refers to the Scutari Barrack Hospital which served as Nightingale's headquarters while the General Hospital was smaller and had fewer patients. The existence of these two hospitals outside of the military zone of conflict was one of the reasons given by Herbert to permit women to serve as nurses in the war. Later, however, Nightingale also assumed responsibility for female nurses in military hospitals located in the Crimea.

[3]The reference is to a cavalry battle in Balaclava, which is located on the Black Sea in the Crimea.

good. And I can truly say, like St. Peter, 'it is good for us to be here' tho' I doubt whether, if St. Peter had been here, he would have said so. As I went my night-rounds among the newly wounded that first night, there was not one murmur, not one groan, the strictest discipline, the most absolute silence & quiet prevailed, only the step of the sentry and I heard one man say, I was dreaming of my friends at home, & another said And I was thinking of them. These poor fellows bear pain and mutilation with unshrinking heroism, and die or are cut up without a complaint. Not so the Officers, but we have nothing to do with the Officers. The wounded are now lying up to our very door, and we are landing 540 men from the "Andes"—I take rank in the army as Brigadier-General, because 40 British females, whom I have with me, are more difficult to manage than 4000 men. Let no lady come out here who is not used to fatigue & privation—for the Devonport Sisters,[4] who ought to know what self-denial is, do nothing but complain. Occasionally the roof is torn off our quarters, or the windows blown in—and we are flooded and under water for the night. We have all the Sick Cookery now to do, and have got in four men for the purpose, for the prophet Mahomet does not allow us a female. And we are now able to supply these poor fellows with something besides Gov*t.* rations. The climate is very good for the healing of wounds.

I wish you would recall me to Dr. Bence Jones's remembrance, when you see him, and tell him that I have had but too much occasion to remember him in the constant use of his dreadful presents. Now comes the time of haemorrhage and Hospital Gangrene, and every ten minutes an orderly runs, and we have to go and cram lint into the wound till a Surgeon can be sent for, and stop the bleeding as well as we can. In all our Corridors I think we have not an

average of three limbs per man—and there are two ships more "loading" at the Crimea with wounded, this is our phraseology. Then come the operations and a melancholy, not an encouraging list is this. They are all performed in the wards—no time to move them. One poor fellow, exhausted with haemorrhage, has his leg amputated as a last hope and dies ten minutes after the surgeons have left him. Almost before the breath has left his body, it is sewn up in its blanket and carried away—buried the same day. We have no room for corpses in the wards. The Surgeons pass on to the next, an excision of the shoulder-joint—beautifully performed and going on well—ball lodged just in the head of the joint, and fracture starred all round. The next poor fellow has two stumps for arms—and the next has lost an arm and leg. As for the balls, they go in where they like, and do as much harm as they can in passing. That is the only rule they have. The next case has one eye put out, and paralysis of the iris of the other. He can neither see nor understand. But all who can walk come into us for Tobacco, but I tell them that we have not a bit to put into our own mouths. Not a sponge, nor a rag of linen, not anything have I left. Everything is gone to make slings and stump pillows and shirts. These poor fellows have not had a clean shirt nor been washed for two months before they came here, and the state in which they arrive from the transport is literally *crawling.* I hope in a few days we shall establish a little cleanliness. But we have not a basin nor a towel nor a bit of soap nor a broom—I have ordered 300 scrubbing brushes. But one half the Barrack is so sadly out of repair that it is impossible to use a drop of water on the stone floors, which are all laid upon rotten wood, and would give our men fever in no time. The next case is a poor fellow [*sic*] where the ball went in at the side of the head, put out one eye, made a hole in his tongue and came out in the neck. The wound was doing very nicely when he was seized with agonizing pain and died suddenly, without convulsion or paralysis. At the P[ost] M[ortem], an abscess in the anterior part of the

---

[4]The Devonport Sisters included Sisters from the Sisterhood of the Holy Cross and the Society of the Most Holy Trinity. They gained most of their nursing experience in the cholera epidemics of the late 1840s and early 1850s.

head was found as big as my fist—yet the man kept his reasoning faculties till the last. And nature had thrown out a false coat all round it.

I am getting a screen now for the Amputations, for when one poor fellow, who is to be amputated tomorrow, sees his comrade today die under the knife it makes impression—and diminishes his chance. But, anyway, among these exhausted frames the mortality of the operations is frightful. We have Erysipelas Fever and Gangrene. And the Russian wounded are the worst. We are getting on nicely though in many ways. They were so glad to see us. The Senior Chaplain is a sensible man, which is a remarkable providence. I have not been out of the hospital wards yet. But the most beautiful view in the world lies outside. If you ever see Mr. Whitfield, the House Apothecary of St. Thomas's, will you tell him that the nurse he sent me, Mrs. Roberts,[5] is worth her weight in gold. There was another engagement on the 8th, and more wounded, who are coming down to us. The text which heads my letter was expounded thus—Mrs. Lawfield was recommended to return home and set her cap, vulgarly speaking, at some one elsewhere than here, but on begging for mercy, was allowed to make another trial. Mrs. Drake[6] is a treasure—the four others are not fit to take care of themselves nor of others in a Military Hosp*l*. This is my first impression. But it may modify, if I can convince them of the absolute necessity of discipline and propriety in a drunken Garrison.

Continued on inside flap of envelope.

Believe me, dear Sir, yours very truly and gratefully,
This is only the beginning of things. We are still expecting the assault.

---

[5]Mrs. Roberts was a nurse from St. Thomas Hospital who provided care for Nightingale when she later became ill in the Crimea.

[6]Mrs. Elizabeth Drake was one of the nurses from St. John's House. She never returned to England but died in Balaclava in 1855.

119. Sophia Jex-Blake
"The Medical Education of Women"

Sophia Jex-Blake (1840–1912) pioneered women's medical education in Great Britain. In the 1880s, when Jex-Blake sought to become a doctor, Elizabeth Blackwell and Elizabeth Garret-Anderson were the only two women who served on the Medical Register and the means by which they had obtained this status were now blocked. Blackwell, the first Englishwoman doctor, had entered the medical profession in the United States. Although she was allowed to practice medicine when she returned to England, the Medical Act of 1858 subsequently required all future medical candidates to possess a British university degree. In 1865, Garret-Anderson became the first woman to qualify in England by passing the examinations of the Society of Apothecaries, but the regulations were subsequently altered to prevent other women from qualifying in the same manner. Undaunted by these obstacles, Jex-Blake entered Edinburgh University in 1869 along with several other female students who sought to become doctors. After several years of study, however, the women were informed that they had been

Originally written by Sophia Jex-Blake in 1873.

improperly admitted and were not allowed to take their final examinations. While some female students then sought qualification in France and Switzerland, Jex-Blake attempted to gain admission to the medical profession in Britain by taking an examination in Midwifery at the College of Surgeons from which female candidates could not be barred. But the entire Board of Examiners resigned when she and another woman presented themselves as candidates. In her 1873 article, "The Medical Education of Women," reproduced here, Jex-Blake presents a detailed case for opening up the medical profession to women. She subsequently focused on organizing the London School of Medicine for Women which was established in 1874. It first granted medical degrees three years later after the Medical Act of 1876 removed all restrictions based on sex. Jex-Blake returned to Edinburgh to practice medicine and founded another medical school for women which existed for a short period. In 1924, at the time of the fifty-year anniversary celebration of the London School of Medicine for Women, there were over 2,000 women doctors out of a total of 40,000 medical practitioners and half of the women had received their training at the school.

In the very short time at my command I shall not attempt any exhaustive discussion of the primary question which lies at the root of my subject—the question, namely, whether it is, or is not, desirable that women should be educated for, and admitted to, the medical profession. I can indicate only in the briefest outline the reasons which lead me to answer this question in the affirmative.

In approaching the subject, the first point seems to be to divest the mind as completely as possible of all conventional pre-existing theories respecting it, and to consider it as if now presented for the first time. Let it be supposed that no doctors as yet existed, that society now for the first time awoke to the great want of skilled medical aid, as distinguished from the empiricism of domestic treatment, and that it was resolved at once to set aside persons to acquire a scientific knowledge of the human frame in health and in disease, and of all remedies available as curative agents, with the object that such knowledge should be used for the benefit of all the rest of the race. In such case, would the natural idea be that members of each sex should be so set apart for the benefit of their own sex respectively—that men should fit themselves to minister to the maladies of men, and women to those of women—or that one sex

only should undertake the care of the health of all, under the circumstances? For myself, I have no hesitation in saying that the former seems to me the *natural* course, and that to civilised society, if unaccustomed to the idea, the proposal that persons of one sex should in every case be consulted about every disease incident to those of the other, would be very repugnant; nay, that were every other condition of society the same as now, it would probably be held wholly inadmissible. Indeed, I will even go a step further, and say that if any question arose respecting the relative fitness of men and women for attendance on the sick, the experience of daily life would go far to prove that, of the two, women have more love of medical work, and are naturally more inclined, and more fitted, for it than most men. If a child falls downstairs and is more or less seriously hurt, is it the father or the mother, the sister or the brother, (where all are without medical training) who is most equal to the emergency, and who first applies the needful remedies? Or again, in the heart of the country, where doctors are not readily accessible, is it the squire and the parson, or their respective wives, who minister to the ailments of half the parish? And if women are thus naturally inclined for medical practice, I do not know who has the right to say that they shall not be allowed to make their work scientific when they desire it, but shall be limited to merely the mechanical details and wearisome routine of nursing, while to men is reserved all intelligent knowledge of disease, and all study of the laws by which health may be preserved or restored. I confess that, as regards natural instincts and social propriety alike, it seems to me that their evidence, such as it is, is wholly for, and not against, the cause of women as physicians of their own sex, and it is for this alone that I am pleading.

To glance at another aspect of the question, I must further say that I believe few people are aware of the very widespread desire existing among women for the services of doctors of their own sex; and yet there are probably few present who have not known individual cases where severe suffering has been borne, and danger perhaps incurred, in consequence of the excessive reluctance felt by some young girl, or woman even of maturer years, to consult a doctor in certain circumstances, and more especially to tell him all the facts of her case with the absolute frankness which is essentially necessary. Doctors have often told me themselves of the extreme pity they have felt in such cases, and yet I believe that doctors know, as a rule, less of the facts to which I am referring than anyone else is likely to do.... In this case, of all others, there should at least be a *choice*. If I am wrong in my belief on this point, at least no harm will be done; if I am right, a quite incalculable benefit will have been conferred.

Even more briefly must I glance at the historical evidence in favour of the medical education of women. Those learned in Greek literature will remember that Homer speaks of medical women both in the *Iliad* and in the *Odyssey,* and that Euripides represents the nurse as reminding Queen Phædra that if her disease is 'such as may not be told to men' there are skilled women at hand to whom she may have recourse. Everyone is aware how common it seems to have been in the days of chivalry for women to be adepts in medicine and surgery, and how frequently the knights wounded in battle were healed in the adjacent nunneries. The midwives and wise women of the middle ages held at least an equal position with the apothecaries and barber-surgeons of the time, and their competition seems to have been so formidable as to drive the male practitioners to seek protection from them, for it appears that in 1421 a petition was presented to Henry V praying that 'No woman use the practise of fysik under the payne of long emprisonment.' Again, in the reign of Henry VIII, we find an Act passed for the relief of irregular practitioners of medicine, 'both men and women,' who were being 'sued, vexed, and troubled by the Company and Fellowship of Surgeons of London, mynding onlie their own lucres,' the said surgeons having but recently obtained an Act of Incorporation, which they now endeavoured to employ as an engine of oppressive monopoly. Maitland, in his *History of Edinburgh,* also mentions that at the time of the foundation of the Edinburgh College of Physicians in 1681, 'both men and women' were engaged in the 'practice of physic' in that city. I have not time even to enumerate the women whose medical skill has obtained record in history during the past few centuries, but I may mention that no less than three women have held chairs in the Medical Faculty of the University of Bologna, namely, Dorotea Bucca, Professor of Medicine, early in the fifteenth century; Anna Mazzolini, Professor of Anatomy, in 1760; and Maria Della Donne, who was appointed Professor of Midwifery by Napoleon Buonaparte at the beginning of the present century. The Italian Universities have always been freely open to women, and a large number seem to have studied medicine there during the middle ages. In Germany also several such instances have occurred, and considerable numbers of women are now studying in the universities of France and Switzerland.

But even if we had no such historical precedents, and if the special suitability of women for the study of medicine were totally denied, I should still confidently rest my case on my last and most comprehensive argument—namely, the right possessed by every intelligent human being to choose out his or her own life work, and to decide what is and what is not calculated to conduce to his or her personal benefit and happiness. In the words of the late Mrs. J. S. Mill,[1] 'We deny the right of any portion of the species to decide for another portion, or any individual for another individual, what is and what is not their "proper sphere." The proper sphere for all human beings is the largest and highest which they are able to attain to. What this is, cannot be ascertained without complete liberty of choice.'...If women do on the whole make worse doctors than men, they will simply fail in competition with them; if their sister women do not desire their services, they will not obtain practice, and the thing will die out of itself. Where, then, is the need of opposition and prohibition?...

If, then, it may be assumed that women are at least not to be prevented from entering the medical profession, it is clearly to be desired that they should aim at the same standard, and comply with the same requirements, as all other practitioners of medicine. This is manifestly for the good of the community, and women themselves ask for nothing else. They are quite ready to undertake the ordinary studies, and to rest their claims to practice on their ability to pass the ordinary examinations. They ask for no exceptions and no indulgence; they simply ask that no exceptional hindrances shall be thrown in their way. Two things, then, are essentially necessary.

(1) That they should obtain such an education as shall make them thoroughly competent to take their share of responsibility in the care of the national health; and

(2) That this education should be secured in accordance with the regulations prescribed by authority, that they may be recognised by the State as having complied with its

requirements, and may thus enter upon practice on terms of perfect quality with other practitioners.

The first condition is strictly natural and inevitable; the second, though equally imperative, is in a certain sense arbitrary, or at least artificial.

...A positive law exists in England to limit the recognised practice of medicine to those whose names are entered on the Government Register, admission to that register being obtainable only through certain specified examining boards. It cannot be too often repeated, as it cannot be too clearly understood, that it is to the existence of this law that all the fundamental difficulties in the way of the medical education of women are due; and that, not because the Medical Act contains any clause excluding women from the profession, but because it vests all power of examination in bodies who seem only too much inclined to wrest and misuse that power for the purpose of effecting such exclusion. The Medical Act says not a word respecting the sex of the practitioner of medicine; it simply lays down certain conditions and requirements, and all who comply with these are *ipso facto* entitled to registration....

...[S]uch registration is obtainable only after examination by certain specified boards; these boards have laid down as a positive condition of examination that students must have studied in certain recognised public schools—it being, moreover, defined...that it is not sufficient to be taught even by the recognised public teacher, unless such teaching be received publicly and in the ordinary class; and lastly, women are not, so far as I know, allowed admission to any public school in Great Britain, and yet they cannot obtain permission to organise a separate medical school for themselves, or, at least, they are told that the teaching in any such school, however excellent it may be, shall be refused recognition by the examining boards.

I trust that I have given a clear view of the somewhat complicated difficulties of the case; in the few minutes that remain I desire to direct at-

---

[1]The reference is to Harriet Taylor Mill whose essay "On Marriage" appears as Source 121 in this chapter.

tention to the possible modes of remedy. The difficulties are threefold; depending, that is,

(1) on the law,
(2) on the examining boards,
(3) on the medical schools.

If the conditions were different in any one of these three directions, the present problem might admit of solution. If women were admitted to the ordinary medical schools on the same terms as men, the examining boards could hardly refuse to examine them; if the regulations of the boards were relaxed to the extent of admitting candidates from a new school formed expressly for women, no more would be necessary, for such a school could readily be organised, and might be in complete working order in the course of six months; or, thirdly, if the Medical Act permitted the examination of women by a special board, and their subsequent registration as practitioners, the existing schools and boards might maintain their attitude of exclusion, and yet women could enter the profession on equal terms with men.

At any rate, we confidently trust that whenever a new Medical Bill is introduced, its provisions will be such as to make the present exclusionist policy impossible. It is sufficiently plain that the sole object of legislation on this subject ought to be to give some guarantee to the public that the medical practitioners recognised by law have received an adequate education, and are not unfit to be trusted with the care of the sick, so far, indeed, as such fitness or unfitness can be ascertained by examinations ....I feel every confidence that in a very short time the needful remedy will be found and applied by the Legislature.

Indeed, this assurance is redoubled by the fact that the proposed action of Government with reference to the medical profession is exactly of the kind which will of itself remove all our difficulties. Let but the 'One Portal System' be fairly established; let the Government appoint an examining board whose composition shall secure universal confidence; let all candidates who desire to enter the medical profession be admitted to the same examinations; let their papers be identified by numbers instead of names, so that this vexed question of sex may be so far absolutely excluded, and then let all students who are able successfully to pass the ordeal be admitted to registration; and the whole problem which has so long troubled and perplexed us will be solved at once and for ever. In the great scheme of national education the principle of payment by results is becoming daily more and more supreme; let it prevail in this department also....

I hope even in my lifetime that the day will come when the axiom that 'Mind is of no sex' shall cease to be a mere dead letter; when equal facilities of study shall be afforded to all earnest students in connection with the national universities, and all other places of education supported in any degree by public money; but it is not for this that I am now pleading. Whether it is just and fair that women should be taxed in order to supply educational advantages which they are not allowed to share, I leave it to others to decide. I only ask that, if women are to be excluded from all the medical schools that at present exist, they may at least be allowed by their own labour and at their own expense— with such help as may be voluntarily vouchsafed to them—to form and organise a medical school of their own; with the certainty that, on the completion of their studies, due provision shall be made for their admission to the recognised examinations, and for their subsequent registration as legally qualified practitioners, on exactly the same footing as all others who have pursued the same course of study and attained the same standard of professional acquirement.

This is all that is absolutely necessary, and I leave it to the justice of my countrymen to say whether this is too much to ask.

120. Marie Curie
Obtaining Radium

Marie Curie (1867–1934), the first woman to win a Nobel Prize, was born in Warsaw, Poland as Marie Sklodowska. As women were not admitted into universities in Warsaw, then under Russian rule, she arrived in Paris in 1891 to study at the Sorbonne and soon excelled in classes in physics, chemistry, and mathematics. Here, Marie met Pierre Curie whom she married in 1895. Her research interests in radioactivity, a term which she coined, were inspired by French physicist Henri Becquerel's demonstration that uranium compounds spontaneously emitted radiation similar to those of x-rays. Marie hypothesized that uranium ore contained two unknown elements that were more radioactive than uranium itself. Pierre soon joined Marie's research and they announced the existence of polonium and radium in 1898. The Curies aimed to offer actual proof by isolating the two elements from pitchblende, a uranium ore. Marie carried out the final stages of research and achieved the isolation of radium in 1902. She describes these experiences in this selection from her book *Pierre Curie* (1923) which contains extensive autobiographical information. The Curies shared a Nobel Prize in physics with Becquerel in 1903. Three years later, Pierre was killed in a traffic accident in Paris. Marie was offered his chair in physics at the Sorbonne and became the first woman to teach at the university. In 1911, she was awarded the Nobel Prize in chemistry for the preparation of pure radium and became the first recipient of two Nobel Prizes. Nonetheless, the all-male Academy of Sciences voted against her candidacy for election to the Academy. Just prior to the outbreak of World War I, Marie was designated director of the research division of the newly established Radium Institute in Paris. Marie and Pierre's first daughter, Irène, became a physicist and continued their commitment to science. During the war, Marie organized radiology services for military hospitals, trained women in radiology, and equipped special ambulances in which she and Irène traveled to provide x-rays to wounded soldiers at the front. In 1922, Marie was elected to the Academy of Medicine. Shortly before Marie's death in 1934 from leukemia caused by exposure to radiation, Irène and her husband Frédéric Joliot discovered artificial radioactivity for which they received a Nobel Prize in 1935. Eve, a second daughter, an internationally renowned journalist, commemorated Marie's life in a biography, *Madame Curie* (1937). In 1995, French President François Mitterand enshrined Marie and Pierre Curie in the Pantheon in Paris.

[I]t was only after several years of most arduous labor that we finally succeeded in completely separating the new substance, now known to everybody as radium. Here is, briefly, the story of the search and discovery.

As we did not know, at the beginning, any of the chemical properties of the unknown substance, but only that it emits rays, it was by these rays that we had to search. We first undertook the analysis of a pitchblende from St.

From *Pierre Curie* by Marie Curie, translated by Charlotte and Vernon Kellogg. © 1923 by The Macmillan Company.

Joachimsthal. Analyzing this ore by the usual chemical methods, we added an examination of its different parts for radioactivity, by the use of our delicate electrical apparatus. This was the foundation of a new method of chemical analysis which, following our work, has been extended, with the result that a large number of radioactive elements have been discovered.

In a few weeks we could be convinced that our prevision had been right, for the activity was concentrating in a regular way. And, in a few months, we could separate from the pitchblende a substance accompanying the bismuth, much more active than uranium, and having well defined chemical properties. In July, 1898, we announced the existence of this new substance, to which I gave the name of polonium, in memory of my native country.

While engaged in this work on polonium, we had also discovered that, accompanying the barium separated from the pitchblende, there was another new element. After several months more of close work we were able to separate this second new substance, which was afterwards shown to be much more important than polonium. In December, 1898, we could announce the discovery of this new and now famous element, to which we gave the name of radium.

However, the greatest part of the material work had yet to be done. We had, to be sure, discovered the existence of the remarkable new elements, but it was chiefly by their radiant properties that these new substances were distinguished from the bismuth and barium with which they were mixed in minute quantities. We had still to separate them as pure elements. On this work we now started.

We were very poorly equipped with facilities for this purpose. It was necessary to subject large quantities of ore to careful chemical treatment. We had no money, no suitable laboratory, no personal help for our great and difficult undertaking. It was like creating something out of nothing, and if my earlier studying years had once been called by my brother-in-law the heroic period of my life, I can say without exaggeration that the period on which my husband

and I now entered was truly the heroic one of our common life.

We knew by our experiments that in the treatment of pitchblende at the uranium plant of St. Joachimsthal, radium must have been left in the residues, and, with the permission of the Austrian government, which owned the plant, we succeeded in securing a certain quantity of these residues, then quite valueless,—and used them for extraction of radium. How glad I was when the sacks arrived, with the brown dust mixed with pine needles, and when the activity proved even greater than that of the primitive ore! It was a stroke of luck that the residues had not been thrown far away or disposed of in some way, but left in a heap in the pine wood near the plant. Some time later, the Austrian government, on the proposition of the Academy of Science of Vienna, let us have several tons of similar residues at a low price. With this material was prepared all the radium I had in my laboratory up to the date when I received the precious gift from the American women.

The School of Physics could give us no suitable premises, but for lack of anything better, the Director permitted us to use an abandoned shed which had been in service as a dissecting room of the School of Medicine. Its glass roof did not afford complete shelter against rain; the heat was suffocating in summer, and the bitter cold of winter was only a little lessened by the iron stove, except in its immediate vicinity. There was no question of obtaining the needed proper apparatus in common use by chemists. We simply had some old pine-wood tables with furnaces and gas burners. We had to use the adjoining yard for those of our chemical operations that involved producing irritating gases; even then the gas often filled our shed. With this equipment we entered on our exhausting work.

Yet it was in this miserable old shed that we passed the best and happiest years of our life, devoting our entire days to our work. Often I had to prepare our lunch in the shed, so as not to interrupt some particularly important operation. Sometimes I had to spend a whole day mixing a boiling mass with a heavy iron rod nearly as large as myself. I would be broken with fatigue at the day's end. Other days, on the contrary, the work would be a most minute and delicate fractional crystallization, in the effort to concentrate the radium. I was then annoyed by the floating dust of iron and coal from which I could not protect my precious products. But I shall never be able to express the joy of the untroubled quietness of this atmosphere of research and the excitement of actual progress with the confident hope of still better results. The feeling of discouragement that sometimes came after some unsuccessful toil did not last long and gave way to renewed activity. We had happy moments devoted to a quiet discussion of our work, walking around our shed.

One of our joys was to go into our workroom at night; we then perceived on all sides the feebly luminous silhouettes of the bottles or capsules containing our products. It was really a lovely sight and one always new to us. The glowing tubes looked like faint, fairy lights.

Thus the months passed, and our efforts, hardly interrupted by short vacations, brought forth more and more complete evidence. Our faith grew ever stronger, and our work being more and more known, we found means to get new quantities of raw material and to carry on

some of our crude processes in a factory, allowing me to give more time to the delicate finishing treatment.

At this stage I devoted myself especially to the purification of the radium, my husband being absorbed by the study of the physical properties of the rays emitted by the new substances. It was only after treating one ton of pitchblende residues that I could get definite results. Indeed we know to-day that even in the best minerals there are not more than a few decigrammes of radium in a ton of raw material.

At last the time came when the isolated substances showed all the characters of a pure chemical body. This body, the radium, gives a characteristic spectrum, and I was able to determine for it an atomic weight much higher than that of the barium. This was achieved in 1902. I then possessed one decigramme of very pure radium chloride. It had taken me almost four years to produce the kind of evidence which chemical science demands, that radium is truly a new element. One year would probably have been enough for the same purpose, if reasonable means had been at my disposal. The demonstration that cost so much effort was the basis of the new science of radioactivity.

QUESTIONS FOR ANALYSIS

1. What were Florence Nightingale's initial reactions on arriving at the Scutari Barrack Hospital? How did she challenge existing nursing and administrative practices?

2. Discuss Sophia Jex-Blake's historical analysis of the medical education of women and their roles in medicine. On what basis does she call for admitting women into the medical profession? What prevented women from pursuing careers in medicine and how does Jex-Blake propose to eliminate these obstacles?

3. What was the theoretical basis of Marie Curie's research on radioactivity? Under what conditions did Marie and Pierre Curie carry out their research? How do these conditions compare with those found in contemporary scientific laboratories?

4. With reference to the introductory materials and their own writings, compare the medical and scientific careers pursued by Florence Nightingale, Sophia Jex-Blake, and Marie Curie. What common issues and obstacles did they confront as women? What circumstances and initiatives enabled them to overcome these obstacles? Consider women's current status in medicine and science. What has changed and what remains the same?

CRITIQUES OF MARRIAGE, SEXUAL NORMS, AND WOMEN'S FASHION

Women's critiques of marriage and sexuality in the middle and last decades of the nineteenth century encompassed a broad range of issues that ranged from women's unequal economic and legal status in the family to dress fashions that were injurious to women's health. The sexual double standard became a particular target of women's opposition to the Contagious Diseases Acts passed in Britain in the 1860s. Enforced in garrison towns, the acts compelled prostitutes as well as women suspected of prostitution to submit to medical examinations for venereal disease. Until the latter part of the nineteenth century, however, women's writings on sexuality typically focused on the necessity to control male sexuality. This began to change following the publication of various studies by the German and English sexologists Richard von Krafft-Ebing, Magnus Hirschfeld, and Havelock Ellis. Although Krafft-Ebing stipulated that his *Psychopathia Sexualis* (1886) was to be limited to a medical audience, the sexologists' studies soon became common knowledge and provided an unprecedented legitimacy for public discussion of sexuality, including the experiences of homosexuals. Krafft-Ebing's view that homosexuals were "congenitally defective" and Hirschfeld's concept that homosexuals constituted a "third" or "intermediate" sex provided a new, "scientific" basis for the homosexual rights movement.

121. Harriet Taylor
"On Marriage"

Harriet Taylor's (1808–1858) writings on marriage and divorce were inspired by her personal experiences and political convictions. She was unhappily married and lived apart from her husband for fifteen years, but could not obtain a divorce. Beginning in the 1830s, Harriet Taylor carried out a romantic attachment with John Stuart Mill (1806–1873). In 1832, they both wrote essays entitled, "On Marriage," and hers is reproduced here. She also contributed to the Unitarian journal the *Monthly Repository.* The editor, W. J. Fox, was a radical Unitarian minister and the journal served as a vehicle for the discussion of such issues as divorce, civil marriage, and women's suffrage. Mill met Harriet in the radical Unitarian circles to which she and her husband John Taylor belonged. Only in 1851, following her husband's death in 1849, were they finally married. In addition to the right to divorce, Harriet Taylor Mill advocated women's suffrage, education, and employment opportunities. In her article, "Enfranchisement of Women," published anonymously in the *Westminister Review* in 1851, she described the ideas and actions of the women's suffrage movement in the United States and systematically refuted arguments against women's suffrage. Harriet and John Stuart Mill's marriage was an intellectually collaborative one and she

Originally written by Harriet Taylor in 1832–33.

strongly influenced the ideas on individual and women's rights contained in Mill's publications. Following Harriet's sudden death, her daughter Helen Taylor (1831–1907) and Mill continued to work together on women's rights and suffrage. It was at this time, too, that Mill published his essay *On Liberty* (1859) and, with Helen's encouragement, completed *The Subjugation of Women* (1869).

If I could be providence to the world for a time, for the express purpose of raising the condition of women, I should come to you to know the *means*—the *purpose* would be to remove all interference with affection, or with any thing which is, or which even might be supposed to be, demonstrative of affection—In the present state of women's minds, perfectly uneducated, and with whatever of timidity and dependence is natural to them increased a thousand fold by their habits of utter dependence, it would probably be mischievous to remove at once all restraints, they would buy themselves protectors at a dearer cost than even at present—but without raising their natures at all, it seems to me, that once give women the desire to raise their social condition, and they have a power which in the present state of civilization and of men's characters, might be made of tremendous effect. Whether nature made a difference in the nature of men and women or not, it seems now that all men, with the exception of a few lofty minded, are sensualists more or less—Women on the contrary are quite exempt from this trait, however it may appear otherwise in the cases of some—It seems strange that it should be so, unless it was meant to be a source of power in demi-civilized states such as the present—or it may not be so—it may be only that the habits of freedom and low indulgence in which boys grow up and the contrary notion of what is called purity in girls may have produced the appearance of different natures in the two sexes—As certain it is that there is equality in nothing, now—all the pleasures such as there are being men's, and all the disagreables and pains being women's, as that every pleasure would be infinitely heightened both in kind and degree by the perfect equality of the sexes. Women are educated for one single object, to

gain their living by marrying—(some poor souls get it without the churchgoing in the same way—they do not seem to me a bit worse than their honoured sisters)—To be married is the object of their existence and that object being gained they do really cease to exist as to anything worth calling life or any useful purpose. One observes very few marriages where there is any real sympathy or enjoyment of companionship between the parties—The woman knows what her power is, and gains by it what she has been taught to consider "proper" to her state—The woman who would gain power by such means is unfit for power, still they do use this power for paltry advantages and I am astonished it has never occurred to them to gain some large purpose: but their minds are degenerated by habits of dependance—I should think that 500 years hence none of the follies of their ancestors will so excite wonder and contempt as the fact of legislative restraint as to matters of feeling—or rather in the expressions of feeling. When once the law undertakes to say which demonstration of feeling shall be given to which, it seems quite inconsistent not to legislate for *all,* and say how many shall be seen, how many heard, and what kind and degree of feeling allows of shaking hands—The Turks is the only consistent mode—

I have no doubt that when the whole community is really educated, tho' the present laws of marriage were to continue they would be perfectly disregarded, because no one would marry—The widest and perhaps the quickest means to do away with its evils is to be found in promoting education—as it is the means of all good—but meanwhile it is hard that those who suffer most from its evils and who are always the best people, should be left without remedy.

Would not the best plan be divorce which could be attained by *any, without any reason assigned,* and at small expence, but which could only be finally pronounced after a long period? not *less* time than two years should elapse between suing for divorce and permission to contract again—but what the decision will be *must* be certain at the moment of asking for it—*unless* during that time the suit should be withdrawn—

(I feel like a lawyer in talking of it only! O how absurd and little it all is!)

In the present system of habits and opinions, girls enter into what is called a contract perfectly ignorant of the conditions of it, and that they should be so is considered absolutely essential to their fitness for it!—But after all the one argument of the matter which I think might be said so as to strike both high and low natures is—Who would wish to have the person without the inclination? Whoever would take the benefit of a law of divorce must be those whose inclination is to separate and who on earth would wish another to remain with them against their inclination? I should think no one—people sophisticate about the matter now and will not believe that one "*really* would *wish to go.*" Suppose instead of calling it a "law of divorce" it were to be called "Proof of affection"—They would like it better then—

At this present time, in this state of civilization, what evil would be caused by, first placing women on the most entire equality with men, as to all rights and privileges, civil and political, and then doing away with all laws whatever relating to marriage? Then if a woman had children she must take the charge of them, women would not then have children without considering how to maintain them. Women would have no more reason to barter person for bread, or for any thing else, than men have—public offices being open to them alike, all occupations would be divided between the sexes in their natural arrangement. Fathers would provide for their daughters in the same manner as for their sons—

All the difficulties about divorce seem to be in the consideration for the children—but on this plan it would be the women's *interest* not to have children—*now* it is thought to be the woman's interest to have children as so many *ties* to the man who feeds her.

Sex in its true and finest meaning, seems to be the way in which is manifested all that is highest best and beautiful in the nature of human beings—none but poets have approached to the perception of the beauty of the material world—still less of the spiritual—and there never yet existed a poet, except by the inspiration of that feeling which is the perception of beauty in all forms and by all the means which are given us, as well as by *sight.* Are we not born with the *five* senses, merely as a foundation for others which we may make by them—and who extends and refines those material senses to the highest—into infinity—best fulfils the end of creation—That is only saying—*Who enjoys most, is most virtuous*—It is for *you*—the most worthy to be the apostle of all loftiest virtue—to teach, such as may be taught, that the higher the *kind* of enjoyment, the *greater* the *degree*—perhaps there is but one class to whom this *can* be *taught*—the poetic nature struggling with superstition: *you* are fitted to be the saviour of such—

122. Harriet Martineau
"Dress and Its Victims"

Harriet Martineau (1802–1876) was a leading feminist and one of the first professional female journalists in nineteenth-century England. She was born into a Unitarian middle-class manufacturer's family in Norwich. Martineau was educated in a Unitarian girls' school and at home. Martineau published her first article, "Female Writers of Practical Divinity," in 1822 in the Unitarian *Monthly Repository.* As a single woman who never married, Martineau decided to support herself by writing after the collapse of her father's investments in the mid-1820s and his subsequent death left her without financial resources. She achieved immediate fame with her *Illustrations of Political Economy.* Published as a monthly series from 1832 to 1834, it explained the theories of Adam Smith, David Ricardo, Thomas Malthus, James Mill, and Jeremy Bentham in the form of stories accessible to the general reading public. Some twenty years later, in 1852, Martineau translated and published Auguste Comte's *Positive Philosophy* which diffused Comte's ideas throughout Great Britain. A prolific and influential writer, she wrote hundreds of articles and editorials on such topics as the abolition of slavery in the United States, poor law reform, women's education, domestic violence, divorce, and women's health. Martineau also published her *Autobiography* in 1855. She took up the cause of suffrage in the 1860s and was one of the signatories of the 1866 petition to Parliament advocating the extension of suffrage to women on the same basis as men. Martineau's last campaign concerned the Contagious Diseases Acts.[1] Among the first to sign a petition drawn up by the Ladies National Association for the Repeal of the Contagious Diseases Acts, she also wrote numerous articles opposing the Acts. Martineau's article "Dress and Its Victims," excerpted here, originally appeared in *Once a Week* as part of a series on health issues and was later anthologized in her *Health, Husbandry, and Handicraft* (1861).

There are a good many people who cannot possibly believe that dress can have any share in the deaths of the 100,000 persons who go needlessly to the grave every year in our happy England, where there are more means of comfort for everybody than in any other country in Europe.

How can people be killed by dress, now-a-days? they ask....

It will not seem so wonderful that the familiar clothing of our neighbours and ourselves may be of such importance when we remember the explanations of physicians—that dress may and usually does, affect the condition and action of almost every department of the human frame;—the brain and nervous system, the lungs, the stomach, and other organs of the trunk, the eyes, the skin, the muscles, the glandular system, the nutritive system, and even the bony frame, the skeleton on which all hangs. If dress can meddle mischievously with the action, or affect the condition of all these, it can be no marvel that it is responsible for a good many of the hundred thousand needless deaths which are happening around us this year.

---

[1]See Source 123 by Josephine Butler for an explanation of these Acts.

Putting aside the ordinary associations, as far as we can, and trying for the moment to consider what is to be desired in the clothing of the human body,—what is requisite to make dress good and beautiful,—let us see what is essential.

Dress should be a covering to all the parts of the body which need warmth or coolness, as the case may be. It should be a shelter from the evils of the atmosphere, whether these be cold, or heat, or wet, or damp, or glare. This is the first requisite; for such shelter is the main purpose of clothing. In our own country the dress should easily admit of the necessary changes in degrees of warmth demanded by our changeable climate.

Dress should bear a close relation to the human form. No other principle can be permanent; no other can be durably sanctioned by sense and taste, because no other has reality in it....

Where it follows the outline of the frame it should fit accurately enough to fulfil its intention, but so easily as not to embarrass action. It should neither compress the internal structure nor impede the external movement. An easy fit, in short, is the requisite. It is a part of this easy fit that the weight of the clothes should be properly hung and distributed....

Next; dress ought to be agreeable to wear: and this includes something more than warmth and a good fit. It should be light, and subject to as few dangers and inconveniences as possible.

These conditions being observed, it follows of course that the costume will be modest, and that it will be graceful. Grace and beauty are flowers from the root of utility....

If we consider the female dress of 1859..., what can we say of it? Does the costume, as a whole, follow the outline of the form? Does it fit accurately and easily? Is the weight made to hang from the shoulders? Are the garments of to-day convenient and agreeable in use? Is the mode modest and graceful? So far from it, that all these conditions are conspicuously violated by those who think they dress well. Here and there we may meet a sensible woman, or a girl who has no money to spend in new clothes,

whose appearance is pleasing—in a straw bonnet that covers the head, in a neat gown which hangs gracefully and easily from the natural waist, and which does not sweep up the dirt: but the spectacle is now rare; for bad taste in the higher classes spreads very rapidly downwards, corrupting the morals as it goes.

The modern dress perverts the form very disagreeably. The evil still begins with the stays, in too many instances, though there is less tight-lacing than formerly....The ribs are pressed out of their places, down upon the soft organs within, or overlapping one another: the heart is compressed, so that the circulation is irregular: the stomach and liver are compressed, so that they cannot act properly: and then parts which cannot be squeezed are thrust out of their places, and grave ailments are the consequence. At the very best, the complexion loses more than the figure can be supposed to gain. It is painful to see what is endured by some young women in shops and factories, as elsewhere. They cannot stoop for two minutes over their work without gasping and being blue, or red, or white in the face. They cannot go upstairs without stopping to take breath every few steps. Their arms are half numb, and their hands red or chilblained; and they must walk as if they were all-of-a-piece, without the benefit and grace of joints in the spine and limbs. A lady had the curiosity to feel what made a girl whom she knew so like a wooden figure, and found a complete palisade extending round the body. On her remonstrating, the girl pleaded that she had "only six-and-twenty whalebones!"

Any visitor of a range of factories will be sure to find that girls are dropping in fainting-fits, here and there, however pure the air and proper the temperature; and here and there may be seen a vexed and disgusted proprietor, seeking the warehouse-woman, or some matron, to whom he gives a pair of large scissors, with directions to cut open the stays of some silly woman who had fainted. Occasional inquests afford a direct warning of the fatal effects which may follow the practice of tight-lacing; but slow and painful disease is much more common; and the register exhibits, not the stays, but the malady created by the stays as the cause of death. That such cases are common, any physician who practises among the working-classes will testify....

What is to be done? Will anything ever be done? or is feminine wilfulness and slavishness to fashion to kill off hundreds and thousands of the race, as at present? I see, with much satisfaction, that the Messrs. Courtauld, the great silk manufacturers in Essex, have put up a notice in their factories, that a fine is imposed on the wearing of crinoline by their workwomen. The ground of the regulation is, that the work cannot be done with either decency or safety in that kind of dress. I hope this example will be followed in all mills and factories where the same reason can be assigned. There are whole societies in America who do not see the necessity for such mischief, and who hope to put an end to it—in their own country at least. The Dress-Reform Association of the United States was instituted some years since by women who refused the inconvenience of Paris fashions in American homesteads: and they have been aided, not only by physicians, but by other men, on the ground of the right of women to wear what suits their occupations and their taste, without molestation. The dress which was long ago agreed upon, after careful consideration—the so-called Bloomer costume (not as we see it in caricature, but in its near resemblance to the most rational English fashion of recent times)—is extensively worn, not only in rural districts, but in many towns. It seems to fulfil the various conditions of rational, modest, and graceful dress better than any other as yet devised for temperate climates; and if so, it will spread, in spite of all opposition.

123. Josephine Butler
The Contagious Diseases Acts

Josephine Butler (1828–1906) led a crusade against the Contagious Diseases Acts. She was involved in rescue work among prostitutes in Liverpool in the 1850s, but first became politically engaged around educational issues. With her husband, the Reverend George Butler, she participated in the North of England Council for the Higher Education of Women and became its president in 1867. She petitioned Cambridge to provide examinations for women and published a pamphlet, *The Education and Employment of Women,* in 1868. Butler relinquished much of her educational work, fearful that it would be adversely affected by her activism around the Contagious Diseases Acts.

The first two Contagious Diseases Acts were passed in 1864 and 1866, but the third and harshest act of 1869 provoked the most concerted opposition. Enforced in fifteen garrison towns, it subjected prostitutes and women suspected of prostitution to examination and treatment, including treatment in confinement. Butler became the leader of the Ladies National Association for the Repeal of the Contagious Diseases Acts formed in January 1870. Florence Nightingale, Harriet Martineau, Mary Carpenter, and Elizabeth Blackwell were among those who signed an initial petition against the Acts. Later, in 1871, the Association presented a petition to Parliament with 250,000 signatures. In this same year, Butler testified before

Originally written by Josephine Butler in 1871.

the Royal Commission on the Contagious Diseases Acts and published the article reproduced here, "Letter to My Countrywomen, Dwelling in the Farmsteads and Cottages of England." The Ladies National Association faced a formidable task. The medical establishment and the press generally supported the Acts and the great majority of middle-class men and women either approved or were indifferent to how they affected working-class women and prostitutes. Butler therefore focused her efforts on religious organizations and the working class in the hope that working-class men would vote against Parliamentary members who supported the Acts, noting that the law would not even exist if women could vote. She addressed large meetings in numerous cities, including Birmingham, Liverpool, Manchester, and Glasgow. Butler sustained her seventeen years of activism against the Acts with an evangelical faith committed to social justice which she shared with her husband. The Acts were suspended in 1883 and then repealed in 1886. Butler then turned her attention to the traffic in women between Britain and the Continent and the repeal of Contagious Diseases Acts in India which occupied the Ladies National Association from 1886 to 1915. Butler's writings include *On the Moral Reclaimability of Prostitutes* (1870) and *Personal Reminiscences of a Great Crusade* (1896).

LETTER TO MY COUNTRYWOMEN, DWELLING IN THE FARMSTEADS AND COTTAGES OF ENGLAND

My Dear Friends,

There is a law now in force in this country which concerns you all, and yet of which many among you have never even heard. I want you to listen to me for a little, whilst I try to explain it to you, and when you have heard and understood, you will, I think, feel about it as I do.

I daresay you all know that there are women, alas, thousands of women, in England who live by sin. Sometimes, when you have been late at your market-town, you may have passed one such in the street, and have shrunk aside, feeling it shame even to touch her; or perhaps, instead of scorn, a deep pity has filled your heart, and you have longed to take her hand, and to lead her back to a better and happier life. Now, it is the pity and not the scorn which I would fain have you feel towards these poor women, and when you have read what I have to tell you about them, I think it will not be hard for you to be merciful to them in your thoughts.

In the first place, you must understand that very few ever begin to lead a bad life from choice. Thousands of the miserable creatures whom we call *fallen* have really not fallen at all, for they never stood upon any height of virtue or knowledge from which it was possible for them to descend, and if they love darkness, it is because no light ever shone upon them; no tender mother ever spoke to *them* of God or Christ; no kind father ever shielded *them* from temptation; no pure examples ever encouraged *them* to resist evil, and to seek after that which is good; rather, they have been sold—yes, sold—into their life of bondage by those who ought to have died to save them from such misery.

I daresay you will hardly believe it, yet it is but too true. It is said that there were in one large seaport town, only a few years ago, 1500 prostitutes under fifteen years of age, some of them mere children of eleven or twelve; and of these many had been sent upon the streets by their own parents, who lived upon the wages of their sin and shame.

Again, many girls are led astray when very young—sometimes most shamefully deceived and betrayed—then, finding the doors of their relatives and friends shut against them, they are driven by despair into recklessness and vice.

A still larger number, in fact far more than half of all the women who live by prostitution, fall into it through lack of food, and clothes, and shelter. Are we sure, you and I, that, in like case, we should not have done the same? Hunger and cold are hard to bear; it needs the courage of a martyr to die rather than to sin— and not only a martyr's courage, but a martyr's faith also; and how should those who find this world so cruel and so sad, place their trust in the God who made it?

To show you how much want has to do with prostitution, I will just mention here that a French doctor, who inquired carefully into the histories of 3,000 fallen women in Paris, found that of those 3,000 only thirty-five had had any chance of earning their bread honestly.

Now, if you were to add together all the women who are trained to sin from their cradle, all those who are betrayed into it by deceit, or driven into it by despair, and all those who sink into it through real starvation, you would find very few left whom you could justly call bad— doing evil because they love evil; very few, therefore, to whom you have any right to deny your pity and your help.

Perhaps you will say, 'if all that this lady tells us about these women is true, we are very sorry for them, but we don't see how we can help them; the Government to whom we pay so much money that it may take care of the people ought to do that. Surely it is trying to save some of the poor creatures.'

To this I answer: the Parliament and the Government of England have done nothing *for* these women, what it has done *against* them you now shall hear.

I have taken it for granted that you all know that there is such an evil as prostitution in the land; but perhaps some of you are not aware that those who lead vicious lives are liable to certain painful and dangerous diseases, called *contagious,* because it is supposed that one person can only take them from another by *contact,* or *touch.* A healthy man, by merely walking down a street where there is small-pox or fever, may sicken and die; but with these contagious diseases it is quite different, all men are safe from them so long as they live virtuous lives. Therefore, as I daresay you will think, every man can preserve himself from them if he chooses, and if he does not choose, he deserves to suffer, and ought not to be saved from suffering, because through it he may, perhaps, learn to be wiser and better.

But as regards certain men, at least—I mean soldiers and sailors—our Government judges otherwise.

You know it costs a great deal of money to train a man to be a soldier or sailor, and if he is often ill and unfit for service that money is as good as lost. Now, when the Government found that very many soldiers and sailors were constantly in hospital, owing to contagious disorders, they asked themselves what they could do to prevent such a waste of the public funds, and such a weakening of the forces on which the country has to depend for safety in time of war, which was quite right; but instead of teaching or helping them to lead pure lives, they resolved to protect them against contagious diseases—in other words, to make it safe for them to sin....

The particular measure of which I am writing was passed in 1869. There had been laws to protect soldiers and sailors against contagious diseases since 1864, but as they were milder in their provisions, and their operations confined to a very few places, little had been thought or said about them.

Now, the men who drew up these different Acts say that they kept them so quiet because they thought the matter with which they had to do indecent and disgusting, and unfit to be talked of or read about. But when I have told you what was written in the Act of 1869—for which the Acts passed in 1864 and 1866 prepared the way—I think you will agree with me that they were afraid to let Englishmen and Englishwomen understand the new laws lest they should cry out against them.

The Contagious Diseases Act of 1869 provides that in fifteen towns where there are always many soldiers and sailors—such as

Canterbury, Aldershot, Portsmouth, and Plymouth—there shall be surgeons appointed to examine with certain instruments the persons of all prostitutes, to see in what state they are, and those whom they find to be healthy they are to allow to go away, having given them a notice of when they are to come to be examined again, and so long as a woman appears on the stated days for examination, the police do not arrest or otherwise interfere with her; but those whom the surgeons find to be diseased, they are to send to what are called hospitals, but are really prisons, to be cured, and when they are well, they are dismissed to follow their former pursuits, their certificates of health being put into the hands of the police.

Further, this Act gives power to certain police inspectors, to watch all women and girls. The Act does not specify the age. As a matter of fact, girls of most tender years are brought under it, and the police-officers say that they especially watch seamstresses, labourers' daughters, and domestic servants, in those towns, and for fifteen miles round each, and if one of these paid spies—for since they wear plain clothes, so that people cannot tell they belong to the police-force, they are really spies—*suspects* any woman of being a prostitute, or of *intending* prostitution, he can go to her and say, 'Come to Mr. So and So', naming the Examining Surgeon, 'and sign the paper'. Now this paper is called the Voluntary Submission, and by signing it a woman agrees to submit for a whole year (it might, under the Act, be a shorter time, but the police have *always* filled it up for a year) to be examined whenever she is called upon to be so, but it is so carefully worded that an ignorant girl might put her name to it without knowing that by so doing she was signing away her liberty for twelve months, her character for ever.

If, however, she does understand this, and refuses to go to the surgeon, then the police spy can take her before a single magistrate, and that one magistrate, on the oath of that one policeman, who only needs to swear that he *suspects*, not that he knows her to be leading a vicious

life, may order her to submit to the examination, periodically for twelve months, and if she does not obey can send her to prison for three months, and when the first three months are over, if she still refuses to be examined by the surgeon, the magistrate can commit her for three more, and so on again and again.

So, if a woman is ignorant or frightened, and goes to the surgeon, she is put on the list of common women, and if she is well informed and brave, and goes before the magistrate, she may be imprisoned for life, without ever having been properly tried; and the examination to which she is ordered to submit is so cruel and indecent, that it is shameful even to speak of it; and those who have undergone it a few times, become so hardened and degraded, that almost all hope of saving them is lost. The object for which they are thus brutalised, is that the men who share their guilt and often tempt them to it, may go on sinning as much as they like, without any danger to their health; for in the eyes of *generals and admirals,* the souls of English women are of less consequence than the bodies of English men—if those bodies happen to be clothed in a red coat or a blue jacket.

Now, I beg you to think first with what awful dangers such a law surrounds innocent women. Remember, a policeman can accuse a girl, or even a child, of prostitution merely on suspicion, and there is no rule laid down to fix the signs which shall be held to give him a right to think evil of her. Each policeman is to judge of them for himself; and it is clear that they do not all agree....But if to be out of doors alone at night, or to be seen talking to men in the streets, is to be looked upon as a sign of a bad character, and to give a policeman the right to accuse a girl of prostitution, what woman will not fear to leave her house after dark, or to exchange a greeting with a friend?

And if the oath of one policeman is enough to condemn a girl, how will she be safe even in her own house? For supposing that a policeman has a spite against her, or any of her people, what is there to prevent his accusing her falsely, since he cannot be punished, even if the accusation be proved false, if he swears that he believed it to be true?

And the danger is not only from the paid spy; every man if he is angry with a woman, perhaps because he has tried to seduce her, and she has resisted him, can write to the police, not even signing his name to the letter, since he will not be called upon to appear as a witness, saying—'such a person is a prostitute', and without even knowing who her accuser is, she can be brought before the examining surgeon, or the magistrate, and so may be doomed to sin and shame for life, because some wretch envied her her innocence. Think, too, how the threat of such a false accusation might be used to extort money from the timid and weak.

I do not say that women would often be falsely accused, nor yet that money would often be extorted from them by the threat of such false accusation; but I do assert that whilst the Contagious Diseases Acts continue the law of the land, such things might happen, and that no innocent woman, however poor or ignorant, ought for a single day to be exposed to so frightful a risk; and I want you also to understand that it is just the poor and ignorant who are endangered by these Acts, since no police spy, or any other man, would dare to accuse a woman of wealth and position, able to protect herself, or with friends strong enough to protect her, unless he had the most certain proofs of her guilt, and perhaps not even then.

I feel sure you will see directly how cruel these laws are to good women, but I want you to see, too, how unjust they are to sinful ones. For when a woman sins, does she sin alone? Rather for one sinful woman are there not fifty—ay, a hundred—sinful men? And which of them, the ignorant, half-starved prostitutes, or the men, often well-taught and well-fed, who consort with them, are they that carry disease to virtuous wives, and transmit it to unconscious infants? Surely the men. Then they are not only more guilty, but also more dangerous in their guilt. Yet the punishment of the sin of *two* is made by these laws to fall upon *one*

alone, and that one the least to be blamed, and the most to be pitied.

But even if a woman be utterly vile—the tempter not the tempted—ought she to be deprived of all her rights and liberties? It has always been the boast of Englishmen that the law of England treats every accused person as innocent until he has been proved to be guilty.... All these safeguards against injustice and oppression the Law of England gives to the man accused of the darkest crime, but for the last two years Parliament has denied every one of them to women only charged with sin, whose accomplices it not only does not punish, but even tries to protect against those penalties which God himself has attached to vice.

And this brings me to the third charge which I have to make against the Contagious Diseases Acts, namely, that they tend to encourage men in vicious habits.

This they do in two ways; directly, by affording them opportunities of sinning safely; appointing for them, as it were, cities of refuge, to which they can flee, and in which they can indulge their evil desires without fear; and, indirectly, by accustoming them to think that vice is so natural, so necessary, that no one expects them to be virtuous; for they must see that those who framed these Acts despaired of ever improving mankind, since if they had had any hope of being able to reform it, instead of building hospitals in which to cure prostitutes, so that there may be always healthy women for soldiers and sailors (and of course all other men living in or near the subjected towns), to consort with, they would surely have founded schools and reading rooms, and clubs, and workshops, and other like institutions, where they might spend their idle time in learning good, instead of in doing evil....

And that these Acts have really increased vice among men is proved by the fact that on the nights after the prostitutes have been examined by the surgeons, and when it is *supposed* that there is no danger in approaching them, since all who were found to be diseased had been sent to the hospitals, the bad houses in the fifteen towns are crowded, especially by young lads and married men.

Supposed, I say, for the promise of safety contained in the fact of a woman being at large just after having been examined, often proves false, in truth it has not been shown that contagious disorders have diminished in England among soldiers and sailors, since the Contagious Diseases Act became law; and in France, where like laws have been in force for many years past, they abound so frightfully that persons, capable of forming an opinion on the subject, believe that it is to them that the French owe the terrible defeats of the late war, for they could neither march nor fight like the healthier and more virtuous Germans; and if this opinion is correct, it deserves to be carefully considered, since it is for the avowed purpose of keeping the army and navy healthy that our Contagious Diseases Acts were framed and carried....

Now, I think, I have made you understand what the Contagious Diseases Acts are, and the purpose for which they were passed; and I hope I have succeeded in convicting you of their cruelty, injustice, and immorality. In a second letter I will tell you what has been done towards forcing Parliament to undo its evil work, and will show you how you too can help to rid the country of these bad laws. But before concluding this I must just point out that all these police spies, and examining surgeons, and hospitals for diseased women, who are to be cured, not that they may be saved from their vicious life, but that soldiers and sailors may share it without risk, cost a great deal of money, and that all that money comes out of the pockets of the people.

I am, my dear Friends,
Yours very faithfully,
An English Lady

124. Anna Rueling
The Women's Movement and the Homosexual Question

Anna Rueling was an activist in the homosexual and women's movements in Berlin. Although the homosexual movement in Germany flourished from the last decades of the nineteenth century until its suppression by the rise of Nazism, exclusively women's organizations did not exist until the 1920s. Homosexual women, like men, often referred to themselves as "Uranians" and worked in such organizations as the Scientific Humanitarian Committee founded in Berlin in 1897 by the sexologist Magnus Hirschfeld. His writings include *Sex in Human Relations* and *Berlin's Third Sex* which describes the homosexual, mainly male, subculture in Berlin. Hirschfeld actively promoted women's participation in the Scientific Humanitarian Committee, and articles by and about women appeared regularly in its *Yearbook of Intermediate Sexual Types* published annually from 1899 to 1923. The committee aimed to achieve societal tolerance for homosexuals and to abolish discriminatory laws. These included Paragraph 175 of the German Penal Code, legislated in 1871, which defined homosexual acts between men as a crime. Public discussions of female homosexuality led to a backlash and, in 1910, a new draft

Speech given by Anna Rueling at the annual conference of the Scientific Humanitarian Committee, October 8, 1904, Prinz Albrecht Hotel, Berlin.

penal code proposed to include women in Paragraph 175. This provoked opposition from several women's organizations in Germany, including the League for the Protection of Mothers, which argued that it would make women vulnerable to informers and blackmail and cast suspicion on all single women who lived together. The organized opposition to the code and its subsequent defeat marked an important turning point in the relations between the women's and homosexual movements in Germany. Prior to this time, homosexual women had assumed an important, but often unacknowledged, role in the women's organizations because they were silent about their sexuality. These organizations, in turn, had generally failed to support the cause of homosexual rights. These issues are the subject of Anna Rueling's speech, "What Interest Does the Women's Movement Have in the Homosexual Question?" which she gave in 1904 at an annual conference of the Scientific Humanitarian Committee in Berlin.

WHAT INTEREST DOES THE WOMEN'S MOVEMENT HAVE IN THE HOMOSEXUAL QUESTION?

The women's movement is necessary to the history of civilization. Homosexuality is a necessity in terms of natural history, representing the bridge, the natural and obvious link between men and women. This, now, is a scientific fact against which ignorance and intolerance struggle in vain. Nevertheless, some will ask why I mention the history of civilization and natural history in one breath, two fields which upon cursory examination seem to be diametrically opposed. There is a basis for this broader view.

In general, when homosexuality is discussed, one thinks only of the Uranian men and overlooks the many homosexual women who exist and about whom much less is said because—I would almost like to say "unfortunately"—they don't have to fight an unjust penal code which resulted from false moral views. Women are not threatened with painful trials and imprisonment when they follow their inborn drive for love. But the mental stress that Uranian women endure is just as great, or greater, than the burden under which Uranian men suffer. To the world which bases its judgment on outward appearances, these women are much more obvious than even the most effeminate man. Only too often, misdirected morality exposes them to scorn and mockery.

Uranian women, even if they are not discussed, are important for our entire social structure because they influence it in many ways. Upon consideration of the facts, one must conclude that homosexuality and the women's movement are not opposed to each other, but rather that they are destined to help each other find justice and recognition and to abolish the injustice against which they now struggle.

The homosexual movement fights for the rights of all homosexuals, men as well as women. The Scientific-Humanitarian Committee is distinguished from the other movement groups which have or should have an interest in this struggle, in that it has dedicated itself enthusiastically to the Uranian woman as well as the Uranian man.

The women's movement strives for long-neglected women's rights; it is fighting especially for the greatest possible independence for women and their legal equality with men both in and out of marriage. The latter is of particular importance, first of all because of present economic conditions, and second, because the statistically proven surplus of women in the population of our country means that a large number of women simply cannot get married. Since only 10 percent of these women inherit sufficient means to live, the other 90 percent are forced to enter the labor market to earn their living in some sort of occupation. The position and participation of homosexual women in the women's movement and the movement's attempts to solve these problems are significant and deserve extensive, universal attention.

One must distinguish between two facets of the homosexual woman, her general personality and her sexual proclivity. Her overall personality is of primary importance; of secondary importance is the direction of her sex drive, which must be considered in all its complexity before it can be completely understood, since the physical love drive is generally an overflow, a natural result of psychological qualities, i.e., in people with primarily masculine characteristics, it naturally directs itself toward women and vice versa, regardless of the actual physical sex of the person. The homosexual woman possesses many qualities, inclinations, and capacities which we ordinarily consider masculine. She particularly deviates from the feminine norm in her emotional life. While emotion is almost always—exceptions prove the rule—the predominant and deciding trait in the heterosexual woman, clear reason rules the Uranian woman. She is, like the average man, more objective, energetic, and goal oriented than the feminine woman; her thoughts and feelings are those of a man; she does not imitate man, she is inherently similar to him....

Since the homosexual woman with her masculine proclivities will never suitably complement a masculine man, it is clear that the Uranian woman is not suited for marriage. Uranian women usually are aware of this fact, at least subconsciously, and accordingly refuse to go to the altar. But often they must deal with parents, cousins, aunts, and all the other dear friends and relatives, who tell them day in and day out about the necessity of marriage, and with this wise advice make life hell for them. Thanks to the poor education we provide for young girls, Uranian women often stumble blindly into marriage, without clear views and concepts of sexuality and sex life. As long as so-called "society" views spinsterhood as something unpleasant, even inferior, Uranian women will all too often allow outer circumstances to drive them into marriages in which they will neither give nor receive happiness. Aren't such marriages far more immoral than the love pact of two people who are drawn to each other by a powerful force?

The women's movement wants to reform marriage. It wants to bring about legal changes so that present conditions will cease to exist, so that discord and injustice, arbitrariness and slavish subjection, will disappear from the family, so that future generations will be healthier and stronger.

In connection with these attempts to reform, the women's movement must not forget the degree to which absurd attitudes toward homosexual women are responsible for tragic marriages. I specifically say "the degree to which" because naturally I do not attribute total blame to those absurd attitudes. But because even part of the blame lies there, the women's movement cannot dismiss its responsibility for informing society by spoken and written work how pernicious it is to force homosexual women into marriage....Because feminists care about the moral fiber and health of our people, they must wholeheartedly combat the pressures used to force homosexuals to marry. The women's movement can spread enlightenment which will enable society to see that the marriage of homosexuals is a triple crime; it is a crime against the state, against society itself, and against an unborn generation, for experience teaches us that the offspring of Uranians are seldom healthy and strong. The unhappy creatures who are conceived and born without love, or even desire, represent a large percentage of the mentally disturbed, retarded, epileptics, tuberculars, and degenerates of all kinds. Morbid sexual drives such as sadism and masochism are often the legacy from Uranians who procreated against their nature. The state and society should have an urgent interest in preventing Uranians from marrying, since it is the state and society who bear the burden of the care of these sick and weak beings who are unable to make any contribution in return.

It seems to me that an essentially practical point for heterosexual women to remember is that if homosexuals would remain single without damage to their social status, there would be more husbands available for those women whose natural inclinations are satisfied by the role of wife, housekeeper, and mother. Unfortunately, we still lack valid statistics regarding the number of homosexual women, but according to my intense work in this area, I believe that the statistics which resulted from Dr. Hirschfeld's studies of male homosexuality can be applied also to women. Therefore, there are approximately as many Uranian women as single women in Germany. To be more precise: say there are two million unmarried women and two million homosexual women; among these probably around 50 percent (one million) of the single women are Uranians, and around 50 percent (one million) of female homosexuals are married. They married because of social pressure, and thus blocked marriage for one million single heterosexual women. The conclusion is obvious. If more Uranian women remained single, more heterosexual women would have the opportunity to marry....

If the women's movement would focus on homosexuality as it relates to the marriage question, the original conception of marriage as a union between a man and woman who love each other might again come into its own. Today's too frequent marriages for money or "sensible" matches are in direct contrast to the ethical demand that people marry only for love.

I have observed that many homosexual women marry because they recognize their natural inclinations too late, and as a result they create unhappiness for themselves and others. Here too the women's movement can help, since it is very concerned with the education of the young; it can point out the importance of explaining the nature of homosexuality rationally and sensibly to older children and young people in whom homosexual tendencies have been detected by loving parents and honest, understanding friends. In this way, immense tortures and misery caused by attempts to force homosexual children into heterosexual paths could be prevented....

Now, I personally want to reiterate a point frequently made by Dr. Hirschfeld, and that is that homosexuals do not belong exclusively to any particular social class; that is, homosexuality does not occur more frequently in the upper class than in the lower class, or vice versa. No father or mother—not even those among you—can safely assume that there is no Uranian child among his or her offspring. There is a strange belief prevalent in the middle class that homosexuality does not exist in their circles, and from this group comes the greatest opposition to Uranian liberation. I myself remember that once in my parents' home when homosexuality became a topic of conversation, my father declared with conviction: "This sort of thing can't happen in my family!" The facts prove the opposite! Nothing else need be said!...

In order to obtain for homosexuals and all women generally the opportunity to live according to their natures, it is necessary to actively aid the women's movement's efforts to expand educational opportunities and new professions for women....One must conclude that it is wrong to value one sex more highly than the other—to speak of a first-class sex, man, a second-class sex, woman, and a third-class sex, the Uranian. The sexes are not of different value, they are merely of different kind. Because of this, it is clear that men, women, and Uranians are not equally suited for all professions. This is a fact the women's movement cannot change, nor does it want to....

The combinations of masculine and feminine characteristics vary so much from one person to another that all children, whether masculine or feminine, should be educated for independence in the name of simple justice. This will enable the adults to decide for themselves whether their nature suits them for the home or the world, marriage or no marriage. There must be free choice available to enable women to make their own decisions to pursue an artistic or learned profession, or to feel that they are not strong enough to do so. It is the sacred responsibility of parents to avoid stereotyped upbringing for their children and to see that each is educated according to his or her individual characteristics. Of course the schools now follow established patterns, but in the future they must provide equal education for both boys and girls and discard the notion that girls have less intellectual capacity than boys.

One need not fear that equal education for boys and girls will cause increasing competition in the professions—particularly in academia, as our opponents claim. It is true that homosexual women are specially suited for the sciences because they have those qualities lacking in feminine women: greater objectivity, energy, and perseverance. Naturally, this observation does not preclude the fact that there are extremely capable heterosexual women who are doctors, lawyers, etc. Nevertheless, I must maintain that under favorable conditions most heterosexual women choose marriage. They seek a broader, more comprehensive education in order to be esteemed companions for their husbands, not just sensual love-objects, and to be wives who are respected by their husbands as intellectual equals, and accordingly granted equal rights and responsibilities in the marriage.

Therefore, men, women, and homosexuals all benefit from a more equitable upbringing and education. Men will gain rational, sensitive companions for their lives; women will gradually gain a more worthy and legally protected position; and the Uranians will be free to dedicate themselves to their chosen professions.

Just as homosexual men often prefer professions that have a feminine quality, such as women's fashions, nursing, cooking, or being servants, homosexual women also lean toward certain professions. Thus there are many homosexual women in the medical, judicial, and agricultural professions, as well as in the creative arts....

Contrary to the belief of the anti-feminists that women are inferior and that only those with strong masculine characteristics are to be valued, I believe that women in general are equal to men. I am convinced, however, that the homosexual woman is particularly capable of playing a leading role in the international women's rights movement for equality. And indeed, from the beginning of the women's movement until the present day, a significant number of homosexual women assumed the leadership in the numerous struggles and, through their energy, awakened the naturally indifferent and submissive average women to an awareness of their human dignity and rights. I am unable and unwilling to name names because as long as many consider homosexuality criminal and unnatural, at best sick, ladies I could call homosexual might feel insulted. Above all, decency and duty forbid indiscretion, and neither the noble love of a Uranian suffragette nor the feelings of a heterosexual need be aired in a public forum. But anyone with the slightest bit of familiarity with homosexual traits who has been following the women's movement at all or who knows any of its leading women personally or by pictures, will find the Uranians among the suffragettes and recognize that Uranians are often noble and fine.

Considering the contributions made to the women's movement by homosexual women for decades, it is amazing that the large and influential organizations of the movement have never lifted a finger to improve the civil rights and social standing of their numerous Uranian members. It is amazing that they have done nothing, absolutely nothing, to protect many of their best-known and most deserving leaders from ridicule and scorn by enlightening the general public about the true nature of Uranianism....

I understand the reason for the reluctance of the women's movement to deal with this problem, although it deals very frankly with other purely sexual matters. It stems from its fear that the movement could suffer in the eyes of the still blind and ignorant masses if it took up the homosexual question by energetically supporting the human rights of Uranians. I'll frankly admit that this fear was justified during the early days of the women's movement when it had to carefully avoid losing converts, and this fear was a credible excuse for temporarily ignoring the homosexual question. Today, however, when the movement is advancing unimpeded, when no bureaucratic wisdom, no philistinism, can block its triumphant march, this failure to deal with an important question is an injustice, an injustice which the women's movement inflicts on itself....

...[I]t is not necessary to preach about the injustices done to the Uranians from every rooftop—this would only harm the movement—I understand this aspect well; it need only speak objectively about the homosexual question when addressing the sexual, official, economic, and purely human relationships of the sexes to each other. That it can do, and in so doing, it can slowly and quietly bring about enlightenment.

I am now getting to a topic with which the women's movement has been specially concerned in recent years—prostitution. From an ethical standpoint, one can think of it as one wishes; in any case, it will have to be dealt with for a long time to come. Personally, I regard prostitution as an unfortunate but necessary evil which will be impossible to eradicate as long as there are human passions, but which, if we are fortunate, we can lessen—a goal worth striving for.

An important fact that has been completely ignored by the women's movement's struggle against the increase of prostitution and its destructive companion, venereal disease, is that 20 percent of all prostitutes are homosexual. At first this may seem odd because of the contradiction between homosexuality and constant sexual intercourse with the male. This situation has been explained to me more than once by a "girl of the streets" who told me that she considered her said trade as business—completely divorced from her sexual drive, which was satisfied by her woman lover. Adverse domestic and economic conditions had driven these girls into the street.

...In a certain sense, the struggle of the homosexual woman for social acceptance is also a struggle against prostitution, although I must emphasize the fact that this struggle could only result in a lessening of prostitution, not its eradication.

One must not forget that justice for Uranians in general would mean that a great number of homosexual men who are driven to prostitutes by their fear of detection would no longer find this necessary. This, of course, would result in a reduction of venereal disease, which though small would be valuable, for each single case in which syphilis or other venereal infection is avoided is a victory for the health of the people and the coming generation on whom rests the well-being and greatness of our fatherland.

The women's movement fights for the right of individuality and self-determination. It must admit that the alienating ban that society still places on Uranians suppresses this right; and therefore its responsibility is to join the homosexuals in their struggles, just as it actively assists unwed mothers, working women, and many others as they fight for freedom and right, battling against old, false, traditional concepts of a morality which is in actuality an immorality of the worst kind.

...Without the active support of the Uranian women, the women's movement would not be where it is today—this is an undisputable fact.

The women's movement and the homosexual rights movement have long traveled a dark path filled with innumerable obstacles. Now the light is gradually being turned on in human hearts, and it is becoming brighter for us....[B]oth movements will reach the point when they will recognize that they have many mutual interests, when they will peacefully join hands in order to join forces in battle where it is necessary.

And if at first we find serious and difficult hours in store for us, we must not give up in cowardice, but must move courageously through the hostile forces, onward to the victory which is assured us. For the sun of understanding and truth has risen in the east, and no power of darkness can alter its radiant course—slowly it will rise higher and higher! Not today or tomorrow, but in the not too distant future, the women's movement and Uranians will raise their flags of victory!

Per aspera ad astra![1]

---

[1]Through adversity, we will reach the stars!

QUESTIONS FOR ANALYSIS

1. How does Harriet Taylor evaluate the institution of marriage? On what basis does she argue for the right to divorce? How does she envision the transformation of women's lives if they were granted equality with men and all laws concerning marriage were abolished?

2. Discuss the main points of Harriet Martineau's critique of women's dress. What does she propose as an alternative? Why did dress reform become a feminist issue?

3. What were Josephine Butler's views on prostitution and how were they informed by her religious convictions? On what basis did she argue for the repeal of the Contagious Diseases Acts?

4. According to Anna Rueling, what are the main attributes of "Uranian women"? What contributions have they made to German society in general and the women's movement in particular? What common goals did she believe were shared by the homosexual and women's movements?

5. Compare the writings in this section with contemporary feminist writings and discussions on marriage, divorce, prostitution, and women's sexuality.

MASS, SOCIALIST PARTIES, TRADE UNIONS, AND PROTECTIVE LEGISLATION

By the end of the nineteenth century, mass, working-class socialist parties existed throughout most of Europe. This prompted the formation of the Second International in 1889. With a strong parliamentary base, socialist parties were able to effect legislation on behalf of workers. Socialist women mobilized women into trade unions and socialist parties where they raised demands specific to working women's work and lives, including the issue of suffrage. Protective legislation became an important source of debate. Such legislation was generally supported by socialist women but opposed by middle-class feminists who argued that it undermined the struggle for women's equality.

125. Clara Zetkin
Women and the Trade Unions

Clara Zetkin's (1857–1933) political activism as a socialist, trade unionist, and advocate of women workers spanned the earliest years of the German Social Democratic Party (SPD) to the founding and consolidation of the German Communist Party[1] in 1918. Following the expiration of Bismark's Anti-Socialist Law in 1890, the SPD grew into the largest socialist party in Europe. Until 1908, however, laws of association enforced throughout most of Germany prohibited women from officially joining and attending political meetings of electoral parties. On principle, and not only because of these laws, Zetkin advocated separate women's groups, meetings, conferences, and delegate assembly meetings. She also edited and contributed to the biweekly journal *Die Gleichheit (Equality)* founded in 1892 in Stuttgart where

---

[1]On the origins of the German Communist Party, see Rosa Luxemburg, Source 142 in Chapter 12.

Originally written by Clara Zetkin in 1893.

Zetkin was active in the trade union movement. From a left-wing socialist perspective, the journal published articles on such issues as child labor; suffrage; women's working conditions, including that of domestic workers; and labor strikes in Germany and abroad. Zetkin's article "Women's Work and the Organization of Trade Unions," reproduced here, was published in the November 1, 1893, issue of *Die Gleichheit* in preparation for an SPD congress in Cologne. At the first International Socialist Women's Congress held in Stuttgart in 1907, Zetkin was chosen secretary of the International Women's Bureau, and *Die Gleichheit* was designated its official journal. While recruiting working women to the SPD, Zetkin also opposed its reformism and support for German imperialism which culminated in the party's infamous vote for war credits in August 1914. She organized an International Conference of Socialist Women held in Berne, Switzerland, in 1915. Zetkin became a member of the Independent Social-Democratic Party (USPD) formed in 1917 by antiwar members of the SPD and soon afterward was removed as editor of *Die Gleichheit*. In 1918, Zetkin led rank-and-file members of the USPD into the German Communist Party which she subsequently represented in the Reichstag during the Weimar Republic.

WOMEN'S WORK AND THE ORGANIZATION OF TRADE UNIONS

The Party Congress of Cologne will have to address itself to the question of trade union organization, i.e., the relationship between the political and trade union movement. The question will be dealt with because of the urgings of trade union circles. Recently the trade unions have declined; within the trade union movement there is a tendency to blame, among other factors, the attitude of the political movement for this phenomenon....

There remains the indubitable fact that in all capitalist countries, women's work in industry plays an ever larger role. The number of industrial branches in which women nowadays toil and drudge from morning till night increases with every year. [see "Employed in 1882" table on p. 373.] Factories which have traditionally employed women, employ more and more women workers. It is not only that the number of all industrially employed women is constantly growing, but their number in relation to the men who are working in industry and trade is also on the increase.

Some branches of industry (one has only to think of clothing) are virtually dominated by women's labor which constantly reduces and replaces men's labor.

For understandable reasons, particularly during periods of recession (like the one we are experiencing right now),[1] the number of women workers has increased in both relative and absolute terms whereas the number of employed male laborers has decreased. As we have already reported, during 1892 in Saxony, the number of male workers over 16 years of age decreased by 1,633 whereas the number of female workers of similar age increased by 2,466.

According to the Viennese university instructor, J. Singer, five million women were working in Germany's industry during the last few years.

The business survey of 1882 points out that out of 7,340,789 individually employed persons in Germany, 1,509,167 (20.6%) were women. Thus there were 21 women for every 100 persons involved in industrial production....

These statistics give only an approximate idea of the extent to which female labor is being used since the myriad of women who work in factories which are not "under the protection of the law" and do not, therefore, come under factory inspection, have not been included. How extensive is just the number of women who slave away as domestic servants!

---

[1]Zetkin is referring to the worldwide economic depression that occurred in 1893.

| | Employed in 1882 | | Women as % of |
Industry	Men	Women	All Employed
Lace Production	5,676	30,204	84.1
Clothing, Linen, Finery	279,978	440,870	61.2
Spinning	69,272	100,459	59.2
Haberdashery Items	13,526	17,478	56.4
Service and Restaurants	172,841	141,407	45.0
Tobacco Production	64,477	48,919	43.1
Embroidery and Weaving	42,819	31,010	42.0
Paper Manufacturing	37,685	20,847	35.6
Textiles	336,400	155,396	31.6
Messenger Service	9,212	3,265	26.2
Commerce	536,221	181,296	25.3
Bookbinding and Carton Making	31,312	10,409	24.9

The reasons for the constantly growing use of female laborers have been repeatedly pointed out: their cheapness and the improvement of the mechanical means and methods of production. The automatic machine, which in many cases does not even stand in need of having to be regulated, works with the powers of a giant, possesses unbelievable skill, speed and exactness and renders muscle power and acquired skills superfluous. The capitalist entrepreneur can employ only female labor at those places where he previously had to use male employees. And he just loves to hire women because female labor is cheap, much cheaper than male labor.

Even though the productive capacity of female workers does not lag behind that of male workers, the difference between men's and women's wages is very significant. The latter is often only half of the former and often only a third.

According to the Leipzig Chamber of Commerce, the following weekly wages were paid:

In 1892, the Leipzig Health Insurance Office made a statistical analysis of wages which determined that 60% of the women workers have weekly earnings of below or up to 9 marks, 32% up to 12 marks and only 7% up to 15 or 19 to 21 marks. As far as earnings are concerned, men, too, do not fare well but they do better than their female counterparts; 37% of the men earn up to 15 marks, 30% up to 19 marks and 33% up to 21 marks.

The women laborers of Berlin's chemical industry earn highly unfavorable wages; 74% of them have a weekly wage of only up to 10 marks and 50 pfennigs. Of the remaining 26%, only 2% have a weekly salary of up to 24 marks.

From Hessen, Bavaria, Saxony, Thuringia, Württemberg, i.e., from all of God's little German fatherlands, the factory inspectors report that the wages of women workers are far below those of male laborers. Factory Inspector Worrishoffer of Baden undertook a very thorough

	Men	Women
	(Marks)	(Marks)
Fabrication of Lace	20–35	7–15
Factory for Paper Lanterns	16–22	7.50–10
Woolen Industry	15–27	7.20–10.20
Cloth Glove Factory	12–30	6–15
Fabrication of Leather and Leather Goods	12–28	7–18
Linen and Jute Factory	12–27	5–10
Sugar Factory	10.50–31	7.50–10
Rubber Factory	9–27	6–17

investigation of the social situation of factory workers. It, too, demonstrates very clearly the miserable earnings of women who work in industry. Worrishoffer divided male and female workers according to their earnings into three wage groups: a low one with a weekly salary of less than 13 marks, a medium one whose weekly wages amount to 15 to 24 marks and a high one whose weekly salary is more than 24 marks. Of the female workers of Mannheim, 99.2% belong to the low category, 0.7% to the medium and 0.1% to the higher wage group. In other words, of 100 women workers in Mannheim, 99 have a weekly salary of below 15 marks and 27 [of these] have a salary of up to 10 marks. These statistics amply illustrate the fact that the living conditions of these female workers correspond to their miserable earnings. It is easily understandable that these customary starvation wages for female laborers push thousands of them from the proletariat into the lumpenproletariat. Their dire straits force some of them to take up part-time or temporary prostitution so that by selling their bodies, they may earn the piece of bread that they cannot secure by the sale of their labor.

But it is not just the women workers who suffer because of the miserable payment of their labor. The male workers, too, suffer because of it. As a consequence of their low wages, the women are transformed from mere competitors into unfair competitors who push down the wages of men. Cheap women's labor eliminates the work of men and if the men want to continue to earn their daily bread, they must put up with low wages. Thus women's work is not only a cheap form of labor, it also cheapens the work of men and for that reason it is doubly appreciated by the capitalist, who craves profits. An entire branch of industry—the textile business—is living proof of how women's work is used to depress wages. The low salaries paid to textile workers is in part the result of the extensive use of female labor in that industry. The wool and cotton barons have used the cheap work of women in order to lower the working and living conditions of an entire category of the proletariat to a level that defies culture.

The transfer of hundreds of thousands of female laborers to the modernized means of production that increase productivity ten or even a hundredfold should have resulted (and did result in some cases) in a higher standard of living for the proletariat, given a rationally organized society. But as far as the proletariat is concerned, capitalism has changed blessing into curse and wealth into bitter poverty. The economic advantages of the industrial activity of proletarian women only aid the tiny minority of the sacrosanct guild of coupon clippers and extortionists of profit.

Frightened by the economic consequences of women's work and the abuses connected with it, organized labor demanded for a while the prohibition of female labor. It was viewing this question merely from the narrow viewpoint of the wage question. Thanks to Socialist propaganda, the class-conscious proletariat has learned to view this question from another angle, from the angle of its historical importance for the liberation of women and the liberation of the proletariat. It understands now how impossible it is to abolish the industrial labor of women. Thus it has dropped its former demand and it attempts to lessen the bad economic consequences of women's work within capitalist society (and only within it!) by two other means; by the legal protection of female workers and by their inclusion in trade union organizations....

Given the fact that many thousands of female workers are active in industry, it is vital for the trade unions to incorporate them into their movement. In individual industries where female labor plays an important role, any movement advocating better wages, shorter working hours, etc., would be doomed from the start because of the attitude of those women workers who are not organized. Battles which began propitiously enough, ended up in failure because the employers were able to play off non-union female workers against those that are organized in unions. These non-union workers continued to work (or took up work) under any conditions, which transformed them from competitors in dirty work to scabs.

It is not only because of the successful economic battles of trade unions that women should be included in them. The improvement of the starvation wages of female workers and the limitation of competition among them requires their organization into unions.

The fact that the pay for female labor is so much lower than that of male labor has a variety of causes. Certainly one of the reasons for these poor wages for women is the circumstance that female workers are practically unorganized. They lack the strength which comes with unity. They lack the courage, the feeling of power, the spirit of resistance and the ability to resist which is produced by the strength of an organization in which the individual fights for everybody and everybody fights for the individual. Furthermore, they lack the enlightenment and the training which an organization provides. Without an understanding of modern economic life in whose machinery they are inextricably caught up, they will neither be able to take advantage of periods of boom through conscious, calculating and unified conduct nor will they be able to protect themselves against the disadvantages occurring during periods of economic recession. If, under the pressure of unbearable conditions they finally fight back, they usually do so at an inopportune moment and in a disorganized fashion.

This situation exercises a great influence upon the miserable state of women's work and is further reflected by the bitterness that male workers feel about women's competition. Thus in the interest of both men and women workers, it is urgently recommended that the latter be included in the trade unions. The larger the number of organized female workers who fight shoulder to shoulder with their comrades from the factory or workshop for better working conditions, the sooner and the greater will women's wages rise so that soon there may be the realization of the principle: Equal pay for equal work regardless of the difference in sex. The organized female worker who has become the equal of the male worker ceases to be his scab competitor.

The unionized male workers realize more and more just how important it is that the female workers are accepted into the ranks of their organization. During these past few years, there was no lack of effort on the part of the unions in regard to this endeavor. And yet how little has been accomplished and how incredibly much remains to be done in this respect.

According to the Report of the General Commission of the Trade Unions of Germany, out of fifty-two organizations, there are only fourteen that have a membership of both male and female workers. Then there are two organizations that consist only of women and girls. What does all this mean given the large and steadily growing number of industries which employ women?

Even in those industrial branches in which the trade union organization of women began, these organizations are still in their infancy....

As far as the percentage of female membership is concerned, the Tobacco Workers rank first, and yet these women workers do not even constitute a fourth of its entire membership. In 1882, 43.1% of all tobacco industry workers were women. In the other four trade unions which come next, as far as the percentage of women that work in the industries they represent are concerned, women workers do not even constitute 10% of the membership. The Organization of Gold and Silver Workers does not have a female membership of even 5% even though there are large numbers of women workers who are employed by the gold and silver industry. In 1882, 60% of all laborers in spinning mills and 30% of all laborers in weaving mills happened to be women, yet the percentage of them who were unionized amounted to only 9½%. These numbers, in conjunction with the slave wages which generally prevail in the textile industry, speak whole volumes about the necessity of unionizing women.

In recognition of this necessity, the trade unions should use all of their energies to work for the inclusion of women in their organizations.

We certainly do not fail to recognize the difficulties raised by women workers which are detrimental to the solution of this problem.

Stupid resignation, lack of a feeling of solidarity, shyness, prejudices of all kinds and fear of the factory tyrant keep many women from joining unions. Even more than the just mentioned factors, the lack of time on the part of female workers represents a major obstacle against their mass organization because women are house as well as factory slaves and are forced to bear a double workload. The economic developments, however, as well as the increasing acuteness of the class struggle, educate both male and female laborers and force them to overcome the above-mentioned difficulties.

We certainly recognize the fact that during the past few years, the trade unions have made a serious effort to enroll female workers alongside their male colleagues. But what has been accomplished and aimed for does not come up to the urgency and the importance of the task. Theoretically, most male union members admit that the common unionization of both male and female workers of the same trade has become an unavoidable necessity. In practice, however, many of them do not make the effort that they could be making. Rather there are only a few unions and within them only certain individuals who pursue with energy and perseverance the organization of female workers. The majority of trade union members give them precious little support. They treat such endeavors as a hobby which should be tolerated but not supported "as long as there are still so many indifferent non-union male workers." This point of view is totally wrong.

The unionization of women workers will make significant progress only when it is no longer merely aided by the few, but by every single union member making every effort to enlist their female colleagues from factory and workshop. In order to fulfill this task, two things are necessary. The male workers must stop viewing the female worker primarily as a woman to be courted if she is young, beautiful, pleasant and cheerful (or not). They must stop (depending on their degree of culture or lack of it) molesting them with crude and fresh sexual advances. The workers must rather get accustomed to treat female laborers primarily as female proletarians, as working-class comrades fighting class slavery and as equal and indispensable co-fighters in the class struggle. The unions make such a big thing out of having all of the members and followers of the political party become members of the unions. It seems to us that it would be much more important to put the emphasis on enrolling the broad, amorphous masses in the labor movement. In our opinion, the main task of the unions is the enlightenment, disciplining and education [of all workers] for the class struggle. In view of the increasing use of female labor and the subsequent results, the labor movement will surely commit suicide if, in its effort to enroll the broad masses of the proletariat, it does not pay the same amount of attention to female workers as it does to male ones.

126. Louise Kautsky
A Socialist Platform for the Protection of Women Workers

International socialist congresses provided an important forum for women to address common issues across national lines. Protective legislation was one such issue. Two proposals were adopted at the International Socialist Workers' Congress held in Zurich in 1893. Anna Kuliscioff, a founding member of the Italian Socialist Party, proposed that women should receive equal pay for equal work and that protective legislation should not restrict employment opportunities for

Louise Kautsky (Austria), International Socialist Workers' Congress. Zurich, 1893.

women. The second resolution, reproduced here, was introduced by Louise Kautsky of the German Social Democratic Party.

Considering that the bourgeois women's movement rejects all special legislation for the protection of women as interfering with the rights of woman and her equality with man, that on the one hand it ignores the character of our present-day society which rests on the exploitation of the working class by the capitalist class, women as well as men, and on the other hand fails to recognize the special role which has fallen to women through the differentiation of the sexes, namely their role as mother of children which is so important for the future, the International Congress of Zürich declares:

It is the duty of the labor representatives of all countries to advocate energetically the passage of the following measures which will offer legal protection to women workers:

1. Introduction of a maximum eight-hour day for women and a six-hour day for girls under eighteen.
2. Fixing an uninterrupted period of time off once a week of thirty-six hours.

3. Prohibition of night work.
4. Prohibition of women working in all jobs which are detrimental to their health.
5. Prohibition of work by pregnant women, two weeks before and four weeks after childbirth.
6. Appointment of women factory inspectors in sufficient numbers in all industries where women are employed.
7. Application of the above measures to all working women whether it be in factories, workshops, in home industry or as women agricultural workers.[1]

---

[1]Kautsky's resolution was passed with the exception of this article concerning the application of protective legislation. The Congress subsequently affirmed "equal pay for equal work" in these terms: "In order to protect women workers and, above all, to forcibly inhibit the exploitation of women, it is necessary not only to reduce working hours but to accept the principle of 'equal pay for equal work.'"

127. Beatrice Webb
"Women and the Factory Acts"

Beatrice Webb (1858–1943) was a leading member of the Fabian Society and an appointee on the Royal Commission on the Poor Law which met from 1905 to 1909. Of upper-middle-class origins, she carried out philanthropic work and investigated the causes of poverty in the East End of London in the 1880s and contributed to Charles Booth's *Life and Labour of the People of London*. These experiences led her to critique unregulated capitalism. She embraced socialism after reading Fabian Society publications and joined the Fabians after her 1892 marriage to Sidney Webb. Established in 1884 in London and primarily middle-class and professional in its composition, the Fabian Society promoted municipal socialism and a gradual, evolutionist transformation of society through state intervention. The Fabians carried out extensive investigations of factory conditions and working-class life, many of which were summarized in published

Fabian Tract No. 67 (February 1896).

tracts distributed to the public. Although the Fabian Women's Group was not formed until 1908, a Women's Tract Committee existed within the society from 1892 to 1896 which advocated publication of Beatrice Webb's "Women and the Factory Acts" in 1896.[1] Here, she presents her views in favor of the protective legislation contained in the Factory and Workshop Act of 1895. In addition to her diaries, Beatrice Webb's writings, many co-authored with Sidney Webb, include *The History of Trade Unionism* (1894), *Minority Report* (1909), the report of the socialist and progressive members of the Royal Commission on the Poor Law; *The Prevention of Destitution* (1911); *A Constitution for the Socialist Commonwealth of Great Britain* (1902); *My Apprenticeship* (1926); *Soviet Communism. A New Civilisation* (1935); and *Our Partnership* (1948).

The discussions on the Factory Act of 1895 raised once more all the old arguments about Factory legislation, but with a significant new cleavage. This time legal regulation was demanded, not only by all the organizations of working women whose labor was affected,[1] but also by, practically, all those actively engaged in Factory Act administration. The four women Factory Inspectors unanimously confirmed the opinion of their male colleagues. Of all the classes[2] having any practical experience of Factory legislation, only one—that of the employers—was ranged against the Bill, and that not unanimously. But the employers had the powerful aid of most of the able and devoted ladies who have usually led the cause of women's enfranchisement, and whose strong theoretic objection to Factory legislation caused many of the most important clauses in the Bill to be rejected.

The ladies who resist further legal regulation of women's labor usually declare that their objection is to special legislation applying only to women. They regard it as unfair, they say, that women's power to compete in the labor market should be 'hampered' by any regulation from which men are free. Any such restriction, they assert, results in the lowering of women's wages, and in diminishing the aggregate demand for women's work. I shall, later on, have something to say about this assumed competition between men and women. But it is curious that we seldom find these objectors to unequal laws coming forward to support even those regulations which apply equally to men and to women. Nearly all the clauses of the 1895 Bill, for instance, and nearly all the amendments proposed to it, applied to men and women alike. The sanitary provisions; the regulations about fire-escapes; the pre-eminently important clause making the giver-out of work responsible for the places where his work is done; the power to regulate unhealthy trades or processes; all these made no distinction between the sexes. Yet the ladies who declared that they objected

---

[1]Petitions were sent in, and meetings held in support of the Bill by, I believe, all the Trade Unions of Women, as well as by the Women's Co-operative Guild, which is mainly composed of women textile workers, whose hours of labor have, for nearly forty years, been rigidly fixed by law.

[2]See the *Report of the Chief Inspector of Factories* for 1894, C. 7745, price 5s. 3d.; also the *Opinions on Overtime,* published by the Women's Trade Union League (Club Union Buildings, Clerkenwell Road, London). The evidence before the Royal Commission on Labor was decidedly in favor of an extension of, and the more rigid enforcement of Factory legislation: see, in particular, the Minority Report (published separately, price 2d., by the Manchester Labor Press, Tib Street, Manchester).

---

[1]The Fabian Executive Committee commissioned the Women's Tract Committee to write a tract on women and socialism that was to address such issues as trade unionism, health, law, technical education, and factory legislation. But the tract was never produced as the author, Harriot Stanton Blatch, refused to revise her work. She subsequently published an article in *Gunton's Magazine* which instead proposed paying housework and childcare as specialized professions. Beatrice Webb's "Women and the Factory Acts" appears to have fulfilled, at least in part, the committee's goals. Published as Fabian Tract No. 67, it was based on papers she gave to the Nottingham Conference of the National Union of Women Workers (October, 1895) and the Fabian Society (January, 1896).

only to inequality of legislation, gave no effective aid to the impartial sections of the Bill. If we believe that legal regulation of the hours and conditions of labor is found, in practice, to promote the economic independence and positively to add to the industrial efficiency of the workers concerned, why should we not help women workers in unregulated trades to gain this superior economic position, even if Parliament persists in denying it to the men? It is clear that there lurks behind the objection of inequality an inveterate scepticism as to the positive advantages of Factory legislation. Indeed, the most energetic and prominent opponents of women's Factory Acts openly avow as much.... Therefore before discussing whether any particular Factory Act is good for women or not, we had better make up our minds on the general question. Does State regulation of the hours and conditions of labor increase or decrease the economic independence and industrial efficiency of the workers concerned?

Now those who object to further Factory legislation are right in asserting that the issue cannot be decided by harrowing accounts of factory tyranny, or particular cases of cruelty or hardship. I shall not trouble you with the long list of calamities in the unregulated trades, on which the official report of the Chief Inspector of Factories lays so much stress—the constitutions ruined by long hours in dressmakers' workrooms or insanitary laundries, the undermining of family life by the degradation of the home into a workshop, the diseases and deaths caused by white lead and lucifer matches. And, I hope, no one in the discussion will think it any argument against Factory Acts that some poor widow might find it more difficult to get bread for her starving children if she were forbidden to work at the white lead factory; that some sick man's daughter would not be allowed to earn the doctor's fee by taking extra work home after her factory day; or that some struggling laundress might find it impossible to make a living if she could not employ her girls for unlimited hours. Either way there must be hard cases, and individual grievances. The question is whether, taking the whole population and all considerations into account, the evils will be greater under regulation or under free competition.

Let us concede to the opponents of Factory legislation that we must do nothing to impair or limit the growing sense of personal responsibility in women; that we must seek, in every way, to increase their economic independence, and their efficiency as workers and citizens, not less than as wives and mothers; and that the best and only real means of attaining these ends is the safeguarding and promoting of women's freedom. The only question at issue is how best to obtain this freedom. When we are concerned with the propertied classes—when, for instance, it is sought to open up to women higher education or the learned professions—it is easy to see that freedom is secured by abolishing restrictions. But when we come to the relations between capital and labor an entirely new set of considerations come into play. In the life of the wage-earning class, absence of regulation does not mean personal freedom. Fifty years' experience shows that Factory legislation, far from diminishing individual liberty, greatly increases the personal freedom of the workers who are subject to it. Everyone knows that the Lancashire woman weaver, whose hours of labor and conditions of work are rigidly fixed by law enjoys, for this very reason, more personal liberty than the unregulated laundry-woman in Notting Hill. She is not only a more efficient producer, and more capable of associating with her fellows in Trade Unions, Friendly Societies, and Co-operative Stores, but an enormously more independent and self-reliant citizen. It is the law, in fact, which is the mother of freedom.

... No sensible person can really assert that the individual operative seeking a job has either the knowledge or the opportunity to ascertain what the conditions are, or to determine what they should be, even if he could bargain about them at all. On these matters, at any rate, there can be no question of free contract. We may, indeed, leave them to be determined by the employer himself: that is to say, by the competition between employers as to who can most reduce the expenses of production. What this means, we know from the ghastly experience of the early factory system;

when whole generations of our factory hands were stunted and maimed, diseased and demoralized, hurried into early graves by the progressive degeneration of conditions imposed on even the best employers by the reckless competition of the worst.[3] The only alternative to this disastrous reliance on a delusive freedom is the settlement, by expert advice, of standard conditions of health, safety, and convenience, to which all employers, good and bad alike, are compelled by law to conform....

We can now understand why it is that the representative wage earner declares, to the astonishment of the professional man or the journalist, that a rule fixing his hours of labor, or defining conditions of sanitation or safety, is not a restriction in his personal liberty. The workman knows by experience that there is no question of his ever settling these matters for himself. There are only two alternatives to their decision by the employer. One is their settlement by a conference between the representatives of the employers and the representatives of the organized workmen; both sides, of course, acting through their expert salaried officials. This is the method of collective bargaining—in short, Trade Unionism. The other method is the settlement by the whole community of questions which affect the health and industrial efficiency of the race. Then we get expert investigation as to the proper conditions, which are enforced by laws binding on all. This is the method of Factory legislation.

...Rates of wages, for instance, are best settled by collective bargaining; and sanitation, safety, and the prevention of overwork by fixed hours of labor are best secured by legal enactment.

But this question of the relative advantages of legislative regulation and Trade Unionism has unhappily no bearing on the women employed in the sweated industries. Before we can have Trade Union regulation we must build up strong Trade Unions; and the unfortunate women workers whose overtime it was proposed to curtail, and whose health and vigor it was proposed to improve, by Mr. Asquith's Bill of 1895, are without any effective organization. The Lancashire women weavers and card-room hands were in the same predicament before the Factory Acts. It was only when they were saved from the unhealthy conditions and excessive hours of the cotton mills of that time that they began to combine in Trade Unions, to join Friendly Societies, and to form Co-operative Stores. This, too, is the constant experience of men's trades. Where effective Trade Unions have grown up, legal protection of one kind or another has led the way.[4] And it is easy to see why this is so. Before wage-earners can exercise the intelligence, the deliberation, and the self-denial that are necessary for Trade Unionism, they must enjoy a certain standard of physical health, a certain surplus of energy, and a reasonable amount of leisure. It is cruel mockery to preach Trade Unionism, and Trade Unionism alone, to the seamstress sewing day and night in her garret for a bare subsistence; to the laundrywoman standing at the tub eighteen hours at a stretch; or to the woman whose health is undermined with 'Wrist-drop,' 'Potter's-rot,' or 'Phossy-jaw.' If we are really in earnest in wanting Trade Unions for women, the way is unmistakable. If we wish to see the capacity for organization, the self-reliance, and the personal independence of the Lancashire cotton weaver spread to other trades, we must give the women workers in these trades the same legal fixing of hours, the same effective prohibition of overtime, the same legal security against accident and disease, the same legal standard of sanitation and health as is now enjoyed by the women in the Lancashire cotton mills.

So much for the general theory of Factory legislation. We have still to deal with the special arguments directed against those clauses of the 1895 Bill which sought to restrict the overtime worked by women in the sweated trades.

---

[3]Some account of this development is given in the first chapter of my *Co-operative Movement in Great Britain*. See also Engels' *Condition of the English Working Classes in 1844,* or Arnold Toynbee's *The Industrial Revolution.*

[4]For proof of this see *The History of Trade Unionism,* by Sidney and Beatrice Webb, particularly the first chapter.

If, however, we have fully realized the advantages, both direct and indirect, which the workers obtain from the legal regulation of their labor, we shall regard with a good deal of suspicion any special arguments alleged in opposition to any particular Factory Acts. The student of past Factory agitations sees the same old bogeys come up again and again. Among these bogeys the commonest and most obstructive has always been that of foreign competition, that is to say, the risk that the regulated workers will be supplanted by 'free labor'— whether of other countries or of other classes at home. At every step forward in legal regulation the miner and the textile worker have been solemnly warned that the result of any raising of their standard of sanitation, safety, education or leisure would be the transference of British capital to China or Peru. And to my mind it is only another form of the same fallacy when capitalists' wives and daughters seek to alarm working women by prophesying, as the result of further Factory legislation, the dismissal of women and girls from employment, and their replacement by men. The opposition to Factory legislation never comes from workers who have any practical experience of it. Every existing organization of working women in the kingdom has declared itself in favor of Factory legislation. Unfortunately, working women have less power to obtain legislation than middle-class women have. Unfortunately, too, not a few middle-class women have allowed their democratic sympathies and Collectivist principles to be overborne by this fear of handicapping women in their struggle for employment. Let us, therefore, consider, as seriously as we can, this terror lest the capitalist employing women and girls at from five to twelve shillings a week, should, on the passage of a new Factory Act, replace them by men at twenty or thirty shillings.

First let us realize the exact amount of the inequality between the sexes in our Factory Acts. All the regulations with respect to safety, sanitation, employers' liability, and age apply to men and women alike. The only restriction of any importance in our Labor Code which bears unequally on men and women is that relating to the hours of labor.[5] Up to now there has been sufficient influence among the employers, and sufficient prejudice and misunderstanding among legislators, to prevent them expressly legislating, in so many words, about the hours of labor of adult men. That better counsels are now prevailing is shown by the fact that Parliament in 1892 gave power to the Board of Trade to prevent excessive hours of work among railway servants, and that the Home Secretary has now a similar power in respect of any kind of manual labor which is injurious to health or dangerous to life and limb. I need hardly say that I am heartily in favor of regulating, by law, the hours of adult men, wherever and whenever possible.[6] But although the prejudice is breaking down, it is not likely that the men in the great staple industries will be able to secure for themselves the same legal limitation of hours and prohibition of overtime that the women in the textile manufactures have enjoyed for nearly forty years. And thus it comes about that some of the most practical proposals for raising the condition of the

---

[5] *The Law relating to Factories and Workshops,* by May Abraham and A. Llewelyn Davies (Eyre and Spottiswoode, 1896, 5/-), contains a convenient summary of all the Acts. With regard to hours, the main provisions are as follows: Textile factories employing women or children, may work only between 6 a.m. and 6 p.m. (or 7 a.m. and 7 p.m.), only 56½ hours net per week, and overtime is absolutely prohibited. In non-textile factories and in ordinary workshops, women may be worked 60 hours per week, overtime is (usually) permitted under certain conditions, and the day's work may (except on Saturdays) range over a period from 6 a.m. to 8 p.m., or, if no children or young persons are employed, even from 6 a.m. to 10 p.m. This absence of a precisely determined legal working-day makes it practically impossible to enforce the law. In 'domestic workshops' there is no restriction on women's hours, and in laundries the only limit is a general one of sixty hours per week (or fourteen in any one day), without regulation of the hours of beginning or ending, or of mealtimes. This is quite illusory.

[6] See Fabian Tract, No. 48, *Eight Hours by Law: a Practicable Solution.*

women in the sweated trades must take the form of regulations applying to women only.

It is frequently asserted as self-evident that any special limitation of women's labor must militate against their employment. If employers are not allowed to make their women work overtime, or during the night, they will, it is said, inevitably prefer to have men. Thus, it is urged, any extension of Factory legislation to trades at present unregulated must diminish the demand for women's labor. But this conclusion, which seems so obvious, really rests on a series of assumptions which are not borne out by facts.

The first assumption is, that in British industry to-day, men and women are actively competing for the same employment. I doubt whether any one here has any conception of the infinitesimal extent to which this is true. We are so accustomed, in the middle-class, to see men and women engaged in identical work, as teachers, journalists, authors, painters, sculptors, comedians, singers, musicians, medical practitioners, clerks, or what not, that we almost inevitably assume the same state of things to exist in manual labor and manufacturing industry. But this is very far from being the case. To begin with, in over nine-tenths of the industrial field there is no such thing as competition between men and women: the men do one thing, and the women do another. There is no more chance of our having our houses built by women than of our getting our floors scrubbed by men. And even in those industries which employ both men and women, we find them sharply divided in different departments, working at different processes, and performing different operations. In the tailoring trade, for instance, it is often assumed that men and women are competitors. But in a detailed investigation of that trade I discovered that men were working at entirely separate branches to those pursued by the women....

I do not wish to imply that there are absolutely no cases in British industry in which men and women are really competing with each other. It is, I believe, easy to pick out an instance here and there in which it might be prophesied that the removal of an existing legal restriction might, in the first instance, lead to some women being taken on in place of men. In the book and printing trade of London, for instance, it has been said that if women were allowed by law to work all through the night, a certain number of exceptionally strong women might oust some men in book-folding and even in compositors' work.[7] We must not overlook these cases; but we must learn to view them in their proper proportion to the whole field of industry. It would clearly be a calamity to the cause of women's advancement if we were to sacrifice the personal liberty and economic independence of three or four millions of wage-earning women in order to enable a few hundreds or a few thousands to supplant men in certain minor spheres of industry.[8]

The second assumption is, that in the few cases in which men and women may be supposed really to compete with each other for employment, the effect of any regulation of women's hours is pure loss to them, and wholly in favor of their assumed competitors who are unrestricted. This, I believe, is simply a delusion. Any investigator of women's work knows full well that what most handicaps women is their general deficiency in industrial capacity and technical skill. Where the average woman fails is in being too much of an amateur at her work, and too little of a professional. Doubtless it may be said that the men are to blame here: it is they who induce women to marry, and thus divert their attention from professional life. But though we cannot cut at the root of

---

[7]With regard to the employment of women as compositors, an article by Amy Linnett, in the *Economic Review* for January, 1892, should be referred to.

[8]Looked at from the point of view of the whole community, and not merely from that of one sex, it would, of course, be a matter for further consideration whether, and in what directions, it is socially desirable that men should be replaced by women as industrial operatives. Throughout this paper I have abstained from discussing this consideration.

this, by insisting, as I once heard it gravely suggested, on 'three generations of unmarried women,' we can do a great deal to encourage the growth of professional spirit and professional capacity among women workers, if we take care to develop our industrial organization along the proper lines. The first necessity is the exclusion of illegitimate competitors. The real enemies of the working woman are not the men, who always insist on higher wages, but the 'amateurs' of her own sex. So long as there are women, married or unmarried, eager and able to take work home, and do it in the internals of another profession, domestic service, we shall never disentangle ourselves from that vicious circle in which low wages lead to bad work, and bad work compels low wages. The one practical remedy for this disastrous competition is the extension of Factory legislation, with its strict limitation of women's hours, to all manufacturing work wherever carried on.[9] It is no mere coincidence that the only great industry in which women get the same wages as men Lancashire cotton weaving—is the one in which precise legal regulation of women's hours has involved the absolute exclusion of the casual amateur. No woman will be taken on at a cotton mill unless she is prepared to work the full factory hours, to come regularly every day, and put her whole energy into her task.... If we want to bring the women wage-earners all over England up the level of the Lancashire cotton weavers, we must subject them to the same conditions of exclusively professional work.

There is another way in which the extension of the Factory Acts to the unregulated trades is certain to advance women's industrial position....

If there is one result more clearly proved by experience than another, it is that the legal fixing of definite hours of labor, the requirement of a high standard of sanitation, and the prohibition of overtime, all favor production on a large scale. It has been the employers' constant complaint against the Factory Acts that they

inevitably tend to squeeze out the 'little master.' The evidence taken by the House of Lords' Committee on Sweating conclusively proved that any effective application of factory regulations to the workplaces of East London and the Black Country would quickly lead to the substitution of large factories.... Those well-meaning ladies who, by resisting the extension of Factory legislation, are keeping alive the domestic workshop and the sweaters' den, are thus positively curtailing the sphere of women's employment. The 'freedom' of the poor widow to work, in her own bedroom, 'all the hours that God made'; and the wife's privilege to supplement a drunken husband's wages by doing work at her own fireside, are, in sober truth, being purchased at the price of the exclusion from regular factory employment of thousands of 'independent women.'

We can now sum up the whole argument. The case for Factory legislation does not rest on harrowing tales of exceptional tyranny, though plenty of these can be furnished in support of it. It is based on the broad facts of the capitalist system, and the inevitable results of the Industrial Revolution.[10] A whole century of experience proves that where the conditions of the wage-earner's life are left to be settled by 'free competition' and individual bargaining between master and man, the worker's 'freedom' is delusive....

All these general considerations apply more forcibly to women wage-earners than to men. Women are far more helpless in the labor market, and much less able to enforce their own common rule by Trade Unionism. The only chance of getting Trade Unions among women workers lies through the Factory Acts. We have before us nearly forty years' actual experience of the precise limitation of hours and the absolute prohibition of overtime for women workers in the cotton manufacture; and they teach us nothing that justifies us in refusing to extend the like protection to the women slaving for

---

[9]See Fabian Tract, No. 50, *Sweating: its Cause and Remedy.*

[10]See Fabian Tract, No. 23, *The Case for an Eight Hours Bill.*

irregular and excessive hours in laundries, dressmakers' workrooms, and all the thousand and one trades in which women's hours of work are practically unlimited....

Finally, we have seen that the fear of women's exclusion from industrial employment is wholly unfounded.... The real enemy of the woman worker is not the skilled male operative, but the unskilled and half-hearted female 'amateur' who simultaneously blacklegs both the workshop and the home. The legal regulation of women's labor is required to protect the independent professional woman worker against these enemies of her own sex. Without this regulation it is futile to talk to her of the equality of men and women. With this regulation, experience teaches us that women can work their way in certain occupations to a man's skill, a man's wages, and a man's sense of personal dignity and independence.

QUESTIONS FOR ANALYSIS

1. With reference to Clara Zetkin's article and accompanying tables, discuss the role of women workers in German industry. To what factors does Zetkin attribute women's low wages? What were the economic and political consequences of these low wages? What is Zetkin's critique of the trade union movement? What solutions does she propose for improving the conditions faced by women workers?

2. Discuss the premises concerning the nature of women's work under capitalism that inform Louise Kautsky's statement and resolution on protective legislation. Why would the Congress have refused to adopt article seven of the resolution?

3. How does Beatrice Webb characterize the opponents of protective legislation? How does her definition of individual freedom and liberty differ from theirs? Discuss the distinction that she draws between collective bargaining and protective legislation and the implications of this distinction for women workers. How does she address the concern that protective legislation will disadvantage women in relation to male workers and exclude them from industry?

4. Compare the perspectives on women's work expressed in these writings by Clara Zetkin, Louise Kautsky, and Beatrice Webb. What do they reveal about certain patterns that characterized European industrialization, trade unionism, and the organization of women workers? How does the nineteenth-century debate on protective legislation compare with contemporary debates on this issue?

SUFFRAGE MOVEMENTS ON THE EVE OF WORLD WAR I

Prior to World War I, women could vote in only two European countries. Women first obtained the vote in Finland in 1906 and then in Norway, partially in 1907 and fully in 1913. The failure of previous movements compelled the formation of new suffrage organizations in England and France such as the Women's Social and Political Union (1903) and the French Union for Women's Suffrage (1909). Socialist women also raised the issue of suffrage on an international level. At the second International Socialist Women's Conference held in Copenhagen in 1910, Clara Zetkin introduced a resolution for an annual celebration of International Women's Day on March 8 to demand universal women's suffrage without property restrictions that would exclude working women. This day was chosen to coincide with a demonstration called in New York City on March 8, 1907, by women workers in the needle trades to demand the vote and the formation of a union. The conference accepted Zetkin's proposal and the first International Women's Day was commemorated in 1911.

128. The French Union for Women's Suffrage "The Question of the Vote for Women"

The Union Française pour le suffrage des femmes (French Union for Women's Suffrage) was formed in 1909 by Jeanne Schmahl (1846–1915). She was the leader of L'Avant Courrière (The Forerunner) which focused on women's economic and legal issues. Departing from previous feminist strategies, the organization attacked the Napoleonic Code[1] in a piecemeal fashion rather than in its entirety. The Forerunner achieved the passage of the first married women's property act in France, which granted women the right to their own wages, and it also won legal rights for women in court cases, including their right to sit on a jury and to testify in court. Schmahl then took up the cause of women's suffrage. The approach of the French Union for Women's Suffrage is illustrated in this excerpt from "The Question of the Vote for Women," a report it presented to the Besançon Municipal Council in 1913. Following the outbreak of World War I, the organization joined with other feminist groups to issue a "Manifesto to the Women of the Neutral and Allied Countries" which declared the commitment of French women to endure the sacrifices required to see the war to its conclusion. Women in France were not granted the right to vote until 1944 when the Consultative Assembly of the French Committee of National Liberation, held in Algiers, voted in favor of women's suffrage. Women first voted in municipal elections in 1945 and then in 1946, in the first post-World War II national elections.

Some Arguments in Favor of the Vote for Women

We are going to try to prove that the vote for women is a just, possible and desirable reform.

It is just that a woman vote.

A woman is subject to the law, pays direct and indirect taxes just as a man does; in a country of universal suffrage, laws ought to be established by all the taxpayers, men and women.

A woman possesses her own property; she can inherit, make a will; she has an interest in having her say in the laws relating to property.

A woman has responsibility in the family; she ought to be consulted about the laws establishing her rights and duties with respect to her husband, her children, her parents.

Women work—and in ever greater numbers; a statistic of 1896 established that at that date 6,400,000 French women were gainfully employed, that the proportion of female workers was 42 per cent of the women over thirteen

---

[1] On the Napoleonic Code, see Source 98 in Chapter 8.

The French Union for Women's Suffrage, France, 1913.

years of age, and that the number of women workers was 35 per cent of the total number of workers, both male and female.

If she is in business, she, like any businessman, has interests to protect; it would be unjust for her not to be represented in the Chambers of Commerce, and in the regulatory bodies and courts dealing with commercial matters. Many questions pertaining to business can be decided only by the Municipal Councils and regulatory bodies.

If a woman is a worker or a domestic, she ought to participate as a man does in voting on unionization laws, laws covering workers' retirement, social security, the limitation and regulation of work hours, weekly days off, labor contracts, etc.

If she is a civil servant (postal employee, schoolteacher, professor), should she not have the right to give her opinions on the questions of her salary, of her service, of vacations, of the special rules to which she is subject? If she is a doctor, lawyer, writer, artist, she must fight to make her rights recognized. And the others, who do not yet work but who will have to work in view of the increasing costs of the necessities of life, will also have to fight to assure that new careers will be open to them as well as to men.

Certainly many beneficial reforms have been made on behalf of women, in the name of justice, by a legislature composed of men. But in order for them to correspond to the real rights of women it is necessary that the latter participate in their establishment.

A woman is from this day on capable of voting.

Her education has improved considerably; the elementary school curriculum for boys and girls is the same, and in coeducational schools girls profit from the instruction at least as much as boys; higher education is available to young women, secondary education is as serious in girls' secondary schools as it is in boys' secondary schools.

Woman's importance in the family is greater and greater, her moral authority and economic power are increasing; new legislation on marriage and divorce and on paternity suits tend to make her independent and allow her to develop her personality.

Finally, her special characteristics of order, economy, patience and resourcefulness will be as useful to society as the characteristics of man and will favor the establishment of laws too often overlooked until now.

The women's vote will assure the establishment of important social laws.

All women will want:

To fight against alcoholism, from which they suffer much more than men;

To establish laws of health and welfare;

To obtain the regulation of female and child labor;

To defend young women against prostitution;

Finally, to prevent wars and to submit conflicts among nations to courts of arbitration.

We will see, by studying what has been accomplished by the women's vote in countries where women have obtained it, that it is legitimate to expect that all these urgent reforms will be realized in France, too, when French women vote.

129. Emmeline Pankhurst
"Why We Are Militant"

Emmeline Pankhurst (1858–1928) attended her first suffrage rally in her native city of Manchester at the age of fourteen. In *My Own Story* (1914), Pankhurst relates that she left the rally with her mother as a "conscious and confirmed suffragist," but it was only later experiences as a Poor Law Guardian and Manchester School Board member that convinced her of the vital necessity of women's vote for improving their status in society. In 1903, Emmeline established the Women's Social and Political Union (WSPU) with her daughters, Christabel and Sylvia,[1] and several women from the Independent Labor Party to which all three Pankhursts belonged. Recruiting from Manchester factories, the WSPU focused on suffrage as a single issue and the key to social reform and pledged its members not to work within existing political parties until the vote was won. The WSPU achieved national prominence in 1905 when, on the eve of the general elections, Christabel and Annie Kenney, a textile worker, disrupted a Manchester political rally of Liberal Party member Sir Edward Grey to raise the issue of votes for women. They were thrown out of the meeting, arrested on charges of obstruction, and served time in prison after refusing to pay their fines. Shifting its headquarters to London in 1906, the WSPU began publication of a new, monthly newspaper, *Votes for Women,* opened branches throughout England, and became a national organization. When the victorious Liberal Party refused to consider female suffrage, the WSPU stepped up its militant campaign with huge pageants and processions, disruptions of political meetings, heckling of Cabinet ministers, confrontations with parliamentary members and police, the convening of "Women's Parliaments," window smashings, and arson attacks on property. In prison, WSPU members, including all three Pankhursts, engaged in hunger strikes, endured forced feedings, and insisted on being treated as political offenders. Prison treatment ruined the health of several WSPU members and even caused the death of some. With the outbreak of World War I, the WSPU suspended its suffrage campaign and committed its members to the war effort which Emmeline and Christabel enthusiastically supported. All suffragists in prison were released in August 1914 in exchange for terminating their suffrage activities. Only in 1918, at the end of the war, did women over thirty obtain the right to vote in England and Ireland; and not until 1928, the year of Pankhurst's death, were all adult women enfranchised. "Why We Are Militant" is a speech she gave in New York City in 1913.

I know that in your minds there are questions like these; you are saying, 'Woman Suffrage is sure to come; the emancipation of humanity is an evolutionary process, and how is it that some women, instead of trusting to that evolution, instead of educating the masses of people of their country, instead of educating their own sex to prepare them for citizenship, how is it that these militant women are using violence and upsetting the business arrangements of the country in their undue impatience to attain their end?'

---

[1] On Sylvia Pankhurst, see Source 133 in Chapter 11.

A speech delivered in New York on October 21, 1913.

Let me try to explain to you the situation.

Although we have a so-called democracy, and so called representative government there, England is the most conservative country on earth....

...Nothing ever has been got out of the British Parliament without something very nearly approaching a revolution. You need something dynamic in order to force legislation through the House of Commons; in fact, the whole machinery of government in England may almost be said to be an elaborate arrangement for not doing anything.

The extensions of the franchise to the men of my country have been preceded by very great violence, by something like a revolution, by something like civil war. In 1832, you know we were on the edge of a civil war and on the edge of revolution, and it was at the point of the sword—no, not at the point of the sword— it was after the practice of arson on so large a scale that half the city of Bristol was burned down in a single night, it was because more and greater violence and arson were feared that the Reform Bill of 1832 was allowed to pass into law.[2] In 1867, John Bright urged the people of London to crowd the approaches to the Houses of Parliament in order to show their determination, and he said that if they did that no Parliament, however obdurate, could resist their just demands. Rioting went on all over the country, and as the result of that rioting, as the result of that unrest, which resulted in the pulling down of the Hyde Park railings, as a result of the fear of more rioting and violence the Reform Act of 1867 was put upon the statute books.

In 1884 came the turn of the agricultural labourer. Joseph Chamberlain, who afterwards became a very conservative person, threatened that, unless the vote was given to the agricultural labourer, he would march 100,000 men from Birmingham to know the reason why. Rioting was threatened and feared, and so the agricultural labourers got the vote.

Meanwhile, during the '80's, women, like men, were asking for the franchise. Appeals, larger and more numerous than for any other reform, were presented in support of Woman's Suffrage. Meetings of the great corporations, great town councils, and city councils, passed resolutions asking that women should have the vote. More meetings were held, and larger, for Woman Suffrage than were held for votes for men, and yet the women did not get it. Men got the vote because they were and would be violent. The women did not get it because they were constitutional and law-abiding. Why, is it not evident to everyone that people who are patient where mis-government is concerned may go on being patient! Why should anyone trouble to help them? I take to myself some shame that through all those years, at any rate from the early '80's, when I first came into the Suffrage movement, I did not learn my political lessons.

I believed, as many women still in England believe, that women could get their way in some mysterious manner, by purely peaceful methods. We have been so accustomed, we women, to accept one standard for men and another standard for women, that we have even applied that variation of standard to the injury of our political welfare....

Well, we in Great Britain, on the eve of the General Election of 1905, a mere handful of us—why, you could almost count us on the fingers of both hands—set out on the wonderful adventure of forcing the strongest Government of modern times to give the women the vote. Only a few in number; we were not strong in influence, and we had hardly any money, and yet we quite gaily made our little banners with the words 'Votes for Women' upon them, and we set out to win the enfranchisement of the women of our country.

---

[2]For more information on this bill, see Source 101 in Chapter 8, "Chartist Women Demand Equal Political Rights."

The Suffrage movement was almost dead. The women had lost heart. You could not get a Suffrage meeting that was attended by members of the general public. We used to have about 24 adherents in the front row. We carried our resolutions and heard no more about them.

Two women changed that in a twinkling of an eye at a great Liberal demonstration in Manchester, where a Liberal leader, Sir Edward Grey, was explaining the programme to be carried out during the Liberals' next turn of office. The two women put the fateful question, 'When are you going to give votes to women?' and refused to sit down until they had been answered. These two women were sent to gaol, and from that day to this the women's movement, both militant and constitutional, has never looked back. We had little more than one moribund society for Woman Suffrage in those days. Now we have nearly 50 societies for Woman Suffrage, and they are large in membership, they are rich in money, and their ranks are swelling every day that passes. That is how militancy has put back the clock of Woman Suffrage in Great Britain....

I want to say here and now that the only justification for violence, the only justification for damage to property, the only justification for risk to the comfort of other human beings is the fact that you have tried all other available means and have failed to secure justice, and as a law-abiding person—and I am by nature a law-abiding person, as one hating violence, hating disorder—I want to say that from the moment we began our militant agitation to this day I have felt absolutely guiltless in this matter.

I tell you that in Great Britain there is no other way....

...The women who are waging this war are women who would fight, if it were only for the idea of liberty—if it were only that they might be free citizens of a free country—I myself would fight for that idea alone. But we have, in addition to this love of freedom, intolerable grievances to redress.

We do not feel the weight of those grievances in our own persons. I think it is very true that people who are crushed by personal wrongs are not the right people to fight for reform. The people who can fight best are the people who have happy lives themselves, the fortunate ones. At any rate, in our revolution it is the happy women, the fortunate women, the women who have drawn prizes in the lucky bag of life, in the shape of good fathers, good husbands and good brothers, they are the women who are fighting this battle. They are fighting it for the sake of others more helpless than themselves, and it is of the grievances of those helpless ones that I want to say a few words to-night to make you understand the meaning of our militant campaign.

Those grievances are so pressing that, so far from it being a duty to be patient and to wait for evolution, in thinking of those grievances the idea of patience is intolerable. We feel that patience is something akin to crime when our patience involves continued suffering on the part of the oppressed....

Our marriage and divorce laws are a disgrace to civilization. I sometimes wonder, looking back from the serenity of past middle age, at the courage of women. I wonder that women have the courage to take upon themselves the responsibilities of marriage and motherhood when I see how little protection the law of my country affords them. I wonder that a woman will face the ordeal of childbirth with the knowledge that after she has risked her life to bring a child into the world she has absolutely no parental rights over the future of that child. Think what trust women have in men when a woman will marry a man, knowing, if she has knowledge of the law, that if that man is not all she in her love for him thinks him, he may even bring a strange woman into the house, bring his mistress into the house to live with her, and she cannot get legal relief from such a marriage as that.

How often is women's trust misplaced, and yet how whole-hearted and how touching that trust must be when a woman, in order to get love and companionship will run such terrible risks in entering into marriage! Yet women have done it, and as we get to know more of life we militant Suffragists have nerved ourselves and forced ourselves to learn something of how other people live. As we get that knowledge we realise how political power, how political influence, which would enable us to get better laws, would make it possible for thousands upon thousands of unhappy women to live happier lives.

Well, you may say, the laws may be inadequate, the laws may be bad, but human nature, after all, is not much influenced by laws, and upon the whole, people are fairly happy. Now, for those who are fortunate it is very comfortable to have that idea, but if you will really look at life as we see it in our centralised civilisation in Europe, you will find that after all the law is a great educator, and if men are brought up to think the law allows them to behave badly to those who should be nearest and dearest to them, the worst kind of man is very apt to take full advantage of all the laxity of the law.

What have we been hearing of so much during the last few years! It is a very remarkable thing, ladies, and gentlemen, that along with this woman's movement, along with this woman's revolt, you are having a great uncovering of social sores. We are having light let into dark places, whether it is in the United States or whether it is in the old countries of Europe, you find the social ills from which humanity suffers, are very much the same. Every civilised country has been discussing how to deal with that most awful slavery, the white slave traffic.

...We have now a White Slave Act, and in that Act of Parliament they have put a flogging clause. Certain men are to be flogged if they are convicted and found guilty under that Act of Parliament, and the British House of Commons, composed of men of varying moral standard, waxed highly eloquent on the need of flogging these tigers of the human race, men engaged in the white slave traffic.

Well, we women looked on and we read their speeches, but in our hearts we said, 'Why don't they decide to go to the people for whom the white slave traffic exists? What is the use of dealing with the emissaries, with the slave hunters, with the purveyors? Why don't they go to the very foundation of the evil; why don't they attack the customers? If there was no demand there would be no traffic, because business does not exist if there is no demand for it?' And so we women said, 'It's no use, gentlemen, trying to put us off with sentimental legislation on the white slave traffic. We don't trust you to settle it; we want to have a hand in settling it ourselves, because we think we know how.' And we have a right to distrust that legislation. They passed the Act very, very quickly; they put it on the Statute Book, and we have seen it in operation, and we know that the time of Parliament and the time of the nation was wasted on a piece of legislation which I fear was never intended to be taken very seriously; something to keep the women quiet, something to lull us into a sense of security, something to make us believe that now, at least, the Government were really grappling with the situation.

And so we attacked this great evil. We said, 'How can we expect real legislation to deal with the white slave traffic on a small scale when the Government of the country is the biggest white slave trading firm that we have got?

And it is true, because you know, although we have suppressed such regulation of vice in England, we have got it in full swing in the great dependencies that we own all over the world, and we have only to turn to India and look to other places where our Army is stationed to find the Government, which is in no way responsible to women, actually taking part in that awful trade, in absolute cold bloodedness where native women are concerned, all, forsooth, in the name of the health of the men of our forces.

Well, we have been speaking out, ladies and gentlemen; we have been saying to our nation and the rulers of our nation, 'We will not have the health of one-half of the community, their pretended health, maintained at the expense of the degradation and sorrow and misery of the other half.'

I want to ask you whether, in all the revolutions of the past, in your own revolt against British rule, you had deeper or greater reasons for revolt than women have to-day?

Take the industrial side of the question: have men's wages for a hard day's work ever been so low and inadequate as are women's wages to-day? Have men ever had to suffer from the laws, more injustice than women suffer? Is there a single reason which men have had for demanding liberty that does not also apply to women?

Why, if you were talking to the *men* of any other nation you would not hesitate to reply in the affirmative. There is not a man in this meeting who has not felt sympathy with the uprising of the men of other lands when suffering from intolerable tyranny, when deprived of all representative rights. You are full of sympathy with men in Russia. You are full of sympathy with nations that rise against the domination of the Turk. You are full of sympathy with all struggling people striving for independence. How is it, then, that some of you have nothing but ridicule and contempt and reprobation for women who are fighting for exactly the same thing? . . .

Our hearts burn within us when we read the great mottoes which celebrate the liberty of your country; when we go to France and we read the words, liberty, fraternity and equality, don't you think that we appreciate the meaning of those words? And then when we wake to the knowledge that these things are not for us, they are only for our brothers, then there comes a sense of bitterness into the hearts of some women, and they say to themselves, 'Will men never understand?' But so far as we in England are concerned, we have come to the conclusion that we are not going to leave men any illusions upon the question.

When we were patient, when we believed in argument and persuasion, they said, 'You don't really want it because, if you did, you would do something unmistakable to show you were de-termined to have it.' And then when we did something unmistakable they said, 'You are behaving so badly that you show you are not fit for it.' . . .

Well, in Great Britain, we have tried persuasion, we have tried the plan of showing (by going upon public bodies, where they allowed us to do work they hadn't much time to do themselves) that we are capable people. We did it in the hope that we should convince them and persuade them to do the right and proper thing. But we had all our labour for our pains, and now we are fighting for our rights, and we are growing stronger and better women in the process. We are getting more fit to use our rights because we have such difficulty in getting them. . . .

We know the joy of battle. When we have come out of the gates of Holloway at the point of death, battered, starved, forcibly fed as some of our women have been—their mouths forced open by iron gags—their bodies bruised, they have felt when the prison bars were broken and the doors have opened, even at the point of death, they have felt the joy of battle and the exultation of victory.

People have said that women could never vote, never share in the government, because government rests upon force. We have proved that is not true. Government rests not upon force; government rests upon the consent of the governed; and the weakest woman, the very poorest woman, if she withholds her consent cannot be governed.

They sent me to prison, to penal servitude for three years. I came out of prison at the end of nine days. I broke my prison bars. Four times they took me back again; four times I burst the prison door open again. And I left England openly to come and visit America, with only three or four weeks of the three years' sentence of penal servitude served. Have we not proved, then, that they cannot govern human beings who withhold their consent?

And so we are glad we have had the fighting experience, and we are glad to do all the fighting for all the women all over the world. All that we ask of you is to back us up. We ask you to show that although, perhaps, you may not mean to fight as we do, yet you understand the meaning of our fight; that you realise we are women fighting for a great idea; that we wish the betterment of the human race, and that we believe this betterment is coming through the emancipation and uplifting of women.

130. British Suffrage Art

The suffrage movement provided new opportunities for female artists who experienced hostility from the established Royal Academy as well as avant-garde modernists. Two new organizations of women artists were formed in the years immediately preceding World War I, the Artists Suffrage League (1907) and the Suffrage Atelier (1909). Art was commissioned by all three main suffrage organization at this time: the National Union of Women's Suffrage Societies, the Women's Social and Political Union (WSPU), and the Women's Freedom League, an organization that broke away from the WSPU. Suffrage artists transformed mainstream images of Edwardian culture into bearers of profeminist messages. They designed and produced graphics, posters, banners, and costumes for pageants, such as the Pageant of Women's Trades and Professions held in April 1909. Suffrage artists drew on artistic models provided by the Arts and Crafts movement as well as the new genres of magazine covers, which often resembled posters, and advertising.

STUDIES IN NATURAL HISTORY.
The Antysuffragyst or Prejudicidon.

The Antysuffragyst or Prejudicidon. This curious animal has the smallest brain capacity of any living creature. Its sight is so imperfect that it cannot see further than the end of its nose; but it has a wonderful capacity for discovering the stupefying plant called "Humbugwort," on which it feeds voraciously. It is closely allied to the Lunaticodon, and it is a fierce enemy of the Justiceidon.

A. "The Antysuffragyst or Prejudicidon" appeared in the September 26, 1913, issue of the *Vote,* the newspaper of the Women's Freedom League. (*"The Antysuffragyst" cartoon from* Vote, *September 26, 1913.*)

B. "The Cat and Mouse Act" In 1913, the Liberal Party passed the Prisoners' Temporary Discharge for Ill-Health Act for which Home Secretary Reginald McKenna became responsible. The law, which came to be known as the Cat and Mouse Act, stipulated that hunger strikers should be released when their lives were endangered, but reimprisoned immediately after recovery. Hunger strikers were force-fed in prison, with a tube placed in their nose, which often caused its own health problems. Over a thousand suffragettes endured forcible feeding. This poster, "The Cat and Mouse Act," was produced in 1914 for the Women's Social and Political Union. (*Museum of London © 1913.*)

QUESTIONS FOR ANALYSIS

1. How does the resolution of the French Union for Women's Suffrage argue for women's right to vote? How are the issues of female suffrage and social legislation linked?
2. What reasons does Emmeline Pankhurst give to justify the militant tactics of the Women's Social and Political Union? What lessons does she draw from the means by which men obtained suffrage? Why does she believe that it is essential for women to obtain the vote? How do her views in this regard compare with those of the French Union for Women's Suffrage? What is the purpose of Pankhurst's speech? Compare Pankhurst's arguments for women's suffrage with those presented by Barbara Leigh-Smith Bodichon, Source 107 in Chapter 9.
3. Discuss the image and text of the "Studies in Natural History" drawing. To what scientific theory does it allude? Discuss the imagery and symbolism of the "The Cat and Mouse Act" poster. Why did the Women's Social and Political Union commission this poster? Evaluate the effectiveness of these two different forms of political art. Would they appeal to the same or different audiences?

CHAPTER 11

World War I and the Russian Revolution

The threat of war from colonial rivalries, the Balkan Wars of 1912 and 1913, and the increasing militarism of the European powers inspired the international peace movements of the late nineteenth and early twentieth centuries. Proponents of peace ranged from pacifists who opposed the use of force on principle to socialists who condemned imperialism as the main source of war. When the war did break out in 1914, however, the socialist parties failed to fulfill their prewar pledges not to wage war against workers in other countries. The Second International dissolved as a consequence. Only a minority of socialists adhered to internationalism and to the Leninist call to "turn the imperialist war into a civil war" against the bourgeoisie of each country. The majority of women, like the population as a whole, participated in the war enthusiasm that characterized the initial phase of World War I. But women who opposed the war continued to organize for peace across national lines, as in the peace conference of socialist women organized by Clara Zetkin in March 1915 in Switzerland and the International Women's Congress convened one month later in neutral Holland. Contrary to popular expectations on both sides, a quick victory proved elusive. The absurdity and horrors of the war reached their height in the trench warfare waged between French and German soldiers on the Western Front. Here, the war dragged on with enormous casualties but few military gains for either side. Demands for peace gained momentum as conditions deteriorated on the battlefront and homefront, especially in Russia and Germany where food shortages provoked food riots and massive protests led by women.

The war created the conditions for the Easter Rising of 1916 in Ireland and two revolutions, the Bolshevik Revolution of October 1917 and the German Revolution of November 1918. These revolutions, in turn, proved decisive in ending World War I. In Russia, Tsar Nicholas II abdicated in the face of the deteriorating Russian war effort. While the Provisional Government formed in March 1917 pledged to continue the war until victory, the Bolsheviks demanded an end to Russian participation in what they denounced as an imperialist war. The Bolshevik slogan "Peace, Land, and Bread" gained increasing mass support in the face of the hopelessness of the Russian war effort, spontaneous land seizures by deserting soldiers, and food shortages experienced at home. In November 1917, the Bolsheviks seized the Winter Palace and began the process of suing for peace. In Germany, the mutiny of the Kiel sailors in October 1918 precipitated the November 1918 revolution which resulted in the removal of the kaiser from power and the declaration of the Weimar Republic. In Germany, in contrast to Russia, the left-wing forces were not strong enough to carry out a socialist revolution and the republic remained bourgeois and capitalist throughout its short-lived existence from 1918 to January 1933 when Hitler became chancellor of Germany.

WORLD WAR I AND THE HOMEFRONT

In the first months of the war, thousands of women became unemployed as industries that were not related to the war effort sharply reduced production. Soon, however, the war created new work opportunities for middle-class as well as working-class women. In the rapidly expanding munitions industry, women

"These women are doing their bit. Learn to make munitions." Great Britain, c. 1917. Septimus E. Scott. (*Hoover Institution on War, Revolution & Peace.*)

found steady work but often experienced poor and even dangerous conditions. Women also replaced men at the front in heavy industry, public transportation, and agriculture. Various forms of volunteer activities, such as nursing and the provision of assistance to refugees, also engaged women, especially those of the middle class. Although there were certain common elements, the homefront was not the same in each country, but varied in accordance with the existing forms of political rule. In Ireland, often described as the first and last colony of England, Irish republicans took advantage of England's vulnerability during the war to launch the Easter Rising of 1916 for Irish independence. In Germany and Russia, women led food protests and street demonstrations which undermined political support for monarchical rule and played a crucial role in the revolutions of 1917 and 1918. Throughout Europe, however, the horrific losses of soldier's lives inspired peace initiatives and vows to prevent such a war from ever occurring again.

131. Vera Brittain
Diary Entries and Poems on the War

Vera Brittain (1893–1970) was a feminist, pacifist, and international peace activist. Best known for her *Testament of Youth* (1933), she was a prolific author of journalistic articles, novels, memoirs, and diaries. Initially enthusiastic about Britain's entry into the war, Brittain changed her views as the war's horrors and futility became ever more apparent, the effects of which she directly witnessed when she temporarily left Oxford in 1915 to serve as a nurse in the Voluntary Aid Detachment (V.A.D.). Brittain also experienced deep personal losses, as her fiancé Roland Leighton, brother Edward, and two close friends were all killed in the war. She first attempted to publish her diary of World War I in 1922, in response to a publishing competition, but it was rejected and only seven years later did Brittain begin to write *Testament of Youth*. Although its discussion of World War I draws extensively on her diary, she continued to believe that it should appear separately. The diary was finally published in 1939 on the eve of World War II. Explaining the crucial difference between the diary and *Testament of Youth*, Brittain noted: "But diaries give one effect which *Testament of Youth*, because of its very perspective & the long period it covered, could only partially convey—the cumulative effect of day-by-day suspense & anguish, the unillumined length of the night which had done so much permanent quenching of youthful hope & energy before it was finished."[1] Brittain wrote these diary entries during a three-week period from July 29, 1914, when Austria declared war on Serbia, to August 25. The second selection, "The German Ward," is from the fall of 1917 when she was treating German prisoners of war in France. It is one of several poems Brittain published in 1918 in *Verses of a V.A.D. and Other War Poems*.

---

[1]See Brittain's notes for "Introduction to War Diaries," written around 1939, which are included in the editor's introduction to *War Diary, 1913–1917: Chronicle of Youth*, (London, Gollancz 1981), p. 15. Her diary of World War II was published posthumously as *Wartime Chronicle: Diary 1939–1945* (London, Gollancz 1989).

From *Vera Brittain: War Diary, 1913–1917: Chronicle of Youth*, edited by Alan Bishop with Terry Smart. © 1981 by Victor Gollancz, Ltd.

A. War Diaries

Wednesday July 29th

War was declared to-day between Austria and Servia. Of course it is feared that the whole of Europe will be involved. This critical state seems to have arisen so suddenly out of the long-time hostile attitude of Teuton & Slav, represented in the feeling of Servia towards Austria-Hungary, that it is difficult to follow each step. But Germany & Italy in case of Russia's intervention to protect Servia are bound to assist Austria by the Triple Alliance, while France is allied to Russia & we are connected with both by an Entente Cordiale, though without a definite alliance. The Irish question seems quite to have sunk into the background beside these momentous issues.

Saturday August 1st

According to the paper this morning, the last hope of peace is about to be abandoned, and Germany is mobilising....

Sunday August 2nd

Mother tried to get a paper when she went down to the Baths, but they were all sold up; however the Ellingers lent us theirs. But it was not so much what was in the papers that caused excitement as the rumours that were spreading about all day. It is said the Germans have declared war against Russia & that also they have attacked Luxembourg. The situation is very grave indeed.

Monday August 3rd

To-day has been far too exciting to enable me to feel at all like sleep—in fact it is one of the most thrilling I have ever lived through, though without doubt there are many more to come. That which has been so long anticipated by some & scoffed at by others has come to pass at last—Armageddon in Europe! On Saturday evening Germany declared war upon Russia & also started advancing towards the French frontier. The French, in order to make it evident that they were not the aggressors, wasted some hours & then the order to mobilise was given. Great excitement in France continued throughout the night & yesterday the Germans attacked France without declaring war. Unconfirmed rumour says that in one place they have been repulsed with heavy losses. They also broke a treaty in occupying the neutral Duchy of Luxembourg. Luxembourg's neutrality was guaranteed in 1807 by England, France & Germany, & thus Germany's attack upon it is said to be a direct challenge to Great Britain. Some of the papers seem to think that the Austrian-Servian war was only a blind & that Germany was at the bottom of the whole affair—the "mailed fist" anxious to strike. At any rate Germany has destroyed the tottering hopes of peace and has plunged Europe into a situation the like of which, *The Times* says, has never been known since the fall of the Roman Empire. The great fear now is that our bungling Government will declare England's neutrality. If we at this critical juncture were to refuse to help our friend France, we should be guilty of the grossest treachery & sacrifice our credit for ever. Besides we should gain nothing, for if we were to stand aside & let France be wiped out, a terrible retribution would fall upon us from a strengthened & victorious Germany.

I sat this morning after breakfast reading various newspapers for about two hours. A rumour is going round to-night that England has declared to Germany that if a German sets foot in Belgian territory her (England's) navy will immediately act. There are many who think that this policy of vacillation is losing us the opportunity to strike a telling blow—that we should send troops to prevent the Germans getting into Belgium instead of waiting till they *are* in.

I should think this must be the blackest Bank Holiday within memory. Pandemonium reigned in the town. What with holiday-trippers, people struggling for papers, trying to lay in stores of food & dismayed that the price of everything had gone up, there was confusion everywhere. Mother met Mrs. Whitehead in the town; she is in great anxiety because she has one son in Russia, one—Jack—in Servia, and another on his way from India. Marjorie Briggs, who was to have been married on Saturday, was married in a hurry on Friday as her husband had to have joined his regiment on Saturday. The papers are full of stories of tourists in hopeless plights trying to get back to England. Paper money is useless & the majority of the trains are cut off. It is rumoured that there is fear in Paris that a fleet of German Zeppelins are going to destroy Paris from above in the night.

Tuesday August 4th

Late as it is & almost too excited to write as I am, I must make some effort to chronicle the stupendous events of this remarkable day. The situation is absolutely unparalleled in the history of the world. Never before has the war strength of each individual nation been of such great extent, even though all the nations of Europe, the dominant continent, have been armed before. It is estimated that when the war begins *14 millions* of men will be engaged in the conflict. Attack is possible by earth, water & air, & the destruction attainable by the modern war machines used by the armies is unthinkable & past imagination.

This morning at breakfast we learnt that war is formally declared between France & Germany, that the German ambassador has left Paris & the French ambassador Berlin. Germany has declared to Belgium that if her troops are allowed to pass unmolested through Belgian territory she will protect her interests in the Treaty at the end of the war. Belgium has indignantly refused any such violation of international honour, and the King of the Belgians has appealed to King George for aid....

All day long rumours kept coming that a naval engagement had been fought off the coast of Yorkshire. I went up to the tennis club this afternoon, more to see if I could hear anything than to play, as it kept on pouring with rain. No one knew any further definite news, but we all discussed the situation. I mentioned Edward's & Maurice's keenness to do something

definite & Bertram Spafford said they ought either to apply to Mr Heathcote or Mr Goodman, who were the chief Territorials here, or to go to the Territorial headquarters in Manchester. I told him yesterday that the fact of a strong healthy man like himself being absolutely ignorant of military tactics was a proof that our military system was at fault somewhere. He said that at the Manchester Grammar School, where he went, they had no corps, & that many men were in the same case as himself.

The war will alter everything &, even if I pass my exam., there would probably be no means to send both Edward & me to Oxford at the same time. There is nothing to do now but wait. When I got in I found Edward had procured an evening paper with the startling news that England had sent an ultimatum to Germany, to expire at midnight to-night, demanding the immediate withdrawal of her troops from Belgium. Germany declared earlier in the day that if it became necessary to her tactics to treat Belgium as an enemy she would do so. German troops are said to have crossed the Belgian frontier & reached Verviers. Sir E. Grey's speech has caused great satisfaction in France.

Immediately after dinner I had to go to a meeting of the University Extension Lectures Committee. Small groups of people, especially men, were standing about talking, & in front of the Town Hall was quite a large crowd, as on the door was posted up the mobilisation order, in large black letters, ordering all army recruits to take up the colours & all Territorials to go to their headquarters. Edward has been reading the papers carefully & says that at present only the trained army & the Territorials are wanted & there is no demand for untrained volunteers. Though anxious to fight he says he will wait until he hears that people like himself are needed; he is of course very young & not overexperienced....

I could not rest indoors so got Mother & Daddy to come out with me to look for further news. In the town the groups of people had increased, and suppressed excitement was everywhere in the air. There was a crowd round the

Post Office; at first I thought they were attracted by the mobilisation order up in the window, but it turned out that Mr Heathcote & his motor car were there, with Mrs Heathcote inside driving. He was in his uniform, which looked as if he were going to-night, & was very busy sending telegrams off. We next went to the station & found there that a last edition extra of the *Chronicle* had been issued but all the copies were sold. However Smith the foreman, who told us his son had gone to the front, gave us his copy. It contained the thrilling news that Germany has formally declared war on Belgium! This looks like an answer to our ultimatum, & will perhaps free us from the necessity of waiting until midnight for our answer....

Mrs Kay told us that her son Tom, Mrs Johnson's chauffeur, has been told he is required as he belongs to the Army Medical Corps. His wife, who had a baby only last Friday, is terribly excited because he has to leave her, as she only gets about 2/- a day when he is gone, & has three tiny children on her hands. Mrs Kay says she will have to look after them. She thinks all her four sons who are Reservists will have to go abroad as they have all volunteered for foreign service. Luckily she takes it all calmly & philosophically, though she seems to think she will never see them again.

To sum up the situation in any way is impossible, every hour brings fresh & momentous events & one must stand still & await catastrophes each even more terrible than the last. All the nations of this continent are ready with their swords drawn, & Germany the aggressor with her weaker ally Austria stands alone facing an armed Europe united against her. She has broken treaty after treaty & disregarded every honourable tie with other nations. Italy, her old ally, has reaffirmed her neutrality, & thus assists our side by remaining out of the conflict. This conflict is a mortal struggle between herself & France; life to the one will mean death to the other. Indeed this war is a matter of life & death to us, & Daddy says the key to the whole situation is the British navy & that as that

stands or falls the fate of Europe will be decided.

Wednesday August 5th
All the news of last night was confirmed this morning, and it is further announced that the time limit given by Britain for an answer to her ultimatum expired without a reply coming from Germany, and that war between England & Germany is formally declared. Papers seemed to differ as to whether England must be said to have made war on Germany or Germany on England. Some say that the Germans have started hostilities against us by sinking a British mining ship and chasing a British cruiser. Thus, as the papers point out, Germany has declared war on four powers—Russia, France, Belgium and Britain, within 3 days. Nothing like it, they say, has been known since the time of Napoleon, and even Napoleon did not make war on his neighbours at so mad a rate....

The French have sunk a German cruiser, & have seized a dreadnought and another cruiser. Heavy firing has been heard from Margate, but nothing has been seen. Mrs Kay brought the pleasing information this afternoon that Mrs Johnson is going to keep the chauffeur son's place open for him & has promised to pay the 8/- a week rent for his wife & children. All decent firms are keeping open the Reservists' places for them, & in addition to this Harrods Stores have promised 10/- each family to the families of those who have been called out from assisting in the shop.

I showed Edward an appeal in *The Times* & the *Chronicle* for young unmarried men between the ages of 18 & 30 to join the army. He suddenly got very keen & after dinner he & Maurice wandered all round Buxton trying to find out what to do in order to volunteer for home service....

Thursday August 6th
To-day has principally been one of the weary waiting kind. Nothing very definite has happened.... England's policy & the papers are

naturally secretive about the position of her Fleet. The chief news is that the Germans have been repulsed with heavy losses while trying to storm the Belgian fortification of Liège. The Belgians are said to have behaved magnificently & while the defenders of the forts were engaged in keeping the invaders at bay, a Belgian brigade arrived & crowned the splendid efforts of the defenders with success.

To-day I started the only work it seems possible as yet for women to do—the making of garments for the soldiers. I started knitting sleeping-helmets, and as I have forgotten how to knit, & was never very brilliant when I knew, I seemed to be an object of some amusement. But even when one is not skillful it is better to proceed slowly than to do *nothing* to help.

Friday August 7th
This morning came the somewhat depressing news that the British cruiser H.M.S. *Amphion,* which sank the German liner *Königin Luise,* ran into a mine & was blown up. 131 men were drowned. The information cast a gloom over breakfast, during which meal Daddy worked himself into a thorough temper, raved away at us, & said he would not allow Edward to go abroad whatever happened—"Whatever you do, don't volunteer until you're *quite* sure there's no danger," sort of thing. Edward replied quite calmly that no one could prevent him serving his country in any way he wanted to.

The Belgian fortress of Liège is still holding out, though it is very hard pressed. 25,000 Belgians are holding it against a reinforcement of 100,000 Germans. The opposition is a serious hindrance to Germany, who reckoned on storming Liège with scarcely any trouble. All day long I knitted away. Various reports kept coming in of battles, different dreadnoughts being sunk, multitudes of Germans being killed, but none of them were confirmed.

Maurice & Edward wandered about all day waiting for an answer & at last they got it. Just at dinner-time Mrs Goodman came to see us &

said that she had heard from her husband—who is Territorial officer at Chesterfield—that there was no room for more recruits in the 6th Derby & Notts, Regt. but that the adjutant was delighted with their letter & had selected it from many hundred other applications & sent it to the War Office. Edward's & Maurice's qualifications are considered excellent & vastly superior to the majority of those who volunteer. Daddy was quite angry about the letter being sent to the War Office, but E. said that Daddy, not being a public school man or having had any training, could not possibly understand the impossibility of his remaining in inglorious safety while others, scarcely older than he, were offering their all. E. is of course rather young to volunteer really, being only eighteen. Maurice was nineteen to-day. E. faces the prospect of whatever he may have to do with perfect tranquillity, & says that even death can only come once. We spoke of the entire absence of future prospects which war seems to produce; E. said that but for this he would have been eagerly speculating about Oxford, but now he scarcely thought of it at all. Intellect, except in very high places, seems scarcely to count at all in time of war—the ordinary average soldier fights just as well for his ignorance as any cultured man for his knowledge. And then the value of human life becomes so cheap, so that while the loss of ten men under tragic circumstances amid ordinary conditions would fill the whole country with horror, the news of the loss of thousands is now regarded with a philosophical calm and an unmoved countenance. My beloved brother! What will become of him? But as I told him this evening, dreary as life is without his presence here, dreary as are the prospects of what may lie before him, yet I would not have his decision back, or keep him here.

Saturday August 8th
No news has arrived for Edward or Maurice from the War Office, but as it is so inundated with requests, doubtless we shall not hear for several days.

Liège has kept up its gallant defence & still holds out. A rumour that arrived last night is confirmed this morning, that is, 25,000 Germans have been killed before the Liège forts. The Germans asked for a truce so that they could bury their dead, but so far the truce has not been granted. I am incapable of feeling glad at such a wholesale slaughter of the Germans, whatever use it may be to us. I can only think of the 25,000 mothers who bore & reared those men with toil, & of the wives & families, never ardent for war or for a quarrel with us, which they leave behind them.

The two Germans cruisers which took refuge in Messina harbour have courageously left the neutral waters, & drew the British ships in hot pursuit. They prefer death to disarmament; it *is* a splendid instance of German patriotism.

This afternoon I went to the St. John's bandaging class. Of course I have never been to one before, never having taken a real interest, but I managed to take in quite a lot & learnt how to do 3 different kinds of bandages. With the greatest industry in the world however one cannot get a certificate & therefore cannot volunteer as a nurse for six weeks.

Edward received a letter from Roland saying that he had applied for a commission in a Norfolk regiment but as yet had heard nothing. He hopes very much to get it. I cannot but think it a terrible waste of good material that such an intellect as that should put itself readily in danger, but E. says Roland has an excellent military brain, & is quite worth a commission. R. says that several timid people in Lowestoft, because they are at the nearest point to Germany in England, fear every moment to be suddenly attacked by the enemy's fleet. A good deal of excitement exists there.

Monday August 10th
No special news has come through from the front to-day. Probably momentous happenings are going on, but the blinds are drawn down and we know nothing of them as yet. Certainly great bodies of French & German troops seem to be approaching one another round about Al-

sace Lorraine. The Japanese are stirring themselves & there are rumours of their joining the war on the side of England....

Tuesday August 11th
Very little news has come through to-day, probably owing to the censorship of the War news....

Friday August 14th
These days I seem to wake up weary; they are both long & full of work of a somewhat tiring nature. But one must not grumble at that; the great thing to be thankful for is the having something to do, for without that life would be unbearable. This morning as it happened the knitted helmets had to be given in to Mrs Heathcote, & we have no more materials in the house at present, so I had no sewing for the War to do at the time. I occupied myself in learning up parts of the First Aid book, and practising what bandages I could do single handed. This afternoon we had the bandaging class; it was very hot, but luckily not quite such a crowd of cackling women was present.

About lunch-time Maurice rang up to say he goes to Oxford tomorrow. He expects to have a month's drilling & then be given a commission & go to fight. When Edward & Daddy came in we told them about it. Edward said little, but thought a great deal, I've no doubt; he said to me later that this Oxford business could not be entered into without the written consent of their parents. Mother in her heart of hearts does not feel altogether happy about his not going, but can do nothing when Daddy, not possessing the requisite courage, refuses to let him undertake any military duties whatsoever.

Just after dinner Eirene Dodd rang up, wanting me to go & help to move some things at the Red Cross headquarters, which is the old Royal Hotel. I walked home with Eirene & then said goodbye to Maurice—though I am going to see him off in the morning, & walked a little way down the road with Edward & him. I told him to come back "couvert de gloire" & that then I would show him off with great pride to all the community in general.

I could have written a much more interesting account of to-day than this, but am altogether too weary for literary efforts—the war seems to swallow up even one's best gifts...

Tuesday August 18th

The paper officially announces this morning that the British Army—the Expeditionary Force—has all been landed in France without a single casualty. The French seem to welcome with great joy the nation which for the first time is fighting with [them] in a European war. Meanwhile very little fresh news comes of the conflict. Japan has sent Germany an ultimatum with regard to certain territory seized by the Germans in China.

Thursday August 20th

The paper announced this morning that the British army was in the fighting line, but the statement was denied by the War Office in the evening. At night Mother, E. & I went to the Upper Circle at the Opera House to see Martin Harvey in *The Breed of the Treshams,* one of those melodramatic plays which seem to be dying out but which are rather a relief after the problem & rather plotless pieces that hold the stage just now.

Friday August 21st

...I heard this morning, to my joy, that Roland Leighton, owing to his defective eyesight, has not passed the necessary exam, for serving in the army, & therefore cannot go. I am glad because I did not want that brilliant intellect to be wasted, & that most promising career to be spoilt at its outset.

Saturday August 22nd

News came to-day that the Germans have pushed back the Belgians and occupied Brussels, which surrendered without resistance for the sake of the safety of its large population of unarmed civilians. The Germans appear to have entered the city in an arrogant & bullying spirit, offering the inhabitants every kind of insult imaginable. The leading article in the *Daily Mail* this morning tried to point out to us the horrors that the poor gallant little nation is undergoing. The article was entitled "The Agony of Belgium". It seems wrong to play tennis when such terrors are convulsing Europe—but if one is used to regular exercise, the cessation of it only leads to weariness, morbidness, and general unfitness.

Sunday August 23rd

I had a letter from Roland this morning, longer than he usually sends me, & according to him written for the pure pleasure of writing to me rather than because he had anything to say. He seems very distressed because his eyesight is keeping him out of the army.

Monday August 24th

Rather more news than usual came from the front to-day. The Russians have obtained a brilliant victory over the Germans in Eastern Prussia, & are advancing quickly in their myriads. Also the Servians have so thoroughly defeated the Austrians on the banks of the Drina that the Austrian Campaign against Servia is practically abandoned.

But, while the news from one frontier is so reassuring, the news from the other, in which we are of course most concerned, is scarcely encouraging. This evening comes the report that Namur, which was expected to hold out at least as long as Liège, has surrendered to the enemy. This is admitted to be a great German military achievement. Some British troops were engaged all yesterday in battle. No list of casualties has appeared as yet.

Tuesday August 25th

Very grave news from the front this morning,—so much so that all faces look grave, & there are vague rumours in the paper about sudden conclaves at the War Ministry & audiences given by Lord Kitchener in the early morning. The report of the fall of Namur is confirmed; it seems to have been taken without a struggle. Without doubt, the heavy punishment inflicted on brave Liège, the fall of Brussels, & the feeling in Belgium that she has had to bear the whole shock of the war in the north without much assistance from her allies, all tended to weaken the resistance of the defenders of Namur. In consequence of this unexpected blow, the Allies have been obliged to withdraw from the Meuse towards the French frontier. At the other end of the battle the Germans in Lorraine have forced the French back, & taken Luneville, which is some miles over the French frontier. The French admit that their plan of attack has been a failure & that they were better prepared for a defensive movement. All the papers are very pessimistic & no one talks about scare-mongering *now;* the scare has become only too real, & the warning is lost in the actual.

Two British Army Corps—the 1st & the 2nd—have fought their first battle against the Germans at Mons & have held their ground. No actual casualty lists are yet issued, but the number of them this evening is estimated by Sir John French at 2,000. Eight German Zeppelins, the existence of which no one suspected, are said to be intending to sail over England, dropping dynamite on our ports & probably on our rich cities like London. Truly we of this generation are born to a youth very different from anything we ever supposed or imagined for ourselves. Trouble & disasters are menacing us the nature of which we cannot even guess at....

B. "The German Ward"

("Inter arma caritas")

When the years of strife are over and my recollection fades
　Of the wards wherein I worked the weeks away,

I shall still see, as a vision rising 'mid the Wartime shades,
　The ward in France where German wounded lay.

I shall see the pallid faces and the half-suspicious eyes,
　I shall hear the bitter groans and laboured breath,
And recall the loud complaining and the weary tedious cries,
　And sights and smells of blood and wounds and death.

I shall see the convoy cases, blanket-covered on the floor,
　And watch the heavy stretcher-work begin,
And the gleam of knives and bottles through the open theatre door,
　And the operation patients carried in.

I shall see the Sister standing, with her form of youthful grace,
　And the humour and the wisdom of her smile,
And the tale of three years' warfare on her thin expressive face—
　The weariness of many a toil-filled while.

I shall think of how I worked for her with nerve and heart and mind,
　And marvelled at her courage and her skill,
And how the dying enemy her tenderness would find
　Beneath her scornful energy of will.

And I learnt that human mercy turns alike to friend or foe
　When the darkest hour of all is creeping nigh,
And those who slew our dearest, when their lamps were burning low,
　Found help and pity ere they came to die.

So, though much will be forgotten when the sound of War's alarms
　And the days of death and strife have passed away,
I shall always see the vision of love working amidst arms
　In the ward wherein the wounded prisoners lay.

From *Verses of a V.A.D. and Other War Poems* by Vera Brittain, with a new introduction by Paul Berry and Mark Bostridge. © 1995 by Imperial War Museum Department of Printed Books, London.

132. Flora Shaw/Lady Lugard
Belgian War Refugees

During World War I, Flora Shaw/Lady Lugard (1852–1929) organized a War Refugees Committee in London to assist thousands of Belgium refugees to find housing accommodations in England. For this work, she was awarded a Dame of the British Empire award in 1916. The award was especially fitting as it was preceded by many years of activity on behalf of the British Empire. Described as a "Crusader for Empire,"[1] Shaw served as the first Colonial Editor of *The Times* from 1893 until 1900. While at the newspaper, she met Sir Frederick Lugard, later appointed High Commissioner of Northern Nigeria, whom she married in 1902. The organizational network of the War Refugees Committee had its origins in the Ulster Relief formed prior to World War I in anticipation of a possible civil war in Ireland. An Irish Protestant, Lady Lugard opposed the Irish Home Rule Bill of 1912 because it left Protestants in a minority in politics and she believed it was an opening wedge to the dissolution of the British Empire. Fearful that the bill's passage would provoke violence in Ulster, Lady Lugard helped to organize the Ulster Relief to arrange for the evacuation of thousands of Protestant women and children from Ulster to England. The outbreak of World War I temporarily eclipsed the Irish question. As the German army swept through Belgium, Lady Lugard activated the Ulster Relief network to assist Belgian refugees. She obtained support from the British and Belgian governments as well as the Roman Catholic Church in England as most of the refugees would be Catholic. In forming the War Refugees Committee, which received its first refugees on August 24, 1914, Lady Lugard specified that it "should have no politics and no religious distinctions." She became especially concerned with finding appropriate housing for middle-class and professional refugees. The willingness of British families to house refugees began to wane by the end of the year and donated homes were converted into hostels as an alternative. The first document reproduced here is a solicitation for these hostels. The second, "The Work of the War Refugees Committee," is an address given by Lady Lugard to the Royal Society of Arts on March 24, 1915, and published in the same year. The committee did not conclude its work with the Armistice but continued until the end of January 1919 to assist refugees with all the necessary arrangements for their return to Belgium.

---

[1]Helen Callaway and Dorothy O. Helly, "Crusader for Empire: Flora Shaw/Lady Lugard," in Nupur Chaudhuri and Margaret Strobel, *Western Women and Imperialism. Complicity and Resistance* (Ind.: 1992).

From *The Work of The War Refugees Committee, an address given by Lady Lugard to the Royal Society of Arts, March 24th, 1915.* © 1915 by G. Bell and Sons, Ltd., London.

A. "Lady Lugard's Hostels for Belgian Refugees"

Committee—Lady Lugard.
The Hon. Mrs. Roland Leigh.
Stuart Hogg, Esq.

These Hostels have been instituted by Lady Lugard for the reception of Belgians who have hitherto lived on their private means but have come to the end of their resources. Also for some of a poorer class who have received hospitality offered for a definite period which has now come to an end.

There are at present eleven houses, accommodating a total of about 400 people. Two of these are more in the nature of hospitals, the rest are carried on like private hotels or boarding-houses. Care is taken to make the life as pleasant as possible. Guests are placed in the different houses according to their social rank; there is a capable manageress in every house, a Belgian cook, and to a large extent the other servants are Belgian.

In many cases, where the refugees have some small means of their own, it has been found desirable to assist in payment of the rent of flats, or by direct contributions. At present 125 are helped in these ways.

Lady Lugard's aim has been to make each house a "little corner of Belgium," as one of the guests happily expressed it. There is a committee of ladies, who visit these houses regularly and see that the inmates are as happy and comfortable as possible.

All expenditure is accounted for to the Central Committee, and care is taken that there is no waste.

The scheme has a certain amount of financial help from the War Refugees Committee, but all expenses of furnishing, rent, lighting, and general upkeep are borne by Lady Lugard's Committee.

Your help is asked to carry on this undertaking, which is one of the attempts to repay a small portion of the immense debt we owe to the unhappy Belgian nation.

Cheques and postal orders should be made out to Lady Lugard, and addressed to her at

51, Rutland Gate, S.W.[1]

---

[1]This was Lady Lugard's home address.

B. "The Work of the War Refugees Committee"

I have been asked to speak to-day about the work of the War Refugees Committee.

The work of the War Refugees Committee is intimately associated with what will, I believe, hereafter be regarded as one of the most acutely pathetic chapters of our island history. Because we are an island, because a stretch of sea lies between us and Europe, because, above all, we have a Navy which for a thousand years has known how to defend that strip of sea, we have been able, not for the first time in our history, to offer refuge to a people stricken and driven out from their proper home.

There is no need for me to speak now of what Belgium has done—we all have the knowledge in our hearts. In the Titanic struggle in which we are engaged Belgium bore for a time the burden of the world, and the world can never forget, and never repay.

...An appeal was sent to the papers on Sunday night,[1] and as a net result of our exertions we were enabled on the following Monday morning to take possession as a Committee of the empty offices which have since developed into the well-known headquarters of the War Refugees Committee at Aldwych. That first morning we had hardly pens and ink, we had not chairs to sit upon, the offices were almost entirely without furniture, and while we were trying to organize our immediate plan of operations, the response to our Appeal, which had appeared only in that morning's papers, took the embarrassing if at the same time encouraging form of no less than 1,000 letters, all containing offers of hospitality and help.

The response of the country to the movement was absolutely phenomenal. The 1,000 letters of that day became 2,000 on the following day, then 3,000, then 4,000, then 5,000, and on the day on which we received 5,000 letters there

---

[1]The date was August 23, 1914.

were also 1,200 callers at the Office. Every letter and every visitor brought proposals of help in one form or another. Within a fortnight we had at our disposal hospitality for 100,000 persons. Cheques, clothing, food, offers of personal service flowed in upon us. I could spend hours rather than minutes in telling you the details of that first outpouring of public generosity. The sense of the country was made absolutely clear that if it could not share the acute suffering caused to the people of Belgium by the war it desired to diminish that suffering by every means that it possessed. These offers came not from one class nor from one place, but from all classes and from all places. Catholic and Protestant, Jew and Nonconformist, high and low, rich and poor united, all unaware, in a spontaneous tribute of sympathy and respect. Nations, like individuals, have their moments of unconscious self-revelation. It was a moment which unmistakably revealed the heart of England.

The enthusiasm and volume of the movement were cheering, and no offers touched us more deeply than the hundreds we received, often on postcards, from the very poor....

Among the offers which had been made to us was one from the Army and Navy Stores proposing to lend us an empty shirt factory conveniently situated just opposite Victoria Station. ... Willing help came from every side, and the result was achieved that before three on the following afternoon the shirt factory had been converted into a hostel, where 250 beds were made up with clean sheets and pillow-cases; a kitchen was arranged down-stairs with eight cooking-stoves; dining-tables were ready laid; and a hot dinner for several hundred people awaited the arrival of the refugees. Our first batch of 250 arrived there that afternoon. We disposed of the others in different places, and from that day, though we continued to receive refugees in London at the rate of several hundreds per day, and were often at our wits' end what to do, not one who reached our hands was ever left without food and lodging.

The experience of this first week gave us the formation of the principal Departments of the War Refugees Committee....

Our first need was obviously a Card Index and Correspondence Department....

We needed a Transport Department to meet refugees at the stations to convey them to and from the Refugees....

Our next obvious need was an organized system of fitting the refugees into the offers of hospitality which were received for them. This has remained from the beginning the most complicated and difficult work we have had to do. A Department, afterwards known as our Allocation Department, was organized at once under Lady Gladstone, Mrs. Alfred Lyttelton, and Mrs. Gilbert Samuel, who have been assisted in the work by an army of willing volunteers. The work of this Department, of which a beginning had been made in the Belgian Consulate even before the War Refugees Committee came into existence, has since been carried on in four main divisions. There has been our Central Allocation Department, of which the direction has remained in the hands of Mrs. Gilbert Samuel.... There has been the Allocation of the Belgian Consulate, also carried on at Aldwych, under the direction of the Misses Rothschild and a group of helpers, and there has been the Allocation of the Catholic Women's League, under the direction of Miss Streeter, working always in co-operation with Aldwych, but carried on from their own headquarters in Victoria Street. In addition to these there has been also the Allocation, carried on independently of Aldwych, by the Jewish community, who from their own private offers have provided for upwards of 6,000 people. The Catholic ladies have allocated upwards of 6,000. In Miss Rothschild's room at Aldwych some 30,000 have been either allocated or helped in other ways. Our own two branches of Allocation have since the beginning of the movement arranged for the placing of between 50,000 and 60,000 persons. In all its branches the War Refugees Committee has found homes for about 100,000 persons.

A Department taking its rise in the same necessities as the Allocation Department proper is the Department of Local Committees, which early in the movement formed themselves throughout the country for the better management of local offers of hospitality, while working in correspondence with Aldwych....

To these Departments one other of great importance was added in the first days. It was our Clothing Department, with headquarters at 23, Warwick Square. Here Lady Emmott, ably assisted by Lady MacDonnell and other devoted ladies, has been enabled by the generosity of the public to distribute nearly a million garments, including much-needed boots and shoes....

It is as a task of consolation that we have from the beginning conceived of our work. I regret to have detained you so long with a description of the machinery by which the work was done. I take you back now to the days when the first refugees, fleeing from the terror of fire and sword, began to reach our shores. These refugees were different from the refugees who are now arriving. They had actually borne the first onslaught of German fury. Men had seen their wives and daughters shot, and worse than shot, before their eyes. Fathers and mothers had seen their little children trampled to death under German feet. Old and young had alike been driven before the bayonet and placed as shields to protect the enemy from Belgian bullets. Some had been forced to dig graves, and even to bury men who were not yet dead. All had been smoked and burned out of their pillaged homes, holding themselves lucky if they were not forced back to be consumed in the funeral pyres of their domestic possessions. It has become the fashion now to cast doubt upon the authenticity of deeds fit only for the annals of the Middle Ages. Those of us who helped at that time nightly to receive the refugees as they arrived can never forget the tales of inconceivable horror which were poured into our ears, nor the convincing simplicity of narration which made it impossible to doubt their general truth. I remember the first refugee with whom I happened to speak about herself. It was not a horrible case—on the contrary, quite simple—but it brought home to me with shock of realization what was happening within an ordinary day's journey of London. It was only a mother feeding her child with a basin of bread and milk in one of our Refuges. I asked her where she came from. She said "Charleroi." "Then you have seen the fighting?" "Oh, yes, I carried him—indicating the baby—out under the German guns." It was nothing. She had had the luck to escape, but the contrast between the peacefulness of her actual occupation and her words brought home what she had escaped from. In the same Refuge on a later day there was a man whose face was like the face of a tragic fate. He did not speak, he did not move. The ladies who were working in the Refuge approached him for some time in vain. One reminded him that he had his wife, while many had lost their wives, and at last he spoke. "Yes," he said, "I have my wife! But we had five children, and we have not one left. Four of the little ones were trampled to death under the feet of a German regiment, and my little girl, my eldest, fourteen years old, was given to the German soldiery, who misused her before my eyes. Afterwards they took her away with the regiment." And he fell back to the only thing he seemed able to say, "We had five children—we have not one left." The stories which we heard at that time, daily and nightly, from not one alone, but from practically every refugee who reached us, were such as surpass all imagination of horror and brutality....It was also abundantly evident that they were not the isolated acts of brutal or drunken individuals. Evidence was unanimous, and to our minds conclusive, that the crimes were committed in pursuance of a general order from above.

I will not hold your imagination in this atmosphere. Let it be placed to the credit of twentieth-century civilization that the universal abhorrence aroused by the conduct of the German army towards civilians was such as to force German authorities to a recognition of the mistake they had committed. Orders to terrorize the population were apparently withdrawn, and, so far as we are aware, the brutalities of the first weeks of the campaign have for the present ceased....

The first chapter of Government intervention was to relieve the War Refugees Committee of the expense and difficulty of providing Refuges in London.[2] The Government took the Alexandra Palace, and in that and other available public institutions it organized immediately, under the Metropolitan Asylums Board and the Board of Guardians, Refuges which had a total capacity of about 8,000 persons. After the fall of Antwerp, Earl's Court Camp, with a further capacity of 4,000 persons, was added to the government Refuges....

The first refugees arrived usually in a state of absolute destitution. Their constant prayer was that they might be immediately allowed to work and to earn for themselves some portion back of what they had lost. But an opinion was at that time held that no attempt should be made to obtain employment for these refugees in the ordinary labour market of the country, and the lavish hospitality which was offered to them encouraged the hope that they might be amply provided for by private beneficence during the continuance of the war.

The first work of the War Refugees Committee when the refugees arrived in the Government Refuges was, therefore, to supply them as far as possible with immediate necessaries. They needed everything. Besides the substantial requirements of clothes and shoes, they wanted combs, brushes, soap, hair-pins, boot-laces, braces, needles, cotton, thimbles— everything that even the poorest find necessary in daily life. The men, of course, urgently wanted tobacco; the women wanted knitting-needles and wool to knit. We did our best to supply all these, and among the small articles which at that time were distributed freely none were more eagerly accepted than rosaries. We gave them away by thousands. The exodus had been so sudden that they had apparently in many cases been left behind, and men and women alike among the first arrivals from the Walloon country seemed anxious to possess themselves of this usual accompaniment of prayer.

...They seemed themselves to realize, in the tragic extremity of their distress, that they had lost everything except their God, and I cannot easily convey the touching fervour of the prayers in the chapels of the Refuges at which I once or twice incidentally assisted. Piety, courage, extraordinary fortitude, and overflowing heartfelt gratitude for all that was being done for them in England were the principal characteristics that enlisted our sympathy and admiration for our guests....

The refugees were supposed to remain in the London Refuges for a period of only three to five days at the outside. Once rested and refitted it was the work of the War Refugees Committee to pass them on to the permanent homes so cordially offered by the hospitality of the country....

...At the beginning of the movement refugees had to be dealt with only at the rate of 100 or 200 per day. From the date of the public offer of national hospitality made by the Government, the number increased steadily....

...During the stress created by the fall of Antwerp,[3] when upwards of 4,000 refugees arrived in one day in London by trainloads from the Continent, and as many as 2,000 had to be sent in small individual groups to different stations of the British Isles, a total of 6,000 had to be handled every day! It has been estimated that during this period as many as 8,000 and 10,000 refugees crossed the Channel daily to our shores. No warning nor preparation could be given as to the numbers to be dealt with. While the crisis lasted they poured in day and night, taxing the energies of the whole organization almost to breaking-point....

The crisis lasted only a couple of weeks. The occupation of Ostend by the Germans on October 17th closed the Belgian coast and stopped the daily transport service. Since that time refugees have been only able to reach us by way of Holland, and though this country has continued to provide such facilities as are possible

---

[2]Prime Minister Asquith pledged government assistance to Belgian refugees on September 9, 1914.

[3]The fall of Antwerp took place on October 10, 1914.

for their transit, the figures of the daily arrivals have fallen considerably. The total for November was the lowest for any month since the beginning of the war. In December and January the numbers again mounted, giving a total of 12,000 for December and 14,000 for January. Refugees are still, notwithstanding the dangers of mines and submarines and the prohibition of our blockade zone, arriving in numbers which are to be counted daily in three figures. But the rush is over. We are no longer working under the same conditions of pressure.

There are noticeable also some other remarkable differences. We are working now with a different class of refugee. The simple country folk of the first exodus have given place to the urban population of the great towns, and they come to us under different conditions. The early refugees had, as I have told you, suffered in their own persons all the worst horrors of war. Since the fall of Antwerp the flight has been rather—though not of course wholly—from "the wrath to come." Many refugees are fleeing from what they fear may happen rather than from what has actually happened. I speak chiefly for the moment of the working-classes. Many of those now coming have been attracted to this country by the accounts sent back in the first moments of relief and gratitude by the earlier refugees....

The gradual development of the situation which has brought us a different class of refugee has also brought about a very important modification of opinion with regard to the conditions of their reception. It has been decided that the employment of refugees instead of being deprecated should now be encouraged, and that instead of depending for subsistence on the hospitality of the country they should, as far as possible, be enabled to support themselves....Belgian Labour Bureaux working in connection with the Central Labour Bureaux have been established in the Government Refuges, as also in the Rink at Aldwych. Recruiting Bureaux have been established in the Government Refuges, by means of which Belgians of military age are enabled to join

their colours and return to the front at Flanders....

Outside these questions the problem with which since Christmas we have been most acutely pre-occupied is the problem of giving suitable help to the urgent needs of the propertied and professional classes. This is a class with which I have myself been thrown into close and constant touch, and the sorrows and difficulties of their position are very vivid to me. They have suffered, of course, horribly in regard to their material possessions, and the numbers increase daily of persons accustomed to live in the comfort of comparative affluence who are reduced to absolute penury. Such cases call for the sincerest sympathy and for practical help....

In the early part of the movement such cases as these were provided for by private hospitality, and I come now to the greatest change of all which the movement has undergone. The movement of private hospitality, which has provided from first to last for a figure approaching to something like a quarter of a million refugees, has, as was to a certain extent inevitable, exhausted its first impulse. About Christmas time we began to realize that the offers of hospitality had ceased. No fresh offers came, and hosts who had previously had Belgians in their houses wrote that they would shortly be needing this accommodation for other purposes. Our Allocation Department became a Department of Re-Allocation. Gifts of clothing also sensibly diminished.

...The part of private generosity for better-class refugees still remains to bring the bare necessities of life up to the standard which the nation would wish to offer in such cases as those I have just now cited.

There are many obvious ways in which this can be done. Among the most generally successful so far has been the organization of large houses on the basis of gratuitous hotels. I have myself organized two or three such houses....

Another way of meeting the necessities of the class of refugees of whom we are now speaking is by paying the rent of furnished flats, in which a very small grant is sometimes enough to render domestic life a possibility. Among the propertied and professional classes there are some who have still some small resources....

I would like to have been able to do justice to other institutions for the assistance of refugees which have from the beginning of the movement developed as branches of the various Departments at Aldwych....

A Department which is probably doing in its way as much humane work as any other is what we call the "Missing Relatives" Department.... We have, of course, now registered the addresses of many thousands of refugees....In the "Lost Relatives" section all urgent cases, such, for instance, as a father or mother searching for their lost children, or a husband his wife, etc., are handled immediately by our Correspondence Department, who make every effort to trace the missing person. The machinery which is used for tracing the letters is put in operation, and I am glad to say that we frequently succeed in finding and uniting the members of families who have lost each other in the flight.

...I would like to say that our work would have been absolutely impossible had it not been for the devoted, generous, and regular support of hundreds of volunteers who have given every bit as much as we have given, and who have been content to do it—to come early, to stay late, to work day after day unflinchingly at the least interesting tasks, to spend their strength, their emotions, their money, and their time in the background, so to speak, of our organization, without a thought of anything but the help that they could give. These volunteers have come from every rank. I have mentioned 500. Had we wanted 5,000 we could have had them. I am almost ashamed even to speak of thanks or recognition where it has been so little sought. I would only say of many of our unmentioned helpers that their names should be written in letters of gold, were it possible that any true record could be kept of the service which this movement has called forth.

...In the details which I have given you we are simply working out the national resolution that the exiles now in our midst shall be cared for, helped, and protected to the limits of our ability until the day dawns for them, when they may return to the homes they love. We see no end, and we desire to see no end, to our exertions but the day of repatriation. Be that day near or far, it is our hope to continue our work till it is reached, and we look with quiet confidence and absolute assurance to the public we know to give us the full support of its sympathy and its help.

133. E. Sylvia Pankhurst
"How to Meet Industrial Conscription"

With her mother, Emmeline and her older sister, Christabel, Sylvia Pankhurst (1882–1960) founded the Women's Social and Political Union (WSPU) in 1903.[1] For the cause of suffrage, she endured fifteen arrests; nine hunger, sleep, and thirst strikes; and several forced feedings. In 1912, Sylvia Pankhurst formed the East London Federation of the WSPU in the East End. Critical of how the WSPU had lost much of its former working-class base and was now predominantly middle class, she argued that the WSPU should seek the support of, if not an alliance with,

---

[1]On the WSPU's formation, see Emmeline Pankhurst, Source 129 in Chapter 10.

The Woman's Dreadnought, 20 March 1915.

the Labor Party and the trade unions. Following an official separation from the WSPU in 1914, the organization changed its name to the East London Federation of Suffragettes (ELFS) and initiated its own weekly newspaper, the *Woman's Dreadnought,* which Pankhurst edited until its demise in 1924. World War I divided the Pankhurst family anew. Sylvia and Adela, her younger sister in Australia, opposed the war while Emmeline and Christabel suspended the WSPU's suffrage activities and actively supported the war effort. Defying the August 1914 Defense of the Realm Act (DORA) which forbade the diffusion of information that could undermine the war effort, the ELFS combined the campaign for suffrage with antiwar activities. While opposing the war, the ELFS sought to alleviate wartime hardships by establishing "cost price" restaurants, centers for the daily milk distribution, a mother and infant clinic, and day nurseries. The ELFS also acted to improve women's wages and working conditions in war industries. Proposals for protecting women workers are outlined in Pankhurst's article "How to Meet Industrial Conscription," published in the March 20, 1915 issue of the *Woman's Dreadnought* soon after the government's announcement of women's registration for war work. Renamed the Workers Suffrage Federation (WSF) in 1916, the organization moved in an increasingly left direction as it supported the Easter Rising in 1916 and the Russian Revolution in 1917. In the 1930s, Pankhurst opposed the Italian invasion of Ethiopia, supported the Republican forces in the Spanish Civil War, and denounced Britain's appeasement policies toward Nazi Germany. Pankhurst's books include *The Suffragette* (1911), a history commissioned by the WSPU; *Soviet Russia as I Saw It in 1920* (1921); *Save the Mothers* (1930); *The Suffragette Movement* (1931); and *The Home Front* (1932), which recounts her activities in the East End of London during World War I.

To the women whom they have refused to grant the rights of enfranchised citizens, the Government, through the President of the Board of Trade, has issued an appeal to enlist for War service.[1]

The Women's Societies which the Government has so often flouted are urged to lend their aid in marshalling the volunteers.

Registers of women who are prepared to undertake any kind of paid work, industrial, agricultural, clerical, etc., are to be kept at the Labour Exchanges, and registration forms are being sent out to the women's organisations.

Those who register must state their ages and whether they are married, widowed or unmarried; if they have ever done any paid work, and if so, what and when, and in whose employ; if they are free to work whole or part time, or to leave their homes; whether there is any kind of work that they are willing or able to do, and whether they are willing to train for work which they have not previously done.

In view of this appeal, which is being made to women by the Government—appeals by Governments usually tend to become irresistible demands—it is surely time that all the women's organisations, trade union, political, educational and social, should come together to discuss this important matter and formulate their demands to safeguard the position of women of all ranks in the labour army.

The men who signed the Army forms that were sent round to the householders, found

---

[1]Registration for war work was still voluntary in March 1915. Soon afterward, however, the government passed the National Register Act which compelled all persons between fifteen and sixty-five years of age to register their trade and profession with the government. The ELFS organized demonstrations against this act which it regarded as the prelude to national conscription.

themselves called up for service, sometimes much to their surprise. The women who sign their names on the War Service Register will probably find themselves called up too, whether they wish or not. Shall we allow them to go without fair conditions first being assured?

The Government, through Mr Lloyd George and Lord Kitchener, has announced that it is about to take extensive control of industry.[2]

The Government makes it plain that it is determined that the provisions of munitions of war, both for Great Britain and the Allies, shall absorb all our entire national energies, so that all our people may become part of a great war machine engaged either in fighting, supplying the wherewithall to fight, or in providing necessaries of food, clothing, housing and transport for the soldiers or armament makers.

In order to conciliate the British workmen (who, by their votes, have been made the ultimate arbiters of the nation's destiny, though they scarcely realise their power), Mr Lloyd George has held conference with the Great Trade Unions which, as yet, are almost entirely controlled by men. The Government has promised that limits shall be set to the profits of employers, and that good wages and fair conditions of labour shall be ensured.

Various increases in wages have been made, and negotiations are taking place in regard to demands for much larger increases. The Trade Union leaders and Labour Members of Parliament occupy a position of grave and anxious responsibility at this time, for on their handling of the situation the position of millions of workers largely depends.

Perhaps an even vaster responsibility rests on the shoulders of women who are leaders of women at this time. As yet, the working women, the sweated drudges of the world, are but poorly organised, and all the women's suffrage and other political and social organisations must lend their aid at this crisis, in securing the best possible terms for the masses of women workers, on whom the future of our race so largely depends.

It is more urgently imperative than ever that every woman who works for her living should join a Trade Union, in order that she may have a strong organisation to protect her interests, and that she may help to protect the interests of other women.

A national conference of women should be called immediately to formulate demands for the regulation of this industrial enlistment of women. Here are some of the demands which would, undoubtedly, be adopted by such a conference:—

(1) As the Government is already by far and away the largest employer of labour in the country, and may soon be almost the sole employer, it is absolutely imperative that *women who are to be enlisted as recruits in the National War Service shall have the Vote at once.*

(2) That fair wages shall be assured to women. *That where a woman is employed on work hitherto done by men she shall receive the wage hitherto paid to men, in addition to any war bonus or increase in wages which might have been paid for the work now, in the case of men employees. That in no case shall an unskilled woman be employed at a lower wage than the current rate to men unskilled labourers.*

(3) The Government has announced its determination to put an end to industrial disputes, and proposes that, where the parties concerned fail to come to an agreement:—

'The matter shall be referred to an impartial tribunal, nominated by his Majesty's Government, for immediate investigation and report to the Government with a view to a settlement.'

---

[2]Lord Herbert Kitchener was appointed Secretary of State at War in August 1914 in the government of Liberal Party Prime Minister Herbert Asquith. David Lloyd George was Chancellor of the Exchequer (Treasury) at the time of Pankhurst's article. He was chosen to head the Ministry of Munitions following its creation in June 1915. Lloyd George succeeded in promoting passage of the National Register Act and the Munitions of War Act of July 1915 which gave the government extensive powers to regulate war industries and levy fines on workers who went on strike in these industries. The act allowed women to engage in work from which they were excluded before the war but did not affirm equal pay for equal work with men. Lloyd George replaced Asquith as prime minister in December 1916.

The Women's Conference would undoubtedly demand that *women should have strong representation on this tribunal, and that in all disputes in regard to women's employment, a woman of standing and experience, (the nation has many such to draw upon) should be the chairman of the tribunal, or in case of the appointment of a sole arbiter, a woman should be the arbiter of the dis-pute.*

(4) That proper safeguards in regard to hours, wages, and conditions be arranged in conjunction with representatives of the women concerned, and that no woman shall be compelled to work under conditions which the representative of the organisation to which she belongs, reports to be unsatisfactory.

This is a moment of very vital importance to women, calling for all our energy and resource, all our earnestness, all our solidarity.

Let us band ourselves together—sinking our differences—to build up a position of dignity and security for our sisters, in order that as free citizens they may give their services to the nation willingly and with enthusiasm.

134. European Women Suffragists A Call to the Women of All Nations

Dr. Aletta Jacobs (1854–1927) and Dutch suffragists issued a call in the March 1, 1915 issue of *Jus Suffragii (The Right to Suffrage)*, the newspaper of the International Woman Suffrage Alliance, to bring together women from both sides of the war to attend an international conference in neutral Holland. A women's rights and peace activist, Jacobs was Holland's first woman physician and opened the world's first birth control clinic in Amsterdam in 1882. She was cofounder of the Dutch Association for Woman Suffrage, which she chaired, and president of the International Woman Suffrage Alliance. When the German Union for Woman Suffrage announced in October, 1914 that it was canceling plans to hold an international meeting of the Alliance in the spring of 1915 in Berlin, Jacobs proposed convening a conference in Holland to discuss peace proposals outlined by women's peace associations. Invitations were to be extended to all international women's organizations as well as those involved in the campaign for women's suffrage. As the executive committee of the Dutch Association for Woman Suffrage refused to provide official support, Jacobs and other Dutch activists proceeded to organize the conference as individuals.[1] It was attended by some 1,200 delegates from twelve countries even though major suffrage organizations in England and France rejected and denounced the conference invitation. Presided over by the American peace activist Jane Addams, the International Women's Congress was held at The Hague from April 28 to May 1, 1915. The conference adopted a twelve-point resolution, which deputations of delegates were to present to various heads of state, and established the

---

[1]For Jacobs' detailed account of the Hague conference, see her autobiography, originally published in Amsterdam in 1924 as *Herinneringen* (1996 English translation, *Memories: My Life as an International Leader in Health, Suffrage, and Peace*, pp. 82–85).

International Committee of Women for Permanent Peace (ICWPP). National sections of the committee were subsequently established in various countries. Following the conclusion of World War I, at a conference held in Zurich in 1919, the ICWPP was renamed the Women's International League for Peace and Freedom (WILPF) which remains active to this day. Reproduced here are the call to the conference published in the March 1, 1915 issue of *Jus Suffragii* and the resolutions of the Hague conference.

A. "Call to the Women of All Nations"

From many countries appeals have come asking us to call together an International Women's Congress to discuss what the women of the world can do and ought to do in the dreadful times in which we are now living.

We women of the Netherlands, living in a neutral country, accessible to the women of all other nations, therefore take upon ourselves the responsibility of calling together such an International Congress of Women. We feel strongly that at a time when there is so much hatred among nations, we women must show that we can retain our solidarity and that we are able to maintain a mutual friendship.

Women are waiting to be called together. The world is looking to them for their contribution towards the solution of the great problems of to-day.

Women, whatever your nationality, whatever your party, your presence will be of great importance.

The greater the number of those who take part in the Congress, the stronger will be the impression its proceedings will make.

Your presence will testify that you, too, wish to record your protest against this horrible war, and that you desire to assist in preventing a recurrence of it in the future.

Let our call to you not be in vain!

B. "Some Principles of a Peace Settlement"

1. Plea for Definition of Terms of Peace

Considering that the people in each of the countries now at war believe themselves to be

Call to the Women of All Nations, Holland 1915.

fighting not as aggressors, but in self-defence and for their national existence, this International Congress of Women urges the Governments of the belligerent countries publicly to define the terms on which they are willing to make peace, and for this purpose immediately to call a truce.

2. Arbitration and Conciliation

This International Congress of Women, believing that war is the negation of all progress and civilisation, declares its conviction that future international disputes should be referred to arbitration or conciliation, and demands that in future these methods shall be adopted by the Governments of all nations.

3. International Pressure

This International Congress of Women urges the Powers to come to an agreement to unite in bringing pressure to bear upon any country which resorts to arms without having referred its case to arbitration or conciliation.

4. Democratic Control of Foreign Policy

War is brought about not by the peoples of the world, who do not desire it, but by groups of individuals representing particular interests.

This International Congress of Women demands, therefore, that foreign politics shall be subject to democratic control, and at the same time declares that it can only recognise as democratic a system which includes the equal representation of men and women.

5. Transference of Territory

That there should be no transference of territory without the consent of the men and women in it.

6. Protest

War, the ultima ratio of the statesmanship of men, we women declare to be a madness, possible only to a people intoxicated with a false idea; for it destroys everything the constructive powers of humanity have taken centuries to build up.

7. Women's Responsibility

This International Congress of Women is convinced that one of the strongest forces for the prevention of war will be the combined influence of the women of all countries, and that therefore upon women as well as men rests the responsibility for the outbreak of future wars. But as women can only make their influence effective if they have equal political rights with men, this Congress declares that it is the duty of the women of all countries to work with all their force for their political enfranchisement.

8. Women's Sufferings in War

This International Congress of Women protests against the assertion that war means the protection of women. Not forgetting their sufferings as wives, mothers, and sisters, it emphasises the fact that the moral and physical sufferings of many women are beyond description, and are often of such a nature that by the tacit consent of men the least possible is reported. Women raise their voices in commiseration with those women wounded in their deepest sense of womanhood and powerless to defend themselves.

9. Women Delegates to Conference of Powers

Believing that it is essential for the future peace of the world that representatives of the people should take part in the conference of the Powers after the war, this International Women's Congress urges that, among the representatives, women delegates should be included.

10. Woman Suffrage Resolution

This International Women's Congress urges that, in the interests of civilisation, the conference of the Powers after the war should pass a resolution affirming the need in all countries of extending the Parliamentary franchise to women.

11. Promotion of International Good Feeling

This International Women's Congress, which is in itself an evidence of the serious desire of women to bring together mankind in the work of building up our common civilisation, considers that every means should be used for promoting mutual understanding and goodwill between the nations, and for resisting any tendency towards a spirit of hatred and revenge.

12. Education of Children

Realising that for the prevention of the possibility of a future war every individual should be convinced of the inadmissibility of deciding disputes by force of arms, this International Congress of Women urges the necessity of so directing the education of children as to turn their thoughts and desires towards the maintenance of peace, and to give them a moral education so as to enable them to act on this conviction whatever may happen.

135. Ellen Key
"War and the Sexes"

Ellen Key (1849–1926) was a Swedish activist who promoted motherhood as an important social endeavor and as the highest vocation for women as determined by their biology. Key believed in separate spheres for men and women, opposed outside employment for mothers, and was initially against women's suffrage because it distracted women from their "natural" activities. To free mothers from their dependence on men, she called on the government to financially support all women raising children both within and outside of marriage. However, Key opposed public centers for child care because she believed it should be carried out by mothers privately in the home. Widely translated, her publications include *The Century of the Child* (1900), *Love and Marriage* (1904), and *The Renaissance of Motherhood* (1914). Key's ideas were embraced by conservative women's groups in the United States and Europe. She became especially influential in Germany due to the efforts of the German Society for the Protection of Motherhood. World War I led Key to assess how fears of depopulation arising out of the war were transforming public attitudes and policies toward motherhood. She expressed her concerns in the article "War and the Sexes," published in the June 1916 issue of *The Atlantic Monthly*, and in her book, *War, Peace, and the Future* (1916).

I

...After the war, woman's prospects, from the point of view of her natural duty—motherhood—will be dark indeed.

The number of women who will have to dismiss all thought of marriage—already far too large—is destined to become much larger still. The number of those who lead immoral lives and are childless, or who bear illegitimate children, will therefore increase. Others, from a sense of patriotic duty to which appeal has already been made, may marry invalids. How many of these will be disappointed in their most justified wishes for happiness! Those women who have chosen among the men who are rejected from military service quite often have defective children. The possibilities for millions of women who are now at the most favorable age for marriage decrease steadily, for with every day that goes by the number of young men who might return from the war without severe bodily or mental injuries grows less and less—not to mention the millions who will never return. And, lastly, the higher the development of women, the more they chafe under the 'patriotic' mandate to bear many children to replace the nation's losses. For they know that, from the point of view of their personal development as well as that of the race, *fewer* but *better* children are to be preferred.

If, therefore, the future is dark for the women of the warring countries, is not the present much darker? Apart from all the women who, directly or indirectly, have been killed by the war before ever becoming wives or mothers, there are all those who have borne children during the horrors of war—children that died soon after birth; there are those who have been separated from children whom they will probably never recover; there are those who bear the children of the invading enemy. Added to these is

From *The Atlantic Monthly*, June 1916 issue, Vol. 117, no. 6, pp. 837–844. Contributed by Sandi Cooper.

the host of women who have lost their fathers and children; all the widows, all the homeless, that war has created....

II

A considerable number of plans have already been suggested in Europe to relieve the abnormal sex-conditions, which have, of course, met with much formidable opposition.

Some one in London has conceived the idea of founding a 'society for the marrying of wounded heroes'—an appeal to woman's self-sacrifice and patriotism to make the lives of these men bearable and to propagate children who will inherit their fathers' qualities of heroism. These wives, who would, in most cases, have to become the supporters of their families, would, therefore, be paid a man's wages and would, in many cases, also be given a stipend to facilitate their marriage. Moreover, in order to insure suitable mating, it is suggested that recourse be had to selective committees of clergymen and physicians; it is evidently not proposed to let the parties themselves choose. Women who are physically strong will be expected to marry men who need to be carried or pushed in a chair. Blind men, who can still at least enjoy good food, will be married to good cooks, and so forth.

It seems impossible to believe the statement that the society already has hundreds of thousands of female members. Can it be possible that women are willing to offer themselves for such a pitiful purpose—where love is quite out of the question?

In Germany some one has suggested that the government give invalids an opportunity to own their homes. This would enable the heroes of the war to found families—for it is to be expected that thousands of heroic women who are widowed by the war will remarry these invalids. Another thoughtful German has suggested that the government open a marriage department, partly to further early marriages, partly in order to help young men make suitable acquaintances. The young men who survive the war, he thinks, will not have time for the social life that formerly gave them opportunities for becoming acquainted.

At the beginning of the war, before any one suspected either its length or the number of its victims, a German feminist wrote an article decidedly consoling to the German women, pointing out that the greatest percentage of marriages in Germany took place after the War of 1870. This was, however, the result of the great economic boom that this war brought Germany. It gave the young men of between twenty and thirty the chance that they otherwise too often lack, of having a family. The same authoress predicted the duplication of this state of affairs as the result of German victory in the present war; but after twenty months of desperate struggle, such an optimistic view can hardly be sustained. The capital accumulated by the prosperity of the last decades is quickly disappearing. The future of every country is being more deeply mortgaged with every hour that passes. The graves that are now being filled with the bodies of youths of sixteen and seventeen are growing in number. It is not strange, therefore, that here and there the idea of polygamy, which already had its advocates in Germany before the war, should now be considered as tenable from the standpoint of race-hygiene. Those men who return sound from the war know for a fact that young Germans pure-mindedly and seriously consider this idea from patriotic reasons.

And the same idea has been openly expressed by an Indian prince studying sociology and ethnology in Oxford. He points out that even before the war England had 1,200,000 more women than men; and with the present losses of young men between the ages of twenty and thirty, he estimates that every fourth woman in England must remain unmarried. Similar conditions must naturally follow in other countries. Of course, from the point of view of race-hygiene, only those men who are physically, psychically, and morally sound should be allowed to marry two wives. Love must, of course, be sacrificed for the sake of patriotism;

and women (this prince believes) will sooner make this compromise than remain single for life. From the standpoint of the race, to be sure, such marriages are infinitely to be preferred to invalid marriages; but it does not seem probable at present that any state will formally adopt this idea. It is probable, however, that there will actually be a state of polygamy such as existed after the Thirty Years' War. The increase of population will, therefore, probably be greater than a condition of strict monogamy would permit. But it is unlikely that many unmarried self-supporting women will replace marriage with free love. The question is, whether these women will want to become mothers; and if so, whether the community will lend dignity and responsibility to such form of matriarchal law.

In most countries where these questions have been seriously considered, very rational means have been found for increasing the birth-rate. In Germany, for instance, they have done away with the law preventing women with children from becoming teachers, as well as the difficulties attending military marriages, and the red tape attending the remarriage of the divorced; and they have also increased the salaries of the official class.

A question that is causing great anxiety in Germany is the danger to maternity in the increase and spreading of contagious diseases during the war. Another source of anxiety lies in the disastrous effect of nervous shock and life at the battle-front on the potential fatherhood of the race. For these reasons, many women who marry men returning from the war are destined to remain childless.

III

First in the sphere of literature, then in that of social work, and finally from the point of view of race-hygiene, people everywhere during the last few decades have been considering the problem of the unmarried mother and her child. All those who, for humanitarian and social reasons, urged the care and protection of the community for these mothers and their children were considered apostles of immorality....

...The fact that the battlefields swallow up millions of lives makes the birth-rate a national question and revolutionizes ideas of sexual morality. Everything is now looked upon in a Spartan spirit as being a matter of the State. All these facilities for military marriages are being made because the State expects the men to propagate themselves before they die. It is to ensure a good crop of soldiers for the year 1936 that Joffre[1] has, to the greatest possible degree, given the French soldiers four days' leave with free journeys home. It has been proposed in France to tax the unmarried and childless and to reduce the taxes of those who are married or have many children; and similar measures will probably be taken in the other warring countries.

What was formerly considered a sin—loveless marriages contracted simply for the purpose of having offspring—will perhaps, from the national point of view, come to be considered a duty hereafter. The bearing of children outside of marriage, and perhaps other deviations from the ideal of monogamy, will be practiced openly after the war to a far greater extent than was done secretly by people of Europe before the war. Twenty months of war have already dealt heavier blows to the foundations of 'Holy Marriage' than all the 'apostles of immorality' were able to compass. That all new forms of sex-relation will not be *officially* sanctioned is self-evident, but they may have the sanction of custom; and this, in some cases, means more than the approval of the State.

When the German 'Society for the Protection of Motherhood' celebrated its tenth anniversary in 1915, Helena Stocker was able to show that the protection of motherhood, which, ten years ago, was almost considered indecent, had become the watchword of the day.

---

[1]General Joseph Joffre: commander-in-chief of the French armies in World War I.

The 'Society for the Protection of Motherhood,' the German 'Society for the Increase of Population,' and another for the protection and growth of the race, all met in October, 1915. And for each of them the principal question was, how to diminish the mortality of infants, and how best to extend the protection of motherhood. For financial aid during confinement and illness, nursing premiums, and so on, they now turn to the State. The idea that I have so long advocated, that *mothers should be considered the servants of the State,* has already been taken up in Germany. And no difference is made between married and unmarried mothers.

Another moral question that was previously discussed—that of birth-prevention—has come up again during the war. In East Prussia the question has been discussed as to whether the law against abortion should be suspended for those women who fell victims to the Russian soldiers. And in France, where many women have, with great suffering, borne the children of their enemies, some people still advocate preventive measures; some one even suggested killing these children, in order to ensure the purity of the race. Surely one cannot go further from the ideals of Christian morality! And though these suggestions have been rejected, the mere fact that they have been discussed proved what this whole war has so clearly shown: that the religion of Europe is no longer that of *Christianity* but that of *nationalism,* and that everything that is considered good for the nation is assumed to be right.

IV

The question for the future will be whether patriotism will have become to such a degree a religion to women that they will be willing to sacrifice their idea of love—which, to the more advanced modern woman, had also become a religion—and marry for the convenience of the State....

But one thing is certain, and that is that after the war very many women simply *will not have the strength* to undertake the duties of marriage,—at least, not if they are to have large families. Even before the war, many women found the fourfold duties of a wife—to help support the family, to bear and care for children, to be the companion of her husband, and to care for the home—too much of an undertaking. After this war, millions of women will have to become the supporters of their families, even if their invalided husbands are able to contribute. Many women will have to be nurses to the husbands whom the war has returned to them a wreck. With the new taxes, the burden of making both ends meet will be greatly increased. Through the loss of the male members of the family, women have become the sole supporters of the old and helpless of the family. Many of these, to be sure, will not have been able to survive the sufferings and deprivations of the war, but those who are left will be dependent on the arm of a single woman. In some cases, no doubt, women will have become physically and psychically stronger through the work and sacrifices war has brought on them. Many imaginary illnesses will have disappeared, but such cases are, no doubt, comparatively few compared to those where women's health has been ruined by the sorrow and tribulations of war. Therefore they will have to spare themselves in some sphere. And the only possible sphere will be that in which the state will expect most of them: motherhood.

I have never agreed with those feminists who claim that the one way in which the married woman proves her worth is by her ability to earn a livelihood. Her ability to bear and educate her children and build a home is so handicapped by her leaving her home to procure a livelihood that the only way to solve the problem would be to consider her motherhood a state service, and reward it accordingly....

But this ideal way of solving the problem of motherhood and self-support was very distant even before the war, and though now, in the interest of the birth-rate, there is a good deal of talk about different means of helping mothers, when peace comes the people will have to shoulder the mountain of war debt, and there will be hardly any funds left in Europe with which to help women. Therefore, this ideal solution of the problem will be postponed to a still more distant future. Among the nations so heavily oppressed by the war, it will inevitably be necessary to count on a far greater number of women having to become self-supporting than formerly. This will bring about very radical changes in the community, in economic conditions, in family life, and in the increase of population. Family life, during the next generations, will be more sober, more prosaic. The death of so many men will, to a certain extent, do away with competition between the sexes, but also with marriage. The number of illegitimate children will increase, but they will be better cared for. On the whole, the increase of population will be hindered by woman's inability both to bear and provide for children, and to those who look upon woman as the producer of soldiers, this will seem a misfortune. To those, however, who look upon the matter in a more human way, it will, on the contrary, *become a condition for future development that women resolutely refuse mass production of children,* and more consistently seek to improve the quality of humanity, while they, at the same time, try more energetically to procure the right to have a share in dictating the politics on which the lives of their sons and daughters are so dependent....

V

...If women, after the war, willingly comply with the wish for 'national child-bearing,' and 'patriotically' support this competition, they do not deserve anything better than that their sons twenty years hence shall fill new trenches! Let us hope that they will not be willing!

If, for national reasons, woman should become untrue to the highest instincts of her nature, which lead her to give the race only children of love, she will sink so deep that neither the right to vote nor any other rights will be able to help her. Warning voices have already been heard pointing out that, from a biological point of view (that is, the transmission

of hereditary traits), love is necessary. My intuition in this respect seems therefore to be verified. What love means to spiritual happiness every one knows who is truly loved. It may be selfish to think of one's self; but for the good of the race, one may well wish that the women of the generation out of which every fourth must remain single, will sooner bear this sacrifice than submit to bear loveless children for the sake of the nation....

VI

The war has destroyed millions of homes. It has shattered happiness beyond all belief. It has spoiled innumerable lives, and yet we must remember that it has also made unforeseen happiness possible. The literature of the war is full of stories of the heroic women who have braved every danger in order to be able to follow or become united with their lovers. It also tells of unions that have been sundered, and of anguished doubt that has become crushing certainty. Even in the love-life of the community, war brought some slight compensation with its incalculable evil. It has sometimes appeared as the deliverer as well as the enslaver.

The war has called forth a new and pathetic phenomenon in the nation of mothers. From many of these one has heard the cry, 'My son is dead—give me another.' They have heard of some homeless soldier, whom, without knowing him, they have overwhelmed with presents, even offering him a home....

The fact that many little war-children have been adopted by mothers who have lost their own children, or by women who have never known what motherhood means, shows one of the ways in which women have been able to glean some sweetness from the bitterness of war. But how meagre, how artificial are these joys compared to all the natural, life-giving, promising human relationships that have been crushed under the iron hoofs of the black horse of War!

136. Constance Markievicz
A Call to the Women of Ireland

Constance Markievicz (1868–1926) was born Constance Gore-Booth to a Protestant, aristocratic, landowning family in County Sligo in what later became part of the Republic of Ireland. She was a founder and a leader of numerous Republican and women's organizations, including the Dublin-based Irish Citizens Army led by socialist labor leader James Connolly. The Easter Rising of 1916 was carried out jointly by the Irish Citizens Army, which called for an independent, socialist Ireland, and the Irish Volunteers whose aims were limited to freeing Ireland from British rule. The two groups merged to form two wings of the Irish Republican Army. Unlike the Cumann na mban, the women's auxiliary of the Irish Volunteers, of which Markievicz later became president, women in the Citizens Army drilled alongside men and were part of the military action of the Rising. For her role as a company commander in the Irish Citizen Army, Markievicz was sentenced to execution by shooting, but her sentence was commuted to prison on account of her sex. Although released in 1917, she was soon arrested and imprisoned again. During the ten years between the Rising and her death, Markievicz was often on the run to elude frequent police raids of Republican meetings and altogether served five prison sentences which sometimes included solitary confinement. These experiences, recorded in letters to her sister, Eva Gore-Booth, were published as the *Prison Letters of Countess Markievicz* (1934). During her second incarceration, Markievicz became the first woman to be elected to the British Parliament but chose not to take her seat. Instead, she served as minister of labor in the Dail Eireann of 1919, the independent parliament set up by Sinn Fein members who refused to serve in the British Parliament. Markievicz opposed the Treaty of 1921 between Britain and Ireland which divided Ireland into the Republic of Ireland in the south, while the six predominantly Protestant counties of Northern Ireland remained under British rule. She died in 1926 in Dublin following a severe illness, her health undermined by various imprisonments. Over 300,000 supporters and admirers lined her funeral procession. *A Call to the Women of Ireland* was originally a lecture, "Women, Ideals, and the Nation," which Markievicz delivered to the Students' National Literary Society in Dublin in 1907. It was first published as a pamphlet in 1909 and then reprinted in 1918 as a contribution to post-Rising debates about the future of Ireland.

I take it as a great compliment that so many of you, the rising young women of Ireland, who are distinguishing yourselves every day and coming more and more to the front, should give me this opportunity. We older people look to you with great hopes and a great confidence that in your gradual emancipation you are bringing fresh ideas, fresh energies, and above all a great genius for sacrifice into the life of the nation.

In Ireland the women seem to have taken less part in public life, and to have had less share in the struggle for liberty, than in other nations. In Russia, among the people who are working

From *A Call to the Women of Ireland* © 1918 by Fergus O'Connor, Dublin.

to overthrow the tyrannical and unjust government of the Czar and his officials, and in Poland where, to be a nationalist, men and women must take their lives in their hands, women work as comrades, shoulder to shoulder, with their men. No duty is too hard, no act too dangerous for them to undertake. Many a woman has been incarcerated in the dungeons under St. Peter and St. Paul....

Now, England in this twentieth century is much more civilized, and much more subtle than Russia in her methods for subjugating a nation; therefore, more difficult to fight; and much more difficult to realise as an enemy.... Irish people, being very simple and honest themselves, have often taken a very long time to realise that we are being governed—not as we are told for the ultimate good of the Irish nation, but as an alien province that must be prevented from interfering with the commerce of, and whose interest must always be kept subservient to England. She has systematically overtaxed us—for our own good; she has depopulated our country—and it is for our own good; she has tried to kill our language—for our own good; she entices our young men into her armies, to fight her battles for her—and still it is for our own good. She began this policy in 1800 when—entirely for our own good—our Parliament was disbanded, and we were given instead the great privilege of sending to Westminster a small band of representatives to make the best fight they could for Irish rights against an overwhelming majority; which, of course, while causing but a small annoyance to England, brought there with the representatives of Ireland, their families and the whole of the society in which they moved. In fact the wealthiest and most influential section of the Irish nation was, at one fell swoop, transferred to London, there to spend its money, and to learn to talk about the "Empire." The immense privilege of belonging to the "greatest Empire in the world," of being one with "the greatest people in the world," has since been shouted and preached and sung to us, till many of us have been beguiled into believing this

story of faery gold only to be lost—lost to our country in her direst need!

In this desertion our women participated quite as much as our men, they abandoned their Dublin mansions, to hire or buy houses in London, they followed the English Court about and joined the English ranks of toadies and place-hunters, bringing up their daughters in English ways and teaching them to make English ideals their ideals, and when possible marrying them to Englishmen.

Of course this could not go on for ever, and the Irish nation, at last realising that they and their interests had been sold for years, refused to be represented by them any longer...but it was too late—the rich and the aristocratic section of the men and women of Ireland had been lost to their country for years, if not for all eternity.

Now, I am not going to discuss the subtle psychological question of why it was that so few women in Ireland have been prominent in the national struggle, or try to discover how they lost in the dark ages of persecution the magnificent legacy of Maeve, Fleas, Macha and their other great fighting ancestors, True, several women distinguished themselves on the battlefields of '98,[1] and we have the women of the *Nation* newspaper, of the Ladies' Land League, also in our own day the few women who have worked their hardest in the Sinn Féin movement and in the Gaelic League, and we have the woman who won a battle for Ireland, by preventing a wobbly Corporation from presenting King Edward of England with a loyal address. But for the most part our women, though sincere, steadfast Nationalists at heart, have been content to remain quietly at home, and leave all the fighting and striving to the men.

Lately things seem to be changing. As in the last century, during the sixties, a strong tide of liberty swept over the world, so now again a

---

[1]This is a reference to the failed Irish Revolt of 1798, which was inspired by the French Revolution.

strong tide of liberty seems to be coming towards us, swelling and growing and carrying
before it all the outposts that hold women enslaved and bearing them triumphantly into the
life of the nations to which they belong.

We are in a very difficult position here, as so
many Unionist women would fain have us work
together with them for the emancipation of
their sex and votes—obviously to send a member to Westminster. But I would ask every Nationalist woman to pause before she joined a
Suffrage Society or Franchise League that did
not include in their Programme the Freedom of
their Nation. "A Free Ireland with no Sex Disabilities in her Constitution" should be the
motto of all Nationalist women. And a grand
motto it is.

There are great possibilities in the hands and
the hearts of the young women of Ireland—great
possibilities indeed, and great responsibilities.
For as you are born a woman, so you are born an
Irelander, with all the troubles and responsibilities of both. You may shirk or deny them, but
they are there, and some day—as a woman and as
an Irelander—you will have to face the question
of how your life has been spent, and how have
you served your sex and your nation?

The greatest gifts that the young women of
Ireland can bring into public life with them,
are ideals and principles. Ideals, that are but the
Inward Vision, that will show them their nation glorious and free, no longer a reproach to
her sons and daughters; and principles that will
give them courage and strength—the patient
toil of the worker, the brilliant inspiration of
the leader.

Women, from having till very recently stood
so far removed from all politics, should be able
to formulate a much clearer and more incisive
view of the political situation than men. For a
man from the time he is a mere lad is more or
less in touch with politics, and has usually the
label of some party attached to him, long before
he properly understands what it really means.

. . . Men all their lives are so occupied in examining closely, from a narrow party point of view
all the little Bills "relating to Ireland." . . .

Now, here is a chance for our women. Let
them remind their men, that their first duty is
to examine any legislation proposed not from a
party point of view, not from the point of view
of a sex, a trade, or a class, but simply and only
from the standpoint of their Nation. Let them
learn to be statesmen and not merely politicians. . . .

Fix your mind on the ideal of Ireland free,
with her women enjoying the full rights of citizenship in their own nation, and no one will be
able to side-track you, and so make use of you
to use up the energies of the nation in obtaining all sorts of concessions—concessions, too,
that for the most part were coming in the natural course of evolution, and were perhaps just
hastened a few years by the fierce agitations to
obtain them. . . .

Tenant right and peasant proprietorship, extension of the franchise and universal suffrage,
are all but steps in the evolution of the world;
for, as education and with it the knowledge of
the rights of a man or a woman to live, is
gained by the masses of mankind, so gradually
they push their way—individually and collectively—into the life of their nation, and being
in the majority, the moment they realise their
power, the world may be theirs for the taking.

But our national freedom cannot, and must
not, be left to evolution. If we look around us,
we will find that evolution—as far as Ireland is
concerned—is tending rather to annihilate us
as a nation altogether. We seem day by day to
be brought more and more in touch with England, and little by little to be losing all that
distinctiveness which pertains to a nation, and
which may be called nationality. London seems
to be coming nearer and nearer to us till quite
imperceptibly it has become the centre of the
universe to even many good Irishmen.

Of course all modern inventions have helped
England in the task of submerging our interests
in hers—trains, the penny post, the telegraph
system, have all brought her nearer, and given
her more power over us. More especially the
daily papers, forcing England upon us as the
headquarters of our politics, our society, our

stock exchange, our sport, teaching us to regard her foreign policy and her wars from the point of view of one or other of her political parties— all this, I say, wears away the rock of our national pride, and little by little we drift nearer to the conventional English views of life

The educational systems through the country have also, been used to work for the destruction of our nationality from the smart English governess who despised the mere Irish, to the village schoolmaster forced to train up his scholars in ignorance of Ireland's wrongs, in ignorance of Ireland's language and history.

The schools, too, were usually under the patronage of the priest or parson of the district, and therefore very naturally concentrated on developing strong sectarian feelings in the children, instead of the broader creed of nationalism.

Every right granted to us by England has been done in such a way that it helped to split us up into divisions and sub-divisions. That policy is being continued now. We have had landlords and tenants, Catholics and Protestants, North and South, besides the sub-divisions into the different sections of the English political parties. We are rapidly adding graziers and peasants, farmers and labourers, to the list of Irishmen who are all losing sight of their ideal, and sordidly scrambling for what they hope to get. They are curiously blind and inconsistent, too, for no great and real prosperity can be ours, while our interests are always the very last to be considered in an Empire....

In the ready-made clothing trades, England ...with her great manufacturing towns, has always been our worst enemy, and a tax on foreign ready-made clothes would tend to close the markets to all but English goods— the very ones (that once grant we require Protection at all) we require to be protected the most against. England's greatest rival in ready-mades is Germany. We, at present, count for nothing, and, of course, German goods excluded by a prohibitary tariff, England would do practically what she liked with the markets over here. Her firms would be in the position of

a certain English boot factory which established a shop in Limerick in competition to a Limerick boot factory. Being a rich firm, they were able to sell their wares under cost price to the unsuspicious people of Limerick, till the day when the Limerick factory closed its doors, unable to stand up against the competition, and from that day the people of Limerick have had to pay through the boots they wear—the expenses of the fight, and a huge dividend to the English company, as well as having the unemployed from the ruined industry to support, unless they emigrate.

All this points to the one way in which the women of Ireland can help their country; and, indeed, many of them are already doing so; and it is a movement too that all creeds, all classes, and all politics can join in. We have the Irish Industrial Development Associations, and the Sinn Fein organisations both working very hard for this object, but still there is much to be done.

It is not enough just vaguely to buy Irish goods where you can do so without trouble, just in a sort of sentimental way. No; you must make Irish goods as necessary to your daily life, as your bath or your breakfast. Say to yourselves, "We must establish here a Voluntary Protection against foreign goods." By this I mean that we must not resent sometimes having to pay an extra penny for an Irish-made article—which is practically protection—as the result of protection is that native manufacturers are enabled to charge the extra penny that will enable their infant factory to live and grow strong, and finally to compete on absolutely equal terms with the foreign-made article.

Every Irish industry that we manage even to keep in existence is an added wealth to our nation, and therefore indirectly to ourselves. It employs labour which serves to keep down the poor rates, and to check emigration. The large sums of money it turns over benefits every other trade in the country. So you also will benefit, for the richer the country, the better price you will be able to command for your services in the professions and trades, and the more positions will be created to which you may aspire....

If the women of Ireland would organise the movement for buying Irish goods more, they might do a great deal to help their country. If they would make it the fashion to dress on Irish clothes, feed on Irish food—in this as in everything, LIVE REALLY IRISH LIVES, they would be doing something great, and don't let our clever Irish colleens rest content with doing this individually, but let them go out and speak publicly about it, form leagues, of which "No English Goods" is the war-cry. Let them talk, and talk, publicly and privately, never minding how they bore people—till not one even of the peasants in the wilds of Galway but has heard and approved of the movement.

I daresay you will think this all very obvious and very dull, but Patriotism and Nationalism and all great things are made up of much that is obvious and dull, and much that in the beginning is small, but that will be found to lead out into fields that are broader and full of interest. You will go out into the world and get elected on to as many public bodies as possible, and by degrees through your exertions no public institution—whether hospital, workhouse, asylum or any other, and no private house—but will be supporting the industries of your country....

Again, another way of helping Ireland—make this country untenable for the British Army—let them be taught to paraphrase the Cromwellian cry and say when ordered to Ireland, "To Hell or Ireland." Take a leaf from the book of the Italian ladies in their treatment of the armies of the Austrian usurper. Boycott them, men and officers, let them realise what the sight of a red or khaki coat means to a right-thinking Irishman or woman; let them feel that you would force them to leave, that you would fain see in their place the gorgeous uniforms of an Irish army, the brilliant ranks of regiments like the Volunteers of '82. Make public opinion so strong that no Irish lad will ever again join the army of his country's enemies, to be at any moment called upon to "quell sedition" in his own country or to fight against other noble nations in the same plight as themselves.

If Irish boys could realize the contempt the British army is held in abroad, if they heard it talked of as the last relic of barbarism, a "mercenary army," and, as the most immoral army in the world, they would indeed hesitate before they entered it.

Then, again, you can educate your universities, colleges and schools. Don't permit pro-English propaganda....

England is now holding by force three civilised nations—nations whose ideals are Freedom, Justice and Nationhood—Ireland, India and Egypt, not to consider her savage territories and South Africa. Her colonies have to be coaxed into loyalty, and her House of Commons goes into hysterics over the news that Germany is building ships. Does that look like an Empire that is replete with a great National confidence, that is going to last for all eternity?

Wherever Ireland is known in the world, she is known by the great legacy her martyrs have left her, tales of noble deeds, of fearless deaths, of lives of self-denial and renunciation. Her name stands for the emblem of all that is brave and true, while England, her conqueror, has but gained for herself universally among the nations, the sobriquet of "La Perfide Albion."

To sum up in a few words what I want the Young Ireland women to remember from me: Regard yourselves as Irish, believe in yourselves as Irish, as units of a nation distinct from England, your conqueror, and as determined to maintain your distinctiveness and gain your deliverance. Arm yourselves with weapons to fight your nation's cause. Arm your souls with noble and free ideas. Arm your minds with the histories and memories of your country and her martyrs, her language, and a knowledge of her arts, and her industries. And if in your day the call should come for your body to arm, do not shirk that either.

May this aspiration towards life and freedom among the women of Ireland bring forth a Joan of Arc to free our nation!

137. Women's Popular Protests in Berlin

The homefront in Germany was characterized by high prices and shortages of basic necessities. Popular protests, which first took place in 1915, involved strikes, formal political demonstrations, and, above all, food riots. The majority of women protestors, predominantly working class, but also of the lower middle class, insisted on the ability as well as the obligation of authorities at every level to provide for basic needs and thereby prove their legitimacy. By the "hunger winter" of 1916–1917, women who lacked alternatives routinely spent the night in line, often in record freezing temperatures, for a chance to obtain food when shops opened in the morning. Women also directly confronted various authorities in the hope that the military government would recognize their contributions as "soldiers of the home front." But the protestors' hopes were largely unfulfilled and food protests and other demonstrations became increasingly violent and desperate. Many women working in the munitions industries believed that they could stop the war by participating in a general strike; the massive walkout of January 1918 offers insight into women's sentiment and the fearful official response. Women participated in great numbers in the Revolution of November 1918, but the final collapse of the government also owed much to their ceaseless protests during the four years of the war. Official concerns about the political consequences of these protests are revealed in thousands of "reports on morale" drawn up by policy and military personnel. The reports vividly recorded the comments and actions of protestors throughout Germany and were used by the government to assess civilian morale. These accounts of protests, which took place in Berlin, include two written by women protestors and five by the police. They cover nearly a four-year period, from February 1915 to November 8, 1918, the day before the German Revolution. The date preceding each police report indicates when it was written, often a day or two after the events described.

A. February 17, 1915

On the 16th of the month at 5:00 p.m., thousands of women and children gathered at the municipal market hall in Andreas Street to buy a few pounds of potatoes. As the sale commenced, everyone stormed the market stands. The police, who were trying to keep order, were simply overrun and were powerless against the onslaught. A life-threatening press at the stands ensued; each sought to get past the next ...women had their possessions ripped from them and children were trampled on the ground as they pleaded for help.... [W]omen who got away from the crowds with some ten pounds of potatoes each were bathed in sweat and dropped to their knees from exhaustion before they could continue home.

—*Report of Officer Rhein*

B. October 17, 1915

After 25 minutes I entered Edison St. in Öberschoneweide [just outside Berlin] and then turning onto Wilhelm St. came upon a crowd of several thousand men and women, who were loudly howling and pushing the policemen aside. I learned from the sergeant on duty, who had received several head injuries...that the crowd had already stormed several butter shops

From Bundesarchiv Lichterfelde, Provinz Brandenburg Repositur 30, Berlin C. Police Presidium, Nrs. 15809, 15814, 15821, and 15851. Contributed and translated by Belinda Davis.

because of the prices.... [S]everal large display windows were shattered, shop doors destroyed, and entire stocks were simply taken.... I was asked to close off the street as the police and officers were completely helpless against the crowd. We cleared the street with fifteen mounted officers.... Various objects such as flower pots were thrown at us.

—*Report of Officer Krupphausen*

C. October 29, 1915

The demonstration planned by the women of the Social Democratic Party (SPD) has now actually taken place.... About 250 women gathered [before SPD headquarters]. Shortly after 11:00 a.m., a deputation of six women entered the executive offices, where a meeting of the party executive and board was taking place, to communicate their desires and demands. The women's patience was tested, as it took about an hour before the party executive would speak to them. The women became even more restless as the executive declared that the meeting would not be interrupted for any reason. The deputation was therefore not admitted. This was communicated to the women waiting impatiently outside. Like a bolt of lightning, they stormed the steps and forced their way into the meeting room shouting abuses ("traitor to the people"). They took the empty seats that remained in the room. Those without seats blocked the entrance. Now, comrade Haase entered and bade the women to wait in the adjacent room until something definite could be determined in the meeting.... Many women cried out that they could no longer bear their sad existence. Others rebuked the party executive and again called party members traitors to the people... who no longer cared about the people's needs.

—*Report of Officer Schrott*

D. [In early February 1917], despite the state of siege[1] more than 500 of us [factory workers] marched towards the Britz [suburb of Berlin] municipal office. [Inside the office], the police were pushed to one side [by the workers]. Mayor Schmiedigen and the Bread Commission could not get out of the room [as the door was blocked by the workers]. The members of the City Council had to remain in their seats. The women's delegation pointed out the urgent needs of the people. As... Schmiedigen tried to insult me with the epithet "representative of the people," many women responded, "we are all representatives of the people." The demonstration of women from Britz was so effective that not only the [local working-class newspapers], but also *Der Volksfreund* in Braunschweig ... reported on it. We also succeeded in obtaining extra bread ration cards and cards for rolls and barley as a substitute for potatoes.

—*Account of Martha Balzer*

E. July 4, 1917

At today's public market in Viktoria Platz, there was not a single vegetable to be had [due to extreme speculation and the holding back of goods by farmers]. The city magistrate is apparently not in a position to provide a substitute for potatoes. The mood of the populace is therefore extremely agitated. Discussion with totally reasonable women confirms that a portion of the population is now actually suffering from starvation.[2] Many people left the market empty-handed and in small groups headed for the City Hall where they demanded bread and potatoes. Informed by the civil servant that they should get bread where it was sold, a number of women walked down Gürtel Street to the shop of the baker Hans Schwarz... [where] against the will of the clerk, they forcibly took 30 loaves of bread without paying. A riot arose over this, involving about 150 people, in which the shop's awning

---

[1] In August, 1914, the government imposed the Prussian state of siege law which limited assemblies in the street, criticism of the state and/or the war effort, and increased censorship and police controls.

[2] As a result of food shortages caused by the Allied blockade, over 750,000 German civilians died of starvation during the war.

was damaged.... This episode demonstrates that it is high time that efforts were made to bring in sufficient supplies of fruits and vegetables and to distribute them to the broad masses of the population whose patience is evidently exhausted.

—*Report of Officer Kuhlmann to Von Oppen*

F. February 22, 1918

The Independent Social Democratic Party of Germany[1] has worked zealously in secret, particularly among women, to gain support for the mass strike it is planning in the munitions works. The women took part in an outstanding way in the [January 1918 mass] strike itself, as well as in the demonstrations on the streets. Spurred on by the "Independents," the women then let themselves get carried away in several riots. If the political mass strike has not brought about the success the Independents had hoped for, they still believe that the strike put real pressure on the government, and created support in the factories for continuing the strike. Here again, the women are playing an important role. At the same time, if agitated women are often the ones leading the strike and street demonstrations, there is also a

---

[1]The Independent Social Democratic Party (USPD) was formed in January 1917 by members of the antiwar faction of the Social Democratic Party.

large portion of women who have been discouraged from initiating a new strike because of the arrest and conviction of many under war law [state of seige]. These second period women do not want to know anything more about a new strike and, for the time being, are not interested in street demonstrations either.

—*Report of Officer Palm*

G. On November 8, 1918, the revolutionary shop stewards of the Berlin metal shops came together...in the party headquarters of the USPD to finalize plans for a Berlin general strike and demonstration on November 9, 1918. Various representatives of Berlin working-class youth were also present to advise on the participation of youth....I belonged to the group that was supposed to go to Knorr-Bremse in Lichtenberg. After the discussion... we went to Alexanderplatz [a large, public square in Berlin]....We youth comrades possessed various weapons [for possible use the next day]. Erich Habersaath again showed me how to take a revolver apart, clean it, and load it. At first, some didn't want to give us girls any weapons, because they thought such things weren't for women. But in the end I got my revolver.

—*Account of Lucie Gottschar-Heimburger*

138. Käthe Kollwitz
Letters and Diaries from World War I

Käthe Kollwitz (1867–1945) is one of the most important graphic artists of the first half of the twentieth century. The first woman elected to the Prussian Academy of Arts in 1919, she worked in a variety of mediums: woodcuts, posters, charcoal drawings, engravings, lithographs, and sculpture. Kollwitz's art was imbued with her feminist, socialist, and pacifist political views. She became interested in depicting working-class life following her marriage in 1891 to Dr. Karl Kollwitz. Through his practice, she met the industrial workers of Berlin. Her series "The Weavers' Revolt" (1895–1897) consisted of three engravings and lithographs depicting the 1844 revolt of the Silesian weavers. The

From pp. 62–63, 73, 87–90 of *The Diary and Letters of Kaethe Kollwitz,* edited by Hans Kollwitz, translated by Richard and Clara Winston. Copyright © 1988 by Northwestern University Press. Reprinted by permission.

A. **"The Parents"** "The Parents," 1920 charcoal drawing for the "War" series of woodcuts published in 1924. *(Kupferstichkabinett. Staatliche Museen zu Berlin/Preussischer Kulturbesitz. Photo: Jorg P. Anders.)*

prominence that Kollwitz gained from "The Weavers' Revolt" led to her appointment to teach at the Berlin School for Women Artists. In 1913, she founded the Women's Art Union in Berlin. Following the outbreak of World War I, the mother and child theme as well as the general suffering caused by war became central to Kollwitz's work and she is perhaps best known for her works memorializing World Wars I and II. Kollwitz lost her youngest son, Peter, in the first war and her grandson, also named Peter, in the second. In 1924, she published her series "War," comprising seven woodcuts, and produced her three most famous posters: "Germany's Children Starving," "Bread," and "Never Again War." Kollwitz's sculptures of grieving parents, "The Mother" and "The Father," were unveiled in 1932 in the Roggevelde Military Cemetery in Belgium. Modeled on herself and her husband, Karl, the granite sculptures were placed at the cemetery entrance and overlook the field of crosses where their son was buried. The Nazi government forced Kollwitz to resign from the Academy in February 1933, prevented her from exhibiting in museums, and prohibited art galleries from purchasing her art. She completed "Death," her last major series of lithographs, in 1936. Kollwitz's views on the war are also evident in her private writings. The following diary entries and letters, published in *The Diary and Letters of Käthe Kollwitz*, are from August 1914 to January 1919.

B. Diary Entries, August 1914–January 1919

August 27, 1914

In the heroic stiffness of these times of war, when our feelings are screwed to an unnatural pitch, it is like a touch of heavenly music, like sweet, lamenting murmurs of peace, to read that French soldiers spare and actually help wounded Germans, that in the franc-tireur villages German soldiers write on the walls of houses such notices as: Be considerate! An old woman lives here.—These people were kind to me.—Old people only.—Woman in child-bed.—And so on.

A piece by Gabriele Reuter in the *Tag* on the tasks of women today. She spoke of the joy of sacrificing—a phrase that struck me hard. Where do all the women who have watched so carefully over the lives of their beloved ones get the heroism to send them to face the cannon? I am afraid that this soaring of the spirit will be followed by the blackest despair and dejection. The task is to bear it not only during these few weeks, but for a long time—in dreary November as well, and also when spring comes again, in March, the month of young men who wanted to live and are dead. That will be much harder.

Those who now have only small children, like Lise her Maria, seem to me so fortunate. For us, whose sons are going, the vital thread is snapped.

September 30, 1914

Cold, cloudy autumnal weather. The grave mood that comes over one when one knows: there is war, and one cannot hold on to any illusions any more. Nothing is real but the frightfulness of this state, which we almost grow used to. In such times it seems so stupid that the boys must go to war. The whole thing is so ghastly and insane. Occasionally there comes the foolish thought: how can they possibly take part in such madness? And at once the cold shower: they *must, must!* All is leveled by death; down with all the youth! Then one is ready to despair.

Only one state of mind makes it at all bearable: to receive the sacrifice into one's will. But how can one maintain such a state?

[Peter Kollwitz was killed on October 22, 1914.][1]

---

[1]Käthe Kollwitz's diaries and letters were edited and published by her oldest son, Hans, who provided the information that appears in brackets here and in subsequent diary entries. Peter Kollwitz volunteered to serve in the German army in August 1914, the first month of the war. He was eighteen years old when he was killed in Flanders in October.

December 1, 1914

Conceived the plan for a memorial for Peter tonight, but abandoned it again because it seemed to me impossible of execution. In the morning I suddenly thought of having Reike ask the city to give me a place for the memorial. There would have to be a collection taken for it. It must stand on the heights of Schildhorn, looking out over the Havel. To be finished and dedicated on a glorious summer day. Schoolchildren of the community singing, "On the way to pray." The monument would have Peter's form, lying stretched out, the father at the head, the mother at the feet. It would be to commemorate the sacrifice of all the young volunteers.

It is a wonderful goal, and no one has more right than I to make this memorial.

December 9, 1914

My boy! On your memorial I want to have your figure on top, *above* the parents. You will lie outstretched, holding out your hands in answer to the call for sacrifice: "Here I am." Your eyes—perhaps—open wide, so that you see the blue sky above you, and the clouds and birds. Your mouth smiling. And at your breast the pink I gave you.

August 27, 1916

Read an essay on liberalism by Leopold von Wiese. It showed me all the contradictory elements within myself. My untenably contradictory position on the war. How did I come to it? Because Peter sacrificed his life. What I saw so clearly then and what I wanted to preserve in my work now seems to be once more so dubious. I think I can keep Peter only if I do not let anyone take away from me what he taught me then. Now the war has been going on for two years and five million young men are dead, and more than that number again are miserable, their lives wrecked. Is there *anything at all* that can justify that?

And now Wiese speaks of the necessity of "opposing utterly all sacrifice of the living to a lifeless idea." "For a pair of happy eyes means more than all the doctrines of worldly wisdom." Surely that is something different from the joy in the law with which Peter and his fellows marched into the field. And different from what Rupp taught: "Man is not born for happiness, but to do his duty."

September 1916

My work seems so hopeless that I have decided to stop for the time being. My inward feeling is one of emptiness. How shall I find joy outside of the work? Talking to people means nothing at all. Nothing and no one can help me. I see Peter far, far in the distance. Naturally I will not give it up—possibly I cannot—but I shall make a pause. Now I have no joy in it. All day yesterday I took care of a host of things. But what for?

March 19, 1918

Yesterday I kept thinking about this: How can one cherish joy now when there is really nothing that gives joy? And yet the imperative is surely right. For joy is really equivalent to strength. It is possible to have joy within oneself and yet shoulder all the suffering. Or is it really impossible?

If all the people who have been hurt by the war were to exclude joy from their lives, it would almost be as if they had died. Men without joy seem like corpses. They seem to obstruct life.... When someone dies because he has been sick—even if he is still young—the event is so utterly beyond one's powers that one *must* gradually become resigned to it. He is dead because it was not in his nature to live. But it is different in war. There was only one possibility, one point of view from which it could be justified: the free willing of it. And that in turn was possible only because there was the conviction that Germany was in the right and had the duty to defend herself. At the beginning it would have been wholly impossible for me to conceive of letting the boys go as par-

ents *must* let their boys go now, without inwardly affirming it—letting them go simply to the slaughterhouse. That is what changes everything. The feeling that we were betrayed then, at the beginning. And perhaps Peter would still be living had it not been for this terrible betrayal. Peter and millions, many millions of other boys. All betrayed.

That is why I cannot be calm. Within me all is upheaval, turmoil.

Finally I ask myself: What has happened? After the sacrifice of the boys themselves, and our own sacrifice—will not everything be the same?

All is turbulence.

October 1, 1918

Germany is near the end. Wildly contradictory feelings. Germany is losing the war.

What is going to happen now? Will the patriotic emotion flare up once more so powerfully that a last-ditch defense will start? Kerr's poem. I find in myself no agreement with it. Madness not to cut the war short if the game is up, not to save what still may be saved. The young men who are still alive, Germany must keep; otherwise the country will be absolutely impoverished. Therefore, *not another day of war* when it is clear that the war is lost.

October 30, 1918

The *Vorwaerts* printed my reply to Dehmel after all, and the *Vossische Zeitung* has reprinted it.

[Article from *Vorwaerts*][2]

To Richard Dehmel. Reply by Kaethe Kollwitz.

In the *Vorwaerts* of October 22 Richard Dehmel published a manifesto entitled *Sole Salvation*. He appeals to all fit men to volun-

---

[2]Käthe Kollwitz's letter to the poet Richard Dehmel in *Vorwärts* is not part of her original diaries but was added by Hans Kollwitz for the benefit of the reader. It was in the last months of the war, in October of 1918 (four years after the death of Peter Kollwitz), that Dehmel issued his appeal for old men and young boys to voluntarily enlist in the German army.

teer. If the highest defense authorities issued a call, he thinks, after the elimination of the "poltroons" a small and therefore more select band of men ready for death would volunteer, and this band could save Germany's honor.

I herewith wish to take issue with Richard Dehmel's statement. I agree with his assumption that such an appeal to honor would probably rally together a select band. And once more, as in the fall of 1914, it would consist mainly of Germany's youth—what is left of them. The result would most probably be that these young men who are ready for sacrifice would in fact be sacrificed. We have had four years of daily bloodletting—all that is needed is for one more group to offer itself up, and Germany will be bled to death. All the country would have left would be, by Dehmel's own admission, men who are no longer the flower of Germany. For the best men would lie dead on the battlefields. In my opinion such a loss would be worse and more irreplaceable for Germany than the loss of whole provinces.

We have learned many new things in these four years. It seems to me that we have also learned something about the concept of honor. We did not feel that Russia had lost her honor when she agreed to the incredibly harsh peace of Brest-Litovsk. She did so out of a sense of obligation to save what strength she had left for internal reconstruction. Neither does Germany need to feel dishonored if the Entente refuses a just peace and she must consent to an imposed and unjust peace. Then Germany must be proudly and calmly conscious that in so consenting she no more loses her honor than an individual man loses his because he submits to superior force. Germany must make it a point of honor to profit by her hard destiny, to derive inner strength from her defeat, and to face resolutely the tremendous labors that lie before her.

I respect the act of Richard Dehmel in once more volunteering for the front, just as I respect his having volunteered in the fall of 1914. But it must not be forgotten that

Dehmel has already lived the best part of his life. What he had to give—things of great beauty and worth—he has given. A world war did not drain his blood when he was twenty.

But what about the countless thousands who also had much to give—other things beside their bare young lives? That these young men whose lives were just beginning should be thrown into the war to die by legions—can this really be justified?

There has been enough of dying! Let not another man fall! Against Richard Dehmel I ask that the words of an even greater poet be remembered: "Seed for the planting must not be ground."

Kaethe Kollwitz

New Year, 1918[3]

This year we want to go to the Sterns. The past five New Year's Eves were turned backward. Were full of grief, mourning, longing for peace. We do not want to pass this New Year's Eve alone. Hans is here. Together with him we want to be with our dearest friends, the Sterns, to greet the coming year. For now all is future. A future that we resolve to see as bright beyond the immediate darkness. We don't want to be alone today; we want to give ourselves courage, to strengthen and express faith.

This year has brought the end of the war.

There is no peace yet. The peace will probably be very bad. But there is no more war. It might be said that instead we have civil war. But no, there have been troubles, but it has not come to that. The year 1918 ended the war and brought the revolution. The frightful pressure of war that grew steadily more intolerable is lifted now; one can breathe more easily. No one imagines that good times will follow right away. But we have finally crawled through the narrow shaft in which we were prisoned, in which we could not stir. We see light and breathe air.

---

[3]As the German Revolution and the end of the war took place in November 1918, this must be from New Year's 1919.

QUESTIONS FOR ANALYSIS

1. How is women's war work represented in the poster titled "These Women are Doing Their Bit. Learn to Make Munitions."? Why is the main female worker waving to the soldier?

2. What were Vera Brittain's initial reactions to World War I? How do her diaries illuminate how her friends, family, and much of the general public regarded the events that took place during the first weeks of the war? How does Brittain's poem "The German Ward" convey her changing views on the war?

3. How does Lady Lugard's description of the plight of the Belgium refugees help us to understand the mobilization of public support for British intervention in the war? Discuss the organizational structure and activities of the War Refugees Committee and what it reveals about women's voluntarism during the war. How did class, gender, and religion inform the work of the committee?

4. What special measures does Sylvia Pankhurst propose to protect women workers in war industries? How does she link the issue of war work to the issue of women's suffrage? Contrast Pankhurst's discussion of industrial conscription with the depiction of women workers in the munitions poster.

5. On what basis does the "Call to the Women of All Nations" appeal to women to attend an International Congress of Women in Holland? Discuss the underlying principles of the twelve-point resolution adopted at the Congress. How does it define women's relationship to issues of war and peace?

6. In Ellen Key's view, how has World War I transformed marriage and motherhood? Why does she foresee a bleak future for women and mothers in the postwar period? What alternatives does Key propose? What concepts of biology and "race-hygiene" inform her views?

7. Discuss Constance Markievicz's analysis of Ireland's relationship to England. How does Markievicz call on women to participate in the struggle for Irish independence, and what unique contributions does she believe they can make?

8. What was the nature of women's militant protests in Berlin as related in police reports and the accounts of women workers? What opinions did the police express about these protests? How did the women's conditions and attitudes change in the three-year period from 1915 to 1918? What do the reports reveal about how the women viewed their rights as consumers and workers and the obligation of the Social Democratic Party and the government to represent their interests?

9. Discuss the evolution of Käthe Kollwitz's views on World War I as recorded in her diary entries from 1914 to 1919. How did Kollwitz's role as a mother shape these views and how did she seek to memorialize her son Peter? Discuss the imagery and multilayered meanings of her drawing "The Parents."

10. How do the sources in this section reveal different aspects of the World War I homefront as well as the variations in the conditions faced by women in Britain and Germany?

THE RUSSIAN REVOLUTIONS OF 1905 AND 1917

The Russian Revolution took place in three phases: the Revolution of 1905 created a weak Russian Parliament, or Duma; the February Revolution of 1917 followed the abdication of Tsar Nicholas II and established a liberal, democratic Provisional Government which remained committed to the war effort; and the Bolshevik Revolution of November 1917 ended Russian involvement in the war and began the process of building socialism around the principles of the Bolshevik Party led by Lenin.

139. Program of the Women's Progressive Party

Following the Revolution of 1905, Tsar Nicholas II created a new State Duma. Hopeful of the possibilities for social transformation, various groups drew up petitions and presented their demands in the interlude between the October 17, 1905 Manifesto and the convocation of the Duma on July 10, 1906. In the October Manifesto, the tsar made a qualified pledge to broaden the electorate for the new Duma. In December, however, he announced that women would not be granted the right to vote. This provoked protests from Russian feminists who continued to demand women's suffrage. Two new women's groups were also formed in 1905 and 1906. The first, the Union of Women's Equality (1905–1908) was primarily a suffrage organization with a membership of about 8,000 women and eighty branches throughout Russia. Its most radical members supported universal suffrage and labor reform. But the union's influence diminished soon after 1905. The Women's Progressive Party (1906–1917), by contrast, advocated a broad program of women's rights and social reform. It was founded in St. Petersburg by Dr. Mariya Ivanovna Pokrovskaya, a member of the Russian Women's Mutual Philanthropic Society, then in existence for ten years. In 1902, she launched a campaign against state-regulated prostitution and two years later initiated a new journal, the *Women's Messenger,* which focused on medical and social issues concerning children, the poor, and prostitutes. When the Russian Women's Mutual Philanthropic Society rejected her proposal to form a political party, maintaining that it was a charitable rather than a political organization, Pokrovskaya established the Women's Progressive Party. First meeting in January 1906, the party included a Women's Club that discussed women's topics on a regular basis.

Women are the most rightless, dispossessed part of the population. They find themselves in subordination to men and in complete dependence upon their will. They are deprived of their human personality and enslaved by the existing laws and their economic dependence on the ruling sex. Because of their subordination, economic dependence and lack of rights, women cannot develop all their spiritual abilities and render active assistance to the perfection of mankind (in spiritual and physical respects) and improvement of the social order. Because lack of political rights is one of the main causes of the enslavement of women, the Women's Progressive Party makes its immediate goal the attainment of complete political equality of women with men.

The Women's Progressive Party regards it as necessary to struggle with all the shortcomings of contemporary life and to realize general human ideals: truth, equality, fraternity, freedom, justice and humanitarianism. It holds that realization of these ideals is only possible through a peaceful, evolutionary path. To achieve these goals, it deems the following necessary:

1. All citizens of the Russian Empire, without distinction of sex, confession or nationality, are to be equal before the law. Any differences based on social estate [*soslovie*] and all limitations on personal and property rights should be eliminated.

2. All citizens should be granted freedom of conscience and confession, freedom of speech (oral, written and printed), inviolability of person and home, and freedom of assembly, strike, petition, movement (through abolition of passports), and business. No one may be subjected

"Program of the Women's Progressive Party, 30 January 1906."

to persecution and punishment except through judicial authority and on the basis of law.

3. At the present time, a constitutional-democratic monarchy is the most appropriate form of government for Russia. A popular representative assembly should be elected by all citizens who have attained the age of 21 years, without distinction of sex, nationality and confession, on the basis of an equal, direct and secret vote. All citizens are to be accorded a passive electoral right. The popular assembly is to prepare a constitution, to wield the right of legislative initiative, to hold ultimate authority for the promulgation of laws, to exercise broad control over executive authority, to have the right to confirm the state budget, and to exercise control over it. The ministers are to be responsible to the popular assembly.

4. All the peoples inhabiting Russia are to be united in the name of general human ideals. All the peoples of Russia are to enjoy complete freedom in the use of their language in the press, court, school, and various public institutions. Russian is recognized as the state language.

5. The broad development of local self-government [should be achieved] over the entire expanse of the Russian state on the same principals as those of the popular assembly. Self-governing units for the better realization of their goals can form themselves into unions. Satisfaction of all local needs is in the jurisdiction of self-administrative organs. They have the power to tax themselves. Central authorities are only to oversee the activity of local self-government organs and to take them to court in the event they violate the law.

6. The court should be free from any kind of external pressure and is to be guided solely by the law. Justice should be immediate, prompt, merciful, and the same for all citizens. The court should be composed of women (in the capacity of judges, lawyers and jurors). They are to have the right to occupy all judicial positions. Capital punishment is to be abolished. Legal custody, suspended sentences, and parole are to be introduced. Women are to participate in the review of existing laws, both criminal and civil.

Civil marriage is to be established in family law and made obligatory for all. Religious sanctification should be left to the free will of each. Parents hold equal authority over the children. Wives are equal to their husbands in all respects, have the right to one-half of family savings, and in a legal manner should be made economically independent of their husbands if they cannot have their own income because of family circumstances. The procedure for divorce is to be simplified and made easier. A broad defense of children, especially illegitimate children, should be established through legislation. Laws should be promulgated to prohibit the establishment and maintenance of every kind of house of prostitution, "homes of tolerance" and commerce in female bodies, and to establish severe punishment of middlemen, souteneurs, and other people who derive material profit from trade in women or who contribute to their seduction into prostitution. Medical and police supervision of prostitution and prohibition of marriage by military men are to be abolished. Punishment for appearing drunk in public places, more severe punishment for crimes and misdemeanors committed under the influence of alcohol, and limitation (and, in time, a total ban) on the production and sale of spirits are to be promulgated.

7. Indirect taxes are abolished. So far as possible, a progressive income tax and progressive inheritance tax are to be introduced.

In the economic sphere, one should strive to abolish the unequal distribution of wealth and to [introduce] a more just compensation for labor. Each adult capable of working should earn his own means of existence; no one should live off the exploitation of others' labor. Industrial enterprises should be established primarily on collective artisan [*artel'*] principles or, at least, the workers should have the right to a certain percent of the profit. Legislative defense should exist for all forms of hired labor (including house servants). Inspection that is independent of entrepreneurs and that includes participation by elected representatives of the workers should oversee the enforcement of laws concerning the protection of labor. Inspection by women should be established, especially in those industrial enterprises where female labor is employed, but it should also be admitted to other establishments as well. An eight-hour working day should also be introduced. It may be increased or reduced, depending upon the nature of industry: in harmful industries, the working day should be reduced to the minimum feasible level, while in industries that are not harmful it can be increased, but only with the consent of the workers and by no more than two hours [per day]. The workday should include a two-hour break for lunch and rest. Night work is permissible only for those kinds of production where it is [technically] unavoidable. Night work should not exceed six hours. Overtime work is prohibited. Use of child labor of either sex is prohibited for those of school age (under 16); adolescents between 16 and 18 are not to work more than six hours a day. Women are to be freed of work for four weeks before and six weeks after giving birth and are to receive their wages in full during this period. Child-care for infants and small children is to be established in all enterprises where women work or serve. Women who are still nursing infants are to be freed [from work] for one-half hour every three hours. To supervise women working in factories, female supervisors are to be appointed in lieu of male supervisors. Compensation for labor by both sexes is to be made equal, and equality is also to be established in their work relationships.

State insurance is to be established for all (without exception) for old age, illness and disability. Arbitration courts are to be established to resolve misunderstandings between workers and employers. Payment in kind is to be prohibited.

8. Peasant landholding is to be expanded through various means: transfer to peasant ownership or use of state, church and privately owned lands on terms that will be worked out by the popular assembly (according to local conditions) and that will facilitate the transfer of a sufficient quantity to the people so that they can cultivate [the land] by their own family and without using hired labor. Colonization [to Siberia] must be improved. Women have equal right to the use of land. Peasant women should have a land allotment equal to that of peasant men. Preservation of communal ownership or the transition to private household allotments should be left to the judgment of the peasants themselves. A broad development of agricultural cooperatives, unions and other organizations (with the goal of improving the material condition of small landowners) is needed. A rural inspectorate is needed wherever hired labor is used. It is necessary to have a broad development of model farms, experimental fields, and other institutions that can assist in improving agriculture.

9. Representatives of central authorities are only to exercise control over education. They are not to violate freedom of moral development and education. The moral and educational aspects of the schools and other educational institutions are to be in the hands of local self-government organs; moreover, final authority over these matters belongs to people employed in moral development and education. Along with public and state schools, private schools that have been founded by private persons or organizations may also exist. Such schools are under the control of authorities but have freedom in their organization of moral and educational development. Women should have the right to hold all positions in schools. Education of both sexes is to be identical. There is to be equal, free, obligatory, universal and professional education for all children of both sexes until the age of 16 years. Free education in secondary and higher institutions of learning is desirable. Institutions of higher education are to have full autonomy. There is to be joint education of both sexes in all forms of schools, but separate schools for them are permissible. There are to be reforms in institution in order to unite all schools (elementary, secondary and higher) so as to facilitate the transfer of children from one school to the next. Schools should give attention to the intellectual development of pupils; to the improvement in their moral level; to the ennoblement of their souls through a strengthening of their will; to the development of a sense of justice, respect for women, sensitiv-

ity to beauty, love and a devotion to general human ideals; and to physical development. There is to be a broad concern for professional, agricultural and extracurricular education.

10. There is to be a broad concern for public health with the aim of improving the sanitary conditions of life for the urban, rural and factory populations. Rational removal of human waste and refuse from populated areas, good water supplies, healthy and inexpensive housing for workers (and in general for all the unpropertied classes), and other useful institutions (e.g., playgrounds for children, public laundries, bathhouses, etc.) are to be established. There is also to be sanitation supervision over housing, food products and other items of consumption, every kind of hired labor (including house servants, industrial and commercial institutions, schools, etc); in addition, women are to be involved in supervision on the same terms as

men. Criminal liability is to be established for violation of laws on the protection of labor and, in general, laws and binding decrees on the protection of public health. Medical assistance is to be free. But the main effort is to be directed toward sanitary measures, which, by protecting the health of the populace and preventing illness, should reduce medical assistance to a minimum.

11. Militarism must be abolished and, instead of armies, a militia should be established. For defense against external foes, it is necessary to conclude arbitration agreements with all the neighboring countries that surround Russia, and to strive to develop peaceful unions between various groups of the Russian population with various foreign states. It is necessary to repudiate the policy of aggression and to direct efforts toward the peaceful development of the country.

140. Muslim Women and the First All-Russian Muslim Congress

In the wake of the February Revolution of 1917, the new Provisional Government announced immediate preparations for calling a Constituent Assembly to decide the form of Russia's government. Among the groups that organized to formulate positions on key questions of governance and civil rights were representatives of Russia's vast Muslim population. Muslim members of the Fourth Duma convened provincial conferences in April to prepare a program and elect deputies to the first All-Russian Muslim Congress which was held in Moscow from May 1 to May 11. Muslim women also gathered in April at a special conference in Kazan to draft a specific set of demands on women's issues. The general congress in May was attended by more than 770 delegates, including about 200 women. The delegates represented some 30 million Muslims from all regions of the Russian empire. Liberal reformers dominated the Congress and strove for the introduction of a democratic federal republic, the emancipation of Muslim women, and the reorganization of Muslim religious administration. The presence of women and the presentation of their program evoked vigorous objections from conservative clerical deputies at the Congress. Three women nevertheless attained influential positions: to the Congress Presidium, to the Orenburg Spiritual Assembly entrusted with reform of the religious administration, and to the

Contributed by Sagit Fiaritovich Faizov and translated by Faizov (Tatar and other Turkic languages into Russian) and Joan Afferica (Russian into English).

Central Muslim Council. The first document presents excerpts from a speech by Fatima-tutash Kulhamidova,[1] a representative of the Women's Section of the Congress. The second document is the full text of a ten-point resolution prepared for the Congress by the Kazan women's conference.[2] It was introduced by Kulhamidova on behalf of the Women's Section and adopted by the Congress.

A. Fatima-tutash Kulhamidova, Report from the Women's Section

Ladies and Gentlemen! When one speaks of the "Muslim woman," the most dismal, the most unattractive images appear before one's eyes. Rights and freedoms taken away, oppression and coercion, slavery and a deplorable lack of understanding by women themselves of their position, or the absence of even the possibility of understanding it—such are the images that appear one after another when one views the life of our Muslim women.

Muslim women, called upon to aid their husbands along life's path...are regarded as some kind of unnecessary and extraneous element in social existence....Not only are their political rights sacrificed, but their civil rights are as well. Moreover, they are not considered fit for situations that require them to be alongside and act together with men, with their comrades. Because of this, a veil covers [the Muslim woman's] face and from childhood she is taught that it is forbidden to show herself before the eyes of others. She is raised shut up in a dark humid house. She is dominated by her father, husband, and children. They see her as a virtual slave deprived of political, national, and religious rights. Therefore, they draw her to tasks that they deem desirable [for women]. And although the woman is supposed to be mistress of the house, she is denied those rights which even servants possess. For she is considered only purchased property. When they say "she is a woman," they oppress her spirit and deny her right to an inner world. She is accorded no respect. She is considered un-

worthy not only of a politically and socially-oriented upbringing, but a national and religious one as well. They say: "After all she is not a father, she is a mother; she is not a human being, she is a woman...."

To this day, however, there is one secret feeling that has not been extinguished in the soul of the Turco-Tatar woman. She is still alive, she has survived. So! This feeling tells her to look closely at the surrounding reality. She has seen that everywhere life-giving principles have become confused, have been trampled and crushed by the absence of rights, by injustice, by ignorance. Her spirit has been filled with bitterness. She could not consent to the fact that the people around her live such a life. She deemed it necessary to improve their lives, to give them consciousness and knowledge and, in this way, to make our future way of life brighter....

She has become a teacher, she has seen the light. She has had to sacrifice much on this path. But nothing terrified or frightened her.... She drew public attention to the fact that her sisters each year die by the thousands owing to the prejudices—coming from who knows where—that they not show themselves to a male physician. She decided to reform even this side of life. She became a physician herself.... The most respected, the most noble of our sisters were sacrificed along the way. In her struggle for the rights granted by nature, she drew apart from the attitudes of fathers and grandfathers and understood the need to study the law herself. And along this path she came closer to her goal. She was not satisfied by her immediate surroundings alone; she deemed it necessary

---

[1] The *tutash,* which follows the first name Fatima, is a part of speech that indicates an unmarried woman.

[2] These documents appear in the Congress protocols which were published in Tatar and other Turkic languages in Arabic script in 1917 in Petrograd.

to help sisters far away, those who live beyond the Black Sea. She invited them to unite on the path of struggle for the rights of women. She proved in deed, even for those countries, that she indeed wishes to achieve her aims. She showed the way to those Turco-Tatar women living beyond the Black Sea. She extended a helping hand to her Muslim women friends there. Turco-Tatar women consider it their most important task to send their own female representatives or those male representatives deserving of trust to all institutions which prepare civil laws for all Muslims and to judicial institutions connected in many ways with the interests of Muslim women.

The Turco-Tatar woman believes that success awaits her on the path to the posited goal. (Applause.) If she did not believe this, she would not believe that for her will shine the bright star of the future, the future of our nation. This star will begin to shine only when women, as equals, begin to take part in a broad range of activities. . . . (Applause.)

B. The Ten-Point Resolution of the Women's Section

I.1. In accordance with the Shariah,[1] men and women are equal.

2. In accordance with the Shariah, women have the right to take part in political and social life. Therefore, in accordance with the Shariah, women have the right to take part in elections.

3. In the Shariah, there is no requirement that women wear the *hijab* (veil).

II. Only that nation can be strong whose women are granted equal rights, and this equality can be realized only if women are ready to take part in the activity of institutions drafting legislation concerning women. Muslim women, therefore, must have equal political rights with men. Women must elect and be elected to the Constituent Assembly.[2]

III. To elect and be elected is a new experience for Muslim women: in the first place, because women until now have not had experience participating in public activity; in the second place, it is probable that men will oppose women visiting voting places. In order to avoid losing the votes of half the population, the votes of women, the participation of each Muslim woman must be regarded as an obligation at the time of the elections to the Constituent Assembly. Arrangements must be made to give women voting places separate from men.

IV. Among Russian Muslim women, marriages take place very frequently without the consent of the prospective bride. This creates unhappiness in family life for parents as well as children; therefore, the marriage blessing must be given with the consent of the prospective bride and groom.

V. When the husband does not consent to his wife's desire for a divorce, he often holds on to her by force. For the women, it is a form of torture to remain in a family where there is no consent, only oppression. Since this is injurious to the upbringing of children, the Congress, taking into account the rights of women and the interests of the upbringing of the future generation, has adopted the following resolution: the blessing of marriage must include the provision that the wife has the right to divorce if there is no longer mutual good will between wife and husband.

VI. We must take account of the fact that in Turkestan and the Caucasus and also in Kazakhstan it is acceptable to give in marriage unfortunate eleven- and twelve-year-old girls. Having in mind the resulting physical sickness and early death of young women and the sickness of the children, our future generation, the Congress has resolved to ban the giving in marriage of girls who have not reached sixteen years, whether in the north or in the south.

VII. Taking into account the interests of preserving the health and well-being of the children of the nation and the fact that the misfortune of

---

[1]The Shariah is Islamic law grounded in the Koran and the sayings and deeds of the Prophet Mohammed.

[2]The Constituent Assembly met in January 1918.

many families proceeds from the susceptibility of men and women to infectious diseases, and aiming to end these misfortunes, the Congress has resolved that "at the time of marriage the prospective groom and bride must present proof that they have no infectious diseases."

VIII. Considering that men often take as wives more than one woman, regarding this as a kind of game, and divorce their former wives without any cause, and considering this custom the cause of abuses, the Congress has resolved: At the time of marriage the prospective groom must declare that he will not marry a second woman, and if for some reason he takes a second wife, then he will give the first a divorce and guarantee her subsistence."

IX. In Kazakhstan, fathers of girls sell them at a young age for a bride-price *(kalym)*. Later, if the need for divorce arises, the return of *kalym* is required. This *kalym* is not a gift given by the prospective groom to the prospective bride [*mahr*]. *Kalym* is a sum for acquiring the prospective bride from her father.[3] Therefore,

taking into account the rights of Kazakh girls, the Congress has resolved: "*Kalym* must not be paid when Kazakh girls marry. Even if *kalym* has already been paid for a girl as of this day, she has the right not to marry because her husband was chosen without her consent."

X. For various reasons, many women are obliged to enter public houses.[4] And if one of them attempts to tear herself away from there, the law on public houses forces them to remain in that unfortunate situation forever. In view of this, public houses and legislation about them must end.

Following the adoption of these points, Kulhamidova introduced a resolution with two additional subpoints drawn up by the Women's Section at the Congress. These were joined to the original ten points as point XI:

XI.
1. Muslim women must be equal with men in political and civil rights.
2. In view of the fact that polygamy violates humanity and justice, it must be decisively eliminated.

---

[3]This point distinguishes between *kalym* and *mahr*. *Kalym* is the bride-price paid by the prospective husband to the father. *Mahr* is the gift given by the groom to the prospective bride. *Mahr* is consonant with Islamic tradition, while *kalym* is alien to this tradition and so is censured in this resolution.

---

[4]Houses of prostitution.

141. Alexandra Kollontai "Women and the Family in the Communist State"

Alexandra Kollontai (1872–1952) served as the first commissar of public welfare in the Soviet Union, the first woman to hold an official Cabinet position in Europe. From an aristocratic family, she became irrevocably committed to the revolutionary process after witnessing the massacre of hundreds of peaceful petitioners in St. Petersburg on Bloody Sunday, January 9, 1905. An initiator of the Proletarian Women's Movement of 1905 to 1908 which sought to organize the factory women of St. Petersburg, Kollontai was especially critical of bourgeois feminists for diverting working women into all-women's organizations and away from the struggle for socialism. She went into exile in 1908 and returned to Russia soon after the February 1917 revolution. Following the Bolshevik Revolution of November, Kollontai became the first woman elected to the Central Committee of the Communist Party. As commissar of public welfare, she played a key role in drafting the 1918 Family Code[1] which secularized marriage and family laws, allowed for divorce by mutual consent, and eliminated the legal category of "illegitimate" children. With Inessa Armand and Nadezhda Krupskaya,[2] Kollontai organized the first Soviet All-Russian Congress of Working Women which drew over a thousand delegates to Moscow in November 1918. It was here that Kollontai gave her speech "Women and the Family in the Communist State." The Congress established the basis for what became the Zhenotdel, the Women's Department of the Communist Party. It was directed by Armand until her death in 1920 and then by Kollontai who reoriented Zhenotdel policies for the post–Civil War period, initiated work among Muslim women in Central Asia, and sponsored unveilings of Muslim women at conferences in Moscow. In 1922, however, Kollontai was dismissed as commissar of public welfare and head of the Zhenotdel because of her involvement in the Workers' Opposition. She was sent to Norway on a diplomatic, trade mission and later served as the first ambassador to Sweden. Although Kollontai never recovered her previous level of political influence in Russia, the women's movements of the 1960s and 1970s revived interest in her writings on the family, love, and sex. These include *The Social Bases of the Woman Question* (1909), *The New Morality and the Working Class* (1918), *The Autobiography of a Sexually Emancipated Communist Woman* (1926), *Love of Worker Bees* (1923), and *A Great Love* (1927).

1. The Family and Wage Labor by Women

Will the family continue to exist in a communist state? Will it be the same as it is now: This question troubles the heart of many working class women, and worries men also. Working women have had to think about these questions frequently since life began changing before our very eyes. Old customs and habits are disappearing and the entire life of the proletarian family is being recreated on a basis many find different, new, unaccustomed, strange. Additional confusion is introduced by the fact that in Soviet Russia divorce has been made easier.

By a decree of the People's Commissars on December 18 [16], 1917, divorce ceased to be a luxury available only to the wealthy; no longer must a woman worker wait months, even years, to separate from a husband who has subjected her to drunkenness, coarse behavior, beatings. Now divorce by mutual consent can be obtained within two weeks at most. But this very ease of divorce, which is a blessing to those women who were unhappy in their marriage, frightens other women, especially those accustomed to look upon the husband as breadwinner and the sole support of life. These latter do not see that *women must learn to search for support*

---

[1]For later debates on this code, see Source 153 in Chapter 12.
[2]On Krupskaya, see Source 149 in Chapter 12.

From *Documents of Soviet History, Volume 1: The Triumph of Bolshevism, 1917–1919* edited by Rex A. Wade. © 1991 by Academic International Press.

in other places, to seek and receive it not from an individual man but from the collective, from the state.

There is no hiding the truth: the typical former family, where the man was head and breadwinner and the woman existed only in relationship to her husband, without will, time or money of her own, is changing before our eyes. There is no need to fear this....

...In the traditional family the husband worked and provided for his wife and children, while the wife took care of the house and raised the children. But during the past century this traditional family has fallen apart in all countries where capital dominates, where factories and other capitalist enterprises based on wage labor have rapidly developed. Its customs and morals are changing along with the general conditions of life. The first major change in the basic nature of the family has come along with wage labor by women. Formerly the only breadwinner was the man. However, during the last fifty to sixty years in Russia (in other countries earlier) capitalism has forced women to seek work outside the family, outside the home. The wages of the male breadwinner were insufficient, and thus the woman had to seek wage work and to knock on the factory door. Each year the number of working class women working outside the home, at the factory or other daily wage work, as saleswomen, office clerks, servants, washer-women, waitresses, increases. Statistics show that before the world war began in 1914, sixty million women in the countries of Europe and America were earning wages of their own. During the war the number increased. About half of these women were married. But what kind of family life is this now, where the wife-mother is away at work eight hours a day, and away ten hours with travel time! Household affairs are neglected; the children grow up without maternal supervision, spending more time on the streets and exposed to the danger there. The wife, mother-worker, strains herself to fulfill three roles simultaneously: to do her own work at the factory, printshop or office the same as her husband, to

take care of the house, and to look after the children. Capitalism has put an unbearable burden upon the shoulders of the woman; it turns her into a wage worker but does not lighten her household or maternal duties. The woman's shoulders bend under this unbearable triple load, from her breast escapes a stifled groan, and tears are always in her eyes. "Women's life" always was difficult, but never so hard, so cheerless, as it is for millions of working women under capitalism, in the age of industrial production.

The more quickly wage labor by women spreads, the faster the old family breaks down. What kind of family life is there when husband and wife work different shifts! What kind of family life is there if the wife has no time even to fix meals! What kind of parents if, after a full day of hard work, they can find no time for the children! Formerly the mother worked at household tasks and the children grew up around her, under her protective watch. Now, the woman hurries to the factory when the whistle blows, and then again at the whistle hurries home in order to try to take care of her household tasks, and then again, without enough sleep, to work again. For the working married woman, life is more like penal servitude. In such circumstances the family disintegrates, falls apart. Many of the things which held the family together disappear. *The family ceases to be necessary, either for its members or for society.* The old family structure becomes a burden.

What held the former strong, cohesive family together? First, the fact that the family breadwinner was the husband and father. Second, the family economy was essential to all members. Third, that the children were raised by their parents. What remains of this kind of family? We already have noted that the husband has ceased to be the sole breadwinner. The wife-worker earns wages along with him. She has learned to support herself, her children, and sometimes even her husband. There remains only the functions of housekeeping and the care and instruction of young children. Let us look further at whether even these tasks are not to be taken from the family.

2. Household Work Ceasing to Be a Necessity

There was a time when the entire life of women of the poorer class, in the city as well as in the country, was passed in the bosom of the family. Beyond the threshold of her own house, the woman knew nothing and doubtless hardly wished to know anything. To compensate for this, she had within her own house a most varied group of occupations, of a most necessary and useful kind, not only to the family itself but also to the entire society. The woman did everything that is now done by any working woman or peasant woman. She cooked, she washed, she cleaned the house, she went over and mended the family clothing; but she not only did that. She had also to discharge a great number of duties which are no longer fulfilled by the woman of to-day; she spun wool and linen; she wove cloth and garments, she knitted stockings, she made lace, and she took up, as far as her resources permitted, the pickling and smoking of preserved foods; she made beverages for the household; she moulded her own candles. How manifold were the duties of the woman of earlier times!...

3. The Industrial Work of Woman in the Home

In the days of our grandmothers this domestic work was an absolutely necessary and useful thing, on which depended the well-being of the family; the more the mistress of the house applied herself to these duties, the better was life in the house, and the more order and affluence it presented. Even the State was able to draw some profit from this activity of woman as a housekeeper. For, as a matter of fact, the woman of other days did not limit herself to preparing potato soup, but her hands also created many valuable products such as cloth, thread, butter, etc., all of which were things which could serve as commodities on the mar-

ket and which therefore could be considered as merchandise, as things of value....

4. The Married Woman and the Factory

But capitalism has changed all this ancient mode of living. All that was formerly produced in the bosom of the family is now being manufactured in quantity in workshops and factories. The machine has supplanted the active fingers of the wife. What housekeeper would now occupy herself in moulding candles, spinning wool, weaving cloth? All these products can be bought in the shop next door....the fact is that the contemporary family is becoming more and more liberated from all domestic labours, without which concern our grandmothers could hardly have imagined a family....

5. Individual Housekeeping Doomed

The family consumes but no longer produces. The essential labours of the housekeeper are now four in number: matters of cleanliness (cleaning the floors, dusting, heating, care of lamps, etc), cooking (preparation of dinners and suppers), washing, and the care of the linen and clothing of the family (darning and mending).

These are painful and exhausting labours; they absorb all the time and all the energies of the working woman, who must in addition put in her hours of labour in a factory....

The working woman would in vain spend all the day from morning to evening cleaning her home, washing and ironing the linen, using herself up in ceaseless efforts to keep her worn-out clothing in order, she might kill herself preparing with her modest resources such food as might please her, and there would nevertheless at nightfall remain not one material result of all her day's work, and she would have created with her indefatigable hands nothing that could be considered as a commodity on the commercial market. Even if a working woman should live a thousand years, there would never be any change for her. There would always be a new layer of dust to be removed from the mantlepiece, her husband would always come in hungry at night, her little tots would always bring in mud on their shoes. The work of the housekeeping woman is becoming more useless day by day, more unproductive.

6. The Dawn of Collective Housekeeping

The individual household has passed its zenith. It is being replaced more and more by collective housekeeping. The working woman will sooner or later need to take care of her own dwelling no longer; in the communist society of to-morrow this work will be carried on by a special category of working women who will do nothing else. The wives of the rich have long been freed from these annoying and tiring duties. Why should the working woman continue to carry out these painful tasks? In Soviet Russia, the life of the working woman should be surrounded with the same ease, with the same brightness, with the same hygiene, with the same beauty, which has thus far surrounded only the women of the richer classes. In a communist society the working women will no longer have to spend their few, alas too few, hours of leisure in cooking, *there will be in a communist society public restaurants and central kitchens* to which everybody may come to take his meals.

These establishments have already been on the increase in all countries, even under the capitalist regime. In fact, for half a century the number of restaurants and cafes in all the great cities of Europe has increased day by day; they have sprung up like mushrooms after autumn rain. But while under the capitalist system only people with well-lined purses could afford to take their meals in a restaurant, in the communist city anyone who likes may come to eat in the central kitchens and restaurants. The case will be the same with washing and other work: the working woman will no longer be obliged to sink in an ocean of filth or to ruin her eyes in darning her stockings or mending her linen, she will simply carry these things to the *central laundries* each week, and take them out again each week already washed and ironed. The working woman will have one care less to face.

Also, special clothes-mending shops will give the working women the opportunity to devote their evenings to instructive reading, to healthy recreation, instead of spending them as at present in exhausting labour. Therefore, the four last duties still remaining to burden our women, as we have seen above, will soon also disappear under the triumphant communist regime. And the working women will surely have no cause to regret this. Communist society will only have broken the domestic yoke of women in order to render life richer, happier, freer and more complete.

7. The Child's Upbringing Under Capitalism

But what will happen of the family after all these labours of individual housekeeping have disappeared? We still have *the children* to deal with. But here also the State of the working comrades will come to the rescue of the family by creating a substitute for the family. Society will gradually take charge of all that formerly devolved on parents. Under the capitalist regime, *the instruction of the child has ceased to be the duty of the parents.* The children were taught in schools. Once the child had attained school age, the parents breathed more freely. Beginning with this moment, the intellectual development of their child ceased to be their affair. But all the obligations of the family towards the child were not therefore finished. There was still the duty of feeding children, buying them shoes, clothing them, making skilled and honest workers of them, who might be able when the time came to live by themselves and to feed and support their parents in their old age. However, it was very unusual for a worker's family to be able to fulfill entirely all these obligations towards their children; their low wages did not permit them even to give the children enough to eat, while lack of leisure prevented the parents from devoting to the education of the rising generation the full attention which it demanded. The family was supposed to bring up the children. But did it really? As a matter of fact, it is the street which brings up the children of the proletariat. The children of the proletarians are ignorant of the amenities of family life, pleasures which we still shared with our own fathers and mothers.

Furthermore, the low wages of the parents, insecurity, even hunger, frequently bring it about that when hardly ten years of age, the son of the proletarian already becomes in his turn an independent worker. Now, as soon as the child (boy or girl) begins to earn money, he considers himself the master of his own person to such an extent that the words and counsels of his parents cease to have any effect upon him, the authority of the parents weakens and obedience is at an end. As the domestic labours of the family die out one by one, all obligations of support and training will be fulfilled by society in place of the parents. Under the capitalist regime, children were frequently, too frequently, a heavy and unbearable burden on the proletarian family.

8. The Child and the Communist State

Here also the communist society will come to the aid of the parents. In Soviet Russia, owing to the care of the Commissariats of Public Education and of Social Welfare, great advances are being made, and already many things have been done in order to facilitate for the family the task of bringing up and supporting the children. There are homes for the very small babies; day nurseries, kindergartens, children's colonies and homes, infirmaries, and health resorts for sick children, restaurants, free lunches at school, free distribution of textbooks, of warm clothing, of shoes to the pupils of the educational establishments—does not all this sufficiently show that the child is passing out of the confines of the family and being removed from the shoulders of the parents on to those of the community?

The care of children by the parents consisted of three distinct parts; (1) the care necessarily devoted to very young babies; (2) the bringing up of the child; (3) the instruction of the child. As for the instruction of children in primary schools, and later in gymnasiums and universi-

ties, it has become the duty of the State, even in capitalist society. The other occupations of the working class, its conditions of life, imperatively dictated even to capitalist society the creation, for the benefit of the young, of playgrounds, infants' schools, homes, etc.... But bourgeois society was afraid of going too far in this matter of meeting the interests of the working class, lest it contribute in this way to the disintegration of the family. The capitalists themselves are not unaware of the fact that the family of old, with the wife a slave and the man responsible for the support and well-being of the family, that the family of this type is the best weapon to stifle the proletarian effort towards liberty, to weaken the revolutionary spirit of the working man and working woman. Worry for his family takes the backbone out of the worker, obliges him to compromise with capital. The father and the mother, what will they not do when their children are hungry? Contrary to the practice of capitalist society, which has not been able to transform the education of youth into a truly social function, a State task, communist society will consider the social education of the rising generation as the very basis of its laws and customs, as the corner-stone of the new edifice.... Our new man, in our new society, is to be moulded by socialist organizations, such as playgrounds, gardens, homes, and many other such institutions, in which the child will pass the greater part of the day and where intelligent educators will make of him a communist who is conscious of the greatness of this sacred motto: solidarity, comradeship, mutual aid, devotion to the collective life.

9. The Mother's Livelihood Assured

But now, with the bringing up gone and with the instruction gone, what will remain of the obligations of the family towards its children, particularly after it has been relieved also of the greater portion of the material cares involved in having a child, except for the care of a very small baby while it still needs its mother's attention, while it is still learning to walk, clinging to its mother's skirts? Here again the communist State hastens to the aid of the working mother. No longer shall the child-mother be bowed down with a baby in her arms! The Workers' State charges itself with the duty of assuring a livelihood to every mother, whether she be legitimately married or not, as long as she is suckling her child, of creating everywhere maternity homes, of establishing in all the cities and all the villages day nurseries and other similar institutions, in order thus to permit the woman to serve the State in a useful manner and to be a mother at the same time.

10. Marriage No Longer a Chain

Let the working mothers be reassured. The communist society is not intending to take the children away from the parents, nor to tear the baby from its mother's breast; nor has it any intention of resorting to violence in order to destroy the family as such. No such thing! Such are not the aims of the communist society.... The communist society therefore approaches the working woman and the working man and says to them: "You are young, you love each other. Everyone has the right to happiness. Therefore live your life. Do not flee happiness. Do not fear marriage, even though marriage was truly a chain for the working man and woman of capitalist society. Above all, do not fear, young and healthy as you are, to give to your country new workers, new citizen-children. The society of the workers is in need of new working forces; it hails the arrival of every new-born child in the world. Nor should you be concerned because of the future of your child: your child will know neither hunger nor cold. It will not be unhappy nor abandoned to its fate as would have been the case in capitalist society. A subsistence ration and solicitous care are secured to the child and to the mother by the communist society, by the Workers' State, as soon as the child arrives in the world.... The communist society will take upon itself all the duties involved in the education of the child, but the paternal joys, the maternal satisfaction—these will not be taken away from those

who show themselves capable of appreciating and understanding these joys. Can this be called a destruction of the family by means of violence?—or a forcible separation of child and mother?

11. The Family a Union of Affection and Comradeship

There is no escaping the fact: the old type of family has seen its day....But on the ruins of the former family we shall soon see a new form rising which will involve altogether different relations between men and women, and which will be *a union of affection and comradeship, a union of two equal members of the communist society, both of them free, both of them independent, both of them workers.* No more domestic "servitude" for women. No more inequality within the family. No more fear on the part of the woman lest she remain without support or aid with little ones in her arms if her husband should desert her. The woman in the communist city no longer depends on her husband but on her work. It is not her husband but her robust arms which will support her. There will be no more anxiety as to the fate of her children. The State of the Workers will assume responsibility for these. Marriage will be purified of all its material elements, of all money calculations, which constitute a hideous blemish on family life in our days. Marriage is henceforth to be transformed into a sublime union of two souls in love with each other, having faith in each other; this union promises to each working man and to each working woman, simultaneously, the most complete happiness, the maximum of satisfaction which can be the lot of creatures who are conscious of themselves and of the life which surrounds them. *This free union,* which is strong in the comradeship with which it is inspired, *instead of the conjugal slavery of the past—that is what the communist society of to-morrow offers to both men and women.* Once the conditions of labour have been transformed, and the material security of working women has been increased, and after marriage such as was performed by the Church—that so-called indissoluble marriage which was at bottom merely a fraud—after this marriage has given place to the free and honest union of men and women who are lovers and comrades, another shameful scourge will also be seen to disappear, another frightful evil which is a stain on humanity and which falls with all its weight on the hungry working woman: prostitution.

This evil we owe to the economic system now in force, to the institution of private property. Once the latter has been abolished, the trade in women will automatically disappear....

The woman who is called upon to struggle in the great cause of the liberation of the workers—such a woman should know that in the new State there will be no more room for such petty divisions as were formerly understood: "These are my own children; to them I owe all my maternal solicitude, all my affection; those are your children, my neighbour's children; I am not concerned with them. I have enough to do with my own." Henceforth the worker-mother, who is conscious of her social function, will rise to a point where she is no longer differentiates between *yours* and *mine;* she must remember that there are henceforth only *our* children, those of the communist State, the common possession of all the workers.

12. Social Equality of Men and Women

The Workers' State has need of a new form of relation between the sexes. The narrow and exclusive affection of the mother for her own children must expand until it embraces all the children of the great proletarian family. In place of the indissoluble marriage based on the servitude of woman, we shall see rise the free union, fortified by the love and the mutual respect of the two members of the Workers' State, equal in their rights and in their obligations. In place of the individual and egotistic family, there will arise a great universal family of workers, in which all the workers, men and women, will be, above all, workers, comrades. Such will be the relation between men and women in the communist society of to-morrow. This new relation will assure to humanity all the joys of so-called free love ennobled by a true social equality of the mates, joys which were unknown to the commercial society of the capitalist regime.

...The red flag of the social revolution which will shelter, after Russia, other countries of the world also, already proclaims to us the approach of the heaven on earth to which humanity has been aspiring for centuries.

QUESTIONS FOR ANALYSIS

1. Discuss how the program of the Women's Progressive Party analyzes women's status in Russian society. What are the party's goals and how does it propose to achieve them? Overall, what political viewpoint does the program represent?

2. How does Fatima-tutash Kulhamidova describe the status and emerging political and social consciousness of Muslim women in Russia? How does the Ten-Point Resolution interpret the Shariah and what is the significance of this interpretation? On what basis are certain prevailing customs concerning marriage and divorce condemned? What changes are advocated to ensure Muslim women's equal rights with men?

3. How does Alexandra Kollontai depict the traditional family in Russian society and to what factors does she attribute its dissolution? What are the remaining functions of the family and how does Kollontai propose to collectivize these functions? How does she envision the communist family and its relationship to the new socialist economy and state?

4. Compare the political viewpoints and proposals for change elaborated by the Women's Progressive Party, the Women's Section of the First All-Russian Muslim Congress, and Alexandra Kollontai.

CHAPTER 12

The Interwar Period

he Weimar Republic was declared on November 9, 1918, following the mutinies of the Kiel sailors in late October. Kaiser Wilhelm II was removed from power, the monarchy abolished, and Friedrich Ebert, a leader of the Social Democratic Party, appointed president of the new Republic. The Republican government, and not the Kaiser, thus became identified with Germany's defeat, including the punitive terms of the Versailles Peace Treaty which strengthened the anti-Republican sentiments of revanchist, right-wing groups.

From the outset, the Social Democrats and the newly formed German Communist Party (KPD) were divided over their vision of the new Republic. The Social Democrats favored a reformist, bourgeois republic that preserved capitalist property relations. The KPD was formed by the left-wing critics of the Social Democratic Party who blamed its reformism for the August 1914 parliamentary vote for war credits. The KPD advocated a worker's revolution along the lines of the Bolshevik Revolution. It became part of the Third International formed in 1919 under the leadership of the Soviet Union to replace the Second International which had collapsed at the outset of the war. The KPD's revolutionary views gained the support of only a minority of German workers and in 1919 its leaders, Rosa Luxemburg and Karl Liebknecht, were murdered. While defeating the challenge from the left through police repression, the Social Democratic government of the Weimar Republic proved unable to sustain itself in the face of challenges from the right. Less than fifteen years after it was proclaimed, the Weimar Republic collapsed in January 1933 when Hitler became chancellor of Germany.

Although shaped by Germany's defeat in World War I and the conditions of the Versailles Treaty, many aspects of Weimar Germany were shared by other European nations during the interwar period. These included social dislocations caused by the war as well as war demobilization, innovative cultural trends (many of which were also shared by Soviet artists), economic instability due to inflation and rising unemployment, the formation of right-wing groups, and new economic and political gains for women, including the achievement of female suffrage in Britain and Germany in 1918.

In the Soviet Union, socialist construction combined industrialization and the transformation of property relations with revolutionary cultural, legal, and social policies in education, housing, health care, marriage, family, and sexuality. Formed in 1919, the Zhenotdel, or Women's Department of the Communist Party, played a crucial role in securing women's support for the Revolution and advocating for women's interests within the revolutionary process. The Zhenotdel was first led by Inessa Armand and then by Alexandra Kollontai, the commissar of public welfare, following Armand's death in 1920. During the Civil War (1918–1921), the Zhenotdel organized the establishment of child-care centers, communal kitchens, and public dining halls which socialized private household work and facilitated women's participation in the economy. But these gains were soon undermined by the New Economic Policy (NEP; 1921–1928). Not only were women replaced in the work force by demobilized soldiers, but the NEP policy of shrinking the large, state enterprises affected women in especially adverse ways. Funding for public services was also severely diminished.

In the 1930s, by contrast, new economic and career opportunities were created for women by the implementation of the first and second five-year plans of 1928–1932 and 1933–1937. Production grew by some 15 to 20 percent as millions of male and female Soviet workers were mobilized to carry out industrialization and fulfill the goals of the plans. The 1930s saw the greatest expansion of women's entry into the Soviet economy as they comprised 82 percent of workers entering the labor force for the first time. Unprecedented numbers of women became skilled industrial workers and professionals.

At the same time, however, the social conservatism of the Stalin period terminated the artistic and cultural innovations of the early, Soviet period. Stalin abolished the Zhenotdel in 1930, claiming that there was no longer a need for a separate women's department. Writers, musicians, and artists were required to abandon experimental forms, which were denounced as bourgeois, and to adhere to the new requirements of socialist realism. Soviet family policies were also transformed. Motherhood was now exalted and women lost their right to reproductive choice when abortion, first made legal in 1920, was prohibited in 1936.

These reversals and Stalin's political repression notwithstanding, rapid industrialization provided the Soviet Union with the military and industrial basis essential for waging war against Nazi Germany. Within fifteen years, the Soviet Union had created the industrial infrastructure that it had taken France and Britain a century and a half to achieve.

While the Soviet economy expanded, the Great Depression of the 1930s plunged millions of workers in industrialized Europe and the

United States into unemployment, poverty, and despair. The exclusion of women workers, especially married women, was soon promoted as a means of alleviating (male) unemployment and maintaining the integrity of the family. In Germany and France, in particular, the Depression fueled the nationalistic, anti-Semitic, and patriarchal sentiments of right-wing groups. In France, the effects of the Depression were not felt until 1931 but here, too, right-wing groups, typically mobilized into fascist leagues, attacked the parliamentary system of government as ineffectual. Although the right wing never unified, as it did in Germany, its show of strength in a mass demonstration in February 1934 galvanized the Socialist, Communist, and Radical parties to form the Popular Front government of 1936–1938. The government aimed to resolve the economic crisis by improving the living and working conditions of workers. Ultimately, however, the Depression was ended only by the military production and mass conscription that accompanied World War II.

THE GERMAN REVOLUTION AND THE WEIMAR REPUBLIC

The Weimar Republic granted suffrage to women in 1918 and initiated reforms in the area of social welfare and family legislation, although Paragraph 218 of the Constitution prohibited abortion. Pledged to reform rather than social revolution, the Social Democratic government left the capitalist economy intact along with the officer corps and the aristocracy. The Social Democratic vision of the Weimar Republic was challenged by women in the newly formed German Communist Party (KPD) for whom the Republic signified only the beginning of a socialist revolution which alone could guarantee the rights of women and the working class.

142. Rosa Luxemburg "The Beginning"

Rosa Luxemburg (1871–1919) was a founder of the German Communist Party (KPD). Of Polish-Jewish origin, she participated in underground revolutionary activities while a high school student in Warsaw. Luxemburg attended the University of Zurich in Switzerland where she became immersed in the political life of the Russian and Polish socialist exile communities. In 1898, Luxemburg moved to Berlin and joined the German Social Democratic Party (SPD). With Clara Zetkin,[1] she opposed the party's revisionist tendencies and its refusal to condemn German

[1] On Zetkin, see Chapter 10, Source 125, "Women and the Trade Unions."

militarism and imperialism. A staunch opponent of World War I, Luxemburg intended to attend the April 1915 International Women's Conference at The Hague[2] but was arrested prior to the conference for her antiwar activities. From prison, where she spent much of the war years, Luxemburg smuggled out various leaflets and brochures, including *The Junius Pamphlet,* written in 1915. Here, she critiqued the notion of Germany's "defensive" war, condemned all national wars in the era of imperialism, and called for a new International. On New Year's Day 1916, Karl Liebknecht and antiwar members of the Social Democratic Party from throughout Germany met secretly to form the Gruppe International within the SPD and adopted a statement of principles that Luxemburg sent from prison. Following the Revolution of November 9, when she and other political prisoners were released, the Gruppe International transformed itself into the Spartacus League, for which Luxemburg wrote a program. The Spartacus League refused to join the new government and subsequently formed the basis of the German Communist Party. In her speech to the KPD's founding congress, held in December 1918 and January 1919, Luxemburg underlined Marx's prediction that capitalism would descend into barbarism if the socialist revolution was not carried out. Soon afterward, on January 15, 1919, Luxemburg and Liebknecht were arrested by the Social Democratic government and murdered while in police custody. Luxemburg's writings include *Reform or Revolution* (1899), *The Mass Strike, the Political Party, and the Trade Unions* (1906), and *The Accumulation of Capital* (1913). Her article "The Beginning" was published in the November 18, 1918, issue of the newspaper *Rhote Fahn (Red Flag).*

The revolution has begun. What we need now is not rejoicing over its accomplishments, not celebrations of victory over the prostrate foe, but rigorous self-criticism and strict marshaling of our strength so the work now begun can go forward. For little has been attained and the enemy is *not* defeated.

What has been accomplished? The monarchy has been swept away. Supreme governmental power has been handed over to the workers' and soldiers' representatives. But the monarchy was never the real enemy. It was only the cover, the figurehead for imperialism. It was not the Hohenzollern who ignited the World War, spread fire to the four corners of the earth, and brought Germany to the brink of the abyss. Like all bourgeois governments, the monarchy was only an administrator for the ruling classes.

The criminals who must be held responsible for the genocide are the imperialist bourgeoisie, the capitalist ruling class.

Abolition of capital's domination and achievement of a socialist order: that and nothing less is the historic theme of the current revolution. A massive task, this cannot be dispatched in a twinkling by a few decrees from on high, but can be set in motion only through the conscious action of the urban and rural working people. It can be carried through all tempests and brought safely to port only by the highest intellectual maturity and unflagging idealism of the popular masses.

The revolution's goal clearly points out its course, and its tasks indicate the needed methods. *All power to the toiling masses, and to the workers' and soldiers' councils; safeguard the*

---

[2]On this conference, see Source 134 in Chapter 11.

revolution's accomplishments from the enemies that lie in wait for it. These are the guidelines for all measures of the revolutionary government.

Every step, every action of the government must point like a compass in this direction:

- Expand and reelect local workers' and soldiers' councils to replace the chaotic and impulsive character of their initial actions through a conscious process of understanding the revolution's goals, tasks, and course.

- Maintain representative bodies of the masses in permanent session. Real political power should be transferred from the Executive Committee of the councils, a small body, to the broader basis of the workers' and soldiers' councils.

- Immediately convoke a national parliament of the workers and soldiers in order to organize all of Germany's proletariat as a class, a solid political power, the bulwark and driving force of the revolution.

- Immediately organize not the "peasants," but the farm workers and small peasants, a layer that has not participated in the revolution up to now.

- Build a proletarian Red Guard for ongoing defense of the revolution and train a workers' militia in order to organize the entire proletariat to be on guard at all times.

- Expel the surviving organs of the absolutist militaristic police state from the administration, judiciary, and army.

- Immediately confiscate dynastic fortunes and property and large landed estates as an initial, preliminary measure to secure the people's food supply, since hunger is the most dangerous ally of counterrevolution.

- Immediately convene in Germany a world congress of workers to loudly and clearly proclaim the socialist and international character of the revolution, because the future of the German revolution is anchored in the International and in the world proletarian revolution.

We have listed only the first, most necessary steps. What is the present revolutionary government doing?

It simply leaves the state as an administrative organism, from top to bottom, in the hands of yesterday's supporters of Hohenzollern absolutism and tomorrow's tools of the counterrevolution.

It convenes the national constituent assembly, thereby creating a bourgeois counterweight to the workers' and soldiers' power, shunting the revolution onto the rails of a bourgeois revolution, and conjuring away the socialist goals of the revolution.

It does nothing to demolish the continuing power of capitalist class rule.

It does everything to reassure the bourgeoisie, to preach the sacredness of private property, and to ensure the inviolability of capitalist property relations.

It retreats before the constantly advancing counterrevolution without appealing to the masses, without sharply warning the people.

Law and order! Law and order! These words reverberate from all sides and from all government statements and are jubilantly echoed from all wings of the bourgeoisie. The clamor against the specter of "anarchy" and "putschism," the familiar infernal whine of the capitalist worried about his safes, property, and profits: this is the overriding theme song of today, and the revolutionary workers' and soldiers' government calmly tolerates the sounding of the rallying cry for the assault on socialism. Worse—it participates in word and deed.

The results of the revolution's first week are as follows: in the land of the Hohenzollern, basically nothing has changed. The workers' and soldiers' government functions as a stand-in for the bankrupt imperialist government. All its acts—of commission and omission—are based on a fear of the working masses. Before the revolution could develop power and momentum, its life blood, which is its socialist and proletarian character, was drained.

Everything is as you would expect. The most reactionary country in the civilized world does not become a revolutionary people's republic in twenty-four hours. Soldiers who yesterday killed revolutionary proletarians in Finland, Russia, the Ukraine, and in the Baltics—and workers who quietly allowed this to happen—have not in twenty-four hours become conscious fighters for socialism.

The state of the German revolution reflects the maturity of German political conditions. Scheidemann-Ebert are the government befitting the German revolution in its present stage. And the Independents[1], who believe they can build socialism together with Scheidemann-Ebert[2] and who solemnly certify in *Die Freiheit* that with them they are forming an "exclusively socialist govern-ment," thereby become the authorized corporate partners in this first, provisional stage.

But revolutions do not stand still. It is a fundamental law that they constantly move forward and outgrow themselves. The first stage is already pressing against its internal contradictions. The situation is understandable as a beginning, but untenable in the long run. The masses must be on guard if the counterrevolution is not to win across the board.

We have made a beginning. What remains is not in the hands of the petty creatures who want to block the flow of the revolution and stop the wheel of world history. World history's order of the day calls for achieving the final goals of socialism. The German revolution is on the path of this guiding light. Step by step, through storm and stress, through struggle and anguish, misery and victory, the revolution will triumph.

It must!

[1]Members of the Independent Social-Democratic Party.

[2]Philip Scheidemann and Friedrich Ebert, Social Democratic leaders who became Chancellor and President, respectively, of the Weimar Republic.

143. "Manifesto for International Women's Day"

This "Manifesto for International Women's Day" was published in a March 1921, issue of *Die Kommunistin (The Woman Communist)*, a newspaper of the German Communist Party (KPD), edited by Clara Zetkin[1] and Rosa Luxemburg.

TO ALL WORKING WOMEN!

Working women, employed women of all kinds in city and countryside, small-property holders, mothers of the proletariat and the dispossessed.

Come out for International Women's Day!

It must become your day!

Your lives and deeds are dominated by exorbitant price increases with which small and medium incomes cannot keep pace. Exorbitant prices deplete your bread and season it with the bitterest of worries and scalding tears. They tear the shirt from the backs of your children and rob them of their rosy cheeks and happy smiles. Uncounted numbers of you are massed in stifling back rooms, in dark and airless courtyard apartments, in damp, moldy cellars and drafty garrets. Unmanageable fees and

[1]On Zetkin, see Source 125 in Chapter 10.

taxes increase the burden of your worries and add to your privations.

Those who still have work and earnings appear happy. In front of the employment agencies and factory gates, troops of unemployed men and women gather together. That means people without bread, often the homeless too, and those without a country. Millions are no longer earning enough to salt their bread, and they tremble in fear that tomorrow unemployment will be their fate. Among all these unfortunates are countless women. As long as the men were being misused in the trenches as cannon fodder for the rich, *you women* were drawn by the hundreds of thousands into the factories as machine fodder for scant pay, used as exalted "emergency assistants" in banks, department stores, offices, and schoolrooms. Now you are pocketing the promised "gratitude of the fatherland."

Since distant millionaires and billionaires cannot pit you against the men to suppress wages and salaries, you have the doors to the places of work and employment slammed in your faces. It is *the family* you should tend to, *which you don't have* or for which you would need bread. Think of the mass graves in which flourishing male life lies moldering, the endless lines of war cripples and invalids. Of what concern is it to the factory owner, merchant, stock company, local government, state, or nation how you prolong your troubled existence? The spirit of mammon, of exploitative capitalism, is their master. That is why you are cast into the streets as prostitutes, even those half-grown children in need of protection among you!

The house they talk about, *what is it?* The exploitative capitalist economy transforms it from a home—a place of rest, peace, and happiness—into a treadmill whose operation mercilessly crushes you, body and soul. Hard, growing burdens of work and duty press you to the floor; condemned to stay inside your narrow four walls, your *spirit is numbed* into a backwardness alien to our times. It stunts your humanity. But because you cannot be whole, cannot be free persons, neither are you capable of being understanding, helpful comrades to your husbands, nor mothers, raising your children to be beautiful, proud human beings.

And what a mockery of your fate! You and yours are ground down by toil and drudgery, you and yours starve and suffer. Next to you, however, usurers, racketeers, speculators, profiteers of war and revolution, capitalist exploiters of every type and color squander and waste immeasurable riches in absurd, disgusting sensual frenzies. People, many of whom have never lifted a finger in productive work, harvest where you and yours have plowed and sowed. *Working women, is this injustice, this insanity to last forever?*

The political equality of the female sex, the democracy of the new Germany, will overcome this insanity, put an end to this injustice. Do you not hear this chorus daily, working women? Yes, in the new Germany you may vote for democracy. Yes, you may even be voted for. But, despite everything, is not the *equality of women* in political as well as public life, just on *paper?* Does it count that a few women in official positions shape the social conditions of women, make decisions about affairs of inherent concern to women? Just recently the majority of the Reichstag decided on an empty pretext to delay still further the day when women may act as jurors. In reality, teachers, women civil servants on the whole, still remain condemned to celibacy if they do not want to lose their positions.

But above all, working women, employed women, housewives of the dispossessed and those of scant means, *who among you has been made full, free, and happy by the right to vote and be elected?* The economic power of the capitalists to exploit, to take advantage of, and to oppress you and yours persists. The democracy of the bourgeois order offers you the drudgery of work without its fruits—poverty and unfreedom. It defends idleness and excess for some. It has prison and bullets for others. It has *prison and bullets for you, too, working women,* if you "no longer want the lazy...to squander what diligent hands have earned." [...]

You working women, you who make demands and struggle, count in the millions. International Women's Day proves it. In all countries where, demanding their rights, the disinherited surge forward under the sign of communism against the exploiting and subjugating power of capital, on International Women's Day mothers filled with pain, housewives bent with worry, exhausted working women, clerical workers, teachers, and small-property holders flow together. The same need forces the same demands to their lips. Their shared understanding lends them to the common struggle. Across borders, over mountains and seas, they are of a single will. *The demands of the working women of Germany are also the demands of their sisters in Switzerland, in Austria, in Holland, in France and other countries. They are the international demands of women.* The women communists, proletarians, and small-property holders in Russia are taking the lead, setting an example for the women struggling here. The revolution in Russia is also their immortal work. With it demands for rights were fulfilled for which women in other countries must still struggle hard. Your Women's Day can call the manual and mental laborers to work on the construction of the communist order. *Working women, in a front with these gallant ones, with the rising, fighting proletarians of the whole world!*

144. Elsa Herrmann
"This Is the New Woman"

The "New Woman" was a cultural phenomenon which arose from the transformation of women's lives by World War I and the changes which took place in the immediate post-war period. By 1925, over eleven million women in Germany were employed in paid, wage labor and constituted about a third of the entire workforce. The majority of women still worked in agriculture and industry, but increasing numbers were engaged in white-collar, service, and professional employment. In this 1929 article, Elsa Herrmann offers her definition of the characteristics, opportunities, and values of the new generation of German women.

To all appearances, the distinction between women in our day and those of previous times is to be sought only in formal terms because the modern woman refuses to lead the life of a lady and a housewife, preferring to depart from the ordained path and go her own way. In fact, however, the attitude of the new woman toward traditional customs is the expression of a worldview that decisively influences the direction of her entire life. The difference between the way women conceived of their lives today as distinguished from yesterday is most clearly visible in the objectives of this life.

The woman of yesterday lived exclusively for and geared her actions toward the future. Already as a half-grown child, she toiled and stocked her hope chest for her future dowry. In the first years of marriage she did as much of the household work as possible herself to save on expenses, thereby laying the foundation for future prosperity, or at least a worry-free old age. In pursuit of these goals she helped her

From *The Weimer Republic Sourcebook*, edited by Anton Kaes, Martin Jay, and Edward Dimendberg, pp. 206–208. Copyright © 1994 the Regents of the University of California. Reprinted by permission from the University of California Press.

husband in his business or professional activities. She frequently accomplished incredible things by combining her work in the household with this professional work of her own, the success of which she could constantly observe and measure by the progress of their mutual prosperity. She believed she had fulfilled her life's purpose when income deriving from well-placed investments or from one or more houses allowed her and her husband to retire from business. Beyond this, the assets saved and accumulated were valued as the expression of her concern for the future of her children.

The woman of yesterday pursued the same goal of securing the future in all social spheres, varied only according to her specific conditions. The woman defined exclusively by her status as a lady determined the occasions when she would allow herself to be seen in public by considering the possible advantages to herself and her family, a standpoint that would often determine the selection of the places she would frequent and where she would vacation. Less well-off women often kept a so-called "big house." They invited guests and took part in social functions to give the impression in their milieu that all the financial and social requisites for their husbands' career advancement were at hand. For every genuine woman of yesterday it was quite natural to make all manner of sacrifices in a completely selfless fashion, provided they served to advance the social ascent of the family or one of its members.

Her primary task, however, she naturally saw to be caring for the well-being of her children, the ultimate carriers of her thoughts on the future. Thus the purpose of her existence was in principle fulfilled once the existence of these children had been secured, that is, when she had settled the son in his work and gotten the daughter married. Then she collapsed completely, like a good racehorse collapses when it has maintained its exertions up to the very last minute. She changed quickly, succumbing to various physical ailments whose symptoms she had never before noticed or given any mind.

The woman of yesterday was intent on the future; the woman of the day before yesterday was focused on the *past*. For the latter, in other words, there was no higher goal than honoring the achievements of the "good old days." In their name she strove to ward off everything that could somehow have disturbed her accepted and recognized way of life.

In stark contrast, the woman of today is oriented exclusively toward the present. That which is is decisive for her, not that which should be or should have been according to tradition.

She refuses to be regarded as a physically weak being in need of assistance—the role the woman of yesterday continued to adopt artificially—and therefore no longer lives by means supplied to her from elsewhere, whether income from her parents or her husband. For the sake of her economic independence, the necessary precondition for the development of a self-reliant personality, she seeks to support herself through gainful employment. It is only too obvious that, in contrast to earlier times, this conception of life necessarily involves a fundamental change in the orientation of women toward men which acquires its basic tone from concerns of equality and comradeship.

The new woman has set herself the goal of proving in her work and deeds that the representatives of the female sex are not second-class persons existing only in dependence and obedience but are fully capable of satisfying the demands of their positions in life. The proof of her personal value and the proof of the value of her sex are therefore the maxims ruling the life of every single woman of our times, for the sake of herself and the sake of the whole. [...]

The people of yesterday are strongly inclined to characterize the modern woman as unfeminine because she is no longer wrapped up in kitchen work and the chores that have to be done around the house. Such a conception is less informative about the object of the judgment than the ones making it, who have adopted a view about the essence of the sexes based upon various accidental, external features. The concepts *female* and *male* have their ultimate origin in the erotic sphere and do not refer to the ways in which people might engage in activity. A woman is not female because she wields a cooking spoon and turns everything upside down while cleaning, but because she manifests characteristics that the man finds desirable, because she is kind, soft, understanding, appealing in her appearance, and so on. [. . .]

Despite the fact that every war from time immemorial has entailed the liberation of an intellectually, spiritually, or physically fettered social group, the war and postwar period of our recent past has brought women nothing extraordinary in the slightest but only awakened them from their lethargy and laid upon them the responsibility for their own fate. Moreover, the activity of women in our recent time of need represented something new neither to themselves nor to the population as a whole, since people had long been theorizing the independence and equality of woman in her relationship to man.

The new woman is therefore no artificially conjured phenomenon, consciously conceived in opposition to an existing system; rather, she is organically bound up with the economic and cultural developments of the last few decades. Her task is to clear the way for equal rights for women in all areas of life. That does not mean that she stands for the complete equality of the representatives of both sexes. Her goal is much more to achieve recognition for the complete legitimacy of women as human beings, according to each the right to have her particular physical constitution and her accomplishments respected and, where necessary, protected.

QUESTIONS FOR ANALYSIS

1. How does Rosa Luxemburg assess the accomplishments and limitations of the November 1918 Revolution in Germany? What political and economic measures does she propose for the successful completion of the Revolution?
2. On what basis does the "Manifesto for International Women's Day" appeal to working women? What is its analysis of the conditions faced by women in the workplace and the family? How does this manifesto critique bourgeois democracy, the form of government that characterized the Weimar Republic?
3. How does Elsa Herrmann distinguish between the "woman of yesterday" and the "New Woman"? What political and economic factors account for the rise of the "New Woman"? How is she different from the women addressed in the "Manifesto for International Women's Day"?

WOMEN AND MODERNISM
IN THE INTERWAR PERIOD

In the interwar period, modernism affected all forms of artistic activities: writing, art, architecture, and fashion. Although it assumed different national forms, such as the avant-garde Constructivist movement in the Soviet Union and the Bauhaus movement in Weimar Germany, modernists shared in common a celebration of the democratizing effects of mass cultural production and industrial streamlined design in place of ornate, craft-based decorative arts. Thus, the dress designs of the Soviet Constructivist artist Varvara Stepanova have many features in common with those of French couture designer Coco Chanel. The internationalist character of modernism was especially evident in the 1925 Paris Exhibition of Decorative and Industrial Arts which featured Soviet and French designers. Photographers, too, gained a new artistic prominence in the interwar period and experimented with new forms of representation, such as photomontage, which was especially popular in the Soviet Union and Weimar Germany.

145. Lotte Lenya and The Threepenny Opera

Singer and actress Lotte Lenya (1898–1981) starred in *Cabaret* as well as several Bertolt Brecht and Kurt Weill productions, including *Rise and Fall of the City of Mahagonny* and *The Threepenny Opera*. In their collaborations, Brecht wrote the script while Weill, Lenya's first husband, scored the music. *The Threepenny Opera* was an adaptation of John Gay's eighteenth-century *The Beggar's Opera* which Brecht's secretary, Elisabeth Hauptmann, brought to his attention following a successful London revival. It premiered in 1928 in Berlin with Lenya in the role of the prostitute Ginny Jenny. Lenya's fame spread with her 1930 recordings of selections from the opera and appearance in the same year in the film *The Threepenny Opera* directed by G. W. Pabst. By this time, 4,000 performances of the play had been given in 120 German theaters. But this success took place against the backdrop of growing Nazi influence and terror. Between 1928 and 1933, productions of *Mahagonny,* which critiqued the Nazi vision of an "ideal, racial state," were punctuated by Nazi violence. In 1933, Weill, who was Jewish, was warned of his imminent arrest and left Berlin for Paris. Lenya soon followed and in 1935 they both departed for New York City. Following Weill's death in 1950, Lenya became the guardian of his legacy. A 1954 revival in Greenwich Village, in which she appeared, introduced Lenya and *The Threepenny Opera* to a new generation.[1]

A. Photograph of Lotte Lenya

Lotte Jacobi (1896–1990) studied photography in Munich and began working in the Atelier Jacobi, her father's portrait studio in Berlin, in 1927. She was the fourth generation member of her family and the first female to work as a photographer in the studio. Jacobi's career exemplified the new professional opportunities for women as well as the transformation of many artistic genres which characterized the Weimar period. She expanded the work of the studio to include photographs for magazines, advertising, and newspapers. Jacobi was especially interested in photographing directors and performers in the contemporary cinema and theater. She took this photograph of Lotte Lenya in August 1928 during rehearsals for *The Threepenny Opera.* The photograph helped to establish Lenya's reputation and was her favorite. Working conditions became increasingly difficult for Jacobi following the Nazi seizure of power in 1933. Jacobi published photographs under an assumed name for two years before emigrating to New York City where she opened a studio with her sister.

---

[1]For Lenya's account of the original production of *The Threepenny Opera,* see her article "That Was a Time!" published in *Theatre Arts* in May 1956 and reprinted as "August 28, 1928" in the 1960 Grove Press edition of the *Threepenny Opera* translated by Eric Bentley.

Lotte Lenya, Berlin, 1928. Photograph by Lotte Jacobi. (*Lotte Jacobi Archives/University of New Hampshire.*)

B. "Pirate Jenny"

Lotte Lenya's recordings from The Threepenny Opera include "Pirate Jenny" which belonged to the character of Polly Peachum, the daughter of the king of the beggars, in the original theater production. Lenya also sang "Pirate Jenny" in the 1930 film of the opera and so is often associated with it. A prostitute, Polly has a vision of a pirate ship that will take her away, destroy the town, and thereby punish all who have abused her. In *Lenya: A Life,* Donald Spoto describes Lenya's film portrayal: "When she finally sings 'Pirate Jenny' in the bordello, she stands absolutely still, leans against a window, hands on one hip. Her eyelids are half-closed, a rueful smile occasionally appears—but the look is one of infinite weariness, of near-despair.... Had we no other film performance by Lotte Lenya than this, she would still remain one of the most intriguing players to come from the Weimar years."[1]

Gentlemen, today you see me washing up the glasses
And making up the beds and cleaning.
When you give me p'raps a penny, I will curtsey rather well.
When you see my tatty clothing and this tatty old hotel
P'raps you little guess with whom you're dealing.
One fine afternoon there will be shouting from the harbor.
Folk will ask: what's the reason for that shout?
They will see me smiling while I rinse the glasses
And will say: what has she to smile about?
    And a ship with eight sails and
    With fifty great cannon
    Sails in to the quay.

They say: go and wipe your glasses, my girl
And their pennies are thrown to me.
And I thank them for the pennies and I do the beds up right
(Though nobody is going to sleep in them that night)
And they haven't the least idea who I may be.
One fine afternoon there will be roaring from the harbor.

---

[1]Donald Spoto, *Lenya: A Life* (New York: Ballantine, 1990), 87–88.

Folk will ask; what's the reason for that roar?
They will see me standing just beside the window
And will say: now what's she sneering for?
   And the ship with eight sails and
   With fifty great cannon
   Will shoot up the town.

Gentlemen, I fear this puts an end to your
   laughter
For your walls, they will all cave in.
And this whole fair city will be razed to the
   ground.
Just one tatty old hotel will survive safe and
   sound.
Folk will ask what special person dwells therein.
And all night long round this hotel there will
   be shouting.
Folk will ask: why was it this they'd spare?
Folk will see me leave the place the following
   morning
And will say: so that's who was in there!

And the ship with eight sails and
With fifty great cannon
Will run flags up the mast.

And a hundred men will come ashore before
   it's noon
And will go where it's dark and chill.
And every man they find, they will drag along
   the street
And they'll clap him in chains and lay him at
   my feet
And they'll ask: now which of these are we to
   kill?
And when the clock strikes noon it will be still
   down by the harbor.
When folk ask: now just who has got to die?
You will hear me say at that point: All of them!
And when their heads fall, I'll say: Whoopee!
   And the ship with eight sails and
   With fifty great cannon
   With sail off with me.

146. Varvara Stepanova
Soviet Textile and Dress Design

Varvara Stepanova (1894–1958) played a leading role in establishing Soviet artistic culture. She was a theorist and practitioner of the Constructivist movement which included her husband, Alexander Rodchenko; Vladimir Tatlin; Liubov Popova; and Alexandra Exter. Sharing many of the ideals of the European and prerevolutionary Russian avant-garde, the Constructivists sought to eliminate the distinction between the fine and applied arts. Of a multidimensional nature, their work combined construction, texture, and design. Stepanova's artistic productions thus included photomontage film posters, book and magazine covers, woodcuts, paintings, textile and dress designs, and stage sets. The revolutionary needs of the Moscow textile industry provided the Constructivists with an unprecedented opportunity to implement their belief that art should serve society and that design should be fused with mass production. Constructivist artists were among the first to respond to a 1923 *Pravda* call for artists to address industrial issues. Stepanova taught in the textile department of VKhUTEMAS (Higher State Artistic and Technical Workshops) and advocated new methods for designing fabrics and clothing. With Liubov Popova (1889–1924), whose writings include the 1923 article "The Dress of Today Is the Industrial Dress," she became one of the first artists to work in textile design and production at the First State Textile Printing Factory in Moscow. Stepanova's artistic approach is exemplified in this 1924 sample of dress design and in her 1929 article "From Cloth to Pattern and Fabric."

A.

B.

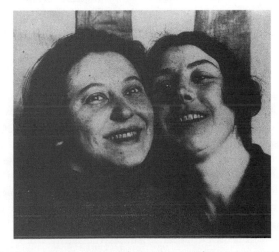

Left (A.), Double portrait of Stepanova and Popova, 1924. Photo by Alexander Rodchenko; right (B.), Stepanova, 1924, design for a professional woman's suit. Left, (*Photo by Alexander Rodchenko from* Varvara Stepanova: The Complete Work *by Alexander Lavrentiev © 1988 MIT Press Edition.*) Right, (*1924 dress design from* Varvara Stephanova: The Complete Work *by Alexander Lavrentiev.*)

C. "From Cloth to Pattern and Fabric"

Until the present, artists have worked in the textile industry along the lines of decoration and the application of decorative patterns to pre-existing fabrics. The artist has taken no part in either the application of new dyeing processes, in working out new fabric structures, or in inventing new materials for fabrics. Though he works in an industrialized factory he retains all the hallmarks of a handicraftsman. He has become the decorative executor of the so-called "demands of the market," which have evolved without any contribution from him. Therefore, the artist's importance to the textile industry is negligible, even compared to that of the artist-constructor in a car factory.

Textile production has preserved the artist in the swamp of petit-bourgeois taste by linking his work to so-called "fashion."

Ultimately the artist is transformed into a trade draftsman vulnerable to every vagary of the trade apparatus. The artist has no independence and has nothing with which to demonstrate the validity of his achievements, his aspirations, and his projects. The principle of expediency, which motivates design in every other sphere of industry, has no standing with him, and he has no incentive to develop that very field where, at first sight, it would seem that artistic design should reign supreme.

The only correct approach would be for the artist to participate, at first, in clothing design. Out of this would develop his participation in the manufacture and dyeing of the fabric.

"Fashion" is directly affected by fabric designs—usually the cut of a garment presupposes existing types of fabrics and fabric designs—which means that printed fabric is subject to more frequent changes in pattern, whereas the facture of the fabric or the manner in which it is manufactured exerts a significantly greater influence on the cut of the garment and is more impervious to changes in fashion.

It is a mistake to think that fashion is an unnecessary appendage. In the final analysis, so-called "fashion" follows in the wake of rationalization, just as our way of life is also becoming gradually more rationalized.

In a planned Socialist economy fashion will depend, not on competition in the market, but on the improvement and rationalization of the garment and textile industries. The textile artist will then occupy an independent, responsible position both in the factory and in the trade and sales apparatus. We may predict that not a single production shop will be able to function without him, but he will have to make significant changes in the way he works. He will have to become more of an artist-technician, and his creative work will be rationalized along with the factory itself.

In the future we won't have stylish clothing that characterizes the life-style of society in every detail.

The decline of graphic design and ornamentation which we are now observing in many fields of art indicates that the fabric printing industry no longer has much of a future. The artist's whole attention should be focused on the processing and coloring of fabric, and on working out new fabric types.

Pattern can be of some importance in the details of a garment, where it will probably continue to be used for decoration. Like everything else pattern will be submitted to a standard and will ultimately be expressed through the structure of the fabric.

The principal task of the textile artist now is to coordinate his work on fabric design with the design of the garment, to refuse to design fabrics in the abstract for an unknown purpose, to eliminate all handicraft working methods, to introduce mechanical devices with the aim of geometricizing his working methods and, most important (and at the moment what is really lacking), to infiltrate the life of the consumer and find out what happens to his fabric after it is shipped from the factory.

The gulf that separates the fabric from the garment made out of it is becoming a great obstacle to improving the quality of our clothing production.

We can no longer speak of compiling a pattern. It is time to move from designing a garment to designing the structure of a fabric. This will immediately allow the textile industry to abandon the excessive assortment it now operates with, and will make it possible to really standardize and improve, at last, the quality of its production.

147. Coco Chanel
The Chanel Look

Coco Chanel (1883–1971) transformed the world of European couture in the interwar period, and her designs continue to influence contemporary fashion. Born as August Gabrielle Chanel, she was raised in a convent in France. Here, all the girls were required to wear black. A transformed version of this "convent style" became the hallmark of the Chanel look: simple, but elegant black dresses that were in marked contrast to the elaborate dress designs of her day. As François Baudot notes of her success, "Thousands of women now began to real-

Coco Chanel, 1935. Photograph by Man Ray.
(*Man Ray's portrait of Coco Chanel, Ca. 1935* ©
Man Ray Trust ADAGP/ARS/Telimage—1998.)

ize that 'poor chic' could be the answer to social snobbery. The Chanel look, with its lines reduced to their simplest expression, shows that *how* clothes are worn is much more important than *what* is worn; that a good line is worth more than a pretty face; that well-dressed is not the same as dressy, and that the acme of social cachet was to be proletarian."[1] Chanel set up her first fashion boutique in Deauville in 1913 and her first couture house in Biarritz in 1915. By 1935, when this photograph of Coco Chanel was taken by Man Ray, Chanel had reached the height of her success. She employed almost 4,000 workers and sold nearly 28,000 designs annually throughout the world.

QUESTIONS FOR ANALYSIS

1. How does Lotte Jacobi's 1928 photograph of Lotte Lenya convey the image of the "New Woman" in the Weimar Republic? Discuss the multilayered meanings and political symbolism of the "Pirate Jenny" song in *The Threepenny Opera*. Consider "Pirate Jenny" and the Jacobi photograph in relation to the issues raised in "Manifesto for International Women's Day" and "This Is the New Woman" (Sources 143 and 144 in this chapter).
2. How does Varvara Stepanova define the role of the artist in the new Soviet textile industry? Discuss her use of fabric and design for a woman's professional suit.
3. How did Coco Chanel's fashion innovations express the "New Woman" in France? Compare her fashion style with that of Varvara Stepanova.
4. What do the sources in this section reveal about how women shaped and experienced modernity?

---

[1]François Baudet, *Universe of Fashion, Chanel* (New York: Universe Publishing, 1996), p. 9.

WOMEN AND SOVIET SOCIALIST CONSTRUCTION

Soviet socialist construction involved cultural as well as economic transformations. The following decrees, debates, and articles illustrate how educators, jurists, and women's organizations aimed to create the new Soviet man and woman by expanding educational opportunities and revising marriage and family laws.

148. Decree on Open Admission to Universities and Higher Education, August 1918

The Soviet government placed a high priority on education as it regarded an educated population as essential for the construction of socialism in its political and economic aspects. The Commissariat of Enlightenment was created to direct new educational policies. It was led by three long-standing Bolsheviks: Commissar Anatole Lunacharsky, M. N. Pokrovsky, and Nadezhda K. Krupskaya. The Commissariat defined education in the broadest sense to include art, film, museums, libraries, music, and drama in addition to more traditional, institutional methods of instruction. As a result of the Revolution, primary education became universal and compulsory and tremendous progress was made in the area of adult literacy. Soviet policies also opened up the universities, once the prerogative of the elite, to the children of workers and peasants and increased the overall numbers of universities.

Admission to Higher Institutions of Learning

1. Any person, regardless of citizenship or sex, who is sixteen years of age may be admitted to any of the higher institutions of learning without presenting a diploma or certificate of graduation from a secondary or any other school.

2. To require that the candidate present any credentials other than an identification card is forbidden.

3. All higher institutions of learning in the Republic, on the basis of the resolution "on the introduction of obligatory coeducation" in all educational institutions, are accessible to all, regardless of sex, in conformity with the degree on coeducation. Violators of this provision are subject to criminal prosecution by the Revolutionary Tribunal.

4. Admissions of first-year students (for 1918–1919) already made on the basis of either school certificates or competitive examinations are hereby declared void. New entrance conditions, conforming to the demands of the general statutes on higher institutions of learning which are now under consideration, will be published not later than September 1 of this year.

5. Tuition fees in higher educational institutions of the R.S.F.S.R. are henceforth abolished. Tuition fees already paid for the first half of the academic year 1918–1919 shall be refunded accordingly.

149. Nadezhda K. Krupskaya
"Adult Education in Soviet Russia"

Nadezhda K. Krupskaya (1869–1939) first became interested in educational reform and pedagogical theory while a young teacher in Petersburg. In the 1890s, Krupskaya joined various revolutionary circles in Petersburg and, in a Marxist circle, met Lenin whom she later married. Krupskaya also taught adult workers at the Evening-Sunday School, one of several established by philanthropic factory owners to provide their workers with an elementary education. As an adult education teacher, Krupskaya gained access to factories and her vivid descriptions of working conditions were included in leaflets distributed by her circle to factory workers. The circle's success in inspiring strikes in 1895 led to the arrests of numerous members, including Krupskaya and Lenin. After serving a year in prison, Krupskaya was exiled and joined Lenin who was then in Siberia. She served as his personal secretary and as the correspondent of the Bolshevik Party. Krupskaya also wrote *The Woman Worker* (1900), the first major writing on Russian women workers from a Marxist perspective, which was smuggled into Russia and circulated widely among factory workers. In 1913, she became an editor of the new journal *Rabotnitsa* (*The Woman Worker*), founded to mobilize women workers into political as well as economic action. Following the Bolshevik Revolution, Krupskaya applied her expertise in pedagogy to shaping Soviet educational policies. She helped to organize the Commissariat of Enlightenment, directed its adult education department, and served on the executive committee. "Adult Education in Soviet Russia," a speech Krupskaya delivered to the first All-Russian Congress of National Education held in Moscow in 1918, appeared in the periodical *Narodnoe Prosvyeschenie* (*Public Enlightenment*). Krupskaya consistently argued for equal educational opportunities for men and women and melded her educational initiatives with work among women. With Alexandra Kollontai[1] and Inessa Armand, she organized the 1918 First All-Russian Congress of Working Women out of which the Zhenotdel evolved. Krupskaya also founded *Kommunistka,* the Zhenotdel's theoretical journal, which she edited until its dissolution in 1930. During the Civil War, she and Kollontai boarded special propaganda trains and distributed educational materials to stiffen the resolve of the population in the Red areas. Krupskaya worked with the Commissariat of Enlightenment until her death. Her writings include *Public Education and Democracy* (1917), **On Project Method** (1931), *Woman, USSR Citizen with Equal Rights* (1937), and the posthumously published anthology, *Pedagogic Works* (1957–1963).

---

[1]On Kollontai, see Source 141 in Chapter 11.

From *Adult Education in Russia* by Mme. Nikolai Lenine. © 1920.

The war has taken millions of people from their ordinary life and placed them in abnormal conditions in which they had to face death. This, of course, compelled them to yearn for and seek a solution to the questions which had arisen in their minds. A great craving for knowledge appeared. Then the Revolution, particularly the October Revolution, created, for the masses of toilers, problems of immense importance and difficulty. The old state of things bequeathed a sad legacy—darkness, ignorance, and the absence of the very elements of knowledge. In the work of reconstruction, the great majority of people felt at every turn their impotence through a lack of knowledge. They learned by bitter experience that knowledge means power, and so they began passionately and irresistibly to crave for it. The sabotage of the intelligentsia showed them most clearly that knowledge had hitherto been a prerogative and monopoly of the ruling classes.

Adult education could not thrive during the autocracy. Hundreds of regulations, circulars and orders, fettered, maimed, and spoiled the work. The adult student was always under supervision. The authorities did all in their power to hinder any living word or thought from reaching the masses. But there has been an end to all this. The work, however, has not passed into full power. That which has been accomplished is no more than a drop in the ocean.

The whole country should be covered with a network of elementary schools for adults who cannot read and write, and for such as can do so only slightly. There must be no illiterates in Communist Russia. Let every one who has knowledge realize that knowledge, just as any material blessing, should not be the possession of the few, but the property of all; but, chiefly, he should use his time as far as he can to give knowledge to others. In the work of giving instruction time must not be wasted—"as much as possible in as short a time as possible," should be the method. In this connection care should be exercised to see in every case if the pupil really needs the instruction which is given. Many professional teachers adopt those methods in schools for adults which they followed in schools for children. They starve their pupils with explaining children's tales, with dictation, with grammar exercises, and so forth. But adults at once pass to the reading of papers and pamphlets, the language of which is not difficult, to the copying out in exercise books of any articles which pleased them, to the writing down of their own ideas, to short original compositions. . . .

The elementary school is an immense problem, but the problem of the practical school is not less important. Hitherto, applied knowledge interested mainly those who wished to get on in the world. The changed conditions, however, have achieved this result: that the most progressive workers and peasants look upon applied knowledge as a fundamental need. Knowledge of a quite special kind is required for the control and management of production, for the establishing of agricultural Communes on the basis of improved management. The workers and peasants feel that without such knowledge they are unable to master the conditions of life. The character of special education must, however, be different from what it used to be. Previously a professional training sought to fit a worker for some mechanical action—to grind, to be a locksmith, to plane, and so on; but now, in addition to all that, a professional training must enable the worker to understand the industry in which he is engaged in its entirety, and its place in the world market. Science must light up its particular nature, the history of the ramifications of the industry must be made known, and that must be connected with the study of the history of labor and of civilization; light must be shed upon it from the side of economics and politics, and so forth. In short, together with purely technical methods, a professional training should give a breadth of outlook, a grasp of the conditions in which the industry developed, such as are essential to a worker who is to be a masterly creator of the commonwealth—but were not of much use to the mere wage earner.

Finally, schools of a higher type, i.e., People's Universities, must be established. The reform of the higher school has opened the door of the university to all who want it. But that reform, as such, does not, of course open the higher school to those who so far have had no education at all. To choose a particular branch of study, which one might more fully pursue, it is necessary to have a more or less clear idea with regard to the branches of knowledge in existence: one must have a general education, and know the methods by which knowledge is acquired. Anyone entering a University without such a preliminary qualification, would soon be obliged to withdraw from it. Higher type schools should therefore give that preliminary general education to such as do not possess it....

In close connection with adult education is the organization of discussions and lectures, of cinematographic gatherings, excursions, and museums.

I will not dwell at length on these necessary complementary activities in adult education, but will only make a few observations about them....

The cinematograph, like the school, may be a great instrument of emancipation, or of enslavement. In the bourgeois system it was a powerful means to instill into the masses bourgeois ideas and feelings. There is a cinematograph section at the Commissariat for Public Instruction.[1] Six million roubles have been assigned to it in order that films may be produced which will suggest quite other ideas and feelings—i.e., feelings of human solidarity, internationalism, the idea of carefully organizing all production in the interest of the masses of the people, and so on. Provincial cinemas will make use of these films—at present they have no suitable films, or such as deprave the soul are circulated, or, at best, such pictures as are not very harmful.

In museums a great deal has been made, so far, of Natural History, Ethnography, Hygiene, and so on. A social section has been absent. Now, however, in Moscow, at the Socialist Academy, a Social Museum has been organized. There are in it, at present, a set of colored, very artistically produced, diagrams, dealing with problems of militarism, of concentration of manufacture, and so on. The program of the museum is dealt with by a special commission of Socialist Communists. Colored copies of these diagrams, pictures, etc., will be prepared and sent out to the museums in the provinces.

The organization of libraries is as important as the establishment of schools for adults. A terrible waste is going on in that sphere at present. Every union, every village, organizes its own library, and a great deal of money is spent in this way. Yet these libraries are poor, and their readers are not satisfied. The scantiness of our cultural forces, and the impoverished state of the book market, should lead us to strict economy of forces and books. Yet there is nowhere so much overlapping as in the department of library organization. For every locality a carefully-planned network of libraries should be arranged, having a central library, or libraries, and a range of points which should be served by travelling libraries in the American style.

... But it must not be forgotten that the most important work in a library is the selection of books. At present people who are not well informed often do the purchasing work.... To help the librarian in such work there should exist "a standard catalogue." A particular Commission of Specialists is now at work at the Commissariat of Instruction to produce such a catalogue (showing the most important books in all branches of knowledge). To help local institutions to purchase books for libraries and schools there is organized a Department of Supply at the Commissariat of Public Instruction. It will supply provincial warehouses as well as separate educational institutions with books, school appurtenances and aids.

I will not deal with the place of Art in adult education. That is a big subject. There are particular departments at the Commissariat of Public Instruction—Music, Drama, Fine Arts—and the department for adult education is closely connected with them....

---

[1]This is another term for the Commissariat of Enlightenment as "Enlightenment" and "Instruction" are often used interchangeably in translation.

All the phases of Adult Education will fully develop only then when the most active and direct participation in them will be exercised by those sections of the population for whom they exist. Every library should have its committee of readers, every school its committee of teachers and pupils, and so on. Then the work will live and last.

And workers and peasants should share not only in the organizing of particular Adult Education institutions. By taking part in the departments of "the Soviet for Public Instruction" they will share in every branch of Adult Education as a whole, and thus will lift it to that summit which will make knowledge the possession of the vast majority of the citizens of the Soviet Republic.

150. Decree on the Liquidation of Illiteracy (1919)

The following decree on illiteracy was issued in January 1919 by the Council of People's Commissars. The issue of literacy was especially important for women. At the time of the Revolution, 85 percent of women were illiterate and they constituted 14 million out of the 17 million who were illiterate in the Soviet population as a whole. In 1920, the All-Russian Extraordinary Commission for the Liquidation of Illiteracy, headed by Anna Kurskaya, was established in the Commissariat of Enlightenment. In the 1920s, local societies were created to assist the work of the commission. Women wrote the majority of educational texts produced by these societies and the commission. The Zhenotdel also assisted in the campaign to abolish illiteracy. In addition, mobile libraries, like the one depicted here, increased access to learning on collective farms. By the 1930s, over two-thirds of Soviet adults had achieved basic literacy skills.

A. Georgy Petrusov, Mobile library on a *kolkhoz* (collective farm), Ukraine, 1934. (*Photo by Georgy Petrusov from the book* Propaganda and Dreams *by Leah Bendavid-Val, Edition Stemmle, 1999.*)

B. For the purpose of giving the entire population of the Republic the opportunity for conscious participation in the country's political life, the Council of People's Commissars has decreed:

1. Everyone in the Republic from ages 8 to 50 who is unable to read or write is obligated to learn how to read and write in Russian, or in their native language, according to their choice. Instruction is given in government schools both existing and those instituted for the illiterate population according to the plan of the People's Commissariat of Enlightenment.

Note. This activity applies to Red Army men, with the corresponding work in military units taking place with the closest participation of the Red Army and Fleet political departments.

2. The date for liquidation of illiteracy is established by the province and city soviets as appropriate. General plans for liquidation of illiteracy locally are to be formulated by organs of the People's Commissariat of Enlightenment within two months after the publication of the current decree.

3. The People's Commissariat of Enlightenment and its local organs are given the right to recruit, for teaching the illiterate, the country's entire literate population which has not been called to war, as a labor responsibility. Payment is to be made according to the educational worker norm.

4. The People's Commissariat of Enlightenment and its local organs recruit all organizations of the working population (namely trade unions, local Russian Communist Party cells, the Communist Youth League, commissions for work among women, and others) to participate directly in the work of liquidating illiteracy.

5. For those learning to read and write who are working at hourly wages, excluding those occupied at militarized enterprises, the work day is abbreviated by two hours for instruction, with the same wages.

6. For the liquidation of illiteracy the organs of the People's Commissariat of Enlightenment are allowed to use People's Houses, churches, clubs, private homes, appropriate rooms at plants, factories and Soviet institutions, etc.

7. Supply departments are charged with the responsibility to satisfy the needs of those institutions whose goal is to liquidate illiteracy before satisfying the needs of other institutions.

8. Those deviating from the obligations established by the current decree and those hindering the illiterate from attending schools are held criminally responsible.

9. The People's Commissariat of Enlightenment is charged with publishing instructions regarding application of the current decree within a two week period.

151. Nadezhda I. Shikshaeva
Model Soviet Teacher

In April 1923, the newspaper *Pravda* initiated a competition for the best teachers in the Soviet Union. Descriptions of model Soviet teachers were then published in *Pravda*. The following account concerns Nadezhda I. Shikshaeva.

Rex A. Wade, *Documents of Soviet History,* Volume 3, Lenin's Heirs, 1923–1925, (Academic International Press, Gulf Breeze, Fla., 1995). Reprinted by permission.

Nadezhda I. Shikshaeva was born in 1883. Her father had been a teacher, and sent her to the Geneva University, where she graduated in history in 1906, and then came back to Russia to teach. At the present time she is headmistress of the First Tver School Commune, unquestionably the best in the province.

The children belong to the working and peasant classes, and are for the most part orphans. They are passionately attached to their principal, who in all her teaching endeavors to inculcate in them habits of self-discipline and love of work and knowledge.

The chief subjects of study at the school are the physical, natural, and social sciences, and labor—its theory and practice. All the children, particularly the older ones, have a general knowledge of the Constitution of the R.S.F.S.R. and to a certain extent of the history of the working class movement in Russia.

Productive processes in school workshops are taught to girls and boys alike. The school, on the initiative of Shikshaeva, has been equipped with carpentry, locksmith, boot, and sewing workrooms in which the children make their own furniture, boots, and clothing. Sometimes they even execute orders for other children's institutions. Nor is agriculture neglected; here, too, theoretical instruction goes side by side with practical work. Thus in 1922 the school ploughed and sowed 21.6 acres providing a supply of food for the school almost for the whole of winter.

The studies of the children include singing, dramatic art, clay modelling, drawing, etc., all based on a study of the best masters, as well as practical work. Theatricals, concerts, whole children's operas are produced. Nearly all the children take part in these entertainments which have a tremendous success with the local population. So highly indeed does that latter value the school and its principal that very often at her request the local peasants and workers repair the school buildings, or do other work for the school free of charge.

The school takes an active part in all special campaigns and "weeks," in all anniversaries and revolutionary holidays. But N. I. Shikshaeva is not only an excellent and highly-esteemed teacher, she is also a keen public and social worker. Since 1918 she has taken a prominent part in the effort to stamp out illiteracy amongst the adult population; she belongs to the provincial educational committees and other local public bodies. Nor has her work passed without notice. In 1920 the Commissariat for Education registered the school as a model school, and the Tver county educational department conferred on N. I. Shikshaeva the honor of being recognized as a "Hero of Labor."

152. Soviet Poster on Women and Socialist Construction

Soviet posters played an important role in mobilizing workers and peasants in the construction of socialism. The caption on this 1926 poster by A. Staakhov-Braslavskii reads: "Emancipated Woman—Build Socialism!" The artist also utilized this striking image for a 1926 March 8 International Women's Day poster.

"Emancipated Woman—Build Socialism!" by A. Staakhov-Braslavskii. (*1926 poster from* Russia 20th Century: History of the Country in Posters *by N. Barburina.*)

153. Debates on the 1926 Family Code

The Family Code of 1918 revolutionized marriage and family relations in the Soviet Union. It removed marriage from the purview of the Church, liberalized divorce, and granted alimony for an unlimited time to women unable to work. Internal debates among jurists as well as the adverse effects of the New Economic Policy on women prompted a reconsideration of the Code, especially with regard to alimony and the legal status of de facto marriages which were not registered. Soviet jurists debated revisions to the Code for nearly two years, from 1923 until October 1925 when the Council of People's Commissars submitted a draft of a new Family Code to the Central Executive Committee for ratification. But only 20 percent of the discussion participants supported the Code without criticism, while some 60 percent opposed it on moral terms. The draft code was then circulated for public debate, which took place between October 1925 and November 1926 when the Code was officially ratified. The debate took place in local organizations, the press, and meetings held in schools, factories, and villages, including some six thousand village discussions. Alexandra Kollontai[1] also joined the debate by writing articles from abroad. Arguing against alimony, which she emphasized could not be paid by all men, Kollontai called instead for taxation to create a general fund that would establish financial support for single mothers, nurseries, and children's homes. Although widely discussed, Kollontai's vision of collective responsibility was rejected. The new Family Code of 1926, which went into effect in

---

[1]On Kollontai, see Source 141 in Chapter 11.

January 1927, extended alimony to women who were in need or unemployed for up to six months but limited payments to the disabled, also to six months. De facto marriages were recognized, but only if they met four conditions: cohabitation, a joint household, mutual upbringing of children, and verification of the marriage by a third party and/or written documentation. Reproduced here are excerpts from reports and discussions on the Code which took place in the October 17, 1925 and November 15, 1926 sessions of the Central Executive Committee. The sessions were chaired by People's Commissar of Justice Dmitri Kursky whose closing speech in the November session, included here, was immediately followed by a committee vote affirming the new Code.

Second Session of the Twelfth Assembly of the Central Executive Committee, October 17, 1925.

Address by People's Commissar of Justice Dmitri Kursky.

...The Family Code of 1918, depriving Church marriage of any significance, granted civil recognition only to the civil—Soviet—marriage. For our country this was, of course, a fundamental revolution....

Here are the main points in which the Code we are bringing before the All-Russian Central Executive Committee for study and approval differs from the old one:

Firstly, the fact of registration has changed its role and its importance.

Secondly, registered and non-registered marriages become equal in their material consequences.

Thirdly, there is increased protection for the children in cases where a marriage is annulled either by divorce or at the unilateral desire of one of the spouses.

Fourthly, we introduce a demand for certain guarantees before a marriage is registered, which will make for greater care in the concluding of marriages.

Lastly, in some cases the property relations between the married parties have been differently construed.

You can see how many points of difference there are between the new project and the old one. I will now deal with these points one by one:

First point:—the change in the role and importance of the fact of registration. The previous Code was enacted at a time when Church marriage still prevailed and there was still no other way of formalizing marriages. To counterbalance the Church ceremony, the Code provided for the setting up of Registrar's Offices and afforded the protection of the Law only to those marriages which had been registered under the order established by the Code.... Even at that time (I myself took part in the session of the All-Russian Central Executive Committee which accepted that Code), even at that time the criticism was voiced that by such limitations *de facto* marriages would be deprived of absolutely all rights, since registered marriages alone enjoyed the State's protection, and that in only one respect, namely in connection with children, were *de facto* marriages protected by the law.

In the Family Code children's rights were based on the facts of parenthood, and were safeguarded irrespective or whether the marriage was registered or not. But the wife in a *de facto* marriage enjoyed no rights. This aspect of the Family Code is fundamentally altered in the present project: registered marriages and *de facto* marriages are now to have equal rights before the Law....

As regards the rights of women and children the most essential point is the so-called question of alimony, the right to support and main-

From *The Family in the USSR: Documents and Readings,* edited with an introduction by Rudolf Schlesinger, Ph.D. Published by Routledge and Kegan Paul. Reprinted by permission of Taylor & Francis.

tenance when a marriage is dissolved for any reason, and in general the married party's claim to support as well as the children's claim to be supported by their parents.

We tackle this problem by laying down that a destitute spouse who is unable to work has a right to support. Here is the text of article 11:

"A destitute spouse who is unable to work has the right to receive support from the other spouse if the court decides that the latter is able to provide such support. The spouse who is able to work may claim the right to receive support during unemployment."

The note to the article says:

"Persons maintaining *de facto* marriage relations, even though they have not registered these, may equally avail themselves of the right to receive support." I have already mentioned this.

This right of the spouses to alimony is exceptionally important in that it clarifies the very meaning of our Code and the rights of women generally.

(*Voice from the floor:* Why women and not men?)

Kursky: Article 11 is concerned with the rights of men also, since by spouse is meant either husband or wife; but the article is primarily intended to safeguard the women. And it is right that this should be so.

The first point that is here enacted, and that was already provided for by the old Code, is that a destitute spouse who is unable to work is entitled to support. Not only needy, but also unable to work: these are the conditions to be met before a married party can claim support. But the Soviet of People's Commissars went considerably further, and to our project, which had this sentence only, they added: "and during unemployment".

I cannot tell what amplification the Commission of the All-Russian Central Executive Committee may yet add, but it appears that in court practice, if this point is accepted in principle, it will be necessary to provide a more precise definition of the term "unemployment", so that it should not end by covering a spouse's intention to live at someone else's expense. This is an important point which requires greater precision.

This year the Moscow District Court made a most interesting experiment. It arranged a meeting of the People's Judges who had previous experience of the procedure in alimony suits. It emerged that the question of alimony among the working masses did not give rise to any particularly serious problems which would require the enactment of specific amendments, and that there was no difficulty in levying forced contributions as these are exacted from wages. And there is a basic principle of the law of civil procedure to the effect that after minimum deductions not more than half the wages earned may be exacted. The problem is here, therefore, more or less regularized.

But our peasant comrades among the members of the All-Russian Central Executive Committee will no doubt have their say as regards the villages. And the judges' meeting heard the main objections and had its attention drawn to the question of alimony where peasants are concerned.

In the Land Code there is a rider to the basic article 66 which says: Persons joining a *Dvor*[1] by marriage or adoption acquire a right to use the land and the communal property which constitute the *Dvor* in question, in accordance with common law; at the same time they lose their rights in any other *Dvor*.[2]

This clause of the Land Code unquestionably entitles the peasant wife to maintenance, even if she is divorced during the first year of her marriage. Officially she has a claim to a certain portion, calculated by the membership and property of the *Dvor*. This point called forth serious objections on the part of the peasants. It was pointed out that this would lead to the impoverishment of the *Dvor*. But in all fairness we must recall that if there have indeed been cases where the payment of alimony went beyond the powers of a *Dvor*, on the other hand our enquiries have shown that among the peasants there are cases of so-called "working wives"—

---

[1]*Dvor:* peasant household.

[2]The "other *Dvor:*" the household to which the woman belonged before marriage.

girl-workers who are taken as wives by registered marriage and subsequently ejected from the *Dvor*. I do not know how widespread this sort of thing may be, but it has been taken up by the press, and recently a journalist, Bragin, published an account of this usage under the headline *A Wife For a Season*.

There is another point to which our meeting paid special attention: as a matter of court practice it has become the rule that where a peasant-wife's claim to support is investigated, the property of both *Dvors* must be borne in mind; if the wife returns to her own *Dvor* and her material conditions become no worse than before, she is not entitled to alimony.

The meeting suggested that the demand for alimony might be satisfied in kind. Our courts pointed out that sometimes decisions like the following are made: a peasant woman who has one child is divorced; she takes the child with her, and with it is given a cow, in consideration of which her rights to alimony are cancelled for two years.

Similar decisions have been given in court. In any case the meeting, having heard the peasant point of view, suggested that in court practice it was desirable to satisfy the claim to alimony in kind by a given amount of rye, flour, etc., but not to divide the *Dvor* or force it to part with its chief asset, land—a course of action by which *Dvors* have been impoverished time and again.

Such are the modifications introduced by court practice. But in any case our law will assure the wife, be she peasant or worker, the basic right to alimony.

In this connection we even go so far as to provide for the woman's protection where during the hearing of a case the defendant alleges several co-habitants....

...Prior to this code our court practice found a way out by declaring all the defendants responsible (*laughter*) and letting the woman exact a contribution from each. By the way, a French visitor of mine called this the judgment of Solomon. (*Laughter.*)

In our project, as it was brought before the Soviet of People's Commissars, we had come to the following decision: we established *en bloc* responsibility; in other words we made all responsible, but we left the woman the right to demand alimony from only one of them—whichever she chose (*laughter*).

(*Voice from the floor:* Whichever earns most!)

Kursky: Yes, whichever earns most (*laughter*)—to protect women's rights. However, this version was not passed and the Soviet of People's Commissars approved instead the ruling formulated in article 27, which says:

"If the court, while examining the question, decides that at the time of conception the child's mother maintained sex relations with persons other than the person specified by article 23 of this Code, the court shall determine which of these persons is to be considered the child's father, and shall place upon that person the responsibilities provided for under article 26 of the Code."

In other words, as a general rule the court will be guided by the evidence of the plaintiff. Whoever is indicated by the plaintiff is recognized by the court to be the father (*laughter*), unless he succeeds in proving that he was not implicated, in which case he will be freed from the responsibility.

Such are the main questions regarding the right to alimony which I thought fit to examine. Of course the comrades who are going to state their opinion of our project may devise a different solution; but I think that they will not discover a new way of tackling this problem until the day when the State undertakes the bringing-up of all children. Vladimir Ilyich (Lenin), in his lecture to the workers of Moscow in 1919, pointed out the road we must travel—the road that leads to the communal rearing of children.

For the towns, the road signs pointing the way have already been erected. But these new methods have not yet found their way into the villages, where we still have some 20,000,000 private households with the smoke daily rising from the family hearth and individual management still in charge. There the problem will not be solved so soon, and we must therefore think hard how to safeguard the rights of women in the villages particularly.

And what is happening in those national republics where the marriage problem has made no progress at all? I have already outlined the fact that in the heart of the R.S.F.S.R. (Russian Socialist Federative Soviet Republic) registration of marriage has become a widespread habit and the rights of women are protected by our law. But in the national republics, as I discovered recently when studying this problem, we have not even got to a proper beginning yet.

(*Comrade Samursky, from the floor:* And we won't get there soon!)

Kursky: In the distant Auls[3] and far-away Nomad camps we still find, untouched and inviolate, the antediluvian custom of formalizing marriage and family under the auspices of the clergy.

Comrade Shurupova [comments on Kursky's address]:

Comrades, I am going to examine the question of divorce. This is a painful question both for us women and for the men. I have noticed that some comrades are laughing and giggling. But during this session this question will have to be discussed in common and seriously. I shall not defend women overmuch. We, too, make mistakes. Nevertheless, in the majority of cases the men are to blame (*laughter*). Don't be angry. I am not so young any more. I won't lie (*laughter*).

There are many of these divorces, but consider: do they benefit us in any way? Every one of us, every man or woman, will agree, they do not. For example: a man and a woman get divorced. They own a little house, a cow, and they have three children. And here they are, splitting their little property in two. The mother of course will not leave the children with the father, for children are always dearer to the mother. What is the woman to do now, with her children?

There is no danger for the husband. He will find another woman to live with. But for the wife life under such conditions is terribly difficult. The result of it all is poverty, and we have too much of that as it is. You yourself have pointed out what we lack. Above all we lack children's homes. If the State had made itself responsible for all this, it would not have managed. This is no matter for laughter, but for tears.

Now I shall turn on the men (*laughter*). I shall not take them under my protection. Husband and wife get divorced. We women are not yet fully educated; we are still in the dark, we were enslaved for centuries. All we know is priest's gossip—which we are only now beginning to forget—about "the wife must fear her husband". But now women are beginning to learn a little....

We are being drawn into all sorts of work. Our men comrades, they know a bit more than we do. You must teach us, you must not just laugh and giggle; that is of no use, particularly on the part of the enlightened comrades, the Party men. I do not consider this the way of comrades. They should be the first to give the women a hand, to teach them, show them the way about which Vladimir Ilyich used to tell us. You must not forget that Vladimir Ilyich was the first to sound the battle-cry on behalf of the oppressed women.

His road should be followed. At any rate there should not be laughing at women. To us that is very insulting.

What is the position of a peasant woman? She looks after the house, she sews, she washes and she helps her husband take in the harvest, while he—forgive me, comrades, for saying so—he will not go to bed alone and she has to obey his pleasure. And if she does not, he kicks her out (*laughter*).

We should think about these problems.

Third Session of the Twelfth Assembly of the Central Executive Committee, November 15, 1926.

Report by Comrade Kursky.

Comrades: May I recall that the Second Session of the Twelfth Assembly of the All-Rus-

---

[3]Villages in the Caucasus.

sian Executive Committee has already examined the project of the Code of Laws relating to Marriage, Family and Guardianship, submitted as drafted by the Soviet of People's Commissars, and that, having accepted this project as a basis, the Committee decided to have it circulated for local discussion at widely-held people's meetings, with a view to ascertaining the attitude of the broad masses to this project of law which vitally concerns the interests of one and all?

The discussion has assumed unusually large proportions. According to calculations made by the People's Commissariat of Justice, the number of village meetings alone amounted to 6,000, and may in fact have been considerably greater.

In paying attention to the trends of opinion that the examination of our project in the various districts has revealed, I must first of all dwell on those which became apparent at the village meetings. Here a tendency showed itself to preserve, to conserve the patriarchal peasant family; the motive being, of course, an economic one: the desire to preserve an economically more powerful household, and so on.

The letter which Comrade Platov, a member of the All-Russian Executive Committee, recently published in *Izvestiya* was more or less in that tone; in it he declared that the project of the People's Commissariat of Justice which had been circulated for local discussion was a project of no marriage or polygamy, that it would bring havoc to the villages, that the villages still cling obstinately to Church marriage, to the patriarchal—the letter does not use the word "patriarchal"—but precisely to the old family, and that every divorce is a misfortune for the village....

I must begin by saying that, on the strength of existing statistical data, this viewpoint does not in the least accord with the economic processes now affecting the villages, which could be slowed down by some legislative measure if anyone so desired....

Even if we remember that on the Volga the rate of division among the peasant *Dvors* has since then slightly decreased, as has the number

of families sharing out their property on account of the famine, and that later these processes for a while completely ceased, as was shown by an investigation, we must still give the average annual percentage of *Dvors* dividing as at least 2 per cent. This means half a million new *Dvors* each year. We thus find nothing less than a definite natural process of disintegration of the old many-membered family....

The facts show that we must completely abandon the reactionary Utopian idea of preserving the patriarchal family and preventing the division of peasant families into smaller units. It is impossible to stop processes which have their roots in the life and morals of the broad peasant masses.

As regards the other argument put forward by the advocates of these views, namely that the villages still adhere strongly to Church marriage, we must admit that this may well be the case....

On the one hand there is a very considerable increase in the number of registered marriages after the end of the imperialist, and particularly after the end of the civil, war (a similar increase in the number of registered marriages can be observed in all countries). On the other hand there is a gradual decrease in the number of marriages, bringing the figure to a norm of more or less 100 for every 10,000 inhabitants.

Reviewing these developments as a whole, we are bound to declare that the registration of marriage has become a custom, that it has become the normal way of formalizing marital relations, and is fully recognized throughout the territory of the R.S.F.S.R.

To speak of the villages as clinging to the Church ritual is therefore wrong. At best we may say that side by side with registration, marriages are still being formalized by a Church ritual which has no legal import. In any case we can assert the complete victory of our way of formalizing marital relations. Thus the second argument put forward by the adherents of this right wing at the village meetings is without foundation.

Another trend of opinion which came to light at the village meetings I would call the middle course, the basic trend. It advocates the registration of marriages, i.e., the preservation of the law as it is at present....

Very typical is the resolution adopted by a general meeting of the citizens of the village of Piremen, in the *Volost* of Trufangorsk, in the Onega district. It says:

(1) Only registered marriages should be considered as legal marriages with all their consequences, because the peasants have not yet fully abandoned Church marriage.

(2) Divorce must be free.

(3) Only such property must be shared between spouses after divorce as they acquired during their life together.

(4) The spouse able to work should not receive assistance from the other spouse, nor should he or she receive assistance during unemployment.

(5) Relatives, whether direct or by marriage, should not share the responsibility for alimony of a member of the family.

(6) Where, in establishing fatherhood, other cohabitants are cited, the latter should be made to contribute towards the child's upkeep.

Note.—Young people present at the general meeting, but out-voted at the count, during the discussion of the question dealt with in paragraph 1 raised their voices for the equalization of *de facto* marriages with registered marriages.

This note which accompanies the minutes of this meeting is characteristic.

We can thus discern three definite trends of opinion in the villages. When we turn to the towns, we shall meet with a completely reversed attitude to the problem. In the towns an overwhelming majority of the meetings, especially those held in the workers' quarters, voted for the extension of legal protection to non-registered marital relations. This can be stated, not as an approximate survey, but as a definite fact....

The first point to be made clear is that side by side with registered marriages we face a considerable spread of marital relationships unformalized by registration. Some highly revealing statistical data are available in this connection....

In terms of percentages we have 36.4 per cent. of marriages non-registered but involving

a comparatively long period of association (23 per cent. of these last for about one year and 13 per cent. last longer than one year, even two years and some even longer than that). The marriage is entered into, children are born, and then the father abandons the mother and the latter brings a suit for alimony.

As we have seen, there is a considerable proportion of *de facto* marriages which without registration last for a more or less considerable period of time and in the main correspond to the outward criteria indicated in article 12 of the project of the Soviet of People's Commissars. Life itself thus creates a clear-cut necessity—that of providing protection for those *de facto* marriages which involve an association of a more or less lasting nature....

The main argument raised against us declared that we were affording protection to every casual liaison without defining what we mean by *de facto* marriages....

Article 12 of the present project lays down the conditions to which this protection is subject. These are the criteria:

Evidence acceptable in court of a marital relationship where the marriage has not been registered comprises: the fact of living together, the existence during this association of a common household, awareness by a third party of these marital relations, evidence thereof in personal correspondence and other documents, as well as reciprocal material support according to the circumstances, joint education of the children, etc....

Comrade Ryazanov (from the floor): Is it worth all this fuss?

Kursky: Yes, it is, because you would otherwise leave hundreds of thousands of women without any protection. That is why it is worth all this fuss, Comrade Ryazanov.

Discussion on Comrade Kursky's Report

Comrade Kasparova (of the Women's Section[1] of the Central Committee of the All-Union Communist Party {Bolshevists})

---

[1]This is the Zhenotdel, also referred to as the Women's Department of the Communist Party.

We are building socialism in our country. Yet, when we face the facts of everyday life, we for some reason allow ourselves to be turned aside; although our task is to go forward towards socialism. The present project is certainly a step forward. And if we alter article 12 of this project, we shall achieve what we had in view, what we are striving for.

We must approach the realities of our life realistically. Our actual economic situation shows that we are not sufficiently strong to give social security to all who need support. We are therefore considering that some of the destitute should be cared for by the State, and the others by the family.

Article 12, in a very concealed fashion, indicates that very many women will be left completely without aid. If you stand by the three points enumerated in article 12 to guide the courts in proving marital cohabitation where the marriage is not registered—the fact of living together, the existence of a joint household and the awareness of a third party of these marital relations—you will tie the judges down and they will have no means of helping the women.

I suggest that article 12 be dropped altogether and that it be left to the discretion of the judge to find marital relations proven, on the basis of the facts of the case. For you know very well, comrades, that during the present period *de facto* marriages do exist. Can you ignore them?

I agree that at the present time it is difficult to provide for all destitute persons at the government's expense. But at the same time we cannot do away with those little channels of socialism through which we guide our aid for destitute workers.[2] If we are to do away with this form of assistance, we shall have taken a reactionary step back. I hold that we must have social security. But article 12, as I see it, indirectly opposes social security and the case for

aiding women who live in a state of marriage but whose marriage is not registered.

Comrades, I approve the project of the Code, but I oppose article 12.

Comrade Rasynkova (Vyatka District)

I do not understand why in this project *de facto* marriages only are put on a par with registered ones, and not casual marriages also. The law makes no mention of the casual marriage. I feel that an amendment putting casual marriages on a par with registered ones should be introduced into the Code. There were some men who said that when husband and wife separate, the children should be left with the wife and the husband should only pay alimony. It was also said here that women are only interested in the alimony. That is not true, comrades! A woman is interested in bringing up her child. If the man does not want the child, the woman finds it difficult to rear the child—and the result is large numbers of Bezprisornye.[3]

I consider that the government should create special children's homes where children can be brought up. And the means for this education should be provided by the father and the mother.

Comrade Gnipova (Kursk District)

I speak from this platform in the wake of lawyers and scholars: it is thus evident that I shall not have much to add. But we women must make use of what rights are granted us under Soviet law. Comrades, all of you are defending woman and child; and yet we must admit that so far the women have received very little; we must not be blamed for the way in which we have been brought up, the old-fashioned way, and that we cannot therefore make proper use of the rights given to us.

Burdening an entire family with the payment of alimony is naturally liable to interfere with our agriculture. I consider it unfair, and I

---

[2]Kasparova is referring to the tendency of Soviet judges to give women the benefit of the doubt in alimony cases.

[3]Children roaming wild for lack of care and supervision.

think it will lead to nothing but a campaign of ill-feeling against women.

If, for example, three brothers live together and possess one cow, and the court decides that alimony is to be paid by the whole family, is the cow to be divided into small pieces? No good will have come from such a decision, but the farm will have been ruined.

One of the comrades here said that the time limit within which paternity must be proven should be extended. If we were with child for eighteen months, then this time limit could be extended; but as we carry our children for nine months, as always, the father should be traced even more quickly and be made to pay.

Then, the question of age has been discussed here, the fact that early marriages are detrimental to health. I must subscribe to this view: what kind of a new generation will we have from 16-year-old mothers? But there is yet another danger. Nowadays even girls of 25 are worn out—and the cause of this is abortions. I do not know who is competent to deal with this; but something must be done, for not only 16-year-olds are ruined by abortions, but women of 25 and 30 too.

Comrade Soltz mentioned our struggle to abolish Muslim polygamy, and yet we find, even among communists, people with four wives.[4] But apart from that, dear ladies, we shall have to set ourselves a rigid and well-defined limit; so that no ill words are spoken of us from this platform. If our men are not to think us cheap, let us show them what we are worth. I forget the name of the comrade who spoke of people getting registered fifteen times. This must be rectified: the law should lay down how many times the same person may register a marriage, whether just once, or twice, or three times. This must be established in law.

Last night I was unable to sleep, trying to think of what to say and how to contradict Comrade Ryazanov. But I lack the strength. Speaking from this platform he reaped a great deal of applause, and this is what he said: Our communism is not worth a kopek.

For him life was made easy, and now nothing will satisfy him. But to us women who have been oppressed throughout the ages, the Communist Party has thrown out a rope which we are clutching with both hands; the Communist Party has freed us from oppression.

A voice from the floor: Did Comrade Ryazanov say that?

Comrade Gnipova: He did! If Comrade Ryazanov intends to abolish *de facto* marriages, why has he not, in the sixty years of his life, arranged matters in such a fashion that we beget children only after registration; for now we beget them before registration, some before and some after. Why did not Comrade Ryazanov alter this?

Comrade Ryazanov (from the floor): One can't keep everyone in line!

Comrade Gnipova: I think that even if Comrade Ryazanov had five heads he could not alter things, nor can anyone else. We must recognize the *de facto* marriage. Who are the women bearing children? A widow, a young girl, a daily woman. But we refuse to recognize *de facto* marriage. We do wrong, comrades!

We must pay the most serious attention to our young generation!

Closing Speech by Comrade Dmitri Kursky

Comrades, the debate on the report of the Code of Laws relating to Marriage, Family and Guardianship has faithfully reproduced the general picture we got when the project was discussed at the workers' meetings in towns and villages. I could not say definitely and exactly how the votes are divided in the Republic as concerns the fundamental problems of compulsory registration. But like the discussions, the present debate convinces me that the majority agree that the time has come for us to afford the protection of the Law to the new shoots of the life around us, and that in this

---

[4]The maximum number of wives permitted under Islam. See Source 140 in Chapter 11, "Muslim Women and the First All-Russian Muslim Congress," for a discussion by Muslim women of marriage and polygamy.

sense our project takes into account the demands of reality....

Let me dwell on the fundamental aspects. There is firstly the question of divorce. An analysis of divorce statistics would provide us with a useful insight into married life. Now at the village meetings people said: Divorce must be restricted because divorces are becoming much too frequent; some sort of check should be imposed; it should not be permissible to divorce very frequently, sufficient reasons should be required, etc....What are the statistics in the connection?

For every 10,000 inhabitants of the R.S.F.S.R. there were in 1922, 10 divorces; in 1923, 9; in 1924, 11; and during nine months of 1925, 8.5, i.e., averaged out over the whole year—11 again. You can see that we have here a fairly stable and definite figure....

...Take a further look. What is the age at which divorce is most frequent? The majority of divorces dissolve a first marriage. This proves that we are not faced with endless series of divorces. The age of the majority of persons divorcing each other lies between 20 and 24....

This shows that these people divorce in order to achieve a new and stable marriage. 53 per cent. of divorces occurred in marriages which had lasted not more than one year. Only in 13 per cent. of cases were marriages of greater duration concerned....

...Very wrong is the attitude of those comrades who ask for a restriction of divorce. The absolute right to divorce is one of the achievements of the October Revolution, and in this respect women cannot make any concessions.

As to the question of alimony in the towns and in the villages, I shall confine myself to replying to the suggestion that alimony suits are more numerous in the towns than on the land. There is quite a number of them in the villages, too: in many regions one-third of all the cases heard in the People's Courts are concerned with alimony and other aspects of family life. The partly landed character of rural economy constitutes the main impediment to any solution of the alimony problem in the villages, simple though it be where the workers are concerned. It has been suggested here that alimony should be "nationalized". The proposal was to levy it for the State treasury so that the State could then rear the children in children's homes. This would be a most inexpedient solution. During the present years of transition this problem must be solved on the basis of existing conditions....

I shall now come to the main question. I do understand the comrades from the villages who propose compulsory registration. The whole system on the land does, as some comrades put it, make life a matter of hard cash and a good wife. But we must not forget that at the village meetings 40 per cent. opposed compulsory registration. They were the spokesmen of the new school of thought in the villages, and they, too, must be considered....

...There is only one real way of curing the evil—to protect property interests in *de facto* marriages by acknowledging both parties' right to the property and by entitling the *de facto* wife (and this is a very serious legal point) to the same alimony benefits as the registered wife.

I am sure that when the project is examined in the Commission we shall take all considerations into account....We shall not only succeed in bringing our Code into line with the dominant plan of bringing conditions nearer to those that will exist in a communist society, but at the same time take fully into account the peripheral phenomena which demand the protection of the law. I think Comrade Terechova was right when she said that 75 per cent. of members of this session will vote for the project.

Comrade Ryazanov (from the floor): Wait for the count!

Comrade Kursky: I propose that a vote be taken on the project. *(Applause.)*

QUESTIONS FOR ANALYSIS

1. How did the 1918 Soviet decree on open admission to universities and higher education transform the educational system?
2. How does Nadezhda Krupskaya define the aims and methods of Soviet adult education? What institutional changes does she propose? How does Krupskaya relate the task of education to that of Soviet socialist construction?
3. Discuss the main points of the 1919 decree on the liquidation of illiteracy. What is the significance of mobile libraries on collective farms, for example, of the library depicted in the photograph?
4. What does the description of Nadezhda Shikshaeva as a "model teacher" reveal about Soviet educational practices and women's role in the new educational system?
5. Discuss the imagery of the 1926 Soviet poster on women. How does it appeal to women to build socialism?
6. What was at stake in the Soviet family code debates concerning registered and de facto marriages? How and why did attitudes toward divorce and alimony vary among urban and peasant households, and what measures did the new code propose to accommodate these differences? What do the debates reveal about the process of Soviet family legislation at this time?
7. Compare the "New Woman" of the Soviet Union with her counterparts in the capitalist West. What accounts for their points of difference and similarity?

THE GREAT DEPRESSION

The Great Depression dramatically transformed family as well as work relations. In 1930s capitalist Europe, women had to defend their right to work against calls for their exclusion from the work force. Prolonged periods of unemployment created psychological as well as economic hardships for adults and children, undermined family relations, and often destroyed families altogether. The crisis of unemployment was greatest in Germany. By 1932, 6 million were out of work, or two out of every five workers, and production had decreased by over 40 percent.

154. Marie Jahoda, Paul F. Lazarsfeld, and Hans Zeisel
Marienthal: The Sociography of an Unemployed Community

Trained as social scientists at the University of Vienna, Marie Jahoda, Paul F. Lazarsfeld, and Hans Zeisel were active members of the Austrian Socialist movement. Marienthal is located near Vienna which had a progressive Social Democratic administration at this time. *Marienthal: The Sociography of an Unemployed Community* (1933) examines how the small textile town of Marienthal coped with a factory closing in 1930 that plunged the majority of its working population into unemployment. The town had been founded a hundred years earlier at the same time as the textile factory. Its total population of 1,486, consisting mainly of Roman Catholics, comprised 478 households averaging three individuals per household. Three-quarters of the families now received relief payments. Jahoda, Lazarsfeld, and Zeisel analyzed data from family files, which were compiled for each family; life histories of thirty-two men and thirty women; time sheets filled out by eighty persons; and meal records maintained by forty families. Data collection was combined with immersion in the life of the town as the sociologists believed that researchers should participate in community activities and not merely act as "outside observers." Several projects were launched that allowed for a more casual form of interaction with the Marienthal residents. These included clothing drives, political meetings, a pattern design course taught to women twice a week, medical services, a girls gymnastics course, and parental guidance sessions with mothers. The Marienthal study was carried out over a one-and-a-half-year period, from the fall of 1931 to January 1932. Jahoda, Lazarsfeld, and Zeisel left Austria in the 1930s and obtained faculty positions in England and the United States. Jahoda became a professor of social psychology at the University of Sussex; Lazarsfeld, a professor of sociology at Columbia University; and Zeisel, a faculty member at the University of Chicago.

Chapter 7, "The Meaning of Time"

...Time in Marienthal has a dual nature: it is different for men and women. For the men, the division of the days into hours has long since lost all meaning. Of one hundred men, eighty-eight were not wearing a watch and only thirty-one of these had a watch at home. Getting up, the midday meal, going to bed, are the only remaining points of reference. In between, time elapses without anyone really knowing what has taken place. The time sheets reveal this in a most graphic manner. A thirty-three year old unemployed man provided the following time sheet for a single day:

A.M.

6–7	Getting up.
7–8	Wake the boys because they have to go to school.
8–9	When they have gone, I go down to the shed to get wood and water.
9–10	When I get back up to the house my wife always asks me what she ought to cook; to avoid the question I go off into the field.
10–11	In the meantime midday comes around.
11–12	Empty.

P.M.

12–1	We eat at one o'clock; the children don't come home from school until then.
1–2	After the meal I take a look at the newspaper.
2–3	Go out.
3–4	Go to Treer's (the shopkeeper's).
4–5	Watch trees being cut down in the park; a pity about the park.
5–6	Go home.
6–7	Then it's time for the evening meal—noodles and semolina pudding.
7–8	Go to bed.

The men, for the most part, passed their time in idleness....

The term "unemployed" applies in the strict sense only to the men, for the women are merely unpaid, not really unemployed. They have the household to run, which fully occupies their day. Their work has a definite purpose, with numerous fixed tasks, functions, and duties that make for regularity. The following tabulation of the principal activities of one hundred women shows how very differently their mornings and afternoons pass compared with those of the men.

	Morning	Afternoon
Housework	75	42
Washing	10	8
Attending the children	6	12
Minor household activities (sewing, etc.) and idleness	9	38
	100	100

Here is a typical account of a woman's day:

A.M.

6–7	Get dressed, light the fire, prepare breakfast.
7–8	Wash and get ready, dress the children and take them to school.
8–9	Wash up, go shopping.
9–10	Tidy the rooms.
10–11	Start preparing the meal.
11–12	Finish cooking the meal and have lunch.

P.M.

12–1	Do the washing-up and clear up the kitchen.
1–2	Take the children to the kindergarten.
2–5	Sewing and darning.
5–6	Fetch the children.
6–7	Have supper.
7–8	Undress the children, wash and put them to bed.
8–10	Sewing.
10–11	Go to bed.

So for the women the day is filled with work. They cook and scrub, stitch, take care of the children, fret over the accounts, and are allowed little leisure by the housework that becomes, if anything, more difficult at a time when resources shrink. The different significance of time for the unemployed man and his wife is brought home by occasional minor domestic conflicts. One woman writes:

Although I now have much less to do than before, I am actually busy the whole day, and have no time to relax. Before, we could buy clothes for the children. Now I have to spend the whole day patching and darning to keep them looking decent. My husband tells me off because I'm never finished; he says he sees other women out chatting on the street while I'm at it all day long without ever getting through. He simply doesn't understand what it means to be always mending clothes for the children so that they don't have to be ashamed of themselves.

Another woman:

Nowadays we are always having rows at lunchtime because my husband can never be in on time, although he was as regular as clockwork before.

Thus the men do not even manage to keep to the few fixed times that remain, for punctuality loses meaning when there is nothing in the world that definitely has to be done.

Watching the women at their work, it is hard to believe that they used to do all this on top of an eight-hour day in the factory. Although housekeeping has become more difficult and time consuming for them, the purely physical effort involved before was nevertheless much greater. The women know this and re-mark on it. Nearly all the accounts of their lives mention the fact that their housework used to keep them up late into the night after a day at the factory. But nearly all of these accounts also contain a sentence such as this: "If only we could get back to work." This wish would be understandable enough on purely financial grounds, but it is repeatedly qualified by the disclaimer that it is not merely because of the money:

> If I could get back to the factory it would be the happiest day of my life. It's not only for the money; stuck here alone between one's own four walls, one isn't really alive (Frau A., age twenty-nine).
>
> The work is easier now than it was in the factory days. I used to be up half the night doing the housework, but I still preferred it (Frau R., age twenty-eight).
>
> It used to be lovely in Marienthal before, just going to the factory made a change (Frau M., age thirty-two).
>
> Since the factory closed down, life is much harder. You have to rack your brains to think what to cook. The money doesn't go far enough. You are shut up inside all day long and never go out anywhere (Frau S., age thirty-seven).
>
> I would go straight back to weaving at once if I could; I really miss the work (Frau P., age seventy-eight).

So the women want to return to work despite the greater work load, and not simply for finan-cial reasons. The factory widened their sphere of existence and provided them with social con-tacts they now miss. But there is no evidence that the women's sense of time has been dis-rupted in the way it happened with the men....

Chapter 8, "Fading Resilience"

We have described the state of affairs in Mari-enthal at the time we were there. But even if the people with their altered sense of time

scarcely notice its progress any longer, the months go by and the foundations on which their life still rests are crumbling gradually and irresistibly away. The question is, how long can this life continue?...

Let us see which of our data can contribute to an answer. To begin with, economic condi-tions are constantly changing, and changing for the worse. This is a direct consequence of the unemployment relief laws. After a certain time, unemployment payments are superseded by emergency relief, which in turn is gradually re-duced, and can eventually be stopped alto-gether. But reduced relief payments are only one cause of the deteriorating economic situa-tion. The process is significantly accelerated by the wear and tear on all personal belongings. There is no allowance for their replacement or even for their repair in the carefully balanced budget of a Marienthal household. The mo-ment will come when shoes and clothes, repeat-edly mended, finally reach the stage where they can no longer be repaired. Crockery breaks and cannot be replaced. A case of illness will plunge a whole family into debt.

That wear and tear did not raise greater havoc earlier is due to the fact that people had unusually large stocks of materials, especially textiles, which they used to get from the fac-tory for next to nothing. Even now, many fami-lies still have some odd pieces of material from which something can be sewn when the need arises. The children's clothes suffered most, of course, and where no more material was avail-able, the parents would take some of their own clothing that was still wearable and decent and have it cut down into coats and other clothing for the youngsters....

In all three families for which we have detailed data, the children were better provided with clothes than the adults. The women have a ready explanation: it is the children who first benefit from any charity action (public or private). Fur-thermore, adult clothes can be converted into children's, but not the other way around; and fi-nally it is the children who are cared for most. As a result, the state of children's clothing is relatively good while the adults often lack the most minimal necessities....

As economic conditions deteriorate, how will the attitude of the people of Marienthal change? It will be useful to refer back here to the four categories into which we divided the population according to their basic attitude. Following is the average income per consumer unit in each of these four attitude categories:

		Schillings per Month
Unbroken		34
Resigned		30
In despair	} Broken	25
Apathetic		19

This table is not only significant for the connection it establishes between a family's attitudes and its economic situation; it also allows us to foresee at approximately what point the deterioration of income will push a family into the next lower category. In Chapter 1, we summarized the basic differences between these four attitude groups, and we know already that this difference of approximately five schillings a month means the difference between being still able to use sugar or having to cook with saccharine, between having the children's shoes repaired or keeping the children at home, between occasionally having a cigarette or having to pick up butts on the street. But this difference means also the difference between being unbroken, resigned, in despair, or apathetic.

These figures may have no validity beyond Marienthal. In places where not all people are pushed out of work at the same time, neglect and despair may set in at an earlier stage, when the level of income is still higher. Comparison with the surrounding world seems to play its part in matters of mood and attitude.

The economic deterioration carries with it an almost calculable change in the prevailing mood....

One of the questions of major import for the future of the unemployed individual is how unemployment affects his personal relationships. We have already seen how political passions subsided and how personal animosity increased, and we have seen also evidence of touching helpfulness, especially toward children.

Our knowledge about the effect of unemployment on relationships within the family came primarily from our conversations with the women. We are aware that this source by itself does not give the whole picture, since their remarks are often engendered by isolated incidents in the family. The evidence would have been better if we ourselves had observed family life over a long period of time; this, however, was not possible. Nevertheless, there is merit to those isolated reports, precisely because they select the incidents that stick in the women's minds as worth reporting. We shall draw no conclusions but simply point to possibilities.

In some cases, unemployment improved the relationship between husband and wife; for example, in one family mentioned earlier, the new situation forced the husband to give up drink. Often, too, where the wife used to feel neglected by her husband, his presence at home is now a source of satisfaction. One woman wrote:

> By and large I get on better with my husband now that he is out of work, because he helps more around the house and keeps an eye on the children. He was not so good with the first children we had as he is with the younger ones.

One has the impression that a tendency to improve the relationship already existed on the woman's side, and their common predicament strengthened a latent inclination by removing obstacles on the man's side.

On the other hand, in some marriages that had developed quite normally before, the new pressures created nervous tension and occasional quarrels. The best example of this was found in the diary that one unemployed worker kept conscientiously from the beginning of his unemployment. It showed how new and unfamiliar tensions and minor conflicts darkened the relationship between the man and his wife without, however, ever really destroying their basic understanding. Both spoke of one another with the greatest respect and affection when talking to a third party. Here is an entry made not many weeks after the onset of unemployment:

Going into the forest with Martha [his wife] to collect some wood. The best, the only real friend one has in life is a good wife.

A few days later:

Martha, that most faithful life companion, has just accomplished a feat worth recommending to all for imitation; she has managed to prepare an evening meal for three adults and four children for only sixty-five groschen.

A few weeks later:

I am condemned to silence but Martha is beginning to waver. Today was pay day; after settling our debts at the shop we simply did not have a penny left. Icy silence at home, petty things disrupt the harmony. She did not say good night.

A few days later:

What strangers we are to each other; we are getting visibly harder. Is it my fault that times are bad, do I have to take all the blame in silence???

Family 178 provides an example of the same development. The wife wrote:

We sometimes have quarrels at home these days, but only minor ones, mainly when the boys go on wild hikes and ruin their shoes.

Finally there are cases where family relationships are seriously impaired as a result of unemployment. Yet a closer examination might reveal that these marriages were not exactly happy to begin with. A tense situation might have decisively deteriorated under the pressure of privation. One woman said:

He had always been fairly quarrelsome, even during the old days at the factory, but his colleagues liked him and simply looked the other way when they saw him fly into a temper. Now things are worse, of course, because he takes it all out on the family.

Another woman:

I often quarrel with my husband because he does not care about a thing any longer and is never at home. Before unemployment it was not so bad because the factory provided a distraction.

A third woman related how her husband used to drink and get on badly with her, adding:

Hardship has made our rows more frequent because our nerves are on edge and we have so little patience left.

On the whole, it seems, improvements in the relationship between husband and wife as a result of unemployment are definitely exceptional. Generally, in happy marriages minor quarrels appear to occur more frequently than before. In marriages that were already unsettled, difficulties have become more acute. Tendencies already latent in a marriage are thus intensified by external circumstances....

But how things will continue, we cannot foresee, even assuming that no unexpected changes occur in the external situation. Two developments are possible. As conditions deteriorate, forces my emerge in the community ushering in totally new events, such as revolt or migration. It is, however, also possible that the feeling of solidarity that binds the people of Marienthal together in the face of adversity will one day dissolve, leaving each individual to scramble after his own salvation.

Events of the first type are entirely out of our range of prediction. But as to the second, we can make some contribution to a question that could become important. How does an individual's life history affect his powers of resistance during unemployment? What connection is there between past experience and present attitude?...

We found in our files a number of...cases, people whose power of resistance, after gradually deteriorating, suffered a sudden collapse. This usually occurred with men whose earlier life had been characterized by ambition and high expectations.

However, those who had been particularly well-off in the past were apt to develop a different reaction to unemployment, as in the following example:

Frau J. K. was born in 1890 in Erlach near Pitten. Her father had been an active member of the Social Democratic Party and consequently had been forced to continually change his place of work. Before reaching the age of six she lived in six different places. The family moving ended in Marienthal, where the political circumstances were quite favorable. The father worked here as a weaver. She was one of five children, enjoyed school very much and learned well without having any preference for a particular subject. She wanted to become a dressmaker, but the other children were still small and she could not leave. She entered the factory as a messenger girl and worked there until 1914. She liked to go out, was passionately fond of dancing, and frequently went to the theater or the movies in Vienna.

She married in 1910; her husband also worked in the factory. It was a very good marriage; she never had a moment's unhappiness. After her second child she stayed at home and thought she would devote herself to her children. When the war came, her husband enlisted; in 1917 he was killed. At that point the children were one and a half, three, and seven years old. She had to go back to earning a living and began in the cannery. Then she worked for a time for the railway in Mitterndorf, returned to Marienthal in 1920 and worked there until 1929. She now draws thirty-nine schillings relief money.

All her sons have made good. The oldest is a gardener in Marchegg, earning forty-four schillings a week. But he is not giving her anything; he is saving up for a motor bike. The second son works in Vienna, earns forty schillings a week, has part of his clothing provided by the firm he works for, and he sends her thirty schillings a week. She still has to support the youngest boy, who is apprenticed in Vienna. She always made all the children's clothes. The youngest was musical; she let him have music lessons, and even when things were at their worst, she paid seven schillings a month for a music teacher.

She has never lost her cheerfulness and, although "an old girl" by now, still likes to dance. After the war she began to take an active part in the Social Democratic movement, first working in its Women's Section and later in its Child Welfare organization, where she is on the committee. Since the committee had to let the theater go, she now runs the nursery school one afternoon a week. Her hardest time had been between 1916 and 1918, and then last year up to August. During the war things had been bad because her husband had been killed and she was left alone with three children. The situation did not improve until 1918 when, working in the cannery, she was able to get food more easily. Last year had been bad because she had been completely dependent on her sons. She had not been starving, but didn't like depriving the boys.

Her best time was the present because she could see that her children had got somewhere. They all were devoted to her, took her to the movies in Vienna, and looked after her generally. She divides her money so that what she receives from her son, together with the unemployment relief, is spent on food; her pension of fifty schillings a month goes on clothes. When no clothes are needed, one eats a little better. She is already buying things for the boys without their knowing it. She likes to think that when one of them gets married he will have something put by.

This woman had always shown a great capacity for organizing her life successfully. This ability did not get lost. In her youth she liked to amuse herself, sought the company of others, went dancing, and often made trips to the theater. Today, she still lets her children take her to the movies once in a while. Work on the Child Welfare Committee, an activity more in keeping with her years, has taken the place of dancing. Her need for contact with other people has remained the same, and she still knows how to satisfy it. One has the impression that she has retained her supply of physical and spiritual energy from her better days, from the time when she used to be at work; her attitude to life is not easily shaken. To be sure, one cannot say what will happen if her relief money is cut or her sons lose their jobs.

On the whole, those who had been particularly well-off in the past either held out for an especially long time or broke down especially quickly. For those who held out especially long, it was hard to determine how much was due to economic advantage and how much to adaptability, since both factors were almost always jointly operative, and had been there before the onset of unemployment. One woman, for example, declared that she had no trouble with clothes because she had been so well provided with them when she was married; today, ten years later, she is still well equipped. And she proudly informed us that she never had to go out to work....

On the other hand, the people who had been well off in the past but now put up particularly poor resistance are above all characterized

by their complete lack of adaptability. Their life broke because they could neither grasp nor bear the enormous difference between past and present. Some of them give the impression that, because of some early developed mental posture, the original shock effect lasted a particularly long time. Eventually the shock will recede and give way to resignation. With others, however, the feeling of being the victims of an unexpected and undeserved catastrophe is so strong that they show not the faintest sign of coming to terms with their predicament. They seem to be heading for individual disaster long before the village as a whole reaches that point. It may well be that this is the psychological condition which, in a large city, culminates in suicide or a similar catastrophic reaction.

People who had been particularly bad off in the past, now either belong again to the broken families or, consoled by the fact that everyone is in the same miserable plight, have joined the class of the resigned....

In some cases, however, where extreme poverty formerly prevailed and (judging from the father's life-history) the family always belonged to the broken group, unemployment brought a certain relaxation. For example, there is the mother of three children who lost her husband soon after they were married. Since early youth she had been comparing her own life with that of others in her age group and now clearly finds this comparison less unfavorable. The double burden of household and factory has been lifted from her; she draws relief money, but most important, now her life is much like that of any other woman in the village. Our files contain a number of such cases.

Finally, there are those who in the past led normal working lives with no particular distinguishing features; they are to be found in all three attitude categories. Just which one depends in each case on such factors as age, income, and character traits, but our rough analysis of the past was not sufficient to disclose such differentiating constellations.

Thus we have endeavored, here at the end, to put before the reader a living picture of some of these people with whom we have had such close contact for a few months. This brings us to the limit of our inquiry and also of our method, aimed as it is at the general and typical. We entered Marienthal as scientists; we leave it with only one desire: that the tragic opportunity for such an inquiry may not recur in our time.

QUESTIONS FOR ANALYSIS

1. Discuss the methodological approach of sociologists Jahoda, Lazarsfeld, and Zeisel; specifically, what categories of analysis framed their research of the unemployed community of Marienthal and what were their sources of information?
2. What did their study reveal about the dual nature of time for men and women? How was family life transformed by unemployment, especially as economic conditions progressively deteriorated? Discuss how women's past experiences and expectations affected the ways in which they coped with unemployment. What conclusions do the sociologists draw from their study?

WOMEN AND THE FASCIST LEAGUES IN FRANCE

The Great Depression fueled the rise of right-wing groups and movements throughout Europe. In France, the effects of the Depression were delayed and did not become apparent until around 1931. Organizing in urban areas as well as among peasants, fascist leagues gained strength as the economic crisis deepened in the 1930s. While many were more successful than the first wave of facist groups formed in the 1920s, these leagues never unified organizationally into a single party like the Nazi Party. Their most dramatic show of strength was on February 6, 1934, when they organized a mass rally to protest parliamentary corruption in front of the Palais Bourbon, the meeting place of the Chamber of Deputies. Violent confrontations subsequently occurred between the fascist leagues and left-wing counterdemonstrators. Fearful of the rise of the right, left forces began to unite in defense of the Republic. The Communist Party took the lead in initiating a broad union of the socialist, communist, and radical parties in accordance with the Soviet Union's new Popular Front strategy which became the official policy of communist parties belonging to the Third Communist International. For the Bastille Day demonstration of July 14, 1935, the left mobilized around the slogan "give the workers bread, give the young work, and give the world peace." Victories in the local elections of 1935 were followed by the formation on the national level of the Popular Front government of 1936–1938. Headed by Leon Blum, the government initiated a number of economic and social reforms but suppressed workers' strikes and actions that challenged the private ownership of the means of production. The fascist leagues' assault on parliamentary democracy now focused on the programs of the Popular Front, and their anti-Semitism found expression in repeated attacks on Leon Blum who was Jewish.

155. Lucienne Blondel
Women and French Fascism

Lucienne Blondel was a member of the French fascist league La Solidarité Française and the secretary-general of its journal by the same name. The league was formed in 1933 by François Coty, a perfume magnate. Coty funded the fascist Action Française and the Croix-de-Feu (Cross of Fire), which were established in the 1920s, and owned several newspapers, including *Le Figaro*. The Solidarité Française was a paramilitary organization whose members organized into squads by region, wore uniforms (blue shirts and berets), and gave the Roman-style salute. The Solidarité Française participated in the February 6, 1934 riots in Paris. In its rhetoric and actions, it consistently targeted the parliament of the Third Republic, which it considered corrupt and ineffective. Like many of the fascist leagues, the Solidarité Française attempted to recruit women to its movement and it created a women's section for this purpose. The following article by Lucienne Blondel, "La Solidarité Française et les Femmes," appeared in October 1935 in *Problèmes Actuels* (*Current Problems*), a special edition of the journal *Solidarité Française*.

Women, who numerically make up the most important part of the Nation, are at the moment the great unknown in the future of politics.

What do women think? What do women want?

On this subject, the only opinions we have heard so far are those of some aspiring female politicians who only lack a mandate to honorably serve in the discordant choir of Parliament.

For National Revolutionaries like us, this Parliament is destined to disappear. The real country of workers, taxpayers, deceived and humiliated voters, has come to regard these politicians as *outlaws* who are manipulated by the lodges[1] and the parties...and by the millions of complicities, profits, hushed-up scandals, and compromises that make up the secret history of the French Third Republic.

...Our legislators of today work a little like the Homeric gods who meddled in the lives of the Greeks without an iota of wisdom or fairness, but only according to their passions and interests. Therefore, they must now cede their place to new men, who are free and responsible to the nation.

And women, what do you say?...

Men vote: therefore we must vote!...

A simple concern of justice is typically the only reason women give for voting.

We propose to offer them some others, for the next epoch when voting will actually mean something.

As with cleaning products, we must examine the question of the vote *before* and *after!*

"Before," that is today, elections represent the triumph of schemes and trickery, the most cynical bluffs, and the influences of money. A woman candidate for Parliament or for *arrondissement* council would necessarily be presented and subsidized by a political party, which means that she would be compromised before even having the possibility of accomplishing anything....

---

[1]Here, the reference is to such groups as the Freemasons.

Contributed and translated by Daniella Sarnoff. Used by permission.

And, finally, were one to grant the vote to women tomorrow, the odds are that the majority of women would abstain, which is significant. The only ones to cast their vote, as a weapon in combat, would be the female red army, Jewish women lawyers, socialist women, and female schoolteachers who have been unleashed against the country and its traditions. . . .

But we, the women whom no political tarantula has stung, and who only want certain institutional reforms and changes, would we obtain what we want by voting for Blum, Frot, or Tartempion?[2] Men's voting experiences have shown the opposite. It is sufficient, and let us not forget, that women, traditional by instinct and conservative in the best sense of the term, are much more practical than their male companions. . . .

To vote today, assuming that there would be a great demand for women's votes, would only succeed in complicating a condemned parliamentary machine, without anything viable, useful, or serious resulting from it. From it would only burst forth the discourses and beliefs of a favored minority. And that is all!

That is not what women who are inclined toward a constructive and durable feminism could possibly want. . . .

But as of today, they could share demands and desires for the new type of voting that we will impose. . . .[I]t will no longer be this shamefully distorted lottery, in which three quarters of the country is uninterested. . . .

What can women do?

To become accustomed to the idea, and it is not easy, of an extra responsibility that will become theirs tomorrow. To spread from one to another the facts of certain problems which await a solution and, also, to repeat a small question which will cause reflection:

—You want to vote? Fine! One could ask you for whom . . . but it is much more important to ask you for what?

A hundred questions await you which are now at a standstill because of the indifference of politicians who are more occupied with the game of shifting majorities than with tragic or difficult problems.

Depopulation is a terrible threat to our future, which cannot be fought effectively with the expenses and fanfare of rewarding all too rare large families.

We need measures that are energetic as well as sensitive which are part of a comprehensive plan instead of encouragement limited to superficial congratulations.

As long as the economic crisis, which affects the daily existence of forty million people is not resolved, as long as the moral redress of the country is a hope and not a reality, as long as real and practical advantages do not favor an increased birth rate, we will not get the French people to be fruitful and multiply according to the word of God and the needs of a powerful nation.

In addition to this absolutely basic problem, there are other pressing issues directly related to it.

There is the question of childhood which daily scandals will never resolve if they are but fodder for journalists and don't in the least influence officials, civil servants, and politicians.

There is the reform of the entire educational system, which we must decimate—in a figurative sense—rather than allow professors and anarchist school teachers to continue to fail in the educational mission.

Let us . . . dissolve the teacher's union, just like all unions of civil servants paid by the state, and we will see if the others do not return to a more modest conception of their role.

Finally, in a different vein, there are the problems specific to women which are as varied as they are complicated.

Let the articles of the civil code be revised in order to give women neither exceptional nor particular rights, but rather simple equality before the law.

---

[2]Eugene Frot was the minister of interior during the February 1934 riots. He was a dissident socialist. *Tartempion* is a pejorative term signifying "anyman" and is often used to connote condescension or even scorn. Used here, it connotes scorn of Blum and Frot.

Let intense, even angry discussions provoke debates about women's work. We will confront the vigorous defenders of the *female proletariat,* as if merely coupling these two words were not a condemnation of a liberal regime that has encouraged the exploitation of a poorly defended workforce with low wages and the disintegration of the home.

Depopulation, even more so than male unemployment, is caused by a change of lifestyle which is all too widespread.

Women have always worked, but never as they do now.

Let us repeat, so that those with eyes and ears can understand: we will always be the defenders of women's interests and rights, whether in the domain of law or economics.

But it is desirable to reduce women's wage work through the combination of a new moral orientation of the nation and a social transformation.

There are enough women who work without pleasure and out of necessity because they are single or supporters of their families.

These women would be the first to benefit from such a reform as they would receive better pay and more protections. But the others, who do not need to work, would be invited to seek other distractions.

Alas, will we have demonstrated in vain, even with the support of abundant proof, that France is dying of depopulation, that in thirty or forty years its population will be half of what it was before the war, that it is on the road to becoming a nation of old people. It will be obvious, and not only in statistics and graphs.

Of course, it is good that articles and films, sketches and books bring to light these appalling pieces of information.

But in addition to documents and figures, it is necessary to create social and moral propaganda which will awaken the masses spiritually. In the twentieth century, the newspapers, the radio, and cinema alone have that possibility, at least with regard to the masses. They are the mirrors, often unfaithful but enticing, in which men today seek their reflection, dreams, and even inspiration for their ideas.

The day when good films celebrate childhood and the family with the same ingenuity and diversity that is used today to increase the number of vaudeville or gangster stories, the night the radio sounds an alarm for the future, and the newspapers print in a thousand different ways the law of our national salvation, a large step will have been taken to direct the attention of the entire nation toward its most dramatic problem: its own existence and endurance.

But this unanimous orchestration of modern propaganda can only be achieved by the authority of an energetic and stable government. The same is true for all the social measures, protections, and measures which will serve as the concrete counterpart to all moral campaigns.

This again exposes the incompetence of our present regime, which always casts aside essential issues in favor of those concerning details and of little importance, whether it be the budget, finances, national security, or the survival of the country.

As for those, the more numerous who work "because they cannot do otherwise," who would prefer to be mistresses at home, to have children, to make health, beauty, and elegance reign in their homes, they will be able to attain this when the crisis passes and men's wages regain their pre-war values. Our mothers and grandmothers of the middle classes lived in this manner and we will no longer see anything wrong with it. Were it not for the harsh economic laws of today, their granddaughters would voluntarily consent to resuming their time-honored role which the war compromised. And they would wear an ironic smile, devoid of regrets, when they filled out "no profession" on official forms, provided that this no longer means, as it does today, "without protection."

QUESTIONS FOR ANALYSIS

1. What is Lucienne Blondel's appraisal of Parliament and women's demands for the right to vote in France? What cultural and economic issues does Blondel identify as women's issues and what solutions does she propose?

2. Compare Blondel's right-wing critique of the Popular Front government in France with the left-wing critique of the Weimar Republic expressed by Rosa Luxemburg and the "Manifesto for International Women's Day," Sources 142 and 143 in this chapter. What do these comparisons reveal about the varying political views held by women concerning the role of the state and women's status in the workplace and the family?

CHAPTER 13

Women and Fascism, World War II, and the Holocaust

Fascists first came to power in Italy in 1922, followed by Germany in 1933, Spain in 1939, and Vichy France in 1940. Initially of a marginal character, right-wing groups and movements in interwar Italy and Germany soon acquired a mass base due to the convergence of several factors: high levels of unemployment, currency inflation, the shock of defeat, a male backlash against the gender dislocations caused by the war, political polarization and middle-class fears of socialism and communism. In Germany, anti-Semitism played an especially important role in the growth and consolidation of the Nazi Party, while anti-Semitic laws were not adopted in fascist Italy until 1938. Conservative elites in both Germany and Italy were at first suspicious and wary of these right-wing movements because they condemned the existing political system as corrupt and ineffectual. But elite fears of the right were soon subsumed by an even greater fear of the left. Fascist movements gained political power in Italy and Germany not only on the basis of popular support and electoral strength, but also because of the political decisions made by these elites. In 1922, King Victor Emmanuel III appointed Mussolini prime minister in response to his march on Rome and, in 1933, President Paul von Hindenburg made Hitler chancellor of Germany.

In Spain, by contrast, the Popular Front government elected in February 1936 was soon challenged militarily by the fascist military forces of General Francisco Franco. Spanish women mobilized against fascism in various ways, including armed defense of the Republic. In December 1936, the Spanish Republican government appealed to the League of Nations for assistance and condemned the prevailing "murderous peace." But Britain, the United States, and France did not waver from their policy of nonintervention, while Italy and Germany continued to provide arms and troops to the Franco forces without hindrance. Thus, it was in Spain that Germany first tested its new military aircraft by bombing cities and villages. The Soviet Union alone supported the Republican forces which also received crucial support from an estimated 40,000 volunteers who comprised the International Brigades. The defeat of the Spanish Republic in the spring of 1939 represented an important turning point, but debates about nonintervention continued, especially around the crisis of Czechoslovakia in 1938. These debates divided feminists as well as the left in general and it was not until the invasion of Poland on September 1, 1939, that Britain and France honored their treaty obligations and declared war on Germany.

In June 1940, however, less than a year after the outbreak of World War II, the French government capitulated to Nazi Germany. On June 22, the French government under the leadership of Marshall Henri Philippe Pétain, a World War I hero, signed an armistice with Germany which divided France into two. The northern portion, which included Paris, was occupied by Germany while the "free zone in the South" remained under French control. On July 10, the National Assembly dissolved itself and granted full powers to Pétain who relocated his government in Vichy. Attributing France's defeat to "too few children, too few arms, too few allies," Pétain collaborated with the Nazis while implementing his own National Revolution in Vichy France, which embodied a long tradition of right-wing, anti-Semitic opposition to the secular ideals of the French Revolution. For the revolutionary slogans of "Liberty, Equality, and Fraternity," Pétain substituted "Work, Family, and Fatherland." In November 1942, following the Allied invasion of North Africa, the whole of France was occupied by Nazi Germany.

Fascist governments throughout Europe implemented many similar policies. These included the prohibition of trade unions and independent political organizations and parties, the brutal repression of political opposition, and legislation which reinforced patriarchal authority in the family, excluded women from paid, wage labor, and denied them reproductive choice. Racial laws were first implemented in Nazi Germany which carried out the most extreme policies of racial persecution that culminated in genocide. Nazi doctors implemented governmental euthanasia policies on those declared "useless eaters": the terminally ill, severely disabled, and mentally retarded. Jews were excluded from civic and economic life by a series of laws and then forcibly moved to and confined in ghettos from which they were later deported to concentration or death camps. Roma (Gypsies) and homosexuals, too, were regarded as subhuman and rounded up for the camps. Moreover, German industrial firms reaped enormous profits from the slave labor of Jews and Gypsies in concentration camps and the forced labor of non-Jews deported to Germany from Poland, Russia, and various Eastern European countries under Nazi occupation. Companies utilizing concentration camp slave labor were required to pay only a minimal fee to the SS (a paramilitary Nazi party organization) which provided the laborers and maintained the camps. Receiving less than subsistence food rations, vast numbers of slave laborers died of malnutrition and disease as they were literally worked to death.

After the first waves of repression, resistance movements sprang up in the occupied countries, in Vichy France, and, to a much smaller extent, in Nazi Germany itself. Women participated in various aspects of these movements, including the partisan networks that operated behind enemy lines. In Britain, the only European homefront outside the fascist orbit from 1939 to 1941, women worked in the munitions industry, joined the Women's Land Army to carry out agricultural work, and participated in various volunteer organizations that provided social services to civilians and assistance to the war effort. Following the German invasion of the Soviet Union in June 1941, Soviet women played a central role in the Great Patriotic War in which the homefront and the battlefront were often inseparable. Women not only carried out essential noncombatant activities, but also served in the military in a variety of ways, including as commanders of tank battalions and as pilots of airplanes in bombing missions. The entry of the United States into the war, combined with the German surrender at Stalingrad in February 1943, signaled the beginning of the end for the German war machine. The conclusion of the war in 1945 left Europe in ruins. Of all the Allied countries, the Soviet population suffered the greatest losses, with 25 million dead out of a total of 50 million killed during the war. But the Cold War began even before the war's conclusion, as the United States and England began to envision a partnership with a new Germany and an adversarial relationship with the Soviet Union.

FASCIST FAMILY, WORK, AND RACIAL POLICIES

Fascist governments prohibited independent women's organizations, formed their own organizations to promote conservative ideals, and passed laws to bolster patriarchal family and work relations. Racial laws passed in Germany between 1933 and 1938 systematically excluded Jews from virtually all aspects of economic, social, and political life within German society. Racial laws were also legislated in Italy in 1938 following Mussolini's rapprochement with Hitler, in France following its defeat in 1940 (in both the occupied and Vichy zones), and in the various countries occupied by Nazi Germany.

156. Teresa Noce
Women and Italian Fascism

Teresa Noce (1900–1980) was a leader in the Italian Communist Party and the international movement of women against war and fascism. Born in Turin, in the industrial north of Italy, Noce began working in its clothing and textile industries in her youth and later became president of the federation of textile employees and factory workers. She participated in strikes, joined the Young Socialist movement, and opposed World War I. In 1921, Noce joined the Communist Party. She led factory occupations during the postwar wave of labor militancy, in the process confronting fascist bands used by employers to intimidate militant workers. In 1926, the year in which Mussolini officially banned all political parties in Italy, Noce went into exile, first in the Soviet Union and then in Paris where the Italian Communist Party had its official headquarters. She wrote numerous journalistic articles, often under the code name of "Estella," which exposed the patriarchal nature of Italian fascism and the complicity of the Catholic Church in supporting various fascist policies, especially those concerning women's role in the family. Noce made a clandestine return to Italy in 1930 and became active in the Resistance. She participated as a member of the Italian delegation and spoke at the International Women's Congress Against War and Fascism held in Paris in 1934. Noce's article "The Situation of the Working Women of Italy and the Struggle Against War," reproduced here, appeared shortly before the congress in the July 20, 1934 English-language edition of the *International Press Correspondence,* a weekly newspaper which reported on resistance movements and opposition to fascism throughout Europe. During the Spanish Civil War, she worked as a journalist in Barcelona and edited *Volontario della Libertà (Volunteer for Liberty),* the newspaper of the Garibaldi Brigades. Noce returned to France in 1939 and worked in a metallurgical factory until she was arrested by French authorities. Noce was freed in 1941, only to be placed in a series of concentration camps by the Germans and was liberated from Ravensbruck in May, 1945. Returning to Italy, she served on the Executive Committee of the Central Committee of the Communist Party (1945–1954), having been a member of the Central Committee since 1933. Noce was elected to the Constituent Assembly which drew up a new, postwar constitution and subsequently served in the Italian Parliament's Chamber of Deputies (1948–1958) where she worked on laws to improve women's status in the family and the workplace.

The position into which fascism has forced working women in Italy is one of such hardship that even the fascist trade unions find themselves obliged to raise demagogic protests. The trade union functionaries "protest" not infrequently in the fascist newspapers against the competition represented by the low wages paid working women, to the disadvantage of the working men, and demand that "measures" be taken. But such measures generally aim at greater profits for the capitalists: A number of the women employed in an undertaking are discharged, and replaced by men, whose wages are, however, considerably reduced in order to

From *International Press Correspondence,* July 20, 1934.

"equalise" them with the wages of the women workers. The remaining women workers are then threatened with dismissal if they do not submit to further curtailments.

According to the collective agreements concluded by the fascist trade unions, and valid at the present time, women workers are being paid wages between two and eight lire for eight to nine working hours. Wages exceeding eight lire are extremely rare, even for skilled women workers. As a rule, however, not even these starvation wages are actually paid out to women workers, because the collective agreements are systematically violated by the employers, especially with regard to wages. This is evidenced by the enormous number of complaints raised collectively and individually against the industrialists yearly (more than 500,000).

If these wages are now to be further reduced by 7 to 10 per cent, as announced by Mussolini in his speech of May 26, then it is obvious that the work of the women can only be regarded as a kind of semi-gratis work, a kind of compulsory labour, differing from what is understood under compulsory labour only in being carried on in the factories for the direct profit of the employers instead of in concentration camps for the benefit of the State.

As a matter of fact the working women of Italy are forced to work at wages which do not suffice to cover the barest necessities of life. The collective agreements are concluded by the fascist trade union functionaries, without the workers involved having any say in the matter. In case of unemployment the majority of the working women do not even receive the unemployment benefit granted to the men for the period of three months.

A few figures from the fascist papers themselves give a clear idea of the want and misery into which the great masses of the working people are plunged, and the effect upon the working women. In Milan a night shelter has been opened for women only. In 1933 no fewer than 51,593 women, destitute of bread or shelter, took refuge here. The majority of these women were between the ages of 30 and 50, that is to say, they were working women who should have been at the height of their powers.

The "Work for Mother and Child," taken up by fascism during the last few years, gave aid in 1933—as pompously announced by the fascist press—to more than two million mothers and children. But those who know that this work confines itself to providing free accommodation only for expectant mothers who are utterly destitute, that its children's homes take in only deserted children, that the dining rooms for expectant mothers are open only to women who can prove that they have absolutely no other means, upon which they are given one meal a day—those who know this see in the figure of two million a clear proof of the poverty and misery of the greater part of the working women of Italy after 12 years of fascism. It must be remembered that the "Mother and Child" action gives no aid to women whose complete destitution is not acknowledged, even if they are out of work, if only one single member of the family has employment, so that the women are forced to pay the costs of their confinement themselves, or to pay back in instalments the maternity aid given them by the municipality. Even those working women who pay the maternity insurance (in Italy there is an obligatory maternity insurance, which has to be paid by all working women between the ages of 15 and 60), and whose dues are fully paid up, receive only 300 lire as "birth premium," though in reality the expenses of medical or hospital aid alone exceed this sum.

Therefore it is not surprising that in Italy the cases of confinements in the open streets, in railway compartments, etc., are increasing in frequency. The newspapers of the great industrial towns (for instance, *Lavoro Fascista*, June, 1934) frequently record such "joyful events."

Although fascism is carrying on a demagogic campaign against women's work, the proportion of female labour is increasing as compared with male labour. In Italy's trade and industry alone the number of women employed at the present time is 1,500,000, and has increased in these branches of gainful work from 26.5 per cent to 27.2 per cent.

This is the result of the fact that fascism is doubly anxious to replace male labour by female: In the first place, women workers performing the same work are paid from one-third to one-half less than male workers. And in the second place the employment of large numbers of women signifies that in case of war the men called up can be rapidly replaced by women.

It is not by accident that in spite of unemployment the number of women employed in various branches of industry, especially the chemical industry, is growing. Now that the preparations for war are being openly supported by fascism, juveniles and women are being given employment in a number of factories manufacturing arms and munitions.

The situation confronts the Communist Party of Italy, and all revolutionary workers, with important and urgent tasks.

The struggle against war must be extended and better popularised among working women. The working women have demonstrated emphatically that they will and can fight against fascism, against exploitation. Again and again they have defied fascism; in innumerable incidents of the class struggle, of the protest movements, of the strikes and demonstrations, they have forced the exploiters to retreat. These day-to-day struggles of the working women must be linked up with the struggle against war. The demands of the working women, especially the demand for equal pay, for equal work, must be supported by the whole of the toiling masses, and working men and women must join their efforts in the struggle to prevent the war which is being prepared by fascism.

The campaign in preparation of the *International Women's Congress against War and Fascism* has commenced to reach broad strata of the working women of Italy. The Italian Initiative Committee which has been formed, and is composed of socialist, anarchist, Communist, and non-party working women, has issued an appeal to all the women of Italy to raise the alarm against the war being prepared by fascism, and to work for the defence of the Soviet Union and for participation in the World Congress. Thousands of copies of this appeal are being secretly circulated in the works and factories, in the rural districts, in the workers' houses....

157. Hanna Schmitt
"The Disfranchisement of Women"

Hanna Schmitt was a Swiss activist in the Women's International League for Peace and Freedom who had worked closely with German feminists during the Weimar period. In her article "Die Entrechtung der Frauen" ("The Disfranchisement of Women"), Schmitt describes how women were being affected by the implementation of Nazi racial and family policies. The article appeared in *Deutsche Frauenschicksale (The Fate of German Women)*, an anthology produced by the Union fur Recht und Freiheit (Union for Rights and Freedom) and published in 1937 by a German press based in London. The book comprises various articles detailing the repressive nature of Nazi policies on Communists, trade unionists, Jews, and women. It also includes a list of women who were arrested and imprisoned or, in a few instances, sent to concentration camps for such "crimes" as high treason, work with the German Communist Party or the German Social Democratic Party, propaganda against the German government, anti-fascist agitation, and being a member of the Jehovah's Witnesses.

In 1918 German women were given the active and passive franchise. From that time on, they were represented in the parliaments of the Reich, the states and the communities. The progressive and socialist parties in particular sent women representatives to these bodies. The only exception was the Nazi party and once the dictatorship was in power, it was this party that expelled women from parliaments and deprived them again of their public functions. Thousands of women who formerly had been active in state and communal positions experienced the hostile attitude of the new regime. In numerous cases they were turned out of hospitals and schools where they were employed as physicians, school principals and valuable teachers.

Mr. Goebbles[1] tried to explain the measure in his own peculiar brand of dishonest pathos: "When we exclude women from public life, we give them back their honor." Apparently the exclusion of women from public life did not suffice to restore their "honor." For the lives of women as individuals were also restricted, or actually destroyed, even if these only involved private organizations. All women's organizations unpopular with the regime were disbanded or "brought into line," that is, placed under the oversight of the National Socialists, if allowed to continue to exist at all. In particular, anti-fascist and pacifist women's organizations were disbanded, for example, the Women's International League for Peace and Freedom, once very active with 280 local groups. Its one crime was that it had wanted to achieve German honor in the world with something other than arms and weapons.

I have observed for years the activities of this organization [Women's International League for Peace and Freedom] and learned to appreciate the labor of those women who, after the dissolution of the League [by the Nazis] had to leave their homeland to escape suffering in German prisons and concentration camps. Frieda Perlen, the chairwoman of the Stuttgart chapter, a tire-less worker especially for German-French rapprochement, did not escape her persecutors. Her detention in a concentration camp broke her health and after her discharge she succumbed to her suffering. Three other friends, genuine and upright personalities, became stateless refugees. Why? Only because they wanted the best for their country, because they stood up for peace and freedom. The destruction of German women's cultural rights is particularly evident in academics. Twenty-five years before the beginning of German fascism women were allowed to study [at the universities] and to train for every type of public service. Today they are restricted to ten percent of the students. Out of 10,500 young women who took their secondary school exams in 1934, only 100 were allowed to study at the universities, and this permission to study offers no guarantees of an academic position in the future. Certain university faculties go even further than state laws in setting barriers. One faculty of medicine gave an almost classical justification for this: "The female physician is a person with two sexes, who is required to give up her healthy and natural instincts."

Women when they are admitted to the universities are subject to work service duty in the so-called *Frauendienst* or "women's service." They are trained in anti-aircraft defense, for the signal corps and ambulance service, and get practice in the use of gas masks; in short, the first stage of their studies is given over to one theme only: readiness for war.

Women have disappeared from editorial offices and are deprived of professorships. There is no room for them in laboratories. If they are suspect for racial and ideological reasons, the regime persecutes them even more ruthlessly than men. For they are out of favor not only because of their convictions but also because of their sex, which, according to National Socialism, places them on a lower level than the male.

This is particularly evident in the economic and social life of women. An official decree issued on April 27, 1933, shortly after Hitler's seizure of power, said: "Management is to see to it that all married women employees ask for

---

[1]Joseph Goebbels, the minister of propaganda in the Nazi government.

their discharge. If they do not comply voluntarily, the employer is free to dismiss them upon ascertaining that they are economically protected some other way."

Thousands of employees lost their positions by this decree, married as well as single women, who were simply told that parents or other members of their families could take care of them. It goes without saying that such a loss of income seriously threatens the existence of a family, considering the low wage level in Germany. Besides, the men who take the place of the women are not paid any better. Thus, poorly paid husbands are supposed to take care of their wives. If they cannot manage, the wives will look for work, but these days married women are treated worse than ever. Their wages are minimal for a maximum amount of work. Wage contracts are ignored. Only in munitions factories are they sought after as a reserve labor force in case of war. Due to the slavedriver system and the unsanitary conditions in the war industry, the health of the women is most seriously endangered. It isn't any better for those women and girls, often only 14 or 15 years old, who are ordered to work on farms. They often return home undernourished, sick, and despairing. Many of them turn to prostitution. In the cities, so-called "retraining institutes" are established, in which women from various occupations receive training as household servants. Girls who have completed their schooling are also sent to such courses, or pressed into work service. Instead of a salary they receive a little bit of pocket money and room and board. Women's work service for those who are 17 to 21 years old is organized militarily; in the 242 work camps organized in 1934, the young women did not learn any occupation except marching and training for war service. Things are the worst for female servants who work in the households of "reliable National Socialists"; these employers gain innumerable employees in this way, whom they can trick to their heart's content. Young women there perform "service" in the truest sense of the word, and not only as household personnel.

And what about the private life of women? The Nazis claim to honor housewives and mothers above all. But this honor is denied to a great number of mothers. It does not cover, for example, the mothers of illegitimate children. They bear the burden of care but are morally disqualified and do not get the benefit of the necessary legal protection.

Marriage is subject to the severest restrictions. Women who have been in a concentration camp must provide proof before marriage that they are in perfect physical and mental health, racially pure and not dependent on public relief. The threat of sterilization hangs over many women like the sword of Damocles. Every woman with a physical defect, be it ever so slight, must make it known to the authorities. The doors are wide open for denunciations and vindictive vengeance.

According to official German statistics, more than 500,000 people were sterilized in the last few years; of these, 30,000 cases resulted in death. While in the U.S.A., which has sterilization laws also, 19,000 people were sterilized in 28 years, 55,000 people were subjected to this dangerous operation in a single year in Germany. With such massive surgical intervention, the necessary precautionary measures are naturally not observed. The magazine *German Justice* [*Die deutsche Justiz*] reports that by the end of 1934, 27,958 women were sterilized, with five per cent of the cases resulting in death.

The most tender and intimate relationship held sacred by all civilized peoples is that between mother and child. But with what brutality does National Socialism often tear asunder these bonds. If, for example, a mother is Jewish and the father Aryan, the child is taken away from the mother in case of divorce to avoid exposure to "Jewish influences." The fate of Jewish mothers is especially tragic in small towns. They must look on as their children grow up in isolation which has a devastating effect on the young minds. How heart-rending is the authentic report of a Jewish mother forced to live with husband and child in such a town. When she asked her little daughter what she wanted for Christmas, she tearfully replied, "Only a little playmate, mommy!" It is not surprising that Jewish families of means send their children abroad to be educated. There is a tragic saying among them, "Our children become letters." Though the hardest tests are those for Jewish or mixed marriages—which are no longer allowed—"Aryan" couples should not be considered lucky. The "blockwatch," trusted people placed by the fascist system in each apartment building (in other words, informers), "politely" make the families of officials, for example, aware of their duty to procreate, for "the Führer needs soldiers."

There are also premiums for families with many children. However, in the past year, 3,563 women who had given birth to their third child applied for such a premium, but only 787 were found worthy of this, for only they could satisfactorily prove their "pure race" back to the great-grandparents on both sides of the family.

The lot of the housewife and mother is not better than that of the woman with a job. Whether at home or on the job, she must depend on a man. Such a system offers no space for her to have her own life. Women are to be kept from thinking, for their thinking could lead to danger. The more the German woman is good hearted and sentimental rather than hard, the more she must keep silent.

Women who are silent, women who serve, are not free women. For only free women can develop their characters to the fullest, be it in pursuit of studies or at work, as housewives and mothers. A nation that honors its women honors itself. However much it may boast of national virtues, a regime that oppresses the women shows contempt for the people and at the same time for those who are responsible for the future, the mothers of its children.

158. Vérine (Marguerite Lebrun) "God, Work, Family, and Fatherland"

Vérine (Marguerite Lebrun) was a journalist and activist in 1930s pro-family and pro-natalist circles who became an outspoken advocate for Pétain's National Revolution. Like many right-wing women, Vérine supported Vichy family policies that reinforced patriarchal and religious values. Along with Andrée Butillard, the president of the large Catholic women's group, the Union féminine civique et sociale (Women's Civil and Social Union), she was one of only two women to serve on the Conseil Consultatif de la famille française, the Vichy advisory council on families. The widow of a prominent physician-legislator, Vérine achieved such prominence in government circles largely because of her involvement in the École des Parents (Parents' School), an organization she founded in 1929 with the aid of a network of right-wing men in the natalist-familialist movement of the interwar period. In cooperation with legislators and propagandists who were members of the Alliance Nationale Contre la Dépopulation (National Alliance Against Depopulation), the Plus Grande Famille (the Largest Family), fascist leagues like the Croix de Feu, as well as mainstream conservative political parties, Vérine and the École des Parents organized conferences and lectures throughout the interwar period aimed at inculcating in French parents a child-rearing philosophy that privileged the ideals of God, Fatherland, Order, and Authority that would later be at the heart of the Vichy regime's "New Order." The following excerpt is from Vér-

Contributed and translated by Cheryl Koos. Used with permission.

ine's article "God, Work, Family, and Fatherland" which appeared in the January 1941 issue of the journal *Education*. Founded in 1935 with right-wing financial support, *Education* became a major advocate of the Pétain government's educational philosophy which instituted separate educational programs and goals for boys and girls and sought to reinstate the Catholic Church at the center of classroom instruction.

From the bottom of the abyss where we have fallen because of the defeat of 1940, all appeared at first glance to be an incomprehensible mystery with unfathomable and insoluble problems....

We have taken inventory of our inner wealth and we have seen for ourselves, in unhappiness and in familial and collective anguish, the passions that must raise man above himself—God—Work—Family—Fatherland....

GOD: Have we not always made self-interested and childish demands on Him, paid Him some pitiful premium in order to pay our insurance against the evils of this world and the other? Is He really OUR FATHER to all? The one who commands love between men so that they can truly be brothers?

Have we lived the Lord's Prayer? Have we asked first of all that His kingdom comes? Haven't we ourselves created our own hell in neglecting the essential, in transgressing the eternal law of love and order? Isn't man himself punished in worshipping Baal, Astarte, and the Golden Calf, founder of pleasure-seeking capitalism? What then? What are we complaining about?

WORK: Work is a task to perform and perform well. Have we sufficiently taught our children to love this task, to give to them the passion of work and work done well? Have we demonstrated to them sufficiently that one is only happy through effort and that man must earn his life and his happiness by the sweat of his brow? Have we shown them that *nothing* replaces effort, neither the gifts of nature, intelligence, goodness, beauty, neither fortune, neither science, nor even the effort of those whom we love? Man, in spite of his deep affections and perhaps because of them, must live alone making effort his companion. If we want to spare our children effort, we sap from them what they can produce permanently; because of us, they will join the category of parasites, life's great failures or losers.

Work also has the great advantage of thrusting youth into reality and the experiential. It regulates sensitivity, the nervous system, mental balance, and protects against stupid vanity, envy, malicious gossip, artificial needs, and false intoxicating pleasures. Work is of key importance at all ages, but especially for the youth because work's virtues—temperance, strength, health, justice, and purity—are found in its nobleness.

FAMILY: Certainly we all have wanted to serve the family, but have we served it well? Indeed, we have not understood nor have made clear the family's social role.... On the board of directors of the main organizations devoted to the family, we haven't had an equitable distribution of different social classes like the current government wants in order to restore an awareness of the family. This is why intimate and fruitful collaboration among various milieus has not been successful. It is high time to repair these past errors.

FATHERLAND: Have we not known, we mothers, how to love the Fatherland with a truly maternal love—this sublime, unbiased, comprehensive, intuitive love capable of sacrificing itself solely to see the happiness and glory of that which it loves?—We withdraw within "Our France" much like we withdraw into "our family." The world's youth did not know each other. How could they then have become united and love each other since there is not love without knowledge. Not being able to love, they are beaten...! This time will soon be past! Thank God! In a New Europe, exhausted and sober, peoples' profound qualities and virtues will finally be spread around the continent and the promotion of vice will be limited. Mothers will protect public health, and when we say "mothers", we don't mean just French mothers. All European mothers must work in yearly international conferences, toward the grand civilizing and human task; every mother who is truly "mother" of the West and East, is an awakener of souls and cultivator of effort.

Every mother hates war because every mother hates hatred. She is unwilling to accept that, to the end of time, Cain will kill Abel and that man will forever be a wolf to one another. For that, it doesn't suffice to perfect man; instead, leave it to God to remake him. This way the Holy Spirit dominates anew the material that invades all. What matters is the shape of future institutions. Their value will depend on the value of man's honesty and integrity.

Also, we mothers are not content to create Men; we know how to raise them. To raise a man is to mount toward the human summit, slowly, progressively and without faltering, because the domain of the Good that doesn't advance, retreats. We teach our sons to hate only two things: Evil and Hate.... We teach them to be French above all and quite simply not to give into their irrational passions, not to be confused by misleading appearances or disoriented by difficult situations. There is greatness in accepting an unavoidable disaster. To rebel is to think of oneself.... Teach them to know how to obey *le Chef*[1] who inherits the double heritage of heroism and wisdom from the great men of Antiquity and our History; teach them how to keep their smile and goodwill during difficult times and how to sometimes laugh at things in order not to cry. This is what it means to be "French."

_____

[1] *Le Chef:* "Our Leader"—a common reference to Pétain.

159. Marta Appel
Memoirs

Marta Appel (1894–1987) and her husband Dr. Ernst Appel, a rabbi, served as co-presidents of the B'nai B'rith Lodge in Dortmund, Germany. The lodge organized social activities and carried out educational and social work. In April 1937, they were arrested by the Gestapo which dissolved the lodge and confiscated its assets. Soon after their release in May, the Appels and their daughters secretly fled Germany, first to Holland and then to the United States. In these excerpts from her *Memoirs,* written in English in 1940—1941, Marta Appel describes how her family and the Jewish community were affected by the implementation of Nazi racial policies and laws.

The children had been advised not to come to school on April 1, 1933, the day of the boycott. Even the principal of the school thought Jewish children's lives were no longer safe. One night they placed big signs on every store or house owned by Jewish people. In front of our temple, on every square and corner, billboards were scoffing at us. Everywhere, and on all occasions, we read and heard that we were vermin and had caused the ruin of the German people. No Jewish store was closed on that day; none was willing to show fear in the face of the boycott. The only

From *Jewish Life in Germany: Memoirs from Three Centuries,* edited by Monika Richarz, pp. 351–361. Copyright © 1991 by Indiana University Press. Reprinted by permission.

building which did not open its door as usual, since it was Saturday, was the temple. We did not want this holy place desecrated by any trouble.

I even went downtown that day to see what was going on in the city. There was no cheering crowd as the Nazis had expected, no running and smashing of Jewish businesses. I heard only words of anger and disapproval. People were massed before the Jewish stores to watch the Nazi guards who were posted there to prevent anyone from entering to buy. And there were many courageous enough to enter, although they were called rude names by the Nazi guards, and their pictures were taken to show them as enemies of the German people in the daily papers. Inside the stores, in the offices of the owners, there was another battle proceeding. Nazis were forcing those Jewish men to send wires abroad to foreign businesses, saying that there was no Jewish boycott and that nothing unusual was happening. Accompanied by two Nazi officials, one of the men was taken even to Holland to convince the foreign customers and businessmen there....

How much our life changed in those days! Often it seemed to me I could not bear it any longer, but thinking of my children, I knew we had to be strong to make it easier for them. From then on I hated to go out, since on every corner I saw signs that the Jews were the misfortune of the people. Wherever I went, when I had to speak to people in a store I imagined how they would turn against me if they knew I was Jewish. When I was waiting for a streetcar I always thought that the driver would not stop if he knew I was Jewish. Never did anything unpleasant happen to me on the street, but I was expecting it at every moment, and it was always bothering me. I did not go into a theater or a movie for a long time before we were forbidden to,[1] since I could not bear to be among people who hated me so much. Therefore, when, later on, all those restrictions came, they

did not take away from me anything that I had not already renounced. Nevertheless, it meant a new shame. Not to go of my own accord was very different from not being allowed to go.

In the evenings we sat at home at the radio listening fearfully to all the new and outrageous restrictions and laws which almost daily brought further suffering to Jewish people. We no longer visited our friends, nor did they come anymore to see us....

Since I had lived in Dortmund, I had met every four weeks with a group of women, all of whom were born in Metz, my beloved home city. We all had been pupils or teachers in the same high school. After the Nazis came, I was afraid to go to the meetings. I did not want the presence of a Jewess to bring any trouble, since we always met publicly in a café. One day on the street, I met one of my old teachers, and with tears in her eyes she begged me: "Come again to us; we miss you; we feel ashamed that you must think we do not want you anymore. Not one of us has changed in her feeling toward you." She tried to convince me that they were still my friends, and tried to take away my doubts. I decided to go to the next meeting. It was a hard decision, and I had not slept the night before. I was afraid for my gentile friends. For nothing in the world did I wish to bring them trouble by my attendance, and I was also afraid for myself. I knew I would watch them, noticing the slightest expression of embarrassment in their eyes when I came. I knew they could not deceive me; I would be aware of every change in their voices. Would they be afraid to talk to me?

It was not necessary for me to read their eyes or listen to the change in their voices. The empty table in the little alcove which always had been reserved for us spoke the clearest language. It was even unnecessary for the waiter to come and say that a lady phoned that morning not to reserve the table thereafter. I could not blame them. Why should they risk losing a position only to prove to me that we still had friends in Germany?

I, personally, did not mind all those disappointments, but when my children had to face them, and were not spared being offended everywhere, my heart was filled with anguish.

---

[1]The ban forbidding Jews to attend the theater, concerts, movie houses, etc. was not effected until November 12, 1938. Three days later Jewish children were expelled from the public schools.

It required a great deal of inner strength, of love and harmony among the Jewish families, to make our children strong enough to bear all that persecution and hatred. [...]

Almost every lesson began to be a torture for Jewish children. There was not one subject anymore which was not used to bring up the Jewish question. And in the presence of Jewish children the teachers denounced all the Jews, without exception, as scoundrels and as the most destructive force in every country where they were living. My children were not permitted to leave the room during such a talk; they were compelled to stay and to listen; they had to feel all the other children's eyes looking and staring at them, the examples of an outcast race....

One day, for the first time in a long while, I saw my children coming back from school with shining eyes, laughing and giggling together. Most of the classes had been gathered that morning in the big hall, since an official of the new *Rasseamt,* the office of races, had come to give a talk about the differences of races. "I asked the teacher if I could go home," my daughter was saying, "but she told me she had orders not to dismiss anyone. You may imagine it was an awful talk. He said that there are two groups of races, a high group and a low one. The high and upper race that was destined to rule the world was the Teutonic, the German race, while one of the lowest races was the Jewish race. And then, Mommy, he looked around and asked one of the girls to come to him." The children again began to giggle about their experience. "First we did not know," my girl continued, "what he intended, and we were very afraid when he picked our Eva. Then he began, and he was pointing at Eva, 'Look here, the small head of this girl, her long forehead, her very blue eyes, and blond hair. 'And look,' he said, 'at her tall and slender figure. These are the unequivocal marks of a pure and unmixed Teutonic race.' Mommy, you should have heard how at this moment all the girls burst into laughter. Even Eva could not help laughing. Then from all sides of the hall there was shouting, 'She is a Jewess!' You should have seen the officer's face! I guess he was lucky that the principal got up so quickly and, with a sign to the

pupils, stopped the laughing and shouting and dismissed the man, thanking him for his interesting and very enlightening talk. At that we began again to laugh, but he stopped us immediately. Oh, I was so glad that the teacher had not dismissed me and I was there to hear it."

When my husband came home, they told him and enjoyed it again and again. And we were thankful to know that they still had not completely forgotten how to laugh and to act like happy children.

"If only I could take my children out of here!" That thought was occupying my mind more and more. I no longer hoped for any change as did my husband. Besides, even a changed Germany could not make me forget that all our friends, the whole nation, had abandoned us in our need. It was no longer the same country for me. Everything had changed, not people alone—the city, the forest, the river— the whole country looked different in my eyes.

[*In the spring of 1935 a Jewish doctor flees from Dortmund, leaving all that he owned behind.*]

A few days after the doctor had left with his family, we were invited to a friend's house. Of course the main subject of the evening was the doctor's flight. The discussion became heated. "He was wrong," most of the men were arguing. "It indicates a lack of courage to leave the country just now when we should stay together, firm against all hatred." "It takes more courage to leave," the ladies protested vigorously. "What good is it to stay and to wait for the slowly coming ruin? Is it not far better to go and to build up a new existence somewhere else in the world, before our strength is crippled by the everlasting strain on our nerves, on our souls? Is not our children's future more important than a fruitless holding out against Nazi cruelties and prejudices?" Unanimously the women felt that way, and took the doctor's side, while the men, with more or less vehemence, were speaking against him.

On our way home I still argued with my husband. He, like all the other men, could not imagine how it was possible to leave our beloved homeland, to leave all the duties which constitute a man's life. "Could you really leave all this behind you to enter nothingness?" From the heavy sound of his voice I realized how the mere thought was stirring him. "I could," I said frankly, and there was not a moment of hesitation on my part. "I could," I said again, "since I would go into a new life." And I really meant it.

Our private life became more and more troublesome. It was not simply that my husband always had difficulties with his sermons. Everywhere they tried to set a trap for him. There were so many regulations always being set up for a large congregation, and the rabbis had to abide by them. Every four weeks from now on we had to send revised typewritten lists of all the members of the different clubs. Whenever some of them were moving, we had to know it and inform "the party" immediately. We had to ask the Gestapo for permission for everything that went on in the temple and in the community house.

Our New Year and the Day of Atonement, the highest Jewish holy days, were approaching [1935]. Carefully my husband and I went through each of his sermons. Word for word we read aloud and considered whether they would pass the scrutiny of the Nazi supervisor who was present at every service....

Two or three days before the holy days the pavement around the temple was besmeared with big white letters, "The Jews are our bad luck," and many similar signs. We cleaned the pavement on the eve of the first holy day, but the next morning it was even worse. That was not the only disturbance; a group of young Hitler Boys in their uniforms were posted at the entrance of the temple to make a deafening noise. When people entered the temple they had to walk though two lines of boys who were beating drums with all their force and sounding their trumpets in dreadful dissonance. A policeman was posted not more than ten steps away at the street corner, but when the ushers asked him to help them send the boys away, his reply was, "There is nothing I can do about it; they are sent on a special order."...

The next holy day passed without interference. Sometime later we learned that numerous anonymous letters had poured in to the Gestapo expressing the indignation of the Christian citizens at the disturbance of our service. At that time it still may have seemed wiser to the Nazis not to arouse too much sympathy for the Jewish people, and not to show their real intentions toward them too openly.

There were still millions of Germans who did not believe that this treatment was accorded us with the official sanction of the Hitler government. We would hear them say constantly, "If Hitler knew about such cruelty, he never would allow it. He would stop them at once. He does not know about what is going on. This is all illegal, and done by these vile Storm Troopers without his knowledge."

Many eyes were opened only when, in the fall of that year, 1935, the *Nuernberger Gesetze*, the laws of Nuremberg, were proclaimed. Only then did they believe, after reading in every paper and listening to each broadcast, that Jewish people were no longer citizens. Furthermore, the Jews remained subjects of the German Reich, with all the duties of a subject but without any of the rights of a citizen. [...]

Whenever we went to a meeting I saw faces which I had never seen before, while many of the old and well known were missing. Our congregation changed its members constantly now. Following the laws of Nuremberg, more and more businesses were forced to close, and the former owners left the country or went to Berlin, where they thought to lose themselves in the great mass of Jews. In spite of that our congregation did not decrease, since, from all smaller places around, Jews came to bigger cities. Life in a small town had become intolerable for Jews. They could not even buy food any longer in an Aryan store, and the Jewish ones had been closed for a long time....

Again it was spring! [1936] Nature, in its own new strength, was bringing new hope. But, for us, there was no hope. The restrictions were drawn tighter and tighter, strangling the life of the Jewish congregations. The budget of our congregation, which had been 800,000 marks when Hitler came to power, had diminished to 80,000 marks. Where we had had less than a quarter of our people on relief before, we now had more than three-fourths of them on our relief roll. The city relief scarcely provided for Jewish people, and now no religious or private institution was permitted to make a charity drive. The only thing to do for our paupers was to have, as the Nazis had, a weekly "pound" collection of food and another of old clothing. Week after week a truck went around and every Jewish household gave at least one pound of some foodstuff. More than once, those of us on the committee did not know what to do when nothing but sacks of dried peas and lentils came in. Even people with money could not buy what they needed. Fruit, vegetables, butter, eggs, and meat were no longer on the market in sufficient quantities. It was clear that people could give only of what they themselves had plenty. We had the same sad experience with our clothing collection. The new substitutes, which looked very nice in the show windows, did not last long. People would alter their old clothes to make them look like new, so that the wardrobe for the poor in our community house gradually became empty.

Never, in my memory, had we had so much activity in our community house. From the basement to the attic there was not a single space where a group of people was not taking a course of some sort. Men and women, boys and girls came to learn new trades and professions which they hoped would enable them to make a living abroad. Ladies who had never touched a thing in their own homes came now to learn to cook, to sew, to become a milliner, a hairdresser, or to prepare in some other way to make a living abroad. Men who had been retired for several

years, or those who had had big businesses and factories of their own, came to learn to be farmers, shoemakers, or carpenters, or to fit themselves for some other vocation. And, mingled with the sounds of all these trades coming out of the various rooms, there was also a mixture of different languages echoing throughout the building. Spanish, French, English, or Hebrew could be heard whenever one passed a room.

The hardest task I had to do was to arrange for the transportation of children to foreign countries: the United States of America, Palestine, England, and Italy. It was most heartbreaking to see them separate from their parents. Yet the parents themselves came to beg and urge us to send their children away as soon as possible, since they could no longer stand to see them suffer from hatred and abuse. The unselfish love of the parents was so great that they were willing to deprive themselves of their most precious possessions so that their children might live in peace and freedom.[2] [...]

In 1937 it became more and more difficult for my husband to perform his duties under the supervision of the Gestapo of the city. All this was wearing on my husband's health, and besides, the uncertain, unpromising future of our girls was upon his mind both day and night. I could not help him. What I wished to say was, "Let us go," but I had promised not to urge him anymore, not to make it still harder for him. So what was ever-present in our thoughts lay unspoken between us.

---

[2] From 1934 to the end of 1939 over 18,000 Jewish children and young people emigrated from Germany without their parents, 8,100 to England and 5,300 to Palestine.

160. Mrs. Miniver
Appeal on Behalf of Jewish Refugees

Mrs. Miniver was a popular English novel serialized in *The Times* in the late 1930s. A celebration of domestic life, the novel is virtually bereft of references to politics. Only in the very latest installments do we obtain a glimpse of how events in Europe are beginning to impinge on the middle-class Miniver family, as in brief references made by Mrs. Miniver to the possibility of war and the arrival of refugee families in England. In this December 1938 letter to the *The Times,* author Jan Struthers/Joyce Anstruther (1901–1953) uses the fictional persona of Carolyn Miniver to call on the British people to provide material assistance to Jewish refugees. *Mrs. Miniver* was first published as a single volume in 1939, shortly after the outbreak of World War II. Struthers also consulted on the 1941 Hollywood film adaptation that starred Greer Garson. Set during the Battle of Britain, the film depicts how the Miniver family and the town in which they lived coped with the German bombings. The letter reproduced here was one of several written by the fictional Mrs. Miniver to *The Times* before as well as during the war. The references to "Vin," "Toby," and "Judy" are to her children in the novel.

FOR THE REFUGEES

CLOTHES, BOOKS AND TOYS

THE 'PUT-AWAY CUPBOARD'

To the editor of *The Times*

Sir, Since the last time I appeared in your columns, on November 29, I have had so many letters from those of your readers who are good enough to take an interest in this family chronicle that I venture to ask whether you will let me send them, through you, a reply and a request.

All the writers of these letters share my feelings about the Jewish tragedy, but some of them question the truth of the sentence, 'to shrink from direct pain was bad enough, but to shrink from vicarious pain was the ultimate cowardice.' If one reads everything that is written about such horrors, they suggest, one may find oneself 'haunted almost beyond endurance'. 'And what is the use of that,' they add beseechingly, 'if there is nothing whatever one can do about it?'

I am afraid I still think that it is a great deal of use. The only force which can be brought to bear upon those who are responsible for such an outrage is the tide of public opinion. But what is this tide, except the aggregate of a million small drops of private opinion? A paragraph in a letter to a friend, a remark passed on the telephone or at the dinner-table, may seem to do little good; but in reality every word, spoken or written, helps to

swell the tide. In such a cause, surely, one should be willing to accept the burden of being haunted?

Besides, is there really 'nothing whatever one can do about it'? I feel sure that the writers of these letters have already turned out their pockets on behalf of the refugees. But have they, I wonder, turned out their cupboards as well? And if not, will they join with me in doing so? I am not suggesting a raid upon their current wardrobe, though if there is anything to spare from that, too, so much the better. What I am talking about is the 'put-away cupboard'—that time-honoured institution which is one of the glories of English family life: the cupboard which contains the spare blankets; and the brown cardigan suit which one bought in a hurry and never really liked; and those flannel shirts of one's husband's which have shrunk in the wash but are still too big for Vin: and a whole collection of Vin's clothes which will probably have the moth in them before Toby can wear them; and Judy's and Toby's last year's winter coats, kept on the off-chance that they would be able to get into them again this year (which they can't); and a rabble of stockings and sweaters and woollen underclothes of various shapes and sizes, all carefully, if rather vaguely, hoarded against a rainy day.

Well, our own rainy day may never actually arrive; but for other people, at this moment, it is raining very hard indeed. There is a desperate

---

need for clothes—especially children's clothes—for the refugees. I feel sure that a concerted raid upon all the 'put-away cupboards' of this country would go far towards meeting the need. And while my fellow 'Mrs. Minivers' (this very week-end, perhaps) are engaged upon that could the Judys and Tobys be persuaded to have yet another go at their toy-cupboards? Probably they have just combed them out for the sake of the hospitals or the unemployed; but if they could manage to spare even one more toy apiece it would help to make the refugee children's Christmas a less forlorn one. Books would be welcome, too, for many of the children already speak English and the rest are eager to learn.

Parcels should be addressed to:—

Lord Baldwin's Appeal Fund (Clothing Dept.), Westbourne Terrace, W.2.

May I take the opportunity of wishing all my unknown friends, on behalf of this family, a very happy Christmas?

I am, Sir, yours faithfully,

CAROLINE MINIVER

(*The Times,* Saturday, 17 December 1938)

QUESTIONS FOR ANALYSIS

1. How does Teresa Noce analyze the main features of fascist policies toward working women in Italy? How were women affected by these policies? What political strategy does Noce advocate?
2. Discuss Hanna Schmitt's description of how the Nazi regime transformed women's political, economic, and social status in Nazi Germany. How does she describe the effects of the racial laws at this time (1937)?
3. Compare and contrast the accounts of women and fascism provided by Teresa Noce and Hanna Schmitt. Why was it important for women in other European countries to learn about the situation of women in Italy and Germany?
4. To what causes does Vérine attribute France's defeat in 1940? What special tasks does she assign to mothers in France and the "New Europe"? Compare Vérine's views with those of Lucienne Blondel, Source 155 in Chapter 12.
5. How were the lives of Marta Appel, her family, and the Jewish community in Dortmund, Germany, transformed by the implementation of Nazi racial laws? What differing views did Marta and her husband initially have on the issue of emigration? What factors might account for these differences?
6. On what basis does the fictional "Mrs. Miniver" appeal to the British to assist Jewish refugees? Why did author Jan Struthers frame this appeal in the voice of Carolyn Miniver?

INTERWAR DEBATES ON WAR AND PEACE

Women acted on an international level to ensure a permanent peace after World War I. At its 1919 founding congress in Zurich, the Women's International League for Peace and Freedom (WILPF)[1] called for immediate, general disarmament, condemned the punitive terms proposed in the Versailles Peace Treaty, and opposed the continuing Allied economic blockade of Germany as harmful to the civilian population. Right-wing movements flourished in the aftermath of defeat, and the consolidation of fascism in Italy and Germany once again threatened Eu-

---

[1]On the origins of WILPF, see Chapter 11, Source 134, "Call to the Women of All Nations."

rope with war. In 1934, an International Women's Congress Against War and Fascism drew 1,500 participants to Paris. Soon afterward, Italy invaded Ethiopia in 1935 and Germany, already in violation of the Versailles Treaty prohibition on rearmament, occupied the Rhineland in 1936. These developments created divisions within the international peace movement that only deepened with the Spanish Civil War (1936–1939) and the crisis in Czechoslovakia provoked by German claims to the Sudetenland which culminated in the Munich Pact of 1938. While some women continued to oppose military intervention on principle, others argued that such intervention was essential to preventing a second world war.

161. Dolores Ibarruri
"Help Us Win!"

Dolores Ibarruri (1895–1981), known as La Pasionaria, was a leader of the Spanish Communist Party and the secretary of its women's section. An organizer of the National Committee of Women Against War and Fascism in Spain, Ibarruri headed the Spanish delegation to the International Women's Congress Against War and Fascism held in Paris in 1934. She became a member of the Executive Committee of the Third Communist International in 1935, the year in which the International initiated its Popular Front strategy. Ibarruri served as a Communist Party deputy in the Parliament of the Popular Front government elected in February 1936 on a program of democracy, land reform, worker's rights, and anti-clericalism. The first fascist revolt against the government broke out on July 18, 1936, in Seville, followed by military risings in Madrid and Barcelona the next day. Throughout the three years of the Spanish Civil War, Ibarruri ceaselessly rallied the Spanish people to resist fascism. In a July 19 radio broadcast from Madrid, she defiantly declared, "The whole country is shocked by the actions of these villains. They want with fire and sword to turn democratic Spain, the Spain of the people, into a hell of terrorism and torture. But they shall not pass!"[1] Ibarruri also criticized the non-interventionist policies of democratic European governments. These policies are the topic of her speech "Help Us Win!" delivered in Paris on December 3, 1937, and later reproduced as a pamphlet. Elected in June 1936, the Popular Front government in France refused to provide assistance to the Republican government because it was fearful of provoking a civil war. Following the defeat of the Republic in 1939, Ibarruri emigrated to the Soviet Union where she wrote her autobiography *El Unico Camino* (*The Only Road*), published in English as *They Shall Not Pass* (1966). It was intended not only to record the past, but also to inspire the ongoing struggle for democracy in Spain. Ibarruri did not return to Spain until after General Francisco Franco's death in 1975. She was elected as a Communist Party deputy in the June 1977 elections and served in the first democratic Parliament in post-Franco Spain.

---

[1] "Danger! To Arms!," in *Dolores Ibarruri, Speeches and Articles, 1936–1938* (Moscow: Foreign Languages Publishing House,

Excerpts, as attached, from "Help Us Win!" from *Dolores Ibarruri: Speeches and Articles, 1936–1938*, pp. 165–174. Copyright © 1938 by Foreign Languages Publishing House, Moscow.

Democrats of France, comrades and friends, it is over a year since I first spoke from this world tribune.[1] At that time every day saw countless hardships, incredible sufferings and frightful privations inflicted on our people, and the blood of the finest sons of Spain flowed in streams in defence of peace, democracy and liberty.

I came here then to recount and explain to the public throughout the world the meaning of our struggle, the significance of the war which had been fomented by the betrayers of our fatherland and which for sixteen months now has been annihilating people, towns and villages of our martyred Spain.

Your protest against the invasion of our territory by German and Italian armies, your demands that aeroplanes and guns be sent to Republican Spain showed then that you understood what danger a victory of the forces of reaction and fascism, which had launched themselves against our country, would entail for France, democratic France, the France of the "rights of man."...

We have come here, as the Spanish comrades who have already spoken have said, to tear down the web of lies and vile calumny that has been woven around our struggle.

We have come to assure you, in the name of genuine democracy, that Spain is not thinking of an armistice, that Spain does not want a compromise, that Spain is imbued with only one thought and one aspiration—to crush fascism. (*Cheers.*)...

True, we have lost the provinces of the North. That was a severe blow to us....

It was not due to lack of courage on the part of our men that the North fell. No! They fought like lions. The North fell because of the policy of so-called non-intervention. This "non-intervention" deprived our brothers who were defending the North of arms and munitions, while the fascists were receiving troops and

munitions from Italy and Germany without let or hindrance. (*Indignant cries from the audience: "Open the Spanish frontier!"*)[2]

Despite all this, Republican Spain is staunchly holding her own....

In vain do the fascists and reactionaries of all countries try to pervert the meaning of our struggle. They will delude nobody by asserting that they are fighting communism. A few days ago de Brouckère, Chairman of the Second International, rightly warned all democrats against this lie. Fascism is pleading a struggle against communism as an excuse for the war. But with the fascists everything that does not belong to fascism is communism.

Has not Mussolini made this quite clear? This butcher of the Italian workers said in a recent speech that the enemies of fascism "are the liberals, the parliamentary democrats, the Bolsheviks, socialists and communists, and also certain Catholics with whom we shall settle accounts sooner or later." Does this leave any room for doubt? The fascist crusade is directed against everything that bears the stamp of humaneness and progress....

The reactionary elements in our country would not reconcile themselves to their defeat at the February elections. They were afraid of the strength and might of the People's Front. They decided to act as your "Cagoulards"[3] are acting in France: they organized assassinations and killed prominent Lefts, formed stores of arms—in a word, they tried by means of innumerable acts of provocation against the working people to terrorize the country. That is how the reactionaries prepared for a coup d'état....

If, dear friends, you compare the activities of the "Cagoulards" with those of the Spanish fas-

---

[1]Ibarruri delivered this speech in the Winter Velodrome, an indoor sports stadium in Paris. Five years later, in July 1942, the stadium was used as a detention center for 12,884 foreign Jews, including children, who were rounded up by the French police and deported from France to Auschwitz.

[2]The Popular Front government in France announced a policy of nonintervention in Spain on August 2, 1936, and closed the border between France and Spain on August 8.

[3]The Cagoule (Hooded Cloak) was a terrorist, anti-communist organization founded shortly after the Popular Front came to power. It attempted to inspire a military revolt against the government. The Cagoule's secret activities were exposed in the French press in France in 1937, the same year as Ibarruri's speech.

cists you will realize that what I say is true. (*Shouts of "Right, quite right! Down with the rebels! To jail with the Cagoulards!"*)

A few days ago *Matin,* the French reactionary newspaper, asserted that we, the delegates of the Spanish People's Front, not satisfied with the rivers of blood shed in Spain, want to open the red floodgates of civil war in France....

As a woman and a mother, and as a Spanish citizen, I declare that this is a vile calumny!

We love peace profoundly. We have come to France as heralds of peace and not as emissaries of war! (*Loud applause.*) We want the French people to avoid the sufferings and incredible sacrifices which the Spanish people have had to bear in the struggle against fascism. We want to spare French women the boundless grief of the mothers, wives and sisters of Spain before whose eyes fascist machine guns mow down their children, brothers and husbands....

And yet we are alone in this titanic conflict between fascism and democracy.... Alone? No, not entirely. For far away lives a people whose example is an inspiration to all who are fighting for a better future, a people in whose midst, as in a protected fortress, all who are persecuted by world reaction find refuge.... There is a land, the Land of Socialism—the Soviet Union—which has extended a hand to us across borders, continents and seas and has helped us, openly declaring to the world that our cause is not the cause of Spain alone but the cause of all advanced and progressive mankind!

In the name of all the fighters in our country who are shedding their blood for freedom, we thank this great nation, which, despite the distance that separates us, inspires us and assists us despite all obstacles. (*The audience rises and loudly cheers. Cries: "Long live the Soviet Union! Long live Stalin!"*)

We also feel the active solidarity of the proletariat of all countries and of all who love democracy. Unfortunately, this assistance is not as effective as it might be because the governments of the democratic countries retreat all too easily in face of fascism....

We want France, whom we deeply love, to correctly appraise the real state of affairs: Spain has an army of over half a million men; and if the youth of pre-recruitment age are reckoned this figure will reach one million. If we win the war, this army will serve as a guarantee of freedom, democracy and peace. But if, owing to the cowardice of those in whose hands the fate of democracy lies, the German and Italian fascists continue to dispatch munitions and troops against us, if at the same time our frontiers remain closed, if the naval blockade continues, if, owing to all this, our Republic is destroyed, then, people of France, do not forget that you too will be in immediate danger. What good will the Maginot line be then? None whatever! Danger will come then not only from Germany but also from the Pyrenees. (*Shouts of "Open the Spanish frontier!"*)

Do not forget that the future of the whole world is at stake.

You must remember the sacrifices borne by the Spanish workers, the anti-fascists of our country, the mothers and wives who are losing children, husbands and fathers in the struggle against fascism....

People of France, help us to win the war, and you will by that act be defending yourselves from your enemies, who are at the same time our enemies.

Let no one have cause to come to you and say what Aiksa said to his son Baodil after the fall of Granada: "Bewail like a woman what you could not defend like a man!" (*The audience rises in a storm of cheers and applause.*)

162. Virginia Woolf
The Three Guineas

Virginia Woolf (1882–1941) is one of the most important writers of the twentieth century. The daughter of literary figure Sir Leslie Stephens and Julia Stephens,[1] she grew up in a upper-middle-class Victorian literary milieu. Along with her sister Vanessa Bell, a painter, Woolf was part of the avant-garde Bloomsbury Group in London which included Maynard Keynes, E. M. Forester, and Leonard Woolf. She was a leading literary critic of her day and with Leonard established the Hogarth Press in 1917. Woolf's political views were feminist, socialist, and pacifist. Her modernist novels include *Jacob's Room* (1922), *Mrs. Dalloway* (1925), *To the Lighthouse* (1927), and *Orlando* (1928), inspired by her relationship with Vita Sackville-West. Woolf also published essays, literary reviews, and political treatises such as *A Room of One's Own* (1929) and *The Three Guineas* (1938), which she actually began writing in 1932. By the time it was published, the political terrain of Europe had changed dramatically, but Woolf's anti-interventionist views were only confirmed when her nephew Clive Bell, a volunteer in the International Brigades, was killed in Spain in 1937. To Forester she wrote, "I have never known anyone of my generation have that feeling about a war. We were all C.O.'s [conscientious objectors] in the Great War.... The moment force is used, it becomes meaningless and unreal to me."[2] In *The Three Guineas*, Virginia Woolf answers the hypothetical question "How are we to prevent war?" with three proposals: a university education for women that would inculcate peaceful values, an independent income to provide the material basis for women's intellectual independence from men, and the formation of an "Outsider's Society" comprised of the "daughters of educated men." In this third portion, excerpted here, Woolf critiques the theoretical premises of anti-fascist cultural organizations in Britain with which she was associated. During World War II, Virginia and Leonard's house in London was severely damaged by a German bomb and enemy planes often flew over Sussex to which they had moved. Woolf's wartime experiences are recorded in her posthumously published diaries and letters and in "Thoughts on Peace in an Air Raid" (1940). Having suffered from depression and periodic mental breakdowns throughout her life, Woolf committed suicide in March 1941.

Let us then draw rapidly in outline the kind of society which the daughters of educated men found and join outside your society but in co-operation with its ends. In the first place, this new society, you will be relieved to learn, would have no honorary treasurer, for it would need no funds. It would have no office, no committee, no secretary; it would call no meetings; it would hold no conferences. If name it must have, it could be called the Outsiders' Society. That is not a resonant name, but it has the advantage that it squares with facts—the facts of history, of

---

[1]See Chapter 9, Source 117, for Julia Margaret Cameron's photograph of Julia Stephens, "Mrs. Herbert Duckworth, 1867."

[2]Hermione Lee, *Virginia Woolf* (New York: Alfred A. Knopf, 1997), p. 658.

law, of biography; even, it may be, with the still hidden facts of our still unknown psychology. It would consist of educated men's daughters working in their own class—how indeed can they work in any other?—and by their own methods for liberty, equality and peace. Their first duty...would be not to fight with arms. This is easy for them to observe, for in fact, as the papers inform us, "the Army Council have no intention of opening recruiting for any women's corps." The country ensures it. Next they would refuse in the event of war to make munitions or nurse the wounded. Since in the last war both these activities were mainly discharged by the daughters of working men, the pressure upon them here too would be slight, though probably disagreeable. On the other hand the next duty to which they would pledge themselves is one of considerable difficulty, and calls not only for courage and initiative, but for the special knowledge of the educated man's daughter. It is, briefly, not to incite their brothers to fight, or to dissuade them, but to maintain an attitude of complete indifference. But the attitude expressed by the word "indifference" is so complex and of such importance that it needs even here further definition. Indifference in the first place must be given a firm footing upon fact. As it is a fact that she cannot understand what instinct compels him, what glory, what interest, what manly satisfaction fighting provides for him— "without war there would be no outlet for the manly qualities which fighting develops"—as fighting thus is a sex characteristic which she cannot share, the counterpart some claim of the maternal instinct which he cannot share, so is it an instinct which she cannot judge. The outsider therefore must leave him free to deal with this instinct by himself, because liberty of opinion must be respected, especially when it is based upon an instinct which is as foreign to her as centuries of tradition and education can make it. This is a fundamental and instinctive distinction upon which indifference may be based. But the outsider will make it her duty not merely to base her indifference upon instinct, but upon reason. When he says, as history proves that he has said, and may say again, "I am fighting to protect our

country" and thus seeks to rouse her patriotic emotion, she will ask herself, "What does 'our country' mean to me an outsider?" To decide this she will analyse the meaning of patriotism in her own case. She will inform herself of the position of her sex and her class in the past. She will inform herself of the amount of land, wealth and property in the possession of her own sex and class in the present—how much of "England" in fact belongs to her. From the same sources she will inform herself of the legal protection which the law has given her in the past and now gives her....She will find that she has no good reason to ask her brother to fight on her behalf to protect "our" country. "'Our country,'" she will say, "throughout the greater part of its history has treated me as a slave; it has denied me education or any share in its possessions. 'Our' country still ceases to be mine if I marry a foreigner. 'Our' country denies me the means of protecting myself, forces me to pay others a very large sum annually to protect me, and is so little able, even so, to protect me that Air Raid precautions are written on the wall. Therefore if you insist upon fighting to protect me, or 'our' country, let it be understood, soberly and rationally between us, that you are fighting to gratify a sex instinct which I cannot share; to procure benefits which I have not shared and probably will not share; but not to gratify my instincts, or to protect myself or my country. For," the outsider will say, "in fact, as a woman, I have no country. As a woman I want no country. As a woman my country is the whole world."...

Such then will be the nature of her "indifference" and from this indifference certain actions must follow. She will bind herself to take no share in patriotic demonstrations; to assent to no form of national self-praise; to make no part of any claque or audience that encourages war; to absent herself from military displays, tournaments, tattoos, prize-givings and all such ceremonies as encourage the desire to impose "our" civilization or "our" dominion upon other people. The psychology of private life, moreover, warrants the belief that this use of indifference by the daughters of educated men would help materially to prevent war. For psychology would seem to show that it is far harder for human beings to take action when other people are indifferent and allow them complete freedom of action, than when their actions are made the centre of excited emotion. The small boy struts and trumpets outside the window: implore him to stop; he goes on; say nothing; he stops. That the daughters of educated men then should give their brothers neither the white feather of cowardice nor the red feather of courage, but no feather at all; that they should shut the bright eyes that rain influence, or let those eyes look elsewhere when war is discussed—that is the duty to which outsiders will train themselves in peace before the threat of death inevitably makes reason powerless.

Such then are some of the methods by which the society, the anonymous and secret Society of Outsiders would help you, Sir, to prevent war and to ensure freedom. Whatever value you may attach to them you will agree that they are duties which your own sex would find it more difficult to carry out than ours; and duties moreover which are specially appropriate to the daughters of educated men. For they would need some acquaintance with the psychology of educated men, and the minds of educated men are more highly trained and their words subtler than those of working men. There are other duties, of course—many have already been outlined in the letters to the other honorary treasurers. But at the risk of some repetition let us roughly and rapidly repeat them, so that they may form a basis for a society of outsiders to take its stand upon. First, they would bind themselves to earn their own livings. The importance of this as a method of ending war is obvious; sufficient stress has already been laid upon the superior cogency of an opinion based upon economic independence over an opinion based upon no income at all or upon a spiritual right to an income to make further proof unnecessary. It follows that an outsider must make it her business to press for a living wage in all the professions now open to her sex; further that she must create new professions in which she can earn the right to an independent opinion. Therefore she must bind herself to

press for a money wage for the unpaid worker in her own class—the daughters and sisters of educated men who, as biographies have shown us, are now paid on the truck system, with food, lodging and a pittance of £40 a year. But above all she must press for a wage to be paid by the State legally to the mothers of educated men. The importance of this to our common fight is immeasurable; for it is the most effective way in which we can ensure that the large and very honourable class of married women shall have a mind and a will of their own, with which, if his mind and will are good in her eyes, to support her husband, if bad to resist him, in any case to cease to be "his woman" and to be her self....

The outsiders...would bind themselves to remain outside any profession hostile to freedom, such as the making or the improvement of the weapons of war. And they would bind themselves to refuse to take office or honour from any society which, while professing to respect liberty, restricts it, like the universities of Oxford and Cambridge....

... [T]he description thus loosely and imperfectly given is enough to show you, Sir, that the Society of Outsiders has the same ends as your society—freedom, equality, peace; but that it seeks to achieve them by the means that a different sex, a different tradition, a different education, and the different values which result from those differences have placed within our reach. Broadly speaking, the main distinction between us who are outside society and you who are inside society must be that whereas you will make use of the means provided by your position—leagues, conferences, campaigns, great names, and all such public measures as your wealth and political influence place within your reach—we, remaining outside, will experiment not with public means in public but with private means in private...

... [A]s this letter has gone on, adding fact to fact, another picture has imposed itself upon the foreground. It is the figure of a man; some say, others deny, that he is Man himself, the quintessence of virility, the perfect type of which all the others are imperfect adumbrations. He is a man certainly. His eyes are glazed; his eyes glare. His body, which is braced in an unnatural position, is tightly cased in a uniform. Upon the breast of that uniform are sewn several medals and other mystic symbols. His hand is upon a sword. He is called in German and Italian Führer or Duce; in our own language tyrant or Dictator. And behind him lie ruined houses and dead bodies—men, women and children. But we have not laid that picture before you in order to excite once more the sterile emotion of hate. . . . A common interest unites us; it is one world, one life. How essential it is that we should realise that unity the dead bodies, the ruined houses prove. For such will be our ruin if you in the immensity of your public abstractions forget the private figure, or if we in the intensity of our private emotions forget the public world. Both houses will be ruined, the public and the private, the material and the spiritual, for they are inseparably connected. . . .

To return then to the form that you have sent and ask us to fill up: for the reasons given we will leave it unsigned. But in order to prove as substantially as possible that our aims are the same as yours, here is the guinea, a free gift, given freely, without any other conditions than you choose to impose upon yourself. It is the third of three guineas; but the three guineas, you will observe, though given to three different treasurers are all given to the same cause, for the causes are the same and inseparable.

QUESTIONS FOR ANALYSIS

1. On what basis does Dolores Ibarruri appeal to the people of France to support the Spanish Republic? How does she critique the view that French support for the Republic will unleash a civil war in France? What does Ibarruri predict will be the consequences of noninterventionism for Spain, France, and Europe as a whole?
2. Discuss Virginia Woolf's proposal for an "Outsiders' Society." Who will comprise the society and what will be its purpose? How does Woolf link the issues of peace, anti-fascism, and feminism? What is the meaning of her statement "As a woman I want no country. As a woman my country is the whole world"?
3. Compare and contrast the views of Dolores Ibarruri and Virginia Woolf. What personal and political factors account for their differences on issues of war and peace?

WOMEN AND THE HOLOCAUST

To create a "pure, Aryan, super race," the Nazi regime first carried out a euthanasia program against the terminally ill, mentally incurable, and severely disabled. Over 100,000 individuals were murdered between 1934 and 1941 when the program was officially terminated, although continued in secret. A conference on racial policy held in September 1939 outlined Nazi plans for the elimination of Jews and Roma (Gypsies) from Germany and German-incorporated territories: the concentration of Jews in urban ghettos and the confinement of the Roma in special encampments, followed by deportations to concentration camps. The mass murder of Jews was first carried out during the 1941 German invasion of the Soviet Union by special SS units under orders to arrest and shoot Communist Party officials and Jews. Although over a million people, mostly Jews, were killed in this manner, it was considered inefficient and the summary executions were then replaced by toxic gassing in transport vans. At the January 1942 Wannsee Conference, plans for the "Final Solution" were systematized and placed under SS control. Deportations of German Jews to concentration and extermination camps began soon afterward. The deportations of Jews from France were initiated in the summer of 1942 with the full cooperation of the Vichy government which required all Jews to register their names and addresses. In Italy, the deportation of Jews to Auschwitz began when Germany occupied Italy immediately after its surrender to the Allies in September 1943. Deportations of Jews were also carried out in the countries occupied by Germany, such as Holland, Greece, and Norway. Six death camps were established in Poland. The largest camp was at Auschwitz which also included a slave labor camp. Over a million Jews and 23,000 Roma from eleven European countries were murdered at Auschwitz. In all, about six million Jews and a quarter million Roma were murdered during World War II.

163. Charlotte Delbo
"Arrivals, Departures"

Charlotte Delbo (1913–1985) was traveling in South America in September 1941 when she learned that the Gestapo had guillotined a communist friend. She returned to France and joined her husband, Georges Dudach, in the Resistance. In March 1942, they were arrested by the French police while producing anti-German leaflets in their apartment. Dudach was executed in May and, after five months of imprisonment in France, Delbo was sent to Birkenau-Auschwitz which comprised both a concentration and a death camp. In *Le Convoi du 24 Janvier (The Convoy of January 24)* (1965), published in English as *Convoy to Auschwitz* (1997), Delbo provides short biographies of the 230 French women with whom she was transported to the camp in January 1943. All of the women were arrested for political reasons and over half were communists, although not all were party members, and two, as it turned out, were collaborators. Only forty-nine women survived the camp, where they were placed in a special section for political prisoners. In January 1944, Delbo and some of the women were sent from Auschwitz to the women's camp of Ravensbruck. Shortly before the end of the war in Europe, she was released to the Red Cross and repatriated to France in June 1945. Delbo has related her experiences and reflections on Auschwitz in the form of poems and short prose. The poem "Arrivals, Departures" appeared in *None of Us Will Return* (1946) which is anthologized with two other volumes in *Auschwitz and After* (1995).

People arrive. They look through the crowd of those who are waiting, those who await them. They kiss them and say the trip exhausted them.

People leave. They say good-bye to those who are not leaving and hug the children.

There is a street for people who arrive and a street for people who leave.

There is a café called "Arrivals" and a café called "Departures."

There are people who arrive and people who leave.

But there is a station where those who arrive are those who are leaving

a station where those who arrive have never arrived, where those who have left never came back.

It is the largest station in the world.

This is the station they reach, from wherever they came.

They get here after days and nights

having crossed many countries

they reach it together with their children, even the little ones who were not to be part of this journey.

They took the children because for this kind of trip you do not leave without them.

Those who had some took gold because they believed gold might come in handy.

All of them took what they loved most because you do not leave your dearest possessions when

you set out for far-distant lands.

Each one brought his life along, since what you must take with you, above all, is your life.

And when they have gotten there

they think they've arrived in Hell

maybe. And yet they did not believe in it.

They had no idea you could take a train to Hell but since they were there they took their courage in their hands ready to face what's coming

together with their children, their wives and their aged parents

with family mementoes and family papers.

They do not know there is no arriving in this station.

They expect the worst—not the unthinkable.

And when the guards shout to line up five by five, the men on one side, women and children on the other, in a language they do not understand, the truncheon blows convey the message so they line up by fives ready for anything.

Mothers keep a tight hold on their children—trembling at the thought they might be taken away—because the children are hungry and thirsty and disheveled by lack of sleep after crossing so many countries. At last they have reached their destination, they will be able to take care of them now.

And when the guards shout to leave their bundles, comforters and keepsakes on the platform, they do so since they are ready for the worst and do not wish to be taken aback by anything. They say: "We shall see." They have already seen so much and they are weary from the journey.

The station is not a railroad station. It is the end of the line. They stare, distressed by the surrounding desolation.

In the morning, the mist veils the marshes.

In the evening floodlights reveal the white barbed wire with the sharpness of astrophotography. They believe this is where they are being taken, and are filled with fear.

At night they wait for the day with the small children heavy in their mothers' arms. They wait and wonder.

With the coming of daylight there is no more waiting. The columns start out at once. Women and children first, they are the most exhausted. After that the men. They are weary too but relieved that their women and children should go first.

For women and children are made to go first.

In the winter they are chilled to the bone. Particularly those who come from Herakleion, snow is new to them.

In the summer the sun blinds them when they step out of the cattle cars locked tight on departure.

Departure from France and Ukraine Albania Belgium Slovakia Italy Hungary Peloponnesos Holland Macedonia Austria Herzegovina from the shores of the Black Sea the shores of the Mediterranean the banks of the Vistula.

They would like to know where they are. They have no idea that this is the center of Europe. They look for the station's name. This is a station that has no name.

A station that will remain nameless for them.

Some of them are traveling for the first time in their lives.

Some of them have traveled in all the countries in the world, businessmen. They were familiar with all manner of landscape, but they do not recognize this one.

They look. Later on they will be able to describe how it was.

All wish to remember the impression they had and how they felt they would never return.

This is a feeling one might have had earlier in one's life. They know you should not trust feelings.

Some came from Warsaw wearing large shawls and with tied-up bundles

some from Zagreb, the women their heads covered by scarves

some from the Danube wearing multicolored woolen sweaters knitted through long night hours

some from Greece, they took with them black olives and loukoums

some came from Monte Carlo

they were in the casino

they are still wearing tails and stiff shirt fronts mangled from the trip

paunchy and bald

fat bankers who played keep the bank

there are married couples who stepped out of the synagogue the bride all in white wrapped in her veil wrinkled from having slept on the floor of the cattle car

the bridegroom in black wearing a top hat his gloves soiled

parents and guests, women holding pearl-embroidered handbags

all of them regretting they could not have stopped home to change into something less dainty

The rabbi holds himself straight, heading the line. He has always been a model for the rest.

There are boarding-school girls wearing identical pleated skirts, their hats trailing blue ribbons. They pull up their knee socks carefully as they clamber down, and walk neatly five by five, holding hands, unaware, as though on a regular Thursday school outing. After all, what can they do to boarding-school girls shepherded by their teacher? She tells them, "Be good, children!" They don't have the slightest desire not to be good.

There are old people who used to get letters from their children in America. Their idea of foreign lands comes from postcards. Nothing ever looked like what they see here. Their children will never believe it.

There are intellectuals: doctors, architects, composers, poets. You can tell them by the way they walk, by their glasses. They too have seen a great deal in their lifetimes, studied much. Many made use of their imagination to write books, yet nothing they imagined ever came close to what they see now.

All the furriers of large cities are gathered here, as well as the men's and women's tailors and the manufacturers of ready-to-wear who had moved to western Europe. They do not recognize in this place the land of their forebears.

There is the inexhaustible crowd of those who live in cities where each one occupies his own cell in the beehive. Looking at the endless lines you wonder how they ever fit into the stacked-up cubicles of a metropolis.

There is a mother who's boxing her five-year-old's ears because he won't hold her by the hand and she expects him to stay quietly by her side. You run the risk of getting lost if you're separated in a strange, crowded place. She hits her child, and we who know cannot forgive her for it. Yet, were she to smother him with kisses, it would all be the same in the end.

There are those who having journeyed for eighteen days lost their minds, murdering one another inside the boxcars and

those who suffocated during the trip when they were tightly packed together

they will not step out.

There's a little girl who hugs her doll against her chest, dolls can be smothered too.

There are two sisters wearing white coats. They went out for a stroll and never got back for dinner. Their parents still await their return anxiously.

Five by five they walk down the street of arrivals. It is actually the street of departures but no one knows it. This is a one-way street.

They proceed in orderly fashion so as not to be faulted for anything.

They reach a building and heave a sigh. They have reached their destination at last.

And when the soldiers bark their orders, shouting for the women to strip, they undress the children first, cautiously, not to wake them all at once. After days and nights of travel the little ones are edgy and cranky

then the women shed their own clothing in front of their children, nothing to be done

and when each is handed a towel they worry whether the shower will be warm because the children could catch cold

and when the men enter the shower room through another door, stark naked, the women hide the children against their bodies.

Perhaps at that moment all of them understand.

But understanding doesn't do any good since they cannot tell those waiting on the railway platform

those riding in the dark boxcars across many countries only to end up here

those held in detention camps who fear leaving, wondering about the climate, the working conditions, or being parted from their few possessions

those hiding in the mountains and forests who have grown weary of concealment. Come what may they'll head home. Why should anyone come looking for them who have harmed no one

those who imagined they found a safe place for their children in a Catholic convent school where the sisters are so kind.

A band will be dressed in the girls' pleated skirts. The camp commandant wishes Viennese waltzes to be played every Sunday morning.

A blockhova will cut homey curtains from the holy vestments worn by the rabbi to celebrate the sabbath no matter what, in whatever place.

A kapo will masquerade by donning the bridegroom's morning coat and top hat, with her girlfriend wrapped in the bride's veil. They'll play "wedding" all night while the prisoners, dead tired, lie in their bunks. Kapos can have fun since they're not exhausted at the end of the day.

Black Calamata olives and Turkish delight cubes will be sent to ailing German hausfrauen who couldn't care less for Calamata olives, nor olives of any kind.

All day all night

every day every night the chimneys smoke, fed by this fuel dispatched from every part of Europe

standing at the mouth of the crematoria men sift through ashes to find gold melted from gold teeth. All those Jews have mouths full of gold, and since there are so many of them it all adds up to tons and tons.

In the spring men and women sprinkle ashes on drained marshland plowed for the first time. They fertilize the soil with human phosphates.

From bags tied round their bellies they draw human bone meal which they sow upon the furrows. By the end of the day their faces are covered with white dust blown back up by the wind. Sweat trickling down their faces over the white powder traces their wrinkles.

They need not fear running short of fertilizer since train after train gets here every day and every night, every hour of every day and every night.

This is the largest station in the world for arrivals and for departures.

Only those who enter the camp find out what happened to the others. They cry at the thought of having parted from them at the station the day an officer ordered the young prisoners to line up separately

people are needed to drain the marshes and cover them with the others' ashes.

They tell themselves it would have been far better never to have entered, never to have found out.

164. Gerda Weissmann Klein
All but My Life

Gerda Weissmann Klein (b. 1924) was fifteen years old when German troops invaded her home town of Bielitz, Poland, on September 3, 1939. Polish troops retreated to nearby Krakow to fight the Germans, but the city surrendered, soon followed by Warsaw. Within eighteen days of crossing the border on September 1, Germany had occupied Poland. In *All but My Life* (1957, 1995), Klein relates her experiences and those of her parents and brother, Arthur, during the occupation. In these excerpts, she describes her experiences in the slave labor textile factory to which she was transported in 1944. Over a thousand women and girls worked at the factory in Grünberg, Poland, near Auschwitz. In December, the workers began to hear air raid sirens indicating the nearby presence of Allied bombers. At the end of January 1945, they learned that Soviet troops had liberated Auschwitz and defeated the German army in Poland and were now marching toward Germany which was being approached from the West by British and American troops. The women from the Grünberg factory as well as other camps in the vicinity, some four thousand in all, were now divided into two columns and forced on a three-month death march toward the West. Of Klein's original column of two thousand, only a hundred and twenty survived. They arrived in Czechoslovakia in early May, just before Germany surrendered to the Allies, and were rescued by the U.S. Army. Gerda Weissmann subsequently married Kurt Klein, a lieutenant in the army, and began a new life in the United States.

When I think of Grünberg I grow very sad. It was cruelty set against a backdrop of beauty. The gentle vineyard-covered hills silhouetted against the sapphire sky seemed to mock us.

The vast camp had been built as part of a textile mill not long before the war. The sun shone through the glass roof. The camp was modern, well scrubbed, clean, and filled with suffering.

That day in 1944, when we arrived, there were approximately a thousand girls there. Some were bursting with health and color, others were half-starved and walked about with bent backs, decaying teeth, the pallor of death already on their faces. Those who worked outdoors looked healthy and fresh, the others seemed to be gray, moving parts of the infernal machines. Though clean, the camp was badly run. The staff, appointed by the factory authorities in the days when the plant had first become a camp, consisted of a particularly evil and stupid group of girls who feasted while the rest starved as they wove and spun. The contrast was sharp; there was no in-between.

There was one SS man who guarded the entrance, one *Lagerführerin* very different from Frau-Kügler, and several helpers, vicious and ignorant. The SS guard and the *Lagerführerin* checked and double-checked us in the yard in front of the building in which we were to be housed. Finally, when they had established to their satisfaction that all were present, we were admitted to the building, which consisted of three enormous halls with concrete floors. Two of the halls contained our bunks. The third was the dining hall. It was isolated from the kitchen by a partition with a window through which the food was handed.

Suse came running to embrace Ilse and me. "I have saved you bunks next to mine," she said. "I met a friend here, a girl I used to know back home. You will like her."

We approached the bunks and there stood a beautiful girl. She held out both her hands.

"I am Liesel Stepper," she said in a silvery voice, and as we clasped hands I knew we would be friends. Liesel, who was from Czechoslovakia, told us about herself. She smiled a great deal and when she did her round eyes, brown and soft, always lit up. She reminded me of a fairy-tale princess. When she walked, her movements were so graceful that she hardly seemed to touch the ground.

The conversation inevitably got around to conditions in camp and so we learned about the horrors of Grünberg, and the part of the camp that spelled disaster and death, the *Spinnerei.* Whoever worked there did not last very long.

In the morning we girls from Landeshut assembled in the courtyard. Again we stood *Appell* for a long time, and were snapped to attention when the *Betriebsleiter,* as the director was called here, arrived. The girls had told us about him already. I looked curiously at the man who was so feared. He was tall and slender. His face was large and pale, with a square look and hollow cheeks. His eyes were deep set and water-blue. They seemed to have no lashes at all. He had grotesquely long arms and large hands; on his right hand he wore a large signet ring.

"That must be the ring," I thought—the girls had told us how he would make a fist and beat his victim with the ring until her face or body was covered with blood. It was not long before I had to witness such a scene....

But this was still our first morning in Grünberg. We stood in the courtyard, waiting to be picked for our first jobs.

"*Spinnerei!*" the *Betriebsleiter* announced, and I felt a shiver go through my body.

Suse was the first one to go. A few more of the tallest and prettiest girls were chosen next. Then he pointed to me. I detached myself from my group, and walked over to where Suse and the others were standing.

A dozen or more girls followed—then Ilse. Our eyes met. The worst had happened, but somehow now I was not afraid. It was always uncertainty that I feared most.

We were marched to the factory. It was a model German factory. In previous years, it was

said, they had brought foreign visitors to see it. The architecture, layout, and machinery were beautiful, I had to agree. The office building in front, with its multitude of spotless windows, overlooked vast star-shaped flower beds, and against the factory walls were hundreds of rose bushes. Along the paths were beds of tulips in full bloom.

We were taken to the spinning room. There I saw the girls, living skeletons with yellowish-gray skin drawn tight over prominent cheekbones; there were gaping holes in their mouths where teeth had either been knocked out or rotted out. These girls ran to and from huge spinning machines, repairing broken threads with nimble fingers. Their tired eyes and sallow jaws seemed to belie the swiftly running feet and dexterous fingers.

I was put to tending one of those monstrous machines. I thought I would never learn to operate it, to tie the knots before the machine, which moved rapidly, could smash my fingers. Wherever I looked, the threads tore. I kept running from one break to another until I reached a point of complete exhaustion. Worst of all, my throat felt dry and itchy from the dust and lint in the air. But time went fast. Before I knew it, we were marching back to camp. After another long roll call we fell on our bunks.

The next day the work went more smoothly, and in less than a week I mastered the machine. However, my throat continued to bother me, as did my ankles, which had been swollen for several days. But I preferred the spinning room to staying in camp, where one did not know what might happen from one minute to the next.

About every two months the spinning-room girls were X-rayed. Each time a number were found to have contracted tuberculosis; they were immediately sent to Auschwitz. The spinning room had the greatest turnover of girls, and these bimonthly examinations came to be dreaded by all of us.

May passed, and June came. Thousands of roses burst open in a marvelous range of colors. In the early morning, when we marched to the factory, the dew was still on the fragrant petals. Sometimes the first rays of the sun would make the drops glitter like diamonds. Day after day I had to resist the desire to run out of line and touch those beautiful blossoms.

The day we all feared most came toward the end of July. We were called to have our X-rays taken. We marched through the factory gates and out into the street. I felt a strange animation, felt color shooting to my cheeks. I remembered the girl in Sosnowitz to whom I had given the bowl of food, the first girl whom I had met from a slave camp. I could still see the two red spots glowing on her cheeks and I felt spots burning on mine.

This must be it, I thought. This is the end! I want to live, I want to live!

In the physician's office we were told to strip to the waist. I removed the little sack from my neck, with the piece of broken glass and charred wood that Arthur and I gathered from our Temple the day before he left Bielitz. It might show as a dark spot on my chest. I kept the sack in my hand.

"Arthur," I whispered, "be with me, I am so afraid."

I shivered in the summer heat. Slowly the line crept forward.

"Your name?" a female voice asked.

"Gerda Weissmann." My voice sounded high and unnatural. I heard it trail off in the dark room and there was the sound of the pencil scratching over the paper, checking my name....

I waited in another line while my plate was developed. Why did it take so long? Then I heard the doctor snap. "Clear!"

I saw him pointing at me. "Clear! Clear!" he said, pointing to others.

We returned to camp. Ilse was sitting on the bunk, her face white.

"Ilse" I called frantically.

She jumped up.

"Are you all right?" she whispered hoarsely.

I nodded. "And you?"

She nodded too.

We embraced. We had a lease on life for another two months!

A week later the tubercular girls departed. There was one among them whom I had gotten to know fairly well. I went to see her before she left. We embraced, there was nothing to say. What does one say to someone who knows that in a few days she won't be alive any more?

It is often said that it is best that we cannot know the future, but this case was an exception. About two years later, when I was working in Munich, I was in the German Museum, looking through some lists of refugees that were published there daily, when I heard a gay voice call to me.

"Isn't that Gerda? It's me, not my ghost," said a rosy girl in a blue sweater.

As she came closer, I remembered our sad farewell in Grünberg before she had left for Auschwitz. As she embraced me, we both started to cry. Then she told me what had happened. When her group reached Auschwitz they were taken to the death house, but traffic was heavy there. Things were so busy, they had to wait for death. As she sat on the ground she idly dug into the soil of which she was soon to become a part. The gesture was her salvation, for she unearthed a handful of gems. To what forgotten soul had they belonged?

The girl was momentarily dazed. Then she ran up to an SS guard—what had she to lose? "Help me," she pleaded, and gave him the bundle. "I want to live."

Somehow he got her and two other girls from her group places to work in the kitchen. That parcel of gems saved three lives, even though the guard could have taken the gift and refused to help the girls. When she finished her story, I asked the girl about her lungs.

She smiled. "That is another strange story. I went to a number of doctors, but they found nothing wrong. Perhaps that is another miracle, or perhaps the doctor in Grünberg made a mistake."

Whatever the explanation was, the girl had been given back her life.

As the hot summer wore on, conditions in Grünberg became worse. Working hours grew longer; there was less food; the transports to Auschwitz were more frequent....

One morning in September I was returning to camp from the factory, looking forward to some sleep. My head constantly felt heavy and dull, my legs hurt from running all night after the devilish spinning machine.

Suddenly, when our guard had his head turned, a piece of bread was thrown over the fence into the courtyard. The girl ahead of me quickly picked up the bread and as she did, I heard her faint, terror-stricken voice whisper, "Don't give me away!"

The SS man saw the incident. He demanded to know who threw the bread. We knew there would be a penalty—head-shaving, beating, or worse. One by one we were questioned.

"I don't know," was each answer.

Then the husky SS guard with his heavy arm would hit the tired face, again and again. I was frightened. I would rather have gone without food, or worked twenty-four hours straight through, than bear physical punishment.

But my turn came, and I muttered an exhausted, "I don't know."

I felt a heavy blow over my eyes.

"Who?" he demanded.

"I don't know," I repeated.

Another blow fell, deafening my left ear. I swayed and staggered toward the building. When I finally reached my bunk I sat there without undressing. My face was puffed and bruised, my skin and lips were moist with blood. I must have cried without knowing it.

Thirty girls were beaten that morning. I don't know what the blows did to the others, but they shattered the wall of strength that I had built for myself. Neither propaganda talks, designed to break our morale, nor hunger, nor work, no matter how hard, had affected my resistance as had the brutal blows of that guard.

If only they don't touch my flesh, I can survive, I thought.

Survive? All of a sudden it did not seem worth while. All the suffering and agony, for what? How could we be free when they would kill us first? How much longer would it last? It was now September, 1944—five years since the Germans had taken Bielitz—and there was still no end in sight.

How long could we go on? When would it be my turn to go to Auschwitz? How many more X-ray examinations could I pass before I would be doomed?

Ilse, who worked on the day shift, came back at noon. She walked quietly past my bunk, thinking I was asleep. I called her. When she turned around she was smiling. I hid my bruised face under the blanket. She turned away from me so that I could not see what she was doing, and dug into her pocket.

"I brought you a present!" she announced triumphantly.

There, on a fresh leaf, was one red, slightly mashed raspberry!

As I sat up, she stared at my face. The raspberry rolled under the bunk. I crawled under to get it. I ate it slowly, dust and all; its sweet juice mingled with the saltiness on my lips.

A few days later fortune smiled at me. The incredible happened.

Suse Kunz and I were chosen for different work. There was an old German in the spinning room who weighed the finished crates of yarn, recorded the weights in a ledger, and put each crate on a conveyor belt which carried it to storage. One day we noticed that the crates remained piled up at the end of the row of machines; the man who kept the production records had not come. The *Obermeister,* or head foreman, went from machine to machine several times that day. In the afternoon he called Suse and me, and asked whether we could write. We both said we could.

"Now," he said, shaking a bony finger, "that requires thinking; I don't know if you would be able to do it."

I wondered if he wanted to hurt us, or if he really thought that we were some type of animal. He showed us the ledger columns in which each machine was listed by number, with spaces to note the weight of material each machine produced. Then he made us enter a few figures.

"Lightly," he cautioned, "so we can erase it, in case of error. Think," he repeated. "Think!"

With serious faces, we both thought. We almost burst into laughter. We passed the tests well, and were told that one of us would work days, the other nights. The foreman then took us to his office and gave us each a folded piece of paper.

"Give them to the *Lagerführerin*," he said.

Out of sight, we looked at the messages. We had been given hard-labor cards, entitling each of us to an extra bowl of food daily. An extra bowl—we were rich! That meant that Isle and I would not be hungry any more and it meant the same for Suse and Liesel.

The work was very hard. We had to lift the heavy crates and carefully place them on the running belt; a crooked crate might tangle with the yarn or cause congestion, and this would have meant sabotage. But I liked the work. I was not confined to one machine. Instead, I went all over the vast halls. Other workers envied me, and stared when they saw me walking around with the ledger.

It was interesting to walk through the spinning halls. Huge machines tore the raw materials to shreds. These shreds were beaten into a loose, cotton-like substance that a carding machine then combed into loose strips that could be spun. Each spinning hall had its own carding machine.

During the long hours of the autumn nights, while the machines hummed busily and the sleepwalking skeletons ran after the threads, I walked through the plant, weighing the heavy crates, entering the amounts of yarn that helped Germany's economy—helped Germany fight us.

Deliveries of old clothes arrived daily from Auschwitz to be shredded up and converted into yarn. A number of the girls who ran the shredding machines insisted that they had recognized their parents' clothes. We had heard that in Auschwitz prisoners were told to undress for showers; that they were then handed bread rations and sent into the "showers"—only when the vents opened it was not water but gas.

Once as I passed the shredder I thought I saw Mama's coat. I turned away, praying, then forced myself to look again. It was just a black coat. It could have been anybody's—hundreds of people wore black coats....

The night wore on. The horizon became lighter. Another day was coming to Grünberg. The machines seemed to go faster. The shredder rolled louder, tearing to bits the clothes from Auschwitz—and I held in my heart the picture of my homecoming.

QUESTIONS FOR ANALYSIS

1. Discuss Gerda Weissmann Klein's experiences in the slave labor textile factory in Grünberg. How was the factory linked to Auschwitz? What do Klein's experiences reveal about the nature of slave labor in Nazi-occupied Europe?
2. How does Charlotte Delbo's poem "Arrivals, Departures" contribute to our understanding of the Holocaust? Discuss her choice of a poem rather than a narrative account to convey what took place at Auschwitz.

WOMEN AND THE RESISTANCE

Women participated in resistance groups in Germany, Italy, France, and the various countries of occupied Europe. They served as couriers, liaison agents, transporters of arms, intelligence agents, saboteurs, propagandists, writers and distributers of clandestine newspapers, and partisan fighters. Members of the

Resistance held a broad spectrum of political views and, for security purposes, resistance groups sometimes comprised only a handful of trusted friends and relatives. Other groups functioned as cells within large networks that received considerable assistance from American, British, and Soviet intelligence services. From the outset, communists were among the most organized and committed political opponents of fascism and they were also the first to be placed in concentration camps in Nazi Germany. But the activities of Communist parties and the Communist International followed the trajectory of Soviet foreign policy which sometimes created confusion. While the Popular Front policy of 1935–1939 led to the creation of a broad European front of anti-fascist and democratic forces, the two-year period of the Nazi-Soviet Pact of 1939–1941 undermined anti-Nazi activities in certain countries, especially in France which capitulated to Germany in June 1940. Following its invasion by Germany in June 1941, however, the Soviet Union declared all-out war on fascism throughout Europe and supplied resistance groups with financial and material assistance to carry out acts of resistance and sabotage behind enemy lines. Between 1942 and 1943, uprisings also took place within at least twenty Jewish ghettos in Eastern Europe, of which the most famous was the Warsaw Ghetto Uprising of 1943. Revolts occurred within five concentration camps, including Auschwitz.

165. The White Rose Munich Student Group, "Leaflets of the White Rose"

Sophie Scholl (1921–1943) was a member of the White Rose resistance group, a small circle of students at the University of Munich. Sophie, her brother, Hans, and other members of the group were executed in 1944 for their opposition to the Nazi government. From the summer of 1942 to February 1943, the group distributed thousands of leaflets in Munich and several other German cities calling on the German people to carry out acts of resistance and sabotage. The White Rose students also painted slogans on walls, such as "Down with Hitler," "Hitler the Mass Murderer," and "Freedom," as well as canceled out swastikas. Reproduced here is the third of four leaflets in the series "Leaflets of the White Rose." The group distributed two additional leaflets in January and February 1943 as part of a new series "Leaflets of the Resistance." The last leaflet, entitled, "Fellow Fighters in the Resistance," focused on the new opportunities for resistance presented by the German defeat at Stalingrad on January 31, 1943. On February 18, Sophie and Hans were arrested after being observed scattering leaflets at the university; they were executed soon afterward. Very little appeared in the German press about the White Rose, but knowledge of the executions spread within Germany as well as abroad. Several hundred people attended a 1944 memorial in New York City at which Eleanor Roosevelt spoke. Inge Scholl, the sister of Hans and Sophie, subsequently documented the group's activities in *Students Against Tyranny: The White Rose, Munich 1942–1943* (1970).

Inge Scholl, pp. 81–84 of *The White Rose: Munich 1942–1943* © 1983 by Inge Aicher-Scholl, Wesleyan University Press by permission of University Press of New England.

Salus publica suprema lex

All ideal forms of government are utopias. A state cannot be constructed on a purely theoretical basis; rather, it must grow and ripen in the way an individual human being matures. But we must not forget that at the starting point of every civilization the state was already there in rudimentary form. The family is as old as man himself, and out of this initial bond man, endowed with reason, created for himself a state founded on justice, whose highest law was the common good. The state should exist as a parallel to the divine order, and the highest of all utopias, the *civitas dei,* is the model which in the end it should approximate. Here we will not pass judgment on the many possible forms of the state—democracy, constitutional monarchy, monarchy, and so on. But one matter needs to be brought out clearly and unambiguously. Every individual human being has a claim to a useful and just state, a state which secures the freedom of the individual as well as the good of the whole. For, according to God's will, man is intended to pursue his natural goal, his earthly happiness, in self-reliance and self-chosen activity, freely and independently within the community of life and work of the nation.

But our present "state" is the dictatorship of evil. "Oh, we've known that for a long time," I hear you object, "and it isn't necessary to bring that to our attention again." But, I ask you, if you know that, why do you not bestir yourselves, why do you allow these men who are in power to rob you step by step, openly and in secret, of one domain of your rights after another, until one day nothing, nothing at all will be left but a mechanized state system presided over by criminals and drunks? Is your spirit already so crushed by abuse that you forget it is your right—or rather, your *moral duty*—to eliminate this system? But if a man no longer can summon the strength to demand his right, then it is absolutely certain that he will perish. We would deserve to be dispersed through the earth like dust before the wind if we do not muster our powers at this late hour and finally find the courage which up to now we have lacked. Do not hide your cowardice behind a cloak of expediency, for with every new day that you hesitate, failing to oppose this offspring of

Hell, your guilt, as in a parabolic curve, grows higher and higher.

Many, perhaps most, of the readers of these leaflets do not see clearly how they can practice an effective opposition. They do not see any avenues open to them. We want to try to show them that everyone is in a position to contribute to the overthrow of this system. It is not possible through solitary withdrawal, in the manner of embittered hermits, to prepare the ground for the overturn of this "government" or bring about the revolution at the earliest possible moment. No, it can be done only by the cooperation of many convinced, energetic people—people who are agreed as to the means they must use to attain their goal. We have no great number of choices as to these means. The only one available is *passive resistance.* The meaning and the goal of passive resistance is to topple National Socialism, and in this struggle we must not recoil from any course, any action, whatever its nature. At *all* points we must oppose National Socialism, wherever it is open to attack. We must soon bring this monster of a state to an end. A victory of fascist Germany in this war would have immeasurable, frightful consequences. The military victory over Bolshevism dare not become the primary concern of the Germans. The defeat of the Nazis must *unconditionally* be the first order of business. The greater necessity of this latter requirement will be discussed in one of our forthcoming leaflets.

And now every convinced opponent of National Socialism must ask himself how he can fight against the present "state" in the most effective way, how he can strike it the most telling blows. Through passive resistance, without a doubt. We cannot provide each man with the blueprint for his acts, we can only suggest them in general terms, and he alone will find the way of achieving this end:

Sabotage in armament plants and war industries, sabotage at all gatherings, rallies, public ceremonies, and organizations of the National Socialist Party. Obstruction of the smooth functioning of the war machine (a machine for war that goes on solely to shore up and perpetuate the National Socialist Party and its dictatorship).

Sabotage in all the areas of science and scholarship which further the continuation of the war—whether in universities, technical schools, laboratories, research institutes, or technical bureaus. *Sabotage* in all cultural institutions which could potentially enhance the "prestige" of the fascists among the people. *Sabotage* in all branches of the arts which have even the slightest dependence on National Socialism or render it service. *Sabotage* in all publications, all newspapers, that are in the pay of the "government" and that defend its ideology and aid in disseminating the brown lie. Do not give a penny to public drives (even when they are conducted under the pretense of charity). For this is only a disguise. In reality the proceeds aid neither the Red Cross nor the needy. The government does not need this money; it is not financially interested in these money drives. After all, the presses run continuously to manufacture any desired amount of paper currency. But the populace must be kept constantly under tension, the pressure of the bit must not be allowed to slacken! Do not contribute to the collections of metal, textiles, and the like. Try to convince all your acquaintances, including those in the lower social classes, of the senselessness of continuing, of the hopelessness of this war; of our spiritual and economic enslavement at the hands of the National Socialists; of the destruction of all moral and religious values; and urge them to *passive resistance!*

Aristotle, *Politics:* "...and further, it is part [of the nature of tyranny] to strive to see to it that nothing is kept hidden of that which any subject says or does, but that everywhere he will be spied upon,...and further, to set man against man and friend against friend, and the common people against the privileged and the wealthy. Also it is part of these tyrannical measures, to keep the subjects poor, in order to pay the guards and soldiers, and so that they will be occupied with earning their livelihood and will have neither leisure nor opportunity to engage in conspiratorial acts....Further, [to levy] such taxes on income as were imposed in Syracuse, for under Dionysius the citizens gladly paid out their whole fortunes in taxes within five years. Also, the tyrant is inclined constantly to foment wars."

Please duplicate and distribute!

166. Bronka Klibanski "In the Ghetto and in the Resistance"

Bronka Klibanski participated in the Jewish resistance movement as a member of the Dror, a socialist-Zionist youth movement. Klibanski's hometown of Grodno, Poland was incorporated by the Soviet Union following the German invasion of Poland in September 1939. After June 1941, when the Soviet Union itself was invaded by Germany, Grodno came under Nazi rule and racial laws were imposed on its Jewish population, which was soon confined in ghettos. Klibanski served as a courier for the underground and fought with the partisans in the forest. Emigrating to Israel after the war, she worked as a researcher at Yad Vashem, the Holocaust museum, library, and archive in Jerusalem. Klibanski is the author of *The Archive of Abraham Silberschein* (Hebrew, 1984) and *The Collection of Testimonies, Diaries and Memoirs in the Yad Vashem Archives* (1990). She also edited and published the papers of Mordechai Tenenbaum with whom she worked in the Resistance.

From "In the Ghetto and in the Resistance: A Personal Narrative" by Bronka Klibanski. From *Women in the Holocaust,* edited by Dalia Ofer and Lenore J. Weitzman, pp. 175–186. Copyright © 1998 by Dalia Ofer and Lenore J. Weitzman. Reprinted by permission from Yale University Press.

The war reached my hometown of Grodno and the entire Bialystok region on June 22, 1941, with a surprise attack by Germany on the Soviet Union. From the outbreak of World War II in 1939 until that point, we had lived under Soviet rule.

With the German invasion and the collapse of the Red Army, the world we had known was shattered. Overnight, our lives became public property. We suddenly found ourselves deprived of basic rights, unprotected by any law. We applied all our energy and ingenuity to the cause of finding food. Parents found it hard to adjust to the situation, leaving the burden of earning a livelihood almost exclusively to the children. To bring home something to ease the gnawing hunger, we sought any kind of work. Each day gave birth to new decrees and new humiliations. Jews were excluded from bread rations. First we were ordered to wear an armband with a blue Star of David and then, a week later, the yellow star. We were forbidden to walk on sidewalks but had to goose-step in the gutter. We saw no logic in these laws, only cruelty for its own sake and a bizarre wish to degrade us.

After the initial shock, we lifted our heads once again. We tried to make light of the situation, but we had hardly become used to it when a new blow fell and the Jews of Grodno were ordered to move into two ghettos, Grodno Ghetto 1 and Grodno Ghetto 2.

I shall omit from my testimony my personal recollections of the life in Grodno Ghetto 2 and shall move to my activities in the resistance movement in Bialystok. These activities grew out of my prewar involvement in Dror, a socialist-Zionist youth movement. Members of the prewar youth movements, Zionists and others, began meeting in the Grodno ghetto. In early January 1942 Mordechai Tenenbaum, a leader of the Dror movement, traveled to Grodno from Vilna on false documents. He told us what happened to the Jews of Vilna—they were taken to the Ponar forest, where fifty thousand out of about sixty thousand of them were massacred over a period of six months. He believed that the Germans intended to murder all the Jews under their control, and he exhorted us to be alert to

everything around us, to put up resistance in any way possible, and to hide and try to escape when the Germans came to expel us from the ghetto. He told us not to cooperate with the Jewish police and not to believe their promises. He mentioned that a similar proclamation had been read at an assembly of Zionist youth organizations held in the Vilna ghetto on January 1, before he left. Now, at long last, we understood why we had been segregated from the rest of the population and enclosed in the ghetto. The Germans had contrived this situation in order to round us up and murder us with greater efficiency. Nevertheless, we found this truly hard to accept; somewhere in our hearts nestled the hope that we would be excluded from this fate....

. . . [I]n November 1942, the Germans surrounded Grodno Ghetto 2 in a surprise move. Everyone in my family, along with all the other Jews in the ghetto, was taken to the death camps. Thus I was not at their side in the last and worst moments of their lives. Since then, I have searched for them in every mound of earth that covers the soil of the former death camps in Polish territory.

I learned what had happened to my family from Mordechai Tenenbaum, who came as an emissary of the Jewish Fighters' Organization and the National Jewish Committee in Warsaw, to organize the resistance movement in the Bialystok ghetto. He found that the ghetto had been closed by the Germans in order to prevent the entry of Jewish fugitives from the ghettos in the Bialystok area, whose liquidation had begun on November 1, 1942. After overcoming many obstacles, Tenenbaum reached Grodno Ghetto 1, where he learned about the liquidation of Ghetto 2 and the deportations of my family. In fact the entire Bialystok area, including Grodno, was being steadily purged of Jews. Spared for the time being was only the Bialystok ghetto.

Day and night the deportation trains raced westward from the Bialystok region, delivering their human cargo to the death camps of Treblinka, Majdanek, and Auschwitz. How much time remained for the more than forty thousand Jews in the ghetto of Bialystok who were still alive? The extermination facilities could not murder all of us at once, so we waited and hoped. Most of the Jews in the Bialystok ghetto were assigned to work for the German army; they sabotaged their output and adopted a business-as-usual pose that camouflaged the fear of death. Perhaps, by some miracle, they might be spared after all....

The acting chairman of the Judenrat, Ephraim Barasz, met repeatedly with the heads of the German administration in Bialystok to try to convince them that the labor of all the ghetto inmates was useful to the war effort. German military and economic committees visited the ghetto frequently to see for themselves how vitally needed the Jews were, to pocket bribes, and to calm the suspicious ghetto inmates. The Judenrat carried out the Germans' instructions faithfully. They encouraged the Jews to obey German orders, to work hard, and not to endanger themselves by taking action or even thinking of resistance.

In this way the web of illusion was spun in the Bialystok ghetto—the myth that hard work and cooperation with the Germans would spare the inmates from the fate of the other ghettos. The Germans, aware of these beliefs, exploited them cynically to calm the Jews' fears and quell the spirit of revolt. This complacency persisted even after the first Aktion, the roundup for deportation in early February 1943, and lasted until August of that year, when the final liquidation of the ghetto began.

By late 1942, most of the ghettos in the region had been liquidated; only the youth movements—a predominant part of the underground, which had become a force to reckon with in the Bialystok ghetto—tried to devise methods of resistance. Concurrently, small groups from the underground were sent or made their way to the forest to establish a base where the survivors of the uprising and downfall of the ghetto would continue the battle as partisans.

Only the young people, such as our group of Dror, felt that we could do something. We knew that we could not stop the mass murder, but we understood that we had the power to sabotage the Germans' efforts if we could overcome the Jews' paralyzing illusions; fight the complacency and silence that the Germans were intent on evoking; preach vengeance; and find ways of taking action and, perhaps, effecting rescue. We had to defend ourselves for the sake of human dignity....

It was my privilege to belong to one of the movements whose members acted in this cause. I was not one of the leaders, but a rank-and-file member of a team of Dror activists. One day Mordechai Tenenbaum suggested that I cross to the Aryan side to serve as a liaison officer for the movement. In late 1942 only one liaison officer, Tema Shneiderman, remained alive. Other experienced officers had been caught by the Germans, and I was to replace them....

It was not by chance that women were chosen as liaisons or couriers. It was easier for us to disguise ourselves. Until the Bialystok ghetto uprising began in August 1943, we worked on the Aryan side, hardly communicating with one another and never collaborating. This situation changed when the uprising commenced. The heavy siege that the Germans imposed on the ghetto rendered us useless to the rebels. Even so, I moved around the outside of the ghetto in the hope that I might find some way to communicate with comrades inside in order to help them. When the uprising broke out, two women from Hashomer Hatsa'ir slipped out of the ghetto. Together we tried to make contact with their comrades who had joined the partisans. We knew that a small group of Jewish partisans—mostly Communists but some from Hashomer Hatsa'ir[1]—had assembled in the forests surrounding Bialystok after having fled the ghetto between January and April 1943. The members of this group had frequently visited the ghetto to maintain communications and receive food and weapons. Few in number and poorly equipped, they often changed hiding places to avoid detection by the Germans and the villagers. I had not had any contact with them before the uprising.

---

[1] A Zionist youth movement.

August 16, 1943, was the first day of the uprising in the Bialystok ghetto. Chaika Grossman (head of Hashomer Hatsa'ir in the Bialystok ghetto and later a member of the Israeli Parliament) and I met in the street with the forest group's liaison, Marylka Rozycka, and through her we established contact with the partisans there. I wandered daily, as if dazed, around the ghetto walls and the surrounding streets, looking for Jews that I could help. At this point, we pledged all our energy to saving fugitives from the deportation trains and helping the group of partisans who sheltered them.

The uprising lasted more than three weeks. Mordechai Tenenbaum was killed, or committed suicide, and with him died the great majority of his fighters. I was unable to assist the fighters and my loved ones, and I suffered for my inability to be with them.... All these events were frustrating, and I thought bitterly that in the end none of us would survive.

Members of the resistance who had escaped from the trains or gone into hiding joined the forest group—some on their own initiative and others with our assistance. Thus the size of the group soon doubled, from fifty to one hundred members. The newcomers were in abysmal physical and psychological condition....

Our group of young women initially numbered six and then five: two Communists, two women from Hashomer Hatsa'ir, and one from Dror. Our sixth member, Rivkele Medajska, lost her life helping Jews who had gone into hiding outside the ghetto and who were in great danger before she was able to lead them to the partisans....

It was not until the spring of 1944 that the first squad of Soviet paratroopers reached the forest surrounding Bialystok and begin to unify the groups operating there, including Jews and Soviet prisoners of war who had escaped their captors. Our partisan group, known as Forojs (Forward), joined them and thus crossed an important watershed. The Jewish partisans' situation improved greatly once they joined the larger partisan camp, although they were not spared humiliations as Jews. We became liaisons at the Soviet headquarters of the large partisan formation. The unit was well organized and worked in coordination with the Red Army; it had all the equipment it needed and communicated directly with partisan centers in Minsk (the capital of Belorussia) and Moscow.

According to instructions, an antifascist committee was set up in Bialystok in May 1944 under one of our Communist comrades, Liza Chapnik....

We five women liaisons effectively constituted the executive core of the antifascist committee. We stayed in touch with non-Jewish comrades who worked in such places as the Bialystok airport and German military installations. They fed us information on the movements of army units, the positioning of anti-aircraft batteries, and the number of warplanes and other craft. One member of our group used this information to prepare detailed sketches of German military and aircraft formations, which were forwarded by partisan headquarters to the Soviet military command....

In the last few weeks of the German occupation, fleeing soldiers—including Belorussians, Ukrainians, and others from the German auxiliary units—passed through our area. We were ordered to convince them that the Soviet authorities would give them amnesty if they defected from the enemy army and turned over their weapons to us. In many cases we were successful, and by liberation day we had a bunker full of weapons....

Liberation came on July 27, 1944, a beautiful, clear summer day. Earlier, the SS men who had lived in the area had fled Bialystok in panic, and one of their special commando units had set the entire downtown area ablaze. Three of us—Ania Rud, Liza Chapnik, and I—had taken shelter in a German bunker; the others had gone to the forest. But now, the flames that had danced over downtown Bialystok had finally subsided. A strange silence hovered in the air, as if the whole city were holding its breath.

The first to come were the Soviet sappers, searching for mines. When we heard Russian being spoken, we cautiously stepped forward to meet them, but only after satisfying ourselves beyond doubt that they were indeed Soviet soldiers. Then came the tanks. We ran to kiss their grimy crew members and thank them for having liberated us from the Germans at long last. Our joy was boundless, our smiles unlimited. We organized a small parade the next day, clutching red flags and marching to the outskirts of the city, where we welcomed the general who led the troops into the town. He was moved to receive this kind of welcome on Polish soil. We were drunk with happiness that day, having witnessed our people's murderers in flight.

But what then? How were we to go on? What followed were days of mourning and bereavement. There were no homes left, no families. I wandered through the streets looking for familiar faces. Once a Russian soldier whispered to me: "Nye platch, dyevushka; lubimiy tvoy vernyetsa"—"Don't cry, my girl; your lover will come back." I did not even feel the tears streaming down my face.

After the liberation of Bialystok, we helped arrest collaborators who were known to have revealed the hiding places of Jews and handed them over to the Germans. . . . Of the Jewish partisan group, which had attained the size of more than one hundred members after the uprising in and liquidation of the ghetto, only fifty or so survived. Most of them went to Israel. . . .

What typified the group of liaisons to which I belonged? Each of us had belonged to a youth movement, either leftist, Zionist, or Communist; the main feature we shared was probably the ideological education our respective movements had provided. The values that informed our education were Jewish and universal. We had learned at an early age to be frugal; to improve ourselves; to transcend trivialities, egotistic inclinations, and cheap temptations; and to work on behalf of the collective. These values, which we internalized along with our hatred of the Germans and our wish to exact revenge, may explain, at least partially, our actions in the underground and after the war. Complementary to these ideals were the education we received at home and our own personality traits, such as courage and audacity.

It is important to note the respectful and admiring attitude the Soviet command and partisans evinced toward our group of liaisons. We were awarded citations and our names and activities were mentioned in newspapers, books, and periodicals (without mention of our Jewishness).

We saw no need to wage a feminist struggle to ensure our human dignity; we earned respect by behaving and acting in ways that spoke for themselves. In comparison to the men, it seems to me that we women were more loyal to the cause, more sensitive to our surroundings, wiser—or perhaps more generously endowed with intuition—and more resolute in our commitment to change reality while accepting that some things are unattainable. We did so without despairing, without giving up, and without ceasing to try.

QUESTIONS FOR ANALYSIS

1. On what basis did the White Rose student group call on the German people to oppose the Nazi government? What forms of resistance did the group propose?
2. How was Bronka Klibanski's life transformed by the 1941 German invasion of Bialystok, Poland which was previously under Soviet rule? What is her critique of the role of the Judenrat in the Jewish ghettos? Discuss the composition and activities of the Resistance groups with which she worked. How was the Resistance affected by the arrival of Soviet paratroopers in 1944?
3. Compare and contrast the resistance activities of the White Rose and Bronka Klibanski.

THE HOMEFRONT IN BRITAIN
AND THE SOVIET UNION

Britain declared war on Germany on September 3, 1939, two days after the German invasion of Poland. For nearly a year and a half, Britain alone mobilized an effective homefront against Nazi Germany. Fighting in the first months of the war took place at sea as the German navy attacked British merchant vessels and the navy ships protecting them. Severe shortages occurred as a result of interrupted trade, and food rationing was implemented. In the spring of 1940, Germany won a series of quick victories in Europe: Norway surrendered on May 7 soon followed by Holland, Belgium, and Luxemburg. On May 10, the anti-appeasement Conservative Party leader Winston Churchill replaced Neville Chamberlain as prime minister, a position he held throughout the war. The Battle of Britain (July 1940 to October 1941) began almost immediately after France surrendered on June 22, 1940, as German control of French airfields allowed German bombers to easily reach southeastern England. Air-raid sirens, civilian evacuations, the issuing of gas masks, and mass destruction from the bombings became a daily feature of life until the Royal Air Force (RAF) achieved victory in the air war. The German bombings killed some 30,000 civilians. The Soviet Union entered World War II following the German invasion of June 1941. Here, the distinctions between the homefront and military front were often blurred as the German army occupied huge areas of Soviet territory in the initial months of the war. A million and a half people died during the Siege of Leningrad which took place between 1941 and 1943. Altogether, about 25 million Soviet people lost their lives during World War II.

167. Nella Last
A Mother's Diary

Nella Last responded to an appeal from the Mass-Observation project for ordinary individuals to maintain daily records of their lives. Initiated in 1937, the project aimed to "record the voice of the people." A middle-aged housewife in the English shipbuilding town of Barrow-in-Furness, located near Liverpool, Last began her diary at the project's outset, and continued it for almost thirty years. Her diaries from World War II, the most extensive of those in the Mass-Observation Archives, were edited and published by Suzie Fleming and Richard Broad as *Nella Last's War: A Mother's Diary, 1939–1945* (1981). Here, Last records the full range of her wartime experiences, including food shortages, the German bombings of Britain, participation in volunteer work, reactions to the events of the war as reported in the newspapers and the BBC, and how the war affected her relationship with her husband, a joiner and shipfitter, and her sons Arthur, a tax inspector apprentice and the eldest, and Cliff, a soldier. Last volunteered at the Women's Volunteer Service (W.V.S.) Center, a canteen for soldiers on leave, and a Red Cross second-hand shop. The W.V.S. was established in 1939 to provide services for evacuated civilians and hospitals in the event of war. During World War II, the center provided clothing and blankets for hospitals and the military forces. The volunteers also raised funds to purchase raw materials, such as wool, which they knit into clothing. These excerpts from Last's diary are drawn from September 1939, the first month of the war; May 1941, during the Battle of Britain; and May 1945, the last month of the war in Europe.

Sunday, 3 September, 1939
Bedtime

Well, we know the worst. Whether it was a kind of incredulous stubbornness or a faith in my old astrological friend who was right in the last crisis when he said 'No war', I *never* thought it would come. Looking back I think it was akin to a belief in a fairy's wand which was going to be waved.

I'm a self-reliant kind of person, but today I've longed for a close woman friend—for the first time in my life. When I heard Mr. Chamberlain's voice, so slow and solemn, I seemed to see Southsea Prom the July before the last crisis. The Fleet came into Portsmouth from Weymouth and there were hundreds of extra ratings walking up and down. There was a sameness about them that was not due to their clothes alone, and it puzzled me. It was the look on their faces—a slightly brooding, faraway look. They all had it—even the jolly-looking boys—and I felt I wanted to rush up and ask them what they could see that I could not. And now I know.

The wind got up and brought rain, but on the Walney shore men and boys worked filling sandbags. I could tell by the dazed look on many faces that I had not been alone in my belief that 'something' would turn up to prevent war....

My younger boy will go in just over a week. His friend who has no mother and is like another son will go soon—he is twenty-six. My elder boy works in Manchester. As a tax inspector he is at present in a 'reserved occupation'.

Tuesday, 5 September, 1939

I went to the W.V.S. Centre today and was amazed at the huge crowd. We have moved into a big room in the middle of town now, but big as it is, every table was crowded uncomfortably with eager workers. Afterwards, huge stacks of wool to be knitted into bedcovers, and dozens of books of tailor's patterns to be machined together, were taken. They average about seventy-seven yards of machining to join each piece with a double row of stitching and a double-stitched hem. I'm on my third big one and have made about a dozen cot quilts....

Tonight I had my first glimpse of a blackout, and the strangeness appalled me. A tag I've heard somewhere, 'The City of Dreadful Night', came into my mind and I wondered however the bus and lorry drivers would manage. I don't think there is much need for the wireless to advise people to stay indoors—I'd need a dog to lead me.

{The first blackouts meant that all street lights were extinguished, no car headlights allowed, and windows were shuttered or curtained with heavy black material or paper. Later, torches and car headlights were allowed, but they were masked to provide only a slit of light.}[1]

Wednesday, 6 September, 1939

Today I was in the company of several women of my own age, and we talked of the beginning weeks of the last war—of the mad stampede of boys and men to rush to the Shipyard and get under 'Vickers Umbrella', making them indispensable on munitions so they

---

[1]Explanatory passages in brackets provided by Suzie Fleming and Richard Broad, the editors of Last's diary.

wouldn't be called up. There is so little of that now that it is not heard of. Instead, there seems a kind of resignation—a 'Well I'll get my turn I suppose', and a look on the faces of the lads who have joined the Territorials and Militia that was not there this time last week.

I looked at my own lad sitting with a paper, and noticed he did not turn a page often. It all came back with a rush—the boys who set off so gaily and lightly and did not come back—and I could have screamed aloud....

There are big 'secret' preparations for hospitals here, and it looks as if a big new grammar school—girls'—has been taken as well. Two Isle of Man boats which lie up for the winter in the docks are nearly finished fitting out as hospital-ships, and I wonder if they will bring patients by water.

Sunday, 4 May, 1941

A night of terror, and there are few windows left in the district—or roof tiles! Land mines, incendiaries and explosives were dropped, and we cowered thankfully under our indoor shelter. I've been so dreadfully sick all day, and I'm sure it's sheer fright, for last night I really thought our end had come. By the time the boys come, I'll be able to laugh about it. Now I've a sick shadow over me as I look at my loved little house that will never be the same again. The windows are nearly all out, the metal frames strained, the ceilings down, the walls cracked and the garage roof showing four inches of daylight where it joins the wall. Doors are splintered and off—and there is the *dirt* from the blast that swept down the chimney. The house rocked, and then the kitchenette door careered down the hall and plaster showered on to the shelter....

All neighbours who have cars, and friends in the country, have fled—a woman opposite brought her key and said, 'Keep an eye on things please,' I said, '*No*, you must do it yourself. You have no right to expect me or anyone else to do it for you. You are strong and well and have no children to think of. I'd not put out an incendiary if I saw it strike your house—unless I thought the flame would be a danger to others.' I think I'm a little mad today. I'd

never have spoken so plainly until now. The damage to houses is very widespread—and all round, there's the circle of hovering balloons. It's not saved the Yard, though, for two shops have been destroyed—one, the pattern shop, burned like a torch with its wood-fed flames.

The birds sang so sweetly at dawning today—just as the all-clear sounded and people timidly went round looking at the damage. I wonder if they will sing as sweetly in the morning—and if we all will hear them. Little sparrows had died as they crouched—from blast possibly. It looked as if they had bent their little heads in prayer, and had died as they did so. I held one in my hand: 'He counteth the sparrow and not one falleth that He does not see'—Poles, Czechs, Greeks, all sparrows....

Our newspapers did not come till 3:30; there must have been a bomb on the line somewhere. Only Merseyside was mentioned on the wireless. We will not want them to know they have got to the Shipyard perhaps....

Bedtime

... There seems to be a lot of damage done, with whole districts roped off. My new neighbours say that nothing they had in London was so bad—poor things, they have so little furniture and all of it obviously second-hand bits. They were bombed out at Harrow, and then again at Liverpool....

My fish swam like silver balloons in a blue-grey bowl, and then, as the sun sank, they turned to a faint blush-pink. So odd how one changes: I have loved the crash and roar of waves all my life, but now I never look at them if I can help it; they make me think too much of shipwrecks and horror. Always I have loved the moon—tonight I felt a detachment, a sense of menace. No 'peaceful', 'benign', 'serene', 'kindly' moon, as she rose to point the way for devil-bombers, but a sneering, detached Puck who delighted in holding a burglar's lantern.

Saturday, 5 May, 1945

I think this changeable weather saps energy. It even makes Mrs. Atkinson say she is tired! All her talk and ideas lately seem to be about 'how the

Germans can be punished'. She does not seem to see the horror that *is* Germany, millions of homeless ones adrift in the very essence of the word, no homes, no work, sanitation, water, light or cooking facilities, untold dead to bury, sick and mad people to care for. I said, 'Would you like to live in Barrow if every house downtown had gone in the blitz, the water, light and transport gone, and all the Shipyard—with idle, desperate men beginning to seek food and shelter from *anywhere*, by *any* means?' Any safe corner or peaceful place with buildings or food of any kind—flocks, fields of growing food, orchards, cows and poultry—will be a target for the half-crazed, hungry people wandering round. Me, I see mad chaos, which we, the Allies, can do little about. She thinks it's 'only right for what they did to Poland and Holland'. I said, 'There is no right and wrong in it. If a man has a gangrened wound, a malignant ulcer, he doesn't speak of right and wrong—he seeks cure or amputation before the poison has spread through the whole of his body.'

Tonight, before I came to bed, it had stopped raining and I went down the garden path to smell the grateful earth and the scent of damp greenery. At the bottom corner, I was conscious of an unpleasant smell and, as I thought, Murphy[2] had left a very dead rat he had caught. As I got the spade, scolding him the while, and buried the loathsome thing in the soft earth, the thought of decay and death under the acres of fallen masonry in Europe set my mind again on 'What *will* happen?'—till my head ticked so badly I could see the beat of a nerve in my throbbing temple when I looked in the glass. People say all round, 'I'm glad the war is over before my son has to go'—not realising the problems of Europe or the Pacific War.

{As the Allied troops moved into Germany and Eastern Europe, they found the concentration camps— among them, Belsen. Horrific newsreel pictures of half-living skeletons, piles of dead, emaciated bodies and the apparatus of death were sent back and shown

in the cinemas. Over six million people had been murdered in these camps.}

Sunday, 6 May, 1945

I'd suggested going to the pictures, for I knew my husband would feel a bit stale after sticking to his paper work all day....Last week, I would not go to see the Belsen horror-camp pictures. I felt the ones in the paper quite dreadful enough. They were shown again tonight, as 'requested' by someone. I looked in such pity, marvelling how human beings could have clung so to life: the poor survivors must have had both a good constitution and a great will to live. What kept them alive so long before they dropped as pitiful skeletons? Did their minds go first, I wonder, their reasoning, leaving nothing but the shell to perish slowly, like a house left untenanted? Did their pitiful cries and prayers rise into the night to a God who seemed as deaf and pitiless as their cruel jailers?

I've a deep aversion to interference, having suffered from it all my life till recent years. I've always said, 'Let every country govern itself, according to its own ways of thought and living. Let them develop their own way and not have standards forced on them, standards so often governed by commercial or political considerations, rather than their own good. Let them reach out in friendly neighbourliness, rather than 'by order of treaties or pacts.' Now I see it would not do. Germany had that creed, developed to a degree of isolationism. People knew about concentration camps, but nothing could seemingly be done about it. This horror is not just one of war. No power can be left so alone that, behind a veil of secrecy, *any*thing can happen.

Odd if V.E. Day comes next Wednesday, for a Naval man in the canteen said it would 'be about next Wednesday'. The tide is running out so swiftly and unnoticeably that the actual cease-fire will be shorn of excitement and any of the wild 'whoopee' of the last armistice. It's a very dreadful thought: I'm not very old, but can clearly remember three wars—and at the rate they have been, could yet see another.

---

[2]Murphy was the family cat.

168. Soviet Poster, 1941

"Fascism—the Most Evil Enemy of Women. Everyone to the Struggle against Fascism!" (*1941 poster from* Russia 20th Century: History of the Country in Posters *by N. Barburina.*)

The production of posters increased dramatically following the German invasion of the Soviet Union in June, 1941. During the course of the war, the Iskusstvo publishing house in Moscow printed 34 million copies of 80 different poster designs. Many posters, like this one by the artist Nina Nikolaevna Vatolina, identified "fascism" rather than Nazi Germany as the enemy. The top portion of the caption reads: "Fascism—the Most Evil Enemy of Women." The lower portion reads: "Everyone to the Struggle Against Fascism."

169. Soviet Women in Defense of the Motherland

Over 800,000 Soviet women served at the front in World War II. Their military service included work as signallers, snipers, gunners, tank drivers, commanders of tank battalions, bomber pilots, anti-aircraft gunners, sailors, submachine-gunners in the infantry, and paratroopers. Women worked in the field as cooks and bakers, drivers, political workers, field engineers, nurses, and medical orderlies. Women also joined underground and partisan military units in Soviet territory occupied by Nazi Germany. Some 60,000 women, for example, participated in the partisan movement in Belorussia. These two accounts of Soviet

From *War's Unwomanly Face* by S. Alexiyevich, pp. 82, 84–88, 193–195, 240–243, 200–202. Copyright © 1988 by Progress Publishing Group. Reprinted by permission.

women's defense of the Motherland are drawn from interviews carried out by S. Alexiyevich and anthologized in *War's Unwomanly Face* (1985, translated 1988). Many of the women who fought together remained friends for over forty years and they are sometimes interviewed together rather than individually. Recalling the significance of the women's wartime experiences, Klara Semyonovna Tikhonovich, an anti-aircraft gunner, wrote, "No matter what our date of birth was, we were all born in 1941."[1]

A. Sergeant Valentina Pavlovna Chudayeva, Anti-aircraft-gun Crew Commander from Siberia

"And now imagine—it was 1941. The farewell party for the school-leavers was under way in our school. We all had our own plans, our own dreams—we were young girls after all. After the party, we sailed down the river Ob to an island. When we left the city, it looked normal, as it always did. We were all so merry, so happy.... We had never yet been kissed by any boys; we didn't even have boy friends. We watched dawn breaking and then sailed back.... But the whole of the city was in commotion; many people were in tears. And from all sides came, 'There's war!' There's war!' All the radios were turned on. But we couldn't grasp it. What war? We had been so happy, we had made such plans—what college we would enter, what we would become.... And then like a bolt from the blue: there's war! Grown-ups were in tears, but we felt no fear; we assured one another that it would take us less than a month 'to smash the Nazis to smithereens', a song we sang before the war, it was a current idea that we would carry the war back onto enemy territory.... We understood everything when people began receiving 'killed in action' notices.

"They refused to call up my father for active duty. But he kept pestering the military registration and enlistment office and soon he went to the front. In spite of his poor health, his grey hair, and his weak lungs—he had tuberculo-sis—and in spite of his age! But he joined the 'Steel Division'—it was also called 'Stalin's Division'—which included many Siberians. We, too, felt that there could be no war without us taking part, that we, too, must fight. We wanted arms at once! So we hurried to the military registration and enlistment office. And on the 10th of February I was ready to leave for the front....

"For two months we travelled in heated goods vans. Two thousand girls—a whole train. We were accompanied by officers who trained us. We were to be signallers. We arrived in the Ukraine and it was there that we were bombed for the first time....

"So I became a signaller in an anti-aircraft unit. I was assigned to a command post to receive and send messages. And I might have remained a signaller right to the end of the war, had I not received a notice that my father had been killed in action. When that happened I asked them to send me to the front line, 'I want to avenge my father, to settle scores with the Nazis.' I wanted to fight, to take revenge, to shoot.... They tried to tell me that a telephone was very important for the artillery. But a telephone receiver does not shoot you know.... I wrote an application to our regimental commander. But he turned down my request. Then, without wasting time, I wrote to the commander of the division. Soon Colonel Krasnykh came, lined us all up, and said, 'Where is the girl who wants to be a gun-crew commander?' And there I stood—a short girl, with a

---

[1]S. Alexiyevich, *War's Unwomanly Face,* trans. Keith Hammond and Lyudmila Lezhneva (Moscow: Progress Publishers, 1988), p. 7

thin neck. And from that neck a submachine gun was hanging—a heavy one, with seventy-one cartridges....I must have been quite a pitiful sight....He looked at me, 'What do you want?' And I said, 'I want to shoot.' I don't know what he thought. He gazed at me for a long time, then turned round and walked off. Well, I thought, that's a flat refusal. Presently our commander came running. 'The colonel has given you permission....'

"I took a short-term training course, very short indeed—just three months—and became a gun-crew commander. I was sent to Anti-Aircraft Regiment 1357. At first my nose and ears bled and my stomach was completely upset.... It wasn't so terrible at night, but in the daytime it was simply awful. The planes seemed to be heading straight for you, right for your gun. In a second they would make mincemeat of you....It was not really a young girl's job.... At first we had '85s; they had acquitted themselves well at the approaches to Moscow. Later they were used as anti-tank guns, and we were given '37s. That was in the Rzhev sector. Heavy fighting was under way there....In the spring the ice on the Volga began to break. ...And one day we saw an ice-floe drifting downstream and on it two or three Germans and one Russian soldier....They had died gripping one another, got frozen into that floe, and the whole floe was splashed with blood. Can you picture it? All the water in Mother Volga was mixed with blood...."

Valentina Pavlovna suddenly paused in the middle of a word and appealed, "I can't....Let me catch my breath....It's so hard....Alexandra has persuaded me to take some sedative; that at least holds my tears back...."

"And I recalled besieged Leningrad, while listening to Valentina," Alexandra Fyodorovna said. "Especially one occasion which stunned all of us. We were told about some elderly lady who opened the window every day and tipped out a saucepan of water into the street, reaching out farther and farther with every throw. We thought at first that perhaps she was insane—

so many things happened during the siege—but then somebody went to her and asked her what she was doing. And just listen to what she said. 'If the Nazis enter Leningrad and step upon my street I'll scald them with boiling water. I'm old and no longer capable of doing anything else, so I'll scald them with boiling water.' And she kept practising....She was an educated lady, and I remember her face even now.

"She chose that method of opposition that was in her power. You have to picture that period....The enemy was already close, battles were fought by the Narva Gates, and the shops of the Kirov Works were being shelled.... There were many people who sought any way to join the fighting even when it seemed that they were not capable of much. That was what impressed us most of all...."

"I returned from the front a cripple," Valentina Pavlovna continued her story. "I was wounded in the back by a fragment. The wound was not large, but I was thrown far away into a snowdrift....So when they found me, my legs had been badly frozen. Evidently I had been buried under the snow; but I was breathing and that had formed a kind of pipe through the snow. I was discovered by dogs. They dug in the snow, and brought my hat to the orderlies. In it was my identification card. Everybody had such cards—with the names and addresses of relatives to be informed in case of death. They dug me out and put me on a waterproof cape. My sheepskin coat was all soaked in blood....But nobody paid any attention to my legs....

"I stayed in hospital for six months. They wanted to amputate one leg above the knee, because gangrene had already set in....It turned out that my ward doctor was against the amputation. He had proposed another method, which was new at the time: to introduce oxygen under the skin with a special syringe. Oxygen feeds....Well, I can't tell you exactly, I'm not a medical specialist....

"And that young lieutenant managed to persuade the hospital head. They did not amputate my leg, but began treating me according to the new method. And two months later I already started walking—on crutches, to be sure.... After I was discharged, I was entitled to convalescence leave. But what kind of leave could I have? Where could I go? To whom? So I went back to my unit, to my gun. There I joined the Communist Party. At the age of nineteen....

"I celebrated Victory Day in Eastern Prussia. There had already been a lull in the fighting for a couple of days; nobody was shooting. Suddenly in the middle of the night, there was an air alarm. We all jumped up. And then we heard, 'It's victory! They've surrendered!' That they had surrendered was very good, but the main thing which we grasped at once was that it was victory. 'The war's over! The war's over!' Everyone started firing whatever was to hand—a submachine gun, a pistol.... And a gun was fired, too.... Some men were wiping away tears, others were dancing. 'We're alive! We're alive!' Afterwards our commander said, 'Well, you won't be demobbed until you've paid for the shells. What have you done? How many shells have you fired?' It seemed to us that peace would reign in the world forever, none would ever want another war, and that all the shells must be destroyed. So that people should never even speak about war.

B. Guards Junior Sergeant Tamara Stepanovna Umnyagina, Medical Orderly

"I remember running to the military registration and enlistment office in a coarse cotton skirt and white plimsolls, buckled-like shoes—they were the very last word. So there I was in that skirt and those shoes, asking to be sent to the front. I arrived at my unit—it was a rifle division, based near Minsk—to be told that there was no need for seventeen-year-old girls to fight; the men would feel ashamed, they said. The enemy would soon be crushed and I should go home to mummy. I was upset at not being allowed to fight, of course. What did I do? I managed to see the chief of staff. Sitting with him was the same colonel who had rejected me. 'Comrade Commander who is more senior,' I said, 'permit me not to obey Comrade Colonel. I shan't go home in any case, I shall retreat with you. Where should I go—the Germans are already close.' That was what everyone called me after that—'Comrade Commander who is more senior'. It was the seventh day of the war. We began to retreat....

"Soon we were bathed in blood. There were very many wounded, but they were so quiet, so patient, they wanted to live so much. No one believed that the war would last so long; we expected it to end at any moment. I remember that everything I had was soaked in blood—through to the skin.... My shoes were torn and I went about barefoot. What did I see? Near Mogilev a station was bombed. A trainload of children was in the station. The adults began to throw them out of the carriage windows, little children—three or four years old. There was a wood nearby and they ran towards it. Immediately tanks began to advance and ran over the children. There was nothing left of them.... Even today that picture could make you go out of your mind.

"But the most terrible experience of all was still to come: Stalingrad, that was most terrible.... There was no battle-field at Stalingrad, the entire city was the battlefield—streets, houses, cellars. You just try to carry a wounded man out of there! My body was one big bruise, my trousers were covered with blood. The sergeant-major told us: 'Girls, there are no more trousers, so don't ask for them'. But our trousers were covered with blood; when they dried they stood up, stiffer from blood than they would have been from starch—you could cut yourself. There wasn't a clean centimeter anywhere and in spring there was nothing to exchange for summer uniforms. Everything was on fire; on the Volga, for example, even the water was on fire. The water did not freeze even in winter, but burned. Everything burned.... In Stalingrad there wasn't a gramme of earth that wasn't soaked in human blood.

"Reinforcements would arrive. Such young, handsome fellows they were. And in a day or two they would all be dead, not one of them left. I was beginning to be afraid of new people. Afraid of remembering them, their faces, their talk. Because no sooner had they arrived than they were gone. This was 1942, you see—the hardest, most difficult period. On one occasion ten of us were left out of three hundred at the end of the day. And when we were down to ten and the battle subsided, we began to kiss each other and weep because we had happened to survive. . . .

"When everything was over in Stalingrad we were given the task of evacuating the most seriously wounded in ships and barges to Kazan and Gorky. It was already spring—March and April. But we found so many more wounded, in the ground, in trenches, in dug-outs, in cellars—there were so many of them, it's impossible to tell you. It was horrible! We had always thought as we carried the wounded out of the battlefields that no more of them were left, that they had all been sent away and there were none in Stalingrad itself, but when the fighting ended there were so many of them that it could be hardly believed. . . . In the ship I sailed on there were people without legs or arms and hundreds who were suffering from tuberculosis. Not only had they lost themselves physically, but they suffered mentally, they agonised inwardly. We had to treat them, persuade them with gentle words, calm them with smiles. . . .

"After the war I could not escape from the smell of blood. It pursued me for a long time. As soon as summer came I thought that war was about to break out. When the sun warmed everything—trees, houses, asphalt,—it had a smell, for me everything smelled of blood. Whatever I ate, whatever I drank, I could not escape from that smell! I would make up the bed with clean linen and to me even that would smell of blood . . ."

170. Marlene Dietrich
Entertaining the Allied Troops

Singer and actress Marlene Dietrich (1901–1993) appeared in several silent films before her film career was launched with *The Blue Angel* (1930). This was also the last film she made in Germany. In this same year, Dietrich came to Hollywood where she became famous for such films as *Morocco* (1930), *Touch of Evil* (1958), *Witness to the Prosecution* (1958), and *Judgment at Nuremberg* (1961). During World War II, Dietrich and Lotte Lenya[1] sang popular American songs in German which were broadcast as part of the American war propaganda effort. As a member of the U.S. Army, Dietrich also entertained American troops in Europe and North Africa. She routinely performed from the back of a jeep when theaters were not available and, on one occasion, even from behind enemy lines. In this September 1944 photograph, Dietrich appears in a canteen in London. She did not return to Germany after the war, preferring to live in the United States and Paris instead. In his documentary *Marlene* (1984), Maximilian Schell asked Dietrich about her decision to join the U.S. Army and subsequent criticisms that she had betrayed Germany. Dietrich replied that the decision was not at all difficult:

---

[1]On Lotte Lenya, see Source 145 in Chapter 12.

Marlene Dietrich, September, 1944. Entertaining in a London canteen. *(Hulton-Getty/Liaison Agency.)*

We wanted to finish the war as quickly as possible. We didn't know a thing about politics. But naturally we were against the Nazis—of course we were. We knew about the concentration camps, children being gassed, etc. We knew all about that. So it wasn't difficult to decide....I couldn't help the fact that it was Germany. You see I didn't know just how much the Germans actually knew. But at any rate they were following Hitler. That's all I knew....And, as I know the Germans, after all, I am a German myself, well, they wanted their Fuhrer and they got their Fuhrer.[2]

_____

[2]*Marlene* (1984), a Karel Dirka/Zev Braun Production directed by Maximilian Schell.

QUESTIONS FOR ANALYSIS

1. How does Nella Last's diary reveal her views and experiences concerning the major turning points of World War II in Britain? For historians, what is the significance of a diary maintained by an "average housewife" in Britain?
2. Discuss the imagery of the 1941 Soviet poster. On what basis does it appeal to women? What is the significance of the caption's identification of fascism as the enemy rather than Germany?
3. Compare and contrast the experiences of Soviet women in World War II. How did they regard their roles in the defense of the Soviet Union?
4. Compare the homefronts in Britain and the Soviet Union. How do Nella Last's diary and the accounts of Soviet women illuminate the relationship between the battlefront and the homefront?
5. What was the role of entertainment in the homefront? How does Marlene Dietrich explain her decision to join the U.S. Army? Discuss her explanation of Hitler's appeal to the German people.

CHAPTER 14

Women and Postwar Europe, 1945–1980

Hopes for a new, peaceful, and just post–World War II order were embodied in the formation of the United Nations in 1945 and in the UN's Universal Declaration of Human Rights of 1948 which defined human rights in the broadest terms to include economic and social as well as political rights.

With the division of Europe, postwar European construction assumed two basic forms: the construction of welfare states in the West and socialism in the East. In both the East and the West, however, domestic as well as foreign policy were shaped by the politics of the Cold War which lasted from 1945 until the collapse of socialism in Eastern Europe in 1989. The division of Germany in 1949, in particular, became the focal point of Cold War policies. The governments of Western Europe aimed to differentiate themselves from socialism and central planning which characterized the Soviet Union and the countries of Eastern Europe. But the exigencies of postwar reconstruction required a continuation and even expansion of the economic and social planning implemented during World War II. Although Winston Churchill had led Britain throughout all but the first months of the war, his Conservative Party was defeated in the general elections held in the summer of 1945, in no small measure due to its refusal to acknowledge popular support for governmental economic and social intervention in the postwar period. Embracing Keynesian economic policies, the Labor Party government of Clement Attlee (1945–1951) nationalized key sectors of British industry and implemented the social welfare proposals detailed in Sir William Beveridge's report, *Social Insurance and Allied Services* (1942). Beveridge envisioned a "New Britain" that was "free, as free as humanely possible, of the five giant evils, of Want, of Disease, of Ignorance, of Squalor and of Idleness."[1]

Decolonization, sometimes carried out under the auspices of the United Nations, also transformed Europe in the immediate postwar period. India first achieved independence from Britain in 1949, following decades of massive protests that included campaigns of civil disobedience. The most violent and protracted war of national liberation took place in Algeria. Inspired by the 1954 defeat of French forces at Dien Bien Phu in Indochina, militants in Algeria initiated terrorist attacks on French forces in their own country and the struggle for independence soon gained mass support. In France, the protracted nature of the Algerian War (1954–1962), combined with revelations concerning the routine use of torture by the French forces, mobilized students and prominent intellectuals in support of Algerian independence. Up to this point, the French Communist Party had enjoyed enormous prestige due to its role in the Resistance, but its failure to support the right of Algeria to full independence created the conditions for the emergence of the French New Left.

Student, peace, and women's movements dramatically altered the political landscape of the United States and Europe in the 1960s and 1970s. The heterogeneous organizations of the New Left were galvanized by their opposition to the Vietnam War, the American government's continuation of the failed French colonial war in Indochina. Anti-imperialist in its broad outlook, the New Left allied with Third World liberation and civil rights movements and challenged the ethical and ecological implications of mass, consumer society. In 1968, massive student and workers' demonstrations and strikes erupted throughout Europe, and most dramatically in France. Second-wave feminism, so-called in recognition of the first wave in the nineteenth century, arose from the failure of the New Left to adequately address women's issues within its overall critique of society. Women formed caucuses within New Left organizations as well as autonomous groups to mobilize around discrimination against women in education and employment; women's roles within the family; the sexual double-standard and reproductive choice; the lack of female political representation; and the marginalization of women in academic programs, both as educators and as subjects of inquiry.

---

[1]Oxford University address given by Beveridge on December 6, 1942 shortly after the publication of his report and anthologized in William H. Beveridge, *The Pillars of Security and Other War-Time Essays and Addresses* (London: George Allen and Unwin, 1943), pp. 80–96.

THE FAMILY ALLOWANCES ACT
AND THE BRITISH WELFARE STATE

The Family Allowances Act of 1945 represented the culmination of a campaign for family allowances, also referred to as endowments, initiated by Eleanor Rathbone and the Family Endowment Council, later called the Family Endowment Society. Family allowances paid by the state were an integral aspect of the post–World War II British welfare state created by Labor Party legislation which also included the National Insurance and National Health Service Acts of 1946 and the National Assistance Act of 1948. Family allowances were paid to mothers on a weekly basis for the support of two or more children. The payments were less than the sums recommended by Rathbone and Sir William Beveridge, who served for a time as president of the Family Endowment Society. The Family Allowances Act did, however, establish the principle of a state-funded system of assistance for the care of children, which Beveridge described as the most innovative aspect of post–World War II welfare legislation.

171. Eleanor Rathbone
The Case for Family Allowances

Eleanor Florence Rathbone (1872–1946) waged a campaign for family allowances for some thirty years. Allowances paid by the British government to soldiers' wives during World War I provided her with a model for family allowances during peacetime. Rathbone formed the Family Endowment Council in 1917 and subsequently published *The Disinherited Family* (1924).[1] As president of the National Union of Societies for Equal Citizenship (NUSEC), she won support for including family allowances and free access to birth control information in the organization's program. Rathbone also lobbied for family allowances as a member of the Liverpool City Council and in Parliament where she served from 1929 to 1945 as one of its first female members. Feminist as well as eugenicist views informed Rathbone's advocacy of family allowances. She argued for the payment of allowances by the government directly to mothers and for the inclusion of all children, both within and outside of wedlock, as recipients. Rathbone emphasized that family allowances paid for each child were essential to eliminating the economic dependence of mothers on husbands or male providers. At the same time, as a member of the Eugenics Council, to which Beveridge also belonged, she aimed to encourage several children among the higher socioeconomic groups, in particular, and the NUSEC also advocated sterilization, if necessary, to prevent the "unfit" from reproducing. Rathbone believed that the Family Endowment Society should include all supporters of

---

[1] The Family Endowment Society published a posthumous edition in 1949 as *Family Allowances*. In his epilogue, Beveridge credited Rathbone's original 1924 edition for his own conversion to the idea of family allowances.

family allowances irrespective of their viewpoints. This approach increased the mass support for allowances, but the issue of ensuring women's economic independence was eclipsed as a result. In *The Case for Family Allowances* (1940), Rathbone reformulated and condensed her earlier ideas in order to reach a popular audience. In these excerpts from Chapter 3, "Remedies," she systematically critiques the alternatives most commonly proposed in place of family allowances.

THE REMEDIES

...What are the remedies? There are four which are commonly put forward and may be considered as possible alternatives to that of direct provision for children through Family Allowances. These are:—

1. To rely on the wage-system, raising wages and salaries through the ordinary methods of collective bargaining aided by labour legislation, so that they shall be sufficient to enable men to keep their families at the standard normal in their occupations and grades. Or, in the precise form put forward by Mr. Rowntree[1] and at one time favoured in labour circles, to secure at least a minimum wage which will cover the "human needs" of a "normal" or "average" family supposed to consist of man, wife and three children.

2. To meet the extra temporary needs of child dependency through an extended system of communal services in kind—school meals, milk, etc.

3. To ask individuals to solve the problem for themselves by producing no more children than they can adequately maintain on their actual incomes.

4. To keep down the cost of living by controlling prices, whether by Government subsidies or by limiting profits. This proposal is usually put forward only as a war-time measure and has only come into general discussion since war became a certainty.

Let us see how far these devices are likely to provide a solution.

1. Why "a Living Wage" Is Not the Cure

...I shall proceed to show:—

First, that even if such a "living wage" were achieved, a large proportion of the families with children would still remain undernourished and in poverty;

Secondly, that it has never yet been achieved in this country even in the most prosperous years, nor—allowing for differences in standards of "human needs"—in any other country;

Thirdly, that it has no prospect of being achieved within measurable distance of time.

The fact is that there is no such thing as a "normal family". At the time of the 1921 Census (that taken in 1931 was in a form which yielded no exactly comparable figures) only 6.2 per cent. of the men over 20 in England and Wales were married with just three children under 16 years of age; 60.6 per cent. had no such children; 26.5 per cent. had fewer and 6.7 per cent. more than three. The average number of children per man was less than one, actually 0.88.

But the families with over three children, though a small group, included 37 per cent. of the children, and Mr. Rowntree calculated that no less than 54 per cent. were members of such families during five or more years of their childhood. During the subsequent twenty years the falling birth rate has reduced still further the proportion of children per man, and the results of a recent survey of York by Mr. Rowntree indicate that a wage based on the needs of a family with three dependent children would result in only 5 per cent. receiving just what is necessary, while 91 per cent. would get more and 3.9 per cent. of the families less than

---

[1]Seebohm Rowntree carried out extensive surveys of the living conditions of working-class households with the aim of determining minimum wages as well as the level at which insufficient income reduced a household to a level of poverty requiring assistance. The "minimum" wage was based on an "average" household's expenses in rent, food, clothing, and miscellaneous expenditures.—Ed.

enough to cover their basic needs. But this last group would include at any one time 23 per cent. of the children and, as before, a much larger percentage for part of their childhood.

Many people will reflect that even if the normality of the five-member (three children) family is a fiction bearing little relationship to the facts, it would be all to the good if adherence to it did produce a surplus for 91 per cent. of the families concerned. No one, in a wealthy country like ours, ought to be compelled to live at subsistence level. This perfectly true reflection probably explains the tenacity with which a section of the Labour movement, and especially of the trade unionists, have clung to the objective we are now considering, ignoring its unpleasant results for families. To put it plainly, some of them have hoped to repeat the success of a hundred years ago, and to win the battle for higher wages, as their forefathers won the battle for shorter hours and better hygiene in factories, from behind the skirts of the women and children. Believing, as I think rightly, that they were in justice entitled to a larger share of the product of industry than they were receiving, they have sought to strengthen their claim by urging the needs of "our wives and families."

But we are entitled to ask how far they have succeeded in this legitimate aim and what greater measure of success may be expected in the future. And if the answer is unsatisfactory, does not that perhaps indicate that there may have been mistakes in the tactics adopted and that it might be better to try another road?

The facts as to the distribution of the product of industry are certainly disquieting to anyone possessing a social conscience, whether he himself belongs to the "haves" or the "have-nots". Consider the position as summarised by Mr. Colin Clark,[2] perhaps the greatest of the younger experts on this subject:—

One tenth of the whole working population (those with incomes over £250) take nearly

half of the national income, and a small class comprising 1½ per cent. of the population (with incomes over £1000) take one quarter.

The share of wages in the national income oscillates with the trade cycle but has shown little change in the last 25 years.

Elsewhere he shows that wages now claim an extra 2 per cent. of the product as compared with twenty-five years ago.

Two per cent. increase in twenty-five years! Yet those years have witnessed the rise of the Parliamentary Labour Party, two short Labour Governments and an immense increase in the membership and influence of the trade unions, the chief of whose many functions is to safeguard and raise standards of living and of remuneration. Have, then, their labours been in vain? That would indeed be a rash conclusion. The position and standard of life of the workers in continuous employment have improved substantially during that period. . . . But we have to remember that the past twenty years have been marked by much more widespread unemployment and short time, continuous though fluctuating in intensity, than anything experienced during the years before 1914. The great majority of the workers have suffered from this themselves for shorter or longer periods, or through relatives whom they have helped to support, and very many have been left with exhausted savings and homes stripped of their plenishings. It remains true, however, that most workers who have been lucky enough to retain their jobs during recent years draw bigger pay than their predecessors for shorter hours worked under less uncomfortable conditions, and can buy more, though at higher prices. They and their families benefit also by much-improved social services, paid for largely by themselves through indirect taxation. But these improvements are mainly the result of the vastly improved productivity of labour, in this and other countries, due to scientific discoveries and improved methods.

Thus the cake to be divided is a larger cake and the share of it which falls to be consumed by the wage-earners is a larger share—actually though not proportionately, except perhaps to

[2]*National Income and Outlay,* pp. xiii and 96. (Macmillan & Co., 1938.)

the extent of that modest 2 per cent. And for that measure of success the efforts of the organised workers and their leaders are no doubt entitled to much of the credit. Without their efforts, the wage-earners might not have been able even to keep their footing on the slope of distribution. They might have slipped farther down. And they might have achieved even more if more of the workers had been active and loyal trade unionists.

But for all that, it is not a very magnificent result, not when we remember that quarter of the child population which is still being reared on 4s. worth of food or so per week. Many of them, it is true, are the offspring of parents who have been rash enough to indulge in more than three children, and so come outside the objective of the theory of the "living wage". But not all; nor has that objective come anywhere near general attainment. Mr. Rowntree himself has always been an obstinate adherent of the theory, while proposing to provide for the children in excess of three, through family allowances. In 1936 he recalculated the minimum cost of the "normal" family at 53s. weekly for urban and 41s. for rural workers (equivalent in April 1940, to 63s. and 49s. respectively)—that at a standard which allowed no fresh milk, no butter, the cheapest margarine, only one egg a week and home-baked bread. He estimated that baker's bread would cost another 1s. to 2s. and that 6s. should be added to bring the diet up to the optimum standard laid down by League of Nations health experts—i.e., at this standard the minimum to-day would be 70s. for urban and 56s. for rural workers. Compare those figures with the following estimates of actual earnings:

Mr. Rowntree himself reckons that four out of every ten adult male urban workers were earning in 1936 less than 55s. and one in every three less than 50s.

The minimum enforced by the Agricultural Wages Board in 1937, varying in different areas, averaged 33s. 4d.

Half the male applicants for unemployment assistance in 1937 were earning less than 50s. in their normal occupations.

The railwaymen's claim for a 50s. minimum was rejected by the employers up to the outbreak of War, but later accepted for London with a minimum of 48s. for the provinces....

Five shillings per family per week certainly would not go very far to abolish poverty. But supposing that, instead, the addition to the income of every family was 5s. for each child under 15—the proposal I am about to defend, with the powerful support of Mr. Maynard Keynes—is that not a much more attractive proposition, as well as being one realisable by a very much less impracticable form of redistribution than that supposed above?...

2. Why Communal Services Will Not Meet the Need

The section of trade-union opinion which has looked askance at Family Allowances has unquestionably been influenced partly by the fear that such allowances, even if paid for wholly by the State would "affect detrimentally negotiations regarding wage-fixing," while communal services in kind are believed to be less likely to have that effect. This was the opinion expressed, though they explicitly refused to argue it, by a minority of three out of twelve members composing a Joint Committee of the T.U.C. and the Labour Party Executive appointed in 1927 to consider Family Allowances and cognate subjects related to the Living Wage. After an exhaustive enquiry lasting nearly a year, the majority of nine had:—

> arrived at the conclusion that the most valuable step that can now be taken to further the welfare of the nation's children is the institution of a scheme of Family Allowances, to be paid in cash to the mother.

They recommended a State-paid scheme, limited to children of families below income-tax level, at the rate of 5s. for the first and 3s. for each subsequent child. But the General Council of the T.U.C., by sixteen votes to eight, preferred the Minority Report, and their verdict was endorsed after a brief debate at the Trade Union Congress and subsequently acquiesced in by the Labour

Party. That Report, be it noted, did not reject cash allowances on principle or outright. In fact, it recommended that these should be paid "for the first year or two after birth"; but for the rest it preferred, partly on the ground of financial stringency, to concentrate what money might be obtainable on social services.

But is there in fact any truth in the view that communal services are less likely than cash allowances to affect unfavourably the wage bargain? Is not just the opposite more probable? What is certain is that the average member of the well-to-do classes is led by the very vagueness of the term "social services" to greatly overestimate—not their general utility—but the extent to which they relieve financial obligations which would otherwise have to be made by the wage-earners. Hence, when his attention is drawn to the facts of poverty, he salves his conscience by reflecting that "the poor have so much done for them".[3]

Let us consider therefore how far social services actually affect the question of child maintenance or could be made to do so if their provision were extended. Improved educational, medical and recreational services are clearly beside the point. They do nothing to lighten the burden of providing children with basic necessaries, but rather increase it, by extending the period of dependency and giving the children higher standards and bigger appetites. Housing subsidies only benefit the small minority dwelling in houses owned by local authorities, and unless the authority is one of the few which have had the wisdom to concentrate the subsidy on giving rebates on rent to tenants according to the number of their children or below a certain income level, the houses are usually too expensive for the poorer or larger families.

The only existing social services relevant to our subject are school meals and milk and the supply of milk by Health authorities to mothers and infants. These services might with advantage be considerably extended and there might perhaps be added the provision of school uniforms and school holidays. Before the War, slightly under 2 per cent. of school children on any one school day were getting free school dinners. With the help of a State subsidy of £660,000 per annum,[4] slightly over half the school population received a third of a pint of milk daily for a halfpenny; a minority of them free on a health and means test. The Children's Nutrition Council, an all-party body of which I am Chairman, has for some years been urging that all school children should get their milk ration free—a bigger one if they can take it. We also asked that the meals should be widely extended and supplied free to those below a certain income level and that canteens should form part of the equipment of all new schools. We have further asked that the present meagre provision of free or cheapened milk to mothers and infants through local authorities, which works with very varied success and in some areas not at all, should be replaced by a much larger and more generous scheme. But though these reforms were supported by nearly every important organisation in the country concerned with child welfare, they were not conceded. . . .[5]

But suppose they were achieved, the provision of one meal on school days does not cover Sundays and holidays and, taking the average value of the meals as fourpence, is equivalent only to an allowance of 2s. every school week. It does not benefit the children under school age, nor those in the expensive period when the child is beginning his or her industrial life but is not self-supporting. And communal meals, if provided by a paid staff in properly equipped buildings instead of, as at present, through all sorts of makeshift arrangements at the cost of much labour and discomfort to the teachers, are a doubtful economy

---

[3]I have scarcely ever spoken to a middle-class audience on Family Allowances without having that remark hurled at me by someone every line of whose body and raiment testified to generous living.

[4]October 1938–September 1939.

[5]Since this was written, the provision of free or cheapened milk has been substantially extended.

compared with money allowances to the parents, though on other grounds there is much to be said for them. After all, the home fire has usually to be kept burning and the home meal prepared by an unpaid mother for herself and the other members of the household. In spite of the greater cheapness of large-scale buying, comparisons between the cost of institutional feeding and what an efficient working-class mother actually spends, when she can afford it, on serving nourishing meals are seldom favourable to the former kind of provision.... In their Penguin Book on *Our Food Problem*, Mr. Le Gros Clark and Mr. Titmuss reckon that if every school child on every school day received an ample and varied meal and also a pint of milk, it would probably cost the country about £40 million a year. For little more than that, according to calculations recently made by a competent statistician, the mother of every working-class child could receive 5s. a week for every child in her family under 15 years of age except the first. Which do you suppose the mother would prefer?

I have not much doubt, for, to say nothing of the one-third of the children who are below school age and the days when school does not meet, it must be remembered that a child needs other meals besides dinner and other things besides food....

3. Limitation of Families Not the Solution

The third solution of our problem—one which finds favour especially with the more unthinking members of the well-to-do and indeed of all classes—is that of later marriages and limitation of births. In the early days of the movement for Family Endowment, as we then called it, perhaps the most formidable argument we had to meet was that based on the belief that Great Britain was overpopulated and that allowances for children would lead to an increased birth rate, especially among the poorer parents and those least able to give their children a desirable environment. Our reply was that this could hardly happen, because this class did not practise birth control and already produced almost as many children as Nature permitted. Restriction of families was already

taking effect, but only in the upper, middle and artisan classes and among the abler, more ambitious and more far-sighted of the parents. Of all this there was ample proof.

Since then the position has changed. Knowledge of contraceptive methods, though not universal nor the methods universally practised, has spread steadily downwards. The birth rate, with only slight fluctuations, continued to fall until 1933, and since then has remained fairly steady. Though this country, or at least the urban part of it, may be said to be overcrowded, the fear of all who have seriously studied the question is now that the population may decline to an extent which will menace our prosperity and our security as a nation and the headquarters of an empire. The danger is still in the future, but it is coming unpleasantly near. Here are some of the facts.

We have already about the lowest birth rate in Europe—lower than that of France, which formerly led the race towards national suicide. In the words of Mr. Carr Saunders, one of the chief experts on the subject and now Director of the London School of Economics, "we are not only not reproducing ourselves but are between 25 and 30 per cent. below replacement rate".... But the War is likely to affect the figures. It is estimated that the war of 1914–18 resulted in about half a million fewer babies being born than would have otherwise seen the light.

Nor do these changes in the character of the population concern only a distant future. The number of children under 15 in England and Wales has already fallen from a peak of about 11 millions in 1911 to just under 9 millions, while the number of people over 65 has almost doubled. The proportion of old to young is so changing that whereas half a century ago children under 15 were 35 per cent. of the population and old people under 5 per cent., to-day the corresponding proportions are 22 and $8\frac{1}{2}$ per cent., and it is estimated that by 1971 the old will probably slightly outnumber the children....[6]

---

[6]These figures are taken from a P.E.P. broadsheet of April 1940.

Some readers may reflect that an England with half or even with a tenth of its present population would be a much pleasanter place. But that is doubtfully true when one remembers how closely the phenomena of decay in a nation resemble those of decay in the individual. A country of disused factories, closed shops, empty houses, many bathchairs and few perambulators might not be a particularly cheerful place. Nor are families with one or two children the happiest or the most wholesome kind of families for the children or the parents. Also, a nation with powerful neighbours who have not chosen to decay casting envious eyes on its empty spaces at home and abroad may not be a very safe nation to belong to, as Sweden—once an Empire—is finding now. Let future generations, you say, look after themselves. But we owe it to posterity at least to reflect that it cannot be good for the world nor for ourselves that the proportion of the white races to the coloured and of the Anglo-Saxon race to all others should be a steadily dwindling proportion. One may admit that without being either an imperialist or a militarist.

But the danger does not lie only in a remote future, as our enemies themselves have noted:—

> The German *Army Year Book* for 1937 pointedly remarks that 'England's position in the world through the way her population is developing is seriously and almost irretrievably threatened' and the writer goes on to comment that by 1950 Germany will have at her disposal 12,994,000 men of military age between twenty and forty-five, whilst Great Britain will have only 8,721,000.[7]

4. Price Control or Higher Wages Not the Solution

I have supplied evidence of the extent of poverty and malnutrition and of the declining population during the past decade.... These conditions are changing too rapidly for exact measurement. But already the effects of the rising cost of living have been brought home to everyone. At the end of the first seven months of the war, retail prices had risen, according to Board of Trade figures, by 17 per cent. There is reason to think that 20 per cent. better represents what the working housewife actually pays. This has happened in spite of a Government subsidy which is costing the State about £60 million a year for the purpose of pinning down the prices of only four commodities—meat, milk, bread and flour—to the level they had reached by the end of 1939....

Moreover, even if price control can be strengthened and extended to all necessaries, whether by Government subsidy or by restricting profits, it has some grave disadvantages. First; the better it effects its own purpose, the more it tends to defeat another purpose which the Government have in mind—namely, the cutting down of all unnecessary buying so as to free more ships, more labour and effort for the importing or production of those things which are necessary to winning the war and for the exporting of commodities which will help to pay for our imports. Secondly, control of prices, though in some forms it may be necessary to stop profiteering, obviously helps most those who can afford to purchase most—namely, the well-to-do. Thirdly, even if limited to necessaries, price control favours the purchase of luxuries by freeing more money for it. If the wealthy housewife pays less for her bit of sirloin, she can better afford a dish of asparagus to follow it.

By contrast with all that, if whatever the Government could afford to spend on encouraging rather than restricting consumption were concentrated on Family Allowances, the benefits would go to those who can least afford to cut down their purchases without incurring socially injurious results, and who normally spend and would continue to spend the highest proportion of their incomes on necessaries rather than luxuries—namely, the families with dependent children.

The same range of objections applies to the other possible method of counteracting higher prices—that is, by increased wages and salaries. That too is likely to work out on the principle of "To him that hath shall be given"....

---

[7]*Our Food Problem,* by F. Le Gros Clark and R. M. Titmuss (Penguin Special).

5. Family Allowances the Solution

While again repudiating exaggerated claims for this reform, this at least we do claim: that Family Allowances are not only desirable both in peace and in war-time, but are achievable here and now, even more achievable in war than in peace because more visibly necessary; further, that when achieved they will go far to drain the morass of extreme poverty which disgraces our land and will help to check most of the other evils which have been touched on in these pages. . . .

QUESTIONS FOR ANALYSIS

1. How does Eleanor Rathbone critique the four commonly proposed alternatives to family allowances: a "living wage," communal services, limitation of families, and price controls or higher wages?
2. Compare Rathbone's views on women and the family with those of Alexandra Kollontai, Source 141 in Chapter 11. What political and economic factors account for their differing viewpoints?

WOMEN AND THE ALGERIAN WAR

The most protracted war for national independence waged in the post–World War II period, the Algerian War (1954–1962), profoundly transformed French politics, brought about the collapse of the Fourth Republic government, and led to the development of a New Left in France. The FLN (National Liberation Front), comprised of militant Arabs and Berbers seeking full independence from France, launched its first armed attack on colonial authorities just six months after the French defeat at Dien Bien Phu, Indochina, in 1954.[1] Unlike the colonial war in Indochina (1946–1954), the Algerian War involved conscripted French soldiers and a large population of over a million European settlers known as the *pied noirs.* The settlers consistently obstructed attempts by the French government to reach a negotiated settlement with the FLN that would undermine French rule over the nine million Muslims in Algeria. In May 1958, settlers and army officers carried out a military coup in Algiers when they suspected the government's intent to negotiate a settlement with the FLN. The coup toppled the Fourth Republic and Charles de Gaulle agreed to form a new government on the condition that the French could write a new constitution granting him sufficient authority to resolve the Algerian War. In 1959, President de Gaulle proposed to allow the Algerian population as a whole to determine its own future, but this was rejected by both the FLN and the settlers. In France, an emerging New Left organized demonstrations in support of full independence for Algeria and self-determination for the Muslim population. Public opinion in

---

[1] By this time, the United States was providing nearly eighty percent of the war materials for the French forces in Indochina. It continued to carry out covert military activities in the region which erupted into full-scale war in the 1960s.

France increasingly turned against the war as a consequence of its reversal of the ten-year postwar economic miracle, constant political crises, the French military's failure to quash the rebellion, and publicized accounts of torture routinely carried out by the military and police against the Muslim population. In 1960, a group of French women formed the Djamila Boupacha Defense Committee in Paris and brought international attention to the case of Djamila Boupacha, a young Algerian girl tortured by the French military.

172. The Case of Djamila Boupacha

The case of Djamila Boupacha, a young Algerian woman tortured by the French military, dramatically exposed the nature of the war France was waging against Algerian independence. A courier for the FLN (National Liberation Front), Boupacha, her brother-in-law, and father were arrested in February 1960. Boupacha was repeatedly tortured in a prison cell where she was held until officially charged in May with "attempted wilful murder and consorting with malefactors." The charges alluded to her association with the FLN and a bomb placed in an Algiers University restaurant several months earlier. Although evidence was never presented to link Boupacha to the bomb, which was defused, she "confessed" to the crime when threatened with additional torture. Boupacha was represented by the Tunisian-born attorney Gisèle Halimi[1] who had previously defended political prisoners in Tunisia. Halimi advised Boupacha to retract her forced confession and to file a civil indictment against the French military authorities. On Halimi's request, Simone de Beauvoir[2] chaired the Djamila Boupacha Defense Committee which launched a campaign to bring Boupacha's case to the attention of the general public in France and abroad as well as high political officials in France, including President Charles de Gaulle and Minister of War Pierre Messmer. The Defense Committee succeeded in moving Boupacha's trial from Algiers to metropolitan France and, in December 1960, the judge of the High Court of Caen began to investigate, often encountering obstruction from French authorities in Algeria. To further publicize the case, de Beauvoir and Halimi co-authored *Djamila Boupacha* (1962), translated in English as *Djamila Boupacha: The Story of the Torture of a Young Algerian Girl Which Shocked Liberal French Opinion* (1962). The Algerian War was ended by the Evian agreement of March 1962, which granted amnesty for all Algerian political prisoners. Boupacha and her family were released in April, but the agreement also terminated her case against the military authorities in Algeria. The two documents reproduced here are the full text of Boupacha's civil indictment and excerpts from de Beauvoir's introduction to *Djamila Boupacha*.

---

[1] For additional information on Halimi, see Source 174 in this chapter.

[2] For background information on Simone de Beauvoir, see Source 175, also in this chapter.

A. The Text of Djamila Boupacha's Civil Indictment

During the night of 10/11th February, 1960, a party of about fifty *gardes mobiles, harkis*[1] and police inspectors drove up to my parent's house at Dely Ibrahim, Algiers, in jeeps and army trucks, and dismounted there. One of them was Captain D——, *en second*[2] at the El Biar Centre. I was resident at my parent's house at the time. I was savagely beaten up there before even being taken away. My brother-in-law Abdelli Ahmed, who was present that evening, suffered a similar ordeal, as did my father Abdelaziz Boupacha, who is seventy years old.

All three of us were removed to the Classification Centre [*Centre de Tri*] at El Biar. There I received a second beating up, so violent that I was knocked off my feet and collapsed. It was then that certain military personnel, including a captain in the paratroops, kicked my ribs in. I still suffer from a costal displacement on my left side.

After four or five days I was transferred to Hussein Dey. This, I had been told, was where I would get a taste of the 'third degree'. I found out what this implied—firstly, torture by electricity. (Since the electrodes would not stay in place when affixed to my nipples, one of my torturers fastened them on with Scotch tape.) I received similar electrical burns on my legs, face, anus, and vagina. This electrical torture was interspersed with cigarette-burns, blows, and the 'bath treatment': I was trussed up and hung over a bath on a stick, and submerged till I nearly choked.

A few days later I was given the most appalling torture of all, the so-called 'bottle treatment'. First they tied me up in a special posture, and then they rammed the neck of a bottle into my vagina. I screamed and fainted. I was unconscious, to the best of my knowledge, for two days.

During the earlier part of my time in El Biar, I was brought into the presence of my brother-in-law Abdelli Ahmed, who also bore the most frightful marks of the beatings and tortures he had undergone. Nor was my father spared, despite his great age.

On 15th May, 1960, I was formally committed and charged with attempted wilful murder and consorting with malefactors. When brought before the examining magistrate I repeated the confession that had been forcibly extracted from me, under torture, by my inquisitors. I was then, and am still, severely shocked and shaken by my terrible ordeal. To my own sufferings must be added my father's experience—a most frightful shock for an old man—of seeing his twenty-year-old daughter still disfigured by the tortures she had endured.

My father is at present interned in the camp at Beni-Messous, but earlier his condition gave rise to such anxiety that he had to spend nearly a week in the Maillot Hospital.

My brother-in-law is under detention in the Civil Prison, Algiers, and his case is being dealt with separately from mine. Yet they are intimately linked: we were arrested on the same day, and the 'malefactors' we stand accused of 'consorting with' are the same men on the run. The reason for this separation is obvious. My brother-in-law and I are each a witness to the fact of the other's torture, and the authorities well might fear that if we were brought into a public courtroom together, we should testify to our common experience.

Though I chose Maître[3] Gisèle Halimi of the Paris Bar as my defending counsel several weeks ago, it is only today that she has been able to come and see me, since her visitor's permit for Algiers was (with singular restrictiveness) made valid for three days, only, ie *from 16th to 19th May, 1960.*

The facts adduced above constitute the crime of wrongful detention of the person, with aggravating circumstances as under, *in that the aforesaid wrongful detention was prolonged for over a month, and accompanied by 'physical torture'.* These crimes are covered, and penalized, by Articles

---

[1]*Harkis*: Algerians who fought on the side of the French military.

[2]*en second*: second in command.

[3]*Maître*: a general form of address for a lawyer.

341, 342, and 344 *in fine* of the new Penal Code. In the circumstances, *Monsieur le Juge d'Instruction,* I have the honour to lodge with you an indictment in respect of the aforesaid crimes, and hereby constitute myself plaintiff in the civil suit arising therefrom. *[Signed]* Boupacha. Detainee [No 1134] in Algiers Prison. Algiers, 17th May, 1960.

B. Simone de Beauvoir, Introduction to *Djamila Boupacha*

An Algerian girl of twenty-three, an FLN liaison agent, illegally imprisoned by French military forces, who subjected her to torture and deflowered her with a bottle: it is a common enough story. From 1954 onwards we have all compounded our consciences with a species of racial extermination that—first in the name of 'subjugating rebellious elements' and later in that of 'pacification'—has claimed over a million victims. Men and women, old folk and children, have been machine-gunned during 'mopping-up operations' [*ratis-sages*], burnt alive in their villages, had their throats slit or their bellies ripped open, died countless sorts of martyrs' deaths. Whole tribes have been bundled off to so-called rehabilitation or 'regroupment' centres, where they were starved, beaten, and decimated by exposure and epidemics. Such places are in fact death camps, though they have a subsidiary function as brothels for the crack regiments. Today more than five hundred thousand Algerians are confined in them. During the last few months even our most circumspect papers have been full of horror-stories: murders, lynchings, *ratonnades*[1] and man-hunts through the streets of Oran, dozens of corpses strung up from trees in the Bois de Boulogne, beside the Seine, in the very heart of Paris; endless cases involving maimed limbs or broken skulls. Algeria has become a second Haiti. Can we still be moved by the sufferings of one young girl? After all—as was delicately hinted by M. Patin, President of the Committee of Public Safety, during an interview at which I was present—Djamila Boupacha is still alive, so her ordeal cannot have been all that frightful....

In telling this story Gisèle Halimi is not attempting to convert those whose hearts still remain stubbornly impervious to the deep shame most of us feel because of it. The paramount interest of her book lies in its detailed exposure of a lying propaganda machine—a machine operated so efficiently that during the past seven years only a few faint glimmers of truth have contrived to slip past it. How many times have I been brought up short by the unanswerable gambit: 'Yes, but if all this was as widespread and ghastly and monstrous a scandal as you assert, surely it would be common knowledge?' But therein lies the whole point: for the scandal to reach such appalling heights, and attain such a degree of prevalence, it was vital that its central element should *not* be common knowledge. The use of torture has been publicly extolled by General Massu,[2] openly recommended to young officers, sanctioned by large numbers of churchmen, applauded by the European minority in Algeria, and systematically practised in prisons, barracks, *djebels*,[3] and the so-called 'transit camps' [*centres de tri*]. Thanks to this unanimity of opinion, it has proved easy to deny every individual allegation of torture. The exceptional thing about the Boupacha case is not the nature of the facts involved, but their publication. A happy conjunction of pride on the defendant's part and dogged persistence in her lawyer, coupled with the great professional courage shown by a certain judge, have made it possible to lift the curtain of darkness and misrepresentation behind which the widespread atrocities of 'war against subversion' have hitherto been concealed. It is true that one stumbling-block has survived all attempts at dislodgment, but at least it has become glaringly conspicuous in the process. By his public utterances General Ailleret, Commander-in-Chief of the Forces in Algeria (a post to which he was appointed by General De Gaulle), has made it quite clear that the Army actively opposes the exposure of Djamila's torturers....

---

[1]*Ratonnades*: summary executions.

[2]General Jacques Massu, a leader of the May, 1958 coup carried out by French settlers and army officers in Algiers when they suspected that the French government intended to negotiate a settlement with the FLN.

[3]*Djebels*: desert areas.

What of the victim who comes through... an ordeal of torture? Though found innocent and released, he remains gagged by threats the effectiveness of which he knows all too well. In the normal course of events, too, and as an extra security measure, his place of residence is laid down for him, and thus his gaolers guarantee that he will hold his tongue. If, on the other hand, he is found guilty, it is generally too late for him to lodge a complaint after the verdict has been pronounced. But, you may ask, does he not get a chance to speak out at the preliminary hearing? Far from it: he knows very well that if his testimony is dismissed, he will be 'interrogated' all over again. Sometimes his torturers are actually waiting for him outside the magistrate's court. In Algeria judges, doctors and barristers alike all regard such defendants as 'the enemy', and it is on this tacit professional collusion that the entire system rests. A verdict of 'guilty' is inevitable. The sentence has been decided in advance, and the only purpose of the trial is to camouflage the arbitrary nature of the proceedings. In this respect Djamila's case is most instructive. Brought up before a hostile magistrate, the burning mementoes of her 'interrogation' still visible on her flesh, Djamila, though in a state of abject traumatic shock after the savage treatment she had received, both stuck to her original testimony and openly declared: 'I have been tortured. I insist on a medical examination.' The judge made no attempt whatsoever to take up this point or cross-question her; he merely recorded her assertion in the transcript of evidence. Then he called in one of those doctors whose business it is to cover up for the judge when the latter wants to cloak his actions in a spurious appearance of legality. A few months later, however, several Paris doctors, when summoned to give expert evidence for the defence, all agreed that Djamila had been subjected to what they termed 'traumatic defloration'. Yet in Algeria [doctors]... are there, invariably, to *deny* that maltreatment has taken place, not to confirm it: they are simply performing their allotted role. In the same way, Algerian lawyers never dream of really *helping* their clients; even

if the idea occurred to them, they would be far too scared to put it into practice.... They ask nothing better than to collaborate with the Army authorities, the police, the courts, and most of the European population in defeating the enemy by fair means or foul. Djamila could expect no help from her Algerian defence counsel, who cheerfully told Gisèle Halimi: 'Open and shut case: ten minutes'll see it through.'

Thus Djamila came within a hair's-breadth of being condemned as so many others had been—that is, on evidence obtained under torture. No proof of her guilt was forthcoming. Those ghastly days she spent at El Biar and Hussein Dey would have had no official existence except in her memory....

Early in 1958 General De Gaulle, on being pressed to denounce the use of torture, proudly replied that this was an integral element of the old 'System', and would be abolished with the fall of the Fourth Republic. Indeed, Malraux[4] later proclaimed that, as from 28th May, it *had* been abolished. Yet here we find—after two and a half years of the Gaulliste régime—De Gaulle's Minister of War, together with the Commander-in-Chief in Algeria also a nominee of De Gaulle's, guaranteeing their subordinates complete immunity from the law, whatever they might have done. This more or less amounts to giving them public *carte blanche* for any excesses they might care to perpetrate, without the least danger to their 'morale', let alone their personal safety. For a long time now we have been in the habit of saying that though torture is not unknown in the Army, the Army as such does not approve the use of torture. But today such pious distinctions are no longer valid....

In such circumstances, moral indignation would be useless. To protest in the name of morality against 'excesses' or 'abuses' is an error which hints at active complicity. There are no 'abuses' or 'excesses' here, simply an all-pervasive system.... *Solitudinem faciunt, pacem appellant,*[5]

---

[4] André Malraux, a novelist and the Minister of Culture.

[5] *Solitudinem faciunt, pacem appellant*: "They create desolation and call it peace."

Tacitus wrote of the Germans. His words apply exactly to the military authorities' so-called 'pacification', which can only be accomplished in areas that have first been reduced to a barren wilderness, and could not be completely enforced until every Algerian was either dead or rotting away behind barbed wire. Victory on any other terms is inconceivable. So if victory is the end in view—as generals, colonels, paratroopers and *légionnaires* all proclaim—why quibble over the means employed? The end more than justifies the worst of them; indeed, the end itself far exceeds them in villainy.

'I am only one among thousands of other detainees,' Djamila told her lawyer the other day, and that is no less than the truth. There are fourteen thousand Algerians confined in French camps and prisons, seventeen thousand in gaol in Algeria itself, and hundreds of thousands more filling the Algerian camps. The efforts made on Djamila's behalf would fail in their purpose if they did not create a general revulsion against the sufferings inflicted on her fellow-prisoners—sufferings of which her own case furnished a by no means extreme example. But any such revulsion will lack concrete reality unless it takes the form of political action. The alternatives are simple and clear-cut. Either—despite your willing and facile grief over such past horrors as the Warsaw ghetto or the death of Anne Frank—you align yourselves with our contemporary butchers rather than their victims, and give your unprotesting assent to the martyrdom which thousands of Djamilas and Ahmeds are enduring in your name, almost, indeed, before your very eyes; or else you reject, not merely certain specific practices, but the greater aim which sanctions them, and for which they are essential. In the latter case you will refuse to countenance a war that dares not speak its true name—not to mention an Army that feeds on war, heart and soul, and a Government that knuckles under to the Army's demands; and you will raise heaven and earth to give this gesture of yours effective force. There is no alternative, and I hope this book will help to convince you of the fact. The truth confronts you on all sides. You can no longer mumble the old excuse 'We didn't know'; and now that you *do* know, can you continue to feign ignorance, or content yourselves with a mere token utterance of horrified sympathy? I hope not.

QUESTIONS FOR ANALYSIS

1. How does Djamila Boupacha describe the events that took place on the night of her arrest in February, 1960, and her subsequent treatment in prison? What legal charges did she bring against the French colonial authorities in her civil indictment?
2. What is Simone de Beauvoir's analysis of the role of torture in the Algerian War? On what basis does de Beauvoir call on the people of France to support Algerian independence? How do her views on French colonialism in Algeria contrast with the nineteenth-century women's writings on British and German colonialism that appear in Chapter 9 (Sources 112–114)?

WOMEN'S MOVEMENTS OF THE 1960S AND 1970S IN WESTERN EUROPE

The women's movements that emerged in the 1960s and 1970s offered a variety of critiques to explain the causes of women's oppression and the means for women to attain equality within the overall context of societal transformation. As in the nineteenth century, feminist movements in any given European nation were shaped by varying conditions and political traditions. In Northern Ireland, for example, women played a crucial role in the civil rights movement which sought to end discrimination against Catholics by the majority Protestant population. Women also organized on an international level. Their efforts to press the United Nations to take up women's issues in a more systematic fashion culminated in the 1979 Convention on the Elimination of All Forms of Discrimination Against Women.

173. Juliet Mitchell
"Women: The Longest Revolution"

Juliet Mitchell (b. 1940) is the author of "Women: The Longest Revolution" (1966), one of the most influential essays of second-wave feminism, as well as numerous works on feminism and psychoanalysis. In the 1960s, Mitchell lectured in English literature at Leeds and Reading universities in England and became involved in Marxist politics around the *New Left Review*. She published "Women: The Longest Revolution" in the journal to provide a socialist feminist perspective on women's issues. Four years later, Mitchell helped to organize the Women's Liberation Conference held at Ruskin College in Oxford in 1970. The conference is regarded as the beginning of the contemporary women's movement in Britain. In the 1970s, Mitchell gave up her university position to devote herself full-time to feminist politics, writing and lecturing on literature and feminism in Europe and the United States. She subsequently trained in psychoanalysis and became a practicing psychoanalyst. Mitchell's publications include *Woman's Estate* (1972), *Psychoanalysis and Feminism* (1974), and *Women, the Longest Revolution: Essays in Feminism, Literature, and Psychoanalysis* (1984). Most recently, she co-edited with Ann Oakley *Who's Afraid of Feminism: Seeing Through the Backlash* (1997).

[Mitchell begins her essay by stating,

The situation of women is different from that of any other social group. This is because they are not one of a number of isolable units, but half a totality: the human species. Women are essential and irreplaceable; they cannot therefore be exploited in the same way as other social groups can. They are fundamental to the human condition, yet in their economic, social and political roles, they are marginal. It is precisely this combination—fundamental and marginal at one and the same time—that has been fatal to them.[1]

Mitchell proceeds to critique the contributions and limitations of socialist theorists on "the woman question," such as Charles Fourier, Karl Marx, Friedrich Engels, August Bebel, and Simone de Beauvoir. She concludes that they have failed to fully take into consideration the interrelated nature of all four structures of women's oppression: production, reproduction, sexuality, and socialization of children. This excerpt is from the last portion of the article in which Mitchell examines women's status concerning these structures and how each must be transformed simultaneously if women are to be truly liberated.]

It is only in the highly developed societies of the West that an authentic liberation of women can

---

[1]"Women: The Longest Revolution," *New Left Review*, November/December 1996, p. 9.

"Women: The Longest Revolution" by Juliet Mitchell. From *New Left Review*. November/December 1996 issue. Reprinted by permission of New Left Review.

be envisaged today. But for this to occur, there must be a transformation of all the structures into which they are integrated, and an *'unité de rupture'*.[1] A revolutionary movement must base its analysis on the uneven development of each, and attack the weakest link in the combination. This may then become the point of departure for a general transformation. What is the situation of the different structures today?

1. **Production:** The long-term development of the forces of production must command any socialist perspective. The hopes which the advent of machine technology raised as early as the 19th century have already been discussed. They proved illusory. Today, automation promises the *technical* possibility of abolishing completely the physical differential between man and woman in production, but under capitalist relations of production, the *social* possibility of this abolition is permanently threatened, and can easily be turned into its opposite, the actual diminution of woman's role in production as the labour force contracts.

This concerns the future, for the present the main fact to register is that woman's role in production is virtually stationary, and has been so for a long time now. In England in 1911 30 per cent of the work-force were women; in the 1960's 34 per cent. The composition of these jobs has not changed decisively either. The jobs are very rarely 'careers'. When they are not in the lowest positions on the factory-floor they are normally white-collar auxiliary positions (such as secretaries)—supportive to masculine roles. They are often jobs with a high 'expressive' content, such as 'service' tasks. Parsons says bluntly: 'Within the occupational organization they are analogous to the wife-mother role in the family.[2] The educational system underpins this role-structure. 75 per cent of 18-year-old girls in England are receiving neither training nor edu-

cation today. The pattern of 'instrumental' father and 'expressive' mother is not substantially changed when the woman is gainfully employed, as her job tends to be inferior to that of the man's, to which the family then adapts.

Thus, in all essentials, work as such—of the amount and type effectively available today—has not proved a salvation for women.

2. **Reproduction:** Scientific advance in contraception could, as we have seen, make involuntary reproduction—which accounts for the vast majority of births in the world today, and for a major proportion even in the West—a phenomenon of the past. But oral contraception—which has so far been developed in a form which exactly repeats the sexual inequality of Western society—is only at its beginnings. It is inadequately distributed across classes and countries and awaits further technical improvements. Its main initial impact is, in the advanced countries, likely to be psychological—it will certainly free women's sexual experience from many of the anxieties and inhibitions which have always afflicted it.[3] It will definitely divorce sexuality from procreation, as necessary complements....

3. **Socialization:** The changes in the composition of the work-force, the size of the family, the structure of education, etc—however limited from an ideal standpoint—have undoubtedly diminished the societal function and importance of the family. As an organization it is not a significant unit in the political power system, it plays little part in economic production and it is rarely the sole agency of integration into the larger society; thus at the macroscopic level it serves very little purpose.

The result has been a major displacement of emphasis on to the family's psycho-social

---

[1]See Louis Althusser, "Contradiction et Surdétermination" in *Pour Marx* (1965).

[2]Talcott Parson's and Robert F. Bales, *Family, Socialization, and Interaction Process* (1956), p. 15n.

[3]Jean Baby records the results of an enquiry carried out into attitudes to marriage, contraception and abortion of 3,191 women in Czechoslovakia in 1959: 80 per cent of the women had limited sexual satisfaction because of fear of conception. *Un Monde Meilleur* (1964), p. 82n.

function, for the infant and for the couple....[4]
The vital nucleus of truth in the emphasis on
socialization of the child has been discussed. It
is essential that socialists should acknowledge
it and integrate it entirely into any programme
for the liberation of women....[T]here is no
doubt that the need for permanent, intelligent
care of children in the initial three or four years
of their lives can (and has been) exploited ideo-
logically to perpetuate the family as a total
unit, when its other functions have been
visibly declining. Indeed, the attempt to focus women's existence exclusively on bringing
up children, is manifestly harmful to children.
Socialization as an exceptionally delicate process requires a serene and mature socializer—
a type which the frustrations of a *purely* familial
role are not liable to produce. Exclusive maternity is often in this sense 'counter-pro-
ductive'. The mother discharges her own
frustrations and anxieties in a fixation on the
child. An increased awareness of the critical im
portance of socialization, far from leading to a
restitution of classical maternal roles, should
lead to a reconsideration of them—of what
makes a good socializing agent, who can genuinely provide security and stability for the
child.

The same arguments apply, *a fortiori,* to the
psycho-social role of the family for the couple.
The beliefs that the family provides an impreg-
nable enclave of intimacy and security in an
atomized and chaotic cosmos assumes the absurd—that the family can be isolated from the
community, and that its internal relationships
will not reproduce in their own terms the external relationships which dominate the society.
The family as refuge in a bourgeois society inevitably becomes a reflection of it.

4. **Sexuality:** It is difficult not to conclude that
the major structure which at present is in rapid

evolution is sexuality. Production, reproduction,
and socialization are all more or less stationary in
the West today, in the sense that they have not
changed for three or more decades. There is
moreover, no widespread *demand* for changes in
them on the part of women themselves—the
governing ideology has effectively prevented
critical consciousness. By contrast, the dominant
sexual ideology is proving less and less successful
in regulating spontaneous behaviour. Marriage
in its classical form is increasingly threatened by
the liberalization of relationships before and
after it which affects all classes today. In this
sense, it is evidently the weak link in the chain—
the particular structure that is the site of the
most contradictions. The progressive potential
of these contradictions has already been emphasized. In a context of juridical equality, the liberation of sexual experience from relations which
are extraneous to it—whether procreation or
property—could lead to true inter-sexual freedom. But it could also lead simply to new forms
of neocapitalist ideology and practice. For one of
the forces behind the current acceleration of sexual freedom has undoubtedly been the conversion of contemporary capitalism from a
production-and-work ethos to a consumption-
and-fun ethos. Riesman commented on this development early in the 1950's: '...there is not
only a growth of leisure, but work itself becomes
both less interesting and less demanding for
many...more than before, as job-mindedness
declines, sex permeates the daytime as well as the
playtime consciousness. It is viewed as a consumption good not only by the old leisure
classes, but by the modern leisure masses....'[5]
Bourgeois society at present can well afford a
play area of premarital *non*-procreative sexuality.
Even marriage can save itself by increasing
divorce and remarriage rates, signifying the importance of the institution itself. These considerations make it clear that sexuality, while it
presently may contain the greatest potential for
liberation—can equally well be organized
against any increase of its human possibilities.

---

[4]See Berger and Kellner: "Marriage and the Construction
of Reality," *Diogenes* (Summer 1964) for analyses of marriage and parenthood 'nomic-building' structure.

[5]Riesman, *The Lonely Crowd* (1950), p. 154.

New forms of reification are emerging which may void sexual freedom of any meaning. This is a reminder that while one structure may be the *weak link* in a unity like that of woman's condition, there can never be a solution through it alone. The utopianism of Fourier or Reich[6] was precisely to think that sexuality could inaugurate such a general solution. Lenin's remark to Clara Zetkin is a salutary if over-stated corrective: 'However wild and revolutionary (sexual freedom) may be, it is still really quite bourgeois. It is, mainly, a hobby of the intellectuals and of the sections nearest them. There is no place for it in the Party, in the class conscious, fighting, proletariat.'[7] For a general solution can only be found in a strategy which affects *all* the structures of women's exploitation. This means a rejection of two beliefs prevalent on the left:

Reformism: This now takes the form of limited ameliorative demands: equal pay for women, more nursery-schools, better retraining facilities, etc. In its contemporary version it is wholly divorced from any fundamental critique of women's condition or any vision of their real liberation (it was not always so). Insofar as it represents a tepid embellishment of the *status quo,* it has very little progressive content left.

Voluntarism: This takes the form of maximalist demands—the abolition of the family, abrogation of all sexual restrictions, forceful separation of parents from children—which have no chance of winning any wide support at present, and which merely serve as a substitute for the job of theoretical analysis or practical persuasion. By pitching the whole subject in totally intransigent terms, voluntarism objectively helps to maintain it outside the framework of normal political discussion.

What, then, is the responsible revolutionary attitude? It must include both immediate and fundamental demands, in a single critique of the *whole* of women's situation, that does not fetishize any dimension of it. Modern industrial development, as has been seen, tends towards the separating out of the originally unified function of the family—procreation, socialization, sexuality, economic subsistence, etc....

In practical terms this means a coherent system of demands. The four elements of women's condition cannot merely be considered each in isolation; they form a structure of specific interrelations. The contemporary bourgeois family can be seen as a triptych of sexual, reproductive and socializatory functions (the woman's world) embraced by production (the man's world)—precisely a structure which in the final instance is determined by the economy. The exclusion of women from production—social human activity—and their confinement to a monolithic condensation of functions in a unity—the family—which is precisely unified in the *natural part* of each function, is the root cause of the contemporary *social* definition of women as *natural* beings. Hence the main thrust of any emancipation movement must still concentrate on the economic element—the entry of women fully into public industry. The error of the old socialists was to see the other elements as reducible to the economic; hence the call for the entry of women into production was accompanied by the purely abstract slogan of the abolition of the family. Economic demands are still primary, but must be accompanied by coherent policies for the other three elements, policies which at particular junctures may take over the primary role in immediate action.

Economically, the most elementary demand is not the right to work or receive equal pay for work—the two traditional reformist demands—but *the right to equal work itself.* At present, women perform unskilled, uncreative, service jobs that can be regarded as 'extensions' of their expressive

---

[6]Charles Fourier, the French utopian socialist. Wilhelm Reich established sexual hygiene clinics and in 1930 founded the German Society of Proletarian Sexual Politics in Berlin. His writings include *The Sexual Revolution* (1930, trans. 1945).—Ed.

[7]Clara Zetkin: *Reminiscences of Lenin* (1925, trans. 1929), pp. 52–53.

familial role. They are overwhelmingly wait-ressess, office-cleaners, hair-dressers, clerks, typ-ists. In the working-class occupational mobility is thus sometimes easier for girls than boys—they can enter the white-collar sector at a lower level. But only two in a hundred women are in admin-istrative or managerial jobs, and less than five in a thousand are in the professions. Women are poorly unionized (25 per cent) and receive less money than men for the manual work they do perform: in 1961 the average industrial wage for women was less than half that for men, which, even setting off part-time work, represents a mas-sive increment of exploitation for the employer.

Education

The whole pyramid of discrimination rests on a solid extra-economic foundation—education. The demand for equal work, in Britain, should above all take the form of a demand for an *equal educational system,* since this is at present the main single filter selecting women for inferior work-roles. At present, there is something like equal education for both sexes up to 15. There-after three times as many boys continue their education as girls. Only one in three 'A'-level entrants, one in four university students is a girl. There is no evidence whatever of progress. The proportion of girl university students is the same as it was in the 1920's. Until these injus-tices are ended, there is no chance of equal work for women. It goes without saying that the con-tent of the educational system, which actually instills limitation of aspiration in girls needs to be changed as much as methods of selection. Education is probably the key area for immedi-ate economic advance at present.

Only if it is founded on equality can produc-tion be truly differentiated from reproduction and the family.... Traditionally, the socialist movement has called for the 'abolition of the bourgeois family'. This slogan must be rejected as incorrect today.... The strategic concern for socialists should be for the equality of the sexes, not the abolition of the family. The conse-quences of this demand are no less radical, but they are concrete and positive, and can be inte-grated into the real course of history. The fam-ily as it exists at present is, in fact, incompati-ble with the equality of the sexes. But this equality will not come from its administrative abolition, but from the historical differentia-tion of its functions. The revolutionary demand should be for the liberation of these functions from a monolithic fusion which oppresses each. Thus dissociation of reproduction from sexual-ity frees sexuality from alienation in unwanted reproduction (and fear of it), and reproduction from subjugation to chance and uncontrollable causality. It is thus an elementary demand to press for free State provision of oral contracep-tion. The legalization of homosexuality—which is one of the forms of non-reproductive sexuality—should be supported for just the same reason, and regressive campaigns against it in Cuba or elsewhere should be unhesitat-ingly criticized. The straightforward abolition of illegitimacy as a legal notion as in Sweden and Russia has a similar implication; it would separate marriage civically from parenthood.

From Nature to Culture

The problem of socialization poses more difficult questions, as has been seen. But the need for in-tensive maternal care in the early years of a child's life does not mean that the present single sanctioned form of socialization—marriage and family—is inevitable. Far from it. The funda-mental characteristic of the present system of marriage and family is in our society its *mono-lithism:* there is only one institutionalized form of inter-sexual or inter-generational relationship possible. It is that or nothing. This is why it is essentially a denial of life. For all human experi-ence shows that intersexual and intergenera-tional relationships are infinitely various—indeed, much of our creative literature is a cele-bration of the fact—while the institutionalized expression of them in our capitalist society is ut-terly simple and rigid. It is the poverty and sim-plicity of the institutions in this area of life which are such an oppression. Any society will require some institutionalized and social recog-nition of personal relationships. But there is absolutely no reason why there should be only one legitimized form—and a multitude of unlegitimized experience. Socialism should properly mean not the abolition of the family, but the diversification of the socially acknowledged relationships which are today forcibly and rigidly compressed into it. This would mean a plural range of institutions—where the family is only one, and its abolition implies none. Couples living together or not living together, long-term unions with children, single parents bringing up children, children socialized by conventional rather than biological parents, extended kin groups, etc—all these could be encompassed in a range of institutions which matched the free invention and variety of men and women.

It would be illusory to try and specify these institutions. Circumstantial accounts of the future are idealist and worse, static. Socialism will be a process of change, of becoming. A fixed image of the future is in the worst sense ahistorical; the form that socialism takes will depend on the prior type of capitalism and the nature of its collapse. As Marx wrote: 'What (is progress) if not the absolute elaboration of (man's) creative dispositions, without any preconditions other than antecedent historical evolution which makes the totality of this evolution—i.e. the evolution of all human powers as such, unmeasured by any *previously established* yardstick—an end in itself? What is this, if not a situation where man does not reproduce himself in any determined form, but produces his totality? Where he does not seek to remain something formed by the past, but is the absolute movement of becoming?'[8] The liberation of women under socialism will not be 'rational' but a human achievement, in the long passage from Nature to Culture which is the definition of history and society.

_____

[8]Karl Marx: *Precapitalist Economic Formations,* in *Early Writings,* trans. T.B. Bottomore (1963), p. 85.

174. The Right to Choose: Abortion on Trial in France

The abortion issue in France received national attention with the publication of the "Manifesto of the 343" in the April 5, 1971, edition of the journal *Le Nouvel Observateur.* Drawn up by the Mouvement de libération des femmes (MLF), the manifesto denounced the existing law which criminalized abortion, except for when the mother's life was in danger, as well as the public silence surrounding the dangerous practice of illegal abortions. The manifesto was signed by such prominent women as Catherine Deneuve, Delphine Seyrig, Marguerite Duras, and Simone de Beauvoir who all declared that they had received an abortion. Soon afterward, the attorney Gisèle Halimi[1] initiated the Association Choisir (literally, "to choose") with Jean Rostand of the French Academy, Christiane Rochefort, Delphine Seyrig, and Simone de Beauvoir who became president. Open in membership to men as well as women, Choisir's purpose was to ensure free and accessible contraception, repeal all repressive legislation against abortion, and defend and assist anyone accused of having, performing, or serving as an accomplice to an abortion.

In 1972, Choisir took up the defense of seventeen-year-old Marie-Claire Chevalier who was charged with having an illegal abortion. Halimi represented Marie-Claire in her October, 1972 trial before the Juvenile Court of Bobigny. For the trial

_____

[1]For additional information on Halimi, see Source 172 in this chapter on the Djamila Boupacha case.

date, Choisir mobilized a rally for Marie-Claire in front of the courthouse, the leaflet for which is reproduced here. Marie-Claire was acquitted on the grounds that she did not freely make her own decision and was instead "subjected to pressures of a moral, family, and social kind that she was not able to resist."[2] However, the case did not end with her acquittal. Tried as accomplices were Michèle Chevalier, her mother, and Renée Sausset and Lucette Duboucheix, two co-workers who directed Chevalier to Micheline Bambuck, a secretary-stenographer, who performed the abortion. During their trial in November 1972, Marie-Claire appeared on the women's behalf along with several physicians, scientists, and politicians who testified as expert witnesses against the law which criminalized abortion. Bambuck received a suspended sentence of one year in prison while Marie-Claire's mother was fined 500 francs, which was also suspended. The full trial proceedings were published by Choisir as *Abortion: The Bobigny Affair: A Law on Trial* (1973, 1975 English translation). As Marie-Claire's own trial took place behind closed doors, the material reproduced here is from the November trial of the four women. It includes transcripts from the hearings of Micheline Bambuck and Michèle Chevalier, Marie-Claire's testimony as a witness on their behalf, and Gisèle Halimi's concluding speech for the defense.

The Bobigny trial and verdicts made the front pages of major newspapers in France and boosted the movement to repeal anti-abortion legislation. In 1975, the National Assembly voted into law the main aspects of a bill drafted by Choisir and guaranteed a women's right to abortion on request up to the end of the tenth week of pregnancy. A detailed account of the Bobigny trial is provided by Halimi in *La Cause des femmes* (1973), published in English as *The Right to Choose* (1977).

A. Text of Choisir Leaflet

LEAFLET FOR MARIE-CLAIRE
A GIRL OF SEVENTEEN
IS TO BE JUDGED
FOR HAVING HAD AN ABORTION

Like a million other women in France each year, Marie-Claire has been through the drama of an illicit abortion.

- —BECAUSE SHE did not have 3000 frs to go and have an abortion in comfort in a clinic in Geneva, London, or even Paris
- —BECAUSE SHE is the natural daughter of an unmarried mother who works on the *Métro* and has brought up her three daughters single-handed

- —BECAUSE THERE IS NO SEX EDUCATION in school and because contraception is sabotaged in France (as admitted by Mr. Neuwirth, UDR Deputy, who is the originator of the law on contraception)
- —BECAUSE, as in all these cases, she was left ALONE to find a way out,

she must today relive this drama and undergo JUDGEMENT BEHIND CLOSED DOORS by a society which itself is basically responsible for this situation.

We, women who have been through this situation and who, each month, could have to go through it again, express our wholehearted support of Marie-Claire.

---

[2] Association Choisir, *Abortion: The Bobigny Affair: A Law on Trial* (Sidney, Australia: Wild and Woolley, Ltd., 1975), p. 7.

EVERYONE, MEN AND WOMEN,
MEET
IN FRONT OF BOBIGNY COURTHOUSE
ADMINISTRATIVE CENTRE
ON TUESDAY, 11 OCTOBER.
AT 9 a.m.
Métro to Eglise de Pantin—then Bus
to Administrative Centre
CHOISIR ASSOCIATION
174 RUE DE L'UNIVERSITÉ
PARIS 7

- —FOR CONTRACEPTION
- —FOR THE ABOLITION OF REPRES-
SIVE LEGISLATION ON ABORTION
- —FOR FREE DEFENCE OF EVERY-
ONE ACCUSED OF ABORTION

Founder Members
Jean ROSTAND of the French Academy
Simone de BEAUVOIR, Gisèle HALIMI
Christiane ROCHEFORT, Delphine SEYRIG

B. Trial Proceedings, November 1972

THE HEARING OF THE ACCUSED MICHELINE BAMBUCK

Presiding Judge—Please give us a brief account of the facts.

(Mme Bambuck [the abortionist] begins her statement but her voice is inaudible. The Judge asks her to move to a place where her voice will carry better.)

Judge—Come forward to the witness box. You received a telephone call, didn't you?

Mme Bambuck—I had a phone call, I think it was in October 1971, from a woman . . .

Judge—From a woman unknown to you?

Mme Bambuck—Yes. She said, 'I am calling on behalf of your friend, Mme Sausset.[1] Come to see me in the Metro, I work there.'

Judge—Did she say why?

Mme Bambuck—No, I don't remember her saying why. . . .

Judge—Did this woman phone you only the once?

Mme Bambuck—There were several calls because I didn't want to do it, your Honour, but I felt sorry for her.

Judge—You knew then what it was about?

Mme Bambuck—I found out. I said I didn't want to do it, but she begged me. She phoned me several times. . . .

Judge—So, you did know what she wanted. You met her the first time at the Metro?

Mme Bambuck—She explained that her daughter was pregnant and that she didn't want the child. I know about these things because of the circumstances of my own life. I didn't want to do it. She called me again and I felt sorry for her. I sincerely regret having broken the law, I won't do it again. It's the last time. But from the human point of view . . . my conscience doesn't tell me I did anything wrong.

Judge—Did you receive a sum of money?

Mme Bambuck—I received some money.

Judge—How much?

Mme Bambuck—1200 francs. I had to pay some taxes. I'm a widow with two children and I used the money to pay the taxes. That is the absolute truth, your Honour.

Judge—How many times did you go back?

Mme Bambuck—I had to go back five times in fifteen days. . . .

Judge—Did you give her something to take?

Mme Bambuck—Certainly not. I used a speculum and a probe.

Judge—You say, 'I went back five times.' Why was that?

Mme Bambuck—Mme Chevalier telephoned me and each time she said that nothing had happened yet. I swear, your Honour, that I am telling the truth.

Judge—Was your so-called operation performed in Mme Chevalier's home?

---

[1]Renée Sausset, a coworker of Michèle Chevalier, charged as an accomplice for helping Marie-Claire to obtain an abortion.

Mme Bambuck—Yes, your Honour.

Judge—Under what conditions? Did Mme Chevalier actually help you perform the operation?

Mme Bambuck—No, I did it on my own. I swear it, your Honour, I am speaking only the truth....

THE HEARING OF THE ACCUSED MICHÈLE CHEVALIER

Presiding Judge—I would like you to tell us in your own words what happened.

Mme Chevalier [the mother of Marie-Claire]—In September 1971 my daughter told me she thought she might be pregnant, but she wasn't sure. We began by consulting a doctor, a gynaecologist. He did a test and then he gave her some injections. They didn't work and he asked 4500 francs for an abortion. It was impossible; I am bringing up my three children by myself and I hadn't any money so I had to try somewhere else. I was very unhappy about it but I had to try somewhere else.

I said to Marie-Claire, 'My child, we will have to go without a lot, but if you want to keep the child, we will manage.' Marie-Claire said to me, 'I've thought it over and I won't keep the child, not for anything in the world.' She had made up her mind; there was nothing else to do.

Then, we hadn't any money but we had to find a way out. I started asking my friends to see if they knew of any names or addresses. I often travelled with Mme Duboucheix[3] and I told her about my worries. She said to me, 'No, I don't know anyone, I can't tell you anything, I know nothing.' Then one day she telephoned me. She said, 'Look, I think I have found someone. I will tell you about it later....'

It was Mme Bambuck. She came to see me one evening, and she told me what she would do. She wanted 1200 francs in advance. I didn't have the money; I had to find some. So I borrowed it. We came to an agreement and Mme Bambuck came to the house. She performed the abortion on Marie-Claire. Between the fifth and the twentieth of November, she had to come back five times.

During the night of 20 November, Marie-Claire began to have horrible pains and she had a haemorrhage. At half-past one that night I went to a public telephone and called a doctor and I took Marie-Claire to a nursing home where she had a curette.

But when we got to the nursing home I had another shock. They wanted 1000 francs deposit. I didn't have it. The next day I went to see my friend again and she lent me another 1000 francs. To get Marie-Claire out of the nursing home two days later I had to pay 900 francs. I paid it with an uncovered cheque.

Judge—How much did it finally cost you?

Mme Chevalier—More than 3000 francs.

Judge—How much do you earn a month?

Mme Chevalier—1500 francs.

Judge—Is that inclusive?

Mme Chevalier—We get a bonus according to the station we work at each day. I also get a family allowance in addition to my single salary.

Judge—How much is that altogether?

Mme Chevalier—Just under 2300 francs at the most.

Judge—The report of the police inquiry speaks well of you. I read it out because the Court should be aware of it. 'Her behaviour, morals and reputation are good. She is a good mother. Her home is well looked after and so are her children....'

Mme Chevalier—My children don't want for anything. I am an unmarried mother. I have devoted my life to bringing up my children. I do it properly....

Mme Chevalier—I didn't have any choice. When you haven't any money, you have to break the law. I've said it before, and I say it again. I am grateful to Mme Bambuck.

Judge—You work as a supervisor in the R.A.T.P.[4] and you are away from home from

---

[3]Lucette Duboucheix, a coworker of Michèle Chevalier charged as an accomplice for helping Marie-Claire to obtain an abortion.

[4]R.A.T.P., *Régie automne des transports parisiens,* the Paris Metro.

eleven in the morning to about half past eight at night.

Mme Chevalier—That's right. We start work at twenty past twelve and we finish at ten to eight.

Judge—What sort of home do you live in?

Mme Chevalier—I have a flat in the H.L.M.[5] The R.A.T.P. recommended me for it.

Judge—How many rooms do you have?

Mme Chevalier—Three rooms, kitchen and bathroom.

Judge—And three children, aged sixteen, fifteen and fourteen?

Mme Chevalier—That's what they were then. Now they are a year older and that means we have been waiting a year for this trial.

Judge—How do you feel about what happened?

Mme Chevalier—I have no regrets, none at all.

Judge—M. Public Prosecutor, have you any questions to ask?

Prosecutor—Do you admit to having boiled the water?

Mme Chevalier—Of course. Mme Bambuck came by car from Paris. She wasn't going to bring a pot of boiling water along with her. I admit I boiled the water. That's right.

Judge—And did you provide the alcohol?

Mme Chevalier—She had told me to get some alcohol. In a house with children you always have some alcohol and some cotton wool.

Judge—Have you any questions, M[e] Halimi?

M[e] Halimi—I would like to ask Mme Chevalier whether her daughter had been given any sex education at the C.E.T.[6] where she went to school, or whether Mme Chevalier herself had given her any instruction at all on the matter.

Mme Chevalier—There is absolutely no sex education at the C.E.T. Marie-Claire is my little girl: as a mother, I still see her as a child.

And then, I had no sex education myself. I don't know what words to use.

Judge—These things come up in the press, on the radio and on television. Didn't you think that one day your daughter would be confronted with the question?

Mme Chevalier—No, I didn't see any urgent need to talk to her about it.

Judge—Did you know that she went out with a young man? Did you know who he was?

Mme Chevalier—Her friendship with that particular boy was the same as with many of her other friends. She went to a coeducational secondary college and the boys and girls all knew each other.

M[e] Halimi—A last question. I would like to know who took the initial decision about the abortion.

Mme Chevalier—It was Marie-Claire herself.

Judge—Nevertheless, you did discuss this matter with your daughter?

Mme Chevalier—Yes.

Judge—Your friend, Mme Duboucheix, who will be the next witness, at one time did advise you to keep this child and put it up for adoption?

Mme Chevalier—First to bring a child into the world and then to abandon it, that was not what Marie-Claire wanted. We had made up our minds....

Judge—During the police inquiry you stated, 'Marie-Claire confessed to me that she was pregnant. There was only one way out—to get her an abortion. I live on my own with three children, since my husband abandoned me. I could not imagine bringing up my daughter's child, especially as she has no occupation.'

Mme Chevalier—That's right. She has left school and she hasn't got a job.

Judge—So you asked yourself whether you could take responsibility for this child and you didn't do it?...

Mme Chevalier—My daughter didn't want the child and I wasn't going to keep it against her will. I have lived through that ordeal myself and I didn't want my daughter to have to go through it too.

---

[5]H.L.M., *Habitations Loyer Modéré,* hi-rise subsidized rental apartments.

[6]C.E.T., College d'Enseignement Technique, a technical school.

THE TESTIMONY OF
MARIE-CLAIRE CHEVALIER
AS A WITNESS FOR THE ACCUSED

(The girl who had the abortion, who was acquitted 11 October 1972 in the Juvenile Court of Bobigny.)

Public Prosecutor—M. President, in view of the witness's minority, I ask that she be heard in camera.

Me Halimi—We are opposed to a hearing in camera.

Judge—In view of the girl's age and the family ties which bind her to her mother, would it not be more appropriate for the defence to forgo hearing this witness, to avoid causing her to relive painful moments without adding much to the clarity of the proceedings?

Me Halimi—M. President, I am very much aware of the concern that the judges are showing today towards Marie-Claire's suffering, but I have been her defence counsel, and because the law required it, I had to defend the case in camera.[7] The present case is for complicity in an abortion. It was Marie-Claire who had the abortion. Everything that she said about it was heard in closed court. This evidence is the major item, not only for the prosecution, but also for the defence, and we cannot allow you to question her in camera. I must say, this already resembles a closed session. By the provision of a room like this one for a case the public wishes to follow because it feels concerned, the principle of public court hearings is already infringed....

[The public gallery held only sixty persons. A request by the defence for a bigger room was rejected by the Minister of Justice. Several hundred people who had come to Bobigny in support of Mme Chevalier and her friends were unable to be present during the proceedings.]

Judge—The Court waives the decision to hold the hearing in camera—in deference to your wishes. *(To Marie-Claire)* Give your surname, your Christian names, age, occupation and address.

Marie-Claire—Marie-Claire Chevalier, no occupation, living at Neuilly-Plaisance, born on 12 July 1955.

Judge—What question do you wish me to ask?

Me Halimi—It is not correct to speak of a question. We are talking about Marie-Claire's abortion and her decision to have an abortion. I want Marie-Claire to tell us how things happened. I want her to be allowed to speak without interruption. Marie-Claire, tell your story, and this time it is for all to hear.

Marie-Claire—It was in August that I met Daniel P. We were both students and I trusted him. He had a car; he asked me to go to his place to listen to some records. I said yes. I went, and he got violent. Then...

Me Halimi—Then what?...

Marie-Claire—I got away. I went home. I stayed indoors. After three weeks I was worried and then that is all. I told my mother. She was very upset. She said to me, 'We will do our best to bring the child up.' I said, 'No, I don't want a hooligan's child. I am still at school. I don't want this child.' Then my mother took me to a gynaecologist to confirm my pregnancy. He told her I was pregnant. He asked us for 4500 francs to... Then, we left.... Later, I met Mme Bambuck. She put a probe into me five times. After the fifth time I had a haemorrhage. I went into the nursing-home the night of the twentieth and I came out two days later.

Judge—You went to see a gynaecologist. Which one?

Me Halimi—M. President, we are not informers at this bar. We do not have to inform against him....

Judge—What did your mother say to you?

Marie-Claire—That we could try and bring it up; I said no again. I did not want to.

Judge—And then what happened? Did you ask her to find someone or did she suggest it to you?

Marie-Claire—It was me. I said at the beginning that I could have an abortion....

Judge—Have you any further questions?

Me Halimi—I would like Marie-Claire to tell us what happened when she told Daniel—the cause of it all—that she was pregnant, that

---

[7]Marie-Claire's closed court trial took place at the Juvenile Court of Bobigny in October 1972.

she was going to have an abortion, and that she did not want to see him again. In particular I would like her to tell us about the threats against her. Marie-Claire stated in the previous hearing—which she was in the dock—that after learning that she was pregnant, young Daniel P continued to annoy her and tried to see her. After that, young Daniel proceeded to threats, particularly against Mme Chevalier....[8]

Marie-Claire—Every night—at three o'clock in the morning—he came with a group of friends and knocked on our shutters. He shouted from his car that if he met my mother he would run her over. He also said that if I was pregnant it was not by him.

M[e] Halimi—Did Marie-Claire have any sex education at the C.E.T. she attended? Or any information about contraception?

Marie-Claire—There was no sex education at my school.

Me Halimi—Could your mother tell you?

Marie-Claire—No, my mother didn't tell me; besides, she didn't have the time.

Judge—Have you any questions, M. Prosecutor?

Prosecutor—No.

Judge—I want the witness to wait in the witness's waiting room.

SPEECH FOR THE DEFENSE BY ATTORNEY GISÈLE HALIMI

Members of the Court, there falls to me today a very rare privilege. I am experiencing, with a satisfaction never known by me until today, a perfect harmony between my job as a defending counsel and my condition as a woman. Never until today have I so strongly felt myself at the same time the accused in the dock and the advocate at the bar.

Since a certain distance between the advocate and her client is prescribed by the very proper requirements of our profession, it has doubtless never been imagined that women advocates, like all women, have had abortions and that they can say it and say it publicly as I am saying it today....[9]

And if I speak today, Gentlemen, only of abortion and of what is done to women by a repressive law, a law from another age, it is not just because of the particular dossier that brings us here, but more importantly because this law, devoid of value, of any relevance, of the least sense today or tomorrow, is the touchstone of the oppression that degrades women....

But what I want to say about abortion is that in the suppression of abortion...discrimination reigns supreme. It is always the same class that is arraigned, the class of women who are poor, economically and socially vulnerable, the class without money or connections. I have been a barrister for twenty years, Gentlemen, and... I have never yet had to defend the wife of a top public servant or the wife of a famous doctor, or of an important barrister, or of a company director, or the mistress of any of these same gentlemen....You always convict the same people—the Mme Chevaliers.

Another example of the Class Justice which is the rule, without exception, where women are concerned, is the 'Manifesto of the 343.'[10] You have heard three of the signatories at this bar. I am one of them myself....

Have we ever been arrested? Has there ever been an inquiry about us?...

According to the official figures, as Professor Palmer has just told us, the number of backyard abortions is four hundred thousand a year. But it is thought the true figure is between eight hundred thousand and a million. The number of convictions for the last five years, according to the I.N.S.E.E.[11]

Year	Convictions
1965	588
1966	720

---

[8]Marie-Claire's abortion became known to the authorities after Daniel P. informed the police during an interrogation for a car theft.

[9]Halimi is referring to her own abortions which she describes in her book, *The Right to Choose,* pp. 21–23.

[10]The "Manifesto of the 343" published in the April 5, 1971 issue of *Le Nouvel Observateur.*

[11]Statistics compiled by the I.N.S.E.E., the Institut National de Statistique et des Études Économiques.

1967	623
1968	698
1969	471
1970	340

If you take an average for these years it is about five hundred for eight hundred thousand abortions. If you take the average for the last two years it is less than four hundred.... When a law no longer hinders the multiplication of the crimes it is supposed to prohibit, it is certain that it is worthless. This is a law to be condemned, everybody says so, our witnesses, the press—I could say public opinion in its entirety....

Let us go back to the beginning. For Marie-Claire, finding herself pregnant at sixteen, to be accused of the offence of abortion, it would have had to be proved that she had the knowledge how not to become pregnant. Here, Gentlemen, I come to the problem of sex education....

You have heard the testimony of Simone Iff, the Vice-President of Family Planning. She has told you about the deliberate sabotage practised by the public authorities against her organisation, which exists to give information to help avoid unwanted pregnancies and that is what we are talking about....

Whatever the case, the fact remains that we cannot leave the entire responsibility of sex education to parents. It needs specialists in education, with the parents co-operating in every way they are able....

Another responsibility. Can the prosecution claim that genuine free contraception, accessible to everybody, exists in France? I am not speaking of do-it-yourself amateurish methods which we all use today. I am speaking of proper contraception....

Contraception, at the present time, is used by about six per cent or eight per cent of women. In what section of the population? Among working-class people it is one per cent. And it is, of course, the women in this part of the population who suffer....

This law, there isn't much left of it.... It could at least be coherent. It should at least have some cohesion with other laws. I felt it

was essential that you hear an unmarried mother testify. I hope the Court was moved by her testimony. There are girls here, young girls who have gone through to the end of their pregnancies for various reasons, but let us say, because they respect the law, this famous clause 317.[12] They go through to the end. And what do we do for them? We treat them as outcasts. We take their children away from them. More often than not we force them to abandon them. We take eighty per cent of their salary. It is a real victimisation that is meted out to unmarried mothers....

I come to what seems to me most important in the condemnation of this law. This law cannot survive, Gentlemen, and not one moment more—if I am listened to. Why? For my part, all that needs to be said is this: because it is fundamentally an infringement of the liberty of woman—that half of humanity that has always been oppressed....

To be brought before you in Court—does not this symbolise our oppression?... This claim of ours, elementary, physical, this birthright—to have this control of ourselves, of our bodies—when we ask for this right, from whom do we ask it? From men. We have to apply to you....

Would you agree, Gentlemen, to be brought to trial before a court of women because you had done as you thought fit with your bodies? ... That would sound crazy to you.

Applause in Court...

I claim the right to put questions to those people who proclaim respect for life, those people who talk about it to us when we talk about our rights in the matter of procreation.

Was life being respected when, without any word of protest, we let six million Jews be massacred in the gas ovens? Was life being respected in Algeria, when torture and napalm were systematically used? Was life respected in the Vietnamese children I myself have seen, their bodies riddled by anti-personnel bombs?

---

[12]Article 317 of the Criminal Code made abortion a crime punishable by a period of imprisonment from six months to one year and a fine of 360 to 7,200 francs.

Or in the colonial wars? Or in the factories with their infernal din? ...

Is it showing respect for the life of the infant, to force the woman to give birth? Everything except the mother and the child seems to have the respect here. I repeat, the foetus is respected by our legislators, but after it ceases to be a foetus, the rest of its life seems to interest them but little.

Do you believe, Gentlemen, that it is good for a child to be born unwanted? Is it good for the person that he will become? ...

Suppose Marie-Claire, last year, at the age of sixteen, had decided to have this child. Do you really think she would have been able to look after it, give it a decent education and make it happy, and be able to develop and to realise her own potential at the same time?

Something that impressed me, Gentlemen, was a survey conducted by S.O.F.R.E.S.[13] last year, which was published in a very devout newspaper *Le Pelerin*.[14] It addressed a series of questions to some young French couples on their reasons for limiting their families. Here is a list of their replies:

Insufficient family income	78 per cent
Difficulties for young people in finding jobs	74 per cent
Insecurity of employment	73 per cent
Difficulties in obtaining accommodation	68 per cent
Insufficient social facilities	61 per cent
Impossibility for low-income families to provide adequate education for their children	72 per cent

When I speak of freedom I mean being in possession of the power to practise that freedom; and the great lack in our system of society is that we do not grant women the power to make a choice in the matter of parenthood. ...

Look, Gentlemen, since the outset of this affair—and this is something a court in a democratic country cannot ignore—a chain of solidarity has been forged around Marie-Claire, Madame Chevalier, Madame Duboucheix, Madame Sausset and Madame Bambuck and all the women whom they represent. This chain of support is the expression of a responsible choice. It began underground in the Metro. I have in my briefcase three hundred signatures of Metro employees from all the stations on the line where Mme Chevalier worked—Montmartre, Drouot, Robespierre, Républic, Charonne, all of them. All her colleagues joined in her support. In that, they took on a responsibility—they added their small personal and material contribution to symbolise their own participation in this arraignment before you, for they could see themselves in the same situation. Their solidarity is complete.

You have our witnesses too. I don't think there was one who would deny support for Mme Chevalier and those accused with her. Each witness has stated his or her share in the same responsibility. One witness came and said to you, 'I myself would have performed this abortion operation.' Another, not a gynaecologist, said, 'I would have sent her to a gynaecologist.' ...

There are all the telegrams you have received. There are a hundred signatures from Seine-Saint-Denis[15] that we call 'the anonymous ones.' The women who say to you, 'You cannot convict these women.' Three hundred signatures from Valence. ...

There has been an overwhelming mail. You have to see it at association *Choisir's* office to believe it. Men and women give their support to Marie-Claire, Mme Chevalier and the others: 'You cannot convict them' ...

For you, for us, for all of us who await your decision, and for your good conscience, it is no use to allege that you only wish to concern yourselves with the facts, those facts spoken of by the Public Prosecutor. And it is of no use

---

[13]S.O.F.R.E.S., Société Française d'Estimation.

[14]*Le Pelerin*, The Pilgrim.

[15]The district in which the Court was located.

just to say that the law, whether good or bad, is the law. This attitude is a refusal to face responsibilities, and also—I must speak frankly—it is unworthy of the magistrature. It would amount to saying, Gentlemen, that you are there merely for the mechanical application of a law... They say you must 'uphold the law.' Quite so. But to 'uphold the law' has never meant to become a robot judge and to ignore the great problems of our life. 'To uphold the law is to do justice.' We learnt it in our law studies, M. Prosecutor. I can still remember the essays of my student days.... We learned how, by prolonged efforts, up to the highest Court of Appeal, a law which was no longer at one with social realities, or was fundamentally unjust or immoral, came to be modified, not by the legislature, but by you, Gentlemen, the magistrates....

I beg you for the sake of all women, to choose courage, to take the road taken by those who preceded you in this very Court less than a month ago. Follow them and declare acquittal for all; Mme Chevalier, Mme Sausset, Mme Bambuck and Mme Duboucheix must all be discharged.

When you have finished your notes and have put them aside, ask yourselves this: Do we have the right in France today, in a country that prides itself on being civilised, to send women to prison for having taken into their own hands the control of their own bodies, or for having helped one of their own number to do the same? I am not evading the difficulty—that is why I speak of courage, but this pronouncement of acquittal will be irreversible, and, following your lead, the legislators will deal with it. We say this to you: it has to be said, for we women, half of humanity, are on the move. We shall no longer accept the continuation of this oppression. Gentlemen, it is up to you to say today, 'An era has ended and a new one has begun.'

Applause in court.

Presiding Judge—The matter will be considered. The judgment will be given on 22 November 1972, that is to say, in a fortnight.

175. Simone de Beauvoir "I Am a Feminist"

Simone de Beauvoir (1908–1986) shaped the literary and political landscape of postwar France and provided the theoretical foundations for second-wave feminism in the United States and Europe. De Beauvoir and Jean-Paul Sartre, her lifelong companion, exemplified the "engaged intellectual" of the postwar period. Their journal *Les Temps Modernes* promoted existentialist and, later, feminist thought. De Beauvoir and Sartre were also active in mobilizing French support for Algerian independence.[1] Best known for *The Second Sex* (1949), in which she emphasized, "one is not born, but rather becomes, a woman," de Beauvoir's publications include a wide array of journalistic articles, novels, and memoirs, such as *The Blood of Others* (1945, English translation 1948), *The Mandarins* (1954, English translation 1960), and *Memoirs of a Dutiful Daughter* (1958, English translation 1959). Not until the rise of second-wave feminism in the 1960s and 1970s, however, did the ideas contained in *The Second Sex* become influential on a mass basis. It was at this time, too, that de Beauvoir became involved in various feminist issues, such as the right to reproductive choice.[2]

---

[1]See in this chapter, Source 172, "The Case of Djamila Boupacha."

[2]See in this chapter, Source 174, "The Right to Choose: Abortion on Trial in France."

Excerpts from pp. 29–48 of *After the Second Sex: Conversations with Simone de Beauvoir* by Alice Schwarzer, translated by Marianne Howarth. © by Pantheon Books. (Also published as *Simone de Beauvoir Today* by Chatto & Windus, Ltd.)

She also helped to found and served as president of the League of Women's Rights which campaigned against sexist legislation. De Beauvoir's insights continue to generate activist as well as scholarly debate. In 1999, 140 scholars from 31 countries presented their ongoing research on de Beauvoir at a conference held in Paris to commemorate the fifty-year anniversary of *The Second Sex*. Although de Beauvoir typically described her work as derivative of Sartre's philosophical insights, stating, "Philosophically I only had the role of a disciple," some scholars now credit her with certain crucial insights of existentialism, such as the concept of "otherness," which de Beauvoir discussed in her unpublished diaries as early as 1927 when she was a philosophy student at the Sorbonne. In this 1972 interview with Alice Schwarzer, originally published in *Le Nouvel Observateur* and then anthologized in *After the Second Sex: Conversations with Simone de Beauvoir* (1984), de Beauvoir reflects on *The Second Sex* and the process by which she became a feminist activist.

ALICE SCHWARZER] Your analysis of the situation of women is still the most radical we have, in that no author has gone further than you have since your book *The Second Sex* came out in 1949, and you have been the main inspiration for the new women's movements. But it is only now, twenty-three years later, that you have involved yourself actively in women's actual, collective struggle. You joined the International Women's March last November. Why?

SIMONE DE BEAUVOIR] Because I realised that the situation of women in France has not really changed in the last twenty years. There have been a few minor things in the legal sphere, such as marriage and divorce law. And the availability of contraception has increased—but it still does not go far enough, given that only seven per cent of all French women take the Pill. Women haven't made any significant progress in the world of work either. There may be a few more women working now than there were, but not very many. But in any case, women are still confined to the low-grade jobs. They are more often secretaries rather than managing directors, nurses rather than doctors. The more interesting careers are virtually barred to them, and even within individual professions their promotion prospects are very limited. This set me thinking. I thought it was necessary for women who really wanted their situation to change to take matters

into their own hands. Also, the women's groups which existed in France before the MLF[1] was founded in 1970 were generally reformist and legalistic. I had no desire to associate myself with them. The new feminism is radical, by contrast. As in 1968, its watchword is: change your life today. Don't gamble on the future, act now, without delay.

When the women in the French women's movement got in touch with me, I wanted to join them in their struggle. They asked me if I would work with them on an abortion manifesto, making public the fact that I, and others, had had an abortion. I thought this was a valid way of drawing attention to a problem which is one of the greatest scandals in France today: the ban on abortion.[2]

So it was quite natural for me to take to the streets and to join the MLF militants in the march [in November 1971] and to adopt their slogans as my own. Free abortion on demand, free contraception, free motherhood! ...

A. S.] The term 'feminism' is much misunderstood. What is your definition of it?

S. de B.] At the end of *The Second Sex* I said that I was not a feminist because I believed that

---

[1] Mouvement de Libération des Femmes.

[2] For a discussion of this manifesto, see "The Right to Choose: Abortion on Trial in France," Source 174 in this chapter.

the problems of women would resolve themselves automatically in the context of socialist development. By feminist, I meant fighting on specifically feminine issues independently of the class struggle. I still hold the same view today. In my definition, feminists are women— or even men too—who are fighting to change women's condition, in association with the class struggle, but independently of it as well, without making the changes they strive for totally dependent on changing society as a whole. I would say that, in that sense, I am a feminist today, because I realised that we must fight for the situation of women, here and now, before our dreams of socialism come true. Apart from that, I realised that even in socialist countries, equality between men and women has not been achieved. Therefore it is absolutely essential for women to take their destiny into their own hands. That is why I have now joined the Women's Liberation Movement.

There is another reason—and I believe that it is one of the reasons why so many women have come together to found the movement— namely, that a profound inequality exists between men and women even in left-wing and revolutionary groups and organisations in France. Women always do the most lowly, most tedious jobs, all the behind-the-scenes things, and the men are always the spokesmen; they write the articles, do all the interesting things and assume the main responsibility. So, even within these groups, whose theoretical aim is to liberate everybody, including women, even there women are still inferior. It goes still further. Many—not all—men on the left are aggressively hostile to women's liberation....

A. S.] You said in a comment on *The Second Sex* that the problem of femininity had never affected you personally, that you found yourself 'in a position of great impartiality'. Do you mean to say that individually a woman can escape her female condition? Professionally, as well as in her relationships with her fellow human beings?

S. de B.] Escape one's female condition completely? No! I have the body of a woman—but clearly I have been very lucky. I have escaped many of the things that enslave a woman, such as motherhood and the duties of a housewife. And professionally as well—in my day there were fewer women who studied than nowadays. And, as the holder of a higher degree in philosophy, I was in a privileged position among women. In short, I made men recognise me: they were prepared to acknowledge in a friendly way a woman who had done as well as they had, because it was so exceptional. Now that many women undertake serious study, men are fearful for their jobs. Admitting, as I have done, that a woman doesn't necessarily have to be a wife and mother to have a fulfilled and happy life, means that there will be a certain number of women who will be able to have fulfilled lives without suffering the enslavement of women. Of course they have to be born into a privileged family or possess certain intellectual abilities.

A. S.] You once said, 'the greatest success of my life is my relationship with Sartre'...

S. de B.] Yes.

A. S.]...yet all your life you have always had a great need for your own independence and a fear of being dominated. Given that it is very difficult to establish relationships between men and women that are based on equality, do you believe that you personally have succeeded?

S. de B.] Yes. Or rather, the problem never arose, because there is nothing of the oppressor about Sartre. If I'd loved someone other than Sartre, I would never have let myself be oppressed. There are some women who escape male domination, mostly by means of their professional autonomy. Some have a balanced relationship with a man. Others have inconsequential affairs.

A. S.] You have described women as an inferior class...

S. de B.] I didn't say class. But in *The Second Sex* I did say that women were an 'inferior caste', a caste being a group one is born into and cannot move out of. In principle, though, one can transfer from one class to another. If you are a woman, you can never become a man. Thus women are genuinely a caste. And the way women are treated in economic, social and political terms makes them into an inferior caste.

A. S.] Some women's movements define women as a class outside the existing classes. They base this on the fact that housework, which has no exchange value, is done exclusively by women for nothing. As they see it, patriarchal oppression is therefore the main contradiction, not a subsidiary one. Do you agree with this analysis?

S. de B.] I find the analysis lacking on this point. I'd like someone to do some serious work on it. In *Women's Estate,* for example, Juliet Mitchell[3] showed how to ask the question, but she doesn't claim to resolve it in that book. I remember it was one of the first questions I put when I first came into contact with the militant feminists in the MLF: what, in your view, is the exact connection between patriarchal oppression and capitalist oppression? At the moment, I still don't know the answer. It's a problem which I'd very much like to work on in the next few years. I'm extremely interested in it. But the analyses which regard patriarchal oppression as the equivalent of capitalist oppression are not correct in my view. Of course, housework doesn't produce any surplus value. It's a different condition than that of the worker who is robbed of the surplus value of his work. I'd like to know exactly what the relationship is between the two. Women's entire future strategy depends upon it.

It's very right to emphasise unpaid housework. But there are many women who earn their own living, and who cannot be considered as exploited in the same way as housewives are....

A. S.] You've also been very active in the class struggle, since May 1968. For instance, you've assumed responsibility for a radical left magazine. You've taken to the streets. In brief, what is the connection between the class struggle and the war between the sexes, in your opinion?

S. de B.] What I have been able to establish is that the class struggle in the strict sense does not emancipate women. That has made me change my mind since *The Second Sex* was published. It does-

n't matter whether you're dealing with Communists, Trotskyists or Maoists, women are always subordinate to men. As a result, I'm convinced of the need for women to be truly feminist, to take their problems into their own hands.

... When I wrote *The Second Sex,* I was very surprised at the bad reception it got from the left. I remember one discussion I had with the Trotskyists who said that the women's problem was not a true problem, and that it simply didn't exist. When the revolution came, women would automatically find their place.

It was the same with the Communists... who exposed me to a great deal of ridicule. They wrote articles saying that the working-class women in Bilancourt[4] really couldn't give a damn about the women's problem. Once the revolution had taken place, women would be equal with men. But they were not interested in what would happen to women in the time it took for the revolution to come....

A. S.] After *The Second Sex* was published, you were often accused of not having developed any tactics for the liberation of women and of having come to a halt in your analysis.

S. de B.] That's right. I admit it was a shortcoming in my book. I finish with vague confidence in the future, the revolution and socialism.

A. S.] And today?

S. de B.] I have changed my views now. As I've been telling you, I really am a feminist.

A. S.] What concrete possibilities do you see for the liberation of women on an individual and on a collective level?

S. de B.] On an individual level, women must work outside the home. And, if possible, they should refuse to get married. I could have married Sartre but I think we were wise not to have done so, because when you are married, people treat you as married and eventually you think of yourself as married. As a married woman, you simply do not have the same rela-

---

[3]On Juliet Mitchell, see Source 173 in this chapter.

[4]The Communist Party had a strong influence in Boulogne Bilancourt, an industrial area outside of Paris and the site of a large Renault plant with an established history of labor militancy.

tionship with society as an unmarried woman. I believe that marriage is dangerous for a woman.

Having said that, there can be reasons for it—if you want to have children, for example. Having children is still very difficult if the parents are not married because the children encounter all sorts of difficulties in life.

What really counts, if one wants to be truly independent, is work, a job. That is my advice to all women who ask me. It is a necessary precondition. If you are married and want a divorce, it means you can leave, and support your children, and have a life of your own. Of course, work is not a miracle cure. Work today does have a liberating side, but it is also alienating. As a result, many women have to choose between two sorts of alienation: the alienation of the housewife and that of the working woman. Work is not a panacea, but all the same, it is the first condition for independence.

A. S.] And what about the women who are already married and have children?

S. de B.] I think there are some women who really don't stand much of a chance. If they are thirty-five, with four children to cope with, married and lacking any professional qualifications—then I don't know what they can do to liberate themselves. You can only talk about the real prospect of liberation for future generations.

A. S.] Can women who are struggling for their liberation do so as individuals, or must they act collectively?

S. de B.] They must act collectively. I myself have not done so up to now because there was no organised movement with which I was in agreement. But all the same, writing *The Second Sex* was an act which went beyond my own liberation. I wrote that book out of concern for the feminine condition as a whole, not just to reach a better understanding of the situation of women, but also to contribute to the struggle and to help other women to understand themselves.

In the last twenty years I have received an enormous number of letters from women telling me that my book has been a great help to them in understanding their situation, in their struggle and in making decisions for themselves. I've always taken the trouble to reply to these women. I've met some of them. I've always tried to help women in difficulties....

A. S.] Now that you describe yourself as a militant feminist and have involved yourself in the active debate, what action do you intend to take in the immediate future?

S. de B.] I'm working on a project with a group of women. We want to organise a kind of public hearing to denounce the crimes committed against women. The first two days will be devoted to questions of motherhood, abortion and contraception, and will take place on 13 and 14 May in the Mutualité hall in Paris. There will be a sort of committee of enquiry composed of about ten women. They will cross-examine the witnesses, who will include biologists, sociologists, psychiatrists, doctors, midwives and, above all, women who have suffered from the conditions society imposes upon women.

We hope to convince the public that women must be assured of the right to procreate freely, of public support for the burdens of motherhood—especially crèches—and of the right to refuse unwanted pregnancies through contraceptive measures and abortion. We are demanding that these be free and that women have the right to choose.

A. S.] The struggle of the women's liberation movements is often linked with the struggle for free abortion. Do you personally want to go beyond this stage?

S. de B.] Of course. I think the women's movements, including me, will have to work together on many things. We are not only fighting for free abortion but also for widespread availability of contraceptives, which will mean that abortion will only play a marginal role. On the other hand, contraception and abortion are only a point of departure for the liberation of women. Later on, we will be organising other meetings at which we will expose the exploitation of female labour, be it as a housewife, a white-collar worker, or a working-class woman.

176. Wages for Housework in Italy

The campaign for "wages for housework" was initiated in the early 1970s by women's groups within the autonomous women's movement in Italy. The campaign was carried out in the context of a new law legalizing divorce as well as the movement for reproductive choice which faced intense opposition from the Catholic Church. These documents from 1973 present two different feminist viewpoints concerning the implications of wages for housework paid by the state. The first, "Introduction to the Debate," was written by the Padua-based group Lotta Femminista (Feminist Struggle) which made wages for housework central to its activities. It organized meetings and conferences and joined with other feminist groups to create the Triveneto Committee on salaries for housework. The second statement was issued by the Rome-based Movimento Femminista Romano (Feminist Movement of Rome) at a meeting with Lotta Femminista in Naples.

A. Lotta Femminista, Introduction to the Debate, 1973

For us women, Marx has never been a myth. 'Our' Marx didn't waste too many words on women and on their work, domestic work. It's also true that at the time he was writing, women workers scarcely had the time to reproduce themselves and their children. But if he was a theoretician of a whole period, he must have perceived the centrality of domestic work. So if we feminists quote some of his work, it is to make our comrades who refer to him all the time go back and re-read more carefully the passages where we can perhaps see him approaching the problem.

Let us begin where he approaches the problem but does not touch it:

> the worker... gives himself means of subsistence to keep up his working strength, just as a steam engine is given water and coal, and a wheel is given oil. So the worker's means of consumption are pure and simple means of consumption of a means of production, and the *individual consumption of the worker is a directly productive consumption.*

Except that, and our man does not see this, this consumption presupposes work of some kind.

This work is housework.

Housework is done by women.

This work has never been seen, precisely because it is not paid.

It seems to us no accident that theoretical obsession with productive work has never even touched on the problem of the productivity of housework. If Marx at least came close to the problem, his followers have always carefully kept their distance.

As for the *workers,* we acknowledge their hard struggle over pay, at the moment of production in the factory.

But the workers' struggle has always failed to include the reproduction of working strength and the absence of pay which mystified that reproduction.

The workers have never turned their attention to that part of the productive cycle which has always been carried out without pay.

This is no accident. This was women's work, the other half of the class which, as a class, has been denied by all.

One part of the class with a salary, the other without. This discrimination has been the basis of a stratification of power between the paid and the non-paid, the root of the class weakness which movements of the left have only increased.

Just to quote some of their commonplace accusations, we are 'interclassist', 'corporative', we 'split the class', and so on, and so on.

Even today more than half of the world's population *works without remuneration.*

So women's demand for pay for housework is today the most revolutionary and strategic demand for the whole class.

B. Movimento Femminista Romano, Statement at a Meeting with Lotta Femminista, 1973

Giving housewives a salary means transforming a mass of enslaved individuals, who work without pay or organized hours, into a salaried workforce.

But what will this salary consist of?

The members of Lotta Femminista rightly compare the demand for salaries for housewives to social security (the so-called 'welfare' of Ango-Saxon countries); and it would in fact be nothing but a concession, a charity on the part of the government. This is because housewives, given the private character of housework and emotional implications connected with it, do not possess any contractual power. By what means could they obtain a salary rise? Would they leave children and husbands, the elderly and the sick, and so on, in order to take part in demonstrations? Would they fold their arms and refuse to do the housework, knowing that, afterwards, they would have to slave twice as hard in order to regain lost time? And, finally, who would pay the social cost of this salary if not the women themselves, through price increases and salary cuts elsewhere in the workforce? Nevertheless, even a small income concession may appear to be an achievement to women who have always worked for free, and without recognition, and who have been forced to be totally dependent on their husbands or fathers. In our opinion the demand for salaried housewives conceals elements which are not only reformist, but also dangerous, because:

1 it confines women to their traditional role.
2 it endorses their social function as private, though it grants it some social, economic and symbolic value.
3 it stands in antithesis to the demand for social services and to the involvement of men in them.
4 it defines as work an activity which cannot be considered as such, because it does not have time limits or precise methods.
5 it presents itself as an obstacle to the inclusion of women in the world of active production, because it will facilitate redundancies and self-exclusions.
6 it recognizes the scientific organization of work based on the division of labour (manual/intellectual, master/servant), and the specific division based on sexual difference, which, as Engels says, is the foundation of all other divisions.

It is, in the end, a dangerous struggle. At this point, we'd like to remind you of the fact that the female section of the Francoist[1] party in Spain is currently trying to introduce salaried housework as its national policy. This demand is based on the principle that it will strengthen the family and, consequently, the state. We should bear in mind that Franco and his party are so conscious of the fundamental role of women in the family as the pillar of the present regime that they have created a special law which forbids writing or saying anything against the family. Our slavery has continued for thousands of years, and we can understand the impatience to find immediate tactical goals to galvanize women into action. In this respect, though, and in order to become politically aware of our exploitation and oppression, the salary demand appears restricting, because it does not question the real power-relations of men and women, and the whole ideological basis on which this power rests, but simply makes of it a question of economic discrimination and unpaid work. We then ask ourselves: what is the actual meaning of a struggle for salary? If it is about a mobilization for its own sake, then we must bear in mind that that would not really be an autonomous fight, in that it would require men as intermediaries, because they are in direct contact with the production of

---

[1] The fascist party of Francisco Franco.

exchange value. They are also possible supporters of this struggle, since it does not question their role in profiting from a series of privileges connected with it.

We think that the fight against the patriarchal and the middle-class ideology of the family is fundamental for the liberation of women and of the whole society. Women must be able to refuse marriage in order to achieve full independence.

A massive request for jobs for all women who are currently housewives would go against the system, which cannot renounce gratuitous housework.

The right to manage our own bodies, the collectivization by the state of all the social ser-vices (canteens, local laundries and so on) currently provided for free by housewives, collective education of small children, equality of work with no sex discrimination—these are our objectives. But to achieve them women must come to realize their objective condition. It is impossible to say today how this will happen or how long it will take, but women's struggle has only just begun. The fact that it is spreading throughout the world shows that it responds to needs so real that it could put a bomb under all the contradictions of bourgeois, patriarchal and male society.

177. The United Nations 1979 Convention on the Elimination of All Forms of Discrimination Against Women

The 1979 United Nations Convention on the Elimination of All Forms of Discrimination Against Women (CEDAW) amplifies the rights enumerated in the UN Universal Declaration of Human Rights of 1948. On the insistence of Eleanor Roosevelt, who played an important role in the UN's establishment, and the UN Commission on the Status of Women, which first met in New York in 1947, the Universal Declaration defined human rights broadly to include social and economic as well as political rights. In the decades subsequent to the Universal Declaration, the UN Commission on the Status of Women and women's groups and movements throughout the world pressured the United Nations to become more responsive to women's issues, to include more women in its organizational and leadership bodies, and to define women's rights as human rights in more specific terms than the Universal Declaration. Adopted by the UN General Assembly during the UN Decade for Women (1976–1985), CEDAW has been ratified by 163 countries with 22 countries remaining to ratify, including the United States. A pledge was made at the Fourth World Congress of Women held in Beijing in 1995 for all UN members to ratify by the year 2000.[1] The implementation of CEDAW is monitored by the UN Committee on the Elimination of Discrimination Against Women. Excerpted here is the introduction and Part 1 of CEDAW.

---

[1]For an assessment of CEDAW's significance and accomplishments, see Patricia Flor, Source 187 in Chapter 15.

The States Parties to the present Convention,

Noting that the Charter of the United Nations re-affirms faith in fundamental human rights, in the dignity and worth of the human person and in the equal rights of men and women,

Noting that the Universal Declaration of Human Rights affirms the principle of the inadmissibility of discrimination and proclaims that all human beings are born free and equal in dignity and rights and that everyone is entitled to all the rights and freedoms set forth therein, without distinction of any kind, including distinction based on sex,

Noting that the States Parties to the International Covenants on Human Rights have the obligation to ensure the equal right of men and women to enjoy all economic, social, cultural, civil and political rights,

Considering the international conventions concluded under the auspices of the United Nations and the specialized agencies promoting equality of rights of men and women,

Noting also the resolutions, declarations and recommendations adopted by the United Nations and the specialized agencies promoting equality of rights of men and women,

Concerned, however, that despite these various instruments extensive discrimination against women continues to exist,

Recalling that discrimination against women violates the principles of equality of rights and respect for human dignity, is an obstacle to the participation of women, on equal terms with men, in the political, social, economic and cultural life of their countries, hampers the growth of the prosperity of society and the family and makes more difficult the full development of the potentialities of women in the service of their countries and of humanity,

Concerned that in situations of poverty women have the least access to food, health, education, training and opportunities for employment and other needs,

Convinced that the establishment of the new international economic order based on equity and justice will contribute significantly towards the promotion of equality between men and women,

Emphasizing that the eradication of apartheid, of all forms of racism, racial discrimination, colonialism, neocolonialism, aggression, foreign occupation and domination and interference in the internal affairs of States is essential to the full enjoyment of the rights of men and women....

Convinced that the full and complete development of a country, the welfare of the world and the cause of peace require the maximum participation of women on equal terms with men in all fields...

Aware that a change in the traditional role of men as well as the role of women in society and in the family is needed to achieve full equality between men and women,

Determined to implement the principles set forth in the Declaration on the Elimination of Discrimination against Women and, for that purpose, to adopt the measures required for the elimination of such discrimination in all its forms and manifestations,

Have agreed on the following:

Part I

Article 1. For the purposes of the present Convention, the term "discrimination against women" shall mean any distinction, exclusion or restriction made on the basis of sex which has the effect or purpose of impairing or nullifying the recognition, enjoyment or exercise by women irrespective of their marital status, on a basis of equality of men and women, of human rights and fundamental freedoms in the political, economic, social, cultural, civil or any other field.

Article 2. States Parties condemn discrimination against women in all its forms, agree to pursue by all appropriate means and without delay a policy of eliminating discrimination against women and, to this end, undertake:

(a) To embody the principle of the equality of men and women in their national constitutions or other appropriate legislation if not yet incorporated therein and to ensure, through law and other appropriate means, the practical realization of this principle;

(b) To adopt appropriate legislative and other measures, including sanctions where appropriate, prohibiting all discrimination against women;

(c) To establish legal protection of the rights of women on an equal basis with men and to ensure through competent national tribunals and other public institutions the effective protection of women against any act of discrimination;

(d) To refrain from engaging in any act or practice of discrimination against women and to ensure that public authorities and institutions shall act in conformity with this obligation;

(e) To take all appropriate measures to eliminate discrimination against women by any person, organization or enterprise;

(f) To take all appropriate measures, including legislation, to modify or abolish existing laws, regulations, customs and practices which constitute discrimination against women;

(g) To repeal all national penal provisions which constitute discrimination against women.

Article 3. States Parties shall take in all fields, in particular in the political, social, economic and cultural fields, all appropriate measures, including legislation, to ensure the full development and advancement of women, for the purpose of guaranteeing them the exercise and enjoyment of human rights and fundamental freedoms on a basis of equality with men.

Article 4. 1. Adoption by States Parties of temporary special measures aimed at accelerating *de facto* equality between men and women shall not be considered discrimination as defined in the present Convention, but shall in no way entail as a consequence the maintenance of unequal or separate standards; these measures shall be discontinued when the objectives of equality of opportunity and treatment have been achieved.

2. Adoption by States Parties of special measures, including those measures contained in the present Convention, aimed at protecting maternity shall not be considered discriminatory.

Article 5. States Parties shall take all appropriate measures:

(a) To modify the social and cultural patterns of conduct of men and women, with a view to achieving the elimination of prejudices and customary and all other practices which are based on the idea of the inferiority or the superiority of either of the sexes or on stereotyped roles for men and women;

(b) To ensure that family education includes a proper understanding of maternity as a social function and the recognition of the common responsibility of men and women in the upbringing and development of their children, it being understood that the interest of the children is the primordial consideration in all cases.

Article 6. States Parties shall take all appropriate measures, including legislation, to suppress all forms of traffic in women and exploitation of prostitution of women.

QUESTIONS FOR ANALYSIS

1. What changes does Juliet Mitchell call for in the four structures of production, reproduction, socialization of children, and sexuality? How does she argue against reformism and voluntarism? What does she propose as an alternative strategy? What is the meaning of her concluding statement?

2. On what basis does the Choisir leaflet call for a demonstration on behalf of Marie-Claire? Based on their court testimonies, discuss the individual circumstances and views on abortion of Marie-Claire Chevalier; Michèle Chevalier, her mother; and Micheline Bambuck, who performed the abortion. How does the attorney Gisèle Halimi place the French abortion law on trial in her speech for the defense? What reasons does she give for a woman's right to reproductive choice?

3. What led Simone de Beauvoir to become active in feminist causes in the 1970s? How does she explain her statement in *The Second Sex* that she "was not a feminist"? Discuss her analysis of the

political differences within the women's movement in France. How does she analyze the relationship between the struggle for women's emancipation and the struggle for socialism?

4. On what basis did Italian feminists argue for wages for housework? How do their views on housework compare with those of Eleanor Rathbone, Source 171 in this chapter, and Alexandra Kollontai, Source 141 in Chapter 11?

5. In what sense is the Convention on the Elimination of All Forms of Discrimination Against Women an international bill of women's rights? How does it define women's rights in political, economic, social, and legal terms?

6. Based on the documents provided in this chapter, compare and contrast the issues and perspectives of the various feminist theories, organizations, and movements that emerged in the 1960s and 1970s.

꧁ CHAPTER 15

The 1980s to the Present

The oil crisis of 1973 dramatically increased the costs of industrial production and signaled the end of the post–World War II boom. In the West, Keynesian economics was first abandoned by Margaret Thatcher who brought the Conservative Party to power in 1979. Thatcher's free market approach ended the postwar Conservative and Labor consensus that the priority of government policies was to reduce unemployment rather than inflation even if this required sustained deficit spending. A staunch supporter of NATO (North Atlantic Treaty Organization), Thatcher embraced NATO plans to station intermediate cruise missiles throughout Western Europe, including Britain. In response, the Soviet Union constructed similar missile sites in former East Germany. Peace protests were organized in both the East and the West. The most sustained opposition took place in Greenham Common, England, where a Women's Peace Camp initiated in 1981 drew supporters from all over the world.

In the 1970s and 1980s, dissident movements in Eastern Europe assumed various forms, such as the Charter 77 movement in Czechoslovakia and the Solidarity movement in Poland. Significant variations existed among these movements, but they generally called for multiparty elections, autonomous women's and trade union movements, and freedom of the press. A key turning point in Soviet relations with Eastern Europe occurred when Mikhail Gorbachev became the general secretary of the Central Committee of the Soviet Communist Party in 1985 and announced a new policy of *glasnost,* or "openness." The policy encouraged dissent as it implied that the Soviet Union would no longer intervene militarily in Eastern Europe, as it had in East Germany in 1956 and in Czechoslovakia during the Prague Spring of 1968 and as it had threatened during the early years of the Solidarity movement in the 1980s.

The fall of the Berlin Wall in 1989 signaled the beginning of the end of the Cold War, but no one could have predicted the consequences of this political development for the countries of Eastern Europe and the Soviet Union. For a brief period in 1990, dissident East German groups, which included an autonomous women's association, participated in the Round Table, a provisional governmental arrangement that aimed to create a new democratic and socialist German Democratic Republic (GDR). But the new process of political transformation was soon co-opted by a movement for the unification of Germany, long a goal of the then ruling Conservative Christian Democratic government of West Germany and the United States. In 1991, the Soviet Union itself began to disintegrate, as one republic after another declared independence in what has been described as "a revolution from the top." The Cold War was now over, but so too were the initial attempts at creating new societies in the Soviet Union and Eastern Europe within the framework of socialism.

The rapid introduction of capitalism affected women in particularly adverse ways. Women experienced the greatest percentage of unemployment, while also losing essential social services such as subsidized child care, education, health care, and housing. In 1999, the United Nations Children's Fund (UNICEF), directed by Carol Bellamy, carried out the first comprehensive examination of how the emergence of market economies has transformed the lives of some 150 million women and 50 million girls in 27 countries in Central and Eastern Europe and the former Soviet Union. The *Women in Transition* report was intended to coincide with the twenty year anniversary of the 1979 UN Convention on the Elimination of All Forms of Discrimination Against Women (CEDAW).[1] It concluded that the situation for women and girls "has spiraled downward since the collapse of Communism."[2] Specifically, of some 25 million jobs lost in the 1990s, over 14 million were held by women; in most countries, women's political representation averaged less than 10 percent, a one-third reduction of their former numbers in office; and women's life expectancy had decreased in sixteen countries, in part due to a dramatic increase in the sexual trafficking of women.[3]

One of the main challenges facing Europe in the twenty-first century is the issue of multiculturalism. The dissolution of Yugoslavia, which began with Slovenia's declaration of independence in 1991, was accompanied by several conflicts and wars revolving around ethnic and religious issues. The war in Bosnia (1992–1995) was the bloodiest European conflict of the last quarter of the twentieth century. In Northern Ireland, by contrast, a peaceful resolution of the long-standing conflicts between the minority Catholic and majority Protestant populations appears to be in sight following an unprecedented round of peace talks car-

---

[1] For an excerpt of the text of CEDAW, see Source 177 in Chapter 14. For a twenty year evaluation of CEDAW, see Patricia Flor, Source 187 in this Chapter.

[2] "Free Markets Leave Women Worse Off, Unicef Says," *New York Times,* September 23, 1999, p. 9.

[3] See *Women in Transition.* Regional Monitoring Reports, No. 6. Florence: UNICEF International Child Development Center (1999). Introduction and Overview, pp. vii–xi.

ried out between 1996 and 1998 under the auspices of British Prime Minister Tony Blair's Labor Party government. Within many countries of Western Europe, the issue of multiculturalism has acquired a new significance as the European Union is establishing new, uniform laws for immigration. Some have criticized Europe for attempting to construct a fortress around itself with new, more restrictive policies. At the same time, the economic crisis and high levels of unemployment in Germany and France, in particular, have fueled the growth of xenophobic right-wing parties and skinhead youth groups. Many openly idealize the Nazi Party and have carried out violent assaults on immigrants and their children, reminiscent of the racist and anti-Semitic attacks of the 1930s.

What will be the future for women in the new Europe of the twenty-first century? This will depend on a number of factors: the as-yet-unknown consequences of European unification; Europe's role in the new global economy; European immigration and citizenship policies; the general configuration of political forces; the strength of European women's organizations, such as the European Women's Lobby; and the influence of various commissions of the United Nations. The majority of European parties in power are Social Democratic in one form or another. With Britain's Labor Party taking the lead, most of these parties advocate centrist positions that combine state intervention and support for education and social services with various free market economic measures traditionally characteristic of Conservative political parties. To meet the standards for European currency union and to compete in the global market, Social Democratic governments are privatizing their economies at an unprecedented rate, diminishing the public sector, cutting social services, and increasing the part-time sector of the work force, which is predominantly comprised of women. As the rule of the market increasingly dictates governmental policies in Europe, European-wide and international organizations will assume a more important role in protecting women's rights. The 1948 United Nations Universal Declaration of Human Rights and the 1979 Convention on the Elimination of All Forms of Discrimination Against Women will provide an especially important framework within which to uphold women's economic, social, and political rights.

THE END OF POSTWAR CONSENSUS

The Conservative Party government of Prime Minister Margaret Thatcher reoriented governmental priorities to focus on the interrelated issues of government debt and inflation. By celebrating the free market, reducing spending for social services, and dismantling key aspects of the state sector, Thatcher shattered the postwar consensus concerning the economic and societal benefits of state intervention. Many features of Thatcherism are embodied in the centrist policies of the Social Democratic parties which currently dominate the European political landscape.

178. Margaret Thatcher
Conservative Party Principles

Margaret Thatcher (b. 1925) became prime minister of Britain in 1979, the first woman in Europe to hold this office. Thatcher began her political career in 1959 as a member of the House of Commons. She rose within the Conservative Party and was chosen as the party leader in 1975. During her eleven years as prime minister, Thatcher profoundly transformed British politics and society. Reorienting the British economy toward competition in the global market, "Thatcherism" was based on three basic principles: privatization, weakening the trade unions, and reducing inflation. Thatcher's government privatized many of the industries, utilities, and transportation systems nationalized by the post–World War II Labor government. It also deregulated industry, reduced state intervention and public funding, and strengthened the free market by undermining the trade unions and promoting low wages and part-time work. Thatcher thus set into motion many of the economic precepts later adopted in modified form by centrist governments throughout Europe, including the "New" Labor Party in Britain. In foreign policy, Thatcher was a firm supporter of NATO and an advocate of a strong British military and nuclear capability. Thatcher's writings include her multivolume memoirs, *The Downing Street Years* (1993) and *The Path to Power* (1995). Reproduced here are excerpts from Thatcher's speech to the 1980 Conservative Party conference in which she staunchly defended her policies of economic restructuring against critics who condemned the resulting sharp increase in unemployment. Thatcher remains active in the Conservative Party today and is more adamant than ever in opposing Britain's participation in European unification, even claiming at a 1999 Conservative Party conference that "all of Britain's problems have come from the Continent."

At our Party Conference last year I said that the task in which the Government was engaged—to change the national attitude of mind—was the most challenging to face any British administration since the war. Challenge is exhilarating. This week we Conservatives have been taking stock, discussing the achievements, the setbacks and the work that lies ahead as we enter our second parliamentary year....

In its first seventeen months this Government has laid the foundations for recovery. We have undertaken a heavy load of legislation, a load we

do not intend to repeat, because we do not share the Socialist fantasy that achievement is measured by the number of laws you pass. But there was a formidable barricade of obstacles that we had to sweep aside. For a start, in his first Budget, Geoffrey Howe began to restore incentives to stimulate the abilities and inventive genius of our people. Prosperity comes not from grand conferences of economists but from countless acts of personal self-confidence and self-reliance.

Under Geoffrey's stewardship Britain has repaid $3,600 million of international debt, debt

Excerpted from 1980 speech to Conservative Party Conference.

which had been run up by our predecessors. And we paid quite a lot of it before it was due. In the past twelve months Geoffrey has abolished exchange controls over which British Governments have dithered for decades. Our great enterprises are now free to seek opportunities overseas. This will help to secure our living standards long after North Sea oil has run out. This Government thinks about the future. We have made the first crucial changes in trade union law to remove the worst abuses of the closed shop, to restrict picketing to the place of work of the parties in dispute, and to encourage secret ballots.

Jim Prior [the Employment Secretary] has carried all these measures through with the support of the vast majority of trade union members. Keith Joseph, David Howell, John Nott and Norman Fowler [Ministers responsible for Industry, Energy, Trade and Transport respectively] have begun to break down the monopoly powers of nationalization. Thanks to them, British Aerospace will soon be open to private investment. The monopoly of the Post Office and British Telecommunications is being diminished. The barriers to private generation of electricity for sale have been lifted. For the first time nationalized industries and public utilities can be investigated by the Monopolies commission—a long overdue reform.

Free competition in road passenger transport promises travellers a better deal. Michael Heseltine [Secretary of State for the Environment] has given to millions—yes, millions—of council tenants the right to buy their own homes. It was Anthony Eden who chose for us the goal of 'a property-owning democracy'. But for all the time that I have been in public affairs that has been beyond the reach of so many, who were denied the right to the most basic ownership of all—the homes in which they live.

They wanted to buy. Many could afford to buy. But they happened to live under the jurisdiction of a Socialist council, which would not sell and did not believe in the independence that comes with ownership. Now Michael Heseltine has given them the chance to turn a dream into reality. And all this and a lot more in seventeen months.

The Left continues to refer with relish to the death of capitalism. Well, if this is the death of capitalism, I must say that it has quite a way to go. But all this will avail us little unless we achieve our prime economic objective—the defeat of inflation. Inflation destroys nations and societies as surely as invading armies do. Inflation is the parent of unemployment. It is the unseen robber of those who have saved. No policy which puts at risk the defeat of inflation—however great its short-term attraction—can be right. Our policy for the defeat of inflation is, in fact, traditional. It existed long before ... 'monetarism' became a convenient term of political invective.

But some people talk as if control of the money supply was a revolutionary policy. Yet it was an essential condition for the recovery of much of continental Europe. Those countries knew what was required for economic stability. Previously, they had lived through rampant inflation; they knew that it led to suitcase money, massive unemployment and the breakdown of society itself. They determined never to go that way again.

Today, after many years of monetary self-discipline, they have stable, prosperous economies better able than ours to withstand the buffeting of world recession. So at international conferences to discuss economic affairs many of my fellow-Heads of Government find our policies not strange, unusual or revolutionary, but normal, sound and honest. And that is what they are. Their only question is: 'Has Britain the courage and resolve to sustain the discipline for long enough to break through to success?'

Yes, Mr. Chairman, we have, and we shall. This government is determined to stay with the policy and see it through to its conclusion. That is what marks this administration as one of the truly radical ministries of post-war Britain. Inflation is falling and should continue to fall.

Meanwhile, we are not heedless of the hardships and worries that accompany the conquest of inflation. Foremost among these is unemployment. Today our country has more than two million unemployed.

Now you can try to soften that figure in a dozen ways. You can point out—and it is quite legitimate to do so—that two million today does not mean what it meant in the 1930s; that the percentage of unemployment is much less now than it was then. You can add that today many more married women go out to work. You can stress that, because of the high birth rate in the early 1960s, there is an unusually large number of school leavers this year looking for work and that the same will be true for the next two years. You can emphasize that about a quarter of a million people find new jobs each month and therefore go off the unemployment register. And you can recall that there are nearly twenty-five million people in jobs compared with only about eighteen million in the 1930s. You can point out that the Labour Party conveniently overlooks the fact that of the two million unemployed, for which they blame us, nearly a million and a half were bequeathed by their Government. But when all that has been said, the fact remains that the level of unemployment in our country today is a human tragedy. Let me make it clear beyond doubt. I am profoundly concerned about unemployment. Human dignity and self-respect are undermined when men and women are condemned to idleness. The waste of a country's most precious assets—the talent and energy of its people—makes it the bounden duty of Government to seek a real and lasting cure.

. . . This government is pursuing the only policy which gives any hope of bringing our people back to real and lasting employment. It is no coincidence that those countries, of which I spoke earlier, which have had lower rates of inflation have also had lower levels of unemployment.

I know that there is another real worry affecting many of our people. Although they accept that our policies are right, they feel deeply that the burden of carrying them out is falling much more heavily on the private, than on the public, sector. They say that the public sector is enjoying advantages but the private sector is taking the knocks and at the same time maintaining those in the public sector with better pay and pensions than they enjoy.

I must tell you that I share this concern and understand the resentment. That is why I and my colleagues say that to add to public spending takes away the very money and resources that industry needs to stay in business, let alone to expand. Higher public spending, far from curing unemployment, can be the very vehicle that loses jobs and causes bankruptcies in trade and commerce. That is why we warned local authorities that since rates are frequently the biggest tax that industry now faces, increases in them can cripple local businesses. Councils must, therefore, learn to cut costs in the same way that companies have to.

That is why I stress that if those who work in public authorities take for themselves large pay increases, they leave less to be spent on equipment and new buildings. That in turn deprives the private sector of the orders it needs, especially some of those industries in the hard-pressed regions. Those in the public sector have a duty to those in the private sector not to take out so much in pay that they cause others unemployment. That is why we point out that every time high wage settlements in nationalized monopolies lead to higher charges for telephones, electricity, coal and water, they can drive companies out of business and cost other people their jobs.

If spending money like water was the answer to our country's problems, we would have no problems now. If ever a nation has spent, spent, spent and spent again, ours has. Today that dream is over. All of that money has got us nowhere, but it still has to come from somewhere. . . .

. . . Without a healthy economy we cannot have a healthy society. Without a healthy society the economy will not stay healthy for long.

But it is not the state that creates a healthy society. When the State grows too powerful, people feel that they count for less and less. The State drains society, not only of its wealth but of initiative, of energy, the will to improve and innovate as well as to preserve what is best. Our aim is to let people feel that they count for more and more. . . .

. . . A healthy society is not created by its institutions, either. Great schools and universities do not make a great nation any more than great armies do. Only a great nation can create and evolve great institutions—of learning, of healing, of scientific advance. And a great nation is the voluntary creation of its people—a people composed of men and women whose pride in themselves is founded on the knowledge of what they can give to a community of which they in turn can be proud.

If our people feel that they are part of a great nation and they are prepared to will the means to keep it great, a great nation we shall be, and shall remain. So, what can stop us from achieving this? What then stands in our way? The prospect of another winter of discontent? I suppose it might.

But I prefer to believe that certain lessons have been learnt from experience, that we are coming, slowly, painfully, to an autumn of understanding. And I hope that it will be followed by a winter of common sense. If it is not, we shall not be diverted from our course.

To those waiting with bated breath for that favourite media catchphrase, the 'U' turn, I have only one thing to say: 'You turn if you want to. The lady's not for turning.' I say that not only to you, but to our friends overseas and also to those who are not our friends.

In foreign affairs we have pursued our national interest robustly while remaining alive to needs and interests of others....

...[T]he hallmarks of Tory policy are, as they have always been, realism and resolve. Not for us the disastrous fantasies of unilateral disarmament, of withdrawal from NATO, of abandoning Northern Ireland.

The irresponsibility of the Left on defence increases as the dangers which we face loom larger. We for our part, under Francis Pym's brilliant leadership [as Defence secretary], have chosen a defence policy which potential foes will respect.

We are acquiring, with the co-operation of the United States Government, the Trident missile system. This will ensure the credibility of our strategic deterrent until the end of the century and beyond, and it is very important for the reputation of Britain abroad that we should keep our independent nuclear deterrent, as well as for our citizens here.

We have agreed to the stationing of Cruise missiles in this country. The unilateralists object, but the recent willingness of the Soviet government to open a new round of arms control negotiations shows the wisdom of our firmness.

We intend to maintain and, where possible, to improve our conventional forces so as to pull our weight in the Alliance. We have no wish to seek a free ride at the expense of our allies. We will play our full part....

This afternoon I have tried to set before you some of my most deeply held convictions and beliefs. This party, which I am privileged to serve, and this Government, which I am proud to lead, are engaged in the massive task of restoring confidence and stability to our people.

I have always known that that task was vital. Since last week it has become even more vital than ever. We close our conference in the aftermath of that sinister Utopia unveiled at Blackpool. Let Labour's Orwellian nightmare of the Left be the spur for us to dedicate with a new urgency our every ounce of energy and moral strength to rebuild the fortunes of this free nation.

If we were to fail, that freedom could be imperilled. So let us resist the blandishments of the faint hearts; let us ignore the howls and threats of the extremists; let us stand together and do our duty, and we shall not fail.

QUESTIONS FOR ANALYSIS

1. What economic measures did Thatcher carry out during the first year and a half of her government? How, in her view, did these measures build the "foundations for recovery"? On what basis does Thatcher critique the economic policies of the Labor Party?
2. Discuss Thatcher's analysis of the causes of unemployment in Britain and her proposed solutions.
3. How does Thatcher envision the relationship of the individual to the government?
4. What are the main elements of Thatcher's foreign policy?

PEACE AND ECOLOGY MOVEMENTS

In December 1979, NATO decided to station U.S. cruise and Pershing II nuclear missiles with first-strike capacity in Western Europe. Cruise missiles, which can be launched from sea, ground, or air, have a maximum range of 1,500 miles and a nuclear warhead with fifteen times the destructive power of the atomic bomb that destroyed Hiroshima. The NATO plan called for the production of 4,000 cruise missiles and the deployment of 464 land-based missiles in Britain, Italy, West Germany, Holland, and Belgium. A series of mass protests took place at the designated base sites, including the U.S. military base in Greenham, England. In West Germany, the Green Party assumed a leading role in opposing the use of nuclear power as well as nuclear weapons and promoting peace and ecology issues on the grassroots and parliamentary levels. Moreover, the Greens critiqued the materialism of German consumer society as a contributing factor to the destruction of the world's environment. Peace actions also took place in former East Germany when the Soviet Union, in response to the NATO decision, determined that missiles should be based in the GDR.

179. The Greenham Common Women's Peace Camp

The largest and most sustained protests against the installation of cruise missiles in Europe took place outside the United States military base in Greenham, England where a Women's Peace Camp was established. The main camp at Greenham Common remained in existence for seven years and the last group of women did not leave until the spring of 2000. The protests began on August 27, 1981 when a small group of anti-nuclear activists began a "Women for Life on Earth" march from Cardiff, South Wales to the Greenham base. On arrival a week later, the group presented the base commander with a letter stating its determination to halt the installation of cruise missiles and the arms race in general. A Women's Peace Camp was subsequently organized at Greenham Common, its numbers continually replenished by new supporters, including women from abroad. The women at the camp organized a wide variety of nonviolent actions. These included an international "Embrace the Base" protest held in December 1982 in which some 30,000 women surrounded the military base, holding candles and lighted flares. In January 1984, shortly after the installation of the first cruise missiles, the women wove a giant web as a symbol of the strength and fragility of life which they floated over the base with helium balloons. The Greenham Common protests were informed by several strands of feminism that emerged in the 1970s. The groundwork for the opposition to the cruise missiles in the 1980s was also prepared by decades of peace activism by such groups as the Women's International League for Peace and Freedom (WILPF)[1] and the

---

[1]On the origins of WILPF, see Source 134 in Chapter 11.

From *Greenham Common: Women at the Wire* by Barbara Harford and Sarah Hopkins. (Women's Press, 1984). Reprinted by permission from Sarah Hopkins.

Campaign for Nuclear Disarmament (CND), as well as by more recent environmentalist opposition to nuclear power and nuclear weapons. In turn, the nonviolent and nonhierachical nature of the Greenham Common direct action protests had a major impact on subsequent social movements in Europe, including Eastern Europe. The dismantling of the main camp began soon after a December 1987 U.S.-Soviet treaty stipulating the destruction within three years of Pershing II and land-based cruise missiles, including the ninety-six missiles at the Greenham base, in exchange for a certain number of Soviet missiles. Three documents are reproduced here: the 1981 "Women for Life on Earth" letter to the base commander, a leaflet calling women to the 1982 "Embrace the Base" protest, and a chronology of protests from 1981 to 1984 which illustrates the creativity and persistence of the Greenham Common women.

A. "Women for Life on Earth" Letter

We are a group of women from all over Britain who have walked one hundred and twenty miles from Cardiff to deliver this letter to you. Some of us have brought our babies with us this entire distance. We fear for the future of all our children, and for the future of the living world which is the basis of all life.

We have undertaken this action because we believe that the nuclear arms race constitutes the greatest threat ever faced by the human race and our living planet. We have chosen Greenham Common as our destination because it is this base which our government has chosen for 96 'Cruise' without our consent. The British people have never been consulted about our government's nuclear defence policy. We know that the arrival of these hideous weapons at this base will place our entire country in the position of a front-line target in any confrontation between the two superpowers, Russia and the United States of America. We in Europe will not accept the sacrificial role offered us by our North Atlantic Treaty Organisation (NATO) allies. We will not be the victims in a war which is not of our making. We wish to be neither the initiators nor the targets of a nuclear holocaust. We have had enough of our military and political leaders who squander vast sums of money and human resources on weapons of mass destruction while we can hear in our hearts the millions of human beings throughout the world whose needs cry out to be met. We are implacably opposed to the siting of US Cruise Missiles in this country. We represent thousands of ordinary people who are opposed to these weapons and we will use all our resources to prevent the siting of these missiles here. We want the arms race to be brought to a halt now—before it is too late to create a peaceful, stable world for our future generations.

B. "Embrace the Base" Leaflet

(1982 leaflet from The Road to Greenham Common *by Jill Liddington © 1989. Syracuse University Press. Photo: Schlesinger Library Radcliffe Institute for Advanced Studies.)*

C. Chronology of the Women's Peace Camp, 1981–1984

1981

27 Aug 'Women for Life on Earth' march leaves Cardiff.

5 Sept March arrives at Greenham.

21 Dec First action: women stop sewage pipes being laid.

1982

18 Jan Women keen outside House of Commons on re-opening of Parliament.

20 Jan Newbury District Council serves notice of intention to evict camp from common land.

Early Feb Camp becomes women only.

21 Mar Equinox Festival of Life.

22 Mar First blockade of base by 250 women. 34 arrests.

Mar–May Spot blockades.

27 May First eviction from common land. Camp re-sited on MOT land. 4 women arrested.

28 May Eviction trial. 4 women imprisoned.

7 June 75 women die symbolically outside London Stock Exchange to point up profits made from arms race as President Reagan visits Britain. 7 women arrested.

6 Aug Hiroshima Day. Women place 10,000 stones on Newbury War Memorial.

9 Aug Nagasaki Day. 8 women enter base to present Commander with the 1000th origami crane, a symbol of peace.

27 Aug 18 women occupy the main gate sentry box until arrested.

29 Sept Eviction from Ministry of Transport land.

5 Oct Obstruction of work to lay sewage pipes. 13 arrested.

15 Nov Sentry box action trial.

17 Nov Sewage pipe action trial. 23 women imprisoned.

12 Dec 30,000 women link hands to encircle the base.

13 Dec 2,000 women blockade the base. Some enter to plant snowdrops.

1983

1 Jan 44 women climb the fence to dance on the Cruise missile silos.

17 Jan At reopening of Parliament 73 women occupy the lobby of the Houses of Parliament to emphasise demands that the issue of Cruise missiles be debated.

20 Jan Second camp set up at Green Gate.

22 Jan First attempt to set up the Blue Gate camp. 3 arrests.

7 Feb Over 100 women enter base as snakes. Secretary of State for Defence, Heseltine, visits Newbury. Action at Council chambers.

15 Feb silo action trials. 36 women imprisoned.

22 Feb High Court injunctions and eviction hearing adjourned when 400 women present affidavits stating that Greenham is their home.

25 Feb 6 women climb on to roof of Holloway women's prison to draw attention to the racial, class and economic injustices for which the imprisoned women are victimised.

8 Mar International Women's Day.

9 Mar High Court injunctions and eviction hearing. Injunctions awarded against 21 women.

24 Mar Visit of NATO Generals to base— women blockade.

28 Mar Holloway action trial. Women give evidence of inhumane treatment in prison. Charges against the 6 women dismissed.

31 Mar CND blockade the base.

1 Apr 200 women enter the base disguised as furry animals to have a picnic. CND link-up - 70,000 form a human chain linking Aldermaston, Burghfield and Greenham.

27 Apr Citadel locks action. All gates padlocked by women.

1 May Children's party on the common and inside the base.

12 May Eviction from common land at Yellow (main) Gate. Women's cars illegally impounded.

24 May International Women's Day for Disarmament. Silent vigil and fasting at Greenham.

31 May Women dressed in black enter base to scatter ashes as US Cruise technicians arrive.

24 June Rainbow Dragon festival. 2000 women sew 4½-mile serpent tail that weaves in and out of the base.

4–9 July Week of blockades. Women enter to make personal rituals. Christian exorcism performed. Culminates in removal of 50 ft of fence. New camps set up at Orange and Blue Gates.

24 July Women die symbolically in front of the public, politicians and military buyers who have come to view war hardware on display at an Air Tattoo at Greenham.

25 July 7 women cut hole in fence and paint women's peace symbols on US spy plane 'Blackbird'.

6 Aug Hiroshima Day. Arrival of 20 'Stop The Arms Race' marches from all over UK. Silent vigil and fasting at Newbury War Memorial.

9 Aug Nagasaki Day. Die-in at Yellow Gate. Women's outlines painted on road.

18 Aug 'Blackbird' trial. Charges withdrawn by MOD. One woman imprisoned for contempt.

5 Sept Second birthday celebrations. Women obstruct laying of fuel pipes into base. 27 arrests.

6 Sept Sabotage of fuel pipe workers' machinery.

29 Oct Halloween. 2,000 women take down 4 miles of fence. 187 arrests with increased violence from police and army.

1 Nov Heseltine tells House of Commons that 'intruders' near missile silos run risk of being shot at.

9 Nov Start of Federal court action in New York against President Reagan and George Bush to ban them from employing Cruise missiles in Britain. 24-hour vigils begin at all 102 US bases in Britain.

13 Nov Demonstration to honour the memory of Karen Silkwood, 9 years dead.

14 Nov First of the missile-carrying transporters arrive at Greenham. Night watches begin to check on Cruise activity. 18 peace activists in Turkey sentenced for 5–8 years' imprisonment.

15 Nov Blockade of all gates leads to 143 arrests.

11 Dec 50,000 women encircle the base, reflecting it back on itself with mirrors. Parts of the fence pulled down. Hundreds of arrests as police violence escalates—beatings, broken bones, concussions.

20 Dec Camps at all gates but one. Violence from soldiers, all-night harassment to prevent sleep, horrendous language amounting to verbal rape. Bricks and stones thrown at benders.

27 Dec 3 women spend 3 hours undetected in air traffic control tower.

31 Dec Final gate (Indigo) camped. New Year's party.

1984

1 Jan Women weave giant web, symbol of strength and fragility, to float over base lifted by helium-filled balloons.

5 Mar MOT, NDC and MOD combine forces to commence programme of evictions, starting at north side of base in preparation for Cruise convoy deployment.

9 Mar 12.30 a.m. 100 policemen with dogs converge on 12 women at Blue Gate to prevent them alerting other camps that part of a Cruise convoy (only 3 vehicles) has left the base. Despite police efforts women manage to clock the convoy's route an hour later.

180. Petra Kelly and the German Green Party

Petra Kelly (1947–1992) was a founder and leader of the Green Party in West Germany. A student in the United States in the 1960s, Kelly's political ideas were shaped by the anti-war and civil rights movements. She became passionately committed to environmental issues when her sister Grace died

from cancer in 1970 at the age of ten. Kelly participated in citizen's action groups formed throughout West Germany in the 1970s. Over 300,000 people participated in the more than one thousand initiatives launched by these groups to tackle such issues as the installations of nuclear power plants and environmental pollution. In 1972, the West German Association of Environmental Protection Action Groups was formed as an umbrella organization and coordinated opposition to the construction of a nuclear power plant at Whyl in Germany. In 1975, citizen groups from West Germany, Switzerland, and France carried out a mass occupation of the site. Petra Kelly joined the German Social Democratic Party when Willy Brandt was chancellor, but she left in 1979 in protest against the party's policies on nuclear defense, women, and health issues. A founder of the Green Party in 1979, Kelly was elected to the Bundestag in March 1983. In this same year, she went to East Germany to discuss the stationing of nuclear missiles in East and West Germany with Erich Honecker, chair of the ruling Socialist Unity Party of Germany (SED), as well as with East German dissidents. She was reelected to the Bundestag in January 1987 and gave her final speech in October 1990. Two years later, in October 1992, Kelly was murdered under mysterious circumstances, but the police hastily concluded a murder/suicide by her companion, former army general Gert Bastian. Kelly's writings include *Let Us Look for the Cranes* (1983), *Hiroshima* (1986), *Love Can Conquer Sorrow: Cancer in Children* (1986), and, with Gert Bastian, *Tibet—A Violated Country* (1988). The two speeches reproduced here are anthologized in Kelly's *Fighting for Hope* (1983, English translation 1984).

A. "The System Is Bankrupt"

'Eco-peace means standing up for life, and it includes resisting any threat to life.... The only people who have really understood what revolution means are those who consider non-violent revolution possible' *Philip Berrigan*

We are living at a time when authoritarian ruling elites are devoting more and more attention to their own prospects and less and less to the future of mankind. We have no option but to take a plunge into greater democracy. This does not mean relieving the established parties, parliament and the law courts of their responsibilities, nor forcing them out of office. Information reaches society via the political parties; and the reverse process is important too: the parties and the trade unions also act as sounding-boards for ideas which first arise within society. But the formation of political opinion within the parliamentary system is undoubtedly a process that needs extending further. It needs to be revitalized by a non-violent and creative ecology and peace movement and an uncompromising anti-party party—the Greens. In a period of crisis, a conveyor-belt system between society and the remote, established parties, is required. Otherwise the real problems are evaded in the endless games of power tactics, until eventually, they become quite unmanageable.

A third to a half of the 2,000 million people in the developing countries are either starving or suffering from malnutrition. Twenty-five per cent of their children die before their fifth birthday. In the last twelve months, 40,000 children under the age of five died every day, and the trend for 1983 indicates that this shocking figure is likely to increase.

A UNICEF survey shows that less than 10 per cent of the 15 million children who died this year had been vaccinated against the six most common and dangerous children's diseases. Vac-

cinating every child in the Third World costs
£3.00 ($4.5) per child. But not doing so costs us
five million lives a year. These are classic examples of 'structural violence'. Conflict and unrest
prevail, because the people live under threat of
premature death at the hands of a brutal social
order. For years now, social scientists studying
the Third World have been saying that growth is
taking place without development. Development has to meet the basic needs of the people in
these societies—needs such as work, food,
health, literacy and housing.

The system is bankrupt when $2.3 million a
minute are spent on perfecting the machinery of
destruction, whilst the means of survival become
increasingly scarce. And the system is bankrupt
when rational names are dreamed up for everything irrational. We are told that we have a ministry responsible for our security. It is called the
'Ministry of Defence'. However, there can be no
defence for our country in the event of war. There
can only be destruction. So why not call it by its
proper name, 'Ministry of Destruction'!

'Security policy' has led us into the most dire
insecurity the world has ever faced. The politics
of nuclear confrontation imposes a brand of insanity upon us which runs, 'In order to defend
freedom, we must be prepared to destroy life itself.' Or put another way, 'If you threaten me,
I'll commit suicide.' Every day the United
States and the Soviet Union add new weapons
of mass destruction to arsenals which already
have the equivalent of three tons of TNT in
store for every man, woman and child in the
world. The system is bankrupt when the politicians and generals gaily term this lunacy
'MAD'—mutual assured destruction.

The system is bankrupt when humanity
shrinks from recognising that it is in the
process of destroying itself. As Sam Keen of
Harvard has put it, while we continue to sit
tight on our stocks of nerve gas and nuclear
weapons, we are all members of a sect committed to irrational violence. The enormous expenditure of energy, scientific sophistication and
wealth on the military is the main cause of
poverty, inflation and despair in the world.

The system is bankrupt when micro-electron-
ics undermines the deterrent model. Missiles
with a target accuracy of fifty metres can launch
a direct attack on the opponent's arsenal. Thus
they no longer operate simply as a deterrent, but
should be regarded as weapons of nuclear attack.
The microchip has made the manufacture of
first-strike technology possible, and made it
seem rewarding. According to the peace researchers of the Club of Rome, the world should
be less alarmed by the overkill potential of the
superpowers, than by the threat resulting from
the sophistication of new control systems.

The system is bankrupt when doctors have to
issue warnings against nuclear war, when they
have to publish statements in full-page adver-
tisements about the consequences, immediate
and long-term, of a nuclear blast. Doctors have
been drawing attention to the fact that there is
no effective remedy for the multiplicity of illnesses and injuries caused by nuclear war, especially where radiation sickness is concerned.
None of the precautions for the event of nuclear
war, whether draft legislation or civil defence,
can change any of that.

The system is bankrupt when more than
3,000 accidents are reported in American power
stations in one year. In our view, there is no such
thing as a minor incident in a nuclear power station. Clearly, 'minor incidents' can add up to a
very grave danger. The system is bankrupt when
it costs approximately £50 million ($75 million)
to shut down a nuclear power station, and the
electricity companies tell us that once a nuclear
power station has been shut down, the residual
radioactivity does nobody any harm, and very
soon green fields will grow where nuclear power
stations used to be.

After ten years of Social Democrat/Liberal
government in West Germany, those of us who
stood outside the Chancellor's office with tears in
our eyes and lighted torches in our hands on the
night of Willy Brandt's election victory are now
having to face facts: not even the smallest step
has been taken towards reform, towards greater
democracy. Our illusions are dead and gone. We
do not believe a word the established parties have

to say any more. We shall not be carrying any more torches for them. We only trust ourselves now. To people who advise us to 'Go East, if you don't like it here,' we say East German principles apply in our country too. There is minimal provision for the poor and less and less of an opportunity to speak out. The system is the same; the differences are only of degree.

The system is bankrupt and nothing is certain. As Marie-Luise Kaschnitz's poem says,

Whether we will escape without being
  tortured
and die a natural death,
or will be starving again
and searching the dustbins for potato peelings,
or will be herded together in packs,
we have seen it all before.
Will we steal away in time
to a clean bed or will we perish in a
hundred nuclear strikes?
Will we succeed in
dying with hope?
It is uncertain,
nothing is sure.

B. E. F. Schumacher Memorial Lecture

'We can best help you prevent war, not by repeating your words and repeating your methods, but by finding new words and creating new methods' *Virginia Woolf*[1]

On 3 November 1983, I read in the German newspapers with great shock about the warnings of the British Defence Minister Michael Heseltine, in which he made clear that the Military Police would shoot at peaceful demonstrators near the American Base of Greenham Common. His warnings led to very strong reactions on the side of the opposition and the peace movement. For it is now clear, very clear, that the laws in the Western democracies protect the bombs and not the people. The warning that the State is ready to kill those engaged in non-violent resistance

against nuclear weapons, shows how criminal this atomic age has become.

Great Britain, I am told, has more nuclear bases and consequently targets per head of population—and per square mile—than any country of the world. And I am coming from a country, the Federal Republic of Germany, that is armed to the teeth with atomic and conventional weapons. I come here on this weekend to hold the E. F. Schumacher Memorial Lecture and would like to dedicate this lecture to the Greenham Common women. I dedicate a poem to them by Joan Cavanagh:

I am a dangerous woman
Carrying neither bombs nor babies,
Flowers nor molotov cocktails.
I confound all your reason, theory, realism
Because I will neither lie in your ditches
Nor dig your ditches for you
Nor join in your armed struggle
For bigger and better ditches.
I will not walk with you nor walk for you,
I won't live with you
And I won't die for you
But neither will I try to deny you
The right to live and die.
I will not share one square foot of this earth
  with you
While you are hell-bent on destruction,
But neither will I deny that we are of the same
  earth,
Born of the same Mother.
I will not permit
You to bind my life to yours
But I will tell you that our lives
Are bound together
And I will demand
That you live as though you understand
This one salient fact.

I am a dangerous woman
Because I will tell you, Sir,
Whether you are concerned or not
Masculinity has made of this world a living hell,
A furnace burning away at hope, love, faith and
  justice.

[1]From Virginia Woolf's *The Three Guineas*, Source 162 in Chapter 13.

A furnace of My Lais, Hiroshimas, Dachaus.
A furnace which burns the babies
You tell us we must make.
Masculinity made femininity,
Made the eyes of our women go dark and cold
Sent our sons—yes Sir, our sons—
To war
Made our children go hungry
Made our mothers whores
Made our bombs, our bullets, our 'food for
    peace',
Our definitive solutions and first-strike poli-
    cies.
Masculinity broke women and men on its knee,
Took away our futures,
Made our hopes, fears, thoughts and good
    instincts
'Irrelevant to the larger struggle',
And made human survival beyond the year 2000
An open question.

I am a dangerous woman
Because I will say all this
Lying neither to you nor with you
Neither trusting nor despising you.
I am dangerous because
I won't give up or shut up
Or put up with your version of reality.
You have conspired to sell my life quite cheaply
And I am especially dangerous
Because I will never forgive nor forget
Or ever conspire
To sell your life in return.

Women all over the world are taking the lead in defending the forces of life—whether by demanding a nuclear-free constitution in the Pacific islands of Belau, campaigning against the chemical industry after Seveso, or developing a new awareness of the rights of animals, plants and children.

We must show that we have the power to change the world and contribute towards the development of an ecological/feminist theory, capable of challenging the threat to life before it is too late. Just a few days ago the American House of Representatives agreed to go ahead with the construction of the first 21 of 100 planned MX inter-continental missiles. I have just returned from a trip to the men at the Kremlin in Moscow and to those men in power in East Germany. And this year I have been to Washington several times to meet those in power there too. And during each trip, whether it was to Moscow or Washington or East Berlin, I tried at the same time to speak with the people at the grass-roots level, those struggling against the military-industrial complex whether it be capitalist or state socialist. And while I sat listening to those men, those many incompetent men in power, I realized that they are all a mirror image of each other. They each threaten the other side and try to explain that they are forced to threaten the other side; that they are forced to plan more evil things to prevent other evil things. And that is the heart of the theory of atomic deterrence....

QUESTIONS FOR ANALYSIS

1. How does the "Women for Life on Earth" letter critique the Thatcher government's decision to install NATO cruise missiles at Greenham Common? Analyze the imagery of the December 1982 "Embrace the Base" leaflet. What does the Women's Peace Camp chronology reveal about the political views and protest methods of its participants?

2. Why does Petra Kelly conclude that "the system is bankrupt"? How does she connect the issues of nuclear weapons and nuclear power? How does Kelly relate peace activities in Germany to the Greenham Common protests? Why does Joan Cavanagh describe herself as a "dangerous woman" in the poem which Kelly includes in her E. F. Schumacher Memorial Lecture?

WOMEN IN EASTERN EUROPE

The end of the monopolistic rule of the Communist parties in the former Soviet Union and Eastern Europe led to the creation of new political parties and the political unfettering of the media and the universities where women academics are now freer to pursue topics specific to women and gender studies. Overall, however, the collapse of socialism and the dissolution of the Soviet Union has had a particularly adverse effect on women's status. Not only do women comprise a disproportionate number of the unemployed, but the elimination of social services, including child care, has further reduced their access to employment opportunities. Thousands of women, including professional women, have turned to prostitution for their livelihood as they cater to the new "business class." This is the exact opposite of the changes envisioned by women who aimed to improve women's situation under socialism and democratize the political process through the formation of autonomous women's and workers' organizations. The complexity of women's roles in these movements is illustrated by the following writings by women in former East Germany and Poland.

181. The "Lila Manifesto" of East German Feminists

The "Lila Manifesto" was drawn up by women in the Lila Offensive (Lila), a Berlin Women's Group in the Autonomous Women's Association (*Unabhängiger Frauenverband*) of the former German Democratic Republic (GDR). The association was formed in December 1989 at a mass meeting of more than a thousand women held in the People's Theater in former East Berlin. Prior to the November 9, 1989 Revolution, which brought down the Berlin Wall and effectively ended the political monopoly of the ruling Socialist Unity Party of Germany (SED), women were not allowed to organize autonomously as the government recognized only the official women's organization, the Democratic Women's Union (DFD or *Demokratischer Frauenband Deutschland*). One of several new organizations created in the wake of the November Revolution, the Autonomous Women's Association sought to ensure women's inclusion in the new political process. It served as an umbrella organization and united more than two dozen groups throughout the GDR around its slogan, "Cooperation, Ecological Awareness, and Multiculturalism." The association gained official status at the Round Table, a forum comprised of various opposition groups, which aimed to influence policy on a wide range of issues, including a new national constitution. The association demanded 50 percent quotas for women at all levels of political decision making and the establishment of a cabinet-level secretary for equal opportunity. It also participated in national elections in coalitions with such parties and organizations as New Forum, Democracy Now, and the East German Greens. The "Lila Manifesto" was published in January 1990 in the pamphlet *Frauen in die Offensive* (*Women Take the Initiative*). A visionary statement of the new GDR, it provides a historical critique of GDR policies on women as well as concrete proposals for reform within a socialist and democratic framework.

Originally from *Frauen in die Offensive,* January 1990, Dietz Verlag, Berlin. Translated by Heide Fehrenbach. Used with permission.

WORKING PAPER DEFINING OUR POSITION: WOMEN'S INITIATIVE "LILA[1] OFFENSIVE"

1. Point of Departure

During this period of radical social change, women are also and increasingly moving "into action." Women, as well as men, have been affected by the deforming effects of centralized, administrative socialism. Women, however, are united by the fact that they experience the crisis in a particularly threatening manner.

For in the GDR the formulated rights relating to the women's question could not be fulfilled. On the contrary, misguided policies on women, especially since the 1970s, have led to the reproduction and solidification of patriarchal structures and to retrogression in the emancipation process of women.

In the GDR the ideal of a socialist society of self-determining and self-creating women, men, and children was sacrificed to a social concept in which people were subordinated to economic premises, prescribed from the outside, and finally, made into objects of politics. Policies regarding women were no longer aimed at the goal of women's equality but became an instrument of administrative population and economic policies. With the aim of raising the birth rate, a state-decreed social policy unilaterally assigned the responsibility for family and housework to women. For the sake of a putatively more effective economy, women remained in underpaid, so-called typical women's occupations, and were rarely admitted to higher levels of management. This led to the solidification of a historically antiquated social division of labor on the basis of gender, with particularly weighty consequences:

- Women are expected to assume a double burden through the unilaterally prescribed "reconcilability of occupation and motherhood" and are thereby enormously overstrained, psychologically and physically.

- Women's orientation to assume primary responsibility for house and family work necessarily relegates them to a secondary position in employment, politics, and social life.

- Women occupy the majority of underpaid and undervalued jobs. They have little access to promising professions and leadership positions.

- The undervaluation and depreciation of female competence and capabilities has resulted in the reproduction and solidification of a patriarchal, discriminatory image of women.

- The representation of women's bodies as sexual objects, as well as an increase in violence against women, appears in all areas of social life and is exaggerated through the media by means of beauty pageants, striptease, and pornography.

- The individuality of women and men is impoverished by an upbringing which prescribes gender and sexual stereotypes from childhood on.

- As a result of discrimination, feelings of self-worth and the ability to articulate and realize one's own interests are less developed among women than they are among men.

The discrimination of women is characterized as sexism. Sexism is an expression of patriarchal relationships and indicates the oppression, degradation, and discrimination of individuals on the basis of sex.

We women of the "Lila Offensive" believe that the women's question in the GDR has not been resolved. This means that the elimination of capitalist means of production is indeed a prerequisite, but no guarantee of the cessation of patriarchal repression. This type of repression possesses a general cultural dimension that transcends class and economic organization. Women must become conscious of this fact. The initiative to change this situation must therefore originate with them.

2. Goals of the Women's Initiative "Lila Offensive"

The goal of "Lila Offensive" is the equality of women and men, which will lead to a qualita-

---

[1]Lila, or lavendar, is the women's color in Germany.

tively new self-awareness between the sexes and therefore to a new kind of social relations. Sexual equality is, for us, one of the most basic values of a socialist society. The alternative-social model we are striving for takes as its fundamental goal the right of all women, men, and children to a self-determined development of individual potential. Economics, politics, and civil society must be employed as instruments to realize this social model, and not the reverse.

We want to participate in the creation of a socialist society:

- which is ecologically minded, democratic, feminist, multicultural, non-totalitarian, and socially just;
- which is not oriented to consumption and competition;
- which is free from social discrimination on the basis of sex, life-style, sexuality, age, skin color, language, and disabilities....

We women of the "Lila Offensive" especially want to identify and eliminate the mechanisms and structures that reproduce and dictate the social inequality of women and men:

1. We want to contribute to, provoke, and problematize awareness regarding the position of women, i.e., to reveal the details of their current experience.

2. We want to demand changes in social conditions aimed at creating actual equality between women and men.

3. We want to assist women in recognizing their abilities and (social) position, in articulating their needs and desires, and finally in realizing and decisively raising the resulting objectives and demands.

3. Philosophy of the Women's Initiative "Lila Offensive"

We consider ourselves feminists. Women are placed at a disadvantage vis-à-vis men in most areas of social life in the GDR (see above, 1. Point of Departure). Feminism means, for us, the protection and representation of women's interests, with the ultimate goal of achieving equality between the sexes, irrespective of life-style and love relations.

Feminism is definable according to two points of view:

- Feminism is a way of viewing social relationships. It consciously analyzes and observes what position, role, and significance women occupy in the various areas of society....
- Feminism also designates a politics that has as its goal the termination of sexual and gender relationships as power relationships....

Women's fight for actual equality requires a double strategy of autonomy and self-reliance, on the one hand, and cooperation and integration, on the other.

Women's independent and autonomous organization and representation of interests is essential

1. because women themselves must be responsible for their own liberation,
2. for consciousness-raising and self-discovery among women in existing structures free from patriarchal models and constraints,
3. to motivate women to champion their own interests confidently,
4. to raise consciousness of their position as the disadvantaged sex.

Cooperation and discriminating collaboration are essential, because

1. the women's question is a general social question and must therefore be an issue for all social forces,
2. the actual equality of women will only be accomplished through cooperation between women and men in all areas of social life.

4. Method of Operation and Structure

- The women's group "Lila Offensive," as well as the issue-oriented working groups and concrete projects undertaken by the group are to be reserved for women.

- "Lila Offensive" is independent of other parties and organizations.
- We see ourselves as an independent women's initiative within the Autonomous Women's Association.

We seek the cooperation and timely collaboration of parties, social organizations, and autonomous groups, which have corresponding or similar views with regard to women's politics and social-strategic goals.

CATALOG OF DEMANDS
Classification:

1. Politics and Economy
2. Individual Reproduction
3. Individual Development of Personality
4. Socialization
5. Legal Rights
6. Sexual Self-Determination
7. Violence Against Women
8. Fostering Sensitivity for the Women's Question

Prefatory Remarks:
In order to realize our conceptions of society, we consider the following basic conditions as essential:

- The sovereignty of the GDR is to be preserved.
- Military expenditure should be radically reduced and these funds redistributed for the benefit of the socially vulnerable.
- International and national conflicts should be resolved by peaceful measures.
- The right to work must be fundamentally guaranteed for every woman and man.
- The reorganization of our society along capitalist lines must not be permitted.
- Our nation must be based upon the principles of antifascism, anti-imperialism, and anti-Stalinism.

1. Politics and Economy
Women's interests must be recognized in all areas of social life. We therefore demand:

- Quotas in all political and economic spheres of influence. Quota plans should be drawn up and implemented in a step-by-step process.
- For this purpose, material and mental preconditions must be created. Among others, the formulation of new evaluative criteria for performance, public discussion of quotas in order to sensitize people to this problem and the necessity of its implementation.
- Accordingly, quotas must be established by law.

1.1 Politics
In the future, political decisions at all levels can no longer be made without women's participation. For this reason, we demand:

- the introduction of a 50 percent quota for the presentation of candidates for all levels of popular representation (national [*Volkskammer*], regional [*Bezirks*], and local assemblies [*Kreistage*]);
- the adoption of quotas regulating the proportion of men and women in the leadership of parties and organizations, affirmation and support for alternative representation for women's interests, recognition of the Autonomous Women's Association;
- the allocation of meeting space and the material-technical means for the organization and coordination of the Women's Association's work;
- parliamentary division of power with respect to women's interests and their representation;
- through May, an Advisor for Women's Issues in the Government, and after May the creation of a State Office [*Staatssekretariat*] on the national level.

1.2 Economy
1.2.1. Ecological Restructuring of Industry and Agriculture
To date, economic development in our country has disregarded the menacing condition of our environment, for which development must take

the blame. Only inadequate measures have been taken for its recovery. Therefore, we demand:

- the reconstruction of industry and agriculture on an ecologically sound basis;
- research and implementation of alternative methods of resource utilization...
- daily information on all environmental data, issued in plain language and at a fixed time.

1.2.2. Social Reorganization of the Economy
We demand:

- the particular protection of the constitutionally guaranteed right to work, above all for women, during the impending economic and administrative reforms;
- that there shall be no social decline for women as a result of the restructuring of the labor force, of dismissals, and replacements;
- that economic agreements with other nations and international firms not contain any provisions which discriminate against women;
- the creation of works councils, which include women's representatives with appropriate authority in the various enterprises;
- the reduction of socially required working hours.

1.2.3. Employment
The actual economic equality of women and men, as well as the unimpeded access of women and men to all occupational sectors must be achieved. Therefore we demand:

- higher valuation and better pay for so-called typical women's occupations;
- encouragement and furtherance of access for women to scientific and technical professions, including through structural changes;
- encouragement and furtherance of access for men to enter professions in the social domain (child education, health and service professions);
- equal opportunities for advancement of women and men within professions;

- quotas for apprenticeships, in order to break down the traditional [gender] composition of certain occupations.

1.2.4. Labor Legislation Beneficial to Parents and Children

- support for the compatibility of motherhood and fatherhood with employment
- expansion of tax breaks or financial allowances for those raising children
- equal opportunities for women and men for part-time work
- expansion of opportunities to care for ailing relatives without resulting financial disadvantages.

2. Reproduction of the Individual
Women and men are to share equally in the responsibility and work of the household. To that end, the following steps are essential:

- an expanded and redefined notion of work that includes reproduction as a general social contribution;
- removal of the social policy measures assigned one-sidedly to mothers;
- rights and duties for social fatherhood...

3. Individual Development of Personality
Freedom of choice in creating life conditions and love relationships:

- the dismantling of all privileges granted to marriage (including adoption);
- protection through civil law of all non-kin households (including lesbian);
- legal guarantees for the economic support of single parents and caregivers;
- the legalization of self-help groups;
- creation of the material and intellectual prerequisites for female pensioners to determine their own lifestyles;
- creation of the material and intellectual prerequisites for the integration of disabled persons into society;

- redefinition of relations to so-called socially marginal groups, eradication of prejudice and criminalization;
- elimination of taboos surrounding death, and the social integration of the dying.

4. Socialization
Destruction of stereotypical norms through the abolition of role-determined socialization:

- critical analysis of school curricula and textbooks with respect to the transmission of sexual/gender stereotypes, and the development of new teaching material;
- elimination of taboos surrounding sexual education;
- topical public discussions in the media;
- education which fosters the acceptance of shared responsibility for birth control, pregnancy, the birth and life of the newborn.

5. Legal Rights

- The constitutionally established right of equality for women and men should be enforceable in court.
- Revision of the penal code to ensure the consistent punishment of every form of violence against women (including within marriage).
- Creation of equal status for men with respect to child custody in the case of divorce, and general recognition of social fatherhood.
- Broadening of the adoption laws and their non-bureaucratic administration.
- Humanization of and the elimination of taboos surrounding criminal sentencing with due consideration of women's special status.
- Modification of the military service law (above all, no inclusion of women in the case of mobilization).

6. Sexual, Self-determination
Self-determination over one's body is one of the most elementary rights of women:

- guarantee of the right to abortion;

- psychosocial support/counseling for women before, after, and during the abortion;
- humanization of health services, and especially of gynecological care;
- permission and legitimization of alternative health care and feminist methods of nature healing;
- strict supervision of and legal limitations to genetic research as well as genetic and reproductive engineering;
- public discussions and the dismantling of structures of thought and education which unilaterally prescribe heterosexuality to be the dominant orientation;
- opportunities for sterilization for both sexes upon request.

7. Violence Against Women
The increasing trend of violence against women and children should be openly acknowledged and discussed without being sensationalized:

- the fact of violence and sexist behavior toward women must be released from its current status as a taboo subject, in the workplace, in public, in the media, in the context of relationships;
- recognition of the scope of sexual violence, and especially of rape; publication of numbers and estimates of obscured statistics (including those crimes resulting in death);
- measures to protect women from violence (emergency telephone numbers, service cars in night trains, taxi service for women);
- the creation of women's centers for women and children in crisis situations;
- sensitization of the public to the verbal marginalization, defamation, and insulting of women;
- public recognition of prostitution in the past and present, and disclosure of its origins and causes, and possibilities for its reduction;
- no criminalization of prostitution;
- battle against pornography.

8. Public Sensitization to the Women's Question
Creation of possibilities for women to meet and communicate, and so to discover and articulate

their own needs (women's cafes, clubs, libraries, communal living arrangements, vacation homes).

- The fostering of a public consciousness and of discussion regarding the women's question/gender question;
- media attention to these problems, permission for new women's journals, broadcasts, etc.;

- publication of and free access to the results of feminist research;
- public discussion, analysis, and eradication of sexist content in the media;
- support for nonsexist language and forms of address;
- support for research on women and an institute for women's research.

182. Małgorzata Tarasiewicz
Women in Solidarity

Małgorzata Tarasiewicz (b. 1960) is a human rights activist in Poland and a former coordinator of the Women's Section of Solidarity. Solidarity was an independent trade union and social movement that achieved official recognition following a wave of strikes in August 1980. The strikes began in the shipyards in Gdańsk and soon encompassed some four hundred workplaces throughout Poland's Baltic region. Two women at the Gdańsk Shipyards, Anna Walentynowicz, a crane operator, and Alina Pieńkowska, a nurse, served on the committee that negotiated the Gdańsk Agreement of 1980. The agreement guaranteed the right of workers to form independent trade unions and to strike. It also pledged the government to improve workers' living and working conditions, including the expansion of child care and health services. In 1981, however, the Polish government declared martial law and Solidarity was forced underground until 1989 when it regained official status as well as the right to elect a portion of the Parliament. Solidarity won all but one of the open seats and Tadeusz Mazowiecki, a Solidarity leader, was elected prime minister. The Women's Section of Solidarity was also formed in 1989 and Tarasiewicz became its coordinator after several years of work in a dissident organization. She was forced to resign in 1991 under pressure from Marian Krzaklewski, the newly elected president of Solidarity, and the Solidarity Board. They opposed the Women's Section's advocacy of legalized abortion and demands for more women in the Solidarity leadership. In this same year, Tarasiewicz was elected the first president of Amnesty International in Poland. She is currently coordinating two of its projects, the International Criminal Court Project and the Gender Network, and serving as a board member of the Network of East-West Women which links advocates for women's issues in Eastern and Central Europe as well as in the United States. In this paper presented in New York City in 1997,[1] Tarasiewicz evaluates women's participation in Solidarity, the work of the Women's Section, and how developments in Polish society in the 1990s have affected women.

---

[1]Originally titled "Recent Developments on Women, Liberalism, and Trade Unionism in Poland," the paper was presented to the "Gender in Transition: Women in Europe" workshop sponsored by New York University's Center for European Studies.

Some people say, let us first establish our democracies and then we can work out the details (meaning the rights of women). But if these two things do not go hand in hand, not only will women remain second-class citizens, but our societies will also remain backward, limiting themselves only to talk about democracy. Without women's involvement in political, social, and economic issues, democracy will never be achieved.

This simple truth is illustrated in Izabella Filipiak's novel, *Absolute Amnesia* (1996). The novel presents the Menstrual Police, whose task is to control menstruation, as a metaphor of women's oppression. Marianna, the main protagonist of *Absolute Amnesia,* complains to her mother about being oppressed. As Maria Janion notes in her analysis of the novel:

> The mother's name is Immaculate. She is a gynecologist and a member of the opposition to the communist government. Marianna asks her mother, who is a "human rights activist," for help. She asks her to write a letter to Amnesty International. But Immaculate does not consider her daughter's request seriously:
>
> *"We have really important matters to deal with,"* she reproaches me. *"We have this anniversary, that anniversary or still another anniversary."*
>
> *"But this can be a form of pressure,"* I explain. *"This is a form of political repression." "You are wrong,"* says Immaculate. *"This is your private problem { ... }"* And she leaves for a secret meeting.

This difference of opinion illustrates the type of discussions characteristic of the opposition in the 1970s and 1980s in Poland. Serious talks dealt with independence. Women's issues were not serious. Those who fought for independence were repressed and became victims of political repression. Women's feelings of repression and lack of control over their lives were considered private issues. "Solidarity" was to win independence and democracy for the whole society and then, after achieving these "higher goals," it was to deal with women's issues and improving their situation.[1]

Women opposition activists were pressed to subordinate their demands for equality to the "larger, more important causes." If they criticized "Solidarity" for being undemocratic, they were blamed for their lack of patriotism. Criticizing "Solidarity" meant weakening its power and, in consequence, betraying it. Someone who criticized "Solidarity" was regarded either as a secret police agent or as a mentally ill person.

After the fall of communism in 1989, a similar situation arose, this time not with regard to "Solidarity," but to the Catholic church which was defended from both sides of the political spectrum, the left as well as the right. Adam Michnik,[2] a left-wing publicist, made a point of calling attention to his participation in a "Catholic male pilgrimage," an annual event in which groups of men travel together to one of the Polish shrines. And in his daily *Gazeta Wyborcza,*[3] Michnik has defended the Catholic church. To understand his motivation, one can only imagine that he believes Poland is unprepared to construct a democratic society without church guidance. Michnik has joined other defenders of the Catholic church who regard it as a major patriotic force that has secured Polish identity throughout the ages. They believe that Polish independence is still too weak to exist without this protective force. On the right side of the political spectrum are the fundamentalists. Wanda Nowicka, the president of the Polish Planned Parenthood federation, was furiously attacked as a traitor, mainly by fundamentalists, when she stated that it was unfortunate to have a Polish

---

[1]Maria Janion, *Kobiety i duch innosci* (Warsaw, 1996). In English, the title is *Women and the Spirit of Otherness.*

[2]Adam Michnik was a leading member of the Workers' Defense Committee (KOR). Comprised primarily of intellectuals, it assisted workers arrested during the strikes and protests that occurred in September 1976 when the government, in an attempt to reduce Poland's soaring debt, ended food subsidies and announced a forty percent increase in consumer prices. KOR subsequently expanded its activities to a more general defense of workers' interests and, in the 1980s, played a key advisory role to the Solidarity movement. Michnik strongly supported the 1989 Round Table discussions between Solidarity and the government which led to the 1989 elections.

[3]*Gazeta Wyborcza* first appeared in 1989 as an election newspaper. Michnik is now its editor-in-chief.

pope at the 1994 United Nations Cairo Conference.[4]

Setting aside their "personal problems," women waited for independence which never seems to have fully arrived. First, because of the constant dangers during the transitional period, then because of communists coming back to power, and thirdly because of the failure to fully exclude former members of the secret police from assuming positions in the government cabinet, court system, and parliament.

The struggle over ideas which took place during the transition period was especially important because whatever ideas emerged as dominant would influence the lives of more than one generation. But women's voices were either absent or silenced. Why? Because women were either too submissive to voice their "private" demands or they failed to obtain positions of power in Solidarity and in the government which were taken over by men.

Some years ago, an unofficial organization in Poland received a letter from someone who owned a paint sprayer and wanted to spray slogans on walls but did not know what to write and so asked for advice. In Poland, it soon became a reality that whenever government positions were at stake, the only groups that could make any gains were well-organized and experienced in applying political pressure. Women were definitely not such a group. Rather, like the person with the paint sprayer, they did not know how to exert their rights.

Over the past forty years, the Communist Party instrumentalized women's organizations. While the legal system in Poland was favorable to women, as in other East European countries, a great discrepancy existed between the legal system and the social reality. Women's front organizations, such as the communist-controlled Women's League, with their conformist memberships, provided destructive and demoralizing examples of women's organizations and created negative stereotypes which predominated after 1989 and remain influential even today. So attempts to establish new women's groups were often discredited because they were regarded as just another version of the Women's League.

Moreover, issues like abortion were used for various political purposes during such political turning points as the parliamentary elections which took place during the spring of 1989. All of the major political forces exploited the issue of abortion for their own ends. Ex-communists, for example, advocated the right to choose. Consequently, in the popular consciousness, pro-choice attitudes became identified with the political corruption of the past forty-five years. The strong influence of the Catholic church led the majority of Polish people to adhere to the traditional vision of the female role as one limited to the stereotypes of wife and mother.

Overall, traditional values dominated the political climate. With the growth of unemployment, governmental plans appeared to send women home in order to increase opportunities for male workers. Moreover, women saw staying at home as new and progressive. The predominant point of reference for Poland became the pre-war period which provided a model of social organization for an independent Poland. It was now viewed with nostalgia for, among other things, what a family should be like and what should constitute the feminine role. However, pre-war Polish society, dominated by Catholic ideals, cannot serve as a model for a completely different post-totalitarian country which is aspiring to a free market economy with all of the setbacks and benefits created by this new situation.

The most visible advantage of post-communism has been the transformation of the meaning of work. Under communism, there was no unemployment but professionals often received low wages that did not reflect their abilities or experience. Due to shortages, it was often better to be employed as a shop assistant, even if one had a university degree, because such work provided access to various goods. Now, the main danger that women face is unemployment. Because of

---

[4]The Cairo Conference, held from September 5–13, 1994, was a United Nations International Conference on Population and Development. The conference adopted the Cairo Program of Action which affirmed women's reproductive rights.

their double work load in the paid workforce and in the family, women have not been able to compete with men on the labor market. Hence, since 1989, 98% of registered unemployed women have not been able to obtain work and many women have been dismissed from their jobs. One reason for this development was cultural, as it seems to be generally accepted in Poland that a man needs a job more than a woman.

Thus, a few years after the communists lost power, women's situation leaves much to be desired. Women who want to present their demands once again find themselves in the position of being considered "traitors." The debate over the abortion law in the 1990s is a case in point. Aleksander Kwaśniewski, the current Polish president, is a former communist and former communists hold positions in his cabinet. In January 1997 the government introduced a law to legalize abortion. Soon afterwards, Romana Cieśla, a feminist and former opposition activist, wrote the following analysis of the situation to the Network of East-West Women:

> This is such a Polish paradox: those women who are rooted in the Solidarity movement and are pro-choice activists now have to cooperate with and support former communists like Kwaśniewski. Barbara Labuda and other women activists like her, who were imprisoned under martial law, now find themselves in a paradoxical situation. Bishop Pieronek (the spokesman for the Polish Episcopate) repeated the earlier church warning about the danger to the Polish nation posed by the new law. Journalists treated the bishop's opinion as if it was the most accurate representation of the nation's interests. The whole abortion campaign was, first of all, a political game in which men fought for power and women could not say anything about an issue concerning themselves.
>
> Next year, we will have parliamentary elections. If the right-wing parties supported by the Catholic church win the election, they will attempt once more to ban abortions. Poles dissatisfied with the post-communist government will vote again for pro-church candidates and for a strict anti-abortion law.

Cieśla also mentioned that Marian Krzaklewski, the president of Solidarity, publicly declared his pro-church stand and was one of the main organizers of pro-life demonstrations held in front of the parliament during the vote on the abortion bill.

Solidarity's opposition to the 1997 bill was consistent with its opposition to abortion over the past several years. Solidarity first became engaged in the debate on abortion legislation in 1990 when a parliamentary bill was drafted to ban abortions which had been legal under Communist rule. The national assembly of Solidarity, held in this same year, took a vote in "defense of life" in support of this bill. This was contrary to what the majority of women present at the assembly wanted. However, they only constituted ten percent of the delegates. The Solidarity assembly vote on abortion only confirmed the validity of the resolution that the Women's Section prepared and presented to this assembly in which it charged that Solidarity's political discussions and decisions were monopolized and manipulated by men. Repression soon followed and the Women's Section was declared illegal.

From the very beginning, Solidarity, which was both a trade union and a socio-political movement, was closely bound to the Catholic church. The symbolism of the church became an inseparable part of Solidarity's image; moreover, Solidarity had close ties to the church hierarchy. Solidarity's relationship with the church severely limited women's chances to succeed in their struggle to improve their situation through Solidarity and within its ranks. It also contributed significantly to the political power of the church and defined Polish nationalism as religious in character.

According to Kristi Long, who has written a comprehensive analysis of women's participation in Solidarity during the strikes in 1980,[5] women

---

[5]Kristi Long wrote *We All Fought for Freedom: Women in Poland's Solidarity Movement* (Boulder, Colo.: Westview Press, 1996). Her study of women in Solidarity is based on extensive interviews with activists, including Małgorzata Tarasiewicz, which she carried out in 1991 and 1992 in the Baltic cities of Gdańsk, Gdynia, and Sopot.

who provided food for the strikers found themselves in the sphere of public activity for the first time in their lives. Some transcended this role and became engaged in political and union activities. Many women therefore began their activism by standing in front of the gate to the shipyards and factories and supporting the strike. In helping the strikers and giving them food, they carried out their maternal role in a political way. Motherhood is a crucial notion for comprehending women's consciousness in Poland, for it means providing food, clothing, and shelter. Women began their patriotic involvement as food-givers and then went on to organizing trade union cells in their workplaces. One of the organizers, as Long notes, first became involved when she received a phone call from the shipyard summoning her to make sandwiches. In 1980, when Solidarity was formed, Polish women achieved a chance to speak for themselves and to express their concerns. For many women, this came as a revelation. Often, it was the first time that they had a chance to formulate their visions and convictions.

Throughout Solidarity's existence, however, women activists have had to deal with all sorts of limitations placed on their activities. Although Solidarity greatly depended on women, especially when it was underground from 1981–1989, its leadership has consisted mostly of men. Even now, women in Solidarity rarely attain leadership positions. During martial law, not a single woman was represented on the board of the union, not even in the sections that represented areas of employment with a preponderant majority of women, such as the textile industry or education. Women have not been able to substantially influence the union's policies. An additional factor limiting women's political activism is their double work load of home and workplace.

The place of Polish women in social and political spheres is to a great extent determined by romantic and pragmatic traditions, no matter how restrictive they may seem. In a speech given at the College de France in 1842, Polish national poet Adam Mickiewicz characterized the ideal of a Polish female. In his view, women are "real patriots" when they combine "feminine values with strength and courage." He

pointed out that during the last uprising (against the Russian occupation) quite a few women of this kind had appeared. Mickiewicz also spoke about national emancipation which he connected to a moral stance. He stated: "This is the way that humanity should proceed. First, you have to sacrifice and then you obtain your rights."

This was the way he proposed for Polish women to gain their rights and freedoms. Mickiewicz also believed that women had more freedom in Poland than anywhere else. He noted: "In Poland, a woman carries out illegal political work alongside her husband and brothers and risks her life to help prisoners." Mickiewicz held up the example of Emilia Plater, who fought for Polish independence in the November 1831 uprising and is regarded as a spiritual descendent of the mythical, medieval Lithuanian Grażyna.[6] Both Plater and Grażyna, however, dressed as men. Mickiewicz emphasizes that women will obtain their freedom on the basis of sacrifice. Women did leave behind their domestic duties because of historical circumstances. Women were the spirit of uprisings. But women carried out their heroic work in anonymity because of the dominant view in Polish society that women should not be leaders.

The failure of armed uprisings against Russia led to the development of the Positivist movement in the second half of the nineteenth century. It was based on the glorification of "organic work" which meant building Polish social institutions and preserving Polish culture under foreign occupation. This movement fashioned the Polish mother as a nationalist hero. National interests again subordinated the individual interests of women.[7] The Positivist period shaped the ideal of Matka-Polka[8] who

---

[6]Grażyna is only known by her first name.

[7]On this point, see Maria Janion, *Women and the Spirit of Otherness*.

[8]Matka-Polka was an ideal of a woman who sacrificed herself for bringing up children in the spirit of patriotism and "keeping the family together" while men fought for the homeland. This ideal originated during the period of Poland's partitions in the eighteenth-century.

taught her children Polish language and culture, which was not taught in the schools, as well as the true history of Poland which was falsified by the foreign occupants of Poland. The idealization of this sort of activity limited the possibilities for women's participation in the underground resistance or in trade unionism. Motherhood was promoted as women's most important mission and the popular vision of femininity was very much modeled on Our Lady, the Virgin Mary. Even women in Solidarity had difficulty in diverging from this model.

Women's activities in Solidarity remain unappreciated even though they played a vital role in the movement by organizing union cells in their workplaces; provisioning and assisting strikers, prisoners, and underground workers; hiding activists; and publishing and distributing underground newspapers. For numerous women, participation in Solidarity represented their first involvement in the social and political spheres of life. This produced a radical change in their lives and gave them a sense of independence and self-reliance.

During its underground period, Solidarity received steady support from trade unions in the West. The International Confederation of Free Trade Unions (ICFTU) was one of Solidarity's main supporters. Solidarity counted on its support and, to some extent, became dependent on it. The idea of creating a Women's Section within Solidarity first came from the ICFTU which made this a condition of its continued support. The Women's Section was formed in 1989. At first, the Solidarity leadership did not believe that women could organize themselves into something that would prove dangerous to its authority. In 1990, the Women's Section organized a conference of women from all over Poland to which it invited women from feminist organizations and trade union women from the West to speak about "raising the consciousness" of women. Women from Solidarity discussed the issues of importance to them. These included attempts to ban abortion in Poland and the lack of women's representation in the Solidarity leadership. Women in Solidarity ex-

pressed regret that meetings of the Solidarity Executive Committee failed to discuss their problems. At the conference, many women spoke about the problems that they faced in their private lives, such as alcoholic husbands, domestic violence, and lack of adequate access to the system of social benefits that includes paid leaves, health insurance, and financial security in retirement.

The national leadership of Solidarity did not support the Women's Section and never granted it sufficient funds. The situation quickly worsened in 1990 when the section began to actively campaign against a bill that would ban abortions and contraceptives. Women organized discussions and referendums. They even sent delegates to read a letter of protest in Parliament. The section was becoming dangerous because it was not following the political line of the union leadership. Consequently, the leadership began to control the section's activities. The situation worsened when Marian Krzaklewski was elected the new president of Solidarity. He saw a chance to strengthen the union's relationship with the Catholic church. The national leadership of Solidarity prevented the Women's Section from carrying out its work by refusing to allow its members access to such office equipment as faxes, phones, and photocopy machines. Krzaklewski personally attacked the Section for being anti-Polish and liquidated it in July 1991 for "politicking," that is, for advancing a political agenda; in this case, undermining Solidarity's relationship with the Catholic church, rather than genuinely working on behalf of women.

Almost immediately afterwards, a new Women's Section was created and Irena Krzyżanowska became its leader. She opposed women's right to abortion and maintained close relations with the Episcopate. This new section only lasted for a few months and women in Solidarity were again deprived of representation. At the time of the dissolution of the Women's Section, not a single woman served on the Executive Committee of Solidarity. The Section had offered women the opportunity to become visible following their anonymous participation in oppositional work. But they now returned to their former position of anonymity.

During the brief existence of the first Women's Section, many women revealed how they had internalized a negative view of themselves. They either did not want to organize or denied the existence of discrimination. An additional factor responsible for women's denial of discrimination in our society is their pride in being a Polish mother and in preserving the patriotic values of the Polish nation. During the Communist period, the private sphere, which was mainly women's domain, made the role of mother essential to all forms of resistance to the Communist system. Now, the uniqueness of the private sphere is much diminished as civil society is being established in Poland. When men join public life, women remain in a less valued sphere of life. Long associated with the home, women are simply not seen as part of the new civil society.

From 1990 to 1993, the two most important issues facing women were the threat of laws banning abortion and economic hardship. The attempt to pass laws restricting abortion occurred in two stages. In the first stage, which took place in 1990, an extremely restrictive law was proposed which, in addition to outlawing contraceptives, would have enforced a total ban on abortions, including a three year prison sentence for patients and doctors. In 1993, a more liberal law banning abortions was introduced and passed which made only the doctors, and not the patients, punishable for performing abortions, and allowed for the distribution of contraceptives. Following years of opposition to this law, a new bill introduced in the Parliament on January 5, 1997 legalized abortions on three conditions: in cases of rape, if the mother's life is in danger, and economic hardship. The new bill passed and overturned the 1993 law.

In concluding, it is important to note that the strike in the Gdańsk shipyard began with the firing of Anna Walentynowicz, a Baltic Free Trade Unions activist.[9] She was an important early leader in Solidarity. But women's role, so indispensable in oppositional work, has never been fully appreciated. Women's contributions as well as their opinions and activities have been marginalized. Kristi Long provides a very revealing example of women's status within Solidarity in her analysis of "Człowiek z żelaza" ("Man of Iron"), a film by the renowned Polish director Andrzej Wajda:

> Wajda's depiction of the female contribution to the union movement is that of women as "nurturers" in keeping with the stereotypical view of women in strongly Catholic Polish society. In this film, as in much of public memory, women take care of men and, paradoxically, memory (which consists mainly of the memory of the actions of men) but stand aside during active protest and decision-making. Women's own memories, especially the memories of female activists, tell a rather different story.... Wajda, among others (especially, for example, Wałesa)[10] relegates women's activism to a second class private memory—in the public memory, the doers and shakers are men....[11]

---

[9] Anna Wałentynowicz began working at the Gdańsk shipyard in the 1950s, first as a welder and then as a crane operator. She joined the Baltic Free Trade Unions which were formed in Gdańsk in 1978 by workers in the shipyards of Poland's Baltic region. For her union activities, Wałentynowicz was arrested and searched numerous times by the authorities. She was also beaten by the militia and secret service. In 1980, the government again aimed to reduce its indebtedness by raising food prices and, as in 1976, encountered mass opposition, including workplace stoppages and strikes. On August 9, Wałentynowicz was fired for "disciplinary reasons." This provoked the Gdańsk shipyard strike, which included a demand for her reinstatement, and set into motion the events that led to the organization of Solidarity, initially a grouping of strike committees in the Baltic shipyards.

[10] Lech Wałesa, an electrician in the Gdańsk shipyard, was elected Chairman of Solidarity's National Coordinating Commission in September 1980 and reelected in October 1981 by the First National Congress of Solidarity. Wałesa also served as President of Poland from 1990 to 1995.

[11] Kristi Long, op. cit. p. 50.

QUESTIONS FOR ANALYSIS

1. What is the overall aim and approach of the "Lila Manifesto"? On what basis does it critique women's status in GDR society? What economic, political, and social changes does the manifesto propose? How does it define feminism? How do the societal transformations envisioned in the manifesto compare with what took place in the GDR following the unification of Germany?

2. How does Małgorzata Tarasiewicz analyze the treatment of women's issues and women activists within the Solidarity movement? What was the role of the Women's Section of Solidarity? How was it transformed by the Solidarity leadership? What role did the Solidarity leadership and former members of the Communist Party in Poland assume during the debates on abortion legislation in the 1990s? How does Tarasiewicz analyze the role of Polish nationalism in influencing women's status in Polish society in the past and present, including within Solidarity? How does she compare the status of women in Polish society before and after 1989?

WOMEN AND THE NEW MULTICULTURAL EUROPE

The "new" Europe of the late twentieth and early twenty-first centuries is increasingly multicultural. For the most part, the ethnic minorities in each European country have emigrated from its former colonies. In West Germany, however, where the largest minority is Turkish, the post–World War II government preferred to recruit largely male Turkish workers to alleviate labor shortages rather than promote German women's industrial employment. Women of minority groups face a double form of discrimination: racism and sexism. Their concerns are often ignored by society at large as well as by white feminists. They are also expected to conform to the traditional, patriarchal norms of their ethnic group. These concerns are being addressed by local and nationally based grassroots women's organizations as well as by such European-wide women's organizations as the European Women's Lobby.

183. Pragna Patel
"Third Wave Feminism and Black Women's Activism in Britain"

Born in Kenya and raised in Britain, Pragna Patel is a founder of the first Black Women's Center in West London and Women Against Fundamentalism. Since 1982, she has been active in Southall Black Sisters, a community organization. Patel's writings on issues arising from her activism include *Against the Grain* (1990) and "Multi-culturalism: Myth and Reality" (1991). Her article "Third Wave Feminism and Black Women's Activism in Britain" appeared in the anthology *Black British Feminism* (1997) edited by Heidi Safia Mirza. In Britain, *black* is a political definition that refers to people from Southeast Asia, Africa, and the

Excerpted from "Third Wave Feminism and Black Women's Activism" by Pragna Patel. From *Black British Feminism: A Reader* edited by Heidi Safia Mirza, pp. 255–257, 260–264, 267–268. Copyright © 1997 by Heidi Safia Mirza. Reprinted by permission from Taylor & Francis Books Ltd.

Caribbean. This use of the term emerged in the 1960s as a means of uniting people with the common experiences of colonialism, imperialism, and racism. In these excerpts, Patel examines the activities of the Southall Black Sisters and its sister organization, the Brent Asian Women's Refuge. She also critiques how the use of "multiculturalism" has served to rationalize patriarchal customs detrimental to women. At the time of this article, Patel was studying law.

The time has come, the Walrus said, to talk of many things...

(L. Carroll, *Alice's Adventures in Wonderland*)

I was born in Kenya and came to Britain at the age of 4. I have known no other landscape, but I never felt that I belonged here. With no other choice but to make my life here, I grew into a politics of resistance; against the racism that I experienced outside my home because I was the wrong colour, and against the injustices I experienced because I was the wrong gender. In this way I fashioned for myself a strong political identity, in struggle with other black men and women. Despite hovering on the margins of British society, this identity is a source of tremendous power and strength, and even, dare I say it, moral righteousness.

It was precisely this sense of belonging, this black identity, which fell apart in December 1992. When militant right-wing Hindu nationalists destroyed the sixteenth century Babri-Masjid mosque in India, I was forced to confront the elements of the 'Hindu' identity within me which I had supposed had all but withered away. By virtue of being a member of that diaspora of Indian-Hindu origin, I was, whether I liked it or not, also part of a Hindu collectivity. This collectivity contained elements which, as part of a majority in India, was embarking in the name of God and religion on a course of annihilation of minorities and dissenters, and attacking the very foundations of democracy in that country. Yet this very same collectivity, as a minority elsewhere in the world, knows what it is like to experience discrimination and hatred. These

painful contradictions compelled me to critically re-examine my own Hindu background in order to be able to understand, and crucially to oppose those who, in the name of Hinduism, were acting in a way which was deeply inhuman and shameful to witness.

The recognition that I may belong at one and the same time to an oppressed minority and to an oppressive majority, with all the contradictions that entails, has found an echo in my experiences in Britain. Many of the struggles we have waged as black people here have rested, sometimes uncritically, upon a white majority/black minority dichotomy. This has been useful in creating the sense of solidarity necessary to mobilize against racist attacks from the state and thugs on our streets, uniformed or otherwise, but in asserting a singular and absolute identity—as 'victims' of racism—we have evaded the need to look critically at the inner dynamics of our communities. This has resulted in a tendency to deny uncomfortable realities and has tended to give us a distorted and partial view of ourselves and the world around us. This tendency has been particularly difficult for black women to deal with, as our struggles often arise out of our experiences *within* our communities, and in fighting to force these onto the wider political agenda we have also often had to fight against the imposition of a singular identity either on ourselves or on our communities.

What follows is an attempt to locate these struggles by retracing some of the campaigns of Southall Black Sisters (SBS) and our sister organization, Brent Asian Women's Refuge. Our struggles have, out of necessity, arisen from the routine experiences of many Asian, African, Caribbean and other women who come to these centres with stories of violence, persecution, imprisonment, poverty and homelessness experienced at the hands of their husbands, families and/or the state. In attempting to meet the challenges they pose in their demands for justice for themselves and for women generally, we have had to organize autonomously. But we have always endeavoured to situate our practice within wider anti-racist and socialist movements, involving alliances and coalitions within and across the minority and majority divides. This has not always been easy, but it is the only way we know in which a new and empowering politics can be forged.

By organizing in women's groups and refuges, many of us have fought for autonomous spaces and for the right for our own voices to be heard in order to break free from the patriarchal stranglehold of the family. In the process we have also had to challenge the attitudes of the wider society, as well as the theory and practice of social policy and legislation which seeks to restrict our freedom to make informed choices about our lives. Our organizations and our practice are critical in unmasking the failures, not only of our communities and the state and wider society, but perhaps more tellingly, of so-called multi-culturalist and anti-racist policies.

Throughout our campaigns on domestic violence, whilst countering racist stereotypes about the 'problematic' nature of South Asian families, SBS has sought to highlight not only the familiar economic and legal obstacles faced by all women struggling to live free of abuse, but also the particular plight of Asian women; language barriers, racism, and the specific role of culture and religion which can be used to sanction their subordinate role and to circumscribe their responses. Culture and religion in all societies act to confer legitimacy upon gender inequalities, but these cultural constraints affect some women more than others in communities where 'culture' carries the burden of protecting minority identities in the face of external hostility. We have had to formulate demands and strategies which recognize the plurality of our experiences, without suppressing anything for the sake of political expediency. Alliances have been crucial in this, not only in gaining wider support, but also in breaking down mutual suspicion and stereotypes, and to ensure that some rights are not gained at the expense of others.

We began our protests in the early 1980s over the murder of Mrs. Dhillon and her three daughters by her husband who burnt them to death. In 1984 we took to the streets in response to the death of Krishna Sharma, who committed suicide as a result of her husband's assaults. Organizing with other women in very public ways, through demonstrations and pickets, we broke the silence of the community. Until that point there had been not a single voice of protest from either progressive or conservative elements within the community. The women who led the demonstrations had themselves fled their own families in Southall, but returned to join us with scarves wrapped around their faces so that they might escape recognition. We demanded and won the support of many white women in the wider feminist movement, although initially they were hesitant in offering support for fear of being labelled 'racist'! One of our slogans—'self-defence is no offence'—was appropriated from the anti-racist 'street-fighting' traditions, but ironically it has now become the much quoted slogan of the wider women's movement against male violence in Britain. The form of our protests drew directly from the varied and positive feminist traditions of the Indian sub-continent. We picketed directly outside Krishna Sharma's house, turning accepted notions of honour and shame on their heads. It is the perpetrators of violence, we shouted, who should be shamed and disrobed of their honour by the rest of the community, not the women who are forced to submit. Another slogan—'black women's tradition, struggle not submission'—was first coined on this demonstration, and that, too, has been adapted to become the rallying cry of feminists against male violence in this country.

The lessons of those early years have ensured that we have understood the importance of campaigns and direct action as an essential means of articulating the needs of the women who turn to us daily. From the murder of Balwant Kaur by her husband at the Brent refuge in 1985, to the life imprisonment of Kiranjit Ahluwalia for killing her violent husband in 1989, our response has been driven by a recognition that those tragedies reflected, albeit in extreme forms, the day-to-day experiences of many Asian women facing violence in the home. Over the years we have managed to retain a campaigning edge to our work, while also providing day-to-day services....

The State and the Family

The state for us has never been an abstract concept. It has a real existence which defines our roles and position in society; it negotiates our existence as women within our families....

It has been left to women to highlight the manner in which immigration law can combine with the institution of the family to construct the women as an appendage to her husband, economically and socially dependent upon him, and a potential prisoner of violence and abuse within the home. When a woman has come from abroad to marry here, should the marriage break down within a year her immigration status is rendered illegitimate should she leave her husband (this is known as the 'one year rule'). In the absence of an immigration status in her own right, a women's option to leave a violent or abusive home becomes virtually non-existent. In such a situation if a women does leave her husband, not only is she ineligible for any form of state assistance in the form of housing or welfare benefits, which are a prerequisite for giving women a real choice about leaving a violent home, but she also risks deportation. As there is no right of appeal in such cases, her fate is then entirely dependent upon political decisions taken by the Home Office. The arbitrary and discriminatory nature of such decisions, underpinned by notions of Third World peoples as 'aliens' or 'undesirables', means that the majority of women in such cases are forcibly deported to countries where their futures may be at risk. Persecution based on gender is not recognized as grounds for asylum in this, as in many other western countries.[1]

In this way, as in many others, we see the state applying double standards to the treatment of families from different communities. The premise of social services intervention, for example, is to preserve the unity of the family as far as possible, whilst the police and immigration services end up dividing and separating many black families. Women in the majority community have, through women's own action, managed to extend their choices to enable women to leave unhappy marriages, but for women from minority communities, particularly those with immigration difficulties, that choice is absent. Our demand is for the right of black and minority families to live undivided when they choose, but for women to have a real option of leaving an unhappy marriage without the state and the community colluding to deny that choice....

Multiculturalism and Religious Fundamentalism

Religion and culture is the terrain on which the politics of multiculturalism and variants of anti-racism are built, often amounting to nothing more than a preservation and celebration of minority culture and religion. Multiculturalism has its roots in past British colonial practices in such countries as India.[2] In Britain it allows the state to mediate between itself and minority communities, using so-called 'community leaders' as power-brokers and middle-men. Needless to say, such leaders are male, from religious, business and other socially conservative backgrounds who, historically, have had little or no interest in promoting an agenda for social justice and equality, least of all the rights of Asian

---

[1]See Southall Black Sisters submission to the Home Affairs Select Committee, *Immigration and Domestic Violence*, HMSO, 1992.

[2]Sahgal, G., 'Secular Spaces: The Experience of Asian Women Organising' in Sahgal, G. and Yuval-Davis, N. (eds) *Refusing Holy Orders*, London, Virago Press, 1992.

women.[3] In return for information and votes, the state concedes some measure of autonomy to the 'community leaders' to govern their communities. In reality, this means control over the family—women and children. Together the state and community leaders define the needs of minority communities, to limit their influence and to separate off the more radical elements by labelling them extremists.

In the name of tolerance of 'cultural differences', the rights of women are dismissed, and many Asian women seeking support to escape from violence are often told by state agencies that such a breach is not an acceptable method of resolving their problems in 'their cultures'. They are denied protection and delivered back to their families and communities....[4]

Multiculturalism has provided the ideological framework for fundamentalist and conservative leaders within the Asian communities to emphasize the primacy of religious identities. In this country the rise of religious fundamentalism is in part a response to the upsurge in European nationalism and racism, and the failure of progressive left politics, coupled with the fallout from the Rushdie affair and the Gulf War....

The resurgence of religious fundamentalism feeds off parallel developments within the majority community. The reassertion of Christianity as the main signifier of 'British' identity in schools, or the 'Back to Basics' campaign, underlined by a Christian morality aimed at preserving the nuclear and heterosexual family unit, are developments that have fuelled reactionary demands for formal recognition of minority religious life-styles. Fundamentalist movements may differ in detail, but they have two major objectives in common: recognition as distinctive (to legitimate the claim for access to resources); and the reclamation of family values, with particular emphasis on control over the sexuality and fertility of women.

---

[3]Ali, Y. 'Muslim Women and the Politics of Ethnicity and Culture in Northern England,' op cit.

[4]Patel, P. "Multi-culturalism: Myth and Reality," *Women, a Cultural Review,* vol. 2, no. 3, winter 1991.

Increasingly the received wisdom in the formulation and implementation of social policy is that minority communities are identified according to their religious backgrounds. Other social divisions of class, caste and gender are hidden beneath this monolithic, characterization. Increasingly references are made not to Asian culture, but to Sikh, Muslim or Hindu culture. Such multicultural norms are also permeating popular perceptions of Asian communities.

Women's minds and bodies are the battleground for the preservation of the 'purity' of religious and communal identities. So the role of women as signifiers and transmitters of identity within the family becomes crucial. There is a growing phenomenon of organized gangs and networks of Asian men who hunt down runaway Asian girls and women who are perceived to have transgressed the mores of their culture and religion, and to have defiled their honour and identity. The family has therefore become a site of struggle for feminists and fundamentalists alike....

... The third wave of feminism has a lot to contend with. The rise of new forms of racism, fascism, nationalism and religious fundamentalism world-wide demands from us a new and visionary politics. We must avoid the pitfalls of the identity politics of the 1970s and 1980s which made it so difficult to share experiences, and we must move beyond the limitations of anti-racism and multiculturalism which equally limit our perspectives and our ability to act. We must reject the vicious and blinkered vision of nationalism and fundamentalism. Our task is to find new ways of resisting and new ways of truly democratic thinking which give us the optimism to go beyond all of these failed forms of politics. Our alliances must cross our different identities, and help us to reconceptualize notions of democracy, human rights and citizenship. Whatever the dividing lines drawn by priests, mullahs, gurus and politicians, we will then be able to reach out to our each other, to support one another in our transgressions and defiance. Above all, we must leave room for doubt and uncertainty in our own orthodoxies. The time has come, in the words of the Walrus in Lewis Carroll's poem, to talk of many things....

184. Iona Zambo
"Gypsy Women: Barriers to Citizenship"

There are some 2 million Gypsies, or Roma, as they prefer to be called, living in the member states of the European Union and 15 million in Europe as a whole. Their status in Eastern Europe has deteriorated sharply since the collapse of communism and the emergence of market economies. In the Czech Republic, for example, three-quarters of Roma were unemployed at the beginning of 1999, a situation specifically criticized by the European Union. Roma throughout Europe face increasing discrimination in housing and education, and many are the object of right-wing violence carried out by skinheads. In response, Roma are increasingly organizing on their own behalf, as in Cologne, Germany, where the Association for the Advancement of Roma opened a center in March 1999 to document the history and culture of the Roma and to challenge negative stereotypes. In this article, Iona Zambo examines the specific issues faced by Roma women in Hungary, how Roma women are affected by the market economy, and the patriarchal structure of Roma families. The article appeared in an anthology edited by Tanya Renne, *Ana's Land: Sisterhood in Eastern Europe* (1997).

All over Europe reemerging nationalism aggravates the situation of minorities. Research shows that in the region, specifically in Hungary, prejudice against Gypsies has grown. Social and economic crisis and a spread of poverty were followed by a dramatic increase in segregation. The prevalent ideology, held by politicians and common people alike, is that Gypsies are culturally, morally, and intellectually inferior. Let me quote a few lines from an article that appeared in the November issue of *Cigakritika* (*Gypsy Critic*). A reporter asked random Hungarians on the street about their opinion of Gypsies. A taxi driver replied: "It's the easiest thing in the world to recognize a Gypsy. I hate them in principle, but I still have to give them a ride because they are the only social groups who can afford taxis. You don't have to ask where their money comes from; just read the police reports." An unemployed skinhead answered: "They should be wiped out—every last one. Hitler's biggest sin was starting this job and never finishing it. We skinheads will do it and defeat all inferior races." Of course, there are a few positive voices as well,

but the forgoing examples describe the general attitude. Out of Europe's Gypsy population of 15 million, eight hundred thousand live in Hungary. The Gypsies are the biggest losers in the transition process. Neither the country nor the Gypsy community was prepared for the effects of the transition. It caught the Gypsies unprepared and without economic resources.

Because of a lack of basic living conditions, Gypsy families that are multiply disadvantaged can't join the market economy. Most Gypsies live in extreme poverty, have large families, and are unemployed and unskilled. Their housing is unhealthy, dark, damp, and crowded. It doesn't meet basic hygienic standards. Gypsy settlements, *putri,* are often located near garbage dumps or swamps far away from infrastructure.

Our state of health is radically deteriorating. There is not much hope that it will improve in the near future. Infant mortality as well as premature birth is very high. Children's physical and mental development is slowed, which presents serious problems in education. They are behind their peers and can't keep up with the rest of the class, a factor that affects their whole lives.

As the very survival of Gypsies is becoming threatened, more Gypsy representatives have begun to speak out, but we are still not participating in the political decisionmaking process. We are tiptoeing in front of the walls surrounding the economic and political system. Our actions remain invisible. It is a fact that Gypsies do not have resources, financial or social, to make their own representation more effective. There is no mother country or politically and financially strong international Gypsy organization to rely on or get support from.

The transition created a new social situation that gravely affected Gypsy women. Most Gypsy families are single-income families. The man's income, which is becoming increasingly deficient, is the sole income. Binding tradition, many children, and lack of training keep women from joining the workforce in spite of the need to do so. Gypsy families are known for having many children (families of ten or twelve are not uncommon). Family planning has never existed; women become mothers as early as age fourteen. Therefore, a sharp decrease in the birthrate cannot be expected; only better living conditions could bring the rate down and ensure an escape from poverty.

The Role and Rights of Gypsy Women in the Family

Families are multigenerational. There are two things that control the community: the *patyi,* or the regulation of traditions, and the *krisz,* or the ethical and moral codes. These are unwritten laws mandatory for everyone. For example, these rules ensure unconditional respect for the elderly and assign women a secondary place within the family. Women are present but can't control their own lives. They are subject to traditions and their husband's arbitrary power. Adultery is a forgivable sin for men but not for women. A woman who cheats on her husband has to face the contempt of the whole community and very often physical violence from the husband.

The Hungarian Gypsy community is characterized by a kind of social stratification that is similar to a caste system. A young person beginning a relationship with someone from a lower caste is scorned. Relationships, however, are based on emotions and not prearranged; eloping is very common for young couples. When a woman joins a man in marriage, she becomes part of his family as property. Before then it was her own family that controlled her; now it is her husband's. She has obligations to fulfill toward them. Women who rebel against family traditions and try to break out of this one-way street most often become prostitutes. Their fate is determined by their lack of education and training and their naiveté. Most of them look for shelter in big cities and end up in the hands of pimps.

The process of transition will be a long one for Gypsy women. We have to break out of a double bind, first by struggling against restrictive traditions in the community and then by proving ourselves as women in society. The key question here is whether Gypsy women have equal opportunities to make free choices in the future. A lot of tolerance is needed for us to define our needs, to learn to take our own lives into our own hands.

QUESTIONS FOR ANALYSIS

1. What led Pragna Patel to reexamine her Hindu background? How did this affect her sense of identity? Describe the aims and activities of the Southall Black Sisters and the Brent Asian Women's Refuge. How are women differentially affected by the immigration laws in Britain? How does Patel critique "multiculturalism" in her discussion of the challenges facing "third wave feminism"?

2. In Iona Zambo's view, how has the growth of ethnic nationalism in postsocialist Europe influenced public attitudes toward Gypsies? How are Gypsies affected by the implementation of market economies? How does Zambo analyze women's status within traditional Gypsy families?

WOMEN AND THE WAR IN BOSNIA

The dissolution of Yugoslavia began when Slovenia and Croatia declared independence in 1991 followed by Bosnia in 1992. While the independence of Slovenia, with its relatively homogeneous population, was uneventful, the Serbian populations in Croatia and Bosnia sought to create independent Serbian states within these newly declared republics. A cease-fire in December 1992 curtailed the conflict between Serbia and Croatia, but violence soon erupted in Bosnia, which is made up of Serbs, Croats, and Muslims. To carry out their territorial aims, the Bosnian Serbs received support from the Serbian government, headed by Slobodan Milosevic, which controlled the federal Yugoslav army. In addition to launching numerous attacks on the civilian population of Bosnia, including those in designated United Nations "safe areas," the Bosnian Serbs laid siege to the capital city of Sarajevo in an attempt to incorporate the city into Serbian territory. The siege continued for several months, but Sarajevo, for centuries a positive example of multiculturalism in the Balkans, refused to capitulate. In 1993 and early 1994, the government of Franjo Tudjman sought to create a Croatian enclave in western Bosnia with a view to its future incorporation into Croatia. The war in Bosnia (1992–1995) was ended by the Dayton Accords which stipulated that Sarajevo would remain as Bosnia's undivided capital city. Bosnia itself was to exist as a loose federal state with a central administration, but it was nonetheless divided into a Moslem-Croat Federation and a Serb Republic. By 1995, the wars in Croatia and Bosnia had resulted in over 200,000 deaths and 3 million refugees. The following documents were written by members of the Serbian women's organization Women in Black Against War and the organizers of Medica Zenica, a multiethnic Women's Therapy Center established in Zenica, Bosnia in 1993, to treat female victims of the war.

185. Women in Black Against War

The Serbian group Women in Black Against War was formed in 1991. It was inspired by the Israeli women's group by the same name which has sought to strengthen the peace process in the Middle East. For seven years, and with only brief interruptions, the organization held demonstrations every Wednesday in front of the National Parliament building in Belgrade to protest the ethnic nationalism of the Serbian government and the violence and war in Kosovo, Croatia, and Bosnia. Although Women in Black opposed the Serbian government's actions in Kosovo, governmental laws on security and fear of public reprisals compelled it to suspend public protests during the NATO bombings of Yugoslavia, including Belgrade, which were carried out in the spring of 1999. The bombings followed the refusal of the Serbian government to sign the Rambouillet Agreement of October 1998 concerning the status of Kosovo and its predominantly Albanian population. While virtually all Serbian opposition groups assumed a "patriotic" position during the period of the bombing and stopped criticizing the government, Women in Black continued to express its opposition to the government's policies in various ways, including Internet communications with the international community. These documents include "Nationalism and Serb Women" by Stasa Zajovic, a peace activist and a Women in Black founder, and a "A Call for Action" issued during the war in Bosnia.

A. Stasa Zajovic, "Nationalism and Serb Women"[1]

The militarization of the former Yugoslavia has meant the imposition of military values, symbols, and militaristic language; a cult of necrophilia (expressed in slogans such as "The frontiers of Serbia are where Serbs are buried"); and an acceptance of political and moral totalitarianism. Along with these ideological shifts have come a rigid separation of masculine and feminine roles—in short, woman as mother and man as warrior—and the political marginalization of women.

Throughout the postwar period, concern over women's social condition in Serbia and Yugoslavia has been reduced to an obsession with keeping up the working-class birthrate. As nationalism replaced class struggle as the basis of political discourse in Serbia, this obsession with reproduction was transferred to the nation. With the 1987 "antibureaucratic revolution" (led by Slobodan Milosevic), a strange sort of sisterhood began to show itself in the streets. Huge crowds throughout the country shouted, "We want arms"—in a collective trance, united in hate and desire for vengeance for Serbia's "offended nationhood."

In tandem with the cult of blood and soil, the new Serbian nationalists also summoned to life the symbolic medieval figure of Mother Jugovic—the long-suffering, brave, stoic mother of nine offering her children up to death in defense of the fatherland. Maternity is now to be seen as an obligation, not as a free option for women; women's sexuality is to be controlled and reduced to procreation.

Kosovo: The Demographic Counterrevolution

Serbia's demographic slump has been described as "one of the greatest tragedies of the Serbian people," particularly since the "demographic counterrevolution," a perceived threat posed by the Albanian women of Kosovo province, who have the highest fecundity rate in Europe. Serbian repression in Kosovo has as one of its objectives a reversal of this demographic gap. This reversal has been seen in increasingly militarist (or "national security") terms. The structures of militarized power now insist that the birthrate must grow so that the nation can defend itself in military conflict with other people.

Women are blamed for any shortfall in this sacred mission: "I call upon all Serbian women to give birth to one more son to carry out their national debt," stated one politician. Another, Rada Trajkovic of the Association of Kosovo Serbs, was even more explicit in viewing mothers' sons as cannon fodder: "For each soldier fallen in the war against Slovenia (June 1991) Serbian women must give birth to 100 more sons."

Political Pimps

The manipulation of women by the military establishment began some time ago. The clearest examples, however, are found in the rallies held by the Women's Movement of Yugoslavia, which is linked to the promilitary parties and formed in 1990. In February 1991 the women of this move-

---

[1]This article is not dated, but was probably written in early 1992 prior to the war in Bosnia. Zajovic discusses the independence of Slovenia in 1991 and the war between Serbia and Croatia following Croatia's declaration of independence, which was also in 1991. The last date referred to is December 1991 and no mention is made of the war in Bosnia that began in 1992.

ment publicly lent their support to the Yugoslav Narodni Armi, saying they considered it the only force capable of saving the country. The women have been used; the military hierarchy carries out the function of "political pimp," putting women on the street to give their approval to ends that are contrary to women's own interests.

Before the massive wave of mobilization for civil war, there were warnings in some parts of the country—such as Montenegro, known for its martial traditions—that men should be prepared to give up their lives for the fatherland and that anything less would be a blow to their masculine dignity. Men were expected to follow national tradition, whereby "in war not one Montenegrin man can be protected by a woman." One parliamentarian stated, "We in Montenegro believe that a man who is fighting at the front and allows himself to be hauled back by a woman should commit suicide at once." This perception that women are preventing men from fulfilling their national duty was reinforced in radio broadcasts from the Dubrovnik front, in which a high percentage of Montenegrins were involved. Soldiers would send greetings to their fathers, brothers, and male friends but not to their wives or girlfriends.

Rejecting Manipulation

Happily, the number of men who wish to free themselves from this macho war culture is growing, and more are not ashamed of being protected by their mothers, wives, and sisters. A committee of women was formed in Montenegro in October 1991: "We raise the voice of protest against the private war which those in charge are making from their offices. They have sent their sons out of the country and onto the tennis courts, while our sons carry on being carried by force to the front and to their graves. We demand that these demented leaders, politicians and members of the military resign immediately in order to save this country."

Women as Soldiers

An increase in the number of women in the army is no indicator of sexual equality, least of all where there has been no corresponding democratization of society. There are women in all the militias and national armies now active in the former Yugoslavia, and these women are appropriating the most brutal patriarchal values.

Women members of the Serbian military in Croatia do not occupy important positions in the military hierarchy. The front line is for the men; the *knidze,* or "female militia," in the words of one member, "occupy the administrative jobs, communication services, health, stores. We are simply replacing the boys who have gone into combat. But we, too, have passed the military entrance exam, and we can help the boys at any moment."

A small number of Croatian and Serbian women fighters have gone to the front and have been rapidly converted into mythical figures. This confirms what patriarchal history teaches us: that women enter history only when they have taken on masculine roles. The media celebrate these women as heroines when they kill the enemy; but when women fighters from the other side are captured, they are denounced as "monstrous women."

The first female battalion of the war was established in Glina (a Serb-majority town in Croatia annexed to the Serbian autonomous region in December 1991). Members took the oath "We will fight against all of Serbia's enemies under the protection of God" and heard male officers declare sentiments like "If our mothers would not have been heroines, they would not have given birth to such valiant children." Patriotic women also fight on the "home front," of course. In Belgrade, women knit socks and gather winter clothing to keep the boys at the front warm. The progovernment media fawn on every evidence of maternal mobilization, at the same time ridiculing the work of women in the antiwar communities and centers.

Clearly, the majority of women are on the side of peace. They are convinced that they can offer the historical female alternative: nonviolence in place of violence, life in place of death, vitality in place of destruction. Stana Pavic, an elderly refugee from a Serbian village in Croatia, told me that "we women...should have united like them [the military]—we could have made a peace accord in no time at all."

The Mothers' Movement

With the end of the war in Slovenia in early July 1991, human rights—above all, the right to life—lay in ruins all over Yugoslavia. In such a situation, women burst onto the political scene demanding the right to live. On July 2 a parliamentary session in Belgrade was interrupted by several hundred parents, mostly conscripts' mothers. This was the first civil society initiative against the war in the federal capitol and the first protest against the abuse of women's reproductive work by the state, nation, army, and party. "Men are the controllers of the war and of our sons. We do not give them permission to push our sons forward to kill one another."

As a result, the very role that marginalizes women in their private lives—reproduction—has had the effect of converting them into active participants in the political life of the nation. So the patriarchal divide between the personal and the political weakens; the personal has become the political in this war. The mothers' movement has contained all the ingredients for a militarist's nightmare. Nevertheless, the mother's movement has been subject to different types of manipulation by political structures. A part of the mother's movement has been used—both in Serbia and Croatia—for patriotic propaganda. It is easy to exploit the sentiments of mothers whose sons' lives are in danger. The soldiers' mothers at times feel confused, internalizing the struggle between the "patriotic" sentiment that underlies official propaganda and the desire to save their own sons.

Yugoslav women have been linked in a feminist network since 1987; they have managed to maintain solidarity and plan joint activities ever since, rejecting the conditions that support policies of divide and rule. As the war continues, normal communication is impossible with the cutting of telephone lines, post, and transport between the republics. In spite of such practical obstacles, feminists—Serbs, Albanians, Croats, Slovenes, Hungarians, and Montenegrins—are

united in organizing against the war. Joint and simultaneous protests, such as the weekly Women in Black Against War demonstrations, are a product of this unity of purpose.

B. Women in Black Against War, "A Call for Action"[1]

They have been ceaselessly killing, torturing, and raping for two and a half years already. They have banished more than 3 million lives. They manipulate women. They blackmail men. They spread hate, destruction, and death. We are left without words to express our horror and anger.

They haven't stopped yet.

Fascist leaders of Serbian politics threaten us with war in Kosovo, Macedonia, and Serbia. Meanwhile, they have stopped all electricity, water, and telephone systems in Bosnia and Herzegovina. People die by the minute. No matter which names they have, they die of the cold, illness, and hunger. And it is only November.

Fascist leaders of Serbian politics continue to destroy all positive interethnic communications. They have separated streets, classrooms, families, and cities. They are drawing lines on mountains and corridors through the countryside.

Since 9 October Women in Black Against War has come out every Wednesday on the streets to express its absolute disapproval of all nationalist politics. Above all, Women in Black accuses the fascist Serbian regime of being responsible for the death and destruction. According to their ideology, not a single life has value, not a thousand lives, not 1 million, not 3 million and counting. In the end it doesn't matter if they are Serbian or not.

---

[1]The "Call for Action" is not dated, but appears to have been written some time in 1993 or early 1994. It refers to two and a half years of killing and to the realities of war in Bosnia and Herzegovina; in particular, the towns of Sarajevo, Bihac, and Mostar. The beginning of the killing could refer to the war between Croatia and Serbia in 1991 or to the war between Bosnia and Serbia that began in 1992.

"A Call for Action: Women in Black Against War." From *Ana's Land: Sisterhood in Eastern Europe,* edited by Tanya Renne.

Since their bullets and the cold will wipe out another one hundred thousand people in Bosnia and Herzegovina, since men will rape thousands of women of every nationality, since war is possible in Belgrade and anywhere—Women in Black calls on all women for all types of civil disobedience.

The misery in which we live should not frighten us but incite us to resistance. It is strange that we have not yet started to scream. Our friends from Sarajevo, Bihac, Mostar, sit every day in darkness and cold without any hope. If we ever see them again, they will be difficult to recognize: Their hair has gone white, they are thinner, and they have aged in this short time. Refusing to know how they live and refusing to confront the government that tortures people are crimes.

186. A Women's Therapy Center in Zenica, Bosnia

In 1993, a women's therapy center was established in Zenica in central Bosnia as a multiethnic project. Staffed by an all-female medical team, Medica Zenica planned to provide medical and psychological assistance to women in Zenica as well as refugees from various parts of Bosnia. In particular, Medica Zenica treated the special traumas associated with rape, a pervasive weapon of war and ethnic cleansing during the war in Bosnia. Resisting nationalist pressures, it provided services to patients from all three groups in Bosnia: Croatian, Serbian, and Muslim. A mobile team also visited women in the refugee camps located near Zenica and Sarajevo. Medica Zenica's activities include ongoing research on the effects of the war on women. It has shared its findings with the United Nations Tribunal investigating war crimes in the former Yugoslavia. Created in 1993, the Tribunal defined rape as a crime against humanity when carried out in armed conflict against a civilian population. It thus set a precedent for the prosecution of rape and sexual assault as violations of international law. The project statement reproduced here outlines the origins and aims of Medica Zenica.

Introduction

The Women's Therapy Center "Medica" in Zenica, Central Bosnia began to take care of war-traumatized women and girls in April, 1993. "Medica" is an independent non-governmental women's organization, officially registered in Zenica. Adequate professional aid for war traumatized women and girls did not exist at this time due to the desolate economic and social conditions of the war situation. We decided to found the Medica project to ensure highly professional aid to the female victims of the war in Bosnia and Herzegovina. The assistance which we offer is a combination of medical and psychological treatment provided by a highly professional local staff of about sixty women.

Organizational Profile

1. Women's Therapy Center (Medica 1) which consists of:

(a) A gynecological outpatient ward for refugee women and inhabitants of Zenica. The all-female professional team consists of two gy-

Reprinted by permission from Network of East-West Women. This statement was posted on an electronic mailing list, Women-East-West@neww.org, on February 23, 1995. Contributed by Sonia Jaffe Robbins.

necologists, one general practitioner, one internist, two anesthetists, and six nurses.

(b) An inpatient ward for psychological and physical rehabilitation. This ward provides ongoing psychological and medical care. Our team consists of three psychologists, one psychiatrist, one sociologist, and a female theologian, all working in shifts.

(c) Other staff include kitchen personnel, administrative staff, a field officer, and kindergarten and pre-school teachers....

Problem Statement

Medica's basic aim is to assist women and girls who have been raped, held in detention camps, and who have become refugees during the war, to find their psychic balance, re-establish family relationships, and participate in normal social life (although at the moment there is no reason to believe that this can be achieved). We decided to found Medica on our own initiative when we realized that none of the main international humanitarian organizations were able to provide concrete help for these women. The idea to provide shelter, gynecological and general medical treatment, and psychological therapy for war-traumatized women and girls first came from Dr. Monika Hauser, a feminist gynecologist living in Germany who was moved by news reports on raped women and girls in Bosnia and Herzegovina.

Zenica in Central Bosnia is one of the last cities of Bosnia-Herzegovina which is still free. Since the beginning of the war in April, 1992, Zenica has provided shelter to a huge number of refugees and displaced persons from all of the occupied areas of Bosnia. At the moment, Zenica accommodates approximately 50,000 refugees, almost half of the city's original population. Since April, 1993, Zenica has been shelled continuously. Central Bosnia has been destroyed economically and can only survive through humanitarian aid. The traumatized women and their children, who form the majority of the refugees, suffer from a catastrophic lack of medical and psychological assistance, food, and nutrition for babies.

The project is based on the recognition that systematic torture, in the form of mass rapes of women and girls in former Yugoslavia, is a war crime. This was confirmed by the Amnesty International Report on Bosnia and Herzegovina ("Rape and Sexual Abuse by the Armed forces," January, 1993). We wish to support the creation of facilities that provide relief to traumatized women. Different forms of psychic and physical violence by men are part of women's reality all over the world. This kind of violence reaches its peak in war. This fact prompted us to take the necessary steps to establish our Women's Therapy Center with the three areas of work described above: provision of shelter, gynecological, and psychological assistance. We expect that our project will become a center for the protection of women from the domestic violence ensuing from war. It will relate to the activities of local women's groups....

Project Beneficiaries

1. Women raped in the war and women held in the detention camps for women, which we have reason to believe still exist, and other victims of all forms of sexual, physical, and psychic torture.
2. Women held in the detention camps who have not been raped but have suffered from malnutrition and illnesses.
3. Displaced women, especially those who are pregnant or with small children, in need of medical treatment.
4. Local women experiencing a psychological crisis.
5. The children of our patients as well as child relatives who need our professional help.

Project Activities
Women's Therapy Center
1. Gynecological Outpatient Ward.

We are providing all kinds of gynecological treatment...The outpatient ward is open five days a week to all women for free, but priority is given to refugee women. We provide treatment to around 500–1,000 women a month, of whom 70% are refugee women and 30% are inhabitants of Zenica....

2. Inpatient Ward.

The inpatient ward offers ongoing psychological and psychiatric therapy. The treatment is intended to enable patients to overcome traumatic experiences related to all forms of the physical and psychological torture which they endured. The psychological treatment consists of intensive individual and group-therapy work, such as language courses, sewing, and other activities.

Our psychological program includes:

a. Acute crisis intervention in our inpatient ward with ongoing individual and group therapy.

b. The continuation of therapy for patients released from the Women's Therapy Center to Medica 2 or 3.

c. Ambulant therapy for traumatized women and girls who cannot stay in our house for lack of space, because they want to remain with their families, or because they live in private accommodations in town.

d. We offer consultation to refugee women whose needs range from a single session to ongoing psychological therapy.

General Care

The warm atmosphere in the Center, created by all members of the team, supports the healing process. The medical staff is also concerned with the general health of our patients. Unlike the refugee camps, we offer nutritious food that is rich in vitamins, heating during the winter, and shelter from disastrous weather conditions. On average, we provide shelter to 25 women and 15 children in Medica 1. Until now, we have provided professional care to approximately 250 women and children (Medica 1, 2, and 3).

Follow-Up Projects

In the two follow-up projects (Medica 2/ Women for women in Zenica, and Medica 3 in Visoko, which is on the way to Sarajevo), we have been realizing that women and girls can re-establish their individuality and integrity by running their own households. We want them to take daily organizational tasks into their own hands. They achieve self-assurance through various forms of education and training in nearby enterprises or in our own small, handicraft workshops. This work also enables them to become financially independent.

1. For these purposes, we opened Medica 2/ Women for women in Zenica in July, 1993. The center consists of three private houses connected by a yard to a little vegetable garden, a playground, and a handicraft workshop. Medica 2 has a capacity of 25 women and 15 children.

2. Medica 3 in Visoko (50 km from Zenica, on the way to Sarajevo), opened in August, 1993. We found a spacious house with 1,000 square meters of land. On the cultivated 500 square meters are vegetables, 150 fruit trees, a poultry yard, and three cows, and two (twin) calves. Medica 3 has a capacity of 25 women and 15 children.

Social Assistance

Regular visits are made by the social and medical staff to the women in the refugee camps in Zenica and other places in Central Bosnia. Our aim is to explain the activities of our center and to bring material support (food, including baby food, drugs, clothes, etc.). We also organize "roundtable discussions" as therapy talk groups. We do this, not only to give the women in the camps the opportunity to talk, but also to identify the heavily traumatized women and girls who need our help. We then encourage them to attend our Center....

Kindergarten and Pre-school Services

We offer kindergarten and pre-school services to 30 children of the Center as well as to children of the nearby collective center. At least for a few hours, we wish to take them out of the desperate conditions of the collective center and to give them the possibility to play and learn in a relaxed atmosphere. They are accompanied by a professional staff (teachers and instructors of various sorts) who try to help them to overcome some of their often neglected traumatic experiences.

Research

When our mobile teams visit the refugee camps they also collect data on refugees in order to understand and analyze the full dimension of war-related violence against women in former Yugoslavia, provide evidence for international law courts against war criminals, and contribute to the world-wide feminist analysis of patriarchal violence in times of peace and war.

Time Frame

Medica started its work in the Women's Therapy Center on April 1, 1993. The follow-up projects were developed later on. We intend to continue our work for as long as we can. We estimate that our research work will take between one and five years. The results will be evaluated as we work.

We have been asked to contribute the results of our research to the United Nations Tribunal.

QUESTIONS FOR ANALYSIS

1. How does Stasa Zajovic analyze the implications for women of Serbian ethnic nationalism and militarism? What is her view of women serving in the armies of the former Yugoslavia? How does she describe the aims of the mothers' movement that emerged in Belgrade in July 1991 (following the end of the war in Slovenia) and those of the committee of women formed in Montenegro in October 1991?
2. In its "Call for Action," how does the Serbian group Women in Black Against War critique the policies of the Serbian government and its role in the war in Bosnia?
3. What are the main aims of the Women's Therapy Center in Zenica, Bosnia? How does the project statement analyze the special traumas experienced by women during the war in Bosnia? What kind of research is Medica Zenica carrying out, and what is its relationship to the United Nations Tribunal investigating war crimes in the former Yugoslavia?

WOMEN AND THE UNITED NATIONS

In the 1990s, women succeeded in pressing the United Nations to assume more responsibility for women's issues, gained greater organizational representation within the UN, and assumed leadership of several special commissions. Women's contemporary roles within the United Nations are illuminated in these two speeches, the first by Patricia Flor, chairperson of the UN Commission on the Status of Women, and the second by Mary Robinson, UN High Commissioner for Human Rights.

187. Patricia Flor
Twenty Years of CEDAW (Convention on the Elimination of All Forms of Discrimination Against Women)

Patricia Flor (b. 1961) is the chairperson of the UN Commission on the Status of Women. Educated in universities in Germany and the United States, Flor has served as an attaché in the Foreign Office of the Federal Republic of Germany, as a political and press officer in the German Embassy in Almaty, Kazakhstan,

"55th United Nations Commission on Human Rights: Special Event on the Occasion of the 20th Anniversary of CEDAW" by Dr. Patricia Flor. April 13,1999, Geneva. Printed by permission from Dr. Patricia Flor.

and as a political officer for the advancement of women and social affairs in the Permanent Mission of the Federal Republic of Germany to the United Nations. In March 1998, Flor was appointed to a two-year term as chairperson of the UN Commission on the Status of Women. Formed in 1947, the Commission played a key role in initiating International Women's Year (1975), the Decade for Women (1976–1985), and the adoption by the UN General Assembly of the 1979 Convention on the Elimination of All Forms of Discrimination Against Women (CEDAW).[1] Flor gave this speech in Geneva in April 1999 to commemorate CEDAW's twenty-year anniversary.

Madam Chairperson, Excellencies, Distinguished Delegates, Representatives of Non-governmental Organizations, Ladies and Gentlemen,

We gather today to commemorate the 20th anniversary of the adoption of the Convention on the Elimination of all Forms of Discrimination against Women by the General Assembly in 1979. CEDAW was a milestone in the promotion of the human rights of women and in the fight against all discrimination on the basis of sex. For the first time, an international human rights instrument addressed the specific and systematic discrimination which women face in many forms around the world. The Convention was a major breakthrough also, because it acknowledged not only women's right to non-discrimination and the obligation of states to respect this right, but also states' responsibility for taking active steps to eliminate existing discrimination in society, in the economy, in education, in human rights, in all areas of life.

Since 1979, 163 states have ratified the Convention making it one of the really powerful instruments of international human rights law. This success of the Convention combined with the fact that many states and governments have indeed taken action to implement the Convention and to accord equality in law and practice to women and men, these facts alone are certainly a reason to celebrate.

Women have even more cause to celebrate, because the Commission on the Status of Women at its 43rd Session adopted a second human rights instrument for women, namely the Optional Protocol under CEDAW which will give a right to petition to women. In order to allow us to appreciate the real significance of the Optional Protocol for making CEDAW even more effective in future, I would like to explore some of the achievements and limitations of the Convention itself.

First of all, what have we achieved and where have we failed? In terms of civil and political rights, women have a right to vote in nearly all countries and enjoy equal civil and political rights in most countries at least on paper. But, if we look at female representation in decision-making, in parliaments, in government executive functions, we know of course where our shortcomings are. In terms of social, economic and cultural rights, discrimination continues in too many countries. In many instances, women still do not enjoy equal rights to land or inheritance, they suffer from discrimination which restricts their access to capital, resources, employment, food, social services etc. to name but a few. That women constitute a majority of the poor and illiterate illustrates the continuing disadvantages women and girls face. On the other hand, progress is visible when looking at enrollment rates for girls, at women's expanding economic activities or at the growing number of countries which adopt special measures for the advancement of women in all areas of life.

Let us look at how this description of reality relates to an instrument like CEDAW. CEDAW

---

[1] For an excerpt of the text of CEDAW and additional background information, see Source 177 in Chapter 14.

like any other such instrument has clear limitations in what it can and what it cannot bring about. One constraint to the immediate implementation of CEDAW is reality itself. Equality between women and men cannot simply be imposed or decreed, although state action and legislation remain of course crucial. Equality demands fundamental societal and attitudinal change and only when this change occurs as a result of a process to which both women and men must contribute actively and with good will, will the vision of equality come true.

Another constraint has to do with the character of international treaties. The Convention on the Elimination of all Forms of Discrimination against Women is an international, legally binding instrument, that is states have an obligation to comply after ratification. However, the international community has no police, no judge and no punishments for those who violate the Convention. While the International Criminal Court will be the first permanent institution on the international level to deal with crimes against humanity like genocide, there is no such institution for our Convention. Essentially, this means that states and governments implement CEDAW primarily because of their own commitment to do so. Nevertheless, additional incentives for implementation are required and that is where the Committee on the Elimination of all Forms of Discrimination against Women and the Optional Protocol come in.

The Committee has the primary responsibility to monitor the implementation of the Convention on the basis of states' reports. The Committee has an essential function, because it provides a public forum to discuss the status of women in countries and the policies adopted by governments. However, the Committee has no power to consider individual cases or to look at the situation in countries which, for whatever reason, do not submit their reports or from which disquieting news about violations of the human rights of women emerge.

There are instances where the specific constraints of the Convention create a paradoxical situation for women who suffer from discrimination. Imagine a country which ratified the Convention, but where women do not yet enjoy full and independent legal standing. That means that an individual woman might need the consent and approval of her husband, her oldest brother or her father to bring a complaint before a domestic court. In this case, despite the fact that the Convention guarantees non-discrimination to women, the woman would be unable to enforce her right through the domestic legal system. Or take a country which does not report under the Convention, perhaps because it lacks the technical means or the expertise to do so. If in that country women are subject to specific kinds of discrimination, the Committee under the Convention cannot discuss the matter nor will the government receive assistance in and recommendations for implementing CEDAW, since this requires a report as a basis.

The necessity of an additional and complementary instrument thus becomes abundantly clear. Women need a Protocol which will increase the effectiveness of CEDAW. The Optional Protocol as adopted on 12 March 1999 by the Commission on the Status of Women in New York does exactly that. After its entry into force, it will allow women as individuals or as groups, on their own or represented by a lawyer, a relative, an NGO or whatever other representation they choose, to bring their grievances directly to the United Nations when they cannot find justice at the national level. The Optional Protocol works independently of the reporting requirement under the Convention. Regardless of whether or not a state has reported, if it has ratified the Protocol, a woman can make her voice heard at the United Nations, can attract the attention of the international community to violations of her rights which, in many cases, will also be violations of the rights of many other women.

By its mere existence, the Protocol should also encourage States Parties to improve domestic remedies for women. The Protocol comes into play only after all legal possibilities have been exhausted at the national level. It is certainly desirable for both parties concerned, the woman and the state, that women can find justice at home and need not embark on the long road to the United Nations.

Experience shows that a major ally of women in the full implementation of CEDAW is civil society, the NGO's, the world's public and the media. It is the moral force of an open debate about the status of women, about brutal human rights violations against women like female genital mutilation or trafficking which helps states and governments to find the courage and the determination to adopt measures to fight them and to grant full equality to women both de jure and de facto. The Optional Protocol, by allowing individual cases to be discussed, will contribute greatly to this moral force of persuasion behind CEDAW and it will heighten awareness of ongoing discrimination against women.

Against this background, the importance of CEDAW and the Optional Protocol for promoting gender equality in real life cannot be underestimated. Let me therefore conclude with 3 appeals:

- I appeal to all states which have not yet done so to ratify CEDAW as soon as possible to reach the goal of universal ratification by the year 2000.
- I appeal to all States Parties of CEDAW to ratify the Optional Protocol as soon as possible to allow for its entry into force five years after Beijing, before the Special Session of the General Assembly "Women 2000" in June 2000 and before the turn of the century.
- I appeal to all women and all like-minded men who strive for equality to make full use of both the Convention and the Optional Protocol as powerful tools for the achievement of gender equality. Any instrument is only as effective as its beneficiaries make it. And the beneficiaries are we, the women of this world who have a common cause to bring before the United Nations, and that is the cause of equal rights for everybody regardless of sex.

I thank you for your attention.

188. Mary Robinson
"The Mortal Power of Affirmation"

Mary Robinson (b. 1944) is the United Nations High Commissioner for Human Rights and a former president of the Republic of Ireland. Prior to her career in politics, Robinson was a professor of law at Trinity College. A senator in the Irish Parliament from 1969 to 1989, Robinson tackled such issues as the reform of family law and the legalization of divorce. She also founded and served as director of the Irish Center for European Law from 1988 to 1990. Elected president of the Republic of Ireland on the Labor Party ticket in 1990, Robinson became the first woman ever to hold the office. During her seven-year term, Robinson played a prominent international role. She was the first head of state to visit Rwanda in the aftermath of its 1994 genocidal war and the only Western leader to visit Somalia during the famine of 1992. A patron of the Irish Famine Museum, she has spoken of the lessons Ireland's history provides for contemporary problems of world

Reprinted by permission from the United Nations Association of Great Britain and Northern Ireland. www.oneworld.org/UNA_UK.

hunger. Robinson also contributed to the peace process in Northern Ireland by facilitating meetings between women's groups from Northern Ireland and the Republic of Ireland and, in 1992, undertook the first official visit by an Irish head of state to Northern Ireland. Robinson was appointed UN High Commissioner for Human Rights in 1997, the second person to hold this position, which was created at the World Conference on Human Rights held in 1993 in Vienna. In 1998, she pledged to "rekindle some of the vision" of the 1948 UN Universal Declaration of Human Rights which defined social and economic as well as political rights as human rights.[1] Robinson gave the following speech, "The Mortal Power of Affirmation: Reflections on Fifty Years of the Universal Declaration of Human Rights," in November 1998 in London.[2]

Anniversaries have a peculiar and powerful attraction for human beings. Some mark personal milestones, of independence, achievement, or significant alliances. As with individuals, so it is with anniversaries of nations and institutions. A century provides a satisfactory historical perspective, while a thousand years fills us with awe at human grandeur and puniness. A fiftieth anniversary, at the human and the institutional level, provides a punctuation mark for drawing up a balance sheet of success and failure in the past, and setting an agenda for the future. In giving—and making—account of fifty years experience of the Universal Declaration of Human Rights, we affirm the centrality of human rights to personal and social development, and address the expectations and anxieties of the planet's inhabitants on the edge of a new Millennium in the Christian era.

The Universal Declaration of Human Rights was drawn up just three years after the end of the second world war which had claimed fifty million lives, including six million Jews murdered in the Holocaust with millions of others whose lives were taken because of their political beliefs, their sexual orientation, or other supposed imperfections. Let us bear in mind that this year also marks the 50th anniversary of the first genocide convention.

In the intervening fifty years the world has witnessed many destructive events. The magnitude of these events, the numbers of deaths, the manner of them, take us to the limits of the human imagination. To deal with them requires the witness of survivors to bring those events to a human scale. One such was the Italian Auschwitz survivor, Primo Levi. Shortly after his liberation from Auschwitz Primo Levi encountered a child who looked about three years of age. He was, Levi said: "A nobody, a child of death, a child of Auschwitz." The nameless child died soon afterwards. He died, Levi said, "free, but not redeemed. Nothing remains of him; he bears witness through these words of mine." Levi referred to "the mortal power of affirmation of the smallest and most harmless among us," a three year old who died at Auschwitz without even a name. If the Universal Declaration is to be made a reality, then the international community needs to respond to the mortal affirmation of the smallest and the weakest voices. We must bring home to States their responsibility under the Declaration; and reaffirm our joint commitment, in the words of the Declaration, to "the inherent dignity" and "the equal and inalienable rights of all members of the human family", recognizing that these rights are "the foundation of freedom,

---

[1]Mary Robinson, "Addressing the Gap Between Rhetoric and Reality," epilogue to *The Universal Declaration of Human Rights: Fifty Years and Beyond* ed. Yael Danieli, Elsa Stamatopoulou, and Clarence J. Dias (New York: Baywood Publishing Company, 1999), p. 424.

[2]Robinson's speech was given before a meeting organized jointly by the UNA-UK and the London UN Information Center.

justice and peace in the world." In doing so we recuperate the memories of those whose rights have been denied in the past.

Speaking on the 40th Anniversary of the Universal Declaration of Human Rights, Judge G W Weeramantry described the spread of human rights consciousness as one of the cultural phenomena of all history. He commented that: "Never in historical experience has there been a *universal* acceptance of minimum norms in relation to human rights comparable to the Universal Declaration." In the ensuing decade, there have been major advances both in the conceptual understanding of rights, and in the practical requirements to deliver them. In examining them, it is essential to emphasize that the Declaration set down *minimum* requirements with the intention that they have *universal effect.*

Acceptance of these norms implies compliance with them. The history of the last fifty years underscores the necessity for these norms, but also the fact that in many respects they remain to be implemented. Indeed, the work that has been done in setting standards against which to judge the compliance of States must be regarded as one of the major successes of the last decade. The coming decade must see us moving towards ways of more effectively applying those standards. Given the need to ensure progress towards a universal application of human rights, this will be a daunting task, and not one to be undertaken by a single organisation. It will require the wholehearted commitment of individuals, groups and nations.

The growth of a culture of human rights has continued apace in the last decade, so that we can refer now to the existence of a human rights community. Significant organs of civil society, such as trade unions and the churches, increasingly couch their recommendations in the language of human rights. Relief and development agencies have long since recognized the importance of incorporating human rights into their plans. The work of these bodies, secular and religious, attests to the belief in an ethic of human behaviour that can be translated into effective practical action. Countless organisa-

tions, of varying sizes and resources, operate in their own countries and internationally, to advance the rights of vulnerable groups: children, ethnic minorities, prisoners, the handicapped; and to highlight the issues that continue to bedevil us: torture, extrajudicial executions, modern forms of slavery, and social and economic discrimination and exclusion.

Thoughtful business leaders are responding to the intensification of economic interdependency by becoming powerful actors in the human rights community. Last month in Dublin, Peter Sutherland, speaking on the theme *Towards a Global Ethic,* said:

> Ensuring that low income countries don't miss out on the benefits of globalization is a crucial test for international economic governance and for developed countries. Poverty remains the world's most urgent moral challenge. Yet particularly following the end of the Cold War, there has been a disturbing tendency to look on the widening gap between rich and poor with indifference.
>
> This is short-sighted. Eliminating poverty is not only the right thing to do; it is essential to fulfilling the world's growth potential. Even if a moral imperative to address human suffering did not exist, it would be in the self interest of developed countries to confront global poverty aggressively.

This identification of a business interest in human rights offers enormous potential.

While recognizing the vibrancy and richness of the contributions of these groups we must also acknowledge that the spread of a rights-based culture remains skewed. For many States, both in the developed and developing world, human rights responses are inadequate or nonexistent. The challenge for the future must be to embed the human rights culture in the operations of Governments, and to make human rights activities central to all decision making.

For this to happen we must reinforce the universality and interdependence of all human rights. There is no ownership in any one region:

they are not Western or Eastern, Northern or Southern human rights. A few weeks ago I was glad to have the opportunity to bring together and listen to experts in Islamic Law discussing among themselves the theme "Enriching the Universality of Human Rights: Islamic perspectives on the Universal Declaration of Human Rights". Of particular relevance to our modern world was the emphasis they placed on the "duties owed to community" referred to in Article 29.

We should also recognize the specific advances that have been made in the field of human rights in recent times, and the extent to which necessary social and political change can be driven by rights concerns. The momentous changes in the Republic of South Africa indicate the importance of applying the philosophy underlying the Universal Declaration, and its explicit statement that respect for and implementation of rights constitutes "the foundation of freedom, justice and peace in the world." In particular, the recently completed work of the Truth and Reconciliation Commission shows that revisiting the pain of the past can help us to imagine a route to a more secure and inclusive future. More recently still the Belfast Agreement, by placing human rights concerns at the heart of the peace process, indicates the capacity of such an approach to effect a means of addressing conflict. In both cases, the setting up of powerful independent Human Rights Commissions ensures redress. But their very presence means that those who take decisions have to take into account questions of equality and rights.

The ending of the Cold War has opened up new opportunities for identifying and addressing human rights stripped of ideological rhetoric. We have the opportunity to face problems and issues more honestly, and the experience of the past gives us the analytical tools required for a New World Order where civil, political and economic rights are recognized....

We can also take courage from the increasing centrality accorded to human rights in foreign policy considerations and relations between States. Accelerating and nurturing this develop-

ment is crucial, particularly in relation to increasing globalisation and its potential impact on economic, social and cultural rights. Such potential has generally been seen as a threat. We must ensure that it is seen as an opportunity to be grasped consistent with the 1986 Declaration on the Right to Development. This declaration recognized that development needed to be conceptualized as a comprehensive economic, social, cultural and political process. Such development aims at the constant improvement of the well-being of the entire population and of all individuals on the basis of their active, free and meaningful participation in development, and in an equitable distribution of its benefits.

This places an obligation upon States to have confidence in civil society. They need to empower it to participate in and influence decision-making, especially in relation to the impact of public decisions on rights and equality. We need to "bring rights home" to those affected by decisions impacting on their social inclusion. Initiating a "conversation" on rights between the powerful and the powerless, on the basis of equality, and using the Declaration as a basis, is a prerequisite for such an undertaking. Two considerations are relevant here. The first is the pressing need to tackle the question of debt. The UN has recognized that the foreign debt burden remains one of the most critical factors adversely affecting living standards for a number of developing countries and their people. Thus far, measures for dealing with the debt problem have not achieved an effective, durable and development oriented solution, particularly towards the poorest and most indebted countries.

In April of this year a resolution of the UN Commission of Human Rights emphasized the need for action in relation to the social impact of measures arising from foreign debt....

Secondly, the problem places an obligation upon both States and Non-State Actors in relation to their international activities. States must ensure that the activities of corporations in their jurisdictions are consistent with international obligations, but they could also reflect their capacity to encourage and reward best practices. They could foster the growing recognition on the part of some sectors of the business community of the need to assess the impact of their activities on the counties where they operate, and that a rights-based approach makes economic sense.

In the past fifty years, we have made considerable progress in elaborating and codifying the minimal standards required for effective implementation of human rights, but the legal instruments are by their nature cumbersome, and consuming of time and resources. Governments can show their transparency and commitment by ensuring domestic applicability of international commitments. In this regard, and particularly in this place, I warmly welcome the fact that the Human Rights Act, incorporating the European Convention on Human Rights, received the Royal assent on November 9th. I hope I may be forgiven for personalizing my feelings by referring to the work of Lord Irvine, the Lord Chancellor, and Lord Lestor in facilitating this far-reaching and farseeing endeavour....

Ultimately the work of the United Nations and the human rights community as a whole will be judged by those who are in a precarious position at the margins. Our work will be judged by outcomes, and by measurable change in people's lives. Dignity and respect constitute the ethic. Freedom from fear of death, violence and wrongful imprisonment; and access to health, shelter, food, education and employment: these represent its delivery. In moving to ensure a more just world the UN must continue to engage with a variety of actors who can contribute to fairer outcomes: States themselves, transnational bodies, regional and international organisations. The nature of this engagement requires a greater focus on outcomes, and how they are to be achieved. In calling for change from others, we can hardly exempt ourselves from reform and renewal. We will learn from changes that have promoted better human rights safeguards. We will learn from those who have suffered from human rights abuses, and we will incorporate those lessons into our own practice. The constituencies of the excluded look to us for support and protection. We dare not fail them, least of all by failure to attend to our own inadequacies.

The Universal Declaration affirmed that the rights enumerated in it should be placed at the centre of all human activity. It recognized that all had a part to play—"every individual and every organ of society"—in promoting respect for these rights, as well as securing their universal and effective observance. Thus the mass media have a role to play, as have educators, and within education itself human rights must be part of our shared cultural heritage and our vision of what it means to be human.

The experience gained in the past half century maps out a difficult route for us, but one that is better signposted. Our endeavours in the future require a greater focus on what is required to achieve better implementation. We know that we can effect change. Increasingly we know what is required to get it. The ghosts of our shared history demand Levi's mortal power of affirmation to redeem the present and secure the future.

QUESTIONS FOR ANALYSIS

1. How does Patricia Flor assess the historical significance of the 1979 UN Convention on the Elimination of All Forms of Discrimination Against Women? How does she evaluate improvements in women's lives and status in the twenty years since the adoption of CEDAW? What is the Optional Protocol and how will it enhance the effectiveness of CEDAW?

2. How does Mary Robinson define human rights and the significance of the 1948 UN Universal Declaration of Human Rights? How has the end of the Cold War transformed human rights issues and campaigns? How does Robinson describe the relevance of human rights issues to globalization and relations between "developed" and "developing" nations? On whom does Robinson place the responsibility for ensuring the fulfillment of human rights? What is the meaning of her title, "The Mortal Power of Affirmation"?